Mastering Algebra
Advanced Level

Book Title: Mastering Algebra - Advanced Level

Author: Said Hamilton

Editor: Pat Eblen

Cover design by: Kathleen Myers

First published in the year 2000

Hamilton Education Guides
P.O. Box 681
Vienna, Va. 22183

Library of Congress Catalog Card Number 99-97605
Library of Congress Cataloging-in-Publication Data

ISBN 0-9649954-3-3

General Contents

Detailed Contents

Chapter 1 - Review of Introductory and Intermediate Algebra

Chapter 3 - Matrices

Chapter 4 - Sequences and Series

Chapter 5 - Differentiation

Appendix - Exercise Solutions

Acknowledgments and Dedication

I would like to acknowledge my wife and children for giving me inspiration and for their understanding and patience in allowing me to take on the task of writing this book. I am grateful to Pat Eblen for his editorial comments. As always, his constructive comments and suggestions on clearer presentation of topics truly elevated the usefulness of this book. His continual support and recommendations to further enhance this book is greatly appreciated. I would also like to acknowledge and give my thanks to many education professionals who provided comments to further enhance this book. Finally, my special thanks to Kathleen Myers for her outstanding cover design. *This book is dedicated to my wife and children for their support and understanding.*

Introduction and Overview

As we start the new century, with anticipation and a great deal of aspiration, a primary hope of our nation is that the new millennium will bring new discoveries and advancements to the fields of science, technology, communication, medicine, and space exploration - to mention a few - that will ultimately enhance human life on this planet. This hope and aspiration can only be fulfilled if the individuals in our society obtain a thorough knowledge of technical concepts. A key to achieving this goal is by providing quality educational materials that can build and develop individuals abilities in solving complex problems.

Similar to the previous books published by the Hamilton Education Guides, the intent of this book is to build a strong foundation by increasing student confidence in solving mathematical problems. To achieve this objective, the author has diligently tried to address each subject in a clear, concise, and easy to understand step-by-step format. A great deal of effort has been made to ensure the subjects presented in each chapter are explained simply, thoroughly, and adequately. It is the authors hope that this book can fulfill these objectives by building a solid foundation in pursuit of more advanced technical concepts.

The scope of this book is intended for educational levels ranging from the 11th grade to adult. The book can also be used by students in home study programs, parents, teachers, special education programs, tutors, high schools, preparatory schools, and adult educational programs including colleges and universities as a main text, a thorough reference, or a supplementary book. A thorough knowledge of algebraic concepts in subject areas such as linear equations and inequalities, fractional operations, exponents, radicals, polynomials, factorization, and quadratic equations is required.

"Mastering Algebra: Advanced Level" is the third and final book in a series on algebra. It addresses subjects such as functions of real and complex variables, matrices, sequences, series, limits, factorials, and differentiation. In the second book *"Mastering Algebra: Intermediate Level"* students are introduced to subjects such as linear and non linear equations, inequalities, factoring, quadratic equations, algebraic fractions, and logarithms. In the first book *"Mastering Algebra: An Introduction"* topics such as integer fractions, exponents, radicals, fractional exponents, scientific notations, and polynomials are introduced.

"Mastering Algebra: Advanced Level" is divided into five chapters. Chapter 1 reviews, in some detail, selected subjects addressed in the *"Mastering Algebra: An Introduction"* and the *"Mastering Algebra: Intermediate Level"* books. Topics such as exponents, radicals, factoring polynomials, quadratic equations and factoring, and operations involving with real and complex algebraic fractions are reviewed. Functions of real and complex variables are addressed in Chapter 2. In this chapter students learn how to identify and solve various math operations involving functions. In addition, the steps as to how composite and inverse functions are computed are also addressed in this chapter. Finally, math operations involving addition, subtraction, multiplication, and division of complex numbers are discussed in Chapter 2. Matrix operations including the steps for computing the determinant and inverse of a matrix are discussed in Chapter 3. Solutions to linear systems using various methods such as the Addition, the Substitution, the Inverse Matrices, the Cramer's Rule, the Gaussian Elimination, and the Gauss-Jordan Elimination method are addressed in Chapter 3. Students are encouraged to gain a thorough understanding of the various solution methods for solving linear

systems of equations. Sequences and series are introduced in Chapter 4. Additionally, how to compute and find the limit of arithmetic and geometric sequences and series including expansion and simplification of factorial expressions are also discussed in this chapter. Derivatives and its applicable differentiation rules using the Prime and $\frac{d}{dx}$ notations are introduced in Chapter 5. Use of the Chain rule in solving different types of equations is also discussed in this chapter. Additionally, the steps for solving higher order equations including the implicit differentiation method is introduced in Chapter 5. Finally, detailed solutions to the exercises are provided in the Appendix. Students are encouraged to solve each problem in the same detail and step-by-step format as shown in the text.

In keeping with our commitment of excellence in providing clear, easy to follow, and concise educational materials to our readers, I believe this book will again add value to the Hamilton Education Guides series for its clarity and special attention to detail. I hope readers of this book will find it valuable as a learning tool and a reference. Any comments or suggestions for improvement of this book will be appreciated.

With best wishes,

Said Hamilton

Chapter 1
Review of Introductory and Intermediate Algebra

Quick Reference to Chapter 1 Problems

1.1 Exponents

1.1a Introduction to Integer Exponents 5

Case I - Real Numbers Raised to Positive Integer Exponents, *p. 5*

$$\boxed{2^3} = ;\quad \boxed{(-3)^5} = ;\quad \boxed{-8^3} =$$

Case II - Real Numbers Raised to Negative Integer Exponents, *p. 8*

$$\boxed{4^{-3}} = ;\quad \boxed{(-8)^{-3}} = ;\quad \boxed{-2^{-4}} =$$

1.1b Operations with Positive Integer Exponents 10

Case I - Multiplying Positive Integer Exponents, *p. 10*

$$\boxed{\left(x^3y^2\right)\cdot\left(x^2y\right)\cdot y^3} = ;\quad \boxed{\left(e^3e^5e\right)\cdot\left(-\frac{4}{32}e^2\right)} = ;\quad \boxed{(2-3)^2\cdot\left(5x^3y^2\right)\cdot(-2xy)} =$$

Case II - Dividing Positive Integer Exponents, *p. 12*

$$\boxed{\frac{2ab}{-4a^3b^4}} = ;\quad \boxed{\left(\frac{u^2v^3w^2}{8u^7v^5}\right)\cdot\left(\frac{u}{v^2}\right)} = ;\quad \boxed{\frac{4k^3lm^2}{kl^2m^3}} =$$

Case III - Adding and Subtracting Positive Integer Exponents, *p. 14*

$$\boxed{x^3+3y^2+2x^3-y^2+5} = ;\quad \boxed{\left(a^3+2a^2+4^3\right)-\left(4a^3+20\right)} = ;\quad \boxed{a^{3b}+2a^{2b}-4a^{3b}+5+3a^{2b}} =$$

1.1c Operations with Negative Integer Exponents 16

Case I - Multiplying Negative Integer Exponents, *p. 16*

$$\boxed{5^{-2}\cdot 5\cdot a^{-3}\cdot b^{-3}\cdot a^{-1}\cdot b} = ;\quad \boxed{\left(x^{-2}y^{-2}z^2\right)\cdot\left(x^{-1}y^3z^{-4}\right)} = ;\quad \boxed{\left(3^{-2}\cdot 2^{-1}\right)\cdot\left(3^{-1}\cdot 2^{-2}\right)\cdot 2} =$$

Case II - Dividing Negative Integer Exponents, *p. 19*

$$\boxed{\frac{2^{-3}u^{-1}v^{-3}}{2^{-1}u^{-2}v}} = ;\quad \boxed{\frac{e^{-5}f^{-2}e^0}{2^{-3}f^5e^{-4}}} = ;\quad \boxed{\frac{2^{-3}\cdot a^{-1}}{(-2)^{-2}a^{-3}}} =$$

Case III - Adding and Subtracting Negative Integer Exponents, *p. 21*

$$\boxed{x^{-1}+x^{-2}+2x^{-1}-4x^{-2}+5^{-2}} = ;\quad \boxed{a^{-1}-b^{-1}+2a^{-1}+3b^{-1}} = ;\quad \boxed{k^{-2n}+k^{-3n}-3k^{-2n}+2^{-2}} =$$

1.2 Radicals

1.2a Introduction to Radicals 24

Case I - Roots and Radical Expressions, *p. 24*

$$\boxed{\sqrt{64}} = ;\quad \boxed{\sqrt[3]{375}} = ;\quad \boxed{-\sqrt[2]{324}} =$$

Case II - Rational, Irrational, Real, and Imaginary Numbers, *p. 26*

$\boxed{\sqrt{25} \quad \textit{is a rational number}}$; $\boxed{\sqrt{7} \quad \textit{is an irrational number}}$; $\boxed{-\sqrt{-3} \quad \textit{is not a real number}}$

Case III - Simplifying Radical Expressions with Real Numbers as a Radicand, *p. 27*

$$\boxed{\frac{-3}{-2}\sqrt{400}} = ; \quad \boxed{\frac{\sqrt[3]{162}}{9}} = ; \quad \boxed{\frac{1}{-5}\sqrt[2]{1000}} =$$

1.2b Operations Involving Radical Expressions..29

Case I - Multiplying Monomial Expressions in Radical Form, with Real Numbers, *p. 29*

$$\boxed{\sqrt{5}\cdot\sqrt{15}} = ; \quad \boxed{\left(-2\sqrt[3]{512}\right)\cdot\left(-5\sqrt[3]{108}\right)} = ; \quad \boxed{\sqrt{6}\cdot\sqrt{48}\cdot\sqrt{45}} =$$

Case II - Multiplying Binomial Expressions in Radical Form, with Real Numbers, *p. 31*

$$\boxed{\left(2+\sqrt{2}\right)\cdot\left(5-\sqrt{8}\right)} = ; \quad \boxed{\left(2\sqrt[4]{162}+3\right)\cdot\left(3\sqrt[4]{2}+5\right)} = ; \quad \boxed{\left(\sqrt{24}+3\sqrt{60}\right)\cdot\left(\sqrt{25}-\sqrt{72}\right)} =$$

Case III - Multiplying Monomial and Binomial Expressions in Radical Form, with Real Numbers, *p. 34*

$$\boxed{\sqrt{5}\cdot\left(\sqrt{50}+2\sqrt{27}\right)} = ; \quad \boxed{-2\sqrt{24}\cdot\left(\sqrt{36}-\sqrt{125}\right)} = ; \quad \boxed{-2\sqrt[4]{4}\cdot\left(\sqrt[4]{64}-\sqrt[4]{162}\right)} =$$

Case IV - Rationalizing Radical Expressions - Monomial Denominators with Real Numbers, *p. 36*

$$\boxed{\frac{-8\sqrt{3}}{32\sqrt{45}}} = ; \quad \boxed{\frac{5\sqrt{(-2)^2\cdot 3}}{-\sqrt{7}}} = ; \quad \boxed{\frac{3\sqrt[5]{8}}{\sqrt[5]{81}}} =$$

Case V - Rationalizing Radical Expressions - Binomial Denominators with Real Numbers, *p. 39*

$$\boxed{\frac{8}{2-\sqrt{2}}} = ; \quad \boxed{\frac{\sqrt{125}}{\sqrt{3}-\sqrt{5}}} = ; \quad \boxed{\frac{3+\sqrt{5}}{\sqrt{3}+\sqrt{5}}} =$$

Case VI - Adding and Subtracting Radical Terms, *p. 42*

$$\boxed{6\sqrt{2}+4\sqrt{2}} = ; \quad \boxed{20\sqrt[5]{3}-8\sqrt[5]{3}+5\sqrt[5]{3}} = ; \quad \boxed{a\sqrt{xy}+b\sqrt[3]{xy}-c^2\sqrt{xy}-d} =$$

1.3 Factoring Polynomials

1.3a Factoring Polynomials Using the Greatest Common Factoring Method..................44

Case I - Factoring the Greatest Common Factor to Monomial Terms, *p. 44*

Find the G.C.F. to: $\boxed{6x^2 \textit{ and } 8x^3}$; $\boxed{8x^2yz^3 \textit{ and } 24xy^3z}$; $\boxed{r^3s^3, 8rs^2, \textit{and } 9r^2s}$

Case II - Factoring the Greatest Common Factor to Binomial and Polynomial Terms, *p. 47*

$$\boxed{6a^3b^2c^2-2a^2bc^2} ; \quad \boxed{12x^3y^2z+36x^2z^2} ; \quad \boxed{8a^2b^3+4ab^2-2a^2b}$$

1.3b Factoring Polynomials Using the Grouping Method..50

$$\boxed{x^2-4x-9x+36} = ; \quad \boxed{60m^2+24m-15m-6} = ; \quad \boxed{5(x+y)^2+15x+15y} =$$

1.3c Factoring Polynomials Using the Trial and Error Method....................................52

Case I - Factoring Trinomials of the Form ax^2+bx+c where $a=1$, *p. 52*

1.5 Algebraic Fractions

$-\frac{-a}{+b}=-\left(-\frac{a}{b}\right)=+\frac{a}{b}$; $\frac{2}{x+1}$ *is not defined when* $x=-1$; $\frac{1+x}{x}=\frac{2+2x}{2x}=\frac{a+ax}{ax}$

$\frac{3a^2b^2c}{15ab^3c^2}=$; $\frac{3x+6}{x^2-x-6}=$; $\frac{x^3-x}{x^3-2x^2-3x}=$

Case I - Addition and Subtraction of Algebraic Fractions with Common Denominators, *p. 99*

$\frac{5x}{x+y}-\frac{3x}{x+y}=$; $\frac{3a+b}{2a^2b^3}+\frac{2a-b}{2a^2b^3}+\frac{a-2b}{2a^2b^3}=$; $\frac{y^3+3y}{(y-2)(y+3)}-\frac{y^3+2y-3}{(y-2)(y+3)}=$

Case II - Addition and Subtraction of Algebraic Fractions without Common Denominators , *p. 101*

$\frac{2}{x-2}+\frac{4}{x+3}=$; $\frac{x}{yz}+\frac{y}{xz}+\frac{z}{xy}=$; $\frac{y}{x^2-xy}-\frac{x}{xy-y^2}=$

Case III - Multiplication of Algebraic Fractions, *p. 103*

$\frac{x^2y}{x}\cdot\frac{y}{x^3}=$; $u^2v^2w\cdot\frac{1}{uv^2}\cdot\frac{w}{v^2w^3}=$; $\frac{ab^2}{2}\cdot\left(\frac{1}{a^2b}\cdot\frac{4a^3}{b}\right)=$

Case IV - Division of Algebraic Fractions, *p. 105*

$\frac{1}{x^3y^3z}\div\frac{3z}{x^2y^2}=$; $a^2b^2\div\frac{ab}{3ab^3}=$; $\left(\frac{1}{xy}\div\frac{3x}{y^2}\right)\div\frac{y^2}{x^2}=$

Case I - Addition and Subtraction of Complex Algebraic Fractions, *p. 107*

$\frac{\frac{1}{2x}+\frac{2}{3y}}{\frac{3}{3y}-\frac{5}{2x}}=$; $\frac{\frac{3x^3y^2}{xy}-1}{\frac{x^2y}{xy^2}+1}=$; $\frac{\frac{1}{a}+\frac{1}{b}}{\frac{1}{a}-\frac{1}{b}}-\frac{a}{b}=$

Case II - Multiplication of Complex Algebraic Fractions, *p. 110*

$\frac{\frac{xy}{x^2}\cdot x}{x^2y^2\cdot\frac{1}{xy}}=$; $\frac{\frac{x^2y^2z}{z^2}\cdot\frac{z}{x^3y^3}}{\frac{1}{xy}}=$; $\frac{\frac{xyz^2}{xy}\cdot\frac{x^2}{y^3}}{\frac{x}{y}\cdot\frac{y^2}{x^3}}=$

Case III - Division of Complex Algebraic Fractions, *p. 112*

$\frac{\frac{a^3b^2c}{b^2c^3}\div a^3}{\frac{ab}{a^3}\div b^2}=$; $\frac{\frac{uv^2w}{w^3}\div\frac{u^2v^3}{u}}{\frac{u}{v}}=$; $3xyz\div\frac{\frac{xy}{z}\div z^2}{\frac{xy^2}{xyz^2}}=$

Chapter 1 – Review of Introductory and Intermediate Algebra

The objective of this chapter is to review in some detail selected subjects that were addressed in the "*Mastering Algebra - An Introduction*" and "*Mastering Algebra - Intermediate Level*" books. Students are expected to be familiar with the subjects presented in these books (Parentheses and Brackets, Integer Fractions, Exponents, Radicals, Fractional Exponents, Scientific Notation, Polynomials, One Variable Linear Equations and Inequalities, Factoring Polynomials, Quadratic Equations and Factoring, Algebraic Fractions, and Logarithms) before proceeding with the subjects introduced in this book. Students who may need to review additional examples, beyond what is shown in this chapter, need to study the "*Mastering Algebra - An Introduction*" and "*Mastering Algebra - Intermediate Level*" books.
Real numbers raised to positive and negative integer exponents are reviewed in Section 1.1. Mathematical operations involving addition, subtraction, multiplication, and division of positive and negative integer exponents is also discussed in Section 1.1. An introduction to radical expressions, the steps in simplifying radicals, as well as how rational, irrational, real, and imaginary numbers are identified is reviewed in Section 1.2. In addition, multiplication, division, addition, and subtraction of radical expressions is addressed in this section. Different methods for factoring polynomials and quadratic equations are reviewed in Sections 1.3 and 1.4, respectively. Algebraic fractions and their simplification to lower terms is addressed in Section 1.5. Addition, subtraction, multiplication, and division of simple and complex algebraic fractions are also reviewed in this section. Additional examples, followed by practice problems, are provided in the cases presented in each section to help meet the objective of this chapter.

1.1 Exponents

In this section an introduction to exponents and how real numbers are raised to positive or negative integer exponents are reviewed in section 1.1a, Cases I and II. In addition, operations involving positive and negative integer exponents are addressed in sections 1.1b, Cases I through III, and 1.1c, Cases I through III, respectively.

1.1a Introduction to Integer Exponents

Integer exponents are defined as a^n where a is referred to as the **base**, and n is the **integer exponent**. Note that the base a can be a real number or a variable. The integer exponent n can be a positive or a negative integer. Real numbers raised to positive and negative integer exponents (Cases I and II) are addressed below:

Case I Real Numbers Raised to Positive Integer Exponents

In general, real numbers raised to positive integer exponents are shown as:

$$a^{+n} = a^n = a \cdot a \cdot a \cdot a \ldots a \qquad \textit{where n is a positive integer and } a \neq 0$$

For example,

$$8^{+4} = 8^4 = 8 \cdot 8 \cdot 8 \cdot 8 = 4096$$

Real numbers raised to a positive integer exponent are solved using the following steps:

Step 1 Multiply the base a by itself as many times as the number specified in the exponent. For example, 2^5 implies that multiply 2 by itself 5 times, i.e., $2^5 = 2 \cdot 2 \cdot 2 \cdot 2 \cdot 2$.

Step 2 Multiply the real numbers to obtain the product, i.e., $2 \cdot 2 \cdot 2 \cdot 2 \cdot 2 = 32$.

Examples with Steps

The following examples show the steps as to how real numbers raised to positive integer exponents are solved:

Example 1.1-1

$$\boxed{2^3} =$$

Solution:

Step 1 $\boxed{2^3} = \boxed{2 \cdot 2 \cdot 2}$

Step 2 $\boxed{2 \cdot 2 \cdot 2} = \boxed{\mathbf{8}}$

Example 1.1-2

$$\boxed{1.2^4} =$$

Solution:

Step 1 $\boxed{1.2^4} = \boxed{(1.2) \cdot (1.2) \cdot (1.2) \cdot (1.2)}$

Step 2 $\boxed{(1.2) \cdot (1.2) \cdot (1.2) \cdot (1.2)} = \boxed{\mathbf{2.074}}$

Example 1.1-3

$$\boxed{(-3)^5} =$$

Solution:

Step 1 $\boxed{(-3)^5} = \boxed{(-3) \cdot (-3) \cdot (-3) \cdot (-3) \cdot (-3)}$

Step 2 $\boxed{(-3) \cdot (-3) \cdot (-3) \cdot (-3) \cdot (-3)} = \boxed{\mathbf{-243}}$

Note that:

- A negative number raised to an even integer exponent such as 2, 4, 6, 8, 10, 12, etc. is always positive. For example,

  $(-3)^6 = (+3)^6 = +729 = 729$ $\quad (-2)^2 = (+2)^2 = +4 = 4$ $\quad (-5)^4 = (+5)^4 = +625 = 625$

- A negative number raised to an odd integer exponent such as 1, 3, 5, 7, 9, 11, etc. is always negative. For example,

  $(-3)^5 = -243$ $\quad (-2)^3 = -8$ $\quad (-3)^7 = -2187$

Additional Examples - Real Numbers Raised to Positive Integer Exponents

The following examples further illustrate how to solve real numbers raised to positive integer exponents:

Example 1.1-4

$\boxed{(-10)^0} = \boxed{1}$ (See the notes on page 9 relative to numbers raised to the zero power.)

Example 1.1-5

$\boxed{-(6)^5} = \boxed{-(6\cdot 6\cdot 6\cdot 6\cdot 6)} = \boxed{-(7776)} = \boxed{-7776}$

Example 1.1-6

$\boxed{(-4.25)^3} = \boxed{(-4.25)\cdot(-4.25)\cdot(-4.25)} = \boxed{-76.77}$

Example 1.1-7

$\boxed{(10.45)^4} = \boxed{(10.45)\cdot(10.45)\cdot(10.45)\cdot(10.45)} = \boxed{11925.19}$

Example 1.1-8

$\boxed{-(-20)^3} = \boxed{-[(-20)\cdot(-20)\cdot(-20)]} = \boxed{-[-8000]} = \boxed{+8000} = \boxed{8000}$

Practice Problems - Real Numbers Raised to Positive Integer Exponents

Section 1.1a Case I Practice Problems - Solve the following exponential expressions with real numbers raised to positive integer exponents:

1. $4^3 =$
2. $(-10)^4 =$
3. $0.25^3 =$
4. $12^5 =$
5. $-(3)^5 =$
6. $489^0 =$

Case II Real Numbers Raised to Negative Integer Exponents

Negative integer exponents are defined as a^{-n} where a is referred to as the **base**, and n is the **integer exponent**. Again, note that the base a can be a real number or a variable. The integer exponent n can be a positive or a negative integer. In this section, real numbers raised to negative integer exponents are addressed.

In general, real numbers raised to negative integer exponents are shown as:

$$a^{-n} = \frac{1}{a^{+n}} = \frac{1}{a^n} = \frac{1}{a \cdot a \cdot a \cdot a \ldots \cdot a} \qquad \textit{where } n \textit{ is a positive integer and } a \neq 0$$

For example,

$$5^{-4} = \frac{1}{5^4} = \frac{1}{5 \cdot 5 \cdot 5 \cdot 5} = \frac{1}{625}$$

Real numbers raised to a negative integer exponent are solved using the following steps:

Step 1 Change the negative integer exponent a^{-n} to a positive integer exponent of the form $\frac{1}{a^n}$. For example, change 3^{-4} to $\frac{1}{3^4}$.

Step 2 Multiply the base a in the denominator by itself as many times as the number specified in the exponent. For example, rewrite $\frac{1}{3^4}$ as $\frac{1}{3 \cdot 3 \cdot 3 \cdot 3}$.

Step 3 Multiply the real numbers in the denominator to obtain the answer, i.e., $\frac{1}{3 \cdot 3 \cdot 3 \cdot 3} = \frac{1}{81}$.

Examples with Steps

The following examples show the steps as to how real numbers raised to negative integer exponents are solved:

Example 1.1-9

$$\boxed{4^{-3}} =$$

Solution:

Step 1 $\boxed{4^{-3}} = \boxed{\frac{1}{4^3}}$

Step 2 $\boxed{\frac{1}{4^3}} = \boxed{\frac{1}{4 \cdot 4 \cdot 4}}$

Step 3 $\boxed{\frac{1}{4 \cdot 4 \cdot 4}} = \boxed{\mathbf{\frac{1}{64}}}$

Example 1.1-10

$$\boxed{(-8)^{-3}} =$$

Solution:

Step 1 $\boxed{(-8)^{-3}} = \boxed{\frac{1}{(-8)^3}}$

Step 2 $\boxed{\frac{1}{(-8)^3}} = \boxed{\frac{1}{(-8) \cdot (-8) \cdot (-8)}}$

Step 3 $\boxed{\dfrac{1}{(-8)\cdot(-8)\cdot(-8)}} = \boxed{\dfrac{1}{-512}} = \boxed{-\dfrac{\mathbf{1}}{\mathbf{512}}}$

Additional Examples - Real Numbers Raised to Negative Integer Exponents

The following examples further illustrate how to solve real numbers raised to negative integer exponents:

Example 1.1-11

$$\boxed{2^{-3}} = \boxed{\frac{1}{2^3}} = \boxed{\frac{1}{2\cdot 2\cdot 2}} = \boxed{\frac{\mathbf{1}}{\mathbf{8}}}$$

Example 1.1-12

$$\boxed{-(6)^{-4}} = \boxed{-\frac{1}{(6)^4}} = \boxed{-\frac{1}{6^4}} = \boxed{-\frac{1}{6\cdot 6\cdot 6\cdot 6}} = \boxed{-\frac{\mathbf{1}}{\mathbf{1296}}}$$

Example 1.1-13

$$\boxed{(5.2)^{-4}} = \boxed{\frac{1}{(5.2)^4}} = \boxed{\frac{1}{5.2^4}} = \boxed{\frac{1}{(5.2)\cdot(5.2)\cdot(5.2)\cdot(5.2)}} = \boxed{\frac{\mathbf{1}}{\mathbf{731.16}}}$$

Example 1.1-14

$$\boxed{(-9)^{-4}} = \boxed{\frac{1}{(-9)^4}} = \boxed{\frac{1}{(-9)\cdot(-9)\cdot(-9)\cdot(-9)}} = \boxed{\frac{1}{+6561}} = \boxed{\frac{\mathbf{1}}{\mathbf{6561}}}$$

Example 1.1-15

$$\boxed{-(-4.5)^{-3}} = \boxed{-\frac{1}{(-4.5)^3}} = \boxed{-\frac{1}{(-4.5)\cdot(-4.5)\cdot(-4.5)}} = \boxed{-\frac{1}{-91.125}} = \boxed{+\frac{1}{91.125}} = \boxed{\frac{\mathbf{1}}{\mathbf{91.125}}}$$

Note 1: Any number or variable raised to the zero power is always equal to 1. For example,

$$55^0 = 1,\ (-15)^0 = 1,\ (5{,}689{,}763)^0 = 1,\ [(5x+2)-8]^0 = 1,\ \left[(axy)^3\right]^0 = 1, \text{ and } \left(\sqrt{x^3y^2z}\right)^0 = 1$$

Note 2: Zero raised to the zero power is not defined, i.e., 0^0 is undefined.

Note 3: Any number or variable divided by zero is not defined, i.e., $\frac{1}{0}$, $\frac{x}{0}$, $\frac{355}{0}$, etc. are undefined.

Note 4: Zero divided by any number or variable is always equal to zero. For example, $\frac{0}{1} = 0$,

$$\frac{0}{2{,}560} = 0,\ \frac{0}{\sqrt{10}} = 0,\ \frac{0}{x^2} = 0, \text{ and } \frac{0}{\sqrt[3]{a^2}} = 0.$$

Practice Problems - Real Numbers Raised to Negative Integer Exponents

Section 1.1a Case II Practice Problems - Solve the following exponential expressions with real numbers raised to negative integer exponents:

1. $4^{-3} =$
2. $(-5)^{-4} =$
3. $0.25^{-3} =$
4. $12^{-5} =$
5. $-(3)^{-4} =$
6. $48^{-2} =$

1.1b Operations with Positive Integer Exponents

To multiply, divide, add, and subtract integer exponents, we need to know the following laws of exponents (shown in Table 1.1-1). These laws are used to simplify the work in solving problems with exponential expressions and should be memorized.

Table 1.1-1: Exponent Laws 1 through 7 (Positive Integer Exponents)

Law	Formula	Description
I. **Multiplication**	$a^m \cdot a^n = a^{m+n}$	When multiplying positive exponential terms, if bases a are the same, add the exponents m and n.
II. **Power of a Power**	$\left(a^m\right)^n = a^{mn}$	When raising an exponential term to a power, multiply the powers (exponents) m and n.
III. **Power of a Product**	$(a \cdot b)^m = a^m \cdot b^m$	When raising a product to a power, raise each factor a and b to the power m.
IV. **Power of a Fraction**	$\left(\frac{a}{b}\right)^m = \frac{a^m}{b^m}$	When raising a fraction to a power, raise the numerator and the denominator to the power m.
V. **Division**	$\frac{a^m}{a^n} = a^m \cdot a^{-n} = a^{m-n}$	When dividing exponential terms, if the bases a are the same, subtract exponents m and n.
VI. **Negative Power**	$a^{-n} = \frac{1}{a^n}$	A non-zero base a raised to the $-n$ power equals 1 divided by the base a to the n power.
VII. **Zero Power**	$a^0 = 1$	A non-zero base a raised to the zero power is always equal to 1.

In this section students learn how to multiply (Case I), divide (Case II), and add or subtract (Case III) positive integer exponents by one another.

Case I Multiplying Positive Integer Exponents

Positive integer exponents are multiplied by each other using the following steps and the exponent laws I through III shown in Table 1.1-1.

Step 1 Group the exponential terms with similar bases.

Step 2 Apply the Multiplication Law (Law I) from Table 1.1-1 and simplify the exponential expressions by adding the exponents with similar bases.

Examples with Steps

The following examples show the steps as to how positive integer exponents are multiplied by one another:

Example 1.1-16

$$\boxed{\left(x^3 y^2\right) \cdot \left(x^2 y\right) \cdot y^3} =$$

Solution:

Step 1 $\left(x^3y^2\right)\cdot\left(x^2y\right)\cdot y^3 = \left(x^3x^2\right)\cdot\left(y^3y^2y\right) = \left(x^3x^2\right)\cdot\left(y^3y^2y^1\right)$

Step 2 $\left(x^3x^2\right)\cdot\left(y^3y^2y^1\right) = \left(x^{3+2}\right)\cdot\left(y^{3+2+1}\right) = x^5\cdot y^6 = \mathbf{x^5y^6}$

Example 1.1-17

$$\left(-\frac{1}{5}a^2\right)\cdot(10ab)\cdot\left(-\frac{1}{4}ab^2\right) =$$

Solution:

Step 1 $\left(-\frac{1}{5}a^2\right)\cdot(10ab)\cdot\left(-\frac{1}{4}ab^2\right) = \left(-\frac{1}{5}\times-\frac{1}{4}\times 10\right)\cdot\left(a^2aa\right)\cdot\left(b^2b\right) = \left(\frac{10}{20}\right)\cdot\left(a^2a^1a^1\right)\cdot\left(b^2b^1\right)$

Step 2 $\left(\frac{10}{20}\right)\cdot\left(a^2a^1a^1\right)\cdot\left(b^2b^1\right) = \frac{\not{1}\not{0}}{\underset{2}{\not{2}\not{0}}}\cdot\left(a^{2+1+1}\right)\cdot\left(b^{2+1}\right) = \frac{1}{2}\cdot a^4\cdot b^3 = \mathbf{\frac{1}{2}a^4b^3}$

Note - Non zero numbers or variables raised to the zero power are always equal to 1, i.e., $10^0 = 1$, $(23456)^0 = 1$, $a^0 = 1$ *for* $a \neq 0$, $(a\cdot b)^0 = 1$ *for* $a\cdot b \neq 0$, $(x\cdot y\cdot z)^0 = 1$ *for* $x\cdot y\cdot z \neq 0$, etc.

Additional Examples - Multiplying Positive Integer Exponents

The following examples further illustrate how positive exponential terms are multiplied:

Example 1.1-18

$2^3\cdot 2^5 = 2^{3+5} = 2^8 = \mathbf{256}$

Example 1.1-19

$3^0\cdot 3\cdot 3^2\cdot 3^3 = 3^0\cdot 3^1\cdot 3^2\cdot 3^3 = 3^{0+1+2+3} = 3^6 = \mathbf{729}$

Example 1.1-20

$x^2\cdot y^2\cdot z^3\cdot x^2\cdot y^2\cdot z^4\cdot x = \left(x^2\cdot x^2\cdot x\right)\cdot\left(y^2\cdot y^2\right)\cdot\left(z^3\cdot z^4\right) = \left(x^{2+2+1}\right)\cdot\left(y^{2+2}\right)\cdot\left(z^{3+4}\right) = \mathbf{x^5y^4z^7}$

Example 1.1-21

$(-3-2)\cdot^2\left(-4k^2p^2\right)\cdot(-5kp) = \left[(-5)^2\cdot(-4\times-5)\right]\cdot\left(k^2k\right)\cdot\left(p^2p\right) = \left[(+25)\cdot(+20)\right]\cdot\left(k^2k^1\right)\cdot\left(p^2p^1\right)$

$= 500\cdot\left(k^{2+1}\right)\cdot\left(p^{2+1}\right) = \mathbf{500k^3p^3}$

Practice Problems - Multiplying Positive Integer Exponents

Section 1.1b Case I Practice Problems - Multiply the following positive integer exponents:

1. $x^2\cdot x^3\cdot x =$
2. $2\cdot a^2\cdot b^0\cdot a^3\cdot b^2 =$
3. $\frac{4}{-6}a^2b^3ab^4b^5 =$
4. $2^3\cdot 2^2\cdot x^{2a}\cdot x^{3a}\cdot x^a =$
5. $\left(x\cdot y^2\cdot z^3\right)^0\cdot w^2z^3zw^4z^2 =$
6. $2^0\cdot 4^2\cdot 4^2\cdot 2^2\cdot 4^1 =$

Case II Dividing Positive Integer Exponents

Positive integer exponents are divided by one another using the exponent laws I through VI shown in Table 1.1-1.

Case II Dividing Positive Integer Exponents

Positive integer exponents are divided by one another using the following steps:

Step 1 a. Apply the Division and/or the Negative Power Laws (Laws V and VI) from Table 1.1-1.
b. Group the exponential terms with similar bases.

Step 2 Apply the Multiplication Law (Law I) from Table 1.1-1 and simplify the exponential expressions by adding the exponents with similar bases.

Examples with Steps

The following examples show the steps as to how positive integer exponents are divided by one another:

Example 1.1-22

$$\boxed{\frac{2ab}{-4a^3b^4}} =$$

Solution:

Step 1

$$\boxed{\frac{2ab}{-4a^3b^4}} = \boxed{-\frac{2}{4}\frac{a^1b^1}{a^3b^4}} = \boxed{-\frac{2}{4}\frac{1}{\left(a^3a^{-1}\right)\cdot\left(b^4b^{-1}\right)}}$$

Step 2

$$\boxed{-\frac{2}{4}\frac{1}{\left(a^3a^{-1}\right)\cdot\left(b^4b^{-1}\right)}} = \boxed{-\frac{\overset{1}{\cancel{2}}}{\underset{2}{\cancel{4}}}\frac{1}{\left(a^{3-1}\right)\cdot\left(b^{4-1}\right)}} = \boxed{-\frac{1}{2}\frac{1}{a^2\cdot b^3}} = \boxed{-\frac{\mathbf{1}}{\mathbf{2}}\left(\frac{\mathbf{1}}{\boldsymbol{a^2b^3}}\right)}$$

Example 1.1-23

$$\boxed{\frac{-3xy^4z^3y}{-15x^2z^2}} =$$

Solution:

Step 1

$$\boxed{\frac{-3xy^4z^3y}{-15x^2z^2}} = \boxed{\frac{-3x^1y^4z^3y^1}{-15x^2z^2}} = \boxed{+\frac{3}{15}\frac{x^1y^4z^3y^1}{x^2z^2}} = \boxed{\frac{3}{15}\frac{\left(y^4y^1\right)\cdot\left(z^3z^{-2}\right)}{x^2x^{-1}}}$$

Step 2

$$\boxed{\frac{3}{15}\frac{\left(y^4y^1\right)\cdot\left(z^3z^{-2}\right)}{x^2x^{-1}}} = \boxed{\frac{\overset{1}{\cancel{3}}}{\underset{5}{\cancel{15}}}\frac{\left(y^{4+1}\right)\cdot\left(z^{3-2}\right)}{x^{2-1}}} = \boxed{\frac{1}{5}\frac{y^5\cdot z^1}{x^1}} = \boxed{\frac{\mathbf{1}}{\mathbf{5}}\left(\frac{\boldsymbol{y^5z}}{\boldsymbol{x}}\right)}$$

Additional Examples - Dividing Positive Integer Exponents

The following examples further illustrate how to divide positive integer exponential terms by one another:

Example 1.1-24

$$\boxed{\frac{x^4y^3z}{x^2y^2}} = \boxed{\frac{\left(x^4x^{-2}\right)\cdot\left(y^3y^{-2}\right)\cdot z}{1}} = \boxed{\frac{\left(x^{4-2}\right)\cdot\left(y^{3-2}\right)\cdot z}{1}} = \boxed{\frac{x^2\cdot y^1\cdot z}{1}} = \boxed{\frac{x^2yz}{1}} = \boxed{\mathbf{x^2yz}}$$

Example 1.1-25

$$\boxed{\frac{5a^3b^5}{15a^2b^0}} = \boxed{\frac{5\cdot a^3b^5}{15\,a^2\cdot 1}} = \boxed{\frac{\overset{1}{\cancel{5}}}{\underset{3}{\cancel{15}}}\frac{\left(a^3a^{-2}\right)\cdot b^5}{1}} = \boxed{\frac{1}{3}\frac{\left(a^{3-2}\right)\cdot b^5}{1}} = \boxed{\frac{1}{3}\frac{a^1\cdot b^5}{1}} = \boxed{\mathbf{\frac{1}{3}ab^5}}$$

Example 1.1-26

$$\boxed{\frac{8u^3w^3z^2}{2u^3w^2z}} = \boxed{\frac{8u^3w^3z^2}{2u^3w^2z^1}} = \boxed{\frac{\overset{4}{\cancel{8}}\cdot\left(w^3w^{-2}\right)\cdot\left(z^2\cdot z^{-1}\right)}{\underset{1}{\cancel{2}}\cdot\left(u^3u^{-3}\right)}} = \boxed{\frac{4\cdot\left(w^{3-2}\right)\cdot\left(z^{2-1}\right)}{1\cdot\left(u^{3-3}\right)}} = \boxed{\frac{4\cdot w^1\cdot z^1}{u^0}} = \boxed{\frac{4wz}{1}} = \boxed{\mathbf{4wz}}$$

Example 1.1-27

$$\boxed{\frac{100p^2t^2u}{5pt^4u^5}} = \boxed{\frac{100\;p^2t^2u^1}{5\;\;p^1t^4u^5}} = \boxed{\frac{\overset{20}{\cancel{100}}}{\underset{1}{\cancel{5}}}\frac{p^2p^{-1}}{\left(t^4t^{-2}\right)\cdot\left(u^5u^{-1}\right)}} = \boxed{\frac{20}{1}\frac{p^{2-1}}{\left(t^{4-2}\right)\cdot\left(u^{5-1}\right)}} = \boxed{\frac{20}{1}\left(\frac{p^1}{t^2\cdot u^4}\right)} = \boxed{\mathbf{\frac{20p}{t^2u^4}}}$$

Example 1.1-28

$$\boxed{\left(\frac{w^2z^2}{z}\right)\cdot\left(\frac{w}{z^3}\right)} = \boxed{\left(\frac{w^2z^2}{z^1}\right)\cdot\left(\frac{w^1}{z^3}\right)} = \boxed{\frac{w^2z^2\cdot w^1}{z^1\cdot z^3}} = \boxed{\frac{w^2w^1}{z^1z^3z^{-2}}} = \boxed{\frac{w^{2+1}}{z^{1+3-2}}} = \boxed{\mathbf{\frac{w^3}{z^2}}}$$

Practice Problems - Dividing Positive Integer Exponents

Section 1.1b Case II Practice Problems - Divide the following positive integer exponents:

1. $\dfrac{x^5}{x^3} =$

2. $\dfrac{a^2b^3}{a} =$

3. $\dfrac{a^3b^3c^2}{a^2b^6c} =$

4. $\dfrac{3^2\cdot\left(rs^2\right)}{(2rs)\cdot r^3} =$

5. $\dfrac{2p^2q^3pr^4}{-6p^4q^2r} =$

6. $\dfrac{\left(k^2l^3\right)\cdot\left(kl^2m^0\right)}{k^4l^3m^5} =$

Case III Adding and Subtracting Positive Integer Exponents

A common mistake among students is dealing with addition and subtraction of exponential expressions. In this section positive integer exponents addressing addition and subtraction of numbers that are raised to positive exponents is introduced. Positive exponential expressions are added and subtracted using the following steps:

Step 1 Group the exponential terms with similar bases.

Step 2 Simplify the exponential expressions by adding or subtracting the like terms.

Note that **like terms** are defined as terms having the same variables raised to the same power. For example, x^3 *and* $2x^3$; y^2 *and* $4y^2$ are like terms of one another.

Examples with Steps

The following examples show the steps as to how exponential expressions having positive integer exponents are added or subtracted:

Example 1.1-29

$$\boxed{x^3+3y^2+2x^3-y^2+5}=$$

Solution:

Step 1

$$\boxed{x^3+3y^2+2x^3-y^2+5}=\boxed{\left(x^3+2x^3\right)+\left(3y^2-y^2\right)+5}$$

Step 2

$$\boxed{\left(x^3+2x^3\right)+\left(3y^2-y^2\right)+5}=\boxed{(1+2)x^3+(3-1)y^2+5}=\boxed{\mathbf{3x^3+2y^2+5}}$$

Example 1.1-30

$$\boxed{\left(2^3+x^2+4y\right)-\left(3x^2+y\right)+2x^2}=$$

Solution:

Step 1

$$\boxed{\left(2^3+x^2+4y\right)-\left(3x^2+y\right)+2x^2}=\boxed{8+x^2+4y-3x^2-y+2x^2}$$

$$=\boxed{\left(x^2+2x^2-3x^2\right)+(4y-y)+8}$$

Step 2

$$\boxed{\left(x^2+2x^2-3x^2\right)+(4y-y)+8}=\boxed{(1+2-3)x^2+(4-1)y+8}=\boxed{0x^2+3y+8}=\boxed{\mathbf{3y+8}}$$

Additional Examples - Adding and Subtracting Positive Integer Exponents

The following examples further illustrate addition and subtraction of exponential terms:

Example 1.1-31

$$\boxed{5x^3+3x^2+2x^3-x^2+5}=\boxed{\left(5x^3+2x^3\right)+\left(3x^2-x^2\right)+5}=\boxed{(5+2)x^3+(3-1)x^2+5}=\boxed{\mathbf{7x^3+2x^2+5}}$$

Example 1.1-32

$$\boxed{\left(-2m^4-3m^2+2m^4+3m-10\right)-\left(5m^2+2m+3\right)}=\boxed{-2m^4-3m^2+2m^4+3m-10-5m^2-2m-3}$$

$$=\boxed{\left(-2m^4+2m^4\right)+\left(-3m^2-5m^2\right)+(3m-2m)+(-10-3)}=\boxed{(-2+2)m^4+(-3-5)m^2+(3-2)m-13}$$

$$=\boxed{0m^4-8m^2+m-13}=\boxed{\mathbf{8m^2+m-13}}$$

Example 1.1-33

$$\boxed{x^2+3x^2+y^2+x-4y^2-5^2+2x^2+6x}=\boxed{\left(x^2+3x^2+2x^2\right)+\left(y^2-4y^2\right)+(x+6x)-25}$$

$$=\boxed{(1+3+2)x^2+(1-4)y^2+(1+6)x-25}=\boxed{\mathbf{6x^2-3y^2+7x-25}}$$

Example 1.1-34

$$\boxed{\left(-5w^3-3w-5\right)-\left(3w^3-w-4\right)+5w+2}=\boxed{\left(-5w^3-3w-5\right)+\left(-3w^3+w+4\right)+5w+2}$$

$$=\boxed{\left(-5w^3-3w^3\right)+(-3w+5w+w)+(-5+4+2)}=\boxed{(-5-3)w^3+(-3+5+1)w+1}=\boxed{\mathbf{-8w^3+3w+1}}$$

Example 1.1-35

$$\boxed{\left(a^2b^3+3a^2-b+2^4\right)+\left(2a^2b^3+2a^2+3^0\right)-3^3}=\boxed{a^2b^3+3a^2-b+16+2a^2b^3+2a^2+1-27}$$

$$=\boxed{\left(a^2b^3+2a^2b^3\right)+\left(3a^2+2a^2\right)-b+(16+1-27)}=\boxed{(1+2)a^2b^3+(3+2)a^2-b-10}=\boxed{\mathbf{3a^2b^3+5a^2-b-10}}$$

Practice Problems - Adding and Subtracting Positive Integer Exponents

Section 1.1b Case III Practice Problems - Add or subtract the following positive integer exponential expressions:

1. $x^2+4xy-2x^2-2xy+z^3=$
2. $\left(a^3+2a^2+4^3\right)-\left(4a^3+20\right)=$
3. $3x^4+2x^2+2x^4-\left(x^4-2x^2+3\right)=$
4. $-\left(-2l^3a^3+2l^2a^2-5^3\right)-\left(4l^3a^3-20\right)=$
5. $\left(m^{3n}-4m^{2n}\right)-\left(2m^{3n}+3m^{2n}\right)+5m=$
6. $\left(-7z^3+3z-5\right)-\left(-3z^3+z-4\right)+5z+20=$

1.1c Operations with Negative Integer Exponents

To proceed with simplification of negative exponents, we need to know the Negative Power Law in addition to the other exponent laws (shown in Table 1.1-2). The Negative Power Law states that a base raised to a negative exponent is equal to one divided by the same base raised to the positive exponent, or vice versa, i.e.,

$$a^{-n} = \frac{1}{a^n}$$

and

$$a^n = \frac{1}{a^{-n}} \qquad \text{since} \qquad a^n = \frac{1}{a^{-n}} = \frac{1}{\frac{1}{a^n}} = \frac{\frac{1}{1}}{\frac{1}{a^n}} = \frac{1 \times a^n}{1 \times 1} = \frac{a^n}{1} = a^n .$$

Note that the objective is to write the final answer without a negative exponent. To achieve this the exponent laws are used when simplifying expressions that have negative integer exponents. These laws are used to simplify the work in solving exponential expressions and should be memorized.

Table 1.1-2: Exponent Laws 1 through 6 (Negative Integer Exponents)

Law	Rule	Description
I. **Multiplication**	$a^{-m} \cdot a^{-n} = a^{-m-n}$	When multiplying negative exponential terms, if bases a are the same, add the negative exponents $-m$ and $-n$.
II. **Power of a Power**	$\left(a^{-m}\right)^{-n} = a^{-m \times -n}$	When raising a negative exponential term to a negative power, multiply the negative powers (exponents) $-m$ and $-n$.
III. **Power of a Product**	$(a \cdot b)^{-m} = a^{-m} \cdot b^{-m}$	When raising a product to a negative power, raise each factor a and b to the negative power $-m$.
IV. **Power of a Fraction**	$\left(\frac{a}{b}\right)^{-m} = \frac{a^{-m}}{b^{-m}}$	When raising a fraction to a negative power, raise the numerator and the denominator to the negative power $-m$.
V. **Division**	$\frac{a^{-m}}{a^{-n}} = a^{-m} \cdot a^{n} = a^{-m+n}$	When dividing negative exponential terms, if the bases a are the same, add exponents $-m$ and n.
VI. **Negative Power**	$a^{-n} = \frac{1}{a^n}$	A non-zero base a raised to the $-n$ power equals 1 divided by the base a to the n power

In this section students learn how to multiply (Case I), divide (Case II), and add or subtract (Case III) negative integer exponents by one another.

Case I Multiplying Negative Integer Exponents

Negative integer exponents are multiplied by one another using the following steps and the exponent laws I through III shown in Table 1.1-2.

Step 1 Group the exponential terms with similar bases.

Step 2 Apply the Multiplication Law (Law I) from Table 1.1-2 and simplify the exponential expressions by adding the exponents with similar bases.

Step 3 Change the negative integer exponents to positive integer exponents.

Examples with Steps

The following examples show the steps as to how negative integer exponents are multiplied by one another:

Example 1.1-36

$$\boxed{3^{-2}\cdot 2\cdot 3^{-1}\cdot 2^{-3}} =$$

Solution:

Step 1
$$\boxed{3^{-2}\cdot 2\cdot 3^{-1}\cdot 2^{-3}} = \boxed{3^{-2}\cdot 2^{1}\cdot 3^{-1}\cdot 2^{-3}} = \boxed{\left(3^{-2}\cdot 3^{-1}\right)\cdot\left(2^{-3}\cdot 2^{1}\right)}$$

Step 2
$$\boxed{\left(3^{-2}\cdot 3^{-1}\right)\cdot\left(2^{-3}\cdot 2^{1}\right)} = \boxed{\left(3^{-2-1}\right)\cdot\left(2^{-3+1}\right)} = \boxed{3^{-3}\cdot 2^{-2}}$$

Step 3
$$\boxed{3^{-3}\cdot 2^{-2}} = \boxed{\frac{1}{3^{3}\cdot 2^{2}}} = \boxed{\frac{1}{(3\cdot 3\cdot 3)\cdot(2\cdot 2)}} = \boxed{\mathbf{\frac{1}{108}}}$$

Example 1.1-37

$$\boxed{5^{-2}\cdot 5\cdot a^{-3}\cdot b^{-3}\cdot a^{-1}\cdot b} =$$

Solution:

Step 1
$$\boxed{5^{-2}\cdot 5\cdot a^{-3}\cdot b^{-3}\cdot a^{-1}\cdot b} = \boxed{5^{-2}\cdot 5^{1}\cdot a^{-3}\cdot b^{-3}\cdot a^{-1}\cdot b^{1}} = \boxed{\left(5^{-2}5^{1}\right)\cdot\left(a^{-3}a^{-1}\right)\cdot\left(b^{1}b^{-3}\right)}$$

Step 2
$$\boxed{\left(5^{-2}5^{1}\right)\cdot\left(a^{-3}a^{-1}\right)\cdot\left(b^{1}b^{-3}\right)} = \boxed{\left(5^{-2+1}\right)\cdot\left(a^{-3-1}\right)\cdot\left(b^{1-3}\right)} = \boxed{5^{-1}\cdot a^{-4}\cdot b^{-2}} = \boxed{5^{-1}a^{-4}b^{-2}}$$

Step 3
$$\boxed{5^{-1}a^{-4}b^{-2}} = \boxed{\frac{1}{5^{1}}\cdot\frac{1}{a^{4}}\cdot\frac{1}{b^{2}}} = \boxed{\frac{1\cdot 1\cdot 1}{5\cdot a^{4}\cdot b^{2}}} = \boxed{\mathbf{\frac{1}{5a^{4}b^{2}}}}$$

Additional Examples - Multiplying Negative Integer Exponents

The following examples further illustrate how to multiply negative exponential terms by one another:

Example 1.1-38

$$\boxed{a^{-2}\cdot a^{3}\cdot a^{-1}} = \boxed{a^{-2+3-1}} = \boxed{a^{0}} = \boxed{\mathbf{1}}$$

Example 1.1-39

$$\boxed{3^{-2}\cdot 3^{-1}\cdot 2^{-2}\cdot 3^{-4}\cdot 2} = \boxed{3^{-2}\cdot 3^{-1}\cdot 2^{-2}\cdot 3^{-4}\cdot 2^{1}} = \boxed{\left(3^{-2}\cdot 3^{-1}\cdot 3^{-4}\right)\cdot\left(2^{1}\cdot 2^{-2}\right)} = \boxed{\left(3^{-2-1-4}\right)\cdot\left(2^{1-2}\right)} = \boxed{3^{-7}\cdot 2^{-1}}$$

$$= \boxed{\frac{1}{3^7} \cdot \frac{1}{2^1}} = \boxed{\frac{1}{3 \cdot 3 \cdot 3 \cdot 3 \cdot 3 \cdot 3 \cdot 3} \cdot \frac{1}{2}} = \boxed{\frac{1}{2187} \cdot \frac{1}{2}} = \boxed{\frac{1 \cdot 1}{2187 \cdot 2}} = \boxed{\mathbf{\frac{1}{4374}}}$$

Example 1.1-40

$$\boxed{a^{-5} \cdot b^{-3} \cdot a^{-1} \cdot b^{-2}} = \boxed{\left(a^{-5} \cdot a^{-1}\right) \cdot \left(b^{-3} \cdot b^{-2}\right)} = \boxed{\left(a^{-5-1}\right) \cdot \left(b^{-3-2}\right)} = \boxed{a^{-6} \cdot b^{-5}} = \boxed{a^{-6} b^{-5}} = \boxed{\mathbf{\frac{1}{a^6 b^5}}}$$

Example 1.1-41

$$\boxed{\left(5^{-3} \cdot 5^{-1}\right) \cdot \left(2^{-3} \cdot 5^{-2}\right)} = \boxed{5^{-3} \cdot 5^{-1} \cdot 2^{-3} \cdot 5^{-2}} = \boxed{\left(5^{-3} \cdot 5^{-1} \cdot 5^{-2}\right) \cdot 2^{-3}} = \boxed{\left(5^{-3-1-2}\right) \cdot 2^{-3}} = \boxed{5^{-6} \cdot 2^{-3}}$$

$$= \boxed{\frac{1}{5^6} \cdot \frac{1}{2^3}} = \boxed{\frac{1}{(5 \cdot 5 \cdot 5 \cdot 5 \cdot 5 \cdot 5)} \cdot \frac{1}{(2 \cdot 2 \cdot 2)}} = \boxed{\frac{1}{15625} \cdot \frac{1}{8}} = \boxed{\frac{1 \cdot 1}{(15625 \cdot 8)}} = \boxed{\mathbf{\frac{1}{125000}}}$$

Example 1.1-42

$$\boxed{(-1+3)^{-2}\left(r^{-2} s^2 t\right) \cdot \left(r^3 s^{-2} t^{-3} s\right)} = \boxed{(2)^{-2} r^{-2} s^2 t^1 \cdot r^3 s^{-2} t^{-3} s^1} = \boxed{2^{-2}\left(r^{-2} r^3\right) \cdot \left(s^2 s^{-2} s^1\right) \cdot \left(t^1 t^{-3}\right)}$$

$$= \boxed{2^{-2}\left(r^{-2+3}\right) \cdot \left(s^{2-2+1}\right) \cdot \left(t^{1-3}\right)} = \boxed{2^{-2} r^1 \cdot s^1 \cdot t^{-2}} = \boxed{2^{-2} \cdot r \cdot s \cdot t^{-2}} = \boxed{\frac{1}{2^2} \cdot \frac{r}{1} \cdot \frac{s}{1} \cdot \frac{1}{t^2}} = \boxed{\frac{1 \cdot r \cdot s \cdot 1}{2 \cdot 2 \cdot 1 \cdot 1 \cdot t^2}} = \boxed{\mathbf{\frac{rs}{4t^2}}}$$

Practice Problems - Multiplying Negative Integer Exponents

Section 1.1c Case I Practice Problems - Multiply the following exponential expressions by one another:

1. $\left(3^{-3} \cdot 2^{-1}\right) \cdot \left(2^{-3} \cdot 3^{-2} \cdot 2\right) =$
2. $a^{-6} \cdot b^{-4} \cdot a^{-1} \cdot b^{-2} \cdot a^0 =$
3. $\left(a^{-2} \cdot b^{-3}\right)^2 \cdot \left(a \cdot b^{-2}\right) =$
4. $(-2)^{-4}\left(r^{-2} s^2 t\right) \cdot \left(r^3 s t^{-2} s^{-1}\right) =$
5. $\left(\frac{4}{5}\right)^{-4} 2^2 v^{-5} 2^{-4} v^3 v^{-2} =$
6. $2^{-1} \cdot 3^2 \cdot 3^{-5} \cdot 2^2 \cdot 2^0 =$

Case II Dividing Negative Integer Exponents

Negative integer exponents are divided by one another using the exponent laws I through VI shown in Table 1.1-2. These laws are used in order to simplify division of negative fractional exponents by each other. Negative integer exponents are divided by one another using the following steps:

Step 1 a. Apply the Division and/or the Negative Power Laws (Laws V and VI) from Table 1.1-2.
b. Group the exponential terms with similar bases.

Step 2 Apply the Multiplication Law (Law I) from Table 1.1-2 and simplify the exponential expressions by adding the exponents with similar bases.

Examples with Steps

The following examples show the steps as to how negative integer exponents are divided by one another:

Example 1.1-43

$$\frac{5^{-2}}{5^{-3}} =$$

Solution:

Step 1 $$\frac{5^{-2}}{5^{-3}} = \frac{5^{3}\cdot 5^{-2}}{1} =$$

Step 2 $$\frac{5^{3}\cdot 5^{-2}}{1} = \frac{5^{3-2}}{1} = \frac{5^{1}}{1} = \boxed{5}$$

Example 1.1-44

$$\frac{a^{-2}b^{-3}}{a^{-1}b^{-2}} =$$

Solution:

Step 1 $$\frac{a^{-2}b^{-3}}{a^{-1}b^{-2}} = \frac{1}{\left(a^{2}a^{-1}\right)\cdot\left(b^{3}b^{-2}\right)}$$

Step 2 $$\frac{1}{\left(a^{2}a^{-1}\right)\cdot\left(b^{3}b^{-2}\right)} = \frac{1}{\left(a^{2-1}\right)\cdot\left(b^{3-2}\right)} = \frac{1}{a^{1}\cdot b^{1}} = \boxed{\frac{1}{ab}}$$

Additional Examples - Dividing Negative Integer Exponents

The following examples further illustrate how to divide negative integer exponents by one another:

Example 1.1-45

$$\frac{a^{-2}c^{-3}}{-ac^{4}} = -\frac{a^{-2}c^{-3}}{a^{1}c^{4}} = -\frac{1}{\left(a^{1}a^{2}\right)\cdot\left(c^{4}c^{3}\right)} = -\frac{1}{\left(a^{1+2}\right)\cdot\left(c^{4+3}\right)} = \boxed{-\frac{1}{a^{3}c^{7}}}$$

Example 1.1-46

$$\boxed{\frac{(-3)^{-3}}{-(3)^{-3}}}=\boxed{-\frac{(-3)^{-3}}{(3)^{-3}}}=\boxed{-\frac{(3)^{3}}{(-3)^{3}}}=\boxed{-\frac{3\cdot 3\cdot 3}{-3\cdot -3\cdot -3}}=\boxed{-\frac{27}{-27}}=\boxed{+\frac{\overset{1}{\cancel{27}}}{\underset{1}{\cancel{27}}}}=\boxed{\frac{1}{1}}=\boxed{\mathbf{1}}$$

Example 1.1-47

$$\boxed{\frac{(-2)^{-4}}{-(-2)^{-3}}}=\boxed{-\frac{(-2)^{-4}}{(-2)^{-3}}}=\boxed{-\frac{(-2)^{3}}{(-2)^{4}}}=\boxed{-\frac{-2\cdot -2\cdot -2}{-2\cdot -2\cdot -2\cdot -2}}=\boxed{-\frac{-8}{16}}=\boxed{+\frac{\overset{1}{\cancel{8}}}{\underset{2}{\cancel{16}}}}=\boxed{\frac{\mathbf{1}}{\mathbf{2}}}$$

Example 1.1-48

$$\boxed{\frac{-2^{-3}\cdot a^{-1}}{(-2)^{-2}a^{-3}}}=\boxed{-\frac{2^{-3}\cdot a^{-1}}{(-2)^{-2}a^{-3}}}=\boxed{-\frac{(-2)^{2}\cdot\left(a^{3}\cdot a^{-1}\right)}{2^{3}}}=\boxed{-\frac{(-2\cdot -2)\cdot\left(a^{3-1}\right)}{2\cdot 2\cdot 2}}=\boxed{-\frac{4\cdot a^{2}}{8}}=\boxed{-\frac{\overset{1}{\cancel{4}}}{\underset{2}{\cancel{8}}}a^{2}}=\boxed{-\frac{\mathbf{1}}{\mathbf{2}}\mathbf{a^{2}}}$$

Example 1.1-49

$$\boxed{\frac{a^{2}}{2^{-3}\cdot a^{-2}}}=\boxed{\frac{2^{3}\cdot a^{2}\cdot a^{2}}{1}}=\boxed{\frac{8\cdot a^{2+2}}{1}}=\boxed{\frac{8a^{4}}{1}}=\boxed{\mathbf{8a^{4}}}$$

Practice Problems - Dividing Negative Integer Exponents

Section 1.1c Case II Practice Problems - Divide the following negative integer exponents:

1. $\frac{x^{-2}x}{x^{3}x^{0}}=$
2. $\frac{-2a^{-2}b^{3}}{-6a^{-1}b^{-2}}=$
3. $\frac{-(-3)^{-4}}{3\cdot(-3)^{-3}}=$
4. $\frac{-3^{3}y^{-3}yw}{(-3)^{-2}y^{2}w^{-3}}=$
5. $\frac{a^{-2}b^{2}a^{-5}y^{-2}}{a^{-3}y}=$
6. $\frac{(x\cdot y\cdot z)^{0}\cdot yx^{-2}}{x^{-4}y^{-1}}=$

Case III Adding and Subtracting Negative Integer Exponents

Negative exponential expressions are added and subtracted using the following steps:

Step 1 Group the exponential terms with similar bases.

Step 2 Apply the Negative Power Law (Law VI) from Table 1.1-2, i.e., change a^{-n} to $\frac{1}{a^n}$.

Step 3 Simplify the exponential expression by:

a. Using the fraction techniques learned in Section 1.1, and

b. Using appropriate exponent laws such as the Multiplication Law (Law I) from Table 1.1-2.

Examples with Steps

The following examples show the steps as to how exponential expressions having negative integer exponents are added or subtracted:

Example 1.1-50

$$\boxed{3^{-3}+3^{-2}} =$$

Solution:

Step 1 *Not Applicable*

Step 2

$$\boxed{3^{-3}+3^{-2}} = \boxed{\frac{1}{3^3}+\frac{1}{3^2}} = \boxed{\frac{1}{27}+\frac{1}{9}}$$

Step 3

$$\boxed{\frac{1}{27}+\frac{1}{9}} = \boxed{\frac{(1\cdot 9)+(1\cdot 27)}{27\cdot 9}} = \boxed{\frac{9+27}{243}} = \boxed{\frac{\overset{4}{\cancel{36}}}{\underset{27}{\cancel{243}}}} = \boxed{\mathbf{\frac{4}{27}}}$$

Example 1.1-51

$$\boxed{x^{-1}+x^{-2}+2x^{-1}-4x^{-2}+5^{-2}} =$$

Solution:

Step 1

$$\boxed{x^{-1}+x^{-2}+2x^{-1}-4x^{-2}+5^{-2}} = \boxed{\left(x^{-1}+2x^{-1}\right)+\left(x^{-2}-4x^{-2}\right)+5^{-2}}$$

$$= \boxed{(1+2)x^{-1}+(1-4)x^{-2}+5^{-2}} = \boxed{3x^{-1}-3\,x^{-2}+5^{-2}}$$

Step 2

$$\boxed{3x^{-1}-3\,x^{-2}+5^{-2}} = \boxed{\frac{3}{x^1}-\frac{3}{x^2}+\frac{1}{5^2}} = \boxed{\frac{3}{x}-\frac{3}{x^2}+\frac{1}{5^2}}$$

Step 3

$$\boxed{\frac{3}{x}-\frac{3}{x^2}+\frac{1}{5^2}} = \boxed{\left(\frac{3}{x}-\frac{3}{x^2}\right)+\frac{1}{25}} = \boxed{\left(\frac{\left(3\cdot x^2\right)-(3\cdot x)}{x\cdot x^2}\right)+\frac{1}{25}} = \boxed{\left(\frac{3x^2-3x}{x^{1+2}}\right)+\frac{1}{25}}$$

$$= \boxed{\frac{3x^2-3x}{x^3}+\frac{1}{25}} = \boxed{\frac{25\cdot\left(3x^2-3x\right)+1\cdot x^3}{x^3\cdot 25}} = \boxed{\frac{75\,x^2-75\,x+x^3}{25\,x^3}} = \boxed{\frac{x^3+75\,x^2-75\,x}{25\,x^3}}$$

$$= \boxed{\frac{x\left(x^2+75\,x-75\right)}{25\,x^3}} = \boxed{\frac{x^2+75x-75}{25\left(x^3x^{-1}\right)}} = \boxed{\frac{x^2+75x-75}{25x^{3-1}}} = \boxed{\mathbf{\frac{x^2+75x-75}{25\,x^2}}}$$

Additional Examples - Adding and Subtracting Negative Integer Exponents

The following examples further illustrate addition and subtraction of negative exponential terms:

Example 1.1-52

$$\boxed{(x+y)^{-5}} = \boxed{\mathbf{\frac{1}{(x+y)^5}}}$$

Note: $\boxed{(x\cdot y)^{-5}} = \boxed{\frac{1}{(x\cdot y)^5}} = \boxed{\mathbf{\frac{1}{x^5\,y^5}}}$

Example 1.1-53

$$\boxed{a^{-1}-b^{-1}+2a^{-1}+3b^{-1}} = \boxed{\left(a^{-1}+2a^{-1}\right)+\left(3b^{-1}-b^{-1}\right)} = \boxed{(1+2)a^{-1}+(3-1)b^{-1}} = \boxed{3a^{-1}+2b^{-1}} = \boxed{\frac{3}{a}+\frac{2}{b}}$$

$$= \boxed{\frac{(3\cdot b)+(2\cdot a)}{a\cdot b}} = \boxed{\mathbf{\frac{2a+3b}{ab}}}$$

Example 1.1-54

$$\boxed{a^{-1}+b^{-2}+c^{-3}+3b^{-2}} = \boxed{a^{-1}+\left(b^{-2}+3b^{-2}\right)+c^{-3}} = \boxed{a^{-1}+(1+3)b^{-2}+c^{-3}} = \boxed{a^{-1}+4b^{-2}+c^{-3}}$$

$$= \boxed{\left(a^{-1}+4b^{-2}\right)+c^{-3}} = \boxed{\left(\frac{1}{a}+\frac{4}{b^2}\right)+\frac{1}{c^3}} = \boxed{\left(\frac{\left(1\cdot b^2\right)+(4\cdot a)}{a\cdot b^2}\right)+\frac{1}{c^3}} = \boxed{\left(\frac{b^2+4a}{a\,b^2}\right)+\frac{1}{c^3}}$$

$$= \boxed{\frac{\left[\left(b^2+4a\right)\cdot c^3\right]+\left(1\cdot ab^2\right)}{a\,b^2\cdot c^3}} = \boxed{\frac{b^2\cdot c^3+4a\cdot c^3+a\,b^2}{a\,b^2\,c^3}} = \boxed{\mathbf{\frac{b^2c^3+4ac^3+ab^2}{ab^2c^3}}}$$

Example 1.1-55

$$\boxed{5x^{-3}+3x^{-2}+2x^{-3}-x^{-2}+5^{-2}} = \boxed{\left(5x^{-3}+2x^{-3}\right)+\left(3x^{-2}-x^{-2}\right)+5^{-2}} = \boxed{(5+2)x^{-3}+(3-1)x^{-2}+5^{-2}}$$

$$= \boxed{7x^{-3}+2x^{-2}+5^{-2}} = \boxed{\frac{7}{x^3}+\frac{2}{x^2}+\frac{1}{5^2}} = \boxed{\left(\frac{7}{x^3}+\frac{2}{x^2}\right)+\frac{1}{25}} = \boxed{\left(\frac{7\cdot x^2+2\cdot x^3}{x^3\cdot x^2}\right)+\frac{1}{25}} = \boxed{\frac{7x^2+2x^3}{x^{3+2}}+\frac{1}{25}}$$

$$= \boxed{\frac{x^2(7+2x)}{x^5}+\frac{1}{25}} = \boxed{\frac{7+2x}{x^5\cdot x^{-2}}+\frac{1}{25}} = \boxed{\frac{7+2x}{x^{5-2}}+\frac{1}{25}} = \boxed{\frac{7+2x}{x^3}+\frac{1}{25}} = \boxed{\frac{[(7+2x)\cdot 25]+\left(1\cdot x^3\right)}{25\cdot x^3}}$$

$$= \boxed{\frac{175+50x+x^3}{25x^3}} = \boxed{\mathbf{\frac{x^3+50x+175}{25x^3}}}$$

Example 1.1-56

$$\boxed{x^{-2}+3x^{-2}+y^{-2}+x^2-4y^{-2}-5^2+2x^2} = \boxed{\left(x^{-2}+3x^{-2}\right)+\left(y^{-2}-4y^{-2}\right)+\left(x^2+2x^2\right)-25}$$

$$= \boxed{(1+3)x^{-2}+(1-4)y^{-2}+(1+2)x^2-25} = \boxed{4x^{-2}-3y^{-2}+3x^2-25} = \boxed{\left(\frac{4}{x^2}-\frac{3}{y^2}\right)+3x^2-25}$$

$$= \boxed{\left(\frac{4\cdot y^2-3\cdot x^2}{x^2\cdot y^2}\right)+\frac{3x^2-25}{1}} = \boxed{\frac{4y^2-3x^2}{x^2y^2}+\frac{3x^2-25}{1}} = \boxed{\frac{\left[\left(4y^2-3x^2\right)\cdot 1\right]+\left[x^2y^2\left(3x^2-25\right)\right]}{x^2y^2}}$$

$$= \boxed{\mathbf{\frac{4y^2-3x^2+3x^4y^2-25x^2y^2}{x^2y^2}}}$$

Practice Problems - Adding and Subtracting Negative Integer Exponents

Section 1.1c Case III Practice Problems - Simplify the following negative integer exponential expressions:

1. $x^{-1}+2x^{-2}+3x^{-1}-6x^{-2} =$
2. $\left(3a^{-4}-b^{-2}\right)+\left(-2a^{-4}+3b^{-2}\right) =$
3. $(xy)^{-1}+y^{-2}+4(xy)^{-1}-3y^{-2}+2^{-3} =$
4. $4x^{-1}+y^{-3}+5y^{-3} =$
5. $m^{-5}-\left(m^{-2}-3m^{-5}+m^0\right)+3m^{-2} =$
6. $\left(a^3\right)^{-2}+\left(a^{-2}b\right)^2-6a^{-6}+3a^{-4}b^2 =$

1.2 Radicals

In this section radical expressions and the steps as to how they are simplified are reviewed in section 1.2a, Cases I through III. In addition, operations involving radical expressions which include multiplication, division (rationalization), addition, and subtraction of radicals are addressed in section 1.2b, Cases I through VI.

1.2a Introduction to Radicals

A description of roots and radicals (Case I), classification of numbers (Case II), and simplification of radical expressions with real numbers as a radicand (Case III) are discussed below:

Case I Roots and Radical Expressions

In the general radical expression $\sqrt[a]{b}=c$, the symbol $\sqrt{\ }$ is called a **radical sign**. The expression under the radical b is called the **radicand,** a is called the **index**, and the positive square root of the number c is called the **principal square root**.

Exponents are a kind of shorthand for multiplication. For example, $5\times 5=25$ can be expressed in exponential form as $5^2=25$. Radical signs are used to reverse this process. For example, to write the reverse of $5^2=25$ we take the square root of the terms on both sides of the equal sign, i.e., we write $\sqrt{25}=\sqrt{5^2}=5$. Note that since $5^2=25$ and $(-5)^2=25$, we use $\sqrt{25}$ to indicate the positive square root of 25 is equal to 5 and $-\sqrt{25}$ to indicate the negative square root of 25 is equal to -5. Table 1.2-1 provides square roots, cube roots, fourth roots, and fifth roots of some common numbers used in solving radical expressions. This table should be used as a reference when simplifying radical terms. Students are not encouraged to memorize this table. Following are a few examples on simplifying radical expressions using Table 1.2-1:

a. $\sqrt{64}=\sqrt{8^2}=8$

b. $-2\sqrt{25}=-2\sqrt{5^2}=-(2\cdot 5)=-10$

c. $5\sqrt[5]{32}=5\sqrt[5]{2^5}=(5\cdot 2)=10$

d. $\sqrt{125}=\sqrt{25\cdot 5}=\sqrt{5^2\cdot 5}=5\sqrt{5}$

e. $\sqrt[2]{147}=\sqrt{49\cdot 3}=\sqrt{7^2\cdot 3}=7\sqrt{3}$

f. $2\sqrt{32}=2\sqrt{16\cdot 2}=2\sqrt{4^2\cdot 2}=(2\cdot 4)\sqrt{2}=8\sqrt{2}$

g. $\sqrt[5]{2048}=\sqrt[5]{1024\cdot 2}=\sqrt[5]{4^5\cdot 2}=4\sqrt[5]{2}$

h. $\sqrt[3]{375}=\sqrt[3]{125\cdot 3}=\sqrt[3]{5^3\cdot 3}=5\sqrt[3]{3}$

i. $2\sqrt{250}=2\sqrt{25\cdot 10}=2\sqrt{5^2\cdot 10}=(2\cdot 5)\sqrt{10}=10\sqrt{10}$

j. $\sqrt[4]{324}=\sqrt[4]{81\cdot 4}=\sqrt[4]{3^4\cdot 4}=3\sqrt[4]{4}$

k. $\sqrt[3]{648}=\sqrt[3]{216\cdot 3}=\sqrt[3]{6^3\cdot 3}=6\sqrt[3]{3}$

l. $-\sqrt[2]{324}=-\sqrt[2]{81\cdot 4}=-\sqrt{9^2\cdot 2^2}=-(9\cdot 2)=-18$

Practice Problems - Roots and Radical Expressions

Section 1.2a Case I Practice Problems - Simplify the following radical expressions by using Table 1.2-1:

1. $\sqrt[2]{98}=$
2. $3\sqrt{75}=$
3. $\sqrt[3]{125}=$
4. $\sqrt[5]{3125}=$
5. $\sqrt[4]{162}=$
6. $\sqrt[2]{192}=$

Table 1.2-1: Square roots, cube roots, fourth roots, and fifth roots

Square Roots	***Cube Roots***
$\sqrt{1}=\sqrt{1^2}=(1)^{\frac{1}{2}}=\left(1^2\right)^{\frac{1}{2}}=1$ *Note:* $\sqrt[2]{a}=\sqrt{a}$	$\sqrt[3]{1}=\sqrt[3]{1^3}=(1)^{\frac{1}{3}}=\left(1^3\right)^{\frac{1}{3}}=1$
$\sqrt{4}=\sqrt{2^2}=(4)^{\frac{1}{2}}=\left(2^2\right)^{\frac{1}{2}}=2$	$\sqrt[3]{8}=\sqrt[3]{2^3}=(8)^{\frac{1}{3}}=\left(2^3\right)^{\frac{1}{3}}=2$
$\sqrt{9}=\sqrt{3^2}=(9)^{\frac{1}{2}}=\left(3^2\right)^{\frac{1}{2}}=3$	$\sqrt[3]{27}=\sqrt[3]{3^3}=(27)^{\frac{1}{3}}=\left(3^3\right)^{\frac{1}{3}}=3$
$\sqrt{16}=\sqrt{4^2}=(16)^{\frac{1}{2}}=\left(4^2\right)^{\frac{1}{2}}=4$	$\sqrt[3]{64}=\sqrt[3]{4^3}=(64)^{\frac{1}{3}}=\left(4^3\right)^{\frac{1}{3}}=4$
$\sqrt{25}=\sqrt{5^2}=(25)^{\frac{1}{2}}=\left(5^2\right)^{\frac{1}{2}}=5$	$\sqrt[3]{125}=\sqrt[3]{5^3}=(125)^{\frac{1}{3}}=\left(5^3\right)^{\frac{1}{3}}=5$
$\sqrt{36}=\sqrt{6^2}=(36)^{\frac{1}{2}}=\left(6^2\right)^{\frac{1}{2}}=6$	$\sqrt[3]{216}=\sqrt[3]{6^3}=(216)^{\frac{1}{3}}=\left(6^3\right)^{\frac{1}{3}}=6$
$\sqrt{49}=\sqrt{7^2}=(49)^{\frac{1}{2}}=\left(7^2\right)^{\frac{1}{2}}=7$	$\sqrt[3]{343}=\sqrt[3]{7^3}=(343)^{\frac{1}{3}}=\left(7^3\right)^{\frac{1}{3}}=7$
$\sqrt{64}=\sqrt{8^2}=(64)^{\frac{1}{2}}=\left(8^2\right)^{\frac{1}{2}}=8$	$\sqrt[3]{512}=\sqrt[3]{8^3}=(512)^{\frac{1}{3}}=\left(8^3\right)^{\frac{1}{3}}=8$
$\sqrt{81}=\sqrt{9^2}=(81)^{\frac{1}{2}}=\left(9^2\right)^{\frac{1}{2}}=9$	$\sqrt[3]{729}=\sqrt[3]{9^3}=(729)^{\frac{1}{3}}=\left(9^3\right)^{\frac{1}{3}}=9$
$\sqrt{100}=\sqrt{10^2}=(100)^{\frac{1}{2}}=\left(10^2\right)^{\frac{1}{2}}=10$	$\sqrt[3]{1000}=\sqrt[3]{10^3}=(1000)^{\frac{1}{3}}=\left(10^3\right)^{\frac{1}{3}}=10$
Fourth Roots	***Fifth Roots***
$\sqrt[4]{1}=\sqrt[4]{1^4}=(1)^{\frac{1}{4}}=\left(1^4\right)^{\frac{1}{4}}=1$	$\sqrt[5]{1}=\sqrt[5]{1^5}=(1)^{\frac{1}{5}}=\left(1^5\right)^{\frac{1}{5}}=1$
$\sqrt[4]{16}=\sqrt[4]{2^4}=(16)^{\frac{1}{4}}=\left(2^4\right)^{\frac{1}{4}}=2$	$\sqrt[5]{32}=\sqrt[5]{2^5}=(32)^{\frac{1}{5}}=\left(2^5\right)^{\frac{1}{5}}=2$
$\sqrt[4]{81}=\sqrt[4]{3^4}=(81)^{\frac{1}{4}}=\left(3^4\right)^{\frac{1}{4}}=3$	$\sqrt[5]{243}=\sqrt[5]{3^5}=(243)^{\frac{1}{5}}=\left(3^5\right)^{\frac{1}{5}}=3$
$\sqrt[4]{256}=\sqrt[4]{4^4}=(256)^{\frac{1}{4}}=\left(4^4\right)^{\frac{1}{4}}=4$	$\sqrt[5]{1024}=\sqrt[5]{4^5}=(1024)^{\frac{1}{5}}=\left(4^5\right)^{\frac{1}{5}}=4$
$\sqrt[4]{625}=\sqrt[4]{5^4}=(625)^{\frac{1}{4}}=\left(5^4\right)^{\frac{1}{4}}=5$	$\sqrt[5]{3125}=\sqrt[5]{5^5}=(3125)^{\frac{1}{5}}=\left(5^5\right)^{\frac{1}{5}}=5$
$\sqrt[4]{1296}=\sqrt[4]{6^4}=(1296)^{\frac{1}{4}}=\left(6^4\right)^{\frac{1}{4}}=6$	$\sqrt[5]{7776}=\sqrt[5]{6^5}=(7776)^{\frac{1}{5}}=\left(6^5\right)^{\frac{1}{5}}=6$
$\sqrt[4]{2401}=\sqrt[4]{7^4}=(2401)^{\frac{1}{4}}=\left(7^4\right)^{\frac{1}{4}}=7$	$\sqrt[5]{16807}=\sqrt[5]{7^5}=(16807)^{\frac{1}{5}}=\left(7^5\right)^{\frac{1}{5}}=7$
$\sqrt[4]{4096}=\sqrt[4]{8^4}=(4096)^{\frac{1}{4}}=\left(8^4\right)^{\frac{1}{4}}=8$	$\sqrt[5]{32768}=\sqrt[5]{8^5}=(32768)^{\frac{1}{5}}=\left(8^5\right)^{\frac{1}{5}}=8$
$\sqrt[4]{6561}=\sqrt[4]{9^4}=(6561)^{\frac{1}{4}}=\left(9^4\right)^{\frac{1}{4}}=9$	$\sqrt[5]{59049}=\sqrt[5]{9^5}=(59049)^{\frac{1}{5}}=\left(9^5\right)^{\frac{1}{5}}=9$
$\sqrt[4]{10000}=\sqrt[4]{10^4}=(10000)^{\frac{1}{4}}=\left(10^4\right)^{\frac{1}{4}}=10$	$\sqrt[5]{100000}=\sqrt[5]{10^5}=(100000)^{\frac{1}{5}}=\left(10^5\right)^{\frac{1}{5}}=10$

Case II Rational, Irrational, Real, and Imaginary Numbers

A **rational number** is a number that **can** be expressed as:

1. An integer fraction $\frac{a}{b}$, where a and b are integer numbers and $b \neq 0$. For example: $\frac{3}{8}$, $-\frac{4}{5}$, $\frac{25}{100}$, and $\frac{2}{7}$ are rational numbers.
2. The square root of a perfect square, the cube root of a perfect cube, etc. For example: $\sqrt{36} = \sqrt{6^2} = 6$, $\sqrt{49} = \sqrt{7^2} = 7$, $-\sqrt[3]{125} = -\sqrt[3]{5^3} = -5$, $\sqrt[4]{81} = \sqrt[4]{3^4} = 3$, and $-\sqrt[5]{1024} = -\sqrt[5]{4^5} = -4$ are rational numbers.
3. An integer (a whole number). For example: $5 = \frac{5}{1}$, 0, $\frac{0}{28} = 0$, and 10 are rational numbers.
4. A terminating decimal. For example: $0.25 = \frac{25}{100}$, -0.75, and $5.5 = -5\frac{1}{2}$ are rational numbers.
5. A repeating decimal. For example: $0.3333333... = \frac{1}{3}$, $0.45454545...$, and are rational numbers.

An **irrational number** is a number that:

1. Can not be expressed as an integer fraction $\frac{a}{b}$, where a and b are integer numbers and $b \neq 0$. For example: π, $\frac{2}{\sqrt{2}}$, and $-\frac{5}{\sqrt{3}}$ are irrational numbers.
2. Can not be expressed as the square root of a perfect square, the cube root of a perfect cube, etc. For example: $\sqrt{5}$, $-\sqrt{7}$, $\sqrt{12}$, $\sqrt[3]{4}$, $-\sqrt[5]{6}$, and $\sqrt{3}$ are irrational numbers.
3. Is not a terminating or repeating decimal. For example: $0.432643...$, $-8.346723...$, and $3.14159...$ are irrational numbers.

The **real numbers** consist of all the rational and irrational numbers. For example: π, $\frac{2}{\sqrt{2}}$, $-\sqrt{7}$, $\sqrt[3]{4}$, $-\sqrt[5]{6}$, $\sqrt{3}$, $\sqrt{36} = \sqrt{6^2} = 6$, $0.25 = \frac{25}{100}$, -0.75, $-5.5 = -5\frac{1}{2}$, $-3.8 = -3\frac{4}{5}$, $5 = \frac{5}{1}$, 0, and $-25 = -\frac{25}{1}$ are real numbers.

The **not real numbers** or **imaginary numbers** are square root of any negative real number. For example: $\sqrt{-15}$, $\sqrt{-9}$, $\sqrt{-45}$, and $\sqrt{-36}$ are imaginary numbers. Note that imaginary numbers are also shown as $\sqrt{-15} = \sqrt{15}\,i$, $\sqrt{-9} = \sqrt{9}\,i = 3i$, $\sqrt{-45} = \sqrt{45}\,i$, and $\sqrt{-36} = \sqrt{36}\,i = 6i$ in advanced math books (See Chapter 2, Sections 2.5 and 2.6).

Practice Problems - Rational, Irrational, Real, and Imaginary Numbers

Section 1.2a Case II Practice Problems - Identify which one of the following numbers are rational, irrational, real, or not real:

1. $\frac{5}{8} =$
2. $\sqrt{45} =$
3. $450 =$
4. $-\frac{2}{\sqrt{10}} =$
5. $-\sqrt{-5} =$
6. $\frac{\sqrt{5}}{-2} =$

Case III Simplifying Radical Expressions with Real Numbers as a Radicand

Radical expressions with a real number as radicand are simplified using the following general rule:

$$\sqrt[n]{a^n} = a^{\frac{n}{n}} = a \qquad \text{The } n^{th} \text{ root of } a^n \text{ is } a$$

Where a is a positive real number and n is an integer.

Radicals of the form $\sqrt[n]{a^n} = a$ are simplified using the following steps:

Step 1 Factor out the radicand a^n to a perfect square, cube, fourth, fifth, etc. term (use Table 1.2-1). Write any term under the radical that exceeds the index n as multiple sum of the index.

Step 2 Use the Multiplication Law for exponents (see Section 1.1) by writing a^{m+n} in the form of $a^m \cdot a^n$.

Step 3 Simplify the radical expression by using the general rule $\sqrt[n]{a^n} = a$. Note that any term under the radical which is less than the index n stays inside the radical.

Examples with Steps

The following examples show the steps as to how radical expressions with real terms are simplified:

Example 1.2-1

$$\boxed{\frac{1}{-8}\sqrt[2]{72}} =$$

Solution:

Step 1

$$\boxed{\frac{1}{-8}\sqrt[2]{72}} = \boxed{-\frac{1}{8}\sqrt{72}} = \boxed{-\frac{1}{8}\sqrt{36\cdot 2}} = \boxed{-\frac{1}{8}\sqrt{(6\cdot 6)\cdot 2}} = \boxed{-\frac{1}{8}\sqrt{\left(6^1\cdot 6^1\right)\cdot 2}}$$

$$= \boxed{-\frac{1}{8}\sqrt{\left(6^{1+1}\right)\cdot 2}} = \boxed{-\frac{1}{8}\sqrt{6^2\cdot 2}}$$

Step 2

Not Applicable

Step 3

$$\boxed{-\frac{1}{8}\sqrt{6^2\cdot 2}} = \boxed{-\frac{1}{8}\cdot 6\sqrt{2}} = \boxed{-\frac{\overset{3}{\cancel{6}}}{\underset{4}{\cancel{8}}}\sqrt{2}} = \boxed{-\frac{3}{4}\sqrt{2}}$$

Example 1.2-2

$$\boxed{\frac{-3}{10}\sqrt[4]{5^7\cdot 4}} =$$

Solution:

Step 1

$$\boxed{\frac{-3}{10}\sqrt[4]{5^7\cdot 4}} = \boxed{-\frac{3}{10}\sqrt[4]{5^{4+3}\cdot 4}}$$

Step 2 $\boxed{-\frac{3}{10}\sqrt[4]{5^{4+3}\cdot4}}=\boxed{-\frac{3}{10}\sqrt[4]{5^4\cdot5^3\cdot4}}$

Step 3 $\boxed{-\frac{3}{10}\sqrt[4]{5^4\cdot5^3\cdot4}}=\boxed{-\frac{3}{10}\cdot5\sqrt[4]{5^3\cdot4}}=\boxed{-\frac{\overset{3}{\cancel{15}}}{\underset{2}{\cancel{10}}}\sqrt[4]{5^3\cdot4}}=\boxed{-\frac{3}{2}\sqrt[4]{125\cdot4}}=\boxed{-\frac{3}{2}\sqrt[4]{500}}$

Additional Examples - Simplifying Radical Expressions with Real Numbers as a Radicand

The following examples further illustrate how to solve radical expressions with real numbers as radicand:

Example 1.2-3

$\boxed{\sqrt[2]{25}}=\boxed{\sqrt{25}}=\boxed{\sqrt{5\cdot5}}=\boxed{\sqrt{5^1\cdot5^1}}=\boxed{\sqrt{5^{1+1}}}=\boxed{\sqrt{5^2}}=\boxed{5}$

Example 1.2-4

$\boxed{\sqrt[2]{18}}=\boxed{\sqrt{18}}=\boxed{\sqrt{9\cdot2}}=\boxed{\sqrt{(3\cdot3)\cdot2}}=\boxed{\sqrt{\left(3^1\cdot3^1\right)\cdot2}}=\boxed{\sqrt{\left(3^{1+1}\right)\cdot2}}=\boxed{\sqrt{3^2\cdot2}}=\boxed{3\sqrt{2}}$

Example 1.2-5

$\boxed{\sqrt[2]{125}}=\boxed{\sqrt{125}}=\boxed{\sqrt{25\cdot5}}=\boxed{\sqrt{(5\cdot5)\cdot5}}=\boxed{\sqrt{\left(5^1\cdot5^1\right)\cdot5}}=\boxed{\sqrt{\left(5^{1+1}\right)\cdot5}}=\boxed{\sqrt{5^2\cdot5}}=\boxed{5\sqrt{5}}$

Example 1.2-6

$\boxed{\sqrt[2]{2400}}=\boxed{\sqrt{2400}}=\boxed{\sqrt{400\cdot6}}=\boxed{\sqrt{(20\cdot20)\cdot6}}=\boxed{\sqrt{\left(20^1\cdot20^1\right)\cdot6}}=\boxed{\sqrt{\left(20^{1+1}\right)\cdot6}}=\boxed{\sqrt{20^2\cdot6}}=\boxed{20\sqrt{6}}$

Example 1.2-7

$\boxed{\sqrt[2]{120}}=\boxed{\sqrt{120}}=\boxed{\sqrt{4\cdot30}}=\boxed{\sqrt{(2\cdot2)\cdot30}}=\boxed{\sqrt{\left(2^1\cdot2^1\right)\cdot30}}=\boxed{\sqrt{\left(2^{1+1}\right)\cdot30}}=\boxed{\sqrt{2^2\cdot30}}=\boxed{2\sqrt{30}}$

Example 1.2-8

$\boxed{\frac{2}{15}\sqrt[2]{50}}=\boxed{\frac{2}{15}\sqrt{50}}=\boxed{\frac{2}{15}\sqrt{25\cdot2}}=\boxed{\frac{2}{15}\sqrt{(5\cdot5)\cdot2}}=\boxed{\frac{2}{15}\sqrt{\left(5^1\cdot5^1\right)\cdot2}}=\boxed{\frac{2}{15}\sqrt{\left(5^{1+1}\right)\cdot2}}=\boxed{\frac{2}{15}\sqrt{5^2\cdot2}}$

$=\boxed{\frac{2}{15}\cdot5\sqrt{2}}=\boxed{\frac{\overset{2}{\cancel{10}}}{\underset{3}{\cancel{15}}}\sqrt{2}}=\boxed{\frac{2}{3}\sqrt{2}}$

Practice Problems - Simplifying Radical Expressions with Real Numbers as a Radicand

Section 1.2a Case III Practice Problems - Simplify the following radical expressions:

1. $-\sqrt{49}=$ 2. $\sqrt{54}=$. $-\sqrt{500}=$

4. $\sqrt[5]{3^5\cdot5}=$ 5. $\sqrt[2]{216}=$. $-\frac{1}{4}\sqrt[4]{4^5\cdot2}=$

1.2b Operations Involving Radical Expressions

In this section multiplication, division, addition, and subtraction of monomial and binomial radical expressions, with real numbers, is addressed in Cases I through VI.

Case I Multiplying Monomial Expressions in Radical Form, with Real Numbers

Radicals are multiplied by each other by using the following general product rule:

$$a\sqrt[n]{x} \cdot b\sqrt[n]{y} \cdot c\sqrt[n]{z} = (a \cdot b \cdot c)\sqrt[n]{x \cdot y \cdot z} = abc\sqrt[n]{xyz}$$

Note that radicals can only be multiplied by each other if they have the same index n.

A monomial expression in radical form is defined as:

$$\sqrt{8x^5}\,,\ \sqrt{y}\,,\ \sqrt{27}\,,\ -3\sqrt[5]{x^6y^7}\,,\ \sqrt[3]{x^2y^5}\,,\ 2\sqrt{125}\,,\ etc.$$

A binomial expression in radical form is defined as:

$$\sqrt{x}+\sqrt{y}\,,\ 1+\sqrt{8x}\,,\ xy+\sqrt{y^3}\,,\ x^3y^3-\sqrt{x^2y}\,,\ 9-\sqrt[5]{x^5y^6}\,,\ \sqrt[3]{64}+\sqrt[3]{x^2y^5}\,,\ etc.$$

Monomial expressions in radical form are multiplied by each other using the above general product rule. Radical expressions with real numbers as radicands are multiplied by each other using the following steps:

Step 1 Simplify the radical terms (see Section 1.2a, Case III).

Step 2 Multiply the radical terms by using the product rule. Repeat Step 1, if necessary.

$$k_1\sqrt[n]{a} \cdot k_2\sqrt[n]{b} \cdot k_3\sqrt[n]{c} = (k_1 \cdot k_2 \cdot k_3)\sqrt[n]{a \cdot b \cdot c} = k_1k_2k_3\sqrt[n]{abc} \qquad a, b, and\ c \geq 0$$

Examples with Steps

The following examples show the steps as to how radical expressions in monomial form are multiplied by one another:

Example 1.2-9

$$\boxed{\sqrt{5} \cdot \sqrt{15}} =$$

Solution:

Step 1 $\boxed{Not\ Applicable}$

Step 2 $\boxed{\sqrt{5} \cdot \sqrt{15}} = \boxed{\sqrt{5 \cdot 15}} = \boxed{\sqrt{75}} = \boxed{\sqrt{25 \cdot 3}} = \boxed{\sqrt{5^2 \cdot 3}} = \boxed{5\sqrt{3}}$

Example 1.2-10

$$\boxed{\sqrt{98} \cdot \sqrt{48} \cdot \sqrt{108}} =$$

Solution:

Step 1 $\boxed{\sqrt{98} \cdot \sqrt{48} \cdot \sqrt{108}} = \boxed{\sqrt{49 \cdot 2} \cdot \sqrt{16 \cdot 3} \cdot \sqrt{36 \cdot 3}} = \boxed{\sqrt{7^2 \cdot 2} \cdot \sqrt{4^2 \cdot 3} \cdot \sqrt{6^2 \cdot 3}}$

$= \boxed{7\sqrt{2} \cdot 4\sqrt{3} \cdot 6\sqrt{3}}$

Step 2 $\boxed{7\sqrt{2} \cdot 4\sqrt{3} \cdot 6\sqrt{3}} = \boxed{(7 \cdot 4 \cdot 6) \cdot \left(\sqrt{2} \cdot \sqrt{3} \cdot \sqrt{3}\right)} = \boxed{168\left(\sqrt{2 \cdot 3 \cdot 3}\right)} = \boxed{168\sqrt{2 \cdot 3^2}}$

$= \boxed{(168\cdot 3)\sqrt{2}} = \boxed{\mathbf{504\sqrt{2}}}$

Additional Examples - Multiplying Monomial Expressions in Radical Form, with Real Numbers

The following examples further illustrate how to multiply radical terms by one another:

Example 1.2-11

$$\boxed{\sqrt{12}\cdot\sqrt{9}} = \boxed{\left(\sqrt{4\cdot 3}\right)\cdot\sqrt{3^2}} = \boxed{\left(\sqrt{2^2\cdot 3}\right)\cdot 3} = \boxed{2\sqrt{3}\cdot 3} = \boxed{(2\cdot 3)\sqrt{3}} = \boxed{\mathbf{6\sqrt{3}}}$$

Example 1.2-12

$$\boxed{\sqrt{20}\cdot\sqrt{50}} = \boxed{\sqrt{4\cdot 5}\cdot\sqrt{25\cdot 2}} = \boxed{\sqrt{2^2\cdot 5}\cdot\sqrt{5^2\cdot 2}} = \boxed{2\sqrt{5}\cdot 5\sqrt{2}} = \boxed{(2\cdot 5)\cdot\left(\sqrt{5\cdot 2}\right)} = \boxed{\mathbf{10\sqrt{10}}}$$

Note that we can also simplify the radical terms in the following way:

$$\boxed{\sqrt{20}\cdot\sqrt{50}} = \boxed{\sqrt{20\cdot 50}} = \boxed{\sqrt{1000}} = \boxed{\sqrt{100\cdot 10}} = \boxed{\sqrt{10^2\cdot 10}} = \boxed{\mathbf{10\sqrt{10}}}$$

Example 1.2-13

$$\boxed{\sqrt{50}\cdot\sqrt{32}\cdot\sqrt{3}} = \boxed{\sqrt{25\cdot 2}\cdot\sqrt{16\cdot 2}\cdot\sqrt{3}} = \boxed{\sqrt{5^2\cdot 2}\cdot\sqrt{4^2\cdot 2}\cdot\sqrt{3}} = \boxed{5\sqrt{2}\cdot 4\sqrt{2}\cdot\sqrt{3}} = \boxed{(5\cdot 4)\cdot\left(\sqrt{2}\cdot\sqrt{2}\cdot\sqrt{3}\right)}$$

$$= \boxed{20\cdot\left(\sqrt{2\cdot 2\cdot 3}\right)} = \boxed{20\cdot\left(\sqrt{2^1\cdot 2^1\cdot 3}\right)} = \boxed{20\cdot\left(\sqrt{2^{1+1}\cdot 3}\right)} = \boxed{20\sqrt{2^2\cdot 3}} = \boxed{(20\cdot 2)\sqrt{3}} = \boxed{\mathbf{40\sqrt{3}}}$$

Example 1.2-14

$$\boxed{\sqrt[3]{128}\cdot\sqrt[3]{500}} = \boxed{\sqrt[3]{2\cdot 64}\cdot\sqrt[3]{4\cdot 125}} = \boxed{\sqrt[3]{2\cdot 4^3}\cdot\sqrt[3]{4\cdot 5^3}} = \boxed{4\sqrt[3]{2}\cdot 5\sqrt[3]{4}} = \boxed{(4\cdot 5)\cdot\left(\sqrt[3]{2}\cdot\sqrt[3]{4}\right)} = \boxed{20\cdot\left(\sqrt[3]{2\cdot 4}\right)}$$

$$= \boxed{20\cdot\sqrt[3]{8}} = \boxed{20\cdot\sqrt[3]{2^3}} = \boxed{20\cdot 2} = \boxed{\mathbf{40}}$$

Example 1.2-15

$$\boxed{\sqrt[4]{54}\cdot\sqrt[4]{48}} = \boxed{\sqrt[4]{2\cdot 27}\cdot\sqrt[4]{3\cdot 16}} = \boxed{\sqrt[4]{2\cdot 3^3}\cdot\sqrt[4]{3\cdot 2^4}} = \boxed{\sqrt[4]{2\cdot 3^3}\cdot 2\sqrt[4]{3^1}} = \boxed{2\sqrt[4]{2\cdot 3^3\cdot 3^1}} = \boxed{2\sqrt[4]{2\cdot 3^{3+1}}} = \boxed{2\sqrt[4]{2\cdot 3^4}}$$

$$= \boxed{(2\cdot 3)\cdot\sqrt[4]{2}} = \boxed{\mathbf{6\sqrt[4]{2}}}$$

Practice Problems - Multiplying Monomial Expressions in Radical Form, with Real Numbers

Section 1.2b Case I Practice Problems - Multiply the following radical expressions:

1. $\sqrt{72}\cdot\sqrt{75} =$
2. $-3\sqrt{20}\cdot 2\sqrt{32} =$
3. $\sqrt[2]{16}\cdot\sqrt[2]{27} =$
4. $\sqrt{64}\cdot\sqrt{100}\cdot\sqrt{54} =$
5. $-\sqrt{125}\cdot -2\sqrt{98} =$
6. $\sqrt[4]{625}\cdot\sqrt[4]{324}\cdot\sqrt[4]{48} =$

Case II Multiplying Binomial Expressions in Radical Form, with Real Numbers

To multiply two binomial radical expressions the following multiplication method known as the **FOIL** method needs to be memorized:

$$(a+b)\cdot(c+d)=(a\cdot c)+(a\cdot d)+(b\cdot c)+(b\cdot d)$$

Multiply the **F**irst two terms, i.e., $(a\cdot c)$.

Multiply the **O**uter two terms, i.e., $(a\cdot d)$.

Multiply the **I**nner two terms, i.e., $(b\cdot c)$.

Multiply the **L**ast two terms, i.e., $(b\cdot d)$.

Examples:

1. $\boxed{\left(\sqrt{u}+\sqrt{v}\right)\cdot\left(\sqrt{u}-\sqrt{v}\right)}=\boxed{\left(\sqrt{u}\cdot\sqrt{u}\right)-\left(\sqrt{u}\sqrt{v}\right)+\left(\sqrt{v}\cdot\sqrt{u}\right)-\left(\sqrt{v}\cdot\sqrt{v}\right)}=\boxed{\left(\sqrt{u\cdot u}\right)-\left(\sqrt{u\cdot v}\right)+\left(\sqrt{u\cdot v}\right)-\left(\sqrt{v\cdot v}\right)}$

   $=\boxed{\sqrt{u^2}-\sqrt{v^2}}=\boxed{\mathbf{u-v}}$

2. $\boxed{\left(3-\sqrt{5}\right)\cdot\left(5+\sqrt{7}\right)}=\boxed{(3\cdot 5)+\left(3\cdot\sqrt{7}\right)-\left(5\cdot\sqrt{5}\right)-\left(\sqrt{5}\cdot\sqrt{7}\right)}=\boxed{15+3\sqrt{7}-5\sqrt{5}-\sqrt{5\cdot 7}}=\boxed{\mathbf{15+3\sqrt{7}-5\sqrt{5}-\sqrt{35}}}$

Binomial radical expressions are multiplied by each other using the following steps:

Step 1 Simplify the radical terms (see Section 1.2a, Case III).

Step 2 Use the FOIL method to multiply each term. Repeat Step 1, if necessary.

$$(a+b)\cdot(c+d)=(a\cdot c)+(a\cdot d)+(b\cdot c)+(b\cdot d)$$

Examples with Steps

The following examples show the steps as to how binomial radical expressions with real numbers as radicands are multiplied by one another:

Example 1.2-16

$$\boxed{\left(2+\sqrt{2}\right)\cdot\left(5-\sqrt{8}\right)}=$$

Solution:

Step 1 $\boxed{\left(2+\sqrt{2}\right)\cdot\left(5-\sqrt{8}\right)}=\boxed{\left(2+\sqrt{2}\right)\cdot\left(5-\sqrt{4\cdot 2}\right)}=\boxed{\left(2+\sqrt{2}\right)\cdot\left(5-\sqrt{2^2\cdot 2}\right)}$

$=\boxed{\left(2+\sqrt{2}\right)\cdot\left(5-2\sqrt{2}\right)}$

Step 2 $\boxed{\left(2+\sqrt{2}\right)\cdot\left(5-2\sqrt{2}\right)}=\boxed{(2\cdot 5)-\left(2\cdot 2\sqrt{2}\right)+\left(5\cdot\sqrt{2}\right)-\left(2\sqrt{2}\cdot\sqrt{2}\right)}$

$=\boxed{10-4\sqrt{2}+5\sqrt{2}-2\sqrt{2\cdot 2}}=\boxed{10-4\sqrt{2}+5\sqrt{2}-2\sqrt{2^2}}=\boxed{10+(-4+5)\sqrt{2}-(2\cdot 2)}$

$$= \boxed{10+\sqrt{2}-4} = \boxed{(10-4)+\sqrt{2}} = \boxed{\mathbf{6+\sqrt{2}}}$$

Example 1.2-17

$$\boxed{\left(\sqrt{24}+3\sqrt{60}\right)\cdot\left(\sqrt{25}-\sqrt{72}\right)} =$$

Solution:

Step 1

$$\boxed{\left(\sqrt{24}+3\sqrt{60}\right)\cdot\left(\sqrt{25}-\sqrt{72}\right)} = \boxed{\left(\sqrt{4\cdot 6}+3\sqrt{4\cdot 15}\right)\cdot\left(\sqrt{5\cdot 5}-\sqrt{36\cdot 2}\right)}$$

$$= \boxed{\left(\sqrt{2^2\cdot 6}+3\sqrt{2^2\cdot 15}\right)\cdot\left(\sqrt{5^2}-\sqrt{6^2\cdot 2}\right)} = \boxed{\left(2\sqrt{6}+(3\cdot 2)\sqrt{15}\right)\cdot\left(5-6\sqrt{2}\right)}$$

$$= \boxed{\left(2\sqrt{6}+6\sqrt{15}\right)\cdot\left(5-6\sqrt{2}\right)}$$

Step 2

$$\boxed{\left(2\sqrt{6}+6\sqrt{15}\right)\cdot\left(5-6\sqrt{2}\right)} = \boxed{\left(5\cdot 2\sqrt{6}\right)-\left(2\sqrt{6}\cdot 6\sqrt{2}\right)+\left(5\cdot 6\sqrt{15}\right)-\left(6\sqrt{15}\cdot 6\sqrt{2}\right)}$$

$$= \boxed{10\sqrt{6}-(2\cdot 6)\sqrt{6\cdot 2}+30\sqrt{15}-(6\cdot 6)\sqrt{15\cdot 2}} = \boxed{10\sqrt{6}-12\sqrt{12}+30\sqrt{15}-36\sqrt{30}}$$

$$= \boxed{10\sqrt{6}-12\sqrt{4\cdot 3}+30\sqrt{15}-36\sqrt{30}} = \boxed{10\sqrt{6}-12\sqrt{2^2\cdot 3}+30\sqrt{15}-36\sqrt{30}}$$

$$= \boxed{10\sqrt{6}-(12\cdot 2)\sqrt{3}+30\sqrt{15}-36\sqrt{30}} = \boxed{\mathbf{10\sqrt{6}-24\sqrt{3}+30\sqrt{15}-36\sqrt{30}}}$$

Additional Examples - Multiplying Binomial Expressions in Radical Form, with Real Numbers

The following examples further illustrate how to multiply radical expressions by one another:

Example 1.2-18

$$\boxed{\left(3+\sqrt{300}\right)\cdot\left(8-\sqrt{50}\right)} = \boxed{\left(3+\sqrt{100\cdot 3}\right)\cdot\left(8-\sqrt{25\cdot 2}\right)} = \boxed{\left(3+\sqrt{10^2\cdot 3}\right)\cdot\left(8-\sqrt{5^2\cdot 2}\right)} = \boxed{\left(3+10\sqrt{3}\right)\cdot\left(8-5\sqrt{2}\right)}$$

$$= \boxed{(3\cdot 8)-(3\cdot 5)\sqrt{2}+(8\cdot 10)\sqrt{3}-(10\cdot 5)\sqrt{3}\cdot\sqrt{2}} = \boxed{24-15\sqrt{2}+80\sqrt{3}-50\sqrt{3\cdot 2}} = \boxed{\mathbf{-15\sqrt{2}+80\sqrt{3}-50\sqrt{6}+24}}$$

Example 1.2-19

$$\boxed{\left(3+\sqrt{12}\right)\cdot\left(\sqrt{75}-\sqrt{2}\right)} = \boxed{\left(3+\sqrt{4\cdot 3}\right)\cdot\left(\sqrt{25\cdot 3}-\sqrt{2}\right)} = \boxed{\left(3+\sqrt{2^2\cdot 3}\right)\cdot\left(\sqrt{5^2\cdot 3}-\sqrt{2}\right)}$$

$$= \boxed{\left(3+2\sqrt{3}\right)\cdot\left(5\sqrt{3}-\sqrt{2}\right)} = \boxed{(3\cdot 5)\sqrt{3}-\left(3\cdot\sqrt{2}\right)+(2\cdot 5)\left(\sqrt{3}\cdot\sqrt{3}\right)-2\left(\sqrt{3}\cdot\sqrt{2}\right)} = \boxed{15\sqrt{3}-3\sqrt{2}+10\sqrt{3\cdot 3}-2\sqrt{3\cdot 2}}$$

$$= \boxed{15\sqrt{3}-3\sqrt{2}+10\sqrt{3^2}-2\sqrt{6}} = \boxed{15\sqrt{3}-3\sqrt{2}+(10\cdot 3)-2\sqrt{6}} = \boxed{\mathbf{15\sqrt{3}-3\sqrt{2}-2\sqrt{6}+30}}$$

Example 1.2-20

$$\boxed{\left(\sqrt[3]{3\cdot 5^3}-\sqrt[3]{4\cdot 2^3}\right)\cdot\left(\sqrt[3]{3}-\sqrt[3]{2\cdot 2^3}\right)} = \boxed{\left(5\sqrt[3]{3}-2\sqrt[3]{4}\right)\cdot\left(\sqrt[3]{3}-2\sqrt[3]{2}\right)}$$

$= \boxed{\left(5\sqrt[3]{3}\cdot\sqrt[3]{3}\right)-(5\cdot 2)\left(\sqrt[3]{3}\cdot\sqrt[3]{2}\right)-\left(2\sqrt[3]{4}\cdot\sqrt[3]{3}\right)+(2\cdot 2)\left(\sqrt[3]{4}\cdot\sqrt[3]{2}\right)} = \boxed{\left(5\sqrt[3]{3\cdot 3}\right)-10\left(\sqrt[3]{3\cdot 2}\right)-\left(2\sqrt[3]{4\cdot 3}\right)+4\left(\sqrt[3]{4\cdot 2}\right)}$

$= \boxed{5\sqrt[3]{9}-10\sqrt[3]{6}-2\sqrt[3]{12}+4\sqrt[3]{8}} = \boxed{5\sqrt[3]{9}-10\sqrt[3]{6}-2\sqrt[3]{12}+4\sqrt[3]{2^3}} = \boxed{5\sqrt[3]{9}-10\sqrt[3]{6}-2\sqrt[3]{12}+(4\cdot 2)}$

$= \boxed{\mathbf{5\sqrt[3]{9}-10\sqrt[3]{6}-2\sqrt[3]{12}+8}}$

Example 1.2-21

$\boxed{\left(6\sqrt{48}+2\right)\cdot\left(2\sqrt{18}-4\right)} = \boxed{\left(6\sqrt{3\cdot 16}+2\right)\cdot\left(2\sqrt{2\cdot 9}-4\right)} = \boxed{\left(6\sqrt{3\cdot 4^2}+2\right)\cdot\left(2\sqrt{2\cdot 3^2}-4\right)}$

$= \boxed{\left(6\cdot 4\sqrt{3}+2\right)\cdot\left(2\cdot 3\sqrt{2}-4\right)} = \boxed{\left(24\sqrt{3}+2\right)\cdot\left(6\sqrt{2}-4\right)} = \boxed{\left(24\sqrt{3}\cdot 6\sqrt{2}\right)-\left(4\cdot 24\sqrt{3}\right)+\left(2\cdot 6\sqrt{2}\right)-(2\cdot 4)}$

$= \boxed{\left(24\cdot 6\sqrt{3\cdot 2}\right)-\left(96\sqrt{3}\right)+\left(12\sqrt{2}\right)-8} = \boxed{\mathbf{144\sqrt{6}-96\sqrt{3}+12\sqrt{2}-8}}$

Example 1.2-22

$\boxed{\left[\left(-\sqrt{3}+2\right)\cdot\left(3-\sqrt{3}\right)\right]\cdot\left(\sqrt{3}-4\right)} = \boxed{\left[\left(-3\cdot\sqrt{3}\right)+\left(\sqrt{3}\cdot\sqrt{3}\right)+(2\cdot 3)-\left(2\cdot\sqrt{3}\right)\right]\cdot\left(\sqrt{3}-4\right)}$

$= \boxed{\left[-3\sqrt{3}+\sqrt{3\cdot 3}+6-2\sqrt{3}\right]\cdot\left(\sqrt{3}-4\right)} = \boxed{\left[\left(-3\sqrt{3}-2\sqrt{3}\right)+\sqrt{3^2}+6\right]\cdot\left(\sqrt{3}-4\right)} = \boxed{\left[-5\sqrt{3}+3+6\right]\cdot\left(\sqrt{3}-4\right)}$

$= \boxed{\left[-5\sqrt{3}+9\right]\cdot\left(\sqrt{3}-4\right)} = \boxed{\left(-5\sqrt{3}\cdot\sqrt{3}\right)+(5\cdot 4)\sqrt{3}+9\cdot\sqrt{3}-(9\cdot 4)} = \boxed{-5\sqrt{3\cdot 3}+20\sqrt{3}+9\sqrt{3}-36}$

$= \boxed{-5\sqrt{3^2}+(20+9)\sqrt{3}-36} = \boxed{-(5\cdot 3)+29\sqrt{3}-36} = \boxed{-15-36+29\sqrt{3}} = \boxed{\mathbf{-51+29\sqrt{3}}}$

Practice Problems - Multiplying Binomial Expressions in Radical Form, with Real Numbers

Section 1.2b Case II Practice Problems - Multiply the following radical expressions:

1. $\left(2\sqrt{3}+1\right)\cdot\left(2+\sqrt{2}\right) =$
2. $\left(1+\sqrt{5}\right)\cdot\left(\sqrt{8}+\sqrt{5}\right) =$
3. $\left(2-\sqrt{2}\right)\cdot\left(3+\sqrt{2}\right) =$
4. $\left(5+\sqrt{5}\right)\cdot\left(5-\sqrt{5^3}\right) =$
5. $\left(2+\sqrt{6}\right)\cdot\left(\sqrt[4]{16}-\sqrt{18}\right) =$
6. $\left(2-\sqrt{5}\right)\cdot\left(\sqrt{45}+\sqrt[4]{81}\right) =$

Case III Multiplying Monomial and Binomial Expressions in Radical Form, with Real Numbers

To multiply monomial and binomial expressions in radical form the following general multiplication rule is used:

$$a \cdot (b+c) = a \cdot b + a \cdot c$$

Monomial and binomial expressions in radical form are multiplied by each other using the following steps:

Step 1 Simplify the radical terms (see Section 1.2a, Case III).

Step 2 Multiply each term using the general multiplication rule, i.e., $a \cdot (b+c) = a \cdot b + a \cdot c$. Repeat Step 1, if necessary.

Examples with Steps

The following examples show the steps as to how monomial and binomial expressions in radical form are multiplied by one another:

Example 1.2-23

$$\sqrt{5} \cdot \left(\sqrt{50} + 2\sqrt{27}\right) =$$

Solution:

Step 1

$$\sqrt{5} \cdot \left(\sqrt{50} + 2\sqrt{27}\right) = \sqrt{5} \cdot \left(\sqrt{25 \cdot 2} + 2\sqrt{9 \cdot 3}\right) = \sqrt{5} \cdot \left(\sqrt{5^2 \cdot 2} + 2\sqrt{3^2 \cdot 3}\right)$$

$$= \sqrt{5} \cdot \left[5\sqrt{2} + (2 \cdot 3)\sqrt{3}\right] = \sqrt{5} \cdot \left[5\sqrt{2} + 6\sqrt{3}\right]$$

Step 2

$$\sqrt{5} \cdot \left[5\sqrt{2} + 6\sqrt{3}\right] = 5\left(\sqrt{2} \cdot \sqrt{5}\right) + 6\left(\sqrt{5} \cdot \sqrt{3}\right) = 5\sqrt{2 \cdot 5} + 6\sqrt{5 \cdot 3} = \mathbf{5\sqrt{10} + 6\sqrt{15}}$$

Example 1.2-24

$$-2\sqrt[4]{4} \cdot \left(\sqrt[4]{64} - \sqrt[4]{162}\right) =$$

Solution:

Step 1

$$-2\sqrt[4]{4} \cdot \left(\sqrt[4]{64} - \sqrt[4]{162}\right) = -2\sqrt[4]{4} \cdot \left(\sqrt[4]{16 \cdot 4} - \sqrt[4]{81 \cdot 2}\right) = -2\sqrt[4]{4} \cdot \left(\sqrt[4]{2^4 \cdot 4} - \sqrt[4]{3^4 \cdot 2}\right)$$

$$= -2\sqrt[4]{4} \cdot \left(2\sqrt[4]{4} - 3\sqrt[4]{2}\right)$$

Step 2

$$-2\sqrt[4]{4} \cdot \left(2\sqrt[4]{4} - 3\sqrt[4]{2}\right) = -(2 \cdot 2) \cdot \left(\sqrt[4]{4} \cdot \sqrt[4]{4}\right) + (2 \cdot 3) \cdot \left(\sqrt[4]{4} \cdot \sqrt[4]{2}\right) = -4\left(\sqrt[4]{4 \cdot 4}\right) + 6\left(\sqrt[4]{4 \cdot 2}\right)$$

$$= -4\sqrt[4]{16} + 6\sqrt[4]{8} = -4\sqrt[4]{2^4} + 6\sqrt[4]{8} = -(4 \cdot 2) + 6\sqrt[4]{8} = -8 + 6\sqrt[4]{8} = \mathbf{2\left(3\sqrt[4]{8} - 4\right)}$$

Additional Examples - Multiplying Monomial and Binomial Expressions in Radical Form, with Real Numbers

The following examples further illustrate how to multiply radical terms by one another:

Example 1.2-25

$$3\sqrt{5}\cdot\left(\sqrt{5}+6\sqrt{10}\right) = \left(3\sqrt{5}\cdot\sqrt{5}\right)+\left(3\sqrt{5}\cdot 6\sqrt{10}\right) = \left(3\sqrt{5\cdot 5}\right)+(3\cdot 6)\sqrt{5\cdot 10} = 3\sqrt{5^2}+18\sqrt{50}$$

$$= (3\cdot 5)+18\sqrt{25\cdot 2} = 15+18\sqrt{5^2\cdot 2} = 15+(18\cdot 5)\sqrt{2} = 15+90\sqrt{2} = \mathbf{15\left(1+6\sqrt{2}\right)}$$

Example 1.2-26

$$-2\sqrt{6}\cdot\left(-\sqrt{5}+\sqrt{50}\right) = \left(+2\sqrt{6}\cdot\sqrt{5}\right)-\left(2\sqrt{6}\cdot\sqrt{50}\right) = 2\sqrt{6\cdot 5}-2\sqrt{6\cdot 50} = 2\sqrt{30}-2\sqrt{300} = 2\sqrt{30}-2\sqrt{3\cdot 100}$$

$$= 2\sqrt{30}-2\sqrt{10^2\cdot 3} = 2\sqrt{30}-(2\cdot 10)\sqrt{3} = 2\sqrt{30}-20\sqrt{3} = \mathbf{2\left(\sqrt{30}-10\sqrt{3}\right)}$$

Example 1.2-27

$$\sqrt{3}\cdot\left(2\sqrt{3}+\sqrt{6}\right) = \left(2\sqrt{3}\cdot\sqrt{3}\right)+\left(\sqrt{3}\cdot\sqrt{6}\right) = \left(2\sqrt{3\cdot 3}\right)+\left(\sqrt{3\cdot 6}\right) = 2\sqrt{3^1\cdot 3^1}+\sqrt{18} = 2\sqrt{3^{1+1}}+\sqrt{9\cdot 2}$$

$$= 2\sqrt{3^2}+\sqrt{3^2\cdot 2} = (2\cdot 3)+3\sqrt{2} = 6+3\sqrt{2} = \mathbf{3\left(2+\sqrt{2}\right)}$$

Example 1.2-28

$$3\sqrt{2}\cdot\left(\sqrt{10}+4\sqrt{20}\right) = \left(3\sqrt{2}\cdot\sqrt{10}\right)+\left(3\sqrt{2}\cdot 4\sqrt{20}\right) = \left(3\sqrt{2\cdot 10}\right)+(3\cdot 4)\sqrt{2\cdot 20} = 3\sqrt{20}+12\sqrt{40}$$

$$= 3\sqrt{4\cdot 5}+12\sqrt{4\cdot 10} = 3\sqrt{2^2\cdot 5}+12\sqrt{2^2\cdot 10} = (3\cdot 2)\sqrt{5}+(12\cdot 2)\sqrt{10} = 6\sqrt{5}+24\sqrt{10} = \mathbf{6\left(\sqrt{5}+4\sqrt{10}\right)}$$

Example 1.2-29

$$\sqrt[3]{5}\cdot\left(\sqrt[3]{25}-\sqrt[3]{216}\right) = \sqrt[3]{5}\cdot\left(\sqrt[3]{5^2}-\sqrt[3]{6^3}\right) = \left(\sqrt[3]{5}\cdot\sqrt[3]{5^2}\right)-\left(\sqrt[3]{5}\cdot\sqrt[3]{6^3}\right) = \left(\sqrt[3]{5\cdot 5^2}\right)-\left(\sqrt[3]{5}\cdot 6\right) = \sqrt[3]{5^1\cdot 5^2}-6\sqrt[3]{5}$$

$$= \sqrt[3]{5^{1+2}}-6\sqrt[3]{5} = \sqrt[3]{5^3}-6\sqrt[3]{5} = \mathbf{5-6\sqrt[3]{5}}$$

Practice Problems - Multiplying Monomial and Binomial Expressions in Radical Form, with Real Numbers

Section 1.2b Case III Practice Problems - Multiply the following radical expressions:

1. $2\sqrt{3}\cdot\left(2+\sqrt{2}\right) =$
2. $\sqrt{5}\cdot\left(\sqrt{8}+\sqrt{5}\right) =$
3. $-\sqrt{8}\cdot\left(3-\sqrt{3}\right) =$
4. $4\sqrt{98}\cdot\left(3-\sqrt{2^3}\right) =$
5. $\sqrt[4]{48}\cdot\left(\sqrt[4]{324}+\sqrt[4]{32}\right) =$
6. $2\sqrt{5}\cdot\left(\sqrt{45}+\sqrt[4]{81}\right) =$

Case IV Rationalizing Radical Expressions - Monomial Denominators with Real Numbers

Radicals are divided by each other using the following general rule:

$$\sqrt[n]{\frac{x}{y}} = \frac{\sqrt[n]{x}}{\sqrt[n]{y}} \qquad x \ge 0,\ y > 0$$

In section 1.2a the difference between rational and irrational numbers was discussed. We learned that the square root of non perfect squares, the cube root of non perfect cubes, etc. are irrational numbers. For example, $\sqrt{3}, \sqrt{7}, \sqrt{10}, \sqrt[3]{4}, \sqrt[5]{7}$, *etc.* are classified as irrational numbers. In division of radicals, if the denominator of a fractional radical expression is not a rational number, we rationalize the denominator by changing the radicand of the denominator to a perfect square, a perfect cube, etc.

Simplification of radical expressions being divided requires rationalization of the denominator. A monomial and irrational denominator is rationalized by multiplying the numerator and the denominator by the irrational denominator. This change the radicand of the denominator to a perfect square.

Examples:

1. $\frac{\sqrt{1}}{\sqrt{7}} = \frac{\sqrt{1}}{\sqrt{7}} \times \frac{\sqrt{7}}{\sqrt{7}} = \frac{\sqrt{1 \cdot 7}}{\sqrt{7 \cdot 7}} = \frac{\sqrt{7}}{\sqrt{7^2}} = \mathbf{\frac{\sqrt{7}}{7}}$

   Note that $\sqrt{7}$ is an irrational number. By multiplying $\sqrt{7}$ by itself the denominator is changed to a rational number, i.e., 7.

2. $\sqrt{\frac{20}{3}} = \frac{\sqrt{20}}{\sqrt{3}} = \frac{\sqrt{4 \cdot 5}}{\sqrt{3}} = \frac{\sqrt{2^2 \cdot 5}}{\sqrt{3}} = \frac{2\sqrt{5}}{\sqrt{3}} = \frac{2\sqrt{5}}{\sqrt{3}} \times \frac{\sqrt{3}}{\sqrt{3}} = \frac{2\sqrt{5 \cdot 3}}{\sqrt{3} \cdot \sqrt{3}} = \frac{2\sqrt{15}}{\sqrt{3^2}} = \mathbf{\frac{2\sqrt{15}}{3}}$

   Again, note that $\sqrt{3}$ is an irrational number. By multiplying $\sqrt{3}$ by itself the denominator is changed to a rational number, i.e., 3.

Radical expressions with monomial denominators are simplified using the following steps:

Step 1 Change the radical expression $\sqrt{\frac{a}{b}}$ to $\frac{\sqrt{a}}{\sqrt{b}}$ and simplify.

Step 2 Rationalize the denominator by multiplying the numerator and the denominator of the radical expression $\frac{\sqrt{a}}{\sqrt{b}}$ by $\sqrt{b}$.

Step 3 Simplify the radical expression (see Section 1.2a, Case III).

Examples with Steps

The following examples show the steps as to how radical expressions with monomial denominators are simplified:

Example 1.2-30

$$\boxed{\frac{-8\sqrt{3}}{32\sqrt{45}}} =$$

Solution:

Step 1 $\frac{-8\sqrt{3}}{32\sqrt{45}} = -\frac{\overset{1}{\cancel{8}}\sqrt{3}}{\underset{4}{\cancel{32}}\sqrt{45}} = -\frac{\sqrt{3}}{4\sqrt{9\cdot5}} = -\frac{\sqrt{3}}{4\sqrt{3^2\cdot5}} = -\frac{\sqrt{3}}{4\cdot3\sqrt{5}} = -\frac{\sqrt{3}}{12\sqrt{5}}$

Step 2 $-\frac{\sqrt{3}}{12\sqrt{5}} = -\frac{\sqrt{3}}{12\sqrt{5}}\times\frac{\sqrt{5}}{\sqrt{5}}$

Step 3 $-\frac{\sqrt{3}}{12\sqrt{5}}\times\frac{\sqrt{5}}{\sqrt{5}} = -\frac{\sqrt{3}\times\sqrt{5}}{12\sqrt{5}\times\sqrt{5}} = -\frac{\sqrt{3\cdot5}}{12\sqrt{5\cdot5}} = -\frac{\sqrt{15}}{12\sqrt{5^1\cdot5^1}} = -\frac{\sqrt{15}}{12\sqrt{5^{1+1}}}$

$= -\frac{\sqrt{15}}{12\sqrt{5^2}} = -\frac{\sqrt{15}}{12\cdot5} = \mathbf{-\frac{\sqrt{15}}{60}}$

Example 1.2-31

$\frac{3\sqrt[5]{8}}{\sqrt[5]{81}} =$

Solution:

Step 1 $\frac{3\sqrt[5]{8}}{\sqrt[5]{81}} = \frac{3\sqrt[5]{8}}{\sqrt[5]{3^4}}$

Step 2 $\frac{3\sqrt[5]{8}}{\sqrt[5]{3^4}} = \frac{3\sqrt[5]{8}}{\sqrt[5]{3^4}}\times\frac{\sqrt[5]{3^1}}{\sqrt[5]{3^1}}$

Note that radical expressions with third, fourth, or higher root in the denominator can also be rationalized by changing the denominator to a perfect third, fourth, or higher power.

Step 3 $\frac{3\sqrt[5]{8}}{\sqrt[5]{3^4}}\times\frac{\sqrt[5]{3^1}}{\sqrt[5]{3^1}} = \frac{3\sqrt[5]{8}\times\sqrt[5]{3^1}}{\sqrt[5]{3^4}\times\sqrt[5]{3^1}} = \frac{3\sqrt[5]{8\cdot3}}{\sqrt[5]{3^4\cdot3^1}} = \frac{3\sqrt[5]{24}}{\sqrt[5]{3^{4+1}}} = \frac{3\sqrt[5]{24}}{\sqrt[5]{3^5}} = \frac{\overset{1}{\cancel{3}}\cdot\sqrt[5]{24}}{\underset{1}{\cancel{3}}} = \frac{1\cdot\sqrt[5]{24}}{1}$

$= \mathbf{\sqrt[5]{24}}$

Additional Examples: Rationalizing Radical Expressions - Monomial Denominators with Real Numbers

The following examples further illustrate how to solve radical expressions with monomial denominators:

Example 1.2-32

$\frac{8\sqrt{3}}{\sqrt{2}} = \frac{8\sqrt{3}}{\sqrt{2}}\times\frac{\sqrt{2}}{\sqrt{2}} = \frac{8\sqrt{3}\times\sqrt{2}}{\sqrt{2}\times\sqrt{2}} = \frac{8\sqrt{3\cdot2}}{\sqrt{2\cdot2}} = \frac{8\sqrt{6}}{\sqrt{2^1\cdot2^1}} = \frac{8\sqrt{6}}{\sqrt{2^{1+1}}} = \frac{8\sqrt{6}}{\sqrt{2^2}} = \frac{\overset{4}{\cancel{8}}\sqrt{6}}{\underset{1}{\cancel{2}}} = \frac{4\sqrt{6}}{1} = \mathbf{4\sqrt{6}}$

Example 1.2-33

$$\sqrt[2]{\frac{5}{7}} = \sqrt{\frac{5}{7}} = \frac{\sqrt{5}}{\sqrt{7}} = \frac{\sqrt{5}}{\sqrt{7}} \times \frac{\sqrt{7}}{\sqrt{7}} = \frac{\sqrt{5}\times\sqrt{7}}{\sqrt{7}\times\sqrt{7}} = \frac{\sqrt{5\cdot 7}}{\sqrt{7\cdot 7}} = \frac{\sqrt{35}}{\sqrt{7^1\cdot 7^1}} = \frac{\sqrt{35}}{\sqrt{7^{1+1}}} = \frac{\sqrt{35}}{\sqrt{7^2}} = \boxed{\frac{\sqrt{35}}{7}}$$

Example 1.2-34

$$\frac{1}{\sqrt[2]{5}} = \frac{1}{\sqrt{5}} = \frac{1}{\sqrt{5}} \times \frac{\sqrt{5}}{\sqrt{5}} = \frac{1\times\sqrt{5}}{\sqrt{5}\times\sqrt{5}} = \frac{\sqrt{5}}{\sqrt{5\cdot 5}} = \frac{\sqrt{5}}{\sqrt{5^1\cdot 5^1}} = \frac{\sqrt{5}}{\sqrt{5^{1+1}}} = \frac{\sqrt{5}}{\sqrt{5^2}} = \boxed{\frac{\sqrt{5}}{5}}$$

Example 1.2-35

$$\frac{40\sqrt{12}}{5\sqrt{6}} = \frac{\overset{8}{\cancel{40}}\sqrt{12}}{\underset{1}{\cancel{5}}\sqrt{6}} = \frac{8\sqrt{12}}{\sqrt{6}} = \frac{8\sqrt{4\cdot 3}}{\sqrt{6}} = \frac{8\sqrt{2^2\cdot 3}}{\sqrt{6}} = \frac{(8\cdot 2)\sqrt{3}}{\sqrt{6}} = \frac{16\sqrt{3}}{\sqrt{6}} = \frac{16\sqrt{3}}{\sqrt{6}} \times \frac{\sqrt{6}}{\sqrt{6}} = \frac{16\sqrt{3}\times\sqrt{6}}{\sqrt{6}\times\sqrt{6}}$$

$$= \frac{16\sqrt{3\cdot 6}}{\sqrt{6\cdot 6}} = \frac{16\sqrt{18}}{\sqrt{6^1\cdot 6^1}} = \frac{16\sqrt{9\cdot 2}}{\sqrt{6^{1+1}}} = \frac{16\sqrt{3^2\cdot 2}}{\sqrt{6^2}} = \frac{(16\cdot 3)\sqrt{2}}{6} = \frac{\overset{8}{\cancel{48}}\sqrt{2}}{\underset{1}{\cancel{6}}} = \frac{8\sqrt{2}}{1} = \boxed{8\sqrt{2}}$$

Example 1.2-36

$$\sqrt{\frac{1000}{36}} = \sqrt{\frac{\overset{250}{\cancel{1000}}}{\underset{9}{\cancel{36}}}} = \sqrt{\frac{250}{9}} = \frac{\sqrt{250}}{\sqrt{9}} = \frac{\sqrt{25\cdot 10}}{\sqrt{3^2}} = \frac{\sqrt{5^2\times 10}}{3} = \boxed{\frac{5\sqrt{10}}{3}}$$

Practice Problems: Rationalizing Radical Expressions - Monomial Denominators with Real Numbers

Section 1.2b Case IV Practice Problems - Solve the following radical expressions:

1. $\sqrt{\frac{1}{8}} =$
2. $\sqrt[2]{\frac{50}{7}} =$
3. $\frac{\sqrt{75}}{-5} =$
4. $\sqrt[3]{\frac{25}{16}} =$
5. $\sqrt[5]{\frac{32}{8}} =$
6. $\frac{-3\sqrt{100}}{-5\sqrt{3000}} =$

Case V Rationalizing Radical Expressions - Binomial Denominators with Real Numbers

Simplification of fractional radical expressions with binomial denominators requires rationalization of the denominator. A binomial denominator is rationalized by multiplying the numerator and the denominator by its conjugate. Two binomials that differ only by the sign between them are called **conjugates** of each other. Note that whenever conjugates are multiplied by each other, the two similar but opposite in sign middle terms drop out.

Examples:

1. The conjugate of $2+\sqrt{3}$ is $2-\sqrt{3}$.
2. The conjugate of $\sqrt{6}-10$ is $\sqrt{6}+10$.
3. The conjugate of $\sqrt{3}-\sqrt{5}$ is $\sqrt{3}+\sqrt{5}$.
4. The conjugate of $\sqrt{7}+\sqrt{2}$ is $\sqrt{7}-\sqrt{2}$.

Radical expressions with binomial denominators are simplified using the following steps:

Step 1 Simplify the radical terms in the numerator and the denominator (see Section 1.2a, Case III).

Step 2 Rationalize the denominator by multiplying the numerator and the denominator by its conjugate.

Step 3 Simplify the radical expression using the FOIL method (see Section 1.2b, Case II).

Examples with Steps

The following examples show the steps as to how radical expressions with two terms in the denominator are simplified:

Example 1.2-37

$$\frac{8}{2-\sqrt{2}}$$

Solution:

Step 1 *Not Applicable*

Step 2

$$\frac{8}{2-\sqrt{2}} = \frac{8}{2-\sqrt{2}} \times \frac{2+\sqrt{2}}{2+\sqrt{2}}$$

Step 3

$$\frac{8}{2-\sqrt{2}} \times \frac{2+\sqrt{2}}{2+\sqrt{2}} = \frac{8\times\left(2+\sqrt{2}\right)}{\left(2-\sqrt{2}\right)\times\left(2+\sqrt{2}\right)} = \frac{8\cdot\left(2+\sqrt{2}\right)}{(2\cdot 2)+\left(2\cdot\sqrt{2}\right)-\left(2\cdot\sqrt{2}\right)-\left(\sqrt{2}\cdot\sqrt{2}\right)}$$

$$= \frac{8\left(2+\sqrt{2}\right)}{4+2\sqrt{2}-2\sqrt{2}-\sqrt{2\cdot 2}} = \frac{8\left(2+\sqrt{2}\right)}{4-\sqrt{2^1\cdot 2^1}} = \frac{8\left(2+\sqrt{2}\right)}{4-\sqrt{2^{1+1}}} = \frac{8\left(2+\sqrt{2}\right)}{4-\sqrt{2^2}} = \frac{8\left(2+\sqrt{2}\right)}{4-2}$$

$$= \boxed{\frac{\overset{4}{\cancel{8}}\left(2+\sqrt{2}\right)}{\cancel{2}}} = \boxed{\frac{4\left(2+\sqrt{2}\right)}{1}} = \boxed{\mathbf{4\left(2+\sqrt{2}\right)}}$$

Example 1.2-38

$$\boxed{\frac{\sqrt{8}+\sqrt{4}}{4-\sqrt{2}}} =$$

Solution:

Step 1

$$\boxed{\frac{\sqrt{8}+\sqrt{4}}{4-\sqrt{2}}} = \boxed{\frac{\sqrt{2^2\cdot 2}+\sqrt{2^2}}{4-\sqrt{2}}} = \boxed{\frac{2\sqrt{2}+2}{4-\sqrt{2}}}$$

Step 2

$$\boxed{\frac{2\sqrt{2}+2}{4-\sqrt{2}}} = \boxed{\frac{2\sqrt{2}+2}{4-\sqrt{2}} \times \frac{4+\sqrt{2}}{4+\sqrt{2}}}$$

Step 3

$$\boxed{\frac{2\sqrt{2}+2}{4-\sqrt{2}} \times \frac{4+\sqrt{2}}{4+\sqrt{2}}} = \boxed{\frac{\left(2\sqrt{2}+2\right)\times\left(4+\sqrt{2}\right)}{\left(4-\sqrt{2}\right)\times\left(4+\sqrt{2}\right)}} = \boxed{\frac{\left(4\cdot 2\sqrt{2}\right)+\left(2\sqrt{2}\cdot\sqrt{2}\right)+\left(2\cdot 4\right)+\left(2\cdot\sqrt{2}\right)}{\left(4\cdot 4\right)+\left(4\cdot\sqrt{2}\right)-\left(4\cdot\sqrt{2}\right)-\left(\sqrt{2}\cdot\sqrt{2}\right)}}$$

$$= \boxed{\frac{8\sqrt{2}+2\sqrt{2\cdot 2}+8+2\sqrt{2}}{16+4\sqrt{2}-4\sqrt{2}-\sqrt{2\cdot 2}}} = \boxed{\frac{8\sqrt{2}+2\sqrt{2^2}+8+2\sqrt{2}}{16-\sqrt{2^2}}} = \boxed{\frac{8\sqrt{2}+\left(2\cdot 2\right)+8+2\sqrt{2}}{16-2}}$$

$$= \boxed{\frac{8\sqrt{2}+4+8+2\sqrt{2}}{14}} = \boxed{\frac{\left(8+2\right)\sqrt{2}+12}{14}} = \boxed{\frac{10\sqrt{2}+12}{14}} = \boxed{\frac{\overset{1}{\cancel{2}}\cdot\left(5\sqrt{2}+6\right)}{\underset{7}{\cancel{14}}}} = \boxed{\frac{1\cdot\left(5\sqrt{2}+6\right)}{7}}$$

$$= \boxed{\mathbf{\frac{5\sqrt{2}+6}{7}}}$$

Additional Examples: Rationalizing Radical Expressions - Binomial Denominators with Real Numbers

The following examples further illustrate how to rationalize radical expressions with binomial denominators:

Example 1.2-39

$$\boxed{\frac{\sqrt{5}}{3+\sqrt{5}}} = \boxed{\frac{\sqrt{5}}{3+\sqrt{5}} \times \frac{3-\sqrt{5}}{3-\sqrt{5}}} = \boxed{\frac{\sqrt{5}\times\left(3-\sqrt{5}\right)}{\left(3+\sqrt{5}\right)\times\left(3-\sqrt{5}\right)}} = \boxed{\frac{\left(3\cdot\sqrt{5}\right)-\left(\sqrt{5\cdot 5}\right)}{\left(3\cdot 3\right)-\left(3\cdot\sqrt{5}\right)+\left(3\cdot\sqrt{5}\right)-\left(\sqrt{5}\cdot\sqrt{5}\right)}}$$

$$= \boxed{\frac{3\sqrt{5}-\sqrt{5^2}}{9-3\sqrt{5}+3\sqrt{5}-\sqrt{5^2}}} = \boxed{\frac{3\sqrt{5}-5}{9-\sqrt{5^2}}} = \boxed{\frac{3\sqrt{5}-5}{9-5}} = \boxed{\mathbf{\frac{3\sqrt{5}-5}{4}}}$$

Example 1.2-40

$$\boxed{\frac{\sqrt{5}+\sqrt{3}}{\sqrt{7}+\sqrt{2}}} = \boxed{\frac{\sqrt{5}+\sqrt{3}}{\sqrt{7}+\sqrt{2}} \times \frac{\sqrt{7}-\sqrt{2}}{\sqrt{7}-\sqrt{2}}} = \boxed{\frac{\left(\sqrt{5}+\sqrt{3}\right)\times\left(\sqrt{7}-\sqrt{2}\right)}{\left(\sqrt{7}+\sqrt{2}\right)\times\left(\sqrt{7}-\sqrt{2}\right)}} = \boxed{\frac{\left(\sqrt{5}\cdot\sqrt{7}\right)-\left(\sqrt{5}\cdot\sqrt{2}\right)+\left(\sqrt{3}\cdot\sqrt{7}\right)-\left(\sqrt{3}\cdot\sqrt{2}\right)}{\left(\sqrt{7}\cdot\sqrt{7}\right)-\left(\sqrt{7}\cdot\sqrt{2}\right)+\left(\sqrt{2}\cdot\sqrt{7}\right)-\left(\sqrt{2}\cdot\sqrt{2}\right)}}$$

$$=\boxed{\frac{\sqrt{5\cdot7}-\sqrt{5\cdot2}+\sqrt{3\cdot7}-\sqrt{3\cdot2}}{\sqrt{7\cdot7}-\sqrt{7\cdot2}+\sqrt{2\cdot7}-\sqrt{2\cdot2}}}=\boxed{\frac{\sqrt{35}-\sqrt{10}+\sqrt{21}-\sqrt{6}}{\sqrt{7^2}-\sqrt{14}+\sqrt{14}-\sqrt{2^2}}}=\boxed{\frac{\sqrt{35}-\sqrt{10}+\sqrt{21}-\sqrt{6}}{7-2}}$$

$$=\boxed{\frac{\mathbf{\sqrt{35}-\sqrt{10}+\sqrt{21}-\sqrt{6}}}{\mathbf{5}}}$$

Example 1.2-41

$$\boxed{\frac{3+\sqrt{27}}{3-\sqrt{3}}}=\boxed{\frac{3+\sqrt{9\cdot3}}{3-\sqrt{3}}}=\boxed{\frac{3+\sqrt{3^2\cdot3}}{3-\sqrt{3}}}=\boxed{\frac{3+3\sqrt{3}}{3-\sqrt{3}}}=\boxed{\frac{3+3\sqrt{3}}{3-\sqrt{3}}\times\frac{3+\sqrt{3}}{3+\sqrt{3}}}=\boxed{\frac{\left(3+3\sqrt{3}\right)\times\left(3+\sqrt{3}\right)}{\left(3-\sqrt{3}\right)\times\left(3+\sqrt{3}\right)}}$$

$$=\boxed{\frac{(3\cdot3)+\left(3\cdot\sqrt{3}\right)+(3\cdot3)\sqrt{3}+\left(3\sqrt{3}\cdot\sqrt{3}\right)}{(3\cdot3)+\left(3\cdot\sqrt{3}\right)-\left(3\cdot\sqrt{3}\right)-\left(\sqrt{3}\cdot\sqrt{3}\right)}}=\boxed{\frac{9+3\sqrt{3}+9\sqrt{3}+3\sqrt{3\cdot3}}{9+3\sqrt{3}-3\sqrt{3}-\sqrt{3\cdot3}}}=\boxed{\frac{9+3\sqrt{3}+9\sqrt{3}+3\sqrt{3^2}}{9-\sqrt{3^2}}}$$

$$=\boxed{\frac{9+3\sqrt{3}+9\sqrt{3}+3\cdot3}{9-3}}=\boxed{\frac{(9+9)+(3+9)\sqrt{3}}{9-3}}=\boxed{\frac{18+12\sqrt{3}}{6}}=\boxed{\frac{\not{6}\left(3+2\sqrt{3}\right)}{\not{6}}}=\boxed{\frac{3+2\sqrt{3}}{1}}=\boxed{\mathbf{3+2\sqrt{3}}}$$

Example 1.2-42

$$\boxed{\frac{\sqrt{5}-1}{\sqrt{5}+1}}=\boxed{\frac{\sqrt{5}-1}{\sqrt{5}+1}\times\frac{\sqrt{5}-1}{\sqrt{5}-1}}=\boxed{\frac{\left(\sqrt{5}-1\right)\times\left(\sqrt{5}-1\right)}{\left(\sqrt{5}+1\right)\times\left(\sqrt{5}-1\right)}}=\boxed{\frac{\left(\sqrt{5}\cdot\sqrt{5}\right)-\left(1\cdot\sqrt{5}\right)-\left(1\cdot\sqrt{5}\right)+(1\cdot1)}{\left(\sqrt{5}\cdot\sqrt{5}\right)-\left(1\cdot\sqrt{5}\right)+\left(1\cdot\sqrt{5}\right)-(1\cdot1)}}=\boxed{\frac{\sqrt{5\cdot5}-\sqrt{5}-\sqrt{5}+1}{\sqrt{5\cdot5}-\sqrt{5}+\sqrt{5}-1}}$$

$$=\boxed{\frac{\sqrt{5^2}-\sqrt{5}-\sqrt{5}+1}{\sqrt{5^2}-1}}=\boxed{\frac{5-\sqrt{5}-\sqrt{5}+1}{5-1}}=\boxed{\frac{(5+1)+(-1-1)\sqrt{5}}{5-1}}=\boxed{\frac{6-2\sqrt{5}}{4}}=\boxed{\frac{2\left(3-\sqrt{5}\right)}{\underset{2}{\not{4}}}}=\boxed{\frac{\mathbf{3-\sqrt{5}}}{\mathbf{2}}}$$

Practice Problems: Rationalizing Radical Expressions - Binomial Denominators with Real Numbers

Section 1.2b Case V Practice Problems - Solve the following radical expressions:

1. $\frac{7}{1+\sqrt{7}}=$
2. $\frac{1-\sqrt{18}}{2+\sqrt{18}}=$
3. $\frac{\sqrt{5}}{\sqrt{5}+\sqrt{2}}=$
4. $\frac{3-\sqrt{5}}{\sqrt{7}-\sqrt{4}}=$
5. $\frac{-3+\sqrt{3}}{4+\sqrt{5}}=$
6. $\frac{3-\sqrt{3}}{3+\sqrt{3}}=$

Case VI Adding and Subtracting Radical Terms

Radicals are added and subtracted using the following general rule:

$$k_1\sqrt[n]{a}+k_2\sqrt[n]{a}+k_3\sqrt[n]{a}=(k_1+k_2+k_3)\sqrt[n]{a}$$

Only similar radicals can be added and subtracted. **Similar radicals** are defined as radical expressions with the same index n and the same radicand a. Note that the distributive property of multiplication is used to group the numbers in front of the similar radical terms. Radicals are added and subtracted using the following steps:

Step 1 Group similar radicals.

Step 2 Simplify the radical expression.

Examples with Steps

The following examples show the steps as to how radical expressions are added and subtracted:

Example 1.2-43

$$\boxed{6\sqrt{2}+4\sqrt{2}}=$$

Solution:

Step 1 $\boxed{6\sqrt{2}+4\sqrt{2}}=\boxed{(6+4)\sqrt{2}}$

Step 2 $\boxed{(6+4)\sqrt{2}}=\boxed{\mathbf{10\sqrt{2}}}$

Example 1.2-44

$$\boxed{20\sqrt[5]{3}-8\sqrt[5]{3}+5\sqrt[5]{3}}=$$

Solution:

Step 1 $\boxed{20\sqrt[5]{3}-8\sqrt[5]{3}+5\sqrt[5]{3}}=\boxed{(20-8+5)\sqrt[5]{3}}$

Step 2 $\boxed{(20-8+5)\sqrt[5]{3}}=\boxed{\mathbf{17\sqrt[5]{3}}}$

Example 1.2-45

$$\boxed{\left(6\sqrt{7}+2\sqrt{7}\right)-2\sqrt[3]{7}}=$$

Solution:

Step 1 $\boxed{\left(6\sqrt{7}+2\sqrt{7}\right)-2\sqrt[3]{7}}=\boxed{(6+2)\sqrt{7}-2\sqrt[3]{7}}$

Step 2 $\boxed{(6+2)\sqrt{7}-2\sqrt[3]{7}}=\boxed{\mathbf{8\sqrt{7}-2\sqrt[3]{7}}}$

Additional Examples - Adding and Subtracting Radical Terms

The following examples further illustrate how to add and subtract radical terms:

Example 1.2-46

$$\boxed{2\sqrt{5}+3\sqrt{5}+6} = \boxed{(2+3)\sqrt{5}+6} = \boxed{\mathbf{5\sqrt{5}+6}}$$

Example 1.2-47

$$\boxed{8\sqrt[3]{4}+2\sqrt[3]{4}+5} = \boxed{(8+2)\sqrt[3]{4}+5} = \boxed{10\sqrt[3]{4}+5} = \boxed{\mathbf{5\left(2\sqrt[3]{4}+1\right)}}$$

Example 1.2-48

$$\boxed{2\sqrt[4]{3}+4\sqrt[4]{3}-3\sqrt[4]{3}+\sqrt[4]{5}} = \boxed{(2+4-3)\sqrt[4]{3}+\sqrt[4]{5}} = \boxed{\mathbf{3\sqrt[4]{3}+\sqrt[4]{5}}}$$

Note that the two radical terms have the same index (4) but have different radicands (3 and 5). Therefore, they can not be combined.

Example 1.2-49

$$\boxed{\sqrt[5]{5}+3\sqrt[5]{5}+a\sqrt[5]{5}-(4+a)\sqrt{2}} = \boxed{(1+3+a)\sqrt[5]{5}-(4+a)\sqrt{2}} = \boxed{(4+a)\sqrt[5]{5}-(4+a)\sqrt{2}} = \boxed{\mathbf{(4+a)\left[\sqrt[5]{5}-\sqrt{2}\right]}}$$

Example 1.2-50

$$\boxed{5\sqrt[3]{2x}+8\sqrt[3]{2x}-2c\sqrt[3]{2x}+4\sqrt{2x}-8\sqrt{2x}} = \boxed{(5+8-2c)\sqrt[3]{2x}+(4-8)\sqrt{2x}} = \boxed{\mathbf{(13-2c)\sqrt[3]{2x}-4\sqrt{2x}}}$$

Example 1.2-51

$$\boxed{a\sqrt{xy}+b\sqrt[3]{xy}-c^2\sqrt{xy}-d} = \boxed{a\sqrt{xy}-c^2\sqrt{xy}+b\sqrt[3]{xy}-d} = \boxed{\mathbf{\left(a-c^2\right)\sqrt{xy}+b\sqrt[3]{xy}-d}}$$

Example 1.2-52

$$\boxed{2\sqrt{75}+3\sqrt{125}+\sqrt{20}+3\sqrt{10}-4\sqrt{10}} = \boxed{2\sqrt{25\cdot 3}+3\sqrt{25\cdot 5}+\sqrt{4\cdot 5}+(3-4)\sqrt{10}}$$

$$= \boxed{2\sqrt{5^2\cdot 3}+3\sqrt{5^2\cdot 5}+\sqrt{2^2\cdot 5}-\sqrt{10}} = \boxed{(2\cdot 5)\sqrt{3}+(3\cdot 5)\sqrt{5}+2\sqrt{5}-\sqrt{10}} = \boxed{10\sqrt{3}+15\sqrt{5}+2\sqrt{5}-\sqrt{10}}$$

$$= \boxed{10\sqrt{3}+(15+2)\sqrt{5}-\sqrt{10}} = \boxed{\mathbf{10\sqrt{3}+17\sqrt{5}-\sqrt{10}}}$$

Example 1.2-53

$$\boxed{8\sqrt[3]{6}+4\sqrt[3]{6}+a\sqrt[3]{6}-\sqrt{5}-4\sqrt{5}} = \boxed{(8+4+a)\sqrt[3]{6}+(-1-4)\sqrt{5}} = \boxed{(12+a)\sqrt[3]{6}+(-5)\sqrt{5}} = \boxed{\mathbf{(12+a)\sqrt[3]{6}-5\sqrt{5}}}$$

Practice Problems - Adding and Subtracting Radical Terms

Section 1.2b Case VI Practice Problems - Simplify the following radical expressions:

1. $5\sqrt{3}+8\sqrt{3} =$
2. $2\sqrt[3]{3}-4\sqrt[3]{3} =$
3. $12\sqrt[4]{5}+8\sqrt[4]{5}+2\sqrt[4]{3} =$
4. $a\sqrt{ab}-b\sqrt{ab}+c\sqrt{ab} =$
5. $3x\sqrt[3]{x}-2x\sqrt[3]{x}+4x\sqrt[3]{x^2} =$
6. $5\sqrt[3]{2}+8\sqrt[3]{5} =$

1.3 Factoring Polynomials

In this section different methods for factoring polynomials are reviewed. The steps in factoring polynomials using the Greatest Common Factoring, the Grouping, and the Trial and Error method are addressed in Sections 1.3a, 1.3b, and 1.3c, respectively. Other factoring methods such as using the Difference of Two Square method, the Sum and Difference of Two Cubes method, as well as the method for factoring Perfect Square Trinomials are discussed in Section 1.3d.

1.3a Factoring Polynomials Using the Greatest Common Factoring Method

Solving algebraic fractions, which are reviewed in Section 1.5, requires a thorough knowledge of the factoring and solution methods that are introduced in this and the following section. Therefore, it is essential that students learn how to factor polynomials of second or higher degrees. For example, simplification of math operations such as:

$$\frac{x-y}{2x-y}\cdot\frac{2x^2+xy-y^2}{2y^2-3y+x^2}\ ; \qquad \frac{x^2-y^2}{2x^2+5xy+3y^2}\cdot\frac{4x^2+4xy-3y^2}{5x-5y}\ ; \qquad \frac{x^2-9}{x^3-5x^2+6x}\cdot\frac{16x^2}{8x+24}$$

require familiarization with various factoring methods. In this section students are introduced to factoring the Greatest Common Factor to: monomial terms (Case I) and binomial and polynomial terms (Case II).

Case I Factoring the Greatest Common Factor to Monomial Terms

Factoring a polynomial means writing the polynomial as a product of two or more simpler polynomials. One method in factoring polynomials is by using the Greatest Common Factoring method where the Greatest Common Factor (G.C.F.) is factored out. The Greatest Common Factor to monomial terms is found using the following steps:

Step 1 a. Write the numbers and the variables in their prime factored form.

b. Identify the prime numbers and variables that are common in monomials.

Step 2 Multiply the common prime numbers and variables to obtain the G.C.F.

The following examples show the steps as to how monomial expressions are factored using the Greatest Common Factoring method:

Example 1.3-1

Find the G.C.F. to $27x^2y^3$, $9x^3y$, and $15xy^2$.

Solution:

Step 1

$\boxed{27x^2y^3} = \boxed{3\cdot 9\cdot x\cdot x\cdot y\cdot y^2} = \boxed{3\cdot 3\cdot 3\cdot x\cdot x\cdot y\cdot y\cdot y}$

$\boxed{9x^3y} = \boxed{3\cdot 3\cdot x\cdot x^2\cdot y} = \boxed{3\cdot 3\cdot x\cdot x\cdot x\cdot y}$ and

$\boxed{15xy^2} = \boxed{3\cdot 5\cdot x\cdot y\cdot y}$

Therefore, the common terms are 3, x and y.

Step 2 G.C.F. = $\boxed{3 \cdot x \cdot y} = \boxed{\mathbf{3xy}}$

Example 1.3-2

Find the G.C.F. to $32a^3b^3$, $46ab^2$, and $56a^2b^4$.

Solution:

Step 1 $\boxed{32a^3b^3} = \boxed{4 \cdot 8 \cdot a \cdot a^2 \cdot b \cdot b^2} = \boxed{2 \cdot 2 \cdot 2 \cdot 4 \cdot a \cdot a \cdot a \cdot b \cdot b \cdot b} = \boxed{2 \cdot 2 \cdot 2 \cdot 2 \cdot 2 \cdot a \cdot a \cdot a \cdot b \cdot b \cdot b}$

$\boxed{46ab^2} = \boxed{3 \cdot 12 \cdot a \cdot b \cdot b} = \boxed{3 \cdot 3 \cdot 4 \cdot a \cdot b \cdot b} = \boxed{3 \cdot 3 \cdot 2 \cdot 2 \cdot a \cdot b \cdot b}$ and

$\boxed{56a^2b^4} = \boxed{7 \cdot 8 \cdot a \cdot a \cdot b^2 \cdot b^2} = \boxed{7 \cdot 2 \cdot 4 \cdot a \cdot a \cdot b \cdot b \cdot b \cdot b} = \boxed{7 \cdot 2 \cdot 2 \cdot 2 \cdot a \cdot a \cdot b \cdot b \cdot b \cdot b}$

Therefore, the common terms are 2, 2, a, b, and b.

Step 2 G.C.F. = $\boxed{2 \cdot 2 \cdot a \cdot b \cdot b} = \boxed{\mathbf{4ab^2}}$

Additional Examples - Factoring the Greatest Common Factor to Monomial Terms

The following examples further illustrate how to find the Greatest Common Factor to monomial terms:

Example 1.3-3

Find the G.C.F. to $48xy^2$, $16x^2y$, $4x^3y^2$, and $12xy$.

1. $\boxed{48xy^2} = \boxed{12 \cdot 4 \cdot x \cdot y \cdot y} = \boxed{6 \cdot 2 \cdot 2 \cdot 2 \cdot x \cdot y \cdot y} = \boxed{3 \cdot 2 \cdot 2 \cdot 2 \cdot 2 \cdot x \cdot y \cdot y}$
2. $\boxed{16x^2y} = \boxed{2 \cdot 8 \cdot x \cdot x \cdot y} = \boxed{2 \cdot 2 \cdot 4 \cdot x \cdot x \cdot y} = \boxed{2 \cdot 2 \cdot 2 \cdot 2 \cdot x \cdot x \cdot y}$
3. $\boxed{4x^3y^2} = \boxed{2 \cdot 2 \cdot x \cdot x^2 \cdot y \cdot y} = \boxed{2 \cdot 2 \cdot x \cdot x \cdot x \cdot y \cdot y}$
4. $\boxed{12xy} = \boxed{3 \cdot 4 \cdot x \cdot y} = \boxed{3 \cdot 2 \cdot 2 \cdot x \cdot y}$

Therefore, the common terms are 2, 2, x, and y. Thus, G.C.F. = $\boxed{2 \cdot 2 \cdot x \cdot y} = \boxed{\mathbf{4xy}}$

Example 1.3-4

Find the G.C.F. to $55u^2w^3z$, $50uw^2z^2$, and $15u^3w$.

1. $\boxed{55u^2w^3z} = \boxed{5 \cdot 11 \cdot u \cdot u \cdot w \cdot w^2 \cdot z} = \boxed{5 \cdot 11 \cdot u \cdot u \cdot w \cdot w \cdot w \cdot z}$
2. $\boxed{50uw^2z^2} = \boxed{5 \cdot 10 \cdot u \cdot w \cdot w \cdot z \cdot z} = \boxed{5 \cdot 5 \cdot 2 \cdot u \cdot w \cdot w \cdot z \cdot z}$
3. $\boxed{15u^3w} = \boxed{5 \cdot 3 \cdot u \cdot u^2 \cdot w} = \boxed{5 \cdot 3 \cdot u \cdot u \cdot u \cdot w}$

Therefore, the common terms are 5, u, and w. Thus, G.C.F. = $\boxed{5 \cdot u \cdot w} = \boxed{\mathbf{5uw}}$

Example 1.3-5

Find the G.C.F. to $27abc$, $36a^2b^2c^3$, and $24ac^2$.

1. $\boxed{27abc} = \boxed{9 \cdot 3 \cdot a \cdot b \cdot c} = \boxed{3 \cdot 3 \cdot 3 \cdot a \cdot b \cdot c}$

2. $\boxed{36a^2b^2c^3} = \boxed{2 \cdot 18 \cdot a \cdot a \cdot b \cdot b \cdot c \cdot c^2} = \boxed{2 \cdot 2 \cdot 9 \cdot a \cdot a \cdot b \cdot b \cdot c \cdot c \cdot c} = \boxed{2 \cdot 2 \cdot 3 \cdot 3 \cdot a \cdot a \cdot b \cdot b \cdot c \cdot c \cdot c}$

3. $\boxed{24ac^2} = \boxed{3 \cdot 8 \cdot a \cdot c \cdot c} = \boxed{3 \cdot 2 \cdot 4 \cdot a \cdot c \cdot c} = \boxed{3 \cdot 2 \cdot 2 \cdot 2 \cdot a \cdot c \cdot c}$

Therefore, the common terms are 3, a, and c. Thus, G.C.F. $= \boxed{3 \cdot a \cdot c} = \boxed{\mathbf{3ac}}$

Example 1.3-6

Find the G.C.F. to $12x$, $60xy^2$, and $63x^2$.

1. $\boxed{12x} = \boxed{3 \cdot 4 \cdot x} = \boxed{3 \cdot 2 \cdot 2 \cdot x}$

2. $\boxed{60xy^2} = \boxed{4 \cdot 15 \cdot x \cdot y \cdot y} = \boxed{2 \cdot 2 \cdot 5 \cdot 3 \cdot x \cdot y \cdot y}$

3. $\boxed{63x^2} = \boxed{3 \cdot 21 \cdot x \cdot x} = \boxed{3 \cdot 3 \cdot 7 \cdot x \cdot x}$

Therefore, the common terms are 3 and x. Thus, G.C.F. $= \boxed{3 \cdot x} = \boxed{\mathbf{3x}}$

Practice Problems - Factoring the Greatest Common Factor to Monomial Terms

Section 1.3a Case I Practice Problems - Find the Greatest Common Factor to the following monomial terms:

1. $5x^3$ and $15x$
2. $18x^2y^3z^4$ and $24xy^4z^5$
3. $16a^2bc^3$, $38ab^4c^2$, and $6a^3bc$
4. r^5s^4, $4r^3s^2$, and $3rs$
5. $10u^2vw^3$, $2uv^3w^2$, and uv^2
6. $19a^3b^3$, $12ab^2$, and $6ab$

Case II Factoring the Greatest Common Factor to Binomial and Polynomial Terms

The concept of obtaining Greatest Common Factor can be extended to binomial expressions by obtaining the greater common monomial factor which is found by using the following steps:

Step 1 a. Write each monomial term in its prime factored form.

b. Identify the prime numbers and variables that are common to monomials.

c. Multiply the common prime numbers and variables to obtain the greatest common monomial factor.

Step 2 Factor out the greatest common monomial factor from the binomial expression.

The following examples show the steps as to how binomial expressions are factored:

Example 1.3-7

Factor $6a^3b^2c^2 - 2a^2bc^2$.

Solution:

Step 1 $\boxed{6a^3b^2c^2} = \boxed{2 \cdot 3 \cdot a \cdot a \cdot a \cdot b \cdot b \cdot c \cdot c}$ and

$\boxed{2a^2bc^2} = \boxed{2 \cdot a \cdot a \cdot b \cdot c \cdot c}$

Therefore, the common terms are 2, a, a, b, c, and c which implies that the greatest common monomial factor is $2 \cdot a \cdot a \cdot b \cdot c \cdot c = 2a^2bc^2$. Thus,

Step 2 $\boxed{6a^3b^2c^2 - 2a^2bc^2} = \boxed{\mathbf{2a^2bc^2(3ab - 1)}}$

Example 1.3-8

Factor $12x^3y^2z + 36x^2z^2$.

Solution:

Step 1 $\boxed{12x^3y^2z} = \boxed{4 \cdot 3 \cdot x \cdot x^2 \cdot y \cdot y \cdot z} = \boxed{2 \cdot 2 \cdot 3 \cdot x \cdot x \cdot x \cdot y \cdot y \cdot z}$ and

$\boxed{36x^2z^2} = \boxed{2 \cdot 18 \cdot x \cdot x \cdot z \cdot z} = \boxed{2 \cdot 2 \cdot 9 \cdot x \cdot x \cdot z \cdot z} = \boxed{2 \cdot 2 \cdot 3 \cdot 3 \cdot x \cdot x \cdot z \cdot z}$

Therefore, the common terms are 2, 2, 3, x, x, and z which implies that the greatest common monomial factor is $2 \cdot 2 \cdot 3 \cdot x \cdot x \cdot z = 12x^2z$. Thus,

Step 2 $\boxed{12x^3y^2z + 36x^2z^2} = \boxed{\mathbf{12x^2z(xy^2 + 3z)}}$

Additional Examples - Factoring the Greatest Common Factor to Binomial and Polynomial Terms

The following examples further illustrate how to find the greatest common monomial factor to binomial terms:

Example 1.3-9

Find the greatest common monomial factor to $35m^2n^3 + 5mn^2$.

1. $\boxed{35m^2n^3} = \boxed{5\cdot 7\cdot m\cdot m\cdot n\cdot n^2} = \boxed{5\cdot 7\cdot m\cdot m\cdot n\cdot n\cdot n}$

2. $\boxed{5mn^2} = \boxed{5\cdot m\cdot n\cdot n}$

Therefore, the common terms are 5, m, n, and n which implies that the greatest common monomial factor is $5\cdot m\cdot n\cdot n = 5mn^2$. Thus, $\boxed{35m^2n^3 + 5mn^2} = \boxed{\mathbf{5mn^2(7mn+1)}}$

Example 1.3-10

Find the greatest common monomial factor to $6a^2b + 66ab^4$.

1. $\boxed{6a^2b} = \boxed{2\cdot 3\cdot a\cdot a\cdot b}$

2. $\boxed{66ab^4} = \boxed{6\cdot 11\cdot a\cdot b^2\cdot b^2} = \boxed{2\cdot 3\cdot 11\cdot a\cdot b\cdot b\cdot b\cdot b}$

Therefore, the common terms are 2, 3, a, and b which implies that the greatest common monomial factor is $2\cdot 3\cdot a\cdot b = 6ab$. Thus, $\boxed{6a^2b + 66ab^4} = \boxed{\mathbf{6ab(a+11b^3)}}$

Example 1.3-11

Find the greatest common monomial factor to $7p^3q^6 - 49p^2q^5$.

1. $\boxed{7p^3q^6} = \boxed{7\cdot p\cdot p^2\cdot q^3\cdot q^3} = \boxed{7\cdot p\cdot p\cdot p\cdot q\cdot q^2\cdot q\cdot q^2} = \boxed{7\cdot p\cdot p\cdot p\cdot q\cdot q\cdot q\cdot q\cdot q\cdot q}$

2. $\boxed{49p^2q^5} = \boxed{7\cdot 7\cdot p\cdot p\cdot q\cdot q^4} = \boxed{7\cdot 7\cdot p\cdot p\cdot q\cdot q^2\cdot q^2} = \boxed{7\cdot 7\cdot p\cdot p\cdot q\cdot q\cdot q\cdot q\cdot q}$

Therefore, the common terms are 7, p, p, q, q, q, q, and q which implies that the greatest common monomial factor is $7\cdot p\cdot p\cdot q\cdot q\cdot q\cdot q\cdot q = 7p^2q^5$. Thus, $\boxed{7p^3q^6 - 49p^2q^5} = \boxed{\mathbf{7p^2q^5(pq-7)}}$

Example 1.3-12

Find the greatest common monomial factor to $48x - 20xy$.

1. $\boxed{48x} = \boxed{4\cdot 12\cdot x} = \boxed{2\cdot 2\cdot 4\cdot 3\cdot x} = \boxed{2\cdot 2\cdot 2\cdot 2\cdot 3\cdot x}$

2. $\boxed{20xy} = \boxed{4\cdot 5\cdot x\cdot y} = \boxed{2\cdot 2\cdot 5\cdot x\cdot y}$

Therefore, the common terms are 2, 2, and x which implies that the greatest common monomial factor is $2\cdot 2\cdot x = 4x$. Thus, $\boxed{48x - 20xy} = \boxed{\mathbf{4x(12-5y)}}$

Example 1.3-13

Find the greatest common monomial factor to $24x^3y^3 + 12x^2y^4$.

1. $\boxed{24x^3y^3} = \boxed{2\cdot 12\cdot x\cdot x^2\cdot y\cdot y^2} = \boxed{2\cdot 4\cdot 3\cdot x\cdot x\cdot x\cdot y\cdot y\cdot y} = \boxed{2\cdot 2\cdot 2\cdot 3\cdot x\cdot x\cdot x\cdot y\cdot y\cdot y}$

2. $\boxed{12x^2y^4} = \boxed{2\cdot 6\cdot x\cdot x\cdot y^2\cdot y^2} = \boxed{2\cdot 2\cdot 3\cdot x\cdot x\cdot y\cdot y\cdot y\cdot y}$

Therefore, the common terms are 2, 2, 3, x, x, y, y and y which implies that the greatest common monomial factor is $2 \cdot 2 \cdot 3 \cdot x \cdot x \cdot y \cdot y \cdot y = 12x^2y^3$. Thus, $\boxed{24x^3y^3 + 12x^2y^4}$ $= \boxed{\mathbf{12x^2y^3(2x+y)}}$

Note 1: As one gains more proficiency in solving this class of problems the need for factoring each monomial term to its prime factored form lessens. Therefore, students may simplify a given binomial expression by mentally factoring out the common terms. For example, we can quickly factor out the expression $24x^4y^2z + 12x^2y^3z^2$ by observing that its greatest common monomial factor is $12x^2y^2z$, e.g., $24x^4y^2z + 12x^2y^3z^2 = 12x^2y^2z(2x^2 + yz)$.

Note 2: The process of factoring binomial expressions can further be expanded to include trinomials and polynomials. Following are few additional examples indicating how the greatest common monomial factor to polynomials is obtained:

1. $5x + 15x^2 + 50x^3 = 5x(1 + 3x + 10x^2)$
2. $24xy^2 + 15xy + 12y = 3y(8xy + 5x + 4)$
3. $8a^2b^3 + 4ab^2 - 2a^2b = 2ab(4ab^2 + 2b - a)$
4. $20u^2w^2 + 15u^3w^2 + 5uw = 5uw(4uw + 3u^2w + 1)$
5. $4x^3 + 6x^2 + 2x = 2x(2x^2 + 3x + 1)$
6. $20x^2y^2 - 5x^3y + 15y = 5y(4x^2y - x^3 + 3)$
7. $40x^3y^4z^2 - 12x^2y^2z^2 + 8x^2yz = 4x^2yz(10xy^3z - 3yz + 2)$

Practice Problems - Factoring the Greatest Common Factor to Binomial and Polynomial Terms

Section 1.3a Case II Practice Problems - Find the Greatest Common Factor to the following binomial and polynomial terms:

1. $18x^3y^3 - 12x^2y =$
2. $3a^2b^3c + 15ab^2c^3 =$
3. $xyz^3 + 4x^2y^2z^5 =$
4. $25p^3 + 5p^2q^3 + pq =$
5. $r^2s^2t - 5rst^2 =$
6. $36x^3yz^3 + 4xy^2z^4 - 12x^3y^3z =$

1.3b Factoring Polynomials Using the Grouping Method

In many instances polynomials with four or more terms have common terms that can be grouped together. This process is called factoring by grouping. The steps as to how polynomials are grouped are shown below:

Step 1 Factor the common variables, or numbers, from the monomial terms.

Step 2 Factor the common binomial factor obtained in step 1 by grouping.

Examples with Steps

The following examples show the steps as to how polynomials are factored using the grouping method:

Example 1.3-14

$$\boxed{u^2 - 2u - 16u + 32} =$$

Solution:

Step 1 $\boxed{u^2 - 2u - 16u + 32} = \boxed{u(u-2) - 16(u-2)}$

Step 2 $\boxed{u(u-2) - 16(u-2)} = \boxed{\mathbf{(u-2)(u-16)}}$

Example 1.3-15

$$\boxed{60m^2 + 24m - 15m - 6} =$$

Solution:

Step 1 $\boxed{60m^2 + 24m - 15m - 6} = \boxed{12m(5m+2) - 3(5m+2)}$

Step 2 $\boxed{12m(5m+2) - 3(5m+2)} = \boxed{(5m+2)(12m-3)} = \boxed{\mathbf{4(5m+2)(3m-1)}}$

Example 1.3-16

$$\boxed{15y^3 + 25y^2 + 9y + 15} =$$

Solution:

Step 1 $\boxed{15y^2 + 25y + 9y + 15} = \boxed{5y(3y+5) + 3(3y+5)}$

Step 2 $\boxed{5y(3y+5) + 3(3y+5)} = \boxed{\mathbf{(3y+5)(5y+3)}}$

Additional Examples - Factoring Polynomials Using the Grouping Method

The following examples further illustrate how to factor polynomials using the grouping method:

Example 1.3-17

$$\boxed{5(x+y)^2 + 15x + 15y} = \boxed{5(x+y)^2 + 15(x+y)} = \boxed{\mathbf{5(x+y)[(x+y)+3]}}$$

Example 1.3-18

$$\boxed{8(a+b)^2 + 4(a+b)^3 + 2(a+b)} = \boxed{2(a+b)\left[4(a+b) + 2(a+b)^2 + 1\right]} = \boxed{\mathbf{2(a+b)\{2(a+b)[2+(a+b)]+1\}}}$$

Example 1.3-19

$$\boxed{-x^2+2x-3x+6} = \boxed{x(-x+2)+3(-x+2)} = \boxed{(-x+2)(x+3)}$$

Example 1.3-20

$$\boxed{3ab-7b-3a+7} = \boxed{b(3a-7)-(3a+7)} = \boxed{(3a-7)(b-1)}$$

Example 1.3-21

$$\boxed{x^3+5x^2+x+5} = \boxed{x^2(x+5)+x+5} = \boxed{x^2(x+5)+(x+5)} = \boxed{(x+5)(x^2+1)}$$

Example 1.3-22

$$\boxed{y^3+3y^2+4y+12} = \boxed{y^2(y+3)+4(y+3)} = \boxed{(y+3)(y^2+4)}$$

Example 1.3-23

$$\boxed{24x^2y-12xy-36x+18} = \boxed{12xy(2x-1)-18(2x-1)} = \boxed{(2x-1)(12xy-18)} = \boxed{6(2x-1)(2xy-3)}$$

Example 1.3-24

$$\boxed{12r^3s-6r^2s+4r-2} = \boxed{6r^2s(2r-1)+2(2r-1)} = \boxed{(2r-1)(6r^2s+2)} = \boxed{2(2r-1)(3r^2s+1)}$$

In the following sections, additional factoring methods are reviewed. These methods are used to present polynomials in their equivalent factored form. Students are encouraged to spend adequate time learning each method.

Practice Problems - Factoring Polynomials Using the Grouping Method

Section 1.3b Practice Problems - Factor the following polynomials using the grouping method:

1. $2ab-5b-6a+15=$
2. $y^3+4y^2+y+4=$
3. $42x^2y+21xy-70x-35=$
4. $(x+y)^3+(x+y)^2+x+y=$
5. $4(a+b)^2+32a+32b=$
6. $36r^3s-6r^2s+18r-3=$

1.3c Factoring Polynomials Using the Trial and Error Method

Expressing trinomials as the product of two binomials is one of the most common ways of factoring. In this section, we will review how to factor trinomials of the form ax^2+bx+c, where $a=1$ (Case I) and where $a\rangle 1$ (Case II), using a factoring method which in this book is referred to as the Trial and Error method.

Case I Factoring Trinomials of the Form ax^2+bx+c where $a=1$

To express a trinomial of the form ax^2+bx+c, where $a=1$, in its factored form $(x+m)(x+n)$, let us consider the product $(x+m)(x+n)$ and use the FOIL method to see how each term of the resulting trinomial is formed , i.e.,

$$(x+m)(x+n) = x\cdot x+n\cdot x+m\cdot x+m\cdot n = x^2+(m+n)x+mn$$

Note that the coefficient of the x term is the **sum** of m and n and the constant term is the **product** of m and n. We use this concept in order to express trinomials of the form x^2+bx+c in their equivalent factored form. In addition, in order to choose the right sign for the two integer numbers m and n, the knowledge of the following general sign rules for the indicated cases is needed:

General Sign Rules

When factoring a trinomial of the form x^2+ax+b to its equivalent factored form of $(x+m)(x+n)$, the sign of the two integer numbers m and n is determined based on the following cases:

Case I. If the sum of the two integer numbers $(a+b)$ is positive $(+)$ and the product of the two integer numbers $(a\cdot b)$ is negative $(-)$, then the two integer numbers m and n must have opposite signs.

Case II. If the sum of the two integer numbers $(a+b)$ is negative $(-)$ and the product of the two integer numbers $(a\cdot b)$ is positive $(+)$, then the two numbers must have the same sign. However, since the sum is negative, the two integer numbers m and n must both be negative.

Case III. If the sum of the two integer numbers $(a+b)$ is positive $(+)$ and the product of the two integer numbers $(a\cdot b)$ is also positive $(+)$, then the two integer numbers m and n must both be positive.

Case IV. If the sum of the two integer numbers $(a+b)$ is negative $(-)$ and the product of the two integer numbers $(a\cdot b)$ is also negative $(-)$, then the two integer numbers m and n must have opposite signs.

General Sign Rules - Summary

If sign of the sum $(a+b)$ is	and sign of the product $(a\cdot b)$ is	then, the two integer numbers m and n must
+	−	have opposite signs
−	+	have negative signs
+	+	have positive signs
−	−	have opposite signs

To factor a trinomial of the form $x^2 + ax + b$ to its equivalent factored form of $(x+m)(x+n)$ use the following steps:

Step 1 Obtain two numbers m and n whose sum equals to a and whose product equals to b.

Step 2 Write the trinomial in its factored form. Check the answer by using the FOIL method.

Examples with Steps

The following examples show the steps as to how trinomials of the form $x^2 + ax + b$ are factored:

Example 1.3-25

Factor $x^2 - 16x + 55$.

Solution:

Step 1 Obtain two numbers whose sum is -16 and whose product is 55. Note that since the sum is negative and the product is positive the two integer numbers must both be negative (Case II). Let's construct a table as follows:

Sum	*Product*
$-15-1=-16$	$(-15)\cdot(-1)=15$
$-14-2=-16$	$(-14)\cdot(-2)=28$
$-13-3=-16$	$(-13)\cdot(-3)=39$
$-12-4=-16$	$(-12)\cdot(-4)=48$
$\mathbf{-11-5=-16}$	$\mathbf{(-11)\cdot(-5)=55}$

Step 2 The last line contains the sum and the product of the two numbers that we need. Therefore, $x^2 - 16x + 55 = \mathbf{(x-11)(x-5)}$

Check: $(x-11)(x-5) = x\cdot x - 5\cdot x - 11\cdot x + (-11)(-5) = x^2 + (-5-11)x + 55 = x^2 - 16x + 55$

Example 1.3-26

Factor $x^2 + 2x - 48$.

Solution:

Step 1 Obtain two numbers whose sum is $+2$ and whose product is -48. Note that since the sum is positive and the product is negative the two integer numbers must have opposite signs (Case I). Let's construct a table as follows:

Sum	*Product*
$10-8=2$	$10\cdot(-8)=-80$
$9-7=2$	$9\cdot(-7)=-63$
$\mathbf{8-6=2}$	$\mathbf{8\cdot(-6)=-48}$

Step 2 The last line contains the sum and the product of the two numbers that we need. Therefore, $x^2 + 2x - 48 = \mathbf{(x+8)(x-6)}$

Check: $(x+8)(x-6) = x\cdot x - 6\cdot x + 8\cdot x + 8\cdot(-6) = x^2 + (-6+8)x - 48 = x^2 + 2x - 48$

Additional Examples - Factoring Trinomials of the Form ax^2+bx+c where $a=1$

The following examples further illustrate how to factor trinomials using the Trial and Error method:

Example 1.3-27: Factor $w^2+9w+20$.

Solution:

To factor the above trinomial we need to obtain two numbers whose sum is 9 and whose product is 20. Let's construct a table as follows:

Sum	*Product*
$1+8=9$	$1\cdot 8=8$
$2+7=9$	$2\cdot 7=14$
$3+6=9$	$3\cdot 6=18$
$\mathbf{4+5=9}$	$\mathbf{4\cdot 5=20}$

The last line contains the sum and the product of the two numbers that we need. Thus, $w^2+9w+20 = \mathbf{(w+5)(w+4)}$

Check: $(w+5)(w+4) = w\cdot w+4\cdot w+5\cdot w+4\cdot 5 = w^2+4w+5w+20 = w^2+(4+5)w+20 = w^2+9w+20$

Example 1.3-28: Factor $x^2-10x+12$.

Solution:

To factor the above trinomial we need to obtain two numbers whose sum is -10 and whose product is 12. Let's construct a table as follows:

Sum	*Product*
$-1-9=-10$	$(-1)\cdot(-9)=9$
$-2-8=-10$	$(-2)\cdot(-8)=16$
$-3-7=-10$	$(-3)\cdot(-7)=21$
$-4-6=-10$	$(-4)\cdot(-6)=24$
$-5-5=-10$	$(-5)\cdot(-5)=25$
$-1-12=-13$	$(-1)\cdot(-12)=12$
$-3-4=-7$	$(-3)\cdot(-4)=12$
$-2-6=-8$	$(-2)\cdot(-6)=12$

Since none of the numbers that add to the sum of -10, when multiplied, has a product of 12 and none of the factors of 12, when added, has a sum of -10. Therefore, we conclude that the trinomial $x^2-10x+12$ is **not factorable** using integers, or it is **prime**.

Note: A prime polynomial is one that is not factorable using integers. For example, $4x^2+6x+9$, $2y^2+y+7$, $6w^2+2w-5$, x^2+7y^2, y^2-6y+2, *and* $4x^2+9$ *are* ***prime*** *polynomials.*

Example 1.3-29: Factor x^2+4x-5.

Solution:

To factor the above trinomial we need to obtain two numbers whose sum is 4 and whose product is -5. Let's construct a table as follows:

Sum	***Product***
$\mathbf{5-1=4}$	$\mathbf{5\cdot(-1)=-5}$

In this case, at first trial we obtained the sum and the product of the two numbers that we need. Thus, $x^2+4x-5 = \mathbf{(x+5)(x-1)}$

Check: $(x+5)(x-1) = x\cdot x-1\cdot x+5\cdot x+5\cdot(-1) = x^2-x+5x-5 = x^2+(5-1)x-5 = x^2+4x-5$

Example 1.3-30: Factor $x^2-19x-66$.

Solution:

To factor the above trinomial we need to obtain two numbers whose sum is -19 and whose product is -66. Let's construct a table as follows:

Sum	***Product***
$10-29=-19$	$10\cdot(-29)=-290$
$9-28=-19$	$9\cdot(-28)=-252$
$8-27=-19$	$8\cdot(-27)=-216$
$7-26=-19$	$7\cdot(-26)=-182$
$6-25=-19$	$6\cdot(-25)=-150$
$5-24=-19$	$5\cdot(-24)=-120$
$4-23=-19$	$4\cdot(-23)=-92$
$\mathbf{3-22=-19}$	$\mathbf{3\cdot(-22)=-66}$

The last line contains the sum and the product of the two numbers that we need. Thus, $x^2-19x-66 = \mathbf{(x+3)(x-22)}$

Check: $(x+3)(x-22) = x\cdot x+(-22)\cdot x+3\cdot x+3\cdot(-22) = x^2-22x+3x-66 = x^2+(-22+3)x-66$
$= x^2-19x-66$

Practice Problems - Factoring Trinomials of the Form ax^2+bx+c where $a=1$

Section 1.3c Case I Practice Problems - Factor the following trinomials using the Trial and Error method:

1. $x^2-2x-15$
2. y^2-9y+8
3. $t^2+2t-15$
4. $y^2-2y+11$
5. $x^2+10x+21$
6. $u^2+4u-32$

Case II Factoring Trinomials of the Form ax^2+bx+c where $a \rangle 1$

To express a trinomial of the form ax^2+bx+c, where $a \rangle 1$, in its factored form $(lx+m)(kx+n)$, let us consider the product $(lx+m)(kx+n)$ and use the FOIL method to see how each term of the resulting trinomial is formed ,e.g.,

$$(lx+m)(kx+n) = (k\cdot l)\cdot x\cdot x+(l\cdot n)\cdot x+(k\cdot m)\cdot x+m\cdot n = (kl)x^2+(ln+km)x+mn$$

Note that the product of the coefficient of the x^2 term and the constant term is $kl\cdot mn$. In addition, the product of the coefficients of x is also $kl\cdot mn$. We use this concept in order to express trinomials of the form ax^2+bx+c, where $a \rangle 1$, in their equivalent factored form. The following show the steps in factoring this class of trinomials:

Step 1 Obtain two numbers m and n whose sum equals to b and whose product equals to $a\cdot c$.

Step 2 Rewrite the middle term of the trinomial as the sum of the two numbers found in Step 1.

Step 3 Write the trinomial in its factored form by grouping the first two terms and the last two terms (see Section 1.3b). Check the answer by using the FOIL method.

Examples with Steps

The following examples further illustrate how to factor trinomials of the form ax^2+bx+c, where $a \rangle 1$, using the Trial and Error method:

Example 1.3-31

Factor $6x^2+23x+20$.

Solution:

Step 1 Obtain two numbers whose sum is 23 and whose product is $6\cdot 20=120$. Let's construct a table as follows:

Sum	*Product*
$12+11=23$	$12\cdot 11=132$
$13+10=23$	$13\cdot 10=130$
$14+9=23$	$14\cdot 9=126$
$\mathbf{15+8=23}$	$\mathbf{15\cdot 8=120}$

The last line contains the sum and the product of the two numbers that we need. Therefore,

Step 2 $6x^2+23x+20 = 6x^2+(15+8)x+20 = 6x^2+15x+8x+20 = 3x(2x+5)+4(2x+5)$

Step 3 $3x(2x+5)+4(2x+5) = \mathbf{(2x+5)(3x+4)}$

Check: $(2x+5)(3x+4) = (2\cdot 3)\cdot x\cdot x+(2\cdot 4)\cdot x+(5\cdot 3)\cdot x+(5\cdot 4) = 6x^2+8x+15x+20$
$= 6x^2+(8+15)x+20 = 6x^2+23x+20$

Example 1.3-32

Factor $10x^2 - 9x - 91$.

Solution:

Step 1 Obtain two numbers whose sum is -9 and whose product is $10 \cdot (-91) = -910$. Let's construct a table as follows:

Sum	*Product*
$20 - 29 = -9$	$20 \cdot (-29) = -580$
$21 - 30 = -9$	$21 \cdot (-30) = -630$
$22 - 31 = -9$	$22 \cdot (-31) = -682$
$23 - 32 = -9$	$23 \cdot (-32) = -736$
$24 - 33 = -9$	$24 \cdot (-33) = -792$
$25 - 34 = -9$	$25 \cdot (-34) = -850$
$\mathbf{26 - 35 = -9}$	$\mathbf{26 \cdot (-35) = -910}$

The last line contains the sum and the product of the two numbers that we need. Therefore,

Step 2 $10x^2 - 9x - 91 = 10x^2 + (26 - 35)x - 91 = 10x^2 + 26x - 35x - 91 = 2x(5x + 13) - 7(5x + 13)$

Step 3 $2x(5x + 13) - 7(5x + 13) = \mathbf{(5x + 13)(2x - 7)}$

Check: $(5x + 13)(2x - 7) = (5 \cdot 2) \cdot x \cdot x + (5 \cdot -7) \cdot x + (13 \cdot 2) \cdot x + (13 \cdot -7) = 10x^2 - 35x + 26x - 91$
$= 10x^2 + (-35 + 26)x - 91 = 10x^2 - 9x - 91$

Additional Examples - Factoring Trinomials of the Form $ax^2 + bx + c$ where $a \rangle 1$

The following examples further illustrate how to factor trinomials using the Trial and Error method:

Example 1.3-33: Factor $6x^2 + 16x + 10$.

Solution:

To factor the above trinomial we need to obtain two numbers whose sum is 16 and whose product is $6 \cdot 10 = 60$. Let's construct a table as follows:

Sum	*Product*
$8 + 8 = 16$	$8 \cdot 8 = 64$
$9 + 7 = 16$	$9 \cdot 7 = 63$
$\mathbf{10 + 6 = 16}$	$\mathbf{10 \cdot 6 = 60}$

The last line contains the sum and the product of the two numbers that we need. Therefore,
$6x^2 + 16x + 10 = 6x^2 + (10 + 6)x + 10 = 6x^2 + 10x + 6x + 10 = 2x(3x + 5) + 2(3x + 5) = \mathbf{(3x + 5)(2x + 2)}$

Check: $(3x + 5)(2x + 2) = (3 \cdot 2) \cdot x \cdot x + (3 \cdot 2) \cdot x + (2 \cdot 5) \cdot x + 5 \cdot 2 = 6x^2 + 6x + 10x + 10$
$= 6x^2 + (6 + 10)x + 10 = 6x^2 + 16x + 10$

Example 1.3-34: Factor $5x^2+8x+3$.

Solution:

To factor the above trinomial we need to obtain two numbers whose sum is 8 and whose product is $5\cdot 3=15$. Let's construct a table as follows:

Sum	*Product*
$1+7=8$	$1\cdot 7=7$
$2+6=8$	$2\cdot 6=12$
$\mathbf{3+5=8}$	$\mathbf{3\cdot 5=15}$
$4+4=8$	$4\cdot 4=16$

The third line contains the sum and the product of the two numbers that we need. Therefore,

$5x^2+8x+3 = 5x^2+(3+5)x+3 = 5x^2+3x+5x+3 = x(5x+3)+(5x+3) = \mathbf{(5x+3)(x+1)}$

Check: $(5x+3)(x+1) = (5\cdot 1)\cdot x\cdot x+(5\cdot 1)\cdot x+(3\cdot 1)\cdot x+3\cdot 1 = 5x^2+5x+3x+3 = 5x^2+(5+3)x+3$

$= 5x^2+8x+3$

Example 1.3-35: Factor $6x^2+19x+10$.

Solution:

To factor the above trinomial we need to obtain two numbers whose sum is 19 and whose product is $6\cdot 10=60$. Let's construct a table as follows:

Sum	*Product*
$1+18=19$	$1\cdot 18=18$
$2+17=19$	$2\cdot 17=34$
$3+16=19$	$3\cdot 16=48$
$\mathbf{4+15=19}$	$\mathbf{4\cdot 15=60}$
$5+14=19$	$5\cdot 14=70$

The fourth line contains the sum and the product of the two numbers that we need. Thus,

$6x^2+19x+10 = 6x^2+(4+15)x+10 = 6x^2+4x+15x+10 = 2x(3x+2)+5(3x+2) = \mathbf{(2x+5)(3x+2)}$

Check: $(2x+5)(3x+2) = (2\cdot 3)\cdot x\cdot x+(2\cdot 2)\cdot x+(5\cdot 3)\cdot x+5\cdot 2 = 6x^2+4x+15x+10$

$= 6x^2+(4+15)x+10 = 6x^2+19x+10$

Example 1.3-36: Factor $2w^2-13w+15$.

Solution:

To factor the above trinomial we need to obtain two numbers whose sum is -13 and whose product is $2\cdot 15=30$. Let's construct a table as follows:

Sum	*Product*
$-1-12=-13$	$(-1)\cdot(-12)=12$
$-2-11=-13$	$(-2)\cdot(-11)=22$
$\mathbf{-3-10=-13}$	$\mathbf{(-3)\cdot(-10)=30}$
$-4-9=-13$	$(-4)\cdot(-9)=36$

The third line contains the sum and the product of the two numbers that we need. Thus,

$2w^2-13w+15 = 2w^2+(-3-10)w+15 = 2w^2-3w-10w+15 = w(2w-3)-5(2w-3) = \mathbf{(2w-3)(w-5)}$

Check: $(2w-3)(w-5) = (2\cdot1)\cdot w\cdot w+(2\cdot-5)\cdot w+(-3\cdot1)\cdot w+(-3\cdot-5) = 2w^2-10w-3w+15$
$= 2w^2+(-10-3)w+15 = 2w^2-13w+15$

Example 1.3-37: Factor $5y^2-16y+3$.

Solution:

To factor the above trinomial we need to obtain two numbers whose sum is -16 and whose product is $5\cdot3=15$. Let's construct a table as follows:

Sum	*Product*
$-5-11=-16$	$(-5)\cdot(-11)=55$
$-4-12=-16$	$(-4)\cdot(-12)=48$
$-3-13=-16$	$(-3)\cdot(-13)=39$
$-2-14=-16$	$(-2)\cdot(-14)=28$
$\mathbf{-1-15=-16}$	$\mathbf{(-1)\cdot(-15)=15}$

The last line contains the sum and the product of the two numbers that we need. Thus,
$5y^2-16y+3 = 5y^2+(-1-15)y+3 = 5y^2-y-15y+3 = y(5y-1)-3(5y-1) = \mathbf{(5y-1)(y-3)}$

Check: $(5y-1)(y-3) = 5\cdot y\cdot y+(5\cdot-3)\cdot y-y+(-1)\cdot(-3) = 5y^2-15y-y+3 = 5y^2+(-15-1)y+3$
$= 5y^2-16y+3$

Practice Problems - Factoring Trinomials of the Form ax^2+bx+c where $a \rangle 1$

Section 1.3c Case II Practice Problems - Factor the following trinomials using the Trial and Error method:

1. $10x^2+11x-35$
2. $6x^2-x-12$
3. $-7x^2+46x+21$
4. $6x^2-11xy+3y^2$ y is variable
5. $6x^2+x-40$
6. $2x^2+3x-27$

1.3d Other Factoring Methods for Polynomials

The key to successful factorization of polynomials is recognition and use of the right factoring method. In this section we will review how to factor binomials of the form a^2-b^2 (Case I), $a^3 \pm b^3$ (case II), including Perfect Square Trinomials (Case III) by using formulas that reduce the binomials to lower product terms.

Case I Factoring Polynomials Using the Difference of Two Squares Method

Binomials of the form a^2-b^2 are factored to product of two first degree binomials using the following factorization method:

$$a^2-b^2=(a-b)(a+b)$$

Note that a^2+b^2 is a prime polynomial and can not be factored. The difference of two square terms can be factored using the following steps:

Step 1 Factor the common terms and write the binomial in the standard form of a^2-b^2.

Step 2 Write the binomial in its equivalent factorable form. Check the answer using the FOIL method.

Examples with Steps

The following examples show the steps as to how binomials of the form a^2-b^2 are factored:

Example 1.3-38 Factor $5k^4-3125$ completely.

Solution:

Step 1 $\boxed{5k^4-3125}=\boxed{5\left(k^4-625\right)}=\boxed{5\left(k^{2^2}-25^2\right)}$

Step 2 $\boxed{5\left(k^{2^2}-25^2\right)}=\boxed{5\left(k^2-25\right)\left(k^2+25\right)}=\boxed{5\left(k^2-5^2\right)\left(k^2+25\right)}=\boxed{\mathbf{5(k-5)(k+5)\left(k^2+25\right)}}$

Check: $5(k-5)(k+5)\left(k^2+25\right) = 5(k\cdot k+5\cdot k-5\cdot k-5\cdot 5)\left(k^2+25\right) = 5\left(k^2+5k-5k-25\right)\left(k^2+25\right)$

$= 5\left(k^2-25\right)\left(k^2+25\right) = 5\left(k^2\cdot k^2+25\cdot k^2-25\cdot k^2-25\cdot 25\right) = 5\left(k^4+25k^2-25k^2-625\right)$

$= 5\left(k^4-625\right) = 5k^4-3125$

Example 1.3-39 Factor $81m^4-n^4$ completely.

Solution:

Step 1 $\boxed{81m^4-n^4}=\boxed{\left(9m^2\right)^2-n^{2^2}}$

Step 2 $\boxed{\left(9m^2\right)^2-n^{2^2}}=\boxed{\left(9m^2-n^2\right)\left(9m^2+n^2\right)}=\boxed{\left(3^2m^2-n^2\right)\left(9m^2+n^2\right)}$

$=\boxed{\left[(3m)^2-n^2\right]\left(9m^2+n^2\right)}=\boxed{\mathbf{(3m-n)(3m+n)\left(9m^2+n^2\right)}}$

Check: $(3m-n)(3m+n)\left(9m^2+n^2\right) = (3m\cdot 3m+3m\cdot n-3m\cdot n-n\cdot n)\left(9m^2+n^2\right)$

$= \left(9m^2+3mn-3mn-n^2\right)\left(9m^2+n^2\right) = \left(9m^2-n^2\right)\left(9m^2+n^2\right)$

$= 9m^2\cdot 9m^2+9m^2\cdot n^2-n^2\cdot 9m^2-n^2\cdot n^2 = 81m^4+9m^2n^2-9m^2n^2-n^4 = 81m^4-n^4$

Additional Examples - Factoring Polynomials Using the Difference of Two Squares Method

The following examples further illustrate how to factor binomials of the form a^2-b^2:

Example 1.3-40:

$\boxed{4x^6-64x^2} = \boxed{4x^2\left(x^4-16\right)} = \boxed{4x^2\left(x^{2^2}-4^2\right)} = \boxed{4x^2\left(x^2-4\right)\left(x^2+4\right)} = \boxed{4x^2\left(x^2-2^2\right)\left(x^2+4\right)}$

$= \boxed{\mathbf{4x^2(x-2)(x+2)\left(x^2+4\right)}}$

Check: $4x^2(x-2)(x+2)\left(x^2+4\right) = 4x^2(x\cdot x+2\cdot x-2\cdot x-2\cdot 2)\left(x^2+4\right) = 4x^2\left(x^2+2x-2x-4\right)\left(x^2+4\right)$

$= 4x^2\left(x^2-4\right)\left(x^2+4\right) = 4x^2\left(x^2\cdot x^2+4\cdot x^2-4\cdot x^2-4\cdot 4\right) = 4x^2\left(x^4+4x^2-4x^2-16\right)$

$= 4x^2\left(x^4-16\right) = 4x^2\cdot x^4-4x^2\cdot 16 = 4x^6-64x^2$

Example 1.3-41:

$\boxed{w^8-256} = \boxed{w^{4^2}-16^2} = \boxed{\left(w^4-16\right)\left(w^4+16\right)} = \boxed{\left(w^{2^2}-4^2\right)\left(w^4+16\right)} = \boxed{\left(w^2-4\right)\left(w^2+4\right)\left(w^4+16\right)}$

$= \boxed{\left(w^2-2^2\right)\left(w^2+4\right)\left(w^4+16\right)} = \boxed{\mathbf{(w-2)(w+2)\left(w^2+4\right)\left(w^4+16\right)}}$

Check: $(w-2)(w+2)\left(w^2+4\right)\left(w^4+16\right) = (w\cdot w+2\cdot w-2\cdot w-2\cdot 2)\left(w^2+4\right)\left(w^4+16\right)$

$= \left(w^2+2w-2w-4\right)\left(w^2+4\right)\left(w^4+16\right) = \left(w^2-4\right)\left(w^2+4\right)\left(w^4+16\right)$

$= \left(w^2\cdot w^2+4\cdot w^2-4\cdot w^2-4\cdot 4\right)\left(w^4+16\right) = \left(w^4+4w^2-4w^2-16\right)\left(w^4+16\right)$

$= \left(w^4-16\right)\left(w^4+16\right) = w^4\cdot w^4+16\cdot w^4-16\cdot w^4-16\cdot 16 = w^8+16w^4-16w^4-256 = w^8-256$

Example 1.3-42:

$\boxed{8d^4-200d^2} = \boxed{8d^2\left(d^2-25\right)} = \boxed{8d^2\left(d^2-5^2\right)} = \boxed{\mathbf{8d^2(d-5)(d+5)}}$

Check: $8d^2(d-5)(d+5) = 8d^2(d\cdot d+5\cdot d-5\cdot d-5\cdot 5) = 8d^2\left(d^2+5d-5d-25\right) = 8d^2\left(d^2-25\right)$

$= 8d^2\cdot d^2-8\cdot 25d^2 = 8d^4-200d^2$

Example 1.3-43:

$\boxed{x^2-(y+5)^2} = \boxed{[x-(y+5)][x+(y+5)]} = \boxed{\mathbf{(x-y-5)\,(x+y+5)}}$

Check: $(x-y-5)\,(x+y+5) = x\cdot x+x\cdot y+5\cdot x-x\cdot y-y\cdot y-5\cdot y-5\cdot x-5\cdot y-5\cdot 5$

$= x^2+xy+5x-xy-y^2-5y-5x-5y-25 = x^2-y^2-5y-5y-25 = x^2-y^2-10y-25$

$$= x^2 - \left(y^2 + 10y + 25\right) = x^2 - (y+5)^2$$

Example 1.3-44:

$$\boxed{(u+3)^2 - v^2} = \boxed{[(u+3)-v][(u+3)+v]} = \boxed{(u+3-v)\,(u+3+v)} = \boxed{\mathbf{(u-v+3)\,(u+v+3)}}$$

Check: $(u-v+3)\,(u+v+3) = u\cdot u + u\cdot v + 3\cdot u - u\cdot v - v\cdot v - 3\cdot v + 3\cdot u + 3\cdot v + 3\cdot 3$

$$= u^2 + \cancel{uv} + 3u - \cancel{uv} - v^2 - \cancel{3v} + 3u + \cancel{3v} + 9 = u^2 - v^2 + 6u + 9 = \left(u^2 + 6u + 9\right) - v^2 = (u+3)^2 - v^2$$

Example 1.3-45:

$$\boxed{x^2 - y^2 - 2y - 1} = \boxed{x^2 - \left(y^2 + 2y + 1\right)} = \boxed{x^2 - (y+1)^2} = \boxed{[x-(y+1)][x+(y+1)]} = \boxed{\mathbf{(x-y-1)(x+y+1)}}$$

Check: $(x-y-1)(x+y+1) = x\cdot x + x\cdot y + 1\cdot x - x\cdot y - y\cdot y - 1\cdot y - 1\cdot x - 1\cdot y - 1\cdot 1$

$$= x^2 + \cancel{xy} + \cancel{x} - \cancel{xy} - y^2 - y - \cancel{x} - y - 1 = x^2 - y^2 - y - y - 1 = x^2 - y^2 - 2y - 1$$

Example 1.3-46:

$$\boxed{(x+2)^2 - (y+4)^2} = \boxed{[(x+2)-(y+4)][(x+2)+(y+4)]} = \boxed{(x+2-y-4)(x+2+y+4)} = \boxed{\mathbf{(x-y-2)(x+y+6)}}$$

Check: $(x-y-2)(x+y+6) = x\cdot x + x\cdot y + 6\cdot x - x\cdot y - y\cdot y - 6\cdot y - 2\cdot x - 2\cdot y - 2\cdot 6$

$$= x^2 + \cancel{xy} + 6x - \cancel{xy} - y^2 - 6y - 2x - 2y - 12 = x^2 + 6x - 2x - y^2 - 6y - 2y - 12$$

$$= x^2 + 4x - y^2 - 8y - 12 = x^2 + 4x - y^2 - 8y + (-16+4) = x^2 + 4x + 4 - y^2 - 8y - 16$$

$$= \left(x^2 + 4x + 4\right) - \left(y^2 + 8y + 16\right) = (x+2)^2 - (y+4)^2$$

Practice Problems - Factoring Polynomials Using the Difference of Two Squares Method

Section 1.3d Case I Practice Problems - Use the Difference of Two Squares method to factor the following polynomials:

1. $x^3 - 16x =$
2. $(x+1)^2 - (y+3)^2 =$
3. $t^5 - 81t =$
4. $\left(x^2 + 10x + 25\right) - y^2 =$
5. $c^4 - 9c^2 =$
6. $p^2 - q^2 - 4q - 4 =$

Case II Factoring Polynomials Using the Sum and Difference of Two Cubes Method

To factor binomials of the form a^3+b^3 or a^3-b^3 we use the following formulas:

$$a^3+b^3=(a+b)(a^2-ab+b^2)$$

$$a^3-b^3=(a-b)(a^2+ab+b^2)$$

Students are encouraged to memorize these two formulas in order to successfully factor this class of polynomials. The sum and difference of two cubed binomial terms can be factored using the following steps:

Step 1 Write the binomial in the standard form of a^3+b^3 or a^3-b^3.

Step 2 Write the binomial in its equivalent factorable form. Check the answer by multiplication.

Examples with Steps

The following examples show the steps as to how binomials of the form a^3+b^3 and a^3-b^3 are factored:

Example 1.3-47 Factor $3x^7+81x^4$.

Solution:

Step 1 $\boxed{3x^7+81x^4}=\boxed{3x^4(x^3+27)}=\boxed{3x^4(x^3+3^3)}$

Step 2 $\boxed{3x^4(x^3+3^3)}=\boxed{3x^4[(x+3)(x^2-3\cdot x+3^2)]}=\boxed{\mathbf{3x^4[(x+3)(x^2-3x+9)]}}$

Check: $3x^4[(x+3)(x^2-3x+9)] = 3x^4(x\cdot x^2-3x\cdot x+9\cdot x+3\cdot x^2-3\cdot 3x+3\cdot 9)$

$= 3x^4(x^3-3x^2+9x+3x^2-9x+27) = 3x^4(x^3+27) = 3x^4\cdot x^3+27\cdot 3x^4 = 3x^{4+3}+81x^4$

$= 3x^7+81x^4$

Example 1.3-48 Factor $2a^3-250$.

Solution:

Step 1 $\boxed{2a^3-250}=\boxed{2(a^3-125)}=\boxed{2(a^3-5^3)}$

Step 2 $\boxed{2(a^3-5^3)}=\boxed{2(a-5)(a^2+5\cdot a+5^2)}=\boxed{\mathbf{2(a-5)(a^2+5a+25)}}$

Check: $2(a-5)(a^2+5a+25) = 2(a\cdot a^2+5a\cdot a+25\cdot a-5\cdot a^2-5\cdot 5a-5\cdot 25)$

$= 2(a^3+5a^2+25a-5a^2-25a-125) = 2(a^3-125) = 2a^3-250$

Additional Examples - Factoring Polynomials Using the Sum and Difference of Two Cubes Method

The following examples further illustrate how to factor binomials of the form $a^3\pm b^3$ using the sum and difference of two cubes method:

Example 1.3-49:

$\boxed{x^3+1}=\boxed{x^3+1^3}=\boxed{(x+1)(x^2-1\cdot x+1^2)}=\boxed{\mathbf{(x+1)(x^2-x+1)}}$

Check: $(x+1)(x^2-x+1) = x\cdot x^2 - x\cdot x + 1\cdot x + 1\cdot x^2 - 1\cdot x + 1\cdot 1 = x^3 - x^2 + x + x^2 - x + 1 = x^3 + 1$

Example 1.3-50:

$$\boxed{x^3-1} = \boxed{x^3-1^3} = \boxed{(x-1)(x^2+1\cdot x+1^2)} = \boxed{\mathbf{(x-1)(x^2+x+1)}}$$

Check: $(x-1)(x^2+x+1) = x\cdot x^2 + x\cdot x + 1\cdot x - 1\cdot x^2 - 1\cdot x - 1\cdot 1 = x^3 + x^2 + x - x^2 - x - 1 = x^3 - 1$

Example 1.3-51:

$$\boxed{x^3y^3-27} = \boxed{x^3y^3-3^3} = \boxed{(xy)^3-3^3} = \boxed{(xy-3)\left[(xy)^2+3\cdot xy+3^2\right]} = \boxed{\mathbf{(xy-3)\left[(xy)^2+3xy+9\right]}}$$

Check: $(xy-3)\left[(xy)^2+3xy+9\right] = xy\cdot(xy)^2 + 3xy\cdot xy + 9\cdot xy - 3\cdot(xy)^2 - 3\cdot 3xy - 3\cdot 9$

$= (xy)^3 + 3(xy)^2 + 9xy - 3(xy)^2 - 9xy - 27 = (xy)^3 - 27 = x^3y^3 - 27$

Example 1.3-52:

$$\boxed{a^4-a} = \boxed{a(a^3-1)} = \boxed{a(a^3-1^3)} = \boxed{a\left[(a-1)(a^2+1\cdot a+1^2)\right]} = \boxed{\mathbf{a\left[(a-1)(a^2+a+1)\right]}}$$

Check: $a\left[(a-1)(a^2+a+1)\right] = a(a\cdot a^2 + a\cdot a + 1\cdot a - 1\cdot a^2 - 1\cdot a - 1\cdot 1) = a(a^3 + a^2 + a - a^2 - a - 1)$

$= a(a^3-1) = a^4 - a$

Example 1.3-53:

$$\boxed{r^3-8s^3} = \boxed{r^3-2^3s^3} = \boxed{r^3-(2s)^3} = \boxed{(r-2s)\left[r^2+r\cdot 2s+(2s)^2\right]} = \boxed{\mathbf{(r-2s)(r^2+2rs+4s^2)}}$$

Check: $(r-2s)(r^2+2rs+4s^2) = r\cdot r^2 + r\cdot 2rs + r\cdot 4s^2 - 2s\cdot r^2 - 2s\cdot 2rs - 2s\cdot 4s^2$

$= r^3 + 2r^2s + 4rs^2 - 2r^2s - 4rs^2 - 8s^3 = r^3 - 8s^3$

Example 1.3-54:

$$\boxed{x^3+27y^3} = \boxed{x^3+3^3y^3} = \boxed{x^3+(3y)^3} = \boxed{(x+3y)\left[x^2-x\cdot 3y+(3y)^2\right]} = \boxed{\mathbf{(x+3y)(x^2-3xy+9y^2)}}$$

Check: $(x+3y)(x^2-3xy+9y^2) = x\cdot x^2 - x\cdot 3xy + x\cdot 9y^2 + 3y\cdot x^2 - 3y\cdot 3xy + 3y\cdot 9y^2$

$= x^3 - 3x^2y + 9xy^2 + 3x^2y - 9xy^2 + 27y^3 = x^3 + 27y^3$

Example 1.3-55:

$$\boxed{c^3-27d^3} = \boxed{c^3-3^3d^3} = \boxed{c^3-(3d)^3} = \boxed{(c-3d)\left[c^2+c\cdot 3d+(3d)^2\right]} = \boxed{\mathbf{(c-3d)(c^2+3cd+9d^2)}}$$

Check: $(c-3d)(c^2+3cd+9d^2) = c\cdot c^2 + c\cdot 3cd + c\cdot 9d^2 - 3d\cdot c^2 - 3d\cdot 3cd - 3d\cdot 9d^2$

$= c^3 + 3c^2d + 9cd^2 - 3c^2d - 9cd^2 - 27d^3 = c^3 - 27d^3$

Practice Problems - Factoring Polynomials Using the Sum and Difference of Two Cubes Method

Section 1.3d Case II Practice Problems - Use the sum and difference of two cubes method to factor the following polynomials:

1. $4x^6+4 =$
2. $x^6y^6+8 =$
3. $(x+2)^3-y^3 =$
4. $2r^6-128 =$
5. $(x-7)^3+y^3 =$
6. $x^6y^5+x^3y^2 =$

Case III Factoring Perfect Square Trinomials

Trinomials of the form $a^2+2ab+b^2$ and $a^2-2ab+b^2$ are called perfect square trinomials. Note that these types of polynomials are easy to recognize because their first and last terms are always square and their middle term is twice the product of the quantities being squared in the first and last terms. Once perfect square trinomials are identified, they can then be represented in their equivalent factored form as shown below:

$$a^2+2ab+b^2=(a+b)^2$$

$$a^2-2ab+b^2=(a-b)^2$$

For example, $36x^2-24x+4$, $1+8y+16y^2$, $25x^2+30xy+9y^2$, and $49m^2-70mn+25n^2$ are perfect square trinomials because:

1. Their first term is a square, i.e., $(6x)^2$; $(1)^2$; $(5x)^2$; $(7m)^2$.
2. Their last term is a square, i. e., $(-2)^2$; $(4y)^2$; $(3y)^2$; $(-5n)^2$, and
3. Their middle term is twice the product of the quantities being squared in the first and last terms, i.e., $2\cdot(6x\cdot-2)$; $2\cdot(1\cdot 4y)$; $2\cdot(5x\cdot 3y)$; $2\cdot(7m\cdot-5n)$.

Therefore, the above examples can be represented in their equivalent factored form as: $(6x-2)^2$; $(1+4y)^2$; $(5x+3y)^2$; and $(7m-5n)^2$, respectively.

The following show the steps as to how perfect square trinomials are represented in their equivalent factored form:

Step 1 Write the trinomial in descending order.

Step 2 Check and see if the trinomial match the general forms $a^2+2ab+b^2$ or $a^2-2ab+b^2$.

Step 3 Write the trinomial in its equivalent form, i.e., $(a+b)^2$ or $(a-b)^2$.

Examples with Steps

The following examples show the steps as to how perfect square trinomials are factored:

Example 1.3-56

Factor $25y^2+16+40y$.

Solution:

Step 1 $\boxed{25y^2+16+40y}=\boxed{25y^2+40y+16}$

Step 2 $\boxed{25y^2+40y+16}=\boxed{5^2y^2+40y+4^2}=\boxed{(5y)^2+2\cdot(5y\cdot 4)+4^2}$

Step 3 $\boxed{(5y)^2+2\cdot(5y\cdot 4)+4^2}=\boxed{\mathbf{(5y+4)^2}}$

Example 1.3-57

Factor $16x^2+24xy+9y^2$.

Solution:

Step 1 $\boxed{\textit{Not Applicable}}$

Step 2 $\boxed{16x^2+24xy+9y^2} = \boxed{4^2x^2+24xy+3^2y^2} = \boxed{(4x)^2+2\cdot(4x\cdot 3y)+(3y)^2}$

Step 3 $\boxed{(4x)^2+2\cdot(4x\cdot 3y)+(3y)^2} = \boxed{\mathbf{(4x+3y)^2}}$

Additional Examples - Factoring Perfect Square Trinomials

The following examples further illustrate how to factor perfect square trinomials:

Example 1.3-58:

$\boxed{x^2+4x+4} = \boxed{x^2+4x+2^2} = \boxed{x^2+2\cdot(x\cdot 2)+2^2} = \boxed{\mathbf{(x+2)^2}}$

Example 1.3-59:

$\boxed{9-6y+y^2} = \boxed{y^2-6y+9} = \boxed{y^2-2\cdot(y\cdot 3)+3^2} = \boxed{\mathbf{(y-3)^2}}$

Example 1.3-60:

$\boxed{12t+9+4t^2} = \boxed{4t^2+12t+9} = \boxed{2^2t^2+12t+3^2} = \boxed{(2t)^2+2\cdot(2t\cdot 3)+3^2} = \boxed{\mathbf{(2t+3)^2}}$

Example 1.3-61:

$\boxed{16x^2-40x+25} = \boxed{4^2x^2-40x+5^2} = \boxed{(4x)^2-2\cdot(4x\cdot 5)+5^2} = \boxed{\mathbf{(4x-5)^2}}$

Example 1.3-62:

$\boxed{10x+1+25x^2} = \boxed{25x^2+10x+1} = \boxed{5^2x^2+10x+1^2} = \boxed{(5x)^2+2\cdot(5x\cdot 1)+1^2} = \boxed{\mathbf{(5x+1)^2}}$

Example 1.3-63:

$\boxed{4+9t^2-12t} = \boxed{9t^2-12t+4} = \boxed{3^2t^2-12t+2^2} = \boxed{(3t)^2-2\cdot(3t\cdot 2)+2^2} = \boxed{\mathbf{(3t-2)^2}}$

Example 1.3-64:

$\boxed{9p^2-30pq+25q^2} = \boxed{3^2p^2-30pq+5^2q^2} = \boxed{(3p)^2-2\cdot(3p\cdot 5q)+(5q)^2} = \boxed{\mathbf{(3p-5q)^2}}$

Example 1.3-65:

$\boxed{121u^4-88u^2v^2+16v^4} = \boxed{11^2u^{2^2}-88u^2v^2+4^2v^{2^2}} = \boxed{\left(11u^2\right)^2-2\cdot\left(11u^2\cdot 4v^2\right)+\left(4v^2\right)^2} = \boxed{\mathbf{\left(11u^2-4v^2\right)^2}}$

Practice Problems - Factoring Perfect Square Trinomials

Section 1.3d Case III Practice Problems - Factor the following trinomials:

1. $x^2+18x+81 =$
2. $9+64p^2-48p =$
3. $9w^2+25+30w =$
4. $25+k^2-10k =$
5. $49x^2-84x+36 =$
6. $1+16z+64z^2 =$

1.4 Quadratic Equations and Factoring

In this section the different methods for factoring quadratic equations are reviewed. The Quadratic Formula and its use for solving quadratic equations is addressed in Section 1.4a. Solving quadratic equations using the Quadratic Formula, the Square Root Property, and Completing the Square method are discussed in Sections 1.4b, 1.4c, and 1.4d, respectively. Selection of the best factoring method for solving polynomials or quadratic equations is discussed in Section 1.4e.

1.4a Quadratic Equations and the Quadratic Formula

A quadratic equation is an equation in which the highest power of the variable is 2. For example, $3x^2-16x+5=0$, $x^2=16$, $w^2+9w=0$, $x^2-4x+3=0$, $x^2=-11x-24$, and $y^2-4=0$ are all examples of quadratic equations. Note that any equation that can be written in the form of $ax^2+bx+c=0$, where a, b, and c are real numbers and $a\neq 0$, is called a **quadratic equation**. A quadratic equation represented in the form of $ax^2+bx+c=0$ is said to be in its **standard form**. In the following sections we will review how to solve and represent the solutions to quadratic equations in factored form. However, in order to solve any quadratic equation we first need to become familiar with the quadratic formula.

The Quadratic Formula

To derive the quadratic formula we start with the standard quadratic equation $ax^2+bx+c=0$, where a, b, and c are real numbers and use the method of completing the square to solve the equation as follows:

Step 1 Add $-c$ to both sides of the equation.

$$ax^2+bx+c-c=-c \; ; \; ax^2+bx=-c$$

Step 2 Divide both sides of the equation by a.

$$\frac{ax^2}{a}+\frac{bx}{a}=-\frac{c}{a} \; ; \; x^2+\frac{bx}{a}=-\frac{c}{a}$$

Step 3 Divide $\frac{b}{a}$, the coefficient of x, by 2 and square the term to obtain $\left(\frac{b}{2a}\right)^2$. Add $\left(\frac{b}{2a}\right)^2$ to both sides of the equation.

$$x^2+\frac{b}{a}x+\left(\frac{b}{2a}\right)^2=-\frac{c}{a}+\left(\frac{b}{2a}\right)^2.$$

Step 4 Write the left hand side of the equation , which is a perfect square trinomial, in its equivalent square form.

$$\left(x+\frac{b}{2a}\right)^2=-\frac{c}{a}+\left(\frac{b}{2a}\right)^2$$

Step 5 Simplify the right hand side of the equation using the fraction techniques.

$$\left(x+\frac{b}{2a}\right)^2=-\frac{c}{a}+\left(\frac{b}{2a}\right)^2 \; ; \; \left(x+\frac{b}{2a}\right)^2=-\frac{c}{a}+\frac{b^2}{4a^2} \; ; \; \left(x+\frac{b}{2a}\right)^2=\frac{\left(4a^2\cdot -c\right)+\left(a\cdot b^2\right)}{4a^2\cdot a}$$

$$; \; \left(x+\frac{b}{2a}\right)^2=\frac{ab^2-4a^2c}{4a^3} \; ; \; \left(x+\frac{b}{2a}\right)^2=\frac{a\left(b^2-4ac\right)}{4a^3} \; ; \; \left(x+\frac{b}{2a}\right)^2=\frac{b^2-4ac}{4a^2}$$

Step 6 Take the square root of both sides of the equation.

$$\sqrt{\left(x+\frac{b}{2a}\right)^2}=\pm\sqrt{\frac{b^2-4ac}{4a^2}} \; ; \; x+\frac{b}{2a}=\pm\frac{\sqrt{b^2-4ac}}{\sqrt{2^2a^2}} \; ; \; x+\frac{b}{2a}=\pm\frac{\sqrt{b^2-4ac}}{2a}$$

Step 7 Solve for x by adding $-\frac{b}{2a}$ to both sides of the equation.

$$x+\frac{b}{2a}-\frac{b}{2a}=-\frac{b}{2a}\pm\frac{\sqrt{b^2-4ac}}{2a} \; ; \; x=-\frac{b}{2a}\pm\frac{\sqrt{b^2-4ac}}{2a} \; ; \; x=\frac{-b\pm\sqrt{b^2-4ac}}{2a}$$

The equation $\boldsymbol{x=\frac{-b\pm\sqrt{b^2-4ac}}{2a}}$ is referred to as the **quadratic formula**. Note that the quadratic formula has two solutions $x=\frac{-b+\sqrt{b^2-4ac}}{2a}$ and $x=\frac{-b-\sqrt{b^2-4ac}}{2a}$. We use these solutions to write the quadratic equation $ax^2+bx+c=0$ in its equivalent factored form, i.e.,

$$\boldsymbol{ax^2+bx+c=0} \text{ is factorable to } \boldsymbol{\left(x-\frac{-b+\sqrt{b^2-4ac}}{2a}\right)\left(x+\frac{b+\sqrt{b^2-4ac}}{2a}\right)=0}$$

Let's check the above factored product using the FOIL method. The result should be equal to $ax^2+bx+c=0$.

Check: $\left(x-\frac{-b+\sqrt{b^2-4ac}}{2a}\right)\left(x+\frac{b+\sqrt{b^2-4ac}}{2a}\right)=0 \; ; \; \left(x+\frac{b-\sqrt{b^2-4ac}}{2a}\right)\left(x+\frac{b+\sqrt{b^2-4ac}}{2a}\right)=0$

$$; \; x\cdot x+\left(\frac{b+\sqrt{b^2-4ac}}{2a}\right)\cdot x+\left(\frac{b-\sqrt{b^2-4ac}}{2a}\right)\cdot x+\left(\frac{b+\sqrt{b^2-4ac}}{2a}\right)\cdot\left(\frac{b-\sqrt{b^2-4ac}}{2a}\right)=0$$

$$; \; x^2+\left(\frac{b+\sqrt{b^2-4ac}}{2a}+\frac{b-\sqrt{b^2-4ac}}{2a}\right)x+\left(\frac{\left(b+\sqrt{b^2-4ac}\right)\left(b-\sqrt{b^2-4ac}\right)}{2a\cdot 2a}\right)=0$$

$$; \; x^2+\left(\frac{b+\sqrt{b^2-4ac}+b-\sqrt{b^2-4ac}}{2a}\right)x+\left(\frac{\left(b^2-b\sqrt{b^2-4ac}+b\sqrt{b^2-4ac}-\sqrt{b^2-4ac}\cdot\sqrt{b^2-4ac}\right)}{4a^2}\right)=0$$

$$; \; x^2+\left(\frac{b+b}{2a}\right)x+\left[\frac{b^2-\left(b^2-4ac\right)}{4a^2}\right]=0 \; ; \; x^2+\left(\frac{2b}{2a}\right)x+\left(\frac{b^2-b^2+4ac}{4a^2}\right)=0 \; ; \; x^2+\frac{b}{a}x+\frac{4ac}{4a^2}=0$$

$$; \; x^2+\frac{b}{a}x+\frac{c}{a}=0 \; ; \; \frac{x^2}{1}+\frac{bx}{a}+\frac{c}{a}=0 \; ; \; \frac{ax^2+bx+c}{a}=0 \; ; \; \frac{ax^2+bx+c}{a}=\frac{0}{1} \; ; \; \left(ax^2+bx+c\right)\cdot 1=a\cdot 0 \text{ which}$$

is the same as $ax^2+bx+c=0$.

The quadratic formula is a powerful formula and should be memorized. In the following sections we will use this formula to solve different types of quadratic equations.

Practice Problems - Quadratic Equations and the Quadratic Formula

Section 1.4a Practice Problems - Given the following quadratic equations identify the coefficients a, b, and c.

1. $3x=-5+2x^2$
2. $2x^2=5$
3. $3w^2-5w=2$
4. $15=-y^2-3$
5. $x^2+3=5x$
6. $-u^2+2=3u$

1.4b Solving Quadratic Equations Using the Quadratic Formula

As was stated earlier, the quadratic formula can be used to solve any quadratic equation by expressing the equation in the standard form of $ax^2+bx+c=0$ and by substituting the equivalent numbers for a, b, and c into the quadratic formula. In this section we will review how to solve quadratic equations of the form $ax^2+bx+c=0$, where $a=1$ (Case I) and where $a>1$ (Case II), using the quadratic formula.

Case I Solving Quadratic Equations of the Form $ax^2+bx+c=0$, where $a=1$, Using the Quadratic Formula

Quadratic equations of the form $ax^2+bx+c=0$, where $a=1$, are solved using the following steps:

Step 1 Write the equation in standard form.

Step 2 Identify the coefficients a, b, and c.

Step 3 Substitute the values for a, b, and c into the quadratic equation $x=\frac{-b\pm\sqrt{b^2-4ac}}{2a}$. Simplify the equation.

Step 4 Solve for the values of x. Check the answers by either substituting the x values into the original equation or by multiplying the factored product using the FOIL method.

Step 5 Write the quadratic equation in its factored form.

Examples with Steps

The following examples show the steps as to how quadratic equations are solved using the quadratic formula:

Example 1.4-1

Solve the quadratic equation $x^2+5x=-4$.

Solution:

Step 1 $\boxed{x^2+5x=-4}$; $\boxed{x^2+5x+4=-4+4}$; $\boxed{x^2+5x+4=0}$

Step 2 Let: $\boxed{a=1}$, $\boxed{b=5}$, and $\boxed{c=4}$. Then,

Step 3 Given: $\boxed{x=\frac{-b\pm\sqrt{b^2-4ac}}{2a}}$; $\boxed{x=\frac{-5\pm\sqrt{5^2-4\times1\times4}}{2\times1}}$; $\boxed{x=\frac{-5\pm\sqrt{25-16}}{2}}$

; $\boxed{x=\frac{-5\pm\sqrt{9}}{2}}$; $\boxed{x=\frac{-5\pm\sqrt{3^2}}{2}}$; $\boxed{x=\frac{-5\pm3}{2}}$

Step 4 Separate $x=\frac{-5\pm3}{2}$ into two equations.

I. $\boxed{x=\frac{-5+3}{2}}$; $\boxed{x=-\frac{2}{2}}$; $\boxed{x=-\frac{1}{1}}$; $\boxed{\mathbf{x=-1}}$

II. $\boxed{x = \frac{-5-3}{2}}$; $\boxed{x = -\frac{\overset{4}{\cancel{8}}}{\cancel{2}}}$; $\boxed{x = -\frac{4}{1}}$; $\boxed{\mathbf{x = -4}}$

Check No. 1: I. *Let* $x = -1$ *in* $x^2 + 5x = -4$; $(-1)^2 + (5 \times -1) \overset{?}{=} -4$; $1 - 5 \overset{?}{=} -4$; $-4 = -4$

II. *Let* $x = -4$ *in* $x^2 + 5x = -4$; $(-4)^2 + (5 \times -4) \overset{?}{=} -4$; $16 - 20 \overset{?}{=} -4$; $-4 = -4$

Check No. 2: $x^2 + 5x + 4 \overset{?}{=} (x+1)(x+4)$; $x^2 + 5x + 4 \overset{?}{=} (x \cdot x) + (4 \cdot x) + (1 \cdot x) + (1 \cdot 4)$

; $x^2 + 5x + 4 \overset{?}{=} x^2 + 4x + x + 4$; $x^2 + 5x + 4 \overset{?}{=} x^2 + (4+1)x + 4$; $x^2 + 5x + 4 = x^2 + 5x + 4$

Step 5 Therefore, the equation $x^2 + 5x + 4 = 0$ can be factored to $\mathbf{(x+1)(x+4) = 0}$.

Example 1.4-2

Solve the quadratic equation $x^2 = -12x - 35$.

Solution:

Step 1 $\boxed{x^2 = -12x - 35}$; $\boxed{x^2 + 12x = -12x + 12x - 35}$; $\boxed{x^2 + 12x = 0 - 35}$; $\boxed{x^2 + 12x = -35}$

; $\boxed{x^2 + 12x + 35 = -35 + 35}$; $\boxed{x^2 + 12x + 35 = 0}$

Step 2 Let: $\boxed{a = 1}$, $\boxed{b = 12}$, and $\boxed{c = 35}$. Then,

Step 3 Given: $\boxed{x = \frac{-b \pm \sqrt{b^2 - 4ac}}{2a}}$; $\boxed{x = \frac{-12 \pm \sqrt{12^2 - 4 \times 1 \times 35}}{2 \times 1}}$; $\boxed{x = \frac{-12 \pm \sqrt{144 - 140}}{2}}$

; $\boxed{x = \frac{-12 \pm \sqrt{4}}{2}}$; $\boxed{x = \frac{-12 \pm \sqrt{2^2}}{2}}$; $\boxed{x = \frac{-12 \pm 2}{2}}$

Step 4 Separate $x = \frac{-12 \pm 2}{2}$ into two equations.

I. $\boxed{x = \frac{-12+2}{2}}$; $\boxed{x = -\frac{\overset{5}{\cancel{10}}}{\cancel{2}}}$; $\boxed{x = -\frac{5}{1}}$; $\boxed{\mathbf{x = -5}}$

II. $\boxed{x = \frac{-12-2}{2}}$; $\boxed{x = -\frac{\overset{7}{\cancel{14}}}{\cancel{2}}}$; $\boxed{x = -\frac{7}{1}}$; $\boxed{\mathbf{x = -7}}$

Check No. 1: I. *Let* $x = -5$ *in* $x^2 = -12x - 35$; $(-5)^2 \overset{?}{=} (-12 \times -5) - 35$; $25 \overset{?}{=} 60 - 35$; $25 = 25$

II. *Let* $x = -7$ *in* $x^2 = -12x - 35$; $(-7)^2 \overset{?}{=} (-12 \times -7) - 35$; $49 \overset{?}{=} 84 - 35$; $49 = 49$

Check No. 2: $x^2 + 12x + 35 \overset{?}{=} (x+5)(x+7)$; $x^2 + 12x + 35 \overset{?}{=} (x \cdot x) + (7 \cdot x) + (5 \cdot x) + (5 \cdot 7)$

; $x^2 + 12x + 35 \overset{?}{=} x^2 + 7x + 5x + 35$; $x^2 + 12x + 35 \overset{?}{=} x^2 + (7+5)x + 35$

; $x^2 + 12x + 35 = x^2 + 12x + 35$

Step 5 Therefore, the equation $x^2 + 12x + 35 = 0$ can be factored to $\mathbf{(x+5)(x+7) = 0}$.

Additional Examples - Solving Quadratic Equations of the Form $ax^2+bx+c=0$, where $a=1$, Using the Quadratic Formula

The following examples further illustrate how to solve quadratic equations using the quadratic formula:

Example 1.4-3

Solve the quadratic equation $x^2=16x-55$.

Solution:

First, write the equation in standard form, i.e., $x^2-16x+55=0$

Next, let: $\boxed{a=1}$, $\boxed{b=-16}$, and $\boxed{c=55}$. Then,

Given: $\boxed{x=\frac{-b\pm\sqrt{b^2-4ac}}{2a}}$; $\boxed{x=\frac{-(-16)\pm\sqrt{(-16)^2-4\times1\times55}}{2\times1}}$; $\boxed{x=\frac{16\pm\sqrt{256-220}}{2}}$; $\boxed{x=\frac{16\pm\sqrt{36}}{2}}$

; $\boxed{x=\frac{16\pm\sqrt{6^2}}{2}}$; $\boxed{x=\frac{16\pm6}{2}}$ Therefore:

I. $\boxed{x=\frac{16+6}{2}}$; $\boxed{x=\frac{\overset{11}{\cancel{22}}}{\cancel{2}}}$; $\boxed{x=\frac{11}{1}}$; $\boxed{\mathbf{x=11}}$ II. $\boxed{x=\frac{16-6}{2}}$; $\boxed{x=\frac{\overset{5}{\cancel{10}}}{\cancel{2}}}$; $\boxed{x=\frac{5}{1}}$; $\boxed{\mathbf{x=5}}$

Check No. 1: I. *Let* $x=11$ *in* $x^2=16x-55$; $11^2\overset{?}{=}16\times11-55$; $121\overset{?}{=}176-55$; $121=121$

II. *Let* $x=5$ *in* $x^2=16x-55$; $5^2\overset{?}{=}16\times5-55$; $25\overset{?}{=}80-55$; $25=25$

Check No. 2: $x^2-16x+55\overset{?}{=}(x-11)(x-5)$; $x^2-16x+55\overset{?}{=}(x\cdot x)+(-5\cdot x)+(-11\cdot x)+(-11\cdot-5)$

; $x^2-16x+55\overset{?}{=}x^2-5x-11x+55$; $x^2-16x+55\overset{?}{=}x^2+(-5-11)x+55$

; $x^2-16x+55=x^2-16x+55$

Therefore, the equation $x^2-16x+55=0$ can be factored to $\mathbf{(x-11)(x-5)=0}$.

Example 1.4-4

Solve the quadratic equation $9=-x^2-6x$.

Solution:

First, write the equation in standard form, i.e., $x^2+6x+9=0$.

Next, let: $\boxed{a=1}$, $\boxed{b=6}$, and $\boxed{c=9}$. Then,

Given: $\boxed{x=\frac{-b\pm\sqrt{b^2-4ac}}{2a}}$; $\boxed{x=\frac{-6\pm\sqrt{6^2-4\times1\times9}}{2\times1}}$; $\boxed{x=\frac{-6\pm\sqrt{36-36}}{2}}$; $\boxed{x=\frac{-6\pm\sqrt{0}}{2}}$; $\boxed{x=\frac{-6\pm0}{2}}$

; $\boxed{x=-\frac{\overset{3}{\cancel{6}}}{\cancel{2}}}$; $\boxed{x=-\frac{3}{1}}$; $\boxed{x=-3}$

In this case the equation has one repeated solution, i.e., $\boxed{\mathbf{x=-3}}$ and $\boxed{\mathbf{x=-3}}$.

Thus, the solution set is $\{-3, -3\}$.

Check No. 1: *Let* $x=-3$ *in* $x^2+6x+9=0$; $(-3)^2+6\times-3+9\overset{?}{=}0$; $9-18+9\overset{?}{=}0$; $18-18\overset{?}{=}0$; $0=0$

Check No. 2: $x^2+6x+9\overset{?}{=}(x+3)(x+3)$; $x^2+6x+9\overset{?}{=}(x\cdot x)+(3\cdot x)+(3\cdot x)+(3\cdot 3)$

; $x^2+6x+9\overset{?}{=}x^2+3x+3x+9$; $x^2+6x+9\overset{?}{=}x^2+(3+3)x+9$;

$x^2+6x+9=x^2+6x+9$

Therefore, the equation $x^2+6x+9=0$ can be factored to $\mathbf{(x+3)(x+3)=0}$.

Note that when $c=0$ the quadratic equation $ax^2+bx+c=0$ reduces to $ax^2+bx=0$. For cases where $a=1$, we can solve equations of the form $x^2+bx=0$ using the quadratic formula in the following way:

Example 1.4-5

Solve the quadratic equation $x^2+5x=0$.

Solution:

The equation is already in standard form.

Let: $\boxed{a=1}$, $\boxed{b=5}$, and $\boxed{c=0}$. Then,

Given: $\boxed{x=\frac{-b\pm\sqrt{b^2-4ac}}{2a}}$; $\boxed{x=\frac{-5\pm\sqrt{5^2-4\times1\times0}}{2\times1}}$; $\boxed{x=\frac{-5\pm\sqrt{25-0}}{2}}$; $\boxed{x=\frac{-5\pm\sqrt{25}}{2}}$; $\boxed{x=\frac{-5\pm\sqrt{5^2}}{2}}$

; $\boxed{x=\frac{-5\pm5}{2}}$ Therefore:

I. $\boxed{x=\frac{-5+5}{2}}$; $\boxed{x=\frac{0}{2}}$; $\boxed{\mathbf{x=0}}$ II. $\boxed{x=\frac{-5-5}{2}}$; $\boxed{x=-\frac{\overset{5}{\cancel{10}}}{\cancel{2}}}$; $\boxed{x=-\frac{5}{1}}$; $\boxed{x=-5}$; $\boxed{\mathbf{x=-5}}$

and the solution set is $\{\mathbf{0, -5}\}$.

Check No. 1: I. *Let* $x=0$ *in* $x^2+5x=0$; $0^2+5\cdot0\overset{?}{=}0$; $0+0\overset{?}{=}0$; $0=0$

II. *Let* $x=-5$ *in* $x^2+5x=0$; $(-5)^2+5\cdot-5\overset{?}{=}0$; $25-25\overset{?}{=}0$; $0=0$

Check No. 2: $x^2+5x\overset{?}{=}(x+0)(x+5)$; $x^2+5x\overset{?}{=}(x\cdot x)+(5\cdot x)+(0\cdot x)+(0\cdot5)$; $x^2+5x\overset{?}{=}x^2+5x+0+0$

; $x^2+5x=x^2+5x$

Therefore, the equation $x^2+5x=0$ can be factored to $(x+0)(x+5)=0$ which is the same as $\mathbf{x(x+5)=0}$.

Practice Problems - Solving Quadratic Equations of the Form ax^2+bx+c, where $a=1$, Using the Quadratic Formula

Section 1.4b Case I Practice Problems - Use the quadratic formula to solve the following quadratic equations.

1. $x^2=-5x-6$
2. $y^2-40y=-300$
3. $-x=-x^2+20$
4. $x^2+3x+4=0$
5. $x^2-80-2x=0$
6. $x^2+4x+4=0$

Case II Solving Quadratic Equations of the Form $ax^2+bx+c=0$, where $a \rangle 1$, Using the Quadratic Formula

Trinomial equations of the form $ax^2+bx+c=0$, where $a \rangle 1$, are solved using the following steps:

Step 1 Write the equation in standard form.

Step 2 Identify the coefficients a, b, and c.

Step 3 Substitute the values for a, b, and c into the quadratic equation $x=\frac{-b\pm\sqrt{b^2-4ac}}{2a}$. Simplify the equation.

Step 4 Solve for the values of x. Check the answers by either substituting the x values into the original equation or by multiplying the factored product using the FOIL method.

Step 5 Write the quadratic equation in its factored form.

Examples with Steps

The following examples show the steps as to how second degree trinomial equations are solved using the quadratic formula:

Example 1.4-6

Solve the quadratic equation $2x^2+5x=-3$.

Solution:

Step 1 $\boxed{2x^2+5x=-3}$; $\boxed{2x^2+5x+3=0}$

Step 2 Let: $\boxed{a=2}$, $\boxed{b=5}$, and $\boxed{c=3}$. Then,

Step 3 Given: $\boxed{x=\frac{-b\pm\sqrt{b^2-4ac}}{2a}}$; $\boxed{x=\frac{-5\pm\sqrt{5^2-4\times2\times3}}{2\times2}}$; $\boxed{x=\frac{-5\pm\sqrt{25-24}}{4}}$

; $\boxed{x=\frac{-5\pm\sqrt{1}}{4}}$; $\boxed{x=\frac{-5\pm1}{4}}$

Step 4 Separate $x=\frac{-5\pm1}{4}$ into two equations:

I. $\boxed{x=\frac{-5+1}{4}}$; $\boxed{x=-\frac{4}{4}}$; $\boxed{x=-\frac{1}{1}}$; $\boxed{\mathbf{x=-1}}$ II. $\boxed{x=\frac{-5-1}{4}}$; $\boxed{x=-\frac{\overset{3}{\cancel{6}}}{\underset{2}{\cancel{4}}}}$; $\boxed{\mathbf{x=-\frac{3}{2}}}$

Thus, the solution set is $\left\{\mathbf{-1,-\frac{3}{2}}\right\}$.

Check No. 1: I. *Let* $x=-1$ *in* $2x^2+5x=-3$; $2(-1)^2+(5\times-1)\overset{?}{=}-3$; $2-5\overset{?}{=}-3$; $-3=-3$

II. *Let* $x=-\frac{3}{2}$ *in* $2x^2+5x=-3$; $2\left(-\frac{3}{2}\right)^2+\left(5\times-\frac{3}{2}\right)\overset{?}{=}-3$; $2\times\frac{9}{4}-\frac{15}{2}\overset{?}{=}-3$

; $\frac{18}{4}-\frac{15}{2}\overset{?}{=}-3$; $\frac{(2\times18)-(4\times15)}{4\times2}\overset{?}{=}-3$; $\frac{36-60}{8}\overset{?}{=}-3$; $-\frac{24}{8}\overset{?}{=}-3$; $-3=-3$

Check No. 2: $2x^2+5x+3\stackrel{?}{=}(x+1)(2x+3)$; $2x^2+5x+3\stackrel{?}{=}(2x\cdot x)+(3\cdot x)+(1\cdot 2x)+(1\cdot 3)$

; $2x^2+5x+3\stackrel{?}{=}2x^2+3x+2x+3$; $2x^2+5x+3\stackrel{?}{=}2x^2+(3+2)x+3$

; $2x^2+5x+3=2x^2+5x+3$

Step 5 Therefore, the equation $2x^2+5x+3=0$ can be factored to $(x+1)\left(x+\frac{3}{2}\right)=0$ which is the same as $\mathbf{(x+1)(2x+3)=0}$

Example 1.4-7

Solve the quadratic equation $15x^2=-7x+2$.

Solution:

Step 1 $\boxed{15x^2=-7x+2}$; $\boxed{15x^2+7x=-7x+7x+2}$; $\boxed{15x^2+7x=0+2}$; $\boxed{15x^2+7x=2}$

; $\boxed{15x^2+7x-2=2-2}$; $\boxed{15x^2+7x-2=0}$

Step 2 Let: $\boxed{a=15}$, $\boxed{b=7}$, and $\boxed{c=-2}$. Then,

Step 3 Given: $\boxed{x=\frac{-b\pm\sqrt{b^2-4ac}}{2a}}$; $\boxed{x=\frac{-7\pm\sqrt{7^2-4\times 15\times -2}}{2\times 15}}$; $\boxed{x=\frac{-7\pm\sqrt{49+120}}{30}}$

; $\boxed{x=\frac{-7\pm\sqrt{169}}{30}}$; $\boxed{x=\frac{-7\pm 13}{30}}$

Step 4 Separate $x=\frac{-7\pm 13}{30}$ into two equations:

I. $\boxed{x=\frac{-7+13}{30}}$; $\boxed{x=\frac{\overset{1}{\cancel{6}}}{\underset{5}{\cancel{30}}}}$; $\boxed{\mathbf{x=\frac{1}{5}}}$ II. $\boxed{x=\frac{-7-13}{30}}$; $\boxed{x=-\frac{\overset{2}{\cancel{20}}}{\underset{3}{\cancel{30}}}}$; $\boxed{\mathbf{x=-\frac{2}{3}}}$

Thus, the solution set is $\left\{-\frac{\mathbf{2}}{\mathbf{3}},\frac{\mathbf{1}}{\mathbf{5}}\right\}$.

Check No. 1: I. *Let* $x=\frac{1}{5}$ *in* $15x^2=-7x+2$; $15\left(\frac{1}{5}\right)^2\stackrel{?}{=}\left(-7\times\frac{1}{5}\right)+2$; $15\times\frac{1}{25}\stackrel{?}{=}-\frac{7}{5}+2$

; $\frac{\overset{3}{\cancel{15}}}{\underset{5}{\cancel{25}}}\stackrel{?}{=}-\frac{7}{5}+\frac{2}{1}$; $\frac{3}{5}\stackrel{?}{=}\frac{(-7\times 1)+(2\times 5)}{5\times 1}$; $\frac{3}{5}\stackrel{?}{=}\frac{-7+10}{5}$; $\frac{3}{5}=\frac{3}{5}$

II. *Let* $x=-\frac{2}{3}$ *in* $15x^2=-7x+2$; $15\left(-\frac{2}{3}\right)^2\stackrel{?}{=}\left(-7\times-\frac{2}{3}\right)+2$; $15\times\frac{4}{9}\stackrel{?}{=}\frac{14}{3}+2$

; $\frac{\overset{20}{\cancel{60}}}{\underset{3}{\cancel{9}}}\stackrel{?}{=}\frac{14}{3}+\frac{2}{1}$; $\frac{20}{3}\stackrel{?}{=}\frac{(14\times 1)+(2\times 3)}{3\times 1}$; $\frac{20}{3}\stackrel{?}{=}\frac{14+6}{3}$; $\frac{20}{3}=\frac{20}{3}$

Check No. 2: $15x^2+7x-2\stackrel{?}{=}(5x-1)(3x+2)$; $15x^2+7x-2\stackrel{?}{=}(5x\cdot 3x)+(2\cdot 5x)+(-1\cdot 3x)+(-1\cdot 2)$

; $15x^2+7x-2\stackrel{?}{=}15x^2+10x-3x-2$; $15x^2+7x-2\stackrel{?}{=}15x^2+(10-3)x-2$

; $15x^2+7x-2=15x^2+7x-2$

Step 5 Therefore, the equation $15x^2+7x-2=0$ can be factored to $\left(x-\frac{1}{5}\right)\left(x+\frac{2}{3}\right)=0$ which is the same as $(\mathbf{5x-1})(\mathbf{3x+2})=\mathbf{0}$

Additional Examples - Solving Quadratic Equations of the Form $ax^2+bx+c=0$, where $a \rangle 1$, Using the Quadratic Formula

The following examples further illustrate how to solve quadratic equations:

Example 1.4-8

Solve the quadratic equation $3x^2+7x-6=0$.

Solution:

The equation is already in standard form. Let: $\boxed{a=3}$, $\boxed{b=7}$, and $\boxed{c=-6}$. Then,

Given: $\boxed{x=\frac{-b\pm\sqrt{b^2-4ac}}{2a}}$; $\boxed{x=\frac{-7\pm\sqrt{7^2-4\times3\times-6}}{2\times3}}$; $\boxed{x=\frac{-7\pm\sqrt{49+72}}{6}}$; $\boxed{x=\frac{-7\pm\sqrt{121}}{6}}$

; $\boxed{x=\frac{-7\pm\sqrt{11^2}}{6}}$; $\boxed{x=\frac{-7\pm11}{6}}$ Therefore:

I. $\boxed{x=\frac{-7+11}{6}}$; $\boxed{x=\frac{\overset{2}{\cancel{4}}}{\underset{3}{\cancel{6}}}}$; $\boxed{\mathbf{x=\frac{2}{3}}}$

II. $\boxed{x=\frac{-7-11}{6}}$; $\boxed{x=-\frac{\overset{3}{\cancel{18}}}{\cancel{6}}}$; $\boxed{x=-\frac{3}{1}}$; $\boxed{\mathbf{x=-3}}$

Thus, the solution set is $\left\{\mathbf{-3,\frac{2}{3}}\right\}$.

Check No. 1: I. *Let* $x=-3$ *in* $3x^2+7x-6=0$; $3\cdot(-3)^2+7\cdot(-3)-6\overset{?}{=}0$; $3\cdot9-21-6\overset{?}{=}0$

; $27-21-6\overset{?}{=}0$; $27-27\overset{?}{=}0$; $0=0$

II. *Let* $x=\frac{2}{3}$ *in* $3x^2+7x-6=0$; $3\cdot\left(\frac{2}{3}\right)^2+7\cdot\left(\frac{2}{3}\right)-6\overset{?}{=}0$; $3\cdot\frac{4}{9}+\frac{14}{3}-6\overset{?}{=}0$

; $\frac{\overset{4}{\cancel{12}}}{\underset{3}{\cancel{9}}}+\frac{14}{3}-6\overset{?}{=}0$; $\frac{4}{3}+\frac{14}{3}-6\overset{?}{=}0$; $\frac{4+14}{3}-6\overset{?}{=}0$; $\frac{\overset{6}{\cancel{18}}}{\cancel{3}}-6\overset{?}{=}0$; $\frac{6}{1}-6\overset{?}{=}0$; $6-6\overset{?}{=}0$

; $0=0$

Check No. 2: $3x^2+7x-6\overset{?}{=}(x+3)(3x-2)$; $3x^2+7x-6\overset{?}{=}(x\cdot3x)+(-2\cdot x)+(3\cdot3x)+(3\cdot-2)$

; $3x^2+7x-6\overset{?}{=}3x^2-2x+9x-6$; $3x^2+7x-6\overset{?}{=}3x^2+(-2+9)x-6$

; $3x^2+7x-6=3x^2+7x-6$

Therefore, the equation $3x^2+7x-6=0$ can be factored to $(x+3)\left(x-\frac{2}{3}\right)=0$ which is the same as $(x+3)\left(\frac{x}{1}-\frac{2}{3}\right)=0$; $(x+3)\left(\frac{(3\cdot x)-(1\cdot2)}{1\cdot3}\right)=0$; $(x+3)\left(\frac{3x-2}{3}\right)=0$; $\left(\frac{x+3}{1}\right)\left(\frac{3x-2}{3}\right)=0$

; $\frac{(x+3)\cdot(3x-2)}{1\cdot3}=0$; $\frac{(x+3)\cdot(3x-2)}{1\cdot3}=\frac{0}{1}$; $[(x+3)\cdot(3x-2)]\cdot1=0\cdot3$; $(\mathbf{x+3})(\mathbf{3x-2})=\mathbf{0}$

Example 1.4-9

Solve the quadratic equation $4x^2 + 9x = -6$.

Solution:

First, write the equation in standard form, i.e., $4x^2 + 9x + 6 = 0$.

Next, let: $\boxed{a = 4}$, $\boxed{b = 9}$, and $\boxed{c = 6}$. Then,

Given: $\boxed{x = \frac{-b \pm \sqrt{b^2 - 4ac}}{2a}}$; $\boxed{x = \frac{-9 \pm \sqrt{9^2 - 4 \times 4 \times 6}}{2 \times 4}}$; $\boxed{x = \frac{-9 \pm \sqrt{81 - 96}}{8}}$; $\boxed{x = \frac{-9 \pm \sqrt{-15}}{8}}$

Since the number under the radical is negative, therefore the quadratic equation does not have any real solutions. We state that **the equation is not factorable.**

Note that when $c = 0$ the quadratic equation $ax^2 + bx + c = 0$ reduces to $ax^2 + bx = 0$. For cases where $a \rangle 1$, we can solve equations of the form $ax^2 + bx = 0$ using the quadratic formula in the following way:

Example 1.4-10

Solve the quadratic equation $2x^2 + 5x = 0$.

Solution:

First write the equation in standard form, i.e., $2x^2 + 5x + 0 = 0$.

Next, let: $\boxed{a = 2}$, $\boxed{b = 5}$, and $\boxed{c = 0}$. Then,

Given: $\boxed{x = \frac{-b \pm \sqrt{b^2 - 4ac}}{2a}}$; $\boxed{x = \frac{-5 \pm \sqrt{5^2 - 4 \times 2 \times 0}}{2 \times 2}}$; $\boxed{x = \frac{-5 \pm \sqrt{25 - 0}}{4}}$; $\boxed{x = \frac{-5 \pm \sqrt{25}}{4}}$

; $\boxed{x = \frac{-5 \pm \sqrt{5^2}}{4}}$; $\boxed{x = \frac{-5 \pm 5}{4}}$ Therefore:

I. $\boxed{x = \frac{-5 + 5}{4}}$; $\boxed{x = \frac{0}{4}}$; $\boxed{\mathbf{x = 0}}$ II. $\boxed{x = \frac{-5 - 5}{4}}$; $\boxed{x = -\frac{5}{2}}$; $\boxed{\mathbf{x = -2.5}}$

Thus, the solution set is $\{\mathbf{0}, -\mathbf{2.5}\}$.

Check: I. *Let* $x = 0$ *in* $2x^2 + 5x = 0$; $2 \cdot 0^2 + 5 \cdot 0 \stackrel{?}{=} 0$; $0 + 0 \stackrel{?}{=} 0$; $0 = 0$

II. *Let* $x = -2.5$ *in* $2x^2 + 5x = 0$; $2 \cdot (-2.5)^2 + 5 \cdot -2.5 \stackrel{?}{=} 0$; $2 \cdot 6.25 - 12.5 \stackrel{?}{=} 0$; $12.5 = 12.5$

Therefore, the equation $2x^2 + 5x = 0$ can be factored to $(x + 0)(x + 2.5) = 0$ which is the same as $\mathbf{x(x + 2.5) = 0}$.

Practice Problems - Solving Quadratic Equations of the Form $ax^2 + bx + c = 0$, where $a \rangle 1$, Using the Quadratic Formula

Section 1.4b Case II Practice Problems - Use the quadratic formula to solve the following quadratic equations.

1. $4u^2 + 6u + 1 = 0$
2. $4w^2 + 10w = -3$
3. $6x^2 + 4x - 2 = 0$
4. $15y^2 + 3 = -14y$
5. $2x^2 - 5x + 3 = 0$
6. $2x^2 + xy - y^2 = 0$ x is variable

1.4c Solving Quadratic Equations Using the Square Root Property Method

Quadratic equations of the form $(ax+b)^2 = c$ are solved using a method known as the Square Root Property method where the square root of both sides of the equation are taken and the terms are simplified. Following show the steps as to how quadratic equations are solved using the Square Root property method:

Step 1 Take the square root of the left and the right hand side of the equation. Simplify the terms on both sides of the equation.

Step 2 Solve for the values of x. Check the answers by substituting the x values into the original equation.

Step 3 Write the equation in its factored form.

Examples with Steps

The following examples show the steps as to how equations of the form $(ax+b)^2 = c$ are solved using the Square Root Property method:

Example 1.4-11

Solve the quadratic equation $(x+4)^2 = 36$.

Solution:

Step 1 $\boxed{(x+4)^2 = 36}$; $\boxed{\sqrt{(x+4)^2} = \pm\sqrt{36}}$; $\boxed{\sqrt{(x+4)^2} = \pm\sqrt{6^2}}$; $\boxed{x+4 = \pm 6}$

Step 2 Separate $x+4 = \pm 6$ into two equations.

I. $\boxed{x+4 = +6}$; $\boxed{x = +6-4}$; $\boxed{\mathbf{x = 2}}$

II. $\boxed{x+4 = -6}$; $\boxed{x = -6-4}$; $\boxed{\mathbf{x = -10}}$

Thus, the solution set is $\{\mathbf{-10, 2}\}$.

Check:

I. *Let* $x = 2$ *in* $(x+4)^2 = 36$; $(2+4)^2 \stackrel{?}{=} 36$; $6^2 \stackrel{?}{=} 36$; $36 = 36$

II. *Let* $x = -10$ *in* $(x+4)^2 = 36$; $(-10+4)^2 \stackrel{?}{=} 36$; $(-6)^2 \stackrel{?}{=} 36$; $36 = 36$

Step 3 Therefore, the equation $(x+4)^2 = 36$ can be factored to $\mathbf{(x-2)(x+10) = 0}$.

Example 1.4-12

Solve the quadratic equation $(x-2)^2 = 25$.

Solution:

Step 1 $\boxed{(x-2)^2 = 25}$; $\boxed{\sqrt{(x-2)^2} = \pm\sqrt{25}}$; $\boxed{\sqrt{(x-2)^2} = \pm\sqrt{5^2}}$; $\boxed{x-2 = \pm 5}$

Step 2 Separate $x-2 = \pm 5$ into two equations.

I. $\boxed{x-2 = +5}$; $\boxed{x = 5+2}$; $\boxed{\mathbf{x = 7}}$

II. $\boxed{x-2 = -5}$; $\boxed{x = -5+2}$; $\boxed{\mathbf{x = -3}}$

Thus, the solution set is $\{-3, 7\}$.

Check: I. *Let* $x = 7$ *in* $(x-2)^2 = 25$; $(7-2)^2 \stackrel{?}{=} 25$; $5^2 \stackrel{?}{=} 25$; $25 = 25$

II. *Let* $x = -3$ *in* $(x-2)^2 = 25$; $(-3-2)^2 \stackrel{?}{=} 25$; $(-5)^2 \stackrel{?}{=} 25$; $25 = 25$

Step 3 Therefore, the equation $(x-2)^2 = 25$ can be factored to $(\mathbf{x-7})(\mathbf{x+3}) = \mathbf{0}$.

Additional Examples - Solving Quadratic Equations Using the Square Root Property Method

The following examples further illustrate how to solve quadratic equations using the Square Root Property method:

Example 1.4-13

Solve the quadratic equation $(5y+3)^2 = 15$ using the Square Root Property method.

Solution:

$\boxed{(5y+3)^2 = 15}$; $\boxed{\sqrt{(5y+3)^2} = \pm\sqrt{15}}$; $\boxed{5y+3 = \pm\sqrt{15}}$ Therefore, the two solutions are:

I. $\boxed{5y+3 = +\sqrt{15}}$; $\boxed{5y = \sqrt{15}-3}$; $\boxed{y = \frac{\sqrt{15}-3}{5}}$ II. $\boxed{5y+3 = -\sqrt{15}}$; $\boxed{5y = -\sqrt{15}-3}$; $\boxed{y = -\frac{\sqrt{15}+3}{5}}$

Thus, the solution set is $\left\{-\frac{\sqrt{15}+3}{5}, \frac{\sqrt{15}-3}{5}\right\}$.

Check: I. *Let* $y = \frac{\sqrt{15}-3}{5}$ *in* $(5y+3)^2 = 15$; $\left(5 \cdot \frac{\sqrt{15}-3}{5} + 3\right)^2 \stackrel{?}{=} 15$; $\left(\sqrt{15}-3+3\right)^2 \stackrel{?}{=} 15$; $\left(\sqrt{15}\right)^2 \stackrel{?}{=} 15$; $15 = 15$

II. *Let* $y = -\frac{\sqrt{15}+3}{5}$ *in* $(5y+3)^2 = 15$; $\left[\left(5 \cdot -\frac{\sqrt{15}+3}{5}\right)+3\right]^2 \stackrel{?}{=} 15$; $\left[\left(-\sqrt{15}-3\right)+3\right]^2 \stackrel{?}{=} 15$; $\left(-\sqrt{15}-3+3\right)^2 \stackrel{?}{=} 15$; $\left(-\sqrt{15}\right)^2 \stackrel{?}{=} 15$; $15 = 15$

Therefore, the equation $(5y+3)^2 = 15$ can be factored to $\left(y - \frac{\sqrt{15}-3}{5}\right)\left(y + \frac{\sqrt{15}+3}{5}\right) = 0$ which is the same as $(\mathbf{y - 0.175})(\mathbf{y + 1.375}) = \mathbf{0}$; $y^2 + 1.2y - 0.24 = 0$; $25y^2 + 30y - 6 = 0$, or $(5y+3)^2 = 15$.

Example 1.4-14

Solve the quadratic equation $(x+5)^2 = 49$ using the Square Root Property and the Quadratic Formula method.

Solution:

First Method - The Square Root Property method:

$\boxed{(x+5)^2 = 49}$; $\boxed{\sqrt{(x+5)^2} = \pm\sqrt{49}}$; $\boxed{\sqrt{(x+5)^2} = \pm\sqrt{7^2}}$; $\boxed{x+5 = \pm 7}$ Therefore, the two solutions are:

I. $\boxed{x+5 = +7}$; $\boxed{x = 7-5}$; $\boxed{\mathbf{x = 2}}$

II. $\boxed{x+5 = -7}$; $\boxed{x = -7-5}$; $\boxed{\mathbf{x = -12}}$

Thus, the solution set is $\{\mathbf{-12, 2}\}$.

Check: I. $Let\ x=2\ in\quad (x+5)^2=49\ ;\ (2+5)^2\overset{?}{=}49\ ;\ 7^2\overset{?}{=}49\ ;\ 49=49$

II. $Let\ x=-13\ in\quad (x+5)^2=49\ ;\ (-12+5)^2\overset{?}{=}49\ ;\ (-7)^2\overset{?}{=}49\ ;\ 49=49$

Therefore, the equation $(x+5)^2=49$ can be factored to $\mathbf{(x-2)(x+12)=0}$.

Second Method - The Quadratic Formula method:
Given the expression $(x+5)^2=49$, expand the left hand side of the equation and write the quadratic equation in its standard form, i.e.,

$\boxed{(x+5)^2=49}$; $\boxed{x^2+25+10x=49}$; $\boxed{x^2+10x+(25-49)=0}$; $\boxed{x^2+10x-24=0}$

Let: $\boxed{a=1}$, $\boxed{b=10}$, and $\boxed{c=-24}$. Then,

Given: $\boxed{x=\frac{-b\pm\sqrt{b^2-4ac}}{2a}}$; $\boxed{x=\frac{-10\pm\sqrt{10^2-(4\times1\times-24)}}{2\times1}}$; $\boxed{x=\frac{-10\pm\sqrt{100+96}}{2}}$; $\boxed{x=\frac{-10\pm\sqrt{196}}{2}}$

; $\boxed{x=\frac{-10\pm\sqrt{14^2}}{2}}$; $\boxed{x=\frac{-10\pm14}{2}}$ Therefore, we can separate x into two equations:

I. $\boxed{x=\frac{-10+14}{2}}$; $\boxed{x=\frac{\overset{2}{\cancel{4}}}{\cancel{2}}}$; $\boxed{x=\frac{2}{1}}$; $\boxed{\mathbf{x=2}}$ II. $\boxed{x=\frac{-10-14}{2}}$; $\boxed{x=-\frac{\overset{12}{\cancel{24}}}{\cancel{2}}}$; $\boxed{x=-\frac{12}{1}}$; $\boxed{\mathbf{x=-12}}$

Thus, the solution set is $\{\mathbf{-12, 2}\}$.

The equation $(x+5)^2=49$ can be factored to $\mathbf{(x-2)(x+12)=0}$.

Note: As you may have already noticed, using the quadratic formula may not be a good choice since it requires more work and takes longer to solve. The key to solving quadratic equations is selection of a method that is easiest to use. Further discussions on selection of a best method is addressed in Section 1.4e.

Note that when $b=0$ the quadratic equation $(ax+b)^2=c$ reduces to $(ax)^2=c$. The following examples show the steps as to how quadratic equations of the form $(ax)^2=c$ are solved for cases where the coefficient of x is equal to or greater than one.

- For cases where $a=1$, we can solve equations of the form $x^2=c$ using the Square Root Property method in the following way:

  Example 1.4-15

  Solve $x^2=16$ using the Square Root Property method.

  Solution:

  First - Take the square root of both sides of the equation, i.e., $\sqrt{x^2}=\pm\sqrt{16}$

  Second - Simplify the terms on both sides to obtain the solutions, i.e., $\mathbf{x=\pm4}$. Therefore, the solution set is $\{-4, 4\}$ and the equation $x^2=16$ can be factored to $\mathbf{(x-4)(x+4)=0}$.

Check: I. $Let\ x = -4\ in \quad x^2 = 16\ ;\ (-4)^2 \stackrel{?}{=} 16\ ;\ 16 = 16$

II. $Let\ x = 4\ in \quad x^2 = 16\ ;\ 4^2 \stackrel{?}{=} 16\ ;\ 16 = 16$

- For cases where $a \rangle 1$, we can solve equations of the form $(ax)^2 = c$ (which is the same as $kx^2 = c$, where $k = a^2$) using the Square Root Property method in the following way:

Example 1.4-16

Solve $3x^2 = 27$ using the Square Root Property method.

Solution:

First - Divide both sides of the equation by the coefficient x, i.e., $\frac{\cancel{3}x^2}{\cancel{3}} = \frac{\overset{9}{\cancel{27}}}{\cancel{3}}$; $x^2 = \frac{9}{1}$; $x^2 = 9$

Second - Take the square root of both sides of the equation, i.e., $\sqrt{x^2} = \pm\sqrt{9}$

Third - Simplify the terms on both sides to obtain the solutions, i.e., $\mathbf{x = \pm 3}$

Therefore, the solution set is $\{-3, 3\}$ and the equation $3x^2 = 27$ can be factored to $\mathbf{(x-3)(x+3) = 0}$.

Check: I. $Let\ x = -3\ in \quad 3x^2 = 27\ ;\ 3 \cdot (-3)^2 \stackrel{?}{=} 27\ ;\ 3 \cdot 9 \stackrel{?}{=} 27\ ;\ 27 = 27$

II. $Let\ x = 3\ in \quad 3x^2 = 27\ ;\ 3 \cdot 3^2 \stackrel{?}{=} 27\ ;\ 3 \cdot 9 \stackrel{?}{=} 27\ ;\ 27 = 27$

Practice Problems - Solving Quadratic Equations Using the Square Root Property Method

Section 1.4c Practice Problems - Solve the following equations using the Square Root Property method:

1. $(2y+5)^2 = 36$
2. $(x+1)^2 = 7$
3. $(2x-3)^2 = 1$
4. $x^2 + 3 = 0$
5. $(y-5)^2 = 5$
6. $16x^2 - 25 = 0$

1.4d Solving Quadratic Equations Using Completing-the-Square Method

One of the methods used in solving quadratic equations is called Completing-the-Square method. Note that this method involves construction of perfect square trinomials which was addressed in Section 1.3 d, Case III. In this section we will review how to solve quadratic equations of the form $ax^2+bx+c=0$, where $a=1$ (case I) and where $a\rangle 1$ (Case II), using Completing-the-Square method.

Case I Solving Quadratic Equations of the Form $ax^2+bx+c=0$, where $a=1$, by Completing the Square

The following show the steps as to how quadratic equations, where the coefficient of the squared term is equal to one, are solved using Completing-the-Square method:

Step 1 Write the equation in the form of $x^2+bx=-c$.

Step 2 a. Divide the coefficient of x by 2, i.e., $\frac{b}{2}$.

b. Square half the coefficient of x obtained in step 2a, i.e., $\left(\frac{b}{2}\right)^2$.

c. Add the square of half the coefficient of x to both sides of the equation, i.e., $x^2+bx+\left(\frac{b}{2}\right)^2=-c+\left(\frac{b}{2}\right)^2$.

d. Simplify the equation.

Step 3 Factor the trinomial on the left hand side of the equation as the square of a binomial, i.e., $\left(x+\frac{b}{2}\right)^2=-c+\left(\frac{b}{2}\right)^2$.

Step 4 Take the square root of both sides of the equation and solve for the x values, i.e., $\sqrt{\left(x+\frac{b}{2}\right)^2}=\pm\sqrt{-c+\left(\frac{b}{2}\right)^2}$; $x+\frac{b}{2}=\pm\sqrt{-c+\left(\frac{b}{2}\right)^2}$; $x=-\frac{b}{2}\pm\sqrt{-c+\left(\frac{b}{2}\right)^2}$.

Step 5 Check the answers by substituting the x values into the original equation.

Step 6 Write the quadratic equation in its factored form.

Examples with Steps

The following examples show the steps as to how quadratic equations, where the coefficient of the squared term is equal to one, are solved using Completing-the-Square method:

Example 1.4-17

Solve the quadratic equation $x^2+x-6=0$ by completing the square.

Solution:

Step 1 $\boxed{x^2+x-6=0}$; $\boxed{x^2+x-6+6=+6}$; $\boxed{x^2+x+0=6}$; $\boxed{x^2+x=6}$

Step 2 $\boxed{x^2+x=6}$; $\boxed{x^2+x+\left(\frac{1}{2}\right)^2=6+\left(\frac{1}{2}\right)^2}$; $\boxed{x^2+x+\frac{1}{4}=6+\frac{1}{4}}$; $\boxed{x^2+x+\frac{1}{4}=\frac{6}{1}+\frac{1}{4}}$

$; \boxed{x^2+x+\frac{1}{4}=\frac{(6\cdot 4)+(1\cdot 1)}{1\cdot 4}} ; \boxed{x^2+x+\frac{1}{4}=\frac{24+1}{4}} ; \boxed{x^2+x+\frac{1}{4}=\frac{25}{4}}$

Step 3 $\boxed{x^2+x+\frac{1}{4}=\frac{25}{4}} ; \boxed{\left(x+\frac{1}{2}\right)^2=\frac{25}{4}}$

Step 4 $\boxed{\left(x+\frac{1}{2}\right)^2=\frac{25}{4}} ; \boxed{\sqrt{\left(x+\frac{1}{2}\right)^2}=\pm\sqrt{\frac{25}{4}}} ; \boxed{x+\frac{1}{2}=\pm\sqrt{\frac{5^2}{2^2}}} ; \boxed{x+\frac{1}{2}=\pm\frac{5}{2}}$ therefore:

I. $\boxed{x+\frac{1}{2}=+\frac{5}{2}} ; \boxed{x=-\frac{1}{2}+\frac{5}{2}} ; \boxed{x=\frac{-1+5}{2}} ; \boxed{x=\frac{\overset{2}{\cancel{4}}}{\cancel{2}}} ; \boxed{\mathbf{x=2}}$

II. $\boxed{x+\frac{1}{2}=-\frac{5}{2}} ; \boxed{x=-\frac{1}{2}-\frac{5}{2}} ; \boxed{x=\frac{-1-5}{2}} ; \boxed{x=-\frac{\overset{3}{\cancel{6}}}{\cancel{2}}} ; \boxed{\mathbf{x=-3}}$

Thus, the solution set is $\{\mathbf{-3,2}\}$.

Step 5 Check: Substitute $x=2$ and $x=-3$ in $x^2+x-6=0$

I. *Let* $x=2$ *in* $x^2+x-6=0 ; 2^2+2-6\overset{?}{=}0 ; 4+2-6\overset{?}{=}0 ; 6-6\overset{?}{=}0 ; 0=0$

II. *Let* $x=-3$ *in* $x^2+x-6=0 ; (-3)^2-3-6\overset{?}{=}0 ; 9-3-6\overset{?}{=}0 ; 9-9\overset{?}{=}0 ; 0=0$

Step 6 Thus, the equation $x^2+x-6=0$ can be factored to $\mathbf{(x-2)(x+3)=0}$

Example 1.4-18

Solve the quadratic equation $x^2+2x+5=0$ by completing the square.

Solution:

Step 1 $\boxed{x^2+2x+5=0} ; \boxed{x^2+2x+5-5=-5} ; \boxed{x^2+2x+0=-5} ; \boxed{x^2+2x=-5}$

Step 2 $\boxed{x^2+2x=-5} ; \boxed{x^2+2x+\left(\frac{2}{2}\right)^2=-5+\left(\frac{2}{2}\right)^2} ; \boxed{x^2+2x+1^2=-5+1^2}$

$; \boxed{x^2+2x+1=-5+1} ; \boxed{x^2+2x+1=-4}$

Step 3 $\boxed{x^2+2x+1=-4} ; \boxed{(x+1)^2=-4}$

Step 4 $\boxed{(x+1)^2=-4} ; \boxed{\sqrt{(x+1)^2}=\pm\sqrt{-4}} ; \boxed{x+1=\pm\sqrt{-4}}$ $\sqrt{-4}$ is not a real number.

Therefore, the equation $x^2+2x+5=0$ **does not have any real solutions.**

Step 5 $\boxed{\textit{Not Applicable}}$

Step 6 $\boxed{\textit{Not Applicable}}$

Additional Examples - Solving Quadratic Equations of the Form $ax^2 + bx + c = 0$, where $a = 1$, by Completing the Square

The following examples further illustrate how to solve quadratic equations using Completing the Square method:

Example 1.4-19

Solve the quadratic equation $x^2 + 3x - 7 = 0$ using Completing-the-Square method.

Solution:

$$\boxed{x^2 + 3x - 7 = 0}\,;\,\boxed{x^2 + 3x = 7}\,;\,\boxed{x^2 + 3x + \left(\frac{3}{2}\right)^2 = 7 + \left(\frac{3}{2}\right)^2}\,;\,\boxed{x^2 + 3x + \frac{9}{4} = 7 + \frac{9}{4}}\,;\,\boxed{\left(x + \frac{3}{2}\right)^2 = \frac{7}{1} + \frac{9}{4}}$$

$$;\,\boxed{\left(x + \frac{3}{2}\right)^2 = \frac{(7\cdot4)+(1\cdot9)}{1\cdot4}}\,;\,\boxed{\left(x + \frac{3}{2}\right)^2 = \frac{28+9}{4}}\,;\,\boxed{\left(x + \frac{3}{2}\right)^2 = \frac{37}{4}}\,;\,\boxed{\sqrt{\left(x + \frac{3}{2}\right)^2} = \pm\sqrt{\frac{37}{4}}}\,;\,\boxed{x + \frac{3}{2} = \pm\frac{\sqrt{37}}{2}}$$

therefore: I. $\boxed{x + \frac{3}{2} = +\frac{\sqrt{37}}{2}}\,;\,\boxed{x = \frac{\sqrt{37}}{2} - \frac{3}{2}}\,;\,\boxed{x = \frac{6.083 - 3}{2}}\,;\,\boxed{x = \frac{3.083}{2}}\,;\,\boxed{\mathbf{x = 1.541}}$

II. $\boxed{x + \frac{3}{2} = -\frac{\sqrt{37}}{2}}\,;\,\boxed{x = -\frac{\sqrt{37}}{2} - \frac{3}{2}}\,;\,\boxed{x = \frac{-6.083 - 3}{2}}\,;\,\boxed{x = -\frac{9.083}{2}}\,;\,\boxed{\mathbf{x = -4.541}}$

and the solution set is $\{\mathbf{-4.541, 1.541}\}$.

Check: I. *Let* $x = 1.541$ *in* $x^2 + 3x - 7 = 0$; $(1.541)^2 + (3 \times 1.541) - 7 \stackrel{?}{=} 0$; $2.38 + 4.62 - 7 \stackrel{?}{=} 0$; $0 = 0$

II. *Let* $x = -4.541$ *in* $x^2 + 3x - 7 = 0$; $(-4.541)^2 + (3 \times -4.541) - 7 \stackrel{?}{=} 0$; $20.62 - 13.62 - 7 \stackrel{?}{=} 0$; $0 = 0$

Therefore, the equation $x^2 + 3x - 7 = 0$ can be factored to $\mathbf{(x - 1.541)(x + 4.541) = 0}$.

Example 1.4-20

Solve the quadratic equation $x^2 + 5x + 6 = 0$ using Completing-the-Square method.

Solution:

$$\boxed{x^2 + 5x + 6 = 0}\,;\,\boxed{x^2 + 5x = -6}\,;\,\boxed{x^2 + 5x + \left(\frac{5}{2}\right)^2 = -6 + \left(\frac{5}{2}\right)^2}\,;\,\boxed{x^2 + 5x + \frac{25}{4} = -6 + \frac{25}{4}}\,;\,\boxed{\left(x + \frac{5}{2}\right)^2 = -\frac{6}{1} + \frac{25}{4}}$$

$$;\,\boxed{\left(x + \frac{5}{2}\right)^2 = \frac{(-6\cdot4)+(1\cdot25)}{1\cdot4}}\,;\,\boxed{\left(x + \frac{5}{2}\right)^2 = \frac{-24+25}{4}}\,;\,\boxed{\left(x + \frac{5}{2}\right)^2 = \frac{1}{4}}\,;\,\boxed{\sqrt{\left(x + \frac{5}{2}\right)^2} = \pm\sqrt{\frac{1}{4}}}\,;\,\boxed{x + \frac{5}{2} = \pm\frac{1}{2}}$$

therefore: I. $\boxed{x + \frac{5}{2} = +\frac{1}{2}}\,;\,\boxed{x = \frac{1}{2} - \frac{5}{2}}\,;\,\boxed{x = \frac{1-5}{2}}\,;\,\boxed{x = -\frac{\overset{2}{\cancel{4}}}{\cancel{2}}}\,;\,\boxed{x = -\frac{2}{1}}\,;\,\boxed{\mathbf{x = -2}}$

II. $\boxed{x + \frac{5}{2} = -\frac{1}{2}}\,;\,\boxed{x = -\frac{1}{2} - \frac{5}{2}}\,;\,\boxed{x = \frac{-1-5}{2}}\,;\,\boxed{x = -\frac{\overset{3}{\cancel{6}}}{\cancel{2}}}\,;\,\boxed{x = -\frac{3}{1}}\,;\,\boxed{\mathbf{x = -3}}$

and the solution set is $\{\mathbf{-3, -2}\}$.

Check: I. *Let* $x=-2$ *in* $x^2+5x+6=0$; $(-2)^2+(5\times-2)+6\overset{?}{=}0$; $4-10+6\overset{?}{=}0$; $10-10\overset{?}{=}0$; $0=0$

II. *Let* $x=-3$ *in* $x^2+5x+6=0$; $(-3)^2+(5\times-3)+6\overset{?}{=}0$; $9-15+6\overset{?}{=}0$; $15-15\overset{?}{=}0$; $0=0$

Therefore, the equation $x^2+5x+6=0$ can be factored to $\mathbf{(x+2)(x+3)=0}$.

Example 1.4-21

Solve the quadratic equation $y^2-9y+11=0$ using Completing-the-Square method.

Solution:

$$\boxed{y^2-9y+11=0} ; \boxed{y^2-9y=-11} ; \boxed{y^2-9y+\left(-\frac{9}{2}\right)^2=-11+\left(-\frac{9}{2}\right)^2} ; \boxed{y^2-9y+\frac{81}{4}=-11+\frac{81}{4}}$$

$$; \boxed{\left(y-\frac{9}{2}\right)^2=-\frac{11}{1}+\frac{81}{4}} ; \boxed{\left(y-\frac{9}{2}\right)^2=\frac{(-11\cdot4)+(1\cdot81)}{1\cdot4}} ; \boxed{\left(y-\frac{9}{2}\right)^2=\frac{-44+81}{4}} ; \boxed{\left(y-\frac{9}{2}\right)^2=\frac{37}{4}}$$

$$; \boxed{\sqrt{\left(y-\frac{9}{2}\right)^2}=\pm\sqrt{\frac{37}{4}}} ; \boxed{y-\frac{9}{2}=\pm\frac{\sqrt{37}}{2}}$$ therefore:

I. $$\boxed{y-\frac{9}{2}=+\frac{\sqrt{37}}{2}} ; \boxed{y=\frac{\sqrt{37}}{2}+\frac{9}{2}} ; \boxed{y=\frac{6.083+9}{2}} ; \boxed{y=\frac{15.083}{2}} ; \boxed{\mathbf{y=7.541}}$$

II. $$\boxed{y-\frac{9}{2}=-\frac{\sqrt{37}}{2}} ; \boxed{y=-\frac{\sqrt{37}}{2}+\frac{9}{2}} ; \boxed{y=\frac{-6.083+9}{2}} ; \boxed{y=\frac{2.917}{2}} ; \boxed{\mathbf{y=1.459}}$$

and the solution set is $\{\mathbf{1.459, 7.541}\}$.

Check: I. *Let* $y=7.541$ *in* $y^2-9y+11=0$; $(7.541)^2+(-9\times7.541)+11\overset{?}{=}0$; $56.87-67.87+11\overset{?}{=}0$

; $67.87-67.87\overset{?}{=}0$; $0=0$

II. *Let* $y=1.459$ *in* $y^2-9y+11=0$; $(1.459)^2+(-9\times1.459)+11\overset{?}{=}0$; $2.13-13.13+11\overset{?}{=}0$

; $13.13-13.13\overset{?}{=}0$; $0=0$

Therefore, the equation $y^2-9y+11=0$ can be factored to $\mathbf{(y-7.541)(y-1.459)=0}$.

Practice Problems - Solving Quadratic Equations of the Form $ax^2+bx+c=0$, where $a=1$, by Completing the Square

Section 1.4d Case I Practice Problems - Solve the following quadratic equations using Completing-the-Square method:

1. $x^2+10x-2=0$
2. $x^2-x-1=0$
3. $x(x+2)=80$
4. $y^2-10y+5=0$
5. $x^2+4x-5=0$
6. $y^2+4y=14$

Case II Solving Quadratic Equations of the Form $ax^2+bx+c=0$, where $a \rangle 1$, by Completing the Square

The following show the steps as to how quadratic equations, where the coefficient of the squared term is not equal to one, are solved using Completing-the-Square method:

Step 1 Write the equation in the form of $ax^2+bx=-c$.

Step 2 Divide both sides of the equation by a, i.e., $\frac{ax^2}{a}+\frac{bx}{a}=-\frac{c}{a}$; $x^2+\frac{b}{a}x=-\frac{c}{a}$.

Step 3 a. Divide the coefficient of x by 2, i.e., $\frac{1}{2}\cdot\frac{b}{a}=\frac{b}{2a}$.

b. Square half the coefficient of x obtained in step 3a, i.e., $\left(\frac{b}{2a}\right)^2$

c. Add the square of half the coefficient of x to both sides of the equation, i.e.,

$$x^2+\frac{b}{a}x+\left(\frac{b}{2a}\right)^2=-\frac{c}{a}+\left(\frac{b}{2a}\right)^2.$$

d. Simplify the equation.

Step 4 Factor the trinomial on the left hand side of the equation as the square of a binomial, i.e., $\left(x+\frac{b}{2a}\right)^2=-\frac{c}{a}+\left(\frac{b}{2a}\right)^2$.

Step 5 Take the square root of both sides of the equation and solve for the x values, i.e.,

$$\sqrt{\left(x+\frac{b}{2a}\right)^2}=\pm\sqrt{-\frac{c}{a}+\left(\frac{b}{2a}\right)^2} \; ; \; x+\frac{b}{2a}=\pm\sqrt{-\frac{c}{a}+\left(\frac{b}{2a}\right)^2} \; ; \; x=-\frac{b}{2a}\pm\sqrt{-\frac{c}{a}+\left(\frac{b}{2a}\right)^2}.$$

Step 6 Check the answers by substituting the x values into the original equation.

Step 7 Write the quadratic equation in its factored form.

Examples with Steps

The following examples show the steps as to how quadratic equations, where the coefficient of the squared term is not equal to one, are solved using completing-the-square method:

Example 1.4-22

Solve the quadratic equation $2x^2+3x-6=0$ by completing the square.

Solution:

Step 1 $\boxed{2x^2+3x-6=0}$; $\boxed{2x^2+3x-6+6=+6}$; $\boxed{2x^2+3x+0=+6}$; $\boxed{2x^2+3x=6}$

Step 2 $\boxed{2x^2+3x=6}$; $\boxed{\frac{2}{2}x^2+\frac{3}{2}x=\frac{\overset{3}{6}}{2}}$; $\boxed{x^2+\frac{3}{2}x=\frac{3}{1}}$; $\boxed{x^2+\frac{3}{2}x=3}$

Step 3 $\boxed{x^2+\frac{3}{2}x=3}$; $\boxed{x^2+\frac{3}{2}x+\left(\frac{3}{4}\right)^2=3+\left(\frac{3}{4}\right)^2}$; $\boxed{x^2+\frac{3}{2}x+\frac{9}{16}=3+\frac{9}{16}}$

; $\boxed{x^2+\frac{3}{2}x+\frac{9}{16}=\frac{3}{1}+\frac{9}{16}}$; $\boxed{x^2+\frac{3}{2}x+\frac{9}{16}=\frac{48+9}{16}}$; $\boxed{x^2+\frac{3}{2}x+\frac{9}{16}=\frac{57}{16}}$

Step 4 $\boxed{x^2+\frac{3}{2}x+\frac{9}{16}=\frac{57}{16}}$; $\boxed{\left(x+\frac{3}{4}\right)^2=\frac{57}{16}}$

Step 5 $\boxed{\left(x+\frac{3}{4}\right)^2=\frac{57}{16}}$; $\boxed{\sqrt{\left(x+\frac{3}{4}\right)^2}=\pm\sqrt{\frac{57}{16}}}$; $\boxed{x+\frac{3}{4}=\pm\sqrt{\frac{57}{4^2}}}$; $\boxed{x+\frac{3}{4}=\pm\frac{\sqrt{57}}{4}}$ therefore:

I. $\boxed{x+\frac{3}{4}=+\frac{\sqrt{57}}{4}}$; $\boxed{x=\frac{7.55}{4}-\frac{3}{4}}$; $\boxed{x=\frac{7.55-3}{4}}$; $\boxed{x=\frac{4.55}{4}}$; $\boxed{\mathbf{x=1.138}}$

II. $\boxed{x+\frac{3}{4}=-\frac{\sqrt{57}}{4}}$; $\boxed{x=-\frac{7.55}{4}-\frac{3}{4}}$; $\boxed{x=\frac{-7.55-3}{4}}$; $\boxed{x=\frac{-10.55}{4}}$; $\boxed{\mathbf{x=-2.638}}$

and the solution set is $\{\mathbf{-2.638, 1.138}\}$.

Step 6 Check: Substitute $x=1.138$ and $x=-2.638$ in $2x^2+3x-6=0$

I. *Let* $x=1.138$ *in* $2x^2+3x-6=0$; $2\cdot(1.138)^2+(3\times1.138)-6\overset{?}{=}0$

; $2.59+3.41-6\overset{?}{=}0$; $6-6\overset{?}{=}0$; $0=0$

II. *Let* $x=-2.638$ *in* $2x^2+3x-6=0$; $2\cdot(-2.638)^2+(3\times-2.638)-6\overset{?}{=}0$

; $13.92-7.92-6\overset{?}{=}0$; $13.92-13.92\overset{?}{=}0$; $0=0$

Step 7 Thus, the equation $2x^2+3x-6=0$ which is equal to $x^2+1.5x-3=0$ can be factored to $\mathbf{(x-1.138)(x+2.638)=0}$

Example 1.4-23

Solve the quadratic equation $2u^2+6u-7=0$ by completing the square.

Solution:

Step 1 $\boxed{2u^2+6u-7=0}$; $\boxed{2u^2+6u-7+7=+7}$; $\boxed{2u^2+6u+0=7}$; $\boxed{2u^2+6u=7}$

Step 2 $\boxed{2u^2+6u=7}$; $\boxed{\frac{2}{2}u^2+\frac{\overset{3}{\cancel{6}}}{2}u=\frac{7}{2}}$; $\boxed{u^2+3u=\frac{7}{2}}$

Step 3 $\boxed{u^2+3u=\frac{7}{2}}$; $\boxed{u^2+3u+\left(\frac{3}{2}\right)^2=\frac{7}{2}+\left(\frac{3}{2}\right)^2}$; $\boxed{u^2+3u+\frac{9}{4}=\frac{7}{2}+\frac{9}{4}}$

; $\boxed{u^2+3u+\frac{9}{4}=\frac{(7\cdot4)+(2\cdot9)}{2\cdot4}}$; $\boxed{u^2+3u+\frac{9}{4}=\frac{28+18}{8}}$; $\boxed{u^2+3u+\frac{9}{4}=\frac{46}{8}}$

Step 4 $\boxed{u^2+3u+\frac{9}{4}=\frac{46}{8}}$; $\boxed{\left(u+\frac{3}{2}\right)^2=\frac{\overset{23}{\cancel{46}}}{\underset{4}{\cancel{8}}}}$; $\boxed{\left(u+\frac{3}{2}\right)^2=\frac{23}{4}}$

Step 5 $\boxed{\left(u+\frac{3}{2}\right)^2=\frac{23}{4}}$; $\boxed{\sqrt{\left(u+\frac{3}{2}\right)^2}=\pm\sqrt{\frac{23}{4}}}$; $\boxed{u+\frac{3}{2}=\pm\sqrt{\frac{23}{2^2}}}$; $\boxed{u+\frac{3}{2}=\pm\frac{\sqrt{23}}{2}}$ therefore:

I. $\boxed{u+\frac{3}{2}=+\frac{\sqrt{23}}{2}}$; $\boxed{u=-\frac{3}{2}+\frac{\sqrt{23}}{2}}$; $\boxed{u=\frac{-3+\sqrt{23}}{2}}$; $\boxed{u=\frac{-3+4.8}{2}}$; $\boxed{u=\frac{1.8}{2}}$; $\boxed{\mathbf{u=0.9}}$

II. $\boxed{u+\frac{3}{2}=-\frac{\sqrt{23}}{2}}$; $\boxed{u=-\frac{3}{2}-\frac{\sqrt{23}}{2}}$; $\boxed{u=\frac{-3-\sqrt{23}}{2}}$; $\boxed{u=\frac{-3-4.8}{2}}$; $\boxed{u=\frac{-7.8}{2}}$; $\boxed{\mathbf{u=-3.9}}$

and the solution set is $\{\mathbf{-3.9, 0.9}\}$.

Step 6 Check: Substitute $u=0.9$ and $u=-3.9$ in $2u^2+6u-7=0$

I. *Let* $u=0.9$ *in* $2u^2+6u-7=0$; $2\cdot 0.9^2+(6\times 0.9)-7\stackrel{?}{=}0$; $1.6+5.4-7\stackrel{?}{=}0$; $7-7\stackrel{?}{=}0$; $0=0$

II. *Let* $u=-3.9$ *in* $2u^2+6u-7=0$; $2\cdot(-3.9)^2+(6\times -3.9)-7\stackrel{?}{=}0$; $2\times 15.2-23.4-7\stackrel{?}{=}0$; $30.4-23.4-7\stackrel{?}{=}0$; $7-7\stackrel{?}{=}0$; $0=0$

Step 7 Thus, the equation $2u^2+6u-7=0$, which is the same as $u^2+3u-3.5=0$,can be factored to $\mathbf{(u-0.9)(u+3.9)=0}$.

Additional Examples - Solving Quadratic Equations of the Form $ax^2+bx+c=0$, where $a \rangle 1$, by Completing the Square

The following examples further illustrate how to solve quadratic equations, where the coefficient of the squared term is not equal to one, using Completing-the-Square method:

Example 1.4-24

Solve the quadratic equation $3x^2+2x-1=0$ using Completing-the-Square method.

Solution:

$\boxed{3x^2+2x-1=0}$; $\boxed{3x^2+2x=1}$; $\boxed{\frac{3}{3}x^2+\frac{2}{3}x=\frac{1}{3}}$; $\boxed{x^2+\frac{2}{3}x=\frac{1}{3}}$; $\boxed{x^2+\frac{2}{3}x+\left(\frac{2}{6}\right)^2=\frac{1}{3}+\left(\frac{2}{6}\right)^2}$

; $\boxed{x^2+\frac{2}{3}x+\left(\frac{1}{3}\right)^2=\frac{1}{3}+\left(\frac{1}{3}\right)^2}$; $\boxed{x^2+\frac{2}{3}x+\frac{1}{9}=\frac{1}{3}+\frac{1}{9}}$; $\boxed{\left(x+\frac{1}{3}\right)^2=\frac{1}{3}+\frac{1}{9}}$; $\boxed{\left(x+\frac{1}{3}\right)^2=\frac{(1\cdot 9)+(1\cdot 3)}{3\cdot 9}}$

; $\boxed{\left(x+\frac{1}{3}\right)^2=\frac{9+3}{27}}$; $\boxed{\left(x+\frac{1}{3}\right)^2=\frac{12}{27}}$; $\boxed{\sqrt{\left(x+\frac{1}{3}\right)^2}=\pm\sqrt{\frac{12}{27}}}$; $\boxed{x+\frac{1}{3}=\pm\sqrt{\frac{12}{27}}}$; $\boxed{x+\frac{1}{3}=\pm\sqrt{0.44}}$

; $\boxed{x+0.33=\pm 0.67}$ therefore:

I. $\boxed{x+0.33=+0.67}$; $\boxed{x=0.67-0.33}$; $\boxed{\mathbf{x=0.34}}$ II. $\boxed{x+0.33=-0.67}$; $\boxed{x=-0.67-0.33}$; $\boxed{\mathbf{x=-1}}$

and the solution set is $\{\mathbf{-1, 0.34}\}$.

Check: I. *Let* $x=-1$ *in* $3x^2+2x-1=0$; $3\cdot(-1)^2+(2\cdot -1)-1\stackrel{?}{=}0$; $3\cdot 1-2-1\stackrel{?}{=}0$; $3-3\stackrel{?}{=}0$; $0=0$

II. *Let* $x = 0.34$ *in* $3x^2 + 2x - 1 = 0$; $3 \cdot 0.34^2 + (2 \cdot 0.34) - 1 \stackrel{?}{=} 0$; $3 \cdot 0.11 + 0.68 - 1 \stackrel{?}{=} 0$

; $0.33 + 0.68 - 1 \stackrel{?}{=} 0$; $1 - 1 \stackrel{?}{=} 0$; $0 = 0$

Therefore, the equation $3x^2 + 2x - 1 = 0$ can be factored to $\mathbf{(x - 0.34)(x + 1) = 0}$.

Example 1.4-25

Solve the quadratic equation $3t^2 + 12t - 4 = 0$ using Completing-the-Square method.

Solution:

$\boxed{3t^2 + 12t - 4 = 0}$; $\boxed{3t^2 + 12t = 4}$; $\boxed{\frac{3}{3}t^2 + \frac{12}{3}x = \frac{4}{3}}$; $\boxed{t^2 + 4t = \frac{4}{3}}$; $\boxed{t^2 + 4t + \left(\frac{4}{2}\right)^2 = \frac{4}{3} + \left(\frac{4}{2}\right)^2}$

; $\boxed{t^2 + 4t + \left(\frac{2}{1}\right)^2 = \frac{4}{3} + \left(\frac{2}{1}\right)^2}$; $\boxed{t^2 + 4t + 4 = \frac{4}{3} + 4}$; $\boxed{(t+2)^2 = \frac{4}{3} + \frac{4}{1}}$; $\boxed{(t+2)^2 = \frac{(4 \cdot 1) + (4 \cdot 3)}{3 \cdot 1}}$

; $\boxed{(t+2)^2 = \frac{4+12}{3}}$; $\boxed{(t+2)^2 = \frac{16}{3}}$; $\boxed{\sqrt{(t+2)^2} = \pm\sqrt{\frac{16}{3}}}$; $\boxed{t + 2 = \pm\sqrt{5.33}}$; $\boxed{t + 2 = \pm 2.31}$ therefore:

I. $\boxed{t + 2 = +2.31}$; $\boxed{t = 2.31 - 2}$; $\boxed{\mathbf{t = 0.31}}$

II. $\boxed{t + 2 = -2.31}$; $\boxed{t = -2.31 - 2}$; $\boxed{\mathbf{t = -4.31}}$

and the solution set is $\{\mathbf{0.31, -4.31}\}$.

Check: I. *Let* $t = 0.31$ *in* $3t^2 + 12t - 4 = 0$; $3 \cdot (0.31)^2 + (12 \cdot 0.31) - 4 \stackrel{?}{=} 0$; $3 \cdot 0.096 - 3.72 - 4 \stackrel{?}{=} 0$

; $0.288 + 3.72 - 4 \stackrel{?}{=} 0$; $4 - 4 \stackrel{?}{=} 0$; $0 = 0$

II. *Let* $t = -4.31$ *in* $3t^2 + 12t - 4 = 0$; $3 \cdot (-4.31)^2 + (12 \cdot -4.31) - 4 \stackrel{?}{=} 0$; $3 \cdot 18.57 - 51.72 - 4 \stackrel{?}{=} 0$

; $55.72 - 51.72 - 4 \stackrel{?}{=} 0$; $55.72 - 55.72 \stackrel{?}{=} 0$; $0 = 0$

Therefore, the equation $3t^2 + 12t - 4 = 0$ can be factored to $\mathbf{(t - 0.31)(t + 4.31) = 0}$.

Practice Problems - Solving Quadratic Equations of the Form $ax^2 + bx + c = 0$, where $a \rangle 1$, by Completing the Square

Section 1.4d Case II Practice Problems - Solve the following quadratic equations using Completing-the-Square method. (Note that these problems are identical to the exercises given in Section 1.4b Case II.)

1. $4u^2 + 6u + 1 = 0$
2. $4w^2 + 10w = -3$
3. $6x^2 + 4x - 2 = 0$
4. $15y^2 + 3 = -14y$
5. $2x^2 - 5x + 3 = 0$
6. $2x^2 + xy - y^2 = 0$ x is variable

1.4e How to Choose the Best Factoring or Solution Method

To factor polynomials and to solve quadratic equations a total of seven basic methods have been reviewed in Sections 1.3 and 1.4. Those methods are:

1. The Greatest Common Factoring method
2. The Grouping method
3. The Trial and Error method
4. Factoring methods for polynomials with square and cubed terms
5. The Quadratic Formula method
6. The Square Root Property method, and
7. Completing-the-Square method

The decision as to which one of the above methods is most suitable in factoring a polynomial or solving an equation is left to the student. For example, in some cases, using the Trial and Error method in solving a quadratic equation may be easier than using the Quadratic Formula or Completing-the-Square method. In certain cases, using the quadratic formula in solving a polynomial may be faster than the Grouping or the Trial and Error method. Note that the key in choosing the best and/or the easiest method is through solving many problems. After sufficient practice, students start to gain confidence on selection of one method over the other.

Assumption *- In many instances, the methods used in factoring polynomials (shown in Section 1.3) can also be used in solving quadratic equations (shown in Section 1.4) by recognizing that the left hand side of the equation* $ax^2+bx+c=0$*, namely* ax^2+bx+c *is a polynomial and can be factored as such, using polynomial factoring methods covered in Section 1.3.*

Note 1 - Any quadratic equation can be solved using the quadratic formula. Once the student has memorized the quadratic formula and has learned how to substitute the equivalent values of a, b, and c into the quadratic formula, then the next steps are merely the process of solving the quadratic equation using mathematical operations.

Note 2 - The quadratic formula can be used as an alternative method in factoring polynomials of the form ax^2+bx+c as is stated in the above assumption.

The following examples are solved using the seven factoring and solution methods shown above:

Example 1.4-26

Use different methods to solve the equation $x^2=25$.

Solution:

First Method: (The Trial and Error Method)

Write the equation in the standard quadratic equation form $ax^2+bx+c=0$, i.e., write $x^2=25$ as $x^2+0x-25=0$. To solve the given equation using the Trial and Error method we only consider the left hand side of the equation which is a second degree polynomial. Next, we need to obtain two numbers whose sum is 0 and whose product is -25 by constructing a table as shown below:

Sum	*Product*
$1-1=0$	$1\cdot(-1)=-1$
$2-2=0$	$2\cdot(-2)=-4$
$3-3=0$	$3\cdot(-3)=-9$
$4-4=0$	$4\cdot(-4)=-16$
$\mathbf{5-5=0}$	$\mathbf{5\cdot(-5)=-25}$

The last line contains the sum and the product of the two numbers that we need. Thus, $x^2=25$ or $x^2+0x-25=0$ can be factored to $(\boldsymbol{x-5})(\boldsymbol{x+5})=\mathbf{0}$.

Check: $(x-5)(x+5)=0$; $x\cdot x+5\cdot x-5\cdot x+5\cdot(-5)=0$; $x^2+5x-5x-25=0$; $x^2+(5-5)x-25=0$; $x^2+0x-25=0$

Second Method: (The Quadratic Formula Method)

First, write the equation in the standard quadratic equation form $ax^2+bx+c=0$, i.e., write $x^2=25$ as $x^2+0x-25=0$. Second, equate the coefficients of $x^2+0x-25=0$ with the standard quadratic equation by letting $a=1$, $b=0$, and $c=-25$. Then,

Given: $x=\frac{-b\pm\sqrt{b^2-4ac}}{2a}$; $x=\frac{-0\pm\sqrt{0^2-(4\times1\times-25)}}{2\times1}$; $x=\frac{\pm\sqrt{0+100}}{2}$; $x=\frac{\pm\sqrt{100}}{2}$; $x=\pm\frac{\sqrt{10^2}}{2}$; $x=\pm\frac{10}{2}$. Therefore:

I. $x=+\frac{\overset{5}{\cancel{10}}}{\cancel{2}}$; $x=\frac{5}{1}$; $\boldsymbol{x=5}$ II. $x=-\frac{\overset{5}{\cancel{10}}}{\cancel{2}}$; $x=-\frac{5}{1}$; $\boldsymbol{x=-5}$

Check: I. *Let* $x=5$ *in* $x^2=25$; $5^2\overset{?}{=}25$; $25=25$

II. *Let* $x=-5$ *in* $x^2=25$; $(-5)^2\overset{?}{=}25$; $25=25$

Therefore, the equation $x^2+0x-25=0$ can be factored to $(\boldsymbol{x+5})(\boldsymbol{x-5})=\mathbf{0}$.

Third Method: (The Square Root Property Method)

Take the square root of both sides of the equation, i.e., write $x^2=25$ as $\sqrt{x^2}=\pm\sqrt{25}$; $x=\pm\sqrt{5^2}$; $x=\pm5$. Thus, $\boldsymbol{x=+5}$ or $\boldsymbol{x=-5}$ are the solution sets to the equation $x^2=25$ which can be represented in its factorable form as $(\boldsymbol{x+5})(\boldsymbol{x-5})=\mathbf{0}$.

Fourth Method: (Completing-the-Square Method) - Is not applicable.

Note that from the above three methods using the Square Root Property method is the fastest and the easiest method to obtain the factored terms. The Trial and Error method is the second easiest method to use, followed by the Quadratic Formula method which is the most difficult way of obtaining the factored terms.

Example 1.4-27

Use different methods to solve the equation $x^2+11x+24=0$.

Solution:

First Method: (The Trial and Error Method)

To solve the given equation using the Trial and Error method we only consider the left hand side of the equation which is a second degree polynomial. Next, we need to obtain two numbers whose sum is 11 and whose product is 24 by constructing a table as shown below:

Sum	***Product***
$6+5=11$	$6\cdot 5=30$
$7+4=11$	$7\cdot 4=28$
$\mathbf{8+3=11}$	$\mathbf{8\cdot 3=24}$
$9+2=11$	$9\cdot 2=18$

The third line contains the sum and the product of the two numbers that we need. Thus, $x^2+11x+24=0$ can be factored to $\mathbf{(x+8)(x+3)=0}$.

Check: $(x+8)(x+3)=0$; $x\cdot x+3\cdot x+8\cdot x+8\cdot 3=0$; $x^2+3x+8x+24=0$; $x^2+(3+8)x+24=0$

; $x^2+11x+24=0$

Second Method: (The Quadratic Formula Method)

Given the standard quadratic equation $ax^2+bx+c=0$, equate the coefficients of $x^2+11x+24=0$ with the standard quadratic equation by letting $a=1$, $b=11$, and $c=24$. Then,

Given: $x=\frac{-b\pm\sqrt{b^2-4ac}}{2a}$; $x=\frac{-11\pm\sqrt{11^2-(4\times 1\times 24)}}{2\times 1}$; $x=\frac{-11\pm\sqrt{121-96}}{2}$; $x=\frac{-11\pm\sqrt{25}}{2}$

; $x=\frac{-11\pm\sqrt{5^2}}{2}$; $x=\frac{-11\pm 5}{2}$. Therefore:

I. $x=\frac{-11+5}{2}$; $x=-\frac{\overset{3}{\cancel{6}}}{\cancel{2}}$; $x=-\frac{3}{1}$; $\mathbf{x=-3}$

II. $x=\frac{-11-5}{2}$; $x=-\frac{\overset{8}{\cancel{16}}}{\cancel{2}}$; $x=-\frac{8}{1}$; $\mathbf{x=-8}$

Check: I. *Let* $x=-3$ *in* $x^2+11x+24=0$; $(-3)^2+11\cdot(-3)+24\overset{?}{=}0$; $9-33+24\overset{?}{=}0$; $33-33\overset{?}{=}0$

; $0=0$

II. *Let* $x=-8$ *in* $x^2+11x+24=0$; $(-8)^2+11\cdot(-8)+24\overset{?}{=}0$; $64-88+24\overset{?}{=}0$; $88-88\overset{?}{=}0$

; $0=0$

Therefore, the equation $x^2+11x+24=0$ can be factored to $\mathbf{(x+8)(x+3)=0}$.

Third Method: (Completing-the-Square Method)

$x^2+11x+24=0$; $x^2+11x=-24$; $x^2+11x+\left(\frac{11}{2}\right)^2=-24+\left(\frac{11}{2}\right)^2$; $x^2+11x+\frac{121}{4}=-24+\frac{121}{4}$

; $\left(x+\frac{11}{2}\right)^2=-\frac{24}{1}+\frac{121}{4}$; $\left(x+\frac{11}{2}\right)^2=\frac{(-24\cdot 4)+(1\cdot 121)}{1\cdot 4}$; $\left(x+\frac{11}{2}\right)^2=\frac{-96+121}{4}$; $\left(x+\frac{11}{2}\right)^2=\frac{25}{4}$

; $x+\frac{11}{2}=\pm\sqrt{\frac{25}{4}}$; $x+\frac{11}{2}=\pm\frac{5}{2}$

Therefore: I. $x+\frac{11}{2}=+\frac{5}{2}$; $x=\frac{5}{2}-\frac{11}{2}$; $x=\frac{5-11}{2}$; $x=-\frac{\overset{3}{\cancel{6}}}{\cancel{2}}$; $x=-\frac{3}{1}$; $\mathbf{x=-3}$

II. $x+\frac{11}{2}=-\frac{5}{2}$; $x=-\frac{5}{2}-\frac{11}{2}$; $x=\frac{-5-11}{2}$; $x=-\frac{\overset{8}{\cancel{16}}}{\cancel{2}}$; $x=-\frac{8}{1}$; $\mathbf{x=-8}$

Check: I. *Let* $x=-3$ *in* $x^2+11x+24=0$; $(-3)^2+(11\times -3)+24\overset{?}{=}0$; $9-33+24\overset{?}{=}0$; $0=0$

II. $Let\ x=-8\ in\quad x^2+11x+24=0\ ;\ (-8)^2+(11\times-8)+24\stackrel{?}{=}0\ ;\ 64-88+24\stackrel{?}{=}0\ ;\ 0=0$

Therefore, the equation $x^2+11x+24=0$ can be factored to $(\mathbf{x+8})(\mathbf{x+3})=\mathbf{0}$.

Fourth Method: (The Square Root Property Method) - Is not applicable

Note that from the above three methods using the Trial and Error method is the fastest and the easiest method to obtain the factored terms. Completing-the-Square method is the second easiest method to use, followed by the Quadratic Formula method which is the longest and most difficult way of obtaining the factored terms.

Example 1.4-28

Use different methods to solve the equation $x^2+5x+2=0$.

Solution:

First Method: (The Trial and Error Method)

To solve the given equation using the Trial and Error method we only consider the left hand side of the equation which is a second degree polynomial. Next, we need to obtain two numbers whose sum is 5 and whose product is 2. However, after few trials, it becomes clear that such a combination of integer numbers is not possible to obtain. Hence, the Trial and Error method is not applicable to this particular example.

Second Method: (The Quadratic Formula Method)

Given the standard quadratic equation $ax^2+bx+c=0$, equate the coefficients of $x^2+5x+2=0$ with the standard quadratic equation by letting $a=1$, $b=5$, and $c=2$. Then,

Given: $x=\frac{-b\pm\sqrt{b^2-4ac}}{2a}\ ;\ x=\frac{-5\pm\sqrt{5^2-(4\times1\times2)}}{2\times1}\ ;\ x=\frac{-5\pm\sqrt{25-8}}{2}\ ;\ x=\frac{-5\pm\sqrt{17}}{2}$. Therefore:

I. $x=\frac{-5+\sqrt{17}}{2}\ ;\ x=\frac{-5+4.12}{2}\ ;\ x=-\frac{0.88}{2}\ ;\ \mathbf{x=-0.44}$

II. $x=\frac{-5-\sqrt{17}}{2}\ ;\ x=\frac{-5-4.12}{2}\ ;\ x=-\frac{9.12}{2}\ ;\ \mathbf{x=-4.56}$

Check: I. $Let\ x=-0.44\ in\quad x^2+5x+2=0\ ;\ (-0.44)^2+5\cdot(-0.44)+2\stackrel{?}{=}0\ ;\ 0.2-2.2+2\stackrel{?}{=}0\ ;\ 2.2-2.2\stackrel{?}{=}0$
$;\ 0=0$

II. $Let\ x=-4.56\ in\quad x^2+5x+2=0\ ;\ (-4.56)^2+5\cdot(-4.56)+2\stackrel{?}{=}0\ ;\ 20.8-22.8+2\stackrel{?}{=}0$
$;\ 22.8-22.8\stackrel{?}{=}0\ ;\ 0=0$

Therefore, the equation $x^2+5x+2=0$ can be factored to $(\mathbf{x+0.44})(\mathbf{x+4.56})=\mathbf{0}$.

Third Method: (Completing-the-Square Method)

$x^2+5x+2=0\ ;\ x^2+5x=-2\ ;\ x^2+5x+\left(\frac{5}{2}\right)^2=-2+\left(\frac{5}{2}\right)^2\ ;\ x^2+5x+\frac{25}{4}=-2+\frac{25}{4}\ ;\ \left(x+\frac{5}{2}\right)^2=-\frac{2}{1}+\frac{25}{4}$

$;\ \left(x+\frac{5}{2}\right)^2=\frac{(-2\cdot4)+(1\cdot25)}{1\cdot4}\ ;\ \left(x+\frac{5}{2}\right)^2=\frac{-8+25}{4}\ ;\ \left(x+\frac{5}{2}\right)^2=\frac{17}{4}\ ;\ x+\frac{5}{2}=\pm\sqrt{\frac{17}{4}}\ ;\ x+\frac{5}{2}=\pm\frac{\sqrt{17}}{2}$

Therefore: I. $x+\frac{5}{2}=+\frac{\sqrt{17}}{2}\ ;\ x=\frac{\sqrt{17}}{2}-\frac{5}{2}\ ;\ x=\frac{\sqrt{17}-5}{2}\ ;\ x=\frac{4.12-5}{2}\ ;\ x=-\frac{0.88}{2}\ ;\ \mathbf{x=-0.44}$

II. $x+\frac{5}{2}=-\frac{\sqrt{17}}{2}$; $x=-\frac{\sqrt{17}}{2}-\frac{5}{2}$; $x=\frac{-\sqrt{17}-5}{2}$; $x=\frac{-4.12-5}{2}$; $x=-\frac{9.12}{2}$; $\mathbf{x=-4.56}$

Check: I. *Let* $x=-0.44$ *in* $x^2+5x+2=0$; $(-0.44)^2+5\cdot(-0.44)+2\overset{?}{=}0$; $0.2-2.2+2\overset{?}{=}0$; $2.2-2.2\overset{?}{=}0$; $0=0$

II. *Let* $x=-4.56$ *in* $x^2+5x+2=0$; $(-4.56)^2+5\cdot(-4.56)+2\overset{?}{=}0$; $20.8-22.8+2\overset{?}{=}0$; $22.8-22.8\overset{?}{=}0$; $0=0$

Therefore, the equation $x^2+5x+2=0$ can be factored to $\mathbf{(x+0.44)(x+4.56)=0}$.

Fourth Method: (The Square Root Property Method) - Is not applicable.

Note that from the above two methods using the Quadratic Formula method may be the faster method, for some, than Completing-the-Square method.

Practice Problems - How to Choose the Best Factoring or Solution Method

Section 1.4e Practice Problems - Choose three methods to solve the following quadratic equations. State the degree of difficulty associated with each method you selected.

1. $x^2=16$
2. $x^2+7x+3=0$
3. $(3x+4)^2=36$
4. $x^2+11x+30=0$

1.5 Algebraic Fractions

In this section math operations involving algebraic fractions are reviewed. Algebraic fractions are introduced in Section 1.5a. The steps as to how algebraic fractions are simplified are addressed in Section 1.5b. Addition, subtraction, multiplication, and division of algebraic fractions are addressed in Section 1.5c. Math operations involving complex algebraic fractions are reviewed in Section 1.5d.

1.5a Introduction to Algebraic Fractions

Arithmetic/integer fractions are fractions where the numerator and the denominator are integer numbers. For example, $\frac{2}{3}$, $\frac{1}{5}$, $-\frac{3}{8}$, and $\frac{2}{1}$ are examples of arithmetic fractions. **Algebraic fractions** are fractions where the numerator or the denominator (or both) are variables. For example, $\frac{a}{3}$, $\frac{x}{y}$, $\frac{3}{x+1}$, and $\frac{1}{x}$ are examples of algebraic fractions. The concepts and procedures learned in simplifying, adding, subtracting, multiplying, and dividing arithmetic fractions can directly be applied to algebraic fractions. (The subject of arithmetic fractions has been addressed in detail in the "*Mastering Fractions*" book. Students are encouraged to review chapters 3 and 9 for an overall review of the fractional operations.) In this section we will review the sign rules for fractions; division of algebraic fractions by zero; and equivalent algebraic fractions.

A. Sign Rules For Fractions

In division, we need to consider two signs. The sign of the numerator and the sign of the denominator. Thus, the sign rules for division are:

1. $\frac{-a}{+b} = -\frac{a}{b}$ For example: $\frac{-8}{+2} = -\frac{8}{2} = -\frac{4}{1} = -4$
2. $\frac{+a}{-b} = -\frac{a}{b}$ For example: $\frac{+8}{-2} = -\frac{8}{2} = -\frac{4}{1} = -4$
3. $\frac{-a}{-b} = +\frac{a}{b}$ For example: $\frac{-8}{-2} = +\frac{8}{2} = +\frac{4}{1} = +4$
4. $\frac{+a}{+b} = +\frac{a}{b}$ For example: $\frac{+8}{+2} = +\frac{8}{2} = +\frac{4}{1} = +4$

In fractions, we need to consider three signs. The fractions sign itself, the numerator sign, and the denominator sign. Therefore, the sign rules for fractions are:

1. $+\frac{-a}{+b} = +\left(-\frac{a}{b}\right) = -\frac{a}{b}$
2. $+\frac{+a}{-b} = +\left(-\frac{a}{b}\right) = -\frac{a}{b}$
3. $+\frac{-a}{-b} = +\left(+\frac{a}{b}\right) = +\frac{a}{b}$
4. $+\frac{+a}{+b} = +\left(+\frac{a}{b}\right) = +\frac{a}{b}$
5. $-\frac{-a}{+b} = -\left(-\frac{a}{b}\right) = +\frac{a}{b}$
6. $-\frac{+a}{-b} = -\left(-\frac{a}{b}\right) = +\frac{a}{b}$
7. $-\frac{-a}{-b} = -\left(+\frac{a}{b}\right) = -\frac{a}{b}$
8. $-\frac{+a}{+b} = -\left(+\frac{a}{b}\right) = -\frac{a}{b}$

For example:

1. $+\frac{-8}{+2} = +\left(-\frac{8}{2}\right) = -\frac{8}{2} = -\frac{4}{1} = -4$
2. $+\frac{+8}{-2} = +\left(-\frac{8}{2}\right) = -\frac{8}{2} = -\frac{4}{1} = -4$

3. $+\frac{-8}{-2} = +\left(+\frac{8}{2}\right) = +\frac{8}{2} = +\frac{4}{1} = +4$

4. $+\frac{+8}{+2} = +\left(+\frac{8}{2}\right) = +\frac{8}{2} = +\frac{4}{1} = +4$

5. $-\frac{-8}{+2} = -\left(-\frac{8}{2}\right) = +\frac{8}{2} = +\frac{4}{1} = +4$

6. $-\frac{+8}{-2} = -\left(-\frac{8}{2}\right) = +\frac{8}{2} = +\frac{4}{1} = +4$

7. $-\frac{-8}{-2} = -\left(+\frac{8}{2}\right) = -\frac{8}{2} = -\frac{4}{1} = -4$

8. $-\frac{+8}{+2} = -\left(+\frac{8}{2}\right) = -\frac{8}{2} = -\frac{4}{1} = -4$

B. Division of Algebraic Fractions by Zero

An algebraic fraction is an expression of the form

$$\frac{A}{B} \quad \text{where } A \text{ and } B \text{ are polynomials}$$

Note that the denominator B in an algebraic fraction can not be equal to zero, since division by zero is not defined. For example,

a. $\frac{1}{x^2}$ is not defined when $x = 0$ because $\frac{1}{0^2} = \frac{1}{0}$ is not defined.

b. $\frac{2}{x+1}$ is not defined when $x = -1$ because $\frac{2}{-1+1} = \frac{2}{0}$ is not defined.

c. $\frac{a-2}{2-a}$ is not defined when $a = 2$ because $\frac{2-2}{2-2} = \frac{0}{0}$ is not defined.

d. $\frac{5x}{x-9}$ is not defined when $x = 9$ because $\frac{5 \times 9}{9-9} = \frac{45}{0}$ is not defined.

e. $\frac{x-1}{x^2+6x+5} = \frac{x-1}{(x+1)(x+5)}$ is not defined when $x = -1$ and $x = -5$ because

when $x = -1$ $\qquad \frac{-1-1}{(-1+1)(-1+5)} = \frac{-2}{0 \times 4} = -\frac{2}{0}$ is not defined, and

when $x = -5$ $\qquad \frac{-5-1}{(-5+1)(-5+5)} = \frac{-6}{-4 \times 0} = -\frac{6}{0}$ is not defined.

f. $\frac{x+5}{x^2+2x-15} = \frac{x+5}{(x-3)(x+5)} = \frac{\cancel{(x+5)}}{(x-3)\cancel{(x+5)}} = \frac{1}{x-3}$ is not defined when $x = 3$ because $\frac{1}{3-3}$

$= \frac{1}{0}$ is not defined.

g. $\frac{3}{-x-3}$ is not defined when $x = -3$ because $\frac{3}{-(-3)-3} = \frac{3}{3-3} = \frac{3}{0}$ is not defined.

h. $\frac{2x+3}{2x-3}$ is not defined when $x = \frac{3}{2}$ because $\frac{2 \cdot \frac{3}{2} + 3}{2 \cdot \frac{3}{2} - 3} = \frac{\frac{3}{1}+3}{\frac{3}{1}-3} = \frac{3+3}{3-3} = \frac{6}{0}$ is not defined.

i. $\frac{x}{4x-3}$ is not defined when $x = \frac{3}{4}$ because $\frac{\frac{3}{4}}{4 \cdot \frac{3}{4} - 3} = \frac{\frac{3}{4}}{\cancel{4} \cdot \frac{3}{\cancel{4}} - 3} = \frac{\frac{3}{4}}{\frac{3}{1} - 3} = \frac{\frac{3}{4}}{3-3} = \frac{\frac{3}{4}}{0} = \frac{\frac{3}{4}}{\frac{0}{1}} = \frac{3 \times 1}{4 \times 0}$

$= \frac{3}{0}$ is not defined.

C. Equivalent Algebraic Fractions

When the numerator and the denominator of an algebraic fraction is multiplied by the same number, sign, or a variable the new algebraic fraction is said to be equivalent to the original algebraic fraction. For example, the following algebraic fractions are equivalent to one another:

a. $\frac{a}{b}=\frac{3a}{3b}=\frac{100a}{100b}=\frac{5a}{5b}=\frac{-a}{-b}$

b. $\frac{1+x}{x}=\frac{2+2x}{2x}=\frac{a+ax}{ax}$

c. $\frac{x-3}{4}=\frac{5x-15}{20}=\frac{3x-9}{12}=\frac{ax-3a}{4a}=\frac{-(x-3)}{-4}=\frac{3-x}{-4}$

d. $\frac{x^2}{1+x^2}=\frac{5x^2}{5+5x^2}=\frac{\frac{1}{2}x^2}{\frac{1}{2}+\frac{1}{2}x^2}=\frac{x^2y}{y+x^2y}$

e. $\frac{a-b}{3a}=\frac{2a-2b}{6a}=\frac{10a-10b}{30a}=\frac{ax-bx}{3ax}$

f. $\frac{5}{x-y}=\frac{-5}{-(x-y)}=\frac{-5}{y-x}=\frac{10}{2x-2y}=\frac{5xy}{x^2y-xy^2}$

g. $\frac{x-3}{2x-7}=\frac{-(x-3)}{-(2x-7)}=\frac{3-x}{7-2x}=\frac{2\cdot(x-3)}{2\cdot(2x-7)}=\frac{2x-6}{4x-14}=\frac{-3\cdot(x-3)}{-3\cdot(2x-7)}=\frac{9-3x}{21-6x}$

h. $-\frac{2}{x-1}=-\frac{-2}{-(x-1)}=\frac{2}{1-x}=-\frac{6}{3x-3}=\frac{6}{3-3x}$

i. $\frac{1-x}{1-y}=\frac{-(1-x)}{-(1-y)}=\frac{x-1}{y-1}=\frac{2\cdot(1-x)}{2\cdot(1-y)}=\frac{2-2x}{2-2y}$

Practice Problems - Introduction to Algebraic Fractions

A. Section 1.5a Practice Problems - Write the correct sign for the following fractions.

1. $-\frac{2}{-5}=$
2. $+\frac{-3}{-6}=$
3. $-\frac{-8}{-4}=$
4. $\frac{10}{-2}=$
5. $-\frac{5}{-15}=$
6. $+\frac{-8}{6}=$

B. Section 1.5a Practice Problems - State the value(s) of the variable for which the following fractions are not defined.

1. $\frac{3}{x-1}$
2. $\frac{x-5}{5-x}$
3. $-\frac{1}{x}$
4. $\frac{x}{x+10}$
5. $\frac{2x}{3x-5}$
6. $\frac{5x-2}{x-7}$

C. Section 1.5a Practice Problems - State which of the following algebraic fractions are equivalent.

1. $\frac{2x}{3y}\overset{?}{=}\frac{4x}{9y}$
2. $\frac{3x+1}{2x}\overset{?}{=}\frac{9x+3}{6x}$
3. $\frac{2}{a-b}\overset{?}{=}-\frac{2}{b-a}$
4. $\frac{x-5}{x+1}\overset{?}{=}\frac{5-x}{x-1}$
5. $-\frac{a}{a-1}\overset{?}{=}\frac{a}{1-a}$
6. $\frac{3-x}{-x}\overset{?}{=}\frac{x+3}{x}$

1.5b Simplifying Algebraic Fractions to Lower Terms

In dealing with integer fractions we learned that integer (arithmetic) fractions are reduced to their lowest terms by dividing both the numerator and the denominator by their common terms. For example, the integer fraction $\frac{14}{7}$ is simplified to its lowest term by dividing both the numerator and the denominator by 7, which is common to both, i.e., $\frac{14}{7} = \frac{\cancel{7}\cdot 2}{\cancel{7}} = \frac{2}{1} = 2$. The same principle holds true when simplifying algebraic fractions. Algebraic fractions are simplified using the following steps:

Step 1 Factor both the numerator and the denominator completely (see Sections 1.3 and 1.4).

Step 2 Simplify the algebraic fraction by eliminating the common terms in both the numerator and the denominator.

Examples with Steps

The following examples show the steps as to how algebraic fractions are simplified to their lowest terms:

Example 1.5-1

$$\frac{2y^2-7y-15}{y^2-25} =$$

Solution:

Step 1

$$\frac{2y^2-7y-15}{y^2-25} = \frac{(2y+3)(y-5)}{y^2-5^2} = \frac{(2y+3)(y-5)}{(y-5)(y+5)}$$

Step 2

$$\frac{(2y+3)(y-5)}{(y-5)(y+5)} = \frac{(2y+3)\cancel{(y-5)}}{\cancel{(y-5)}(y+5)} = \mathbf{\frac{2y+3}{y+5}}$$

Example 1.5-2

$$\frac{x^3-x}{x^3-2x^2-3x} =$$

Solution:

Step 1

$$\frac{x^3-x}{x^3-2x^2-3x} = \frac{x(x^2-1)}{x(x^2-2x-3)} = \frac{x(x-1)(x+1)}{x(x+1)(x-3)}$$

Step 2

$$\frac{x(x-1)(x+1)}{x(x+1)(x-3)} = \frac{\cancel{x}(x-1)\cancel{(x+1)}}{\cancel{x}\cancel{(x+1)}(x-3)} = \mathbf{\frac{x-1}{x-3}}$$

Additional Examples - Simplifying Algebraic Fractions to Lower Terms

The following examples further illustrate how to simplify algebraic fractions to their lowest terms:

Example 1.5-3

$$\frac{x^2+5x}{x^2+2x-15} = \frac{x(x+5)}{(x-3)(x+5)} = \frac{x\cancel{(x+5)}}{(x-3)\cancel{(x+5)}} = \mathbf{\frac{x}{x-3}}$$

Example 1.5-4

$$\frac{x^2-1}{4x+4} = \frac{(x-1)(x+1)}{4(x+1)} = \frac{(x-1)\cancel{(x+1)}}{4\cancel{(x+1)}} = \boxed{\mathbf{\frac{x-1}{4}}}$$

Example 1.5-5

$$\frac{7x}{-14x^2+7x} = \frac{7x}{7x(-2x+1)} = \frac{\cancel{7x}}{\cancel{7x}(-2x+1)} = \frac{1}{-2x+1} = \boxed{\mathbf{-\frac{1}{2x-1}}}$$

Example 1.5-6

$$-\frac{5x-3}{3-5x} = \frac{-(5x-3)}{3-5x} = \frac{3-5x}{3-5x} = \frac{\cancel{3-5x}}{\cancel{3-5x}} = \frac{1}{1} = \boxed{\mathbf{1}}$$

Example 1.5-7

$$\frac{6a^2-6ab}{a^2-b^2} = \frac{6a(a-b)}{(a-b)(a+b)} = \frac{6a\cancel{(a-b)}}{\cancel{(a-b)}(a+b)} = \boxed{\mathbf{\frac{6a}{a+b}}}$$

Example 1.5-8

$$\frac{3x^2+9x}{x^3+x^2-6x} = \frac{3x(x+3)}{x\left(x^2+x-6\right)} = \frac{3x(x+3)}{x(x+3)(x-2)} = \frac{3\cancel{x}\cancel{(x+3)}}{\cancel{x}\cancel{(x+3)}(x-2)} = \boxed{\mathbf{\frac{3}{x-2}}}$$

Example 1.5-9

$$\frac{6x^2+x-1}{3x^2+2x-1} = \frac{(2x+1)(3x-1)}{(3x-1)(x+1)} = \frac{(2x+1)\cancel{(3x-1)}}{\cancel{(3x-1)}(x+1)} = \boxed{\mathbf{\frac{2x+1}{x+1}}}$$

Example 1.5-10

$$\frac{(u+1)^2}{2u^2+3u+1} = \frac{(u+1)(u+1)}{(u+1)(2u+1)} = \frac{\cancel{(u+1)}(u+1)}{\cancel{(u+1)}(2u+1)} = \boxed{\mathbf{\frac{u+1}{2u+1}}}$$

Example 1.5-11

$$\frac{t^2-9}{t^2-2t-15} = \frac{t^2-3^2}{t^2-2t-15} = \frac{(t-3)(t+3)}{(t+3)(t-5)} = \frac{(t-3)\cancel{(t+3)}}{\cancel{(t+3)}(t-5)} = \boxed{\mathbf{\frac{t-3}{t-5}}}$$

Example 1.5-12

$$\frac{6y^2+7y-3}{-3y^3+y^2} = \frac{(2y+3)(3y-1)}{-y^2(3y-1)} = \frac{(2y+3)\cancel{(3y-1)}}{-y^2\cancel{(3y-1)}} = \frac{2y+3}{-y^2} = \boxed{\mathbf{-\frac{2y+3}{y^2}}}$$

Practice Problems - Simplifying Algebraic Fractions to Lower Terms

Section 1.5b Practice Problems - Simplify the following algebraic fractions to their lowest terms:

1. $\dfrac{x^2y^2z^5}{-xy^3z^2} =$
2. $-\dfrac{3a^2bc^3}{-9ab^2c} =$
3. $\dfrac{1+2m}{1-2m} =$
4. $\dfrac{2uvw^3}{10u^2v} =$
5. $\dfrac{y^2-4}{y^2-y-6} =$
6. $\dfrac{x^3-3x^2}{x^2-9} =$

1.5c Math Operations Involving Algebraic Fractions

In this section addition, subtraction, multiplication, and division of algebraic fractions (Cases I through IV) are reviewed.

Case I Addition and Subtraction of Algebraic Fractions with Common Denominators

Algebraic fractions with common denominators are added and subtracted using the following steps:

Step 1 Write the common denominator. Add or subtract the numerators.

Step 2 Simplify the algebraic fraction to its lowest term.

Examples with Steps

The following examples show the steps as to how algebraic fractions with common denominators are added and subtracted:

Example 1.5-13

$$\frac{3x^2+5x+5}{(x+3)(x-1)}-\frac{3x^2+4x+2}{(x+3)(x-1)}=$$

Solution:

Step 1

$$\frac{3x^2+5x+5}{(x+3)(x-1)}-\frac{3x^2+4x+2}{(x+3)(x-1)}=\frac{3x^2+5x+5-\left(3x^2+4x+2\right)}{(x+3)(x-1)}$$

Step 2

$$\frac{3x^2+5x+5-\left(3x^2+4x+2\right)}{(x+3)(x-1)}=\frac{3x^2+5x+5-3x^2-4x-2}{(x+3)(x-1)}$$

$$=\frac{\left(3x^2-3x^2\right)+(5x-4x)+(5-2)}{(x+3)(x-1)}=\frac{(3-3)x^2+(5-4)x+3}{(x+3)(x-1)}=\frac{0x^2+x+3}{(x+3)(x-1)}$$

$$=\frac{\cancel{(x+3)}}{\cancel{(x+3)}(x-1)}=\mathbf{\frac{1}{x-1}}$$

Example 1.5-14

$$\frac{3a+b}{2a^2b^3}+\frac{2a-b}{2a^2b^3}+\frac{a-2b}{2a^2b^3}=$$

Solution:

Step 1

$$\frac{3a+b}{2a^2b^3}+\frac{2a-b}{2a^2b^3}+\frac{a-2b}{2a^2b^3}=\frac{3a+b+(2a-b)+(a-2b)}{2a^2b^3}$$

Step 2

$$\frac{3a+b+(2a-b)+(a-2b)}{2a^2b^3}=\frac{(3a+2a+a)+(b-b-2b)}{2a^2b^3}=\frac{(3+2+1)a+(1-1-2)b}{2a^2b^3}$$

$$= \boxed{\frac{6a-2b}{2a^2b^3}} = \boxed{\frac{2(3a-b)}{2a^2b^3}} = \boxed{\frac{\mathbf{3a-b}}{\mathbf{a^2b^3}}}$$

Additional Examples - Addition and Subtraction of Algebraic Fractions with Common Denominators

The following examples further illustrate how to add or subtract algebraic fractions with common denominators:

Example 1.5-15

$$\boxed{\frac{x-3}{(x+4)(x-4)} - \frac{x-5}{(x+4)(x-4)}} = \boxed{\frac{x-3-(x-5)}{(x+4)(x-4)}} = \boxed{\frac{x-3-x+5}{(x+4)(x-4)}} = \boxed{\frac{(x-x)+(-3+5)}{(x+4)(x-4)}} = \boxed{\frac{(1-1)x+2}{(x+4)(x-4)}}$$

$$= \boxed{\frac{0x+2}{(x+4)(x-4)}} = \boxed{\frac{\mathbf{2}}{\mathbf{(x+4)(x-4)}}}$$

Example 1.5-16

$$\boxed{\frac{x+2}{x-3} + \frac{x+3}{x-3}} = \boxed{\frac{x+2+x+3}{x-3}} = \boxed{\frac{(x+x)+(2+3)}{x-3}} = \boxed{\frac{(1+1)x+5}{x-3}} = \boxed{\frac{\mathbf{2x+5}}{\mathbf{x-3}}}$$

Example 1.5-17

$$\boxed{\frac{x^2+3x+2}{(x+1)(x-5)} - \frac{x^2+2x+1}{(x+1)(x-5)}} = \boxed{\frac{x^2+3x+2-\left(x^2+2x+1\right)}{(x+1)(x-5)}} = \boxed{\frac{x^2+3x+2-x^2-2x-1}{(x+1)(x-5)}}$$

$$= \boxed{\frac{\left(x^2-x^2\right)+(3x-2x)+(2-1)}{(x+1)(x-5)}} = \boxed{\frac{(1-1)x^2+(3-2)x+1}{(x+1)(x-5)}} = \boxed{\frac{0x^2+x+1}{(x+1)(x-5)}} = \boxed{\frac{\cancel{(x+1)}}{\cancel{(x+1)}(x-5)}} = \boxed{\frac{\mathbf{1}}{\mathbf{x-5}}}$$

Example 1.5-18

$$\boxed{\frac{3x}{x+3} + \frac{2x-1}{x+3} + \frac{3x+2}{x+3}} = \boxed{\frac{3x+2x-1+3x+2}{x+3}} = \boxed{\frac{(3x+2x+3x)+(-1+2)}{x+3}} = \boxed{\frac{(3+2+3)x+1}{x+3}} = \boxed{\frac{\mathbf{8x+1}}{\mathbf{x+3}}}$$

Example 1.5-19

$$\boxed{\frac{x^2+2x+4}{x+2} - \frac{x^2-1}{x+2} - \frac{x-3}{x+2}} = \boxed{\frac{x^2+2x+4-\left(x^2-1\right)-(x-3)}{x+2}} = \boxed{\frac{x^2+2x+4-x^2+1-x+3}{x+2}}$$

$$= \boxed{\frac{\left(x^2-x^2\right)+(2x-x)+(4+1+3)}{x+2}} = \boxed{\frac{(1-1)x^2+(2-1)x+8}{x+2}} = \boxed{\frac{0x^2+x+8}{x+2}} = \boxed{\frac{\mathbf{x+8}}{\mathbf{x+2}}}$$

Practice Problems - Addition and Subtraction of Algebraic Fractions with Common Denominators

Section 1.5c Case I Practice Problems - Add or subtract the following algebraic fractions. Reduce the answer to its lowest term.

1. $\frac{x}{7}+\frac{3}{7} =$
2. $\frac{8}{a+b}-\frac{7}{a+b} =$
3. $\frac{3x+1}{2y}+\frac{4x+1}{2y}+\frac{3}{2y} =$
4. $\frac{4x}{x-2}-\frac{8}{x-2} =$
5. $\frac{15a}{5a+b}-\frac{-5b}{5a+b} =$
6. $\frac{6x}{3x^2y^2}+\frac{5y}{3x^2y^2} =$

Case II Addition and Subtraction of Algebraic Fractions without Common Denominators

Algebraic fractions without common denominators are solved using the following steps:

Step 1 Obtain a common denominator by multiplying the denominators of the first and second fractions by one another. Cross multiply the numerator of the first fraction with the denominator of the second fraction. Cross multiply the numerator of the second fraction with the denominator of the first fraction. Add or subtract the two products to each other.

Step 2 Simplify the algebraic fraction to its lowest term.

Examples with Steps

The following examples show the steps as to how algebraic fractions without common denominators are added and subtracted:

Example 1.5-20

$$\boxed{\frac{2}{x-2}+\frac{4}{x+3}}=$$

Solution:

Step 1

$$\boxed{\frac{2}{x-2}+\frac{4}{x+3}}=\boxed{\frac{[2\cdot(x+3)]+[4\cdot(x-2)]}{(x-2)\cdot(x+3)}}=\boxed{\frac{2x+6+4x-8}{(x-2)(x+3)}}$$

Step 2

$$\boxed{\frac{2x+6+4x-8}{(x-2)(x+3)}}=\boxed{\frac{(2x+4x)+(-8+6)}{(x-2)(x+3)}}=\boxed{\frac{6x-2}{(x-2)(x+3)}}=\boxed{\frac{\mathbf{2(3x-1)}}{\mathbf{(x-2)(x+3)}}}$$

Example 1.5-21

$$\boxed{\frac{m+2}{m}-\frac{m}{m-1}}=$$

Solution:

Step 1

$$\boxed{\frac{m+2}{m}-\frac{m}{m-1}}=\boxed{\frac{[(m+2)\cdot(m-1)]-(m\cdot m)}{m\cdot(m-1)}}=\boxed{\frac{m^2-m+2m-2-m^2}{m(m-1)}}$$

Step 2

$$\boxed{\frac{m^2-m+2m-2-m^2}{m(m-1)}}=\boxed{\frac{(m^2-m^2)+(2m-m)-2}{m(m-1)}}=\boxed{\frac{\mathbf{m-2}}{\mathbf{m(m-1)}}}$$

Additional Examples - Addition and Subtraction of Algebraic Fractions without Common Denominators

The following examples further illustrate how to add or subtract algebraic fractions without common denominators:

Example 1.5-22

$$\boxed{\frac{7}{x^2y^2}-\frac{3}{3xy^2}}=\boxed{\frac{(7\cdot 3xy^2)-(3\cdot x^2y^2)}{x^2y^2\cdot 3xy^2}}=\boxed{\frac{21xy^2-3x^2y^2}{3x^3y^4}}=\boxed{\frac{3xy^2(7-x)}{3x^3y^4}}=\boxed{\frac{\cancel{3}\cancel{x}\cancel{y^2}(7-x)}{\cancel{3}\cancel{x^3}\,\cancel{y^4}_{\;x^2\,y^2}}}=\boxed{\frac{\mathbf{7-x}}{\mathbf{x^2y^2}}}$$

Example 1.5-23

$$\frac{x+y}{x-y}+\frac{x}{y}=\frac{[(x+y)\cdot y]+[x\cdot(x-y)]}{y\cdot(x-y)}=\frac{xy+y^2+x^2-xy}{y(x-y)}=\frac{(xy-xy)+x^2+y^2}{y(x-y)}=\mathbf{\frac{x^2+y^2}{y(x-y)}}$$

Example 1.5-24

$$\frac{3}{a^2}+\frac{2}{4a}=\frac{(3\cdot 4a)+(2\cdot a^2)}{4a\cdot a^2}=\frac{12a+2a^2}{4a^3}=\frac{2a(6+a)}{4a^3}=\frac{2a(6+a)}{4a^3}=\mathbf{\frac{6+a}{2a^2}}$$

Example 1.5-25

$$\frac{x}{yz}+\frac{y}{xz}+\frac{z}{xy}=\left(\frac{x}{yz}+\frac{y}{xz}\right)+\frac{z}{xy}=\left(\frac{(x\cdot xz)+(y\cdot yz)}{yz\cdot xz}\right)+\frac{z}{xy}=\left(\frac{x^2z+y^2z}{xyz^2}\right)+\frac{z}{xy}=\frac{z(x^2+y^2)}{xyz^2}+\frac{z}{xy}$$

$$=\frac{x^2+y^2}{xyz}+\frac{z}{xy}=\frac{[xy\cdot(x^2+y^2)]+[(z\cdot xyz)]}{xyz\cdot xy}=\frac{x^3y+xy^3+xyz^2}{x^2y^2z}=\frac{xy(x^2+y^2+z^2)}{x^2y^2z}=\mathbf{\frac{x^2+y^2+z^2}{xyz}}$$

Example 1.5-26

$$\frac{3a-1}{2a}-\frac{2a-3}{3a}=\frac{[(3a-1)\cdot 3a]-[(2a-3)\cdot 2a]}{2a\cdot 3a}=\frac{9a^2-3a-4a^2+6a}{6a^2}=\frac{(9a^2-4a^2)+(6a-3a)}{6a^2}$$

$$=\frac{5a^2+3a}{6a^2}=\frac{a(5a+3)}{6a^2}=\mathbf{\frac{5a+3}{6a}}$$

Example 1.5-27

$$\frac{x-1}{x+2}-\frac{x-2}{x+1}=\frac{[(x-1)\cdot(x+1)]-[(x-2)\cdot(x+2)]}{(x+2)\cdot(x+1)}=\frac{(x^2+x-x-1)-(x^2-2x+2x-4)}{(x+2)(x+1)}=\frac{x^2-1-x^2+4}{(x+2)(x+1)}$$

$$=\frac{(x^2-x^2)+(-1+4)}{(x+2)(x+1)}=\frac{(x^2-x^2)+(-1+4)}{(x+2)(x+1)}=\mathbf{\frac{3}{(x+2)(x+1)}}$$

Example 1.5-28

$$\frac{m}{m-n}-\frac{n}{m+n}=\frac{[m\cdot(m+n)]-[n\cdot(m-n)]}{(m-n)\cdot(m+n)}=\frac{m^2+mn-mn+n^2}{m^2+mn-mn-n^2}=\frac{m^2+mn-mn+n^2}{m^2+mn-mn-n^2}=\mathbf{\frac{m^2+n^2}{m^2-n^2}}$$

Practice Problems - Addition and Subtraction of Algebraic Fractions without Common Denominators

Section 1.5c Case II Practice Problems - Add or subtract the following algebraic fractions. Reduce the answer to its lowest term.

1. $\frac{3}{4x^2}+\frac{5}{2x^3}=$
2. $\frac{x}{x+4}-\frac{2}{x-1}=$
3. $\frac{a-b}{a+b}-\frac{a}{b}=$
4. $\frac{x^2}{x+3}+\frac{5x}{x-5}=$
5. $\frac{1}{4x^2y^2z}-\frac{2}{xy^2z}=$
6. $\frac{3}{x+1}+\frac{2}{x-1}-\frac{5}{x}=$

Case III Multiplication of Algebraic Fractions

To multiply algebraic fractions by one another we only review simple cases of algebraic fractions where the numerator and the denominator are mostly monomials. The more difficult algebraic expressions where the terms in the numerator and/or the denominator are polynomials and need to be factored first before reducing the algebraic fraction to lower terms are addressed in Chapter 5 of the *"Mastering Algebra – Intermediate Level"*. Algebraic fractions are multiplied by one another using the following steps:

Step 1 Write the algebraic expression in fraction form, i.e., write x or $u^2v^2w^3$ as $\frac{x}{1}$ and $\frac{u^2v^2w^3}{1}$, respectively.

Step 2 Multiply the numerator and the denominator of the algebraic fraction terms by one another. Simplify the product to its lowest term.

Examples with Steps

The following examples show the steps as to algebraic fractions are multiplied by one another:

Example 1.5-29

$$\boxed{u^2v^2w\cdot\frac{1}{uv^2}\cdot\frac{w}{v^2w^3}} =$$

Solution:

Step 1

$$\boxed{u^2v^2w\cdot\frac{1}{uv^2}\cdot\frac{w}{v^2w^3}} = \boxed{\frac{u^2v^2w}{1}\cdot\frac{1}{uv^2}\cdot\frac{w}{v^2w^3}}$$

Step 2

$$\boxed{\frac{u^2v^2w}{1}\cdot\frac{1}{uv^2}\cdot\frac{w}{v^2w^3}} = \boxed{\frac{u^2v^2w\cdot 1\cdot w}{1\cdot uv^2\cdot v^2w^3}} = \boxed{\frac{u^2v^2w^2}{uv^4w^3}} = \boxed{\frac{\overset{u}{\cancel{u^2}}\,\cancel{v^2}\,\cancel{w^2}}{\cancel{uv^4}\,\underset{v^2\ \ w}{\cancel{w^3}}}} = \boxed{\mathbf{\frac{u}{v^2w}}}$$

Example 1.5-30

$$\boxed{(x-3)\cdot\frac{x}{x^2-9}\cdot\frac{(x+3)}{2x}\cdot 8x^2} =$$

Solution:

Step 1

$$\boxed{(x-3)\cdot\frac{x}{x^2-9}\cdot\frac{(x+3)}{2x}\cdot 8x^2} = \boxed{\frac{(x-3)}{1}\cdot\frac{x}{x^2-9}\cdot\frac{(x+3)}{2x}\cdot\frac{8x^2}{1}}$$

Step 2

$$\boxed{\frac{(x-3)}{1}\cdot\frac{x}{x^2-9}\cdot\frac{(x+3)}{2x}\cdot\frac{8x^2}{1}} = \boxed{\frac{(x-3)\cdot x\cdot(x+3)\cdot 8x^2}{1\cdot\left(x^2-9\right)\cdot 2x}} = \boxed{\frac{8x^3(x-3)(x+3)}{2x(x-3)(x+3)}}$$

$$= \boxed{\frac{8x^3\cancel{(x-3)}\cancel{(x+3)}}{2x\cancel{(x-3)}\cancel{(x+3)}}} = \boxed{\frac{8x^3}{2x}} = \boxed{\frac{\overset{4x^2}{\cancel{8x^3}}}{\cancel{2x}}} = \boxed{\frac{4x^2}{1}} = \boxed{\mathbf{4x^2}}$$

Additional Examples - Multiplication of Algebraic Fractions

The following examples further illustrate how to multiply algebraic fractions:

Example 1.5-31

$$\boxed{ab^2c\cdot\frac{a}{b^2c^2}}=\boxed{\frac{ab^2c}{1}\cdot\frac{a}{b^2c^2}}=\boxed{\frac{ab^2c\cdot a}{1\cdot b^2c^2}}=\boxed{\frac{a^2b^2c}{b^2c^2}}=\boxed{\frac{a^2\cancel{b^2}\cancel{c}}{\cancel{b^2}\,\cancel{c^2}_{\,c}}}=\boxed{\mathbf{\frac{a^2}{c}}}$$

Example 1.5-32

$$\boxed{\frac{x+1}{2}\cdot\frac{x^2-4}{(x+1)(x-2)}}=\boxed{\frac{x+1}{2}\cdot\frac{(x+2)(x-2)}{(x+1)(x-2)}}=\boxed{\frac{(x+1)\cdot[(x-2)(x+2)]}{2\cdot[(x+1)(x-2)]}}=\boxed{\frac{(\cancel{x+1})\cdot[(\cancel{x-2})(x+2)]}{2\cdot[(\cancel{x+1})(\cancel{x-2})]}}=\boxed{\mathbf{\frac{x+2}{2}}}$$

Example 1.5-33

$$\boxed{\frac{2}{x^2y^2z^2}\cdot\frac{x^2z^3}{4yz}}=\boxed{\frac{2\cdot x^2z^3}{x^2y^2z^2\cdot 4yz}}=\boxed{\frac{2x^2z^3}{4x^2y^3z^3}}=\boxed{\frac{\cancel{2}\cancel{x^2}\cancel{z^3}}{\cancel{4}_{\,2}\cancel{x^2}y^3\cancel{z^3}}}=\boxed{\mathbf{\frac{1}{2y^3}}}$$

Example 1.5-34

$$\boxed{\frac{1}{a^2}\cdot\frac{a}{2}\cdot\frac{3a^2}{5}}=\boxed{\frac{1\cdot a\cdot 3a^2}{a^2\cdot 2\cdot 5}}=\boxed{\frac{3a^3}{10a^2}}=\boxed{\frac{3\cancel{a^3}^{\,a}}{10\cancel{a^2}}}=\boxed{\mathbf{\frac{3a}{10}}}$$

Example 1.5-35

$$\boxed{\frac{uv^2}{w^3}\cdot\frac{w^2}{u^2v^2}\cdot\frac{1}{u}}=\boxed{\frac{uv^2\cdot w^2\cdot 1}{w^3\cdot u^2v^2\cdot u}}=\boxed{\frac{uv^2w^2}{u^3v^2w^3}}=\boxed{\frac{\cancel{u}\cancel{v^2}\cancel{w^2}}{\cancel{u^3}_{\,u^2}\cancel{v^2}\cancel{w^3}_{\,w}}}=\boxed{\mathbf{\frac{1}{u^2w}}}$$

Example 1.5-36

$$\boxed{xyz^2\cdot\left(\frac{1}{x^2y^2z^3}\cdot\frac{xy}{z}\right)}=\boxed{\frac{xyz^2}{1}\cdot\left(\frac{1}{x^2y^2z^3}\cdot\frac{xy}{z}\right)}=\boxed{\frac{xyz^2}{1}\cdot\left(\frac{1\cdot xy}{x^2y^2z^3\cdot z}\right)}=\boxed{\frac{xyz^2}{1}\cdot\left(\frac{xy}{x^2y^2z^4}\right)}=\boxed{\frac{xyz^2}{1}\cdot\frac{xy}{x^2y^2z^4}}$$

$$=\boxed{\frac{xyz^2\cdot xy}{1\cdot x^2y^2z^4}}=\boxed{\frac{x^2y^2z^2}{x^2y^2z^4}}=\boxed{\frac{\cancel{x^2}\cancel{y^2}\cancel{z^2}}{\cancel{x^2}\cancel{y^2}\cancel{z^4}_{\,z^2}}}=\boxed{\mathbf{\frac{1}{z^2}}}$$

Practice Problems - Multiplication of Algebraic Fractions

Section 1.5c Case III Practice Problems - Multiply the following algebraic fractions by one another. Simplify the answer to its lowest term.

1. $\frac{1}{xy}\cdot\frac{x^2y^2}{2}=$
2. $\frac{2a^2}{a^3}\cdot\frac{1}{8a}=$
3. $xyz\cdot\frac{1}{x^2y^2z^2}\cdot\frac{x^2}{y}=$
4. $\frac{5u^2v^2}{uv}\cdot\frac{uv^3}{15v^2}\cdot\frac{1}{u^4}=$
5. $8x\cdot\frac{2}{x^3}\cdot\frac{1}{4x^2}=$
6. $\frac{x-4}{x+2}\cdot\frac{x^2-4}{2}\cdot\frac{1}{x-4}=$

Case IV Division of Algebraic Fractions

To divide algebraic fractions by one another we only review simple cases of algebraic factions where the numerator and the denominator are mostly monomials. The more difficult algebraic expressions where the terms in the numerator and/or the denominator are polynomials and need to be factored first before reducing the algebraic fraction to lower terms are addressed in Chapter 5 of the *"Mastering Algebra – Intermediate Level"*. Algebraic fractions are divided by one another using the following steps:

Step 1 Write the algebraic expression in fraction form, i.e., write x or $u^2v^2w^3$ as $\frac{x}{1}$ and $\frac{u^2v^2w^3}{1}$, respectively.

Step 2 Invert the second fraction and change the division sign to a multiplication sign.

Step 3 Multiply the numerator and the denominator of the algebraic fraction terms by one another. Simplify the product to its lowest term, if possible.

Examples with Steps

The following examples show the steps as to algebraic fractions are divided by one another:

Example 1.5-37

$$\frac{1}{x^3y^3z} \div \frac{3z}{x^2y^2} =$$

Solution:

Step 1 *Not Applicable*

Step 2

$$\frac{1}{x^3y^3z} \div \frac{3z}{x^2y^2} = \frac{1}{x^3y^3z} \cdot \frac{x^2y^2}{3z}$$

Step 3

$$\frac{1}{x^3y^3z} \cdot \frac{x^2y^2}{3z} = \frac{1 \cdot x^2y^2}{x^3y^3z \cdot 3z} = \frac{x^2y^2}{3x^3y^3z^2} = \frac{\cancel{x^2}\cancel{y^2}}{3\cancel{x^3}_{x}\cancel{y^3}_{y}z^2} = \mathbf{\frac{1}{3xyz^2}}$$

Example 1.5-38

$$a^2b^2 \div \frac{ab}{3ab^3} =$$

Solution:

Step 1

$$a^2b^2 \div \frac{ab}{3ab^3} = \frac{a^2b^2}{1} \div \frac{ab}{3ab^3}$$

Step 2

$$\frac{a^2b^2}{1} \div \frac{ab}{3ab^3} = \frac{a^2b^2}{1} \cdot \frac{3ab^3}{ab}$$

Step 3

$$\frac{a^2b^2}{1} \cdot \frac{3ab^3}{ab} = \frac{a^2b^2 \cdot 3ab^3}{1 \cdot ab} = \frac{3a^3b^5}{ab} = \frac{3a^2b^4}{1} = \mathbf{3a^2b^4}$$

Example 1.5-39

$$\frac{3x}{x^3} \div \left(\frac{1}{x^2} \div 2x^3\right) =$$

Solution:

Step 1

$$\frac{3x}{x^3} \div \left(\frac{1}{x^2} \div 2x^3\right) = \frac{3x}{x^3} \div \left(\frac{1}{x^2} \div \frac{2x^3}{1}\right)$$

Step 2

$$\frac{3x}{x^3} \div \left(\frac{1}{x^2} \div \frac{2x^3}{1}\right) = \frac{3x}{x^3} \div \left(\frac{1}{x^2} \cdot \frac{1}{2x^3}\right) = \frac{3x}{x^3} \div \left(\frac{1 \cdot 1}{x^2 \cdot 2x^3}\right) = \frac{3x}{x^3} \div \frac{1}{2x^5} = \frac{3x}{x^3} \cdot \frac{2x^5}{1}$$

Step 3

$$\frac{3x}{x^3} \cdot \frac{2x^5}{1} = \frac{3x \cdot 2x^5}{x^3 \cdot 1} = \frac{6x^6}{x^3} = \frac{6x^3}{1} = \mathbf{6x^3}$$

Additional Examples - Division of Algebraic Fractions

The following examples further illustrate how to divide algebraic fractions by one another:

Example 1.5-40

$$\frac{x^2y^2z^3}{xy^2} \div \frac{x^2y^2}{yz} = \frac{x^2y^2z^3}{xy^2} \cdot \frac{yz}{x^2y^2} = \frac{x^2y^2z^3 \cdot yz}{xy^2 \cdot x^2y^2} = \frac{x^2y^3z^4}{x^3y^4} = \mathbf{\frac{z^4}{xy}}$$

Example 1.5-41

$$\frac{a^2b^2}{bc} \div ab^2 = \frac{a^2b^2}{bc} \div \frac{ab^2}{1} = \frac{a^2b^2}{bc} \cdot \frac{1}{ab^2} = \frac{a^2b^2 \cdot 1}{bc \cdot ab^2} = \frac{a^2b^2}{ab^3c} = \mathbf{\frac{a}{bc}}$$

Example 1.5-42

$$\frac{3}{u^2v^2} \div \frac{9u}{u^3v} = \frac{3}{u^2v^2} \cdot \frac{u^3v}{9u} = \frac{3 \cdot u^3v}{u^2v^2 \cdot 9u} = \frac{3u^3v}{9u^3v^2} = \frac{\cancel{3}\cancel{u^3}\cancel{v}}{\underset{3}{\cancel{9}}\cancel{u^3}\underset{v}{\cancel{v^2}}} = \mathbf{\frac{1}{3v}}$$

Example 1.5-43

$$xyz^3 \div \frac{x^2y^2z}{2xy} = \frac{xyz^3}{1} \div \frac{x^2y^2z}{2xy} = \frac{xyz^3}{1} \cdot \frac{2xy}{x^2y^2z} = \frac{xyz^3 \cdot 2xy}{1 \cdot x^2y^2z} = \frac{2x^2y^2z^3}{x^2y^2z} = \frac{2z^2}{1} = \mathbf{2z^2}$$

Example 1.5-44

$$\frac{a^3b^2c}{ab^2} \div \frac{a^2b}{c^3} = \frac{a^3b^2c}{ab^2} \cdot \frac{c^3}{a^2b} = \frac{a^3b^2c \cdot c^3}{ab^2 \cdot a^2b} = \frac{a^3b^2c^4}{a^3b^3} = \frac{\cancel{a^3}\cancel{b^2}c^4}{\cancel{a^3}\underset{b}{\cancel{b^3}}} = \mathbf{\frac{c^4}{b}}$$

Practice Problems - Division of Algebraic Fractions

Section 1.5c Case IV Practice Problems - Divide the following algebraic fractions. Simplify the answer to its lowest term.

1. $\frac{x^2y}{x^3y^2} \div xy =$

2. $\frac{uv^2w}{vw^2} \div \frac{uv^3}{w} =$

3. $a^2b^2c^4 \div \frac{a^2b}{2ac} =$

4. $\frac{xyz}{x^2z^3} \div \frac{x^2z^2}{yz} =$

5. $\left(\frac{uv^2}{v^3} \div 2u^2\right) \div \frac{uv}{3} =$

6. $\frac{x^2y^2z}{xz} \div \left(x^2y \div \frac{4}{yz^3}\right) =$

1.5d Math Operations Involving Complex Algebraic Fractions

A **simple algebraic fraction** is a fraction in which neither the numerator nor the denominator contains a fraction with variables. For example, $\frac{a}{5}$, $\frac{a-1}{a}$, $\frac{2x-1}{3}$, $\frac{3x}{x^2+2x-1}$, and $-\frac{5}{a^2+2a+1}$ are examples of simple algebraic fractions. A **complex algebraic fraction** is a fraction in which either the numerator or the denominator (or both) contains an algebraic fraction. For example, $\frac{1-\frac{1}{w}}{5}$, $\frac{\frac{2x}{5}-\frac{1}{x}}{\frac{2}{x}}$, $\frac{a+\frac{a}{b}}{1-\frac{a}{b}}$, and $\frac{x}{1+\frac{1}{x}}$ are examples of complex algebraic fractions.

Note that an easy way to change complex algebraic fractions to simple algebraic fractions is by multiplying the outer numerator by the outer denominator and the inner denominator by the inner numerator. For example, given the complex fraction $\frac{\frac{x^2}{5}}{\frac{2x}{9}}$; first obtain the numerator of the simple fraction by multiply x^2 , *the outer numerator*, by 9 , *the outer denominator*. Next, obtain the denominator of the simple fraction by multiply 5 , *the inner denominator*, by $2x$, *the inner numerator*. Therefore, the complex fraction $\frac{\frac{x^2}{5}}{\frac{2x}{9}}$ can be written as $\frac{x^2\cdot 9}{5\cdot 2x} = \frac{9x^2}{10x} = \frac{9x^{\not{2}}\,x}{10\not{x}} = \frac{9x}{10}$ which is a simple fraction. In this section addition, subtraction, multiplication, and division of complex algebraic fractions (Cases I through III) are reviewed.

Case I Addition and Subtraction of Complex Algebraic Fractions

Complex algebraic fractions are added or subtracted using the following steps:

Step 1 Add or subtract the algebraic fractions in both the numerator and the denominator. Note that the same process used in simplifying integer (arithmetic) fractions applies to algebraic fractions.

Step 2 Change the complex algebraic fraction to a simple fraction . Reduce the algebraic fraction to its lowest term, if possible.

Examples with Steps

The following examples show the steps as to how complex algebraic expressions are added and subtracted:

Example 1.5-45

$$\boxed{\frac{\frac{3x^3y^2}{xy}-1}{\frac{x^2y}{xy^2}+1}} =$$

Solution:

Step 1

$$\frac{\frac{3x^3y^2}{xy}-1}{\frac{x^2y}{xy^2}+1} = \frac{\frac{3x^3y^2}{xy}-\frac{1}{1}}{\frac{x^2y}{xy^2}+\frac{1}{1}} = \frac{\frac{(3x^3y^2\cdot 1)+(1\cdot xy)}{xy\cdot 1}}{\frac{(x^2y\cdot 1)+(1\cdot xy^2)}{xy^2\cdot 1}} = \frac{\frac{3x^3y^2+xy}{xy}}{\frac{x^2y+xy^2}{xy^2}}$$

Step 2

$$\frac{\frac{3x^3y^2+xy}{xy}}{\frac{x^2y+xy^2}{xy^2}} = \frac{(3x^3y^2+xy)\cdot \not{x}\,y^{\not{2}}}{\not{x}y\cdot(x^2y+xy^2)} = \frac{(3x^3y^2+xy)y}{x^2y+xy^2} = \frac{\not{xy}(3x^2y+1)y}{\not{xy}(x+y)} = \mathbf{\frac{y(3x^2y+1)}{x+y}}$$

Example 1.5-46

$$\frac{\frac{1}{a}+\frac{1}{b}}{\frac{1}{a}-\frac{1}{b}}-\frac{a}{b} =$$

Solution:

Step 1

$$\frac{\frac{1}{a}+\frac{1}{b}}{\frac{1}{a}-\frac{1}{b}}-\frac{a}{b} = \frac{\frac{(1\cdot b)+(1\cdot a)}{a\cdot b}}{\frac{(1\cdot b)-(1\cdot a)}{a\cdot b}}-\frac{a}{b} = \frac{\frac{b+a}{ab}}{\frac{b-a}{ab}}-\frac{a}{b}$$

Step 2

$$\frac{\frac{b+a}{ab}}{\frac{b-a}{ab}}-\frac{a}{b} = \frac{(b+a)\cdot \not{ab}}{\not{ab}\cdot(b-a)}-\frac{a}{b} = \frac{b+a}{b-a}-\frac{a}{b} = \frac{[(b+a)\cdot b]-[a\cdot(b-a)]}{(b-a)\cdot b}$$

$$= \frac{b^2+\not{ab}-\not{ab}+a^2}{b(b-a)} = \mathbf{\frac{a^2+b^2}{b(b-a)}}$$

Additional Examples - Addition and Subtraction of Complex Algebraic Fractions

The following examples further illustrate addition and subtraction of complex algebraic fractions:

Example 1.5-47

$$\frac{2-\frac{2}{3x}}{4-\frac{2}{9x}} = \frac{\frac{2}{1}-\frac{2}{3x}}{\frac{4}{1}-\frac{2}{9x}} = \frac{\frac{(2\cdot 3x)-(2\cdot 1)}{1\cdot 3x}}{\frac{(4\cdot 9x)-(2\cdot 1)}{1\cdot 9x}} = \frac{\frac{6x-2}{3x}}{\frac{36x-2}{9x}} = \frac{(6x-2)\cdot 9x}{3x\cdot(36x-2)} = \frac{18x(3x-1)}{6x(18x-1)} = \frac{\overset{3}{\not{18}}\not{x}(3x-1)}{\not{6}\not{x}(18x-1)} = \mathbf{\frac{3(3x-1)}{18x-1}}$$

Example 1.5-48

$$\frac{\frac{1}{8}+a}{a+\frac{1}{4}} = \frac{\frac{1}{8}+\frac{a}{1}}{\frac{a}{1}+\frac{1}{4}} = \frac{\frac{(1\cdot 1)+(a\cdot 8)}{8\cdot 1}}{\frac{(a\cdot 4)+(1\cdot 1)}{1\cdot 4}} = \frac{\frac{1+8a}{8}}{\frac{4a+1}{4}} = \frac{(1+8a)\cdot 4}{8\cdot(4a+1)} = \frac{(1+8a)\cdot \not{4}}{\underset{2}{\not{8}}\cdot(4a+1)} = \mathbf{\frac{1+8a}{2(1+4a)}}$$

Example 1.5-49

$$\boxed{\frac{\frac{8a^3b^2}{4a^2b}}{\frac{6a^3}{3a^2}}+\frac{\frac{ab}{1}}{\frac{1}{a}}}=\boxed{\frac{\left(8a^3b^2\right)\cdot\left(3a^2\right)}{\left(4a^2b\right)\cdot\left(6a^3\right)}+\frac{\frac{ab}{1}}{\frac{1}{a}}}=\boxed{\overset{b}{\frac{\cancel{24a^5}\,b^{\cancel{2}}}{\cancel{24a^5}\,\cancel{b}}}+\frac{ab\cdot a}{1\cdot 1}}=\boxed{\frac{b}{1}+\frac{a^2b}{1}}=\boxed{\frac{b+a^2b}{1}}=\boxed{b+a^2b}=\boxed{\mathbf{b\left(1+a^2\right)}}$$

Example 1.5-50

$$\boxed{\frac{\frac{x}{y}+\frac{y}{x}}{\frac{3}{xy}+\frac{5}{xy}}}=\boxed{\frac{\frac{(x\cdot x)+(y\cdot y)}{x\cdot y}}{\frac{3+5}{xy}}}=\boxed{\frac{\frac{x^2+y^2}{xy}}{\frac{8}{xy}}}=\boxed{\frac{\left(x^2+y^2\right)\cdot\cancel{xy}}{8\cdot\cancel{xy}}}=\boxed{\mathbf{\frac{x^2+y^2}{8}}}$$

Example 1.5-51

$$\boxed{\frac{\frac{1}{2x}-\frac{1}{3y}}{\frac{2}{3x}+\frac{5}{2y}}-4}=\boxed{\frac{\frac{(1\cdot 3y)-(1\cdot 2x)}{2x\cdot 3y}}{\frac{(2\cdot 2y)+(5\cdot 3x)}{3x\cdot 2y}}-4}=\boxed{\frac{\frac{3y-2x}{6xy}}{\frac{4y+15x}{6xy}}-4}=\boxed{\frac{(3y-2x)\cdot\cancel{6xy}}{\cancel{6xy}\cdot(4y+15x)}-4}=\boxed{\frac{3y-2x}{4y+15x}-4}=\boxed{\frac{3y-2x}{4y+15x}-\frac{4}{1}}$$

$$=\boxed{\frac{(3y-2x)\cdot 1-4\cdot(4y+15x)}{(4y+15x)\cdot 1}}=\boxed{\frac{3y-2x-16y-60x}{4y+15x}}=\boxed{\frac{(3y-16y)+(-60x-2x)}{4y+15x}}=\boxed{\mathbf{\frac{-13y-62x}{4y+15x}}}$$

Example 1.5-52

$$\boxed{\frac{3-\frac{1}{x+1}}{\frac{2}{x+1}+3}}=\boxed{\frac{\frac{3}{1}-\frac{1}{x+1}}{\frac{2}{x+1}+\frac{3}{1}}}=\boxed{\frac{\frac{[3\cdot(x+1)]-(1\cdot 1)}{1\cdot(x+1)}}{\frac{(2\cdot 1)+[3\cdot(x+1)]}{(x+1)\cdot 1}}}=\boxed{\frac{\frac{3x+3-1}{x+1}}{\frac{2+3x+3}{x+1}}}=\boxed{\frac{\frac{3x+2}{x+1}}{\frac{3x+5}{x+1}}}=\boxed{\frac{(3x+2)\cdot\cancel{(x+1)}}{\cancel{(x+1)}\cdot(3x+5)}}=\boxed{\mathbf{\frac{3x+2}{3x+5}}}$$

Example 1.5-53

$$\boxed{\frac{\frac{x}{x+3}+2}{\frac{4}{x+3}-1}}=\boxed{\frac{\frac{x}{x+3}+\frac{2}{1}}{\frac{4}{x+3}-\frac{1}{1}}}=\boxed{\frac{\frac{(x\cdot 1)+[2\cdot(x+3)]}{(x+3)\cdot 1}}{\frac{(4\cdot 1)-[1\cdot(x+3)]}{(x+3)\cdot 1}}}=\boxed{\frac{\frac{x+2x+6}{x+3}}{\frac{4-x-3}{x+3}}}=\boxed{\frac{\frac{3x+6}{x+3}}{\frac{-x+1}{x+3}}}=\boxed{\frac{(3x+6)\cdot\cancel{(x+3)}}{\cancel{(x+3)}\cdot(-x+1)}}=\boxed{\mathbf{\frac{3(x+2)}{1-x}}}$$

Practice Problems - Addition and Subtraction of Complex Algebraic Fractions

Section 1.5d Case I Practice Problems - Simplify the following complex algebraic fractions. Reduce the answer to its lowest term.

1. $\dfrac{2-\frac{1}{5a}}{3-\frac{2}{15a}}=$

2. $\dfrac{\frac{2x^3y^2z}{4x^2z}}{\frac{2x}{xy^2}}-1=$

3. $\dfrac{\frac{2}{a}+\frac{1}{a^3}}{\frac{2}{a^3}+2}=$

4. $\dfrac{a}{a+\frac{1}{a^2}}=$

5. $\dfrac{\frac{x}{y}-3}{3-\frac{y}{x}}=$

6. $\dfrac{\frac{1}{x}-\frac{1}{x+4}}{\frac{3}{x+4}+1}=$

Case II Multiplication of Complex Algebraic Fractions

Complex algebraic fractions are multiplied by one another using the following steps:

Step 1 Multiply the numerator and the denominator of the algebraic fraction terms by one another.

Step 2 Change the complex algebraic fraction to a simple fraction. Reduce the algebraic fraction to its lowest term, if possible.

Examples with Steps

The following examples show the steps as to how complex algebraic fractions are multiplied:

Example 1.5-54

$$\frac{\frac{x^2y^2z}{z^2}\cdot\frac{z}{x^3y^3}}{\frac{1}{xy}} =$$

Solution:

Step 1

$$\frac{\frac{x^2y^2z}{z^2}\cdot\frac{z}{x^3y^3}}{\frac{1}{xy}} = \frac{\frac{x^2y^2z\cdot z}{z^2\cdot x^3y^3}}{\frac{1}{xy}} = \frac{\frac{x^2y^2z^2}{x^3y^3z^2}}{\frac{1}{xy}}$$

Step 2

$$\frac{\frac{x^2y^2z^2}{x^3y^3z^2}}{\frac{1}{xy}} = \frac{x^2y^2z^2\cdot xy}{x^3y^3z^2\cdot 1} = \frac{x^3y^3z^2}{x^3y^3z^2} = \frac{\cancel{x^3}\cancel{y^3}\cancel{z^2}}{\cancel{x^3}\cancel{y^3}\cancel{z^2}} = \frac{1}{1} = \boxed{1}$$

Example 1.5-55

$$\frac{\frac{u^2v^2}{w}\cdot\frac{w^3}{v^4}}{uw^2} =$$

Solution:

Step 1

$$\frac{\frac{u^2v^2}{w}\cdot\frac{w^3}{v^4}}{uw^2} = \frac{\frac{u^2v^2}{w}\cdot\frac{w^3}{v^4}}{\frac{uw^2}{1}} = \frac{\frac{u^2v^2\cdot w^3}{w\cdot v^4}}{\frac{uw^2}{1}} = \frac{\frac{u^2v^2w^3}{wv^4}}{\frac{uw^2}{1}}$$

Step 2

$$\frac{\frac{u^2v^2w^3}{wv^4}}{\frac{uw^2}{1}} = \frac{u^2v^2w^3\cdot 1}{wv^4\cdot uw^2} = \frac{u^2v^2w^3}{uv^4w^2} = \frac{\overset{u}{\cancel{u^2}}\cancel{v^2}\cancel{w^3}}{\cancel{u}\underset{v^2}{\cancel{v^4}}\cancel{w^3}} = \boxed{\frac{u}{v^2}}$$

Additional Examples - Multiplication of Complex Algebraic Fractions

The following examples further illustrate how to multiply complex algebraic fractions:

Example 1.5-56

$$\frac{xyz\cdot\frac{2x}{y^2z^2}}{x^2} = \frac{\frac{xyz}{1}\cdot\frac{2x}{y^2z^2}}{x^2} = \frac{\frac{xyz\cdot 2x}{1\cdot y^2z^2}}{x^2} = \frac{\frac{2x^2yz}{y^2z^2}}{x^2} = \frac{\frac{2x^2yz}{y^2z^2}}{\frac{x^2}{1}} = \frac{2x^2yz\cdot 1}{y^2z^2\cdot x^2} = \frac{2x^2yz}{x^2y^2z^2} = \frac{2x^2yz}{x^2\,\underset{y}{y^2}\,\underset{z}{z^2}} = \mathbf{\frac{2}{yz}}$$

Example 1.5-57

$$\frac{2}{\frac{x^2y^2}{3x}\cdot\frac{6x}{x^3y^3}} = \frac{2}{\frac{x^2y^2\cdot 6x}{3x\cdot x^3y^3}} = \frac{2}{\frac{6x^3y^2}{3x^4y^3}} = \frac{\frac{2}{1}}{\frac{6x^3y^2}{3x^4y^3}} = \frac{2\cdot 3x^4y^3}{1\cdot 6x^3y^2} = \frac{6x^4y^3}{6x^3y^2} = \frac{6\overset{x}{x^4}\,\overset{y}{y^3}}{6x^3y^2} = \frac{xy}{1} = \mathbf{xy}$$

Example 1.5-58

$$\frac{ab}{a^3b^2\cdot\frac{3a}{ab^3}} = \frac{ab}{\frac{a^3b^2}{1}\cdot\frac{3a}{ab^3}} = \frac{ab}{\frac{a^3b^2\cdot 3a}{1\cdot ab^3}} = \frac{ab}{\frac{3a^4b^2}{ab^3}} = \frac{\frac{ab}{1}}{\frac{3a^4b^2}{ab^3}} = \frac{ab\cdot ab^3}{1\cdot 3a^4b^2} = \frac{a^2b^4}{3a^4b^2} = \frac{a^2\,\overset{b^2}{b^4}}{3\underset{a^2}{a^4}\,b^2} = \mathbf{\frac{b^2}{3a^2}}$$

Example 1.5-59

$$\frac{uv\cdot\frac{1}{u^2v^2}\cdot\frac{2u^3}{v^4}}{\frac{u}{v^3}} = \frac{\frac{uv}{1}\cdot\frac{1}{u^2v^2}\cdot\frac{2u^3}{v^4}}{\frac{u}{v^3}} = \frac{\frac{uv\cdot 1\cdot 2u^3}{1\cdot u^2v^2\cdot v^4}}{\frac{u}{v^3}} = \frac{\frac{2u^4v}{u^2v^6}}{\frac{u}{v^3}} = \frac{2u^4v\cdot v^3}{u^2v^6\cdot u} = \frac{2u^4v^4}{u^3v^6} = \frac{2\overset{u}{u^4}\,v^4}{u^3\,\underset{v^2}{v^6}} = \mathbf{\frac{2u}{v^2}}$$

Example 1.5-60

$$\frac{\frac{xyz^2}{xy}\cdot\frac{x^2}{y^3}}{\frac{x}{y}\cdot\frac{y^2}{x^3}} = \frac{\frac{xyz^2\cdot x^2}{xy\cdot y^3}}{\frac{x\cdot y^2}{y\cdot x^3}} = \frac{\frac{x^3yz^2}{xy^4}}{\frac{xy^2}{x^3y}} = \frac{x^3yz^2\cdot x^3y}{xy^4\cdot xy^2} = \frac{x^6y^2z^2}{x^2y^6} = \frac{\overset{x^4}{x^6}\,y^2z^2}{x^2\,\underset{y^4}{y^6}} = \mathbf{\frac{x^4z^2}{y^4}}$$

Practice Problems - Multiplication of Complex Algebraic Fractions

Section 1.5d Case II Practice Problems - Multiply the following complex algebraic expressions. Simplify the answer to its lowest term.

1. $\dfrac{y^2}{\frac{x^2y^2}{xy}\cdot\frac{2x^3y}{x^4}} =$

2. $\dfrac{\frac{1}{2x^3y^2}\cdot\frac{4x}{y}}{\frac{1}{xy^2}}$

3. $\dfrac{\frac{x^2}{x^3}}{2x\cdot\frac{3x^2}{18x^5}}$

4. $\dfrac{\left(x^2yz^2\cdot\frac{2x}{y^2}\right)\cdot\frac{3x}{x^3z^4}}{\frac{1}{y^2z^3}}$

5. $\dfrac{\left(\frac{2a^2b^2}{a^3}\cdot 4ab^3\right)\cdot\frac{1}{3ab^4}}{\frac{8}{3}}$

6. $\dfrac{\frac{u^3v^3w^2}{w^3}\cdot\frac{1}{3w}}{\frac{1}{6uv^2}}$

Case III Division of Complex Algebraic Fractions

Complex algebraic fractions are divided by one another using the following steps:

Step 1 Write the algebraic expression in fraction form, i.e., write x^2y^3 as $\frac{x^2y^3}{1}$.

Step 2 a. Invert the second fraction in either the numerator or the denominator, or both.

b. Change the division sign to a multiplication sign.

c. Multiply the numerator and the denominator of the algebraic fraction terms by one another.

Step 3 Change the complex algebraic fraction to a simple fraction. Reduce the algebraic fraction to its lowest term, if possible.

Examples with Steps

The following examples show the steps as to how complex algebraic fractions are divided:

Example 1.5-61

$$\frac{\frac{a^3b^2c}{b^2c^3} \div a^3}{\frac{ab}{a^3} \div b^2} =$$

Solution:

Step 1

$$\frac{\frac{a^3b^2c}{b^2c^3} \div a^3}{\frac{ab}{a^3} \div b^2} = \frac{\frac{a^3b^2c}{b^2c^3} \div \frac{a^3}{1}}{\frac{ab}{a^3} \div \frac{b^2}{1}}$$

Step 2

$$\frac{\frac{a^3b^2c}{b^2c^3} \div \frac{a^3}{1}}{\frac{ab}{a^3} \div \frac{b^2}{1}} = \frac{\frac{a^3b^2c}{b^2c^3} \cdot \frac{1}{a^3}}{\frac{ab}{a^3} \cdot \frac{1}{b^2}} = \frac{\frac{a^3b^2c \cdot 1}{b^2c^3 \cdot a^3}}{\frac{ab \cdot 1}{a^3 \cdot b^2}} = \frac{\frac{a^3b^2c}{a^3b^2c^3}}{\frac{ab}{a^3b^2}}$$

Step 3

$$\frac{\frac{a^3b^2c}{a^3b^2c^3}}{\frac{ab}{a^3b^2}} = \frac{a^3b^2c \cdot a^3b^2}{a^3b^2c^3 \cdot ab} = \frac{a^6b^4c}{a^4b^3c^3} = \frac{\overset{a^2\ b}{\cancel{a^6}\,\cancel{b^4}\,\cancel{c}}}{\underset{c^2}{\cancel{a^4}\,\cancel{b^3}\,\cancel{c^3}}} = \frac{a^2b}{c^2}$$

Example 1.5-62

$$3xyz \div \frac{\frac{xy}{z} \div z^2}{\frac{xy^2}{xyz^2}} =$$

Solution:

Step 1

$$3xyz \div \frac{\frac{xy}{z} \div z^2}{\frac{xy^2}{xyz^2}} = \frac{3xyz}{1} \div \frac{\frac{xy}{z} \div \frac{z^2}{1}}{\frac{xy^2}{xyz^2}}$$

Step 2

$$\frac{3xyz}{1} \div \frac{\frac{xy}{z} \div \frac{z^2}{1}}{\frac{xy^2}{xyz^2}} = \frac{3xyz}{1} \div \frac{\frac{xy}{z} \cdot \frac{1}{z^2}}{\frac{xy^2}{xyz^2}} = \frac{3xyz}{1} \div \frac{\frac{xy \cdot 1}{z \cdot z^2}}{\frac{xy^2}{xyz^2}} = \frac{3xyz}{1} \div \frac{\frac{xy}{z^3}}{\frac{xy^2}{xyz^2}}$$

Step 3

$$\frac{3xyz}{1} \div \frac{\frac{xy}{z^3}}{\frac{xy^2}{xyz^2}} = \frac{3xyz}{1} \div \frac{xy \cdot xyz^2}{z^3 \cdot xy^2} = \frac{3xyz}{1} \div \frac{x^2y^2z^2}{xy^2z^3} = \frac{3xyz}{1} \cdot \frac{xy^2z^3}{x^2y^2z^2}$$

$$= \frac{3xyz \cdot xy^2z^3}{1 \cdot x^2y^2z^2} = \frac{3x^2y^3z^4}{x^2y^2z^2} = \frac{3\cancel{x^2}\,\overset{y}{\cancel{y^3}}\,\overset{z^2}{\cancel{z^4}}}{\cancel{x^2}\cancel{y^2}\cancel{z^2}} = \frac{3yz^2}{1} = \boxed{3yz^2}$$

Additional Examples - Division of Complex Algebraic Fractions

The following examples further illustrate how to divide complex algebraic fractions:

Example 1.5-63

$$\frac{xyz^2 \div \frac{x^2z^3}{3x}}{xy} = \frac{\frac{xyz^2}{1} \div \frac{x^2z^3}{3x}}{xy} = \frac{\frac{xyz^2}{1} \cdot \frac{3x}{x^2z^3}}{xy} = \frac{\frac{xyz^2 \cdot 3x}{1 \cdot x^2z^3}}{xy} = \frac{\frac{3x^2yz^2}{x^2z^3}}{xy} = \frac{\frac{3x^2yz^2}{x^2z^3}}{\frac{xy}{1}} = \frac{3x^2yz^2 \cdot 1}{x^2z^3 \cdot xy}$$

$$= \frac{3x^2yz^2}{x^3yz^3} = \frac{3\cancel{x^2}\cancel{y}\cancel{z^2}}{\underset{x}{\cancel{x^3}}\,\cancel{y}\,\underset{z}{\cancel{z^3}}} = \boxed{\frac{3}{xz}}$$

Example 1.5-64

$$\frac{\frac{2}{x}}{\frac{x^2y}{2x} \div \frac{x^2y^2}{4x^3}} = \frac{\frac{2}{x}}{\frac{x^2y}{2x} \cdot \frac{4x^3}{x^2y^2}} = \frac{\frac{2}{x}}{\frac{x^2y \cdot 4x^3}{2x \cdot x^2y^2}} = \frac{\frac{2}{x}}{\frac{4x^5y}{2x^3y^2}} = \frac{2 \cdot 2x^3y^2}{x \cdot 4x^5y} = \frac{4x^3y^2}{4x^6y} = \frac{\cancel{4}\cancel{x^3}\,\overset{y}{\cancel{y^2}}}{\cancel{4}\,\underset{x^3}{\cancel{x^6}}\,\cancel{y}} = \boxed{\frac{y}{x^3}}$$

Example 1.5-65

$$\frac{\frac{xyz}{2x} \div \frac{x^2y^2}{x^3}}{\frac{x}{y}} = \frac{\frac{xyz}{2x} \cdot \frac{x^3}{x^2y^2}}{\frac{x}{y}} = \frac{\frac{xyz \cdot x^3}{2x \cdot x^2y^2}}{\frac{x}{y}} = \frac{\frac{x^4yz}{2x^3y^2}}{\frac{x}{y}} = \frac{x^4yz \cdot y}{2x^3y^2 \cdot x} = \frac{x^4y^2z}{2x^4y^2} = \frac{\cancel{x^4}\cancel{y^2}z}{2\cancel{x^4}\cancel{y^2}} = \boxed{\frac{z}{2}}$$

Example 1.5-66

$$a^2b^2 \div \frac{\frac{a}{b^3}}{\frac{a^2b}{b}\div\frac{a^3}{b^2}} = a^2b^2 \div \frac{\frac{a}{b^3}}{\frac{a^2b}{b}\cdot\frac{b^2}{a^3}} = a^2b^2\div\frac{\frac{a}{b^3}}{\frac{a^2b\cdot b^2}{b\cdot a^3}} = a^2b^2\div\frac{\frac{a}{b^3}}{\frac{a^2b^3}{a^3b}} = a^2b^2\div\frac{a\cdot a^3b}{b^3\cdot a^2b^3}$$

$$= a^2b^2\div\frac{a^4b}{a^2b^6} = a^2b^2\cdot\frac{a^2b^6}{a^4b} = \frac{a^2b^2}{1}\cdot\frac{a^2b^6}{a^4b} = \frac{a^2b^2\cdot a^2b^6}{1\cdot a^4b} = \frac{a^4b^8}{a^4b} = \frac{\overset{b^7}{a^4b^8}}{a^4b} = \frac{b^7}{1} = \boxed{b^7}$$

Example 1.5-67

$$\frac{1}{\frac{a^3bc}{b^2c^3}}\div\frac{a}{\frac{a^2b^2}{b}} = \frac{\frac{1}{1}}{\frac{a^3bc}{b^2c^3}}\div\frac{\frac{a}{1}}{\frac{a^2b^2}{b}} = \frac{1\cdot b^2c^3}{1\cdot a^3bc}\div\frac{a\cdot b}{1\cdot a^2b^2} = \frac{b^2c^3}{a^3bc}\div\frac{ab}{a^2b^2} = \frac{b^2c^3}{a^3bc}\cdot\frac{a^2b^2}{ab} = \frac{b^2c^3\cdot a^2b^2}{a^3bc\cdot ab}$$

$$= \frac{a^2b^4c^3}{a^4b^2c} = \frac{\overset{b^2c^2}{a^2b^4c^3}}{\underset{a^2}{a^4b^2c}} = \boxed{\frac{b^2c^2}{a^2}}$$

Example 1.5-68

$$\frac{\frac{u^2v^2}{uw}\div\frac{uv}{w^2}}{\frac{2uw}{u^2}\div\frac{uv}{w}} = \frac{\frac{u^2v^2}{uw}\cdot\frac{w^2}{uv}}{\frac{2uw}{u^2}\cdot\frac{w}{uv}} = \frac{\frac{u^2v^2\cdot w^2}{uw\cdot uv}}{\frac{2uw\cdot w}{u^2\cdot uv}} = \frac{\frac{u^2v^2w^2}{u^2vw}}{\frac{2uw^2}{u^3v}} = \frac{u^2v^2w^2\cdot u^3v}{u^2vw\cdot 2uw^2} = \frac{u^5v^3w^2}{2u^3vw^3} = \frac{\overset{u^2v^2}{u^5v^3w^2}}{\underset{w}{2u^3vw^3}} = \boxed{\frac{u^2v^2}{2w}}$$

Example 1.5-69

$$\frac{3x\div\frac{x^2}{y^2}}{\frac{x}{2}\div y} = \frac{\frac{3x}{1}\div\frac{x^2}{y^2}}{\frac{x}{2}\div\frac{y}{1}} = \frac{\frac{3x}{1}\cdot\frac{y^2}{x^2}}{\frac{x}{2}\cdot\frac{1}{y}} = \frac{\frac{3x\cdot y^2}{1\cdot x^2}}{\frac{x\cdot 1}{2\cdot y}} = \frac{\frac{3xy^2}{x^2}}{\frac{x}{2y}} = \frac{3xy^2\cdot 2y}{x^2\cdot x} = \frac{6xy^3}{x^3} = \frac{6xy^3}{\underset{x^2}{x^3}} = \boxed{\frac{6y^3}{x^2}}$$

Practice Problems - Division of Complex Algebraic Fractions

Section 1.5d Case III Practice Problems - Divide the following complex algebraic fractions. Simplify the answer to its lowest term.

1. $\dfrac{\frac{xy}{x^3}\div\frac{x^2}{xy^2}}{2x} =$

2. $\dfrac{\frac{1}{a}\div\frac{1}{b}}{\frac{a}{ab^2}\div b} =$

3. $xy\div\left(\dfrac{\frac{1}{xy}\div x}{x^3}\right) =$

4. $\dfrac{\frac{u^2vw}{w^2}\div\frac{u^3v}{w}}{w^3} =$

5. $\dfrac{ab^3\div\frac{a^2b^2}{b}}{a\div\frac{a}{b}} =$

6. $\dfrac{\frac{z^3}{z^2}\div 2z}{z^3\div\frac{z^2}{2}} =$

Chapter 2
Functions of Real and Complex Variables

Quick Reference to Chapter 2 Problems

Chapter 2 – Functions of Real and Complex Variables

The objective of this chapter is to introduce students to operations involving functions of real and complex variables and operations involving complex numbers. Relations and functions are introduced in Sections 2.1. Addition, subtraction, multiplication, and division of functions are addressed in Section 2.2. Composition of two or more functions and its computation is discussed in Section 2.3. Identification of a one-to-one function as well as computation of inverse functions are addressed in Section 2.4. Complex numbers and functions of complex variables are introduced in Section 2.5. Finally, math operations involving addition, subtraction, multiplication, and division of complex numbers are addressed in Section 2.6. Each section is concluded by solving examples with practice problems to further enhance the student ability.

2.1 Introduction to Functions of Real Variables

Recall that real numbers are the numbers on the real number line from $-\infty$ to $+\infty$. These are the "normal" every day numbers that we use to count using addition, subtraction, multiplication, and division methods. We identify these numbers as *real* because, as we will see later in this chapter, the need will arise to define a different set of numbers called *imaginary* numbers with some unique set of properties to perform math operations. Note that the functions we are considering in this section have real numbers as variables.

In order to proceed with the definition of a function we first need to learn a few terms. An **ordered pair** is defined by two numbers each called a **component**. The first pair, usually denoted by x, is referred to as the **first component** and the second pair, usually denoted by y, is referred to as the **second component**. In this section only real number components are addressed. The **domain** is defined as the set of all the first components of the ordered pairs. The **range** is defined as the set of all the second components of the ordered pairs. For example, in the relation defined by $\{(1,2),(2,3),(1,6),(3,8),(5,9)\}$ the domain is $\{1, 2, 3, \textit{and } 5\}$ and the range is $\{2, 3, 6, 8, \textit{and } 9\}$. Note that $f(x)$, which is the same as y, represents an element in the range of a function. For example, in the equation $f(x) = y = x^2 + 2x + 1$ since the y value depends on the value of x, we call y the **dependent variable** and x the **independent variable**. Having addressed these terms we can now proceed with defining a function.

In general, a function is denoted by symbols such as f, g, h, p, q, r, s, and t. A symbol such as $f(x)$, read as "f *of* x" represents the range value associated with the domain which is the x value. To understand functions we first need to know about relations. Relations are often defined by equations in two variables. *In a relation each element of the domain is paired with more than one element of the range.* For example: 1. The ordered pairs $\{(1,3),(1,4),(2,6),(3,7)\}$ belong to a relation because each element of the domain is not paired with only one element of the range, i.e., two ordered pairs, $(1,3),(1,4)$, have the same first components or, 2. The equation $y = \pm\sqrt{x+1}$ defines a relation. This is because for each value of $x \rangle 0$ there are two values of y, i.e., at $x = 0$ the variable $y = \pm 1$ or at $x = 2$ the variable $y = \pm\sqrt{3}$, etc.

On the other hand, *a function is defined as a relation in which each element of the domain is paired with only one element of the range.* For example: 1. The ordered pairs $\{(1,2),(2,3),(3,5),(4,6)\}$ belong to a function because each element of the domain is paired with only one element of the range or, 2. The equation $y = x + 3$ defines a function. This is because for each value of x there is only one value of y, i.e., at $x=0$ the variable $y=3$ or at $x=2$ the variable $y=5$, etc.

Example 2.1-1: Given the following equations, find the corresponding y values (range) for the given x values (domain).

a. $2x+3y=0$ at $x=0, x=-1, x=2, and\ x=10$

b. $y=x^2+4$ at $x=-1, x=0, x=3, and\ x=5$

c. $y=\sqrt{2x+1}$ at $x=0, x=2, x=4, and\ x=10$

d. $x-2y=6$ at $x=0, x=-3, x=-1, and\ x=2$

e. $y=x^2+9$ at $x=0, x=-4, x=-1, and\ x=3$

Solutions:

a. Given $2x+3y=0$, at $x=0$ the y value is equal to $\boxed{(2\times 0)+3y=0}$; $\boxed{0+3y=0}$; $\boxed{3y=0}$; $\boxed{\mathbf{y=0}}$

at $x=-1$ the y value is equal to $\boxed{(2\times -1)+3y=0}$; $\boxed{-2+3y=0}$; $\boxed{3y=2}$; $\boxed{y=\frac{2}{3}}$; $\boxed{\mathbf{y=0.66}}$

at $x=2$ the y value is equal to $\boxed{(2\times 2)+3y=0}$; $\boxed{4+3y=0}$; $\boxed{3y=-4}$; $\boxed{y=-\frac{4}{3}}$; $\boxed{\mathbf{y=-1.33}}$

at $x=10$ the y value is equal to $\boxed{(2\times 10)+3y=0}$; $\boxed{20+3y=0}$; $\boxed{3y=-20}$; $\boxed{y=-\frac{20}{3}}$; $\boxed{\mathbf{y=-6.66}}$

Therefore, the ordered pairs are $\{\mathbf{(0,0), (-1,0.66), (2,-1.33), (10,-6.66)}\}$

b. Given $y=x^2+4$, at $x=-1$ the y value is equal to $\boxed{y=(-1)^2+4}$; $\boxed{y=1+4}$; $\boxed{\mathbf{y=5}}$

at $x=0$ the y value is equal to $\boxed{y=0^2+4}$; $\boxed{y=0+4}$; $\boxed{\mathbf{y=4}}$

at $x=3$ the y value is equal to $\boxed{y=3^2+4}$; $\boxed{y=9+4}$; $\boxed{\mathbf{y=13}}$

at $x=5$ the y value is equal to $\boxed{y=5^2+4}$; $\boxed{y=25+4}$; $\boxed{\mathbf{y=29}}$

Therefore, the ordered pairs are $\{\mathbf{(-1,5), (0,4), (3,13), (5,29)}\}$

c. Given $y=\sqrt{2x+1}$, at $x=0$ the y value is equal to $\boxed{y=\sqrt{(2\times 0)+1}}$; $\boxed{y=\sqrt{0+1}}$; $\boxed{y=\sqrt{1}}$; $\boxed{\mathbf{y=1}}$

at $x=2$ the y value is equal to $\boxed{y=\sqrt{(2\times 2)+1}}$; $\boxed{y=\sqrt{4+1}}$; $\boxed{y=\sqrt{5}}$; $\boxed{\mathbf{y=2.24}}$

at $x=4$ the y value is equal to $\boxed{y=\sqrt{(2\times 4)+1}}$; $\boxed{y=\sqrt{8+1}}$; $\boxed{y=\sqrt{9}}$; $\boxed{\mathbf{y=3}}$

at $x=10$ the y value is equal to $\boxed{y=\sqrt{(2\times 10)+1}}$; $\boxed{y=\sqrt{20+1}}$; $\boxed{y=\sqrt{21}}$; $\boxed{\mathbf{y=4.58}}$

Therefore, the ordered pairs are $\{\mathbf{(0,1), (2,2.24), (4,3), (10,4.58)}\}$

d. Given $x-2y=6$, at $x=0$ the y value is equal to $\boxed{0-2y=6}$; $\boxed{-2y=6}$; $\boxed{y=-\frac{6}{2}}$; $\boxed{\mathbf{y=-3}}$

at $x=-3$ the y value is equal to $\boxed{-3-2y=6}$; $\boxed{-2y=6+3}$; $\boxed{-2y=9}$; $\boxed{y=-\frac{9}{2}}$; $\boxed{\mathbf{y=-4.5}}$

at $x=-1$ the y value is equal to $\boxed{-1-2y=6}$; $\boxed{-2y=6+1}$; $\boxed{-2y=7}$; $\boxed{y=-\frac{7}{2}}$; $\boxed{\mathbf{y=-3.5}}$

at $x=2$ the y value is equal to $\boxed{2-2y=6}$; $\boxed{-2y=6-2}$; $\boxed{-2y=4}$; $\boxed{y=-\frac{4}{2}}$; $\boxed{\mathbf{y=-2}}$

Therefore, the ordered pairs are $\{\mathbf{(0,-3),(-3,-4.5),(-1,-3.5),(2,-2)}\}$

e. Given $y=x^2+9$, at $x=0$ the y value is equal to $\boxed{y=0^2+9}$; $\boxed{y=0+9}$; $\boxed{\mathbf{y=9}}$

at $x=-4$ the y value is equal to $\boxed{y=(-4)^2+9}$; $\boxed{y=16+9}$; $\boxed{\mathbf{y=25}}$

at $x=-1$ the y value is equal to $\boxed{y=(-1)^2+9}$; $\boxed{y=1+9}$; $\boxed{\mathbf{y=10}}$

at $x=3$ the y value is equal to $\boxed{y=3^2+9}$; $\boxed{y=9+9}$; $\boxed{\mathbf{y=18}}$

Therefore, the ordered pairs are $\{\mathbf{(0,9),(-4,25),(-1,10),(3,18)}\}$

Example 2.1-2: 1. Specify the domain and the range for each of the following ordered pairs. 2. State which sets constitute a relation or a function.

a. $\{(1,2),(2,3),(3,5),(6,10),(8,11)\}$

b. $\{(2,-2),(4,5),(4,7),(6,8),(10,12)\}$

c. $\{(-1,2),(0,3),(2,4),(5,8)\}$

d. $\{(0,0),(2,4),(3,6),(4,8),(2,5)\}$

e. $\{(-2,3),(-2,6),(4,7),(8,12)\},(10,14)$

f. $\{(2,1),(3,4),(5,6),(7,12),(9,15)\}$

Solutions:

a. The domain and the range for the ordered pairs $\{(1,2),(2,3),(3,5),(6,10),(8,11)\}$ are: $\{1,2,3,6,8\}$ and $\{2,3,5,10,11\}$. Since each domain value corresponds with only one range value **it is a function**.

b. The domain and the range for the ordered pairs $\{(2,-2),(4,5),(4,7),(6,8),(10,12)\}$ are: $\{2,4,6,10\}$ and $\{-2,5,7,8,12\}$. Since each domain value does not corresponds with only one range value **it is a relation**.

c. The domain and the range values for the ordered pairs $\{(-1,2),(0,3),(2,4),(5,8)\}$ are: $\{-1,0,2,5\}$ and $\{2,3,4,8\}$. Since each domain value corresponds with only one range value **it is a function**.

d. The domain and the range values for the ordered pairs $\{(0,0),(2,4),(3,6),(4,8),(2,5)\}$ are: $\{0,2,3,4\}$ and $\{0,4,5,6,8\}$. Since each domain value does not correspond with only one range value **it is a relation**.

e. The domain and the range values for the ordered pairs $\{(-2,3),(-2,6),(4,7),(8,12)\},(10,14)$ are: $\{-2,4,8,10\}$ and $\{3,6,7,12,14\}$. Since each domain value does not correspond with only one range value **it is a relation**.

f. The domain and the range values for the ordered pairs $\{(2,1),(3,4),(5,6),(7,12),(9,15)\}$ are: $\{2,3,5,7,9\}$ and $\{1,4,6,12,15\}$. Since each domain value corresponds with only one range value **it is a function**.

Example 2.1-3: State which of the following equations define a function.

a. $x-y=2$ b. $x^2+y=25$ c. $x^2+y^2=25$ d. $y=x^2$

e. $y^2=x^3$ f. $y=\pm\sqrt{x^2}$ g. $y=\sqrt{x^2-7}$ h. $y=x^2+6$

Solutions:

a. The equation $x-y=2$; $-y=2-x$; $y=x-2$ defines a **function** because there is only one value of y associated with each value of x.

b. The equation $x^2+y=25$; $y=25-x^2$ defines a **function** because there is only one value of y associated with each value of x.

c. The equation $x^2+y^2=25$; $y^2=25-x^2$; $y=\pm\sqrt{25-x^2}$ defines a **relation** because for each value of x there are two values of y.

d. The equation $y=x^2$ defines a **function** because there is only one value of y associated with each value of x.

e. The equation $y^2=x^3$; $y=\pm\sqrt{x^3}$ defines a **relation** because for each value of x there are two values of y.

f. The equation $y=\pm\sqrt{x^2}$; $y=\pm x$ defines a **relation** because for each value of x there are two values of y.

g. The equation $y=\sqrt{x^2-7}$ defines a **function** because there is only one value of y associated with each value of x.

h. The equation $y=x^2+6$ defines a **function** because there is only one value of y associated with each value of x.

Example 2.1-4: Find the corresponding range values for each of the following functions. Write the ordered pair for each case.

a. $f(x)=-x$ for $f(-1)$, $f(0)$, $f(2)$, $f(a+2)$, and $f(-a)$

b. $f(x)=x^2+3x-1$ for $f(-1)$, $f(0)$, $f(5)$, $f(-a)$, and $f(a+1)$

c. $f(x)=\sqrt{3x+1}$ for $f(0)$, $f(1)$, $f(3)$, $f\left(a^2\right)$, and $f(a)$

d. $f(x)=\dfrac{1}{x+1}$ for $f(-2)$, $f(0)$, $f(3)$, $f(a+1)$, and $f(-a)$

Solutions:

a. Given $f(x)=-x$, then $\boxed{f(-1)}=\boxed{-(-1)}=\boxed{\mathbf{1}}$ $\boxed{f(0)}=\boxed{\mathbf{0}}$ $\boxed{f(2)}=\boxed{\mathbf{-2}}$

$\boxed{f(a+2)}=\boxed{-\mathbf{(a+2)}}$ $\boxed{f(-a)}=\boxed{-(-a)}=\boxed{\mathbf{a}}$.

Therefore, the ordered pairs are equal to $\{(-1,1),(0,0),(2,-2),(a+2,-a-2),(-a,a)\}$

b. Given $f(x)=x^2+3x-1$, then $\boxed{f(-1)}=\boxed{(-1)^2+(3\times -1)-1}=\boxed{1-3-1}=\boxed{\mathbf{-3}}$

$\boxed{f(0)} = \boxed{0^2+(3\times 0)-1} = \boxed{0+0-1} = \boxed{-1}$ $\boxed{f(5)} = \boxed{5^2+(3\times 5)-1} = \boxed{25+15-1} = \boxed{39}$

$\boxed{f(-a)} = \boxed{(-a)^2+(3\times -a)-1} = \boxed{a^2-3a-1}$ $\boxed{f(a+1)} = \boxed{(a+1)^2+3(a+1)-1} = \boxed{a^2+5a+3}$

Therefore, the ordered pairs are equal to $\{(-1,-3),(0,-1),(5,39),(-a,a^2-3a-1),(a+1,a^2+5a+3)\}$

c. Given $f(x)=\sqrt{3x+1}$, then $\boxed{f(0)} = \boxed{\sqrt{(3\times 0)+1}} = \boxed{\sqrt{0+1}} = \boxed{\sqrt{1}} = \boxed{1}$

$\boxed{f(1)} = \boxed{\sqrt{(3\times 1)+1}} = \boxed{\sqrt{3+1}} = \boxed{\sqrt{4}} = \boxed{2}$ $\boxed{f(3)} = \boxed{\sqrt{(3\times 3)+1}} = \boxed{\sqrt{9+1}} = \boxed{\sqrt{10}}$

$\boxed{f(a^2)} = \boxed{\sqrt{(3\times a^2)+1}} = \boxed{\sqrt{3a^2+1}}$ $\boxed{f(a)} = \boxed{\sqrt{(3\times a)+1}} = \boxed{\sqrt{3a+1}}$

Therefore, the ordered pairs are equal to $\left\{(0,1),(1,2),(3,\sqrt{10}),\left(a^2,\sqrt{3a^2+1}\right),(a,\sqrt{3a+1})\right\}$

d. Given $f(x)=\frac{1}{x+1}$, then $\boxed{f(-2)} = \boxed{\frac{1}{-2+1}} = \boxed{-\frac{1}{1}} = \boxed{-1}$

$\boxed{f(0)} = \boxed{\frac{1}{0+1}} = \boxed{\frac{1}{1}} = \boxed{1}$ $\boxed{f(3)} = \boxed{\frac{1}{3+1}} = \boxed{\frac{1}{4}}$

$\boxed{f(a+1)} = \boxed{\frac{1}{(a+1)+1}} = \boxed{\frac{1}{a+2}}$ $\boxed{f(-a)} = \boxed{\frac{1}{-a+1}} = \boxed{\frac{1}{1-a}}$

Therefore, the ordered pairs are equal to $\left\{(-2,-1),(0,1),\left(3,\frac{1}{4}\right),\left(a+1,\frac{1}{a+2}\right),\left(-a,\frac{1}{1-a}\right)\right\}$

In some instances, for a given function, it is often useful to calculate the change in the range value, usually shown by the symbol Δ, due to a given change, h, in the domain value, i.e., as the domain value changes from a to $a+h$, the corresponding range value also changes from $f(a)$ to $f(a+h)$. To calculate this change we use the following equation:

$$\Delta = f(a+h)-f(a)$$

Example 2.1-5: Given the following functions, find $f(a+h)-f(a)$.

a. $f(x)=3x+2$ b. $f(x)=x^2+2x-1$ c. $f(x)=x(2x+1)$

Solutions:

a. Given $\boxed{f(x)=3x+2}$ then $\boxed{f(a+h)-f(a)} = \boxed{[3(a+h)+2]-[3a+2]} = \boxed{3a+3h+2-3a-2} = \boxed{3h}$

b. Given $\boxed{f(x)=x^2+2x-1}$ then $\boxed{f(a+h)-f(a)} = \boxed{\left[(a+h)^2+2(a+h)-1\right]-\left[a^2+2a-1\right]} = \boxed{a^2+h^2+2ah+2a}$

$+\boxed{2h-1-a^2-2a+1} = \boxed{h^2+2ah+2h} = \boxed{h(h+2a+2)}$

c. Given $\boxed{f(x)=x(2x+1)=2x^2+x}$ then $\boxed{f(a+h)-f(a)}=\boxed{\left[2(a+h)^2+(a+h)\right]-\left(2a^2+a\right)}=\boxed{2\left(a^2+h^2+2ah\right)}$

$+\boxed{a+h-2a^2-a}=\boxed{2a^2+2h^2+4ah+a+h-2a^2-a}=\boxed{2h^2+4ah+h}=\boxed{\mathbf{h(2h+4a+1)}}$

Finally, the notation $f(x)|_a^b = f(b)-f(a)$ is used in some applications in calculus where the difference between two specific range values are of interest.

Example 2.1-6: Given the following functions, find $f(x)|_b^a$.

a. $f(x)=3x-1$ where $a=2$ and $b=4$

b. $f(x)=x^2+2x-3$ where $a=0$ and $b=3$

c. $f(x)=3x^3+2$ where $a=-2$ and $b=2$

d. $f(x)=\frac{1}{x}+2x$ where $a=3$ and $b=5$

Solutions:

a. Given $\boxed{f(x)=3x-1}$ then $\boxed{f(x)|_2^4}=\boxed{[(3\cdot4)-1]-[(3\cdot2)-1]}=\boxed{(12-1)-(6-1)}=\boxed{11-5}=\boxed{\mathbf{6}}$

b. Given $\boxed{f(x)=x^2+2x-3}$ then $\boxed{f(x)|_0^3}=\boxed{\left[3^2+(2\cdot3)-3\right]-\left[0^2+(2\cdot0)-3\right]}=\boxed{(9+6-3)+3}=\boxed{\mathbf{15}}$

c. Given $\boxed{f(x)=3x^3+2}$ then $\boxed{f(x)|_{-2}^2}=\boxed{\left[3\cdot2^3+2\right]-\left[3\cdot(-2)^3+2\right]}=\boxed{(24+2)-(-24+2)}=\boxed{26+22}=\boxed{\mathbf{48}}$

d. Given $\boxed{f(x)=\frac{1}{x}+2x}$ then $\boxed{f(x)|_3^5}=\boxed{\left(\frac{1}{5}+2\cdot5\right)-\left(\frac{1}{3}+2\cdot3\right)}=\boxed{(0.2+10)-(0.33+6)}=\boxed{10.2-6.33}=\boxed{\mathbf{3.87}}$

In the next section we will address: a. The basic math operations involving functions and b. learn how to identify odd and even functions.

Section 2.1 Practice Problems – Introduction to Functions of Real Variables

1. Find the corresponding y values.

   a. $x-4y=0$ at $x=0, x=-1, and\ x=3$

   b. $y-x^2+1=0$ at $x=-1, x=3, and\ x=-3$

   c. $y-\sqrt{x^2+1}=0$ at $x=2, x=-2, and\ x=-5$

   d. $x+4y=-5$ at $x=0, x=-2, and\ x=4$

   e. $y=2x^2-6$ at $x=0, x=-2, and\ x=-3$

2. a. Specify the domain and the range for each of the following ordered pairs. b. State which set constitute a relation or a function.

   a. $\{(1,4),(2,5),(3,6),(6,9),(8,12)\}$

   b. $\{(1,-1),(2,3),(4,6),(7,9),(10,11)\}$

   c. $\{(1,2),(1,-2),(2,5),(6,8),(9,12)\}$

   d. $\{(0,0),(1,6),(2,5),(2,7),(5,8)\}$

   e. $\{(-1,3),(-1,6),(2,5),(8,10),(10,12)\}$

   f. $\{(1,3),(2,5),(5,6),(7,10),(8,13)\}$

3. State which of the following equations defines a function.

   a. $x+y=12$

   b. $x^2+y^2=81$

   c. $y^2=15-x^2$

d. $y^2 = x^5$ e. $y = x^2 + 9$ f. $2y - 6x = 10$

g. $y = \sqrt{x^2 - 4}$ h. $x - 4y = 8$ i. $x + y^2 = 9$

4. Find the corresponding range values for each of the following functions.

a. $f(x) = -x^2 + 2x$ for $f(-4)$, $f(-1)$, $f(0)$, and $f(1)$

b. $f(x) = x^3 - 2x^2 + 1$ for $f(-2)$, $f(0)$, $f(2)$, and $f(-a)$

c. $f(x) = \sqrt{x^2 + 1}$ for $f(-2)$, $f(-1)$, $f(0)$, and $f(2)$

d. $f(x) = \dfrac{1}{x^2 + 2}$ for $f(-3)$, $f(-1)$, $f(0)$, and $f(3)$

5. Given the following functions, find $f(a+h) - f(a)$.

a. $f(x) = 2x - 1$ b. $f(x) = 2x^2 - 3$ c. $f(x) = \dfrac{x^2 - a^2}{x}$ d. $f(x) = (x-3)(x+1)$

2.2 Math Operations Involving Functions of Real Variables

Two functions $f(x)$ and $g(x)$ are added, subtracted, multiplied, and divided using the following general operational rules:

$$f(x)+g(x) = (f+g)(x)$$

$$f(x)-g(x) = (f-g)(x)$$

$$f(x)\cdot g(x) = (f\cdot g)(x)$$

$$\frac{f(x)}{g(x)} = \left(\frac{f}{g}\right)(x) \quad \text{where } g(x)\neq 0$$

In addition, note that $\frac{1}{g(x)} = \left(\frac{1}{g}\right)(x)$, $\alpha\, f(x) = (\alpha\, f)(x)$, and $\alpha\, f(x) \pm \beta\, f(x) = (\alpha\, f \pm \beta\, f)(x)$ where α and β are real numbers. The following examples show basic math operations involving functions.

Example 2.2-1: Given $f(x)=x^3+3x^2+1$ and $g(x)=2x^3-1$, find

a. $f(x)+g(x) =$ b. $f(x)-g(x) =$ c. $\frac{2f(x)}{g(x)+3} =$ d. $2f(x)\cdot g(x) =$

e. $(5f)(x)-(2g)(x) =$ f. $\frac{1}{g(x)}+f(x) =$ g. $(2f-3g)(x) =$ h. $\left(\frac{2f}{g}\right)(x)-4 =$

i. $f(x)+5g(x) =$ j. $(3f-g)(x) =$ k. $\left(\frac{f}{g}\right)(x)-\left(\frac{1}{g}\right)(x) =$ l. $\left(\frac{3}{f}\right)(x)-\left(\frac{a+1}{f}\right)(x) =$

Solutions:

a. $f(x)+g(x) = \left(x^3+3x^2+1\right)+\left(2x^3-1\right) = x^3+2x^3+3x^2+1-1 = 3x^3+3x^2 = \mathbf{3x^2(x+1)}$

b. $f(x)-g(x) = \left(x^3+3x^2+1\right)-\left(2x^3-1\right) = x^3+3x^2+1-2x^3+1 = x^3-2x^3+3x^2+1+1 = \mathbf{-x^3+3x^2+2}$

c. $\frac{2f(x)}{g(x)+3} = \frac{2\left(x^3+3x^2+1\right)}{\left(2x^3-1\right)+3} = \frac{2\left(x^3+3x^2+1\right)}{2x^3-1+3} = \frac{2\left(x^3+3x^2+1\right)}{2x^3+2} = \frac{2\left(x^3+3x^2+1\right)}{2\left(x^3+1\right)} = \mathbf{\frac{x^3+3x^2+1}{x^3+1}}$

d. $2f(x)\cdot g(x) = 2\left(x^3+3x^2+1\right)\cdot\left(2x^3-1\right) = 2\left(2x^6-x^3+6x^5-3x^2+2x^3-1\right) = \mathbf{2\left(2x^6+6x^5+x^3-3x^2-1\right)}$

e. $(5f)(x)-(2g)(x) = 5\left(x^3+3x^2+1\right)-2\left(2x^3-1\right) = 5x^3+15x^2+5-4x^3+4 = \mathbf{x^3+15x^2+9}$

f. $\frac{1}{g(x)}+f(x) = \frac{1}{2x^3-1}+x^3+3x^2+1 = \frac{1}{2x^3-1}+\frac{x^3+3x^2+1}{1} = \frac{(1\cdot 1)+\left(x^3+3x^2+1\right)\cdot\left(2x^3-1\right)}{2x^3-1}$

$= \frac{1+2x^6-x^3+6x^5-3x^2+2x^3-1}{2x^3-1} = \frac{2x^6+6x^5+x^3-3x^2}{2x^3-1} = \mathbf{\frac{x^2\left(2x^4+6x^3+x-3\right)}{2x^3-1}}$

g. $(2f-3g)(x) = 2(x^3+3x^2+1)-3(2x^3-1) = 2x^3+6x^2+2-6x^3+3 = \mathbf{-4x^3+6x^2+5}$

h. $\left(\frac{2f}{g}\right)(x)-4 = \frac{2(x^3+3x^2+1)}{2x^3-1}-4 = \frac{2x^3+6x^2+2}{2x^3-1}-\frac{4}{1} = \frac{[(2x^3+6x^2+2)\cdot 1]+[-4\cdot(2x^3-1)]}{(2x^3-1)\cdot 1}$

$= \frac{2x^3+6x^2+2-8x^3+4}{2x^3-1} = \frac{2x^3-8x^3+6x^2+2+4}{2x^3-1} = \frac{-6x^3+6x^2+6}{2x^3-1} = \mathbf{\frac{6(-x^3+x^2+1)}{2x^3-1}}$

i. $f(x)+5g(x) = (x^3+3x^2+1)+5(2x^3-1) = x^3+3x^2+1+10x^3-5 = \mathbf{11x^3+3x^2-4}$

j. $(3f-g)(x) = 3(x^3+3x^2+1)-(2x^3-1) = 3x^3+9x^2+3-2x^3+1 = 3x^3-2x^3+9x^2+3+1 = \mathbf{x^3+9x^2+4}$

k. $\left(\frac{f}{g}\right)(x)-\left(\frac{1}{g}\right)(x) = \frac{x^3+3x^2+1}{2x^3-1}-\frac{1}{2x^3-1} = \frac{x^3+3x^2+1-1}{2x^3-1} = \frac{x^3+3x^2}{2x^3-1} = \mathbf{\frac{x^2(x+3)}{2x^3-1}}$

l. $\left(\frac{3}{f}\right)(x)-\left(\frac{a+1}{f}\right)(x) = \frac{3}{x^3+3x^2+1}-\frac{a+1}{x^3+3x^2+1} = \frac{3-(a+1)}{x^3+3x^2+1} = \frac{3-a-1}{x^3+3x^2+1} = \mathbf{\frac{2-a}{x^3+3x^2+1}}$

Example 2.2-2: Given $f(x)=x^2+5x$ and $g(x)=2x-1$, find

a. $f(x)+3g(x) =$ b. $f(x)-g(x) =$ c. $\frac{3f(x)}{g(x)+1} =$ d. $4f(x)\cdot g(x) =$

e. $(3f)(x)-(2g)(x) =$ f. $\frac{1}{g(x)}+f(x) =$ g. $(2f-5g)(x) =$ h. $(3f-g)(x) =$

Solutions:

a. $f(x)+g(x) = (x^2+5x)+(2x-1) = x^2+5x+2x-1 = \mathbf{x^2+7x-1}$

b. $f(x)-g(x) = (x^2+5x)-(2x-1) = x^2+5x-2x+1 = \mathbf{x^2+3x+1}$

c. $\frac{3f(x)}{g(x)+1} = \frac{x^2+5x}{(2x-1)+1} = \frac{x^2+5x}{2x-1+1} = \frac{\not{x}(x+5)}{2\not{x}} = \mathbf{\frac{x+5}{2}}$

d. $4f(x)\cdot g(x) = 4(x^2+5x)\cdot(2x-1) = 4(2x^3-x^2+10x^2-5x) = 4(2x^3+9x^2-5x) = \mathbf{8x^3+36x^2-20x}$

e. $(3f)(x)-(2g)(x) = 3(x^2+5x)-2(2x-1) = 3x^2+15x-4x+2 = \mathbf{3x^2+11x+2}$

f. $\frac{1}{g(x)}+f(x) = \frac{1}{2x-1}+(x^2+5x) = \frac{1}{2x-1}+\frac{x^2+5x}{1} = \frac{1+(x^2+5x)\cdot(2x-1)}{2x-1} = \frac{1+2x^3-x^2+10x^2-5x}{2x-1}$

$= \boxed{\dfrac{2x^3 + (10-1)x^2 - 5x + 1}{2x-1}} = \boxed{\dfrac{\mathbf{2x^3 + 9x^2 - 5x + 1}}{\mathbf{2x-1}}}$

g. $\boxed{(2f - 5g)(x)} = \boxed{2(x^2 + 5x) - 5(2x-1)} = \boxed{2x^2 + 10x - 10x + 5} = \boxed{\mathbf{2x^2 + 5}}$

h. $\boxed{(3f - g)(x)} = \boxed{3(x^2 + 5x) - (2x-1)} = \boxed{3x^2 + 15x - 2x + 1} = \boxed{\mathbf{3x^2 + 13x + 1}}$

Example 2.2-3: Let $f(x) = x+1$ and $g(x) = x^2 + x$. Find and simplify the following expressions.

a. $(f+g)(-1) =$ b. $(g-f)(2) =$ c. $\left(\frac{f}{g}\right)(-2) =$

d. $(f+g)(a^2) =$ e. $\left(\frac{g}{f}\right)(-10) =$ f. $(f \cdot g)(3) =$

g. $f(2) - g(2) =$ h. $(g-f)(-3) =$ i. $g(2) - f(2) =$

j. $\frac{f}{g}(9) =$ k. $(f \cdot g)(9) =$ l. $\frac{g}{f}(-a) =$

Solutions:

a. $\boxed{(f+g)(x)} = \boxed{(x+1) + (x^2 + x)} = \boxed{x^2 + 2x + 1}$ therefore $\boxed{(f+g)(-1)} = \boxed{(-1)^2 + (2 \times -1) + 1} = \boxed{1-2+1} = \boxed{\mathbf{0}}$

b. $\boxed{(g-f)(x)} = \boxed{(x^2 + x) - (x+1)} = \boxed{x^2 + x - x - 1} = \boxed{x^2 - 1}$ therefore $\boxed{(g-f)(2)} = \boxed{2^2 - 1} = \boxed{4-1} = \boxed{\mathbf{3}}$

c. $\boxed{\left(\frac{f}{g}\right)(x)} = \boxed{\frac{x+1}{x^2 + x}} = \boxed{\frac{x+1}{x(x+1)}} = \boxed{\frac{1}{x}}$ therefore $\boxed{\left(\frac{f}{g}\right)(-2)} = \boxed{-\frac{1}{2}} = \boxed{\mathbf{-0.5}}$

d. $\boxed{(f+g)(x)} = \boxed{(x+1) + (x^2 + x)} = \boxed{x^2 + 2x + 1}$ therefore $\boxed{(f+g)(a^2)} = \boxed{(a^2)^2 + (2 \times a^2) + 1} = \boxed{\mathbf{a^4 + 2a^2 + 1}}$

e. $\boxed{\left(\frac{g}{f}\right)(x)} = \boxed{\frac{x^2 + x}{x+1}} = \boxed{\frac{x(x+1)}{x+1}} = \boxed{\frac{x}{1}} = \boxed{x}$ therefore $\boxed{\left(\frac{g}{f}\right)(-10)} = \boxed{\mathbf{-10}}$

f. $\boxed{(f \cdot g)(x)} = \boxed{(x+1)(x^2 + x)} = \boxed{x^3 + x^2 + x^2 + x} = \boxed{x^3 + 2x^2 + x}$ thus $\boxed{(f \cdot g)(3)} = \boxed{3^3 + 2 \cdot 3^2 + 3} = \boxed{\mathbf{48}}$

g. $\boxed{f(x) - g(x)} = \boxed{(x+1) - (x^2 + x)} = \boxed{x + 1 - x^2 - x} = \boxed{-x^2 + 1}$ thus $\boxed{f(2) - g(2)} = \boxed{-(2)^2 + 1} = \boxed{-4+1} = \boxed{\mathbf{-3}}$

h. $\boxed{(g-f)(x)} = \boxed{(x^2 + x) - (x+1)} = \boxed{x^2 + x - x - 1} = \boxed{x^2 - 1}$ therefore $\boxed{(g-f)(-3)} = \boxed{(-3)^2 - 1} = \boxed{9-1} = \boxed{\mathbf{8}}$

i. $\boxed{g(x) - f(x)} = \boxed{(x^2 + x) - (x+1)} = \boxed{x^2 + x - x - 1} = \boxed{x^2 - 1}$ therefore $\boxed{g(2) - f(2)} = \boxed{2^2 - 1} = \boxed{4-1} = \boxed{\mathbf{3}}$

j. $\boxed{\frac{f}{g}(x)} = \boxed{\frac{x+1}{x^2+x}} = \boxed{\frac{x+1}{x(x+1)}} = \boxed{\frac{1}{x}}$ therefore $\boxed{\frac{f}{g}(9)} = \boxed{\frac{1}{9}} = \boxed{\mathbf{0.11}}$

k. $\boxed{(f \cdot g)(x)} = \boxed{(x+1)(x^2+x)} = \boxed{x^3+x^2+x^2+x} = \boxed{x^3+2x^2+x}$ thus $\boxed{(f \cdot g)(9)} = \boxed{9^3+2\cdot 9^2+9} = \boxed{\mathbf{900}}$

l. $\boxed{\frac{g}{f}(x)} = \boxed{\frac{x^2+x}{x+1}} = \boxed{\frac{x(x+1)}{x+1}} = \boxed{\frac{x}{1}} = \boxed{x}$ therefore $\boxed{\frac{g}{f}(-a)} = \boxed{\mathbf{-a}}$

2.2.1 Determining Odd and Even Functions

A function $f(x)$ is called an **odd function** if and only if:

$$f(-x) = -f(x) \qquad \text{for all } x$$

A function $f(x)$ is called an **even function** if and only if:

$$f(-x) = f(x) \qquad \text{for all } x$$

The following examples show how to identify odd or even functions:

Example 2.2-4: State which of the following functions are odd or even.

a. $f(x) = x^2$ b. $f(x) = x^5 + 1$ c. $f(x) = x(x-1)$

d. $f(x) = x - \frac{1}{x}$ e. $f(x) = x^3$ f. $f(x) = 1 + |x|$

g. $f(x) = \frac{x^3}{1-|x|}$ h. $f(x) = \frac{x^2}{1+|x|}$ i. $f(x) = x^3 + 3x^2 + 1$

Solutions:

a. Given $f(x) = x^2$ compute $f(-x)$ and $-f(x)$, i.e., $f(-x) = (-x)^2 = x^2$ and $-f(x) = -x^2$. Since $f(-x) = f(x)$ the function $f(x) = x^2$ is an **even function**.

b. Given $f(x) = x^5 + 1$ compute $f(-x)$ and $-f(x)$, i.e., $f(-x) = (-x)^5 + 1 = -x^5 + 1$ and $-f(x) = -(x^5+1) = -x^5 - 1$. Since $f(-x) \neq -f(x)$ and $f(-x) \neq f(x)$ the function $f(x) = x^5 + 1$ is **neither an odd nor an even function**.

c. Given $f(x) = x(x-1) = x^2 - x$ compute $f(-x)$ and $-f(x)$, i.e., $f(-x) = (-x)^2 - (-x) = x^2 + x$ and $-f(x) = -(x^2 - x) = -x^2 + x$. Since $f(-x) \neq -f(x)$ and $f(-x) \neq f(x)$ the function $f(x) = x^5 + 1$ is **neither an odd nor an even function**.

d. Given $f(x) = x - \frac{1}{x} = \frac{x^2-1}{x}$ compute $f(-x)$ and $-f(x)$, i.e., $f(-x) = \frac{(-x)^2-1}{-x} = \frac{x^2-1}{-x}$ $= \frac{-x^2+1}{x}$ and $-f(x) = -\left(\frac{x^2-1}{x}\right) = \frac{-x^2+1}{x}$. Since $f(-x) = -f(x)$ the function $f(x) = x - \frac{1}{x}$ is an **odd function**.

e. Given $f(x) = x^3$ compute $f(-x)$ and $-f(x)$, i.e., $f(-x) = (-x)^3 = -x^3$ and $-f(x) = -x^3$. Since

$f(-x) = -f(x)$ the function $f(x) = x^3$ is an **odd function.**

f. Given $f(x) = 1+|x|$ compute $f(-x)$ and $-f(x)$, i.e., $f(-x) = 1+|-x| = 1+|x|$ and $-f(x)$ $= -(1+|x|) = -1-|x|$. Since $f(-x) = f(x)$ the function $f(x) = 1+|x|$ is an **even function.**

g. Given $f(x) = \frac{x^3}{1-|x|}$ compute $f(-x)$ and $-f(x)$, i.e., $f(-x) = \frac{(-x)^3}{1-|-x|} = \frac{-x^3}{1-|x|}$ and $-f(x)$ $= -\left(\frac{x^3}{1-|x|}\right) = \frac{-x^3}{1-|x|}$. Since $f(-x) = -f(x)$ the function $f(x) = \frac{x^3}{1-|x|}$ is an **odd function.**

h. Given $f(x) = \frac{x^2}{1+|x|}$ compute $f(-x)$ and $-f(x)$, i.e., $f(-x) = \frac{(-x)^2}{1+|-x|} = \frac{x^2}{1+|x|}$ and $-f(x)$ $= -\left(\frac{x^2}{1+|x|}\right) = \frac{-x^2}{1+|x|}$. Since $f(-x) = f(x)$ the function $f(x) = \frac{x^2}{1+|x|}$ is an **even function.**

i. Given $f(x) = x^3+3x^2+1$ compute $f(-x)$ and $-f(x)$, i.e., $f(-x) = (-x)^3+3(-x)^2+1 = -x^3+3x^2+1$ and $-f(x) = -(x^3+3x^2+1) = -x^3-3x^2-1$. Since $f(-x) \neq -f(x)$ and $f(-x) \neq f(x)$ the function $f(x) = x^3+3x^2+1$ is **neither an odd nor an even function.**

In the next section we will learn how to find the composition of two or more functions shown as $f \circ g = (f \circ g)(x) = f(g(x))$ and $f \circ g \circ h = (f \circ g \circ h)(x) = f(g(h(x)))$.

Section 2.2 Practice Problems – Math Operations Involving Functions of Real Variables

1. Given $f(x)=x^2-3x+5$ and $g(x)=2x^2$, find

a. $2f(x)+g(x) =$ b. $f(x)-3g(x) =$ c. $\frac{f(x)}{x}+5g(x) =$

d. $3f(x)\cdot g(x) =$ e. $3f(x)-5g(x) =$ f. $\frac{3}{g(x)}+2f(x) =$

g. $(3f-2g)(x) =$ h. $\left(\frac{f}{4g}\right)(x)-2 =$ i. $\frac{f(x)-20}{g(x)}+x =$

2. Let $f(x)=x^2+2$ and $g(x)=2x+5$. Find and simplify the following expressions.

a. $(f+g)(-2) =$ b. $(g-f)(0) =$ c. $\left(\frac{f}{g}\right)(-1) =$

d. $(f+2g)(0) =$ e. $\left(\frac{g}{f}\right)(-2) =$ f. $(f\cdot g)(2) =$

3. State which of the following functions are odd or even.

a. $f(x) = x-1$ b. $f(x) = x^6+1$ c. $f(x) = x^2(x-1)$

d. $f(x) = x^2+\frac{1}{x}$ e. $f(x) = 1+x^3$ f. $f(x) = |x|+3$

g. $f(x) = \frac{x^2}{1-2x}$ h. $f(x) = x-\frac{1}{1+x}$ i. $f(x) = x^4+x^2-2$

2.3 Composite Functions of Real Variables

It is sometimes useful to form a new function from two or more given functions. Composite functions are generally shown by a special symbol "$\circ$" as $(f \circ g)(x)$ or $f \circ g$. Note that in math books the composition of two or three functions represented by $f(x)$, $g(x)$, and $h(x)$ is generally shown in one of the following forms: $C(x) = f(g(x)) = (f \circ g)(x) = f \circ g$ or $C(x) = f(g(h(x))) = (f \circ g \circ h)(x) = f \circ g \circ h$. In this section, we will learn how to find composite functions using two or three functions.

Case I Computing Composite Functions Using $f(x)$ and $g(x)$

Given $f(x)=x^2+2x+1$ and $g(x)=x-3$, then $f(g(x))$ is equal to

$$f(g(x)) = f(x-3) = [g(x)]^2+2g(x)+1 = (x-3)^2+2(x-3)+1 = x^2+9-6x+2x-6+1 = x^2-4x+4$$

The new function $f(g(x))$ is called the composite function. Composite functions are obtained from different classes of $f(x)$ and $g(x)$ functions as shown in the following examples.

Example 2.3-1: Find the composite function $f(g(x)) = (f \circ g)(x) = f \circ g$ for the following $f(x)$ and $g(x)$ functions:

a. $f(x)=x+1$; $g(x)=x^2$

b. $f(x)=x^3$; $g(x)=x-1$

c. $f(x)=\frac{1}{x^2-1}$; $g(x)=x^3+1$

d. $f(x)=2x+6$; $g(x)=x^3$

e. $f(x)=x^2+x$; $g(x)=\sqrt{x+1}$

f. $f(x)=\sqrt{x}$; $g(x)=x^2+1$

g. $f(x)=x^5$; $g(x)=\frac{1}{x+1}$

h. $f(x)=\frac{1}{x}+\frac{1}{x-1}$; $g(x)=x^3$

i. $f(x)=x^2+1$; $g(x)=\sqrt{x}$

Solutions:

a. Given $f(x)=x+1$ and $g(x)=x^2$, substitute $g(x)$ in place of x in $f(x)$, i.e.,

$$\boxed{f(g(x))} = \boxed{g(x)+1} = \boxed{\mathbf{x^2+1}}$$

b. Given $f(x)=x^3$ and $g(x)=x-1$, substitute $g(x)$ in place of x in $f(x)$, i.e.,

$$\boxed{f(g(x))} = \boxed{g(x)^3} = \boxed{\mathbf{(x-1)^3+1}}$$

c. Given $f(x)=\frac{1}{x^2-1}$ and $g(x)=x^3+1$, substitute $g(x)$ in place of x in $f(x)$, i.e.,

$$\boxed{f(g(x))} = \boxed{\frac{1}{[g(x)]^2-1}} = \boxed{\frac{1}{(x^3+1)^2-1}} = \boxed{\frac{1}{x^6+2x^3+1-1}} = \boxed{\frac{1}{x^6+2x^3}} = \boxed{\mathbf{\frac{1}{x^3(x^3+2)}}}$$

d. Given $f(x)=2x+6$ and $g(x)=x^3$, substitute $g(x)$ in place of x in $f(x)$, i.e.,

$$\boxed{f(g(x))} = \boxed{2g(x)+6} = \boxed{2x^3+6} = \boxed{\mathbf{2(x^3+3)}}$$

e. Given $f(x)=x^2+x$ and $g(x)=\sqrt{x+1}$, substitute $g(x)$ in place of x in $f(x)$, i.e.,

$$\boxed{f(g(x))} = \boxed{[g(x)]^2+g(x)} = \boxed{\left(\sqrt{x+1}\right)^2+\sqrt{x+1}} = \boxed{\mathbf{x+1+\sqrt{x+1}}}$$

f. Given $f(x)=\sqrt{x}$ and $g(x)=x^2+1$, substitute $g(x)$ in place of x in $f(x)$, i.e.,

$$\boxed{f(g(x))} = \boxed{\sqrt{g(x)}} = \boxed{\mathbf{\sqrt{x^2+1}}}$$

g. Given $f(x)=x^5$ and $g(x)=\frac{1}{x+1}$, substitute $g(x)$ in place of x in $f(x)$, i.e.,

$$\boxed{f(g(x))} = \boxed{[g(x)]^5} = \boxed{\left(\frac{1}{x+1}\right)^5} = \boxed{\mathbf{\frac{1}{(x+1)^5}}}$$

h. Given $f(x)=\frac{1}{x}+\frac{1}{x-1}$ and $g(x)=x^3$, substitute $g(x)$ in place of x in $f(x)$, i.e.,

$$\boxed{f(g(x))} = \boxed{\frac{1}{g(x)}+\frac{1}{g(x)-1}} = \boxed{\frac{1}{x^3}+\frac{1}{x^3-1}} = \boxed{\frac{\left(x^3-1\right)+x^3}{x^3\left(x^3-1\right)}} = \boxed{\frac{x^3-1+x^3}{x^3\left(x^3-1\right)}} = \boxed{\mathbf{\frac{2x^3-1}{x^3\left(x^3-1\right)}}}$$

i. Given $f(x)=x^2+1$ and $g(x)=\sqrt{x}$, substitute $g(x)$ in place of x in $f(x)$, i.e.,

$$\boxed{f(g(x))} = \boxed{[g(x)]^2+1} = \boxed{\left(\sqrt{x}\right)^2+1} = \boxed{\mathbf{x+1}}$$

Example 2.3-2: Find the composite function $g(f(x)) = (g \circ f)(x) = g \circ f$ for the following $g(x)$ and $f(x)$ functions:

a. $g(x)=x^3+1$; $f(x)=\frac{1}{x}$

b. $g(x)=3x^2+5$; $f(x)=5x+1$

c. $g(x)=x^2-1$; $f(x)=\sqrt{x+1}$

d. $g(x)=3x-1$; $f(x)=\frac{1}{x^3}$

e. $g(x)=\frac{1}{x+1}-\frac{1}{x^2}$; $f(x)=\sqrt{x}$

f. $g(x)=\frac{1}{2x+1}+5x$; $f(x)=x^3$

Solutions:

a. Given $g(x)=x^3+1$ and $f(x)=\frac{1}{x}$, substitute $f(x)$ in place of x in $g(x)$, i.e.,

$$\boxed{g(f(x))} = \boxed{[f(x)]^3+1} = \boxed{\left(\frac{1}{x}\right)^3+1} = \boxed{\mathbf{\frac{1}{x^3}+1}}$$

b. Given $g(x)=3x^2+5$ and $f(x)=5x+1$, substitute $f(x)$ in place of x in $g(x)$, i.e.,

$$\boxed{g(f(x))} = \boxed{3[f(x)]^2+5} = \boxed{3(5x+1)^2+5} = \boxed{3\left(25x^2+10x+1\right)+5} = \boxed{75x^2+30x+3+5} = \boxed{\mathbf{75x^2+30x+8}}$$

c. Given $g(x)=x^2-1$ and $f(x)=\sqrt{x+1}$, substitute $f(x)$ in place of x in $g(x)$, i.e.,

$$\boxed{g(f(x))} = \boxed{[f(x)]^2-1} = \boxed{\left(\sqrt{x+1}\right)^2-1} = \boxed{x+1-1} = \boxed{\mathbf{x}}$$

d. Given $g(x)=3x-1$ and $f(x)=\frac{1}{x^3}$, substitute $f(x)$ in place of x in $g(x)$, i.e.,

$$\boxed{g(f(x))} = \boxed{3f(x)-1} = \boxed{3\cdot\frac{1}{x^3}-1} = \boxed{\mathbf{\frac{3}{x^3}-1}}$$

e. Given $g(x)=\frac{1}{x+1}-\frac{1}{x^2}$ and $f(x)=\sqrt{x}$, substitute $f(x)$ in place of x in $g(x)$, i.e.,

$$\boxed{g(f(x))} = \boxed{\frac{1}{f(x)+1}-\frac{1}{[f(x)]^2}} = \boxed{\frac{1}{\sqrt{x}+1}-\frac{1}{\left(\sqrt{x}\right)^2}} = \boxed{\frac{1}{\sqrt{x}+1}-\frac{1}{x}} = \boxed{\mathbf{\frac{x-\sqrt{x}+1}{x\left(\sqrt{x}+1\right)}}}$$

f. Given $g(x)=\frac{1}{2x+1}+5x$ and $f(x)=x^3$, substitute $f(x)$ in place of x in $g(x)$, i.e.,

$$\boxed{g(f(x))} = \boxed{\frac{1}{2f(x)+1} + 5f(x)} = \boxed{\mathbf{\frac{1}{2x^3+1} + 5x^3}}$$

Note that changing the order in which functions are composed in most cases change the result. For example, given $f(x) = x^2 + 1$ and $g(x) = x^3$, then

$(f \circ g)(x) = f(g(x))$ is equal to $f(g(x)) = [g(x)]^2 + 1 = \left(x^3\right)^2 + 1 = \mathbf{x^6 + 1}$ where as

$(g \circ f)(x) = g(f(x))$ is equal to $g(f(x)) = [f(x)]^3 = \mathbf{\left(x^2 + 1\right)^3}$.

However, given $f(x) = \frac{1}{x^2}$ and $g(x) = \frac{1}{x}$ result in having the same $f(g(x))$ and $g(f(x))$, i.e.,

$(f \circ g)(x) = f(g(x))$ is equal to $f(g(x)) = \frac{1}{[g(x)]^2} = \frac{1}{\left(\frac{1}{x}\right)^2} = \frac{1}{\frac{1}{x^2}} = \frac{\frac{1}{1}}{\frac{1}{x^2}} = \frac{1 \cdot x^2}{1 \cdot 1} = \frac{x^2}{1} = \mathbf{x^2}$ and

$(g \circ f)(x) = g(f(x))$ is equal to $g(f(x)) = \frac{1}{f(x)} = \frac{1}{\frac{1}{x^2}} = \frac{\frac{1}{1}}{\frac{1}{x^2}} = \frac{1 \cdot x^2}{1 \cdot 1} = \frac{x^2}{1} = \mathbf{x^2}$.

The following examples further illustrate this point.

Example 2.3-3: Given the following $f(x)$ and $g(x)$ functions, find $f(g(x))$ and $g(f(x))$.

a. $f(x) = x^3 + \frac{1}{x}$ and $g(x) = \sqrt{x}$

b. $f(x) = x^2 + x + 5$ and $g(x) = x + 1$

c. $f(x) = x^2 - 1$ and $g(x) = \frac{1}{x^3}$

d. $f(x) = \frac{1}{2x+5}$ and $g(x) = \sqrt{x-1}$

Solutions:

a. Given $f(x) = x^3 + \frac{1}{x}$ and $g(x) = \sqrt{x}$ substitute $g(x)$ in place of x in $f(x)$ to obtain $f(g(x))$.

$$\boxed{f(g(x))} = \boxed{[g(x)]^3 + \frac{1}{g(x)}} = \boxed{\left(\sqrt{x}\right)^3 + \frac{1}{\sqrt{x}}} = \boxed{\left(x^{\frac{1}{2}}\right)^3 + \frac{1}{\sqrt{x}}} = \boxed{x^{\frac{1}{2}\times 3} + \frac{1}{\sqrt{x}}} = \boxed{x^{\frac{3}{2}} + \frac{1}{\sqrt{x}}} = \boxed{\mathbf{\sqrt{x^3} + \frac{1}{\sqrt{x}}}}$$

Next, substitute $f(x)$ in place of x in $g(x)$ to obtain $g(f(x))$.

$$\boxed{g(f(x))} = \boxed{\sqrt{f(x)}} = \boxed{\mathbf{\sqrt{x^3 + \frac{1}{x}}}}$$

d. Given $f(x) = x^2 + x + 5$ and $g(x) = x + 1$ substitute $g(x)$ in place of x in $f(x)$ to obtain $f(g(x))$.

$$\boxed{f(g(x))} = \boxed{[g(x)]^2 + g(x) + 5} = \boxed{(x+1)^2 + (x+1) + 5} = \boxed{x^2 + 2x + 1 + x + 1 + 5} = \boxed{\mathbf{x^2 + 3x + 7}}$$

Next, substitute $f(x)$ in place of x in $g(x)$ to obtain $g(f(x))$.

$$\boxed{g(f(x))} = \boxed{f(x) + 1} = \boxed{\left(x^2 + x + 5\right) + 1} = \boxed{\mathbf{x^2 + x + 6}}$$

c. Given $f(x) = x^2 - 1$ and $g(x) = \frac{1}{x^3}$ substitute $g(x)$ in place of x in $f(x)$ to obtain $f(g(x))$.

$$\boxed{f(g(x))} = \boxed{[g(x)]^2 - 1} = \boxed{\left(\frac{1}{x^3}\right)^2 - 1} = \boxed{\mathbf{\frac{1}{x^6} - 1}}$$

Next, substitute $f(x)$ in place of x in $g(x)$ to obtain $g(f(x))$.

$$\boxed{g(f(x))} = \boxed{\frac{1}{[f(x)]^3}} = \boxed{\frac{\mathbf{1}}{\left(\boldsymbol{x^2-1}\right)^3}}$$

d. Given $f(x)=\dfrac{1}{2x+5}$ and $g(x)=\sqrt{x-1}$ substitute $g(x)$ in place of x in $f(x)$ to obtain $f(g(x))$.

$$\boxed{f(g(x))} = \boxed{\frac{1}{2g(x)+5}} = \boxed{\frac{\mathbf{1}}{\mathbf{2\sqrt{x-1}+5}}}$$

Next, substitute $f(x)$ in place of x in $g(x)$ to obtain $g(f(x))$.

$$\boxed{g(f(x))} = \boxed{\sqrt{f(x)-1}} = \boxed{\sqrt{\frac{1}{2x+5}-1}} = \boxed{\sqrt{\frac{1-2x-5}{2x+5}}} = \boxed{\sqrt{\frac{\mathbf{-2x-4}}{\mathbf{2x+5}}}}$$

Example 2.3-4: Given $f(x)=x^3+5x-3$ and $g(x)=\dfrac{1}{x^3}+\dfrac{1}{x^2}-1$, find

a. $f(0)$ and $g(-1)$
b. $f(-1)$ and $g(3)$
c. $f(a^2)$ and $g(10)$
d. $f\left(\frac{1}{k}\right)$ and $g(-2k)$
e. $f(-3)$ and $g(2)$
f. $f(2n)$ and $g(0)$

Solutions:

a. $\boxed{f(0)} = \boxed{0^3+(5\cdot 0)-3} = \boxed{\mathbf{-3}}$ and $\boxed{g(-1)} = \boxed{\frac{1}{(-1)^3}+\frac{1}{(-1)^2}-1} = \boxed{\frac{1}{-1}+\frac{1}{1}-1} = \boxed{-1+1-1} = \boxed{\mathbf{-1}}$

b. $\boxed{f(-1)} = \boxed{(-1)^3+(5\cdot -1)-3} = \boxed{-1-5-3} = \boxed{\mathbf{-9}}$ and $\boxed{g(3)} = \boxed{\frac{1}{3^3}+\frac{1}{3^2}-1} = \boxed{\frac{1}{27}+\frac{1}{9}-1} = \boxed{\mathbf{-0.852}}$

c. $\boxed{f(a^2)} = \boxed{(a^2)^3+(5\cdot a^2)-3} = \boxed{\boldsymbol{a^6+5a^2-3}}$ and $\boxed{g(10)} = \boxed{\frac{1}{10^3}+\frac{1}{10^2}-1} = \boxed{\frac{1}{1000}+\frac{1}{100}-1} = \boxed{\mathbf{-0.989}}$

d. $\boxed{f\left(\frac{1}{k}\right)} = \boxed{\left(\frac{1}{k}\right)^3+\left(5\cdot\frac{1}{k}\right)-3} = \boxed{\boldsymbol{\frac{1}{k^3}+\frac{5}{k}-3}}$ and $\boxed{g(-2k)} = \boxed{\frac{1}{(-2k)^3}+\frac{1}{(-2k)^2}-1} = \boxed{\boldsymbol{-\frac{1}{8k^3}+\frac{1}{4k^2}-1}}$

e. $\boxed{f(-3)} = \boxed{(-3)^3+(5\cdot -3)-3} = \boxed{-27-15-3} = \boxed{\mathbf{-45}}$ and $\boxed{g(2)} = \boxed{\frac{1}{2^3}+\frac{1}{2^2}-1} = \boxed{\frac{1}{8}+\frac{1}{4}-1} = \boxed{\mathbf{-0.625}}$

f. $\boxed{f(2n)} = \boxed{(2n)^3+(5\cdot 2n)-3} = \boxed{\boldsymbol{8n^3+10n-3}}$ and $\boxed{g(0)} = \boxed{\frac{1}{0^3}+\frac{1}{0^2}-1} = \boxed{\boldsymbol{\frac{1}{0}+\frac{1}{0}-1}}$ since division by zero is not defined the function $g(x)$ at $x=0$ approaches infinity.

Example 2.3-5: Given $f(x)=x+3$ and $g(x)=x^2+2x+1$, find

a. $f(g(0))$
b. $g(f(0))$
c. $f(g(-1))$
d. $f(g(a))$
e. $g(f(10a))$
f. $g(f(x+1))$
g. $f(g(x^2))$
h. $g(f(2))$

Solutions:

First - Find $f(g(x))$ and $g(f(x))$, i.e.,

$$\boxed{f(g(x))} = \boxed{g(x)+3} = \boxed{\left(x^2+2x+1\right)+3} = \boxed{x^2+2x+4}$$

$\boxed{g(f(x))} = \boxed{[f(x)]^2 + 2\cdot f(x) + 1} = \boxed{(x+3)^2 + 2(x+3) + 1} = \boxed{x^2 + 6x + 9 + 2x + 6 + 1} = \boxed{x^2 + 8x + 16}$

Second - Find $f(g(x))$ or $g(f(x))$ for the specific values given.

a. $\boxed{f(g(0))} = \boxed{0^2 + (2\cdot 0) + 4} = \boxed{0+0+4} = \boxed{\mathbf{4}}$

b. $\boxed{g(f(0))} = \boxed{0^2 + (8\cdot 0) + 16} = \boxed{0+0+16} = \boxed{\mathbf{16}}$

c. $\boxed{f(g(-1))} = \boxed{(-1)^2 + (2\cdot -1) + 4} = \boxed{1-2+4} = \boxed{\mathbf{3}}$

d. $\boxed{f(g(a))} = \boxed{a^2 + (2\cdot a) + 4} = \boxed{\mathbf{a^2 + 2a + 4}}$

e. $\boxed{g(f(10a))} = \boxed{(10a)^2 + (8\cdot 10a) + 16} = \boxed{\mathbf{100a^2 + 80a + 16}}$

f. $\boxed{g(f(x+1))} = \boxed{(x+1)^2 + 8(x+1) + 16} = \boxed{x^2 + 2x + 1 + 8x + 8 + 16} = \boxed{\mathbf{x^2 + 10x + 25}}$

g. $\boxed{f(g(x^2))} = \boxed{(x^2)^2 + (2\cdot x^2) + 4} = \boxed{\mathbf{x^4 + 2x^2 + 4}}$

h. $\boxed{g(f(2))} = \boxed{2^2 + (8\cdot 2) + 16} = \boxed{4+16+16} = \boxed{\mathbf{36}}$

Note that $f(g(x))$ is defined only for values of x where $g(x)$ is defined. Similarly, $g(f(x))$ is defined only for values of x where $f(x)$ is defined. For example, if $f(x) = \frac{1}{x^2}$ and $g(x) = \frac{1}{x}$, then

$$f(g(x)) = \frac{1}{[g(x)]^2} = \frac{1}{\left(\frac{1}{x}\right)^2} = \frac{1}{\frac{1}{x^2}} = \frac{\frac{1}{1}}{\frac{1}{x^2}} = \frac{1\cdot x^2}{1\cdot 1} = \frac{x^2}{1} = \mathbf{x^2} \text{ and}$$

$$g(f(x)) = \frac{1}{f(x)} = \frac{1}{\frac{1}{x^2}} = \frac{\frac{1}{1}}{\frac{1}{x^2}} = \frac{1\cdot x^2}{1\cdot 1} = \frac{x^2}{1} = \mathbf{x^2}.$$

Therefore, we might expect that $f(g(x)) = x^2$ at $x=0$ be equal to zero, i.e., $f(g(0)) = 0^2 = 0$. However, in fact, $f(g(0))$ is undefined because $g(x) = \frac{1}{x}$ at $x=0$ is undefined. Similarly, we might expect $g(f(x))$ at $x=0$ be equal to zero, i.e., $g(f(0)) = 0^2 = 0$. Whereas, in fact, $g(f(0))$ is again undefined because $f(x) = \frac{1}{x^2}$ at $x=0$ is undefined.

Case II Computing Composite Functions Using $f(x)$, $g(x)$, and $h(x)$

The above process of forming a composite function from two functions can be extended to three functions as shown in the following examples.

Example 2.3-6: Given the following $f(x)$, $g(x)$, and $h(x)$ functions, find the composite function $f(g(h(x))) = (f \circ g \circ h)(x) = f \circ g \circ h$

a. $f(x)=2x$, $g(x)=x+1$, and $h(x)=x^2$

b. $f(x)=x+1$, $g(x)=4x$, and $h(x)=x^3$

c. $f(x)=x^3+2$, $g(x)=\frac{1}{x+1}$, and $h(x)=5$

d. $f(x)=x+1$, $g(x)=x^3$, and $h(x)=x^2+x+1$

e. $f(x)=(x+1)^5$, $g(x)=\frac{1}{x-1}$, and $h(x)=\sqrt{x}$

f. $f(x)=\sqrt{x+1}$, $g(x)=x^3+2$, and $h(x)=x-1$

Solutions:

a. Given $f(x)=2x$, $g(x)=x+1$, and $h(x)=x^2$ first substitute $g(x)$ in place of x in $f(x)$ to obtain $f \circ g = f(g(x))$.

$$\boxed{f \circ g} = \boxed{f(g(x))} = \boxed{2g(x)} = \boxed{2(x+1)}$$

Next, substitute $h(x)$ in place of x in $f(g(x))$ to obtain $f \circ g \circ h = f(g(h(x)))$.

$$\boxed{f \circ g \circ h} = \boxed{f(g(h(x)))} = \boxed{2(h(x)+1)} = \boxed{2(x^2+1)} = \boxed{\mathbf{2x^2+2}}$$

b. Given $f(x)=x+1$, $g(x)=4x$, and $h(x)=x^3$ first substitute $g(x)$ in place of x in $f(x)$ to obtain $f \circ g = f(g(x))$.

$$\boxed{f \circ g} = \boxed{f(g(x))} = \boxed{g(x)+1} = \boxed{4x+1}$$

Next, substitute $h(x)$ in place of x in $f(g(x))$ to obtain $f \circ g \circ h = f(g(h(x)))$.

$$\boxed{f \circ g \circ h} = \boxed{f(g(h(x)))} = \boxed{4h(x)+1} = \boxed{\mathbf{4x^3+1}}$$

c. Given $f(x)=x^3+2$, $g(x)=\frac{1}{x+1}$, and $h(x)=5$ first substitute $g(x)$ in place of x in $f(x)$ to obtain $f \circ g = f(g(x))$.

$$\boxed{f \circ g} = \boxed{f(g(x))} = \boxed{g(x)^3+2} = \boxed{\left(\frac{1}{x+1}\right)^3+2} = \boxed{\frac{1}{(x+1)^3}+2}$$

Next, substitute $h(x)$ in place of x in $f(g(x))$ to obtain $f \circ g \circ h = f(g(h(x)))$.

$$\boxed{f \circ g \circ h} = \boxed{f(g(h(x)))} = \boxed{\frac{1}{[h(x)+1]^3}+2} = \boxed{\frac{1}{(5+1)^3}+2} = \boxed{\frac{1}{6^3}+2} = \boxed{\frac{1}{216}+2} = \boxed{\mathbf{2.005}}$$

d. Given $f(x)=x+1$, $g(x)=x^3$, and $h(x)=x^2+x+1$ first substitute $g(x)$ in place of x in $f(x)$ to obtain $f \circ g = f(g(x))$.

$$\boxed{f \circ g} = \boxed{f(g(x))} = \boxed{g(x)+1} = \boxed{x^3+1}$$

Next, substitute $h(x)$ in place of x in $f(g(x))$ to obtain $f \circ g \circ h = f(g(h(x)))$.

$$\boxed{f \circ g \circ h} = \boxed{f(g(h(x)))} = \boxed{[h(x)]^3+1} = \boxed{\mathbf{(x^2+x+1)^3+1}}$$

e. Given $f(x)=(x+1)^5$, $g(x)=\frac{1}{x-1}$, and $h(x)=\sqrt{x}$ first substitute $g(x)$ in place of x in $f(x)$ to obtain $f\circ g = f(g(x))$.

$$\boxed{f\circ g}=\boxed{f(g(x))}=\boxed{[g(x)+1]^5}=\boxed{\left(\frac{1}{x-1}+1\right)^5}$$

Next, substitute $h(x)$ in place of x in $f(g(x))$ to obtain $f\circ g\circ h = f(g(h(x)))$.

$$\boxed{f\circ g\circ h}=\boxed{f(g(h(x)))}=\boxed{\left(\frac{1}{h(x)-1}+1\right)^5}=\boxed{\mathbf{\left(\frac{1}{\sqrt{x}-1}+1\right)^5}}$$

f. Given $f(x)=\sqrt{x+1}$, $g(x)=x^3+2$, and $h(x)=x-1$, first substitute $g(x)$ in place of x in $f(x)$ to obtain $f\circ g = f(g(x))$.

$$\boxed{f\circ g}=\boxed{f(g(x))}=\boxed{\sqrt{g(x)+1}}=\boxed{\sqrt{(x^3+2)+1}}=\boxed{\sqrt{x^3+3}}$$

Next, substitute $h(x)$ in place of x in $f(g(x))$ to obtain $f\circ g\circ h = f(g(h(x)))$.

$$\boxed{f\circ g\circ h}=\boxed{f(g(h(x)))}=\boxed{\sqrt{[h(x)]^3+3}}=\boxed{\mathbf{\sqrt{(x-1)^3+3}}}$$

Example 2.3-7: Given $f(x)=x+5$, $g(x)=x^2+3x+1$, and $h(x)=x$, find

a. $f(g(h(x)))$ b. $g(f(h(x)))$ c. $h(f(g(x)))$

d. $h(g(f(x)))$ e. $g(h(f(x)))$ f. $f(h(g(x)))$

Solutions:

a. Find $\boxed{g(h(x))}=\boxed{[h(x)]^2+3h(x)+1}=\boxed{x^2+3x+1}$ then $\boxed{f(g(h(x)))}=\boxed{g(h(x))+5}=\boxed{(x^2+3x+1)+5}=\boxed{\mathbf{x^2+3x+6}}$

b. Find $\boxed{f(h(x))}=\boxed{h(x)+5}=\boxed{x+5}$ then $\boxed{g(f(h(x)))}=\boxed{[f(h(x))]^2+3f(h(x))+1}=\boxed{(x+5)^2+3(x+5)+1}$

$=\boxed{x^2+10x+25+3x+15+1}=\boxed{\mathbf{x^2+13x+41}}$

c. Find $\boxed{f(g(x))}=\boxed{g(x)+5}=\boxed{(x^2+3x+1)+5}=\boxed{x^2+3x+6}$ then $\boxed{h(f(g(x)))}=\boxed{f(g(x))}=\boxed{\mathbf{x^2+3x+6}}$

d. Find $\boxed{g(f(x))}=\boxed{[f(x)]^2+3f(x)+1}=\boxed{(x+5)^2+3(x+5)+1}=\boxed{x^2+10x+25+3x+15+1}=\boxed{x^2+13x+41}$

then $\boxed{h(g(f(x)))}=\boxed{g(f(x))}=\boxed{\mathbf{x^2+13x+41}}$

e. Find $\boxed{h(f(x))}=\boxed{f(x)}=\boxed{x+5}$ then $\boxed{g(h(f(x)))}=\boxed{[h(f(x))]^2+3h(f(x))+1}=\boxed{(x+5)^2+3(x+5)+1}$

$=\boxed{x^2+10x+25+3x+15+1}=\boxed{\mathbf{x^2+13x+41}}$

f. Find $\boxed{h(g(x))}=\boxed{g(x)}=\boxed{x^2+3x+1}$ then $\boxed{f(h(g(x)))}=\boxed{h(g(x))+5}=\boxed{(x^2+3x+1)+5}=\boxed{\mathbf{x^2+3x+6}}$

Example 2.3-8: Given the functions $f(x)$, $g(x)$, and $h(x)$ in example 2.3-7, find

a. $f(g(h(0)))$ b. $f(g(h(1)))$ c. $g(f(h(-1)))$

d. $g(f(h(0)))$ e. $h(g(f(-a)))$ f. $h(g(f(0)))$

g. $g(h(f(-2)))$ h. $g(h(f(2)))$ i. $f(h(g(1)))$

Solutions:

a. Given $\boxed{f(g(h(x)))} = \boxed{x^2 + 3x + 6}$ then $\boxed{f(g(h(0)))} = \boxed{0^2 + (3 \times 0) + 6} = \boxed{6}$

b. Given $\boxed{f(g(h(x)))} = \boxed{x^2 + 3x + 6}$ then $\boxed{f(g(h(1)))} = \boxed{1^2 + (3 \times 1) + 6} = \boxed{1 + 3 + 6} = \boxed{\mathbf{10}}$

c. Given $\boxed{g(f(h(x)))} = \boxed{x^2 + 13x + 41}$ then $\boxed{g(f(h(-1)))} = \boxed{(-1)^2 + (13 \times -1) + 41} = \boxed{1 - 13 + 41} = \boxed{\mathbf{29}}$

d. Given $\boxed{g(f(h(x)))} = \boxed{x^2 + 13x + 41}$ then $\boxed{g(f(h(0)))} = \boxed{0^2 + (13 \times 0) + 41} = \boxed{\mathbf{41}}$

e. Given $\boxed{h(g(f(x)))} = \boxed{x^2 + 13x + 41}$ then $\boxed{h(g(f(-a)))} = \boxed{(-a)^2 + (13 \times -a) + 41} = \boxed{\mathbf{a^2 - 13a + 41}}$

f. Given $\boxed{h(g(f(x)))} = \boxed{x^2 + 13x + 41}$ then $\boxed{h(g(f(0)))} = \boxed{0^2 + (13 \times 0) + 41} = \boxed{\mathbf{41}}$

g. Given $\boxed{g(h(f(x)))} = \boxed{x^2 + 13x + 41}$ then $\boxed{g(h(f(-2)))} = \boxed{(-2)^2 + (13 \times -2) + 41} = \boxed{4 - 26 + 41} = \boxed{\mathbf{19}}$

h. Given $\boxed{g(h(f(x)))} = \boxed{x^2 + 13x + 41}$ then $\boxed{g(h(f(2)))} = \boxed{2^2 + (13 \times 2) + 41} = \boxed{4 + 26 + 41} = \boxed{\mathbf{71}}$

i. Given $\boxed{f(h(g(x)))} = \boxed{x^2 + 3x + 6}$ then $\boxed{f(h(g(1)))} = \boxed{1^2 + (3 \times 1) + 6} = \boxed{1 + 3 + 6} = \boxed{\mathbf{10}}$

In order to solve some of the problems presented in calculus students are asked to separate composite functions into two other functions, i.e., $f(x)$ and $g(x)$ functions. The quickest way to solve these types of problems is by:

First - identifying the basic math operations (such as addition, subtraction, division, power, and radical) that are used to form the composite function and

Second - Choosing the correct $f(x)$ and $g(x)$ functions that satisfy the identified operations.

For example, the function $C(x) = f(g(x)) = (x^2 + 1)^3$ consists of two operations, i.e., the addition and the power operations. Therefore, let $f(x)$ be equal to $f(x) = x^3$ and $g(x)$ be equal to $g(x) = (x^2 + 1)$, then $C(x) = f(g(x)) = [g(x)]^3 = (x^2 + 1)^3$. The following examples further illustrate how composite functions can be separated into two functions.

Example 2.3-9: Given the composite functions below, find the $f(x)$ and $g(x)$ functions such that $f \circ g(x) = f(g(x)) = C(x)$.

a. $C(x) = 1 + x^2$ b. $C(x) = \frac{1}{x} + 10$ c. $C(x) = \sqrt{x^3 + 1}$

d. $C(x) = \frac{1}{(x+1)^5}$ e. $C(x) = \frac{a}{|x|+a}$ f. $C(x) = x+\sqrt{x}$

g. $C(x) = x^2+5$ h. $C(x) = \frac{1}{(x-1)^3}$ i. $C(x) = \frac{1}{\sqrt{x+2}-1}$

Solutions:

a. The composite function $C(x)=1+x^2$ consist of addition and power operations. Therefore, let

$\boxed{f(x)=1+x}$ and $\boxed{g(x)=x^2}$ then $\boxed{f(g(x))}=\boxed{1+g(x)}=\boxed{\mathbf{1+x^2}}$

b. The composite function $C(x)=\frac{1}{x}+10$ consist of addition and division operations. Therefore, let

$\boxed{f(x)=x+10}$ and $\boxed{g(x)=\frac{1}{x}}$ then $\boxed{f(g(x))}=\boxed{g(x)+10}=\boxed{\mathbf{\frac{1}{x}+10}}$

c. The composite function $C(x)=\sqrt{x^3+1}$ consist of addition, power, and radical operations.

Therefore, let $\boxed{f(x)=\sqrt{x}}$ and $\boxed{g(x)=x^3+1}$ then $\boxed{f(g(x))}=\boxed{\sqrt{g(x)}}=\boxed{\mathbf{\sqrt{x^3+1}}}$

d. The composite function $C(x)=\frac{1}{(x+1)^5}$ consist of addition, division, and power operations.

Therefore, let $\boxed{f(x)=\frac{1}{x^5}}$ and $\boxed{g(x)=x+1}$ then $\boxed{f(g(x))}=\boxed{\frac{1}{[g(x)]^5}}=\boxed{\mathbf{\frac{1}{(x+1)^5}}}$

e. The composite function $C(x)=\frac{a}{|x|+a}$ consist of addition and division operations. Therefore,

let $\boxed{f(x)=\frac{a}{x}}$ and $\boxed{g(x)=|x|+a}$ then $\boxed{f(g(x))}=\boxed{\frac{a}{g(x)}}=\boxed{\mathbf{\frac{a}{|x|+a}}}$

f. The composite function $C(x)=x+\sqrt{x}$ consist of addition and radical operations. Therefore, let

$\boxed{f(x)=x^2+x}$ and $\boxed{g(x)=\sqrt{x}}$ then $\boxed{f(g(x))}=\boxed{[g(x)]^2+g(x)}=\boxed{(\sqrt{x})^2+\sqrt{x}}=\boxed{\mathbf{x+\sqrt{x}}}$

g. The composite function $C(x)=x^2+5$ consist of addition and power operations. Therefore, let

$\boxed{f(x)=x+5}$ and $\boxed{g(x)=x^2}$ then $\boxed{f(g(x))}=\boxed{g(x)+5}=\boxed{\mathbf{x^2+5}}$

h. The composite function $C(x)=(x-1)^3$ consist of subtraction and power operations. Therefore,

let $\boxed{f(x)=x^3}$ and $\boxed{g(x)=x-1}$ then $\boxed{f(g(x))}=\boxed{[g(x)]^3}=\boxed{\mathbf{(x-1)^3}}$

i. The composite function $C(x)=\frac{1}{\sqrt{x+2}-1}$ consist of addition, subtraction, and power operations.

Therefore, let $\boxed{f(x)=\frac{1}{\sqrt{x}-1}}$ and $\boxed{g(x)=x+2}$ then $\boxed{f(g(x))}=\boxed{\frac{1}{\sqrt{g(x)}-1}}=\boxed{\mathbf{\frac{1}{\sqrt{x+2}-1}}}$

Note that composite functions, in some cases, can be computed by writing different forms of

$f(x)$ or $g(x)$ functions. For example, the above composite function $f(g(x)) = \frac{1}{\sqrt{x+2}-1}$ can also be obtained by selecting the following $f(x)$ and $g(x)$ functions:

1. Let, $f(x)=\frac{1}{x-1}$ and $g(x)=\sqrt{x+2}$, then $f(g(x)) = \frac{1}{g(x)-1} = \frac{\mathbf{1}}{\sqrt{\mathbf{x+2}}-\mathbf{1}}$, or

2. Let, $f(x)=\frac{1}{x}$ and $g(x)=\sqrt{x+2}-1$, then $f(g(x)) = \frac{1}{g(x)} = \frac{\mathbf{1}}{\sqrt{\mathbf{x+2}}-\mathbf{1}}$

Having learned about the composite functions, in the next sections we will learn how to identify one-to-one functions and compute the inverse of a function.

Section 2.3 Practice Problems – Composite Functions of Real Variables

1. Find the composite function $f(g(x)) = (f \circ g)(x)$ for the following $f(x)$ and $g(x)$ functions.

   a. $f(x)=2x-1$; $g(x)=-x^2$ b. $f(x)=2x+5$; $g(x)=x+10$ c. $f(x)=\frac{1}{x+1}$; $g(x)=x^3$

   d. $f(x)=x-3$; $g(x)=-x^2$ e. $f(x)=x^2+1$; $g(x)=\sqrt{x}$ f. $f(x)=\sqrt{x}+10x$; $g(x)=x-3$

2. Find the composite function $g(f(x)) = (g \circ f)(x)$ for the following $g(x)$ and $f(x)$ functions.

   a. $g(x)=x^2$; $f(x)=-\frac{1}{x}$ b. $g(x)=x^2-1$; $f(x)=3x$ c. $g(x)=x+2$; $f(x)=\sqrt{x+5}$

   d. $g(x)=2x+1$; $f(x)=\frac{1}{x^2}$ e. $g(x)=\frac{1}{x+1}$; $f(x)=-\sqrt{x}$ f. $g(x)=\frac{1}{2x-3}+x$; $f(x)=3x$

3. Given $f(x)=-x^2+5$ and $g(x)=x^3+2x-1$, find

   a. $f(0)$ and $g(1)$ b. $f(1)$ and $g(2)$ c. $f(k)$ and $g(-k)$

   d. $f(3)$ and $g(-3)$ e. $f(-n)$ and $g(2n)$ f. $f(n+1)$ and $g(0)$

4. Given $f(x)=x-2$ and $g(x)=x^2-1$, find

   a. $f(g(-1))$ b. $g(f(0))$ c. $f(g(-3))$

   d. $g(f(2))$ e. $f(g(n-1))$ f. $g(f(2n))$

5. Given the functions $f(g(h(x)))=x^3-2x^2+1$, $g(f(h(x)))=x-1$, and $h(g(f(x)))=-x^2+2$, find

   a. $f(g(h(2)))$ b. $f(g(h(-2)))$ c. $h(g(f(a-1)))$

   d. $g(f(h(-1)))$ e. $g(f(h(2n+1)))$ f. $h(g(f(0)))$

2.4 One-to-One and Inverse Functions of Real Variables

In this section we will learn how to identify one-to-one functions. We will then discuss two different methods for finding the inverse of functions.

2.4.1 One-to-One Functions

A function is a one-to-one function if and only if for each value of x there is only one corresponding value of y. In other words, a function having two ordered pairs with different first coordinates and the same second coordinates is not a one-to-one function. To find out if a function is a one-to-one function we consider two cases. In the first case, the function is represented by a set of ordered pairs. In the second case, the function is represented in an equation form. Let's find out how a one-to-one function for each of the stated cases is found.

Case I The Function is Represented by a Set of Ordered Pairs

To state if a function is a one-to-one function, for cases where the function is given by a set of ordered pairs, we simply check to see if for each value of x there is only one corresponding value of y. For example, the function $f(x,y)=[(2,1),(4,3),(5,2),(6,8)]$ is a one-to-one function. However, the function $f(x,y)=[(-1,5),(3,7),(1,5)]$ is not a one-to-one function because of the ordered pairs $(-1,5)$ and $(1,5)$, i.e., for $x=\pm 1$ there is one value of $y=5$.

Example 2.4-1: State which of the following ordered paired functions are one-to-one:

a. $f=[(1,4),(2,6),(7,9)]$

b. $f=[(2,4),(2,6),(3,7),(6,9)]$

c. $f=[(-3,1),(-2,0),(-1,3),(3,6)]$

d. $f=[(2,5),(4,7),(8,10),(9,15)]$

Solutions:

a. The ordered pair function $f=[(1,4),(2,6),(7,9)]$ **is a one-to-one function**. Because each x value correspond to only one y value.

b. The ordered pair function $f=[(2,4),(2,6),(3,7),(6,9)]$ **is not a one-to-one function**. Because each x value does not correspond with a different y value, i.e., when $x=2$ the variable y is equal to 4 and 6.

c. The ordered pair function $f=[(-3,1),(-2,0),(-1,3),(3,6)]$ **is a one-to-one function**. Because each x value correspond to only one y value.

d. The ordered pair function $f=[(2,5),(4,7),(8,10),(9,15)]$ **is a one-to-one function**. Because each x value correspond to only one y value.

Case II The Function is Represented by an Equation

To state if a function is a one-to-one function, for cases where the function is given in an equation form, we can use one of the following two methods.

First Method:

Substitute different x values into the given equation and solve for the y value. If each x value result in having only one y value, then the function is one-to-one. For example, the equations

$f(x)=5x-8$, $f(x)=x+1$, $f(x)=\sqrt{x}+2$, $f(x)=2x^{\frac{1}{5}}+5$, $f(x)=\frac{x+2}{x-3}$, $f(x)=x^3$, and $f(x)=(x-7)^{\frac{1}{3}}+4$ are one-to-one functions. This is because for every value of x there is only one value of y. On the other hand, functions such as $f(x)=x^2+1$, $f(x)=x^4$, $f(x)=\frac{1}{1+|x|}$, $f(x)=1-x^2$, $f(x)=(x-3)^2$, $f(x)=(x-3)(x+6)$, $f(x)=x^2$, $f(x)=\left(1-2x^2\right)^5$, and $f(x)=x^2+5x+6$ are not one-to-one functions. This is because, as an example, at $x=\pm 1$ or at $x=\pm 2$ the function $f(x)=x^2+1$ is equal to $f(x)=(\pm 1)^2+1=1+1 \;=2$ and $f(x)=(\pm 2)^2+1=1+4=5$. Thus, the ordered pairs are equal to $\{(1,2),(-1,2),(2,5),(-2,5)\}$ which indicate that the function $f(x)=x^2+1$ is not a one-to-one function. The following examples further illustrate this point.

Example 2.4-2: State which of the following functions are one-to-one.

a. $f(x)=3x+1$ b. $f(x)=x^2$ c. $f(x)=10+x^2$

d. $f(x)=x^3$ e. $f(x)=|x+1|$ f. $f(x)=\frac{1}{x^2+1}$

g. $f(x)=|x-2|+5$ h. $f(x)=2x^3-5$ i. $f(x)=1+e^x$

Solutions:

a. Given the function $f(x)=y=3x+1$, let's find the y value by substituting few x values into the equation $y=3x+1$, i.e.,

at $x=0$ $y = (3\cdot 0)+1 = 0+1 = 1$ at $x=1$ $y = (3\cdot 1)+1 = 3+1 = 4$

at $x=-1$ $y = (3\cdot -1)+1 = -3+1 = -2$ at $x=3$ $y = (3\cdot 3)+1 = 9+1 = 10$

at $x=5$ $y = (3\cdot 5)+1 = 15+1 = 16$ at $x=7$ $y = (3\cdot 7)+1 = 21+1 = 22$

Therefore, the ordered pairs are equal to $\{\mathbf{(0,1),(1,4),(-1,-2),(3,10),(5,16),(7,22)}\}$. Since for every x value there is only one corresponding y value the function $f(x)=y=3x+1$ **is a one-to-one function**.

b. Given the function $f(x)=y=x^2$, let's find the y value by substituting few x values into the equation $y=x^2$, i.e.,

at $x=0$ $y = 0^2 = 0$ at $x=1$ $y = 1^2 = 1$

at $x=-1$ $y = (-1)^2 = 1$ at $x=4$ $y = 4^2 = 16$

at $x=-4$ $y = (-4)^2 = 16$ at $x=7$ $y = 7^2 = 49$

Therefore, the ordered pairs are equal to $\{\mathbf{(0,0),(1,1),(-1,1),(4,16),(-4,16),(7,49)}\}$. Since each x value does not correspond to only one y value the function $f(x)=y=x^2$ **is not a one-to-one function**.

c. Given the function $f(x)=y=10+x^2$, let's find the y value by substituting few x values into the equation $y=10+x^2$, i.e.,

at $x=0$ $y = 10+0^2 = 10+0 = 10$ at $x=1$ $y = 10+1^2 = 10+1 = 11$

at $x=-1$ $y = 10+(-1)^2 = 10+1 = 11$ at $x=5$ $y = 10+5^2 = 10+25 = 35$

at $x=-5$ $y = 10+(-5)^2 = 10+25 = 35$ at $x=8$ $y = 10+8^2 = 10+64 = 74$

Therefore, the ordered pairs are equal to $\{(0,10),(1,11),(-1,11),(5,35),(-5,35),(8,74)\}$. Since each x value does not correspond to only one y value the function $f(x)=y=10+x^2$ **is not a one-to-one function.**

d. Given the function $f(x)=y=x^3$, let's find the y value by substituting few x values into the equation $y=x^3$, i.e.,

at $x=0$ $y = 0^3 = 0$ at $x=1$ $y = 1^3 = 1$

at $x=-1$ $y = (-1)^3 = -1$ at $x=3$ $y = 3^3 = 27$

at $x=-3$ $y = (-3)^3 = -27$ at $x=4$ $y = 4^3 = 64$

Therefore, the ordered pairs are equal to $\{(0,0),(1,1),(-1,-1),(3,27),(-3,-27),(4,64)\}$. Since for every x value there is only one corresponding y value the function $f(x)=y=x^3$ **is a one-to-one function.**

e. Given the function $f(x)=y=|x+1|$, let's find the y value by substituting few x values into the given equation, i.e.,

at $x=0$ $y = |0+1| = |1| = 1$ at $x=1$ $y = |1+1| = |2| = 2$

at $x=-1$ $y = |-1+1| = |0| = 0$ at $x=2$ $y = |2+1| = |3| = 3$

at $x=-2$ $y = |-2+1| = |-1| = |1| = 1$ at $x=-3$ $y = |-3+1| = |-2| = |2| = 2$

Therefore, the ordered pairs are equal to $\{(0,1),(1,2),(-1,0),(2,3),(-2,1),(-3,2)\}$. Since each x value does not correspond to only one y value the function $f(x)=y=|x+1|$ **is not a one-to-one function.**

f. Given the function $f(x)=y=\frac{1}{x^2+1}$, let's find the y value by substituting few x values into the given equation, i.e.,

at $x=0$ $y = \frac{1}{0^2+1} = \frac{1}{0+1} = \frac{1}{1} = 1$ at $x=1$ $y = \frac{1}{1^2+1} = \frac{1}{1+1} = \frac{1}{2} = 0.5$

at $x=-1$ $y = \frac{1}{(-1)^2+1} = \frac{1}{1+1} = \frac{1}{2} = 0.5$ at $x=5$ $y = \frac{1}{5^2+1} = \frac{1}{25+1} = \frac{1}{26} = 0.04$

at $x=-5$ $y = \frac{1}{(-5)^2+1} = \frac{1}{25+1} = \frac{1}{26} = 0.04$ at $x=10$ $y = \frac{1}{10^2+1} = \frac{1}{100+1} = \frac{1}{101} = 0.009$

Therefore, the ordered pairs are equal to $\{(0,1),(1,0.5),(-1,0.5),(5,0.04),(-5,0.04),(10,0.009)\}$. Since each x value does not correspond to only one y value the function $f(x)=y=\frac{1}{x^2+1}$ **is not a one-to-one function.**

g. Given the function $f(x)=|x-2|+5$, let's find the y value by substituting few x values into the equation $y=|x-2|+5$, i.e.,

at $x=0$ $y = |0-2|+5 = |-2|+5 = |2|+5 = 7$ at $x=1$ $y = |1-2|+5 = |-1|+5 = |1|+5 = 6$

at $x=3$ $y = |3-2|+5 = |1|+5 = 6$ at $x=5$ $y = |5-2|+5 = |3|+5 = 8$

Therefore, the ordered pairs are equal to $\{(0,7),(1,6),(3,6),(5,8)\}$. Since each x value does not corresponde to only one y value the function $f(x)=y=|x-2|+5$ **is not a one-to-one function.**

h. Given the function $f(x)=y=2x^3-5$, let's find the y value by substituting few x values into the equation $y=2x^3-5$, i.e.,

at $x=0$ $y = 2\cdot 0^3-5 = 0-5 = -5$ at $x=1$ $y = 2\cdot 1^3-5 = 2-5 = -3$

at $x=-1$ $y = 2\cdot(-1)^3-5 = -2-5 = -7$ at $x=3$ $y = 2\cdot 3^3-5 = 54-5 = 49$

at $x=-3$ $y = 2\cdot(-3)^3-5 = -54-5 = -59$ at $x=4$ $y = 2\cdot 4^3-5 = 128-5 = 123$

Therefore, the ordered pairs are equal to $\{(\mathbf{0},-\mathbf{5}),(\mathbf{1},-\mathbf{3}),(-\mathbf{1},-\mathbf{7}),(\mathbf{3},\mathbf{49}),(-\mathbf{3},-\mathbf{59}),(\mathbf{4},\mathbf{123})\}$. Since for every x value there is only one corresponding y value the function $f(x)=y=2x^3-5$ **is a one-to-one function.**

i. Given the function $f(x)=y=1+e^x$, let's find the y value by substituting few x values into the equation $y=3x+1$, i.e.,

at $x=0$ $y = 1+e^0 = 1+1 = 2$ at $x=1$ $y = 1+e^1 = 1+2.718 = 3.718$

at $x=-1$ $y = 1+e^{-1} = 1+\frac{1}{e} = 1+\frac{1}{2.718} = 1.37$ at $x=2$ $y = 1+e^2 = 1+7.389 = 8.389$

at $x=-2$ $y = 1+e^{-2} = 1+\frac{1}{e^2} = 1+\frac{1}{7.389} = 1.14$ at $x=3$ $y = 1+e^3 = 1+20.08 = 21.08$

Therefore, the ordered pairs are equal to $\{(\mathbf{0},\mathbf{2}),(\mathbf{1},\mathbf{3.718}),(-\mathbf{1},\mathbf{1.37}),(\mathbf{2},\mathbf{8.389}),(-\mathbf{2},\mathbf{1.14}),(\mathbf{3},\mathbf{21.08})\}$. Since for every x value there is only one corresponding y value the function $f(x)=y=1+e^x$ **is a one-to-one function.**

Second Method:

The second way of determining whether a function is one-to-one is by first graphing the function and then drawing a horizontal line through the graph. If the graph of the functions crosses the horizontal line only once, then the function is a one-to-one function. However, if the graph of the function crosses the horizontal line two or more times, then the function is not a one-to-one function. For example, $x^2+y^2=1$; $y=1-x^2$; $y=\frac{1}{4}x^2$; $y=(x-3)^2$; $y=|x-1|+3$; $y=|x|$ are not one-to-one functions because a horizontal line crosses the graph of each function more than once. In summary, a function is a one-to-one function if and only if:

a. For each value of x there is only one corresponding value of y, i.e., no two values of x result in the same value of y, or

b. The graph of the function intersects a horizontal line only once.

2.4.2 Inverse Functions

To find the inverse of a function we consider two cases. In the first case, the function is represented by a set of ordered pairs. In the second case, the function is represented in an equation form. Let's find out how the inverse of a function for each of the stated cases are found.

Case I The Function is Represented by a Set of Ordered Pairs

To find the inverse of a function, for cases where the function is given by a set of ordered pairs, we simply interchange the coordinates in the ordered pairs. For example, the inverse of the function $f=[(1,5),(2,6),(3,7)]$ is obtained by interchanging the x and y coordinates of each ordered pair of f, i.e., $f^{-1} = [(5,1),(6,2),(7,3)]$. Observe that interchanging the coordinates in the ordered pairs result in having the domain of f^{-1} (read as "f inverse of x") to be the same as the range of f, and the range of f^{-1} to be the same as the domain of f. Also note that invertability of a function is dependent upon the ordered pairs of the function. The ordered pairs of a function is invertible if and only if the function is a one-to-one function. In section 2.1 we stated that *a function is defined as a relation in which each element of the domain is paired with only one element of the range.* Therefore, if a function $f(x)$ has ordered pairs with different first coordinates and the same second coordinate, then that function is not a one-to-one function and is not invertible.

Example 2.4-3: Find the inverse of the following ordered paired functions:

a. $f = [(1,3),(2,5),(8,6)]$
b. $f = [(1,5),(1,8),(3,8),(6,10)]$
c. $f = [(-2,1),(-1,0),(1,3),(2,5)]$
d. $f = [(3,10),(4,12),(8,16),(10,25)]$

Solutions:

a. By interchanging the x and y values we obtain $f^{-1} = [(3,1),(5,2),(6,8)]$

b. The ordered pair function $f = [(1,5),(1,8),(3,8),(6,10)]$ does not have an inverse. Because each x value does not correspond with a different y value, i.e., when $x = 1$ the variable y is equal to 5 and 8.

c. By interchanging the x and y values we obtain $f^{-1} = [(1,-2),(0,-1),(3,1),(5,2)]$

d. By interchanging the x and y values we obtain $f^{-1} = [(10,3),(12,4),(16,8),(25,10)]$

Case II The Function is Represented by an Equation

To find the inverse of a function, for cases where the function is given in an equation form, we can use one of the following two methods.

First Method:

In the first method for a given function $f(x)$ the inverse of $f(x)$ is found by simply interchanging the x and y variables as shown in the following steps:

Step 1: Check to see if the function is a one-to-one function.

Step 2: Replace $f(x)$ by y and interchange the x variable with y and the y variable with x.

Step 3: Solve the equation for y. Replace y by $f^{-1}(x)$.

Second Method:

In the second method since the inverse of $f(x)$, denoted by $f^{-1}(x)$, is the unique function that satisfies the equation

$$f\left[f^{-1}(x)\right] = x \qquad \text{for all } x \qquad (1)$$

we can use this equation to find the inverse of the function. The following show the steps for finding the inverse of a function such as $f(x) = x^2 + 1$, where $x \geq 0$:

Step 1: Check to see if the function $f(x) = x^2 + 1$ is a one-to-one function.

Step 2: Since the inverse of any function satisfies $f\left[f^{-1}(x)\right]=x$, use this equation - labeled as equation (1) - and replace x with $f^{-1}(x)$ wherever x is present in the function $f(x)$, i.e., since $f(x)=x^2+1$, then $f\left[f^{-1}(x)\right]=\left[f^{-1}(x)\right]^2+1$. Label this equation as equation No. (2).

Step 3: Equate the right hand side of the equations (1) and (2) with one another and solve for $f^{-1}(x)$, i.e., since $f\left[f^{-1}(x)\right]=x$ and $f\left[f^{-1}(x)\right]=\left[f^{-1}(x)\right]^2+1$, then $x=\left[f^{-1}(x)\right]^2+1$; $x-1=\left[f^{-1}(x)\right]^2$; $\sqrt{x-1}=\sqrt{\left[f^{-1}(x)\right]^2}$; $\sqrt{x-1}=f^{-1}(x)$ or $f^{-1}(x)=\sqrt{x-1}$.

Note: The symbol f^{-1}, used for denoting the inverse of a function, does not represent a negative exponent and should not be confused as an exponent. For example, the functions $e^{-2x}=\frac{1}{e^{2x}}$, $x^{-3}=\frac{1}{x^3}$, or $(x+2)^{-1}=\frac{1}{x+2}$ where as $f^{-1}\neq\frac{1}{f}$.

The following examples show the steps for finding the inverse of a function.

Example 2.4-4: Use two methods to find the inverse of the following functions:

a. $f(x)=2x+1$	b. $f(x)=x^2-2$	c. $f(x)=-x-8$
d. $f(x)=1+x^3$	e. $f(x)=\sqrt{2x-1}$, $x\geq\frac{1}{2}$	f. $f(x)=\sqrt[5]{x}-5$
g. $f(x)=3x^5-5$	h. $f(x)=\frac{1}{3}x-5$	i. $f(x)=x+100$
j. $f(x)=0.4x$	k. $f(x)=\frac{1}{x}+3$, $x\neq 0$	l. $f(x)=-x$
m. $f(x)=\sqrt[5]{x-7}+3$, $x\geq 7$	n. $f(x)=\frac{2x-5}{x-1}$, $x\neq 1$	o. $f(x)=7-3x$

Solutions:

a. The function $f(x)=2x+1$ is a one-to-one function because for each value of x there is only one corresponding value of y. For example, at $x=0$; $y=1$ or at $x=2$; $y=5$, etc.

First Method: Interchange the x variable with y and the y variable with x in the equation $f(x)=y=2x+1$ and solve for y. Next, replace y with $f^{-1}(x)$ to obtain:

$\boxed{x=2y+1}$; $\boxed{x-1=2y}$; $\boxed{\frac{x-1}{2}=\frac{2y}{2}}$; $\boxed{\frac{x-1}{2}=y}$; $\boxed{y=\frac{x-1}{2}}$; $\boxed{\mathbf{f^{-1}(x)=\frac{x-1}{2}}}$

Second Method: Since $f\left[f^{-1}(x)\right]=x$ replace x with $f^{-1}(x)$ wherever x is present in the function $f(x)=2x+1$, i.e., $f\left[f^{-1}(x)\right]=2\cdot f^{-1}(x)+1$.

Next, solve for $f^{-1}(x)$ by equating the right hand side of the equations $f\left[f^{-1}(x)\right]=x$ and $\boxed{f\left[f^{-1}(x)\right]=2\cdot f^{-1}(x)+1}$, i.e., $\boxed{x=2\cdot f^{-1}(x)+1}$; $\boxed{x-1=2f^{-1}(x)}$; $\boxed{\frac{x-1}{2}=\frac{2f^{-1}(x)}{2}}$; $\boxed{\mathbf{f^{-1}(x)=\frac{x-1}{2}}}$

b. The function $f(x)=x^2-2$ is not a one-to-one function. This is because each x value does not correspond to only one y value, i.e., at $x=\pm1$; $y=-1$ or at $x=\pm3$; $y=7$, etc. Therefore, the function $f(x)=x^2-2$ **is not invertible.**

c. The function $f(x)=-x-8$ is a one-to-one function because for each value of x there is only one corresponding value of y. For example, at $x=1$; $y=-9$ or at $x=2$; $y=-10$, etc.

First Method: Interchange the x variable with y and the y variable with x in the equation $f(x)=y=-x-8$ and solve for y. Next, replace y with $f^{-1}(x)$ to obtain:

$\boxed{x=-y-8}$; $\boxed{x+8=-y}$; $\boxed{-y=x+8}$; $\boxed{y=-x-8}$; $\boxed{\mathbf{f^{-1}(x)=-x-8}}$

Second Method: Since $f\left[f^{-1}(x)\right]=x$ replace x with $f^{-1}(x)$ wherever x is present in the function $f(x)=-x-8$, i.e., $f\left[f^{-1}(x)\right]=-f^{-1}(x)-8$.

Next, solve for $f^{-1}(x)$ by equating the right hand side of the equations $f\left[f^{-1}(x)\right]=x$ and $\boxed{f\left[f^{-1}(x)\right]=-f^{-1}(x)-8}$, i.e., $\boxed{x=-f^{-1}(x)-8}$; $\boxed{x+8=-f^{-1}(x)}$; $\boxed{\mathbf{f^{-1}(x)=-x-8}}$

d. The function $f(x)=1+x^3$ is a one-to-one function because for each value of x there is only one corresponding value of y. For example, at $x=1$; $y=2$ or at $x=-1$; $y=0$, etc.

First Method: Interchange the x variable with y and the y variable with x in the equation $f(x)=y=1+x^3$ and solve for y. Next, replace y with $f^{-1}(x)$ to obtain:

$\boxed{x=1+y^3}$; $\boxed{x-1=y^3}$; $\boxed{(x-1)^{\frac{1}{3}}=\left(y^3\right)^{\frac{1}{3}}}$; $\boxed{(x-1)^{\frac{1}{3}}=y}$; $\boxed{y=\sqrt[3]{x-1}}$; $\boxed{\mathbf{f^{-1}(x)=\sqrt[3]{x-1}}}$

Second Method: Since $f\left[f^{-1}(x)\right]=x$ replace x with $f^{-1}(x)$ wherever x is present in the function $f(x)=1+x^3$, i.e., $f\left[f^{-1}(x)\right]=1+\left[f^{-1}(x)\right]^3$.

Next, solve for $f^{-1}(x)$ by equating the right hand side of the equations $f\left[f^{-1}(x)\right]=x$ and $\boxed{f\left[f^{-1}(x)\right]=1+\left[f^{-1}(x)\right]^3}$, i.e., $\boxed{x=1+\left[f^{-1}(x)\right]^3}$; $\boxed{x-1=\left[f^{-1}(x)\right]^3}$; $\boxed{(x-1)^{\frac{1}{3}}=\left[f^{-1}(x)\right]^{3\times\frac{1}{3}}}$; $\boxed{(x-1)^{\frac{1}{3}}=f^{-1}(x)}$

$\boxed{f^{-1}(x)=(x-1)^{\frac{1}{3}}}$; $\boxed{\mathbf{f^{-1}(x)=\sqrt[3]{(x-1)}}}$

e. The function $f(x)=\sqrt{2x-1}$ is a one-to-one function because for each value of $x\geq\frac{1}{2}$ there is only one corresponding value of y, i.e., at $x=\frac{1}{2}$; $y=0$ or at $x=2$; $y=\sqrt{3}$, etc.

First Method: Interchange the x variable with y and the y variable with x in the equation $f(x)=y=\sqrt{2x-1}$ and solve for y. Next, replace y with $f^{-1}(x)$ to obtain:

$\boxed{x=\sqrt{2y-1}}$; $\boxed{x=(2y-1)^{\frac{1}{2}}}$; $\boxed{x^2=(2y-1)^{\frac{1}{2}\times2}}$; $\boxed{x^2=2y-1}$; $\boxed{x^2+1=2y}$; $\boxed{\frac{x^2+1}{2}=\frac{2y}{2}}$; $\boxed{\frac{x^2+1}{2}=y}$

; $\boxed{y=\frac{x^2+1}{2}}$ = $\boxed{\mathbf{f^{-1}(x)=\frac{x^2+1}{2}}}$

Second Method: Since $f\left[f^{-1}(x)\right]=x$ replace x with $f^{-1}(x)$ wherever x is present in the function $f(x)=2x+1$, i.e., $f\left[f^{-1}(x)\right]=\sqrt{2f^{-1}(x)-1}$.

Next, solve for $f^{-1}(x)$ by equating the right hand side of the equations $f\left[f^{-1}(x)\right]=x$ and $\boxed{f\left[f^{-1}(x)\right]=\sqrt{2f^{-1}(x)-1}}$, i.e., $\boxed{x=\sqrt{2f^{-1}(x)-1}}$; $\boxed{x=\left(2f^{-1}(x)-1\right)^{\frac{1}{2}}}$; $\boxed{x^2=\left(2f^{-1}(x)-1\right)^{\frac{1}{2}\times 2}}$

; $\boxed{x^2=2f^{-1}(x)-1}$; $\boxed{x^2+1=2f^{-1}(x)}$; $\boxed{\frac{x^2+1}{2}=\frac{2f^{-1}(x)}{2}}$; $\boxed{\frac{x^2+1}{2}=f^{-1}(x)}$; $\boxed{\mathbf{f^{-1}(x)=\frac{x^2+1}{2}}}$

f. The function $f(x)=\sqrt[5]{x}-5$ is a one-to-one function because for each value of $x\geq 0$ there is only one corresponding value of y, i.e., at $x=0$; $y=-5$ or at $x=2^5$; $y=-3$, etc.

First Method: Interchange the x variable with y and the y variable with x in the equation $f(x)=y=\sqrt[5]{x}-5$ and solve for y. Next, replace y with $f^{-1}(x)$ to obtain:

$\boxed{x=\sqrt[5]{y}-5}$; $\boxed{x+5=\sqrt[5]{y}}$; $\boxed{x+5=y^{\frac{1}{5}}}$; $\boxed{(x+5)^5=y^{\frac{1}{5}\times 5}}$; $\boxed{(x+5)^5=y}$; $\boxed{y=(x+5)^5}$; $\boxed{\mathbf{f^{-1}(x)=(x+5)^5}}$

Second Method: Since $f\left[f^{-1}(x)\right]=x$ replace x with $f^{-1}(x)$ wherever x is present in the function $f(x)=\sqrt[5]{x}-5$, i.e., $f\left[f^{-1}(x)\right]=\sqrt[5]{f^{-1}(x)}-5$.

Next, solve for $f^{-1}(x)$ by equating the right hand side of the equations $f\left[f^{-1}(x)\right]=x$ and $\boxed{f\left[f^{-1}(x)\right]=\sqrt[5]{f^{-1}(x)}-5}$, i.e., $\boxed{x=\sqrt[5]{f^{-1}(x)}-5}$; $\boxed{x+5=\sqrt[5]{f^{-1}(x)}}$; $\boxed{x+5=\left[f^{-1}(x)\right]^{\frac{1}{5}}}$; $\boxed{(x+5)^5=\left[f^{-1}(x)\right]^{\frac{1}{5}\times 5}}$

; $\boxed{(x+5)^5=f^{-1}(x)}$; $\boxed{\mathbf{f^{-1}(x)=(x+5)^5}}$

g. The function $f(x)=3x^5-5$ is a one-to-one function because for each value of x there is only one corresponding value of y. For example, at $x=0$; $y=-5$ or at $x=1$; $y=-2$, etc.

First Method: Interchange the x variable with y and the y variable with x in the equation $f(x)=y=3x^5-5$ and solve for y. Next, replace y with $f^{-1}(x)$ to obtain:

$\boxed{x=3y^5-5}$; $\boxed{x+5=3y^5}$; $\boxed{\frac{x+5}{3}=\frac{3y^5}{3}}$; $\boxed{\frac{x+5}{3}=y^5}$; $\boxed{y^{5\times\frac{1}{5}}=\left(\frac{x+5}{3}\right)^{\frac{1}{5}}}$; $\boxed{y=\sqrt[5]{\frac{x+5}{3}}}$; $\boxed{\mathbf{f^{-1}(x)=\sqrt[5]{\frac{x+5}{3}}}}$

Second Method: Since $f\left[f^{-1}(x)\right]=x$ replace x with $f^{-1}(x)$ wherever x is present in the function $f(x)=3x^5-5$, i.e., $f\left[f^{-1}(x)\right]=3\cdot\left[f^{-1}(x)\right]^5-5$.

Next, solve for $f^{-1}(x)$ by equating the right hand side of the equations $f\left[f^{-1}(x)\right]=x$ and

$\boxed{f\left[f^{-1}(x)\right]=3\cdot\left[f^{-1}(x)\right]^5-5}$, i.e., $\boxed{x=3\cdot\left[f^{-1}(x)\right]^5-5}$; $\boxed{x+5=3\cdot\left[f^{-1}(x)\right]^5}$; $\boxed{\frac{x+5}{3}=\frac{3\cdot\left[f^{-1}(x)\right]^5}{3}}$; $\boxed{\frac{x+5}{3}}$

$=\boxed{\left[f^{-1}(x)\right]^5}$; $\boxed{\left(\frac{x+5}{3}\right)^{\frac{1}{5}}=\left[f^{-1}(x)\right]^{5\times\frac{1}{5}}}$; $\boxed{\left(\frac{x+5}{3}\right)^{\frac{1}{5}}=f^{-1}(x)}$; $\boxed{\mathbf{f^{-1}(x)=\sqrt[5]{\frac{x+5}{3}}}}$

h. The function $f(x)=\frac{1}{3}x-5$ is a one-to-one function because for each value of x there is only one corresponding value of y. For example, at $x=0$; $y=-5$ or at $x=3$; $y=-4$, etc.

First Method: Interchange the x variable with y and the y variable with x in the equation $f(x)=y=\frac{1}{3}x-5$ and solve for y. Next, replace y with $f^{-1}(x)$ to obtain:

$\boxed{x=\frac{1}{3}y-5}$; $\boxed{x+5=\frac{1}{3}y}$; $\boxed{3\times(x+5)=3\times\frac{1}{3}y}$; $\boxed{3x+15=y}$; $\boxed{y=3x+15}$; $\boxed{\mathbf{f^{-1}(x)=3x+15}}$

Second Method: Since $f\left[f^{-1}(x)\right]=x$ replace x with $f^{-1}(x)$ wherever x is present in the function $f(x)=\frac{1}{3}x-5$, i.e., $f\left[f^{-1}(x)\right]=\frac{1}{3}\cdot f^{-1}(x)-5$.

Next, solve for $f^{-1}(x)$ by equating the right hand side of the equations $f\left[f^{-1}(x)\right]=x$ and

$\boxed{f\left[f^{-1}(x)\right]=\frac{1}{3}\cdot f^{-1}(x)-5}$, i.e., $\boxed{x=\frac{1}{3}\cdot f^{-1}(x)-5}$; $\boxed{x+5=\frac{f^{-1}(x)}{3}}$; $\boxed{3(x+5)=f^{-1}(x)}$; $\boxed{\mathbf{f^{-1}(x)=3x+15}}$

i. The function $f(x)=x+100$ is a one-to-one function because for each value of x there is only one corresponding value of y, i.e., at $x=0$; $y=100$ or at $x=10$; $y=110$, etc.

First Method: Interchange the x variable with y and the y variable with x in the equation $f(x)=y=x+100$ and solve for y. Next, replace y with $f^{-1}(x)$ to obtain:

$\boxed{x=y+100}$; $\boxed{x-100=y}$; $\boxed{y=x-100}$; $\boxed{\mathbf{f^{-1}(x)=x-100}}$

Second Method: Since $f\left[f^{-1}(x)\right]=x$ replace x with $f^{-1}(x)$ wherever x is present in the function $f(x)=x+100$, i.e., $f\left[f^{-1}(x)\right]=f^{-1}(x)+100$.

Next, solve for $f^{-1}(x)$ by equating the right hand side of the equations $f\left[f^{-1}(x)\right]=x$ and

$\boxed{f\left[f^{-1}(x)\right]=f^{-1}(x)+100}$, i.e., $\boxed{x=f^{-1}(x)+100}$; $\boxed{x-100=f^{-1}(x)}$; $\boxed{\mathbf{f^{-1}(x)=x-100}}$

j. The function $f(x)=0.4x$ is a one-to-one function because for each value of x there is only one corresponding value of y. For example, at $x=0$; $y=0$ or at $x=2$; $y=0.8$, etc.

First Method: Interchange the x variable with y and the y variable with x in the equation $f(x)=y=0.4x$ and solve for y. Next, replace y with $f^{-1}(x)$ to obtain:

$\boxed{x=0.4y}$; $\boxed{\frac{x}{0.4}=\frac{0.4y}{0.4}}$; $\boxed{\frac{x}{0.4}=y}$; $\boxed{y=2.5x}$; $\boxed{\mathbf{f^{-1}(x)=2.5x}}$

Second Method: Since $f\left[f^{-1}(x)\right] = x$ replace x with $f^{-1}(x)$ wherever x is present in the function $f(x)=0.4x$, i.e., $f\left[f^{-1}(x)\right]=0.4\cdot f^{-1}(x)$.

Next, solve for $f^{-1}(x)$ by equating the right hand side of the equations $f\left[f^{-1}(x)\right]=x$ and

$\boxed{f\left[f^{-1}(x)\right]=0.4\cdot f^{-1}(x)}$, i.e., $\boxed{x=0.4\cdot f^{-1}(x)}$; $\boxed{\frac{x}{0.4}=\frac{0.4f^{-1}(x)}{0.4}}$; $\boxed{2.5x=f^{-1}(x)}$; $\boxed{\mathbf{f^{-1}(x)=2.5x}}$

k. The function $f(x)=\frac{1}{x}+3$ is a one-to-one function because for each value of x (except at $x=0$) there is only one corresponding value of y. For example, at $x=1$; $y=4$ or at $x=2$; $y=3.5$, etc.

First Method: Interchange the x variable with y and the y variable with x in the equation $f(x)=y=\frac{1}{x}+3$ and solve for y. Replace y with $f^{-1}(x)$ to obtain:

$\boxed{x=\frac{1}{y}+3}$; $\boxed{x-3=\frac{1}{y}}$; $\boxed{y(x-3)=1}$; $\boxed{y=\frac{1}{x-3}}$; $\boxed{\mathbf{f^{-1}(x)=\frac{1}{x-3}}}$

Second Method: Since $f\left[f^{-1}(x)\right] = x$ replace x with $f^{-1}(x)$ wherever x is present in the function $f(x)=\frac{1}{x}+3$, i.e., $f\left[f^{-1}(x)\right]=\frac{1}{f^{-1}(x)}+3$.

Next, solve for $f^{-1}(x)$ by equating the right hand side of the equations $f\left[f^{-1}(x)\right]=x$ and

$\boxed{f\left[f^{-1}(x)\right]=\frac{1}{f^{-1}(x)}+3}$, i.e., $\boxed{x=\frac{1}{f^{-1}(x)}+3}$; $\boxed{x-3=\frac{1}{f^{-1}(x)}}$; $\boxed{(x-3)\cdot f^{-1}(x)=1}$; $\boxed{\mathbf{f^{-1}(x)=\frac{1}{x-3}}}$

l. The function $f(x)=-x$ is a one-to-one function because for each value of x there is only one corresponding value of y. For example, at $x=0$; $y=0$ or at $x=1$; $y=-1$, etc.

First Method: Interchange the x variable with y and the y variable with x in the equation $f(x)=y=-x$ and solve for y. Replace y with $f^{-1}(x)$ to obtain: $\boxed{x=-y}$; $\boxed{y=-x}$; $\boxed{\mathbf{f^{-1}(x)=-x}}$

Second Method: Since $f\left[f^{-1}(x)\right] = x$ replace x with $f^{-1}(x)$ wherever x is present in the function $f(x)=-x$, i.e., $f\left[f^{-1}(x)\right]=-f^{-1}(x)$.

Next, solve for $f^{-1}(x)$ by equating the right hand side of the equations $f\left[f^{-1}(x)\right]=x$ and

$\boxed{f\left[f^{-1}(x)\right]=-f^{-1}(x)}$, i.e., $\boxed{x=-f^{-1}(x)}$; $\boxed{\mathbf{f^{-1}(x)=-x}}$

m. The function $f(x)=\sqrt[5]{x-7}+3$ is a one-to-one function because for each value of $x\geq 7$ there is only one corresponding value of y, i.e., at $x=7$; $y=3$ or at $x=10$; $y=\sqrt[5]{3}+3=4.25$, etc.

First Method: Interchange the x variable with y and the y variable with x in the equation $f(x)=y=\sqrt[5]{x-7}+3$ and solve for y. Next, replace y with $f^{-1}(x)$ to obtain:

$\boxed{x=\sqrt[5]{y-7}+3}$; $\boxed{x-3=\sqrt[5]{y-7}}$; $\boxed{x-3=(y-7)^{\frac{1}{5}}}$; $\boxed{(x-3)^5=(y-7)^{\frac{1}{5}\times 5}}$; $\boxed{(x-3)^5=y-7}$; $\boxed{(x-3)^5+7=y}$

$= \boxed{y=(x-3)^5+7} = \boxed{\mathbf{f^{-1}(x)=(x-3)^5+7}}$

Second Method: Since $f\left[f^{-1}(x)\right] = x$ replace x with $f^{-1}(x)$ wherever x is present in the function $f(x)=\sqrt[5]{x-7}+3$, i.e., $f\left[f^{-1}(x)\right]=\sqrt[5]{f^{-1}(x)-7}+3$.

Next, solve for $f^{-1}(x)$ by equating the right hand side of the equations $f\left[f^{-1}(x)\right]=x$ and

$\boxed{f\left[f^{-1}(x)\right]=\sqrt[5]{f^{-1}(x)-7}+3}$, i.e., $\boxed{x=\sqrt[5]{f^{-1}(x)-7}+3}$; $\boxed{x-3=\sqrt[5]{f^{-1}(x)-7}}$; $\boxed{x-3=\left[f^{-1}(x)-7\right]^{\frac{1}{5}}}$

; $\boxed{(x-3)^5=\left[f^{-1}(x)-7\right]^{\frac{1}{5}\times 5}}$; $\boxed{(x-3)^5=f^{-1}(x)-7}$; $\boxed{(x-3)^5+7=f^{-1}(x)}$; $\boxed{\mathbf{f^{-1}(x)=(x-3)^5+7}}$

n. The function $f(x)=\frac{2x-5}{x-1}$ is a one-to-one function because for each value of x (except at $x=1$) there is only one corresponding value of y. For example, at $x=2$; $y=-1$ or at $x=3$; $y=0.5$, etc.

First Method: Interchange the x variable with y and the y variable with x in the equation $f(x)=y=\frac{2x-5}{x-1}$ and solve for y. Replace y with $f^{-1}(x)$ to obtain:

$\boxed{x=\frac{2y-5}{y-1}}$; $\boxed{x(y-1)=2y-5}$; $\boxed{xy-x=2y-5}$; $\boxed{xy-2y=x-5}$; $\boxed{y(x-2)=x-5}$; $\boxed{y=\frac{x-5}{x-2}}$; $\boxed{\mathbf{f^{-1}(x)=\frac{x-5}{x-2}}}$

Second Method: Since $f\left[f^{-1}(x)\right] = x$ replace x with $f^{-1}(x)$ wherever x is present in the function $f(x)=\frac{2x-5}{x-1}$, i.e., $f\left[f^{-1}(x)\right]=\frac{2f^{-1}(x)-5}{f^{-1}(x)-1}$.

Next, solve for $f^{-1}(x)$ by equating the right hand side of the equations $f\left[f^{-1}(x)\right]=x$ and

$\boxed{f\left[f^{-1}(x)\right]=\frac{2f^{-1}(x)-5}{f^{-1}(x)-1}}$, i.e., $\boxed{x=\frac{2f^{-1}(x)-5}{f^{-1}(x)-1}}$; $\boxed{x\cdot\left[f^{-1}(x)-1\right]=2f^{-1}(x)-5}$; $\boxed{x f^{-1}(x)-x=2f^{-1}(x)-5}$

$\boxed{x f^{-1}(x)-2f^{-1}(x)=x-5}$; $\boxed{f^{-1}(x)\cdot(x-2)=x-5}$; $\boxed{\mathbf{f^{-1}(x)=\frac{x-5}{x-2}}}$

o. The function $f(x)=7-3x$ is a one-to-one function because for each value of x there is only one corresponding value of y. For example, at $x=0$; $y=7$ or at $x=\frac{1}{3}$; $y=6$, etc.

First Method: Interchange the x variable with y and the y variable with x in the equation $f(x)=y=7-3x$ and solve for y. Next, replace y with $f^{-1}(x)$ to obtain:

$\boxed{x=7-3y}$; $\boxed{x-7=-3y}$; $\boxed{\frac{x-7}{-3}=\frac{-3y}{-3}}$; $\boxed{\frac{7-x}{3}=y}$; $\boxed{y=\frac{7-x}{3}}$; $\boxed{\mathbf{f^{-1}(x)=\frac{7-x}{3}}}$

Second Method: Since $f\left[f^{-1}(x)\right] = x$ replace x with $f^{-1}(x)$ wherever x is present in the function $f(x)=7-3x$, i.e., $f\left[f^{-1}(x)\right]=7-3\cdot f^{-1}(x)$.

Next, solve for $f^{-1}(x)$ by equating the right hand side of the equations $f\left[f^{-1}(x)\right]=x$ and

$\boxed{f\left[f^{-1}(x)\right] = 7-3\cdot f^{-1}(x)}$, i.e., $\boxed{x = 7-3\cdot f^{-1}(x)}$; $\boxed{x-7 = -3f^{-1}(x)}$; $\boxed{\frac{x-7}{-3} = \frac{-3f^{-1}(x)}{-3}}$; $\boxed{\mathbf{f^{-1}(x) = \frac{7-x}{3}}}$

Note that by limiting the domain of a function, a non-invertible function can become an invertible function. For example, the function $f(x)=x^2$ is not invertible. However, by defining $f(x)=x^2$ where $x \ge 0$ the function $f(x)=x^2$ becomes an invertible function. The following are additional examples of non-invertible functions where by restricting the domain result in having invertible functions:

$f(x)=x^2+3$ where $x \ge 0$

$f(x)=x^2$ where $x \le 0$

$f(x)=|x-1|+3$ where $x \ge 1$

$f(x)=-\frac{1}{2}|x|+4$ where $x \le 0$

$f(x)=\frac{1}{2}|x|$ where $x \ge 0$

$f(x)=2(x+3)^2-4$ where $x \ge -3$

$f(x)=(x-2)^2$ where $x \ge 2$

$f(x)=1-x^2$ where $x \le 0$

In the next section we will learn how to write complex numbers in standard form and perform math operations involving complex numbers.

Section 2.4 Practice Problems – One-to-One and Inverse Functions of Real Variables

1. State which of the following functions are one-to-one.

a. $f(x)=2x+5$ b. $f(x)=-5+\frac{2}{3}x$ c. $f(x)=\frac{1}{5}x-1$

d. $f(x)=x^2-25$ e. $f(x)=\sqrt{6x-5}$ f. $f(x)=\frac{1}{4}(16-3x)$

g. $f(x)=x^3-2$ h. $f(x)=2|x|$ i. $f(x)=x^4$

j. $f(x)=1+e^{2x}$ k. $f(x)=x^8+1$ l. $f(x)=x^2+4$

2. Given the following functions are one-to-one, use the first method to find their inverse.

a. $f(x)=x+3$ b. $f(x)=5x$ c. $f(x)=\sqrt{5x-1}$

d. $f(x)=1-2x^3$ e. $f(x)=\sqrt{2x}+1$, $x \ge 0$ f. $f(x)=0.2x+10$

g. $f(x)=\sqrt[3]{x-2}+1$, $x \ge 2$ h. $f(x)=\frac{2x-3}{x}$, $x \ne 0$ i. $f(x)=2-5x$

j. $f(x)=\frac{2x+5}{x+1}$, $x \ne -1$ k. $f(x)=\frac{2x-3}{x-5}$, $x \ne 5$ l. $f(x)=2x^3-9$

2.5 Complex Numbers and Functions of Complex Variables

In Chapters 2 and 4 of the "*Mastering Algebra – Intermediate Level*" we solved equations of the form $ax+b=0$ and $ax^2+bx+c=0$ and learned that:

1. Any linear equation such as $3x+2=-4$ and $3x-15=0$ has one real solution $(x=-2)$ and $(x=5)$, respectively and
2. Any quadratic equation such as $x^2-x-6=0$ and $x^2=16$ has two real solutions $(3\ and\ -2)$ and $(4\ and\ -4)$, respectively.

However, quadratic equations of the form $x^2=-5$ have no real solutions because the square of every real number is positive. Therefore, in order for equations such as $x^2=-5$ to have a solution imaginary numbers are introduced. **Imaginary numbers** are generally shown by the letters "i" or "j". Imaginary numbers are combined with real numbers to form complex numbers. Numbers of the form $a+bi$, where a and b are real numbers, are defined as **complex numbers**. Note that in the complex number $a+bi$, a is referred to as the **real part** and b is referred to as the **imaginary part**. The imaginary number i is defined as

$$i = \sqrt{-1} \text{ and } i^2 = \left(\sqrt{-1}\right)^2 = -1^{\frac{1}{2}\times 2} = -1$$

The form $a+bi$ is referred to as the **standard form** of a complex number. Furthermore, two complex numbers $a+bi$ and $c+di$ are equal to one another if and only if $a=c$ and $b=d$. The following examples show how higher order imaginary numbers are simplified:

Example 2.5-1: Simplify the following imaginary numbers.

a. $i^3 =$ b. $i^4 =$ c. $i^5 =$ d. $i^8 =$

e. $i^7 =$ f. $i^{14} =$ g. $i^{15} =$ h. $i^{20} =$

i. $i^{18} =$ j. $i^{22} =$ k. $i^{25} =$ l. $i^{40} =$

Solutions:

a. $\boxed{i^3} = \boxed{i^{2+1}} = \boxed{i^2 \cdot i} = \boxed{-1 \cdot i} = \boxed{-i}$

b. $\boxed{i^4} = \boxed{\left(i^2\right)^2} = \boxed{(-1)^2} = \boxed{1}$

c. $\boxed{i^5} = \boxed{i^{4+1}} = \boxed{i^4 \cdot i} = \boxed{\left(i^2\right)^2 \cdot i} = \boxed{(-1)^2 \cdot i} = \boxed{1 \cdot i} = \boxed{i}$

d. $\boxed{i^8} = \boxed{\left(i^2\right)^4} = \boxed{(-1)^4} = \boxed{1}$

e. $\boxed{i^7} = \boxed{i^{6+1}} = \boxed{i^6 \cdot i^1} = \boxed{\left(i^2\right)^3 \cdot i} = \boxed{(-1)^3 \cdot i} = \boxed{-1 \cdot i} = \boxed{-i}$

f. $\boxed{i^{14}} = \boxed{\left(i^2\right)^7} = \boxed{(-1)^7} = \boxed{-1}$

g. $\boxed{i^{15}} = \boxed{i^{14+1}} = \boxed{i^{14} \cdot i} = \boxed{\left(i^2\right)^7 \cdot i} = \boxed{(-1)^7 \cdot i} = \boxed{-1 \cdot i} = \boxed{-i}$

h. $\boxed{i^{20}} = \boxed{\left(i^2\right)^{10}} = \boxed{(-1)^{10}} = \boxed{1}$

i. $\boxed{i^{18}} = \boxed{\left(i^2\right)^9} = \boxed{(-1)^9} = \boxed{-1}$

j. $\boxed{i^{22}} = \boxed{\left(i^2\right)^{11}} = \boxed{(-1)^{11}} = \boxed{-1}$

k. $\boxed{i^{25}} = \boxed{i^{24+1}} = \boxed{i^{24} \cdot i} = \boxed{\left(i^2\right)^{12} \cdot i} = \boxed{(-1)^{12} \cdot i} = \boxed{1 \cdot i} = \boxed{i}$

l. $\boxed{i^{40}} = \boxed{\left(i^2\right)^{20}} = \boxed{(-1)^{20}} = \boxed{1}$

There are many ways to simplify higher powers of i. However, to simplify the higher powers of i, it is much easier to use $i^2 = -1$ and the exponent laws as shown in the above examples (review Section 1.1). Note that all powers of i can be simplified to 1, -1, $i = \sqrt{-1}$, or $-i = -\sqrt{-1}$.

Note 1: The number "negative one" raised to an even number is always equal to plus one. For Example: $(-1)^2, (-1)^8, (-1)^{10}, (-1)^{18}, (-1)^{24}, (-1)^{56}, (-1)^{100}, etc.$ are all equal to 1.

Note 2: The number "negative one" raised to an odd number is always equal to negative one. For Example: $(-1)^3, (-1)^7, (-1)^{13}, (-1)^{19}, (-1)^{23}, (-1)^{45}, (-1)^{101}, etc.$ are all equal to -1.

Example 2.5-2: Determine whether the following complex numbers are real or imaginary. Write the complex number in the standard form $a + bi$.

a. $5i =$ b. $-2i + 5 =$ c. $\frac{1+2i}{5} =$ d. $\frac{3\pi}{2} =$

e. $1 - \sqrt{3}\,i =$ f. $7 =$ g. $-i =$ h. $2 + i^8 =$

Solutions:

a. The complex number $5i$ is **imaginary**. The standard form of $5i$ is $\mathbf{0 + 5i}$.

b. The complex number $-2i + 5$ is **imaginary**. The standard form of $-2i + 5$ is $\mathbf{5 - 2i}$.

c. The complex number $\frac{1+2i}{5}$ is **imaginary**. The standard form of $\frac{1+2i}{5}$ is $\frac{1}{5} + \frac{2}{5}i = \mathbf{0.2 + 0.4i}$.

d. The complex number $\frac{3\pi}{2}$ is **real** (an irrational number). The standard form of $\frac{3\pi}{2}$ is $\mathbf{\frac{3\pi}{2} + 0i}$.

e. The complex number $1 - \sqrt{3}\,i$ is **imaginary**. The standard form of $1 - \sqrt{3}\,i$ is $\mathbf{1 + \left(-\sqrt{3}\,i\right)}$.

f. The complex number 7 is **real** (a real number). The standard form of 7 is $\mathbf{7 + 0i}$.

g. The complex number $-i$ is **imaginary**. The standard form of $-i$ is $\mathbf{0 + (-i)}$.

h. The complex number $2 + i^8 = 2 + \left(i^2\right)^4 = 2 + (-1)^4 = 2 + 1 = 3$ is **real** (a real number). The standard form of 3 is $\mathbf{3 + 0i}$.

Note 3: In the real number system, $\sqrt{-6}$, $\sqrt{-9}$, and $\sqrt{-23}$ are undefined. However, in the complex number system they are defined as $\sqrt{-6} = \sqrt{6}\,i$, $\sqrt{-9} = \sqrt{9}\,i$, and $\sqrt{-23} = \sqrt{23}\,i$. In addition, note that **all math operations involving complex numbers must be performed after converting to the standard form** $a + bi$. For example, to add $\sqrt{-5}$ to $\sqrt{-125}$ we must first replace the square roots of the negative numbers by square roots represented by i, i.e., $\sqrt{-5} = \sqrt{5} \cdot \sqrt{-1} = \sqrt{5}\,i$ and $\sqrt{125} = \sqrt{125} \cdot \sqrt{-1} = \sqrt{125}\,i$ before performing the addition. The following examples further illustrate this point.

Example 2.5-3: Write the following expressions in the standard form $a + bi$.

a. $\sqrt{-5} + \sqrt{-125} =$ b. $\left(\sqrt{-3} + \sqrt{-27}\right) - \sqrt{-9} =$ c. $\frac{\sqrt{-4} - \sqrt{-2}}{4} =$

d. $\sqrt{-9} + \sqrt{-16} =$ e. $\sqrt{-2} - \sqrt{-5} =$ f. $\left(\sqrt{-5}\right)^2 =$

g. $\left(\sqrt{-3}\right)^3 =$ h. $\sqrt{-25}-\sqrt{4} =$ i. $\sqrt{-4}\left(\sqrt{-2}+\sqrt{-9}\right) =$

j. $\dfrac{\sqrt{-3}+\sqrt{-2}}{3} =$ k. $\sqrt{-5}\cdot\sqrt{-2} =$ l. $\left(\sqrt{-4}-\sqrt{-36}\right)^2 =$

m. $\dfrac{3-\sqrt{-4}}{-2} =$ n. $\sqrt{-3}\left(\sqrt{-5}-\sqrt{-16}\right) =$ o. $3+\sqrt{(-1)^2-5} =$

Solutions:

a. $\boxed{\sqrt{-5}+\sqrt{-125}} = \boxed{\left(\sqrt{5}\cdot\sqrt{-1}\right)+\left(\sqrt{125}\cdot\sqrt{-1}\right)} = \boxed{\left(\sqrt{5}\cdot i\right)+\left(\sqrt{125}\cdot i\right)} = \boxed{\sqrt{5}\,i+5\sqrt{5}\,i} = \boxed{2.24i+11.18i} = \boxed{\mathbf{0+13.42\,i}}$

b. $\boxed{\left(\sqrt{-3}+\sqrt{-27}\right)-\sqrt{-9}} = \boxed{\left(\sqrt{3}\cdot\sqrt{-1}+\sqrt{27}\cdot\sqrt{-1}\right)-\left(\sqrt{9}\cdot\sqrt{-1}\right)} = \boxed{\sqrt{3}\;i+\sqrt{27}\,i-\sqrt{9}\,i} = \boxed{1.73i+5.2i-3i}$

$= \boxed{3.93i} = \boxed{\mathbf{0+3.93\,i}}$

c. $\boxed{\dfrac{\sqrt{-4}-\sqrt{-2}}{4}} = \boxed{\dfrac{\left(\sqrt{4}\cdot\sqrt{-1}\right)-\left(\sqrt{2}\cdot\sqrt{-1}\right)}{4}} = \boxed{\dfrac{\sqrt{4}\,i-\sqrt{2}\,i}{4}} = \boxed{\dfrac{2i-1.41i}{4}} = \boxed{\dfrac{0.59}{4}i} = \boxed{0.15i} = \boxed{\mathbf{0+0.15\,i}}$

d. $\boxed{\sqrt{-9}+\sqrt{-16}} = \boxed{\left(\sqrt{9}\cdot\sqrt{-1}\right)+\left(\sqrt{16}\cdot\sqrt{-1}\right)} = \boxed{(3\cdot i)+(4\cdot i)} = \boxed{3i+4i} = \boxed{7i} = \boxed{\mathbf{0+7i}}$

e. $\boxed{\sqrt{-2}-\sqrt{-5}} = \boxed{\left(\sqrt{2}\cdot\sqrt{-1}\right)-\left(\sqrt{5}\cdot\sqrt{-1}\right)} = \boxed{(1.41\cdot i)-(2.24\cdot i)} = \boxed{1.41i-2.24i} = \boxed{-0.83i} = \boxed{\mathbf{0+(-0.83i)}}$

f. $\boxed{\left(\sqrt{-5}\right)^2} = \boxed{\left(\sqrt{5}\cdot\sqrt{-1}\right)^2} = \boxed{\left(\sqrt{5}\cdot i\right)^2} = \boxed{5i^2} = \boxed{-5} = \boxed{\mathbf{-5+0i}}$

g. $\boxed{\left(\sqrt{-3}\right)^3} = \boxed{\left(\sqrt{3}\cdot\sqrt{-1}\right)^3} = \boxed{(1.73\cdot i)^3} = \boxed{5.18\cdot i^3} = \boxed{5.18\cdot -i} = \boxed{-5.18i} = \boxed{\mathbf{0+(-5.18i)}}$

h. $\boxed{\sqrt{-25}-\sqrt{4}} = \boxed{\left(\sqrt{25}\cdot\sqrt{-1}\right)\;-2} = \boxed{(5\cdot i)-2} = \boxed{5i-2} = \boxed{\mathbf{-2+5i}}$

i. $\boxed{\sqrt{-4}\left(\sqrt{-2}+\sqrt{-9}\right)} = \boxed{\left(\sqrt{4}\cdot\sqrt{-1}\right)\left[\left(\sqrt{2}\cdot\sqrt{-1}\right)+\left(\sqrt{9}\cdot\sqrt{-1}\right)\right]} = \boxed{\left(\sqrt{4}\cdot i\right)\left[\left(\sqrt{2}\cdot i\right)+\left(\sqrt{9}\cdot i\right)\right]} = \boxed{2i\,(1.41i+3i)}$

$= \boxed{2.82i^2+6i^2} = \boxed{-2.82-6} = \boxed{-8.82} = \boxed{\mathbf{-8.82+0\,i}}$

j. $\boxed{\dfrac{\sqrt{-3}+\sqrt{-2}}{3}} = \boxed{\dfrac{\sqrt{3}\cdot\sqrt{-1}+\sqrt{2}\cdot\sqrt{-1}}{3}} = \boxed{\dfrac{\sqrt{3}\cdot i+\sqrt{2}\cdot i}{3}} = \boxed{\dfrac{1.73i+1.41i}{3}} = \boxed{\dfrac{3.14i}{3}} = \boxed{1.05i} = \boxed{\mathbf{0+1.05i}}$

k. $\boxed{\sqrt{-5}\cdot\sqrt{-2}} = \boxed{\left(\sqrt{5}\cdot\sqrt{-1}\right)\cdot\left(\sqrt{2}\cdot\sqrt{-1}\right)} = \boxed{\sqrt{5}\,i\cdot\sqrt{2}\,i} = \boxed{(2.24\cdot i)\cdot(1.41\cdot i)} = \boxed{3.16i^2} = \boxed{-3.16} = \boxed{\mathbf{-3.16+0i}}$

Again note that math operations involving complex numbers must be performed after converting to the standard form $a+bi$, i.e., in the above example we must first convert $\sqrt{-5}$ and $\sqrt{-2}$ to $\sqrt{5}\,i$ and $\sqrt{2}\,i$ before multiplying the two number by one another. (We can not multiply $\sqrt{-5}$ by $\sqrt{-2}$

to obtain $\sqrt{-5\cdot-2}=\sqrt{10}=3.16=3.16+0i$. This is the wrong answer. The product rule $\sqrt{a}\cdot\sqrt{b}=\sqrt{ab}$ can be used only when a and b are positive numbers.)

l. $\boxed{\left(\sqrt{-4}-\sqrt{-36}\right)^2}=\boxed{\left(\sqrt{4}\cdot\sqrt{-1}-\sqrt{36}\cdot\sqrt{-1}\right)^2}=\boxed{(2i-6i)^2}=\boxed{(-4i)^2}=\boxed{16i^2}=\boxed{-16}=\boxed{\mathbf{-16+0i}}$

m. $\boxed{\frac{3-\sqrt{-4}}{-2}}=\boxed{\frac{3-\sqrt{4}\cdot\sqrt{-1}}{-2}}=\boxed{\frac{3-\sqrt{4}\cdot i}{-2}}=\boxed{\frac{3-2i}{-2}}=\boxed{\frac{3}{-2}+\frac{-2i}{-2}}=\boxed{-\frac{3}{2}+\frac{-2}{-2}i}=\boxed{\mathbf{-1.5+i}}$

n. $\boxed{\sqrt{-3}\left(\sqrt{-5}-\sqrt{-16}\right)}=\boxed{\left(\sqrt{3}\cdot\sqrt{-1}\right)\left[\left(\sqrt{5}\cdot\sqrt{-1}\right)-\left(\sqrt{16}\cdot\sqrt{-1}\right)\right]}=\boxed{\sqrt{3}\cdot i\left[\left(\sqrt{5}\cdot i\right)-\left(\sqrt{16}\cdot i\right)\right]}=\boxed{\sqrt{3}\,i\left(\sqrt{5}\,i-4\,i\right)}$

$=\boxed{1.73i\,(2.24i-4i)}=\boxed{1.73i\times-1.76i}=\boxed{3.04i^2}=\boxed{-3.04}=\boxed{\mathbf{-3.04+0i}}$

o. $\boxed{3+\sqrt{(-1)^2-5}}=\boxed{3+\sqrt{1-5}}=\boxed{3+\sqrt{-4}}=\boxed{3+\sqrt{4}\cdot\sqrt{-1}}=\boxed{3+\sqrt{4}\,i}=\boxed{\mathbf{3+2i}}$

In Chapter 4, Section 4.2, of the "*Mastering Algebra – Intermediate Level*" we learned that one of the methods for solving quadratic equations of the form $ax^2+bx+c=0$ (where a, b, and c are real numbers and $a\neq 0$) is by using the quadratic formula defined as $x_{1,2}=\frac{-b\pm\sqrt{b^2-4ac}}{2a}$. However, to find the solutions the term under the radical, i.e., b^2-4ac had to be real and positive. The introduction of complex numbers enables us to solve quadratic equations even though the term under the radical, i.e., b^2-4ac is negative. The following examples illustrate this point.

Example 2.5-4: Given the quadratic formula $x_{1,2}=\frac{-b\pm\sqrt{b^2-4ac}}{2a}$, solve for x for the following values of a, b, and c.

a. $a=1$, $b=2$, and $c=6$

b. $a=2$, $b=0$, and $c=4$

c. $a=1$, $b=-1$, and $c=3$

d. $a=3$, $b=2$, and $c=5$

Solutions:

a. Substituting $a=1$, $b=2$, and $c=6$ into $x_{1,2}=\frac{-b\pm\sqrt{b^2-4ac}}{2a}$ we obtain

$$\boxed{x_1}=\boxed{\frac{-2+\sqrt{2^2-(4\cdot1\cdot6)}}{2\cdot1}}=\boxed{\frac{-2+\sqrt{4-24}}{2}}=\boxed{\frac{-2+\sqrt{-20}}{2}}=\boxed{\frac{-2+\sqrt{20}\cdot\sqrt{-1}}{2}}=\boxed{\frac{-2+\sqrt{4\cdot5}\cdot i}{2}}=\boxed{\frac{-2+2\sqrt{5}\,i}{2}}$$

$$=\boxed{\frac{2\left(-1+\sqrt{5}\,i\right)}{2}}=\boxed{-1+\sqrt{5}\,i}=\boxed{\mathbf{-1+2.24i}}$$ and

$$\boxed{x_2}=\boxed{\frac{-2-\sqrt{2^2-(4\cdot1\cdot6)}}{2\cdot1}}=\boxed{\frac{-2-\sqrt{4-24}}{2}}=\boxed{\frac{-2-\sqrt{-20}}{2}}=\boxed{\frac{-2-\sqrt{20}\cdot\sqrt{-1}}{2}}=\boxed{\frac{-2-\sqrt{4\cdot5}\cdot i}{2}}=\boxed{\frac{-2-2\sqrt{5}\,i}{2}}$$

$$= \boxed{\frac{-2\left(1+\sqrt{5}\,i\right)}{2}} = \boxed{-1-\sqrt{5}\,i} = \boxed{\mathbf{-1-2.24}\boldsymbol{i}}$$

b. Substituting $a=2$, $b=0$, and $c=4$ into $x_{1,2}=\frac{-b\pm\sqrt{b^2-4ac}}{2a}$ we obtain

$$\boxed{x_1} = \boxed{\frac{-0+\sqrt{0^2-(4\cdot 2\cdot 4)}}{2\cdot 2}} = \boxed{\frac{-0+\sqrt{-32}}{4}} = \boxed{\frac{\sqrt{32}\cdot\sqrt{-1}}{4}} = \boxed{\frac{\sqrt{16\cdot 2}\cdot i}{4}} = \boxed{\frac{4\sqrt{2}\,i}{4}} = \boxed{\sqrt{2}\,i} = \boxed{\mathbf{0+1.41}\boldsymbol{i}} \text{ and}$$

$$\boxed{x_2} = \boxed{\frac{-0-\sqrt{0^2-(4\cdot 2\cdot 4)}}{2\cdot 2}} = \boxed{\frac{-0-\sqrt{-32}}{4}} = \boxed{\frac{-\sqrt{32}\cdot\sqrt{-1}}{4}} = \boxed{\frac{-\sqrt{16\cdot 2}\cdot i}{4}} = \boxed{-\frac{4\sqrt{2}\,i}{4}} = \boxed{-\sqrt{2}\,i} = \boxed{\mathbf{0-1.41}\boldsymbol{i}}$$

c. Substituting . $a=1$, $b=-1$, and $c=3$ into $x_{1,2}=\frac{-b\pm\sqrt{b^2-4ac}}{2a}$ we obtain

$$\boxed{x_1} = \boxed{\frac{-(-1)+\sqrt{(-1)^2-(4\cdot 1\cdot 3)}}{2\cdot 1}} = \boxed{\frac{1+\sqrt{1-12}}{2}} = \boxed{\frac{1+\sqrt{-11}}{2}} = \boxed{\frac{1+\sqrt{11}\cdot\sqrt{-1}}{2}} = \boxed{\frac{1+\sqrt{11}\cdot i}{2}} = \boxed{\frac{1+3.32i}{2}}$$

$$= \boxed{\frac{1}{2}+\frac{3.32}{2}i} = \boxed{\mathbf{0.5+1.66}\boldsymbol{i}} \text{ and}$$

$$\boxed{x_2} = \boxed{\frac{-(-1)-\sqrt{(-1)^2-(4\cdot 1\cdot 3)}}{2\cdot 1}} = \boxed{\frac{1-\sqrt{1-12}}{2}} = \boxed{\frac{1-\sqrt{-11}}{2}} = \boxed{\frac{1-\sqrt{11}\cdot\sqrt{-1}}{2}} = \boxed{\frac{1-\sqrt{11}\cdot i}{2}} = \boxed{\frac{1-3.32i}{2}}$$

$$= \boxed{\frac{1}{2}-\frac{3.32}{2}i} = \boxed{\mathbf{0.5-1.66}\boldsymbol{i}}$$

d. Substituting $a=3$, $b=2$, and $c=5$ into $x_{1,2}=\frac{-b\pm\sqrt{b^2-4ac}}{2a}$ we obtain

$$\boxed{x_1} = \boxed{\frac{-2+\sqrt{2^2-(4\cdot 3\cdot 5)}}{2\cdot 3}} = \boxed{\frac{-2+\sqrt{4-60}}{6}} = \boxed{\frac{-2+\sqrt{-56}}{6}} = \boxed{\frac{-2+\sqrt{56}\cdot\sqrt{-1}}{6}} = \boxed{\frac{-2+\sqrt{4\cdot 14}\cdot i}{6}}$$

$$= \boxed{\frac{-2+2\sqrt{14}\,i}{6}} = \boxed{\frac{2\left(-1+\sqrt{14}\,i\right)}{\cancel{6}\,3}} = \boxed{\frac{-1+3.74i}{3}} = \boxed{-\frac{1}{3}+\frac{3.74}{3}i} = \boxed{\mathbf{-0.33+1.25}\boldsymbol{i}} \text{ and}$$

$$\boxed{x_2} = \boxed{\frac{-2-\sqrt{2^2-(4\cdot 3\cdot 5)}}{2\cdot 3}} = \boxed{\frac{-2-\sqrt{4-60}}{6}} = \boxed{\frac{-2-\sqrt{-56}}{6}} = \boxed{\frac{-2-\sqrt{56}\cdot\sqrt{-1}}{6}} = \boxed{\frac{-2-\sqrt{4\cdot 14}\cdot i}{6}}$$

$$= \boxed{\frac{-2-2\sqrt{14}\,i}{6}} = \boxed{\frac{2\left(-1-\sqrt{14}\,i\right)}{\cancel{6}\,3}} = \boxed{\frac{-1-3.74i}{3}} = \boxed{-\frac{1}{3}-\frac{3.74}{3}i} = \boxed{\mathbf{-0.33-1.25}\boldsymbol{i}}$$

Note that the solutions to the quadratic equations, for cases where the radicand is negative, are complex conjugates of one another.

Similar to real numbers (see Section 2.1), complex numbers can be substituted in place of a variable in functions of one variable as shown in the following examples:

Example 2.5-5: Given the functions $f(x)=x^2+2x+3$ and $g(x)=2x+7$, find.

a. $f(-1+i)=$ b. $g(2-i)=$ c. $f(1+i)=$ d. $f\left(\sqrt{2}\,i\right)=$

e. $g\left(1-\sqrt{3}\,i\right)=$ f. $f(3+4i)=$ g. $g(-i)=$ h. $f(i)=$

Solutions:

a. $\boxed{f(-1+i)}=\boxed{(-1+i)^2+2(-1+i)+3}=\boxed{1+i^2-2i-2+2i+3}=\boxed{1-\cancel{1}-\cancel{2i}-2+\cancel{2i}+3}=\boxed{3-2}=\boxed{1}=\boxed{\mathbf{1+0i}}$

b. $\boxed{g(2-i)}=\boxed{2(2-i)+7}=\boxed{4-2i+7}=\boxed{(4+7)-2i}=\boxed{\mathbf{11-2i}}$

c. $\boxed{f(1+i)}=\boxed{(1+i)^2+2(1+i)+3}=\boxed{1+i^2+2i+2+2i+3}=\boxed{1-\cancel{1}+2i+2+2i+3}=\boxed{(2+3)+(2i+2i)}=\boxed{\mathbf{5+4i}}$

d. $\boxed{f\left(\sqrt{2}\,i\right)}=\boxed{\left(\sqrt{2}\,i\right)^2+2\left(\sqrt{2}\,i\right)+3}=\boxed{2i^2+2\sqrt{2}\,i+3}=\boxed{-2+2\sqrt{2}\,i+3}=\boxed{(3-2)+2.83i}=\boxed{\mathbf{1+2.83i}}$

e. $\boxed{g\left(1-\sqrt{3}\,i\right)}=\boxed{2\left(1-\sqrt{3}\,i\right)+7}=\boxed{2-2\sqrt{3}\,i+7}=\boxed{(2+7)-2\sqrt{3}\,i}=\boxed{\mathbf{9-3.46i}}$

f. $\boxed{f(3+4i)}=\boxed{(3+4i)^2+2(3+4i)+3}=\boxed{9+16i^2+24i+6+8i+3}=\boxed{9-16+24i+6+8i+3}=\boxed{(9-16+6+3)}$

$\boxed{+(24i+8i)}=\boxed{\mathbf{2+32i}}$

g. $\boxed{g(-i)}=\boxed{2(-i)+7}=\boxed{-2i+7}=\boxed{\mathbf{7-2i}}$

h. $\boxed{f(i)}=\boxed{i^2+2i+3}=\boxed{-1+2i+3}=\boxed{(3-1)+2i}=\boxed{\mathbf{2+2i}}$

Again, similar to real numbers (see Section 2.3), complex numbers can also be substituted in place of a variable in composite functions as shown in the following examples.

Example 2.5-6: Given $f(x)=x+5$ and $g(x)=x^2-2x+1$, find

a. $f(g(2+3i))=$ b. $g(f(1+i))=$ c. $f(g(1-i))=$

d. $g(f(i))=$ e. $f\left(g\left(\sqrt{3}+2i\right)\right)=$ f. $g\left(f\left(2+i^3\right)\right)=$

g. $g\left(f\left(i^2+i\right)\right)=$ h. $f\left(g\left(i^5\right)\right)=$ i. $f\left(g\left(1-i^7\right)\right)=$

Solutions:

First - Find $f(g(x))$ and $g(f(x))$, i.e.,

$\boxed{f(g(x))}=\boxed{g(x)+5}=\boxed{\left(x^2-2x+1\right)+5}=\boxed{x^2-2x+6}$

$\boxed{g(f(x))}=\boxed{[f(x)]^2-2\cdot f(x)+1}=\boxed{(x+5)^2-2(x+5)+1}=\boxed{x^2+10x+25-2x-10+1}=\boxed{x^2+8x+16}$

Second – Find $f(g(x))$ or $g(f(x))$ for the specific values given.

a. $\boxed{f(g(2+3i))} = \boxed{(2+3i)^2 - 2(2+3i)+6} = \boxed{4+9i^2+12i-4-6i+6} = \boxed{(4-9-4+6)+(12i-6i)} = \boxed{\mathbf{-3+6i}}$

b. $\boxed{g(f(1+i))} = \boxed{(1+i)^2+8(1+i)+16} = \boxed{1+i^2+2i+8+8i+16} = \boxed{1-1+2i+8+8i+16} = \boxed{\mathbf{24+10i}}$

c. $\boxed{f(g(1-i))} = \boxed{(1-i)^2-2(1-i)+6} = \boxed{1+i^2-2i-2+2i+6} = \boxed{1-1-2i-2+2i+6} = \boxed{4} = \boxed{\mathbf{4+0i}}$

d. $\boxed{g(f(i))} = \boxed{i^2+8i+16} = \boxed{-1+8i+16} = \boxed{(16-1)+8i} = \boxed{\mathbf{15+8i}}$

e. $\boxed{f\left(g\left(\sqrt{3}+2i\right)\right)} = \boxed{\left(\sqrt{3}+2i\right)^2-2\left(\sqrt{3}+2i\right)+6} = \boxed{3+4i^2+4\sqrt{3}\,i-2\sqrt{3}-4i+6} = \boxed{3-4+6.93i-3.46-4i+6}$

$= \boxed{(3-4-3.46+6)+(6.93i-4i)} = \boxed{\mathbf{1.54+2.93i}}$

f. $\boxed{g\left(f\left(2+i^3\right)\right)} = \boxed{g\left(f\left(2+i^{2+1}\right)\right)} = \boxed{g\left(f\left(2+i^2\cdot i\right)\right)} = \boxed{g(f(2-i))} = \boxed{(2-i)^2+8(2-i)+16} = \boxed{4+i^2-4i+16-8i+16}$

$= \boxed{4-1-4i+16-8i+16} = \boxed{(4-1+16+16)+(-4i-8i)} = \boxed{\mathbf{35-12i}}$

g. $\boxed{g\left(f\left(i^2+i\right)\right)} = \boxed{g(f(-1+i))} = \boxed{(-1+i)^2+8(-1+i)+16} = \boxed{1+i^2-2i-8+8i+16} = \boxed{1-1-2i-8+8i+16}$

$= \boxed{(-8+16)+(-2i+8i)} = \boxed{\mathbf{8+6i}}$

h. $\boxed{f\left(g\left(i^5\right)\right)} = \boxed{f\left(g\left(i^{4+1}\right)\right)} = \boxed{f\left(g\left(i^4\cdot i\right)\right)} = \boxed{f(g(i))} = \boxed{i^2-2i+6} = \boxed{-1-2i+6} = \boxed{(-1+6)-2i} = \boxed{\mathbf{5-2i}}$

i. $\boxed{f\left(g\left(1-i^7\right)\right)} = \boxed{f\left(g\left(1-i^{6+1}\right)\right)} = \boxed{f\left(g\left(1-i^6\cdot i\right)\right)} = \boxed{f(g(1+i))} = \boxed{(1+i)^2-2(1+i)+6} = \boxed{1+i^2+2i-2-2i+6}$

$= \boxed{1-1-2+6} = \boxed{4} = \boxed{\mathbf{4+0i}}$

Example 2.5-7: Given $f(g(h(x))) = x^2+2x+10$ and $g(f(h(x))) = 3x+5$, find

a. $f\left(g\left(h\left(1-\sqrt{-2}\right)\right)\right) =$

b. $f(g(h(-i))) =$

c. $g\left(f\left(h\left(i^3\right)\right)\right) =$

d. $g\left(f\left(h\left(\sqrt{-4}\right)\right)\right) =$

e. $f\left(g\left(h\left(1-i^3\right)\right)\right) =$

f. $g(f(h(3+5i))) =$

Solutions:

a. $\boxed{f\left(g\left(h\left(1-\sqrt{-2}\right)\right)\right)} = \boxed{f\left(g\left(h\left(1-\sqrt{2}\cdot\sqrt{-1}\right)\right)\right)} = \boxed{f\left(g\left(h\left(1-\sqrt{2}\,i\right)\right)\right)} = \boxed{\left(1-\sqrt{2}\,i\right)^2+2\left(1-\sqrt{2}\,i\right)+10}$

$= \boxed{1+2i^2-2\sqrt{2}\,i+2-2\sqrt{2}\,i+10} = \boxed{1-2-2\sqrt{2}\,i+2-2\sqrt{2}\,i+10} = \boxed{(1-2+2+10)+\left(-2\sqrt{2}\,i-2\sqrt{2}\,i\right)}$

$= \boxed{11-4\sqrt{2}\,i} = \boxed{11-(4\cdot 1.414)i} = \boxed{\mathbf{11-5.66i}}$

b. $\boxed{f(g(h(-i)))} = \boxed{i^2+2i+10} = \boxed{-1+2i+10} = \boxed{(-1+10)+2i} = \boxed{\mathbf{9+2i}}$

c. $\boxed{g(f(h(i^3)))} = \boxed{g(f(h(i^{2+1})))} = \boxed{g(f(h(i^2 \cdot i)))} = \boxed{g(f(h(-i)))} = \boxed{(3\cdot -i)+5} = \boxed{-3i+5} = \boxed{\mathbf{5-3i}}$

d. $\boxed{g(f(h(\sqrt{-4})))} = \boxed{g(f(h(\sqrt{4}\,i)))} = \boxed{g(f(h(2i)))} = \boxed{(3\cdot 2i)+5} = \boxed{6i+5} = \boxed{\mathbf{5+6i}}$

e. $\boxed{f(g(h(1-i^3)))} = \boxed{f(g(h(1+i)))} = \boxed{(1+i)^2+2(1+i)+10} = \boxed{1+i^2+2i+2+2i+10} = \boxed{1-1+2i+2+2i+10}$

$= \boxed{2i+2+2i+10} = \boxed{(2+10)+(2i+2i)} = \boxed{\mathbf{12+4i}}$

f. $\boxed{g(f(h(3+5i)))} = \boxed{3(3+5i)+5} = \boxed{9+15i+5} = \boxed{(9+5)+15i} = \boxed{\mathbf{14+15i}}$

In the following section students are introduced to operations involving the addition, subtraction, multiplication, and division of complex numbers.

Practice Problems – Complex Numbers and Functions of Complex Variables

1. Simplify the following imaginary numbers.

a. $i^{10} =$ b. $i^{13} =$ c. $i^{17} =$ d. $i^{21} =$

e. $i^{50} =$ f. $i^{100} =$ g. $i^{47} =$ h. $i^{29} =$

2. Write the following expressions in the standard form $a+bi$.

a. $\sqrt{-6}+\sqrt{-12} =$ b. $\left(\sqrt{-2}\cdot\sqrt{-3}\right)^4 =$ c. $\sqrt{-3}\cdot\sqrt{-5} =$

d. $\left(\sqrt{-2}-\sqrt{-25}\right)^3 =$ e. $\sqrt{-1}\left(\sqrt{-2}+\sqrt{-3}\right) =$ f. $5-\sqrt{(-1)^3-2} =$

3. Given $f(x)=x^2+1$ and $g(x)=x^2-2x+1$, find

a. $f(1-i) =$ b. $f(-i) =$ c. $g\left(1-\sqrt{2}\,i\right) =$

d. $g\left(1+i^3\right) =$ e. $f(1+i) =$ f. $f(2+3i) =$

$g\left(-\sqrt{-1}\right) =$ h. $g(2+i) =$ i. $f\left(3+\sqrt{-3}\right) =$

4. Given $f(x)=x^2-1$ and $g(x)=x-5$, find

a. $f\left(g\left(1-\sqrt{-1}\right)\right) =$ b. $g(f(1-i)) =$ c. $g\left(f\left(-i^4\right)\right) =$

d. $f(g(2+5i)) =$ e. $f\left(g\left(i^6\right)\right) =$ f. $g\left(f\left(1+i^5\right)\right) =$

5. Given the quadratic formula $x_{1,2} = \dfrac{-b\pm\sqrt{b^2-4ac}}{2a}$, solve for x for the following values of a, b, and c.

a. $a=2$, $b=3$, and $c=5$ b. $a=3$, $b=4$, and $c=6$ c. $a=4$, $b=1$, and $c=10$

2.6 Math Operations Involving Complex Numbers

In this section addition, subtraction, multiplication, division, and mixed operations involving complex numbers are addressed in Cases I through IV. Several examples showing the math operations with complex numbers are presented. Students are encouraged to learn how to solve the following cases involving complex numbers for future use in calculus.

Case I Addition and Subtraction of Complex Numbers

In general, complex numbers are added and subtracted by grouping the real and imaginary parts together as shown below.

$$(a+bi)+(c+di) = (a+c)+(b+d)i$$

$$(a+bi)-(c+di) = (a-c)+(b-d)i$$

Note that the commutative and associative properties for addition and subtraction of real numbers are also valid for complex numbers as long as i is treated as a variable. In addition, note that addition and subtraction of complex numbers is similar to addition and subtraction of like terms in exponents (see Section 1.1b, Case III)

Example 2.6-1

$$\boxed{(3+2i)+(5-4i)} = \boxed{(3+5)+(2i-4i)} = \boxed{(3+5)+(2-4)i} = \boxed{\mathbf{8-2i}}$$

Example 2.6-2

$$\boxed{(-3-4i)-(3-6i)} = \boxed{(-3-4i)+(-3+6i)} = \boxed{(-3-3)+(-4i+6i)} = \boxed{-6+(-4+6)i} = \boxed{\mathbf{-6+2i}}$$

Example 2.6-3

$$\boxed{(5-8i)-2i} = \boxed{5-8i-2i} = \boxed{5+(-8-2)i} = \boxed{\mathbf{5-10i}}$$

Example 2.6-4

$$\boxed{(4-2i)-(-2+5i)} = \boxed{(4-2i)+(2-5i)} = \boxed{(4+2)-(-2i-5i)} = \boxed{6+(-2-5)i} = \boxed{\mathbf{6-7i}}$$

Example 2.6-5

$$\boxed{(2+3i)+[(1-5i)-(2-8i)]} = \boxed{(2+3i)+[(1-2)+(-5i+8i)]} = \boxed{(2+3i)+(-1+3i)} = \boxed{(2-1)+(3i+3i)} = \boxed{\mathbf{1+6i}}$$

Example 2.6-6

$$\boxed{[5.8i+(3-3.6i)]+(2-3i)} = \boxed{[3+(5.8-3.6)i]+(2-3i)} = \boxed{(3+2.2i)+(2-3i)} = \boxed{(3+2)+(2.2i-3i)} = \boxed{\mathbf{5-0.8i}}$$

Example 2.6-7

$$\boxed{[3i+(3.5+2.1i)]-(4-1.5i)} = \boxed{[3.5+(3i+2.1i)]+(-4+1.5i)} = \boxed{(3.5+5.1i)+(-4+1.5i)} = \boxed{(3.5-4)+(5.1i+1.5i)}$$

$$= \boxed{(3.5-4)+(5.1+1.5)i} = \boxed{\mathbf{-0.5+6.6i}}$$

Example 2.6-8

$$\boxed{\left(2\sqrt{5}+\sqrt{-4}\right)-\left(3-\sqrt{-9}\right)} = \boxed{\left(2\sqrt{5}+2\sqrt{-1}\right)+\left(-3+3\sqrt{-1}\right)} = \boxed{\left(2\sqrt{5}+2i\right)+(-3+3i)} = \boxed{(4.47+2i)+(-3+3i)}$$

$$= \boxed{(4.47-3)+(2i+3i)} = \boxed{\mathbf{1.47+5i}}$$

Example 2.6-9

$$\boxed{\left(5-\sqrt{8}\,i\right)-\left(1-\sqrt{3}\,i\right)} = \boxed{\left(5-\sqrt{8}\,i\right)+\left(-1+\sqrt{3}\,i\right)} = \boxed{(5-1)+\left(-2\sqrt{2}\,i+\sqrt{3}\,i\right)} = \boxed{4+(-2.83+1.73)i} = \boxed{\mathbf{4-1.1}\boldsymbol{i}}$$

Example 2.6-10

$$\boxed{\left[\left(2+\sqrt{27}\,i\right)+\sqrt{3}\,i\right]-\left(1-5\sqrt{3}\,i\right)} = \boxed{\left[\left(2+3\sqrt{3}\,i\right)+\sqrt{3}\,i\right]+\left(-1+5\sqrt{3}\,i\right)} = \boxed{\left(2+4\sqrt{3}\,i\right)+\left(-1+5\sqrt{3}\,i\right)}$$

$$= \boxed{(2-1)+\left(4\sqrt{3}\,i+5\sqrt{3}\,i\right)} = \boxed{1+9\sqrt{3}\,i} = \boxed{\mathbf{1+15.59}\boldsymbol{i}}$$

Case II Multiplication of Complex Numbers

In general, two complex numbers are multiplied by one another in the following way:

$$(a+bi)\cdot(c+di) = ac+adi+bci+bdi^2 = ac+adi+bci-bd = (ac-bd)+(ad+bc)i$$

Note that the distributive property of multiplication is also valid for complex numbers. In addition, note that multiplication of two complex numbers is similar to the multiplication of two binomials using the Foil Method (see Section 1.2b, Case II).

Example 2.6-11

$$\boxed{(4+5i)(3-2i)} = \boxed{(4\times3)+(4\times-2i)+(5i\times3)+(5i\times-2i)} = \boxed{12-8i+15i-10i^2} = \boxed{12-8i+15i+(-10\times-1)}$$

$$= \boxed{12-8i+15i+10} = \boxed{(12+10)+(-8+15)i} = \boxed{\mathbf{22+7}\boldsymbol{i}}$$

Example 2.6-12

$$\boxed{(-2-3i)(-5+i)} = \boxed{(-2\times-5)+(-2\times i)+(-3i\times-5)+(-3i\times i)} = \boxed{10-2i+15i-3i^2} = \boxed{10-2i+15i+(-3\times-1)}$$

$$= \boxed{10-2i+15i+3} = \boxed{(10+3)+(-2i+15i)} = \boxed{\mathbf{13+13}\boldsymbol{i}}$$

Example 2.6-13

$$\boxed{(4-7i)(0+3i)} = \boxed{(4-7i)\times3i} = \boxed{4\times3i-7i\times3i} = \boxed{12i-21i^2} = \boxed{12i+21} = \boxed{\mathbf{21+12}\boldsymbol{i}}$$

Example 2.6-14

$$\boxed{(5-3i)(2-i)} = \boxed{(5\times2)+(5\times-1)+(2\times-3i)+(-3i\times-i)} = \boxed{10-5i-6i+3i^2} = \boxed{10-5i-6i-3} = \boxed{(10-3)+(-5-6)i}$$

$$= \boxed{\mathbf{7-11}\boldsymbol{i}}$$

Example 2.6-15

$$\boxed{(2+5i)(2-5i)} = \boxed{(2\times2)+(2\times-5i)+(2\times-5i)+(5i\times-5i)} = \boxed{4-10i+10i-25i^2} = \boxed{4+(-10+10)i-25\times-1}$$

$$= \boxed{4+0+25} = \boxed{\mathbf{29}}$$

Example 2.6-16

$$\boxed{\left(3+\sqrt{5}\,i\right)\left(5-\sqrt{2}\,i\right)} = \boxed{(3\times5)+\left(3\times-\sqrt{2}\,i\right)+\left(\sqrt{5}\,i\times5\right)+\left(\sqrt{5}\,i\times-\sqrt{2}\,i\right)} = \boxed{15-3\sqrt{2}\,i+5\sqrt{5}\,i-\sqrt{10}\,i^2}$$

$$= \boxed{\left(15+\sqrt{10}\right)+\left(-3\sqrt{2}+5\sqrt{5}\right)i} = \boxed{(15+3.16)+(-4.24+11.18)i} = \boxed{\mathbf{18.16+6.94}\boldsymbol{i}}$$

Example 2.6-17

$\boxed{\left[5i^3(-5+i)\right](2-i)} = \boxed{[-5i(-5+i)](2-i)} = \boxed{\left(25i-5i^2\right)(2-i)} = \boxed{(5+25i)(2-i)} = \boxed{(5\times 2)+(5\times -i)+(25i\times 2)}$

$\boxed{+(25i\times -i)} = \boxed{10-5i+50i-25i^2} = \boxed{(10+25)+(-5+50)i} = \boxed{\mathbf{35+45i}}$

Example 2.6-18

$\boxed{\left(4-\sqrt{8}\,i\right)\left(2+\sqrt{2}\,i\right)} = \boxed{\left(4-2\sqrt{2}\,i\right)\left(2+\sqrt{2}\,i\right)} = \boxed{(4\times 2)+\left(4\times\sqrt{2}\,i\right)+\left(-2\sqrt{2}\,i\times 2\right)+\left(-2\sqrt{2}\,i\times\sqrt{2}\,i\right)} = \boxed{8+4\sqrt{2}\,i}$

$\boxed{-4\sqrt{2}\,i-4i^2} = \boxed{8-4i^2} = \boxed{8+4} = \boxed{\mathbf{12}}$

Example 2.6-19

$\boxed{\left(\sqrt{6}-3i\right)\left(-2\sqrt{2}-i\right)} = \boxed{\left(\sqrt{6}\times -2\sqrt{2}\right)+\left(\sqrt{6}\times -i\right)+\left(-3i\times -2\sqrt{2}\right)+(-3i\times -i)} = \boxed{-2\sqrt{12}-\sqrt{6}\,i+6\sqrt{2}\,i+3i^2}$

$= \boxed{-4\sqrt{3}-\sqrt{6}\,i+6\sqrt{2}\,i-3} = \boxed{\left(-4\sqrt{3}-3\right)+\left(-\sqrt{6}+6\sqrt{2}\right)i} = \boxed{(-6.93-3)+(-2.45+8.49)i} = \boxed{\mathbf{-9.93+6.04i}}$

Example 2.6-20

$\boxed{i^2[(2-5i)(3-5i)]} = \boxed{-(2-5i)(3-5i)} = \boxed{(-2+5i)(3-5i)} = \boxed{(-2\times 3)+(-2\times -5i)+(5i\times 3)+(5i\times -5i)}$

$= \boxed{-6+10i+15i-25i^2} = \boxed{-6+10i+15i+25} = \boxed{(-6+25)+(10+15)i} = \boxed{\mathbf{19+25i}}$

Case III Division of Complex Numbers

In general, two complex numbers are divided by one another in the following way:

$$\frac{a+bi}{c+di} = \frac{a+bi}{c+di}\times\frac{a-bi}{c-di} = \frac{(a+bi)(c-di)}{(c+di)(c-di)} = \frac{ac-adi+bci-bdi^2}{c^2+cdi-cdi-d^2i^2} = \frac{ac-adi+bci+bd}{c^2+d^2} = \frac{(ac+bd)+(bc-ad)i}{c^2+d^2}$$

$$= \frac{ac+bd}{c^2+d^2}+\frac{(ad+bc)i}{c^2+d^2}$$

Note that division of two complex numbers that contain i in the denominator is similar to rationalizing the denominator for radical expressions where the numerator and the denominator are multiplied by the conjugate of the denominator (see Section 1.2b, Case V).

Example 2.6-21

$\boxed{\frac{2+3i}{i}} = \boxed{\frac{2+3i}{0+i}} = \boxed{\frac{2+3i}{0+i}\times\frac{0-i}{0-i}} = \boxed{\frac{2+3i}{i}\times\frac{-i}{-i}} = \boxed{\frac{(2+3i)\times -i}{-i\times i}} = \boxed{\frac{-2i-3i^2}{-i^2}} = \boxed{\frac{-2i+(-3\times -1)}{-(-1)}} = \boxed{\mathbf{3-2i}}$

Example 2.6-22

$\boxed{\frac{-2+i}{-i}} = \boxed{\frac{-2+i}{0-i}\times\frac{0+i}{0+i}} = \boxed{\frac{-2+i}{-i}\times\frac{i}{i}} = \boxed{\frac{(-2+i)\times i}{-i\times i}} = \boxed{\frac{-2i+i^2}{-i^2}} = \boxed{\frac{-2i-1}{-(-1)}} = \boxed{\frac{-1-2i}{1}} = \boxed{\mathbf{-1-2i}}$

Example 2.6-23

$\boxed{\frac{2-5i}{3+2i}} = \boxed{\frac{2-5i}{3+2i}\times\frac{3-2i}{3-2i}} = \boxed{\frac{6-4i-15i+10i^2}{9-6i+6i-4i^2}} = \boxed{\frac{6-19i-10}{9+4}} = \boxed{\frac{-4-19i}{13}} = \boxed{-\frac{4}{13}-\frac{19}{13}i} = \boxed{\mathbf{-0.31-1.46i}}$

Example 2.6-24

$$\boxed{\frac{i}{1+4i}}=\boxed{\frac{i}{1+4i}\times\frac{1-4i}{1-4i}}=\boxed{\frac{i-4i^2}{1-4i+4i-4i^2}}=\boxed{\frac{i+4}{1-4i^2}}=\boxed{\frac{4+i}{1+4}}=\boxed{\frac{4+i}{5}}=\boxed{\frac{4}{5}+\frac{1}{5}i}=\boxed{\mathbf{0.8+0.2}i}$$

Example 2.6-25

$$\boxed{\frac{4+i}{2+3i}}=\boxed{\frac{4+i}{2+3i}\times\frac{2-3i}{2-3i}}=\boxed{\frac{8-12i+2i-3i^2}{4-6i+6i-9i^2}}=\boxed{\frac{(8+3)-10i}{4+9}}=\boxed{\frac{11-10i}{13}}=\boxed{\frac{11}{13}-\frac{10}{13}i}=\boxed{\mathbf{0.85-0.77}i}$$

Example 2.6-26

$$\boxed{\frac{2+i}{\sqrt{2}+5i}}=\boxed{\frac{2+i}{\sqrt{2}+5i}\times\frac{\sqrt{2}-5i}{\sqrt{2}-5i}}=\boxed{\frac{2\sqrt{2}-10i+\sqrt{2}\,i-5i^2}{2-5\sqrt{2}\,i+5\sqrt{2}\,i-25i^2}}=\boxed{\frac{2\sqrt{2}-10i+\sqrt{2}\,i+5}{2+25}}=\boxed{\frac{\left(2\sqrt{2}+5\right)+\left(-10+\sqrt{2}\right)i}{27}}$$

$$=\boxed{\frac{(2.83+5)+(-10+1.41)i}{27}}=\boxed{\frac{7.83-8.59i}{27}}=\boxed{\frac{7.83}{27}-\frac{8.59}{27}i}=\boxed{\mathbf{0.29-0.32}i}$$

Example 2.6-27

$$\boxed{\frac{-3+\sqrt{2}\,i}{\sqrt{3}-\sqrt{25}\,i}}=\boxed{\frac{-3+\sqrt{2}\,i}{\sqrt{3}-5i}}=\boxed{\frac{-3+\sqrt{2}\,i}{\sqrt{3}-5i}\times\frac{\sqrt{3}+5i}{\sqrt{3}+5i}}=\boxed{\frac{-3\sqrt{3}-15i+\sqrt{6}\,i+5\sqrt{2}\,i^2}{3+5\sqrt{3}\,i-5\sqrt{3}\,i-25i^2}}=\boxed{\frac{-3\sqrt{3}-15i+\sqrt{6}\,i-5\sqrt{2}}{3+25}}$$

$$=\boxed{\frac{\left(-3\sqrt{3}-5\sqrt{2}\right)+\left(-15+\sqrt{6}\right)i}{28}}=\boxed{\frac{(-5.2-7.07)+(-15+2.45)i}{28}}=\boxed{\frac{-12.27-12.55i}{28}}=\boxed{\mathbf{-0.44-0.45}i}$$

Example 2.6-28

$$\boxed{\frac{1-\sqrt{5}\,i}{3-8i^3}}=\boxed{\frac{1-\sqrt{5}\,i}{3+8i}}=\boxed{\frac{1-\sqrt{5}\,i}{3+8i}\times\frac{3-8i}{3-8i}}=\boxed{\frac{3-8i-3\sqrt{5}\,i+8\sqrt{5}\,i^2}{9-24i+24i-64i^2}}=\boxed{\frac{3-8i-3\sqrt{5}\,i-8\sqrt{5}}{9+64}}=\boxed{\frac{\left(3-8\sqrt{5}\right)-\left(8+3\sqrt{5}\right)i}{73}}$$

$$=\boxed{\frac{(3-17.89)-(8+6.71)i}{73}}=\boxed{\frac{-14.89-14.71i}{73}}=\boxed{-\frac{14.89}{73}-\frac{14.71}{73}i}=\boxed{\mathbf{-0.2-0.2}i}$$

Example 2.6-29

$$\boxed{\frac{3-5i}{i^5}}=\boxed{\frac{3-5i}{i^2\cdot i^2\cdot i}}=\boxed{\frac{3-5i}{-1\times-1\times i}}=\boxed{\frac{3-5i}{i}}=\boxed{\frac{3-5i}{0+i}}=\boxed{\frac{3-5i}{0+i}\times\frac{0-i}{0-i}}=\boxed{\frac{3-5i}{i}\times\frac{-i}{-i}}=\boxed{\frac{-3i+5i^2}{-i^2}}=\boxed{\frac{-3i-5}{-(-1)}}$$

$$=\boxed{\frac{-5-3i}{1}}=\boxed{\mathbf{-5-3}i}$$

Example 2.6-30

$$\boxed{\frac{\sqrt{4}+i}{2+\sqrt{3}\,i}}=\boxed{\frac{2+i}{2+\sqrt{3}\,i}\times\frac{2-\sqrt{3}\,i}{2-\sqrt{3}\,i}}=\boxed{\frac{4-2\sqrt{3}\,i+2i-\sqrt{3}\,i^2}{4-2\sqrt{3}\,i+2\sqrt{3}\,i-\sqrt{9}\,i^2}}=\boxed{\frac{4-2\sqrt{3}\,i+2i+\sqrt{3}}{4-3i^2}}=\boxed{\frac{\left(4+\sqrt{3}\right)+\left(-2\sqrt{3}+2\right)i}{4+3}}$$

$$=\boxed{\frac{\left(4+\sqrt{3}\right)+\left(-2\sqrt{3}+2\right)i}{7}}=\boxed{\frac{(4+1.73)+(-3.46+2)i}{7}}=\boxed{\frac{5.73-1.46i}{7}}=\boxed{\frac{5.73}{7}-\frac{1.46}{7}i}=\boxed{\mathbf{0.82-0.21}i}$$

Case IV Mixed Operations Involving Complex Numbers

In this case combining addition, subtraction, multiplication, and division of complex numbers are addressed. Note that math properties discussed in Cases I through III are also applicable here. The following show examples of mixed operations involving complex numbers.

Example 2.6-31

$$\boxed{\frac{2+3i}{1+i}\times\frac{2+i}{3+i}}=\boxed{\frac{(2+3i)\times(2+i)}{(1+i)\times(3+i)}}=\boxed{\frac{4+2i+6i+3i^2}{3+i+3i+i^2}}=\boxed{\frac{4+2i+6i-3}{3+i+3i-1}}=\boxed{\frac{(4-3)+(2i+6i)}{(3-1)+(i+3i)}}=\boxed{\frac{(4-3)+(2+6)i}{(3-1)+(1+3)i}}$$

$$=\boxed{\frac{1+8i}{2+4i}}=\boxed{\frac{(1+8i)\times(2-4i)}{(2+4i)\times(2-4i)}}=\boxed{\frac{2-4i+16i-32i^2}{4-8i+8i-16i^2}}=\boxed{\frac{2-4i+16i+32}{4+16}}=\boxed{\frac{34+12i}{20}}=\boxed{\frac{34}{20}+\frac{12}{20}i}=\boxed{\mathbf{1.7+0.6}\boldsymbol{i}}$$

Example 2.6-32

$$\boxed{\frac{1-2i}{i}+\frac{2-3i}{i}}=\boxed{\frac{(1-2i)+(2-3i)}{i}}=\boxed{\frac{(1+2)+(-2i-3i)}{i}}=\boxed{\frac{3-5i}{i}}=\boxed{\frac{3-5i}{0+i}}=\boxed{\frac{3-5i}{0+i}\times\frac{0-i}{0-i}}=\boxed{\frac{(3-5i)\times -i}{i\times -i}}$$

$$=\boxed{\frac{-3i+5i^2}{-i^2}}=\boxed{\frac{-3i-5}{1}}=\boxed{\mathbf{-5-3}\boldsymbol{i}}$$

Example 2.6-33

$$\boxed{\frac{2+i}{1+3i}\div\frac{1+i}{1-i}}=\boxed{\frac{2+i}{1+3i}\times\frac{1-i}{1+i}}=\boxed{\frac{(2+i)\times(1-i)}{(1+3i)\times(1+i)}}=\boxed{\frac{2-2i+i-i^2}{1+i+3i+3i^2}}=\boxed{\frac{2-2i+i+1}{1+i+3i-3}}=\boxed{\frac{(2+1)+(-2+1)i}{(1-3)+(1+3)i}}=\boxed{\frac{3-i}{-2+4i}}$$

$$=\boxed{\frac{3-i}{-2+4i}\times\frac{-2-4i}{-2-4i}}=\boxed{\frac{(3-i)(-2-4i)}{(-2+4i)(-2-4i)}}=\boxed{\frac{-6-12i+2i+4i^2}{4+8i-8i-16i^2}}=\boxed{\frac{-6-12i+2i-4}{4+16}}=\boxed{\frac{(-6-4)+(-12+2)i}{20}}$$

$$=\boxed{\frac{-10-10i}{20}}=\boxed{-\frac{10}{20}-\frac{10}{20}i}=\boxed{\mathbf{-0.5-0.5}\boldsymbol{i}}$$

Example 2.6-34

$$\boxed{[(2+3i)(1-i)]+(1-2i)}=\boxed{(2-2i+3i-3i^2)+(1-2i)}=\boxed{(2-2i+3i+3)+(1-2i)}=\boxed{[(2+3)+(-2+3)i]+(1-2i)}$$

$$=\boxed{(5+i)+(1-2i)}=\boxed{(5+1)+(i-2i)}=\boxed{\mathbf{6}-\boldsymbol{i}}$$

Example 2.6-35

$$\boxed{i^2(1+i)-i^3(1+2i)}=\boxed{-(1+i)+i(1+2i)}=\boxed{(-1-i)+(i+2i^2)}=\boxed{(-1-i)+(i-2)}=\boxed{(-1-2)+(-i+i)}=\boxed{\mathbf{-3}}$$

Example 2.6-36

$$\boxed{[(5-2i)-(2-i)]\times i^5}=\boxed{[(5-2i)+(-2+i)]\times(i^2\cdot i^2\cdot i)}=\boxed{[(5-2)+(-2i+i)]\times i}=\boxed{(3-i)\times i}=\boxed{3i-i^2}=\boxed{\mathbf{1+3}\boldsymbol{i}}$$

Example 2.6-37

$$\boxed{\frac{2-i}{1+i}\div\frac{1}{1+2i}}=\boxed{\frac{2-i}{1+i}\times\frac{1+2i}{1}}=\boxed{\frac{(2-i)\times(1+2i)}{(1+i)\times 1}}=\boxed{\frac{2+4i-i-2i^2}{1+i}}=\boxed{\frac{2+4i-i+2}{1+i}}=\boxed{\frac{(2+2)+(4-1)i}{1+i}}=\boxed{\frac{4+3i}{1+i}}$$

$$= \boxed{\frac{4+3i}{1+i}\times\frac{1-i}{1-i}} = \boxed{\frac{(4+3i)(1-i)}{(1+i)(1-i)}} = \boxed{\frac{4-4i+3i-3i^2}{1-i+i-i^2}} = \boxed{\frac{4-4i+3i+3}{1+1}} = \boxed{\frac{(4+3)+(-4+3)i}{2}} = \boxed{\frac{7-i}{2}} = \boxed{\mathbf{3.5-0.5}i}$$

Example 2.6-38

$$\boxed{\frac{i^4}{1-i}\times\frac{2+i}{i}} = \boxed{\frac{i^2\times i^2}{1-i}\times\frac{2+i}{i}} = \boxed{\frac{-1\times-1}{1-i}\times\frac{2+i}{i}} = \boxed{\frac{1}{1-i}\times\frac{2+i}{i}} = \boxed{\frac{1\times(2+i)}{(1-i)\times i}} = \boxed{\frac{2+i}{i-i^2}} = \boxed{\frac{2+i}{1+i}} = \boxed{\frac{2+i}{1+i}\times\frac{1-i}{1-i}}$$

$$= \boxed{\frac{(2+i)\times(1-i)}{(1+i)\times(1-i)}} = \boxed{\frac{2-2i+i-i^2}{1-i+i-i^2}} = \boxed{\frac{2-2i+i+1}{1+1}} = \boxed{\frac{(2+1)+(-2+1)i}{2}} = \boxed{\frac{3-i}{2}} = \boxed{\frac{3}{2}-\frac{1}{2}i} = \boxed{\mathbf{1.5-0.5}i}$$

Example 2.6-39

$$\boxed{\frac{(3-4i)(2+i)}{1-2i}} = \boxed{\frac{6+3i-8i-4i^2}{1-2i}} = \boxed{\frac{6+3i-8i+4}{1-2i}} = \boxed{\frac{(6+4)+(3i-8i)}{1-2i}} = \boxed{\frac{10-5i}{1-2i}} = \boxed{\frac{10-5i}{1-2i}\times\frac{1+2i}{1+2i}}$$

$$= \boxed{\frac{(10-5i)\times(1+2i)}{(1-2i)\times(1+2i)}} = \boxed{\frac{10+20i-5i-10i^2}{1+2i-2i-4i^2}} = \boxed{\frac{10+20i-5i+10}{1+4}} = \boxed{\frac{(10+10)+(20i-5i)}{5}} = \boxed{\frac{20+15i}{5}} = \boxed{\mathbf{4+3}i}$$

Example 2.6-40

$$\boxed{\frac{i^3}{4-i}\div i^2} = \boxed{\frac{-i}{4-i}\div\frac{i^2}{1}} = \boxed{\frac{-i}{4-i}\times\frac{1}{i^2}} = \boxed{\frac{-i}{4-i}\times\frac{1}{-1}} = \boxed{\frac{-i\times 1}{(4-i)\times-1}} = \boxed{\frac{-i}{-(4-i)}} = \boxed{\frac{i}{4-i}} = \boxed{\frac{i}{4-i}\times\frac{4+i}{4+i}}$$

$$= \boxed{\frac{i(4+i)}{(4-i)(4+i)}} = \boxed{\frac{4i+i^2}{16+4i-4i-i^2}} = \boxed{\frac{4i-1}{16-i^2}} = \boxed{\frac{-1+4i}{16+1}} = \boxed{\frac{-1+4i}{17}} = \boxed{-\frac{1}{17}+\frac{4}{17}i} = \boxed{\mathbf{-0.06+0.24}i}$$

Example 2.6-41

$$\boxed{(1+3i)\times\frac{2+i}{3i}} = \boxed{\frac{(1+3i)}{1}\times\frac{2+i}{3i}} = \boxed{\frac{(1+3i)\times(2+i)}{1\times 3i}} = \boxed{\frac{2+i+6i+3i^2}{3i}} = \boxed{\frac{2+i+6i-3}{3i}} = \boxed{\frac{(2-3)+(1+6)i}{3i}} = \boxed{\frac{-1+7i}{0+3i}}$$

$$= \boxed{\frac{-1+7i}{0+3i}\times\frac{0-3i}{0-3i}} = \boxed{\frac{(-1+7i)\times-3i}{3i\times-3i}} = \boxed{\frac{3i-21i^2}{-9i^2}} = \boxed{\frac{3i+21}{9}} = \boxed{\frac{21}{9}+\frac{3}{9}i} = \boxed{\mathbf{2.33+0.33}i}$$

Example 2.6-42

$$\boxed{\frac{(4-3i)(2+i)}{(2-i)(1+i)}} = \boxed{\frac{8+4i-6i-3i^2}{2+2i-i-i^2}} = \boxed{\frac{8+4i-6i+3}{2+2i-i+1}} = \boxed{\frac{(8+3)+(4i-6i)}{(2+1)+(2i-i)}} = \boxed{\frac{11-2i}{3+i}} = \boxed{\frac{11-2i}{3+i}\times\frac{3-i}{3-i}}$$

$$= \boxed{\frac{(11-2i)\times(3-i)}{(3+i)\times(3-i)}} = \boxed{\frac{33-11i-6i+2i^2}{9-3i+3i-i^2}} = \boxed{\frac{33-11i-6i-2}{9+1}} = \boxed{\frac{(33-2)+(-11i-6i)}{10}} = \boxed{\frac{31-17i}{10}} = \boxed{\mathbf{3.1-1.7}i}$$

Example 2.6-43

$$\boxed{\frac{3}{1+i}\times\frac{1}{2-3i}} = \boxed{\frac{3\times 1}{(1+i)(2-3i)}} = \boxed{\frac{3}{2-3i+2i-3i^2}} = \boxed{\frac{3}{2-3i+2i+3}} = \boxed{\frac{3}{(2+3)+(-3+2)i}} = \boxed{\frac{3}{5-i}} = \boxed{\frac{3}{5-i}\times\frac{5+i}{5+i}}$$

$$\boxed{\frac{3(5+i)}{(5-i)(5+i)}} = \boxed{\frac{15+3i}{25+5i-5i-i^2}} = \boxed{\frac{15+3i}{25-i^2}} = \boxed{\frac{15+3i}{25+1}} = \boxed{\frac{15+3i}{26}} = \boxed{\frac{15}{26}+\frac{3}{26}i} = \boxed{\mathbf{0.58+0.12}i}$$

Example 2.6-44

$$\boxed{\frac{2}{a-bi}+\frac{5}{a-bi}} = \boxed{\frac{2+5}{a-bi}} = \boxed{\frac{7}{a-bi}\times\frac{a+bi}{a+bi}} = \boxed{\frac{7(a+bi)}{(a-bi)(a+bi)}} = \boxed{\frac{7(a+bi)}{a^2+abi-abi-b^2i^2}} = \boxed{\frac{7(a+bi)}{a^2-b^2i^2}}$$

$$= \boxed{\frac{7(a+bi)}{a^2+b^2}} = \boxed{\frac{\mathbf{7a}}{\mathbf{a^2+b^2}}+\frac{\mathbf{7b}}{\mathbf{a^2+b^2}}\boldsymbol{i}}$$

Example 2.6-45

$$\boxed{\frac{1}{2+3i}-\frac{4}{1-i}} = \boxed{\frac{[1\times(1-i)]-[4\times(2+3i)]}{(2+3i)(1-i)}} = \boxed{\frac{(1-i)+(-8-12i)}{2-2i+3i-3i^2}} = \boxed{\frac{(1-8)+(-i-12i)}{2-2i+3i+3}} = \boxed{\frac{-7-13i}{(2+3)+(-2i+3i)}}$$

$$= \boxed{\frac{-7-13i}{5+i}} = \boxed{\frac{-7-13i}{5+i}\times\frac{5-i}{5-i}} = \boxed{\frac{(-7-13i)(5-i)}{(5+i)(5-i)}} = \boxed{\frac{-35+7i-65i+13i^2}{25-5i+5i-i^2}} = \boxed{\frac{-35+7i-65i-13}{25+1}}$$

$$= \boxed{\frac{(-35-13)+(7i-65i)}{26}} = \boxed{\frac{-48-58i}{26}} = \boxed{-\frac{48}{26}-\frac{58}{26}i} = \boxed{\mathbf{-1.85-2.23}\,\boldsymbol{i}}$$

Example 2.6-46

$$\boxed{\frac{4}{1+i}+\frac{2+3i}{3-2i}} = \boxed{\frac{[4\times(3-2i)]+[(2+3i)\times(1+i)]}{(1+i)(3-2i)}} = \boxed{\frac{[4\times(3-2i)]+[(2+3i)\times(1+i)]}{(1+i)(3-2i)}} = \boxed{\frac{(12-8i)+(2+2i+3i+3i^2)}{3-2i+3i-2i^2}}$$

$$= \boxed{\frac{(12+2-3)+(-8i+2i+3i)}{(3+2)+(-2i+3i)}} = \boxed{\frac{11-3i}{5+i}} = \boxed{\frac{11-3i}{5+i}\times\frac{5-i}{5-i}} = \boxed{\frac{(11-3i)(5-i)}{(5+i)(5-i)}} = \boxed{\frac{55-11i-15i+3i^2}{25-5i+5i-i^2}}$$

$$= \boxed{\frac{55-11i-15i-3}{25+1}} = \boxed{\frac{(55-3)+(-11i-15i)}{26}} = \boxed{\frac{52-26i}{26}} = \boxed{\frac{52}{26}-\frac{26}{26}i} = \boxed{\mathbf{2-}\boldsymbol{i}}$$

Example 2.6-47

$$\boxed{\left(i^3\div\frac{1}{2i}\right)+\frac{1}{i}} = \boxed{\left(\frac{i^3}{1}\times\frac{2i}{1}\right)+\frac{1}{i}} = \boxed{\frac{i^3\times 2i}{1\times 1}+\frac{1}{i}} = \boxed{\frac{2i^4}{1}+\frac{1}{i}} = \boxed{\frac{2\times -1\times -1}{1}+\frac{1}{i}} = \boxed{\frac{2}{1}+\frac{1}{i}} = \boxed{\frac{(2\times i)+(1\times 1)}{1\times i}} = \boxed{\frac{2i+1}{i}}$$

$$= \boxed{\frac{2i+1}{0+i}} = \boxed{\frac{2i+1}{0+i}\times\frac{0-i}{0-i}} = \boxed{\frac{(2i+1)\times -i}{i\times -i}} = \boxed{\frac{-2i^2-i}{-i^2}} = \boxed{\frac{2-i}{1}} = \boxed{\mathbf{2-}\boldsymbol{i}}$$

Example 2.6-48

$$\boxed{\frac{i^2}{1-i}\times\frac{1}{3-i^3}} = \boxed{\frac{i^2}{1-i}\times\frac{1}{3+i}} = \boxed{\frac{i^2\times 1}{(1-i)(3+i)}} = \boxed{\frac{-1}{3+i-3i-i^2}} = \boxed{-\frac{1}{3+i-3i+1}} = \boxed{-\frac{1}{(3+1)+(1-3)i}} = \boxed{-\frac{1}{4-2i}}$$

$$= \boxed{-\frac{1}{4-2i}\times\frac{4+2i}{4+2i}} = \boxed{\frac{-1\times(4+2i)}{(4-2i)(4+2i)}} = \boxed{\frac{-4-2i}{16+8i-8i-4i^2}} = \boxed{\frac{-4-2i}{16+4}} = \boxed{\frac{-4-2i}{20}} = \boxed{-\frac{4}{20}-\frac{2}{20}i} = \boxed{\mathbf{-0.2-0.1}\boldsymbol{i}}$$

Example 2.6-49

$$\boxed{\frac{1+i}{i}\div\frac{1}{2+3i}} = \boxed{\frac{1+i}{i}\times\frac{2+3i}{1}} = \boxed{\frac{(1+i)\times(2+3i)}{i\times 1}} = \boxed{\frac{2+3i+2i+3i^2}{i}} = \boxed{\frac{2+3i+2i-3}{i}} = \boxed{\frac{(2-3)+(3+2)i}{i}}$$

$$= \boxed{\frac{-1+5i}{0+i}} = \boxed{\frac{-1+5i}{0+i} \times \frac{0-i}{0-i}} = \boxed{\frac{(-1+5i)\times -i}{i \times -i}} = \boxed{\frac{i-5i^2}{-i^2}} = \boxed{\frac{i+5}{-(-1)}} = \boxed{\frac{i+5}{1}} = \boxed{5+i}$$

Example 2.6-50

$$\boxed{[(4-5i)(2+i)]+\frac{1}{i^2}} = \boxed{(8+4i-10i-5i^2)+\frac{1}{-1}} = \boxed{(8+4i-10i+5)-1} = \boxed{(8+5-1)+(4i-10i)} = \boxed{12-6i}$$

Practice Problems – Math Operations Involving Complex Numbers

Section 2.6 Case I Practice Problems – Add or subtract the following complex numbers:

a. $(4+2i)+(8-5i) =$ b. $(7-3i)-(-5+4i) =$ c. $(4+7i)+[(2-5i)-(6-i)] =$

d. $(2\sqrt{5}+\sqrt{-3})-(4-\sqrt{-25}) =$ e. $[(2+5i)+\sqrt{5}\,i^3]-(1+3\sqrt{5}\,i) =$ f. $(3-\sqrt{6}\,i)-(1+\sqrt{2}\,i) =$

Section 2.6 Case II Practice Problems – Multiply the following complex numbers by one another:

a. $(5+2i)(3-6i) =$ b. $(-6-2i)(-7+i) =$ c. $[3i^6(5-i)](3+i) =$

d. $(3-\sqrt{9}\,i)(5+\sqrt{3}\,i) =$ e. $(\sqrt{5}-i)(-3\sqrt{5}+i) =$ f. $i^3[(2+4i)(4-2i)] =$

Section 2.6 Case III Practice Problems – Divide the following complex numbers by one another:

a. $\frac{1-4i}{5+3i} =$ b. $\frac{i^3}{1-8i} =$ c. $\frac{2-i}{\sqrt{3}+4i} =$

d. $\frac{-2+\sqrt{2}\,i}{\sqrt{5}-\sqrt{36}\,i} =$ e. $\frac{3-\sqrt{3}\,i}{1-8i^5} =$ f. $\frac{3+2i}{1-i^7} =$

Section 2.6 Case IV Practice Problems – Simplify the following expressions involving complex numbers:

a. $i^3(1-2i)+i^4(2-5i) =$ b. $\frac{4+5i}{1-i} \div \frac{1}{1+3i} =$ c. $\frac{1-i}{2+3i} \times \frac{1}{1+i} =$

d. $i^8 \div \frac{2-4i}{1+2i} =$ e. $\frac{(2+5i)(1-i)}{(1+4i)(2+3i)} =$ f. $(5+8i)-\frac{1}{1-i} =$

g. $(4+2i)\div\frac{2}{1-3i} =$ h. $(2-5i)+\frac{1-3i}{1+i} =$ i. $\frac{2+3i}{i} \div \frac{2-4i}{1+i} =$

Chapter 3
Matrices

Quick Reference to Chapter 3 Problems

$$\boxed{A_{2\times3}=\begin{bmatrix}1&2&3\\-1&3&1\end{bmatrix}=}\ ;\quad \boxed{B_{3\times3}=\begin{bmatrix}1&0&3\\1&-1&1\\0&1&0\end{bmatrix}=}\ ;\quad \boxed{C_{3\times4}=\begin{bmatrix}1&2&-3&5\\0&1&-1&3\\-2&0&4&6\end{bmatrix}=}\ .$$

$$\boxed{\begin{bmatrix}1&2\\3&5\end{bmatrix}+\begin{bmatrix}2&4\\-3&1\end{bmatrix}=}\ ;\quad \boxed{\begin{bmatrix}10&3&-1\\1&5&6\\2&3&8\end{bmatrix}+\begin{bmatrix}1&7&9\\2&-3&0\\1&5&-3\end{bmatrix}=}\ ;\quad \boxed{\begin{bmatrix}1&3&6\\-1&0&2\\5&-2&4\end{bmatrix}-\begin{bmatrix}-3&5&2\\0&-4&8\\5&-3&6\end{bmatrix}=}$$

$$\boxed{\begin{bmatrix}-1&0&1\\3&1&0\\1&0&2\end{bmatrix}\begin{bmatrix}0&1&2\\3&0&1\\-1&1&0\end{bmatrix}=}\ ;\quad \boxed{\begin{bmatrix}1&0&2\\3&-1&4\end{bmatrix}\begin{bmatrix}1&2\\-1&3\\2&5\end{bmatrix}=}\ ;\quad \boxed{\begin{bmatrix}1&0&-1\\2&3&-4\\5&1&-1\end{bmatrix}\begin{bmatrix}x\\y\\z\end{bmatrix}=\begin{bmatrix}1\\2\\5\end{bmatrix}=}$$

$$\boxed{\delta(A)=\begin{vmatrix}0&0&3\\1&0&0\\2&0&-1\end{vmatrix}=}\ ;\quad \boxed{\delta(B)=4\times\begin{vmatrix}-1&3\\2&5\end{vmatrix}=}\ ;\quad \boxed{\delta(C)=\begin{vmatrix}0&0&3\\1&3&0\\2&4&-1\end{vmatrix}=}$$

$$\boxed{\left(\begin{bmatrix}3&2\\1&2\end{bmatrix}\begin{bmatrix}1&-1\\3&1\end{bmatrix}\right)^{-1}=}\ ;\quad \boxed{\begin{bmatrix}1&0\\2&3\end{bmatrix}^{-1}\cdot\begin{bmatrix}1&3\\0&-1\end{bmatrix}^{-1}=}\ ;\quad \boxed{\left(\begin{bmatrix}1&3\\0&-1\end{bmatrix}\cdot\begin{bmatrix}1&0\\2&3\end{bmatrix}\right)^{-1}=}$$

$$\boxed{\begin{array}{l}-2x+3y=-1\\x-5y=1\end{array}}=\ ;\quad \boxed{\begin{array}{l}x+y=-2\\2x-2z=-1\\x+2y+2z=1\end{array}}=\ ;\quad \boxed{\begin{array}{l}x+2y-z=1\\-x+3y+z=0\\x+y-3z=2\end{array}}=$$

$$\boxed{\begin{array}{l}-2x+3y=1\\x-2y=0\end{array}}=\ ;\quad \boxed{\begin{array}{l}x+y=-1\\3x-2z=1\\y+4z=0\end{array}}=\ ;\quad \boxed{\begin{array}{l}x-2y+3z=3\\x+z=-4\\2x+2y+z=-1\end{array}}=$$

Case III - Solving Linear Systems Using the Inverse Matrices Method, *p. 218*

$$\boxed{\begin{array}{l} x+3y-3z=-1 \\ 2x-3y+2z=3 \\ x+2y-z=-2 \end{array}} = ; \quad \boxed{\begin{array}{l} x+y=-3 \\ 2x-z=-4 \\ 2y+2z=1 \end{array}} = ; \quad \boxed{\begin{array}{l} x+3y-z=-1 \\ -x+2y+3z=0 \\ x+y-2z=2 \end{array}} =$$

Case IV - Solving Linear Systems Using Cramer's Rule, *p. 224*

$$\boxed{\begin{array}{l} -x+2y=1 \\ x-4y=0 \end{array}} = ; \quad \boxed{\begin{array}{l} x+y=0 \\ 2x-z=1 \\ y+2z=0 \end{array}} = ; \quad \boxed{\begin{array}{l} x-y+3z=2 \\ x-z=-3 \\ 2x-2y+6z=-1 \end{array}} =$$

Case V - Solving Linear Systems Using the Gaussian Elimination Method, *p. 228*

$$\boxed{\begin{array}{r} 2x+3y=\ \ 6 \\ x-4y=-2 \end{array}} = ; \quad \boxed{\begin{array}{l} 2x-3y+z=-1 \\ 3x+2z=0 \\ \ x-2y=1 \end{array}} = ; \quad \boxed{\begin{array}{l} 2x+3z=-1 \\ x+3y=0 \\ 2x-2y+3z=-2 \end{array}} =$$

Case VI - Solving Linear Systems Using the Gauss-Jordan Elimination Method, *p. 234*

$$\boxed{\begin{array}{l} 2x-y=2 \\ 3x-\frac{2}{3}y=0 \end{array}} = ; \quad \boxed{\begin{array}{l} 3x-2z=-1 \\ x-y+z=0 \\ 2x+3y=-2 \end{array}} = ; \quad \boxed{\begin{array}{l} x-z=0 \\ x-3y=-1 \\ x+y=0 \end{array}} =$$

Chapter 3 - Matrices

The objective of this chapter is to improve the student's ability to solve problems involving matrices. Matrices are introduced in Section 3.1. How to add, subtract, and multiply matrices are introduced in Section 3.2. Calculating minors and cofactors and their use in finding the determinant of a matrix are addressed in Section 3.3. Computing the inverse of a square matrix and the steps in finding an inverse matrix are addressed in Section 3.4. In Section 3.5, solving linear systems using various methods such as the Addition, Substitution, Inverse Matrix, Cramer's Rule, Gaussian Elimination, and Gauss-Jordan Elimination methods are discussed in Section 3.6. Each section is concluded by solving examples with practice problems to further enhance the student's ability. Students are encouraged to gain a thorough knowledge of the matrix operations and learn how to find the determinant and inverse of a matrix. An in depth knowledge of matrix properties will greatly simplify solutions to linear systems of equations and introduce students to various methods used in solving these systems.

3.1 Introduction to Matrices

A matrix is defined as a rectangular array of numbers which are called the entries or elements of the matrix. In general, a matrix A with m rows and n columns is represented as

$$A = \begin{bmatrix} a_{11} & a_{12} & a_{13} & \cdots & a_{1n} \\ a_{21} & a_{22} & a_{23} & \cdots & a_{2n} \\ a_{31} & a_{32} & a_{33} & \cdots & a_{3n} \\ \vdots & \vdots & \vdots & \vdots & \vdots \\ a_{m1} & a_{m2} & a_{m3} & \cdots & a_{mn} \end{bmatrix} \begin{matrix} 1^{st}\ row \\ 2^{nd}\ row \\ 3^{rd}\ row \\ \\ m^{th}\ row \end{matrix}$$

$$\begin{matrix} 1^{st} & 2^{nd} & 3^{rd} & & n^{th} \\ col. & col. & col. & & col. \end{matrix}$$

Note that each matrix entry is read first by its location relative to the row and second by its location relative to the column. For example, a_{11} is read as the entry in the first row and the first column, a_{32} is read as the entry in the third row and the second column, a_{m3} is read as the entry in the mth row and the third column, a_{mn} is read as the entry in the mth row and the nth column, etc. The order or dimension of a matrix is given as $m \times n$ (reads as "m by n"). For example,

$$\begin{bmatrix} 1 & 0 \\ 2 & 1 \end{bmatrix}, \begin{bmatrix} 1 & 2 & 3 \\ -1 & 3 & 1 \end{bmatrix}, \begin{bmatrix} 1 & 0 & 3 \\ 1 & -1 & 1 \\ 0 & 1 & 0 \end{bmatrix}, \begin{bmatrix} 1 \\ 2 \\ -3 \end{bmatrix}, \begin{bmatrix} 1 & 2 & -3 & 5 \\ 0 & 1 & -1 & 3 \\ -2 & 0 & 4 & 6 \end{bmatrix}, \begin{bmatrix} 1 & 0 \\ 3 & 6 \\ -1 & 4 \\ 2 & 1 \end{bmatrix}, \text{ and } \begin{bmatrix} 1 & -1 & 3 & 5 \\ 0 & 3 & 2 & 5 \\ 1 & -1 & 5 & 0 \\ -2 & 3 & 4 & 1 \end{bmatrix}$$

are a 2×2 (reads as a "2 by 2"), a 2×3 (reads as a "2 by 3"), a 3×3 (reads as a "3 by 3"), a 3×1 (reads as a "3 by 1"), a 3×4 (reads as a "3 by 4"), a 4×2 (reads as a "4 by 2"), and a 4×4 (reads as a "4 by 4") matrices, respectively. In addition, a matrix consisting of a single row is referred to as a row matrix and a matrix consisting of a single column is referred to as a column matrix. Finally, note that in general matrices are represented by capital letters. For example,

matrices $\begin{bmatrix} 1 & 0 \\ -3 & 5 \end{bmatrix}$ and $\begin{bmatrix} 3 \\ 5 \end{bmatrix}$ are shown as A or $A_{2\times 2}$ and B or $B_{2\times 1}$ matrices.

In the following sections, we will learn about various matrix operations, determinants, inverse matrices, as well as different methods for solving linear systems of equations. However, we first need to learn about several types of matrices known as: equal, transpose, zero, square, diagonal, identity, coefficient, and augmented matrices.

Definition 3.1-1: Equal Matrices

Two matrices are said to be equal if and only if 1) they both are of the same order and 2) their corresponding entries are equal.

Example 3.1-1: The following matrices are equal to each other.

a. $\begin{bmatrix} 1 & 5 \\ 2 & 3 \end{bmatrix} = \begin{bmatrix} 1 & \frac{15}{3} \\ \sqrt{4} & 3 \end{bmatrix}$ b. $\begin{bmatrix} 2 & 0 \\ -1 & \sqrt{9} \end{bmatrix} = \begin{bmatrix} \frac{8}{4} & 0 \\ -\frac{2}{2} & 3 \end{bmatrix}$ c. $\begin{bmatrix} \frac{3}{4} & -3 \\ -0.6 & 1 \end{bmatrix} = \begin{bmatrix} 0.75 & -\frac{9}{3} \\ -\frac{3}{5} & \frac{2}{2} \end{bmatrix}$ d. $\begin{bmatrix} 3 \\ 5 \end{bmatrix} = \begin{bmatrix} \frac{6}{2} \\ \sqrt{25} \end{bmatrix}$

Example 3.1-2: The following matrices are not equal to each other.

a. $\begin{bmatrix} 3 & 5 \\ 0 & 1 \end{bmatrix} \neq \begin{bmatrix} 3 & 0 \\ 5 & 1 \end{bmatrix}$ b. $\begin{bmatrix} -3 & 2 \\ 0 & 5 \end{bmatrix} \neq \begin{bmatrix} -3 & 2 \\ 5 & 0 \end{bmatrix}$ c. $\begin{bmatrix} 3 & 1 \\ 2 & 4 \end{bmatrix} \neq \begin{bmatrix} \sqrt{9} & \frac{5}{5} \\ \frac{4}{2} & \frac{8}{4} \end{bmatrix}$ d. $\begin{bmatrix} 0 \\ 5 \end{bmatrix} \neq \begin{bmatrix} 0 & 5 \end{bmatrix}$

Example 3.1-3: Given the following equal matrices solve for the unknowns.

a. $\begin{bmatrix} x & -2 \\ y-3 & 5 \end{bmatrix} = \begin{bmatrix} 3 & z \\ -2 & 5 \end{bmatrix}$ b. $\begin{bmatrix} 2 & -2a & 2b \\ c-3 & 3 & d+2 \end{bmatrix} = \begin{bmatrix} e & 6 & -b+5 \\ 3 & 3+f & 4 \end{bmatrix}$

Solution:

a. Equating each corresponding element we obtain:

$x = 3$ $-2 = z$ or $z = -2$ $y - 3 = -2$ or $y = 1$

b. Equating each corresponding element we obtain:

$2 = e$ or $e = 2$ $-2a = 6$ or $a = -3$ $2b = -b + 5$ or $3b = 5$; $b = \frac{5}{3}$

$c - 3 = 3$ or $c = 6$ $3 = 3 + f$ or $f = 0$ $d + 2 = 4$ or $d = 2$

Definition 3.1-2: Transpose of a Matrix

The transpose of a matrix is a matrix in which the rows and columns are interchanged. The transpose of a matrix is denoted by A^t.

Example 3.1-4: Find the transpose of the following matrices.

a. $\begin{bmatrix} 1 & 5 \\ 2 & 3 \end{bmatrix}^t = \begin{bmatrix} 1 & 2 \\ 5 & 3 \end{bmatrix}$ b. $\begin{bmatrix} 1 & 5 & 6 \\ 2 & 0 & 3 \end{bmatrix}^t = \begin{bmatrix} 1 & 2 \\ 5 & 0 \\ 6 & 3 \end{bmatrix}$ c. $\begin{bmatrix} 1 & 0 \\ 0 & 1 \end{bmatrix}^t = \begin{bmatrix} 1 & 0 \\ 0 & 1 \end{bmatrix}$ d. $\begin{bmatrix} 3 \\ 5 \\ 2 \end{bmatrix}^t = \begin{bmatrix} 3 & 5 & 2 \end{bmatrix}$

Definition 3.1-3: Zero Matrix

A matrix with zero entries is referred to as a zero matrix. A zero matrix is generally denoted by the zero symbol 0 or a zero with subscript indicating the number of rows (m) and columns (n), i.e., $0_{m \times n}$. For example, 2×2, 3×1, and 2×3 zero matrices are represented as:

$$0_{2\times2} = \begin{bmatrix} 0 & 0 \\ 0 & 0 \end{bmatrix} \qquad 0_{3\times1} = \begin{bmatrix} 0 \\ 0 \\ 0 \end{bmatrix} \qquad 0_{2\times3} = \begin{bmatrix} 0 & 0 & 0 \\ 0 & 0 & 0 \end{bmatrix}$$

Definition 3.1-4: Square Matrix

A matrix having the same number of rows as columns is referred to as a square matrix. For example, $\begin{bmatrix} 1 & -2 \\ -4 & 0 \end{bmatrix}$, $\begin{bmatrix} -1 & 0 & 5 \\ 3 & 4 & 2 \\ 0 & -2 & 7 \end{bmatrix}$, and $\begin{bmatrix} -3 & 4 & 7 & 2 \\ 1 & 0 & 3 & 5 \\ -2 & 5 & 7 & 10 \\ 1 & 0 & 1 & 4 \end{bmatrix}$ are 2×2, 3×3, and 4×4 matrices.

Definition 3.1-5: Diagonal Matrix

A diagonal matrix is a square matrix in which only the diagonal matrix entries are not equal to zero. For example, $A_{2\times2} = \begin{bmatrix} 1 & 0 \\ 0 & 5 \end{bmatrix}$ and $B_{3\times3} = \begin{bmatrix} -1 & 0 & 0 \\ 0 & 4 & 0 \\ 0 & 0 & 7 \end{bmatrix}$ are diagonal matrices.

Definition 3.1-6: Identity Matrix

An identity matrix is a matrix in which only the diagonal matrix entries are equal to one. An identity matrix is generally denoted by the symbol I. $I_{2\times2} = \begin{bmatrix} 1 & 0 \\ 0 & 1 \end{bmatrix}$, $I_{3\times3} = \begin{bmatrix} 1 & 0 & 0 \\ 0 & 1 & 0 \\ 0 & 0 & 1 \end{bmatrix}$ are examples of the identity matrix.

Definition 3.1-7: Coefficient Matrix

A coefficient matrix is a matrix in which its first, second, third, fourth, etc. columns are formed from the coefficients of unknown variables x, y, z, w, etc. presented as a system of linear equations. For example, given the system of linear equations $\begin{matrix} 2x+5y=3 \\ 3x-2y=10 \end{matrix}$ the matrix $\begin{bmatrix} 2 & 5 \\ 3 & -2 \end{bmatrix}$ is called the coefficient matrix.

Definition 3.1-8: Augmented Matrix

An augmented matrix is a coefficient matrix which includes the columns consisting the right hand sides of the linear equations separated by a dashed line. For example, given the system of linear equations $\begin{matrix} 2x+5y=3 \\ 3x-2y=10 \end{matrix}$ the matrix $\left[\begin{matrix} 2 & 5 \\ 3 & -2 \end{matrix} \;\vdots\; \begin{matrix} 3 \\ 10 \end{matrix}\right]$ is called the augmented matrix.

Having identified different types of matrices, in the following section we will address various matrix operations which includes the addition, subtraction, and multiplication of matrices.

Section 3.1 Practice Problems - Introduction to Matrices

1. State the order and find the transpose of each matrix.

a. $\begin{bmatrix} 1 & 0 \\ 2 & 3 \end{bmatrix}$ b. $\begin{bmatrix} 1 & 4 \\ 0 & 2 \end{bmatrix}$ c. $\begin{bmatrix} 1 & 2 & -3 \\ -1 & 0 & 1 \end{bmatrix}$ d. $\begin{bmatrix} 1 & 0 \\ 2 & 1 \\ -1 & -3 \end{bmatrix}$

e. $\begin{bmatrix} 2 & 3 & 5 \\ 1 & 4 & -1 \\ 3 & 0 & 2 \end{bmatrix}$ f. $\begin{bmatrix} 1 \\ 2 \\ 3 \end{bmatrix}$ g. $\begin{bmatrix} 1 & -1 & 2 \end{bmatrix}$ h. $\begin{bmatrix} 1 & 0 & -1 & 2 \\ 2 & 3 & 1 & 0 \\ -1 & 0 & -2 & 3 \end{bmatrix}$

2. Given $A = \begin{bmatrix} 1 & 2 & -3 \\ 0 & 1 & 1 \\ -2 & 3 & 0 \end{bmatrix}$, find a_{12}, a_{23}, a_{33}, a_{22}, a_{31}, a_{21}, and a_{11}.

3. Given $B = \begin{bmatrix} 1 & -1 & 2 & 3 \\ 0 & 1 & 1 & 3 \\ 3 & -1 & -2 & 5 \end{bmatrix}$, find b_{12}, b_{21}, b_{32}, b_{33}, b_{23}, b_{24}, b_{34}, and b_{14}.

4. Given A_{2x3}, B_{3x1}, A_{1x2}, A_{3x3}, B_{2x4}, B_{1x4}, B_{4x1}, and A_{3x4} write a matrix that corresponds to the order given.

5. State if the given paired matrices are equal to each other.

a. $\begin{bmatrix} 1 & 0 \\ -1 & 3 \end{bmatrix} \stackrel{?}{=} \begin{bmatrix} \sqrt{4} & 0 \\ -\frac{2}{2} & \frac{6}{2} \end{bmatrix}$ b. $\begin{bmatrix} -1 & 2 \\ 3 & 6 \end{bmatrix} \stackrel{?}{=} \begin{bmatrix} -\frac{2}{2} & \frac{4}{2} \\ 6 & \frac{8}{12} \end{bmatrix}$ c. $\begin{bmatrix} -1 & 0 & 2 \\ 1 & 2 & 3 \end{bmatrix} \stackrel{?}{=} \begin{bmatrix} -1 & 0 & \frac{16}{8} \\ \frac{\sqrt{25}}{5} & \frac{6}{3} & \sqrt{9} \end{bmatrix}$

d. $\begin{bmatrix} \frac{3}{4} & -4 \\ -0.8 & 1 \end{bmatrix} \stackrel{?}{=} \begin{bmatrix} 0.75 & -\frac{8}{2} \\ \frac{8}{10} & \frac{4}{4} \end{bmatrix}$ e. $\begin{bmatrix} 1 & 0 \\ -2 & 3 \end{bmatrix} \stackrel{?}{=} \begin{bmatrix} \frac{2}{2} & \frac{0}{2} \\ -\sqrt{4} & \frac{9}{3} \end{bmatrix}$ f. $\begin{bmatrix} 1 & 4 \\ 3 & 5 \\ -1 & 2 \end{bmatrix} \stackrel{?}{=} \begin{bmatrix} \sqrt{1} & \sqrt{16} \\ \frac{27}{9} & \frac{25}{5} \\ -\frac{6}{3} & 2 \end{bmatrix}$

3.2 Matrix Operations

In the previous section we defined different types of matrices. In this section, we will learn about matrix addition and subtraction (Case I) and discuss how matrices are multiplied by one another (Case II).

Case I Matrix Addition and Subtraction

To add matrices, we simply add the corresponding entries of the two matrices. Note that to add matrices the matrices must have the same order, i.e., we can only add a 2×2 or a 3×3 matrix with another 2×2 or 3×3 matrix. Therefore, the sum of two matrices of different orders is not defined. In general, two matrices are added or subtracted in the following way:

$$\begin{bmatrix} a_{11} & a_{12} & \cdots & a_{1n} \\ a_{21} & a_{22} & \cdots & a_{2n} \\ \vdots & \vdots & & \vdots \\ a_{m1} & a_{m2} & \cdots & a_{mn} \end{bmatrix} + \begin{bmatrix} b_{11} & b_{12} & \cdots & b_{1n} \\ b_{21} & b_{22} & \cdots & b_{2n} \\ \vdots & \vdots & & \vdots \\ b_{m1} & b_{m2} & \cdots & b_{mn} \end{bmatrix} = \begin{bmatrix} a_{11}+b_{11} & a_{12}+b_{12} & \cdots & a_{1n}+b_{1n} \\ a_{21}+b_{21} & a_{22}+b_{22} & \cdots & a_{2n}+b_{2n} \\ \vdots & \vdots & & \vdots \\ a_{m1}+b_{m1} & a_{m2}+b_{m2} & \cdots & a_{mn}+b_{mn} \end{bmatrix} = \begin{bmatrix} \mathbf{c_{11}} & \mathbf{c_{12}} & \cdots & \mathbf{c_{1n}} \\ \mathbf{c_{21}} & \mathbf{c_{22}} & \cdots & \mathbf{c_{2n}} \\ \vdots & \vdots & & \vdots \\ \mathbf{c_{m1}} & \mathbf{c_{m2}} & \cdots & \mathbf{c_{mn}} \end{bmatrix}$$

$$\begin{bmatrix} a_{11} & a_{12} & \cdots & a_{1n} \\ a_{21} & a_{22} & \cdots & a_{2n} \\ \vdots & \vdots & & \vdots \\ a_{m1} & a_{m2} & \cdots & a_{mn} \end{bmatrix} - \begin{bmatrix} b_{11} & b_{12} & \cdots & b_{1n} \\ b_{21} & b_{22} & \cdots & b_{2n} \\ \vdots & \vdots & & \vdots \\ b_{m1} & b_{m2} & \cdots & b_{mn} \end{bmatrix} = \begin{bmatrix} a_{11} & a_{12} & \cdots & a_{1n} \\ a_{21} & a_{22} & \cdots & a_{2n} \\ \vdots & \vdots & & \vdots \\ a_{m1} & a_{m2} & \cdots & a_{mn} \end{bmatrix} + \begin{bmatrix} -b_{11} & -b_{12} & \cdots & -b_{1n} \\ -b_{21} & -b_{22} & \cdots & -b_{2n} \\ \vdots & \vdots & & \vdots \\ -b_{m1} & -b_{m2} & \cdots & -b_{mn} \end{bmatrix}$$

$$= \begin{bmatrix} a_{11}-b_{11} & a_{12}-b_{12} & \cdots & a_{1n}-b_{1n} \\ a_{21}-b_{21} & a_{22}-b_{22} & \cdots & a_{2n}-b_{2n} \\ \vdots & \vdots & & \vdots \\ a_{m1}-b_{m1} & a_{m2}-b_{m2} & \cdots & a_{mn}-b_{mn} \end{bmatrix} = \begin{bmatrix} \mathbf{d_{11}} & \mathbf{d_{12}} & \cdots & \mathbf{d_{1n}} \\ \mathbf{d_{21}} & \mathbf{d_{22}} & \cdots & \mathbf{d_{2n}} \\ \vdots & \vdots & & \vdots \\ \mathbf{d_{m1}} & \mathbf{d_{m2}} & \cdots & \mathbf{d_{mn}} \end{bmatrix}$$

The following examples shows how two matrices are added or subtracted with one another:

Example 3.2-1: Add or subtract the following square matrices.

a. $\begin{bmatrix} 1 & 2 \\ 3 & 5 \end{bmatrix} + \begin{bmatrix} 2 & -2 \\ 3 & 4 \end{bmatrix} =$

b. $\begin{bmatrix} 1 & 2 & 5 \\ 3 & 1 & -4 \end{bmatrix} + \begin{bmatrix} 1 & 3 & -3 \\ -1 & 0 & 5 \end{bmatrix} =$

c. $\begin{bmatrix} 3 & -3 & -2 \\ 1 & 0 & 5 \end{bmatrix} - \begin{bmatrix} 2 & 5 & 1 \\ -3 & -2 & 0 \end{bmatrix} =$

d. $\begin{bmatrix} 1 & 3 & 6 \\ -1 & 0 & 2 \\ 5 & -2 & 4 \end{bmatrix} - \begin{bmatrix} -3 & 5 & 2 \\ 0 & -4 & 8 \\ 5 & -3 & 6 \end{bmatrix} =$

Solutions:

a. $\begin{bmatrix} 1 & 2 \\ 3 & 5 \end{bmatrix} + \begin{bmatrix} 2 & -2 \\ 3 & 4 \end{bmatrix} = \begin{bmatrix} 1+2 & 2-2 \\ 3+3 & 5+4 \end{bmatrix} = \begin{bmatrix} \mathbf{3} & \mathbf{0} \\ \mathbf{6} & \mathbf{9} \end{bmatrix}$

b. $\begin{bmatrix} 1 & 2 & 5 \\ 3 & 1 & -4 \end{bmatrix} + \begin{bmatrix} 1 & 3 & -3 \\ -1 & 0 & 5 \end{bmatrix} = \begin{bmatrix} 1+1 & 2+3 & 5-3 \\ 3-1 & 1+0 & -4+5 \end{bmatrix} = \begin{bmatrix} \mathbf{2} & \mathbf{5} & \mathbf{2} \\ \mathbf{2} & \mathbf{1} & \mathbf{1} \end{bmatrix}$

c. $\begin{bmatrix} 3 & -3 & -2 \\ 1 & 0 & 5 \end{bmatrix} - \begin{bmatrix} 2 & 5 & 1 \\ -3 & -2 & 0 \end{bmatrix} = \begin{bmatrix} 3 & -3 & -2 \\ 1 & 0 & 5 \end{bmatrix} + \begin{bmatrix} -2 & -5 & -1 \\ 3 & 2 & 0 \end{bmatrix} = \begin{bmatrix} 3-2 & -3-5 & -2-1 \\ 1+3 & 0+2 & 5+0 \end{bmatrix} = \begin{bmatrix} \mathbf{1} & \mathbf{-8} & \mathbf{-3} \\ \mathbf{4} & \mathbf{2} & \mathbf{5} \end{bmatrix}$

d. $\begin{bmatrix} 1 & 3 & 6 \\ -1 & 0 & 2 \\ 5 & -2 & 4 \end{bmatrix} - \begin{bmatrix} -3 & 5 & 2 \\ 0 & -4 & 8 \\ 5 & -3 & 6 \end{bmatrix} = \begin{bmatrix} 1 & 3 & 6 \\ -1 & 0 & 2 \\ 5 & -2 & 4 \end{bmatrix} + \begin{bmatrix} 3 & -5 & -2 \\ 0 & 4 & -8 \\ -5 & 3 & -6 \end{bmatrix} = \begin{bmatrix} 1+3 & 3-5 & 6-2 \\ -1+0 & 0+4 & 2-8 \\ 5-5 & -2+3 & 4-6 \end{bmatrix} = \boxed{\begin{bmatrix} \mathbf{4} & \mathbf{-2} & \mathbf{4} \\ \mathbf{-1} & \mathbf{4} & \mathbf{-6} \\ \mathbf{0} & \mathbf{1} & \mathbf{-2} \end{bmatrix}}$

Properties of Matrix Addition

a. Matrix addition is both associative and commutative. This implies that matrices of the same dimension can be added in any order. For example, if A, B, and C are $m \times n$ matrices, then:

$(A+B)+C = A+(B+C)$ Associative property of addition

$A+B = B+A$ Commutative property of addition

b. An $m \times n$ A matrix added to any $m \times n$ zero matrix is equal to itself, i.e.,

$$A_{m\times n} + 0_{m\times n} = A_{m\times n}$$

c. An $m \times n$ A matrix added to the negative of an $m \times n$ A matrix is equal to the $m \times n$ zero matrix, i.e.,

$$A_{m\times n} + (-A_{m\times n}) = 0_{m\times n} \text{ and } -A_{m\times n} + A_{m\times n} = 0_{m\times n}$$

The following show examples of matrix properties:

Example 3.2-2: Given $A = \begin{bmatrix} 1 & -3 \\ 2 & 5 \end{bmatrix}$ find.

a. $A+(-A) =$ b. $A+A^t =$ c. $A+I =$ d. $A+0 =$

Solutions:

a. $\boxed{A+(-A)} = \begin{bmatrix} 1 & -3 \\ 2 & 5 \end{bmatrix} - \begin{bmatrix} 1 & -3 \\ 2 & 5 \end{bmatrix} = \begin{bmatrix} 1 & -3 \\ 2 & 5 \end{bmatrix} + \begin{bmatrix} -1 & 3 \\ -2 & -5 \end{bmatrix} = \begin{bmatrix} 1-1 & -3+3 \\ 2-2 & 5-5 \end{bmatrix} = \boxed{\begin{bmatrix} \mathbf{0} & \mathbf{0} \\ \mathbf{0} & \mathbf{0} \end{bmatrix}} = \boxed{\mathbf{0_{2\times 2}}}$

b. $\boxed{A+A^t} = \begin{bmatrix} 1 & -3 \\ 2 & 5 \end{bmatrix} + \begin{bmatrix} 1 & 2 \\ -3 & 5 \end{bmatrix} = \begin{bmatrix} 1+1 & -3+2 \\ 2-3 & 5+5 \end{bmatrix} = \boxed{\begin{bmatrix} \mathbf{2} & \mathbf{-1} \\ \mathbf{-1} & \mathbf{10} \end{bmatrix}}$

c. $\boxed{A+I} = \begin{bmatrix} 1 & -3 \\ 2 & 5 \end{bmatrix} + \begin{bmatrix} 1 & 0 \\ 0 & 1 \end{bmatrix} = \begin{bmatrix} 1+1 & -3+0 \\ 2+0 & 5+1 \end{bmatrix} = \boxed{\begin{bmatrix} \mathbf{2} & \mathbf{-3} \\ \mathbf{2} & \mathbf{6} \end{bmatrix}}$

d. $\boxed{A+0} = \begin{bmatrix} 1 & -3 \\ 2 & 5 \end{bmatrix} + \begin{bmatrix} 0 & 0 \\ 0 & 0 \end{bmatrix} = \begin{bmatrix} 1+0 & -3+0 \\ 2+0 & 5+0 \end{bmatrix} = \boxed{\begin{bmatrix} \mathbf{1} & \mathbf{-3} \\ \mathbf{2} & \mathbf{5} \end{bmatrix}}$

Example 3.2-3: Given $A = \begin{bmatrix} 1 & 2 \\ 3 & 5 \end{bmatrix}$, $B = \begin{bmatrix} 2 & 4 \\ -3 & 1 \end{bmatrix}$, and $C = \begin{bmatrix} 1 & 3 \\ 0 & 5 \end{bmatrix}$ find.

a. $A+B =$ b. $B+A =$ c. $A+(B+C) =$ d. $(A+B)+C =$

e. $A+\left(B^t + C^t\right) =$ f. $A^t + B + C^t =$ g. $A+(-A)+C =$

Solutions:

a. $\boxed{A+B} = \begin{bmatrix} 1 & 2 \\ 3 & 5 \end{bmatrix} + \begin{bmatrix} 2 & 4 \\ -3 & 1 \end{bmatrix} = \begin{bmatrix} 1+2 & 2+4 \\ 3-3 & 5+1 \end{bmatrix} = \boxed{\begin{bmatrix} \mathbf{3} & \mathbf{6} \\ \mathbf{0} & \mathbf{6} \end{bmatrix}}$

b. $\boxed{B+A} = \boxed{\begin{bmatrix} 2 & 4 \\ -3 & 1 \end{bmatrix} + \begin{bmatrix} 1 & 2 \\ 3 & 5 \end{bmatrix}} = \boxed{\begin{bmatrix} 2+1 & 4+2 \\ -3+3 & 1+5 \end{bmatrix}} = \boxed{\begin{bmatrix} \mathbf{3} & \mathbf{6} \\ \mathbf{0} & \mathbf{6} \end{bmatrix}}$

Note that since matrix addition is commutative, the results in parts a and b are the same.

c. $\boxed{A+(B+C)} = \boxed{\begin{bmatrix} 1 & 2 \\ 3 & 5 \end{bmatrix} + \left(\begin{bmatrix} 2 & 4 \\ -3 & 1 \end{bmatrix} + \begin{bmatrix} 1 & 3 \\ 0 & 5 \end{bmatrix} \right)} = \boxed{\begin{bmatrix} 1 & 2 \\ 3 & 5 \end{bmatrix} + \begin{bmatrix} 2+1 & 4+3 \\ -3+0 & 1+5 \end{bmatrix}} = \boxed{\begin{bmatrix} 1 & 2 \\ 3 & 5 \end{bmatrix} + \begin{bmatrix} 3 & 7 \\ -3 & 6 \end{bmatrix}}$

$= \boxed{\begin{bmatrix} 1+3 & 2+7 \\ 3-3 & 5+6 \end{bmatrix}} = \boxed{\begin{bmatrix} \mathbf{4} & \mathbf{9} \\ \mathbf{0} & \mathbf{11} \end{bmatrix}}$

d. $\boxed{(A+B)+C} = \boxed{\left(\begin{bmatrix} 1 & 2 \\ 3 & 5 \end{bmatrix} + \begin{bmatrix} 2 & 4 \\ -3 & 1 \end{bmatrix} \right) + \begin{bmatrix} 1 & 3 \\ 0 & 5 \end{bmatrix}} = \boxed{\begin{bmatrix} 3 & 6 \\ 0 & 6 \end{bmatrix} + \begin{bmatrix} 1 & 3 \\ 0 & 5 \end{bmatrix}} = \boxed{\begin{bmatrix} 3+1 & 6+3 \\ 0+0 & 6+5 \end{bmatrix}} = \boxed{\begin{bmatrix} \mathbf{4} & \mathbf{9} \\ \mathbf{0} & \mathbf{11} \end{bmatrix}}$

Note that since matrix addition is associative, the results in parts c and d are the same.

e. $\boxed{A+\left(B^t+C^t\right)} = \boxed{\begin{bmatrix} 1 & 2 \\ 3 & 5 \end{bmatrix} + \left(\begin{bmatrix} 2 & -3 \\ 4 & 1 \end{bmatrix} + \begin{bmatrix} 1 & 0 \\ 3 & 5 \end{bmatrix} \right)} = \boxed{\begin{bmatrix} 1 & 2 \\ 3 & 5 \end{bmatrix} + \begin{bmatrix} 2+1 & -3+0 \\ 4+3 & 1+5 \end{bmatrix}} = \boxed{\begin{bmatrix} 1 & 2 \\ 3 & 5 \end{bmatrix} + \begin{bmatrix} 3 & -3 \\ 7 & 6 \end{bmatrix}}$

$= \boxed{\begin{bmatrix} 1+3 & 2-3 \\ 3+7 & 5+6 \end{bmatrix}} = \boxed{\begin{bmatrix} \mathbf{4} & \mathbf{-1} \\ \mathbf{10} & \mathbf{11} \end{bmatrix}}$

f. $\boxed{A^t+B+C^t} = \boxed{\begin{bmatrix} 1 & 3 \\ 2 & 5 \end{bmatrix} + \begin{bmatrix} 2 & 4 \\ -3 & 1 \end{bmatrix} + \begin{bmatrix} 1 & 0 \\ 3 & 5 \end{bmatrix}} = \boxed{\begin{bmatrix} 1+2 & 3+4 \\ 2-3 & 5+1 \end{bmatrix} + \begin{bmatrix} 1 & 0 \\ 3 & 5 \end{bmatrix}} = \boxed{\begin{bmatrix} 3 & 7 \\ -1 & 6 \end{bmatrix} + \begin{bmatrix} 1 & 0 \\ 3 & 5 \end{bmatrix}} = \boxed{\begin{bmatrix} \mathbf{4} & \mathbf{7} \\ \mathbf{2} & \mathbf{11} \end{bmatrix}}$

g. $\boxed{A+(-A)+C} = \boxed{\begin{bmatrix} 1 & 2 \\ 3 & 5 \end{bmatrix} - \begin{bmatrix} 1 & 2 \\ 3 & 5 \end{bmatrix} + \begin{bmatrix} 1 & 3 \\ 0 & 5 \end{bmatrix}} = \boxed{\begin{bmatrix} 1-1 & 2-2 \\ 3-3 & 5-5 \end{bmatrix} + \begin{bmatrix} 1 & 3 \\ 0 & 5 \end{bmatrix}} = \boxed{\begin{bmatrix} 0 & 0 \\ 0 & 0 \end{bmatrix} + \begin{bmatrix} 1 & 3 \\ 0 & 5 \end{bmatrix}} = \boxed{\begin{bmatrix} \mathbf{1} & \mathbf{3} \\ \mathbf{0} & \mathbf{5} \end{bmatrix}}$

Example 3.2-4: Given $A = \begin{bmatrix} 1 & 2 & 3 \\ 4 & 3 & 1 \\ 1 & -1 & 0 \end{bmatrix}$ and $B = \begin{bmatrix} -1 & -3 & 1 \\ 4 & 5 & 6 \\ 1 & 0 & -3 \end{bmatrix}$ find.

a. $A+B =$ b. $B+A =$ c. $(A+B)+I =$ d. $(A+B)^t+B =$

Solutions:

a. $\boxed{A+B} = \boxed{\begin{bmatrix} 1 & 2 & 3 \\ 4 & 3 & 1 \\ 1 & -1 & 0 \end{bmatrix} + \begin{bmatrix} -1 & -3 & 1 \\ 4 & 5 & 6 \\ 1 & 0 & -3 \end{bmatrix}} = \boxed{\begin{bmatrix} 1-1 & 2-3 & 3+1 \\ 4+4 & 3+5 & 1+6 \\ 1+1 & -1+0 & 0-3 \end{bmatrix}} = \boxed{\begin{bmatrix} \mathbf{0} & \mathbf{-1} & \mathbf{4} \\ \mathbf{8} & \mathbf{8} & \mathbf{7} \\ \mathbf{2} & \mathbf{-1} & \mathbf{-3} \end{bmatrix}}$

b. $\boxed{B+A} = \boxed{\begin{bmatrix} -1 & -3 & 1 \\ 4 & 5 & 6 \\ 1 & 0 & -3 \end{bmatrix} + \begin{bmatrix} 1 & 2 & 3 \\ 4 & 3 & 1 \\ 1 & -1 & 0 \end{bmatrix}} = \boxed{\begin{bmatrix} -1+1 & -3+2 & 1+3 \\ 4+4 & 5+3 & 6+1 \\ 1+1 & 0-1 & -3+0 \end{bmatrix}} = \boxed{\begin{bmatrix} \mathbf{0} & \mathbf{-1} & \mathbf{4} \\ \mathbf{8} & \mathbf{8} & \mathbf{7} \\ \mathbf{2} & \mathbf{-1} & \mathbf{-3} \end{bmatrix}}$

c. $$\boxed{(A+B)+I} = \boxed{\left(\begin{bmatrix}1&2&3\\4&3&1\\1&-1&0\end{bmatrix}+\begin{bmatrix}-1&-3&1\\4&5&6\\1&0&-3\end{bmatrix}\right)+\begin{bmatrix}1&0&0\\0&1&0\\0&0&1\end{bmatrix}} = \boxed{\begin{bmatrix}1-1&2-3&3+1\\4+4&3+5&1+6\\1+1&-1+0&0-3\end{bmatrix}+\begin{bmatrix}1&0&0\\0&1&0\\0&0&1\end{bmatrix}}$$

$$= \boxed{\begin{bmatrix}0&-1&4\\8&8&7\\2&-1&-3\end{bmatrix}+\begin{bmatrix}1&0&0\\0&1&0\\0&0&1\end{bmatrix}} = \boxed{\begin{bmatrix}0+1&-1+0&4+0\\8+0&8+1&7+0\\2+0&-1+0&-3+1\end{bmatrix}} = \boxed{\begin{bmatrix}\mathbf{1}&\mathbf{-1}&\mathbf{4}\\\mathbf{8}&\mathbf{9}&\mathbf{7}\\\mathbf{2}&\mathbf{-1}&\mathbf{-2}\end{bmatrix}}$$

d. $$\boxed{(A+B)^t+B} = \boxed{\left(\begin{bmatrix}1&2&3\\4&3&1\\1&-1&0\end{bmatrix}+\begin{bmatrix}-1&-3&1\\4&5&6\\1&0&-3\end{bmatrix}\right)^t+\begin{bmatrix}-1&-3&1\\4&5&6\\1&0&-3\end{bmatrix}} = \boxed{\begin{bmatrix}0&-1&4\\8&8&7\\2&-1&-3\end{bmatrix}^t+\begin{bmatrix}-1&-3&1\\4&5&6\\1&0&-3\end{bmatrix}}$$

$$= \boxed{\begin{bmatrix}0&8&2\\-1&8&-1\\4&7&-3\end{bmatrix}+\begin{bmatrix}-1&-3&1\\4&5&6\\1&0&-3\end{bmatrix}} = \boxed{\begin{bmatrix}0-1&8-3&2+1\\-1+4&8+5&-1+6\\4+1&7+0&-3-3\end{bmatrix}} = \boxed{\begin{bmatrix}\mathbf{-1}&\mathbf{5}&\mathbf{3}\\\mathbf{3}&\mathbf{13}&\mathbf{5}\\\mathbf{5}&\mathbf{7}&\mathbf{-6}\end{bmatrix}}$$

Example 3.2-5: Solve the following matrix operations.

a. $$\left(\begin{bmatrix}2&3\\1&0\end{bmatrix}+\begin{bmatrix}2&3\\1&0\end{bmatrix}\right)-\begin{bmatrix}1&3\\2&8\end{bmatrix} =$$

b. $$\begin{bmatrix}2&3\\5&1\\0&2\end{bmatrix}-\left(\begin{bmatrix}8&1\\3&5\\-1&-2\end{bmatrix}-\begin{bmatrix}1&-1\\2&3\\-3&5\end{bmatrix}\right) =$$

Solutions:

a. $$\boxed{\left(\begin{bmatrix}2&3\\1&0\end{bmatrix}+\begin{bmatrix}2&3\\1&0\end{bmatrix}\right)-\begin{bmatrix}1&3\\2&8\end{bmatrix}} = \boxed{\begin{bmatrix}2+2&3+3\\1+1&0+0\end{bmatrix}-\begin{bmatrix}1&3\\2&8\end{bmatrix}} = \boxed{\begin{bmatrix}4&6\\2&0\end{bmatrix}-\begin{bmatrix}1&3\\2&8\end{bmatrix}} = \boxed{\begin{bmatrix}4-1&6-3\\2-2&0-8\end{bmatrix}} = \boxed{\begin{bmatrix}\mathbf{3}&\mathbf{3}\\\mathbf{0}&\mathbf{-8}\end{bmatrix}}$$

b. $$\boxed{\begin{bmatrix}2&3\\5&1\\0&2\end{bmatrix}-\left(\begin{bmatrix}8&1\\3&5\\-1&-2\end{bmatrix}-\begin{bmatrix}1&-1\\2&3\\-3&5\end{bmatrix}\right)} = \boxed{\begin{bmatrix}2&3\\5&1\\0&2\end{bmatrix}-\begin{bmatrix}8-1&1+1\\3-2&5-3\\-1+3&-2-5\end{bmatrix}} = \boxed{\begin{bmatrix}2&3\\5&1\\0&2\end{bmatrix}-\begin{bmatrix}7&2\\1&2\\2&-7\end{bmatrix}} = \boxed{\begin{bmatrix}2-7&3-2\\5-1&1-2\\0-2&2+7\end{bmatrix}}$$

$$= \boxed{\begin{bmatrix}\mathbf{-5}&\mathbf{1}\\\mathbf{4}&\mathbf{-1}\\\mathbf{-2}&\mathbf{9}\end{bmatrix}}$$

Section 3.2 Case I Practice Problems - Matrix Addition and Subtraction

1. Add or subtract the following matrices.

a. $\begin{bmatrix}1&3&5\\3&-1&2\end{bmatrix}+\begin{bmatrix}-1&-5&2\\0&6&-3\end{bmatrix} =$

b. $\begin{bmatrix}1&-8\\-3&5\end{bmatrix}-\begin{bmatrix}1&-3\\-2&4\end{bmatrix} =$

c. $\begin{bmatrix}10&3&-1\\1&5&6\\2&3&8\end{bmatrix}+\begin{bmatrix}1&7&9\\2&-3&0\\1&5&-3\end{bmatrix} =$

d. $\begin{bmatrix}2&3\\1&5\\-3&1\end{bmatrix}-\begin{bmatrix}1&8\\3&0\\1&-2\end{bmatrix} =$

e. $\begin{bmatrix}1&3&6\end{bmatrix}-\begin{bmatrix}0&5&-3\end{bmatrix} =$

f. $\begin{bmatrix}1&-5\\-3&8\end{bmatrix}+\begin{bmatrix}0&0\\0&0\end{bmatrix} =$

g. $\begin{bmatrix} 6 & 3 & -1 \\ 1 & 5 & 0 \\ 0 & 0 & 1 \end{bmatrix} + \begin{bmatrix} 1 & 0 & 0 \\ 0 & 1 & 0 \\ 0 & 0 & 1 \end{bmatrix} =$

h. $\begin{bmatrix} 1 & 5 \\ 2 & -3 \\ -1 & 0 \end{bmatrix} + \begin{bmatrix} -1 & -5 \\ -2 & 3 \\ 1 & 0 \end{bmatrix} =$

2. Given $A = \begin{bmatrix} 1 & 3 & 5 \\ 2 & -1 & 0 \\ -1 & 2 & 4 \end{bmatrix}$ and $B = \begin{bmatrix} -1 & 0 & 1 \\ 2 & 3 & 0 \\ 1 & 5 & -3 \end{bmatrix}$ find.

a. $A + B^t =$ b. $A^t + B^t =$ c. $(A+B) - B^t =$ d. $(A-B) + B^t =$

e. $(A^t + B^t) - (A+B) =$ f. $(A-B) - A^t =$ g. $2A - 3B =$

3. Given $A = [1 \;\; 0 \;\; -3]$, $B = [3 \;\; 1 \;\; 5]$, $C = \begin{bmatrix} 1 & 5 & -1 \\ 2 & 0 & 3 \end{bmatrix}$, and $D = \begin{bmatrix} 10 & 8 & 7 \\ 3 & -5 & 2 \end{bmatrix}$ perform the following operations, if possible.

a. $A + C =$ b. $A - B =$ c. $(C+D) - D^t =$

4. Add or subtract the following matrices.

a. $\left(\begin{bmatrix} 1 & 2 \\ 3 & 0 \end{bmatrix} - \begin{bmatrix} 1 & 3 \\ -5 & 2 \end{bmatrix}\right) + \begin{bmatrix} 2 & 3 \\ -1 & 6 \end{bmatrix} =$

b. $\begin{bmatrix} 1 & 3 \\ 5 & 0 \\ 1 & -1 \end{bmatrix} - \left(\begin{bmatrix} 2 & 3 \\ -1 & 0 \\ 1 & 3 \end{bmatrix} - \begin{bmatrix} 1 & 5 \\ -1 & 2 \\ 4 & -3 \end{bmatrix}\right) =$

c. $\begin{bmatrix} 1 & 3 & 5 \\ 2 & -1 & 3 \\ -1 & 4 & 0 \end{bmatrix} + \left(\begin{bmatrix} 1 & 5 & 0 \\ 0 & -1 & 1 \\ 1 & 2 & 3 \end{bmatrix} + \begin{bmatrix} 1 & 0 & 0 \\ 0 & 1 & 0 \\ 0 & 0 & 1 \end{bmatrix}\right) =$

d. $\left(\begin{bmatrix} 2 & 3 \\ -1 & 0 \end{bmatrix} + \begin{bmatrix} -2 & -3 \\ 1 & 0 \end{bmatrix}\right) - \begin{bmatrix} 3 & 5 \\ 1 & -2 \end{bmatrix} =$

5. Given $A = \begin{bmatrix} 2 & 3 & -1 \\ 5 & 0 & 2 \\ 8 & -3 & 1 \end{bmatrix}$ show that $\left[A^t\right]^t = A$.

6. Given $A = \begin{bmatrix} 1 & 2 \\ 0 & 3 \end{bmatrix}$ and $B = \begin{bmatrix} 3 & -5 \\ -3 & 2 \end{bmatrix}$ show that $[A+B]^t = A^t + B^t$.

7. Given the equal matrices solve for the unknowns.

a. $\begin{bmatrix} u-1 & v & 3 \\ 8 & 6 & 16 \\ 0 & 6 & \frac{z}{3} \end{bmatrix} = \begin{bmatrix} 5 & 2v+3 & \sqrt{9} \\ w+1 & 6 & 2x \\ 0 & 2y-3 & 3 \end{bmatrix}$

b. $\begin{bmatrix} 3x & 5 \\ 3 & y-8 \end{bmatrix} + \begin{bmatrix} 3 & 6 \\ -5 & 4 \end{bmatrix} = \begin{bmatrix} 6 & 11 \\ -2 & -4 \end{bmatrix}$

c. $\begin{bmatrix} 3 & 5 \\ -3 & 8 \end{bmatrix} - \begin{bmatrix} 7 & 1 \\ 2 & -2x \end{bmatrix} = \begin{bmatrix} -4 & 4 \\ -5 & 10 \end{bmatrix}$

d. $\begin{bmatrix} x \\ 0 \\ 0 \end{bmatrix} + \begin{bmatrix} 3y \\ y \\ 0 \end{bmatrix} + \begin{bmatrix} z \\ -2z \\ 3z \end{bmatrix} = \begin{bmatrix} 6 \\ 3 \\ 9 \end{bmatrix}$

Case II Matrix Multiplication

Two matrices A and B can be multiplied by one another only if the number of columns in the A matrix equals the number of rows in the B matrix, i.e.,

$$A_{no.\ of\ rows\ in\ A\ \times\ no.\ of\ cols.\ in\ A} \times B_{no.\ of\ rows\ in\ B\ \times\ no.\ of\ cols.\ in\ B} = C_{no.\ of\ rows\ in\ A\ \times\ no.\ of\ cols.\ in\ B}$$

no. of columns in A must match the no. of rows in B

For example, we can only multiply a 3×2 $(A_{3\times2})$, a 4×3 $(A_{4\times3})$ or a 3×5 $(A_{3\times5})$ matrix by a 2×2 $(B_{2\times2})$, 3×5 $(B_{3\times5})$ or a 5×7 $(B_{5\times7})$ matrix, respectively. The order of the product matrices would then be equal to:

$$A_{3\times2}\times B_{2\times2}=C_{3\times2},\ A_{4\times3}\times B_{3\times5}=C_{4\times5},\ \text{and}\ A_{3\times5}\times B_{5\times7}=C_{3\times7}$$

Multiplication of matrices requires calculating sums of products. In general, a 2×3 $(A_{2\times3})$ matrix is multiplied by a 3×3 $(B_{3\times3})$ matrix in the following way:

$$\begin{bmatrix} a_{11} & a_{12} & a_{13} \\ a_{21} & a_{22} & a_{23} \end{bmatrix} \times \begin{bmatrix} b_{11} & b_{12} & b_{13} \\ b_{21} & b_{22} & b_{23} \\ b_{31} & b_{32} & b_{33} \end{bmatrix} = \begin{bmatrix} \mathbf{c_{11}} & \mathbf{c_{12}} & \mathbf{c_{13}} \\ \mathbf{c_{21}} & \mathbf{c_{22}} & \mathbf{c_{23}} \end{bmatrix}$$

where the entries for the C matrix are calculated in the following way:

$c_{11}=a_{11}\cdot b_{11}+a_{12}\cdot b_{21}+a_{13}\cdot b_{31}$ $\quad c_{12}=a_{11}\cdot b_{12}+a_{12}\cdot b_{22}+a_{13}\cdot b_{32}$ $\quad c_{13}=a_{11}\cdot b_{13}+a_{12}\cdot b_{23}+a_{13}\cdot b_{33}$

$c_{21}=a_{21}\cdot b_{11}+a_{22}\cdot b_{21}+a_{23}\cdot b_{31}$ $\quad c_{22}=a_{21}\cdot b_{12}+a_{22}\cdot b_{22}+a_{23}\cdot b_{32}$ $\quad c_{23}=a_{21}\cdot b_{13}+a_{22}\cdot b_{23}+a_{23}\cdot b_{33}$

The following examples show how two matrices are multiplied by one another:

Example 3.2-6: Multiply the following matrices.

a. $\begin{bmatrix} 1 & 2 & 3 \end{bmatrix}\begin{bmatrix} 2 \\ 0 \\ 1 \end{bmatrix}=$ $\qquad$ b. $\begin{bmatrix} 2 & 4 \\ -3 & 1 \end{bmatrix}\begin{bmatrix} 1 & 2 \\ 0 & -1 \end{bmatrix}=$ $\qquad$ c. $\begin{bmatrix} 1 & 0 & 2 \\ 3 & -1 & 4 \end{bmatrix}\begin{bmatrix} 1 & 2 \\ -1 & 3 \\ 2 & 5 \end{bmatrix}=$

Solutions:

a. $\begin{bmatrix} 1 & 2 & 3 \end{bmatrix}\begin{bmatrix} 2 \\ 0 \\ 1 \end{bmatrix}=\begin{bmatrix}(1\times2)+(2\times0)+(3\times1)\end{bmatrix}=\begin{bmatrix}2+0+3\end{bmatrix}=\boxed{5}$

From the above example it should be clear that in general the product of a matrix with *one* row and *n* columns $(A_{1\times n})$ multiplied by another matrix with *n* rows and *one* column $(B_{n\times1})$ is always a real number - not a matrix, i.e., $A_{1\times n}\cdot B_{n\times1}=k$.

b. $\begin{bmatrix} 2 & 4 \\ -3 & 1 \end{bmatrix}\begin{bmatrix} 1 & 2 \\ 0 & -1 \end{bmatrix}=\begin{bmatrix} (2\times1)+(4\times0) & (2\times2)+(4\times-1) \\ (-3\times1)+(1\times0) & (-3\times2)+(1\times-1) \end{bmatrix}=\begin{bmatrix} 2+0 & 4-4 \\ -3+0 & -6-1 \end{bmatrix}=\begin{bmatrix} \mathbf{2} & \mathbf{0} \\ \mathbf{-3} & \mathbf{-7} \end{bmatrix}$

c. $\begin{bmatrix} 1 & 0 & 2 \\ 3 & -1 & 4 \end{bmatrix}\begin{bmatrix} 1 & 2 \\ -1 & 3 \\ 2 & 5 \end{bmatrix}=\begin{bmatrix} (1\times1)+(0\times-1)+(2\times2) & (1\times2)+(0\times3)+(2\times5) \\ (3\times1)+(-1\times-1)+(4\times2) & (3\times2)+(-1\times3)+(4\times5) \end{bmatrix}=\begin{bmatrix} 1+0+4 & 2+0+10 \\ 3+1+8 & 6-3+20 \end{bmatrix}=\begin{bmatrix} \mathbf{5} & \mathbf{12} \\ \mathbf{12} & \mathbf{23} \end{bmatrix}$

Properties of Matrix Multiplication

a. Matrix multiplication is not commutative. This implies that, for most matrices, $AB \neq BA$.

Example 3.2-7: Given $A=\begin{bmatrix}2&3\\-1&5\end{bmatrix}$ and $B=\begin{bmatrix}1&2\\0&1\end{bmatrix}$, show that $AB \neq BA$.

Solution:

$$\boxed{AB}=\boxed{\begin{bmatrix}2&3\\-1&5\end{bmatrix}\cdot\begin{bmatrix}1&2\\0&1\end{bmatrix}}=\boxed{\begin{bmatrix}(2\times1)+(3\times0)&(2\times2)+(3\times1)\\(-1\times1)+(5\times0)&(-1\times2)+(5\times1)\end{bmatrix}}=\boxed{\begin{bmatrix}2+0&4+3\\-1+0&-2+5\end{bmatrix}}=\boxed{\begin{bmatrix}\mathbf{2}&\mathbf{7}\\\mathbf{-1}&\mathbf{3}\end{bmatrix}}$$

$$\boxed{BA}=\boxed{\begin{bmatrix}1&2\\0&1\end{bmatrix}\cdot\begin{bmatrix}2&3\\-1&5\end{bmatrix}}=\boxed{\begin{bmatrix}(1\times2)+(2\times-1)&(1\times3)+(2\times5)\\(0\times2)+(1\times-1)&(0\times3)+(1\times5)\end{bmatrix}}=\boxed{\begin{bmatrix}2-2&3+10\\0-1&0+5\end{bmatrix}}=\boxed{\begin{bmatrix}\mathbf{0}&\mathbf{13}\\\mathbf{-1}&\mathbf{5}\end{bmatrix}}$$

b. If A, B, and C are square matrices and a and b are real number, then

1. $(AB)C = A(BC)$

Example 3.2-8: Given $A=\begin{bmatrix}0&1\\-1&3\end{bmatrix}$, $B=\begin{bmatrix}-1&2\\5&1\end{bmatrix}$, and $C=\begin{bmatrix}-3&4\\-1&1\end{bmatrix}$ show that $(AB)C = A(BC)$.

Solution:

$$\boxed{(AB)C}=\boxed{\left(\begin{bmatrix}0&1\\-1&3\end{bmatrix}\cdot\begin{bmatrix}-1&2\\5&1\end{bmatrix}\right)\cdot\begin{bmatrix}-3&4\\-1&1\end{bmatrix}}=\boxed{\left(\begin{bmatrix}(0\times-1)+(1\times5)&(0\times2)+(1\times1)\\(-1\times-1)+(3\times5)&(-1\times2)+(3\times1)\end{bmatrix}\right)\cdot\begin{bmatrix}-3&4\\-1&1\end{bmatrix}}$$

$$=\boxed{\begin{bmatrix}0+5&0+1\\1+15&-2+3\end{bmatrix}\cdot\begin{bmatrix}-3&4\\-1&1\end{bmatrix}}=\boxed{\begin{bmatrix}5&1\\16&1\end{bmatrix}\cdot\begin{bmatrix}-3&4\\-1&1\end{bmatrix}}=\boxed{\begin{bmatrix}(5\times-3)+(1\times-1)&(5\times4)+(1\times1)\\(16\times-3)+(1\times-1)&(16\times4)+(1\times1)\end{bmatrix}}$$

$$=\boxed{\begin{bmatrix}-15-1&20+1\\-48-1&64+1\end{bmatrix}}=\boxed{\begin{bmatrix}\mathbf{-16}&\mathbf{21}\\\mathbf{-49}&\mathbf{65}\end{bmatrix}}$$

$$\boxed{A(BC)}=\boxed{\begin{bmatrix}0&1\\-1&3\end{bmatrix}\cdot\left(\begin{bmatrix}-1&2\\5&1\end{bmatrix}\cdot\begin{bmatrix}-3&4\\-1&1\end{bmatrix}\right)}=\boxed{\begin{bmatrix}0&1\\-1&3\end{bmatrix}\cdot\left(\begin{bmatrix}(-1\times-3)+(2\times-1)&(-1\times4)+(2\times1)\\(5\times-3)+(1\times-1)&(5\times4)+(1\times1)\end{bmatrix}\right)}$$

$$=\boxed{\begin{bmatrix}0&1\\-1&3\end{bmatrix}\cdot\begin{bmatrix}3-2&-4+2\\-15-1&20+1\end{bmatrix}}=\boxed{\begin{bmatrix}0&1\\-1&3\end{bmatrix}\cdot\begin{bmatrix}1&-2\\-16&21\end{bmatrix}}=\boxed{\begin{bmatrix}(0\times1)+(1\times-16)&(0\times-2)+(1\times21)\\(-1\times1)+(3\times-16)&(-1\times-2)+(3\times21)\end{bmatrix}}$$

$$=\boxed{\begin{bmatrix}0-16&0+21\\-1-48&2+63\end{bmatrix}}=\boxed{\begin{bmatrix}\mathbf{-16}&\mathbf{21}\\\mathbf{-49}&\mathbf{65}\end{bmatrix}}$$

2. $A(B+C) = AB + AC$

Example 3.2-9: Given $A=\begin{bmatrix}-4&3\\1&-3\end{bmatrix}$, $B=\begin{bmatrix}1&-2\\4&1\end{bmatrix}$, and $C=\begin{bmatrix}0&5\\3&0\end{bmatrix}$ show that $A(B+C) = AB + AC$.

Solution:

$$\boxed{A(B+C)}=\boxed{\begin{bmatrix}-4&3\\1&-3\end{bmatrix}\left(\begin{bmatrix}1&-2\\4&1\end{bmatrix}+\begin{bmatrix}0&5\\3&0\end{bmatrix}\right)}=\boxed{\begin{bmatrix}-4&3\\1&-3\end{bmatrix}\cdot\begin{bmatrix}1+0&-2+5\\4+3&1+0\end{bmatrix}}=\boxed{\begin{bmatrix}-4&3\\1&-3\end{bmatrix}\cdot\begin{bmatrix}1&3\\7&1\end{bmatrix}}$$

$$=\boxed{\begin{bmatrix}(-4\times1)+(3\times7)&(-4\times3)+(3\times1)\\(1\times1)+(-3\times7)&(1\times3)+(-3\times1)\end{bmatrix}}=\boxed{\begin{bmatrix}-4+21&-12+3\\1-21&3-3\end{bmatrix}}=\boxed{\begin{bmatrix}\mathbf{17}&\mathbf{-9}\\\mathbf{-20}&\mathbf{0}\end{bmatrix}}$$

$$\boxed{AB} = \begin{bmatrix} -4 & 3 \\ 1 & -3 \end{bmatrix} \cdot \begin{bmatrix} 1 & -2 \\ 4 & 1 \end{bmatrix} = \begin{bmatrix} (-4\times 1)+(3\times 4) & (-4\times -2)+(3\times 1) \\ (1\times 1)+(-3\times 4) & (1\times -2)+(-3\times 1) \end{bmatrix} = \begin{bmatrix} -4+12 & 8+3 \\ 1-12 & -2-3 \end{bmatrix} = \begin{bmatrix} 8 & 11 \\ -11 & -5 \end{bmatrix}$$

$$\boxed{AC} = \begin{bmatrix} -4 & 3 \\ 1 & -3 \end{bmatrix} \cdot \begin{bmatrix} 0 & 5 \\ 3 & 0 \end{bmatrix} = \begin{bmatrix} (-4\times 0)+(3\times 3) & (-4\times 5)+(3\times 0) \\ (1\times 0)+(-3\times 3) & (1\times 5)+(-3\times 0) \end{bmatrix} = \begin{bmatrix} 0+9 & -20+0 \\ 0-9 & 5+0 \end{bmatrix} = \begin{bmatrix} 9 & -20 \\ -9 & 5 \end{bmatrix}$$

$$\boxed{AB+AC} = \begin{bmatrix} 8 & 11 \\ -11 & -5 \end{bmatrix} + \begin{bmatrix} 9 & -20 \\ -9 & 5 \end{bmatrix} = \begin{bmatrix} 8+9 & 11-20 \\ -11-9 & -5+5 \end{bmatrix} = \begin{bmatrix} \mathbf{17} & \mathbf{-9} \\ \mathbf{-20} & \mathbf{0} \end{bmatrix}$$

3. $(\boldsymbol{B}+\boldsymbol{C})\boldsymbol{A} = \boldsymbol{BA}+\boldsymbol{CA}$

Example 3.2-10: Given $A = \begin{bmatrix} -5 & -2 \\ 0 & 2 \end{bmatrix}$, $B = \begin{bmatrix} 1 & 0 \\ 0 & -1 \end{bmatrix}$, and $C = \begin{bmatrix} 2 & 0 \\ 3 & -4 \end{bmatrix}$ show that $(B+C)A = BA+CA$.

Solution:

$$\boxed{(B+C)A} = \left(\begin{bmatrix} 1 & 0 \\ 0 & -1 \end{bmatrix} + \begin{bmatrix} 2 & 0 \\ 3 & -4 \end{bmatrix}\right)\begin{bmatrix} -5 & -2 \\ 0 & 2 \end{bmatrix} = \begin{bmatrix} 1+2 & 0+0 \\ 0+3 & -1-4 \end{bmatrix}\begin{bmatrix} -5 & -2 \\ 0 & 2 \end{bmatrix} = \begin{bmatrix} 3 & 0 \\ 3 & -5 \end{bmatrix}\begin{bmatrix} -5 & -2 \\ 0 & 2 \end{bmatrix}$$

$$= \begin{bmatrix} (3\times -5)+(0\times 0) & (3\times -2)+(0\times 2) \\ (3\times -5)+(-5\times 0) & (3\times -2)+(-5\times 2) \end{bmatrix} = \begin{bmatrix} -15+0 & -6+0 \\ -15+0 & -6-10 \end{bmatrix} = \begin{bmatrix} \mathbf{-15} & \mathbf{-6} \\ \mathbf{-15} & \mathbf{-16} \end{bmatrix}$$

$$\boxed{BA} = \begin{bmatrix} 1 & 0 \\ 0 & -1 \end{bmatrix} \cdot \begin{bmatrix} -5 & -2 \\ 0 & 2 \end{bmatrix} = \begin{bmatrix} (1\times -5)+(0\times 0) & (1\times -2)+(0\times 2) \\ (0\times -5)+(-1\times 0) & (0\times -2)+(-1\times 2) \end{bmatrix} = \begin{bmatrix} -5+0 & -2+0 \\ 0+0 & 0-2 \end{bmatrix} = \begin{bmatrix} -5 & -2 \\ 0 & -2 \end{bmatrix}$$

$$\boxed{CA} = \begin{bmatrix} 2 & 0 \\ 3 & -4 \end{bmatrix} \cdot \begin{bmatrix} -5 & -2 \\ 0 & 2 \end{bmatrix} = \begin{bmatrix} (2\times -5)+(0\times 0) & (2\times -2)+(0\times 2) \\ (3\times -5)+(-4\times 0) & (3\times -2)+(-4\times 2) \end{bmatrix} = \begin{bmatrix} -10+0 & -4+0 \\ -15+0 & -6-8 \end{bmatrix} = \begin{bmatrix} -10 & -4 \\ -15 & -14 \end{bmatrix}$$

$$\boxed{BA+CA} = \begin{bmatrix} -5 & -2 \\ 0 & -2 \end{bmatrix} + \begin{bmatrix} -10 & -4 \\ -15 & -14 \end{bmatrix} = \begin{bmatrix} -5-10 & -2-4 \\ 0-15 & -2-14 \end{bmatrix} = \begin{bmatrix} \mathbf{-15} & \mathbf{-6} \\ \mathbf{-15} & \mathbf{-16} \end{bmatrix}$$

4. $\boldsymbol{a}(\boldsymbol{AB}) = (\boldsymbol{aA})\boldsymbol{B} = \boldsymbol{A}(\boldsymbol{aB})$

Example 3.2-11: Given $A = \begin{bmatrix} 1 & 0 & -1 \\ 2 & 3 & -1 \\ 0 & 5 & 2 \end{bmatrix}$, $B = \begin{bmatrix} -3 & 2 & -1 \\ -1 & 0 & 1 \\ 0 & 3 & -6 \end{bmatrix}$, and $a = -5$ show that the equalities $a(AB) = (aA)B = A(aB)$ are true.

Solution:

$$\boxed{AB = \begin{bmatrix} 1 & 0 & -1 \\ 2 & 3 & -1 \\ 0 & 5 & 2 \end{bmatrix} \cdot \begin{bmatrix} -3 & 2 & -1 \\ -1 & 0 & 1 \\ 0 & 3 & -6 \end{bmatrix}} = \begin{bmatrix} (1\times -3)+(0\times -1)+(-1\times 0) & (1\times 2)+(0\times 0)+(-1\times 3) & (1\times -1)+(0\times 1)+(-1\times -6) \\ (2\times -3)+(3\times -1)+(-1\times 0) & (2\times 2)+(3\times 0)+(-1\times 3) & (2\times -1)+(3\times 1)+(-1\times -6) \\ (0\times -3)+(5\times -1)+(2\times 0) & (0\times 2)+(5\times 0)+(2\times 3) & (0\times -1)+(5\times 1)+(2\times -6) \end{bmatrix}$$

$$= \begin{bmatrix} -3+0+0 & 2+0-3 & -1+0+6 \\ -6-3+0 & 4+0-3 & -2+3+6 \\ 0-5+0 & 0+0+6 & 0+5-12 \end{bmatrix} = \begin{bmatrix} -3 & -1 & 5 \\ -9 & 1 & 7 \\ -5 & 6 & -7 \end{bmatrix} \text{ thus, } \boxed{-5.(AB)} = -5.\begin{bmatrix} -3 & -1 & 5 \\ -9 & 1 & 7 \\ -5 & 6 & -7 \end{bmatrix} = \begin{bmatrix} \mathbf{15} & \mathbf{5} & \mathbf{-25} \\ \mathbf{45} & \mathbf{-5} & \mathbf{-35} \\ \mathbf{25} & \mathbf{-30} & \mathbf{35} \end{bmatrix}$$

$$\boxed{(-5A)B = \left(-5\cdot\begin{bmatrix}1 & 0 & -1\\ 2 & 3 & -1\\ 0 & 5 & 2\end{bmatrix}\right)\cdot\begin{bmatrix}-3 & 2 & -1\\ -1 & 0 & 1\\ 0 & 3 & -6\end{bmatrix}} = \boxed{\begin{bmatrix}-5\times 1 & 0\times -5 & -1\times -5\\ -5\times 2 & -5\times 3 & -5\times -1\\ 0-5 & -5\times -5 & -5\times 2\end{bmatrix}\cdot\begin{bmatrix}-3 & 2 & -1\\ -1 & 0 & 1\\ 0 & 3 & -6\end{bmatrix}} = \boxed{\begin{bmatrix}-5 & 0 & 5\\ -10 & -15 & 5\\ 0 & -25 & -10\end{bmatrix}\cdot\begin{bmatrix}-3 & 2 & -1\\ -1 & 0 & 1\\ 0 & 3 & -6\end{bmatrix}}$$

$$= \boxed{\begin{bmatrix}(-5\times -3)+(0\times -1)+(5\times 0) & (-5\times 2)+(0\times 0)+(5\times 3) & (-5\times -1)+(0\times 1)+(5\times -6)\\ (-10\times -3)+(-15\times -1)+(5\times 0) & (-10\times 2)+(-15\times 0)+(5\times 3) & (-10\times -1)+(-15\times 1)+(5\times -6)\\ (0\times -3)+(-25\times -1)+(-10\times 0) & (0\times 2)+(-25\times 0)+(-10\times 3) & (0\times -1)+(-25\times 1)+(-10\times -6)\end{bmatrix}}$$

$$= \boxed{\begin{bmatrix}15+0+0 & -10+0+15 & 5+0-30\\ 30+15+0 & -20+0+15 & 10-15-30\\ 0+25+0 & 0+0-30 & 0-25+60\end{bmatrix}} = \boxed{\begin{bmatrix}\mathbf{15} & \mathbf{5} & \mathbf{-25}\\ \mathbf{45} & \mathbf{-5} & \mathbf{-35}\\ \mathbf{25} & \mathbf{-30} & \mathbf{35}\end{bmatrix}} \text{ therefore } \boxed{-5(AB) = (-5A)B}$$

$$\boxed{A(-5B) = \begin{bmatrix}1 & 0 & -1\\ 2 & 3 & -1\\ 0 & 5 & 2\end{bmatrix}\left(-5\cdot\begin{bmatrix}-3 & 2 & -1\\ -1\cdot & 0 & 1\\ 0 & 3 & -6\end{bmatrix}\right)} = \boxed{\begin{bmatrix}1 & 0 & -1\\ 2 & 3 & -1\\ 0 & 5 & 2\end{bmatrix}\begin{bmatrix}-5\times -3 & -5\times 2 & -5\times -1\\ -5\times -1 & -5\times 0 & -5\times 1\\ -5\times 0 & -5\times 3 & -5\times -6\end{bmatrix}} = \boxed{\begin{bmatrix}1 & 0 & -1\\ 2 & 3 & -1\\ 0 & 5 & 2\end{bmatrix}\begin{bmatrix}15 & -10 & 5\\ 5 & 0 & -5\\ 0 & -15 & 30\end{bmatrix}}$$

$$= \boxed{\begin{bmatrix}(1\times 15)+(0\times 5)+(-1\times 0) & (1\times -10)+(0\times 0)+(-1\times -15) & (1\times 5)+(0\times -5)+(-1\times 30)\\ (2\times 15)+(3\times 5)+(-1\times 0) & (2\times -10)+(3\times 0)+(-1\times -15) & (2\times 5)+(3\times -5)+(-1\times 30)\\ (0\times 15)+(5\times 5)+(2\times 0) & (0\times -10)+(5\times 0)+(2\times -15) & (0\times 5)+(5\times -5)+(2\times 30)\end{bmatrix}}$$

$$= \boxed{\begin{bmatrix}15+0+0 & -10+0+15 & 5+0-30\\ 30+15+0 & -20+0+15 & 10-15-30\\ 0+25+0 & 0+0-30 & 0-25+60\end{bmatrix}} = \boxed{\begin{bmatrix}\mathbf{15} & \mathbf{5} & \mathbf{-25}\\ \mathbf{45} & \mathbf{-5} & \mathbf{-35}\\ \mathbf{25} & \mathbf{-30} & \mathbf{35}\end{bmatrix}} \text{ therefore } \boxed{-5(AB) = (-5A)B = A(-5B)}$$

5. $a(bA) = (ab)A$

Example 3.2-12: Given $A = \begin{bmatrix}-5 & 4\\ -1 & 5\end{bmatrix}$, $a = 3$, and $b = -1$ show that $a(bA) = (ab)A$.

Solution:

$$\boxed{a(bA)} = \boxed{3\cdot\left(-1\cdot\begin{bmatrix}-5 & 4\\ -1 & 5\end{bmatrix}\right)} = \boxed{3\cdot\begin{bmatrix}-5\times -1 & 4\times -1\\ -1\times -1 & 5\times -1\end{bmatrix}} = \boxed{3\cdot\begin{bmatrix}5 & -4\\ 1 & -5\end{bmatrix}} = \boxed{\begin{bmatrix}5\times 3 & -4\times 3\\ 1\times 3 & -5\times 3\end{bmatrix}} = \boxed{\begin{bmatrix}\mathbf{15} & \mathbf{-12}\\ \mathbf{3} & \mathbf{-15}\end{bmatrix}}$$

$$\boxed{(ab)A} = \boxed{(3\times -1)\begin{bmatrix}-5 & 4\\ -1 & 5\end{bmatrix}} = \boxed{-3\cdot\begin{bmatrix}-5 & 4\\ -1 & 5\end{bmatrix}} = \boxed{\begin{bmatrix}-5\times -3 & 4\times -3\\ -1\times -3 & 5\times -3\end{bmatrix}} = \boxed{\begin{bmatrix}\mathbf{15} & \mathbf{-12}\\ \mathbf{3} & \mathbf{-15}\end{bmatrix}}$$

6. $a(A+B) = aA + aB$

Example 3.2-13: Given $A = \begin{bmatrix}-6 & 10\\ -2 & 3\end{bmatrix}$, $B = \begin{bmatrix}3 & 2\\ -1 & 0\end{bmatrix}$, and $a = 4$ show that $a(A+B) = aA + aB$.

Solution:

$$\boxed{a(A+B)} = \boxed{4\cdot\left(\begin{bmatrix}-6 & 10\\ -2 & 3\end{bmatrix}+\begin{bmatrix}3 & 2\\ -1 & 0\end{bmatrix}\right)} = \boxed{4\cdot\begin{bmatrix}-6+3 & 10+2\\ -2-1 & 3+0\end{bmatrix}} = \boxed{4\cdot\begin{bmatrix}-3 & 12\\ -3 & 3\end{bmatrix}} = \boxed{\begin{bmatrix}-3\times 4 & 12\times 4\\ -3\times 4 & 3\times 4\end{bmatrix}} = \boxed{\begin{bmatrix}\mathbf{-12} & \mathbf{48}\\ \mathbf{-12} & \mathbf{12}\end{bmatrix}}$$

$$\boxed{aA+aB} = \boxed{4\cdot\begin{bmatrix}-6 & 10\\ -2 & 3\end{bmatrix}+4\cdot\begin{bmatrix}3 & 2\\ -1 & 0\end{bmatrix}} = \boxed{\begin{bmatrix}-6\times 4 & 10\times 4\\ -2\times 4 & 3\times 4\end{bmatrix}+\begin{bmatrix}3\times 4 & 2\times 4\\ -1\times 4 & 0\times 4\end{bmatrix}} = \boxed{\begin{bmatrix}-24 & 40\\ -8 & 12\end{bmatrix}+\begin{bmatrix}12 & 8\\ -4 & 0\end{bmatrix}}$$

$$= \boxed{\begin{bmatrix}-24+12 & 40+8\\ -8-4 & 12+0\end{bmatrix}} = \boxed{\begin{bmatrix}\mathbf{-12} & \mathbf{48}\\ \mathbf{-12} & \mathbf{12}\end{bmatrix}}$$

7. **The following are additional properties of matrix multiplication.**
   a. One multiplied by a matrix A is always equal to A, i.e., $1\cdot A = A$.
   b. Minus one multiplied by a matrix A is equal to $-A$, i.e., $-1\cdot A = -A$.
   c. Zero multiplied by a matrix A is equal to the zero matrix, i.e., $0\cdot A = 0$.
   d. A constant a multiplied by the zero matrix is equal to the zero matrix, i.e., $a\cdot 0 = 0$.
   e. Matrix A multiplied by the identity matrix is equal to the A matrix, i.e., $I\cdot A = A\cdot I = A$.
   f. A constant a multiplied by a matrix A is equal to the B matrix, i.e., $a\cdot A = B$, where the entries of the B matrix are obtained by multiplying the constant a with the entries of the matrix A.

Example 3.2-14: Given $A=\begin{bmatrix}7 & 3\\ 4 & 0\end{bmatrix}$, $I=\begin{bmatrix}1 & 0\\ 0 & 1\end{bmatrix}$, and $a=-6$ show that:

a. $1\cdot A = A$ b. $-1\cdot A = -A$ c. $0\cdot A_{2\times 2} = 0_{2\times 2}$ d. $a\cdot 0_{2\times 2} = 0_{2\times 2}$

e. $A\times I = I\times A = A$. In addition, find matrix B by multiplying a with the matrix A.

Solutions:

a. $\boxed{1\cdot A} = \boxed{1\cdot\begin{bmatrix}7 & 3\\ 4 & 0\end{bmatrix}} = \boxed{\begin{bmatrix}1\times 7 & 1\times 3\\ 1\times 4 & 1\times 0\end{bmatrix}} = \boxed{\begin{bmatrix}7 & 3\\ 4 & 0\end{bmatrix} = A}$ therefore $\boxed{\mathbf{1\cdot A = A}}$

b. $\boxed{-1\cdot A} = \boxed{-1\cdot\begin{bmatrix}7 & 3\\ 4 & 0\end{bmatrix}} = \boxed{\begin{bmatrix}-1\times 7 & -1\times 3\\ -1\times 4 & -1\times 0\end{bmatrix}} = \boxed{\begin{bmatrix}-7 & -3\\ -4 & -1\end{bmatrix} = -A}$ therefore $\boxed{\mathbf{-1\cdot A = -A}}$

c. $\boxed{0\cdot A_{2\times 2}} = \boxed{0\cdot\begin{bmatrix}7 & 3\\ 4 & 0\end{bmatrix}} = \boxed{\begin{bmatrix}0\times 7 & 0\times 3\\ 0\times 4 & 0\times 0\end{bmatrix}} = \boxed{\begin{bmatrix}0 & 0\\ 0 & 0\end{bmatrix} = 0_{2\times 2}}$ therefore $\boxed{\mathbf{0\cdot A_{2\times 2} = 0_{2\times 2}}}$

d. $\boxed{-6\cdot 0_{2\times 2}} = \boxed{-6\cdot\begin{bmatrix}0 & 0\\ 0 & 0\end{bmatrix}} = \boxed{\begin{bmatrix}-6\times 0 & -6\times 0\\ -6\times 0 & -6\times 0\end{bmatrix}} = \boxed{\begin{bmatrix}0 & 0\\ 0 & 0\end{bmatrix} = 0_{2\times 2}}$ therefore $\boxed{\mathbf{-6\cdot 0_{2\times 2} = 0_{2\times 2}}}$

e. $\boxed{I\cdot A} = \boxed{\begin{bmatrix}1 & 0\\ 0 & 1\end{bmatrix}\cdot\begin{bmatrix}7 & 3\\ 4 & 0\end{bmatrix}} = \boxed{\begin{bmatrix}1\times 7+0\times 4 & 1\times 3+0\times 0\\ 0\times 7+1\times 4 & 0\times 3+1\times 0\end{bmatrix}} = \boxed{\begin{bmatrix}7+0 & 3+0\\ 0+4 & 0+0\end{bmatrix}} = \boxed{\begin{bmatrix}\mathbf{7} & \mathbf{3}\\ \mathbf{4} & \mathbf{0}\end{bmatrix}}$

$\boxed{A\cdot I} = \boxed{\begin{bmatrix}7 & 3\\ 4 & 0\end{bmatrix}\cdot\begin{bmatrix}1 & 0\\ 0 & 1\end{bmatrix}} = \boxed{\begin{bmatrix}7\times 1+3\times 0 & 7\times 0+3\times 1\\ 4\times 1+0\times 0 & 4\times 0+0\times 1\end{bmatrix}} = \boxed{\begin{bmatrix}7+0 & 0+3\\ 4+0 & 0+0\end{bmatrix}} = \boxed{\begin{bmatrix}\mathbf{7} & \mathbf{3}\\ \mathbf{4} & \mathbf{0}\end{bmatrix}}$ thus $\boxed{\mathbf{I\cdot A = A\cdot I = A}}$

f. $\boxed{a \cdot A} = \boxed{-6 \cdot \begin{bmatrix} 7 & 3 \\ 4 & 0 \end{bmatrix}} = \boxed{\begin{bmatrix} -6\times 7 & -6\times 3 \\ -6\times 4 & -6\times 0 \end{bmatrix}} = \boxed{\begin{bmatrix} \mathbf{-42} & \mathbf{-18} \\ \mathbf{-24} & \mathbf{0} \end{bmatrix}}$ which is equal to a new matrix B.

Matrix multiplication is used in representing linear systems in matrix form. For example, Instead of showing a linear system in the form of $\begin{array}{llll} 3x & -y & +3z & =5 \\ 2x & +3y & -z & =-1 \\ -x & +4y & +2z & =4 \end{array}$ the system can be represented in its equivalent matrix form of $AX = B$ which is equal to $\begin{bmatrix} 3 & -1 & 3 \\ 2 & 3 & -1 \\ -1 & 4 & 2 \end{bmatrix}\begin{bmatrix} x \\ y \\ z \end{bmatrix} = \begin{bmatrix} 5 \\ -1 \\ 4 \end{bmatrix}$ where $A = \begin{bmatrix} 3 & -1 & 3 \\ 2 & 3 & -1 \\ -1 & 4 & 2 \end{bmatrix}$, $X = \begin{bmatrix} x \\ y \\ z \end{bmatrix}$, and $B = \begin{bmatrix} 5 \\ -1 \\ 4 \end{bmatrix}$

The following provides further examples of linear systems in the form of $AX = B$.

Example 3.2-15: Given $A = \begin{bmatrix} 1 & 2 \\ -1 & 3 \end{bmatrix}$, $X = \begin{bmatrix} x \\ y \end{bmatrix}$, and $B = \begin{bmatrix} 10 \\ -2 \end{bmatrix}$ write the linear system $AX = B$.

Solution:

$\boxed{AX = B}$; $\boxed{\begin{bmatrix} 1 & 2 \\ -1 & 3 \end{bmatrix}\begin{bmatrix} x \\ y \end{bmatrix} = \begin{bmatrix} 10 \\ -2 \end{bmatrix}}$; $\boxed{\begin{bmatrix} 1\times x + 2\times y \\ -1\times x + 3\times y \end{bmatrix} = \begin{bmatrix} 10 \\ -2 \end{bmatrix}}$; $\boxed{\begin{bmatrix} x + 2y \\ -x + 3y \end{bmatrix} = \begin{bmatrix} 10 \\ -2 \end{bmatrix}}$; $\boxed{\begin{array}{c} \mathbf{x + 2y = 10} \\ \mathbf{-x + 3y = -2} \end{array}}$

Example 3.2-16: Represent the given linear system $\begin{bmatrix} 1 & 0 & -1 \\ 2 & 3 & -4 \\ 5 & 1 & -1 \end{bmatrix}\begin{bmatrix} x \\ y \\ z \end{bmatrix} = \begin{bmatrix} 1 \\ 2 \\ 5 \end{bmatrix}$ by three separate equations.

Solution:

$\boxed{\begin{bmatrix} 1 & 0 & -1 \\ 2 & 3 & -4 \\ 5 & 1 & -1 \end{bmatrix}\begin{bmatrix} x \\ y \\ z \end{bmatrix} = \begin{bmatrix} 1 \\ 2 \\ 5 \end{bmatrix}}$; $\boxed{\begin{bmatrix} (1\times x)+(0\times y)+(-1\times z) \\ (2\times x)+(3\times y)+(-4\times z) \\ (5\times x)+(1\times y)+(-1\times z) \end{bmatrix} = \begin{bmatrix} 1 \\ 2 \\ 5 \end{bmatrix}}$; $\boxed{\begin{bmatrix} x + 0y - z \\ 2x + 3y - 4z \\ 5x + y - z \end{bmatrix} = \begin{bmatrix} 1 \\ 2 \\ 5 \end{bmatrix}}$; $\boxed{\begin{array}{c} \mathbf{x - z = 1} \\ \mathbf{2x + 3y - 4z = 2} \\ \mathbf{5x + y - z = 5} \end{array}}$

Having learned the basics of the matrix operations, in the next section, we will learn how to find the determinant of an $n \times n$ matrix.

Section 3.2 Case II Practice Problems - Matrix Multiplication

1. Find the product of the following matrix operations.

a. $\begin{bmatrix} 1 & 5 & 3 \\ 0 & -3 & 1 \end{bmatrix}\begin{bmatrix} 1 & -1 \\ 2 & 0 \\ 1 & 3 \end{bmatrix} =$ b. $\begin{bmatrix} 1 & -2 & 2 \end{bmatrix}\begin{bmatrix} 1 \\ 3 \\ -1 \end{bmatrix} =$ c. $\begin{bmatrix} 1 & -1 \\ 2 & 3 \end{bmatrix}\begin{bmatrix} 1 & -5 \\ 3 & 1 \end{bmatrix} =$

d. $\begin{bmatrix} 4 & -3 \\ 7 & 3 \end{bmatrix}\begin{bmatrix} 5 & 1 \\ -2 & 3 \end{bmatrix} =$ e. $\begin{bmatrix} -1 & 0 & 1 \\ 3 & 1 & 0 \\ 1 & 0 & 2 \end{bmatrix}\begin{bmatrix} 0 & 1 & 2 \\ 3 & 0 & 1 \\ -1 & 1 & 0 \end{bmatrix} =$ f. $\begin{bmatrix} 2 & -3 & -1 \\ 0 & 1 & -2 \\ 0 & 1 & 0 \end{bmatrix}\begin{bmatrix} 1 & 0 & 0 \\ 0 & 1 & 0 \\ 0 & 0 & 1 \end{bmatrix} =$

g. $3\begin{bmatrix}1 & 2 & 3 & 5\\0 & 1 & -1 & 3\\1 & 0 & 1 & 2\end{bmatrix}=$ h. $-5\begin{bmatrix}0 & 1 & -2\\-3 & -1 & 2\end{bmatrix}=$

2. Given $A=\begin{bmatrix}1 & 2\\-1 & 0\end{bmatrix}$ and $B=\begin{bmatrix}-1 & 2\\0 & 1\end{bmatrix}$ find.

a. $AB=$ b. $BA=$ c. $(AB)^t A=$ d. $(BA)^t B=$

e. $A^t(BA)=$ f. $AB^t=$ g. $A^tB^t=$ h. $(A^tB^t)A=$

3. Given the following matrix equations find the matrix Y.

a. $2Y+\begin{bmatrix}1 & 0\\0 & 3\end{bmatrix}=\begin{bmatrix}-3 & 2\\1 & 3\end{bmatrix}$ b. $2Y+3\begin{bmatrix}1 & 1\\-1 & 2\end{bmatrix}=\begin{bmatrix}7 & -1\\5 & 0\end{bmatrix}$

c. $3Y-2I=2\begin{bmatrix}5 & 3\\1 & -2\end{bmatrix}$ d. $Y+4I=\begin{bmatrix}8 & 0 & 0\\10 & 0 & 1\\-1 & 2 & 3\end{bmatrix}$

4. Given $A=\begin{bmatrix}1 & 2\\0 & -1\end{bmatrix}$, $B=\begin{bmatrix}1 & 0\\-1 & 2\end{bmatrix}$, and $C=\begin{bmatrix}2 & -1\\0 & 1\end{bmatrix}$ show that:

a. $(AB)C=A(BC)$ b. $A(B+C)=AB+AC$

c. $(B+C)A=BA+CA$ d. $3(A+B)=3A+3B$

5. Given $A=\begin{bmatrix}1 & 2\\0 & 1\end{bmatrix}$ and $B=\begin{bmatrix}2 & -1\\3 & 0\end{bmatrix}$ show that:

a. $(AB)^t=B^tA^t$ b. $(A+B)(A+B)\neq A^2+2AB+B^2=$

c. $(A+B)(A-B)\neq A^2-B^2$ d. $AB\neq BA$

6. Multiply the following matrices.

a. $3\begin{bmatrix}1 & 2\\0 & 1\end{bmatrix}\begin{bmatrix}-1 & 0\\2 & -2\end{bmatrix}=$ b. $\begin{bmatrix}1 & 2\\-1 & 0\end{bmatrix}\begin{bmatrix}1 & -1\\2 & -3\end{bmatrix}+\begin{bmatrix}1 & 0\\0 & -3\end{bmatrix}\begin{bmatrix}-1 & 2\\0 & 3\end{bmatrix}=$

c. $-2\begin{bmatrix}-1 & 0\\-2 & 3\end{bmatrix}\begin{bmatrix}2 & 0\\3 & -1\end{bmatrix}+2I=$ d. $2\begin{bmatrix}1 & 0\\-2 & 3\end{bmatrix}-\left(\begin{bmatrix}1 & 0\\-1 & 2\end{bmatrix}\begin{bmatrix}-2 & 3\\-1 & 1\end{bmatrix}\right)=$

e. $\begin{bmatrix}1 & 0 & 0\\-1 & 0 & -2\\0 & 1 & 2\end{bmatrix}\left(\begin{bmatrix}1 & 0 & -1\\1 & 1 & 0\\0 & 0 & 1\end{bmatrix}+\begin{bmatrix}-3 & 1 & 0\\0 & -2 & 1\\1 & -1 & 0\end{bmatrix}\right)=$ f. $\left(\begin{bmatrix}1 & 2 & -1\\0 & 1 & 0\\-1 & 0 & 2\end{bmatrix}-\begin{bmatrix}1 & 2 & -3\\1 & 0 & 1\\0 & -2 & -1\end{bmatrix}\right)\begin{bmatrix}0 & -1 & 2\\-3 & 2 & -1\\0 & 1 & -3\end{bmatrix}=$

7. Given $B=\begin{bmatrix}1 & -2 & 3\\5 & -3 & 4\\1 & 2 & 3\end{bmatrix}$, $U=\begin{bmatrix}u\\v\\w\end{bmatrix}$, and $C=\begin{bmatrix}1\\-2\\4\end{bmatrix}$ write the linear system $BU=C$.

3.3 Determinants

In this section we will learn how to calculate the minors and the cofactors of a square matrix which leads to the calculation of determinants. Students are encouraged to spend adequate time learning how to compute minors, cofactors, and determinants. Knowing how to compute determinants will greatly simplify evaluation of the inverse of a matrix - a subject which is addressed in the next section.

Calculating Minors:

Given A is an $n \times n$ matrix, the minor M_{ij} of an entry a_{ij} is equal to the determinant of the A matrix $\delta(A)$ after deleting the ith row and the jth column in which the entry a_{ij} appears. The minor of a matrix is denoted by the symbol M_{ij}.

Example 3.3-2: Given the matrix $A = \begin{bmatrix} -1 & 3 \\ 5 & 2 \end{bmatrix}$, find the following minors.

a. M_{11} b. M_{12} c. M_{21} d. M_{22}

Solutions:

a. To find M_{11} cross out the first row and first column in matrix A to obtain $\mathbf{M_{11} = 2}$.

b. To find M_{12} cross out the first row and second column in matrix A to obtain $\mathbf{M_{12} = 5}$.

c. To find M_{21} cross out the second row and first column in matrix A to obtain $\mathbf{M_{21} = 3}$.

d. To find M_{22} cross out the second row and second column in matrix A to obtain $\mathbf{M_{22} = -1}$.

Example 3.3-3: Given the matrix $A = \begin{bmatrix} 1 & 2 & 3 \\ 5 & 1 & 0 \\ 2 & -1 & 3 \end{bmatrix}$, compute the following minors.

a. M_{11} b. M_{12} c. M_{13} d. M_{21} e. M_{22} f. M_{23}
g. M_{31} h. M_{32} i. M_{33}

Solutions:

a. To calculate M_{11} cross out the first row and first column in matrix A.

$$\boxed{M_{11}} = \boxed{\begin{vmatrix} 1 & 0 \\ -1 & 3 \end{vmatrix}} = \boxed{(1 \times 3) - (0 \times -1)} = \boxed{3 - 0} = \boxed{3}$$

b. To calculate M_{12} cross out the first row and second column in matrix A.

$$\boxed{M_{12}} = \boxed{\begin{vmatrix} 5 & 0 \\ 2 & 3 \end{vmatrix}} = \boxed{(5 \times 3) - (0 \times 2)} = \boxed{15 - 0} = \boxed{15}$$

c. To calculate M_{13} cross out the first row and third column in matrix A.

$$\boxed{M_{13}} = \boxed{\begin{vmatrix} 5 & 1 \\ 2 & -1 \end{vmatrix}} = \boxed{(5 \times -1) - (1 \times 2)} = \boxed{-5 - 2} = \boxed{-7}$$

d. To calculate M_{21} cross out the second row and first column in matrix A.

$$\boxed{M_{21}} = \boxed{\begin{vmatrix} 2 & 3 \\ -1 & 3 \end{vmatrix}} = \boxed{(2 \times 3) - (3 \times -1)} = \boxed{6 + 3} = \boxed{9}$$

e. To calculate M_{22} cross out the second row and second column in matrix A.

$\boxed{M_{22}} = \begin{Vmatrix} 1 & 3 \\ 2 & 3 \end{Vmatrix} = \boxed{(1 \times 3) - (3 \times 2)} = \boxed{3 - 6} = \boxed{-3}$

f. To calculate M_{23} cross out the second row a nd third column in matrix A.

$\boxed{M_{23}} = \begin{Vmatrix} 1 & 2 \\ 2 & -1 \end{Vmatrix} = \boxed{(1 \times -1) - (2 \times 2)} = \boxed{-1 - 4} = \boxed{-5}$

g. To calculate M_{31} cross out the third row and first column in matrix A.

$\boxed{M_{31}} = \begin{Vmatrix} 2 & 3 \\ 1 & 0 \end{Vmatrix} = \boxed{(2 \times 0) - (3 \times 1)} = \boxed{0 - 3} = \boxed{-3}$

h. To calculate M_{32} cross out the third row and second column in matrix A.

$\boxed{M_{32}} = \begin{Vmatrix} 1 & 3 \\ 5 & 0 \end{Vmatrix} = \boxed{(1 \times 0) - (3 \times 5)} = \boxed{0 - 15} = \boxed{-15}$

i. To calculate M_{33} cross out the third row and third column in matrix A.

$\boxed{M_{33}} = \begin{Vmatrix} 1 & 2 \\ 5 & 1 \end{Vmatrix} = \boxed{(1 \times 1) - (2 \times 5)} = \boxed{1 - 10} = \boxed{-9}$

Having learned how to evaluate the minors of an $n \times n$ matrix, next we will learn how to compute the cofactors of an $n \times n$ matrix.

Calculating Cofactors:

Given A is an $n \times n$ matrix, the cofactor A_{ij} of an entry a_{ij} is the product of the minor M_{ij} of the entry a_{ij} multiplied by $(-1)^{i+j}$, i.e.,

$$A_{ij} = (-1)^{i+j} M_{ij}$$

The cofactor of a matrix is denoted by the symbol A_{ij}.

Note that when $i + j$ is an even number $(-1)^{i+j}$ is equal to $+1$, i. e., $(-1)^2 = (-1)^4 = (-1)^6 = (-1)^8 \ldots = 1$ and when $i + j$ is an odd number $(-1)^{i+j}$ is equal to -1, i.e., $(-1)^3 = (-1)^5 = (-1)^7 = (-1)^9 \ldots = -1$. Therefore, the cofactor A_{ij} is simply the minor M_{ij} but with $+$ or $-$ sign attached to it. Hence, in an $n \times n$ matrix, the sign pattern the cofactor assigns to the minor is as follows:

$$\begin{bmatrix} + & - & + & - & \cdots \\ - & + & - & + & \cdots \\ + & - & + & - & \cdots \\ - & + & - & + & \cdots \\ \vdots & \vdots & \vdots & \vdots & \vdots \end{bmatrix}$$

Example 3.3-4: Given the matrix $A = \begin{bmatrix} -1 & 3 \\ 5 & 2 \end{bmatrix}$, find the following cofactors.

a. A_{11} b. A_{12} c. A_{21} d. A_{22}

Solutions:

From example 3.3-2, we have $M_{11} = 2$, $M_{12} = 5$, $M_{21} = 3$, and $M_{22} = -1$. Therefore, using the general cofactor equation, $A_{ij} = (-1)^{i+j} M_{ij}$, we can compute the following cofactors.

a. $\boxed{A_{11}} = \boxed{(-1)^{1+1} M_{11}} = \boxed{(-1)^2 \times 2} = \boxed{1 \times 2} = \boxed{\mathbf{2}}$

b. $\boxed{A_{12}} = \boxed{(-1)^{1+2} M_{12}} = \boxed{(-1)^3 \times 5} = \boxed{-1 \times 5} = \boxed{\mathbf{-5}}$

c. $\boxed{A_{21}} = \boxed{(-1)^{2+1} M_{21}} = \boxed{(-1)^3 \times 3} = \boxed{-1 \times 3} = \boxed{\mathbf{-3}}$

d. $\boxed{A_{22}} = \boxed{(-1)^{2+2} M_{22}} = \boxed{(-1)^4 \times -1} = \boxed{1 \times -1} = \boxed{\mathbf{-1}}$

Example 3.3-5: Given the matrix $A = \begin{bmatrix} 1 & 2 & 3 \\ 5 & 1 & 0 \\ 2 & -1 & 3 \end{bmatrix}$, compute the following cofactors.

a. A_{11} b. A_{12} c. A_{13} d. A_{21} e. A_{22} f. A_{23}

g. A_{31} h. A_{32} i. A_{33}

Solutions:

From example 3.3-3, we have $M_{11} = 3$, $M_{12} = 15$, $M_{13} = -7$, $M_{21} = 9$, $M_{22} = -3$, $M_{23} = -5$, $M_{31} = -3$, $M_{32} = -15$, and $M_{33} = -9$. Therefore, using the general cofactor equation, $A_{ij} = (-1)^{i+j} M_{ij}$, we can compute the cofactors as follows:

a. $\boxed{A_{11}} = \boxed{(-1)^{1+1} M_{11}} = \boxed{(-1)^2 \times 3} = \boxed{1 \times 3} = \boxed{\mathbf{3}}$

b. $\boxed{A_{12}} = \boxed{(-1)^{1+2} M_{12}} = \boxed{(-1)^3 \times 15} = \boxed{-1 \times 15} = \boxed{\mathbf{-15}}$

c. $\boxed{A_{13}} = \boxed{(-1)^{1+3} M_{13}} = \boxed{(-1)^4 \times -7} = \boxed{1 \times -7} = \boxed{\mathbf{-7}}$

d. $\boxed{A_{21}} = \boxed{(-1)^{2+1} M_{21}} = \boxed{(-1)^3 \times 9} = \boxed{-1 \times 9} = \boxed{\mathbf{-9}}$

e. $\boxed{A_{22}} = \boxed{(-1)^{2+2} M_{22}} = \boxed{(-1)^4 \times -3} = \boxed{1 \times -3} = \boxed{\mathbf{-3}}$

f. $\boxed{A_{23}} = \boxed{(-1)^{2+3} M_{23}} = \boxed{(-1)^5 \times -5} = \boxed{-1 \times -5} = \boxed{\mathbf{5}}$

g. $\boxed{A_{31}} = \boxed{(-1)^{3+1} M_{31}} = \boxed{(-1)^4 \times -3} = \boxed{1 \times -3} = \boxed{\mathbf{-3}}$

h. $\boxed{A_{32}} = \boxed{(-1)^{3+2} M_{32}} = \boxed{(-1)^5 \times -15} = \boxed{-1 \times -15} = \boxed{\mathbf{15}}$

i. $\boxed{A_{33}} = \boxed{(-1)^{3+3} M_{33}} = \boxed{(-1)^6 \times -9} = \boxed{1 \times -9} = \boxed{\mathbf{-9}}$

Having learned how to evaluate the minors and the cofactors of an $n \times n$ matrix, next we will learn how to compute the determinant of an $n \times n$ matrix.

Calculating Determinants:

Associated with each square matrix A is a real number called the determinant of A. The determinant of A is denoted by the symbol δA or $\delta(A)$ which reads as "the determinant of A". Note that the determinant of a square matrix is generally denoted by two vertical bars "| |" instead of the brackets "[ ]" which is used for matrix identification.

The determinant of an $n \times n$ square matrix is the sum of n products obtained by multiplying each entry in any selected row or column by its cofactor. For example, given the A matrix

$$A_{n\times n}=\begin{bmatrix} a_{11} & a_{12} & a_{13} & \cdots & a_{1n} \\ a_{21} & a_{22} & a_{23} & \cdots & a_{2n} \\ a_{31} & a_{32} & a_{33} & \cdots & a_{3n} \\ \vdots & \vdots & \vdots & & \vdots \\ a_{n1} & a_{n2} & a_{n3} & \cdots & a_{nn} \end{bmatrix}$$

the determinant by expanding the first row, second row, third row, nth row, first column, second column, third column, and nth column is calculated in the following way:

a. The determinant calculated by expanding the first row.

$$\boxed{\delta(A)}=\boxed{a_{11}\cdot A_{11}+a_{12}\cdot A_{12}+a_{13}\cdot A_{13}+\cdots+a_{1n}\cdot A_{1n}}$$

$$=\boxed{a_{11}\cdot(-1)^{1+1}M_{11}+a_{12}\cdot(-1)^{1+2}M_{12}+a_{13}\cdot(-1)^{1+3}M_{13}+\cdots+a_{1n}\cdot(-1)^{1+n}M_{1n}}$$

b. The determinant calculated by expanding the second row.

$$\boxed{\delta(A)}=\boxed{a_{21}\cdot A_{21}+a_{22}\cdot A_{22}+a_{23}\cdot A_{23}+\cdots+a_{2n}\cdot A_{2n}}$$

$$=\boxed{a_{21}\cdot(-1)^{2+1}M_{21}+a_{22}\cdot(-1)^{2+2}M_{22}+a_{23}\cdot(-1)^{2+3}M_{23}+\cdots+a_{2n}\cdot(-1)^{2+n}M_{2n}}$$

c. The determinant calculated by expanding the third row.

$$\boxed{\delta(A)}=\boxed{a_{31}\cdot A_{31}+a_{32}\cdot A_{32}+a_{33}\cdot A_{33}+\cdots+a_{3n}\cdot A_{3n}}$$

$$=\boxed{a_{31}\cdot(-1)^{3+1}M_{31}+a_{32}\cdot(-1)^{3+2}M_{32}+a_{33}\cdot(-1)^{3+3}M_{33}+\cdots+a_{3n}\cdot(-1)^{3+n}M_{3n}}$$

d. The determinant calculated by expanding the nth row.

$$\boxed{\delta(A)}=\boxed{a_{n1}\cdot A_{n1}+a_{n2}\cdot A_{n2}+a_{n3}\cdot A_{n3}+\cdots+a_{nn}\cdot A_{nn}}$$

$$=\boxed{a_{n1}\cdot(-1)^{n+1}M_{n1}+a_{n2}\cdot(-1)^{n+2}M_{n2}+a_{n3}\cdot(-1)^{n+3}M_{n3}+\cdots+a_{nn}\cdot(-1)^{n+n}M_{nn}}$$

e. The determinant calculated by expanding the first column.

$$\boxed{\delta(A)}=\boxed{a_{11}\cdot A_{11}+a_{21}\cdot A_{21}+a_{31}\cdot A_{31}+\cdots+a_{n1}\cdot A_{n1}}$$

$$=\boxed{a_{11}\cdot(-1)^{1+1}M_{11}+a_{21}\cdot(-1)^{2+1}M_{21}+a_{31}\cdot(-1)^{3+1}M_{31}+\cdots+a_{n1}\cdot(-1)^{n+1}M_{n1}}$$

f. The determinant calculated by expanding the second column.

$$\boxed{\delta(A)}=\boxed{a_{12}\cdot A_{12}+a_{22}\cdot A_{22}+a_{32}\cdot A_{32}+\cdots+a_{n2}\cdot A_{n2}}$$

$$=\boxed{a_{12}\cdot(-1)^{1+2}M_{12}+a_{22}\cdot(-1)^{2+2}M_{22}+a_{32}\cdot(-1)^{3+2}M_{32}+\cdots+a_{n2}\cdot(-1)^{n+2}M_{n2}}$$

g. The determinant calculated by expanding the third column.

$$\boxed{\delta(A)}=\boxed{a_{13}\cdot A_{13}+a_{23}\cdot A_{23}+a_{33}\cdot A_{33}+\cdots+a_{n3}\cdot A_{n3}}$$

$$=\boxed{a_{13}\cdot(-1)^{1+3}M_{13}+a_{23}\cdot(-1)^{2+3}M_{23}+a_{33}\cdot(-1)^{3+3}M_{33}+\cdots+a_{n3}\cdot(-1)^{n+3}M_{n3}}$$

h. Determinant expanded by the nth column.

$$\boxed{\delta(A)} = \boxed{a_{1n} \cdot A_{1n} + a_{2n} \cdot A_{2n} + a_{3n} \cdot A_{3n} + \cdots + a_{nn} \cdot A_{nn}}$$

$$= \boxed{a_{1n} \cdot (-1)^{1+n} M_{1n} + a_{2n} \cdot (-1)^{2+n} M_{2n} + a_{3n} \cdot (-1)^{3+n} M_{3n} + \cdots + a_{nn} \cdot (-1)^{n+n} M_{nn}}$$

- **Calculating the Determinant of a 2×2 Matrix**

The determinant of a 2×2 matrix is a scalar number which is equal to the difference of the products of the entries on the two diagonals and is presented in the following general form as:

$$\delta(A) = \delta\begin{bmatrix} a_{11} & a_{12} \\ a_{21} & a_{22} \end{bmatrix} = \begin{vmatrix} a_{11} & a_{12} \\ a_{21} & a_{22} \end{vmatrix} = \mathbf{a_{11} \times a_{22} - a_{12} \times a_{21}}$$

Note that this is a quick way of finding the determinant of any 2×2 matrix. We can also find the determinant of a 2×2 matrix by using the minors and the cofactors. Since we now know how to obtain the minors and the cofactors of a 2×2 matrix , let's use these principals to calculate the determinant by expanding about the first row, the second row, the first column, and the second column as follows:

a. Expanding about the first row the determinant of A is equal to:

$$\boxed{\delta(A)} = \boxed{a_{11} \cdot A_{11} + a_{12} \cdot A_{12}} = \boxed{a_{11} \cdot (-1)^{1+1} M_{11} + a_{12} \cdot (-1)^{1+2} M_{12}} = \boxed{a_{11} \cdot (-1)^{2} M_{11} + a_{12} \cdot (-1)^{3} M_{12}}$$

$$= \boxed{a_{11} \cdot 1 \cdot M_{11} + a_{12} \cdot (-1) \cdot M_{12}} = \boxed{a_{11} \cdot M_{11} - a_{12} \cdot M_{12}} = \boxed{\mathbf{a_{11} \times a_{22} - a_{12} \times a_{21}}}$$

b. Expanding about the second row the determinant of A is equal to:

$$\boxed{\delta(A)} = \boxed{a_{21} \cdot A_{21} + a_{22} \cdot A_{22}} = \boxed{a_{21} \cdot (-1)^{2+1} M_{21} + a_{22} \cdot (-1)^{2+2} M_{22}} = \boxed{a_{21} \cdot (-1)^{3} M_{21} + a_{22} \cdot (-1)^{4} M_{22}}$$

$$= \boxed{a_{21} \cdot (-1) \cdot M_{21} + a_{22} \cdot 1 \cdot M_{22}} = \boxed{-a_{21} \cdot M_{21} + a_{22} \cdot M_{22}} = \boxed{-a_{21} \times a_{12} + a_{22} \times a_{11}} = \boxed{\mathbf{a_{11} \times a_{22} - a_{12} \times a_{21}}}$$

c. Expanding about the first column the determinant of A is equal to:

$$\boxed{\delta(A)} = \boxed{a_{11} \cdot A_{11} + a_{21} \cdot A_{21}} = \boxed{a_{11} \cdot (-1)^{1+1} M_{11} + a_{21} \cdot (-1)^{2+1} M_{21}} = \boxed{a_{11} \cdot (-1)^{2} M_{11} + a_{21} \cdot (-1)^{3} M_{21}}$$

$$= \boxed{a_{11} \cdot 1 \cdot M_{11} + a_{21} \cdot (-1) \cdot M_{21}} = \boxed{a_{11} \cdot M_{11} - a_{21} \cdot M_{21}} = \boxed{a_{11} \times a_{22} - a_{21} \times a_{12}} = \boxed{\mathbf{a_{11} \times a_{22} - a_{12} \times a_{21}}}$$

d. Expanding about the second column the determinant of A is equal to:

$$\boxed{\delta(A)} = \boxed{a_{12} \cdot A_{12} + a_{22} \cdot A_{22}} = \boxed{a_{12} \cdot (-1)^{1+2} M_{12} + a_{22} \cdot (-1)^{2+2} M_{22}} = \boxed{a_{12} \cdot (-1)^{3} M_{12} + a_{22} \cdot (-1)^{4} M_{22}}$$

$$= \boxed{a_{12} \cdot (-1) \cdot M_{12} + a_{22} \cdot 1 \cdot M_{22}} = \boxed{-a_{12} \cdot M_{12} + a_{22} \cdot M_{22}} = \boxed{-a_{12} \times a_{21} + a_{22} \times a_{11}} = \boxed{\mathbf{a_{11} \times a_{22} - a_{12} \times a_{21}}}$$

Note that no matter which row or column the determinant is expanded about the value of the determinant remains the same.

Example 3.3-6: Compute the determinant of the following matrices.

a. $A = \begin{bmatrix} 4 & 2 \\ 0 & -1 \end{bmatrix}$ b. $B = \begin{bmatrix} 3 & -5 \\ 1 & -2 \end{bmatrix}$ c. $C = \begin{bmatrix} 4 & 6 \\ 2 & 3 \end{bmatrix}$ d. $D = \begin{bmatrix} 6 & 2 \\ -1 & 3 \end{bmatrix}$

Solutions:

a. $\boxed{\delta(A)} = \boxed{\delta\begin{bmatrix} 4 & 2 \\ 0 & -1 \end{bmatrix}} = \boxed{\begin{vmatrix} 4 & 2 \\ 0 & -1 \end{vmatrix}} = \boxed{(4\times -1)-(2\times 0)} = \boxed{-4+0} = \boxed{\mathbf{-4}}$

b. $\boxed{\delta(B)} = \boxed{\delta\begin{bmatrix} 3 & -5 \\ 1 & -2 \end{bmatrix}} = \boxed{\begin{vmatrix} 3 & -5 \\ 1 & -2 \end{vmatrix}} = \boxed{(3\times -2)-(-5\times 1)} = \boxed{-6+5} = \boxed{\mathbf{-1}}$

c. $\boxed{\delta(C)} = \boxed{\delta\begin{bmatrix} 4 & 6 \\ 2 & 3 \end{bmatrix}} = \boxed{\begin{vmatrix} 4 & 6 \\ 2 & 3 \end{vmatrix}} = \boxed{(4\times 3)-(6\times 2)} = \boxed{12+12} = \boxed{\mathbf{0}}$

d. $\boxed{\delta(D)} = \boxed{\delta\begin{bmatrix} 6 & 2 \\ -1 & 3 \end{bmatrix}} = \boxed{\begin{vmatrix} 6 & 2 \\ -1 & 3 \end{vmatrix}} = \boxed{(6\times 3)-(2\times -1)} = \boxed{18+2} = \boxed{\mathbf{20}}$

In this book, in order to reinforce the learning process in computing the minors and the cofactors, in most cases, we calculate the determinant of a 2×2 matrix the long way, i. e., by computing the minors and the cofactors and by expanding about a selected row or column. Students may use the quicker method for computing the determinant, i.e., by obtaining the difference of the products of the entries on the two diagonals $\delta(A) = a_{11}\times a_{22} - a_{12}\times a_{21}$ after sufficient practice in using the minors and the cofactors methods.

Note 1 - Determinant is always a scalar number such as $5, -3, -1, 0, -8, 100, etc.$. A determinant can not be represented as a matrix.

Note 2 - Even though determinant is denoted by two vertical bars, it should not be confused with the absolute value notation which is also denoted by two vertical bars. Absolute value of a number is always positive, i.e., $|-15| = |15| = 15, |20| = |-20| = 20, |-3| = |3| = 3$. However, as is shown in the above example, the determinant of a matrix can be positive, negative, or zero.

- **Calculating the Determinant of a 3×3 Matrix**

Using the minors and the cofactors methods, the steps in computing the determinant of a 3×3 matrix is similar to computing the determinant of a 2×2 matrix. The following example shows the process in finding the determinant of a 3×3 matrix:

Example 3.3-7: Given the matrix $A = \begin{bmatrix} 1 & 2 & 3 \\ 5 & 1 & 0 \\ 2 & -1 & 3 \end{bmatrix}$, compute the determinant by expanding the following rows and columns.

a. Expand about the first row. b. Expand about the second row. c. Expand about the third row.

d. Expand about the first column. e. Expand about the second column. f. Expand about the third column.

Solutions:

The entry elements of the A matrix are $a_{11} = 1$, $a_{12} = 2$, $a_{13} = 3$, $a_{21} = 5$, $a_{22} = 1$, $a_{23} = 0$, $a_{31} = 2$, $a_{32} = -1$, and $a_{33} = 3$. In addition, from example 3.3-5 we have $A_{11} = 3$, $A_{12} = -15$, $A_{13} = -7$, $A_{21} = -9$, $A_{22} = -3$, $A_{23} = 5$, $A_{31} = -3$, $A_{32} = 15$, and $A_{33} = -9$. Thus, determinant about the indicated rows and columns can be calculated as follows:

a. $\boxed{\delta(A)} = \boxed{a_{11}\cdot(-1)^{1+1}M_{11} + a_{12}\cdot(-1)^{1+2}M_{12} + a_{13}\cdot(-1)^{1+3}M_{13}} = \boxed{a_{11}\cdot A_{11} + a_{12}\cdot A_{12} + a_{13}\cdot A_{13}}$

$= \boxed{(1\cdot 3)+(2\cdot -15)+(3\cdot -7)} = \boxed{3-30-21} = \boxed{\mathbf{-48}}$

b. $\boxed{\delta(A)} = \boxed{a_{21}\cdot(-1)^{2+1}M_{21} + a_{22}\cdot(-1)^{2+2}M_{22} + a_{23}\cdot(-1)^{2+3}M_{23}} = \boxed{a_{21}\cdot A_{21} + a_{22}\cdot A_{22} + a_{23}\cdot A_{23}}$

$= \boxed{(5\cdot -9)+(1\cdot -3)+(0\cdot 5)} = \boxed{-45-3+0} = \boxed{\mathbf{-48}}$

c. $\boxed{\delta(A)} = \boxed{a_{31}\cdot(-1)^{3+1}M_{31} + a_{32}\cdot(-1)^{3+2}M_{32} + a_{33}\cdot(-1)^{3+3}M_{33}} = \boxed{a_{31}\cdot A_{31} + a_{32}\cdot A_{32} + a_{33}\cdot A_{33}}$

$= \boxed{(2\cdot -3)+(-1\cdot 15)+(3\cdot -9)} = \boxed{-6-15-27} = \boxed{\mathbf{-48}}$

d. $\boxed{\delta(A)} = \boxed{a_{11}\cdot(-1)^{1+1}M_{11} + a_{21}\cdot(-1)^{2+1}M_{21} + a_{31}\cdot(-1)^{3+1}M_{31}} = \boxed{a_{11}\cdot A_{11} + a_{21}\cdot A_{21} + a_{31}\cdot A_{31}}$

$= \boxed{(1\cdot 3)+(5\cdot -9)+(2\cdot -3)} = \boxed{3-45-6} = \boxed{\mathbf{-48}}$

e. $\boxed{\delta(A)} = \boxed{a_{12}\cdot(-1)^{1+2}M_{12} + a_{22}\cdot(-1)^{2+2}M_{22} + a_{32}\cdot(-1)^{3+2}M_{32}} = \boxed{a_{12}\cdot A_{12} + a_{22}\cdot A_{22} + a_{32}\cdot A_{32}}$

$= \boxed{(2\cdot -15)+(1\cdot -3)+(-1\cdot 15)} = \boxed{-30-3-15} = \boxed{\mathbf{-48}}$

f. $\boxed{\delta(A)} = \boxed{a_{13}\cdot(-1)^{1+3}M_{13} + a_{23}\cdot(-1)^{2+3}M_{23} + a_{33}\cdot(-1)^{3+3}M_{33}} = \boxed{a_{13}\cdot A_{13} + a_{23}\cdot A_{23} + a_{33}\cdot A_{33}}$

$= \boxed{(3\cdot -7)+(0\cdot 5)+(3\cdot -9)} = \boxed{-21+0-27} = \boxed{\mathbf{-48}}$

Note that the determinant $\delta(A)$ is equal to -48 no matter which row or column the determinant is expanded about. However, to expedite the computation of determinant, we should expand about a row or a column with the most number of zero entries. Thus, in the above example expansion about the second row or the third column is the quickest way of calculating the determinant. The following example of a 4×4 matrix further illustrates this point.

- **Calculating the Determinant of a 4×4 Matrix**

The steps in computing the determinant of a 4×4 matrix is similar to computing the determinant of a 3×3 matrix. The following example shows the process in finding the determinant of a 4×4 matrix:

Example 3.3-8: Let $A = \begin{bmatrix} 2 & 0 & 3 & 5 \\ 1 & -1 & 0 & 0 \\ -2 & 1 & 1 & 0 \\ 1 & 0 & -3 & 0 \end{bmatrix}$. Compute $\delta(A)$.

Solution:

Note that since the fourth column has the most number of zero entries expanding about the fourth column is the best choice for calculating the determinant of the above 4×4 matrix.

$$\boxed{\delta(A)} = \boxed{a_{14} \cdot A_{14} + a_{24} \cdot A_{24} + a_{34} \cdot A_{34} + a_{44} \cdot A_{44}} = \boxed{a_{14} \cdot (-1)^{1+4} M_{14} + a_{24} \cdot (-1)^{2+4} M_{24} + a_{34} \cdot (-1)^{3+4} M_{34}}$$

$$\boxed{+a_{44} \cdot (-1)^{4+4} M_{4+4}} = \boxed{5 \cdot (-1)^{1+4} M_{14} + 0 \cdot (-1)^{2+4} M_{24} + 0 \cdot (-1)^{3+4} M_{34} + 0 \cdot (-1)^{4+4} M_{4+4}}$$

$$= \boxed{5 \cdot (-1)^{1+4} M_{14} + 0 + 0 + 0} = \boxed{5 \cdot (-1)^5 M_{14}} = \boxed{-5M_{14}} = -5 \cdot \begin{bmatrix} 1 & -1 & 0 \\ -2 & 1 & 1 \\ 1 & 0 & -3 \end{bmatrix}$$

Using the cofactors and the minors expand the reduced 3×3 matrix by either the third column or the third row. Let's select the third row. Note that $a_{31} = 1$, $a_{32} = 0$, and $a_{33} = -3$.

$$\boxed{\delta(A)} = \boxed{-5 \cdot [a_{31} \cdot A_{31} + a_{32} \cdot A_{32} + a_{33} \cdot A_{33}]} = \boxed{-5 \cdot \left[a_{31} \cdot (-1)^{3+1} M_{31} + a_{32} \cdot (-1)^{3+2} M_{32} + a_{33} \cdot (-1)^{3+3} M_{33}\right]}$$

$$= \boxed{-5 \cdot \left[1 \cdot (-1)^{3+1} M_{31} + 0 \cdot (-1)^{3+2} M_{32} + (-3) \cdot (-1)^{3+3} M_{33}\right]} = \boxed{-5 \cdot [1 \cdot M_{31} + 0 + (-3 \cdot M_{33})]} = \boxed{-5 \cdot [M_{31} - 3 \cdot M_{33}]}$$

$$= \boxed{-5M_{31} + 15M_{33}} = \boxed{-5 \cdot \begin{vmatrix} -1 & 0 \\ 1 & 1 \end{vmatrix} + 15 \cdot \begin{vmatrix} 1 & -1 \\ -2 & 1 \end{vmatrix}} = \boxed{-5 \cdot [(-1 \times 1) - (0 \times 1)] + 15 \cdot [(1 \times 1) - (-1 \times -2)]}$$

$$= \boxed{-5 \cdot [-1 - 0] + 15 \cdot [1 - 2]} = \boxed{-5 \cdot (-1) + 15 \cdot (-1)} = \boxed{5 - 15} = \boxed{\mathbf{-10}} \text{ therefore } \boxed{\delta(A) = \mathbf{-10}}$$

For additional practice, as well as checking our answer, let's recalculate the determinant by expanding the reduced 3×3 matrix by the third column. Note that $a_{13} = 0$, $a_{23} = 1$, and $a_{33} = -3$.

$$\boxed{\delta(A)} = \boxed{-5 \cdot [a_{13} \cdot A_{13} + a_{23} \cdot A_{23} + a_{33} \cdot A_{33}]} = \boxed{-5 \cdot \left[a_{13} \cdot (-1)^{1+3} M_{13} + a_{23} \cdot (-1)^{2+3} M_{23} + a_{33} \cdot (-1)^{3+3} M_{33}\right]}$$

$$= \boxed{-5 \cdot \left[0 \cdot (-1)^{1+3} M_{13} + 1 \cdot (-1)^{2+3} M_{23} + (-3) \cdot (-1)^{3+3} M_{33}\right]} = \boxed{-5 \cdot [0 + (-1) \cdot M_{23} + (-3) \cdot M_{33}]}$$

$$= \boxed{-5 \cdot [-M_{23} - 3M_{33}]} = \boxed{5M_{23} + 15M_{33}} = \boxed{5 \cdot \begin{vmatrix} 1 & -1 \\ 1 & 0 \end{vmatrix} + 15 \cdot \begin{vmatrix} 1 & -1 \\ -2 & 1 \end{vmatrix}} = \boxed{5 \cdot [(1 \times 0) - (-1 \times 1)] + 15 \cdot [(1 \times 1) - (-1 \times -2)]}$$

$$= \boxed{5 \cdot [0 + 1] + 15 \cdot [1 - 2]} = \boxed{5 - 15} = \boxed{\mathbf{-10}} \text{ therefore } \boxed{\delta(A) = \mathbf{-10}}$$

Having learned how to find the determinant of a matrix, let's now look at some properties of determinants.

Properties of Determinants

Evaluation of determinants, particularly when the dimensions of the matrix is more than 3×3, becomes very time consuming. Knowledge of the properties of determinant, in many instances, reduces the number of operations involved in calculating determinants. The following are six properties of determinants that students should become familiar with:

1. *If the entries in two rows or two columns of a determinant are identical, then the determinant is equal to zero.*

   For example, a. $\delta(A)=\begin{vmatrix}-1 & 2\\ -1 & 2\end{vmatrix}=0$ b. $\delta(B)=\begin{vmatrix}2 & 3 & 1\\ 0 & 1 & 0\\ 2 & 3 & 1\end{vmatrix}=0$ c. $\delta(C)=\begin{vmatrix}1 & 4 & 3 & 0\\ -1 & 0 & 0 & 1\\ 2 & 0 & 1 & -2\\ -1 & 0 & 0 & 1\end{vmatrix}=0$

2. *If the entries in an entire row or an entire column of a determinant is zero, then the determinant is equal to zero.*

   For example, a. $\delta(A)=\begin{vmatrix}0 & 0\\ 1 & -2\end{vmatrix}=0$ b. $\delta(B)=\begin{vmatrix}0 & 0 & 3\\ 1 & 0 & 0\\ 2 & 0 & -1\end{vmatrix}=0$ c. $\delta(C)=\begin{vmatrix}0 & 0 & 0 & 1\\ 1 & 0 & 0 & 1\\ 0 & 0 & 0 & 2\\ 0 & 0 & 0 & 0\end{vmatrix}=0$

3. *If the entries in one row or one column of a determinant is multiplied by a constant, then the determinant is multiplied by the same constant.*

   For example, a. $\begin{vmatrix}2 & 3 & -1\\ 1\times 3 & 0\times 3 & 3\times 3\\ 2 & -3 & 4\end{vmatrix}=3\begin{vmatrix}2 & 3 & -1\\ 1 & 0 & 3\\ 2 & -3 & 4\end{vmatrix}$ b. $\begin{vmatrix}1 & 2\times 4 & 5\\ -1 & 1\times 4 & 2\\ 0 & -1\times 4 & 8\end{vmatrix}=4\begin{vmatrix}1 & 2 & 5\\ -1 & 1 & 2\\ 0 & -1 & 8\end{vmatrix}$

Note that the result obtained by multiplying the entries of one row or one column of a determinant by a constant should not be confused with multiplication of a matrix by a constant. In case of the determinant, only the entries to a single row or a single column are multiplied by the constant. Whereas, in case of a matrix, each entry in the matrix is multiplied by the constant. For example, $4\times\begin{bmatrix}-1 & 3\\ 2 & 5\end{bmatrix}=\begin{bmatrix}-1\times 4 & 3\times 4\\ 2\times 4 & 5\times 4\end{bmatrix}=\begin{bmatrix}-4 & 12\\ 8 & 20\end{bmatrix}$. However,

$$\boxed{4\times\begin{vmatrix}-1 & 3\\ 2 & 5\end{vmatrix}}=\boxed{\begin{vmatrix}-1\times 4 & 3\times 4\\ 2 & 5\end{vmatrix}}=\boxed{\begin{vmatrix}-4 & 12\\ 2 & 5\end{vmatrix}}=\boxed{(-4\times 5)-(12\times 2)}=\boxed{-20-24}=\boxed{\mathbf{-44}}\text{ or}$$

$$\boxed{4\times\begin{vmatrix}-1 & 3\\ 2 & 5\end{vmatrix}}=\boxed{\begin{vmatrix}-1 & 3\\ 2\times 4 & 5\times 4\end{vmatrix}}=\boxed{\begin{vmatrix}-1 & 3\\ 8 & 20\end{vmatrix}}=\boxed{(-1\times 20)-(3\times 8)}=\boxed{-20-24}=\boxed{\mathbf{-44}}\text{ or}$$

$$\boxed{4\times\begin{vmatrix}-1 & 3\\ 2 & 5\end{vmatrix}}=\boxed{\begin{vmatrix}-1\times 4 & 3\\ 2\times 4 & 5\end{vmatrix}}=\boxed{\begin{vmatrix}-4 & 3\\ 8 & 5\end{vmatrix}}=\boxed{(-4\times 5)-(3\times 8)}=\boxed{-20-24}=\boxed{\mathbf{-44}}\text{ or}$$

$$\boxed{4\times\begin{vmatrix}-1 & 3\\ 2 & 5\end{vmatrix}}=\boxed{\begin{vmatrix}-1 & 3\times 4\\ 2 & 5\times 4\end{vmatrix}}=\begin{vmatrix}-1 & 12\\ 2 & 20\end{vmatrix}=\boxed{(-1\times 20)-(12\times 2)}=\boxed{-20-24}=\boxed{\mathbf{-44}}$$

4. *Interchanging any two rows or any two columns of a determinant result in having a determinant which is opposite in sign to the original determinant.*

   For example, given the matrix $A=\begin{bmatrix}1 & 3\\ 4 & 5\end{bmatrix}$, determinant of the matrix A is equal to:

$$\boxed{\delta(A)} = \boxed{\begin{vmatrix} 1 & 3 \\ 4 & 5 \end{vmatrix}} = \boxed{(1\times 5)-(3\times 4)} = \boxed{5-12} = \boxed{-7}$$

However, if we interchange the first row with the second row the determinant is equal to:

$$\boxed{\delta(A)} = \boxed{\begin{vmatrix} 4 & 5 \\ 1 & 3 \end{vmatrix}} = \boxed{(4\times 3)-(5\times 1)} = \boxed{12-5} = \boxed{7}$$

or, if we interchange the first column with the second column the determinant is equal to:

$$\boxed{\delta(A)} = \boxed{\begin{vmatrix} 3 & 1 \\ 5 & 4 \end{vmatrix}} = \boxed{(3\times 4)-(1\times 5)} = \boxed{12-5} = \boxed{7}$$

5. *Determinant of a square matrix* A *is equal to the determinant of the transpose of* A*, i.e.,* $|A| = |A^t|$. *For example,*

a. $\boxed{\delta(A) = \begin{vmatrix} -1 & 3 \\ 4 & 7 \end{vmatrix}} = \boxed{(-1\times 7)-(3\times 4)} = \boxed{-7-12} = \boxed{\mathbf{-19}}$ and

$$\boxed{\delta(A^t) = \begin{vmatrix} -1 & 4 \\ 3 & 7 \end{vmatrix}} = \boxed{(-1\times 7)-(4\times 3)} = \boxed{-7-12} = \boxed{\mathbf{-19}}$$

b. $\boxed{\delta(B) = \begin{vmatrix} 0 & 0 & 3 \\ 1 & 3 & 0 \\ 2 & 4 & -1 \end{vmatrix}} = \boxed{b_{11}\cdot B_{11} + b_{12}\cdot B_{12} + b_{13}\cdot B_{13}} = \boxed{0\cdot B_{11} + 0\cdot B_{12} + 3\cdot B_{13}} = \boxed{3\cdot B_{13}} = \boxed{3\cdot(-1)^{1+3}\cdot M_{13}}$

$= \boxed{3\cdot(-1)^4\cdot M_{13}} = \boxed{3\cdot 1\cdot M_{13}} = \boxed{3\cdot M_{13}} = \boxed{3\cdot\begin{vmatrix} 1 & 3 \\ 2 & 4 \end{vmatrix}} = \boxed{3\cdot[(1\times 4)-(3\times 2)]} = \boxed{3\cdot(4-6)} = \boxed{3\cdot -2} = \boxed{\mathbf{-6}}$ and

$$\boxed{\delta(B^t) = \begin{vmatrix} 0 & 1 & 2 \\ 0 & 3 & 4 \\ 3 & 0 & -1 \end{vmatrix}} = \boxed{b_{11}\cdot B_{11} + b_{21}\cdot B_{21} + b_{31}\cdot B_{31}} = \boxed{0\cdot B_{11} + 0\cdot B_{21} + 3\cdot B_{31}} = \boxed{3\cdot B_{31}} = \boxed{3\cdot(-1)^{3+1}\cdot M_{31}}$$

$$= \boxed{3\cdot(-1)^4\cdot M_{31}} = \boxed{3\cdot 1\cdot M_{31}} = \boxed{3\cdot M_{31}} = \boxed{3\cdot\begin{vmatrix} 1 & 2 \\ 3 & 4 \end{vmatrix}} = \boxed{3\cdot[(1\times 4)-(2\times 3)]} = \boxed{3\cdot(4-6)} = \boxed{3\cdot -2} = \boxed{\mathbf{-6}}$$

6. *If the entries of one row or column of a determinant are multiplied by a constant and the product is added to the entries of another row or column, then the resulting determinant is equal to the original determinant.*

Example 3.3-9: Given $\begin{vmatrix} 2 & 3 \\ 1 & -1 \end{vmatrix} = \begin{vmatrix} 2+(1\cdot -2) & 3+(-1\cdot -2) \\ 1 & -1 \end{vmatrix} = \begin{vmatrix} 2-2 & 3+2 \\ 1 & -1 \end{vmatrix} = \begin{vmatrix} 0 & 5 \\ 1 & -1 \end{vmatrix}$, Verify that the determinant $\delta(A) = \begin{vmatrix} 2 & 3 \\ 1 & -1 \end{vmatrix}$ is equal to the determinant $\delta(B) = \begin{vmatrix} 0 & 5 \\ 1 & -1 \end{vmatrix}$

Solution:

To verify that the two determinats are equal, find the determinant $\delta(A) = \begin{vmatrix} 2 & 3 \\ 1 & -1 \end{vmatrix}$ and compare it with the determinant $\delta(B)=\begin{vmatrix}0 & 5\\1 & -1\end{vmatrix}$, i. e.,

$$\boxed{\begin{vmatrix}2 & 3\\1 & -1\end{vmatrix}}=\boxed{(2\cdot-1)-(3\cdot1)}=\boxed{-5} \text{ and } \boxed{\begin{vmatrix}0 & 5\\1 & -1\end{vmatrix}}=\boxed{(0\cdot-1)-(1\cdot5)}=\boxed{-5} \text{ therefore } \boxed{\delta(A)=\delta(B)}$$

Example 3.3-10: Given $\begin{vmatrix}1 & 3 & 4\\-1 & 1 & 2\\3 & 1 & -1\end{vmatrix}=\begin{vmatrix}1 & 3 & 4\\(-1+1\times1) & (1+3\times1) & (2+4\times1)\\3 & 1 & -1\end{vmatrix}=\begin{vmatrix}1 & 3 & 4\\0 & 4 & 6\\3 & 1 & -1\end{vmatrix}$

$=\begin{vmatrix}1 & 3 & 4\\0 & 4 & 6\\3+(1\times-3) & 1+(3\times-3) & -1+(4\times-3)\end{vmatrix}=\begin{vmatrix}1 & 3 & 4\\0 & 4 & 6\\3-3 & 1-9 & -1-12\end{vmatrix}=\begin{vmatrix}1 & 3 & 4\\0 & 4 & 6\\0 & -8 & -13\end{vmatrix}$, verify that the determinant

$\delta(A)=\begin{vmatrix}1 & 3 & 4\\-1 & 1 & 2\\3 & 1 & -1\end{vmatrix}$ is equal to the determinant $\delta(B)=\begin{vmatrix}1 & 3 & 4\\0 & 4 & 6\\0 & -8 & -13\end{vmatrix}$

Solution:

Again, to verify that the two determinats are equal, find the determinant $\delta(A)=\begin{vmatrix}1 & 3 & 4\\-1 & 1 & 2\\3 & 1 & -1\end{vmatrix}$ and

compare it with the determinant $\delta(B)=\begin{vmatrix}1 & 3 & 4\\0 & 4 & 6\\0 & -8 & -13\end{vmatrix}$, i. e.,

First find $\delta(A)$ by expanding the first row.

$$\boxed{\delta(A)}=\boxed{a_{11}\cdot A_{11}+a_{12}\cdot A_{12}+a_{13}\cdot A_{13}}=\boxed{1\cdot A_{11}+3\cdot A_{12}+4\cdot A_{13}}=\boxed{A_{11}+3A_{12}+4A_{13}}$$

$$=\boxed{(-1)^{1+1}\cdot M_{11}+3\cdot(-1)^{1+2}\cdot M_{12}+4\cdot(-1)^{1+3}\cdot M_{13}}=\boxed{M_{11}-3M_{12}+4M_{13}}=\boxed{\begin{vmatrix}1 & 2\\1 & -1\end{vmatrix}-3\begin{vmatrix}-1 & 2\\3 & -1\end{vmatrix}+4\begin{vmatrix}-1 & 1\\3 & 1\end{vmatrix}}$$

$$=\boxed{[(1\times-1)-(2\times1)]-3[(-1\times-1)-(2\times3)]+4[(-1\times1)-(1\times3)]}=\boxed{(-1-2)-3[1-6]+4[-1-3]}=\boxed{-3+15-16}$$

$$=\boxed{-4}$$

Second find $\delta(B)$ by expanding the first column.

$$\boxed{\delta(B)}=\boxed{b_{11}\cdot B_{11}+b_{21}\cdot B_{21}+b_{31}\cdot B_{31}}=\boxed{1\cdot B_{11}+0\cdot B_{21}+0\cdot B_{31}}=\boxed{B_{11}}=\boxed{(-1)^{1+1}\cdot M_{11}}=\boxed{M_{11}}$$

$$=\boxed{\begin{vmatrix}4 & 6\\-8 & -13\end{vmatrix}}=\boxed{(4\times-13)-(6\times-8)}=\boxed{-52+48}=\boxed{-4} \text{ therefore } \boxed{\delta(A)=\delta(B)}.$$

Note that the primary reason for using this property is to produce an equal matrix with a row or a column that contains all - except one - zero entries. This will then make it easier to calculate the determinant by expanding the row or column with most zero entries.

Section 3.3 Practice Problems - Determinants

1. Given $A = \begin{bmatrix} 1 & 3 & 4 & -1 \\ 0 & 1 & -2 & 0 \\ 1 & -1 & -3 & 5 \\ -3 & 0 & 6 & 8 \end{bmatrix}$, write the values of the following matrix entries.

a. $a_{23} =$ b. $a_{11} =$ c. $a_{33} =$ d. $a_{41} =$

e. $a_{43} =$ f. $a_{44} =$ g. $a_{34} =$ h. $a_{32} =$

i. $a_{24} =$ j. $a_{31} =$ k. $a_{13} =$ l. $a_{42} =$

2. Let $A = \begin{bmatrix} 1 & 2 \\ 3 & -1 \end{bmatrix}$. Compute the following minors and cofactors.

a. $M_{11} =$ b. $M_{12} =$ c. $M_{21} =$ d. $M_{22} =$

e. $A_{11} =$ f. $A_{12} =$ g. $A_{21} =$ h. $A_{22} =$

3. Let $A = \begin{bmatrix} 1 & 3 & -1 \\ 2 & 3 & -2 \\ 1 & 0 & 4 \end{bmatrix}$. Compute the indicated minors and cofactors.

a. $M_{23} =$ b. $M_{13} =$ c. $M_{32} =$ d. $M_{11} =$

e. $A_{23} =$ f. $A_{13} =$ g. $A_{32} =$ h. $A_{11} =$

4. Let $A = \begin{bmatrix} 1 & 0 & 0 \\ 1 & -1 & 0 \\ 2 & 1 & 1 \end{bmatrix}$. Compute the indicated minors and cofactors.

a. $M_{31} =$ b. $M_{33} =$ c. $M_{12} =$ d. $M_{22} =$

e. $A_{31} =$ f. $A_{33} =$ g. $A_{12} =$ h. $A_{22} =$

5. Find the determinant of the following matrices.

a. $A = \begin{bmatrix} -1 & 2 \\ 3 & 0 \end{bmatrix}$ b. $B = \begin{bmatrix} 4 & 0 \\ 0 & 3 \end{bmatrix} =$ c. $C = \begin{bmatrix} 1 & -3 \\ 0 & 2 \end{bmatrix} =$

d. $D = \begin{bmatrix} 3 & -2 \\ 5 & -3 \end{bmatrix} =$ e. $E = \begin{bmatrix} 1 & 0 \\ 0 & 1 \end{bmatrix} =$

6. Let $A = \begin{bmatrix} 1 & 0 & -1 \\ 2 & 3 & 1 \\ 0 & 0 & 4 \end{bmatrix}$. Compute the determinant of A.

a. Expand about the second row. b. Expand about the third row.

c. Expand about the first column. d. Expand about the second column.

7. Find the determinant of the following A matrices. Expand about the indicated rows and columns.

a. $\begin{bmatrix} 2 & 1 & 4 \\ 0 & -1 & 3 \\ 0 & 0 & 5 \end{bmatrix} =$ 1. Expand about the third column. 2. Expand about the third row.

b. $\begin{bmatrix} 2 & -1 & 1 \\ -1 & 3 & 0 \\ 0 & 4 & 0 \end{bmatrix} =$ 1. Expand about the third column. 2. Expand about the first row.

c. $\begin{bmatrix} 3 & 1 & 1 \\ 2 & -3 & 2 \\ 9 & 3 & 3 \end{bmatrix} =$ 1. Expand about the second row. 2. Expand about the second column.

d. $\begin{bmatrix} -2 & 0 & 0 \\ -1 & 2 & 0 \\ 5 & 7 & 0 \end{bmatrix} =$ 1. Expand about the third column. 2. Expand about the first column.

8. Determine the determinant of the following matrices by observation only.

a. $\begin{vmatrix} 1 & 3 & -1 \\ 0 & 0 & 0 \\ -3 & 5 & 8 \end{vmatrix} =$ b. $\begin{vmatrix} 5 & 1 & 5 \\ 0 & -1 & 0 \\ 2 & 1 & 2 \end{vmatrix} =$ c. $\begin{vmatrix} 1 & 5 & 6 & 8 \\ 2 & 3 & -1 & 0 \\ 1 & 5 & 6 & 8 \\ 0 & 1 & 2 & -3 \end{vmatrix} =$ d. $\begin{vmatrix} 1 & 0 & 1 & 2 \\ 2 & 0 & -1 & -3 \\ -3 & 0 & 3 & 5 \\ 4 & 0 & 5 & 6 \end{vmatrix}$

9. Solve for the unknown.

a. $\begin{vmatrix} x & 2 \\ 6 & 4 \end{vmatrix} = 4$ b. $\begin{vmatrix} -2 & 3 \\ 4 & x \end{vmatrix} = -4$ c. $\begin{vmatrix} 3 & y \\ 2 & 5 \end{vmatrix} = 1$ d. $\begin{vmatrix} 1 & 3 \\ w & -9 \end{vmatrix} = 12$

3.4 Inverse Matrices

In this section we will discuss three methods for finding the inverse of a square matrix. The first method, which is most commonly used, is referred to as *the minor and the cofactor* method. The second method, which is easy to learn but has a limited practical use, is referred to as *the substitution* method. The third method, which is computation intensive but has broader practical use, is referred to as *the elementary row operations* method. Students are encouraged to learn all three methods and use the most practical method for finding the inverse of any given matrix.

- **First Method - The Minor and the Cofactor Method**

Given A is an $n \times n$ square matrix and the determinant of A is not equal to zero, $\delta(A) \neq 0$, then the inverse of the A matrix, denoted by A^{-1}, is equal to $\frac{1}{\delta(A)}$ times the transpose of the matrix obtained by replacing each entry of the A matrix with its cofactors, i.e.,

$$A^{-1} = \frac{1}{\delta(A)} C^t = \frac{1}{\delta(A)} \begin{bmatrix} A_{11} & A_{12} & \cdots & A_{1n} \\ A_{21} & A_{22} & \cdots & A_{2n} \\ \vdots & \vdots & \vdots & \vdots \\ A_{n1} & A_{n2} & \cdots & A_{nn} \end{bmatrix}^t = \frac{1}{\delta(A)} \begin{bmatrix} A_{11} & A_{21} & \cdots & A_{n1} \\ A_{12} & A_{22} & \cdots & A_{n2} \\ \vdots & \vdots & \vdots & \vdots \\ A_{1n} & A_{2n} & \cdots & A_{nn} \end{bmatrix}$$

The following shows the steps for finding the inverse of a square matrix:

Step 1 Calculate the determinant of the square matrix A. If the determinant of A is equal to zero, $\delta(A) = 0$, the matrix A does not have an inverse. If determinant of A is not equal to zero, $\delta(A) \neq 0$, proceed to the next step.

Step 2 a. Calculate the cofactor matrix C by replacing the entries of the A matrix with its cofactors.

b. Obtain the transpose of the C matrix, i.e., C^t by interchanging its rows and columns.

Step 3 Calculate the inverse of the A matrix by multiplying the inverse of the determinant by C^t, i.e., use the above equation $A^{-1} = \frac{1}{\delta(A)} C^t$.

Step 4 Check the answer by multiplying the matrix A with its inverse. The resultant matrix should be equal to the identity matrix I.

The following examples show the steps for finding the inverse of a square matrix using the minor and the cofactor method.

Example 3.4-1: Given $A = \begin{bmatrix} 1 & 2 \\ -3 & 4 \end{bmatrix}$, find A^{-1}.

Solution:

Step 1 Find $\delta(A)$ by expanding about the first row. Note that $a_{11} = 1$ and $a_{12} = 2$.

$$\boxed{\delta(A)} = \boxed{a_{11} \cdot A_{11} + a_{12} \cdot A_{12}} = \boxed{1 \cdot A_{11} + 2 \cdot A_{12}} = \boxed{A_{11} + 2A_{12}} = \boxed{(-1)^{1+1} M_{11} + 2 \cdot (-1)^{1+2} M_{12}}$$

$$= \boxed{(-1)^2 M_{11} + 2 \cdot (-1)^3 M_{12}} = \boxed{M_{11} - 2M_{12}} = \boxed{4 - 2 \cdot (-3)} = \boxed{4 + 6} = \boxed{10}$$

Since the determinant is not equal to zero matrix A has an inverse.

Step 2 Replace the entries of the A matrix with its cofactors.

$$\boxed{A_{11}} = \boxed{(-1)^{1+1} M_{11}} = \boxed{(-1)^2 \times 4} = \boxed{4} \qquad A_{12} = \boxed{(-1)^{1+2} M_{12}} = \boxed{(-1)^3 \times -3} = \boxed{3}$$

$$\boxed{A_{21}} = \boxed{(-1)^{2+1} M_{21}} = \boxed{(-1)^3 \times 2} = \boxed{-2} \qquad A_{22} = \boxed{(-1)^{2+2} M_{22}} = \boxed{(-1)^4 \times 1} = \boxed{1}$$

Therefore, the cofactor matrix is equal to $C = \begin{bmatrix} A_{11} & A_{12} \\ A_{21} & A_{22} \end{bmatrix} = \begin{bmatrix} 4 & 3 \\ -2 & 1 \end{bmatrix}$ and $C^t = \begin{bmatrix} 4 & -2 \\ 3 & 1 \end{bmatrix}$

Step 3 Compute $\boxed{A^{-1}} = \boxed{\frac{1}{\delta(A)} C^t} = \boxed{\frac{1}{10} \cdot \begin{bmatrix} 4 & -2 \\ 3 & 1 \end{bmatrix}} = \boxed{\begin{bmatrix} 4 \times \frac{1}{10} & -2 \times \frac{1}{10} \\ 3 \times \frac{1}{10} & 1 \times \frac{1}{10} \end{bmatrix}} = \boxed{\begin{bmatrix} \frac{4}{10} & -\frac{2}{10} \\ \frac{3}{10} & \frac{1}{10} \end{bmatrix}} = \boxed{\begin{bmatrix} \mathbf{\frac{2}{5}} & \mathbf{-\frac{1}{5}} \\ \mathbf{\frac{3}{10}} & \mathbf{\frac{1}{10}} \end{bmatrix}}$

Step 4 Check the answer by multiplying the A matrix with A^{-1}. The result should be equal to the identity matrix.

$$\boxed{A \times A^{-1}} = \boxed{\begin{bmatrix} 1 & 2 \\ -3 & 4 \end{bmatrix} \times \begin{bmatrix} \frac{2}{5} & -\frac{1}{5} \\ \frac{3}{10} & \frac{1}{10} \end{bmatrix}} = \boxed{\begin{bmatrix} \left(1 \times \frac{2}{5}\right) + \left(2 \times \frac{3}{10}\right) & \left(1 \times -\frac{1}{5}\right) + \left(2 \times \frac{1}{10}\right) \\ \left(-3 \times \frac{2}{5}\right) + \left(4 \times \frac{3}{10}\right) & \left(-3 \times -\frac{1}{5}\right) + \left(4 \times \frac{1}{10}\right) \end{bmatrix}}$$

$$= \boxed{\begin{bmatrix} \frac{2}{5} + \frac{6}{10} & -\frac{1}{5} + \frac{2}{10} \\ -\frac{6}{5} + \frac{12}{10} & \frac{3}{5} + \frac{4}{10} \end{bmatrix}} = \boxed{\begin{bmatrix} \frac{(2 \times 10) + (6 \times 5)}{5 \times 10} & \frac{(-1 \times 10) + (2 \times 5)}{5 \times 10} \\ \frac{(-6 \times 10) + (12 \times 5)}{5 \times 10} & \frac{(3 \times 10) + (4 \times 5)}{5 \times 10} \end{bmatrix}} = \boxed{\begin{bmatrix} \frac{20 + 30}{50} & \frac{-10 + 10}{50} \\ \frac{-60 + 60}{50} & \frac{30 + 20}{50} \end{bmatrix}}$$

$$= \boxed{\begin{bmatrix} \frac{50}{50} & \frac{0}{50} \\ \frac{0}{50} & \frac{50}{50} \end{bmatrix}} = \boxed{\begin{bmatrix} 1 & 0 \\ 0 & 1 \end{bmatrix}}$$

Since $A \times A^{-1}$ is equal to the identity matrix A^{-1} is computed correctly.

Example 3.4-2: Given $A = \begin{bmatrix} 2 & 4 \\ 5 & 10 \end{bmatrix}$, find A^{-1}.

Solution:

Step 1 Find $\delta(A)$ by expanding about the first row. Note that $a_{11} = 2$, and $a_{12} = 4$.

$$\boxed{\delta(A)} = \boxed{a_{11} \cdot A_{11} + a_{12} \cdot A_{12}} = \boxed{2 \cdot A_{11} + 4 \cdot A_{12}} = \boxed{2 \cdot (-1)^{1+1} \cdot M_{11} + 4 \cdot (-1)^{1+2} \cdot M_{12}}$$

$$= \boxed{2 \cdot (-1)^2 \cdot M_{11} + 4 \cdot (-1)^3 \cdot M_{12}} = \boxed{2 \cdot 1 \cdot M_{11} + 4 \cdot (-1) \cdot M_{12}} = \boxed{2 \cdot M_{11} - 4 \cdot M_{12}} = \boxed{(2 \cdot 10) - (4 \cdot 5)}$$

$$= \boxed{20 - 20} = \boxed{0}$$

Since the determinant is equal to zero matrix A does not have an inverse.

Example 3.4-3: Given $A=\begin{bmatrix}1 & 0 & 2\\ -2 & 3 & 0\\ 1 & 2 & 0\end{bmatrix}$, find A^{-1}.

Solution:

Step 1 Find $\delta(A)$ by expanding about the third column. Note that $a_{13}=2$, $a_{23}=0$, and $a_{33}=0$.

$$\boxed{\delta(A)}=\boxed{a_{13}\cdot A_{13}+a_{23}\cdot A_{23}+a_{33}\cdot A_{33}}=\boxed{2\cdot A_{13}+0\cdot A_{23}+0\cdot A_{33}}=\boxed{2A_{13}+0+0}=\boxed{2A_{13}}$$

$$=\boxed{2\cdot(-1)^{1+3}\cdot M_{13}}=\boxed{2\cdot 1\cdot M_{13}}=\boxed{2M_{13}}=\boxed{2\cdot\begin{bmatrix}-2 & 3\\ 1 & 2\end{bmatrix}}=\boxed{2\cdot[(-2\times 2)-(3\times 1)]}=\boxed{2\cdot(-4-3)}$$

$$=\boxed{2\cdot -7}=\boxed{-14}$$

Since the determinant is not equal to zero matrix A has an inverse.

Step 2 Replace the entries of the A matrix with its cofactors.

$$\boxed{A_{11}}=\boxed{(-1)^{1+1}\cdot M_{11}}=\boxed{(-1)^2\cdot M_{11}}=\boxed{M_{11}}=\boxed{\begin{vmatrix}3 & 0\\ 2 & 0\end{vmatrix}}=\boxed{(3\times 0)-(0\times 2)}=\boxed{0-0}=\boxed{0}$$

$$\boxed{A_{12}}=\boxed{(-1)^{1+2}\cdot M_{12}}=\boxed{(-1)^3\cdot M_{12}}=\boxed{-M_{12}}=\boxed{-\begin{vmatrix}-2 & 0\\ 1 & 0\end{vmatrix}}=\boxed{-[(-2\times 0)-(0\times 1)]}=\boxed{-(0-0)}=\boxed{0}$$

$$\boxed{A_{13}}=\boxed{(-1)^{1+3}\cdot M_{13}}=\boxed{(-1)^4\cdot M_{13}}=\boxed{M_{13}}=\boxed{\begin{vmatrix}-2 & 3\\ 1 & 2\end{vmatrix}}=\boxed{(-2\times 2)-(3\times 1)}=\boxed{-4-3}=\boxed{-7}$$

$$\boxed{A_{21}}=\boxed{(-1)^{2+1}\cdot M_{21}}=\boxed{(-1)^3\cdot M_{21}}=\boxed{-M_{21}}=\boxed{-\begin{vmatrix}0 & 2\\ 2 & 0\end{vmatrix}}=\boxed{-[(0\times 0)-(2\times 2)]}=\boxed{-(0-4)}=\boxed{4}$$

$$\boxed{A_{22}}=\boxed{(-1)^{2+2}\cdot M_{22}}=\boxed{(-1)^4\cdot M_{22}}=\boxed{M_{22}}=\boxed{\begin{vmatrix}1 & 2\\ 1 & 0\end{vmatrix}}=\boxed{(1\times 0)-(2\times 1)}=\boxed{0-2}=\boxed{-2}$$

$$\boxed{A_{23}}=\boxed{(-1)^{2+3}\cdot M_{23}}=\boxed{(-1)^5\cdot M_{23}}=\boxed{-M_{23}}=\boxed{-\begin{vmatrix}1 & 0\\ 1 & 2\end{vmatrix}}=\boxed{-[(1\times 2)-(0\times 1)]}=\boxed{-(2-0)}=\boxed{-2}$$

$$\boxed{A_{31}}=\boxed{(-1)^{3+1}\cdot M_{31}}=\boxed{(-1)^4\cdot M_{31}}=\boxed{M_{31}}=\boxed{\begin{vmatrix}0 & 2\\ 3 & 0\end{vmatrix}}=\boxed{(0\times 0)-(2\times 3)}=\boxed{0-6}=\boxed{-6}$$

$$\boxed{A_{32}}=\boxed{(-1)^{3+2}\cdot M_{32}}=\boxed{(-1)^5\cdot M_{32}}=\boxed{-M_{32}}=\boxed{-\begin{vmatrix}1 & 2\\ -2 & 0\end{vmatrix}}=\boxed{-[(1\times 0)-(2\times -2)]}=\boxed{-(0+4)}=\boxed{-4}$$

$$\boxed{A_{33}}=\boxed{(-1)^{3+3}\cdot M_{33}}=\boxed{(-1)^6\cdot M_{33}}=\boxed{M_{33}}=\boxed{\begin{vmatrix}1 & 0\\ -2 & 3\end{vmatrix}}=\boxed{(1\times 3)-(0\times -2)}=\boxed{3+0}=\boxed{3}$$

Therefore, the cofactor matrix is equal to:

$$\boxed{C} = \begin{bmatrix} A_{11} & A_{12} & A_{13} \\ A_{21} & A_{22} & A_{23} \\ A_{31} & A_{32} & A_{33} \end{bmatrix} = \begin{bmatrix} 0 & 0 & -7 \\ 4 & -2 & -2 \\ -6 & -4 & 3 \end{bmatrix} \text{ and } \boxed{C^t} = \begin{bmatrix} 0 & 4 & -6 \\ 0 & -2 & -4 \\ -7 & -2 & 3 \end{bmatrix}$$

Step 3 Compute $\boxed{A^{-1}} = \boxed{\frac{1}{\delta(A)} C^t} = \boxed{\frac{1}{-14} \cdot \begin{bmatrix} 0 & 4 & -6 \\ 0 & -2 & -4 \\ -7 & -2 & 3 \end{bmatrix}} = \begin{bmatrix} 0 \times -\frac{1}{14} & 4 \times -\frac{1}{14} & -6 \times -\frac{1}{14} \\ 0 \times -\frac{1}{14} & -2 \times -\frac{1}{14} & -4 \times -\frac{1}{14} \\ -7 \times -\frac{1}{14} & -2 \times -\frac{1}{14} & 3 \times -\frac{1}{14} \end{bmatrix}$

$$= \begin{bmatrix} 0 & -\frac{4}{14} & \frac{6}{14} \\ 0 & \frac{2}{14} & \frac{4}{14} \\ \frac{7}{14} & \frac{2}{14} & -\frac{3}{14} \end{bmatrix} = \boxed{\begin{bmatrix} \mathbf{0} & -\frac{\mathbf{2}}{\mathbf{7}} & \frac{\mathbf{3}}{\mathbf{7}} \\ \mathbf{0} & \frac{\mathbf{1}}{\mathbf{7}} & \frac{\mathbf{2}}{\mathbf{7}} \\ \frac{\mathbf{1}}{\mathbf{2}} & \frac{\mathbf{1}}{\mathbf{7}} & -\frac{\mathbf{3}}{\mathbf{14}} \end{bmatrix}}$$

Step 4 Check the answer by multiplying the A matrix with A^{-1}. The result should be equal to the identity matrix.

$$\boxed{A \times A^{-1}} = \begin{bmatrix} 1 & 0 & 2 \\ -2 & 3 & 0 \\ 1 & 2 & 0 \end{bmatrix} \times \begin{bmatrix} 0 & -\frac{2}{7} & \frac{3}{7} \\ 0 & \frac{1}{7} & \frac{2}{7} \\ \frac{1}{2} & \frac{1}{7} & -\frac{3}{14} \end{bmatrix} = \begin{bmatrix} \frac{2}{2} & -\frac{2}{7}+\frac{2}{7} & \frac{3}{7}-\frac{6}{14} \\ 0 & \frac{4}{7}+\frac{3}{7} & -\frac{6}{7}+\frac{6}{7} \\ 0 & -\frac{2}{7}+\frac{2}{7} & \frac{3}{7}+\frac{4}{7} \end{bmatrix} = \begin{bmatrix} 1 & 0 & 0 \\ 0 & 1 & 0 \\ 0 & 0 & 1 \end{bmatrix}$$

Since $A \times A^{-1}$ is equal to the identity matrix A^{-1} is computed correctly.

Example 3.4-4: Verify that $B = \begin{bmatrix} -\frac{5}{7} & \frac{3}{7} \\ \frac{4}{7} & -\frac{1}{7} \end{bmatrix}$ matrix is the inverse of the $A = \begin{bmatrix} 1 & 3 \\ 4 & 5 \end{bmatrix}$ matrix.

Solution:

first find the determinant of A by expanding about the first row. Note that $a_{11} = 1$, and $a_{12} = 3$.

$$\boxed{\delta(A)} = \boxed{a_{11}A_{11} + a_{12}A_{12}} = \boxed{1 \cdot A_{11} + 3 \cdot A_{12}} = \boxed{A_{11} + 3A_{12}} = \boxed{(-1)^{1+1} M_{11} + 3 \cdot (-1)^{1+2} M_{12}} = \boxed{M_{11} - 3M_{12}}$$

$$= \boxed{5 - (3 \cdot 4)} = \boxed{5 - 12} = \boxed{-7}$$

Since $\delta(A) \neq 0$ matrix A has an inverse.

To verify $A \times B \overset{?}{=} I$ we multiply the two matrices by one another. The result should be equal to the identity matrix.

$$\begin{bmatrix} 1 & 3 \\ 4 & 5 \end{bmatrix} \times \begin{bmatrix} -\frac{5}{7} & \frac{3}{7} \\ \frac{4}{7} & -\frac{1}{7} \end{bmatrix} = \begin{bmatrix} 1 \times -\frac{5}{7} + 3 \times \frac{4}{7} & 1 \times \frac{3}{7} + 3 \times -\frac{1}{7} \\ 4 \times -\frac{5}{7} + 5 \times \frac{4}{7} & 4 \times \frac{3}{7} + 5 \times -\frac{1}{7} \end{bmatrix} = \begin{bmatrix} -\frac{5}{7}+\frac{12}{7} & \frac{3}{7}-\frac{3}{7} \\ -\frac{20}{7}+\frac{20}{7} & \frac{12}{7}-\frac{5}{7} \end{bmatrix} = \begin{bmatrix} \frac{7}{7} & \frac{0}{7} \\ \frac{0}{7} & \frac{7}{7} \end{bmatrix} = \begin{bmatrix} 1 & 0 \\ 0 & 1 \end{bmatrix}$$

Therefore, the B matrix is inverse of the A matrix.

- **Second Method - The Substitution Method**

To use the substitution method we first need to know about the following definition.

Definition - If an $n \times n$ matrix A has an inverse, then the product of the A matrix multiplied by A^{-1} matrix is equal to an $n \times n$ identity matrix, i. e.,

$$A \times A^{-1} = I \text{ and } A^{-1} \times A = I$$

Given the above statement we can find the inverse of the A matrix by using the substitution method as shown in the following two examples:

Example 3.4-5: Given $A = \begin{bmatrix} 1 & 7 \\ 0 & 3 \end{bmatrix}$ and $I = \begin{bmatrix} 1 & 0 \\ 0 & 1 \end{bmatrix}$, find A^{-1} matrix using $A \times A^{-1} = I$.

Solution:

first find the determinant of A by expanding about the first column. Note that $a_{11} = 1$, and $a_{21} = 0$.

$$\boxed{\delta(A)} = \boxed{a_{11}A_{11} + a_{21}A_{21}} = \boxed{1 \cdot A_{11} + 0 \cdot A_{21}} = \boxed{A_{11} + 0} = \boxed{A_{11}} = \boxed{(-1)^{1+1} M_{11}} = \boxed{(-1)^2 M_{11}} = \boxed{M_{11}} = \boxed{3}$$

Note that we could have used the quicker method for finding $\delta(A)$ by computing the difference of the products of the entries on the two diagonals, i. e., $\delta(A) = (1 \times 3) - (7 \times 0) = 3 - 0 = 3$. Since $\delta(A) \neq 0$ the matrix A has an inverse.

Let $A^{-1} = \begin{bmatrix} a & b \\ c & d \end{bmatrix}$. Substituting A^{-1} in the equality $A \times A^{-1} = I$ we obtain:

$$\begin{bmatrix} 1 & 7 \\ 0 & 3 \end{bmatrix} \times \begin{bmatrix} a & b \\ c & d \end{bmatrix} = \begin{bmatrix} a+7c & b+7d \\ 3c & 3d \end{bmatrix} = \begin{bmatrix} 1 & 0 \\ 0 & 1 \end{bmatrix}$$

Equating the entries on both sides of the equality we have

1. $a + 7c = 1$ $\quad a = 1 - 7c$ $\qquad$ 2. $b + 7d = 0$ $\quad b = -7d$

3. $3c = 0$ $\quad c = 0$ $\qquad$ 4. $3d = 1$ $\quad d = \frac{1}{3}$

Substitution of $c = 0$ and $d = \frac{1}{3}$ into $a = 1 - 7c$ and $b = -7d$ result in having $a = 1$ and $b = -\frac{7}{3}$.

Therefore, $A^{-1} = \begin{bmatrix} a & b \\ c & d \end{bmatrix} = \begin{bmatrix} 1 & -\frac{7}{3} \\ 0 & \frac{1}{3} \end{bmatrix}$

As a routine check, multiplication of A by A^{-1} should result in the identity matrix.

$$A \times A^{-1} = \begin{bmatrix} 1 & 7 \\ 0 & 3 \end{bmatrix} \times \begin{bmatrix} 1 & -\frac{7}{3} \\ 0 & \frac{1}{3} \end{bmatrix} = \begin{bmatrix} 1 \times 1 + 7 \times 0 & 1 \times -\frac{7}{3} + 7 \times \frac{1}{3} \\ 0 \times 1 + 3 \times 0 & 0 \times -\frac{7}{3} + 3 \times \frac{1}{3} \end{bmatrix} = \begin{bmatrix} 1+0 & -\frac{7}{3} + \frac{7}{3} \\ 0+0 & 0 + \frac{3}{3} \end{bmatrix} = \begin{bmatrix} 1 & \frac{-7+7}{3} \\ 0 & \frac{3}{3} \end{bmatrix} = \begin{bmatrix} 1 & 0 \\ 0 & 1 \end{bmatrix}$$

Example 3.4-6: Given $A = \begin{bmatrix} 1 & 0 & 2 \\ -2 & 3 & 0 \\ 1 & 2 & 0 \end{bmatrix}$ and $I = \begin{bmatrix} 1 & 0 & 0 \\ 0 & 1 & 0 \\ 0 & 0 & 1 \end{bmatrix}$, find A^{-1} matrix using $A \times A^{-1} = I$.

Solution:

first find the determinant of A by expanding about the third column. Note that $a_{13}=2$, $a_{23}=0$ and $a_{33}=0$.

$$\boxed{\delta(A)}=\boxed{a_{13}A_{13}+a_{23}A_{23}+a_{33}A_{33}}=\boxed{2\cdot A_{13}+0\cdot A_{23}+0\cdot A_{33}}=\boxed{2A_{13}+0+0}=\boxed{2A_{13}}=\boxed{2\cdot(-1)^{1+3}M_{13}}$$

$$=\boxed{2M_{13}}=\boxed{2\cdot\begin{vmatrix}-2 & 3\\ 1 & 2\end{vmatrix}}=\boxed{2\cdot[(-2\times 2)-(3\times 1)]}=\boxed{2\cdot(-4-3)}=\boxed{2\cdot(-7)}=\boxed{-14}$$

Since $\delta(A)\neq 0$ matrix A has an inverse.

Let $A^{-1}=\begin{bmatrix}a & b & c\\ d & e & f\\ g & h & i\end{bmatrix}$. Substituting A^{-1} in the equality $A\times A^{-1}=I$ we obtain:

$$\begin{bmatrix}1 & 0 & 2\\ -2 & 3 & 0\\ 1 & 2 & 0\end{bmatrix}\times\begin{bmatrix}a & b & c\\ d & e & f\\ g & h & i\end{bmatrix}=\begin{bmatrix}a+2g & b+2h & c+2i\\ -2+3d & -2b+3e & -2c+3f\\ a+2d & b+2e & c+2f\end{bmatrix}=\begin{bmatrix}1 & 0 & 0\\ 0 & 1 & 0\\ 0 & 0 & 1\end{bmatrix}$$

Equating the entries on both sides of the equality we have

1. $\boxed{a+2g=1}$; $\boxed{a=1-2g}$
2. $\boxed{b+2h=0}$; $\boxed{b=-2h}$
3. $\boxed{c+2i=0}$; $\boxed{c=-2i}$
4. $\boxed{-2a+3d=0}$; $\boxed{d=\frac{2a}{3}}$
5. $\boxed{-2b+3e=1}$; $\boxed{e=\frac{1+2b}{3}}$
6. $\boxed{-2c+3f=0}$; $\boxed{f=\frac{2c}{3}}$
7. $\boxed{a+2d=0}$; $\boxed{d=\frac{-a}{2}}$
8. $\boxed{b+2e=0}$; $\boxed{e=\frac{-b}{2}}$
9. $\boxed{c+2f=1}$; $\boxed{f=\frac{1-c}{2}}$

Equating 4 and 7; 5 and 8; 6 and 9 result in having $a=0$, $b=-\frac{2}{7}$, and $c=\frac{3}{7}$. Substitution of these values into the equations 1 through 6 result in having $d=0$, $e=\frac{1}{7}$, $f=\frac{2}{7}$, $g=\frac{1}{2}$, $h=\frac{1}{7}$ and $i=-\frac{3}{14}$. Therefore, $\mathbf{A^{-1}}=\begin{bmatrix}\mathbf{a} & \mathbf{b} & \mathbf{c}\\ \mathbf{d} & \mathbf{e} & \mathbf{f}\\ \mathbf{g} & \mathbf{h} & \mathbf{i}\end{bmatrix}=\begin{bmatrix}\mathbf{0} & \mathbf{-\frac{2}{7}} & \mathbf{\frac{3}{7}}\\ \mathbf{0} & \mathbf{\frac{1}{7}} & \mathbf{\frac{2}{7}}\\ \mathbf{\frac{1}{2}} & \mathbf{\frac{1}{7}} & \mathbf{-\frac{3}{14}}\end{bmatrix}$

Again, as a routine check, multiplication of A by A^{-1} should result in the identity matrix.

$$A\times A^{-1}=\begin{bmatrix}1 & 0 & 2\\ -2 & 3 & 0\\ 1 & 2 & 0\end{bmatrix}\times\begin{bmatrix}0 & -\frac{2}{7} & \frac{3}{7}\\ 0 & \frac{1}{7} & \frac{2}{7}\\ \frac{1}{2} & \frac{1}{7} & -\frac{3}{14}\end{bmatrix}=\begin{bmatrix}1\times 0+0\times 0+2\times\frac{1}{2} & 1\times-\frac{2}{7}+0\times\frac{1}{7}+2\times\frac{1}{7} & 1\times\frac{3}{7}+0\times\frac{2}{7}-2\times\frac{3}{14}\\ -2\times 0+3\times 0+0\times\frac{1}{2} & 2\times\frac{2}{7}+3\times\frac{1}{7}+0\times\frac{1}{7} & -2\times\frac{3}{7}+3\times\frac{2}{7}+0\times-\frac{3}{14}\\ 1\times 0+2\times 0+0\times\frac{1}{2} & 1\times-\frac{2}{7}+2\times\frac{1}{7}+0\times\frac{1}{7} & 1\times\frac{3}{7}+2\times\frac{2}{7}+0\times-\frac{3}{14}\end{bmatrix}$$

$$=\begin{bmatrix}\frac{2}{2} & -\frac{2}{7}+\frac{2}{7} & \frac{3}{7}-\frac{3}{7}\\ 0 & \frac{4}{7}+\frac{3}{7} & -\frac{6}{7}+\frac{6}{7}\\ 0 & -\frac{2}{7}+\frac{2}{7} & \frac{3}{7}+\frac{4}{7}\end{bmatrix}=\begin{bmatrix}\frac{2}{2} & \frac{-2+2}{7} & \frac{3-3}{7}\\ 0 & \frac{4+3}{7} & \frac{-6+6}{7}\\ 0 & \frac{-2+2}{7} & \frac{3+4}{7}\end{bmatrix}=\begin{bmatrix}\frac{2}{2} & \frac{0}{7} & \frac{0}{7}\\ 0 & \frac{7}{7} & \frac{0}{7}\\ 0 & \frac{0}{7} & \frac{7}{7}\end{bmatrix}=\begin{bmatrix}1 & 0 & 0\\ 0 & 1 & 0\\ 0 & 0 & 1\end{bmatrix}$$

Third Method - The Elementary Row Operations

We can use the elementary row operations approach to find the inverse of a matrix by letting $[A \vdots I]$ be transformed to $[I \vdots A^{-1}]$. The following show the steps in using this method:

Step 1 Write the matrix in the form of $[A \vdots I]$.

Step 2 Perform the elementary row operations to transform the matrix to the form of $[I \vdots A^{-1}]$.

Step 3 Check the answer by multiplying $A \times A^{-1} = I$ to obtain the identity matrix.

The following examples show the steps for using this method.

Example 3.4-7: Given $A = \begin{bmatrix} 1 & -1 \\ 0 & 2 \end{bmatrix}$, find A^{-1} by letting $[A \vdots I]$ be transformed by row operations to $[I \vdots A^{-1}]$.

Solution:

First - Write the matrix A in the form of $[A \vdots I]$, i.e., $\begin{bmatrix} 1 & -1 & \vdots & 1 & 0 \\ 0 & 2 & \vdots & 0 & 1 \end{bmatrix}$

Second - Perform the elementary row operations, i.e.,

1. Divide the second row by 2. $\begin{bmatrix} 1 & -1 & \vdots & 1 & 0 \\ 0 & \frac{2}{2} & \vdots & 0 & \frac{1}{2} \end{bmatrix} = \begin{bmatrix} 1 & -1 & \vdots & 1 & 0 \\ 0 & 1 & \vdots & 0 & \frac{1}{2} \end{bmatrix}$

2. Multiply each element of the second row by 1 and add the result to each element of the first row. $\begin{bmatrix} 1+0 & -1+1 & \vdots & 1+0 & 0+\frac{1}{2} \\ 0 & 1 & \vdots & 0 & \frac{1}{2} \end{bmatrix} = \begin{bmatrix} 1 & 0 & \vdots & 1 & \frac{1}{2} \\ 0 & 1 & \vdots & 0 & \frac{1}{2} \end{bmatrix}$

Note that the matrix on the left hand side is an identity matrix. Therefore, inverse of the A matrix is equal to $\mathbf{A^{-1} = \begin{bmatrix} 1 & \frac{1}{2} \\ 0 & \frac{1}{2} \end{bmatrix}}$

Third - Check the answer by multiplying A by A^{-1}.

$$A \times A^{-1} \overset{?}{=} I \; ; \; \begin{bmatrix} 1 & -1 \\ 0 & 2 \end{bmatrix} \times \begin{bmatrix} 1 & \frac{1}{2} \\ 0 & \frac{1}{2} \end{bmatrix} \overset{?}{=} I \; ; \; \begin{bmatrix} (1\times 1)+(-1\times 0) & \left(1\times\frac{1}{2}\right)+\left(-1\times\frac{1}{2}\right) \\ (0\times 1)+(2\times 0) & \left(0\times\frac{1}{2}\right)+\left(2\times\frac{1}{2}\right) \end{bmatrix} \overset{?}{=} I \; ; \; \begin{bmatrix} 1+0 & \frac{1}{2}-\frac{1}{2} \\ 0+0 & 0+1 \end{bmatrix} \overset{?}{=} I \; ; \; \begin{bmatrix} 1 & 0 \\ 0 & 1 \end{bmatrix} = I$$

Since $A \times A^{-1}$ is equal to the identity matrix A^{-1} is computed correctly.

Example 3.4-8: Given $A = \begin{bmatrix} 1 & 2 & 3 \\ -2 & 1 & 0 \\ 3 & -1 & 1 \end{bmatrix}$, find A^{-1} by letting $[A \vdots I]$ be transformed by row operations to $[I \vdots A^{-1}]$.

First - Write the matrix A in the form of $[A \vdots I]$, i.e., $\left[\begin{array}{ccccccc} 1 & 2 & 3 & \vdots & 1 & 0 & 0 \\ -2 & 1 & 0 & \vdots & 0 & 1 & 0 \\ 3 & -1 & 1 & \vdots & 0 & 0 & 1 \end{array}\right]$

Second - Perform the elementary row operations, i.e.,

1. Multiply each element of the first row by 2 and add the result to each element of the second row.

$$\left[\begin{array}{ccccccc} 1 & 2 & 3 & \vdots & 1 & 0 & 0 \\ -2+2 & 1+4 & 0+6 & \vdots & 0+2 & 1+0 & 0+0 \\ 3 & -1 & 1 & \vdots & 0 & 0 & 1 \end{array}\right] = \left[\begin{array}{ccccccc} 1 & 2 & 3 & \vdots & 1 & 0 & 0 \\ 0 & 5 & 6 & \vdots & 2 & 1 & 0 \\ 3 & -1 & 1 & \vdots & 0 & 0 & 1 \end{array}\right]$$

2. Multiply each element of the first row by -3 and add the result to each element of the third

$$\left[\begin{array}{ccccccc} 1 & 2 & 3 & \vdots & 1 & 0 & 0 \\ 0 & 5 & 6 & \vdots & 2 & 1 & 0 \\ 3-3 & -1-6 & 1-9 & \vdots & 0-3 & 0+0 & 1+0 \end{array}\right] = \left[\begin{array}{ccccccc} 1 & 2 & 3 & \vdots & 1 & 0 & 0 \\ 0 & 5 & 6 & \vdots & 2 & 1 & 0 \\ 0 & -7 & -8 & \vdots & -3 & 0 & 1 \end{array}\right]$$

3. Divide the second row by 5.

$$\left[\begin{array}{ccccccc} 1 & 2 & 3 & \vdots & 1 & 0 & 0 \\ 0 & \frac{5}{5} & \frac{6}{5} & \vdots & \frac{2}{5} & \frac{1}{5} & 0 \\ 0 & -7 & -8 & \vdots & -3 & 0 & 1 \end{array}\right] = \left[\begin{array}{ccccccc} 1 & 2 & 3 & \vdots & 1 & 0 & 0 \\ 0 & 1 & \frac{6}{5} & \vdots & \frac{2}{5} & \frac{1}{5} & 0 \\ 0 & -7 & -8 & \vdots & -3 & 0 & 1 \end{array}\right]$$

4. Multiply each element of the second row by -2 and add the result to each element of the first row.

$$\left[\begin{array}{ccccccc} 1+0 & 2-2 & 3-\frac{12}{5} & \vdots & 1-\frac{4}{5} & 0-\frac{2}{5} & 0+0 \\ 0 & 1 & \frac{6}{5} & \vdots & \frac{2}{5} & \frac{1}{5} & 0 \\ 0 & -7 & -8 & \vdots & -3 & 0 & 1 \end{array}\right] = \left[\begin{array}{ccccccc} 1 & 0 & \frac{3}{5} & \vdots & \frac{1}{5} & -\frac{2}{5} & 0 \\ 0 & 1 & \frac{6}{5} & \vdots & \frac{2}{5} & \frac{1}{5} & 0 \\ 0 & -7 & -8 & \vdots & -3 & 0 & 1 \end{array}\right]$$

5. Multiply each element of the second row by 7 and add the result to each element of the third row.

$$\left[\begin{array}{ccccccc} 1 & 0 & \frac{3}{5} & \vdots & \frac{1}{5} & -\frac{2}{5} & 0 \\ 0 & 1 & \frac{6}{5} & \vdots & \frac{2}{5} & \frac{1}{5} & 0 \\ 0 & -7+7 & -8+\frac{42}{5} & \vdots & -3+\frac{14}{5} & 0+\frac{7}{5} & 1+0 \end{array}\right] = \left[\begin{array}{ccccccc} 1 & 0 & \frac{3}{5} & \vdots & \frac{1}{5} & -\frac{2}{5} & 0 \\ 0 & 1 & \frac{6}{5} & \vdots & \frac{2}{5} & \frac{1}{5} & 0 \\ 0 & 0 & \frac{2}{5} & \vdots & -\frac{1}{5} & \frac{7}{5} & 1 \end{array}\right]$$

6. Multiply each element of the third row by $\frac{5}{2}$.

$$\left[\begin{array}{ccccccc} 1 & 0 & \frac{3}{5} & \vdots & \frac{1}{5} & -\frac{2}{5} & 0 \\ 0 & 1 & \frac{6}{5} & \vdots & \frac{2}{5} & \frac{1}{5} & 0 \\ 0 & 0 & \frac{2}{5}\times\frac{5}{2} & \vdots & -\frac{1}{5}\times\frac{5}{2} & \frac{7}{5}\times\frac{5}{2} & 1\times\frac{5}{2} \end{array}\right] = \left[\begin{array}{ccccccc} 1 & 0 & \frac{3}{5} & \vdots & \frac{1}{5} & -\frac{2}{5} & 0 \\ 0 & 1 & \frac{6}{5} & \vdots & \frac{2}{5} & \frac{1}{5} & 0 \\ 0 & 0 & 1 & \vdots & -\frac{1}{2} & \frac{7}{2} & \frac{5}{2} \end{array}\right]$$

7. Multiply each element of the third row by $-\frac{3}{5}$ and add the result to each element of the

first row. $\begin{bmatrix} 1+0 & 0+0 & \frac{3}{5}-\frac{3}{5} & \vdots & \frac{1}{5}+\frac{3}{10} & -\frac{2}{5}-\frac{21}{10} & 0-\frac{3}{2} \\ 0 & 1 & \frac{6}{5} & \vdots & \frac{2}{5} & \frac{1}{5} & 0 \\ 0 & 0 & 1 & \vdots & -\frac{1}{2} & \frac{7}{2} & \frac{5}{2} \end{bmatrix} = \begin{bmatrix} 1 & 0 & 0 & \vdots & \frac{1}{2} & -\frac{5}{2} & -\frac{3}{2} \\ 0 & 1 & \frac{6}{5} & \vdots & \frac{2}{5} & \frac{1}{5} & 0 \\ 0 & 0 & 1 & \vdots & -\frac{1}{2} & \frac{7}{2} & \frac{5}{2} \end{bmatrix}$

8. Multiply each element of the third row by $-\frac{6}{5}$ and add the result to each element of the second row. $\begin{bmatrix} 1 & 0 & 0 & \vdots & \frac{1}{2} & -\frac{5}{2} & -\frac{3}{2} \\ 0+0 & 1+0 & \frac{6}{5}-\frac{6}{5} & \vdots & \frac{2}{5}+\frac{3}{5} & \frac{1}{5}-\frac{21}{5} & 0-\frac{3}{1} \\ 0 & 0 & 1 & \vdots & -\frac{1}{2} & \frac{7}{2} & \frac{5}{2} \end{bmatrix} = \begin{bmatrix} 1 & 0 & 0 & \vdots & \frac{1}{2} & -\frac{5}{2} & -\frac{3}{2} \\ 0 & 1 & 0 & \vdots & 1 & -4 & -3 \\ 0 & 0 & 1 & \vdots & -\frac{1}{2} & \frac{7}{2} & \frac{5}{2} \end{bmatrix}$

Note that the matrix on the left hand side is an identity matrix. Therefore, inverse of the A matrix is equal to $A^{-1} = \begin{bmatrix} \frac{1}{2} & -\frac{5}{2} & -\frac{3}{2} \\ 1 & -4 & -3 \\ -\frac{1}{2} & \frac{7}{2} & \frac{5}{2} \end{bmatrix}$

Third - Check the answer by multiplying A by A^{-1}.

$$A \times A^{-1} \stackrel{?}{=} I \text{ ; } A = \begin{bmatrix} 1 & 2 & 3 \\ -2 & 1 & 0 \\ 3 & -1 & 1 \end{bmatrix} \times \begin{bmatrix} \frac{1}{2} & -\frac{5}{2} & -\frac{3}{2} \\ 1 & -4 & -3 \\ -\frac{1}{2} & \frac{7}{2} & \frac{5}{2} \end{bmatrix} \stackrel{?}{=} I \text{ ; } \begin{bmatrix} \frac{1}{2}+2-\frac{3}{2} & -\frac{5}{2}-8+\frac{21}{2} & -\frac{3}{2}-6+\frac{15}{2} \\ -1+1+0 & 5-4+0 & 3-3+0 \\ \frac{3}{2}-1-\frac{1}{2} & -\frac{15}{2}+4+\frac{7}{2} & -\frac{9}{2}+3+\frac{5}{2} \end{bmatrix} \stackrel{?}{=} I$$

$$\text{; } \begin{bmatrix} 2-1 & 8-8 & 6-6 \\ -1+1 & 1 & 3-3 \\ \frac{3}{2}-\frac{3}{2} & -4+4 & -2+3 \end{bmatrix} \stackrel{?}{=} I \text{ ; } \begin{bmatrix} 1 & 0 & 0 \\ 0 & 1 & 0 \\ 0 & 0 & 1 \end{bmatrix} = I$$

Since $A \times A^{-1}$ is equal to the identity matrix A^{-1} is computed correctly.

Having learned the different methods of finding the inverse of a matrix, let's now look at the property of inverse matrices.

- **The Property of Inverse Matrices**

If A and B are $n \times n$ matrices, where the inverse of each matrix exist, then their product AB has an inverse which is equal to

$$(AB)^{-1} = B^{-1}A^{-1}$$

Note that the above inverse property can be extended to any number of matrices, as long as the inverse for each matrix exists, i.e.,

$$(ABC)^{-1} = [(AB)C]^{-1} = C^{-1}(AB)^{-1} = C^{-1}B^{-1}A^{-1}$$

$$(ABCD)^{-1} = [(AB)(CD)]^{-1} = (CD)^{-1}(AB)^{-1} = D^{-1}C^{-1}B^{-1}A^{-1}$$

Example 3.4-9: Show that $\left(\begin{bmatrix} 3 & 2 \\ 1 & 2 \end{bmatrix}\begin{bmatrix} 1 & -1 \\ 3 & 1 \end{bmatrix}\right)^{-1} = \begin{bmatrix} 1 & -1 \\ 3 & 1 \end{bmatrix}^{-1}\begin{bmatrix} 3 & 2 \\ 1 & 2 \end{bmatrix}^{-1}$

Solution:

Let A be equal to the product of the two matrices in the left hand side of the equality, i.e., let

$$A = \begin{bmatrix} 3 & 2 \\ 1 & 2 \end{bmatrix} \times \begin{bmatrix} 1 & -1 \\ 3 & 1 \end{bmatrix}$$

First - Find the inverse of the A matrix by using the minor and the cofactor method.

Step 1 - Multiply the two matrices. Let the product be equal to the A matrix.

$$\boxed{\begin{bmatrix} 3 & 2 \\ 1 & 2 \end{bmatrix} \times \begin{bmatrix} 1 & -1 \\ 3 & 1 \end{bmatrix}} = \boxed{\begin{bmatrix} (3\times 1)+(2\times 3) & (3\times -1)+(2\times 1) \\ (1\times 1)+(2\times 3) & (1\times -1)+(2\times 1) \end{bmatrix}} = \boxed{\begin{bmatrix} 3+6 & -3+2 \\ 1+6 & -1+2 \end{bmatrix}} = \boxed{\begin{bmatrix} 9 & -1 \\ 7 & 1 \end{bmatrix}}$$

Step 2 - Find the determinant by finding the difference of the products of the entries on the two sides of the diagonals in the A matrix, i.e.,

$\delta(A) = (9\cdot 1)-(-1\cdot 7) = 9+7 = 16$ since the determinant is not equal to zero A has an inverse.

Step 3 - Replace each entry in the A matrix with its cofactor, i.e.,

$$\boxed{A_{11}} = \boxed{(-1)^{1+1} M_{11}} = \boxed{(-1)^2 \times 1} = \boxed{1} \qquad \boxed{A_{12}} = \boxed{(-1)^{1+2} M_{12}} = \boxed{(-1)^3 \times 7} = \boxed{-7}$$

$$\boxed{A_{21}} = \boxed{(-1)^{2+1} M_{21}} = \boxed{(-1)^3 \times -1} = \boxed{1} \qquad \boxed{A_{22}} = \boxed{(-1)^{2+2} M_{22}} = \boxed{(-1)^4 \times 9} = \boxed{9}$$

Therefore, the cofactor matrix is equal to $C = \begin{bmatrix} A_{11} & A_{12} \\ A_{21} & A_{22} \end{bmatrix} = \begin{bmatrix} 1 & -7 \\ 1 & 9 \end{bmatrix}$ and $C^t = \begin{bmatrix} 1 & 1 \\ -7 & 9 \end{bmatrix}$

Step 4 - Compute $\boxed{A^{-1}} = \boxed{\frac{1}{\delta(A)} C^t} = \boxed{\frac{1}{16} \cdot \begin{bmatrix} 1 & 1 \\ -7 & 9 \end{bmatrix}} = \boxed{\begin{bmatrix} \mathbf{\frac{1}{16}} & \mathbf{\frac{1}{16}} \\ \mathbf{-\frac{7}{16}} & \mathbf{\frac{9}{16}} \end{bmatrix}}$

Step 5 - Check the answer by multiplying the A matrix with A^{-1}. The result should be equal to the identity matrix.

$$\boxed{A \times A^{-1}} = \boxed{\begin{bmatrix} 9 & -1 \\ 7 & 1 \end{bmatrix} \times \begin{bmatrix} \frac{1}{16} & \frac{1}{16} \\ -\frac{7}{16} & \frac{9}{16} \end{bmatrix}} = \boxed{\begin{bmatrix} \left(9\times\frac{1}{16}\right)+\left(-1\times-\frac{7}{16}\right) & \left(9\times\frac{1}{16}\right)+\left(-1\times\frac{9}{16}\right) \\ \left(7\times\frac{1}{16}\right)+\left(1\times-\frac{7}{16}\right) & \left(7\times\frac{1}{16}\right)+\left(1\times\frac{9}{16}\right) \end{bmatrix}} = \boxed{\begin{bmatrix} \frac{9}{16}+\frac{7}{16} & \frac{9}{16}-\frac{9}{16} \\ \frac{7}{16}-\frac{7}{16} & \frac{7}{16}+\frac{9}{16} \end{bmatrix}}$$

$$= \boxed{\begin{bmatrix} \frac{9+7}{16} & \frac{9-9}{16} \\ \frac{7-7}{16} & \frac{7+9}{16} \end{bmatrix}} = \boxed{\begin{bmatrix} \frac{16}{16} & \frac{0}{16} \\ \frac{0}{16} & \frac{16}{16} \end{bmatrix}} = \boxed{\begin{bmatrix} 1 & 0 \\ 0 & 1 \end{bmatrix}}$$

Since $A \times A^{-1}$ is equal to the identity matrix A^{-1} is computed correctly.

Second - Let $B = \begin{bmatrix} 1 & -1 \\ 3 & 1 \end{bmatrix}$ and $D = \begin{bmatrix} 3 & 2 \\ 1 & 2 \end{bmatrix}$. Next, find the inverse of the B and D matrix using the same steps outlined above.

$$\boxed{B^{-1}} = \boxed{\frac{1}{\delta(B)} C^t} = \boxed{\frac{1}{4} \cdot \begin{bmatrix} 1 & 1 \\ -3 & 1 \end{bmatrix}} = \boxed{\begin{bmatrix} \frac{1}{4} & \frac{1}{4} \\ -\frac{3}{4} & \frac{1}{4} \end{bmatrix}} \text{ and } \boxed{D^{-1}} = \boxed{\frac{1}{\delta(D)} C^t} = \boxed{\frac{1}{4} \cdot \begin{bmatrix} 2 & -2 \\ -1 & 3 \end{bmatrix}} = \boxed{\begin{bmatrix} \frac{1}{2} & -\frac{1}{2} \\ -\frac{1}{4} & \frac{3}{4} \end{bmatrix}}$$

Third - Find the product of the two inverse matrices B and D.

$$\boxed{B^{-1} \times D^{-1}} = \boxed{\begin{bmatrix} \frac{1}{4} & \frac{1}{4} \\ -\frac{3}{4} & \frac{1}{4} \end{bmatrix} \times \begin{bmatrix} \frac{1}{2} & -\frac{1}{2} \\ -\frac{1}{4} & \frac{3}{4} \end{bmatrix}} = \boxed{\begin{bmatrix} \left(\frac{1}{4} \times \frac{1}{2}\right) + \left(\frac{1}{4} \times -\frac{1}{4}\right) & \left(\frac{1}{4} \times -\frac{1}{2}\right) + \left(\frac{1}{4} \times \frac{3}{4}\right) \\ \left(-\frac{3}{4} \times \frac{1}{2}\right) + \left(\frac{1}{4} \times -\frac{1}{4}\right) & \left(-\frac{3}{4} \times -\frac{1}{2}\right) + \left(\frac{1}{4} \times \frac{3}{4}\right) \end{bmatrix}} = \boxed{\begin{bmatrix} \frac{1}{8} - \frac{1}{16} & -\frac{1}{8} + \frac{3}{16} \\ -\frac{3}{8} - \frac{1}{16} & \frac{3}{8} + \frac{3}{16} \end{bmatrix}}$$

$$= \boxed{\begin{bmatrix} \frac{16-8}{128} & \frac{-16+24}{128} \\ \frac{-48-8}{128} & \frac{48+24}{128} \end{bmatrix}} = \boxed{\begin{bmatrix} \frac{8}{128} & \frac{8}{128} \\ -\frac{56}{128} & \frac{72}{128} \end{bmatrix}} = \boxed{\begin{bmatrix} \mathbf{\frac{1}{16}} & \mathbf{\frac{1}{16}} \\ \mathbf{-\frac{7}{16}} & \mathbf{\frac{9}{16}} \end{bmatrix}}$$

Fourth - Compare the results obtained from the first and third steps. Since $A^{-1} = B^{-1} \times D^{-1}$ we have shown that the two sides are equal.

In the next section, we will learn how to solve linear systems using several methods, including a methods which involve the use of inverse matrices.

Section 3.4 Practice Problems - Inverse Matrices

1. Use the minor and cofactor method to find the inverse of the following 2×2 matrices, if it exists. Verify each answer by showing that $A \times A^{-1} = I$.

a. $A = \begin{bmatrix} 1 & 2 \\ 3 & 6 \end{bmatrix}$ b. $A = \begin{bmatrix} -3 & -1 \\ 9 & 3 \end{bmatrix}$ c. $A = \begin{bmatrix} 2 & 3 \\ 2 & 4 \end{bmatrix}$

d. $A = \begin{bmatrix} 3 & 1 \\ -1 & 2 \end{bmatrix}$ e. $A = \begin{bmatrix} -2 & -1 \\ 2 & 3 \end{bmatrix}$ f. $A = \begin{bmatrix} 3 & 15 \\ 1 & 5 \end{bmatrix}$

2. Use the minor and cofactor method to find the inverse of the following 3×3 matrices, if it exists. Verify each answer by showing that $A \times A^{-1} = I$.

a. $A = \begin{bmatrix} 1 & 0 & 2 \\ 2 & -3 & 0 \\ 3 & 4 & 0 \end{bmatrix}$ b. $A = \begin{bmatrix} 0 & 4 & 2 \\ 1 & 0 & 3 \\ 1 & -1 & 2 \end{bmatrix}$ c. $A = \begin{bmatrix} 3 & 1 & 1 \\ 1 & 0 & 3 \\ 6 & 2 & 2 \end{bmatrix}$

d. $A = \begin{bmatrix} 0 & 0 & 1 \\ 0 & 1 & 0 \\ 1 & 0 & 0 \end{bmatrix}$ e. $A = \begin{bmatrix} 1 & 0 & 0 \\ 0 & 0 & 1 \\ 1 & 0 & 0 \end{bmatrix}$ f. $A = \begin{bmatrix} 1 & 2 & -3 \\ 3 & -1 & 2 \\ 5 & 2 & 0 \end{bmatrix}$

3. Find the inverse of the following matrices, if it exists, using $A \times A^{-1} = I$.

a. $A=\begin{bmatrix}-3 & \frac{1}{3}\\ 2 & 1\end{bmatrix}$ b. $A=\begin{bmatrix}2 & 0\\ 1 & 0\end{bmatrix}$ c. $A=\begin{bmatrix}1 & 0\\ -2 & 3\end{bmatrix}$

d. $A=\begin{bmatrix}6 & 0\\ 0 & -\frac{1}{2}\end{bmatrix}$ e. $A=\begin{bmatrix}1 & 0 & 0\\ 0 & 0 & 1\\ 1 & 0 & 0\end{bmatrix}$ f. $A=\begin{bmatrix}1 & 0 & 2\\ 2 & 1 & 4\\ 2 & 0 & 4\end{bmatrix}$

4. Show that $\left(\begin{bmatrix}1 & 3\\ 0 & -1\end{bmatrix}\cdot\begin{bmatrix}1 & 0\\ 2 & 3\end{bmatrix}\right)^{-1}=\begin{bmatrix}1 & 0\\ 2 & 3\end{bmatrix}^{-1}\cdot\begin{bmatrix}1 & 3\\ 0 & -1\end{bmatrix}^{-1}$

5. Show that the matrix $A=\begin{bmatrix}1 & 2 & 3\\ 0 & 0 & 0\\ -3 & 4 & 5\end{bmatrix}$ has no inverse.

3.5 Solving Linear Systems

In this section several methods for solving linear systems are introduced. Solutions to systems using the addition and the substitution method are addressed in Cases I and II. These two methods are generally taught at high school level and have a limited practical use with systems having more than three unknowns. Cases III through VI are generally taught at college level and after students are introduced to matrices. Students are encouraged to learn all six methods introduced in this section. It is the author's hope that through enough practice students will learn how to choose one method over the other.

Case I Solving Linear Systems Using the Addition Method

- **Linear Systems with Two Variables**

Linear systems with two unknowns are solved by applying the following steps:

Step 1 a. Eliminate either x or y by multiplying the coefficient of x or y from the other equation. (Note that if the coefficient of x or y are opposite of one another like $-2y$ and $+2y$ simply add the two equations in order to eliminate y.)

b. Add the two equations and solve for x or y.

Step 2 Substitute the x or y value into one of the equations to solve for the second variable.

Step 3 Check the answers by substituting the solution set into one of the original equations.

The following examples show the steps for solving linear systems, with two variables, using the addition method.

Example 3.5-1 Solve the linear system $\begin{matrix} 2x-4y=-3 \\ x+y=5 \end{matrix}$ using the addition method.

Solution:

First - Let's eliminate x from the two equations by multiplying the second equation by -2, i.e.,

$$\begin{matrix} 2x-4y=-3 \\ -2\cdot(x+y)=-2\cdot 5 \end{matrix} \; ; \; \begin{matrix} 2x-4y=-3 \\ \underline{-2x-2y=-10} \end{matrix}$$

$$-6y=-13 \text{ therefore } y=\frac{13}{6}$$

Second - Substitute the y value into $x+y=5$ and solve for x, i.e.,

$$x+\frac{13}{6}=5 \; ; \; x=5-\frac{13}{6} \; ; \; x=\frac{(5\cdot 6)-(13\cdot 1)}{6} \; ; \; x=\frac{30-13}{6} \; ; \; x=\frac{17}{6}$$

Therefore, $\mathbf{x=\frac{17}{6}}$ and $\mathbf{y=\frac{13}{6}}$ and the solution set is $\left\{\left(\mathbf{\frac{17}{6},\frac{13}{6}}\right)\right\}$

Third - Check the answers by substituting the x and y values into one of the original equations. Let's substitute the x and y values into $2x-4y=-3$.

$$\left(2\times\frac{17}{6}\right)+\left(-4\times\frac{13}{6}\right)\stackrel{?}{=}-3 \; ; \; \frac{34}{6}-\frac{52}{6}\stackrel{?}{=}-3 \; ; \; \frac{34-52}{6}\stackrel{?}{=}-3 \; ; \; -\frac{18}{6}\stackrel{?}{=}-3 \; ; \; -\frac{3}{1}\stackrel{?}{=}-3 \; ; \; -3=-3$$

Example 3.5-2 Solve the linear system $\begin{matrix} 3x-2y=-1 \\ x-3y=3 \end{matrix}$ using the addition method.

Solution:

First - Let's eliminate x from the two equations by multiplying the second equation by -3, i.e.,

$$\begin{matrix} 3x-2y=-1 \\ -3\cdot(x-3y)=-3\cdot 3 \end{matrix} \; ; \; \begin{matrix} 3x-2y=-1 \\ \underline{-3x+9y=-9} \end{matrix}$$

$$7y=-10 \qquad \text{therefore } y=-\frac{10}{7}$$

Second - Substitute the y value into $x-3y=3$ and solve for x, i.e.,

$$x+\left(-3\times-\frac{10}{7}\right)=3 \; ; \; x=3-\frac{30}{7} \; ; \; x=\frac{3}{1}-\frac{30}{7} \; ; \; x=\frac{(3\cdot 7)-(30\cdot 1)}{7\cdot 1} \; ; \; x=\frac{21-30}{7} \; ; \; x=-\frac{9}{7}$$

Therefore, $\mathbf{x=-\frac{9}{7}}$ and $\mathbf{y=-\frac{10}{7}}$ and the solution set is $\left\{\left(-\frac{9}{7},-\frac{10}{7}\right)\right\}$

Third - Check the answers by substituting the x and y values into one of the original equations. Let's substitute the x and y values into $3x-2y=-1$.

$$\left(3\times-\frac{9}{7}\right)+\left(-2\times-\frac{10}{7}\right)\stackrel{?}{=}-1 \; ; \; -\frac{27}{7}+\frac{20}{7}\stackrel{?}{=}-1 \; ; \; \frac{-27+20}{7}\stackrel{?}{=}-1 \; ; \; -\frac{7}{7}\stackrel{?}{=}-1 \; ; \; -1=-1$$

- **Linear Systems with Three Variables**

Linear systems with three unknowns are solved by applying the following steps:

Step 1 Reduce the three variable system to a two variable system by choosing a variable and eliminating it.

Step 2 Solve the remaining two equations containing two unknowns.

Step 3 Substitute the values from Step 2 into one of the original equations to find the third value.

Step 4 Check the answers by substituting the solution set into one of the original equations.

The following examples show the steps for solving linear systems, with three variables, using the addition method.

Example 3.5-3 Solve the linear system $\begin{matrix} x+3y-3z=-1 \\ 2x-3y+2z=3 \\ x+2y-z=-2 \end{matrix}$ using the addition method.

Solution:

First - Let's eliminate y from the first and second equations by adding the two, i.e.,

$$\begin{matrix} x+3y-3z=-1 \\ \underline{2x-3y+2z=3} \\ 3x+0y-z=2 \end{matrix} \qquad (1)$$

then, eliminate y from the second and third equation in the following way

$$\begin{matrix} 2\cdot(2x-3y+2z)=3\cdot 2 \\ 3\cdot(x+2y-z)=-2\cdot 3 \end{matrix} \; ; \; \begin{matrix} 4x-6y+4z=6 \\ \underline{3x+6y-3z=-6} \\ 7x+0y+z=0 \end{matrix} \qquad (2)$$

Second - Solve for x and z using the equations (1) and (2), i.e., $\begin{array}{r} 3x-z=2 \\ \underline{7x+z=0} \end{array}$

$$10x=2 \text{ thus } x=\frac{1}{5} \text{ and } z=-\frac{7}{5}$$

Third - Substitute the x and z values into $x+3y-3z=-1$ and solve for y, i.e.,

$$\frac{1}{5}+3y-3\times-\frac{7}{5}=-1 \;;\; \frac{1}{5}+3y+\frac{21}{5}=-1 \;;\; 3y=-1-\frac{22}{5} \;;\; 3y=-\frac{27}{5} \;;\; \frac{3y}{3}=-\frac{27}{5}\times\frac{1}{3} \;;\; y=-\frac{9}{5}$$

Therefore, $\mathbf{x=\frac{1}{5}}$, $\mathbf{y=-\frac{9}{5}}$, $\mathbf{z=-\frac{7}{5}}$ and the solution set is $\mathbf{\left\{\left(\frac{1}{5},-\frac{9}{5},-\frac{7}{5}\right)\right\}}$

Fourth - Check the answers by substituting the x, y, and z values into one of the original equation. Let's substitute the x, y, and z values into $x+2y-z=-2$.

$$\frac{1}{5}+\left(2\times-\frac{9}{5}\right)+\frac{7}{5}\overset{?}{=}-2 \;;\; \frac{1}{5}-\frac{18}{5}+\frac{7}{5}\overset{?}{=}-2 \;;\; \frac{1-18+7}{5}\overset{?}{=}-2 \;;\; -\frac{10}{5}\overset{?}{=}-2 \;;\; -\frac{2}{1}\overset{?}{=}-2 \;;\; -2=-2$$

Note that problems 3.5-1 through 3.5-3 are solved again in this section using the inverse matrices method (see Case III).

Example 3.5-4 Solve the linear system $\begin{array}{l} 2x+y-3z=-4 \\ 2x-3y+z=1 \\ 2y-4z=-3 \end{array}$ using the addition method.

Solution:

First - Let's eliminate x from the first and second equation by multiplying the first equation by -1.

$$\begin{array}{l} -1\cdot(2x+y-3z)=-1\cdot-4 \\ 2x-3y+z=1 \end{array} \;;\; \begin{array}{r} -2x-\;y+3z=4 \\ \underline{2x-3y+\;z=1} \\ -4y+4z=5 \end{array} \quad (1)$$

Since the third equation is already in terms of x and z, let's label $2y-4z=-3$ as equation no. (2).

Second - Solve for x and z using the equations (1) and (2), i.e., $\begin{array}{r} -4y+4z=5 \\ \underline{2y-4z=-3} \end{array}$

$$-2y=2 \text{ thus } y=-1 \text{ and } z=\frac{1}{4}$$

Third - Substitute the y and z values into $2x+y-3z=-4$ and solve for x, i.e.,

$$2x-1-3\times\frac{1}{4}=-4 \;;\; 2x-1-\frac{3}{4}=-4 \;;\; 2x=-3+\frac{3}{4} \;;\; 2x=-\frac{3}{1}+\frac{3}{4} \;;\; 2x=\frac{-12+3}{4} \;;\; x=-\frac{9}{8}$$

Therefore, $\mathbf{x=-\frac{9}{8}}$, $\mathbf{y=-1}$, $\mathbf{z=\frac{1}{4}}$ and the solution set is $\mathbf{\left\{\left(-\frac{9}{8},-1,\frac{1}{4}\right)\right\}}$

Fourth - Check the answers by substituting the x, y, and z values into one of the original equation. Let's substitute the x, y, and z values into $2x-3y+z=1$.

$$\left(2\times-\frac{9}{8}\right)+(-3\times-1)+\frac{1}{4}\overset{?}{=}1 \;;\; -2.25+3+0.25\overset{?}{=}1 \;;\; -2.25+3.25\overset{?}{=}1 \;;\; 1=1$$

• Independent, Dependent, and Inconsistent Systems of Equations

Independent System - If through the addition, substitution, or any other process (method), a solution for each variable is found, then the system is referred to as an independent system. For example, the systems in examples 5.3-1 through 5.3-4 are independent systems.

Dependent System - If through the addition, substitution, or any other process (method), all the variables are eliminated and both sides of the resulting equation are equal, then there are an infinite number of solutions (no unique solutions) to the system and the equations are dependent (see example 3.5-5).

Inconsistent System – If through the addition, substitution, or any other process (method) all the variables are eliminated and both sides of the resulting equation are not equal, then the system has no solutions and is inconsistent (see example 3.5-6).

Example 3.5-5: Solve the linear system $\begin{matrix} 3x+4y=2 \\ 6x+8y=4 \end{matrix}$ using the addition method.

Solution:

Let's eliminate x from the first and third equations by multiplying the first equation by -4, i.e.,

$$\begin{matrix} -2\cdot(3x+4y)=-2\cdot 2 \\ 1\cdot(6x+8y)=1\cdot 4 \end{matrix} \; ; \; \begin{matrix} -6x-8y=-4 \\ \underline{\;6x+8y=\;4\;} \\ 0x+0y=0 \end{matrix} \; ; \; 0=0$$

Since both sides of the resulting equation are equal the linear system has an infinite number of solutions (no unique solution) and the equations are dependent.

Example 3.5-6: Solve the linear system $\begin{matrix} x+2y=1 \\ 4x+8y=0 \end{matrix}$ using the addition method.

Solution:

Let's eliminate x from the first and third equations by multiplying the first equation by -4, i.e.,

$$\begin{matrix} -4\cdot(x+2y)=-4\cdot 1 \\ 1\cdot(4x+8y)=1\cdot 0 \end{matrix} \; ; \; \begin{matrix} -4x-8y=-4 \\ \underline{\;4x+8y=0\;} \\ 0x+0y=-4 \end{matrix} \; ; \; 0\neq -4$$

Since 0 can not be equal to -4 the linear system has no solution and is an inconsistent system.

Section 3.5 Case I Practice Problems - Solving Linear Systems Using the Addition Method

1. Find the solution set of the given systems by using the addition method.

a. $\begin{matrix} x+3y=2 \\ 2x+2y=-1 \end{matrix}$ b. $\begin{matrix} x+2y=0 \\ -x+y=2 \end{matrix}$ c. $\begin{matrix} 2x+4y=-1 \\ 4x+8y=5 \end{matrix}$ d. $\begin{matrix} 4x-2y=-3 \\ x-2y=1 \end{matrix}$

e. $\begin{matrix} 2x+y=3 \\ 4x+2y=6 \end{matrix}$ f. $\begin{matrix} x+y=2 \\ 2x-z=-1 \\ 2y+2z=3 \end{matrix}$ g. $\begin{matrix} x-y+3z=2 \\ x-z=-3 \\ 2x-2y+6z=-1 \end{matrix}$ h. $\begin{matrix} x+3y-z=-2 \\ -x+2y+3z=1 \\ x+y-2z=0 \end{matrix}$

Case II Solving Linear Systems Using the Substitution Method

- **Linear Systems with Two Variables**

Linear systems with two unknowns are solved by applying the following steps:

Step 1 Solve one equation for x or y.

Step 2 Substitute the resulting equation from Step 1 into the second equation and solve for the unknown variable.

Step 3 Substitute the solution from Step 2 into the first equation and solve for the second unknown variable.

Step 4 Check the answers by substituting the solution set into one of the original equations.

The following examples show the steps for solving linear systems, with two variables, using the substitution method.

Example 3.5-7 Solve the linear system $\begin{matrix} x-4y=6 \\ 2x+3y=1 \end{matrix}$ using the substitution method.

Solution:

First - Let's solve the first equation for x in terms of y, i.e., given $x-4y=6$, then $x=4y+6$

Second - Substitute the x value into the second equation $2x+3y=1$ and solve for y

$$2x+3y=1 \; ; \; 2(4y+6)+3y=1 \; ; \; 8y+12+3y=1 \; ; \; 8y+3y=1-12 \; ; \; 11y=-11 \; ; \; y=-1$$

Third - Substitute the y value into the first equation $x-4y=6$ and solve for x

$$x-4y=6 \; ; \; x+(-4\times-1)=6 \; ; \; x+5=6 \; ; \; x=6-5 \; ; \; x=1$$

Therefore, $\mathbf{x=1}$ and $\mathbf{y=-1}$ and the solution set is $\{(\mathbf{1},\mathbf{-1})\}$

Fourth - Check the answers by substituting the x and y values into one of the original equations. Let's substitute the x and y values into $2x+3y=1$.

$$(2\times1)+(3\times-1)\stackrel{?}{=}-1 \; ; \; 2-3\stackrel{?}{=}-1 \; ; \; -1=-1$$

Example 3.5-8 Solve the linear system $\begin{matrix} x+2y=3 \\ 3x+y=-2 \end{matrix}$ using the substitution method.

Solution:

First - Let's solve the second equation for y in terms of x, i.e., given $3x+y=-2$, then $y=-3x-2$

Second - Substitute the y value into the first equation $x+2y=3$ and solve for x

$$x+2y=3 \; ; \; x+2(-3x-2)=3 \; ; \; x-6x-4=3 \; ; \; -5x=3+4 \; ; \; -5x=7 \; ; \; x=-\frac{7}{5}$$

Third - Substitute the x value into the second equation $3x+y=-2$ and solve for y

$$3x+y=-2 \; ; \; \left(3\times-\frac{7}{5}\right)+y=-2 \; ; \; -\frac{21}{5}+y=-2 \; ; \; y=-2+\frac{21}{5} \; ; \; y=\frac{-10+21}{5} \; ; \; y=\frac{11}{5}$$

Therefore, $x=-\frac{7}{5}$ and $y=\frac{11}{5}$ and the solution set is $\left\{\left(-\frac{7}{5},\frac{11}{5}\right)\right\}$

Fourth - Check the answers by substituting the x and y values into one of the original equations. Let's substitute the x and y values into $x+2y=3$.

$$-\frac{7}{5}+\left(2\times\frac{11}{5}\right)\overset{?}{=}3\ ;\ -\frac{7}{5}+\frac{22}{5}\overset{?}{=}3\ ;\ \frac{-7+22}{5}\overset{?}{=}3\ ;\ \frac{15}{5}\overset{?}{=}3\ ;\ 3=3$$

Example 3.5-9 Solve the linear system $\begin{array}{l}2x+3y=1\\4x+6y=2\end{array}$ using the substitution method.

Solution:

First - Let's solve the first equation for x in terms of y, i.e., given $2x+3y=1$, then $x=-1.5y+0.5$

Second - Substitute the x value into the second equation $4x+6y=2$ and solve for y

$$4x+6y=2\ ;\ 4(-1.5y+0.5)+6y=2\ ;\ -6y+2+6y=2\ ;\ 2=2$$

Since the variable y is eliminated and both sides of the equation are equal, thus the linear system has infinite number of solutions (no unique solution) and the equations are dependent.

- **Linear Systems with Three Variables**

Linear systems with three unknowns are solved by applying the following steps:

Step 1 Solve one equation for any one variable x, y, or z, in terms of the other two.

Step 2 Substitute the resulting equation from Step 1 into the other two equations to obtain a two variable system.

Step 3 Repeat the steps used in the substitution method to solve systems with two variables as stated above.

Step 4 Use the back substitution method to solve for the remaining variable.

Step 5 Check the answers by substituting the x, y, and z values into one of the original equations.

The following examples show the steps for solving linear systems, with three variables, using the substitution method.

Example 3.5-10 Solve the linear system $\begin{array}{l}x+2y-z=2\\3x+4y+2z=1\\-2x+y-3z=3\end{array}$ using the substitution method.

Solution:

First - Let's solve the first equation in terms of y and z, i.e., $x+2y-z=2$; $x=-2y+z+2$

Second - Substitute the x value into the second and third equations

$$3x+4y+2z=1\ ;\ 3(-2y+z+2)+4y+2z=1\ ;\ -6y+3z+6+4y+2z=1\ ;\ -2y+5z=-5 \quad (1)$$

$$-2x+y-3z=3\ ;\ -2(-2y+z+2)+y-3z=3\ ;\ 4y-2z-4+y-3z=3\ ;\ 5y-5z=7 \quad (2)$$

Third - Solve equation (1) in terms of z and substitute the z value into the second equation (2).

$-2y+5z=-5$; $-2y=-5z-5$; $y=\frac{5}{2}z+\frac{5}{2}$ substituting the y into (2) we obtain

$5y-5z=7$; $5\left(\frac{5}{2}z+\frac{5}{2}\right)-5z=7$; $\frac{25}{2}z+\frac{25}{2}-5z=7$; $\left(\frac{25}{2}-5\right)z=7-\frac{25}{2}$; $\frac{15}{2}z=-\frac{11}{2}$; $z=-\frac{11}{15}$

Third - Substitute the z value in equation (1) and solve for y

$-2y+5z=-5$; $-2y+\left(5\times-\frac{11}{15}\right)=-5$; $-2y-\frac{55}{15}=-5$; $-2y=-5+\frac{11}{3}$; $-2y=-\frac{4}{3}$; $y=\frac{2}{3}$

Fourth - Substitute the y and z values into the first equation and solve for x

$x+2y-z=2$; $x+\left(2\times\frac{2}{3}\right)+\frac{11}{15}=2$; $x+\frac{4}{3}+\frac{11}{15}=2$; $x=-\frac{4}{3}-\frac{11}{15}+2$; $x=-\frac{31}{15}+2$; $x=\frac{-31+30}{15}$; $x=-\frac{1}{15}$

Therefore, $\mathbf{x=-\frac{1}{15}}$, $\mathbf{y=\frac{2}{3}}$, and $\mathbf{z=-\frac{11}{15}}$ and the solution set is $\left\{\left(-\frac{1}{15},\frac{2}{3},-\frac{11}{15}\right)\right\}$

Fifth - Check the answers by substituting the x, y, and z values into one of the original equations. Let's substitute the x, y, and z values into $-2x+y-3z=3$.

$\left(-2\times-\frac{1}{15}\right)+\frac{2}{3}+\left(-3\times-\frac{11}{15}\right)\overset{?}{=}3$; $\frac{2}{15}+\frac{2}{3}+\frac{33}{15}\overset{?}{=}3$; $\frac{2+33}{15}+\frac{2}{3}\overset{?}{=}3$; $\frac{35}{15}+\frac{2}{3}\overset{?}{=}3$; $\frac{7}{3}+\frac{2}{3}\overset{?}{=}3$; $\frac{9}{3}\overset{?}{=}3$; $3=3$

Example 3.5-11 Solve the linear system $\begin{array}{l}3x-y+2z=1\\x+y+z=3\\2x-3y+z=-1\end{array}$ using the substitution method.

Solution:

First - Let's solve the second equation in terms of y and z, i.e., $x+y+z=3$; $x=-y-z+3$

Second - Substitute the x value into the first and third equations

$3x-y+2z=1$; $3(-y-z+3)-y+2z=1$; $-3y-3z+9-y+2z=1$; $4y+z=8$ (1)

$2x-3y+z=-1$; $2(-y-z+3)-3y+z=-1$; $-2y-2z+6-3y+z=-1$; $5y+z=7$ (2)

Third - Solve equation (1) in terms of z and substitute the z value into the second equation (2).

$4y+z=8$; $4y=-z+8$; $y=-\frac{1}{4}z+2$ substituting the y into (2) we obtain

$5y+z=7$; $5\left(-\frac{1}{4}z+2\right)+z=7$; $-\frac{5}{4}z+10+z=7$; $\left(-\frac{5}{4}+1\right)z=-3$; $-\frac{1}{4}z=-3$; $z=12$

Third - Substitute the z value in equation (1) and solve for y

$4y+z=8$; $4y+12=8$; $4y=-12+8$; $4y=-4$; $y=-1$

Fourth - Substitute the y and z values into the first equation and solve for x

$x+y+z=3$; $x-1+12=3$; $x=3-11$; $x=-8$

Therefore, $\mathbf{x=-8}$, $\mathbf{y=-1}$, and $\mathbf{z=12}$ and the solution set is $\{(-8,-1,12)\}$

Fifth - Check the answers by substituting the x, y, and z values into one of the original equations.

Let's substitute the x, y, and z values into $2x-3y+z=-1$.

$(2\times-8)+(-3\times-1)+12\stackrel{?}{=}-1$; $-16+3+12\stackrel{?}{=}-1$; $-16+15\stackrel{?}{=}-1$; $-1=-1$

Example 3.5-12 Solve the linear system $\begin{array}{l} x-y+z=1 \\ 2x-2y+2z=2 \\ 3x-3y+3z=5 \end{array}$ using the substitution method.

Solution

First - Let's solve the first equation in terms of y and z, i.e., $x-y+z=1$; $x=y-z+1$

Second - Substitute the x value into the second and third equations

$2x-2y+2z=2$; $2(y-z+1)-2y+2z=2$; $2y-2z+2-2y+2z=2$; $2=2$ (1)

$3x-3y+3z=5$; $2(y-z+1)-3y+3z=5$; $2y-2z+2-3y+3z=5$; $-y+z=3$ (2)

Since both sides of the equation (1) are equal to 2 the linear system has an infinite number of solutions (no unique solutions) and the equations are dependent.

Section 3.5 Case II Practice Problems - Solving Linear Systems Using the Substitution Method

1. Find the solution set of the given systems by using the substitution method.

a. $\begin{array}{l} 3x-4y=2 \\ 5x-3y=1 \end{array}$ b. $\begin{array}{l} x-3y=2 \\ y=3x-5 \end{array}$ c. $\begin{array}{l} x+4y=-3 \\ 2x-3y=1 \end{array}$ d. $\begin{array}{l} x+y=-5 \\ 2x-5y=1 \end{array}$

e. $\begin{array}{l} x-2y=3 \\ 2x-4y=5 \end{array}$ f. $\begin{array}{l} 2x+3y=3 \\ 6x+9y=9 \end{array}$ g. $\begin{array}{l} 2x+3y-z=3 \\ x-y+2z=1 \\ x-y+z=-2 \end{array}$ h. $\begin{array}{l} x=2y-z \\ x+3y+z=1 \\ 2x-y+z=3 \end{array}$

Case III Solving Linear Systems Using the Inverse Matrices Method

The linear system of n equations and n unknowns

$$\begin{array}{cccccc} a_{11}x_1 + & a_{12}x_2 + & a_{13}x_3 + & \cdots & +a_{1n}x_n = & b_1 \\ a_{21}x_1 + & a_{22}x_2 + & a_{23}x_3 + & \cdots & +a_{2n}x_n = & b_2 \\ a_{31}x_1 + & a_{32}x_2 + & a_{33}x_3 + & \cdots & +a_{3n}x_n = & b_3 \\ \vdots & \vdots & \vdots & & \vdots & \vdots \\ a_{n1}x_1 + & a_{n2}x_2 + & a_{n3}x_3 + & \cdots & +a_{nn}x_n = & b_n \end{array}$$

can either be written in the matrix form of

$$\begin{bmatrix} a_{11} & a_{12} & a_{13} & \cdots & a_{1n} \\ a_{21} & a_{22} & a_{23} & \cdots & a_{2n} \\ a_{31} & a_{32} & a_{33} & \cdots & a_{3n} \\ \vdots & \vdots & \vdots & & \vdots \\ a_{n1} & a_{n2} & a_{n3} & \cdots & a_{nn} \end{bmatrix} \begin{bmatrix} x_1 \\ x_2 \\ x_3 \\ \vdots \\ x_n \end{bmatrix} = \begin{bmatrix} b_1 \\ b_2 \\ b_3 \\ \vdots \\ b_n \end{bmatrix}$$

or in a more concise matrix notation form of

$$\boldsymbol{AX = B} \qquad (1)$$

where A is an $n \times n$ coefficient matrix and X and B are $n \times 1$ column matrices.
Assuming that matrix A has an inverse we can multiply both sides of $AX = B$ by A^{-1} to obtain the following equality.

$$AX = B$$
$$A^{-1}AX = A^{-1}B$$
$$\left(A^{-1}A\right)X = A^{-1}B$$
$$IX = A^{-1}B$$
$$\boldsymbol{X = A^{-1}B} \qquad (2)$$

where $A^{-1}B$ is an $n \times 1$ column matrix. Note that since X is equal to $A^{-1}B$, this implies that each entry in X is equal to the corresponding entry in $A^{-1}B$. Therefore, equation (2) can be used to solve the unknown variables in a linear system. In addition, note that since matrix multiplication is not commutative, the solution to the system $AX = B$ can only be written as $X = A^{-1}B$ and not as $X = BA^{-1}$.

Linear systems can be solved using the inverse matrix method by applying the following steps:

Step 1 Write the linear system in the form of $AX = B$.

Step 2 Find the determinant of the coefficient matrix A. If $\delta(A) \neq 0$ go to Step 3. If $\delta(A) = 0$ the linear system is either a dependent or an inconsistent system.

Step 3 Find the inverse of the A matrix. Check to ensure $A \times A^{-1} = I$.

Step 3 Use $X = A^{-1}B$ to solve for the unknowns x, y, z, etc. which are also referred to as the solution set.

Step 5 Check the answers by substituting the solution set into one of the original equations.

The following examples show the steps for solving linear systems using the inverse matrix method.

Example 3.5-13 Solve the system $\begin{matrix} 2x-4y=-3 \\ x+y=5 \end{matrix}$ by finding the inverse of the coefficient matrix.

Solution:

First - Write the given system in the form of $AX=B$, i.e., $\begin{bmatrix} 2 & -4 \\ 1 & 1 \end{bmatrix}\begin{bmatrix} x \\ y \end{bmatrix}=\begin{bmatrix} -3 \\ 5 \end{bmatrix}$

Second - Find determinant of the coefficient matrix A. Let's expand about the first row

$$\boxed{\delta(A)}=\boxed{a_{11}A_{11}+a_{12}A_{12}}=\boxed{2A_{11}-4A_{12}}=\boxed{2\cdot(-1)^{1+1}M_{11}-4\cdot(-1)^{1+2}M_{12}}=\boxed{(2\cdot 1)+(4\cdot 1)}=\boxed{6}$$

Third - Find inverse of the A matrix. Note that

$$\boxed{A_{11}=(-1)^{1+1}M_{11}=1} \quad \boxed{A_{12}=(-1)^{1+2}M_{12}=-1} \quad \boxed{A_{21}=(-1)^{2+1}M_{21}=4} \text{ and } \boxed{A_{22}=(-1)^{22}M_{22}=2}$$

thus $\boxed{C=\begin{bmatrix} A_{11} & A_{12} \\ A_{21} & A_{22} \end{bmatrix}=\begin{bmatrix} 1 & -1 \\ 4 & 2 \end{bmatrix}}$ and $\boxed{C^t=\begin{bmatrix} 1 & 4 \\ -1 & 2 \end{bmatrix}}$ therefore

$$\boxed{A^{-1}}=\boxed{\frac{1}{\delta(A)}C^t}=\boxed{\frac{1}{6}\cdot\begin{bmatrix} 1 & 4 \\ -1 & 2 \end{bmatrix}}=\boxed{\begin{bmatrix} \frac{1}{6} & \frac{4}{6} \\ -\frac{1}{6} & \frac{2}{6} \end{bmatrix}}=\boxed{\begin{bmatrix} \frac{1}{6} & \frac{2}{3} \\ -\frac{1}{6} & \frac{1}{3} \end{bmatrix}}$$

Check the answer by multiplying the A matrix by A^{-1}.

$$\boxed{A\times A^{-1}}=\boxed{\begin{bmatrix} 2 & -4 \\ 1 & 1 \end{bmatrix}\times\begin{bmatrix} \frac{1}{6} & \frac{2}{3} \\ -\frac{1}{6} & \frac{1}{3} \end{bmatrix}}=\boxed{\begin{bmatrix} \frac{2}{6}+\frac{4}{6} & \frac{4}{3}-\frac{4}{3} \\ \frac{1}{6}-\frac{1}{6} & \frac{2}{3}+\frac{1}{3} \end{bmatrix}}=\boxed{\begin{bmatrix} 1 & 0 \\ 0 & 1 \end{bmatrix}}$$

Since $A\times A^{-1}$ is equal to the identity matrix A^{-1} is computed correctly.

Fourth - Use $X=A^{-1}B$ to solve for x and y.

$$\boxed{\begin{bmatrix} x \\ y \end{bmatrix}}=\boxed{\begin{bmatrix} \frac{1}{6} & \frac{2}{3} \\ -\frac{1}{6} & \frac{1}{3} \end{bmatrix}\begin{bmatrix} -3 \\ 5 \end{bmatrix}}=\boxed{\begin{bmatrix} -\frac{3}{6}+\frac{10}{3} \\ \frac{3}{6}+\frac{5}{3} \end{bmatrix}}=\boxed{\begin{bmatrix} \frac{17}{6} \\ \frac{13}{6} \end{bmatrix}}$$

Therefore, $\mathbf{x=\frac{17}{6}}$ and $\mathbf{y=\frac{13}{6}}$ and the solution set is $\left\{\left(\mathbf{\frac{17}{6},\frac{13}{6}}\right)\right\}$

Fifth - Check the answers by substituting the x and y values into one of the original equations.

Let's substitute the x and y values into $2x-4y=-3$.

$$\left(2\times\frac{17}{6}\right)+\left(-4\times\frac{13}{6}\right)\stackrel{?}{=}-3\ ;\ \frac{34}{6}-\frac{52}{6}\stackrel{?}{=}-3\ ;\ \frac{34-52}{6}\stackrel{?}{=}-3\ ;\ -\frac{18}{6}\stackrel{?}{=}-3\ ;\ -\frac{3}{1}\stackrel{?}{=}-3\ ;\ -3=-3$$

Example 3.5-14 Solve the system $\begin{matrix} 3x-2y=-1 \\ x-3y=3 \end{matrix}$ by finding the inverse of the coefficient matrix.

Solution:

First - Write the given system in the form of $AX = B$, i.e., $\begin{bmatrix} 3 & -2 \\ 1 & -3 \end{bmatrix}\begin{bmatrix} x \\ y \end{bmatrix} = \begin{bmatrix} -1 \\ 3 \end{bmatrix}$

Second - Find determinant of the coefficient matrix A. Let's expand about the first row

$$\boxed{\delta(A)} = \boxed{a_{11}A_{11} + a_{12}A_{12}} = \boxed{3A_{11} - 2A_{12}} = \boxed{3\cdot(-1)^{1+1}M_{11} - 2\cdot(-1)^{1+2}M_{12}} = \boxed{(3\cdot -3)+(-2\cdot -1)} = \boxed{-9+2} = \boxed{-7}$$

Third - Find inverse of the A matrix. Note that

$\boxed{A_{11} = (-1)^{1+1}M_{11} = -3}$ $\boxed{A_{12} = (-1)^{1+2}M_{12} = -1}$ $\boxed{A_{21} = (-1)^{2+1}M_{21} = 2}$ and $\boxed{A_{22} = (-1)^{22}M_{22} = 3}$

$$\text{thus } \boxed{C = \begin{bmatrix} A_{11} & A_{12} \\ A_{21} & A_{22} \end{bmatrix} = \begin{bmatrix} -3 & -1 \\ 2 & 3 \end{bmatrix}} \text{ and } \boxed{C^t = \begin{bmatrix} -3 & 2 \\ -1 & 3 \end{bmatrix}} \text{ therefore}$$

$$\boxed{A^{-1}} = \boxed{\frac{1}{\delta(A)}C^t} = \boxed{\frac{1}{-7}\cdot\begin{bmatrix} -3 & 2 \\ -1 & 3 \end{bmatrix}} = \boxed{\begin{bmatrix} \frac{3}{7} & -\frac{2}{7} \\ \frac{1}{7} & -\frac{3}{7} \end{bmatrix}}$$

Check the answer by multiplying the A matrix by A^{-1}.

$$\boxed{A \times A^{-1}} = \boxed{\begin{bmatrix} 3 & -2 \\ 1 & -3 \end{bmatrix} \times \begin{bmatrix} \frac{3}{7} & -\frac{2}{7} \\ \frac{1}{7} & -\frac{3}{7} \end{bmatrix}} = \boxed{\begin{bmatrix} \frac{9}{7}-\frac{2}{7} & -\frac{6}{7}+\frac{6}{7} \\ \frac{3}{7}-\frac{3}{7} & -\frac{2}{7}+\frac{9}{7} \end{bmatrix}} = \boxed{\begin{bmatrix} \frac{9-2}{7} & \frac{-6+6}{7} \\ \frac{3-3}{7} & \frac{-2+9}{7} \end{bmatrix}} = \boxed{\begin{bmatrix} \frac{7}{7} & \frac{0}{7} \\ \frac{0}{7} & \frac{7}{7} \end{bmatrix}} = \boxed{\begin{bmatrix} 1 & 0 \\ 0 & 1 \end{bmatrix}}$$

Since $A \times A^{-1}$ is equal to the identity matrix A^{-1} is computed correctly.

Fourth - Use $X = A^{-1}B$ to solve for x and y.

$$\boxed{\begin{bmatrix} x \\ y \end{bmatrix}} = \boxed{\begin{bmatrix} \frac{3}{7} & -\frac{2}{7} \\ \frac{1}{7} & -\frac{3}{7} \end{bmatrix}\begin{bmatrix} -1 \\ 3 \end{bmatrix}} = \boxed{\begin{bmatrix} -\frac{3}{7}-\frac{6}{7} \\ -\frac{1}{7}-\frac{9}{7} \end{bmatrix}} = \boxed{\begin{bmatrix} -\frac{9}{7} \\ -\frac{10}{7} \end{bmatrix}}$$

Therefore, $\mathbf{x = -\frac{9}{7}}$ and $\mathbf{y = -\frac{10}{7}}$ and the solution set is $\left\{\left(-\frac{9}{7}, -\frac{10}{7}\right)\right\}$

Fifth - Check the answers by substituting the x and y values into one of the original equations. Let's substitute the x and y values into $3x - 2y = -1$.

$$\left(3\times -\frac{9}{7}\right) + \left(-2\times -\frac{10}{7}\right) \stackrel{?}{=} -1 \; ; \; -\frac{27}{7}+\frac{20}{7} \stackrel{?}{=} -1 \; ; \; \frac{-27+20}{7} \stackrel{?}{=} -1 \; ; \; -\frac{7}{7} \stackrel{?}{=} -1 \; ; \; -1 = -1$$

Example 3.5-15 Solve the system $\begin{matrix} x+3y-3z=-1 \\ 2x-3y+2z=3 \\ x+2y-z=-2 \end{matrix}$ by finding the inverse of the coefficient matrix.

Solution:

First - Write the given system in the form of $AX = B$, i.e., $\begin{bmatrix} 1 & 3 & -3 \\ 2 & -3 & 2 \\ 1 & 2 & -1 \end{bmatrix}\begin{bmatrix} x \\ y \\ z \end{bmatrix} = \begin{bmatrix} -1 \\ 3 \\ -2 \end{bmatrix}$

Second - Find determinant of the coefficient matrix A. Let's expand about the first row

$$\boxed{\delta(A)} = \boxed{a_{11}A_{11} + a_{12}A_{12} + a_{13}A_{13}} = \boxed{A_{11} + 3A_{12} - 3A_{13}} = \boxed{(-1)^{1+1}M_{11} + 3\cdot(-1)^{1+2}M_{12} - 3\cdot(-1)^{1+3}M_{13}}$$

$$= \boxed{M_{11} - 3M_{12} - 3M_{13}} = \boxed{\begin{vmatrix} -3 & 2 \\ 2 & -1 \end{vmatrix} - 3\begin{vmatrix} 2 & 2 \\ 1 & -1 \end{vmatrix} - 3\begin{vmatrix} 2 & -3 \\ 1 & 2 \end{vmatrix}} = \boxed{(3-4) - 3(-2-2) - 3(4+3)} = \boxed{-1+12-21} = \boxed{-10}$$

Third - Find inverse of the A matrix. Note that

$$\boxed{A_{11} = (-1)^{1+1}M_{11} = M_{11}} \quad \boxed{A_{12} = (-1)^{1+2}M_{12} = -M_{12}} \quad \boxed{A_{13} = (-1)^{1+3}M_{13} = M_{13}}$$

$$\boxed{A_{21} = (-1)^{2+1}M_{21} = -M_{21}} \quad \boxed{A_{22} = (-1)^{2+2}M_{22} = M_{22}} \quad \boxed{A_{23} = (-1)^{2+3}M_{23} = -M_{23}}$$

$$\boxed{A_{31} = (-1)^{3+1}M_{31} = M_{31}} \quad \boxed{A_{32} = (-1)^{3+2}M_{32} = -M_{32}} \quad \boxed{A_{33} = (-1)^{3+3}M_{33} = M_{33}}$$

$$\boxed{A_{11} = \begin{vmatrix} -3 & 2 \\ 2 & -1 \end{vmatrix} = 3-4 = -1} \quad \boxed{A_{12} = -\begin{vmatrix} 2 & 2 \\ 1 & -1 \end{vmatrix} = -(-2-2) = 4} \quad \boxed{A_{13} = \begin{vmatrix} 2 & -3 \\ 1 & 2 \end{vmatrix} = 4+3 = 7}$$

$$\boxed{A_{21} = -\begin{vmatrix} 3 & -3 \\ 2 & -1 \end{vmatrix} = -(-3+6) = -3} \quad \boxed{A_{22} = \begin{vmatrix} 1 & -3 \\ 1 & -1 \end{vmatrix} = -1+3 = 2} \quad \boxed{A_{23} = -\begin{vmatrix} 1 & 3 \\ 1 & 2 \end{vmatrix} = -(2-3) = 1}$$

$$\boxed{A_{31} = \begin{vmatrix} 3 & -3 \\ -3 & 2 \end{vmatrix} = 6-9 = -3} \quad \boxed{A_{32} = -\begin{vmatrix} 1 & -3 \\ 2 & 2 \end{vmatrix} = -(2+6) = -8} \quad \boxed{A_{33} = \begin{vmatrix} 1 & 3 \\ 2 & -3 \end{vmatrix} = -3-6 = -9}$$

thus $\boxed{C = \begin{bmatrix} A_{11} & A_{12} & A_{13} \\ A_{21} & A_{22} & A_{23} \\ A_{31} & A_{32} & A_{33} \end{bmatrix} = \begin{bmatrix} -1 & 4 & 7 \\ -3 & 2 & 1 \\ -3 & -8 & -9 \end{bmatrix}}$ and $\boxed{C^t = \begin{bmatrix} -1 & -3 & -3 \\ 4 & 2 & -8 \\ 7 & 1 & -9 \end{bmatrix}}$ therefore

$$\boxed{A^{-1}} = \boxed{\frac{1}{\delta(A)}C^t} = \boxed{\frac{1}{-10}\begin{bmatrix} -1 & -3 & -3 \\ 4 & 2 & -8 \\ 7 & 1 & -9 \end{bmatrix}} = \boxed{\begin{bmatrix} \frac{1}{10} & \frac{3}{10} & \frac{3}{10} \\ -\frac{2}{5} & -\frac{1}{5} & \frac{4}{5} \\ -\frac{7}{10} & -\frac{1}{10} & \frac{9}{10} \end{bmatrix}}$$

Check the answer by multiplying the A matrix by A^{-1}.

$$\boxed{A \times A^{-1}} = \boxed{\begin{bmatrix} 1 & 3 & -3 \\ 2 & -3 & 2 \\ 1 & 2 & -1 \end{bmatrix} \times \begin{bmatrix} \frac{1}{10} & \frac{3}{10} & \frac{3}{10} \\ -\frac{2}{5} & -\frac{1}{5} & \frac{4}{5} \\ -\frac{7}{10} & -\frac{1}{10} & \frac{9}{10} \end{bmatrix}} = \boxed{\begin{bmatrix} \frac{1}{10} - \frac{6}{5} + \frac{21}{10} & \frac{3}{10} - \frac{3}{5} + \frac{3}{10} & \frac{3}{10} + \frac{12}{5} - \frac{27}{10} \\ \frac{2}{10} + \frac{6}{5} - \frac{14}{10} & \frac{6}{10} + \frac{3}{5} - \frac{2}{10} & \frac{6}{10} - \frac{12}{5} + \frac{18}{10} \\ \frac{1}{10} - \frac{4}{5} + \frac{7}{10} & \frac{3}{10} - \frac{2}{5} + \frac{1}{10} & \frac{3}{10} + \frac{8}{5} - \frac{9}{10} \end{bmatrix}}$$

$$= \begin{bmatrix} \frac{1-12+21}{10} & \frac{3-6+3}{10} & \frac{3+24-27}{10} \\ \frac{2+12-14}{10} & \frac{6+6-2}{10} & \frac{6-24+18}{10} \\ \frac{1-8+7}{10} & \frac{3-4+1}{10} & \frac{3+16-9}{10} \end{bmatrix} = \begin{bmatrix} \frac{22-12}{10} & \frac{6-6}{10} & \frac{27-27}{10} \\ \frac{14-14}{10} & \frac{12-2}{10} & \frac{24-24}{10} \\ \frac{8-8}{10} & \frac{4-4}{10} & \frac{19-9}{10} \end{bmatrix} = \begin{bmatrix} \frac{10}{10} & \frac{0}{10} & \frac{0}{10} \\ \frac{0}{10} & \frac{10}{10} & \frac{0}{10} \\ \frac{0}{10} & \frac{0}{10} & \frac{10}{10} \end{bmatrix} = \begin{bmatrix} 1 & 0 & 0 \\ 0 & 1 & 0 \\ 0 & 0 & 1 \end{bmatrix}$$

Since $A \times A^{-1}$ is equal to the identity matrix A^{-1} is computed correctly.

Fourth - Use $X = A^{-1}B$ to solve for x, y, and z.

$$\begin{bmatrix} x \\ y \\ z \end{bmatrix} = \begin{bmatrix} \frac{1}{10} & \frac{3}{10} & \frac{3}{10} \\ -\frac{2}{5} & -\frac{1}{5} & \frac{4}{5} \\ -\frac{7}{10} & -\frac{1}{10} & \frac{9}{10} \end{bmatrix} \begin{bmatrix} -1 \\ 3 \\ -2 \end{bmatrix} = \begin{bmatrix} -\frac{1}{10}+\frac{9}{10}-\frac{6}{10} \\ \frac{2}{5}-\frac{3}{5}-\frac{8}{5} \\ \frac{7}{10}-\frac{3}{10}-\frac{18}{10} \end{bmatrix} = \begin{bmatrix} \frac{2}{10} \\ -\frac{9}{5} \\ -\frac{14}{10} \end{bmatrix} = \begin{bmatrix} \frac{1}{5} \\ -\frac{9}{5} \\ -\frac{7}{5} \end{bmatrix}$$

Therefore, $x = \frac{1}{5}$, $y = -\frac{9}{5}$, $z = -\frac{7}{5}$ and the solution set is $\left\{\left(\frac{1}{5}, -\frac{9}{5}, -\frac{7}{5}\right)\right\}$

Fifth - Check the answers by substituting the x, y, and z values into one of the original equation. Let's substitute the x, y, and z values into $x + 2y - z = -2$.

$$\frac{1}{5} + \left(2 \times -\frac{9}{5}\right) + \frac{7}{5} \stackrel{?}{=} -2 \; ; \; \frac{1}{5} - \frac{18}{5} + \frac{7}{5} \stackrel{?}{=} -2 \; ; \; \frac{1-18+7}{5} \stackrel{?}{=} -2 \; ; \; -\frac{10}{5} \stackrel{?}{=} -2 \; ; \; -\frac{2}{1} \stackrel{?}{=} -2 \; ; \; -2 = -2$$

Example 3.5-16 Solve the system $\begin{matrix} x + 2z = 1 \\ 3x + y + 6z = -2 \\ 4x + 8z = 0 \end{matrix}$ by finding the inverse of the coefficient matrix.

Solution:

First - Write the given system in the form of $AX = B$, i.e., $\begin{bmatrix} 1 & 0 & 2 \\ 3 & 1 & 6 \\ 4 & 0 & 8 \end{bmatrix} \begin{bmatrix} x \\ y \\ z \end{bmatrix} = \begin{bmatrix} 1 \\ -2 \\ 0 \end{bmatrix}$

Second - Find determinant of the coefficient matrix A. Let's expand about the second column.

$$\boxed{\delta(A)} = \boxed{a_{12}A_{12} + a_{22}A_{22} + a_{32}A_{32}} = \boxed{0 \times A_{12} + 1 \times A_{22} + 0 \times A_{32}} = \boxed{A_{22}} = \boxed{(-1)^{2+2} M_{22}} = \boxed{M_{22}} = \begin{vmatrix} 1 & 2 \\ 4 & 8 \end{vmatrix}$$

$$= \boxed{(1 \times 8) - (2 \times 4)} = \boxed{8 - 8} = \boxed{0}$$

Since $\delta(A) = 0$ the inverse of the A matrix does not exist. The linear system does not have a solution and is an inconsistent system. Note that we could have immediately stated that $\delta(A) = 0$ by observing that in the determinant matrix the elements of the third row are the same as the elements in the first row, except that they are multiplied by 4 (see the properties of determinant).

Section 3.5 Case III Practice Problems - Solving Linear Systems Using the Inverse Matrices Method

1. Find the solution set of the given systems by using the inverse matrices method.

a. $\begin{matrix} x+2y=3 \\ -2x+5y=-1 \end{matrix}$ b. $\begin{matrix} x+3y=0 \\ -x+4y=-1 \end{matrix}$ c. $\begin{matrix} 3x+4y=-2 \\ 6x+8y=10 \end{matrix}$ d. $\begin{matrix} 3x-2y=-3 \\ x-2y=0 \end{matrix}$

e. $\begin{matrix} 2x+y=1 \\ 4x+2y=8 \end{matrix}$ f. $\begin{matrix} 2x-y=-5 \\ 3x-4y=-4 \end{matrix}$ g. $\begin{matrix} x+y=2 \\ 2x-z=-1 \\ 2y+2z=3 \end{matrix}$ h. $\begin{matrix} x-y+3z=2 \\ x-z=-3 \\ 2x-2y+6z=-1 \end{matrix}$

i. $\begin{matrix} x+3y-z=-2 \\ -x+2y+3z=1 \\ x+y-2z=0 \end{matrix}$ j. $\begin{matrix} x+y=0 \\ 2x-z=1 \\ 4x+4y=-1 \end{matrix}$

2. Use the result of exercise number 1-g above to find the solution set for the following system of linear equations.

a. $\begin{matrix} x+y=-3 \\ 2x-z=-4 \\ 2y+2z=1 \end{matrix}$ b. $\begin{matrix} x+y=0 \\ 2x-z=-1 \\ 2y+2z=1 \end{matrix}$

3. Use the result of exercise number 1-i above to find the solution set for the following system of linear equations.

a. $\begin{matrix} x+3y-z=1 \\ -x+2y+3z=3 \\ x+y-2z=-1 \end{matrix}$ b. $\begin{matrix} x+3y-z=-1 \\ -x+2y+3z=0 \\ x+y-2z=2 \end{matrix}$

Case IV Solving Linear Systems Using Cramer's Rule

Cramer's rule is a formula which represent the solution to system of linear equations in determinant form. For example,

a. The solution to a 2×2 linear system $\begin{matrix} a_{11}x + a_{12}y = b_1 \\ a_{21}x + a_{22}y = b_2 \end{matrix}$ using Cramer's rule is given by

$$x = \frac{|A_1|}{|A|} = \frac{\begin{vmatrix} b_1 & a_{12} \\ b_2 & a_{22} \end{vmatrix}}{\begin{vmatrix} a_{11} & a_{12} \\ a_{21} & a_{22} \end{vmatrix}} \qquad y = \frac{|A_2|}{|A|} = \frac{\begin{vmatrix} a_{11} & b_1 \\ a_{21} & b_2 \end{vmatrix}}{\begin{vmatrix} a_{11} & a_{12} \\ a_{21} & a_{22} \end{vmatrix}}$$

Note that $|A|$ denotes the determinant of the coefficient matrix. $|A_1|$ denotes the determinant of the A_1 matrix, where A_1 is obtained by replacing the entries in the first column of A with the entries in the right hand side column of the augmented matrix. $|A_2|$ denotes the determinant of the A_2 matrix, where A_2 is obtained by replacing the entries in the second column of A with the entries in the right hand side of the augmented matrix. Similarly,

b. The solution to a 3×3 linear system $\begin{matrix} a_{11}x + a_{12}y + a_{13}z = b_1 \\ a_{21}x + a_{22}y + a_{23}z = b_2 \\ a_{31}x + a_{32}y + a_{33}z = b_3 \end{matrix}$ using Cramer's rule is given by

$$x = \frac{|A_1|}{|A|} = \frac{\begin{vmatrix} b_1 & a_{12} & a_{13} \\ b_2 & a_{22} & a_{23} \\ b_3 & a_{32} & a_{33} \end{vmatrix}}{\begin{vmatrix} a_{11} & a_{12} & a_{13} \\ a_{21} & a_{22} & a_{23} \\ a_{31} & a_{32} & a_{33} \end{vmatrix}} \qquad y = \frac{|A_2|}{|A|} = \frac{\begin{vmatrix} a_{11} & b_1 & a_{13} \\ a_{21} & b_2 & a_{23} \\ a_{31} & b_3 & a_{33} \end{vmatrix}}{\begin{vmatrix} a_{11} & a_{12} & a_{13} \\ a_{21} & a_{22} & a_{23} \\ a_{31} & a_{32} & a_{33} \end{vmatrix}} \qquad z = \frac{|A_3|}{|A|} = \frac{\begin{vmatrix} a_{11} & a_{12} & b_1 \\ a_{21} & a_{22} & b_2 \\ a_{31} & a_{32} & b_3 \end{vmatrix}}{\begin{vmatrix} a_{11} & a_{12} & a_{13} \\ a_{21} & a_{22} & a_{23} \\ a_{31} & a_{32} & a_{33} \end{vmatrix}}$$

Linear systems can be solved using Cramer's rule by applying the following steps:

Step 1 Write the coefficient and the augmented matrix by first ensuring each equation of the linear system is in the form of $ax + by + cz + dw + \ldots = k$.

Step 2 Find the determinant of the coefficient matrix A. If $\delta(A) \neq 0$ go to Step 3. If $\delta(A) = 0$ the linear system is either a dependent or an inconsistent system.

Step 3 Find the determinant of the A_1 matrix after replacing the entries in the first column of the coefficient matrix A with the entries in the right hand side column of the augmented matrix.

Step 4 Find the determinant of the A_2 matrix after replacing the entries in the second column of the coefficient matrix A with the entries in the right hand side column of the augmented matrix. Repeat Step 4 by replacing third, fourth, fifth, etc. columns of the coefficient matrix with the entries in the right hand side of the augmented matrix.

Step 5 Solve for the unknowns x, y, z, w, etc. by dividing the determinants obtained in Steps 3 and 4 by the determinant of the coefficient matrix obtained in Step 2, i.e.,

$$x = \frac{|A_1|}{|A|} \qquad y = \frac{|A_2|}{|A|} \qquad z = \frac{|A_3|}{|A|} \qquad w = \frac{|A_4|}{|A|}, \text{ etc.}$$

Therefore, the solution set is equal to $\{(x, y, z, w, \ldots)\}$.

Step 6 Check the answers by substituting the solution set into one of the original equations.

The following examples show the steps in solving linear systems using Cramer's rule:

Example 3.5-17 Use Cramer's rule to solve the linear system $\begin{array}{l} -x+2y=1 \\ x-4y=0 \end{array}$

Solution:

Step 1 The coefficient matrix is equal to $A = \begin{bmatrix} -1 & 2 \\ 1 & -4 \end{bmatrix}$ and the augmented matrix is equal to $\begin{bmatrix} -1 & 2 & \vdots & 1 \\ 1 & -4 & \vdots & 0 \end{bmatrix}$

Step 2 Let's find $\delta(A)$ by expanding about the first row. Note that $a_{11} = -1$, $a_{12} = 2$.

$$\boxed{\delta(A)} = \boxed{|A|} = \boxed{a_{11}A_{11} + a_{12}A_{12}} = \boxed{-A_{11} + 2A_{12}} = \boxed{-(-1)^{1+1} M_{11} + 2 \cdot (-1)^{1+2} M_{12}}$$

$$= \boxed{(-1 \cdot -4) + 2 \cdot (-1 \cdot 1)} = \boxed{4-2} = \boxed{2}$$

Since determinant is not equal to zero we can proceed to the next step.

Step 3 1. Replace the entries in the first column of the coefficient matrix with the entries in the right hand side of the augmented matrix to obtain matrix $A_1 = \begin{bmatrix} 1 & 2 \\ 0 & -4 \end{bmatrix}$

2. Find $\delta(A_1)$ by expanding about the first column. Note that $a_{11} = 1$, $a_{21} = 0$.

$$\boxed{\delta(A_1)} = \boxed{|A_1|} = \boxed{a_{11}A_{11} + a_{21}A_{21}} = \boxed{A_{11}} = \boxed{(-1)^{1+1} M_{11}} = \boxed{1 \cdot -4} = \boxed{-4}$$

Step 4 1. Replace the entries in the second column of the coefficient matrix with the entries in the right hand side of the augmented matrix to obtain matrix $A_2 = \begin{bmatrix} -1 & 1 \\ 1 & 0 \end{bmatrix}$

2. Find $\delta(A_2)$ by expanding about the second column. Note that $a_{12} = 1$, $a_{22} = 0$.

$$\boxed{\delta(A_2)} = \boxed{|A_2|} = \boxed{a_{12}A_{12} + a_{22}A_{22}} = \boxed{A_{12}} = \boxed{(-1)^{1+2} M_{12}} = \boxed{-1 \cdot 1} = \boxed{-1}$$

Step 5 Solve for x and y using the Cramer's rule, i.e., $x = \frac{|A_1|}{|A|}$ and $y = \frac{|A_2|}{|A|}$. Therefore,

$$\boxed{x = \frac{|A_1|}{|A|}}; \boxed{x = \frac{-4}{2}}; \boxed{\mathbf{x = -2}} \quad \text{and} \quad \boxed{y = \frac{|A_2|}{|A|}}; \boxed{y = \frac{-1}{2}}; \boxed{\mathbf{y = -\frac{1}{2}}}$$

and the solution set is equal to $\left\{\left(-2, -\frac{1}{2}\right)\right\}$

Step 6 Let's check the answer by substituting the x and y values into $x - 4y = 0$.

$$-2-4\cdot\left(-\frac{1}{2}\right)\overset{?}{=}0\ ;\ -2+\frac{4}{2}\overset{?}{=}0\ ;\ -2+2\overset{?}{=}0\ ;\ 0=0$$

Example 3.5-18 Use Cramer's rule to solve the linear system $\begin{matrix} x+y=0 \\ 2x-z=1 \\ y+2z=0 \end{matrix}$

Solution:

Step 1 The coefficient matrix is equal to $A=\begin{bmatrix} 1 & 1 & 0 \\ 2 & 0 & -1 \\ 0 & 1 & 2 \end{bmatrix}$ and the augmented matrix is equal to $\begin{bmatrix} 1 & 1 & 0 & \vdots & 0 \\ 2 & 0 & -1 & \vdots & 1 \\ 0 & 1 & 2 & \vdots & 0 \end{bmatrix}$

Step 2 Let's find $\delta(A)$ by expanding about the first row.

$$\boxed{\delta(A)}=\boxed{|A|}=\boxed{a_{11}A_{11}+a_{12}A_{12}+a_{13}A_{13}}=\boxed{A_{11}+A_{12}+0\cdot A_{13}}=\boxed{(-1)^{1+1}M_{11}+(-1)^{1+2}M_{12}}$$

$$=\boxed{M_{11}-M_{12}}=\boxed{\begin{vmatrix} 0 & -1 \\ 1 & 2 \end{vmatrix}-\begin{vmatrix} 2 & -1 \\ 0 & 2 \end{vmatrix}}=\boxed{(0+1)-(4+0)}=\boxed{1-4}=\boxed{-3}$$

Since determinant is not equal to zero we can proceed to the next step.

Step 3 1. Replace the entries in the first column of the coefficient matrix with the entries in the right hand side of the augmented matrix to obtain the matrix $A_1=\begin{bmatrix} 0 & 1 & 0 \\ 1 & 0 & -1 \\ 0 & 1 & 2 \end{bmatrix}$

2. Find $\delta(A_1)$ by expanding about the first row.

$$\boxed{\delta(A_1)}=\boxed{|A_1|}=\boxed{a_{11}A_{11}+a_{12}A_{12}+a_{13}A_{13}}=\boxed{0\cdot A_{11}+A_{12}+0\cdot A_{13}}=\boxed{A_{12}}=\boxed{(-1)^{1+2}M_{12}}$$

$$=\boxed{-M_{12}}=\boxed{-\begin{vmatrix} 1 & -1 \\ 0 & 2 \end{vmatrix}}=\boxed{-[(1\times 2)-(0\times -1)]}=\boxed{-2}$$

Step 4 1. Replace the entries in the second column of the coefficient matrix with the entries in the right hand side of the augmented matrix to obtain the matrix $A_2=\begin{bmatrix} 1 & 0 & 0 \\ 2 & 1 & -1 \\ 0 & 0 & 2 \end{bmatrix}$

2. Find $\delta(A_2)$ by expanding about the first row.

$$\boxed{\delta(A_2)}=\boxed{|A_2|}=\boxed{a_{11}A_{11}+a_{12}A_{12}+a_{13}A_{13}}=\boxed{A_{11}+0\cdot A_{12}+0\cdot A_{13}}=\boxed{A_{11}}=\boxed{(-1)^{1+1}M_{11}}$$

$$=\boxed{M_{11}}=\boxed{\begin{vmatrix} 1 & -1 \\ 0 & 2 \end{vmatrix}}=\boxed{(1\times 2)-(0\times -1)}=\boxed{2}$$

Step 5 1. Replace the entries in the third column of the coefficient matrix with the entries in the right hand side of the augmented matrix to obtain the matrix

$$A_3 = \begin{bmatrix} 1 & 1 & 0 \\ 2 & 0 & 1 \\ 0 & 1 & 0 \end{bmatrix}$$

2. Find $\delta(A_3)$ by expanding about the third row.

$$\boxed{\delta(A_3)} = \boxed{|A_3|} = \boxed{a_{31}A_{31} + a_{32}A_{32} + a_{33}A_{33}} = \boxed{0 \cdot A_{31} + A_{32} + 0 \cdot A_{33}} = \boxed{A_{32}} = \boxed{(-1)^{3+2} M_{32}}$$

$$= \boxed{-M_{32}} = \boxed{-\begin{vmatrix} 1 & 0 \\ 2 & 1 \end{vmatrix}} = \boxed{-[(1 \times 1) - (0 \times 2)]} = \boxed{-1}$$

Step 6 Solve for x and y using Cramer's rule, i.e., $x = \frac{|A_1|}{|A|}$, $y = \frac{|A_2|}{|A|}$, and $z = \frac{|A_3|}{|A|}$.

Therefore, $\boxed{x = \frac{|A_1|}{|A|}}$; $\boxed{x = \frac{\mathbf{2}}{\mathbf{3}}}$ $\boxed{y = \frac{|A_2|}{|A|}}$; $\boxed{y = -\frac{\mathbf{2}}{\mathbf{3}}}$ $\boxed{z = \frac{|A_3|}{|A|}}$; $\boxed{z = \frac{\mathbf{1}}{\mathbf{3}}}$

and the solution set is equal to $\left\{\left(\frac{\mathbf{2}}{\mathbf{3}}, -\frac{\mathbf{2}}{\mathbf{3}}, \frac{\mathbf{1}}{\mathbf{3}}\right)\right\}$

Step 7 Let's check the answer by substituting the x, y, and z values into $2x - z = 1$.

$$2 \times \frac{2}{3} - \frac{1}{3} \stackrel{?}{=} 1 \; ; \; \frac{4}{3} - \frac{1}{3} \stackrel{?}{=} 1 \; ; \; \frac{4-1}{3} \stackrel{?}{=} 1 \; ; \; \frac{3}{3} \stackrel{?}{=} 1 \; ; \; 1 = 1$$

Section 3.5 Case IV Practice Problems - Solving Linear Systems Using Cramer's Rule

1. Find the solution set to each of the following linear systems using Cramer's rule. Note that Problems 1-e through 1-i are identical to the exercises 1-e through 1-i in Section 3.5 Case III.

a. $\frac{1}{3}x - \frac{1}{2}y = 1$, $\frac{1}{2}x - \frac{1}{3}y = 0$

b. $x - 4y = 1$, $y = 2$

c. $x + y = -2a$, $x - y = 2b$

d. $x - 3y = 1$, $2x - y = 3$

e. $2x + y = 1$, $4x + 2y = 8$

f. $2x - y = -5$, $3x - 4y = -4$

g. $x + y = 2$, $2x - z = -1$, $2y + 2z = 3$

h. $x - y + 3z = 2$, $x - z = -3$, $2x - 2y + 6z = -1$

i. $x + 3y - z = -2$, $-x + 2y + 3z = 1$, $x + y - 2z = 0$

2. Use the result of exercise number 1-g above to find the solution set for the following linear equations. Note that the answers should agree with practice problems 2a and 2b in Section 3.5 Case III.

a. $x + y = -3$, $2x - z = -4$, $2y + 2z = 1$

b. $x + y = 0$, $2x - z = -1$, $2y + 2z = 1$

Case V Solving Linear Systems Using the Gaussian Elimination Method

The Gaussian Elimination method is used on any system of n linear equations with n unknowns that has a unique solution. The objective in using the Gaussian Elimination method is to transfer the augmented matrix into a matrix for which $a_{ij} = 0$ when $i \rangle j$. To transfer an augmented matrix into a matrix with 1's in the main diagonal entries and zeros in the lower triangle we use the following elementary row operations as many times as required:

1. Interchange any two rows.
2. Multiply each element of a row by a constant k, where $k \neq 0$.
3. Replace elements of a row by the sum of itself plus k times the corresponding elements of another row.

In general, the transferred augmented matrix for a system with four linear equations and four unknowns has the following appearance.

$$\begin{bmatrix} 1 & * & * & * & \vdots & * \\ 0 & 1 & * & * & \vdots & * \\ 0 & 0 & 1 & * & \vdots & * \\ 0 & 0 & 0 & 1 & \vdots & * \end{bmatrix}$$

The following show the steps in solving linear systems using the Gaussian Elimination method:

Step 1 Write the linear system in the matrix form $AX = B$.

Step 2 Write the augmented matrix.

Step 3 Transfer the augmented matrix into a matrix with 1's in the main diagonal entries and zeros in the lower triangle using the elementary row operations.

Step 4 Change the matrix obtained in Step 3 into a system of linear equations. Solve for the variables using the back substitution method.

Step 5 Check the answers by substituting the solutions back into one of the original equations.

The following examples show the steps for solving linear systems using the Gaussian Elimination method:

Example 3.5-19 Solve the following linear systems by applying the Gaussian Elimination method to the augmented matrix.

a. $\begin{matrix} 2x + 3y = 6 \\ x - 4y = -2 \end{matrix}$

b. $\begin{matrix} 2x - 3y + z = -1 \\ 3x + 2z = 0 \\ x - 2y = 1 \end{matrix}$

c. $\begin{matrix} -x + 4z + 2w = 0 \\ 3x - 2y = 1 \\ y - 3w = -1 \\ 2x + 3y - 2z + w = -2 \end{matrix}$

Solution (a):

First - Write the linear system in the form of $AX = B$.

$$\begin{bmatrix} 2 & 3 \\ 1 & -4 \end{bmatrix} \begin{bmatrix} x \\ y \end{bmatrix} = \begin{bmatrix} 6 \\ -2 \end{bmatrix}$$

Second - Write the augmented matrix.

$$\begin{bmatrix} 2 & 3 & \vdots & 6 \\ 1 & -4 & \vdots & -2 \end{bmatrix}$$

Third - Perform the elementary row operations.

1. Replace the second row with the first row. $\begin{bmatrix} 1 & -4 & \vdots & -2 \\ 2 & 3 & \vdots & 6 \end{bmatrix}$

2. Multiply each elements of the first row by -2 and add the result to each element of the second row. $\begin{bmatrix} 1 & -4 & \vdots & -2 \\ 2-2 & 3+8 & \vdots & 6+4 \end{bmatrix} = \begin{bmatrix} 1 & -4 & \vdots & -2 \\ 0 & 11 & \vdots & 10 \end{bmatrix}$

3. Divide the second row by 11. $\begin{bmatrix} 1 & -4 & \vdots & -2 \\ 0 & \frac{11}{11} & \vdots & \frac{10}{11} \end{bmatrix} = \begin{bmatrix} 1 & -4 & \vdots & -2 \\ 0 & 1 & \vdots & \frac{10}{11} \end{bmatrix}$

Note that the augmented matrix has 1's in its main diagonal entries and zero in the lower triangle.

Fourth - Write the augmented matrix in its equivalent linear system form.

$$x-4y=-2$$
$$y=\frac{10}{11}$$

Since $y=\frac{10}{11}$, we can solve for x by back substitution. $x-4y=-2$; $x+\left(-4\times\frac{10}{11}\right)=-2$; $x-\frac{40}{11}=-2$; $x=\frac{40}{11}-2$; $x=\frac{40-22}{11}$; $x=\frac{18}{11}$. Therefore, $\mathbf{x=\frac{18}{11}}$ and $\mathbf{y=\frac{10}{11}}$ and the solution set is $\left\{\left(\mathbf{\frac{18}{11}}, \mathbf{\frac{10}{11}}\right)\right\}$

Fifth - Check the answers by substituting the x and y values into one of the original equations. Let's substitute the x and y values into $2x+3y=6$.

$\left(2\times\frac{18}{11}\right)+\left(3\times\frac{10}{11}\right)\overset{?}{=}6$; $\frac{36}{11}+\frac{30}{11}\overset{?}{=}6$; $\frac{36+30}{11}\overset{?}{=}6$; $\frac{66}{11}\overset{?}{=}6$; $6=6$

Solution (b):

First - Write the linear system in the form of $AX=B$.

$$\begin{bmatrix} 2 & -3 & 1 \\ 3 & 0 & 2 \\ 1 & -2 & 0 \end{bmatrix}\begin{bmatrix} x \\ y \\ z \end{bmatrix} = \begin{bmatrix} -1 \\ 0 \\ 1 \end{bmatrix}$$

Second - Write the augmented matrix.

$$\begin{bmatrix} 2 & -3 & 1 & \vdots & -1 \\ 3 & 0 & 2 & \vdots & 0 \\ 1 & -2 & 0 & \vdots & 1 \end{bmatrix}$$

Third - Perform the elementary row operations.

1. Replace the third row with the first row. $\begin{bmatrix} 1 & -2 & 0 & \vdots & 1 \\ 3 & 0 & 2 & \vdots & 0 \\ 2 & -3 & 1 & \vdots & -1 \end{bmatrix}$

2. Multiply each elements of the first row by -3 and add the result to each element of the second row. $\begin{bmatrix} 1 & -2 & 0 & \vdots & 1 \\ 3-3 & 0+6 & 2+0 & \vdots & 0-3 \\ 2 & -3 & 1 & \vdots & -1 \end{bmatrix} = \begin{bmatrix} 1 & -2 & 0 & \vdots & 1 \\ 0 & 6 & 2 & \vdots & -3 \\ 2 & -3 & 1 & \vdots & -1 \end{bmatrix}$

3. Divide the second row by 6. $\begin{bmatrix} 1 & -2 & 0 & \vdots & 1 \\ 0 & \frac{6}{6} & \frac{2}{6} & \vdots & \frac{-3}{6} \\ 2 & -3 & 1 & \vdots & -1 \end{bmatrix} = \begin{bmatrix} 1 & -2 & 0 & \vdots & 1 \\ 0 & 1 & \frac{1}{3} & \vdots & -\frac{1}{2} \\ 2 & -3 & 1 & \vdots & -1 \end{bmatrix}$

4. Multiply each elements of the first row by -2 and add the result to each element of the third row. $\begin{bmatrix} 1 & -2 & 0 & \vdots & 1 \\ 0 & 1 & \frac{1}{3} & \vdots & -\frac{1}{2} \\ 2-2 & -3+4 & 1+0 & \vdots & -1-2 \end{bmatrix} = \begin{bmatrix} 1 & -2 & 0 & \vdots & 1 \\ 0 & 1 & \frac{1}{3} & \vdots & -\frac{1}{2} \\ 0 & 1 & 1 & \vdots & -3 \end{bmatrix}$

5. Multiply each element of the second row by -1 and add the result to each element of the third row. $\begin{bmatrix} 1 & -2 & 0 & \vdots & 1 \\ 0 & 1 & \frac{1}{3} & \vdots & -\frac{1}{2} \\ 0 & 1-1 & 1-\frac{1}{3} & \vdots & -3+\frac{1}{2} \end{bmatrix} = \begin{bmatrix} 1 & -2 & 0 & \vdots & 1 \\ 0 & 1 & \frac{1}{3} & \vdots & -\frac{1}{2} \\ 0 & 0 & \frac{2}{3} & \vdots & -\frac{5}{2} \end{bmatrix}$

6. Multiply each element of the third row by $\frac{3}{2}$

$$\begin{bmatrix} 1 & -2 & 0 & \vdots & 1 \\ 0 & 1 & \frac{1}{3} & \vdots & -\frac{1}{2} \\ 0 & 0 & \frac{2}{3}\times\frac{3}{2} & \vdots & -\frac{5}{2}\times\frac{3}{2} \end{bmatrix} = \begin{bmatrix} 1 & -2 & 0 & \vdots & 1 \\ 0 & 1 & \frac{1}{3} & \vdots & -\frac{1}{2} \\ 0 & 0 & 1 & \vdots & -\frac{15}{4} \end{bmatrix}$$

Note that the augmented matrix has 1's in its main diagonal entries and zeros in the lower triangle.

Fourth - Write the augmented matrix in its equivalent linear system form.

$$x-2y=1$$
$$y+\frac{1}{3}z=-\frac{1}{2}$$
$$z=-\frac{15}{4}$$

Since $z=-\frac{15}{4}$, we can solve for x and y by back substitution.

$y+\frac{1}{3}z=-\frac{1}{2}$; $y+\left(\frac{1}{3}\times-\frac{15}{4}\right)=-\frac{1}{2}$; $y-\frac{15}{12}=-\frac{1}{2}$; $y=\frac{15}{12}-\frac{1}{2}$; $y=\frac{(15\times2)-(1\times2)}{12\times2}$; $y=\frac{30-2}{24}$; $y=\frac{3}{4}$

$x-2y=1$; $x-2\times\frac{3}{4}=1$; $x-\frac{6}{4}=1$; $x=1+\frac{6}{4}$; $x=\frac{4+6}{4}$; $x=\frac{10}{4}$; $x=\frac{5}{2}$

Therefore, $\mathbf{x=\frac{5}{2}}$, $\mathbf{y=\frac{3}{4}}$, and $\mathbf{z=-\frac{15}{4}}$ and the solution set is $\left\{\left(\frac{5}{2},\frac{3}{4},-\frac{15}{4}\right)\right\}$

Fifth - Check the answers by substituting the x, y, and z values into one of the original equations. Let's substitute the x, y, and z values into $2x-3y+z=-1$.

$\left(2\times\frac{5}{2}\right)+\left(-3\times\frac{3}{4}\right)-\frac{15}{4}\stackrel{?}{=}-1$; $\frac{10}{2}-\frac{9}{4}-\frac{15}{4}\stackrel{?}{=}-1$; $\frac{10}{2}+\left(\frac{-9-15}{4}\right)\stackrel{?}{=}-1$; $\frac{10}{2}-\frac{24}{4}\stackrel{?}{=}-1$; $\frac{10}{2}-\frac{12}{2}\stackrel{?}{=}-1$

; $\frac{10-12}{2}\stackrel{?}{=}-1$; $-\frac{2}{2}\stackrel{?}{=}-1$; $-1=-1$

Solution (c):

First - Write the linear system in the form of $AX = B$.

$$\begin{bmatrix} -1 & 0 & 4 & 2 \\ 3 & -2 & 0 & 0 \\ 0 & 1 & 0 & -3 \\ 2 & 3 & -2 & 1 \end{bmatrix} \begin{bmatrix} x \\ y \\ z \\ w \end{bmatrix} = \begin{bmatrix} 0 \\ 1 \\ -1 \\ -2 \end{bmatrix}$$

Second - Write the augmented matrix.

$$\begin{bmatrix} -1 & 0 & 4 & 2 & \vdots & 0 \\ 3 & -2 & 0 & 0 & \vdots & 1 \\ 0 & 1 & 0 & -3 & \vdots & -1 \\ 2 & 3 & -2 & 1 & \vdots & -2 \end{bmatrix}$$

Third - Perform the elementary row operations.

1. Divide the elements of the first row by -1.

$$\begin{bmatrix} 1 & 0 & -4 & -2 & \vdots & 0 \\ 3 & -2 & 0 & 0 & \vdots & 1 \\ 0 & 1 & 0 & -3 & \vdots & -1 \\ 2 & 3 & -2 & 1 & \vdots & -2 \end{bmatrix}$$

2. Multiply each elements of the first row by -3 and add the result to each element of the second row.

$$\begin{bmatrix} 1 & 0 & -4 & -2 & \vdots & 0 \\ 3-3 & -2+0 & 0+12 & 0+6 & \vdots & 1+0 \\ 0 & 1 & 0 & -3 & \vdots & -1 \\ 2 & 3 & -2 & 1 & \vdots & -2 \end{bmatrix} = \begin{bmatrix} 1 & 0 & -4 & -2 & \vdots & 0 \\ 0 & -2 & 12 & 6 & \vdots & 1 \\ 0 & 1 & 0 & -3 & \vdots & -1 \\ 2 & 3 & -2 & 1 & \vdots & -2 \end{bmatrix}$$

3. Multiply each elements of the first row by -2 and add the result to each element of the fourth row.

$$\begin{bmatrix} 1 & 0 & -4 & -2 & \vdots & 0 \\ 0 & -2 & 12 & 6 & \vdots & 1 \\ 0 & 1 & 0 & -3 & \vdots & -1 \\ 2-2 & 3+0 & -2+8 & 1+4 & \vdots & -2+0 \end{bmatrix} = \begin{bmatrix} 1 & 0 & -4 & -2 & \vdots & 0 \\ 0 & -2 & 12 & 6 & \vdots & 1 \\ 0 & 1 & 0 & -3 & \vdots & -1 \\ 0 & 3 & 6 & 5 & \vdots & -2 \end{bmatrix}$$

4. Divide the second row by -2.

$$\begin{bmatrix} 1 & 0 & -4 & -2 & \vdots & 0 \\ 0 & \frac{-2}{-2} & \frac{12}{-2} & \frac{6}{-2} & \vdots & \frac{1}{-2} \\ 0 & 1 & 0 & -3 & \vdots & -1 \\ 0 & 3 & 6 & 5 & \vdots & -2 \end{bmatrix} = \begin{bmatrix} 1 & 0 & -4 & -2 & \vdots & 0 \\ 0 & 1 & -6 & -3 & \vdots & -\frac{1}{2} \\ 0 & 1 & 0 & -3 & \vdots & -1 \\ 0 & 3 & 6 & 5 & \vdots & -2 \end{bmatrix}$$

5. Multiply each elements of the second row by -1 and add the result to each element of the third row.

$$\begin{bmatrix} 1 & 0 & -4 & -2 & \vdots & 0 \\ 0 & 1 & -6 & -3 & \vdots & -\frac{1}{2} \\ 0 & 1-1 & 0+6 & -3+3 & \vdots & -1+\frac{1}{2} \\ 0 & 3 & 6 & 5 & \vdots & -2 \end{bmatrix} = \begin{bmatrix} 1 & 0 & -4 & -2 & \vdots & 0 \\ 0 & 1 & -6 & -3 & \vdots & -\frac{1}{2} \\ 0 & 0 & 6 & 0 & \vdots & -\frac{1}{2} \\ 0 & 3 & 6 & 5 & \vdots & -2 \end{bmatrix}$$

6. Multiply each element of the second row by -3 and add the result to each element of the

fourth row.
$$\begin{bmatrix} 1 & 0 & -4 & -2 & \vdots & 0 \\ 0 & 1 & -6 & -3 & \vdots & -\frac{1}{2} \\ 0 & 0 & 6 & 0 & \vdots & -\frac{1}{2} \\ 0 & 3-3 & 6+18 & 5+9 & \vdots & -2+\frac{3}{2} \end{bmatrix} = \begin{bmatrix} 1 & 0 & -4 & -2 & \vdots & 0 \\ 0 & 1 & -6 & -3 & \vdots & -\frac{1}{2} \\ 0 & 0 & 6 & 0 & \vdots & -\frac{1}{2} \\ 0 & 0 & 24 & 14 & \vdots & -\frac{1}{2} \end{bmatrix}$$

7. Divide each element of the third row by 6.
$$\begin{bmatrix} 1 & 0 & -4 & -2 & \vdots & 0 \\ 0 & 1 & -6 & -3 & \vdots & -\frac{1}{2} \\ \frac{0}{6} & \frac{0}{6} & \frac{6}{6} & \frac{0}{6} & \vdots & -\frac{1}{12} \\ 0 & 0 & 24 & 14 & \vdots & -\frac{1}{2} \end{bmatrix} = \begin{bmatrix} 1 & 0 & -4 & -2 & \vdots & 0 \\ 0 & 1 & -6 & -3 & \vdots & -\frac{1}{2} \\ 0 & 0 & 1 & 0 & \vdots & -\frac{1}{12} \\ 0 & 0 & 24 & 14 & \vdots & -\frac{1}{2} \end{bmatrix}$$

8. Multiply each element of the third row by -24 and add the result to each element of the fourth row.
$$\begin{bmatrix} 1 & 0 & -4 & -2 & \vdots & 0 \\ 0 & 1 & -6 & -3 & \vdots & -\frac{1}{2} \\ 0 & 0 & 1 & 0 & \vdots & -\frac{1}{12} \\ 0 & 0 & 24-24 & 14+0 & \vdots & -\frac{1}{2}+2 \end{bmatrix} = \begin{bmatrix} 1 & 0 & -4 & -2 & \vdots & 0 \\ 0 & 1 & -6 & -3 & \vdots & -\frac{1}{2} \\ 0 & 0 & 1 & 0 & \vdots & -\frac{1}{12} \\ 0 & 0 & 0 & 14 & \vdots & \frac{3}{2} \end{bmatrix}$$

9. Divide each element of the fourth row by 14.
$$\begin{bmatrix} 1 & 0 & -4 & -2 & \vdots & 0 \\ 0 & 1 & -6 & -3 & \vdots & -\frac{1}{2} \\ 0 & 0 & 1 & 0 & \vdots & -\frac{1}{12} \\ 0 & 0 & 0 & \frac{14}{14} & \vdots & \frac{3}{2}\times\frac{1}{14} \end{bmatrix} = \begin{bmatrix} 1 & 0 & -4 & -2 & \vdots & 0 \\ 0 & 1 & -6 & -3 & \vdots & -\frac{1}{2} \\ 0 & 0 & 1 & 0 & \vdots & -\frac{1}{12} \\ 0 & 0 & 0 & 1 & \vdots & \frac{3}{28} \end{bmatrix}$$

Note that the augmented matrix has 1's in its main diagonal entries and zeros in the lower triangle.

Fourth - Write the augmented matrix in its equivalent linear system form.

$$x-4z-2w=0$$
$$y-6z-3w=-\frac{1}{2}$$
$$z=-\frac{1}{12}$$
$$w=\frac{3}{28}$$

Since $w=\frac{3}{28}$, and $z=-\frac{1}{12}$ we can solve for x and y by back substitution.

$y-6z-3w=-\frac{1}{2}$; $y+\left(-6\times-\frac{1}{12}\right)+\left(-3\times\frac{3}{28}\right)=-\frac{1}{2}$; $y+\frac{6}{12}-\frac{9}{28}=-\frac{1}{2}$; $y+\frac{1}{2}-\frac{9}{28}=-\frac{1}{2}$

; $y=-\frac{1}{2}-\frac{1}{2}+\frac{9}{28}$; $y=\frac{-1-1}{2}+\frac{9}{28}$ $y=-\frac{2}{2}+\frac{9}{28}$; $y=-\frac{1}{1}+\frac{9}{28}$; $y=\frac{(-1\times 28)+(1\times 9)}{1\times 28}$; $y=\frac{-28+9}{28}$

; $y=-\frac{19}{28}$

$$x-4z-2w=0\ ;\ x+\left(-4\times-\frac{1}{12}\right)+\left(-2\times\frac{3}{28}\right)=0\ ;\ x+\frac{4}{12}-\frac{6}{28}=0\ ;\ x+\frac{1}{3}-\frac{3}{14}=0\ ;\ x=-\frac{1}{3}+\frac{3}{14}$$

$$;\ x=\frac{(-1\times14)+(3\times3)}{3\times14}\ ;\ x=\frac{-14+9}{42}\ ;\ x=-\frac{5}{42}$$

Therefore, $x=-\frac{5}{42}$, $y=-\frac{19}{28}$, $z=-\frac{1}{12}$, $w=\frac{3}{28}$ and the solution set is $\left\{\left(-\frac{5}{42},-\frac{19}{28},-\frac{1}{12},\frac{3}{28}\right)\right\}$

Fifth - Check the answers by substituting the x, y, z, and w values into one of the original equations. Let's substitute the x, y, z and w values into $-x+4z+2w=0$.

$$\frac{5}{42}+\left(4\times-\frac{1}{12}\right)+\left(2\times\frac{3}{28}\right)\stackrel{?}{=}0\ ;\ \frac{5}{42}-\frac{4}{12}+\frac{6}{28}\stackrel{?}{=}0\ ;\ 0.119-0.333+0.214\stackrel{?}{=}0\ ;\ -0.214+0.214\stackrel{?}{=}0\ ;\ 0=0$$

Section 3.5 Case V Practice Problems - Solving Linear Systems Using the Gaussian Elimination Method

1. Solve the following linear systems by applying the Gaussian Elimination method to each augmented matrix.

a. $\begin{matrix} x-2y=-3 \\ 2x+3y=4 \end{matrix}$ b. $\begin{matrix} 2x+y=-2 \\ 3x-y=0 \end{matrix}$ c. $\begin{matrix} x-3y=1 \\ 2x+5y=0 \end{matrix}$

d. $\begin{matrix} 4x-3y=1 \\ 3x+y=2 \end{matrix}$ e. $\begin{matrix} 2x+3z=-1 \\ x+3y=0 \\ 2x-2y+3z=-2 \end{matrix}$ f. $\begin{matrix} 3x-z=0 \\ 2x-y-z=1 \\ 3x+2y=-1 \end{matrix}$

2. In the following exercises write the linear system whose augmented matrix is given.

a. $\begin{bmatrix} \frac{1}{2} & 1 & \vdots & -1 \\ -2 & 3 & \vdots & 4 \end{bmatrix}$ b. $\begin{bmatrix} 2 & 0 & \vdots & 2 \\ -1 & 3 & \vdots & 5 \end{bmatrix}$ c. $\begin{bmatrix} 1 & 2 & -3 & \vdots & 1 \\ 0 & 1 & 2 & \vdots & 0 \\ -1 & 3 & 1 & \vdots & 2 \end{bmatrix}$ d. $\begin{bmatrix} 2 & 3 & 5 & \vdots & 10 \\ 1 & -1 & 3 & \vdots & 11 \\ 0 & 2 & -1 & \vdots & -2 \end{bmatrix}$

3. Find the solution set to the following augmented matrices which have been transformed by elementary row operations.

a. $\begin{bmatrix} 1 & 2 & 0 & \vdots & -5 \\ 0 & 1 & -2 & \vdots & 3 \\ 0 & 0 & 1 & \vdots & 2 \end{bmatrix}$ b. $\begin{bmatrix} 1 & -2 & 3 & \vdots & 4 \\ 0 & 1 & 3 & \vdots & 2 \\ 0 & 0 & 1 & \vdots & -3 \end{bmatrix}$ c. $\begin{bmatrix} 1 & -2 & -1 & \vdots & 1 \\ 0 & 1 & 2 & \vdots & 2 \\ 0 & 0 & 1 & \vdots & 3 \end{bmatrix}$

Case VI Solving Linear Systems Using the Gauss - Jordan Elimination Method

The Gauss-Jordan Elimination method is also used on any system of n linear equations with n unknowns that has a unique solution. The objective in using the Gauss-Jordan Elimination method is to transfer the augmented matrix into a matrix for which $a_{ij} = 0$ when $i \rangle j$ and $i \langle j$. To transfer an augmented matrix into a matrix with 1's in the main diagonal entries and zeros in the lower and upper triangles we use the following elementary row operations as many times as required:

1. Interchange any two rows.
2. Multiply each element of a row by a constant k, where $k \neq 0$.
3. Replace elements of a row by the sum of itself plus k times the corresponding elements of another row.

In general, the transferred augmented matrix for a system with four linear equations and four unknowns has the following appearance.

$$\begin{bmatrix} 1 & 0 & 0 & 0 & \vdots & * \\ 0 & 1 & 0 & 0 & \vdots & * \\ 0 & 0 & 1 & 0 & \vdots & * \\ 0 & 0 & 0 & 1 & \vdots & * \end{bmatrix}$$

The following show the steps in solving linear systems using the Gauss-Jordan Elimination method

Step 1 Write the linear system in the matrix form $AX = B$.

Step 2 Write the augmented matrix.

Step 3 a. Transfer the augmented matrix into a matrix with 1's in the main diagonal entries and zeros in the lower and upper triangles using the elementary row operations.

b. Write the solution set to the linear system.

Step 4 Check the answers by substituting the solutions back into one of the original equations.

Note that the steps in solving a linear system using the Gauss-Jordan Elimination method is similar to the Gaussian Elimination method. However, the major difference in the Gauss-Jordan Elimination method is that the augmented matrix is transferred into a matrix that yields to a solution without back-substitution. The following examples show the steps for solving linear systems using the Gauss-Jordan Elimination method:

Example 3.5-20 Solve the following linear systems by applying the Gauss-Jordan Elimination method to the augmented matrix.

a. $\begin{array}{l} 2x - 4y = -3 \\ x + y = 5 \end{array}$

b. $\begin{array}{l} 2x + 4y - z = 4 \\ x - z = 0 \\ y + 2z = -1 \end{array}$

c. $\begin{array}{l} 3x - 4y + 3z - w = -6 \\ 2y + z - 3w = 0 \\ x - 6w = -1 \\ 2x - 3y - z = -2 \end{array}$

Solution (a):

First - Write the linear system in the form of $AX = B$.

$$\begin{bmatrix} 2 & -4 \\ 1 & 1 \end{bmatrix}\begin{bmatrix} x \\ y \end{bmatrix} = \begin{bmatrix} -3 \\ 5 \end{bmatrix}$$

Second - Write the augmented matrix.

$$\begin{bmatrix} 2 & -4 & \vdots & -3 \\ 1 & 1 & \vdots & 5 \end{bmatrix}$$

Third - Perform the elementary row operations.

1. Replace the second row with the first row. $\begin{bmatrix} 1 & 1 & \vdots & 5 \\ 2 & -4 & \vdots & -3 \end{bmatrix}$

2. Multiply each elements of the first row by -2 and add the result to each element of the second row. $\begin{bmatrix} 1 & 1 & \vdots & 5 \\ 2-2 & -4-2 & \vdots & -3-10 \end{bmatrix} = \begin{bmatrix} 1 & 1 & \vdots & 5 \\ 0 & -6 & \vdots & -13 \end{bmatrix}$

3. Divide the second row by -6. $\begin{bmatrix} 1 & 1 & \vdots & 5 \\ 0 & \frac{-6}{-6} & \vdots & \frac{-13}{-6} \end{bmatrix} = \begin{bmatrix} 1 & 1 & \vdots & 5 \\ 0 & 1 & \vdots & \frac{13}{6} \end{bmatrix}$

4. Multiply each element of the second row by -1 and add the result to each element of the first row. $\begin{bmatrix} 1+0 & 1-1 & \vdots & 5-\frac{13}{6} \\ 0 & 1 & \vdots & \frac{13}{6} \end{bmatrix} = \begin{bmatrix} 1 & 0 & \vdots & \frac{17}{6} \\ 0 & 1 & \vdots & \frac{13}{6} \end{bmatrix}$

Note that the augmented matrix has 1's in its main diagonal entries and zeros in the lower and upper triangles. Therefore, $x = \frac{17}{6}$ and $y = \frac{13}{6}$ and the solution set is $\left\{\left(\frac{17}{6}, \frac{13}{6}\right)\right\}$

Fourth - Check the answers by substituting the x and y values into one of the original equations. Let's substitute the x and y values into $2x - 4y = -3$.

$$\left(2\times\frac{17}{6}\right)+\left(-4\times\frac{13}{6}\right) \stackrel{?}{=} -3 \; ; \; \frac{34}{6}-\frac{52}{6} \stackrel{?}{=} -3 \; ; \; \frac{34-52}{6} \stackrel{?}{=} -3 \; ; \; -\frac{18}{6} \stackrel{?}{=} -3 \; ; \; -\frac{3}{1} \stackrel{?}{=} -3 \; ; \; -3 = -3$$

Solution (b):

First - Write the linear system in the form of $AX = B$.

$$\begin{bmatrix} 2 & 4 & -1 \\ 1 & 0 & -1 \\ 0 & 1 & 2 \end{bmatrix}\begin{bmatrix} x \\ y \\ z \end{bmatrix} = \begin{bmatrix} 4 \\ 0 \\ -1 \end{bmatrix}$$

Second - Write the augmented matrix.

$$\begin{bmatrix} 2 & 4 & -1 & \vdots & 4 \\ 1 & 0 & -1 & \vdots & 0 \\ 0 & 1 & 2 & \vdots & -1 \end{bmatrix}$$

Third - Perform the elementary row operations.

1. Replace the second row with the first row. $\begin{bmatrix} 1 & 0 & -1 & \vdots & 0 \\ 2 & 4 & -1 & \vdots & 4 \\ 0 & 1 & 2 & \vdots & -1 \end{bmatrix}$

2. Multiply each elements of the first row by -2 and add the result to each element of the

second row. $\begin{bmatrix} 1 & 0 & -1 & \vdots & 0 \\ 2-2 & 4+0 & -1+2 & \vdots & 4+0 \\ 0 & 1 & 2 & \vdots & -1 \end{bmatrix} = \begin{bmatrix} 1 & 0 & -1 & \vdots & 0 \\ 0 & 4 & 1 & \vdots & 4 \\ 0 & 1 & 2 & \vdots & -1 \end{bmatrix}$

3. Divide the second row by 4. $\begin{bmatrix} 1 & 0 & -1 & \vdots & 0 \\ 0 & \frac{4}{4} & \frac{1}{4} & \vdots & \frac{4}{4} \\ 0 & 1 & 2 & \vdots & -1 \end{bmatrix} = \begin{bmatrix} 1 & 0 & -1 & \vdots & 0 \\ 0 & 1 & \frac{1}{4} & \vdots & 1 \\ 0 & 1 & 2 & \vdots & -1 \end{bmatrix}$

4. Multiply each elements of the second row by -1 and add the result to each element of the third row.

$$\begin{bmatrix} 1 & 0 & -1 & \vdots & 0 \\ 0 & 1 & \frac{1}{4} & \vdots & 1 \\ 0 & 1-1 & 2-\frac{1}{4} & \vdots & -1-1 \end{bmatrix} = \begin{bmatrix} 1 & 0 & -1 & \vdots & 0 \\ 0 & 1 & \frac{1}{4} & \vdots & 1 \\ 0 & 0 & \frac{7}{4} & \vdots & -2 \end{bmatrix}$$

5. Multiply each element of the third row by $\frac{4}{7}$.

$$\begin{bmatrix} 1 & 0 & -1 & \vdots & 0 \\ 0 & 1 & \frac{1}{4} & \vdots & 1 \\ 0 & 0 & \frac{7}{4}\times\frac{4}{7} & \vdots & -2\times\frac{4}{7} \end{bmatrix} = \begin{bmatrix} 1 & 0 & -1 & \vdots & 0 \\ 0 & 1 & \frac{1}{4} & \vdots & 1 \\ 0 & 0 & 1 & \vdots & -\frac{8}{7} \end{bmatrix}$$

6. Multiply each element of the third row by $-\frac{1}{4}$ and add the result to each element of the second row.

$$\begin{bmatrix} 1 & 0 & -1 & \vdots & 0 \\ 0+0 & 1+0 & \frac{1}{4}-\frac{1}{4} & \vdots & 1+\frac{2}{7} \\ 0 & 0 & 1 & \vdots & -\frac{8}{7} \end{bmatrix} = \begin{bmatrix} 1 & 0 & -1 & \vdots & 0 \\ 0 & 1 & 0 & \vdots & \frac{9}{7} \\ 0 & 0 & 1 & \vdots & -\frac{8}{7} \end{bmatrix}$$

7. Multiply each element of the third row by 1 and add the result to each element of the first row.

$$\begin{bmatrix} 1+0 & 0+0 & -1+1 & \vdots & 0-\frac{8}{7} \\ 0 & 1 & 0 & \vdots & \frac{9}{7} \\ 0 & 0 & 1 & \vdots & -\frac{8}{7} \end{bmatrix} = \begin{bmatrix} 1 & 0 & 0 & \vdots & -\frac{8}{7} \\ 0 & 1 & 0 & \vdots & \frac{9}{7} \\ 0 & 0 & 1 & \vdots & -\frac{8}{7} \end{bmatrix}$$

Note that the augmented matrix has 1's in its main diagonal entries and zeros in the lower and upper triangles. Therefore, $x = -\frac{8}{7}$, $y = \frac{9}{7}$, and $z = -\frac{8}{7}$ and the solution set is $\left\{\left(-\frac{8}{7}, \frac{9}{7}, -\frac{8}{7}\right)\right\}$

Fourth - Check the answers by substituting the x, y, and z values into one of the original equations. Let's substitute the x, y, and z values into $2x+4y-z=4$.

$$\left(2\times-\frac{8}{7}\right)+\left(4\times\frac{9}{7}\right)+\frac{8}{7}\overset{?}{=}4 \;;\; -\frac{16}{7}+\frac{36}{7}+\frac{8}{7}\overset{?}{=}4 \;;\; \frac{-16+36+8}{7}\overset{?}{=}4 \;;\; \frac{28}{7}\overset{?}{=}4 \;;\; \frac{4}{1}\overset{?}{=}4 \;;\; 4=4$$

Solution (c):

First - Write the linear system in the form of $AX = B$.

$$\begin{bmatrix} 3 & -4 & 3 & -1 \\ 0 & 2 & 1 & -3 \\ 1 & 0 & 0 & -6 \\ 2 & -3 & -1 & 0 \end{bmatrix} \begin{bmatrix} x \\ y \\ z \\ w \end{bmatrix} = \begin{bmatrix} -6 \\ 0 \\ -1 \\ -2 \end{bmatrix}$$

Second - Write the augmented matrix.

$$\begin{bmatrix} 3 & -4 & 3 & -1 & \vdots & -6 \\ 0 & 2 & 1 & -3 & \vdots & 0 \\ 1 & 0 & 0 & -6 & \vdots & -1 \\ 2 & -3 & -1 & 0 & \vdots & -2 \end{bmatrix}$$

Third - Perform the elementary row operations.

1. Replace the third row with the first row.
$$\begin{bmatrix} 1 & 0 & 0 & -6 & \vdots & -1 \\ 0 & 2 & 1 & -3 & \vdots & 0 \\ 3 & -4 & 3 & -1 & \vdots & -6 \\ 2 & -3 & -1 & 0 & \vdots & -2 \end{bmatrix}$$

2. Multiply each elements of the first row by -3 and add the result to each element of the third row.
$$\begin{bmatrix} 1 & 0 & 0 & -6 & \vdots & -1 \\ 0 & 2 & 1 & -3 & \vdots & 0 \\ 3-3 & -4+0 & 3+0 & -1+18 & \vdots & -6+3 \\ 2 & -3 & -1 & 0 & \vdots & -2 \end{bmatrix} = \begin{bmatrix} 1 & 0 & 0 & -6 & \vdots & -1 \\ 0 & 2 & 1 & -3 & \vdots & 0 \\ 0 & -4 & 3 & 17 & \vdots & -3 \\ 2 & -3 & -1 & 0 & \vdots & -2 \end{bmatrix}$$

3. Multiply each elements of the first row by -2 and add the result to each element of the fourth row.
$$\begin{bmatrix} 1 & 0 & 0 & -6 & \vdots & -1 \\ 0 & 2 & 1 & -3 & \vdots & 0 \\ 0 & -4 & 3 & 17 & \vdots & -3 \\ 2-2 & -3+0 & -1+0 & 0+12 & \vdots & -2+2 \end{bmatrix} = \begin{bmatrix} 1 & 0 & 0 & -6 & \vdots & -1 \\ 0 & 2 & 1 & -3 & \vdots & 0 \\ 0 & -4 & 3 & 17 & \vdots & -3 \\ 0 & -3 & -1 & 12 & \vdots & 0 \end{bmatrix}$$

4. Divide the second row by 2.
$$\begin{bmatrix} 1 & 0 & 0 & -6 & \vdots & -1 \\ 0 & \frac{2}{2} & \frac{1}{2} & -\frac{3}{2} & \vdots & \frac{0}{2} \\ 0 & -4 & 3 & 17 & \vdots & -3 \\ 0 & -3 & -1 & 12 & \vdots & 0 \end{bmatrix} = \begin{bmatrix} 1 & 0 & 0 & -6 & \vdots & -1 \\ 0 & 1 & \frac{1}{2} & -\frac{3}{2} & \vdots & 0 \\ 0 & -4 & 3 & 17 & \vdots & -3 \\ 0 & -3 & -1 & 12 & \vdots & 0 \end{bmatrix}$$

5. Multiply each elements of the second row by 4 and add the result to each element of the third row.
$$\begin{bmatrix} 1 & 0 & 0 & -6 & \vdots & -1 \\ 0 & 1 & \frac{1}{2} & -\frac{3}{2} & \vdots & 0 \\ 0 & -4+4 & 3+2 & 17-6 & \vdots & -3+0 \\ 0 & -3 & -1 & 12 & \vdots & 0 \end{bmatrix} = \begin{bmatrix} 1 & 0 & 0 & -6 & \vdots & -1 \\ 0 & 1 & \frac{1}{2} & -\frac{3}{2} & \vdots & 0 \\ 0 & 0 & 5 & 11 & \vdots & -3 \\ 0 & -3 & -1 & 12 & \vdots & 0 \end{bmatrix}$$

6. Multiply each element of the second row by 3 and add the result to each element of the fourth row.
$$\begin{bmatrix} 1 & 0 & 0 & -6 & \vdots & -1 \\ 0 & 1 & \frac{1}{2} & -\frac{3}{2} & \vdots & 0 \\ 0 & 0 & 5 & 11 & \vdots & -3 \\ 0 & -3+3 & -1+\frac{3}{2} & 12-\frac{9}{2} & \vdots & 0+0 \end{bmatrix} = \begin{bmatrix} 1 & 0 & 0 & -6 & \vdots & -1 \\ 0 & 1 & \frac{1}{2} & -\frac{3}{2} & \vdots & 0 \\ 0 & 0 & 5 & 11 & \vdots & -3 \\ 0 & 0 & \frac{1}{2} & \frac{15}{2} & \vdots & 0 \end{bmatrix}$$

7. Divide each element of the third row by 5.

$$\begin{bmatrix} 1 & 0 & 0 & -6 & \vdots & -1 \\ 0 & 1 & \frac{1}{2} & -\frac{3}{2} & \vdots & 0 \\ 0 & 0 & \frac{5}{5} & \frac{11}{5} & \vdots & -\frac{3}{5} \\ 0 & 0 & \frac{1}{2} & \frac{15}{2} & \vdots & 0 \end{bmatrix} = \begin{bmatrix} 1 & 0 & 0 & -6 & \vdots & -1 \\ 0 & 1 & \frac{1}{2} & -\frac{3}{2} & \vdots & 0 \\ 0 & 0 & 1 & \frac{11}{5} & \vdots & -\frac{3}{5} \\ 0 & 0 & \frac{1}{2} & \frac{15}{2} & \vdots & 0 \end{bmatrix}$$

8. Multiply each element of the third row by $-\frac{1}{2}$ and add the result to each element of the second row.

$$\begin{bmatrix} 1 & 0 & 0 & -6 & \vdots & -1 \\ 0+0 & 1+0 & \frac{1}{2}-\frac{1}{2} & -\frac{3}{2}-\frac{11}{10} & \vdots & 0+\frac{3}{10} \\ 0 & 0 & 1 & \frac{11}{5} & \vdots & -\frac{3}{5} \\ 0 & 0 & \frac{1}{2} & \frac{15}{2} & \vdots & 0 \end{bmatrix} = \begin{bmatrix} 1 & 0 & 0 & -6 & \vdots & -1 \\ 0 & 1 & 0 & -\frac{13}{5} & \vdots & \frac{3}{10} \\ 0 & 0 & 1 & \frac{11}{5} & \vdots & -\frac{3}{5} \\ 0 & 0 & \frac{1}{2} & \frac{15}{2} & \vdots & 0 \end{bmatrix}$$

9. Multiply each element of the third row by $-\frac{1}{2}$ and add the result to each element of the fourth row.

$$\begin{bmatrix} 1 & 0 & 0 & -6 & \vdots & -1 \\ 0 & 1 & 0 & -\frac{13}{5} & \vdots & \frac{3}{10} \\ 0 & 0 & 1 & \frac{11}{5} & \vdots & -\frac{3}{5} \\ 0+0 & 0+0 & \frac{1}{2}-\frac{1}{2} & \frac{15}{2}-\frac{11}{10} & \vdots & 0+\frac{3}{10} \end{bmatrix} = \begin{bmatrix} 1 & 0 & 0 & -6 & \vdots & -1 \\ 0 & 1 & 0 & -\frac{13}{5} & \vdots & \frac{3}{10} \\ 0 & 0 & 1 & \frac{11}{5} & \vdots & -\frac{3}{5} \\ 0 & 0 & 0 & \frac{32}{5} & \vdots & \frac{3}{10} \end{bmatrix}$$

10. Multiply each element of the fourth row by $\frac{5}{32}$.

$$\begin{bmatrix} 1 & 0 & 0 & -6 & \vdots & -1 \\ 0 & 1 & 0 & -\frac{13}{5} & \vdots & \frac{3}{10} \\ 0 & 0 & 1 & \frac{11}{5} & \vdots & -\frac{3}{5} \\ 0 & 0 & 0 & \frac{32}{5}\times\frac{5}{32} & \vdots & \frac{3}{10}\times\frac{5}{32} \end{bmatrix} = \begin{bmatrix} 1 & 0 & 0 & -6 & \vdots & -1 \\ 0 & 1 & 0 & -\frac{13}{5} & \vdots & \frac{3}{10} \\ 0 & 0 & 1 & \frac{11}{5} & \vdots & -\frac{3}{5} \\ 0 & 0 & 0 & 1 & \vdots & \frac{3}{64} \end{bmatrix}$$

11. Multiply each element of the fourth row by $-\frac{11}{5}$ and add the result to each element of the third row.

$$\begin{bmatrix} 1 & 0 & 0 & -6 & \vdots & -1 \\ 0 & 1 & 0 & -\frac{13}{5} & \vdots & \frac{3}{10} \\ 0+0 & 0+0 & 1+0 & \frac{11}{5}-\frac{11}{5} & \vdots & -\frac{3}{5}-\frac{33}{320} \\ 0 & 0 & 0 & 1 & \vdots & \frac{3}{64} \end{bmatrix} = \begin{bmatrix} 1 & 0 & 0 & -6 & \vdots & -1 \\ 0 & 1 & 0 & -\frac{13}{5} & \vdots & \frac{3}{10} \\ 0 & 0 & 1 & 0 & \vdots & -\frac{45}{64} \\ 0 & 0 & 0 & 1 & \vdots & \frac{3}{64} \end{bmatrix}$$

12. Multiply each element of the fourth row by $\frac{13}{5}$ and add the result to each element of the second row.

$$\begin{bmatrix} 1 & 0 & 0 & -6 & \vdots & -1 \\ 0+0 & 1+0 & 0+0 & -\frac{13}{5}+\frac{13}{5} & \vdots & \frac{3}{10}+\frac{39}{320} \\ 0 & 0 & 1 & 0 & \vdots & -\frac{45}{64} \\ 0 & 0 & 0 & 1 & \vdots & \frac{3}{64} \end{bmatrix} = \begin{bmatrix} 1 & 0 & 0 & -6 & \vdots & -1 \\ 0 & 1 & 0 & 0 & \vdots & \frac{27}{64} \\ 0 & 0 & 1 & 0 & \vdots & -\frac{45}{64} \\ 0 & 0 & 0 & 1 & \vdots & \frac{3}{64} \end{bmatrix}$$

13. Multiply each element of the fourth row by 6 and add the result to each element of the first row.

$$\begin{bmatrix} 1+0 & 0+0 & 0+0 & -6+6 & \vdots & -1+\frac{9}{32} \\ 0 & 1 & 0 & 0 & \vdots & \frac{27}{64} \\ 0 & 0 & 1 & 0 & \vdots & -\frac{45}{64} \\ 0 & 0 & 0 & 1 & \vdots & \frac{3}{64} \end{bmatrix} = \begin{bmatrix} 1 & 0 & 0 & 0 & \vdots & -\frac{23}{32} \\ 0 & 1 & 0 & 0 & \vdots & \frac{27}{64} \\ 0 & 0 & 1 & 0 & \vdots & -\frac{45}{64} \\ 0 & 0 & 0 & 1 & \vdots & \frac{3}{64} \end{bmatrix}$$

Note that the augmented matrix has 1's in its main diagonal entries and zeros in the lower and upper triangles.

Thus, $\mathbf{x=-\frac{23}{32}}$, $\mathbf{y=\frac{27}{64}}$, $\mathbf{z=-\frac{45}{64}}$, $\mathbf{w=\frac{3}{64}}$ and the solution set is $\left\{\left(-\frac{23}{32},\frac{27}{64},-\frac{45}{64},\frac{3}{64}\right)\right\}$

Fourth - Check the answers by substituting the x, y, z, and w values into one of the original equations. Let's substitute the x, y, z, and w values into $x-6w=-1$ or $2y+z-3w=0$.

$-\frac{23}{32}+\left(-6\times\frac{3}{64}\right)\overset{?}{=}-1$; $-\frac{23}{32}-\frac{18}{64}\overset{?}{=}-1$; $-\frac{23}{32}-\frac{9}{32}\overset{?}{=}-1$; $\frac{-23-9}{32}\overset{?}{=}-1$; $-\frac{32}{32}\overset{?}{=}-1$; $-\frac{1}{1}\overset{?}{=}-1$; $-1=-1$

or, $2\times\frac{27}{64}-\frac{45}{64}-3\times\frac{3}{64}\overset{?}{=}0$; $\frac{54}{64}-\frac{45}{64}-\frac{9}{64}\overset{?}{=}0$; $\frac{54-45-9}{64}\overset{?}{=}0$; $\frac{45-45}{64}\overset{?}{=}0$; $\frac{0}{64}\overset{?}{=}0$; $0=0$

Section 3.5 Case VI Practice Problems - Solving Linear Systems Using the Gauss-Jordan Elimination Method

1. Solve the following linear systems by applying Gauss-Jordan Elimination method to each augmented matrix

a. $x-3y=-2$
$2x-y=-3$

b. $2x-y=2$
$3x-\frac{2}{3}y=0$

c. $-x+2y=-2$
$\frac{1}{2}x-\frac{1}{3}y=-\frac{1}{2}$

d. $x-y=0$
$2x-3y=-\frac{1}{4}$

e. $3x-2z=-1$
$x-y+z=0$
$2x+3y=-2$

f. $x-z=0$
$x-3y=-1$
$x+y=0$

2. Write the coefficient matrix and the augmented matrix for the following linear systems.

a. $x-2y=-1$
$2x+4y=-3$

b. $x+2y=-3$
$x-y=0$

c. $x+2y-z=1$
$y-3z=-3$
$x-4z=-2$

d. $x-2y=-4$
$x+2y+z=-1$
$y-3z=3$

e. $x+y-2z+w=-1$
$2y-4w=0$
$x-2w=-1$
$x+y-4w=0$

f. $x+y-2w=-1$
$y-3w=-2$
$x-y+2z-w=3$
$y-3z=-2$

Chapter 4
Sequences and Series

Quick Reference to Chapter 4 Problems

Chapter 4 - Sequences and Series

The objective of this chapter is to improve the student's ability to solve problems involving sequences and series. Sequences and series are introduced in Sections 4.1 and 4.2. How to solve arithmetic sequences and arithmetic series are discussed in Section 4.3. Solutions to geometric sequences and geometric series are addressed in Section 4.4. The process of identifying convergence or divergence of a sequence or a series, for large values of n, is discussed in Section 4.5. Finally, the factorial notation and its use in expanding binomial expressions are addressed in Section 4.6. Each section is concluded by solving examples with practice problems to further enhance the student's ability.

4.1 Sequences

A **sequence** is a function whose domain contains a set of positive integer terms such as $(1, 2, 3, 4, \cdots)$. Functions generate sequences. For example, the function $s(n) = s_n = n - 2$ whose domain is $(1, 2, 3, 4, 5, 6)$ generates the sequence

$s(1) = s_1 = 1 - 2 = -1$ $\quad$ $s(2) = s_2 = 2 - 2 = 0$ $\quad$ $s(3) = s_3 = 3 - 2 = 1$

$s(4) = s_4 = 4 - 2 = 2$ $\quad$ $s(5) = s_5 = 5 - 2 = 3$ $\quad$ $s(6) = s_6 = 6 - 2 = 4$

where the first six terms of the sequence are $(s_1, s_2, s_3, s_4, s_5, s_6) = (-1, 0, 1, 2, 3, 4)$.

In general, a function $f(x)$ whose domain is the set of positive integers $(1, 2, 3, \cdots, n)$ including a fixed value for n is called a **finite sequence function**. On the other hand, a function whose domain is the set of $(1, 2, 3, \cdots)$ is called an **infinite sequence function**. The elements of the range of a sequence function are called the **terms** of the sequence function. In some instances a sequence is given by presenting its first few terms, followed by its nth term, $s_n = s(n)$, which is commonly referred to as the **general term** of a sequence. For example, the sequence $4, 3, \frac{16}{5}, \frac{25}{9}, \cdots, \frac{(n+1)^2}{2n-1}$ shows the first four terms and the general term of the sequence. In the following examples we will learn how the various terms of a sequence are found:

Example 4.1-1 List the first six terms of the given sequence.

a. $a_n = \frac{(-3)^n}{n^3}$ $\quad$ b. $b_k = \frac{(-1)^k}{k+1}$ $\quad$ c. $d_n = \frac{5}{n(2n-1)}$ $\quad$ d. $c_n = \left(\frac{1}{2}\right)^n \cdot \frac{(-1)^n}{n}$

Solutions:

a. $\boxed{a_1} = \boxed{\frac{(-3)^1}{1^3}} = \boxed{\frac{-3}{1}} = \boxed{-3}$ $\qquad$ $\boxed{a_2} = \boxed{\frac{(-3)^2}{2^3}} = \boxed{\frac{9}{8}} = \boxed{\mathbf{1.125}}$

$\boxed{a_3} = \boxed{\frac{(-3)^3}{3^3}} = \boxed{\frac{-27}{27}} = \boxed{-1}$ $\qquad$ $\boxed{a_4} = \boxed{\frac{(-3)^4}{4^3}} = \boxed{\frac{81}{64}} = \boxed{\mathbf{1.265}}$

$\boxed{a_5} = \boxed{\frac{(-3)^5}{5^3}} = \boxed{\frac{-243}{125}} = \boxed{\mathbf{-1.944}}$ $\qquad$ $\boxed{a_6} = \boxed{\frac{(-3)^6}{6^3}} = \boxed{\frac{729}{216}} = \boxed{\mathbf{3.375}}$

b. $\boxed{b_1} = \boxed{\frac{(-1)^1}{1+1}} = \boxed{\frac{-1}{2}} = \boxed{\mathbf{-0.5}}$ $\boxed{b_2} = \boxed{\frac{(-1)^2}{2+1}} = \boxed{\frac{1}{3}} = \boxed{\mathbf{0.333}}$

$\boxed{b_3} = \boxed{\frac{(-1)^3}{3+1}} = \boxed{\frac{-1}{4}} = \boxed{\mathbf{-0.25}}$ $\boxed{b_4} = \boxed{\frac{(-1)^4}{4+1}} = \boxed{\frac{1}{5}} = \boxed{\mathbf{0.2}}$

$\boxed{b_5} = \boxed{\frac{(-1)^5}{5+1}} = \boxed{\frac{-1}{6}} = \boxed{\mathbf{-0.167}}$ $\boxed{b_6} = \boxed{\frac{(-1)^6}{6+1}} = \boxed{\frac{1}{7}} = \boxed{\mathbf{0.143}}$

c. $\boxed{d_1} = \boxed{\frac{5}{1\cdot(2\cdot1-1)}} = \boxed{\frac{5}{2-1}} = \boxed{\frac{5}{1}} = \boxed{\mathbf{5}}$ $\boxed{d_2} = \boxed{\frac{5}{2\cdot(2\cdot2-1)}} = \boxed{\frac{5}{2\cdot3}} = \boxed{\frac{5}{6}} = \boxed{\mathbf{0.833}}$

$\boxed{d_3} = \boxed{\frac{5}{3\cdot(2\cdot3-1)}} = \boxed{\frac{\not{5}}{3\cdot\not{5}}} = \boxed{\frac{1}{3}} = \boxed{\mathbf{0.333}}$ $\boxed{d_4} = \boxed{\frac{5}{4\cdot(2\cdot4-1)}} = \boxed{\frac{5}{4\cdot7}} = \boxed{\frac{5}{28}} = \boxed{\mathbf{0.178}}$

$\boxed{d_5} = \boxed{\frac{5}{5\cdot(2\cdot5-1)}} = \boxed{\frac{\not{5}}{\not{5}\cdot9}} = \boxed{\frac{1}{9}} = \boxed{\mathbf{0.111}}$ $\boxed{d_6} = \boxed{\frac{5}{6\cdot(2\cdot6-1)}} = \boxed{\frac{5}{6\cdot11}} = \boxed{\frac{5}{66}} = \boxed{\mathbf{0.076}}$

d. $\boxed{c_1} = \boxed{\left(\frac{1}{2}\right)^1\cdot\frac{(-1)^1}{1}} = \boxed{\frac{1}{2}\cdot-1} = \boxed{-\frac{1}{2}} = \boxed{\mathbf{-0.5}}$ $\boxed{c_2} = \boxed{\left(\frac{1}{2}\right)^2\cdot\frac{(-1)^2}{2}} = \boxed{\frac{1}{4}\cdot\frac{1}{2}} = \boxed{\frac{1}{8}} = \boxed{\mathbf{0.125}}$

$\boxed{c_3} = \boxed{\left(\frac{1}{2}\right)^3\cdot\frac{(-1)^3}{3}} = \boxed{\frac{1}{8}\cdot-\frac{1}{3}} = \boxed{-\frac{1}{24}} = \boxed{\mathbf{-0.042}}$ $\boxed{c_4} = \boxed{\left(\frac{1}{2}\right)^4\cdot\frac{(-1)^4}{4}} = \boxed{\frac{1}{16}\cdot\frac{1}{4}} = \boxed{\frac{1}{64}} = \boxed{\mathbf{0.016}}$

$\boxed{c_5} = \boxed{\left(\frac{1}{2}\right)^5\cdot\frac{(-1)^5}{5}} = \boxed{\frac{1}{32}\cdot-\frac{1}{5}} = \boxed{-\frac{1}{160}} = \boxed{\mathbf{-0.006}}$ $\boxed{c_6} = \boxed{\left(\frac{1}{2}\right)^6\cdot\frac{(-1)^6}{6}} = \boxed{\frac{1}{64}\cdot\frac{1}{6}} = \boxed{\frac{1}{384}} = \boxed{\mathbf{0.003}}$

Example 4.1-2 Find the indicated terms for the following sequences.

a. Write the third and sixth terms of $s_n = \frac{(-2)^{n+1}}{n^3}$

b. Write the tenth term of $a_i = (i-1)^3 \cdot 2^{i-4}$

c. Write the third and fourth terms of $a_n = (-1)^{n-2}$

d. Write the seventh term of $a_k = (0.2)^{k-1}$

e. Write the third and twelfth terms of $a_i = (-i)^3$

f. Write the eleventh term of $s_n = (-1)^{n-1} \cdot 2^{n+1}$

Solutions:

a. $\boxed{s_3} = \boxed{\frac{(-2)^{3+1}}{3^3}} = \boxed{\frac{(-2)^4}{27}} = \boxed{\frac{16}{27}} = \boxed{\mathbf{0.593}}$ $\boxed{s_6} = \boxed{\frac{(-2)^{6+1}}{6^3}} = \boxed{\frac{(-2)^7}{216}} = \boxed{\frac{-128}{216}} = \boxed{\mathbf{-0.593}}$

b. $\boxed{a_{10}} = \boxed{(10-1)^3\cdot2^{10-4}} = \boxed{9^3\cdot2^6} = \boxed{729\cdot64} = \boxed{\mathbf{4.6656\times10^4}}$

c. $\boxed{a_3} = \boxed{(-1)^{3-2}} = \boxed{(-1)^1} = \boxed{\mathbf{-1}}$ $\boxed{a_4} = \boxed{(-1)^{4-2}} = \boxed{(-1)^2} = \boxed{\mathbf{1}}$

d. $\boxed{a_7} = \boxed{(0.2)^{7-1}} = \boxed{(0.2)^6} = \boxed{0.000064} = \boxed{\mathbf{6.4 \times 10^{-5}}}$

e. $\boxed{a_3} = \boxed{(-3)^3} = \boxed{-3^3} = \boxed{\mathbf{-27}}$ $\boxed{a_{12}} = \boxed{(-12)^3} = \boxed{-12^3} = \boxed{\mathbf{-1728}}$

f. $\boxed{s_{11}} = \boxed{(-1)^{11-1} \cdot 2^{11+1}} = \boxed{(-1)^{10} \cdot 2^{12}} = \boxed{1 \cdot 2^{12}} = \boxed{\mathbf{4096}}$

Example 4.1-3 Write s_5, s_6, s_7, and s_{15} for the following sequences.

a. $s_n = \frac{(n+1)\pi}{2}$ b. $s_n = \frac{2^n}{2^{n-1}}$ c. $s_n = 2n(n-1)(n-2)$

Solution:

a. $\boxed{s_5} = \boxed{\frac{(5+1)\pi}{2}} = \boxed{\frac{6\pi}{2}} = \boxed{\mathbf{3\pi}}$ $\boxed{s_6} = \boxed{\frac{(6+1)\pi}{2}} = \boxed{\frac{7\pi}{2}} = \boxed{\mathbf{3.5\pi}}$

$\boxed{s_7} = \boxed{\frac{(7+1)\pi}{2}} = \boxed{\frac{8\pi}{2}} = \boxed{\mathbf{4\pi}}$ $\boxed{s_{15}} = \boxed{\frac{(15+1)\pi}{2}} = \boxed{\frac{16\pi}{2}} = \boxed{\mathbf{8\pi}}$

b. $\boxed{s_5} = \boxed{\frac{2^5}{2^{5-1}}} = \boxed{\frac{2^5}{2^4}}\,;\,\boxed{2^5 \cdot 2^{-4}}\,;\,\boxed{2^{5-4}}\,;\,\boxed{2^1} = \boxed{\mathbf{2}}$

$\boxed{s_6} = \boxed{\frac{2^6}{2^{6-1}}} = \boxed{\frac{2^6}{2^5}} = \boxed{2^6 \cdot 2^{-5}} = \boxed{2^{6-5}} = \boxed{2^1} = \boxed{\mathbf{2}}$

$\boxed{s_7} = \boxed{\frac{2^7}{2^{7-1}}} = \boxed{\frac{2^7}{2^6}} = \boxed{2^7 \cdot 2^{-6}} = \boxed{2^{7-6}} = \boxed{2^1} = \boxed{\mathbf{2}}$

$\boxed{s_{15}} = \boxed{\frac{2^{15}}{2^{15-1}}} = \boxed{\frac{2^{15}}{2^{14}}} = \boxed{2^{15} \cdot 2^{-14}} = \boxed{2^{15-14}} = \boxed{2^1} = \boxed{\mathbf{2}}$

c. $\boxed{s_5} = \boxed{2 \cdot 5 \cdot (5-1)(5-2)} = \boxed{10 \cdot 4 \cdot 3} = \boxed{\mathbf{120}}$ $\boxed{s_6} = \boxed{2 \cdot 6 \cdot (6-1)(6-2)} = \boxed{12 \cdot 5 \cdot 4} = \boxed{\mathbf{240}}$

$\boxed{s_7} = \boxed{2 \cdot 7 \cdot (7-1)(7-2)} = \boxed{14 \cdot 6 \cdot 5} = \boxed{\mathbf{420}}$ $\boxed{s_{15}} = \boxed{2 \cdot 15 \cdot (15-1)(15-2)} = \boxed{30 \cdot 14 \cdot 13} = \boxed{\mathbf{5460}}$

Example 4.1-4 Find the twelfth term of the following sequences:

a. $0.5, 0.25, 0.125, \cdots, \frac{1^{n+1}}{2^n}$ b. $8, 5.063, 4.214, \cdots, \left(1+\frac{1}{n}\right)^{n+2}$ c. $4, -12, 32, \cdots, \frac{2^n \cdot (n+1)}{(-1)^{n+1}}$

Solutions:

a. $\boxed{s_{12}} = \boxed{\frac{1^{12+1}}{2^{12}}} = \boxed{\frac{1^{13}}{2^{12}}} = \boxed{\frac{1}{4096}} = \boxed{0.000244} = \boxed{\mathbf{2.44 \times 10^{-4}}}$

b. $\boxed{s_{12}} = \boxed{\left(1+\frac{1}{12}\right)^{12+2}} = \boxed{\left(\frac{12+1}{12}\right)^{14}} = \boxed{\left(\frac{13}{12}\right)^{14}} = \boxed{1.0833^{14}} = \boxed{\mathbf{3.066}}$

c. $\boxed{s_{12}} = \boxed{\frac{2^{12}\cdot(12+1)}{(-1)^{12+1}}} = \boxed{\frac{2^{12}\cdot 13}{(-1)^{13}}} = \boxed{\frac{4096\cdot 13}{-1}} = \boxed{-53248} = \boxed{\mathbf{-5.3248\times 10^4}}$

Example 4.1-5 Given the general term of the sequence $s(n) = s_n = n(n-2)+5$, write its kth and $k+1$ term.

Solutions:

a. To write the kth term of the sequence simply substitute k in place of n in the general term of the sequence, i.e., $\boxed{s(k)} = \boxed{s_k} = \boxed{k(k-2)+5} = \boxed{\mathbf{k^2-2k+5}}$

b. To write the $k+1$ term of the sequence simply substitute $k+1$ in place of n in the general term of the sequence, i.e., $\boxed{s(k+1)} = \boxed{s_{k+1}} = \boxed{(k+1)[(k+1)-2]+5} = \boxed{(k+1)[k+1-2]+5} = \boxed{(k+1)(k-1)+5}$

$= \boxed{k^2 - \not{k} + \not{k} - 1 + 5} = \boxed{\mathbf{k^2+4}}$

Example 4.1-6 For the given domain $(1, 2, 3, 4)$, write the first four terms of the following functions:

a. $f(x) = x^2 + 2x + 1$

b. $s(x) = 3x - 5$

c. $g(x) = \frac{2}{3}x$

d. $h(x) = x^{-1} + 1$

Solutions:

a. $\boxed{f(1)} = \boxed{f_1} = \boxed{1^2+(2\cdot 1)+1} = \boxed{1+2+1} = \boxed{\mathbf{4}}$

$\boxed{f(2)} = \boxed{f_2} = \boxed{2^2+(2\cdot 2)+1} = \boxed{4+4+1} = \boxed{\mathbf{9}}$

$\boxed{f(3)} = \boxed{f_3} = \boxed{3^2+(2\cdot 3)+1} = \boxed{9+6+1} = \boxed{\mathbf{16}}$

$\boxed{f(4)} = \boxed{f_4} = \boxed{4^2+(2\cdot 4)+1} = \boxed{16+8+1} = \boxed{\mathbf{25}}$

Therefore, the first four terms of the sequence are $(f_1, f_2, f_3, f_4) = \mathbf{(4, 9, 16, 25)}$

b. $\boxed{s(1)} = \boxed{s_1} = \boxed{(3\cdot 1)-5} = \boxed{3-5} = \boxed{\mathbf{-2}}$

$\boxed{s(2)} = \boxed{s_2} = \boxed{(3\cdot 2)-5} = \boxed{6-5} = \boxed{\mathbf{1}}$

$\boxed{s(3)} = \boxed{s_3} = \boxed{(3\cdot 3)-5} = \boxed{9-5} = \boxed{\mathbf{4}}$

$\boxed{s(4)} = \boxed{s_4} = \boxed{(3\cdot 4)-5} = \boxed{12-5} = \boxed{\mathbf{7}}$

Therefore, the first four terms of the sequence are $(s_1, s_2, s_3, s_4) = \mathbf{(-2, 1, 4, 7)}$.

c. $\boxed{g(1)} = \boxed{g_1} = \boxed{\frac{2}{3}\times 1} = \boxed{\mathbf{\frac{2}{3}}}$

$\boxed{g(2)} = \boxed{g_2} = \boxed{\frac{2}{3}\times 2} = \boxed{\mathbf{\frac{4}{3}}}$

$\boxed{g(3)} = \boxed{g_3} = \boxed{\frac{2}{\not{3}}\times \not{3}} = \boxed{\frac{2}{1}} = \boxed{\mathbf{2}}$

$\boxed{g(4)} = \boxed{g_4} = \boxed{\frac{2}{3}\times 4} = \boxed{\mathbf{\frac{8}{3}}}$

Therefore, the first four terms of the sequence are $(g_1, g_2, g_3, g_4) = \mathbf{\left(\frac{2}{3}, \frac{4}{3}, 2, \frac{8}{3}\right)}$

d. $\boxed{h(1)} = \boxed{h_1} = \boxed{1^{-1}+1} = \boxed{\frac{1}{1}+1} = \boxed{\frac{1+1}{1}} = \boxed{\mathbf{2}}$

$\boxed{h(2)} = \boxed{h_2} = \boxed{2^{-1}+1} = \boxed{\frac{1}{2}+1} = \boxed{\frac{1+2}{2}} = \boxed{\mathbf{\frac{3}{2}}}$

$\boxed{h(3)} = \boxed{h_3} = \boxed{3^{-1}+1} = \boxed{\frac{1}{3}+1} = \boxed{\frac{1+3}{4}} = \boxed{\frac{4}{3}}$ $\boxed{h(4)} = \boxed{h_4} = \boxed{4^{-1}+1} = \boxed{\frac{1}{4}+1} = \boxed{\frac{1+4}{4}} = \boxed{\frac{5}{4}}$

Therefore, the first four terms of the sequence are $(h_1, h_2, h_3, h_4) = \left(2, \frac{3}{2}, \frac{4}{3}, \frac{5}{4}\right)$

In the following section we will discuss series and identify its relation with sequences.

Section 4.1 Practice Problems - Sequences

1. List the first four and tenth terms of the given sequences.

a. $a_n = \frac{2n+1}{-2n}$ b. $b_k = \frac{k(k+1)}{k^2}$ c. $d_n = 3-(-2)^n$ d. $k_n = \left(-\frac{1}{2}\right)^n \frac{(-1)^{n+1}}{n+2}$

2. Write s_3, s_4, s_5, and s_8 for the following sequences.

a. $s_n = \frac{n(n+1)}{2n^{-1}}$ b. $s_n = (-1)^{n+1}2^{n-2}$ c. $s_n = \frac{(-2)^{n+1}(n-2)}{2n}$

3. Write the first five terms of the following sequences.

a. $a_n = (-1)^{n+1}(n+2)$ b. $a_i = 3\left(\frac{1}{100}\right)^{i-2}$ c. $c_i = 3\left(-\frac{1}{5}\right)^{i-1}$

d. $a_n = (3n-5)^2$ e. $u_k = ar^{k-2}+2$ f. $b_k = -3\left(\frac{2}{3}\right)^{k-2}$

g. $c_j = \frac{j}{j+1}+j$ h. $y_n = \left(1-\frac{1}{n+2}\right)^{n+1}$ i. $u_k = 1-(-1)^{k+1}$

j. $y_k = \frac{k}{2^{k-1}}$ k. $y_n = 9^{\frac{1}{k}}(k-2)$ l. $c_n = \frac{n^2-2}{n+1}$

4. Given $n!$ read as "n factorial" which is defined as $n! = n(n-1)(n-2)(n-3)\cdots 5\cdot 4\cdot 3\cdot 2\cdot 1$, find

a. The first eight terms of $n!$.

b. The first four terms of $a_n = \frac{2n+1}{n!}$.

c. The tenth and twelfth terms of the $c_n = \frac{1+3^{n-1}}{(n!)^2}$.

d. The first, fifth, tenth, and fifteenth terms of $y_n = \frac{n!(n-1)}{2+n!}$.

5. Write the first three terms of the following sequences.

a. $c_n = \frac{(2n-3)(n+1)}{(n-4)n}$ b. $a_n = \left(\frac{1}{n-1}\right)\left(\frac{n-2}{2+n}\right)$ c. $s_n = (-1)^{n+1}2^{n+1}$

d. $y_k = (-1)^{k+1}\frac{k(k-1)}{2}$ e. $b_n = n^2\left(\frac{n-1}{2+n}\right)$ f. $x_a = (5-a)^{a+1}2^a$

4.2 Series

Addition of the terms in any finite sequence result in having the sum of the sequence. The sum of the sequence is referred to as a **series**. For example, the sequence $y_k = \frac{1}{2^{k-1}}$ for $k = 1, 2, 3, 4, 5, and\ 6$ can be summed and expressed in the following way:

$$y_1 + y_2 + y_3 + y_4 + y_5 + y_6 = \frac{1}{2^0} + \frac{1}{2^1} + \frac{1}{2^2} + \frac{1}{2^3} + \frac{1}{2^4} + \frac{1}{2^5} = 1 + \frac{1}{2} + \frac{1}{4} + \frac{1}{8} + \frac{1}{16} + \frac{1}{32} = \mathbf{1.9687}$$

The sum of a sequence is generally shown by the Greek letter "$\sum$" (sigma) which is also called summation. Thus, using the sigma notation, the above example can be expressed in the following way $\sum_{i=1}^{6} y_k$ *where* $y_k = \frac{1}{2^{k-1}}$. Note that the variable i is referred to as the *index of summation* and the integer range over which the summation occurs is referred to as the *range of summation*. The following are three properties of summation that students should be familiar with:

$$\sum_{i=1}^{n} (a_i + b_i) = \sum_{i=1}^{n} a_i + \sum_{i=1}^{n} b_i$$

$$\sum_{i=1}^{n} k a_i = k \sum_{i=1}^{n} a_i$$

$$\sum_{i=1}^{n} k = nk$$

These properties are used extensively in solving the sum of sequences over a specified range as shown in the following examples:

Example 4.2-1 Given $\sum_{i=1}^{n} a_i = 20$ and $\sum_{i=1}^{n} b_i = 40$, find the solution to the following problems using the summation properties.

a. $\sum_{i=1}^{n} (2a_i + 3b_i) =$ b. $\sum_{i=1}^{n} (a_i - b_i) =$ c. $\sum_{i=1}^{n} (-5a_i + 2b_i) =$ d. $\sum_{i=1}^{n} \left(\frac{1}{2}a_i - \frac{1}{4}b_i\right) =$

Solutions:

a. $\boxed{\sum_{i=1}^{n} (2a_i + 3b_i)} = \boxed{\sum_{i=1}^{n} 2a_i + \sum_{i=1}^{n} 3b_i} = \boxed{2\sum_{i=1}^{n} a_i + 3\sum_{i=1}^{n} b_i} = \boxed{(2 \times 20) + (3 \times 40)} = \boxed{40 + 120} = \boxed{\mathbf{160}}$

b. $\boxed{\sum_{i=1}^{n} (a_i - b_i)} = \boxed{\sum_{i=1}^{n} a_i - \sum_{i=1}^{n} b_i} = \boxed{20 - 40} = \boxed{\mathbf{-20}}$

c. $\boxed{\sum_{i=1}^{n} (-5a_i + 2b_i)} = \boxed{\sum_{i=1}^{n} -5a_i + \sum_{i=1}^{n} 2b_i} = \boxed{-5\sum_{i=1}^{n} a_i + 2\sum_{i=1}^{n} b_i} = \boxed{(-5 \times 20) + (2 \times 40)} = \boxed{-100 + 80} = \boxed{\mathbf{-20}}$

d. $\boxed{\sum_{i=1}^{n}\left(\frac{1}{2}a_i-\frac{1}{4}b_i\right)}=\boxed{\sum_{i=1}^{n}\frac{1}{2}a_i-\sum_{i=1}^{n}\frac{1}{4}b_i}=\boxed{\frac{1}{2}\sum_{i=1}^{n}a_i-\frac{1}{4}\sum_{i=1}^{n}b_i}=\boxed{\left(\frac{1}{2}\times 20\right)+\left(-\frac{1}{4}\times 40\right)}=\boxed{10-10}=\boxed{0}$

Example 4.2-2 Solve the following series:

a. Find $\sum_{n=1}^{6}a_n$ *where* $a_n=\frac{2n+1}{n}$.

b. Find $\sum_{i=1}^{7}x_i$ *where* $x_i=\left(1+i^2\right)(-2)^i$.

c. Find $\sum_{j=0}^{5}\left(x_j-1\right)^2$ *where* $x_j=\frac{1}{1+j}$.

d. Find $\sum_{k=0}^{4}(u_k)^2$ *where* $u_k=k+1$.

e. Find $\sum_{n=1}^{5}(y_n-2)^{n+1^2}$ *where* $y_n=\frac{n}{1+n}$.

f. Find $\sum_{a=0}^{5}(u_a+a)^2$ *where* $u_a=a^2-1$.

Solutions:

a. $\boxed{\sum_{n=1}^{6}a_n \text{ where } a_n=\frac{2n+1}{n}}=\boxed{a_1+a_2+a_3+a_4+a_5+a_6}=\boxed{\frac{2+1}{1}+\frac{4+1}{2}+\frac{6+1}{3}+\frac{8+1}{4}+\frac{10+1}{5}+\frac{12+1}{6}}$

$=\boxed{\frac{3}{1}+\frac{5}{2}+\frac{7}{3}+\frac{9}{4}+\frac{11}{5}+\frac{13}{6}}=\boxed{3+2.5+2.33+2.25+2.2+2.17}=\boxed{\mathbf{14.45}}$

b. $\boxed{\sum_{i=1}^{7}x_i \text{ where } x_i=\left(1+i^2\right)(-2)^i}=\boxed{x_1+x_2+x_3+x_4+x_5+x_6+x_7}=\boxed{\left(1+1^2\right)\left(-2\right)^1+\left(1+2^2\right)\left(-2\right)^2}$

$\boxed{+\left(1+3^2\right)(-2)^3+\left(1+4^2\right)(-2)^4+\left(1+5^2\right)(-2)^5+\left(1+6^2\right)(-2)^6+\left(1+7^2\right)(-2)^7}=\boxed{(2\cdot-2)+(5\cdot 4)+(10\cdot-8)}$

$\boxed{+(17\cdot 16)+(26\cdot-32)+(37\cdot 64)+(50\cdot-128)}=\boxed{-4+20-80+272-832+2368-6400}=\boxed{\mathbf{-4656}}$

c. $\boxed{\sum_{j=0}^{5}\left(x_j-1\right)^2 \text{ where } x_j=\frac{1}{1+j}}=\boxed{(x_0-1)^2+(x_1-1)^2+(x_2-1)^2+(x_3-1)^2+(x_4-1)^2+(x_5-1)^2}$

$=\boxed{\left(\frac{1}{1}-1\right)^2+\left(\frac{1}{2}-1\right)^2+\left(\frac{1}{3}-1\right)^2+\left(\frac{1}{4}-1\right)^2+\left(\frac{1}{5}-1\right)^2+\left(\frac{1}{6}-1\right)^2}=\boxed{0+\left(-\frac{1}{2}\right)^2+\left(-\frac{2}{3}\right)^2+\left(-\frac{3}{4}\right)^2}$

$\boxed{+\left(\frac{4}{5}\right)^2+\left(-\frac{5}{6}\right)^2}=\boxed{\frac{1}{4}+\frac{4}{9}+\frac{9}{16}+\frac{16}{25}+\frac{25}{36}}=\boxed{0.25+0.4444+0.5625+0.64+0.6944}=\boxed{\mathbf{2.5913}}$

d. $\boxed{\sum_{k=0}^{4}(u_k)^2 \text{ where } u_k=k+1}=\boxed{(u_0)^2+(u_1)^2+(u_2)^2+(u_3)^2+(u_4)^2}=\boxed{(0+1)^2+(1+1)^2+(2+1)^2}$

$\boxed{+(3+1)^2+(4+1)^2}=\boxed{1^2+2^2+3^2+4^2+5^2}=\boxed{1+4+9+16+25}=\boxed{\mathbf{55}}$

e. $\boxed{\sum_{n=1}^{5}(y_n-2)^{n+1} \text{ where } y_n=\frac{n}{1+n}} = \boxed{(y_1-2)^{1+1}+(y_2-2)^{2+1}+(y_3-2)^{3+1}+(y_4-2)^{4+1}+(y_5-2)^{5+1}}$

$= \boxed{\left(\frac{1}{1+1}-2\right)^2+\left(\frac{2}{1+2}-2\right)^3+\left(\frac{3}{1+3}-2\right)^4+\left(\frac{4}{1+4}-2\right)^5+\left(\frac{5}{1+5}-2\right)^6} = \boxed{(-1.5)^2+(-1.33)^3+(-1.25)^4}$

$\boxed{+(-1.2)^5+(-1.17)^6} = \boxed{2.25-2.35+2.44-2.49+2.56} = \boxed{\mathbf{2.41}}$

f. $\boxed{\sum_{a=0}^{5}(u_a+a)^2 \text{ where } u_a=a^2-1} = \boxed{(u_0+0)^2+(u_1+1)^2+(u_2+2)^2+(u_3+3)^2+(u_4+4)^2+(u_5+5)^2}$

$= \boxed{(-1+0)^2+(0+1)^2+(3+2)^2+(8+3)^2+(15+4)^2+(24+5)^2} = \boxed{1+1+25+121+361+841} = \boxed{\mathbf{1350}}$

Example 4.2-3 Solve the following series.

a. $\sum_{a=1}^{5}a_n(2a-1) =$

b. $\sum_{a=1}^{5}\frac{2a+1}{a} =$

c. $\sum_{i=1}^{5}\frac{(-1)^{i+1}}{2i} =$

d. $\sum_{n=0}^{4}(n-1)^2(n+1) =$

e. $\sum_{j=-3}^{3}\frac{2^j}{j+5} =$

f. $\sum_{k=1}^{5}\frac{(1-k)^{k-1}}{k} =$

Solutions:

a. $\boxed{\sum_{a=1}^{5}a(2a-1)} = \boxed{[1\cdot(2\cdot1-1)]+[2\cdot(2\cdot2-1)]+[3\cdot(2\cdot3-1)]+[4\cdot(2\cdot4-1)]+[5\cdot(2\cdot5-1)]} = \boxed{1+6+15}$

$\boxed{+28+45} = \boxed{\mathbf{95}}$

b. $\boxed{\sum_{a=1}^{5}\frac{2a+1}{a}} = \boxed{\frac{(2\cdot1)+1}{1}+\frac{(2\cdot2)+1}{2}+\frac{(2\cdot3)+1}{3}+\frac{(2\cdot4)+1}{4}+\frac{(2\cdot5)+1}{5}} = \boxed{\frac{3}{1}+\frac{5}{2}+\frac{7}{3}+\frac{9}{4}+\frac{11}{5}} = \boxed{3+2.5}$

$\boxed{+2.33+2.25+2.2} = \boxed{\mathbf{12.28}}$

c. $\boxed{\sum_{i=1}^{5}\frac{(-1)^{i+1}}{2i}} = \boxed{\frac{(-1)^{1+1}}{2\cdot1}+\frac{(-1)^{2+1}}{2\cdot2}+\frac{(-1)^{3+1}}{2\cdot3}+\frac{(-1)^{4+1}}{2\cdot4}+\frac{(-1)^{5+1}}{2\cdot5}} = \boxed{\frac{(-1)^2}{2}+\frac{(-1)^3}{4}+\frac{(-1)^4}{6}+\frac{(-1)^5}{8}+\frac{(-1)^6}{10}}$

$= \boxed{\frac{1}{2}-\frac{1}{4}+\frac{1}{6}-\frac{1}{8}+\frac{1}{10}} = \boxed{0.5-0.25+0.167-0.125+0.1} = \boxed{\mathbf{0.392}}$

d. $\boxed{\sum_{n=0}^{4}(n-1)^2(n+1)} = \boxed{\left[(0-1)^2(0+1)\right]+\left[(1-1)^2(1+1)\right]+\left[(2-1)^2(2+1)\right]+\left[(3-1)^2(3+1)\right]+\left[(4-1)^2(4+1)\right]}$

$= \boxed{(1\cdot1)+(0\cdot2)+(1\cdot3)+(4\cdot4)+(9\cdot5)} = \boxed{1+0+3+16+45} = \boxed{\mathbf{65}}$

e. $\sum_{j=-3}^{3} \frac{2^j}{j+5} = \frac{2^{-3}}{-3+5} + \frac{2^{-2}}{-2+5} + \frac{2^{-1}}{-1+5} + \frac{2^0}{0+5} + \frac{2^1}{1+5} + \frac{2^2}{2+5} + \frac{2^3}{3+5} = \frac{2^{-3}}{2} + \frac{2^{-2}}{3} + \frac{2^{-1}}{4} + \frac{2^0}{5} + \frac{2^1}{6} + \frac{2^2}{7}$

$+\frac{2^3}{8} = \frac{0.125}{2} + \frac{0.25}{3} + \frac{0.5}{4} + \frac{1}{5} + \frac{2}{6} + \frac{4}{7} + \frac{8}{8} = 0.0625 + 0.0833 + 0.125 + 0.2 + 0.3333 + 0.5714 + 1 = \mathbf{2.3755}$

f. $\sum_{k=1}^{5} \frac{(1-k)^k}{k} = \frac{(1-1)^1}{1} + \frac{(1-2)^2}{2} + \frac{(1-3)^3}{3} + \frac{(1-4)^4}{4} + \frac{(1-5)^5}{5} = \frac{0}{1} + \frac{1}{2} - \frac{8}{3} + \frac{81}{4} - \frac{1024}{5} = 0.5 - 2.67$

$+20.25 - 204.8 = \mathbf{-186.72}$

Example 4.2-4 Prove that both sides of the following series are equal to one another.

a. $\sum_{i=1}^{n} 2x_i + \sum_{i=1}^{n} 4y_i = 2\sum_{i=1}^{n}(x_i + 2y_i)$

b. $\sum_{i=1}^{n} a\, y_i^2 = a \sum_{i=1}^{n} y_i^2$

c. $\sum_{i=1}^{n} a = n\, a$

d. $\sum_{i=1}^{n}(x_i + a) = \sum_{i=1}^{n} x_i + na$

Solutions:

a. $\sum_{i=1}^{n} 2x_i + \sum_{i=1}^{n} 4y_i = (2x_1 + 2x_2 + 2x_3 + \cdots + 2x_n) + (4y_1 + 4y_2 + 4y_3 + \cdots + 4y_n) = (2x_1 + 4y_1) + (2x_2 + 4y_2)$

$+(2x_3 + 4y_3) + \cdots + (2x_n + 4y_n) = 2(x_1 + 2y_1) + 2(x_2 + 2y_2) + 2(x_3 + 2y_3) + \cdots + 2(x_n + 2y_n)$

$= \mathbf{2\sum_{i=1}^{n}(x_i + 2y_i)}$

b. $\sum_{i=1}^{n} a\, y_i^2 = ay_1^2 + ay_2^2 + ay_3^2 + ay_4^2 + \cdots + ay_n^2 = a\left(y_1^2 + y_2^2 + y_3^2 + y_4^2 + \cdots + y_n^2\right) = \mathbf{a \sum_{i=1}^{n} y_i^2}$

c. $\sum_{i=1}^{n} a = \underbrace{a + a + a + a + \cdots + a}_{n \text{ terms}} = \mathbf{n\, a}$

d. $\sum_{i=1}^{n}(x_i + a) = (x_1 + a) + (x_2 + a) + (x_3 + a) + \cdots + (x_n + a) = (x_1 + x_2 + x_3 + \cdots + x_n) + (a + a + a + \cdots + a)$

$= \mathbf{\sum_{i=1}^{n} x_i + na}$

Example 4.2-5 Use the properties of summation to evaluate the following series.

a. $\sum_{i=1}^{6} 2k =$

b. $\sum_{i=1}^{7}(4k - 3) =$

c. $\sum_{k=1}^{4}\left(k^3 - 2k\right) =$

d. $\sum_{k=1}^{5}\left(k^2+a\right)=$ e. $\sum_{k=1}^{4}2\cdot\left(-\frac{2}{3}\right)^{k+1}=$ f. $\sum_{k=1}^{5}\left(2^k+k\right)=$

Solutions:

a. $\boxed{\sum_{i=1}^{6}2k}=\boxed{2\sum_{i=1}^{6}k}=\boxed{2\cdot 6k}=\boxed{\mathbf{12k}}$

b. $\boxed{\sum_{i=1}^{7}(4k-3)}=\boxed{\sum_{i=1}^{7}4k+\sum_{i=1}^{7}-3}=\boxed{4\sum_{i=1}^{7}k-\sum_{i=1}^{7}3}=\boxed{4\cdot 7k-7\cdot 3}=\boxed{\mathbf{28k-21}}$

c. $\boxed{\sum_{k=1}^{4}\left(k^3-2k\right)}=\boxed{\sum_{k=1}^{4}k^3+\sum_{k=1}^{4}-2k}=\boxed{\sum_{k=1}^{4}k^3-2\sum_{k=1}^{4}k}=\boxed{\left(1^3+2^3+3^3+4^3\right)-2(1+2+3+4)}=\boxed{1+8+27+64}$

$\boxed{-(2\cdot 10)}=\boxed{100-20}=\boxed{\mathbf{80}}$

d. $\boxed{\sum_{k=1}^{5}\left(k^2+a\right)}=\boxed{\sum_{k=1}^{5}k^2+\sum_{k=1}^{5}a}=\boxed{\sum_{k=1}^{5}k^2+5a}=\boxed{\left(1^2+2^2+3^2+4^2+5^2\right)+5a}=\boxed{(1+4+9+16+25)+5a}$

$=\boxed{55+5a}=\boxed{\mathbf{5(11+a)}}$

e. $\boxed{\sum_{k=1}^{4}2\cdot\left(-\frac{2}{3}\right)^{k+1}}=\boxed{2\sum_{k=1}^{4}\left(-\frac{2}{3}\right)^{k+1}}=\boxed{2\cdot\left[\left(-\frac{2}{3}\right)^2+\left(-\frac{2}{3}\right)^3+\left(-\frac{2}{3}\right)^4+\left(-\frac{2}{3}\right)^5\right]}=\boxed{2\cdot\left(\frac{4}{9}-\frac{8}{27}+\frac{16}{81}-\frac{32}{243}\right)}$

$=\boxed{2\cdot(0.4444-0.2963+0.1975-0.1317)}=\boxed{2\cdot 0.2139}=\boxed{\mathbf{0.4278}}$

f. $\boxed{\sum_{k=1}^{5}\left(2^k+k\right)}=\boxed{\sum_{k=1}^{5}2^k+\sum_{k=1}^{5}k}=\boxed{\left(2^1+2^2+2^3+2^4+2^5\right)+(1+2+3++4+5)}=\boxed{(2+4+8+16+32)+15}$

$=\boxed{62+15}=\boxed{77}$

Section 4.2 Practice Problems - Series

1. Given $\sum_{i=1}^{n}a_i=10$ and $\sum_{i=1}^{n}b_i=25$, find

a. $\sum_{i=1}^{n}(2a_i+4b_i)=$ b. $\sum_{i=1}^{n}(-a_i+b_i)=$ c. $\sum_{i=1}^{n}(3a_i+5b_i)=$ d. $\sum_{i=1}^{n}\left(\frac{1}{2}a_i+\frac{1}{5}b_i\right)=$

2. Evaluate each of the following series.

a. $\sum_{k=1}^{5}y_k\ where\ y_k=2+k$ b. $\sum_{n=0}^{6}x_n\ where\ x_n=\frac{1}{(-2)^{n+1}}$ c. $\sum_{n=0}^{4}x_n\ where\ x_n=(-1)^{n+1}$

d. $\sum_{j=-3}^{3} u_j$ *where* $u_j = j - 3j^2$ e. $\sum_{a=3}^{5} y^a$ *where* $y = a + 2$ f. $\sum_{i=0}^{5} x_i$ *where* $x_i = \frac{(-1)^{i+1}}{2^i}$

g. $\sum_{k=-2}^{3} y^{k+2}$ *where* $y = 2k - 3$ h. $\sum_{m=1}^{5} (x_m - 1)^2$ *where* $x_m = \frac{1}{m}$

3. Find the sum of the following series within the specified range.

a. $\sum_{i=-3}^{3} 10^i =$ b. $\sum_{n=0}^{6} \frac{n-1}{2^n} =$ c. $\sum_{a=0}^{4} \frac{1}{10^a} =$

d. $\sum_{n=1}^{5} \left(n^2 - n\right) =$ e. $\sum_{m=0}^{6} (-1)^{m+1} =$ f. $\sum_{k=0}^{5} \frac{1+(-1)^k}{2^k} =$

g. $\sum_{a=1}^{6} 5(a-1) + 3 =$ h. $\sum_{k=0}^{5} \left(-\frac{1}{3}\right)^{k-1} =$ i. $\sum_{j=1}^{5} \left(j - 3j^2\right) =$

j. $\sum_{n=1}^{4} \frac{n+1}{n} - \sum_{n=1}^{4} \frac{n^2}{n+1} =$ k. $\sum_{k=1}^{5} 5k^{-1} =$ l. $\sum_{i=1}^{4} (-0.1)^{2i-5} =$

4. Rewrite the following terms using the sigma notation.

a. $\frac{1}{2} + \frac{1}{3} + \frac{1}{4} + \frac{1}{5} + \frac{1}{6} + \frac{1}{7} =$ b. $\frac{1}{2} + \frac{2}{3} + \frac{3}{4} + \frac{4}{5} + \frac{5}{6} + \frac{6}{7} =$ c. $2 + 4 + 8 + 16 + 32 + 64 =$

d. $1 + \frac{1}{2} + \frac{1}{3} + \frac{1}{4} + \frac{1}{5} + \frac{1}{6} =$ e. $0 + \frac{1}{2} + \frac{2}{3} + \frac{3}{4} + \frac{4}{5} + \frac{5}{6} =$ f. $1 - \frac{1}{2} + \frac{1}{3} - \frac{1}{4} + \frac{1}{5} - \frac{1}{6} =$

4.3 Arithmetic Sequences and Arithmetic Series

An arithmetic sequence is a sequence in which each term, after the first term, is obtained by adding a common number to the preceding term. The common number added to each term can be found by taking the **common difference**, denoted by d, of two successive terms. For example, the sequences $3, 6, 9, 12, 15, \cdots$ and $\frac{1}{2}, 1, \frac{3}{2}, 2, \frac{5}{2}, 3, \frac{7}{2}, \cdots$ are arithmetic sequences because the common difference that is added to each term in order to obtain the next term is $6-3 = 3$ and $1-\frac{1}{2} = \frac{1}{2}$, respectively. Note that the nth term in both examples can easily be stated as $s_n = 3n$ and $s_n = \frac{1}{2}n$. Therefore, the two arithmetic sequences can be written as $3, 6, 9, 12, 15, \cdots, 3n$ and $\frac{1}{2}, 1, \frac{3}{2}, 2, \frac{5}{2}, 3, \frac{7}{2}, \cdots, \frac{1}{2}n$.

To obtain the nth term of an arithmetic sequence, a general form can be developed by letting s_n and d be the nth term and the common difference of an arithmetic sequence. Thus, the first terms can be written as:

$s_1 = a$

$s_2 = s_1 + d$ where a and d are real numbers and n is a positive integer

$s_3 = s_2 + d = (s_1 + d) + d = s_1 + 2d$

$s_4 = s_3 + d = (s_1 + 2d) + d = s_1 + 3d$

$s_5 = s_4 + d = (s_1 + 3d) + d = s_1 + 4d$

$\vdots$

$s_n = s_{n-1} + d = [s_1 + (n-2)d] + d = s_1 + nd - 2d + d = s_1 + nd - d = s_1 + (n-1)d$

$s_{n+1} = s_n + d = [s_1 + (n-1)d] + d = s_1 + nd - d + d = s_1 + nd$

Thus, the nth and $n+1$ term of an arithmetic sequence is equal to

$$\mathbf{s_n = s_1 + (n-1)d} \qquad (1)$$

$$\mathbf{s_{n+1} = s_1 + nd} \qquad (2)$$

In the following examples the above equations (1) and (2) are used in order to find several terms of arithmetic sequences.

Example 4.3-1 Find the next five terms of the following arithmetic sequences.

a. $s_1 = 5$, $d = 3$ b. $s_1 = -5$, $d = 2$ c. $s_1 = 20$, $d = 0.4$

Solutions:

a. The nth term for an arithmetic sequence is equal to $s_n = s_1 + (n-1)d$. Substituting $s_1 = 5$ and $d = 3$ into the general arithmetic expression for $n = 2, 3, 4, 5, and\ 6$ we obtain

$\boxed{s_2} = \boxed{s_1 + (2-1)d} = \boxed{s_1 + d} = \boxed{5+3} = \boxed{\mathbf{8}}$

$\boxed{s_3} = \boxed{s_1 + (3-1)d} = \boxed{s_1 + 2d} = \boxed{5 + (2 \times 3)} = \boxed{5+6} = \boxed{\mathbf{11}}$

$\boxed{s_4} = \boxed{s_1 + (4-1)d} = \boxed{s_1 + 3d} = \boxed{5 + (3 \times 3)} = \boxed{5+9} = \boxed{\mathbf{14}}$

$\boxed{s_5} = \boxed{s_1 + (5-1)d} = \boxed{s_1 + 4d} = \boxed{5 + (4 \times 3)} = \boxed{5+12} = \boxed{\mathbf{17}}$

$\boxed{s_6} = \boxed{s_1 + (6-1)d} = \boxed{s_1 + 5d} = \boxed{5 + (5 \times 3)} = \boxed{5+15} = \boxed{\mathbf{20}}$

Thus, the first six terms of the arithmetic sequence are $(\mathbf{5, 8, 11, 14, 17, 20})$.

b. Substituting $s_1 = -5$ and $d = 2$ into the general arithmetic expression for $n = 2, 3, 4, 5, and\ 6$ we obtain

$\boxed{s_2} = \boxed{s_1 + (2-1)d} = \boxed{s_1 + d} = \boxed{-5+2} = \boxed{\mathbf{-3}}$

$\boxed{s_3} = \boxed{s_1 + (3-1)d} = \boxed{s_1 + 2d} = \boxed{-5 + (2 \times 2)} = \boxed{-5+4} = \boxed{\mathbf{-1}}$

$\boxed{s_4} = \boxed{s_1 + (4-1)d} = \boxed{s_1 + 3d} = \boxed{-5 + (3 \times 2)} = \boxed{-5+6} = \boxed{\mathbf{1}}$

$\boxed{s_5} = \boxed{s_1 + (5-1)d} = \boxed{s_1 + 4d} = \boxed{-5 + (4 \times 2)} = \boxed{-5+8} = \boxed{\mathbf{3}}$

$\boxed{s_6} = \boxed{s_1 + (6-1)d} = \boxed{s_1 + 5d} = \boxed{-5 + (5 \times 2)} = \boxed{-5+10} = \boxed{\mathbf{5}}$

Thus, the first six terms of the arithmetic sequence are $(\mathbf{-5, -3, -1, 1, 3, 5})$.

c. Substituting $s_1 = 20$ and $d = 0.4$ into the general arithmetic expression for $n = 2, 3, 4, 5, and\ 6$ we obtain

$\boxed{s_2} = \boxed{s_1 + (2-1)d} = \boxed{s_1 + d} = \boxed{20 + 0.4} = \boxed{\mathbf{20.4}}$

$\boxed{s_3} = \boxed{s_1 + (3-1)d} = \boxed{s_1 + 2d} = \boxed{20 + (2 \times 0.4)} = \boxed{20 + 0.8} = \boxed{\mathbf{20.8}}$

$\boxed{s_4} = \boxed{s_1 + (4-1)d} = \boxed{s_1 + 3d} = \boxed{20 + (3 \times 0.4)} = \boxed{20 + 1.2} = \boxed{\mathbf{21.2}}$

$\boxed{s_5} = \boxed{s_1 + (5-1)d} = \boxed{s_1 + 4d} = \boxed{20 + (4 \times 0.4)} = \boxed{20 + 1.6} = \boxed{\mathbf{21.6}}$

$\boxed{s_6} = \boxed{s_1 + (6-1)d} = \boxed{s_1 + 5d} = \boxed{20 + (5 \times 0.4)} = \boxed{20 + 2} = \boxed{\mathbf{22}}$

Thus, the first six terms of the arithmetic sequence are $(\mathbf{20, 20.4, 20.8, 21.2, 21.6, 22})$.

Example 4.3-2 Find the general term and the fiftieth term of the following arithmetic sequences.

a. $s_1 = 3$, $d = 5$ b. $s_1 = -2$, $d = 4$ c. $s_1 = 10$, $d = -2.5$

Solutions:

a. The nth term for an arithmetic sequence is equal to $s_n = s_1 + (n-1)d$. Substituting $s_1 = 3$ and $d = 5$ into the general arithmetic expression we obtain

$\boxed{s_n} = \boxed{3 + (n-1)5} = \boxed{3 + 5n - 5} = \boxed{5n + (3-5)} = \boxed{\mathbf{5n - 2}}$

substituting $n = 50$ into the general equation $s_n = 5n - 2$ we find

$\boxed{s_{50}} = \boxed{(5 \times 50) - 2} = \boxed{250 - 2} = \boxed{\mathbf{248}}$

b. Substituting $s_1 = -2$ and $d = 4$ into the general arithmetic expression $s_n = s_1 + (n-1)d$ we obtain

$\boxed{s_n} = \boxed{-2 + (n-1)4} = \boxed{-2 + 4n - 4} = \boxed{4n + (-2 - 4)} = \boxed{\mathbf{4n - 6}}$

substituting $n = 50$ into the general equation $s_n = 4n - 6$ we find

$\boxed{s_{50}} = \boxed{(4 \times 50) - 6} = \boxed{200 - 6} = \boxed{\mathbf{194}}$

c. Substituting $s_1 = 10$ and $d = -2.5$ into the general arithmetic expression $s_n = s_1 + (n-1)d$ we obtain $\boxed{s_n} = \boxed{10 + (n-1) \times -2.5} = \boxed{10 - 2.5n + 2.5} = \boxed{-2.5n + (10 + 2.5)} = \boxed{\mathbf{-2.5n + 12.5}}$

substituting $n = 50$ into the general equation $s_n = -2.5n + 12.5$ we find

$\boxed{s_{50}} = \boxed{(-2.5 \times 50) + 12.5} = \boxed{-125 + 12.5} = \boxed{\mathbf{-112.5}}$

Example 4.3-3 Find the next four terms in each of the following arithmetic sequences.

a. $6, 10, \cdots$ b. $x, x+2, \cdots$ c. $2x+1, 2x+5, \cdots$ d. $x, x-29, \cdots$

Solutions:

a. The first term s_1 and the common difference d are equal to $s_1 = 6$ and $d = 10 - 6 = 4$. Thus, using the general arithmetic equation $s_n = s_1 + (n-1)d$ or $s_{n+1} = s_n + d$ the next four terms are as follows: Let's use $s_{n+1} = s_n + d$. Then,

$\boxed{s_3} = \boxed{s_2 + d} = \boxed{10 + 4} = \boxed{\mathbf{14}}$ $\boxed{s_4} = \boxed{s_3 + d} = \boxed{14 + 4} = \boxed{\mathbf{18}}$

$\boxed{s_5} = \boxed{s_4 + d} = \boxed{18 + 4} = \boxed{\mathbf{22}}$ $\boxed{s_6} = \boxed{s_5 + d} = \boxed{22 + 4} = \boxed{\mathbf{26}}$

b. The first term s_1 and the common difference d are equal to $s_1 = x$ and $d = x + 2 - x = 2$. Thus,

$\boxed{s_3} = \boxed{s_2 + d} = \boxed{(x+2)+2} = \boxed{\mathbf{x+4}}$ $\boxed{s_4} = \boxed{s_3 + d} = \boxed{(x+4)+2} = \boxed{\mathbf{x+6}}$

$\boxed{s_5} = \boxed{s_4 + d} = \boxed{(x+6)+2} = \boxed{\mathbf{x+8}}$ $\boxed{s_6} = \boxed{s_5 + d} = \boxed{(x+8)+2} = \boxed{\mathbf{x+10}}$

c. The first term s_1 and the common difference d are equal to $s_1 = 2x + 1$ and $d = (2x+5) - (2x+1)$. $= 2x + 5 - 2x - 1 = 4$. Thus,

$\boxed{s_3} = \boxed{s_2 + d} = \boxed{(2x+1)+4} = \boxed{\mathbf{2x+5}}$ $\boxed{s_4} = \boxed{s_3 + d} = \boxed{(2x+5)+4} = \boxed{\mathbf{2x+9}}$

$\boxed{s_5} = \boxed{s_4 + d} = \boxed{(2x+9)+4} = \boxed{\mathbf{2x+13}}$ $\boxed{s_6} = \boxed{s_5 + d} = \boxed{(2x+13)+4} = \boxed{\mathbf{2x+17}}$

d. The first term s_1 and the common difference d are equal to $s_1 = x$ and $d = (x - 29) - x = -29$. Thus,

$\boxed{s_3} = \boxed{s_2 + d} = \boxed{(x-29)-29} = \boxed{\mathbf{x-58}}$ $\boxed{s_4} = \boxed{s_3 + d} = \boxed{(x-58)-29} = \boxed{\mathbf{x-87}}$

$\boxed{s_5} = \boxed{s_4 + d} = \boxed{(x-87)-29} = \boxed{\mathbf{x-116}}$ $\boxed{s_6} = \boxed{s_5 + d} = \boxed{(x-116)-29} = \boxed{\mathbf{x-145}}$

Example 4.3-4 The first term of an arithmetic sequence is -5 and the fourth term is 10. Find the twentieth term.

Solution:

Since $s_1 = -5$ and $s_4 = 10$ we use the general formula $s_n = s_1 + (n-1)d$ in order to solve for d.

$\boxed{s_4 = s_1 + (4-1)d}$; $\boxed{10 = -5 + (4-1)d}$; $\boxed{10 = -5 + 3d}$; $\boxed{10 + 5 = 3d}$; $\boxed{15 = 3d}$; $\boxed{d = \frac{15}{3}}$; $\boxed{d = 5}$. Then,

$\boxed{s_{20} = s_1 + (20-1)d}$; $\boxed{s_{20} = -5 + 19d}$; $\boxed{s_{20} = -5 + (19 \times 5)}$; $\boxed{s_{20} = -5 + 95}$; $\boxed{\mathbf{s_{20} = 90}}$

Having learned about arithmetic sequences and the steps for finding the terms of an arithmetic sequence, we will next learn about arithmetic series and the steps for finding the sum of arithmetic series over a given range.

Addition of the terms in an arithmetic sequence result in having an arithmetic series. To obtain the arithmetic series formula let $s_k = s_1 + (k-1)d$ be an arithmetic sequence and denote the sum of the first n terms by

$$S_n = \sum_{k=1}^{n} s_1 + (k-1)d$$

then,

$$S_n = s_1 + (s_1 + d) + \cdots + [s_1 + (n-2)d] + [s_1 + (n-1)d] \qquad (a)$$

Let's write the sum in reverse order and add the two series (a) and (b) together.

$$S_n = [s_1 + (n-1)d] + [s_1 + (n-2)d] + \cdots + (s_1 + d) + s_1 \qquad (b)$$

$$S_n + S_n = \{s_1 + [s_1 + (n-1)d]\} + \{(s_1 + d) + [s_1 + (n-2)d]\} + \cdots + \{[s_1 + (n-2)d] + (s_1 + d)\} + \{[s_1 + (n-1)d] + s_1\}$$

$$2S_n = [s_1 + s_1 + (n-1)d] + [s_1 + d + s_1 + (n-2)d] + \cdots + [s_1 + (n-2)d + s_1 + d] + [s_1 + (n-1)d + s_1]$$

$$2S_n = [2s_1 + (n-1)d] + [s_1 + s_1 + (n-2)d + d] + \cdots + [s_1 + s_1 + (n-2)d + d] + [s_1 + s_1 + (n-1)d]$$

$$2S_n = [2s_1 + (n-1)d] + [2s_1 + nd - 2d + d] + \cdots + [2s_1 + nd - 2d + d] + [2s_1 + (n-1)d]$$

$$2S_n = [2s_1 + (n-1)d] + [2s_1 + nd - d] + \cdots + [2s_1 + nd - d] + [2s_1 + (n-1)d]$$

$$2S_n = [2s_1 + (n-1)d] + [2s_1 + (n-1)d] + \cdots + [2s_1 + (n-1)d] + [2s_1 + (n-1)d]$$

$$2S_n = n[2s_1 + (n-1)d] \; ; \; S_n = \frac{n[2s_1 + (n-1)d]}{2} = \frac{n}{2}[2s_1 + (n-1)d]$$

Therefore, the arithmetic series can be written in the following two forms:

$$\mathbf{S_n = \sum_{k=1}^{n} s_1 + (k-1)d} \qquad (1)$$

$$\mathbf{S_n = \frac{n}{2}[2s_1 + (n-1)d]} \qquad (2)$$

Note that equation (2), similar to the nth term of the arithmetic sequence $[s_n = s_1 + (n-1)d]$, is given in terms of s_1, n, and d.

In the following examples the above equations (1) and (2) are used in order to find the sum of arithmetic series.

Example 4.3-5 Find the sum of the following arithmetic series.

a. $\sum_{i=1}^{20}(2i+1) =$ b. $\sum_{i=1}^{15}(3i-2) =$ c. $\sum_{j=3}^{15}(5j-1) =$

Solutions:

a. **First** - Write the first three terms of the arithmetic series in expanded form, i.e.,

$$\boxed{\sum_{i=1}^{20}(2i+1)} = \boxed{(2+1)+(4+1)+(6+1)+\cdots} = \boxed{3+5+7+\cdots}$$

Second - Identify the first term, s_1, the difference between the two terms, d, and n, i.e., $s_1 = 3$, $d = 5-3 = 2$, and $n = 20$.

Third - Use the arithmetic series formula to obtain the sum of the twenty terms.

$$S_n = \frac{n}{2}[2s_1 + (n-1)d]$$

$$\boxed{S_{20}} = \boxed{\frac{20}{2}[2s_1 + (20-1)d]} = \boxed{10[2s_1 + 19d]} = \boxed{10[(2\times 3)+(19\times 2)]} = \boxed{10[6+38]} = \boxed{10\times 44} = \boxed{\mathbf{440}}$$

Note that prior to learning the arithmetic series formula the only method that we could use was by summing each term as shown below:

$$\boxed{\sum_{i=1}^{20}(2i+1)} = \boxed{(2+1)+(4+1)+(6+1)+(8+1)+(10+1)+(12+1)+(14+1)+(16+1)+(18+1)+(20+1)}$$

$$\boxed{+(22+1)+(24+1)+(26+1)+(28+1)+(30+1)+(32+1)+(34+1)+(36+1)+(38+1)+(40+1)}$$

$$= \boxed{3+5+7+9+11+13+15+17+19+21+23+25+27+29+31+33+35+37+39+41} = \boxed{\mathbf{440}}$$

As you note, it is much easier to use the arithmetic series formula as opposed to the summation of each term which is fairly long and time consuming.

b. **First** - Write the first three terms of the arithmetic series in expanded form, i.e.,

$$\boxed{\sum_{i=1}^{15}(3i-2)} = \boxed{(3-2)+(6-2)+(9-2)+\cdots} = \boxed{1+4+7+\cdots}$$

Second - Identify the first term, s_1, the difference between the two terms, d, and n, i.e., $s_1 = 1$, $d = 4-1 = 3$, and $n = 15$.

Third - Use the arithmetic series formula to obtain the sum of the fifteen terms.

$$S_n = \frac{n}{2}[2s_1 + (n-1)d]$$

$$\boxed{S_{15}} = \boxed{\frac{15}{2}[2s_1 + (15-1)d]} = \boxed{7.5[2s_1 + 14d]} = \boxed{7.5[(2\times 1)+(14\times 3)]} = \boxed{7.5[2+42]} = \boxed{7.5\times 44} = \boxed{\mathbf{330}}$$

or, we can obtain the answer by summing up the first fifteen terms of the series, i.e.,

$$\boxed{\sum_{i=1}^{15}(3i-2)} = \boxed{1+4+7+10+13+16+19+22+25+28+31+34+37+40+43} = \boxed{\mathbf{330}}$$

c. **First** - Write the first three terms of the arithmetic series in expanded form, i.e.,

$$\boxed{\sum_{j=3}^{15}(5j-1)} = \boxed{(15-1)+(20-1)+(25-1)+\cdots} = \boxed{14+19+24+\cdots}$$

Second - Identify the first term, s_1, the difference between the two terms, d, and n, i.e., $s_1 = 14$, $d = 19-14 = 5$, and $n = 13$.

Third - Use the arithmetic series formula to obtain the sum of the thirteen terms.

$$S_n = \frac{n}{2}[2s_1 + (n-1)d]$$

$$\boxed{S_{15}} = \boxed{\frac{13}{2}[2s_1 + (13-1)d]} = \boxed{6.5[2s_1 + 12d]} = \boxed{6.5[(2\times 14)+(12\times 5)]} = \boxed{6.5[28+60]} = \boxed{6.5\times 88} = \boxed{\mathbf{572}}$$

or, we can obtain the answer by summing up the first thirteen terms of the series, i.e.,

$$\boxed{\sum_{j=3}^{15}(5j-1)} = \boxed{14+19+24+29+34+39+44+49+54+59+64+69+74} = \boxed{\mathbf{572}}$$

Example 4.3-6 Given the first term s_1 and d, find S_{80} for each of the following arithmetic sequences.

a. $s_1 = 5$, $d = 2$ b. $s_1 = -10$, $d = 3$ c. $s_1 = 500$, $d = 25$

Solutions:

a. The nth term for an arithmetic series is equal to $S_n = \frac{n}{2}[2s_1 + (n-1)d]$. Substituting $s_1 = 5$ and $d = 2$ into the general arithmetic expression we obtain

$$\boxed{S_{80}} = \boxed{\frac{80}{2}[2s_1 + (80-1)d]} = \boxed{40[2s_1 + 79d]} = \boxed{40[(2\times 5)+(79\times 2)]} = \boxed{40[10+158]} = \boxed{40\times 168} = \boxed{\mathbf{6720}}$$

b. Substituting $s_1 = -10$ and $d = 3$ into $S_n = \frac{n}{2}[2s_1 + (n-1)d]$ we obtain

$$\boxed{S_{80}} = \boxed{\frac{80}{2}[2s_1 + (80-1)d]} = \boxed{40[2s_1 + 79d]} = \boxed{40[(2\times -10)+(79\times 3)]} = \boxed{40[-20+237]} = \boxed{\mathbf{8680}}$$

c. Substituting $s_1 = 500$ and $d = 25$ into $s_n = \frac{n}{2}[2s_1 + (n-1)d]$ we obtain

$$\boxed{S_{80}} = \boxed{\frac{80}{2}[2s_1 + (80-1)d]} = \boxed{40[2s_1 + 79d]} = \boxed{40[(2\times 500)+(79\times 25)]} = \boxed{40[1000+1975]} = \boxed{\mathbf{119000}}$$

Example 4.3-7 Find the sum of the following sequences for the indicated values.

a. S_{35} for the sequence $-5, 3, \cdots$ b. S_{200} for the sequence $-10, 10, \cdots$

Solutions:

a. The first term s_1 and the common difference d are equal to $s_1 = -5$ and $d = 3 - (-5) = 3 + 5 = 8$.

Thus, using the general arithmetic series $S_n = \frac{n}{2}[2s_1 + (n-1)d]$ we obtain

$$\boxed{S_{35}} = \boxed{\frac{35}{2}[2s_1 + (35-1)d]} = \boxed{17.5[2s_1 + 34d]} = \boxed{17.5[(2 \times -5) + (34 \times 8)]} = \boxed{17.5[-10 + 272]} = \boxed{\mathbf{4585}}$$

b. The first term s_1 and the common difference d are equal to $s_1 = -10$ and $d = 10 - (-10) = 20$.

Thus, using the general arithmetic series $S_n = \frac{n}{2}[2s_1 + (n-1)d]$ we obtain

$$\boxed{S_{200}} = \boxed{\frac{200}{2}[2s_1 + (200-1)d]} = \boxed{100[2s_1 + 199d]} = \boxed{100[(2 \times -10) + (199 \times 20)]} = \boxed{100[-20 + 3980]}$$

$$= \boxed{\mathbf{396000}}$$

Section 4.3 Practice Problems - Arithmetic Sequences and Arithmetic Series

1. Find the next seven terms of the following arithmetic sequences.

a. $s_1 = 3$, $d = 2$ b. $s_1 = -3$, $d = 2$ c. $s_1 = 10$, $d = 0.8$

2. find the general term and the eighth term of the following arithmetic sequences.

a. $s_1 = 3$, $d = 4$ b. $s_1 = -3$, $d = 5$ c. $s_1 = 8$, $d = -1.2$

3. find the next six terms in each of the following arithmetic sequences.

a. $5, 8, \cdots$ b. $x, x+4, \cdots$ c. $3x+1, 3x+4, \cdots$ d. $w, w-10, \cdots$

4. Find the sum of the following arithmetic series.

a. $\sum_{i=10}^{20}(2i-4) =$ b. $\sum_{k=1}^{1000} k =$ c. $\sum_{k=1}^{100}(2k-3) =$

d. $\sum_{i=1}^{15} 3i =$ e. $\sum_{i=1}^{10}(i+1) =$ f. $\sum_{k=5}^{15}(2k-1) =$

g. $\sum_{i=4}^{10}(3i+4) =$ h. $\sum_{j=5}^{13}(3j+1) =$ i. $\sum_{k=7}^{18}(4k-3) =$

5. The first term of an arithmetic sequence is 6 and the third term is 24. Find the tenth term.

6. Given the first term s_1 and d, find S_{50} for each of the following arithmetic sequences.

a. $s_1 = 2$, $d = 5$ b. $s_1 = -5$, $d = 6$ c. $s_1 = 30$, $d = 10$

7. Find the sum of the following sequences for the indicated values.

a. S_{15} for the sequence $-8, 6, \cdots$ b. S_{100} for the sequence $-20, 20, \cdots$

4.4 Geometric Sequences and Geometric Series

A geometric sequence is a sequence in which each term, after the first term, is obtained by multiplying the preceding term by a common multiplier. This common multiplier is also called the **common ratio** and is denoted by r. The common ratio r is obtained by division of two successive terms in a sequence. For example, the sequences $3, 6, 12, 24, 48, \cdots$ and $\frac{1}{2}, \frac{1}{4}, \frac{1}{8}, \frac{1}{16}, \frac{1}{32}, \cdots$ are geometric sequences because the common ratio that each term is multiplied by in order to obtain the next term is equal to $\frac{6}{3} = 2$ and $\frac{\frac{1}{4}}{\frac{1}{2}} = \frac{1 \times 2}{4 \times 1} = \frac{2}{4} = \frac{1}{2}$, respectively.

To obtain the nth term of a geometric sequence, a general form can be developed by letting s_n and r be the nth term and the common multiplier (common ratio) of a geometric sequence. Thus, the first terms can be written as:

$s_1 = a$ where a is a real number and n is a positive integer

$s_2 = s_1 r$

$s_3 = s_2 r = (s_1 r) \cdot r = s_1 r^2$

$s_4 = s_3 r = (s_1 r^2) \cdot r = s_1 r^3$

$s_5 = s_4 r = (s_1 r^3) \cdot r = s_1 r^4$

$\vdots$

$s_n = s_{n-1} r = (s_1 r^{n-2}) \cdot r = s_1 r^{n-2+1} = s_1 r^{n-1}$

$s_{n+1} = s_n r = (s_1 r^{n-1}) \cdot r = s_1 r^{n-1+1} = s_1 r^n$

Thus, the nth and $n+1$ term of an arithmetic sequence is equal to

$$\mathbf{s_n = s_1 r^{n-1}} \qquad (1)$$

$$\mathbf{s_{n+1} = s_1 r^n} \qquad (2)$$

In the following examples the above equations (1) and (2) are used in order to find several terms of geometric sequences.

Example 4.4-1 Find the next five terms of the following geometric sequences.

a. $s_1 = 5$, $r = 2$ b. $s_1 = -3$, $r = 3$ c. $s_1 = 10$, $r = 0.5$

Solutions:

a. The nth term for an geometric sequence is equal to $s_n = s_1 r^{n-1}$ Substituting $s_1 = 5$ and $r = 2$ into the general geometric expression for $n = 2, 3, 4, 5, and\ 6$ we obtain

$\boxed{s_2} = \boxed{s_1 r^{2-1}} = \boxed{s_1 r} = \boxed{5 \times 2} = \boxed{\mathbf{10}}$

$\boxed{s_3} = \boxed{s_1 r^{3-1}} = \boxed{s_1 r^2} = \boxed{5 \times 2^2} = \boxed{5 \times 4} = \boxed{\mathbf{20}}$

$\boxed{s_4} = \boxed{s_1 r^{4-1}} = \boxed{s_1 r^3} = \boxed{5 \times 2^3} = \boxed{5 \times 8} = \boxed{\mathbf{40}}$

$\boxed{s_5} = \boxed{s_1 r^{5-1}} = \boxed{s_1 r^4} = \boxed{5 \times 2^4} = \boxed{5 \times 16} = \boxed{\mathbf{80}}$

$\boxed{s_6} = \boxed{s_1 r^{6-1}} = \boxed{s_1 r^5} = \boxed{5 \times 2^5} = \boxed{5 \times 32} = \boxed{\mathbf{160}}$

Thus, the first six terms of the geometric sequence are $(\mathbf{5, 10, 20, 40, 80, 160})$.

b. Substituting $s_1 = -3$ and $r = 3$ into the general geometric expression for $n = 2, 3, 4, 5, and\ 6$ we obtain

$\boxed{s_2} = \boxed{s_1 r^{2-1}} = \boxed{s_1 r} = \boxed{-3 \times 3} = \boxed{\mathbf{-9}}$

$\boxed{s_3} = \boxed{s_1 r^{3-1}} = \boxed{s_1 r^2} = \boxed{-3 \times 3^2} = \boxed{-3 \times 9} = \boxed{\mathbf{-27}}$

$\boxed{s_4} = \boxed{s_1 r^{4-1}} = \boxed{s_1 r^3} = \boxed{-3 \times 3^3} = \boxed{-3 \times 27} = \boxed{\mathbf{-81}}$

$\boxed{s_5} = \boxed{s_1 r^{5-1}} = \boxed{s_1 r^4} = \boxed{-3 \times 3^4} = \boxed{-3 \times 81} = \boxed{\mathbf{-243}}$

$\boxed{s_6} = \boxed{s_1 r^{6-1}} = \boxed{s_1 r^5} = \boxed{-3 \times 3^5} = \boxed{-3 \times 243} = \boxed{\mathbf{-729}}$

Thus, the first six terms of the geometric sequence are $(\mathbf{-3, -9, -27, -81, -243, -729})$.

c. Substituting $s_1 = 10$ and $r = 0.5$ into the general geometric expression for $n = 2, 3, 4, 5, and\ 6$ we obtain

$\boxed{s_2} = \boxed{s_1 r^{2-1}} = \boxed{s_1 r} = \boxed{10 \times 0.5} = \boxed{\mathbf{5}}$

$\boxed{s_3} = \boxed{s_1 r^{3-1}} = \boxed{s_1 r^2} = \boxed{10 \times 0.5^2} = \boxed{10 \times 0.25} = \boxed{\mathbf{2.5}}$

$\boxed{s_4} = \boxed{s_1 r^{4-1}} = \boxed{s_1 r^3} = \boxed{10 \times 0.5^3} = \boxed{10 \times 0.125} = \boxed{\mathbf{1.25}}$

$\boxed{s_5} = \boxed{s_1 r^{5-1}} = \boxed{s_1 r^4} = \boxed{10 \times 0.5^4} = \boxed{10 \times 0.0625} = \boxed{\mathbf{0.625}}$

$\boxed{s_6} = \boxed{s_1 r^{6-1}} = \boxed{s_1 r^5} = \boxed{10 \times 0.5^5} = \boxed{10 \times 0.03125} = \boxed{\mathbf{0.3125}}$

Thus, the first six terms of the geometric sequence are $(\mathbf{10, 5, 2.5, 1.25, 0.625, 0.3125})$.

Example 4.4-2 find the general term and the tenth term of the following geometric sequences.

a. $s_1 = 3$, $r = 1.2$ b. $s_1 = -2$, $r = 0.8$ c. $s_1 = 10$, $r = -0.5$

Solutions:

a. The nth term for a geometric sequence is equal to $s_n = s_1 r^{n-1}$. Substituting $s_1 = 3$ and $r = 1.2$ into the general geometric expression we obtain

$\boxed{s_n} = \boxed{3 \times r^{n-1}} = \boxed{\mathbf{3 \times 1.2^{n-1}}}$

substituting $n = 10$ into the general equation $s_n = 3 \times 1.2^{n-1}$ we have

$\boxed{s_{10}} = \boxed{3 \times 1.2^{10-1}} = \boxed{3 \times 1.2^9} = \boxed{3 \times 5.1598} = \boxed{\mathbf{15.479}}$

b. Substituting $s_1 = -2$ and $r = 0.8$ into the general geometric expression $s_n = s_1 r^{n-1}$ we obtain

$\boxed{s_n} = \boxed{-2 \times r^{n-1}} = \boxed{\mathbf{-2 \times 0.8^{n-1}}}$

substituting $n = 10$ into the general equation $s_n = -2 \times 0.8^{n-1}$ we have

$\boxed{s_{10}} = \boxed{-2 \times 0.8^{10-1}} = \boxed{-2 \times 0.8^9} = \boxed{-2 \times 0.1342} = \boxed{\mathbf{-0.2684}}$

c. Substituting $s_1 = 10$ and $r = -0.5$ into the general geometric expression $s_n = s_1 r^{n-1}$ we obtain

$\boxed{s_n} = \boxed{10 \times r^{n-1}} = \boxed{\mathbf{10 \times (-0.5)^{n-1}}}$

substituting $n = 10$ into the general equation $s_n = 10 \times (-0.5)^{n-1}$ we have

$\boxed{s_{10}} = \boxed{10 \times (-0.5)^{10-1}} = \boxed{10 \times (-0.5)^9} = \boxed{10 \times (-0.0019)} = \boxed{\mathbf{-0.019}}$

Example 4.4-3 Find the next four terms and the nth term in each of the following geometric sequences.

a. $1, \frac{1}{2}, \cdots$ b. $-\frac{1}{3}, \frac{1}{9}, \cdots$ c. $\frac{1}{2}x, -\frac{1}{4}x, \cdots$

Solutions:

a. The first term s_1 and the common ratio r are equal to $s_1 = 1$ and $r = \frac{\frac{1}{2}}{1} = \frac{\frac{1}{2}}{\frac{1}{1}} = \frac{1 \times 1}{2 \times 1} = \frac{1}{2}$. Thus,

using the general geometric equation $s_n = s_1 r^{n-1}$ the next four terms are:

$\boxed{s_3} = \boxed{s_1 r^2} = \boxed{1 \cdot \left(\frac{1}{2}\right)^2} = \boxed{\frac{1}{2^2}} = \boxed{\mathbf{\frac{1}{4}}}$ $\boxed{s_4} = \boxed{s_1 r^3} = \boxed{1 \cdot \left(\frac{1}{2}\right)^3} = \boxed{\frac{1}{2^3}} = \boxed{\mathbf{\frac{1}{8}}}$

$\boxed{s_5} = \boxed{s_1 r^4} = \boxed{1 \cdot \left(\frac{1}{2}\right)^4} = \boxed{\frac{1}{2^4}} = \boxed{\mathbf{\frac{1}{16}}}$ $\boxed{s_6} = \boxed{s_1 r^5} = \boxed{1 \cdot \left(\frac{1}{2}\right)^5} = \boxed{\frac{1}{2^5}} = \boxed{\mathbf{\frac{1}{32}}}$

Thus, the first six terms of the geometric sequence are $\left(\mathbf{1, \frac{1}{2}, \frac{1}{4}, \frac{1}{8}, \frac{1}{16}, \frac{1}{32}}\right)$ and the nth term is

equal to $\boxed{s_n} = \boxed{1 \cdot \left(\frac{1}{2}\right)^{n-1}} = \boxed{\frac{1^{n-1}}{2^{n-1}}} = \boxed{\mathbf{\frac{1}{2^{n-1}}}}$

b. The first term s_1 and the common ratio r are equal to $s_1 = -\frac{1}{3}$ and $r = \frac{\frac{1}{9}}{-\frac{1}{3}} = -\frac{1 \times 3}{9 \times 1} = -\frac{1}{3}$. Thus,

using the general geometric equation $s_n = s_1 r^{n-1}$ the next four terms are:

$\boxed{s_3} = \boxed{s_1 r^2} = \boxed{-\frac{1}{3} \cdot \left(-\frac{1}{3}\right)^2} = \boxed{-\frac{1}{3^3}} = \boxed{\mathbf{-\frac{1}{27}}}$ $\boxed{s_4} = \boxed{s_1 r^3} = \boxed{-\frac{1}{3} \cdot \left(-\frac{1}{3}\right)^3} = \boxed{\frac{1}{3^4}} = \boxed{\mathbf{\frac{1}{81}}}$

$\boxed{s_5} = \boxed{s_1 r^4} = \boxed{-\frac{1}{3} \cdot \left(-\frac{1}{3}\right)^4} = \boxed{-\frac{1}{3^5}} = \boxed{\mathbf{-\frac{1}{243}}}$ $\boxed{s_6} = \boxed{s_1 r^5} = \boxed{-\frac{1}{3} \cdot \left(-\frac{1}{3}\right)^5} = \boxed{\frac{1}{3^6}} = \boxed{\mathbf{\frac{1}{729}}}$

Thus, the first six terms of the geometric sequence are $\left(\mathbf{-\frac{1}{3}, \frac{1}{9}, -\frac{1}{27}, \frac{1}{81}, -\frac{1}{243}, \frac{1}{729}}\right)$ and the

n^{th} term is equal to $\boxed{s_n} = \boxed{-\frac{1}{3}\cdot\left(-\frac{1}{3}\right)^{n-1}} = \boxed{-\frac{(-1)^{n-1}}{3\cdot 3^{n-1}}} = \boxed{-\frac{(-1)^{n-1}}{3^n}}$

c. The first term s_1 and the common ratio r are equal to $s_1 = \frac{1}{2}x$ and $r = -\frac{1}{2}$. Thus, using the general geometric equation $s_n = s_1 r^{n-1}$ the next four terms are:

$\boxed{s_3} = \boxed{s_1 r^2} = \boxed{\frac{1}{2}x\cdot\left(-\frac{1}{2}\right)^2} = \boxed{\frac{1}{2^3}x} = \boxed{\frac{1}{8}x}$ $\boxed{s_4} = \boxed{s_1 r^3} = \boxed{\frac{1}{2}x\cdot\left(-\frac{1}{2}\right)^3} = \boxed{-\frac{1}{2^4}x} = \boxed{-\frac{1}{16}x}$

$\boxed{s_5} = \boxed{s_1 r^4} = \boxed{\frac{1}{2}x\cdot\left(-\frac{1}{2}\right)^4} = \boxed{\frac{1}{2^5}x} = \boxed{\frac{1}{32}x}$ $\boxed{s_6} = \boxed{s_1 r^5} = \boxed{\frac{1}{2}x\cdot\left(-\frac{1}{2}\right)^5} = \boxed{-\frac{1}{2^6}x} = \boxed{-\frac{1}{64}x}$

Thus, the first six terms of the geometric sequence are $\left(\frac{1}{2}x, -\frac{1}{4}x, \frac{1}{8}x, -\frac{1}{16}x, \frac{1}{32}x, -\frac{1}{64}x\right)$

and the nth term is equal to $\boxed{s_n} = \boxed{\frac{1}{2}x\cdot\left(-\frac{1}{2}\right)^{n-1}} = \boxed{\frac{x(-1)^{n-1}}{2\cdot 2^{n-1}}} = \boxed{\frac{x(-1)^{n-1}}{2^{n-1+1}}} = \boxed{\frac{x(-1)^{n-1}}{2^n}}$

Example 4.4-4 Given the following terms of a geometric sequence, find the common ratio r.

a. $s_1 = 32$ and $s_7 = \frac{1}{2}$ b. $s_1 = 3$ and $s_5 = \frac{1}{27}$ c. $s_1 = 5$ and $s_8 = 1$

Solutions:

a. Substitute $s_1 = 32$ and $s_7 = \frac{1}{2}$ into $s_n = s_1 r^{n-1}$ and solve for r.

$\boxed{s_7 = s_1 r^{7-1}}$; $\boxed{\frac{1}{2} = 32r^6}$; $\boxed{\frac{\frac{1}{2}}{32} = r^6}$; $\boxed{r^6 = \frac{\frac{1}{2}}{2^5}}$; $\boxed{r^6 = \frac{\frac{1}{2}}{\frac{2^5}{1}}}$; $\boxed{r^6 = \frac{1\times 1}{2\times 2^5}}$; $\boxed{r^6 = \frac{1}{2^6}}$; $\boxed{r = \frac{1}{2}}$

b. Substitute $s_1 = 3$ and $s_5 = \frac{1}{27}$ into $s_n = s_1 r^{n-1}$ and solve for r.

$\boxed{s_5 = s_1 r^{5-1}}$; $\boxed{\frac{1}{27} = 3r^4}$; $\boxed{\frac{\frac{1}{27}}{3} = r^4}$; $\boxed{r^4 = \frac{\frac{1}{3^3}}{3}}$; $\boxed{r^4 = \frac{\frac{1}{3^3}}{\frac{3}{1}}}$; $\boxed{r^4 = \frac{1\times 1}{3^3\times 3}}$; $\boxed{r^4 = \frac{1}{3^4}}$; $\boxed{r = \frac{1}{3}}$

c. Substitute $s_1 = 5$ and $s_8 = 1$ into $s_n = s_1 r^{n-1}$ and solve for r.

$\boxed{s_8 = s_1 r^{8-1}}$; $\boxed{1 = 5r^7}$; $\boxed{\frac{1}{5} = r^7}$; $\boxed{r^7 = \frac{1}{5}}$; $\boxed{r = \frac{1}{\sqrt[7]{5}}}$

Example 4.4-5 Write the first six terms and the nth term of the following geometric sequences.

a. $s_n = \left(\frac{1}{3}\right)^{n-1}$ b. $s_n = \left(\frac{1}{2}\right)^{n+2}$ c. $s_n = \left(-\frac{1}{3}\right)^{2n}$ d. $s_n = \left(-\frac{1}{2}\right)^{n+1}$

Solutions:

a. $\boxed{s_1 = \left(\frac{1}{3}\right)^{1-1}} = \boxed{\left(\frac{1}{3}\right)^0} = \boxed{1}$ $\boxed{s_2 = \left(\frac{1}{3}\right)^{2-1}} = \boxed{\left(\frac{1}{3}\right)^1} = \boxed{\frac{1}{3}}$

$s_3 = \left(\frac{1}{3}\right)^{3-1} = \left(\frac{1}{3}\right)^2 = \frac{1}{3^2} = \mathbf{\frac{1}{9}}$ $s_4 = \left(\frac{1}{3}\right)^{4-1} = \left(\frac{1}{3}\right)^3 = \frac{1}{3^3} = \mathbf{\frac{1}{27}}$

$s_5 = \left(\frac{1}{3}\right)^{5-1} = \left(\frac{1}{3}\right)^4 = \frac{1}{3^4} = \mathbf{\frac{1}{81}}$ $s_6 = \left(\frac{1}{3}\right)^{6-1} = \left(\frac{1}{3}\right)^5 = \frac{1}{3^5} = \mathbf{\frac{1}{243}}$

Thus, the first six terms of the geometric sequence are $\left(\mathbf{1, \frac{1}{3}, \frac{1}{9}, \frac{1}{27}, \frac{1}{81}, \frac{1}{243}}\right)$.

b. $s_1 = \left(\frac{1}{2}\right)^{1+2} = \left(\frac{1}{2}\right)^3 = \frac{1}{2^3} = \mathbf{\frac{1}{8}}$ $s_2 = \left(\frac{1}{2}\right)^{2+2} = \left(\frac{1}{2}\right)^4 = \frac{1}{2^4} = \mathbf{\frac{1}{16}}$

$s_3 = \left(\frac{1}{2}\right)^{3+2} = \left(\frac{1}{2}\right)^5 = \frac{1}{2^5} = \mathbf{\frac{1}{32}}$ $s_4 = \left(\frac{1}{2}\right)^{4+2} = \left(\frac{1}{2}\right)^6 = \frac{1}{2^6} = \mathbf{\frac{1}{64}}$

$s_5 = \left(\frac{1}{2}\right)^{5+2} = \left(\frac{1}{2}\right)^7 = \frac{1}{2^7} = \mathbf{\frac{1}{128}}$ $s_6 = \left(\frac{1}{2}\right)^{6+2} = \left(\frac{1}{2}\right)^8 = \frac{1}{2^8} = \mathbf{\frac{1}{256}}$

Thus, the first six terms of the geometric sequence are $\left(\mathbf{\frac{1}{8}, \frac{1}{16}, \frac{1}{32}, \frac{1}{64}, \frac{1}{128}, \frac{1}{256}}\right)$.

c. $s_1 = \left(-\frac{1}{3}\right)^2 = \frac{1}{3^2} = \mathbf{\frac{1}{9.0 \times 10^0}}$ $s_2 = \left(-\frac{1}{3}\right)^4 = \frac{1}{3^4} = \mathbf{\frac{1}{8.1 \times 10^1}}$

$s_3 = \left(-\frac{1}{3}\right)^6 = \frac{1}{3^6} = \mathbf{\frac{1}{7.29 \times 10^2}}$ $s_4 = \left(-\frac{1}{3}\right)^8 = \frac{1}{3^8} = \mathbf{\frac{1}{6.56 \times 10^3}}$

$s_5 = \left(-\frac{1}{3}\right)^{10} = \frac{1}{3^{10}} = \mathbf{\frac{1}{5.9 \times 10^4}}$ $s_6 = \left(-\frac{1}{3}\right)^{12} = \frac{1}{3^{12}} = \mathbf{\frac{1}{5.31 \times 10^5}}$

Thus, the six terms are $\left(\mathbf{\frac{1}{9.0 \times 10^0}, \frac{1}{8.1 \times 10^1}, \frac{1}{7.29 \times 10^2}, \frac{1}{6.56 \times 10^3}, \frac{1}{5.9 \times 10^4}, \frac{1}{5.31 \times 10^5}}\right)$.

d. $s_1 = \left(-\frac{1}{2}\right)^2 = \frac{1}{2^2} = \mathbf{\frac{1}{4}}$ $s_2 = \left(-\frac{1}{2}\right)^3 = -\frac{1}{2^3} = \mathbf{-\frac{1}{8}}$

$s_3 = \left(-\frac{1}{2}\right)^4 = \frac{1}{2^4} = \mathbf{\frac{1}{16}}$ $s_4 = \left(-\frac{1}{2}\right)^5 = -\frac{1}{2^5} = \mathbf{-\frac{1}{32}}$

$s_5 = \left(-\frac{1}{2}\right)^6 = \frac{1}{2^6} = \mathbf{\frac{1}{64}}$ $s_6 = \left(-\frac{1}{2}\right)^7 = -\frac{1}{2^7} = \mathbf{-\frac{1}{128}}$

Thus, the first six terms of the geometric sequence are $\left(\mathbf{\frac{1}{4}, -\frac{1}{8}, \frac{1}{16}, -\frac{1}{32}, \frac{1}{64}, -\frac{1}{128}}\right)$.

Having learned about geometric sequences and the steps for finding the terms of a geometric sequence, we will next learn about geometric series and the steps for finding the sum of geometric series over a given range.

Similar to arithmetic series, addition of the terms in a geometric sequence result in having a geometric series. To obtain the geometric series formula let $s_k = s_1 r^{k-1}$ be a geometric sequence and denote the sum of the first n terms by

$$S_n = \sum_{k=1}^{n} s_1 r^{k-1}$$

then,

$$S_n = s_1 + s_1 r + s_1 r^2 + \cdots + s_1 r^{n-2} + s_1 r^{n-1} \qquad (a)$$

Let's multiply both sides of the equation (a) by r and subtract (b) from (a).

$$S_n \cdot r = s_1 \cdot r + s_1 r \cdot r + s_1 r^2 \cdot r + \cdots + s_1 r^{n-2} \cdot r + s_1 r^{n-1} \cdot r$$

$$rS_n = s_1 r + s_1 r^2 + s_1 r^3 + \cdots + s_1 r^{n-1} + s_1 r^n \qquad (b)$$

$$S_n - rS_n = \left(s_1 + s_1 r + s_1 r^2 + s_1 r^3 + \cdots + s_1 r^{n-1}\right) - \left(s_1 r + s_1 r^2 + s_1 r^3 + \cdots + s_1 r^{n-1} + s_1 r^n\right)$$

$$S_n(1-r) = \left(s_1 + s_1 r + s_1 r^2 + s_1 r^3 + \cdots + s_1 r^{n-1}\right) + \left(-s_1 r - s_1 r^2 - s_1 r^3 - \cdots - s_1 r^{n-1} - s_1 r^n\right)$$

$$S_n(1-r) = s_1 + \left(s_1 r - s_1 r\right) + \left(s_1 r^2 - s_1 r^2\right) + \left(s_1 r^3 - s_1 r^3\right) + \cdots + \left(s_1 r^{n-1} - s_1 r^{n-1}\right) - s_1 r^n$$

$$S_n(1-r) = s_1 - s_1 r^n \; ; \; S_n = \frac{s_1 - s_1 r^n}{1-r} \; ; \; S_n = \frac{s_1\left(1-r^n\right)}{1-r} \qquad r \neq 1$$

Therefore, the geometric series can be written in the following two forms:

$$\mathbf{S_n = \sum_{k=1}^{n} s_1 r^{k-1}} \qquad (1)$$

$$\mathbf{S_n = \frac{s_1\left(1-r^n\right)}{1-r} \qquad r \neq 1} \qquad (2)$$

Note that equation (2), similar to the nth term of a geometric sequence $\left(s_n = s_1 r^{n-1}\right)$, is given in terms of s_1, n, and r.

A third alternative way of expressing the geometric series is by substituting $s_1 r^n$ with its equivalent value $s_1 r^n = s_1 r^{n-1} \cdot r = r\left(s_1 r^{n-1}\right) = rs_n$ which result in having

$$S_n = \frac{s_1 - s_1 r^n}{1-r} = \frac{s_1 - r\left(s_1 r^{n-1}\right)}{1-r} = \frac{s_1 - r\left(s_1 r^{n-1}\right)}{1-r} = \mathbf{\frac{s_1 - rs_n}{1-r}} \qquad (3)$$

where the geometric series is given in terms of s_1, s_n (the geometric sequence), and r.

In the following examples we will use the above equations (1), (2), and (3) in order to find the sum of geometric series.

Example 4.4-6 Evaluate the sum of the following geometric series.

a. $\sum_{k=1}^{10} 3^{k-2} =$ b. $\sum_{k=1}^{10} (-3)^{k-2} =$ c. $\sum_{k=2}^{6} 8\left(-\frac{1}{2}\right)^{k+1} =$

Solutions:

a. **First** - Write the first few terms of the geometric series in expanded form, i.e.,

$$\boxed{\sum_{k=1}^{10} 3^{k-2}} = \boxed{3^{1-2}+3^{2-2}+3^{3-2}+3^{4-2}+\cdots} = \boxed{3^{-1}+3^{0}+3^{1}+3^{2}+\cdots} = \boxed{3^{-1}+1+3+6+\cdots}$$

Second - Identify the first term, s_1, the common ratio between the two terms, r, and n, i.e.,

$s_1 = 3^{-1}$, $r = \frac{1}{3^{-1}} = \frac{1}{\frac{1}{3}} = \frac{\frac{1}{1}}{\frac{1}{3}} = \frac{1\times 3}{1\times 1} = 3$, and $n = 10$.

Third - Use the geometric series formula to obtain the sum of the ten terms.

$$S_n = \frac{s_1\left(1-r^n\right)}{1-r}$$

$$\boxed{S_{10}} = \boxed{\frac{\frac{1}{3}\left(1-3^{10}\right)}{1-3}} = \boxed{\frac{\frac{1}{3}(1-59049)}{-2}} = \boxed{\frac{-\frac{59048}{3}}{-2}} = \boxed{\frac{-\frac{59048}{3}}{-\frac{2}{1}}} = \boxed{\frac{59048\times 1}{3\times 2}} = \boxed{\frac{59048}{6}} = \boxed{\mathbf{9841.333}}$$

Note that prior to learning the geometric series formula the only method that we could use was by summing each term as shown below:

$$\boxed{\sum_{k=1}^{10} 3^{k-2}} = \boxed{3^{-1}+3^{0}+3^{1}+3^{2}+3^{3}+3^{4}+3^{5}+3^{6}+3^{7}+3^{8}} = \boxed{3^{-1}+1+3+9+27+81+243+729+2187}$$

$$\boxed{+6561} = \boxed{3^{-1}+9841} = \boxed{0.333+9841} = \boxed{\mathbf{9841.333}}$$

As you note, it is much easier to use the geometric series formula as opposed to the summation of each term which is somewhat long and time consuming.

b. **First** - Write the first few terms of the geometric series in expanded form, i.e.,

$$\boxed{\sum_{k=1}^{10} (-3)^{k-2}} = \boxed{(-3)^{1-2}+(-3)^{2-2}+(-3)^{3-2}+(-3)^{4-2}+\cdots} = \boxed{(-3)^{-1}+(-3)^{0}+(-3)^{1}+(-3)^{2}+\cdots}$$

$$= \boxed{-3^{-1}+1-3+9+\cdots}$$

Second - Identify the first term, s_1, the common ratio between the two terms, r, and n, i.e.,

$s_1 = -3^{-1}$, $r = \frac{1}{-3^{-1}} = \frac{1}{-\frac{1}{3}} = -\frac{\frac{1}{1}}{\frac{1}{3}} = -\frac{1\times 3}{1\times 1} = -3$, and $n = 10$.

Third - Use the arithmetic series formula to obtain the sum of the ten terms.

$$S_n = \frac{s_1\left(1-r^n\right)}{1-r}$$

$$\boxed{S_{10}} = \boxed{\frac{-\frac{1}{3}\left[1-(-3)^{10}\right]}{1-(-3)}} = \boxed{\frac{-\frac{1}{3}(1-59049)}{4}} = \boxed{\frac{\frac{59048}{3}}{4}} = \boxed{\frac{\frac{59048}{3}}{\frac{4}{1}}} = \boxed{\frac{59048\times 1}{3\times 4}} = \boxed{\frac{59048}{12}} = \boxed{\mathbf{4920.666}}$$

or, we can obtain the answer by summing up the first ten terms of the series, i.e.,

$$\boxed{\sum_{k=1}^{10}(-3)^{k-2}} = \boxed{(-3)^{-1}+(-3)^{0}+(-3)^{1}+(-3)^{2}+(-3)^{3}+(-3)^{4}+(-3)^{5}+(-3)^{6}+(-3)^{7}+(-3)^{8}}$$

$$= \boxed{-3^{-1}+1-3+9-27+81-243+729-2187+6561} = \boxed{-3^{-1}+4921} = \boxed{-0.333+4921} = \boxed{\mathbf{4920.666}}$$

c. **First** - Write the first few terms of the geometric series in expanded form, i.e.,

$$\boxed{\sum_{k=2}^{6}8\left(-\frac{1}{2}\right)^{k+1}} = \boxed{8\left(-\frac{1}{2}\right)^{2+1}+8\left(-\frac{1}{2}\right)^{3+1}+8\left(-\frac{1}{2}\right)^{4+1}+8\left(-\frac{1}{2}\right)^{5+1}+\cdots} = \boxed{8\left(-\frac{1}{2}\right)^{3}+8\left(-\frac{1}{2}\right)^{4}+8\left(-\frac{1}{2}\right)^{5}}$$

$$\boxed{+8\left(-\frac{1}{2}\right)^{6}+\cdots} = \boxed{8\times-\frac{1}{8}+8\times\frac{1}{16}-8\times\frac{1}{32}+8\times\frac{1}{64}+\cdots} = \boxed{-\frac{8}{8}+\frac{8}{16}-\frac{8}{32}+\frac{8}{64}+\cdots} = \boxed{-1+\frac{1}{2}-\frac{1}{4}+\frac{1}{8}+\cdots}$$

Second - Identify the first term, s_1, the common ratio between the two terms, r, and n, i.e.,

$s_1 = -1$, $r = \frac{\frac{1}{2}}{-1} = -\frac{\frac{1}{2}}{\frac{1}{1}} = -\frac{1\times1}{2\times1} = -\frac{1}{2}$, and $n = 5$.

Third - Use the geometric series formula to obtain the sum of the five terms.

$$S_n = \frac{s_1\left(1-r^n\right)}{1-r}$$

$$\boxed{S_5} = \boxed{\frac{-1\cdot\left[1-\left(-\frac{1}{2}\right)^5\right]}{1-\left(-\frac{1}{2}\right)}} = \boxed{\frac{-\left(1+\frac{1}{2^5}\right)}{1+\frac{1}{2}}} = \boxed{\frac{-\left(1+\frac{1}{32}\right)}{\frac{3}{2}}} = \boxed{\frac{-(1+0.03125)}{1.5}} = \boxed{-\frac{1.03125}{1.5}} = \boxed{\mathbf{-0.6875}}$$

or, we can obtain the answer by summing up the first five terms of the series, i.e.,

$$\boxed{\sum_{k=2}^{6}8\left(-\frac{1}{2}\right)^{k+1}} = \boxed{8\cdot\left[\left(-\frac{1}{2}\right)^{3}+\left(-\frac{1}{2}\right)^{4}+\left(-\frac{1}{2}\right)^{5}+\left(-\frac{1}{2}\right)^{6}+\left(-\frac{1}{2}\right)^{7}\right]} = \boxed{8\cdot\left[-\frac{1}{8}+\frac{1}{16}-\frac{1}{32}+\frac{1}{64}-\frac{1}{128}\right]}$$

$$= \boxed{8\cdot(-0.125+0.0625-0.03125+0.01563-0.00781)} = \boxed{8\times-0.08593} = \boxed{\mathbf{-0.6875}}$$

Example 4.4-7 Given the first term s_1 and r, find S_{10} for each of the following geometric sequences.

a. $s_1 = 5$, $r = 2$ b. $s_1 = -10$, $r = 3$ c. $s_1 = 50$, $r = -2$

Solutions:

a. The nth term for a geometric series is equal to $S_n = \frac{s_1\left(1-r^n\right)}{1-r}$. Substituting $s_1 = 5$ and $r = 2$ into the general geometric expression we obtain

$$\boxed{S_{10}} = \boxed{\frac{5\left(1-2^{10}\right)}{1-2}} = \boxed{\frac{5(1-1024)}{-1}} = \boxed{\mathbf{5115}}$$

b. Substituting $s_1 = -10$ and $r = 3$ into $S_n = \frac{s_1\left(1-r^n\right)}{1-r}$ we obtain

$$\boxed{S_{10}} = \boxed{\frac{-10\left(1-3^{10}\right)}{1-3}} = \boxed{\frac{-10(1-59049)}{-2}} = \boxed{5\times-59048} = \boxed{\mathbf{-295240}}$$

c. Substituting $s_1 = 50$ and $r = -2$ into $S_n = \frac{s_1\left(1-r^n\right)}{1-r}$ we obtain

$$\boxed{S_{10}} = \boxed{\frac{50\left[1-(-2)^{10}\right]}{1-(-2)}} = \boxed{\frac{50[1-1024]}{1+2}} = \boxed{\frac{50\times-1023}{3}} = \boxed{\mathbf{-17050}}$$

Example 4.4-8 Find the x and y values to the following problems.

a. x if $\sum_{i=1}^{6} ix = 20$

b. x and y if $\sum_{i=1}^{5}(ix+2y) = 30$ and $\sum_{i=2}^{5}(ix+2y) = 10$

c. x if $\sum_{i=1}^{5}(ix+5) = 15$

d. x and y if $\sum_{i=3}^{6}(x+iy) = 50$ and $\sum_{i=4}^{8}(x+iy) = 24$

Solutions:

a. Expanding $\sum_{i=1}^{6} ix = 20$ we obtain $\boxed{x+2x+3x+4x+5x+6x=20}$; $\boxed{21x=20}$; $\boxed{x=\frac{20}{21}}$; $\boxed{\mathbf{x=0.952}}$

b. Expanding $\sum_{i=1}^{5}(ix+2y) = 30$ we obtain $\boxed{(x+2y)+(2x+2y)+(3x+2y)+(4x+2y)+(5x+2y)=30}$

; $\boxed{(x+2x+3x+4x+5x)+(2y+2y+2y+2y+2y)=30}$; $\boxed{15x+10y=30}$

Expanding $\sum_{i=2}^{5}(ix+2y) = 10$ we obtain $\boxed{(2x+2y)+(3x+2y)+(4x+2y)+(5x+2y)=10}$; $\boxed{14x+8y=10}$

The two linear equations with two unknowns x and y are solved using the substitution method

to obtain $\boxed{\mathbf{x=-7}}$ and $\boxed{\mathbf{y=13.5}}$

c. Expanding $\sum_{i=1}^{5}(ix+5) = 15$ we obtain $\boxed{(x+5)+(2x+5)+(3x+5)+(4x+5)+(5x+5)}$; $\boxed{15x+25=15}$

; $\boxed{x=-\frac{10}{15}}$; $\boxed{\mathbf{x=-0.667}}$

d. Expanding $\sum_{i=3}^{6}(x+iy) = 50$ we obtain $\boxed{(x+3y)+(x+4y)+(x+5y)+(x+6y)=50}$; $\boxed{4x+18y=50}$

Expanding $\sum_{i=4}^{8}(x+iy) = 24$ we obtain $\boxed{(x+4y)+(x+5y)+(x+6y)+(x+7y)+(x+8y)=24}$; $\boxed{5x+30y=24}$

The two linear equations with two unknowns x and y are solved using the substitution method

to obtain $\boxed{\mathbf{x=35.6}}$ and $\boxed{\mathbf{y=-5.133}}$

Example 4.4-9 Find the value of x for the following geometric sequences.

a. $2, 4x, 16$. b. $2^{-1}, 2^{-1}x, 2^{-3}$ c. $5, 5x, 125$

Solutions:

a. Since the common ratio r of a geometric sequence is defined as the ratio of the $(n+1)$ *st* term to the nth term, we can use this principal to solve for x, i.e.,

$$\boxed{r} = \boxed{\frac{4x}{2} = \frac{16}{4x}} \text{ therefore } \boxed{4x \times 4x = 16 \times 2}; \boxed{16x^2 = 32}; \boxed{x^2 = \frac{32}{16}}; \boxed{x^2 = 2}; \boxed{\mathbf{x = \pm\sqrt{2}}}$$

b. Using the common ratio principal we can solve for x in the following way:

$$\boxed{r} = \boxed{\frac{2^{-1}x}{2^{-1}} = \frac{2^{-3}}{2^{-1}x}} \text{ therefore } \boxed{2^{-1}x \times 2^{-1}x = 2^{-3} \times 2^{-1}}; \boxed{2^{-2}x^2 = 2^{-4}}; \boxed{x^2 = \frac{2^{-4}}{2^{-2}}}; \boxed{x^2 = 2^{-4} \cdot 2^2}$$

$$; \boxed{x^2 = 2^{-4+2}}; \boxed{x^2 = 2^{-2}}; \boxed{x^2 = \frac{1}{2^2}}; \boxed{\mathbf{x = \pm\frac{1}{2}}}$$

c. Using the common ratio principal we can solve for x in the following way:

$$\boxed{r} = \boxed{\frac{5x}{5} = \frac{125}{5x}} \text{ therefore } \boxed{5x \times 5x = 125 \times 5}; \boxed{25x^2 = 625}; \boxed{x^2 = \frac{625}{25}}; \boxed{x^2 = 25}; \boxed{\mathbf{x = \pm 5}}$$

Section 4.4 Practice Problems - Geometric Sequences and Geometric Series

1. Find the next four terms of the following geometric sequences.

   a. $s_1 = 3$, $r = 0.5$ b. $s_1 = -5$, $r = 2$ c. $s_1 = 5$, $r = 0.75$

2. Find the eighth and the general term of the following geometric sequences.

   a. $s_1 = 2$, $r = \sqrt{3}$ b. $s_1 = -4$, $r = 1.2$ c. $s_1 = 4$, $r = -2.5$

3. Find the next six terms and the nth term in each of the following geometric sequences.

   a. $1, \frac{1}{4}, \cdots$ b. $-\frac{1}{2}, \frac{1}{4}, \cdots$ c. $\frac{1}{3}p, -3p, \cdots$

4. Given the following terms of a geometric sequence, find the common ratio r.

   a. $s_1 = 25$ and $s_4 = \frac{1}{5}$ b. $s_1 = 4$ and $s_5 = \frac{1}{64}$ c. $s_1 = 3$ and $s_8 = 1$

5. Write the first five terms of the following geometric sequences.

   a. $s_n = \left(-\frac{1}{3}\right)^{2n-1}$ b. $s_n = \left(\frac{1}{3}\right)^{2n+2}$ c. $s_n = \left(-\frac{1}{5}\right)^{2n-3}$ d. $s_n = \left(-\frac{1}{2}\right)^{n}$

6. Evaluate the sum of the following geometric series.

   a. $\sum_{k=1}^{6} 3^{k-1} =$ b. $\sum_{k=3}^{10} (-2)^{k-3} =$ c. $\sum_{j=4}^{8} 4\left(-\frac{1}{2}\right)^{j+1} =$

d. $\sum_{m=1}^{4}(-2)^{m-3} =$ e. $\sum_{n=5}^{10}(-3)^{n-4} =$ f. $\sum_{k=1}^{5}(-3)^{k-1} =$

g. $\sum_{m=1}^{5}4^{m} =$ h. $\sum_{j=1}^{4}\frac{3^j}{27} =$ i. $\sum_{k=3}^{6}6\left(\frac{1}{2}\right)^{k+1} =$

7. Given the first term s_1 and r, find S_8 for each of the following geometric sequences.

a. $s_1 = 3$, $r = 3$ b. $s_1 = -8$, $r = 0.5$ c. $s_1 = 2$, $r = -2.5$

8. Solve for x and y.

a. $\sum_{i=3}^{7}(ix+2) = 30$ b. $\sum_{i=1}^{4}(ix+y) = 20$ and $\sum_{i=2}^{6}(ix+y) = 10$

4.5 Limits of Sequences and Series

A sequence $s_1, s_2, s_3, s_4, \cdots, s_n, \cdots$ is said to **converge** to the constant K,

$$\lim_{n\to\infty} s_n = K$$

if and only if, for a large value of n, the absolute value of the difference between the nth term and the constant K is very small. For example, the sequence $2, \frac{5}{4}, \frac{10}{9}, \frac{17}{16}, \cdots, \left(1+\frac{1}{n^2}\right), \cdots$ converges to 1. This is because, the absolute value of the difference between $\left(1+\frac{1}{n^2}\right)$, for large n, and 1 is very small. On the other hand, the sequence $s_1, s_2, s_3, s_4, \cdots, s_n, \cdots$ is said to **diverge,** if and only if, for a large value of n, the sequence approaches to infinity (∞). For example, the sequence $4, 8, 16, 32, \cdots, 2^{n+1}, \cdots$ does not converge. This is because, as n increases, the nth term increases without bound, i.e., it approaches to infinity. In the following examples we will learn how to identify a convergent or a divergent sequence:

Example 4.5-1 State which of the following sequences are convergent.

a. $1, 2, 3, 4, 5, \cdots, n, \cdots =$

b. $2, \frac{5}{8}, \frac{10}{27}, \frac{17}{64}, \cdots, \frac{n^2+1}{n^3}, \cdots =$

c. $2, \frac{8}{3}, \frac{26}{9}, \frac{80}{27}, \cdots, \frac{3^n-1}{3^{n-1}}, \cdots =$

d. $\frac{1}{16}, \frac{1}{64}, \frac{1}{256}, \frac{1}{1024}, \cdots, \frac{1}{4^{n+1}}, \cdots =$

e. $3, \frac{9}{4}, \frac{19}{9}, \frac{33}{16}, \cdots, 2+\frac{1}{n^2}, \cdots =$

f. $6, \frac{7}{4}, \frac{8}{9}, \frac{9}{16}, \cdots, \frac{n+5}{n^2}, \cdots =$

g. $\frac{1}{2}, \frac{1}{3}, \frac{1}{4}, \frac{1}{5}, \cdots, \frac{1}{n+1}, \cdots =$

h. $3, 6, 9, 12, 15, \cdots, 3n, \cdots =$

Solutions:

In solving this class of problems write the nth term and observe if it converges or diverges as n approaches to infinity.

a. The sequence $1, 2, 3, 4, 5, \cdots, n, \cdots$ continues to increase. $\lim_{n\to\infty} n = \infty$ which is undefined.

Hence, **the sequence diverges or is divergent.**

b. $\boxed{\lim_{n\to\infty} \frac{n^2+1}{n^3}} = \boxed{\lim_{n\to\infty}\left(\frac{n^2}{n^3}+\frac{1}{n^3}\right)} = \boxed{\lim_{n\to\infty}\left(\frac{1}{n}+\frac{1}{n^3}\right)} = \boxed{\frac{1}{\infty}+\frac{1}{\infty^3}} = \boxed{0+0} = \boxed{0}$

The sequence converges to 0

c. $\boxed{\lim_{n\to\infty} \frac{3^n-1}{3^{n-1}}} = \boxed{\lim_{n\to\infty}\left(\frac{3^n}{3^{n-1}}-\frac{1}{3^{n-1}}\right)} = \boxed{\lim_{n\to\infty}\left(\frac{3^n 3^{-(n-1)}}{1}-\frac{1}{3^{n-1}}\right)} = \boxed{\lim_{n\to\infty}\left(\frac{3^{\cancel{n}-\cancel{n}+1}}{1}-\frac{1}{3^{n-1}}\right)}$

$= \boxed{\lim_{n\to\infty}\left(3-\frac{1}{3^{n-1}}\right)} = \boxed{3-\frac{1}{3^{\infty-1}}} = \boxed{3-\frac{1}{3^{\infty}}} = \boxed{3-\frac{1}{\infty}} = \boxed{3-0} = \boxed{3}$ **The sequence converges to 3**

d. $\boxed{\lim_{n\to\infty}\frac{1}{4^{n+1}}} = \boxed{\frac{1}{4^{\infty+1}}} = \boxed{\frac{1}{4^{\infty}}} = \boxed{\frac{1}{\infty}} = \boxed{0}$ **The sequence converges to 0**

e. $\boxed{\lim_{n\to\infty}\left(2+\frac{1}{n^2}\right)} = \boxed{2+\frac{1}{\infty^2}} = \boxed{2+\frac{1}{\infty}} = \boxed{2+0} = \boxed{2}$ **The sequence converges to 2**

f. $\boxed{\lim_{n\to\infty}\frac{n+5}{n^2}} = \boxed{\lim_{n\to\infty}\left(\frac{n}{n^2}+\frac{5}{n^2}\right)} = \boxed{\lim_{n\to\infty}\left(\frac{1}{n}+\frac{5}{n^2}\right)} = \boxed{\frac{1}{\infty}+\frac{5}{\infty^2}} = \boxed{\frac{1}{\infty}+\frac{5}{\infty}} = \boxed{0+0} = \boxed{0}$

The sequence converges to 0

g. $\boxed{\lim_{n\to\infty}\frac{1}{n+1}} = \boxed{\frac{1}{\infty+1}} = \boxed{\frac{1}{\infty}} = \boxed{0}$

The sequence converges to 0

h. The sequence $3, 6, 9, 12, 15, \cdots, 3n, \cdots$ continues to increase. $\lim_{n\to\infty} 3n = 3\cdot\infty = \infty$ which is undefined. Hence, **the sequence diverges or is divergent.**

Example 4.5-2 State which of the following geometric sequences are convergent.

a. $\frac{1}{3}, \frac{1}{9}, \frac{1}{27}, \frac{1}{81}, \cdots, \frac{1}{3^n}, \cdots =$

b. $2, 4, 8, 16, 32, \cdots, 2^n, \cdots =$

c. $1, -1, 1, -1, \cdots, (-1)^{n+1}, \cdots =$

d. $10, 1, \frac{1}{10}, \frac{1}{100}, \frac{1}{1000}, \cdots, 10\cdot\left(\frac{1}{10}\right)^{n-1}, \cdots =$

e. $0.2, 0.02, 0.0002, \cdots, 2(0.1)^n, \cdots =$

f. $1, \frac{4}{3}, \frac{16}{9}, \frac{64}{27}, \frac{256}{81}, \cdots, \left(\frac{4}{3}\right)^{n-1}, \cdots =$

g. $-2, 4, -8, 16, -32, \cdots, (-1)^n 2^n, \cdots =$

h. $27, 3, \frac{1}{3}, \frac{1}{27}, \cdots, 27\cdot\left(\frac{1}{9}\right)^{n-1}, \cdots =$

Solutions:

a. The sequence $\frac{1}{3}, \frac{1}{9}, \frac{1}{27}, \frac{1}{81}, \cdots, \frac{1}{3^n}, \cdots$ **converges to 0** since, for large value of n, the absolute value of the difference between $\frac{1}{3^n}$ and 0 is very small.

b. The sequence $2, 4, 8, 16, 32, \cdots, 2^n, \cdots$ **diverges** since, as n increases, the nth term increases without bound.

c. The sequence $1, -1, 1, -1, \cdots, (-1)^{n+1}, \cdots$ **diverges** since, as n increases, the nth term oscillates back and forth from $+1$ to -1.

d. The sequence $10, 1, \frac{1}{10}, \frac{1}{100}, \frac{1}{1000}, \cdots, 10\cdot\left(\frac{1}{10}\right)^{n-1}, \cdots$ **converges to 0** since, for large value of n, the absolute value of the difference between $10\cdot\left(\frac{1}{10}\right)^{n-1}$ and 0 is very small.

e. The sequence $0.2, 0.02, 0.0002, \cdots, 2(0.1)^n, \cdots$ **converges to 0** since, for large value of n, the absolute value of the difference between $2(0.1)^n$ and 0 is very small.

f. The sequence $1, \frac{4}{3}, \frac{16}{9}, \frac{64}{27}, \frac{256}{81}, \cdots, \left(\frac{4}{3}\right)^{n-1}, \cdots$ **diverges** since, as n increases, the nth term increases without bound.

g. The sequence $-2, 4, -8, 16, -32, \cdots, (-1)^n 2^n, \cdots$ **diverges** since, as n increases, the nth term oscillates back and forth from a large positive number to a large negative number.

h. The sequence $27, 3, \frac{1}{3}, \frac{1}{27}, \cdots, 27\cdot\left(\frac{1}{9}\right)^{n-1}, \cdots$ **converges to 0** since, for large value of n, the absolute value of the difference between $27\cdot\left(\frac{1}{9}\right)^{n-1}$ and 0 is very small.

Example 4.5-3 Discuss the limiting behavior of the following sequences as n approaches ∞.

a. $\frac{1}{n^2} =$ b. $1-\frac{1}{n^2} =$ c. $\frac{n+5}{n^2} =$ d. $\frac{n^2+5}{n^2} =$

e. $(1)^{-n}\frac{1}{n} =$ f. $\left(1+\frac{1}{2}\right)^{-n} =$ g. $2+\frac{1}{n^2} =$ h. $100n =$

i. $\frac{5n+10}{n} =$ j. $\frac{2^n-1}{2^n} =$

Solutions:

a. $\boxed{\lim_{n\to\infty}\frac{1}{n^2}} = \boxed{\frac{1}{\infty^2}} = \boxed{\frac{1}{\infty}} = \boxed{0}$ **converges to 0**

b. $\boxed{\lim_{n\to\infty}1-\frac{1}{n^2}} = \boxed{1-\frac{1}{\infty^2}} = \boxed{1-\frac{1}{\infty}} = \boxed{1-0} = \boxed{1}$ **converges to 1**

c. $\boxed{\lim_{n\to\infty}\frac{n+5}{n^2}} = \boxed{\lim_{n\to\infty}\left(\frac{\not{n}}{n^{2=1}}+\frac{5}{n^2}\right)} = \boxed{\lim_{n\to\infty}\left(\frac{1}{n}+\frac{5}{n^2}\right)} = \boxed{\frac{1}{\infty}+\frac{5}{\infty^2}} = \boxed{\frac{1}{\infty}+\frac{5}{\infty}} = \boxed{0+0} = \boxed{0}$

converges to 0

d. $\boxed{\lim_{n\to\infty}\frac{n^2+5}{n^2}} = \boxed{\lim_{n\to\infty}\left(\frac{n^2}{n^2}+\frac{5}{n^2}\right)} = \boxed{\lim_{n\to\infty}\left(1+\frac{5}{n^2}\right)} = \boxed{1+\frac{5}{\infty^2}} = \boxed{1+0} = \boxed{1}$

converges to 1

e. $\boxed{\lim_{n\to\infty}(1)^{-n}\frac{1}{n}} = \boxed{\lim_{n\to\infty}\left(\frac{1}{1^n}\cdot\frac{1}{n}\right)} = \boxed{\frac{1}{1^\infty}\cdot\frac{1}{\infty}} = \boxed{\frac{1}{1}\cdot\frac{1}{\infty}} = \boxed{1\cdot 0} = \boxed{0}$ **converges to 0**

f. $\boxed{\lim_{n\to\infty}\left(1+\frac{1}{2}\right)^{-n}} = \boxed{\lim_{n\to\infty}\left(\frac{3}{2}\right)^{-n}} = \boxed{\lim_{n\to\infty}(1.5)^{-n}} = \boxed{\lim_{n\to\infty}\left(\frac{1}{1.5^n}\right)} = \boxed{\frac{1}{1.5^\infty}} = \boxed{\frac{1}{\infty}} = \boxed{0}$

converges to 0

g. $\boxed{\lim_{n\to\infty} 2+\frac{1}{n^2}} = \boxed{2+\frac{1}{\infty^2}} = \boxed{2+\frac{1}{\infty}} = \boxed{2+0} = \boxed{2}$ **converges to 2**

h. $\boxed{\lim_{n\to\infty} 100n} = \boxed{100\cdot\infty} = \boxed{\infty}$ **diverges**

i. $\boxed{\lim_{n\to\infty} \frac{5n+10}{n}} = \boxed{\lim_{n\to\infty}\left(\frac{5\not{n}}{\not{n}}+\frac{10}{n}\right)} = \boxed{\lim_{n\to\infty}\left(5+\frac{10}{n}\right)} = \boxed{5+\frac{10}{\infty}} = \boxed{5+0} = \boxed{5}$

converges to 5

j. $\boxed{\lim_{n\to\infty} \frac{2^n-1}{2^n}} = \boxed{\lim_{n\to\infty}\left(\frac{2^{\not{n}}}{2^{\not{n}}}-\frac{1}{2^n}\right)} = \boxed{\lim_{n\to\infty}\left(1-\frac{1}{2^n}\right)} = \boxed{1-\frac{1}{2^\infty}} = \boxed{1-\frac{1}{\infty}} = \boxed{1-0} = \boxed{1}$

converges to 1

Note that an easier way of finding the answer to sequences as $n\to\infty$ is by rewriting the sequence in its "almost equivalent" form. This approach is only applicable to cases where n is approaching to infinity. For example, $\lim_{n\to\infty} n+8$ is almost the same as $\lim_{n\to\infty} n$. (This is because $\infty+8$ is the same as ∞. Addition of the number eight to a very large number such as infinity does not significantly change the final answer.) Let's use this approach to solve few of the above problems.

$\boxed{\lim_{n\to\infty} \frac{n+5}{n^2}} \approx \boxed{\lim_{n\to\infty} \frac{n}{n^2}} = \boxed{\lim_{n\to\infty} \frac{1}{n}} = \boxed{\frac{1}{\infty}} = \boxed{0}$ which is the same answer as in 4.5-3c.

$\boxed{\lim_{n\to\infty} \frac{n^2+5}{n^2}} \approx \boxed{\lim_{n\to\infty} \frac{n^2}{n^2}} = \boxed{\lim_{n\to\infty} \frac{1}{1}} = \boxed{1}$ which is the same answer as in 4.5-3d.

$\boxed{\lim_{n\to\infty} \frac{5n+10}{n}} \approx \boxed{\lim_{n\to\infty} \frac{5n}{n}} = \boxed{\lim_{n\to\infty} \frac{5}{1}} = \boxed{5}$ which is the same answer as in 4.5-3i.

$\boxed{\lim_{n\to\infty} \frac{2^n-1}{2^n}} \approx \boxed{\lim_{n\to\infty} \frac{2^n}{2^n}} = \boxed{\lim_{n\to\infty} \frac{1}{1}} = \boxed{1}$ which is the same answer as in 4.5-3j.

Example 4.5-4 State whether or not the following sequences are convergent or divergent as n approaches infinity. If the sequence does converge, find its limit.

a. $\frac{n^2}{n-1} =$ b. $\frac{5n}{\sqrt{n^2+1}} =$ c. $\frac{8^n}{2^n+10^5} =$ d. $\frac{2^n}{8^n+10^5} =$

e. $\frac{8^n}{2^n+1} =$ f. $\frac{10^5\sqrt{n}}{1+n} =$ g. $10^n =$ h. $\frac{05^n}{n^2+1} =$

i. $\left(\frac{1}{4}\right)^n =$ j. $\frac{n^3+1}{n^3+n+1} =$ k. $\sqrt{25-\frac{1}{n}} =$ l. $\frac{3^n+1}{3^n} =$

m. $\frac{n^6}{12n^4+5} =$ n. $\frac{n^4+3n}{n^5+3} =$ o. $\frac{n^2}{\sqrt{8n^4+1}} =$ p. $\frac{1}{n}-\frac{1}{n+3} =$

q. $(0.5)^n =$ r. $(0.5)^{-n} =$ s. $\frac{\sqrt{n-1}}{2\sqrt{n}} =$ t. $\frac{2\sqrt{n}}{\sqrt{n+1}} =$

Solutions:

a. $\boxed{\lim_{n\to\infty}\frac{n^2}{n-1}} \approx \boxed{\lim_{n\to\infty}\frac{n^2}{n}} = \boxed{\lim_{n\to\infty}\frac{n}{1}} = \boxed{\lim_{n\to\infty} n} = \boxed{\infty}$ **diverges**

b. $\boxed{\lim_{n\to\infty}\frac{5n}{\sqrt{n^2+1}}} \approx \boxed{\lim_{n\to\infty}\frac{5n}{\sqrt{n^2}}} = \boxed{\lim_{n\to\infty}\frac{5n}{n}} = \boxed{\lim_{n\to\infty} 5} = \boxed{5}$ **converges to 5**

c. $\boxed{\lim_{n\to\infty}\frac{8^n}{2^n+10^5}} \approx \boxed{\lim_{n\to\infty}\frac{8^n}{2^n}} = \boxed{\lim_{n\to\infty}\frac{2^{3n}}{2^n}} = \boxed{\lim_{n\to\infty} 2^{3n}\cdot 2^{-n}} = \boxed{\lim_{n\to\infty} 2^{3n-n}} = \boxed{\lim_{n\to\infty} 2^{2n}}$

$= \boxed{2^\infty} = \boxed{\infty}$ **diverges**

d. $\boxed{\lim_{n\to\infty}\frac{2^n}{8^n+10^5}} \approx \boxed{\lim_{n\to\infty}\frac{2^n}{8^n}} = \boxed{\lim_{n\to\infty}\frac{2^n}{2^{3n}}} = \boxed{\lim_{n\to\infty}\frac{1}{2^{3n}.2^{-n}}} = \boxed{\lim_{n\to\infty}\frac{1}{2^{3n-n}}} = \boxed{\lim_{n\to\infty}\frac{1}{2^{2n}}}$

$= \boxed{\frac{1}{2^\infty}} = \boxed{\frac{1}{\infty}} = \boxed{0}$ **converges to 0**

e. $\boxed{\lim_{n\to\infty}\frac{8^n}{2^n+1}} \approx \boxed{\lim_{n\to\infty}\frac{8^n}{2^n}} = \boxed{\lim_{n\to\infty}\frac{2^{3n}}{2^n}} = \boxed{\lim_{n\to\infty} 2^{3n}\cdot 2^{-n}} = \boxed{\lim_{n\to\infty} 2^{3n-n}} = \boxed{\lim_{n\to\infty} 2^{2n}}$

$= \boxed{2^\infty} = \boxed{\infty}$ **diverges**

f. $\boxed{\lim_{n\to\infty}\frac{10^5\sqrt{n}}{1+n}} \approx \boxed{\lim_{n\to\infty}\frac{10^5\sqrt{n}}{n}} = \boxed{\lim_{n\to\infty}\frac{10^5 n^{\frac{1}{2}}}{n}} = \boxed{\lim_{n\to\infty}\frac{10^5}{n\cdot n^{-\frac{1}{2}}}} = \boxed{\lim_{n\to\infty}\frac{10^5}{n^{1-\frac{1}{2}}}} = \boxed{\lim_{n\to\infty}\frac{10^5}{n^{\frac{1}{2}}}}$

$= \boxed{\frac{10^5}{\infty^{\frac{1}{2}}}} = \boxed{\frac{10^5}{\infty}} = \boxed{0}$ **converges to 0**

g. $\boxed{\lim_{n\to\infty} 10^n} = \boxed{10^\infty} = \boxed{\infty}$ **diverges**

h. $\boxed{\lim_{n\to\infty}\frac{0.5^n}{n^2+1}} \approx \boxed{\lim_{n\to\infty}\frac{0.5^n}{n^2}} = \boxed{\frac{0.5^\infty}{\infty^2}} = \boxed{\frac{0}{\infty}} = \boxed{0}$ **converges to 0**

i. $\boxed{\lim_{n\to\infty}\left(\frac{1}{4}\right)^n} = \boxed{\left(\frac{1}{4}\right)^\infty} = \boxed{0.25^\infty} = \boxed{0}$ **converges to 0**

j. $\boxed{\lim_{n\to\infty}\frac{n^3+1}{n^3+n+1}} \approx \boxed{\lim_{n\to\infty}\frac{n^3}{n^3}} = \boxed{\lim_{n\to\infty}\frac{1}{1}} = \boxed{1}$ **converges to 1**

k. $\boxed{\lim_{n\to\infty}\sqrt{25-\frac{1}{n}}} \approx \boxed{\sqrt{25-\frac{1}{\infty}}} = \boxed{\sqrt{25-0}} = \boxed{\sqrt{5^2}} = \boxed{5}$ **converges to 5**

l. $\boxed{\lim_{n\to\infty}\frac{3^n+1}{3^n}} \approx \boxed{\lim_{n\to\infty}\frac{3^n}{3^n}} = \boxed{\lim_{n\to\infty}\frac{1}{1}} = \boxed{1}$ **converges to 1**

m. $\boxed{\lim_{n\to\infty}\frac{n^6}{12n^4+5}} \approx \boxed{\lim_{n\to\infty}\frac{n^6}{n^4}} = \boxed{\lim_{n\to\infty}\frac{n^6 n^{-4}}{1}} = \boxed{\lim_{n\to\infty} n^2} = \boxed{\infty^2} = \boxed{\infty}$ **diverges**

n. $\boxed{\lim_{n\to\infty}\frac{n^4+3n}{n^5+3}} \approx \boxed{\lim_{n\to\infty}\frac{n^4}{n^5}} = \boxed{\lim_{n\to\infty}\frac{1}{n^5 n^{-4}}} = \boxed{\lim_{n\to\infty}\frac{1}{n}} = \boxed{\frac{1}{\infty}} = \boxed{0}$ **converges to 0**

o. $\boxed{\lim_{n\to\infty}\frac{n^2}{\sqrt{8n^4+1}}} \approx \boxed{\lim_{n\to\infty}\frac{n^2}{\sqrt{8n^4}}} = \boxed{\lim_{n\to\infty}\frac{n^2}{2n^2\sqrt{2}}} = \boxed{\lim_{n\to\infty}\frac{1}{2\sqrt{2}}} = \boxed{\frac{1}{2\sqrt{2}}} = \boxed{0.353}$ **converges to 0.353**

p. $\boxed{\lim_{n\to\infty}\left(\frac{1}{n}-\frac{1}{n+3}\right)} \approx \boxed{\lim_{n\to\infty}\left(\frac{1}{n}-\frac{1}{n}\right)} = \boxed{\lim_{n\to\infty}\left(\frac{1-1}{n}\right)} = \boxed{\frac{0}{\infty}} = \boxed{0}$ **converges to 0**

q. $\boxed{\lim_{n\to\infty}(0.5)^n} = \boxed{(0.5)^\infty} = \boxed{0}$ **converges to 0**

r. $\boxed{\lim_{n\to\infty}(0.5)^{-n}} = \boxed{\frac{1}{(0.5)^n}} = \boxed{\frac{1}{(0.5)^\infty}} = \boxed{\frac{1}{0}} = \boxed{\infty}$ **diverges**

s. $\boxed{\lim_{n\to\infty}\frac{\sqrt{n-1}}{2\sqrt{n}}} \approx \boxed{\lim_{n\to\infty}\frac{\sqrt{\not n}}{2\sqrt{\not n}}} = \boxed{\lim_{n\to\infty}\frac{1}{2}} = \boxed{\frac{1}{2}}$ **converges to $\frac{1}{2}$**

t. $\boxed{\lim_{n\to\infty}\frac{2\sqrt{n}}{\sqrt{n+1}}} \approx \boxed{\lim_{n\to\infty}\frac{2\sqrt{\not n}}{\sqrt{\not n}}} = \boxed{\lim_{n\to\infty}\frac{2}{1}} = \boxed{2}$ **converges to 2**

Infinite Geometric Series

In section 4.4 we stated that the geometric series can be written in the following two forms:

$$S_n = \sum_{k=1}^{n} s_1 r^{k-1} \qquad (1)$$

$$S_n = \frac{s_1(1-r^n)}{1-r} \quad \text{where } r \neq 1 \qquad (2)$$

In equation (2), let's consider the criteria where $|r|$ is less than one and n is considerably large.

Then, under these conditions r^n approaches to zero and $S_n = \frac{s_1(1-r^n)}{1-r}$ reduces to $S_n = \frac{s_1}{1-r}$, i.e.,

$$S_\infty = \sum_{n=1}^{\infty} s_1 r^{n-1} = \lim_{n\to\infty}\frac{s_1(1-r^n)}{1-r} = \frac{s_1(1-0)}{1-r} = \frac{s_1}{1-r}.$$

Thus, the sum of an **infinite geometric series** as $n \to \infty$ and if $|r| \langle 1$ is equal to

$$S_\infty = \sum_{n=0}^{\infty} s_1 r^n = \sum_{n=1}^{\infty} s_1 r^{n-1} = \frac{s_1}{1-r} \qquad (3)$$

In the following examples we will use the above equation in order to find the sum of infinite geometric series.

Example 4.5-5 Find the sum of the following geometric series.

a. $\sum_{j=0}^{\infty} 2\left(\frac{1}{4}\right)^j =$ b. $\sum_{j=0}^{\infty} 5\left(-\frac{1}{2}\right)^j =$ c. $\sum_{k=1}^{\infty} 5\left(\frac{5}{3}\right)^{k-1} =$

d. $\sum_{k=1}^{\infty} \frac{18}{10^{k+1}} =$ e. $\sum_{j=0}^{\infty} \left(\frac{2}{3}\right)^j =$ f. $\sum_{j=0}^{\infty} \left(-\frac{1}{8}\right)^j =$

Solutions:

a. $s_1 = 2$ and $r = \frac{1}{4}$. Since $|r| \langle 1$ we can use equation (3) to obtain the sum

$$\boxed{\sum_{j=0}^{\infty} 2\left(\frac{1}{4}\right)^j} = \boxed{\frac{2}{1-\frac{1}{4}}} = \boxed{\frac{2}{\frac{4-1}{4}}} = \boxed{\frac{2}{\frac{3}{4}}} = \boxed{\frac{\frac{2}{1}}{\frac{3}{4}}} = \boxed{\frac{2 \times 4}{1 \times 3}} = \boxed{\mathbf{\frac{8}{3}}}$$

b. $s_1 = 5$ and $r = -\frac{1}{2}$. Since $|r| \langle 1$ we can use equation (3) to obtain the sum

$$\boxed{\sum_{j=0}^{\infty} 5\left(-\frac{1}{2}\right)^j} = \boxed{\frac{5}{1-\left(-\frac{1}{2}\right)}} = \boxed{\frac{5}{1+\frac{1}{2}}} = \boxed{\frac{5}{\frac{2+1}{2}}} = \boxed{\frac{5}{\frac{3}{2}}} = \boxed{\frac{\frac{5}{1}}{\frac{3}{2}}} = \boxed{\frac{5 \times 2}{1 \times 3}} = \boxed{\mathbf{\frac{10}{3}}}$$

c. $s_1 = 5$ and $r = \frac{5}{3}$. Since $|r| \rangle 1$ the geometric series $\sum_{k=1}^{\infty} 5\left(\frac{5}{3}\right)^{k-1}$ **has no finite sum.**

d. $$\boxed{\sum_{k=1}^{\infty} \frac{18}{10^{k+1}}} = \boxed{\sum_{k=1}^{\infty} \frac{18}{10^2 \cdot 10^{k-1}}} = \boxed{\sum_{k=1}^{\infty} \frac{18}{100}\left(\frac{1}{10^{k-1}}\right)} = \boxed{\sum_{k=1}^{\infty} \frac{18}{100}\left(\frac{1}{10}\right)^{k-1}}$$

$s_1 = \frac{18}{100}$ and $r = \frac{1}{10}$. Since $|r| \langle 1$ we can use equation (3) to obtain the sum

$$\boxed{\sum_{k=1}^{\infty} \frac{18}{100}\left(\frac{1}{10}\right)^{k-1}} = \boxed{\frac{\frac{18}{100}}{1-\frac{1}{10}}} = \boxed{\frac{\frac{18}{100}}{\frac{10-1}{10}}} = \boxed{\frac{\frac{18}{100}}{\frac{9}{10}}} = \boxed{\frac{18 \times 10}{100 \times 9}} = \boxed{\frac{180}{900}} = \boxed{\mathbf{\frac{1}{5}}}$$

e. $s_1 = 1$ and $r = \frac{2}{3}$. Since $|r| \langle 1$ we can use equation (3) to obtain the sum

$$\boxed{\sum_{j=0}^{\infty} \left(\frac{2}{3}\right)^j} = \boxed{\frac{1}{1-\frac{2}{3}}} = \boxed{\frac{1}{\frac{3-2}{3}}} = \boxed{\frac{1}{\frac{1}{3}}} = \boxed{\frac{\frac{1}{1}}{\frac{1}{3}}} = \boxed{\frac{1 \times 3}{1 \times 1}} = \boxed{\frac{3}{1}} = \boxed{\mathbf{3}}$$

f. $s_1 = 1$ and $r = -\frac{1}{8}$. Since $|r| \langle 1$ we can use equation (3) to obtain the sum

$$\boxed{\sum_{j=0}^{\infty}\left(-\frac{1}{8}\right)^{j}} = \boxed{\frac{1}{1-\left(-\frac{1}{8}\right)}} = \boxed{\frac{1}{1+\frac{1}{8}}} = \boxed{\frac{1}{\frac{8+1}{8}}} = \boxed{\frac{1}{\frac{9}{8}}} = \boxed{\frac{\frac{1}{1}}{\frac{9}{8}}} = \boxed{\frac{1\times 8}{1\times 9}} = \boxed{\mathbf{\frac{8}{9}}}$$

Example 4.5-6 Find the sum of the following infinite geometric series.

a. $2-1+\frac{1}{2}-\frac{1}{4}+\cdots=$

b. $-\frac{1}{3}+1-3+9+\cdots=$

c. $1+\frac{1}{9}+\frac{1}{81}+\frac{1}{729}+\cdots=$

d. $1+\frac{1}{2}+\frac{1}{4}+\frac{1}{16}+\cdots=$

Solutions:

a. Given $2-1+\frac{1}{2}-\frac{1}{4}+\cdots$, then $s_1=2$ and $r=-\frac{1}{2}$. Since $|r|\langle 1$ we can use equation (3) to obtain the sum, i.e.,

$$\boxed{2-1+\frac{1}{2}-\frac{1}{4}+\cdots} = \boxed{\frac{2}{1-\left(-\frac{1}{2}\right)}} = \boxed{\frac{2}{1+\frac{1}{2}}} = \boxed{\frac{2}{\frac{2+1}{2}}} = \boxed{\frac{2}{\frac{3}{2}}} = \boxed{\frac{\frac{2}{1}}{\frac{3}{2}}} = \boxed{\frac{2\times 2}{1\times 3}} = \boxed{\frac{4}{3}} = \boxed{1.333}$$

b. Given $-\frac{1}{3}+1-3+9+\cdots$, then $s_1=-\frac{1}{3}$ and $r=\frac{1}{-\frac{1}{3}}=-3$. Since $|r|=|-3|=3$ is greater than one the geometric sequence $-\frac{1}{3}+1-3+9+\cdots$ **has no finite sum.**

c. Given $1+\frac{1}{9}+\frac{1}{81}+\frac{1}{729}+\cdots$, then $s_1=1$ and $r=\frac{\frac{1}{9}}{1}=\frac{1}{9}$. Since $|r|\langle 1$ we can use equation (3) to obtain the sum, i.e.,

$$\boxed{1+\frac{1}{9}+\frac{1}{81}+\frac{1}{729}+\cdots} = \boxed{\frac{1}{1-\frac{1}{9}}} = \boxed{\frac{1}{\frac{9-1}{9}}} = \boxed{\frac{1}{\frac{8}{9}}} = \boxed{\frac{\frac{1}{1}}{\frac{8}{9}}} = \boxed{\frac{1\times 9}{1\times 8}} = \boxed{\mathbf{\frac{9}{8}}}$$

d. Given $1+\frac{1}{2}+\frac{1}{4}+\frac{1}{16}+\cdots$, then $s_1=1$ and $r=\frac{\frac{1}{2}}{1}=\frac{1}{2}$. Since $|r|\langle 1$ we can use equation (3) to obtain the sum, i.e.,

$$\boxed{1+\frac{1}{2}+\frac{1}{4}+\frac{1}{16}+\cdots} = \boxed{\frac{1}{1-\frac{1}{2}}} = \boxed{\frac{1}{\frac{2-1}{2}}} = \frac{1}{\frac{1}{2}} = \boxed{\frac{\frac{1}{1}}{\frac{1}{2}}} = \boxed{\frac{1\times 2}{1\times 1}} = \boxed{\frac{2}{1}} = \boxed{\mathbf{2}}$$

Repeating Decimals

An application of infinite geometric series is in representation of repeating decimals as the quotient of two integers. For example, $0.131313\overline{13}\cdots$ and $0.6666\overline{66}\cdots$ are repeating decimals. The bar above the repeating numbers denotes that the numbers appearing under it are repeated endlessly. The following examples show the steps as to how repeating decimals are converted to fractional forms:

Example 4.5-7 Write the following repeating decimals as the quotient of two positive integers.

a. $0.1313\overline{13}\cdots =$ b. $5.510510\overline{510}\cdots =$ c. $0.1245\overline{1245}\cdots =$

Solutions:

a. **First** - write the decimal number $0.1313\overline{13}\cdots$ in its equivalent form of

$0.1313\overline{13}\cdots = 0.13 + 0.0013 + 0.000013 + \cdots$

Second - Since this is a geometric series, write the first term of the series and its ratio, i.e.,

$s_1 = 0.13$ and $r = \frac{0.0013}{0.13} = 0.01$

Third - Since the ratio r is less than one, use the infinite geometric series equation $s_\infty = \frac{s_1}{1-r}$

to obtain the sum of the infinite series $0.13 + 0.0013 + 0.000013 + \cdots$, i.e.,

$$\boxed{s_\infty} = \boxed{\frac{s_1}{1-r}} = \boxed{\frac{0.13}{1-0.01}} = \boxed{\frac{0.13}{0.99}} = \boxed{\frac{13}{99}} \text{ thus } \boxed{\mathbf{0.1313\overline{13}\cdots = \frac{13}{99}}}$$

b. **First** - Consider the decimal portion of the number $5.510510\overline{510}\cdots$ and write it in its equivalent form of $0.510510\overline{510}\cdots = 0.510 + 0.000510 + 0.000000510 + \cdots$

Second - Since this is a geometric series, write the first term of the series and its ratio, i.e.,

$s_1 = 0.510$ and $r = \frac{0.000510}{0.510} = 0.001$

Third - Since the ratio r is less than one, use the infinite geometric series equation $s_\infty = \frac{s_1}{1-r}$

to obtain the sum of the infinite series $0.510 + 0.000510 + 0.000000510 + \cdots$, i.e.,

$$\boxed{s_\infty} = \boxed{\frac{s_1}{1-r}} = \boxed{\frac{0.510}{1-0.001}} = \boxed{\frac{0.510}{0.999}} = \boxed{\frac{510}{999}} \text{ thus } \boxed{\mathbf{5.510510\overline{510}\cdots = 5\frac{510}{999}}}$$

c. **First** - write the decimal number $0.1245\overline{1245}\cdots$ in its equivalent form of

$0.1245\overline{1245}\cdots = 0.1245 + 0.00001245 + \cdots$

Second - Since this is a geometric series, write the first term of the series and its ratio, i.e.,

$s_1 = 0.1245$ and $r = \frac{0.00001245}{0.1245} = 0.0001$

Third - Since the ratio r is less than one, use the infinite geometric series equation $s_\infty = \frac{s_1}{1-r}$

to obtain the sum of the infinite series $0.1245 + 0.00001245 + \cdots$, i.e.,

$$\boxed{s_\infty} = \boxed{\frac{s_1}{1-r}} = \boxed{\frac{0.1245}{1-0.0001}} = \boxed{\frac{0.1245}{0.9999}} = \boxed{\frac{1245}{9999}} \text{ thus } \boxed{\mathbf{0.1245\overline{1245}\cdots = \frac{1245}{9999}}}$$

Section 4.5 Practice Problems - Limits of Sequences and Series

1. State which of the following sequences are convergent.

a. $1, \frac{3}{2}, 2, \frac{5}{2}, 3, \cdots, \frac{n+1}{2}, \cdots =$

b. $0, \frac{3}{2}, \frac{8}{3}, \frac{15}{4}, \cdots, \frac{n^2-1}{n}, \cdots =$

c. $4, 8, 16, 32, \cdots, 2^{n+1}, \cdots =$

d. $\frac{1}{16}, \frac{1}{64}, \frac{1}{256}, \frac{1}{1024}, \cdots, \frac{1}{4^{n+1}}, \cdots =$

e. $0, \frac{1}{4}, \frac{2}{9}, \frac{3}{16}, \cdots, \frac{n-1}{n^2}, \cdots =$

f. $\frac{1}{25}, \frac{1}{125}, \frac{1}{625}, \frac{1}{3125}, \cdots, \left(\frac{1}{5}\right)^{n+1}, \cdots =$

2. State which of the following geometric sequences are convergent.

a. $\frac{1}{4}, \frac{1}{16}, \frac{1}{64}, \frac{1}{256}, \cdots, \frac{1}{4^n}, \cdots =$

b. $-5, 25, -125, 625, -3125, \cdots, (-5)^n, \cdots =$

c. $2, -2, 2, -2, \cdots, 2(-1)^{n+1}, \cdots =$

d. $1, \frac{1}{2}, \frac{1}{4}, \frac{1}{8}, \cdots, \left(\frac{1}{2}\right)^{n-1}, \cdots =$

e. $-9, 27, -81, 243, \cdots, (-1)^n 3^{n+1}, \cdots =$

f. $1, \frac{1}{3}, \frac{1}{9}, \frac{1}{27}, \frac{1}{81}, \cdots, \left(\frac{1}{3}\right)^{n-1}, \cdots =$

3. State whether or not the following sequences are convergent or divergent as $n \to \infty$. If the sequence does converge, find its limit.

a. $\frac{n^2}{n^3-4} =$

b. $\frac{5n+1}{\sqrt{n^2+1}} =$

c. $\frac{25^n}{5^{n+1}} =$

d. $\frac{5^n+25}{125^n} =$

e. $\frac{(n+2)^2}{n^2} =$

f. $\frac{2^n}{2^n+1} =$

g. $\frac{\sqrt{n^2+2n}}{\sqrt{n^4+1}} =$

h. $\frac{5}{n^2+1} =$

i. $\frac{n+1}{n-1} =$

j. $\frac{n}{n^3-1} =$

k. $10^{\frac{1}{n}} =$

l. $\frac{(n-1)^2}{(1-n)(1+n)} =$

m. $100^{-\frac{1}{n}} =$

n. $3^{\frac{3}{n}} =$

o. $\frac{n+100}{n^3-10} =$

p. $\frac{100^{\frac{1}{n}}}{n^2+3} =$

q. $\frac{1}{n+1}-1 =$

r. $(0.25)^{-n} =$

s. $\frac{\sqrt{n+1}}{\sqrt{n}+1} =$

t. $\frac{\sqrt{n}}{\sqrt{n+1}}+2 =$

4. Find the sum of the following geometric series.

a. $\sum_{j=0}^{\infty} 3\left(\frac{1}{8}\right)^j =$

b. $\sum_{j=0}^{\infty} 3\left(-\frac{1}{4}\right)^j =$

c. $\sum_{k=1}^{\infty} 3\left(\frac{3}{2}\right)^{k-1} =$

d. $\sum_{k=1}^{\infty} \frac{5}{100^{k+1}} =$

e. $\sum_{j=0}^{\infty} \left(\frac{1}{3}\right)^j =$

f. $\sum_{j=0}^{\infty} \left(-\frac{1}{5}\right)^j =$

5. Find the sum of the following infinite geometric series.

a. $5-1+\frac{1}{5}-\frac{1}{25}+\cdots =$

b. $-\frac{1}{2}+2-8+32+\cdots =$

c. $1+\frac{1}{6}+\frac{1}{36}+\frac{1}{216}+\cdots =$

d. $1+\frac{1}{10}+\frac{1}{100}+\frac{1}{1000}+\cdots =$

6. Write the following repeating decimals as the quotient of two positive integers.

a. $0.6666\overline{66}\cdots =$

b. $3.027027\overline{027}\cdots =$

c. $0.1111\overline{11}\cdots =$

4.6 The Factorial Notation

As was stated earlier, the product of several consecutive positive integers is usually written using a special symbol $n!$, read as "n factorial," which is defined by

$$n! = n(n-1)(n-2)(n-3)(n-4)\ldots(4)(3)(2)(1)$$

For example, $1!$ (read as "one factorial") through $10!$ (read as "ten factorial") are written in their equivalent form as:

$\boxed{1!} = \boxed{1}$

$\boxed{2!} = \boxed{2\cdot 1} = \boxed{2}$

$\boxed{3!} = \boxed{3\cdot 2\cdot 1} = \boxed{6}$

$\boxed{4!} = \boxed{4\cdot 3\cdot 2\cdot 1} = \boxed{24}$

$\boxed{5!} = \boxed{5\cdot 4\cdot 3\cdot 2\cdot 1} = \boxed{120}$

$\boxed{6!} = \boxed{6\cdot 5\cdot 4\cdot 3\cdot 2\cdot 1} = \boxed{720}$

$\boxed{7!} = \boxed{7\cdot 6\cdot 5\cdot 4\cdot 3\cdot 2\cdot 1} = \boxed{5{,}040}$

$\boxed{8!} = \boxed{8\cdot 7\cdot 6\cdot 5\cdot 4\cdot 3\cdot 2\cdot 1} = \boxed{40{,}320}$

$\boxed{9!} = \boxed{9\cdot 8\cdot 7\cdot 6\cdot 5\cdot 4\cdot 3\cdot 2\cdot 1} = \boxed{362{,}880}$

$\boxed{10!} = \boxed{10\cdot 9\cdot 8\cdot 7\cdot 6\cdot 5\cdot 4\cdot 3\cdot 2\cdot 1} = \boxed{3{,}628{,}800}$

Note that, for $n \rangle 1$, since $n! = n(n-1)(n-2)(n-3)\ldots 4\cdot 3\cdot 2\cdot 1$ and $(n-1)! = (n-1)(n-2)(n-3)\cdots 4\cdot 3\cdot 2\cdot 1$

we can rewrite the recursive $n!$ relationship in the following way:

$$n! = n\underbrace{(n-1)(n-2)(n-3)(n-4)\ldots 4\cdot 3\cdot 2\cdot 1}_{(n-1)!} = \mathbf{n(n-1)!}$$

$$n(n-1)! = n(n-1)\underbrace{(n-2)(n-3)(n-4)\ldots 4\cdot 3\cdot 2\cdot 1}_{(n-2)!} = \mathbf{n(n-1)(n-2)!}$$

$$n(n-1)(n-2)! = n(n-1)(n-2)\underbrace{(n-3)(n-4)\ldots 4\cdot 3\cdot 2\cdot 1}_{(n-3)!} = \mathbf{n(n-1)(n-2)(n-3)!}$$

$$n(n-1)(n-2)(n-3)! = n(n-1)(n-2)(n-3)\underbrace{(n-4)\ldots 4\cdot 3\cdot 2\cdot 1}_{(n-4)!} = \mathbf{n(n-1)(n-2)(n-3)(n-4)!}$$

For example,

$\boxed{7!} = \boxed{7\cdot 6!} = \boxed{7\cdot 6\cdot 5!} = \boxed{7\cdot 6\cdot 5\cdot 4!} = \boxed{7\cdot 6\cdot 5\cdot 4\cdot 3!} = \boxed{\mathbf{7\cdot 6\cdot 5\cdot 4\cdot 3\cdot 2!}}$

$\boxed{10!} = \boxed{10\cdot 9!} = \boxed{10\cdot 9\cdot 8!} = \boxed{10\cdot 9\cdot 8\cdot 7!} = \boxed{10\cdot 9\cdot 8\cdot 7\cdot 6!} = \boxed{\mathbf{10\cdot 9\cdot 8\cdot 7\cdot 6\cdot 5!}}$

$\boxed{35!} = \boxed{35\cdot 34!} = \boxed{35\cdot 34\cdot 33!} = \boxed{35\cdot 34\cdot 33\cdot 32!} = \boxed{35\cdot 34\cdot 33\cdot 32\cdot 31!} = \boxed{\mathbf{35\cdot 34\cdot 33\cdot 32\cdot 31\cdot 30!}}$

$\boxed{(n+4)!} = \boxed{(n+4)(n+3)!} = \boxed{(n+4)(n+3)(n+2)!} = \boxed{\mathbf{(n+4)(n+3)(n+2)(n+1)!}}$

$\boxed{(n+8)!} = \boxed{(n+8)(n+7)!} = \boxed{(n+8)(n+7)(n+6)!} = \boxed{\mathbf{(n+8)(n+7)(n+6)(n+5)!}}$

$\boxed{(2n+2)!} = \boxed{(2n+2)(2n+1)!} = \boxed{(2n+2)(2n+1)(2n)!} = \boxed{(2n+2)(2n+1)(2n)(2n-1)!} = \boxed{(2n+2)(2n+1)(2n)}$

$\boxed{(2n-1)(2n-2)!} = \boxed{\mathbf{(2n+2)(2n+1)(2n)(2n-1)(2n-2)(2n-3)!}}$

$$\boxed{(2n+5)!} = \boxed{(2n+5)(2n+4)!} = \boxed{(2n+5)(2n+4)(2n+3)!} = \boxed{(2n+5)(2n+4)(2n+3)(2n+2)!} = \boxed{(2n+5)(2n+4)}$$

$$\boxed{(2n+3)(2n+2)(2n+1)!} = \boxed{\mathbf{(2n+5)(2n+4)(2n+3)(2n+2)(2n+1)(2n)!}}$$

The above principal can be used to prove that $0!=1$. Since $1! = 1\cdot(1-1)! = 1\cdot 0!$ in order for equality on both sides of the equation to hold true $0!$ must be equal to one. Hence, we state that $0!=1$.

Example 4.6-1 Expand and simplify the following factorial expressions.

a. $13! =$ b. $(6-2)! =$ c. $\frac{8!}{4!} =$ d. $\frac{12!}{11!} =$

e. $\frac{10!}{5!4!} =$ f. $\frac{8!}{3!(8-2)!} =$ g. $\frac{11!8!}{16!} =$ h. $\frac{(4-2)!8!}{11!(5-3)!} =$

Solutions:

a. $\boxed{13!} = \boxed{13\cdot12\cdot11\cdot10\cdot9\cdot8\cdot7\cdot6\cdot5\cdot4\cdot3\cdot2\cdot1} = \boxed{\mathbf{6{,}227{,}020{,}800}}$

b. $\boxed{(6-2)!} = \boxed{4!} = \boxed{4\cdot3\cdot2\cdot1} = \boxed{\mathbf{24}}$

c. $\boxed{\frac{8!}{4!}} = \boxed{\frac{8\cdot7\cdot6\cdot5\cdot4!}{4!}} = \boxed{\frac{8\cdot7\cdot6\cdot5\cdot\cancel{4!}}{\cancel{4!}}} = \boxed{8\cdot7\cdot6\cdot5} = \boxed{\mathbf{1680}}$

d. $\boxed{\frac{12!}{11!}} = \boxed{\frac{12\cdot11!}{11!}} = \boxed{\frac{12\cdot\cancel{11!}}{\cancel{11!}}} = \boxed{\mathbf{12}}$

e. $\boxed{\frac{10!}{5!4!}} = \boxed{\frac{10\cdot9\cdot8\cdot7\cdot6\cdot5!}{5!4!}} = \boxed{\frac{10\cdot9\cdot8\cdot7\cdot6\cdot\cancel{5!}}{\cancel{5!}4!}} = \boxed{\frac{10\cdot9\cdot8\cdot7\cdot6}{4!}} = \boxed{\frac{10\cdot9\cdot\overset{2}{\cancel{8}}\cdot7\cdot\cancel{6}}{\cancel{4}\cdot\cancel{3}\cdot\cancel{2}\cdot1}} = \boxed{\frac{10\cdot9\cdot2\cdot7}{1}} = \boxed{\mathbf{1260}}$

f. $\boxed{\frac{8!}{3!(8-2)!}} = \boxed{\frac{8!}{3!6!}} = \boxed{\frac{8\cdot7\cdot\cancel{6!}}{3!\cancel{6!}}} = \boxed{\frac{8\cdot7}{3!}} = \boxed{\frac{\overset{4}{\cancel{8}}\cdot7}{3\cdot\cancel{2}\cdot1}} = \boxed{\frac{4\cdot7}{3}} = \boxed{\mathbf{\frac{28}{3}}}$

g. $\boxed{\frac{11!8!}{16!}} = \boxed{\frac{11!8!}{16\cdot15\cdot14\cdot13\cdot12\cdot11!}} = \boxed{\frac{\cancel{11!}8\cdot7\cdot6\cdot5\cdot4\cdot3\cdot2\cdot1}{16\cdot15\cdot14\cdot13\cdot12\cdot\cancel{11!}}} = \boxed{\frac{\cancel{8}\cdot\cancel{7}\cdot\cancel{6}\cdot\cancel{5}\cdot4\cdot3\cdot2\cdot1}{\underset{2}{\cancel{16}}\cdot\underset{3}{\cancel{15}}\cdot\underset{2}{\cancel{14}}\cdot13\cdot\underset{2}{\cancel{12}}}} = \boxed{\frac{\cancel{4}\cdot\cancel{3}\cdot\cancel{2}\cdot1}{2\cdot\cancel{3}\cdot2\cdot13\cdot2}} = \boxed{\mathbf{\frac{1}{13}}}$

h. $\boxed{\frac{(4-2)!8!}{11!(5-3)!}} = \boxed{\frac{2!8!}{11!2!}} = \boxed{\frac{8!}{11!}} = \boxed{\frac{8!}{11\cdot10\cdot9\cdot8!}} = \boxed{\frac{\cancel{8!}}{11\cdot10\cdot9\cdot\cancel{8!}}} = \boxed{\mathbf{\frac{1}{990}}}$

Example 4.6-2 Write the following products in factorial form.

a. $4\cdot3\cdot2\cdot1 =$ b. $10\cdot11\cdot12\cdot13 =$ c. $20\cdot21\cdot22\cdot23\cdot24 =$

d. $9\cdot8\cdot7\cdot6\cdot5 =$ e. $3\cdot4\cdot5\cdot6\cdot7 =$ f. $20 =$

g. $15\cdot14\cdot13\cdot12\cdot11 =$ h. $8 =$ i. $1\cdot2\cdot3\cdot4\cdot5\cdot6 =$

Solutions:

a. $\boxed{4\cdot3\cdot2\cdot1} = \boxed{\mathbf{4!}}$

b. $\boxed{10\cdot11\cdot12\cdot13} = \boxed{\mathbf{\frac{13!}{9!}}}$

c. $\boxed{20\cdot21\cdot22\cdot23\cdot24} = \boxed{\mathbf{\frac{24!}{19!}}}$

d. $\boxed{9\cdot8\cdot7\cdot6\cdot5} = \boxed{\mathbf{\frac{9!}{4!}}}$

e. $\boxed{3\cdot4\cdot5\cdot6\cdot7} = \boxed{\mathbf{\frac{7!}{2!}}}$

f. $\boxed{20} = \boxed{\mathbf{\frac{20!}{19!}}}$

g. $\boxed{15\cdot14\cdot13\cdot12\cdot11} = \boxed{\mathbf{\frac{15!}{10!}}}$

h. $\boxed{8} = \boxed{\mathbf{\frac{8!}{7!}}}$

i. $\boxed{1\cdot2\cdot3\cdot4\cdot5\cdot6} = \boxed{\mathbf{6!}}$

Example 4.6-3 Expand the following factorial expressions.

a. $n! =$

b. $(n-1)! =$

c. $(n-3)! =$

d. $(n-5)! =$

e. $(n-8)! =$

f. $(2n)! =$

g. $(2n-1)! =$

h. $(3n-5)! =$

i. $(2n+1)! =$

j. $(n+1)! =$

k. $(n+3)! =$

l. $(n+8)! =$

m. $(2n+2)! =$

n. $(2n+5)! =$

Solutions:

a. $\boxed{n!} = \boxed{(n)(n-1)(n-2)(n-3)(n-4)(n-5)(n-6)(n-7)(n-8)(n-9)\cdots4\cdot3\cdot2\cdot1}$

b. $\boxed{(n-1)!} = \boxed{(n-1)(n-2)(n-3)(n-4)(n-5)(n-6)(n-7)(n-8)(n-9)\cdots4\cdot3\cdot2\cdot1}$

c. $\boxed{(n-3)!} = \boxed{(n-3)(n-4)(n-5)(n-6)(n-7)(n-8)(n-9)\cdots4\cdot3\cdot2\cdot1}$

d. $\boxed{(n-5)!} = \boxed{(n-5)(n-6)(n-7)(n-8)(n-9)(n-10)(n-11)\cdots4\cdot3\cdot2\cdot1}$

e. $\boxed{(n-8)!} = \boxed{(n-8)(n-9)(n-10)(n-11)\cdots4\cdot3\cdot2\cdot1}$

f. $\boxed{(2n)!} = \boxed{(2n)(2n-1)(2n-2)(2n-3)(2n-4)\cdots4\cdot3\cdot2\cdot1}$

g. $\boxed{(2n-1)!} = \boxed{(2n-1)(2n-2)(2n-3)(2n-4)\cdots4\cdot3\cdot2\cdot1}$

h. $\boxed{(3n-5)!} = \boxed{(3n-5)(3n-6)(3n-7)(3n-8)(3n-9)(3n-10)(3n-11)\cdots4\cdot3\cdot2\cdot1}$

i. $\boxed{(2n+1)!} = \boxed{(2n+1)(2n)(2n-1)(2n-2)(2n-3)(2n-4)\cdots4\cdot3\cdot2\cdot1}$

j. $\boxed{(n+1)!} = \boxed{(n+1)(n)(n-1)(n-2)(n-3)(n-4)(n-5)(n-6)(n-7)(n-8)\cdots4\cdot3\cdot2\cdot1}$

k. $\boxed{(n+3)!} = \boxed{(n+3)(n+2)(n+1)(n)(n-1)(n-2)(n-3)(n-4)(n-5)(n-6)(n-7)(n-8)\cdots4\cdot3\cdot2\cdot1}$

l. $\boxed{(n+8)!} = \boxed{(n+8)(n+7)(n+6)(n+5)(n+4)(n+3)(n+2)(n+1)(n)(n-1)(n-2)(n-3)\cdots4\cdot3\cdot2\cdot1}$

m. $\boxed{(2n+2)!} = \boxed{\mathbf{(2n+2)(2n+1)(2n)(2n-1)(2n-2)(2n-3)(2n-4)\cdots 4\cdot 3\cdot 2\cdot 1}}$

n. $\boxed{(2n+5)!} = \boxed{\mathbf{(2n+5)(2n+4)(2n+3)(2n+2)(2n+1)(2n)(2n-1)(2n-2)(2n-3)(2n-4)\cdots 4\cdot 3\cdot 2\cdot 1}}$

Example 4.6-4 Expand and simplify the following factorial expressions.

a. $\frac{(n-1)!}{(n-3)!} =$ b. $\frac{(n+2)!}{n!} =$ c. $\frac{(n+3)!}{(n-1)!} =$ d. $\frac{(n+2)(n+2)!}{(n+3)!} =$

e. $\frac{(3n+1)!(3n-1)!}{(3n)!(3n-3)!} =$ f. $\frac{(n!)^2}{(n+1)!(n-1)!} =$ g. $\frac{(2n-2)!2(n!)}{(2n)!(n-1)!} =$

Solutions:

a. $\boxed{\frac{(n-1)!}{(n-3)!}} = \boxed{\frac{(n-1)!}{(n-3)(n-2)(n-1)!}} = \boxed{\frac{\cancel{(n-1)!}}{(n-3)(n-2)\cancel{(n-1)!}}} = \boxed{\frac{\mathbf{1}}{\mathbf{(n-3)(n-2)}}}$

b. $\boxed{\frac{(n+2)!}{n!}} = \boxed{\frac{(n+2)(n+1)n!}{n!}} = \boxed{\frac{(n+2)(n+1)\cancel{n!}}{\cancel{n!}}} = \boxed{\mathbf{(n+2)(n+1)}}$

c. $\boxed{\frac{(n+3)!}{(n-1)!}} = \boxed{\frac{(n+3)(n+2)(n+1)(n)(n-1)!}{(n-1)!}} = \boxed{\frac{(n+3)(n+2)(n+1)(n)\cancel{(n-1)!}}{\cancel{(n-1)!}}} = \boxed{\mathbf{(n+3)(n+2)(n+1)n}}$

d. $\boxed{\frac{(n+2)(n+2)!}{(n+3)!}} = \boxed{\frac{(n+2)(n+2)!}{(n+3)(n+2)!}} = \boxed{\frac{(n+2)\cancel{(n+2)!}}{(n+3)\cancel{(n+2)!}}} = \boxed{\frac{\mathbf{n+2}}{\mathbf{n+3}}}$

e. $\boxed{\frac{(3n+1)!(3n-1)!}{(3n)!(3n-3)!}} = \boxed{\frac{[(3n+1)(3n)!][(3n-1)(3n-2)(3n-3)!]}{(3n)!(3n-3)!}} = \boxed{\frac{[(3n+1)\cancel{(3n)!}][(3n-1)(3n-2)\cancel{(3n-3)!}]}{\cancel{(3n)!}\cancel{(3n-3)!}}}$

$= \boxed{\frac{(3n+1)(3n-1)(3n-2)}{1}} = \boxed{\mathbf{(3n+1)(3n-1)(3n-2)}}$

f. $\boxed{\frac{(n!)^2}{(n+1)!(n-1)!}} = \boxed{\frac{(n!)(n!)}{[(n+1)(n)!](n-1)!}} = \boxed{\frac{(n!)\cancel{(n!)}}{[(n+1)\cancel{(n)!}](n-1)!}} = \boxed{\frac{n!}{(n+1)(n-1)!}} = \boxed{\frac{n\cancel{(n-1)!}}{(n+1)\cancel{(n-1)!}}} = \boxed{\frac{\mathbf{n}}{\mathbf{n+1}}}$

g. $\boxed{\frac{(2n-2)!2(n!)}{(2n)!(n-1)!}} = \boxed{\frac{(2n-2)![2n(n-1)!]}{[(2n)(2n-1)(2n-2)!](n-1)!}} = \boxed{\frac{\cancel{(2n-2)!}[2n\cancel{(n-1)!}]}{[(2n)(2n-1)\cancel{(2n-2)!}]\cancel{(n-1)!}}} = \boxed{\frac{2n}{(2n)(2n-1)}}$

$= \boxed{\frac{\cancel{2n}}{\cancel{(2n)}(2n-1)}} = \boxed{\frac{\mathbf{1}}{\mathbf{2n-1}}}$

The factorial notation is used in expansion of the $\binom{n}{r}$ read as "the binomial coefficient n, r" in the following way:

$$\binom{n}{r} = \frac{n!}{r!(n-r)!}$$

Example 4.6-5 Write the following expressions in factorial notation form. Simplify the answer.

a. $\binom{6}{4} =$ b. $\binom{6}{2} =$ c. $\binom{5}{0} =$ d. $\binom{5}{5} =$ e. $\binom{7}{5} =$

f. $\binom{10}{5} =$ g. $\binom{8}{4} =$ h. $\binom{n}{n} =$ i. $\binom{n}{n-1} =$ j. $\binom{n}{n-4} =$

Solutions:

a. $\binom{6}{4} = \frac{6!}{4!(6-4)!} = \frac{6!}{4!\cdot 2!} = \frac{6\cdot5\cdot4\cdot3\cdot2\cdot1}{(4\cdot3\cdot2\cdot1)\cdot(2\cdot1)} = \frac{\overset{3}{\cancel{6}}\cdot5\cdot\cancel{4}\cdot\cancel{3}\cdot\cancel{2}\cdot\cancel{1}}{(\cancel{4}\cdot\cancel{3}\cdot\cancel{2}\cdot\cancel{1})\cdot(\cancel{2}\cdot1)} = \frac{3\cdot5}{1} = \boxed{15}$

b. $\binom{6}{2} = \frac{6!}{2!(6-2)!} = \frac{6!}{2!\cdot 4!} = \frac{6\cdot5\cdot4\cdot3\cdot2\cdot1}{(2\cdot1)\cdot(4\cdot3\cdot2\cdot1)} = \frac{\overset{3}{\cancel{6}}\cdot5\cdot\cancel{4}\cdot\cancel{3}\cdot\cancel{2}\cdot\cancel{1}}{(\cancel{2}\cdot1)\cdot(\cancel{4}\cdot\cancel{3}\cdot\cancel{2}\cdot\cancel{1})} = \frac{3\cdot5}{1} = \boxed{15}$

c. $\binom{5}{0} = \frac{5!}{0!(5-0)!} = \frac{5!}{0!\cdot 5!} = \frac{\cancel{5!}}{1\cdot\cancel{5!}} = \frac{1}{1} = \boxed{1}$

d. $\binom{5}{5} = \frac{5!}{5!(5-5)!} = \frac{5!}{5!\cdot 0!} = \frac{\cancel{5!}}{\cancel{5!}\cdot 1} = \frac{1}{1} = \boxed{1}$

e. $\binom{7}{5} = \frac{7!}{5!(7-5)!} = \frac{7!}{5!\cdot 2!} = \frac{7\cdot6\cdot5!}{5!\cdot2!} = \frac{7\cdot\overset{3}{\cancel{6}}\cdot\cancel{5!}}{\cancel{5!}\cdot\cancel{2}\cdot1} = \frac{7\cdot3}{1} = \boxed{21}$

f. $\binom{10}{5} = \frac{10!}{5!(10-5)!} = \frac{10!}{5!\cdot 5!} = \frac{10\cdot9\cdot8\cdot7\cdot6\cdot5\cdot4\cdot3\cdot2\cdot1}{(5\cdot4\cdot3\cdot2\cdot1)\cdot(5\cdot4\cdot3\cdot2\cdot1)} = \frac{\overset{2}{\cancel{10}}\cdot9\cdot\overset{2}{\cancel{8}}\cdot7\cdot\cancel{6}\cdot\cancel{5}\cdot\cancel{4}\cdot\cancel{3}\cdot\cancel{2}\cdot\cancel{1}}{(\cancel{5}\cdot\cancel{4}\cdot\cancel{3}\cdot\cancel{2}\cdot\cancel{1})\cdot(\cancel{5}\cdot\cancel{4}\cdot\cancel{3}\cdot\cancel{2}\cdot\cancel{1})} = \frac{2\cdot9\cdot2\cdot7}{1} = \boxed{252}$

g. $\binom{8}{4} = \frac{8!}{4!(8-4)!} = \frac{8!}{4!\cdot 4!} = \frac{8\cdot7\cdot6\cdot5\cdot4!}{4!\cdot4\cdot3\cdot2\cdot1} = \frac{\cancel{8}\cdot7\cdot\overset{2}{\cancel{6}}\cdot5\cdot\cancel{4!}}{\cancel{4!}\cdot\cancel{4}\cdot\cancel{3}\cdot\cancel{2}\cdot1} = \frac{7\cdot2\cdot5}{1} = \boxed{70}$

h. $\binom{n}{n} = \frac{n!}{n!(n-n)!} = \frac{n!}{n!0!} = \frac{\cancel{n}!}{\cancel{n}!\cdot1} = \frac{1}{1} = \boxed{1}$

i. $\binom{n}{n-1} = \frac{n!}{(n-1)![n-(n-1)]!} = \frac{n!}{(n-1)!(\cancel{n}-\cancel{n}+1)!} = \frac{n!}{(n-1)!1!} = \frac{n\cdot(n-1)!}{(n-1)!} = \frac{n\cdot(\cancel{n}-\cancel{1})!}{(\cancel{n}-\cancel{1})!} = \boxed{n}$

j. $\binom{n}{n-4} = \frac{n!}{(n-4)![n-(n-4)]!} = \frac{n!}{(n-4)!(\cancel{n}-\cancel{n}+4)!} = \frac{n!}{(n-4)!4!} = \frac{n\cdot(n-1)\cdot(n-2)\cdot(n-3)\cdot(n-4)!}{(n-4)!4!}$

$= \frac{n\cdot(n-1)\cdot(n-2)\cdot(n-3)\cdot(\cancel{n-4})!}{(\cancel{n-4})!4!} = \frac{n\cdot(n-1)\cdot(n-2)\cdot(n-3)}{4\cdot3\cdot2\cdot1} = \boxed{\frac{n\cdot(n-1)\cdot(n-2)\cdot(n-3)}{24}}$

An application of the binomial coefficient n, r, i.e., $\binom{n}{r}$ is in its use for expansion of binomials of the form $(a+b)^n$, where n is a positive integer. In general, the binomial equation of order n can be expanded in the following form:

$$(a+b)^n = \binom{n}{0}a^n + \binom{n}{1}a^{n-1}b + \binom{n}{2}a^{n-2}b^2 + \cdots + \binom{n}{r-1}a^{n-r+1}b^{r-1} + \cdots + \binom{n}{n}b^n \qquad (1)$$

note that the above equation is used for expanding binomial expressions that are raised to the second, third, fourth, or higher powers. For example, $(x-2)^2$, $(x+3)^3$, $(x-1)^5$, $(x-3)^6$, $(2x+y)^4$, $\left(2x-\frac{y}{2}\right)^5$, $\left(x-\sqrt{3}\right)^4$, $(x-2y)^6$, $(x-y)^{16}$, etc. can all be expanded using the above equation. The following examples show the steps as to how binomial coefficients are used in expanding binomial equations:

Example 4.6-6 Expand the following binomial expressions.

a. $(x-1)^3 =$ b. $(x+2)^5 =$ c. $(x-3)^4 =$

Solutions:

a. **First** - Identify the $a, b,$ and n terms, i.e., $a = x$, $b = -1$, $n = 3$.

Second - Use the general binomial expansion formula, i.e., equation (1) above to expand $(x-1)^3$.

$$\boxed{(x-1)^3} = \boxed{\binom{3}{0}x^3 + \binom{3}{1}x^{3-1}\cdot(-1) + \binom{3}{2}x^{3-2}(-1)^2 + \binom{3}{3}x^{3-3}(-1)^3} = \boxed{\binom{3}{0}x^3 - \binom{3}{1}x^2 + \binom{3}{2}x - \binom{3}{3}x^0}$$

$$= \boxed{\binom{3}{0}x^3 - \binom{3}{1}x^2 + \binom{3}{2}x - \binom{3}{3}\cdot 1} = \boxed{\frac{3!}{0!(3-0)!}x^3 - \frac{3!}{1!(3-1)!}x^2 + \frac{3!}{2!(3-2)!}x - \frac{3!}{3!(3-3)!}}$$

$$= \boxed{\frac{3!}{3!}x^3 - \frac{3!}{2!}x^2 + \frac{3!}{2!}x - \frac{3!}{3!}} = \boxed{\frac{\cancel{3!}}{\cancel{3!}}x^3 - \frac{3\cdot\cancel{2!}}{\cancel{2!}}x^2 + \frac{3\cdot\cancel{2!}}{\cancel{2!}}x - \frac{\cancel{3!}}{\cancel{3!}}} = \boxed{\mathbf{x^3 - 3x^2 + 3x - 1}}$$

b. **First** - Identify the $a, b,$ and n terms, i.e., $a = x$, $b = 2$, $n = 5$.

Second - Expand $(x+2)^5$ using equation (1).

$$\boxed{(x+2)^5} = \boxed{\binom{5}{0}x^5 + \binom{5}{1}x^{5-1}\cdot 2 + \binom{5}{2}x^{5-2}\cdot 2^2 + \binom{5}{3}x^{5-3}\cdot 2^3 + \binom{5}{4}x^{5-4}\cdot 2^4 + \binom{5}{5}x^{5-5}\cdot 2^5}$$

$$= \boxed{\binom{5}{0}x^5 + 2\binom{5}{1}x^4 + 4\binom{5}{2}x^3 + 8\binom{5}{3}x^2 + 16\binom{5}{4}x + 32\binom{5}{5}x^0} = \boxed{\binom{5}{0}x^5 + 2\binom{5}{1}x^4 + 4\binom{5}{2}x^3 + 8\binom{5}{3}x^2}$$

$$\boxed{+16\binom{5}{4}x + 32\binom{5}{5}} = \boxed{\frac{5!}{0!(5-0)!}x^5 + 2\frac{5!}{1!(5-1)!}x^4 + 4\frac{5!}{2!(5-2)!}x^3 + 8\frac{5!}{3!(5-3)!}x^2 + 16\frac{5!}{4!(5-4)!}x}$$

$$\boxed{+32\frac{5!}{5!(5-5)!}} = \boxed{\frac{5!}{5!}x^5 + 2\frac{5!}{4!}x^4 + 4\frac{5!}{2!3!}x^3 + 8\frac{5!}{3!2!}x^2 + 16\frac{5!}{4!}x + 32\frac{5!}{5!}} = \boxed{\frac{\cancel{5!}}{\cancel{5!}}x^5 + 2\frac{5\cdot 4!}{4!}x^4}$$

$$\boxed{+4\frac{5\cdot 4\cdot \cancel{3!}}{2!\cancel{3!}}x^3 + 8\frac{5\cdot 4\cdot \cancel{3!}}{\cancel{3!}2!}x^2 + 16\frac{5\cdot \cancel{4!}}{\cancel{4!}}x + 32\frac{\cancel{5!}}{\cancel{5!}}} = \boxed{\mathbf{x^5 + 10\,x^4 + 40\,x^3 + 80x^2 + 80x + 32}}$$

c. **First** - Identify the $a, b,$ and n terms, i.e., $a = x$, $b = -3$, $n = 4$.

Second - Expand $(x-3)^4$ using equation (1).

$$\boxed{(x-3)^4} = \boxed{\binom{4}{0}x^4 + \binom{4}{1}x^{4-1}\cdot(-3) + \binom{4}{2}x^{4-2}(-3)^2 + \binom{4}{3}x^{4-3}(-3)^3 + \binom{4}{4}x^{4-4}(-3)^4} = \boxed{\binom{4}{0}x^4 - 3\binom{4}{1}x^3}$$

$$\boxed{+9\binom{4}{2}x^2 - 27\binom{4}{3}x + 81\binom{4}{4}x^0} = \boxed{\frac{4!}{0!(4-0)!}x^4 - 3\frac{4!}{1!(4-1)!}x^3 + 9\frac{4!}{2!(4-2)!}x^2 - 27\frac{4!}{3!(4-3)!}x}$$

$$\boxed{+81\frac{4!}{4!(4-4)!}} = \boxed{\frac{4!}{4!}x^4 - 3\frac{4!}{3!}x^3 + 9\frac{4!}{2!2!}x^2 - 27\frac{4!}{3!}x + 81\frac{4!}{4!0!}} = \boxed{\frac{4!}{4!}x^4 - 3\frac{4\cdot \cancel{3!}}{\cancel{3!}}x^3 + 9\frac{4\cdot 3\cdot \cancel{2!}}{2!\cancel{2!}}x^2}$$

$$\boxed{-27\frac{4\cdot \cancel{3!}}{\cancel{3!}}x + 81\frac{4!}{4!0!}} = \boxed{\mathbf{x^4 - 12\,x^3 + 54\,x^2 - 108x + 81}}$$

Example 4.6-7 Use the general equation for binomial expansion to solve the following exponential numbers to the nearest hundredth.

a. $(0.83)^6 =$ b. $(1.05)^8 =$ c. $(1.21)^{10} =$

Solutions:

a. **First** - Write the exponential expression in the form of $(0.83)^6 = (1-0.17)^6$.

Second - Identify the $a, b,$ and n terms, i.e., $a = 1$, $b = -0.17$, $n = 6$.

Third - Use the general binomial expansion formula, i.e., equation (1) above to expand $(1-0.17)^6$.

$$\boxed{(1-0.17)^6} = \boxed{\binom{6}{0}1^6 + \binom{6}{1}1^5\cdot(-0.17) + \binom{6}{2}1^4\cdot(-0.17)^2 + \binom{6}{3}1^3\cdot(-0.17)^3 + \binom{6}{4}1^2\cdot(-0.17)^4 + \binom{6}{5}1\cdot(-0.17)^5}$$

$$\boxed{+\binom{6}{6}(-0.17)^6} = \boxed{\binom{6}{0} - 0.17\binom{6}{1} + 0.0289\binom{6}{2} - 0.0049\binom{6}{3} + \cdots} = \boxed{\frac{6!}{0!(6-0)!} - 0.17\frac{6!}{1!(6-1)!}}$$

$$\boxed{+0.0289\frac{6!}{2!(6-2)!} - 0.0049\frac{6!}{3!(6-3)!} + \cdots} = \boxed{\frac{6!}{6!} - 0.17\frac{6!}{5!} + 0.0289\frac{6!}{2!4!} - 0.0049\frac{6!}{3!3!} + \cdots}$$

$$= \boxed{\frac{\cancel{6!}}{\cancel{6!}} - 0.17\frac{6\cdot \cancel{5!}}{\cancel{5!}} + 0.0289\frac{6\cdot 5\cdot \cancel{4!}}{2!\cancel{4!}} - 0.0049\frac{6\cdot 5\cdot 4\cdot \cancel{3!}}{3!\cancel{3!}} + \cdots} = \boxed{1 - 1.02 + 0.4335 - 0.098 + \cdots} \approx \boxed{0.3155}$$

Therefore, $(1-0.17)^6$, to the nearest hundredth, is equal to $\boxed{0.32}$

b. **First** - Write the exponential expression in the form of $(1.05)^8 = (1+0.05)^8$.

Second - Identify the $a, b,$ and n terms, i.e., $a=1$, $b=0.05$, $n=8$.

Third - Use the general binomial expansion formula, i.e., equation (1) above to expand $(1+0.05)^8$.

$$\boxed{(1+0.05)^8} = \boxed{\binom{8}{0}1^8+\binom{8}{1}1^7\cdot(0.05)+\binom{8}{2}1^6\cdot(0.05)^2+\binom{8}{3}1^3\cdot(0.05)^3+\binom{8}{4}1^2\cdot(0.05)^4+\binom{8}{5}1\cdot(0.05)^5+\cdots}$$

$$= \boxed{\binom{8}{0}+0.05\binom{8}{1}+0.0025\binom{8}{2}+0.000125\binom{8}{3}+\cdots} = \boxed{\frac{8!}{0!(8-0)!}+0.05\frac{8!}{1!(8-1)!}+0.0025\frac{8!}{2!(8-2)!}}$$

$$\boxed{+0.000125\frac{8!}{3!(8-3)!}+\cdots} = \boxed{\frac{8!}{8!}+0.05\frac{8!}{7!}+0.0025\frac{8!}{2!6!}+0.000125\frac{8!}{3!5!}+\cdots}$$

$$= \boxed{\frac{8!}{8!}+0.05\frac{8\cdot 7!}{7!}+0.0025\frac{8\cdot 7\cdot 6!}{2!6!}+0.000125\frac{8\cdot 7\cdot 6\cdot 5!}{3!5!}+\cdots} = \boxed{1+0.4+0.07+0.007+\cdots} \approx \boxed{1.477}$$

Therefore, $(1+0.05)^8$, to the nearest hundredth, is equal to $\boxed{1.48}$

c. **First** - Write the exponential expression in the form of $(1.21)^5 = (1+0.21)^5$.

Second - Identify the $a, b,$ and n terms, i.e., $a=1$, $b=0.21$, $n=5$.

Third - Use the general binomial expansion formula, i.e., equation (1) above to expand $(1+0.21)^5$.

$$\boxed{(1+0.21)^5} = \boxed{\binom{5}{0}1^5+\binom{5}{1}1^4\cdot(0.21)+\binom{5}{2}1^3\cdot(0.21)^2+\binom{5}{3}1^2\cdot(0.21)^3+\binom{5}{4}1\cdot(0.21)^4+\binom{5}{5}\cdot(0.21)^5}$$

$$= \boxed{\binom{5}{0}+0.21\binom{5}{1}+0.0441\binom{5}{2}+0.00926\binom{5}{3}+0.0019\binom{5}{4}+0.0004\binom{5}{5}} = \boxed{\frac{5!}{0!(5-0)!}+0.21\frac{5!}{1!(5-1)!}+0.0441\frac{5!}{2!(5-2)!}}$$

$$\boxed{+0.00926\frac{5!}{3!(5-3)!}+0.0019\frac{5!}{5!(5-4)!}+0.0004\frac{5!}{5!(5-5)!}} = \boxed{\frac{5!}{5!}+0.21\frac{5!}{4!}+0.0441\frac{5!}{2!3!}+0.00926\frac{5!}{3!2!}+0.0019\frac{5!}{4!1!}+0.0004\frac{5!}{5!}}$$

$$= \boxed{\frac{5!}{5!}+0.21\frac{5\cdot 4!}{4!}+0.0441\frac{5\cdot 4\cdot 3!}{2!3!}+0.00926\frac{5\cdot 4\cdot 3!}{3!2!}+0.0019\frac{5\cdot 4!}{4!}+0.0004\frac{5!}{5!}} = \boxed{1+1.05+0.441+0.0926+0.0095+0.0004}$$

$$= \boxed{2.594}$$

Therefore, $(1+0.21)^5$, to the nearest hundredth, is equal to $\boxed{2.59}$

Note that in equation (1) the rth term in a binomial expansion is given by

$$\binom{n}{r-1} a^{n-r+1} b^{r-1} = \frac{n!}{(r-1)!\,(n-r+1)!} a^{n-r+1} b^{r-1} \qquad (2)$$

this implies that we can use the above equation to find any specific term of a binomial. For example, the sixth term of $(x-3)^8$ is equal to

$$\boxed{\binom{8}{5} x^3(-3)^5} = \boxed{\frac{8!}{5!(8-5)!} x^3(-3)^5} = \boxed{\frac{8!}{5!3!} x^3 \cdot (-243)} = \boxed{-243 \cdot \frac{8 \cdot 7 \cdot 6 \cdot \cancel{5!}}{\cancel{5!} \cdot 3 \cdot 2 \cdot 1} x^3} = \boxed{\mathbf{-13{,}608\, x^3}}$$

and the fourth term of $(x-3)^4$ is equal to

$$\boxed{\binom{4}{3} x^{4-4+1}(-3)^3} = \boxed{\frac{4!}{3!(4-3)!} x(-3)^3} = \boxed{\frac{4!}{3!1!} x \cdot (-27)} = \boxed{-27 \cdot \frac{4 \cdot \cancel{3!}}{\cancel{3!}} x} = \boxed{\mathbf{-108\, x}}$$

Example 4.6-8 Find the stated term of the following binomial expressions.

a. The sixth term of $(x+2)^{10}$

b. The eighth term of $(x-y)^{12}$

c. The fifth term of $(w-a)^{13}$ =

d. The tenth term of $(x+1)^{20}$

Solutions:

a. **First** - Identify the a, b, r and n terms, i.e., $a = x$, $b = 2$, $r = 6$, and $n = 10$.

Second - Use equation (2) above to find the sixth term of $(x+2)^{10}$.

$$\boxed{\frac{10!}{(6-1)!\,(10-6+1)!} x^{10-6+1} \cdot 2^{6-1}} = \boxed{\frac{10!}{5!5!} x^5 2^5} = \boxed{\frac{10 \cdot 9 \cdot 8 \cdot 7 \cdot 6 \cdot \cancel{5!}}{5!\cancel{5!}} 32x^5} = \boxed{\frac{\overset{2}{\cancel{10}} \cdot \overset{3}{\cancel{9}} \cdot \cancel{8} \cdot 7 \cdot 6}{\cancel{5} \cdot \cancel{4} \cdot \cancel{3} \cdot \cancel{2} \cdot 1} 32x^5} = \boxed{\frac{6 \cdot 7 \cdot 6}{1} 32x^5}$$

$$= \boxed{\mathbf{8064\, x^5}}$$

b. **First** - Identify the a, b, r and n terms, i.e., $a = x$, $b = -y$, $r = 8$, and $n = 12$.

Second - Use equation (2) above to find the eighth term of $(x-y)^{12}$.

$$\boxed{\frac{12!}{(8-1)!\,(12-8+1)!} x^{12-8+1} \cdot b^{8-1}} = \boxed{\frac{12!}{7!5!} x^5(-y)^7} = \boxed{-\frac{12 \cdot 11 \cdot 10 \cdot 9 \cdot 8 \cdot \cancel{7!}}{\cancel{7!}5!} x^5 y^7} = \boxed{-\frac{\cancel{12} \cdot 11 \cdot \cancel{10} \cdot 9 \cdot 8}{\cancel{5} \cdot \cancel{4} \cdot \cancel{3} \cdot \cancel{2} \cdot 1} x^5 y^7}$$

$$= \boxed{-(11 \cdot 9 \cdot 8) x^5 y^7} = \boxed{\mathbf{-792\, x^5 y^7}}$$

c. **First** - Identify the a, b, r and n terms, i.e., $a = w$, $b = -a$, $r = 5$, and $n = 13$.

Second - Use equation (2) above to find the fifth term of $(w-a)^{13}$.

$$\boxed{\frac{13!}{(5-1)!(13-5+1)!} w^{13-5+1}(-a)^{5-1}} = \boxed{\frac{13!}{4!9!} w^9(-a)^4} = \boxed{\frac{13 \cdot 12 \cdot 11 \cdot 10 \cdot \cancel{9!}}{4!\cancel{9!}} w^9 a^4} = \boxed{\frac{13 \cdot \cancel{12} \cdot 11 \cdot \overset{5}{\cancel{10}}}{\cancel{4} \cdot \cancel{3} \cdot \cancel{2} \cdot 1} w^9 a^4}$$

$$= \boxed{\frac{13 \cdot 11 \cdot 5}{1} w^9 a^4} = \boxed{\mathbf{715\, w^9 a^4}}$$

d. **First** - Identify the a, b, r and n terms, i.e., $a = x$, $b = 1$, $r = 10$, and $n = 20$.

Second - Use equation (2) above to find the tenth term of $(x+1)^{20}$.

$$\boxed{\frac{20!}{(10-1)!(20-10+1)!}x^{20-10+1} \cdot 1^{10-1}} = \boxed{\frac{20!}{9!11!}x^{11} \cdot 1^{9}} = \boxed{\frac{20 \cdot 19 \cdot 18 \cdot 17 \cdot 16 \cdot 15 \cdot 14 \cdot 13 \cdot 12 \cdot \cancel{11!}}{9!\cancel{11!}}x^{11}}$$

$$= \boxed{\frac{\overset{2}{\cancel{20}} \cdot 19 \cdot \cancel{18} \cdot 17 \cdot \cancel{16} \cdot 15 \cdot \cancel{14} \cdot 13 \cdot 12}{9 \cdot \cancel{8} \cdot \cancel{7} \cdot \cancel{6} \cdot \cancel{5} \cdot \cancel{4} \cdot \cancel{3} \cdot 2 \cdot 1}x^{11}} = \boxed{\frac{19 \cdot 17 \cdot 2 \cdot 15 \cdot 13 \cdot 12}{9}x^{11}} = \boxed{\mathbf{167{,}960\ x^{11}}}$$

Section 4.6 Practice Problems - The Factorial Notation

1. Expand and simplify the following factorial expressions.

a. $11! =$ b. $(10-3)! =$ c. $\dfrac{12!}{5!} =$ d. $\dfrac{14!}{10!} =$

e. $\dfrac{15!}{8!4!} =$ f. $\dfrac{10!}{4!(10-2)!} =$ g. $\dfrac{12!6!}{14!} =$ h. $\dfrac{(7-3)!\,9!}{12!(7-2)!} =$

2. Write the following products in factorial form.

a. $7 \cdot 6 \cdot 5 \cdot 4 \cdot 3 \cdot 2 \cdot 1 =$ b. $10 \cdot 11 \cdot 12 \cdot 13 \cdot 14 \cdot 15 =$ c. $22 \cdot 23 \cdot 24 \cdot 25 =$

d. $8 \cdot 7 \cdot 6 \cdot 5 \cdot 4 =$ e. $4 \cdot 5 \cdot 6 \cdot 7 \cdot 8 \cdot 9 =$ f. $35 =$

3. Expand the following factorial expressions.

a. $5(n!) =$ b. $(n-7)! =$ c. $(n+10)! =$ d. $(5n-5)! =$

e. $(2n-8)! =$ f. $(2n+6)! =$ g. $(2n-5)! =$ h. $(3n+3)! =$

4. Expand and simplify the following factorial expressions.

a. $\dfrac{(n-2)!}{(n-4)!} =$ b. $\dfrac{(n+4)!}{n!} =$ c. $\dfrac{(n+5)!}{(n-2)!} =$ d. $\dfrac{(n-1)(n+1)!}{(n+2)!} =$

e. $\dfrac{(3n)!(3n-2)!}{(3n+1)!(3n-4)!} =$ f. $\dfrac{(n-1)!}{(n+2)!(n!)^2} =$ g. $\dfrac{(2n-3)!\,2(n!)}{(2n)!(n-2)!} =$

5. Write the following expressions in factorial notation form. Simplify the answer.

a. $\dbinom{5}{3} =$ b. $\dbinom{10}{6} =$ c. $\dbinom{8}{0} =$ d. $\dbinom{8}{8} =$ e. $\dbinom{6}{3} =$

f. $\dbinom{5}{1} =$ g. $\dbinom{n}{n-5} =$ h. $\dbinom{2n}{2n-1} =$ i. $\dbinom{3n}{3n-3} =$ j. $\dbinom{n}{n-6} =$

6. Expand the following binomial expressions.

a. $(x-2)^4 =$ b. $(u+2)^7 =$ c. $(y-3)^5 =$

7. Use the general equation for binomial expansion to solve the following exponential numbers to the nearest hundredth.

a. $(0.95)^5 =$ b. $(2.25)^7 =$ c. $(1.05)^4 =$

8. Find the stated term of the following binomial expressions.

a. The eighth term of $(x+3)^{12}$ b. The ninth term of $(x-y)^{10}$

c. The seventh term of $(u-2a)^{11}$ d. The twelfth term of $(x-1)^{18}$

Chapter 5
Differentiation

Quick Reference to Chapter 5 Problems

Chapter 5 - Differentiation

The objective of this chapter is to improve the student's ability to solve problems involving derivatives. In Section 5.1 derivatives are computed using the Difference Quotient method. Various differentiation rules using the prime and $\frac{d}{dx}$ notation are introduced in Sections 5.2 and 5.3, respectively. The use of the Chain Rule in finding the derivative of functions that are being added, subtracted, multiplied, and divided are addressed in Section 5.4. Implicit differentiation is discussed in Section 5.5. Finding the derivative of exponential and radical expressions is addressed in Sections 5.6 and 5.7, respectively. Finally, computation of second, third, fourth, or higher order derivatives are discussed in Section 5.8. Each section is concluded by solving examples with practice problems to further enhance the student's ability.

5.1 The Difference Quotient Method

In this section students learn how to differentiate functions using the difference quotient equation as the limit h approaches zero. The expression $\frac{f(x+h)-f(x)}{h}$ is referred to as the difference quotient equation. A function $f(x)$ is said to be differentiable at x if and only if $\lim_{h\to 0}$ $\frac{f(x+h)-f(x)}{h}$ exists. If the limit exists, then the result is referred to as the derivative of $f(x)$ at x which is denoted by $f'(x)$. It should be noted that this approach is rather long and time consuming and is merely presented in order to show the usefulness of the differentiation rules which are addressed in the subsequent sections. The following examples show the steps in finding derivatives of functions using the Difference Quotient method:

Example 5.1-1: Use the Difference Quotient method to find the derivative of the following functions.

a. $f(x)=3x+1$ b. $f(x)=4x^2$ c. $f(x)=\sqrt{x+3}$

d. $f(x)=(2x-7)^2$ e. $f(x)=-\sqrt{x+1}$ f. $f(x)=\frac{1}{\sqrt{x+1}}$

Solutions:

a. To find the derivative of the function $f(x) = 3x+1$

First - Substitute $\boxed{f(x+h)} = \boxed{3(x+h)+1} = \boxed{3x+3h+1}$ and $\boxed{f(x)=3x+1}$ into the difference quotient equation, i.e., $\boxed{\frac{f(x+h)-f(x)}{h}} = \boxed{\frac{(3x+3h+1)-(3x+1)}{h}} = \boxed{\frac{\cancel{3x}+3h+\cancel{1}-\cancel{3x}-\cancel{1}}{h}} = \boxed{\frac{3\cancel{h}}{\cancel{h}}} = \boxed{3}$

Second - Compute $f'(x)$ as the $\lim_{h\to 0}$ in the difference quotient equation, i.e.,

$$\boxed{f'(x)} = \boxed{\lim_{h\to 0}\frac{f(x+h)-f(x)}{h}} = \boxed{\lim_{h\to 0} 3} = \boxed{3}$$

b. To find the derivative of the function $f(x) = 4x^2$

First - Substitute $\boxed{f(x+h)} = \boxed{4(x+h)^2} = \boxed{4(x^2+h^2+2xh)} = \boxed{4x^2+4h^2+8xh}$ and $\boxed{f(x)} = \boxed{4x^2}$ into the difference quotient equation, i.e., $\frac{f(x+h)-f(x)}{h} = \frac{(4x^2+4h^2+8xh)-4x^2}{h} = \frac{\cancel{4x^2}+4h^2+8xh-\cancel{4x^2}}{h}$

$= \frac{4\cancel{h}(h+2x)}{\cancel{h}} = \frac{4(h+2x)}{1} = 4h+8x$

Second - Compute $f'(x)$ as the $\lim_{h\to 0}$ in the difference quotient equation, i.e.,

$f'(x) = \lim_{h\to 0}\frac{f(x+h)-f(x)}{h} = \lim_{h\to 0}(4h+8x) = (4\cdot 0)+8x = 0+8x = \mathbf{8x}$

c. To find the derivative of the function $f(x)=\sqrt{x+3}$

First - Substitute $f(x+h) = \sqrt{(x+h)+3} = \sqrt{x+h+3}$ and $f(x)=\sqrt{x+3}$ into the difference quotient equation, i.e., $\frac{f(x+h)-f(x)}{h} = \frac{\sqrt{x+h+3}-\sqrt{x+3}}{h}$

To remove the radical from the numerator multiply both the numerator and the denominator by $\sqrt{x+h+3}+\sqrt{x+3}$ to obtain the following:

$\frac{f(x+h)-f(x)}{h} = \frac{\sqrt{x+h+3}-\sqrt{x+3}}{h} = \frac{\sqrt{x+h+3}-\sqrt{x+3}}{h}\cdot\frac{\sqrt{x+h+3}+\sqrt{x+3}}{\sqrt{x+h+3}+\sqrt{x+3}} = \frac{1}{\sqrt{x+h+3}+\sqrt{x+3}}$

Second - Compute $f'(x)$ as the $\lim_{h\to 0}$ in the difference quotient equation, i.e.,

$f'(x) = \lim_{h\to 0}\frac{f(x+h)-f(x)}{h} = \lim_{h\to 0}\frac{1}{\sqrt{x+h+3}+\sqrt{x+3}} = \frac{1}{\sqrt{x+3}+\sqrt{x+3}} = \mathbf{\frac{1}{2\sqrt{x+3}}}$

d. To find the derivative of the function $f(x) = (2x-7)^2$

First - Substitute $f(x+h) = [2(x+h)-7]^2 = 4(x+h)^2+49-28(x+h) = 4x^2+4h^2+8xh+49-28x-28h$ and $f(x) = (2x-7)^2 = 4x^2+49-28x$ into the difference quotient equation, i.e., $\frac{f(x+h)-f(x)}{h}$

$= \frac{(4x^2+4h^2+8xh+49-28x-28h)-(4x^2+49-28x)}{h} = \frac{\cancel{4x^2}+4h^2+8xh+\cancel{49}-\cancel{28x}-28h-\cancel{4x^2}-\cancel{49}+\cancel{28x}}{h}$

$= \frac{4h^2+8xh-28h}{h} = \frac{\cancel{h}(4h+8x-28)}{\cancel{h}} = 4h+8x-28$

Second - Compute $f'(x)$ as the $\lim_{h\to 0}$ in the difference quotient equation, i.e.,

$f'(x) = \lim_{h\to 0}\frac{f(x+h)-f(x)}{h} = \lim_{h\to 0}(4h+8x-28) = (4\cdot 0)+8x-28 = 0+8x-28 = \mathbf{8x-28}$

e. To find the derivative of the function $f(x) = -\sqrt{x+1}$

First - Substitute $f(x+h) = -\sqrt{(x+h)+1} = -\sqrt{x+h+1}$ and $f(x) = -\sqrt{x+1}$ into the difference quotient equation, i.e., $\boxed{\frac{f(x+h)-f(x)}{h}} = \boxed{-\frac{\sqrt{x+h+1}+\sqrt{x+1}}{h}}$

To remove the radical from the numerator multiply both the numerator and the denominator by $\sqrt{x+h+1}-\sqrt{x+1}$ to obtain the following:

$$\boxed{\frac{f(x+h)-f(x)}{h}} = \boxed{-\frac{\sqrt{x+h+1}+\sqrt{x+1}}{h}} = \boxed{-\frac{\sqrt{x+h+1}+\sqrt{x+1}}{h}\cdot\frac{\sqrt{x+h+1}-\sqrt{x+1}}{\sqrt{x+h+1}-\sqrt{x+1}}} = \boxed{-\frac{1}{\sqrt{x+h+1}+\sqrt{x+1}}}$$

Second - Compute $f'(x)$ as the $\lim_{h\to 0}$ in the difference quotient equation, i.e.,

$$\boxed{f'(x)} = \boxed{\lim_{h\to 0}\frac{f(x+h)-f(x)}{h}} = \boxed{-\lim_{h\to 0}\frac{1}{\sqrt{x+h+1}+\sqrt{x+1}}} = \boxed{-\frac{1}{\sqrt{x+1}+\sqrt{x+1}}} = \boxed{-\frac{\mathbf{1}}{\mathbf{2\sqrt{x+1}}}}$$

f. To find the derivative of the function $f(x) = \frac{1}{\sqrt{x+1}}$

First - Substitute $\boxed{f(x+h)} = \boxed{\frac{1}{\sqrt{x+h+1}}}$ and $\boxed{f(x)} = \boxed{\frac{1}{\sqrt{x+1}}}$ into the difference quotient equation,

i.e., $\boxed{\frac{f(x+h)-f(x)}{h}} = \boxed{\frac{\frac{1}{\sqrt{x+h+1}}-\frac{1}{\sqrt{x+1}}}{h}} = \boxed{\frac{\sqrt{x+1}-\sqrt{x+h+1}}{h\sqrt{x+1}\cdot\sqrt{x+h+1}}}$

To remove the radical from the numerator multiply both the numerator and the denominator by $\sqrt{x+1}+\sqrt{x+h+1}$ to obtain the following:

$$\boxed{\frac{f(x+h)-f(x)}{h}} = \boxed{\frac{\sqrt{x+1}-\sqrt{x+h+1}}{h\sqrt{x+1}\cdot\sqrt{x+h+1}}\cdot\frac{\sqrt{x+1}+\sqrt{x+h+1}}{\sqrt{x+1}+\sqrt{x+h+1}}} = \boxed{\frac{(x+1)-(x+h+1)}{h(x+1)\sqrt{x+h+1}+h(x+h+1)\sqrt{x+1}}}$$

$$= \boxed{\frac{\cancel{x}+\cancel{1}-\cancel{x}-h-\cancel{1}}{h(x+1)\sqrt{x+h+1}+h(x+h+1)\sqrt{x+1}}} = \boxed{\frac{-\cancel{h}}{\cancel{h}\left[(x+1)\sqrt{x+h+1}+(x+h+1)\sqrt{x+1}\right]}} = \boxed{\frac{-1}{(x+1)\sqrt{x+h+1}+(x+h+1)\sqrt{x+1}}}$$

Second - Compute $f'(x)$ as the $\lim_{h\to 0}$ in the difference quotient equation, i.e.,

$$\boxed{f'(x)} = \boxed{\lim_{h\to 0}\frac{f(x+h)-f(x)}{h}} = \boxed{\lim_{h\to 0}\frac{-1}{(x+1)\sqrt{x+h+1}+(x+h+1)\sqrt{x+1}}} = \boxed{\frac{-1}{(x+1)\sqrt{x+1}+(x+1)\sqrt{x+1}}}$$

$$= \boxed{-\frac{\mathbf{1}}{\mathbf{2(x+1)\sqrt{x+1}}}}$$

Example 5.1-2: Given the derivative of the functions in example 5.1-1, find:

a. $f'(2)$ b. $f'(3)$ c. $f'(1)$

d. $f'(0)$ e. $f'(15)$ f. $f'(0)$

Solutions:

a. Given $f'(x) = 3$ then, $\boxed{f'(2)} = \boxed{\mathbf{3}}$

Note that since the derivative is constant $f'(x)$ is independent of the x value. $f'(2)$ can also be calculated directly by using $\boxed{f'(x)} = \boxed{\lim_{h\to 0}\frac{f(x+h)-f(x)}{h}} = \boxed{\lim_{h\to 0}\frac{[3(x+h)+1]-(3x+1)}{h}}$

and by replacing x with 2, i.e., $\boxed{f'(2)} = \boxed{\lim_{h\to 0}\frac{[3(2+h)+1]-(6+1)}{h}} = \boxed{\lim_{h\to 0}\frac{6+3h+1-7}{h}}$

$= \boxed{\lim_{h\to 0}\frac{3h}{h}} = \boxed{\lim_{h\to 0} 3} = \boxed{3}$

b. Given $f'(x) = 8x$ then, $\boxed{f'(3)} = \boxed{8\cdot 3} = \boxed{\mathbf{24}}$

$f'(3)$ can also be calculated directly by using $\boxed{f'(x)} = \boxed{\lim_{h\to 0}\frac{f(x+h)-f(x)}{h}} = \boxed{\lim_{h\to 0}\frac{4(x+h)^2-4x^2}{h}}$

and by replacing x with 3, i.e., $\boxed{f'(3)} = \boxed{\lim_{h\to 0}\frac{4(3+h)^2-4\cdot 3^2}{h}} = \boxed{\lim_{h\to 0}\frac{4(9+h^2+6h)-36}{h}}$

$= \boxed{\lim_{h\to 0}\frac{36+4h^2+24h-36}{h}} = \boxed{\lim_{h\to 0}\frac{h(4h+24)}{h}} = \boxed{\lim_{h\to 0} 4h+24} = \boxed{0+24} = \boxed{\mathbf{24}}$

c. Given $f'(x) = \frac{1}{2\sqrt{x+3}}$ then, $\boxed{f'(1)} = \boxed{\frac{1}{2\sqrt{1+3}}} = \boxed{\frac{1}{2\sqrt{4}}} = \boxed{\frac{1}{2\cdot 2}} = \boxed{\mathbf{\frac{1}{4}}}$

Again, $f'(1)$ can also be calculated directly by using the equation $\boxed{f'(x)} = \boxed{\lim_{h\to 0}\frac{f(x+h)-f(x)}{h}}$

$= \boxed{\lim_{h\to 0}\frac{\sqrt{(x+h)+3}-\sqrt{x+3}}{h}}$ and by replacing x with 1, i.e., $\boxed{f'(1)} = \boxed{\lim_{h\to 0}\frac{\sqrt{(1+h)+3}-\sqrt{1+3}}{h}}$

$= \boxed{\lim_{h\to 0}\frac{\sqrt{h+4}-\sqrt{4}}{h}} = \boxed{\lim_{h\to 0}\frac{\sqrt{h+4}-\sqrt{4}}{h}\cdot\frac{\sqrt{h+4}+\sqrt{4}}{\sqrt{h+4}+\sqrt{4}}} = \boxed{\lim_{h\to 0}\frac{h}{h\left(\sqrt{h+4}+2\right)}} = \boxed{\lim_{h\to 0}\frac{1}{\sqrt{h+4}+2}}$

$= \boxed{\frac{1}{\sqrt{0+4}+2}} = \boxed{\frac{1}{\sqrt{4}+2}} = \boxed{\frac{1}{2+2}} = \boxed{\mathbf{\frac{1}{4}}}$

d. Given $f'(x) = 8x-28$ then, $\boxed{f'(0)} = \boxed{(8\times 0)-28} = \boxed{\mathbf{-28}}$

e. Given $f'(x) = -\frac{1}{2\sqrt{x+1}}$ then, $\boxed{f'(15)} = \boxed{-\frac{1}{2\sqrt{15+1}}} = \boxed{-\frac{1}{2\sqrt{16}}} = \boxed{-\frac{1}{2\cdot 4}} = \boxed{\mathbf{-\frac{1}{8}}}$

f. Given $f'(x) = -\frac{1}{2\sqrt{(x+1)^3}}$ then, $\boxed{f'(0)} = \boxed{-\frac{1}{2\sqrt{(0+1)^3}}} = \boxed{-\frac{1}{2\cdot 1}} = \boxed{\mathbf{-\frac{1}{2}}}$

In problems 5.1-2 d, e, and f students may want to practice finding $f'(x)$ for the specific values of

x by using the general equation $f'(x) = \lim_{h \to 0} \frac{f(x+h)-f(x)}{h}$. The answers should agree with the above stated solutions.

Finally, it should be noted that every differentiable function is continuous. However, not every continuous function is differentiable. The proof of this statement is beyond the scope of this book and can be found in a calculus book. In the following section we will learn simpler methods of finding derivative of functions using various differentiation rules.

Section 5.1 Practice Problems – The Difference Quotient Method

1. Find the derivative of the following functions by using the Difference Quotient method.

a. $f(x)=x^2-1$ b. $f(x)=x^3+2x-1$ c. $f(x)=\frac{x}{x-1}$ d. $f(x)=-\frac{1}{x^2}$

e. $f(x)=20x^2-3$ f. $f(x)=\sqrt{x^3}$ g. $f(x)=\frac{10}{\sqrt{x-5}}$ h. $f(x)=\frac{ax+b}{cx}$

2. Compute $f'(x)$ for the specified values by using the difference quotient equation as the $\lim_{h \to 0}$.

a. $f(x)=x^3$ at $x=1$ b. $f(x)=1+2x$ at $x=0$ c. $f(x)=x^3+1$ at $x=-1$

d. $f(x)=x^2(x+2)$ at $x=2$ e. $f(x)=x^{-2}+x^{-1}+1$ at $x=1$ f. $f(x)=\sqrt{x}+2$ at $x=10$

5.2 Differentiation Rules Using the Prime Notation

In the previous section we differentiated several functions by writing the difference quotient equation and taking the limit as h approaches to zero. This process, however - as was mentioned earlier, is rather long and time consuming. Instead, we can establish some general rules that make the calculation of derivatives simpler. These rules are as follows:

Rule No. 1 - *The derivative of a constant function is equal to zero, i.e.,*

$$\text{if } f(x)=k, \qquad \text{then } f'(x)=0$$

For example, the derivative of the functions $f(x)=10$, $g(x)=-100$, and $s(x)=250$ is equal to $f'(x)=0$, $g'(x)=0$, and $s'(x)=0$, respectively.

Rule No. 2 - *The derivative of the identity function is equal to one, i.e.,*

$$\text{if } f(x)=x, \qquad \text{then } f'(x)=1$$

For example, the derivative of the functions $f(x)=5x$, $g(x)=-10x$, and $s(x)=\sqrt{5}x$ is equal to $f'(x)=5\cdot1=5$, $g'(x)=-10\cdot1=-10$, and $s'(x)=\sqrt{5}\cdot1=\sqrt{5}$, respectively.

Rule No. 3A - *The derivative of the function* $f(x)=x^n$ *is equal to* $f'(x)=nx^{n-1}$, *where* n *is a positive integer.*

For example, the derivative of the functions $f(x)=x$, $f(x)=x^2$, $f(x)=x^3$, and $f(x)=x^4$ is equal to $f'(x)=1\cdot x^{1-1}=x^0=1$, $f'(x)=2\cdot x^{2-1}=2x^1=2x$, $f'(x)=3\cdot x^{3-1}=3x^2$, and $f'(x)=4\cdot x^{4-1}=4x^3$, respectively.

Rule No. 3B - *The derivative of the function* $f(x)=x^n$ *is equal to* $f'(x)=nx^{n-1}$, *where* n *is a negative integer.*

For example, the derivative of the functions $f(x)=x^{-1}$, $f(x)=x^{-2}$, $f(x)=x^{-3}$, $f(x)=x^{-4}$, and $f(x)=x^{-\frac{1}{8}}$ is equal to $f'(x)=-1\cdot x^{-1-1}=-x^{-2}$, $f'(x)=-2\cdot x^{-2-1}=-2x^{-3}$, $f'(x)=-3\cdot x^{-3-1}=-3x^{-4}$, and $f'(x)=-\frac{1}{8}\cdot x^{-\frac{1}{8}-1}=-\frac{1}{8}x^{-\frac{9}{8}}$, respectively.

Note that this rule can also be used to obtain the derivative of functions that are in the form of $f(x)=\frac{1}{x^n}$ by rewriting the function in its equivalent form of $f(x)=x^{-n}$. For example, the derivative of the functions $f(x)=\frac{1}{x}=x^{-1}$, $f(x)=\frac{1}{x^2}=x^{-2}$, $f(x)=-\frac{2}{x^3}=-2x^{-3}$, and $f(x)=\frac{1}{x^8}=x^{-8}$ is equal to $f'(x)=-1\cdot x^{-1-1}=-x^{-2}$, $f'(x)=-2\cdot x^{-2-1}=-2x^{-3}$, $f'(x)=(-2\cdot-3)\cdot x^{-3-1}=6x^{-4}$, and $f'(x)=-8\cdot x^{-8-1}=-8x^{-9}$, respectively.

Rule No. 4 - *If the function* $f(x)$ *is differentiable at* x, *then a constant* k *multiplied by* $f(x)$ *is also differentiable at* x, *i.e.,*

$$(kf)'x = kf'(x)$$

Note that this rule is referred to as the **scalar rule.**

For example, the derivative of the functions $f(x)=5x$, $g(x)=-10x$, and $s(x)=\sqrt{5}x$ is equal to $f'(x)=5$, $g'(x)=-10$, and $s'(x)=\sqrt{5}$, respectively.

Rule No. 5 - *If the function* $f(x)$ *and* $g(x)$ *are differentiable at* x, *then their sum is also differentiable at* x, *i.e.,*

$$(f+g)'(x) = f'(x)+g'(x)$$

In other words, the derivative of the sum of two differentiable functions, $(f+g)'(x)$, is equal to the derivative of the first function, $f'(x)$, plus the derivative of the second function, $g'(x)$. Note that this rule is referred to as the **summation rule.**

For example, the derivative of the functions $h(x)=\underbrace{(5x-3)}_{f(x)}+\underbrace{\left(2x^2-1\right)}_{g(x)}$ and $s(x)=\underbrace{6x^3}_{f(x)}+\underbrace{(3x+2)}_{g(x)}$ is equal to $h'(x)=f'(x)+g'(x)=5+4x$ and $s'(x)=f'(x)+g'(x)=18x^2+3$.

Rule No. 6 - *If the function* $f(x)$ *and* $g(x)$ *are differentiable at* x, *then their product is also differentiable at* x, *i.e.,*

$$(f\cdot g)'(x) = f'(x)g(x)+g'(x)f(x)$$

In other words, the derivative of the product of two differentiable functions, $(f\cdot g)'(x)$, is equal to the derivative of the first function multiplied by the second function, $f'(x)\cdot g(x)$, plus the derivative of the second function multiplied by the first function, $g'(x)\cdot f(x)$. Note that this rule is referred to as the **product rule**.

For example, the derivative of the functions $f(x)=(3x-5)(6x+1)$ and $g(x)=-10x^2\left(5x^3-2\right)$ is equal to $f'(x) = [3\cdot(6x+1)]+[6\cdot(3x-5)] = 18x+3+18x-5 = (18x+18x)+(-5+3) = 36x-2$ and

$$g'(x) = \left[-20x\cdot\left(5x^3-2\right)\right]+\left[15x^2\cdot-10x^2\right] = -100x^4+40x-150x^4 = -250x^4+40x$$

Rule No. 7 - Using the rules 1, 4, 5, and 6 we can write the formula for differentiating polynomials, i.e.,

$$\textit{if } f(x) = a_n x^n + a_{n-1}x^{n-1} + a_{n-2}x^{n-2} + \cdots + a_3x^3 + a_2x^2 + a_1x^1 + a_0 \text{ , } \textit{then}$$

$$f'(x) = n a_n x^{n-1} + (n-1)a_{n-1}x^{n-2} + (n-2)a_{n-2}x^{n-3} + \cdots + 3a_3x^2 + 2a_2x + a_1$$

For example, the derivative of the polynomials $f(x)=6x^4+5x^3-3$, $g(x)=2x^5-3x^2-4x$, and $h(x)=\frac{1}{3}x^2-2x+5$ is equal to $f'(x) = (6\cdot4)x^{4-1}+(5\cdot3)x^{3-1} = 24x^3+15x^2$, $g'(x) = (5\cdot2)x^4-(2\cdot3)x-4$ $= 10x^4-6x-4$, and $h'(x) = \left(2\cdot\frac{1}{3}\right)x-2 = \frac{2}{3}x-2$, respectively.

Rule No. 8 - *If the function* $f(x)$ *and* $g(x)$ *are differentiable at* x, *then their quotient is also differentiable at* x, *i.e.,*

$$\left(\frac{f}{g}\right)'(x) = \frac{f'(x)g(x) - g'(x)f(x)}{[g(x)]^2}$$

In other words, the derivative of the quotient of two differentiable functions, $\left(\frac{f}{g}\right)'(x)$, is equal to the derivative of the function in the numerator multiplied by the function in the denominator, $f'(x)\cdot g(x)$, minus the derivative of the function in the denominator multiplied by the function in the numerator, $g'(x)\cdot f(x)$, all divided by the square of the denominator, $[g(x)]^2$. Note that this rule is referred to as the **quotient rule**.

For example, the derivative of the functions $f(x)=\frac{1+3x}{1+x}$ and $g(x)=\frac{3x^2+5}{x^3+1}$ is equal to $f'(x)$

$$= \frac{[3\cdot(1+x)]-[1\cdot(1+3x)]}{(1+x)^2} = \frac{3+3x-1-3x}{(1+x)^2} = \frac{2}{(1+x)^2} \text{ and } g'(x) = \frac{\left[6x\cdot\left(x^3+1\right)\right]-\left[3x^2\cdot\left(3x^2+5\right)\right]}{\left(x^3+1\right)^2}$$

$$= \frac{6x^4+6x-9x^4-15x^2}{\left(x^3+1\right)^2} = \frac{-3x^4-9x^2}{\left(x^3+1\right)^2}$$

In the following examples the above rules are used in order to find the derivative of various functions:

Example 5.2-1: Differentiate the following functions.

a. $f(x)=5+x$ b. $f(x)=x^3+3x-1$ c. $f(x)=2-x$

d. $f(x)=10x^3+5x^2+5$ e. $f(x)=3\left(x^2+2x\right)$ f. $f(x)=ax^3+bx^2+c$

g. $f(x)=ax^2+b$ h. $f(x)=\frac{x^5}{10}-\frac{x^4}{4}+\frac{x^3}{6}-\frac{x^2}{4}$ i. $f(x)=\frac{10}{x^2}$

j. $f(x)=\frac{1}{x}-\frac{1}{x^3}$ k. $f(x)=x^{-5}+3x^{-3}-2x^{-1}+10$ l. $f(x)=\frac{x^2+2}{x^4}$

m. $f(x)=\frac{x^4}{1-x}$ n. $f(x)=\frac{x^4+10}{x^2+1}$ o. $f(x)=\frac{3+x}{x^3-5}$

p. $f(x)=\frac{2}{x}\left(\frac{2+x^3}{x+1}\right)$ q. $f(x)=\left(x^2+1\right)(x+5)$ r. $f(x)=(x+1)(x+2)$

s. $f(x)=\left(1+\frac{1}{x}\right)\left(1-\frac{1}{x^3}\right)$ t. $f(x)=\left(\frac{3x^2+5}{x}\right)(x+1)$ u. $f(x)=\frac{1+x^2}{1-x^2}$

v. $f(x)=\frac{2x^2+3x+1}{x^2+1}$ w. $f(x)=\frac{ax^2+bx+c}{ax^2-b}$ x. $f(x)=\frac{3-x}{\frac{1}{x}-5}$

Solutions:

a. Given $\boxed{f(x)=5+x}$ then $\boxed{f'(x)} = \boxed{0+x^{1-1}} = \boxed{x^0} = \boxed{1}$

b. Given $\boxed{f(x)=x^3+3x-1}$ then $\boxed{f'(x)}=\boxed{3x^{3-1}+3\cdot x^{1-1}-0}=\boxed{3x^2+3\cdot x^0}=\boxed{3x^2+3\cdot 1}=\boxed{\mathbf{3x^2+3}}$

c. Given $\boxed{f(x)=2-x}$ then $\boxed{f'(x)}=\boxed{0-x^{1-1}}=\boxed{-x^0}=\boxed{\mathbf{-1}}$

d. Given $\boxed{f(x)=10x^3+5x^2+5}$ then $\boxed{f'(x)}=\boxed{10\cdot 3x^{3-1}+5\cdot 2x^{2-1}+0}=\boxed{\mathbf{30x^2+10x}}$

e. Given $\boxed{f(x)=3\left(x^2+2x\right)}$ then $\boxed{f'(x)}=\boxed{3\left(2x^{2-1}+2\cdot x^{1-1}\right)}=\boxed{3\left(2x+2\cdot x^0\right)}=\boxed{3(2x+2\cdot 1)}=\boxed{\mathbf{6x+6}}$

f. Given $\boxed{f(x)=ax^3+bx^2+c}$ then $\boxed{f'(x)}=\boxed{a\cdot 3x^{3-1}+b\cdot 2x^{2-1}+0}=\boxed{\mathbf{3ax^2+2bx}}$

g. Given $\boxed{f(x)=ax^2+b}$ then $\boxed{f'(x)}=\boxed{a\cdot 2x^{2-1}+0}=\boxed{\mathbf{2a\,x}}$

h. Given $\boxed{f(x)=\frac{x^5}{10}-\frac{x^4}{4}+\frac{x^3}{6}-\frac{x^2}{4}}$ then $\boxed{f'(x)}=\boxed{\frac{1}{10}\cdot 5x^{5-1}-\frac{1}{4}\cdot 4x^{4-1}+\frac{1}{6}\cdot 3x^{3-1}-\frac{1}{4}\cdot 2x^{2-1}}$

$=\boxed{\frac{5}{10}x^4-\frac{4}{4}x^3+\frac{3}{6}x^2-\frac{2}{4}x}=\boxed{\mathbf{\frac{x^4}{2}-\frac{x^3}{1}+\frac{x^2}{2}-\frac{x}{2}}}$ or, using the quotient rule we obtain

$$\boxed{f'(x)}=\boxed{\frac{\left(5x^{5-1}\cdot 10\right)-\left(0\cdot x^5\right)}{10^2}-\frac{\left(4x^{4-1}\cdot 4\right)-\left(0\cdot x^4\right)}{4^2}+\frac{\left(3x^{3-1}\cdot 6\right)-\left(0\cdot x^3\right)}{6^2}-\frac{\left(2x^{2-1}\cdot 4\right)-\left(0\cdot x^2\right)}{4^2}}$$

$$=\boxed{\frac{50x^4}{100}-\frac{16x^3}{16}+\frac{18x^2}{36}-\frac{8x}{16}}=\boxed{\mathbf{\frac{x^4}{2}-\frac{x^3}{1}+\frac{x^2}{2}-\frac{x}{2}}}$$

i. Given $\boxed{f(x)=\frac{10}{x^2}}$ then $\boxed{f'(x)}=\boxed{\frac{\left(0\cdot x^2\right)-\left(2x^{2-1}\cdot 10\right)}{\left(x^2\right)^2}}=\boxed{\frac{0-(2x\cdot 10)}{x^4}}=\boxed{\frac{-20\cancel{x}}{x^{4=3}}}=\boxed{\mathbf{-\frac{20}{x^3}}}$

j. Given $\boxed{f(x)=\frac{1}{x}-\frac{1}{x^3}}$ then $\boxed{f'(x)}=\boxed{\frac{(0\cdot x)-(1\cdot 1)}{x^2}-\frac{\left(0\cdot x^3\right)-\left(3x^{3-1}\cdot 1\right)}{\left(x^3\right)^2}}=\boxed{-\frac{1}{x^2}+\frac{3x^{\cancel{2}}}{x^{6=4}}}=\boxed{\mathbf{-\frac{1}{x^2}+\frac{3}{x^4}}}$

k. Given $\boxed{f(x)=x^{-5}+3x^{-3}-2x^{-1}+10}$ then $\boxed{f'(x)}=\boxed{-5x^{-5-1}+3\cdot -3x^{-3-1}-2\cdot -1x^{-1-1}+0}$

$=\boxed{\mathbf{-5x^{-6}-9x^{-4}+2x^{-2}}}$ or, we can rewrite $f(x)$ as $f(x)=\frac{1}{x^5}+\frac{3}{x^3}-\frac{2}{x}+10$ and then find $f'(x)$ using the quotient rule.

$$\boxed{f'(x)}=\boxed{\frac{\left(0\cdot x^5\right)-\left(5x^4\cdot 1\right)}{\left(x^5\right)^2}+\frac{\left(0\cdot x^3\right)-\left(3x^2\cdot 3\right)}{\left(x^3\right)^2}-\frac{(0\cdot x)-(1\cdot 2)}{x^2}+0}=\boxed{-\frac{5x^4}{x^{10}}-\frac{9x^2}{x^6}+\frac{2}{x^2}}=\boxed{-\frac{5x^{\cancel{4}}}{x^{10=6}}-\frac{9x^{\cancel{2}}}{x^{6=4}}+\frac{2}{x^2}}$$

$$= \boxed{-\frac{5}{x^6}-\frac{9}{x^4}+\frac{2}{x^2}} = \boxed{\mathbf{-5x^{-6}-9x^{-4}+2x^{-2}}}$$

l. Given $\boxed{f(x)=\frac{x^2+2}{x^4}}$ then $\boxed{f'(x)} = \boxed{\frac{\left[\left(2x^{2-1}+0\right)\cdot x^4\right]-\left[4x^{4-1}\cdot\left(x^2+2\right)\right]}{\left(x^4\right)^2}} = \boxed{\frac{\left[2x\cdot x^4\right]-\left[4x^3\cdot\left(x^2+2\right)\right]}{x^8}}$

$$= \boxed{\frac{2x^5-4x^5-8x^3}{x^8}} = \boxed{\frac{-2x^5-8x^3}{x^8}} = \boxed{\frac{-2x^3\left(x^2+4\right)}{x^{8=5}}} = \boxed{\frac{\mathbf{-2\left(x^2+4\right)}}{\mathbf{x^5}}}$$

m. Given $\boxed{f(x)=\frac{x^4}{1-x}}$ then $\boxed{f'(x)} = \boxed{\frac{\left[4x^{4-1}\cdot(1-x)\right]-\left[(0-1)\cdot x^4\right]}{(1-x)^2}} = \boxed{\frac{4x^3(1-x)+x^4}{(1-x)^2}} = \boxed{\frac{4x^3-4x^4+x^4}{(1-x)^2}}$

$$= \boxed{\frac{4x^3-3x^4}{(1-x)^2}} = \boxed{\frac{\mathbf{4x^3-3x^4}}{\mathbf{(1-x)^2}}}$$

n. Given $\boxed{f(x)=\frac{x^4+10}{x^2+1}}$ then $\boxed{f'(x)} = \boxed{\frac{\left[\left(4x^{4-1}+0\right)\left(x^2+1\right)\right]-\left[\left(2x^{2-1}+0\right)\left(x^4+10\right)\right]}{\left(x^2+1\right)^2}} = \boxed{\frac{4x^3\left(x^2+1\right)-2x\left(x^4+10\right)}{\left(x^2+1\right)^2}}$

$$= \boxed{\frac{4x^5+4x^3-2x^5-20x}{\left(x^2+1\right)^2}} = \boxed{\frac{2x^5+4x^3-20x}{\left(x^2+1\right)^2}} = \boxed{\frac{\mathbf{2x\left(x^4+2x^2-10\right)}}{\mathbf{\left(x^2+1\right)^2}}}$$

o. Given $\boxed{f(x)=\frac{3+x}{x^3-5}}$ then $\boxed{f'(x)} = \boxed{\frac{\left[(0+1)\left(x^3-5\right)\right]-\left[\left(3x^{3-1}-0\right)(3+x)\right]}{\left(x^3-5\right)^2}} = \boxed{\frac{\left(x^3-5\right)-3x^2(3+x)}{\left(x^3-5\right)^2}}$

$$= \boxed{\frac{x^3-5-9x^2-3x^3}{\left(x^3-5\right)^2}} = \boxed{\frac{\mathbf{-2x^3-9x^2-5}}{\mathbf{\left(x^3-5\right)^2}}}$$

p. Given $\boxed{f(x)=\frac{2}{x}\left(\frac{2+x^3}{x+1}\right)}$ then $\boxed{f'(x)} = \boxed{\left[\frac{(0\cdot x)-(1\cdot 2)}{x^2}\right]\left(\frac{2+x^3}{x+1}\right)+\left\{\frac{\left[\left(0+3x^{3-1}\right)(x+1)\right]-(1+0)\left(2+x^3\right)}{(x+1)^2}\right\}\left(\frac{2}{x}\right)}$

$$= \boxed{\left\{\frac{\left[3x^2(x+1)\right]-\left(2+x^3\right)}{(x+1)^2}\right\}} = \boxed{-\frac{2}{x^2}\left(\frac{2+x^3}{x+1}\right)+\frac{2}{x}\left[\frac{3x^3+3x^2-x^3-2}{(x+1)^2}\right]} = \boxed{-\frac{2}{x^2}\left(\frac{2+x^3}{x+1}\right)+\frac{2}{x}\left[\frac{2x^3+3x^2-2}{(x+1)^2}\right]}$$

q. Given $\boxed{f(x)=\left(x^2+1\right)(x+5)}$ then $\boxed{f'(x)} = \boxed{\left[\left(2x^{2-1}+0\right)(x+5)\right]+\left[(1+0)\left(x^2+1\right)\right]} = \boxed{2x(x+5)+\left(x^2+1\right)}$

$$= \boxed{2x^2+10x+x^2+1} = \boxed{\mathbf{3x^2+10x+1}}$$

A second method would be to multiply the binomial terms by one another, using the FOIL method, and then taking the derivative of $f(x)$ as follows:

$$\boxed{f(x)} = \boxed{(x^2+1)(x+5)} = \boxed{x^3+5x^2+x+5} \text{ then } \boxed{f'(x)} = \boxed{3x^{3-1}+(5\cdot 2)x^{2-1}+x^{1-1}+0} = \boxed{\mathbf{3x^2+10x+1}}$$

r. Given $\boxed{f(x)=(x+1)(x+2)}$ then $\boxed{f'(x)} = \boxed{[(1+0)(x+2)]+[(1+0)(x+1)]} = \boxed{(x+2)+(x+1)} = \boxed{\mathbf{2x+3}}$

An alternative way is by multiplying the terms in parenthesis together and then taking the derivative of the product, i.e.,

$$\boxed{f(x)} = \boxed{(x+1)(x+2)} = \boxed{x^2+2x+x+2} \text{ then } \boxed{f'(x)} = \boxed{2x^{2-1}+2\cdot 1+1+0} = \boxed{2x+2+1} = \boxed{\mathbf{2x+3}}$$

s. Given $\boxed{f(x)=\left(1+\frac{1}{x}\right)\left(1-\frac{1}{x^3}\right)}$ then $\boxed{f'(x)} = \boxed{\left[0+\frac{(0\cdot x)-(1\cdot 1)}{x^2}\right]\left(1-\frac{1}{x^3}\right)+\left[0-\frac{(0\cdot x^3)-(3x^{3-1}\cdot 1)}{x^6}\right]\left(1+\frac{1}{x}\right)}$

$$= \boxed{-\frac{1}{x^2}\left(1-\frac{1}{x^3}\right)+\frac{3x^2}{x^6}\left(1+\frac{1}{x}\right)} = \boxed{-\frac{1}{x^2}+\frac{1}{x^5}+\frac{3x^2}{x^6}+\frac{3x^2}{x^7}} = \boxed{-\frac{1}{x^2}+\frac{1}{x^5}+\frac{3x^{\not 2}}{x^{6=4}}+\frac{3x^{\not 2}}{x^{7=5}}}$$

$$= \boxed{-\frac{1}{x^2}+\frac{1}{x^5}+\frac{3}{x^4}+\frac{3}{x^5}} = \boxed{\mathbf{\frac{4}{x^5}+\frac{3}{x^4}-\frac{1}{x^2}}}$$

A perhaps simpler way is to write $f(x)$ in the following form:

$$\boxed{f(x)} = \boxed{\left(1+\frac{1}{x}\right)\left(1-\frac{1}{x^3}\right)} = \boxed{\left(1+x^{-1}\right)\left(1-x^{-3}\right)} \text{ then } \boxed{f'(x)} = \boxed{\left[\left(0-x^{-1-1}\right)\cdot\left(1-x^{-3}\right)\right]+\left[\left(0+3x^{-3-1}\right)\cdot\left(1+x^{-1}\right)\right]}$$

$$= \boxed{\left[-x^{-2}\cdot\left(1-x^{-3}\right)\right]+\left[3x^{-4}\cdot\left(1+x^{-1}\right)\right]} = \boxed{\left(-x^{-2}+x^{-2-3}\right)+\left(3x^{-4}+3x^{-4-1}\right)} = \boxed{-x^{-2}+x^{-5}+3x^{-4}+3x^{-5}}$$

$$= \boxed{4x^{-5}+3x^{-4}-x^{-2}} = \boxed{\mathbf{\frac{4}{x^5}+\frac{3}{x^4}-\frac{1}{x^2}}}$$

t. Given $\boxed{f(x)=\left(\frac{3x^2+5}{x}\right)(x+1)}$ then $\boxed{f'(x)} = \boxed{\left\{\frac{\left[\left(3\cdot 2x^{2-1}+0\right)x\right]-\left[1\cdot\left(3x^2+5\right)\right]}{x^2}\right\}(x+1)+\left[1\cdot\left(\frac{3x^2+5}{x}\right)\right]}$

$$= \boxed{\left[\left(\frac{6x^2-3x^2-5}{x^2}\right)(x+1)\right]+\left(\frac{3x^2+5}{x}\right)} = \boxed{\mathbf{\left[\left(\frac{3x^2-5}{x^2}\right)(x+1)\right]+\left(\frac{3x^2+5}{x}\right)}}$$

u. Given $\boxed{f(x)=\frac{1+x^2}{1-x^2}}$ then $\boxed{f'(x)} = \boxed{\frac{\left[\left(0+2x^{2-1}\right)\left(1-x^2\right)\right]-\left[\left(0-2x^{2-1}\right)\left(1+x^2\right)\right]}{\left(1-x^2\right)^2}} = \boxed{\frac{2x\left(1-x^2\right)+2x\left(1+x^2\right)}{\left(1-x^2\right)^2}}$

$$= \boxed{\frac{2x-2x^3+2x+2x^3}{x^4-2x^2+1}} = \boxed{\frac{(2x+2x)-\cancel{2x^3}+\cancel{2x^3}}{\left(1-x^2\right)^2}} = \boxed{\mathbf{\frac{4x}{\left(1-x^2\right)^2}}}$$

v. Given $\boxed{f(x)=\frac{2x^2+3x+1}{x^2+1}}$ then $\boxed{f'(x)} = \boxed{\frac{\left[\left(4x^{2-1}+3+0\right)\left(x^2+1\right)\right]-\left[\left(2x^{2-1}+0\right)\left(2x^2+3x+1\right)\right]}{\left(x^2+1\right)^2}}$

$$= \boxed{\frac{\left[(4x+3)\left(x^2+1\right)\right]-\left[2x\left(2x^2+3x+1\right)\right]}{\left(x^2+1\right)^2}} = \boxed{\frac{\cancel{4x^3}+4x+3x^2+3-\cancel{4x^3}-6x^2-2x}{\left(x^2+1\right)^2}} = \boxed{\mathbf{\frac{-3x^2+2x+3}{\left(x^2+1\right)^2}}}$$

w. Given $\boxed{f(x)=\frac{ax^2+bx+c}{ax^2-b}}$ then $\boxed{f'(x)} = \boxed{\frac{\left[\left(2ax^{2-1}+b\right)\left(ax^2-b\right)\right]-\left[2ax^{2-1}\left(ax^2+bx+c\right)\right]}{\left(ax^2-b\right)^2}}$

$$= \boxed{\frac{\left[(2ax+b)\left(ax^2-b\right)\right]-\left[2ax\left(ax^2+bx+c\right)\right]}{\left(ax^2-b\right)^2}} = \boxed{\frac{\cancel{2a^2x^3}-2abx+abx^2-b^2-\cancel{2a^2x^3}-2abx^2-2acx}{\left(ax^2-b\right)^2}}$$

$$= \boxed{\frac{-abx^2-2abx-2acx-b^2}{\left(ax^2-b\right)^2}} = \boxed{\mathbf{\frac{-abx^2-2a(b+c)x-b^2}{\left(ax^2-b\right)^2}}}$$

x. Given $\boxed{f(x)=\frac{3-x}{\frac{1}{x}-5}}$ then $\boxed{f'(x)} = \boxed{\frac{\left[-1\cdot\left(\frac{1}{x}-5\right)\right]-\left\{\left[\frac{(0\cdot x)-(1\cdot 1)}{x^2}\right](3-x)\right\}}{\left(\frac{1}{x}-5\right)^2}} = \boxed{\frac{\left(-\frac{1}{x}+5\right)+\frac{1}{x^2}(3-x)}{\left(\frac{1}{x}-5\right)^2}}$

$$= \boxed{\frac{-\frac{1}{x}+\frac{3-x}{x^2}+5}{\left(\frac{1}{x}-5\right)^2}} = \boxed{\frac{-\frac{1}{x}+\frac{3}{x^2}-\frac{1}{x}+5}{\left(\frac{1}{x}-5\right)^2}} = \boxed{\frac{\frac{3}{x^2}-\frac{2}{x}+5}{\frac{1}{x^2}-\frac{10}{x}+25}} = \boxed{\frac{\frac{3x-2x^2}{x^3}+\frac{5}{1}}{\frac{x-10x^2}{x^3}+\frac{25}{1}}} = \boxed{\frac{\frac{3x-2x^2+5x^3}{x^3}}{\frac{x-10x^2+25x^3}{x^3}}}$$

$$= \boxed{\frac{5x^3-2x^2+3x}{25x^3-10x^2+x}} = \boxed{\frac{\cancel{x}\left(5x^2-2x+3\right)}{\cancel{x}\left(25x^2-10x+1\right)}} = \boxed{\mathbf{\frac{5x^2-2x+3}{25x^2-10x+1}}}$$

Example 5.2-2: Find $f'(0)$, $f'(1)$, and $f'(-2)$ for the following functions.

a. $f(x)=(x+5)x^2$ b. $f(x)=3x^2+1$ c. $f(x)=x^{-5}-2x^{-4}-3x^{-2}+1$

d. $f(x)=x^{-1}(x+2)$ e. $f(x)=x^3+2x+\frac{1}{x}$ f. $f(x)=x^2(x+1)$

g. $f(x)=\dfrac{x^2+4}{3x^2+1}$ h. $f(x)=x^5-2x^2+3x+10$ i. $f(x)=\left(x^2+3\right)(x-1)$

Solutions:

a. Given $f(x)=(x+5)x^2$, then

$\boxed{f'(x)}=\boxed{\left[(1+0)\cdot x^2\right]+\left[2x^{2-1}\cdot(x+5)\right]}=\boxed{x^2+2x(x+5)}=\boxed{x^2+2x^2+10x}=\boxed{3x^2+10x}$ and

$\boxed{f'(0)}=\boxed{\left(3\cdot 0^2\right)+(10\cdot 0)}=\boxed{\mathbf{0}}$

$\boxed{f'(1)}=\boxed{\left(3\cdot 1^2\right)+(10\cdot 1)}=\boxed{3+10}=\boxed{\mathbf{13}}$

$\boxed{f'(-2)}=\boxed{\left[3\cdot(-2)^2\right]+(10\cdot -2)}=\boxed{(3\cdot 4)-20}=\boxed{12-20}=\boxed{\mathbf{-8}}$

b. Given $f(x)=3x^2+1$, then

$\boxed{f'(x)}=\boxed{(3\cdot 2)x^{2-1}+0}=\boxed{6x}$ and

$\boxed{f'(0)}=\boxed{6\cdot 0}=\boxed{\mathbf{0}}$ $\boxed{f'(1)}=\boxed{6\cdot 1}=\boxed{\mathbf{6}}$ $\boxed{f'(-2)}=\boxed{6\cdot -2}=\boxed{\mathbf{-12}}$

c. Given $f(x)=x^{-5}-2x^{-4}-3x^{-2}+1$, then

$\boxed{f'(x)}=\boxed{-5x^{-5-1}+(-2\cdot -4)x^{-4-1}+(-3\cdot -2)x^{-2-1}+0}=\boxed{-5x^{-6}+8x^{-5}+6x^{-3}}=\boxed{-\frac{5}{x^6}+\frac{8}{x^5}+\frac{6}{x^3}}$ and

$\boxed{f'(0)}=\boxed{-\frac{5}{0^6}+\frac{8}{0^5}+\frac{6}{0^3}}=\boxed{\mathbf{-\frac{5}{0}+\frac{8}{0}+\frac{6}{0}}}$ $f'(0)$ is undefined due to division by zero

$\boxed{f'(1)}=\boxed{-\frac{5}{1^6}+\frac{8}{1^5}+\frac{6}{1^3}}=\boxed{-\frac{5}{1}+\frac{8}{1}+\frac{6}{1}}=\boxed{\mathbf{9}}$

$\boxed{f'(-2)}=\boxed{-\frac{5}{(-2)^6}+\frac{8}{(-2)^5}+\frac{6}{(-2)^3}}=\boxed{-\frac{5}{64}+\frac{8}{-32}+\frac{6}{-8}}=\boxed{-0.078-0.25-0.75}=\boxed{\mathbf{-1.078}}$

d. Given $f(x)=x^{-1}(x+2)$, then

$\boxed{f'(x)}=\boxed{\left[-x^{-1-1}(x+2)\right]+\left(1\cdot x^{-1}\right)}=\boxed{\left[-x^{-2}(x+2)\right]+x^{-1}}=\boxed{\left[-\frac{x+2}{x^2}\right]+\frac{1}{x}}=\boxed{-\frac{1}{x}-\frac{2}{x^2}+\frac{1}{x}}=\boxed{-\frac{2}{x^2}}$ and

$\boxed{f'(0)}=\boxed{-\frac{2}{0^2}}=\boxed{\mathbf{-\frac{2}{0}}}$ $f'(0)$ is undefined due to division by zero

$\boxed{f'(1)}=\boxed{-\frac{2}{1^2}}=\boxed{-\frac{2}{1}}=\boxed{\mathbf{-2}}$

$$f'(-2) = -\frac{2}{(-2)^2} = -\frac{2}{4} = \mathbf{-\frac{1}{2}}$$

e. Given $f(x) = x^3 + 2x + \frac{1}{x} = x^3 + 2x + x^{-1}$, then

$f'(x) = 3x^{3-1} + 2x^{1-1} - x^{-1-1} = 3x^2 + 2 - x^{-2}$ and

$f'(0) = 3x^2 + 2 - x^{-2} = 3x^2 + 2 - \frac{1}{x^2} = 3 \cdot 0^2 + 2 - \frac{1}{0^2} = \mathbf{2 - \frac{1}{0}}$ $f'(0)$ is undefined due to division by zero

$$f'(1) = 3x^2 + 2 - x^{-2} = 3x^2 + 2 - \frac{1}{x^2} = 3 \cdot 1^2 + 2 - \frac{1}{1^2} = 3 + 2 - \frac{1}{1} = 3 + 2 - 1 = \mathbf{4}$$

$$f'(-2) = 3x^2 + 2 - x^{-2} = 3 \cdot (-2)^2 + 2 - \frac{1}{(-2)^2} = 3 \cdot 4 + 2 - \frac{1}{4} = 12 + 2 - 0.25 = \mathbf{13.75}$$

f. Given $f(x) = x^2(x+1)$, then

$f'(x) = \left[2x^{2-1}(x+1)\right] + \left(1 \cdot x^2\right) = \left[2x(x+1)\right] + x^2 = 2x^2 + 2x + x^2 = 3x^2 + 2x$ and

$$f'(0) = \left(3 \cdot 0^2\right) + (2 \cdot 0) = \mathbf{0}$$

$$f'(1) = \left(3 \cdot 1^2\right) + (2 \cdot 1) = 3 + 2 = \mathbf{5}$$

$$f'(-2) = \left[3 \cdot (-2)^2\right] + (2 \cdot -2) = (3 \cdot 4) - 4 = 12 - 4 = \mathbf{8}$$

g. Given $f(x) = \frac{x^2 + 4}{3x^2 + 1}$, then

$$f'(x) = \frac{\left[\left(3x^2 + 1\right) \cdot 2x\right] - \left[\left(x^2 + 4\right) \cdot 6x\right]}{\left(3x^2 + 1\right)^2} = \frac{\left(6x^3 + 2x\right) - \left(6x^3 + 24x\right)}{\left(3x^2 + 1\right)^2} = \frac{6x^3 + 2x - 6x^3 - 24x}{\left(3x^2 + 1\right)^2} = -\frac{22x}{\left(3x^2 + 1\right)^2}$$

$$f'(0) = -\frac{22 \cdot 0}{\left(3 \cdot 0^2 + 1\right)^2} = -\frac{0}{(0+1)^2} = -\frac{0}{1^2} = -\frac{0}{1} = \mathbf{0}$$

$$f'(1) = -\frac{22 \cdot 1}{\left(3 \cdot 1^2 + 1\right)^2} = -\frac{22}{(3+1)^2} = -\frac{22}{4^2} = -\frac{22}{16} = \mathbf{-1.375}$$

$$f'(-2) = -\frac{22 \cdot -2}{\left[3 \cdot (-2)^2 + 1\right]^2} = -\frac{-44}{(12+1)^2} = \frac{44}{13^2} = \frac{44}{169} = \mathbf{0.26}$$

h. Given $f(x)=x^5-2x^2+3x+10$, then

$\boxed{f'(x)}=\boxed{5x^{5-1}-2\cdot 2x^{2-1}+3+0}=\boxed{5x^4-4x+3}$

$\boxed{f'(0)}=\boxed{5\cdot 0^4-4\cdot 0+3}=\boxed{0-0+3}=\boxed{3}$

$\boxed{f'(1)}=\boxed{5\cdot 1^4-4\cdot 1+3}=\boxed{5-4+3}=\boxed{4}$

$\boxed{f'(-2)}=\boxed{5\cdot(-2)^4+(-4\cdot -2)+3}=\boxed{(5\cdot 16)+8+3}=\boxed{80+11}=\boxed{\mathbf{91}}$

i. Given $f(x)=(x^2+3)(x-1)$, then

$\boxed{f'(x)}=\boxed{[2x\cdot(x-1)]+[1\cdot(x^2+3)]}=\boxed{2x^2-2x+x^2+3}=\boxed{3x^2-2x+3}$

$\boxed{f'(0)}=\boxed{3\cdot 0^2-2\cdot 0+3}=\boxed{0-0+3}=\boxed{3}$

$\boxed{f'(1)}=\boxed{3\cdot 1^2-2\cdot 1+3}=\boxed{3-2+3}=\boxed{4}$

$\boxed{f'(-2)}=\boxed{3\cdot(-2)^2+(-2\cdot -2)+3}=\boxed{3\cdot 4+4+3}=\boxed{12+7}=\boxed{\mathbf{19}}$

Example 5.2-3: Given $g(x)=\frac{1}{x}+1$ and $h(x)=x$, find $f(x)$, $f'(x)$ and $f'(0)$.

a. $f(x)=x\,g(x)$

b. $f(x)=2x^2-5h(x)$

c. $f(x)=g(x)+\frac{x}{h(x)}$

d. $h(x)=3x\,f(x)$

e. $h(x)=1-f(x)$

f. $3h(x)=2x\,f(x)-1$

Solutions:

a. Given $g(x)=\frac{1}{x}+1$ and $f(x)=x\,g(x)$, then

$\boxed{f(x)}=\boxed{xg(x)}=\boxed{x\cdot\left(\frac{1}{x}+1\right)}=\boxed{\mathbf{1+x}}$ therefore $\boxed{f'(x)}=\boxed{1}$ and $\boxed{f'(0)}=\boxed{1}$

b. Given $h(x)=x$ and $f(x)=2x^2-5h(x)$, then

$\boxed{f(x)}=\boxed{2x^2-5h(x)}=\boxed{2x^2-5\cdot x}=\boxed{\mathbf{2x^2-5x}}$ therefore $\boxed{f'(x)}=\boxed{(2\cdot 2)x^{2-1}-5}=\boxed{\mathbf{4x-5}}$ and

$\boxed{f'(0)}=\boxed{(4\cdot 0)-5}=\boxed{\mathbf{-5}}$

c. Given $g(x)=\frac{1}{x}+1$, $h(x)=x$, and $f(x)=g(x)+\frac{x}{h(x)}$, then

$\boxed{f(x)}=\boxed{g(x)+\frac{x}{h(x)}}=\boxed{\left(\frac{1}{x}+1\right)+\frac{x}{x}}=\boxed{\left(\frac{1}{x}+1\right)+1}=\boxed{\frac{1}{x}+2}=\boxed{\mathbf{x^{-1}+2}}$ therefore $\boxed{f'(x)}=\boxed{-x^{-1-1}+0}$

$=\boxed{-x^{-2}}=\boxed{-\frac{\mathbf{1}}{\mathbf{x^2}}}$ and $\boxed{f'(0)}=\boxed{-\frac{1}{0^2}}=\boxed{-\frac{\mathbf{1}}{\mathbf{0}}}$ which is undefined due to division by zero.

d. Given $h(x)=x$ and $h(x)=3x\,f(x)$, then

$$\boxed{f(x)}=\boxed{\frac{h(x)}{3x}}=\boxed{\frac{x}{3x}}=\boxed{\frac{1}{3}}$$ therefore, $\boxed{f'(x)}=\boxed{0}$ and $\boxed{f'(0)}=\boxed{0}$

e. Given $h(x)=x$ and $h(x)=1-f(x)$, then

$\boxed{f(x)}=\boxed{1-h(x)}=\boxed{1-x}$ therefore, $\boxed{f'(x)}=\boxed{-1}$ and $\boxed{f'(0)}=\boxed{-1}$

f. Given $h(x)=x$ and $3h(x)=2x\,f(x)-1$, then

$$\boxed{f(x)}=\boxed{\frac{3h(x)+1}{2x}}=\boxed{\frac{3x+1}{2x}}$$ therefore, $$\boxed{f'(x)}=\boxed{\frac{[2x\cdot 3]-[2\cdot(3x+1)]}{(2x)^2}}=\boxed{\frac{\not{6x}-\not{6x}-2}{4x^2}}=\boxed{-\frac{2}{4x^2}}$$

$$=\boxed{-\frac{1}{2x^2}}$$ and $$\boxed{f'(0)}=\boxed{-\frac{1}{2\cdot 0^2}}=\boxed{-\frac{1}{0}}$$ which is undefined due to division by zero.

Section 5.2 Practice Problems - Differentiation Rules Using the Prime Notation

1. Find the derivative of the following functions. Compare your answers with Practice Problem number 1 in Section 5.1.

a. $f(x)=x^2-1$ b. $f(x)=x^3+2x-1$ c. $f(x)=\frac{x}{x-1}$ d. $f(x)=-\frac{1}{x^2}$

e. $f(x)=20x^2-3$ f. $f(x)=\sqrt{x^3}$ g. $f(x)=\frac{10}{\sqrt{x-5}}$ h. $f(x)=\frac{ax+b}{cx}$

2. Differentiate the following functions:

a. $f(x)=x^2+10x+1$ b. $f(x)=x^8+3x^2-1$ c. $f(x)=3x^4-2x^2+5$

d. $f(x)=2(x^5+10x^4+5x)$ e. $f(x)=a^2x^3+b^2x+c^2$ f. $f(x)=x^2(x-1)+3x$

g. $f(x)=(x^3+1)(x^2-5)$ h. $f(x)=(3x^2+x-1)(x-1)$ i. $f(x)=x(x^3+5x^2)-4x$

j. $f(x)=\frac{x^3+1}{x}$ k. $f(x)=\frac{x^5+2x^2-1}{3x^2}$ l. $f(x)=\frac{x^2}{(x-1)+3x}$

m. $f(x)=x^2\left(2+\frac{1}{x}\right)$ n. $f(x)=(x+1)\cdot\frac{2x}{x-1}$ o. $f(x)=\frac{x^3+3x-1}{x^4}$

p. $f(x)=(x^2-1)\left(\frac{2x^3+5}{x}\right)$ q. $f(x)=\frac{3x^4+x^2+2}{x-1}$ r. $f(x)=x^{-1}+\frac{1}{x^{-2}}$

3. Compute $f'(x)$ at the specified value of x. Compare your answers with the practice problem number 2 in Section 5.1.

a. $f(x)=x^3$ at $x=1$ b. $f(x)=1+2x$ at $x=0$ c. $f(x)=x^3+1$ at $x=-1$

d. $f(x)=x^2(x+2)$ at $x=2$ e. $f(x)=x^{-2}+x^{-1}+1$ at $x=1$ f. $f(x)=\sqrt{x}+2$ at $x=10$

4. Find $f'(0)$ and $f'(2)$ for the following functions:

a. $f(x)=x^3-3x^2+5$
b. $f(x)=(x^3+1)(x-1)$
c. $f(x)=x(x^2+1)$
d. $f(x)=2x^5+10x^4-4x$
e. $f(x)=2x^{-2}-3x^{-1}+5x$
f. $f(x)=x^{-2}(x^5-x^3)+x$
g. $f(x)=\dfrac{x}{1+x^2}$
h. $f(x)=\dfrac{1}{x}+x^3$
i. $f(x)=\dfrac{ax^2+bx}{cx-d}$

5. Given $f(x)=x^2+1$ and $g(x)=2x-5$, find $h(x)$ and $h'(x)$.

a. $h(x)=x^3 f(x)$
b. $f(x)=3+h(x)$
c. $2g(x)=h(x)-1$
d. $3h(x)=2x\,g(x)-1$
e. $3[f(x)]^2-2h(x)=1$
f. $h(x)=g(x)\cdot 3f(x)$
g. $3h(x)-f(x)=0$
h. $2g(x)+h(x)=f(x)$
i. $f(x)=x^3+5x^2+h(x)$
j. $h(x)=\dfrac{x^3+1}{x}-f(x)$
k. $h(x)=2f(x)+g(x)$
l. $[h(x)]^2-f(x)=10$
m. $f(x)=\dfrac{2g(x)}{h(x)}$
n. $\dfrac{3f(x)}{h(x)}=\dfrac{1}{x}$
o. $f(x)=\dfrac{1}{h(x)+4}$

5.3 Differentiation Rules Using the $\frac{d}{dx}$ Notation

In the previous section the prime notation was used as a means to show the derivative of a function. For example, derivative of the functions $y = f(x) = x^2 + 3x + 1$ was represented as $y' = f'(x) = 2x + 3$. However, derivatives can also be represented by what is referred to as the "double-d" notation. For example, the derivative of the function $y = f(x) = x^2 + 3x + 1$ can be shown as $\frac{dy}{dx} = \frac{d}{dx} f(x) = 2x + 3$. Following are the differentiation rules in the double-d notation form:

Rule No. 1 - *The derivative of a constant function is equal to zero, i.e.,*

$$\text{if } f(x) = k, \qquad \text{then } \frac{d}{dx} f(x) = 0$$

Rule No. 2 - *The derivative of the identity function is equal to one, i.e.,*

$$\text{if } f(x) = x, \qquad \text{then } \frac{d}{dx} f(x) = 1$$

Rule No. 3 - *The derivative of the function* $f(x) = x^n$ *is equal to* $\frac{d}{dx} f(x) = n x^{n-1}$, *where* n *is a positive or negative integer.*

Rule No. 4 (scalar rule) - *If the function* $f(x)$ *is differentiable at* x, *then a constant* k *multiplied by* $f(x)$ *is also differentiable at* x, *i.e.,*

$$\frac{d}{dx}[(k f)x] = k\left[\frac{d}{dx} f(x)\right]$$

Rule No. 5 (summation rule) - *If the function* $f(x)$ *and* $g(x)$ *are differentiable at* x, *then their sum is also differentiable at* x, *i.e.,*

$$\frac{d}{dx}[(f+g)x] = \frac{d}{dx} f(x) + \frac{d}{dx} g(x)$$

Rule No. 6 (product rule) - *If the function* $f(x)$ *and* $g(x)$ *are differentiable at* x, *then their product is also differentiable at* x, *i.e.,*

$$\left[\frac{d}{dx}(f \cdot g)\right](x) = \left[\frac{d}{dx} f(x)\right] g(x) + \left[\frac{d}{dx} g(x)\right] f(x)$$

Rule No. 7 - Using the rules 1, 4, 5, and 6 we can write the formula for differentiating polynomials, i.e.,

$$\text{if } f(x) = a_n x^n + a_{n-1} x^{n-1} + a_{n-2} x^{n-2} + \cdots + a_3 x^3 + a_2 x^2 + a_1 x^1 + a_0 \text{ then,}$$

$$\frac{d}{dx} f(x) = n a_n x^{n-1} + (n-1) a_{n-1} x^{n-2} + (n-2) a_{n-2} x^{n-3} + \cdots + 3a_3 x^2 + 2a_2 x + a_1$$

Rule No. 8 (quotient rule) - *If the function* $f(x)$ *and* $g(x)$ *are differentiable at* x, *then their quotient is also differentiable at* x, *i.e.,*

$$\left[\frac{d}{dx}\left(\frac{f}{g}\right)\right](x) = \frac{\left[\frac{d}{dx}f(x)\right]g(x)-\left[\frac{d}{dx}g(x)\right]f(x)}{[g(x)]^2}$$

Note 1 - depending on the letter used to express the terms of a function, the double-d notation of a derivative is then shown as $\frac{da}{db}\ \frac{(\textit{where a is equal to the letter used in the left hand side of the equation})}{(\textit{where b is equal to the letter used in the right hand side of the equation})}$.

For example,

- if the function y is represented by $f(x)$, i.e., $y = f(x) = x^2 + 2x$, then its derivative is shown as $\frac{dy}{dx} = \frac{d}{dx}f(x) = \frac{d}{dx}\left(x^2 + 2x\right) = 2x + 2$.
- if the function y is represented by $f(t)$, i.e., $y = f(t) = t^3 + 2t^2 + 4$, then its derivative is shown as $\frac{dy}{dt} = \frac{d}{dt}f(t) = \frac{d}{dt}\left(t^3 + 2t^2 + 4\right) = 3t^2 + 4t$.
- if the function u is represented by $f(v)$, i.e., $u = f(v) = v^3 + 3v$, then its derivative is shown as $\frac{du}{dv} = \frac{d}{dv}f(v) = \frac{d}{dv}\left(v^3 + 3v\right) = 3v^2 + 3$.
- if the function p is represented by $f(r)$, i.e., $p = f(r) = 2r^3 - 2r^2 - 3$, then its derivative is shown as $\frac{dp}{dr} = \frac{d}{dr}f(r) = \frac{d}{dr}\left(2r^3 - 2r^2 - 3\right) = 3r^2 - 4r$.
- if the function y is represented by $f(z)$, i.e., $y = f(z) = z^5 + 3z^2 + 1$, then its derivative is shown as $\frac{dy}{dz} = \frac{d}{dz}f(z) = \frac{d}{dz}\left(z^5 + 3z^2 + 1\right) = 5z^4 + 6z$.
- if the function v is represented by $f(x)$, i.e., $v = f(x) = x^8 + 4$, then its derivative is shown as $\frac{dv}{dx} = \frac{d}{dx}f(x) = \frac{d}{dx}\left(x^8 + 4\right) = 8x^7$, etc.

In the following examples the above rules are used in order to find the derivative of various functions:

Example 5.3-1: Find $\frac{dy}{dx}$ for the following functions.

a. $y = x^3 - 2x^2 + 5$ b. $y = 4x^5 - 3x^2 - 1$ c. $y = x^2 + \frac{1}{x}$

d. $y = \frac{3x^2}{1+x}$ e. $y = 5x + \frac{2x}{x^2+1}$ f. $y = x^3\left(x^2 + 1\right)$

g. $y = (x+1)\left(x^2 - 3\right)$ h. $y = 5x(x+1)$ i. $y = 5 + \frac{1-x}{x}$

j. $y = x(x+1)(x-2)$ k. $y = x^2\left(\frac{x-3}{5}\right)$ l. $y = (x+1)(x-1)^{-2}$

Solutions:

a. $\boxed{\frac{dy}{dx}} = \boxed{\frac{d}{dx}\left(x^3 - 2x^2 + 5\right)} = \boxed{\frac{d}{dx}\left(x^3\right) + \frac{d}{dx}\left(-2x^2\right) + \frac{d}{dx}(5)} = \boxed{3x^2 + (-2\cdot 2)x + 0} = \boxed{\mathbf{3x^2 - 4x}}$

b. $\boxed{\frac{dy}{dx}} = \boxed{\frac{d}{dx}\left(4x^5 - 3x^2 - 1\right)} = \boxed{\frac{d}{dx}\left(4x^5\right) + \frac{d}{dx}\left(-3x^2\right) + \frac{d}{dx}(-1)} = \boxed{(4\cdot 5)x^4 + (-3\cdot 2)x + 0} = \boxed{\mathbf{20x^4 - 6x}}$

c. $\boxed{\frac{dy}{dx}} = \boxed{\frac{d}{dx}\left(x^2 + \frac{1}{x}\right)} = \boxed{\frac{d}{dx}\left(x^2 + x^{-1}\right)} = \boxed{\frac{d}{dx}\left(x^2\right) + \frac{d}{dx}\left(x^{-1}\right)} = \boxed{2x + \left(-x^{-2}\right)} = \boxed{2x - x^{-2}} = \boxed{\mathbf{2x - \frac{1}{x^2}}}$

d. $\boxed{\frac{dy}{dx}} = \boxed{\frac{d}{dx}\left(\frac{3x^2}{1+x}\right)} = \boxed{\frac{\left[(1+x)\frac{d}{dx}\left(3x^2\right)\right] - \left[\left(3x^2\right)\frac{d}{dx}(1+x)\right]}{(1+x)^2}} = \boxed{\frac{[(1+x)\cdot 6x] - \left[\left(3x^2\right)\cdot 1\right]}{(1+x)^2}} = \boxed{\mathbf{\frac{3x^2 + 6x}{(1+x)^2}}}$

e. $\boxed{\frac{dy}{dx}} = \boxed{\frac{d}{dx}\left(5x + \frac{2x}{x^2+1}\right)} = \boxed{\frac{d}{dx}(5x) + \frac{d}{dx}\left(\frac{2x}{x^2+1}\right)} = \boxed{\frac{d}{dx}(5x) + \frac{\left[\left(x^2+1\right)\frac{d}{dx}(2x)\right] - \left[2x\frac{d}{dx}\left(x^2+1\right)\right]}{\left(x^2+1\right)^2}}$

$= \boxed{5 + \frac{\left[\left(x^2+1\right)\cdot 2\right] - [2x\cdot 2x]}{\left(x^2+1\right)^2}} = \boxed{5 + \frac{2x^2 + 2 - 4x^2}{\left(x^2+1\right)^2}} = \boxed{\mathbf{5 + \frac{-2x^2 + 2}{\left(x^2+1\right)^2}}}$

f. $\boxed{\frac{dy}{dx}} = \boxed{\frac{d}{dx}\left[x^3\left(x^2+1\right)\right]} = \boxed{\left[\left(x^2+1\right)\frac{d}{dx}\left(x^3\right)\right] + \left[x^3\frac{d}{dx}\left(x^2+1\right)\right]} = \boxed{\left(x^2+1\right)\cdot 3x^2 + x^3\cdot 2x} = \boxed{\mathbf{5x^4 + 3x^2}}$ or,

$\boxed{\frac{dy}{dx}} = \boxed{\frac{d}{dx}\left[x^3\left(x^2+1\right)\right]} = \boxed{\frac{d}{dx}\left(x^5 + x^3\right)} = \boxed{\frac{d}{dx}x^5 + \frac{d}{dx}x^3} = \boxed{5x^{5-1} + 3x^{3-1}} = \boxed{\mathbf{5x^4 + 3x^2}}$

g. $\boxed{\frac{dy}{dx}} = \boxed{\frac{d}{dx}\left[(x+1)\left(x^2-3\right)\right]} = \boxed{\left[\left(x^2-3\right)\frac{d}{dx}(x+1)\right] + \left[(x+1)\frac{d}{dx}\left(x^2-3\right)\right]} = \boxed{\left[\left(x^2-3\right)\cdot 1\right] + [(x+1)\cdot 2x]}$

$= \boxed{x^2 - 3 + 2x^2 + 2x} = \boxed{\mathbf{3x^2 + 2x - 3}}$ or,

$\boxed{\frac{dy}{dx}} = \boxed{\frac{d}{dx}\left[(x+1)\left(x^2-3\right)\right]} = \boxed{\frac{d}{dx}\left(x^3 - 3x + x^2 - 3\right)} = \boxed{\frac{d}{dx}x^3 + \frac{d}{dx}x^2 + \frac{d}{dx}(-3x) + \frac{d}{dx}(-3)} = \boxed{\mathbf{3x^2 + 2x - 3}}$

h. $\boxed{\frac{dy}{dx}} = \boxed{\frac{d}{dx}[5x(x+1)]} = \boxed{\left[(x+1)\frac{d}{dx}(5x)\right] + \left[(5x)\frac{d}{dx}(x+1)\right]} = \boxed{[(x+1)\cdot 5] + [5x\cdot 1]} = \boxed{5x + 5 + 5x} = \boxed{\mathbf{10x + 5}}$ or,

$\boxed{\frac{dy}{dx}} = \boxed{\frac{d}{dx}[5x(x+1)]} = \boxed{\frac{d}{dx}\left(5x^2 + 5x\right)} = \boxed{\frac{d}{dx}5x^2 + \frac{d}{dx}5x} = \boxed{(5\cdot 2)x^{2-1} + 5x^{1-1}} = \boxed{\mathbf{10x + 5}}$

i. $\boxed{\frac{dy}{dx}} = \boxed{\frac{d}{dx}\left(5 + \frac{1-x}{x}\right)} = \boxed{\frac{d}{dx}(5) + \frac{d}{dx}\left(\frac{1-x}{x}\right)} = \boxed{\frac{d}{dx}(5) + \frac{\left[x\frac{d}{dx}(1-x)\right] - \left[(1-x)\frac{d}{dx}(x)\right]}{x^2}} = \boxed{0 + \frac{[x\cdot -1] - [(1-x)\cdot 1]}{x^2}}$

$$= \boxed{\frac{-x-1+x}{x^2}} = \boxed{-\frac{\mathbf{1}}{\mathbf{x^2}}}$$

j. $\boxed{\frac{dy}{dx}} = \boxed{\frac{d}{dx}[x(x+1)(x-2)]} = \boxed{\frac{d}{dx}\left[\left(x^2+x\right)(x-2)\right]} = \boxed{\left[(x-2)\frac{d}{dx}\left(x^2+x\right)\right]+\left[\left(x^2+x\right)\frac{d}{dx}(x-2)\right]}$

$$= \boxed{\left[(x-2)\cdot(2x+1)\right]+\left[\left(x^2+x\right)\cdot 1\right]} = \boxed{2x^2+x-4x-2+x^2+x} = \boxed{\mathbf{3x^2-2x-2}} \text{ or,}$$

$$\boxed{\frac{dy}{dx}} = \boxed{\frac{d}{dx}[x(x+1)(x-2)]} = \boxed{\frac{d}{dx}\left[\left(x^2+x\right)(x-2)\right]} = \boxed{\frac{d}{dx}\left(x^3-2x^2+x^2-2x\right)} = \boxed{\frac{d}{dx}\left(x^3-x^2-2x\right)}$$

$$= \boxed{\frac{d}{dx}x^3+\frac{d}{dx}\left(-x^2\right)+\frac{d}{dx}(-2x)} = \boxed{3x^{3-1}+-2x^{2-1}-2x^{1-1}} = \boxed{3x^2-2x-2x^0} = \boxed{\mathbf{3x^2-2x-2}}$$

k. $\boxed{\frac{dy}{dx}} = \boxed{\frac{d}{dx}\left[x^2\left(\frac{x-3}{5}\right)\right]} = \boxed{\left[\left(\frac{x-3}{5}\right)\frac{d}{dx}\left(x^2\right)\right]+\left[\left(x^2\right)\frac{d}{dx}\left(\frac{x-3}{5}\right)\right]} = \boxed{\left[\left(\frac{x-3}{5}\right)\cdot 2x\right]+\left[\left(x^2\right)\cdot\left(\frac{1\cdot 5-0}{5^2}\right)\right]}$

$$= \boxed{\left(\frac{2x^2-6x}{5}\right)+\left(x^2\cdot\frac{1}{5}\right)} = \boxed{\frac{2x^2-6x}{5}+\frac{x^2}{5}} = \boxed{\frac{2x^2-6x+x^2}{5}} = \boxed{\frac{3x^2-6x}{5}} = \boxed{\frac{\mathbf{3x(x-2)}}{\mathbf{5}}} \text{ or,}$$

$$\boxed{\frac{dy}{dx}} = \boxed{\frac{d}{dx}\left[x^2\left(\frac{x-3}{5}\right)\right]} = \boxed{\frac{d}{dx}\left(\frac{x^3-3x^2}{5}\right)} = \boxed{\frac{\left[5\frac{d}{dx}\left(x^3-3x^2\right)\right]-\left[\left(x^3-3x^2\right)\frac{d}{dx}(5)\right]}{5^2}}$$

$$= \boxed{\frac{\left[5\left(3x^2-6x\right)\right]-\left[\left(x^3-3x^2\right)\cdot 0\right]}{5^2}} = \boxed{\frac{5\left(3x^2-6x\right)-0}{5^2}} = \boxed{\frac{\cancel{5}\left(3x^2-6x\right)}{5^{\cancel{2}=1}}} = \boxed{\frac{3x^2-6x}{5}} = \boxed{\frac{\mathbf{3x(x-2)}}{\mathbf{5}}}$$

l. $\boxed{\frac{dy}{dx}} = \boxed{\frac{d}{dx}\left[(x+1)(x-1)^{-2}\right]} = \boxed{\left[(x-1)^{-2}\frac{d}{dx}(x+1)\right]+\left[(x+1)\frac{d}{dx}(x-1)^{-2}\right]} = \boxed{\left[(x-1)^{-2}\cdot 1\right]+\left[(x+1)\cdot -2(x-1)^{-3}\right]}$

$$= \boxed{(x-1)^{-2}-2(x+1)(x-1)^{-3}} = \boxed{\frac{\mathbf{1}}{\mathbf{(x-1)^2}}-\frac{\mathbf{2(x+1)}}{\mathbf{(x-1)^3}}} \text{ or,}$$

$$\boxed{\frac{dy}{dx}} = \boxed{\frac{d}{dx}\left[(x+1)(x-1)^{-2}\right]} = \boxed{\frac{d}{dx}\left[\frac{x+1}{(x-1)^2}\right]} = \boxed{\frac{\left[(x-1)^2\frac{d}{dx}(x+1)\right]-\left[(x+1)\frac{d}{dx}(x-1)^2\right]}{(x-1)^4}} = \boxed{\frac{\left[(x-1)^2\cdot 1\right]}{(x-1)^4}}$$

$$- \boxed{\frac{\left[(x+1)\cdot 2(x-1)\right]}{(x-1)^4}} = \boxed{\frac{(x-1)^2-2(x+1)(x-1)}{(x-1)^4}} = \boxed{\frac{(x-1)^{\cancel{2}}}{(x-1)^{\cancel{4}=2}}-\frac{2(x+1)\cancel{(x-1)}}{(x-1)^{\cancel{4}=3}}} = \boxed{\frac{\mathbf{1}}{\mathbf{(x-1)^2}}-\frac{\mathbf{2(x+1)}}{\mathbf{(x-1)^3}}}$$

Example 5.3-2: Find the derivative of the following functions.

a. $\frac{d}{dx}\left(3x^2+5x-1\right)=$

b. $\frac{d}{dx}\left(8x^4+3x^2+x\right)=$

c. $\frac{d}{du}\left[\left(u^3+5\right)(u+1)\right]=$

d. $\frac{d}{dt}\left(\frac{2t^2+3t+1}{t^3}\right)=$

e. $\frac{d}{dt}\left[\left(1-t^2\right)(1+t)+t\right]=$

f. $\frac{d}{dt}\left(\frac{t^2+1}{t^2-1}\right)=$

g. $\frac{d}{dt}\left[t^3\left(\frac{-2t}{4}\right)\right]=$

h. $\frac{d}{du}\left(\frac{u}{1-u}+\frac{u^2}{1+u}\right)=$

i. $\frac{d}{ds}\left(\frac{s^3+3s^2+1}{s^3}\right)=$

j. $\frac{d}{dw}\left[\left(w^3+1\right)\left(\frac{1}{w}\right)\right]=$

k. $\frac{d}{dx}\left(\frac{2x}{1+2x}\right)=$

l. $\frac{d}{ds}\left(\frac{s^2}{1+s^2}\right)=$

m. $\frac{d}{dt}\left[t^3\left(t^2+1\right)(3t-1)\right]=$

n. $\frac{d}{du}\left[\frac{4u^3+2}{u^2}\right]=$

o. $\frac{d}{dx}\left[\frac{x^3}{x^2+1}\right]=$

Solutions:

a. $\frac{d}{dx}\left(3x^2+5x-1\right)=\frac{d}{dx}\left(3x^2\right)+\frac{d}{dx}(5x)+\frac{d}{dx}(-1)=(3\cdot 2)x^{2-1}+5x^{1-1}+0=6x+5x^0=\mathbf{6x+5}$

b. $\frac{d}{dx}\left(8x^4+3x^2+x\right)=\frac{d}{dx}\left(8x^4\right)+\frac{d}{dx}\left(3x^2\right)+\frac{d}{dx}(x)=(8\cdot 4)x^{4-1}+(3\cdot 2)x^{2-1}+x^{1-1}=\mathbf{32x^3+6x+1}$

c. $\frac{d}{du}\left[\left(u^3+5\right)(u+1)\right]=\left[(u+1)\frac{d}{du}\left(u^3+5\right)\right]+\left[\left(u^3+5\right)\frac{d}{du}(u+1)\right]=\left[(u+1)\cdot 3u^2\right]+\left[\left(u^3+5\right)\cdot 1\right]$

$=3u^3+3u^2+u^3+5=\mathbf{4u^3+3u^2+5}$

d. $\frac{d}{dt}\left(\frac{2t^2+3t+1}{t^3}\right)=\frac{\left[t^3\frac{d}{dt}\left(2t^2+3t+1\right)\right]-\left[\left(2t^2+3t+1\right)\frac{d}{dt}\left(t^3\right)\right]}{t^6}=\frac{\left[t^3\cdot(4t+3)\right]-\left[\left(2t^2+3t+1\right)\cdot 3t^2\right]}{t^6}$

$=\frac{\left(4t^4+3t^3\right)-\left(6t^4+9t^3+3t^2\right)}{t^6}=\frac{4t^4+3t^3-6t^4-9t^3-3t^2}{t^6}=\frac{-2t^4-6t^3-3t^2}{t^6}=\frac{-t^2\left(2t^2+6t+3\right)}{t^6}$

$=-\frac{t^2\left(2t^2+6t+3\right)}{t^{6=4}}=-\mathbf{\frac{2t^2+6t+3}{t^4}}$

e. $\frac{d}{dt}\left[\left(1-t^2\right)(1+t)+t\right]=\frac{d}{dt}\left[\left(1-t^2\right)(1+t)\right]+\frac{d}{dt}t=\left\{\left[(1+t)\frac{d}{dt}\left(1-t^2\right)\right]+\left[\left(1-t^2\right)\frac{d}{dt}(1+t)\right]\right\}+1$

$=\left\{\left[(1+t)\cdot -2t\right]+\left[\left(1-t^2\right)\cdot 1\right]\right\}+1=\left(-2t-2t^2+1-t^2\right)+1=-2t-2t^2+1-t^2+1=\mathbf{-3t^2-2t+2}$

f. $\boxed{\frac{d}{dt}\left(\frac{t^2+1}{t^2-1}\right)} = \boxed{\frac{\left[\left(t^2-1\right)\frac{d}{dt}\left(t^2+1\right)\right]-\left[\left(t^2+1\right)\frac{d}{dt}\left(t^2-1\right)\right]}{\left(t^2-1\right)^2}} = \boxed{\frac{\left[\left(t^2-1\right)\cdot 2t\right]-\left[\left(t^2+1\right)\cdot 2t\right]}{\left(t^2-1\right)^2}} = \boxed{\frac{\left(2t^3-2t\right)-\left(2t^3+2t\right)}{\left(t^2-1\right)^2}}$

$= \boxed{\frac{2t^3-2t-2t^3-2t}{\left(t^2-1\right)^2}} = \boxed{\frac{\mathbf{-4t}}{\left(\mathbf{t^2-1}\right)^{\mathbf{2}}}}$

g. $\boxed{\frac{d}{dt}\left[t^3\left(-\frac{2t}{4}\right)\right]} = \boxed{\left[\left(\frac{-2t}{4}\right)\frac{d}{dt}t^3\right]+\left[t^3\frac{d}{dt}\left(\frac{-2t}{4}\right)\right]} = \boxed{\left[\left(\frac{-2t}{4}\right)\cdot 3t^2\right]+\left[t^3\left(\frac{4\frac{d}{dt}(-2t)-(-2t)\frac{d}{dt}(4)}{4^2}\right)\right]}$

$= \boxed{-\frac{6t^3}{4}+\left[t^3\left(\frac{[4\cdot -2]-[(-2t)\cdot 0]}{16}\right)\right]} = \boxed{-\frac{6t^3}{4}+\left[t^3\left(\frac{-8}{16}\right)\right]} = \boxed{-\frac{6t^3}{4}-\frac{t^3}{2}} = \boxed{\frac{-12t^3-4t^3}{8}} = \boxed{-\frac{16t^3}{8}} = \boxed{\mathbf{-2t^3}}$

A second way of solving this problem is to simplify $\frac{d}{dt}\left[t^3\left(-\frac{2t}{4}\right)\right]$ as follows:

$\boxed{\frac{d}{dt}\left[t^3\left(-\frac{2t}{4}\right)\right]} = \boxed{\frac{d}{dt}\left(-\frac{2t^4}{4}\right)} = \boxed{\frac{d}{dt}\left(-\frac{2t^4}{\underset{2}{4}}\right)} = \boxed{-\frac{1}{2}\frac{d}{dt}t^4} = \boxed{-\frac{1}{2}\cdot 4t^{4-1}} = \boxed{-\frac{4t^3}{2}} = \boxed{\mathbf{-2t^3}}$

h. $\boxed{\frac{d}{du}\left(\frac{u}{1-u}+\frac{u^2}{1+u}\right)} = \boxed{\frac{d}{du}\left(\frac{u}{1-u}\right)+\frac{d}{du}\left(\frac{u^2}{1+u}\right)} = \boxed{\frac{\left[(1-u)\frac{d}{du}u\right]-\left[u\frac{d}{du}(1-u)\right]}{(1-u)^2}+\frac{\left[(1+u)\frac{d}{du}u^2\right]-\left[u^2\frac{d}{du}(1+u)\right]}{(1+u)^2}}$

$= \boxed{\frac{[(1-u)\cdot 1]-[u\cdot -1]}{(1-u)^2}+\frac{[(1+u)\cdot 2u]-\left[u^2\cdot 1\right]}{(1+u)^2}} = \boxed{\frac{1-u+u}{(1-u)^2}+\frac{2u+u^2-u^2}{(1+u)^2}} = \boxed{\frac{\mathbf{1}}{(\mathbf{1-u})^{\mathbf{2}}}+\frac{\mathbf{2u}}{(\mathbf{1+u})^{\mathbf{2}}}}$

i. $\boxed{\frac{d}{ds}\left(\frac{s^3+3s^2+1}{s^3}\right)} = \boxed{\frac{\left[s^3\frac{d}{ds}\left(s^3+3s^2+1\right)\right]-\left[\left(s^3+3s^2+1\right)\frac{d}{ds}s^3\right]}{s^6}} = \boxed{\frac{\left[s^3\cdot\left(3s^2+6s\right)\right]-\left[\left(s^3+3s^2+1\right)\cdot 3s^2\right]}{s^6}}$

$= \boxed{\frac{\left(3s^5+6s^4\right)-\left(3s^5+9s^4+3s^2\right)}{s^6}} = \boxed{\frac{3s^5+6s^4-3s^5-9s^4-3s^2}{s^6}} = \boxed{\frac{-3s^4-3s^2}{s^6}} = \boxed{\frac{-3s^2\left(s^2+1\right)}{s^{6=4}}} = \boxed{\frac{\mathbf{-3\left(s^2+1\right)}}{\mathbf{s^4}}}$

j. $\boxed{\frac{d}{dw}\left[\left(w^3+1\right)\left(\frac{1}{w}\right)\right]} = \boxed{\frac{d}{dw}\left[\left(w^3+1\right)w^{-1}\right]} = \boxed{\frac{d}{dw}\left(w^{3-1}+w^{-1}\right)} = \boxed{\frac{d}{dw}w^2+\frac{d}{dw}w^{-1}} = \boxed{2w-w^{-2}} = \boxed{\mathbf{2w-\frac{1}{w^2}}}$

k. $\boxed{\frac{d}{dx}\left(\frac{2x}{1+2x}\right)} = \boxed{\frac{\left[(1+2x)\frac{d}{dx}(2x)\right]-\left[(2x)\frac{d}{dx}(1+2x)\right]}{(1+2x)^2}} = \boxed{\frac{[(1+2x)\cdot 2]-[2x\cdot 2]}{(1+2x)^2}} = \boxed{\frac{2+4x-4x}{(1+2x)^2}} = \boxed{\frac{\mathbf{2}}{(\mathbf{1+2x})^{\mathbf{2}}}}$

l. $\boxed{\frac{d}{ds}\left(\frac{s^2}{1+s^2}\right)} = \boxed{\frac{\left[\left(1+s^2\right)\frac{d}{ds}s^2\right]-\left[s^2\frac{d}{ds}\left(1+s^2\right)\right]}{\left(1+s^2\right)^2}} = \boxed{\frac{\left[\left(1+s^2\right)\cdot 2s\right]-\left[s^2\cdot 2s\right]}{\left(1+s^2\right)^2}} = \boxed{\frac{2s+2s^3-2s^3}{\left(1+s^2\right)^2}} = \boxed{\mathbf{\frac{2s}{\left(1+s^2\right)^2}}}$

m. $\boxed{\frac{d}{dt}\left[t^3\left(t^2+1\right)(3t-1)\right]} = \boxed{\frac{d}{dt}\left[\left(t^5+t^3\right)(3t-1)\right]} = \boxed{\left[(3t-1)\frac{d}{dt}\left(t^5+t^3\right)\right]+\left[\left(t^5+t^3\right)\frac{d}{dt}(3t-1)\right]}$

$= \boxed{\left[(3t-1)\cdot\left(5t^4+3t^2\right)\right]+\left[\left(t^5+t^3\right)\cdot 3\right]} = \boxed{\left(15t^5+9t^3-5t^4-3t^2\right)+\left(3t^5+3t^3\right)} = \boxed{\mathbf{18t^5-5t^4+12t^3-3t^2}}$

Another way of solving this problem is by multiplication of the binomial terms using the FOIL method prior to taking the derivative of the function as follows.

$\boxed{\frac{d}{dt}\left[t^3\left(t^2+1\right)(3t-1)\right]} = \boxed{\frac{d}{dt}\left[t^3\left(3t^3-t^2+3t-1\right)\right]} = \boxed{\frac{d}{dt}\left(3t^6-t^5+3t^4-t^3\right)} = \boxed{\mathbf{18t^5-5t^4+12t^3-3t^2}}$

n. $\boxed{\frac{d}{du}\left[\frac{4u^3+2}{u^2}\right]} = \boxed{\frac{\left[u^2\frac{d}{du}\left(4u^3+2\right)\right]-\left[\left(4u^3+2\right)\frac{d}{du}u^2\right]}{u^4}} = \boxed{\frac{\left(u^2\cdot 12u^2\right)-\left(4u^3+2\right)\cdot 2u}{u^4}} = \boxed{\frac{12u^4-8u^4-4u}{u^4}}$

$= \boxed{\frac{4u^4-4u}{u^4}} = \boxed{\frac{4\not{u}\left(u^3-1\right)}{u^{4-3}}} = \boxed{\frac{4\left(u^3-1\right)}{u^3}} = \boxed{4\left(\frac{u^3-1}{u^3}\right)} = \boxed{4\left(\frac{u^3}{u^3}-\frac{1}{u^3}\right)} = \boxed{\mathbf{4\left(1-\frac{1}{u^3}\right)}}$

o. $\boxed{\frac{d}{dx}\left[\frac{x^3}{x^2+1}\right]} = \boxed{\frac{\left[\left(x^2+1\right)\frac{d}{du}x^3\right]-\left[x^3\frac{d}{du}\left(x^2+1\right)\right]}{\left(x^2+1\right)^2}} = \boxed{\frac{\left[\left(x^2+1\right)\cdot 3x^2\right]-\left(x^3\cdot 2x\right)}{\left(x^2+1\right)^2}} = \boxed{\frac{3x^4+3x^2-2x^4}{\left(x^2+1\right)^2}}$

$= \boxed{\frac{x^4+3x^2}{\left(x^2+1\right)^2}} = \boxed{\mathbf{\frac{x^2\left(x^2+3\right)}{\left(x^2+1\right)^2}}}$

Example 5.3-3: Find the derivative of the following functions at the specified value.

a. $\frac{d}{dt}\left[(t+1)(t-2)+3t\right]$ at $t=1$

b. $\frac{d}{du}\left(\frac{u^2+1}{u^3-1}\right)$ at $u=2$

c. $\frac{d}{dx}\left[\frac{x-1}{(x+1)(2x+1)}\right]$ at $x=0$

d. $\frac{d}{dx}\left[\frac{\left(x^3+1\right)(x-1)}{2x^2}\right]$ at $x=2$

e. $\frac{d}{ds}\left(\frac{s^2+3s}{s^2+1}\right)$ at $s=-1$

f. $\frac{d}{dz}\left(\frac{z^2+3z-5}{z}\right)$ at $z=2$

Solutions:

a. $\boxed{\frac{d}{dt}\left[(t+1)(t-2)+3t\right]} = \boxed{\frac{d}{dt}\left[(t+1)(t-2)\right]+\frac{d}{dt}(3t)} = \boxed{\left[(t-2)\frac{d}{dt}(t+1)+(t+1)\frac{d}{dt}(t-2)\right]+\frac{d}{dt}(3t)}$

$$= \boxed{[(t-2)\cdot 1+(t+1)\cdot 1]+3} = \boxed{(t-2)+(t+1)+3} = \boxed{t-2+t+1+3} = \boxed{2t+2}$$

Therefore, at $t=1$ $\boxed{\frac{d}{dt}[(t+1)(t-2)+3t]} = \boxed{2t+2} = \boxed{(2\cdot 1)+2} = \boxed{2+2} = \boxed{\mathbf{4}}$

b. $\boxed{\frac{d}{du}\left(\frac{u^2+1}{u^3-1}\right)} = \boxed{\frac{\left[(u^3-1)\frac{d}{du}(u^2+1)\right]-\left[(u^2+1)\frac{d}{du}(u^3-1)\right]}{(u^3-1)^2}} = \boxed{\frac{\left[(u^3-1)\cdot 2u\right]-\left[(u^2+1)\cdot 3u^2\right]}{(u^3-1)^2}}$

$$= \boxed{\frac{2u^4-2u-3u^4-3u^2}{(u^3-1)^2}} = \boxed{\frac{-u^4-3u^2-2u}{(u^3-1)^2}} = \boxed{-\frac{u(u^3+3u+2)}{(u^3-1)^2}}$$

Therefore, at $u=2$ $\boxed{\frac{d}{du}\left(\frac{u^2+1}{u^3-1}\right)} = \boxed{-\frac{u(u^3+3u+2)}{(u^3-1)^2}} = \boxed{-\frac{2\cdot\left[2^3+(3\cdot 2)+2\right]}{(2^3-1)^2}} = \boxed{-\frac{2\cdot 16}{7^2}} = \boxed{-\frac{32}{49}} = \boxed{\mathbf{-0.653}}$

c. $\boxed{\frac{d}{dx}\left[\frac{x-1}{(x+1)(2x+1)}\right]} = \boxed{\frac{d}{dx}\left[\frac{x-1}{2x^2+3x+1}\right]} = \boxed{\frac{\left[(2x^2+3x+1)\frac{d}{dx}(x-1)\right]-\left[(x-1)\frac{d}{dx}(2x^2+3x+1)\right]}{(2x^2+3x+1)^2}}$

$$= \boxed{\frac{\left[(2x^2+3x+1)\cdot 1\right]-\left[(x-1)\cdot(4x+3)\right]}{(2x^2+3x+1)^2}} = \boxed{\frac{(2x^2+3x+1)-(4x^2+3x-4x-3)}{(2x^2+3x+1)^2}} = \boxed{\frac{(2x^2+3x+1)-(4x^2-x-3)}{(2x^2+3x+1)^2}}$$

$$= \boxed{\frac{2x^2+3x+1-4x^2+x+3}{(2x^2+3x+1)^2}} = \boxed{\frac{-2x^2+4x+4}{(2x^2+3x+1)^2}}$$

Therefore, at $x=0$ $\boxed{\frac{d}{dx}\left[\frac{x-1}{(x+1)(2x+1)}\right]} = \boxed{\frac{-2x^2+4x+4}{(2x^2+3x+1)^2}} = \boxed{\frac{(-2\cdot 0^2)+(4\cdot 0)+4}{\left[(2\cdot 0^2)+[3\cdot 0]+1\right]^2}} = \boxed{\frac{4}{1}} = \boxed{\mathbf{4}}$

d. $\boxed{\frac{d}{dx}\left[\frac{(x^3+1)(x-1)}{2x^2}\right]} = \boxed{\frac{d}{dx}\left[\frac{x^4-x^3+x-1}{2x^2}\right]} = \boxed{\frac{\left[(2x^2)\frac{d}{dx}(x^4-x^3+x-1)\right]-\left[(x^4-x^3+x-1)\frac{d}{dx}(2x^2)\right]}{(2x^2)^2}}$

$$\boxed{\frac{\left[2x^2(4x^3-3x^2+1)\right]-\left[(x^4-x^3+x-1)\cdot 4x\right]}{4x^4}} = \boxed{\frac{(8x^5-6x^4+2x^2)-(4x^5-4x^4+4x^2-4x)}{4x^4}}$$

$$= \boxed{\frac{8x^5 - 6x^4 + 2x^2 - 4x^5 + 4x^4 - 4x^2 + 4x}{4x^4}} = \boxed{\frac{4x^5 - 2x^4 - 2x^2 + 4x}{4x^4}} = \boxed{\frac{2x\left(2x^4 - x^3 - x + 2\right)}{4x^{4-3}}} = \boxed{\frac{2x^4 - x^3 - x + 2}{2x^3}}$$

Therefore, at $x = 2$ $\boxed{\frac{d}{dx}\left[\frac{\left(x^3+1\right)(x-1)}{2x^2}\right]} = \boxed{\frac{2x^4 - x^3 - x + 2}{2x^3}} = \boxed{\frac{\left(2\cdot 2^4\right) - 2^3 - 2 + 2}{2\cdot 2^3}} = \boxed{\frac{24}{16}} = \boxed{\mathbf{1.5}}$

e. $\boxed{\frac{d}{ds}\left(\frac{s^2+3s}{s^2+1}\right)} = \boxed{\frac{\left[\left(s^2+1\right)\frac{d}{ds}\left(s^2+3s\right)\right] - \left[\left(s^2+3s\right)\frac{d}{ds}\left(s^2+1\right)\right]}{\left(s^2+1\right)^2}} = \boxed{\frac{\left[\left(s^2+1\right)(2s+3)\right] - \left[\left(s^2+3s\right)\cdot 2s\right]}{\left(s^2+1\right)^2}}$

$$= \boxed{\frac{\left(2s^3 + 3s^2 + 2s + 3\right) - \left(2s^3 + 6s^2\right)}{\left(s^2+1\right)^2}} = \boxed{\frac{2s^3 + 3s^2 + 2s + 3 - 2s^3 - 6s^2}{\left(s^2+1\right)^2}} = \boxed{\frac{-3s^2 + 2s + 3}{\left(s^2+1\right)^2}}$$

Therefore, at $s = -1$ $\boxed{\frac{d}{ds}\left(\frac{s^2+3s}{s^2+1}\right)} = \boxed{\frac{-3s^2 + 2s + 3}{\left(s^2+1\right)^2}} = \boxed{\frac{\left[-3\cdot(-1)^2\right] + (2\cdot -1) + 3}{\left[\left(-1^2\right)+1\right]^2}} = \boxed{\frac{-3-2+3}{2^2}} = \boxed{-\frac{\mathbf{1}}{\mathbf{2}}}$

f. $\boxed{\frac{d}{dz}\left(\frac{z^2+3z-5}{z}\right)} = \boxed{\frac{\left[z\frac{d}{dz}\left(z^2+3z-5\right)\right] - \left[\left(z^2+3z-5\right)\frac{d}{dz}(z)\right]}{z^2}} = \boxed{\frac{\left[z\cdot(2z+3)\right] - \left[\left(z^2+3z-5\right)\cdot 1\right]}{z^2}}$

$$= \boxed{\frac{2z^2 + 3z - z^2 - 3z + 5}{z^2}} = \boxed{\frac{z^2+5}{z^2}}$$

Therefore, at $z = 2$ $\boxed{\frac{d}{dz}\left(\frac{z^2+3z-5}{z}\right)} = \boxed{\frac{z^2+5}{z^2}} = \boxed{\frac{2^2+5}{2^2}} = \boxed{\frac{9}{4}} = \boxed{\mathbf{2.25}}$

Example 5.3-4: Given the functions below, find their derivatives at the specified value.

a. $\frac{dy}{dx}$, given $y = \frac{x^2+2x-1}{x^3}$ at $x = 2$

b. $\frac{dv}{du}$, given $v = \frac{u^2}{1-u}$ at $u = 4$

c. $\frac{dv}{dx}$, given $v = \left(x^3+1\right)\left(3x^2+5\right)$ at $x = 5$

d. $\frac{du}{dx}$, given $u = \frac{3x}{(x-1)^2}$ at $x = 2$

Solutions:

a. $\boxed{\frac{dy}{dx}} = \boxed{\frac{d}{dx}\left(\frac{x^2+2x-1}{x^3}\right)} = \boxed{\frac{\left[x^3\frac{d}{dx}\left(x^2+2x-1\right)\right] - \left[\left(x^2+2x-1\right)\frac{d}{dx}\left(x^3\right)\right]}{\left(x^3\right)^2}} = \boxed{\frac{\left[x^3(2x+2)\right] - \left[\left(x^2+2x-1\right)\left(3x^2\right)\right]}{x^6}}$

$$= \frac{2x^4 + 2x^3 - \left(3x^4 + 6x^3 - 3x^2\right)}{x^6} = \frac{2x^4 + 2x^3 - 3x^4 - 6x^3 + 3x^2}{x^6} = \frac{-x^4 - 4x^3 + 3x^2}{x^6}$$

Therefore, at $x = 2$ $\frac{dy}{dx} = \frac{d}{dx}\left(\frac{x^2 + 2x - 1}{x^3}\right) = \frac{-x^4 - 4x^3 + 3x^2}{x^6} = \frac{-2^4 - 4\cdot 2^3 + 3\cdot 2^2}{2^6} = \frac{-36}{64} = \mathbf{-0.56}$

b. $\frac{dv}{du} = \frac{d}{du}\left(\frac{u^2}{1-u}\right) = \frac{\left[(1-u)\frac{d}{du}\left(u^2\right)\right] - \left[u^2 \frac{d}{du}(1-u)\right]}{(1-u)^2} = \frac{[(1-u)\cdot 2u] - \left[u^2 \cdot -1\right]}{(1-u)^2} = \frac{2u - 2u^2 + u^2}{(1-u)^2} = \frac{-u^2 + 2u}{(1-u)^2}$

Therefore, at $u = 4$ $\frac{dv}{du} = \frac{d}{du}\left(\frac{u^2}{1-u}\right) = \frac{-u^2 + 2u}{(1-u)^2} = \frac{-(4)^2 + (2\cdot 4)}{(1-4)^2} = \frac{-16+8}{(-3)^2} = \frac{-8}{9} = \mathbf{-0.889}$

c. $\frac{dv}{dx} = \frac{d}{dx}\left[\left(x^3 + 1\right)\left(3x^2 + 5\right)\right] = \left(3x^2 + 5\right)\frac{dv}{dx}\left(x^3 + 1\right) + \left(x^3 + 1\right)\frac{dv}{dx}\left(3x^2 + 5\right) = \left(3x^2 + 5\right)\cdot 3x^2 + \left(x^3 + 1\right)\cdot 6x$

$$= 9x^4 + 15x^2 + 6x^4 + 6x = 15x^4 + 15x^2 + 6x$$

Thus, at $x = 5$ $\frac{dv}{dx} = \frac{d}{dx}\left(\frac{x^2 + 2x - 1}{x^3}\right) = 15x^4 + 15x^2 + 6x = 15\cdot 5^4 + 15\cdot 5^2 + 6\cdot 5 = 9375 + 375 + 30 = \mathbf{9780}$

d. $\frac{du}{dx} = \frac{d}{dx}\left[\frac{3x}{(x-1)^2}\right] = \frac{\left[(x-1)^2 \frac{d}{du}(3x)\right] - \left[3x \frac{d}{du}(x-1)^2\right]}{(x-1)^4} = \frac{\left[(x-1)^2 \cdot 3\right] - [3x \cdot 2(x-1)]}{(x-1)^4} = \frac{3(x-1)^2 - 6x(x-1)}{(x-1)^4}$

$$= \frac{3\left(x^2 - 2x + 1\right) - 6x^2 + 6x}{(x-1)^4} = \frac{3x^2 - 6x + 3 - 6x^2 + 6x}{(x-1)^4} = \frac{-3x^2 + 3}{(x-1)^4}$$

Therefore, at $x = 2$ $\frac{du}{dx} = \frac{d}{dx}\left[\frac{3x}{(x-1)^2}\right] = \frac{-3x^2 + 3}{(x-1)^4} = \frac{\left(-3\cdot 2^2\right) + 3}{(2-1)^4} = \frac{\mathbf{-12+3}}{\mathbf{1}} = -\frac{9}{1} = \mathbf{-9}$

Section 5.3 - Differentiation Rules Using the $\frac{d}{dx}$ Notation

1. Find $\frac{dy}{dx}$ for the following functions:

a. $y = x^5 + 3x^2 + 1$ b. $y = 3x^2 + 5$ c. $y = x^3 - \frac{1}{x}$

d. $y = \frac{x^2}{1 - x^3}$ e. $y = 4x^2 + \frac{1}{x-1}$ f. $y = \frac{x^2 + 2x}{x^3 + 1}$

g. $y = x^3\left(x^2 + 5x - 2\right)$ h. $y = x^2(x+3)(x-1)$ i. $y = 5x - \frac{1}{x^3}$

j. $y = \frac{(x-1)(x+3)}{x^2}$ k. $y = x\left(\frac{x-1}{3}\right)$ l. $y = x^2(x+3)^{-1}$

m. $y = \left(\frac{x}{1+x}\right)\left(\frac{x-3}{5}\right)$ n. $y = x^3\left(1+\frac{1}{x-1}\right)$ o. $y = \frac{1}{x}\left(\frac{2x-1}{3x+1}\right)$

p. $y = \frac{ax^2+bx+c}{bx}$ q. $y = \frac{x^3-2}{x^4-3}$ r. $y = \frac{5x}{(1+x)^2}$

2. Find the derivative of the following functions:

a. $\frac{d}{dt}\left(3t^2+5t\right) =$ b. $\frac{d}{dx}\left(6x^3+5x-2\right) =$ c. $\frac{d}{du}\left(u^3+2u^2+5\right) =$

d. $\frac{d}{dt}\left(\frac{t^2+2t}{5}\right) =$ e. $\frac{d}{ds}\left(\frac{s^3+3s-1}{s^2}\right) =$ f. $\frac{d}{dw}\left(w^3+\frac{w^2}{1+w}\right) =$

g. $\frac{d}{dt}\left[t^2(t+1)\left(t^2-3\right)\right] =$ h. $\frac{d}{dx}\left[(x+1)\left(x^2+5\right)\right] =$ i. $\frac{d}{du}\left(\frac{u^2}{1-u}-\frac{u}{1+u}\right) =$

j. $\frac{d}{dr}\left(\frac{3r^3-2r^2+1}{r}\right) =$ k. $\frac{d}{ds}\left(\frac{3s^2}{s^3+1}-\frac{1}{s^2}\right) =$ l. $\frac{d}{du}\left(\frac{u^3}{1-u}-\frac{u+1}{u^2}\right) =$

3. Find the derivative of the following functions at the specified value.

a. $\frac{d}{dx}\left(x^3+3x^2+1\right)$ at $x=2$ b. $\frac{d}{dx}\left[(x+1)\left(x^2-1\right)\right]$ at $x=1$ c. $\frac{d}{ds}\left[3s^2(s-1)\right]$ at $s=0$

d. $\frac{d}{dt}\left(\frac{t^2+1}{t-1}\right)$ at $t=-1$ e. $\frac{d}{du}\left[\frac{u^3}{(u+1)^2}\right]$ at $u=1$ f. $\frac{d}{dw}\left[\frac{w\left(w^2+1\right)}{3w^2}\right]$ at $w=2$

g. $\frac{d}{dv}\left[\left(v^2+1\right)v^3\right]$ at $v=-2$ h. $\frac{d}{dx}\left(\frac{x^3}{x^2+1}\right)$ at $x=0$ i. $\frac{d}{du}\left[u^3\left(\frac{u^2}{1-u}\right)\right]$ at $u=0$

4. Given the functions below, find their derivatives at the specified value.

a. $\frac{ds}{dt}$, given $s=\left(t^2-1\right)+(3t+2)^2$ at $t=2$ b. $\frac{dy}{dt}$, given $y=\frac{t^3+3t^2+1}{2t}$ at $t=1$

c. $\frac{dw}{dx}$, given $w=\left(x^2+1\right)^2+3x$ at $x=-1$ d. $\frac{dy}{dx}$, given $y=x^2\left(x^3+2x+1\right)^2+3x$ at $x=0$

5.4 The Chain Rule

The chain rule is used for finding the derivative of the composition of functions. In general, the chain rule for two and three differentiable functions are defined in the following way:

a. The chain rule for two differentiable functions $f(x)$ and $g(x)$ is defined as:

$$(f \circ g)'(x) = \frac{d}{dx}\{f[g(x)]\} = f'[g(x)] \cdot g'(x)$$

b. The chain rule for three differentiable functions $f(x)$, $g(x)$ and $h(x)$ is defined as:

$$(f \circ g \circ h)'(x) = \frac{d}{dx}\left[f\{g[h(x)]\}\right] = f'\{g[h(x)]\} \cdot g'[h(x)] \cdot h'(x)$$

The derivative of four or higher differentiable functions using the chain rule involves addition of additional link(s) to the chain. Note that the pattern in finding the derivative of higher order functions is similar to obtaining the derivative of two or three functions, given that they are differentiable.

One of the most common applications of the chain rule is in taking the derivative of functions that are raised to a power. In general, the derivative of a function to the power of n is defined as:

$$\frac{d}{dx}[f(x)]^n = n[f(x)]^{n-1} \cdot \frac{d}{dx} f(x) = n[f(x)]^{n-1} \cdot f'(x)$$

which means, the derivative of a function raised to an exponent, $[f'(x)]^n$, is equal to the exponent times the function raised to the exponent reduced by one, $n[f(x)]^{n-1}$, multiplied by the derivative of the function, $f'(x)$, i.e., $n[f(x)]^{n-1} \cdot f'(x)$.

Note that the key in using the chain rule is that we always take the derivative of the functions by working our way from outside toward inside. The following examples show in detail the use of chain rule in differentiating different types of functions. Students are encouraged to spend adequate time working these examples.

Example 5.4-1: Find the derivative of the following functions. (It is not necessary to simplify the answer to its lowest level. The objective is to learn how to differentiate using the chain rule.)

a. $f(x) = (3-5x)^{-2}$

b. $f(x) = (1+x)^6$

c. $f(x) = \left(1+2x^2\right)^3$

d. $f(x) = \left(x^3 - x^5\right)^8$

e. $f(x) = \left(\frac{1}{x^2} + x\right)^3$

f. $f(x) = \left(1 + \frac{1}{x}\right)^3$

g. $f(x) = \left(x + x^3\right)^4$

h. $f(x) = \left(\frac{1}{1+x^2}\right)^2$

i. $f(x) = \left(\frac{ax+b}{cx-d}\right)^2$

j. $f(t) = \left(\frac{1}{1+t^2}\right)^3$

k. $r(\theta) = \left(\frac{\theta^2}{1+\theta}\right)^3$

l. $p(r) = \left(\frac{r^2+r}{1+r}\right)^3$

m. $g(u) = \left(u^3 + 3u^2\right)^3$

n. $h(t) = \left(\frac{t^3}{t^4-1} + t^2\right)^2$

o. $s(t) = \left[\left(1+t^3\right)^{-1}\right]^3$

p. $f(x)=\left(\frac{x^2}{3}-\frac{2x}{5}\right)^{-1}$ q. $r(\theta)=\left(\frac{\theta}{1+\theta^2}\right)^{-1}$ r. $f(t)=\left(\frac{t^3}{1+t^2}\right)^4$

s. $f(x)=\left[\left(x^3+2x\right)^2-x^2\right]^4$ t. $f(x)=\left[\left(x^{-1}+x^{-3}\right)^2+x\right]^3$ u. $f(x)=\left(2-x^{-1}\right)^{-3}$

v. $f(x)=\left[\left(1+2x^2\right)^3-x^{-2}\right]^5$ w. $f(x)=\left[\left(2-x^{-1}\right)^{-3}+2x^3\right]^{-1}$

Solutions:

a. Given $\boxed{f(x)=(3-5x)^{-2}}$ then $\boxed{f'(x)}=\boxed{-2(3-5x)^{-2-1}\cdot\left(0-5x^{1-1}\right)}=\boxed{-2(3-5x)^{-3}\cdot(-5)}=\boxed{\mathbf{10(3-5x)^{-3}}}$

b. Given $\boxed{f(x)=(1+x)^6}$ then $\boxed{f'(x)}=\boxed{6(1+x)^{6-1}\cdot\left(0+x^{1-1}\right)}=\boxed{6(1+x)^5\cdot 1}=\boxed{\mathbf{6(1+x)^5}}$

c. Given $\boxed{f(x)=\left(1+2x^2\right)^3}$ then $\boxed{f'(x)}=\boxed{3\left(1+2x^2\right)^{3-1}\cdot\left(0+4x^{2-1}\right)}=\boxed{3\left(1+2x^2\right)^2\cdot 4x}=\boxed{\mathbf{12x\left(1+2x^2\right)^2}}$

d. Given $\boxed{f(x)=\left(x^3-x^5\right)^8}$ then $\boxed{f'(x)}=\boxed{8\left(x^3-x^5\right)^{8-1}\cdot\left(3x^{3-1}-5x^{5-1}\right)}=\boxed{\mathbf{8\left(x^3-x^5\right)^7\left(3x^2-5x^4\right)}}$

e. Given $\boxed{f(x)=\left(\frac{1}{x^2}+x\right)^3}=\boxed{\left(x^{-2}+x\right)^3}$ then $\boxed{f'(x)}=\boxed{3\left(x^{-2}+x\right)^{3-1}\cdot\left(-2x^{-2-1}+1\right)}=\boxed{\mathbf{3\left(x^{-2}+x\right)^2\left(-2x^{-3}+1\right)}}$

f. Given $\boxed{f(x)=\left(1+\frac{1}{x}\right)^3}=\boxed{f(x)=\left(1+x^{-1}\right)^3}$ then $\boxed{f'(x)}=\boxed{3\left(1+x^{-1}\right)^{3-1}\cdot\left(0-x^{-1-1}\right)}=\boxed{\mathbf{-3x^{-2}\left(1+x^{-1}\right)^2}}$

g. Given $\boxed{f(x)=\left(x+x^3\right)^4}$ then $\boxed{f'(x)}=(f\circ g)'(x)=\boxed{\mathbf{4\left(x+x^3\right)^3\left(1+3x^2\right)}}$

h. Given $\boxed{f(x)=\left(\frac{1}{1+x^2}\right)^2}$ then $\boxed{f'(x)}=\boxed{2\left(\frac{1}{1+x^2}\right)^{2-1}\cdot\left\{\frac{\left[0\cdot\left(1+x^2\right)\right]-[2x\cdot 1]}{\left(1+x^2\right)^2}\right\}}=\boxed{\left(\frac{2}{1+x^2}\right)\cdot\frac{-2x}{\left(1+x^2\right)^2}}$

$=\boxed{\frac{2\cdot -2x}{\left(1+x^2\right)\cdot\left(1+x^2\right)^2}}=\boxed{\frac{-4x}{\left(1+x^2\right)^{1+2}}}=\boxed{\mathbf{\frac{-4x}{\left(1+x^2\right)^3}}}$

A second method is to rewrite $\boxed{f(x)}=\boxed{\left(\frac{1}{1+x^2}\right)^2}=\boxed{\left[\left(1+x^2\right)^{-1}\right]^2}=\boxed{\left(1+x^2\right)^{-2}}$ and then take the derivative of $f(x)$. Hence, $\boxed{f'(x)}=\boxed{-2\left(1+x^2\right)^{-2-1}\cdot(0+2x)}=\boxed{-2\left(1+x^2\right)^{-3}\cdot(2x)}=\boxed{\mathbf{\frac{-4x}{\left(1+x^2\right)^3}}}$

i. Given $\boxed{f(x)=\left(\frac{ax+b}{cx-d}\right)^2}$ then $\boxed{f'(x)}=\boxed{2\left(\frac{ax+b}{cx-d}\right)^{2-1}\cdot\left[\frac{[a\cdot(cx-d)]-[c\cdot(ax+b)]}{(cx-d)^2}\right]}$

$=\boxed{2\left(\frac{ax+b}{cx-d}\right)\left[\frac{(acx-ad)-(acx+bc)}{(cx-d)^2}\right]}=\boxed{2\left(\frac{ax+b}{cx-d}\right)\left[\frac{\cancel{acx}-ad-\cancel{acx}-bc}{(cx-d)^2}\right]}=\boxed{2\left(\frac{ax+b}{cx-d}\right)\left[\frac{-ad-bc}{(cx-d)^2}\right]}$

$=\boxed{\frac{2(ax+b)(-ad-bc)}{(cx-d)(cx-d)^2}}=\boxed{\frac{-2(ax+b)(ad+bc)}{(cx-d)^{1+2}}}=\boxed{\mathbf{\frac{-2(ax+b)(ad+bc)}{(cx-d)^3}}}$

j. Given $\boxed{f(t)=\left(\frac{1}{1+t^2}\right)^3}$ then $\boxed{f'(t)}=\boxed{3\left(\frac{1}{1+t^2}\right)^{3-1}\cdot\left[\frac{0\cdot(1+t^2)-2t\cdot 1}{(1+t^2)^2}\right]}=\boxed{3\left(\frac{1}{1+t^2}\right)^2\left[\frac{-2t}{(1+t^2)^2}\right]}$

$=\boxed{\frac{3}{(1+t^2)^2}\cdot\frac{-2t}{(1+t^2)^2}}=\boxed{\frac{3\cdot -2t}{(1+t^2)^2\cdot(1+t^2)^2}}=\boxed{\frac{-6t}{(1+t^2)^{2+2}}}=\boxed{\mathbf{\frac{-6t}{(1+t^2)^4}}}$

A second method is to rewrite $\boxed{f(t)}=\boxed{\left(\frac{1}{1+t^2}\right)^3}=\boxed{\left[(1+t^2)^{-1}\right]^3}=\boxed{(1+t^2)^{-3}}$ and then take the derivative of $f(t)$. Hence, $\boxed{f'(t)}=\boxed{-3(1+t^2)^{-3-1}\cdot(0+2t)}=\boxed{-3(1+t^2)^{-4}2t}=\boxed{\mathbf{\frac{-6t}{(1+t^2)^4}}}$

k. Given $\boxed{r(\theta)=\left(\frac{\theta^2}{1+\theta}\right)^3}$ then $\boxed{r'(\theta)}=\boxed{3\left(\frac{\theta^2}{1+\theta}\right)^{3-1}\cdot\left\{\frac{[2\theta\cdot(1+\theta)]-[1\cdot\theta^2]}{(1+\theta)^2}\right\}}=\boxed{3\left(\frac{\theta^2}{1+\theta}\right)^2\left[\frac{2\theta+2\theta^2-\theta^2}{(1+\theta)^2}\right]}$

$=\boxed{\frac{3\theta^4}{(1+\theta)^2}\cdot\frac{\theta^2+2\theta}{(1+\theta)^2}}=\boxed{\frac{3\theta^4\cdot(\theta^2+2\theta)}{(1+\theta)^2\cdot(1+\theta)^2}}=\boxed{\frac{3\theta^4\cdot\theta(\theta+2)}{(1+\theta)^{2+2}}}=\boxed{\frac{3\theta^{4+1}(\theta+2)}{(1+\theta)^4}}=\boxed{\mathbf{\frac{3\theta^5(\theta+2)}{(1+\theta)^4}}}$

l. Given $\boxed{p(r)=\left(\frac{r^2+r}{1+r}\right)^3}$ then $\boxed{p'(r)}=\boxed{3\left(\frac{r^2+r}{1+r}\right)^{3-1}\cdot\left\{\frac{[(2r+1)(1+r)]-[1\cdot(r^2+r)]}{(1+r)^2}\right\}}$

$=\boxed{3\left(\frac{r^2+r}{(1+r)}\right)^2\left[\frac{2r+2r^2+1+r-r^2-r}{(1+r)^2}\right]}=\boxed{3\left[\frac{r(\cancel{1+r})}{(\cancel{1+r})}\right]^2\left[\frac{r^2+2r+1}{(1+r)^2}\right]}=\boxed{3r^2\left[\frac{\cancel{(1+r)^2}}{\cancel{(1+r)^2}}\right]}=\boxed{3r^2\cdot 1}=\boxed{\mathbf{3r^2}}$

A simpler way is to note that $\boxed{p(r)} = \boxed{\left(\frac{r^2+r}{1+r}\right)^3} = \boxed{\left[\frac{r(1+r)}{(1+r)}\right]^3} = \boxed{\left[\frac{r(\cancel{1+r})}{(\cancel{1+r})}\right]^3} = \boxed{r^3}$ then $\boxed{p'(r)} = \boxed{\mathbf{3r^2}}$

m. Given $\boxed{g(u) = \left(u^3+3u^2\right)^3}$ then $\boxed{g'(u)} = \boxed{3\left(u^3+3u^2\right)^{3-1}\cdot\left[3u^2+(3\cdot 2)u\right]} = \boxed{\mathbf{3\left(u^3+3u^2\right)^2\left(3u^2+6u\right)}}$

n. Given $\boxed{h(t) = \left(\frac{t^3}{t^4-1}+t^2\right)^2}$ then $\boxed{h'(t)} = \boxed{2\left(\frac{t^3}{t^4-1}+t^2\right)^{2-1}\cdot\left\{\frac{\left[\left(t^4-1\right)\cdot 3t^2\right]-\left[t^3\cdot\left(4t^3-0\right)\right]}{\left(t^4-1\right)^2}+2t\right\}}$

$= \boxed{2\left(\frac{t^3}{t^4-1}+t^2\right)\left[\frac{3t^6-3t^2-4t^6}{\left(t^4-1\right)^2}+2t\right]} = \boxed{\mathbf{2\left(\frac{t^3}{t^4-1}+t^2\right)\left[\frac{-t^6-3t^2}{\left(t^4-1\right)^2}+2t\right]}}$

o. Given $\boxed{s(t) = \left[\left(1+t^3\right)^{-1}\right]^3 = \left(1+t^3\right)^{-3}}$ then $\boxed{s'(t)} = \boxed{-3\left(1+t^3\right)^{-3-1}\cdot\left(0+3t^2\right)} = \boxed{-3\left(1+t^3\right)^{-4}\cdot 3t^2}$

$= \boxed{-9t^2\left(1+t^3\right)^{-4}} = \boxed{\mathbf{-\frac{9t^2}{\left(1+t^3\right)^4}}}$

p. Given $\boxed{f(x) = \left(\frac{x^2}{3}-\frac{2x}{5}\right)^{-1}}$ then $\boxed{f'(x)} = \boxed{-\left(\frac{x^2}{3}-\frac{2x}{5}\right)^{-1-1}\cdot\left(\frac{2}{3}x-\frac{2}{5}\right)} = \boxed{\mathbf{-\left(\frac{x^2}{3}-\frac{2x}{5}\right)^{-2}\left(\frac{2}{3}x-\frac{2}{5}\right)}}$

q. Given $\boxed{r(\theta) = \left(\frac{\theta}{1+\theta^2}\right)^{-1}}$ then $\boxed{r'(\theta)} = \boxed{-\left(\frac{\theta}{1+\theta^2}\right)^{-1-1}\cdot\left[\frac{1\cdot\left(1+\theta^2\right)-2\theta\cdot\theta}{\left(1+\theta^2\right)^2}\right]} = \boxed{-\left(\frac{\theta}{1+\theta^2}\right)^{-2}\left[\frac{1+\theta^2-2\theta^2}{\left(1+\theta^2\right)^2}\right]}$

$= \boxed{-\left(\frac{\theta}{1+\theta^2}\right)^{-2}\cdot\frac{1-\theta^2}{\left(1+\theta^2\right)^2}} = \boxed{-\left(\frac{1+\theta^2}{\theta}\right)^2\cdot\frac{1-\theta^2}{\left(1+\theta^2\right)^2}} = \boxed{-\frac{\left(1+\theta^2\right)^{\cancel{2}}}{\theta^2}\cdot\frac{1-\theta^2}{\left(1+\theta^2\right)^{\cancel{2}}}} = \boxed{-\frac{1-\theta^2}{\theta^2}} = \boxed{\mathbf{\frac{\theta^2-1}{\theta^2}}}$

r. Given $\boxed{f(t) = \left(\frac{t^3}{1+t^2}\right)^4}$ then $\boxed{f'(t)} = \boxed{4\left(\frac{t^3}{1+t^2}\right)^{4-1}\cdot\left[\frac{3t^2\cdot\left(1+t^2\right)-2t\cdot t^3}{\left(1+t^2\right)^2}\right]} = \boxed{4\left(\frac{t^3}{1+t^2}\right)^3\left[\frac{3t^2+3t^4-2t^4}{\left(1+t^2\right)^2}\right]}$

$= \boxed{\frac{4t^9}{\left(1+t^2\right)^3}\cdot\frac{3t^2+t^4}{\left(1+t^2\right)^2}} = \boxed{\frac{4t^9}{\left(1+t^2\right)^3}\cdot\frac{t^2\left(3+t^2\right)}{\left(1+t^2\right)^2}} = \boxed{\frac{4t^{9+2}\left(3+t^2\right)}{\left(1+t^2\right)^{3+2}}} = \boxed{\mathbf{\frac{4t^{11}\left(3+t^2\right)}{\left(1+t^2\right)^5}}}$

s. Given $\boxed{f(x)=\left[\left(x^3+2x\right)^2-x^2\right]^4}$ then $\boxed{f'(x)}=\boxed{4\left[\left(x^3+2x\right)^2-x^2\right]^{4-1}\cdot\left\{\left[2\left(x^3+2x\right)^{2-1}\cdot\left(3x^2+2\right)\right]-2x\right\}}$

$=\boxed{\mathbf{4\left[\left(x^3+2x\right)^2-x^2\right]^3\cdot\left\{\left[2\left(x^3+2x\right)\left(3x^2+2\right)\right]-2x\right\}}}$

t. Given $\boxed{f(x)=\left[\left(x^{-1}+x^{-3}\right)^2+x\right]^3}$ then $\boxed{f'(x)}=\boxed{3\left[\left(x^{-1}+x^{-3}\right)^2+x\right]^{3-1}\left\{\left[2\left(x^{-1}+x^{-3}\right)^{2-1}\left(-x^{-2}-3x^{-4}\right)\right]+1\right\}}$

$=\boxed{\mathbf{3\left[\left(x^{-1}+x^{-3}\right)^2+x\right]^2\left\{\left[2\left(x^{-1}+x^{-3}\right)\left(-x^{-2}-3x^{-4}\right)\right]+1\right\}}}$

u. Given $\boxed{f(x)=\left(2-x^{-1}\right)^{-3}}$ then $\boxed{f'(x)}=\boxed{-3\left(2-x^{-1}\right)^{-3-1}\cdot\left(0+x^{-2}\right)}=\boxed{-3\left(2-x^{-1}\right)^{-4}x^{-2}}=\boxed{\frac{-3x^{-2}}{\left(2-x^{-1}\right)^4}}$

$=\boxed{\mathbf{-\frac{3}{x^2\left(2-x^{-1}\right)^4}}}$

v. Given $\boxed{f(x)=\left[\left(1+2x^2\right)^3-x^{-2}\right]^5}$ then $\boxed{f'(x)}=\boxed{5\left[\left(1+2x^2\right)^3-x^{-2}\right]^{5-1}\cdot\left[3\left(1+2x^2\right)^{3-1}\cdot4x+2x^{-2-1}\right]}$

$=\boxed{\mathbf{5\left[\left(1+2x^2\right)^3-x^{-2}\right]^4\left[12x\left(1+2x^2\right)^2+2x^{-3}\right]}}$

w. Given $\boxed{f(x)=\left[\left(2-x^{-1}\right)^{-3}+2x^3\right]^{-1}}$ then $\boxed{f'(x)}=\boxed{-\left[\left(2-x^{-1}\right)^{-3}+2x^3\right]^{-1-1}\cdot\left[-3\left(2-x^{-1}\right)^{-3-1}\cdot x^{-2}+6x^2\right]}$

$=\boxed{\mathbf{-\left[\left(2-x^{-1}\right)^{-3}+2x^3\right]^{-2}\left[-3x^{-2}\left(2-x^{-1}\right)^{-4}+6x^2\right]}}$

Example 5.4-2: Find the derivative at $x=0$, $x=-1$, and $x=1$ in example 5.4-1 for problems a - g.

Solutions:

a. Given $f'(x)=10(3-5x)^{-3}$, then

$\boxed{f'(0)}=\boxed{10[3-(5\cdot0)]^{-3}}=\boxed{10[3-0]^{-3}}=\boxed{10\cdot3^{-3}}=\boxed{10\cdot\frac{1}{3^3}}=\boxed{\frac{10}{27}}=\boxed{\mathbf{0.37}}$

$\boxed{f'(-1)}=\boxed{10[3-(5\cdot-1)]^{-3}}=\boxed{10[3+5]^{-3}}=\boxed{10\cdot8^{-3}}=\boxed{10\cdot\frac{1}{8^3}}=\boxed{\frac{10}{512}}=\boxed{\mathbf{0.019}}$

$$\boxed{f'(1)} = \boxed{10[3-(5\cdot 1)]^{-3}} = \boxed{10[3-5]^{-3}} = \boxed{10\cdot(-2)^{-3}} = \boxed{10\cdot\frac{1}{(-2)^3}} = \boxed{\frac{10}{-8}} = \boxed{-\frac{10}{8}} = \boxed{\mathbf{-1.25}}$$

b. Given $f'(x) = 6(1+x)^5$, then

$$\boxed{f'(0)} = \boxed{6(1+0)^5} = \boxed{6\cdot 1^5} = \boxed{6\cdot 1} = \boxed{\mathbf{6}}$$

$$\boxed{f'(-1)} = \boxed{6(1-1)^5} = \boxed{6\cdot 0^5} = \boxed{6\cdot 0} = \boxed{\mathbf{0}}$$

$$\boxed{f'(1)} = \boxed{6(1+1)^5} = \boxed{6\cdot 2^5} = \boxed{6\cdot 32} = \boxed{\mathbf{192}}$$

c. Given $f'(x) = 12x\left(1+2x^2\right)^2$, then

$$\boxed{f'(0)} = \boxed{(12\cdot 0)\left[1+\left(2\cdot 0^2\right)\right]^2} = \boxed{0\cdot(1+0)^2} = \boxed{0\cdot 1^2} = \boxed{0\cdot 1} = \boxed{\mathbf{0}}$$

$$\boxed{f'(-1)} = \boxed{(12\cdot -1)\left[1+2\cdot(-1)^2\right]^2} = \boxed{-12(1+2)^2} = \boxed{-12\cdot 3^2} = \boxed{-12\cdot 9} = \boxed{\mathbf{-108}}$$

$$\boxed{f'(1)} = \boxed{(12\cdot 1)\left[1+\left(2\cdot 1^2\right)\right]^2} = \boxed{12(1+2)^2} = \boxed{12\cdot 3^2} = \boxed{12\cdot 9} = \boxed{\mathbf{108}}$$

d. Given $f'(x) = 8\left(x^3-x^5\right)^7\left(3x^2-5x^4\right)$, then

$$\boxed{f'(0)} = \boxed{8\left(0^3-0^5\right)^7\left[3\cdot 0^2-5\cdot 0^4\right]} = \boxed{8\cdot 0^7\cdot 0} = \boxed{8\cdot 0\cdot 0} = \boxed{\mathbf{0}}$$

$$\boxed{f'(-1)} = \boxed{8\left[(-1)^3-(-1)^5\right]^7\left[3\cdot(-1)^2-5\cdot(-1)^4\right]} = \boxed{8[-1+1]^7[3\cdot 1-5\cdot 1]} = \boxed{8\cdot 0^7\cdot -2} = \boxed{8\cdot 0\cdot -2} = \boxed{\mathbf{0}}$$

$$\boxed{f'(1)} = \boxed{8\left(1^3-1^5\right)^7\left[3\cdot 1^2-5\cdot 1^4\right]} = \boxed{8(1-1)^7(3\cdot 1-5\cdot 1)} = \boxed{8\cdot 0^7(3-5)} = \boxed{8\cdot 0\cdot -2} = \boxed{\mathbf{0}}$$

e. Given $f'(x) = 3\left(x^{-2}+x\right)^2\left(-2x^{-3}+1\right) = 3\left(\frac{1}{x^2}+x\right)^2\left(-\frac{2}{x^3}+1\right)$, then

$$\boxed{f'(0)} = \boxed{3\left(\frac{1}{0^2}+0\right)^2\left(-\frac{2}{0^3}+1\right)} = \boxed{\mathbf{3\left(\frac{1}{0}+0\right)^2\left(-\frac{2}{0}+1\right)}}$$

$f'(0)$ is undefined due to division by zero.

$$\boxed{f'(-1)} = \boxed{3\left(\frac{1}{(-1)^2}-1\right)^2\left(-\frac{2}{(-1)^3}+1\right)} = \boxed{3(1-1)^2(2+1)} = \boxed{3\cdot 0^2\cdot 3} = \boxed{\mathbf{0}}$$

$$\boxed{f'(1)} = \boxed{3\left(\frac{1}{1^2}+1\right)^2\left(-\frac{2}{1^3}+1\right)} = \boxed{3(1+1)^2(-2+1)} = \boxed{3\cdot 2^2\cdot -1} = \boxed{3\cdot 4\cdot -1} = \boxed{\mathbf{-12}}$$

f. Given $f'(x) = -3x^{-2}\left(1+x^{-1}\right)^2 = -\frac{3\left(1+x^{-1}\right)^2}{x^2} = -\frac{3\left(1+\frac{1}{x}\right)^2}{x^2}$, then

$$\boxed{f'(0)} = \boxed{-\frac{3\left(1+\frac{1}{0}\right)^2}{0^2}} = \boxed{-\frac{\mathbf{3}\left(\mathbf{1}+\frac{\mathbf{1}}{\mathbf{0}}\right)^{\mathbf{2}}}{\mathbf{0}}}$$ $f'(0)$ is undefined due to division by zero.

$$\boxed{f'(-1)} = \boxed{-\frac{3\left(1+\frac{1}{-1}\right)^2}{(-1)^2}} = \boxed{-\frac{3(1-1)^2}{1}} = \boxed{-\frac{3\cdot 0^2}{1}} = \boxed{-\frac{0}{1}} = \boxed{\mathbf{0}}$$

$$\boxed{f'(1)} = \boxed{-\frac{3\left(1+\frac{1}{1}\right)^2}{1^2}} = \boxed{-\frac{3(1+1)^2}{1}} = \boxed{-\frac{3\cdot 2^2}{1}} = \boxed{-\frac{3\cdot 4}{1}} = \boxed{\mathbf{-12}}$$

g. Given $f'(x) = 4\left(x+x^3\right)^3\left(1+3x^2\right)$, then

$$\boxed{f'(0)} = \boxed{4\left(0+0^3\right)^3\left(1+3\cdot 0^2\right)} = \boxed{4\cdot 0^3\cdot 1} = \boxed{4\cdot 0\cdot 1} = \boxed{\mathbf{0}}$$

$$\boxed{f'(-1)} = \boxed{4\left[-1+(-1)^3\right]^3\left[1+3\cdot(-1)^2\right]} = \boxed{4\cdot(-1-1)^3\cdot(1+3)} = \boxed{4\cdot(-2)^3\cdot 4} = \boxed{4\cdot(-8)\cdot 4} = \boxed{\mathbf{-128}}$$

$$\boxed{f'(1)} = \boxed{4\left(1+1^3\right)^3\left(1+3\cdot 1^2\right)} = \boxed{4\cdot(1+1)^3\cdot(1+3)} = \boxed{4\cdot 2^3\cdot 4} = \boxed{4\cdot 8\cdot 4} = \boxed{\mathbf{128}}$$

Example 5.4-3: Use the chain rule to differentiate the following functions. Do not simplify the answer to its lowest term.

a. $\frac{d}{dx}\left(x^2+3\right)^5 =$

b. $\frac{d}{dx}\left[\left(x^2+5\right)^3+1\right]^4 =$

c. $\frac{d}{du}\left[\left(u^2+1\right)^3(u+5)\right] =$

d. $\frac{d}{dt}\left[\frac{\left(t^2+3\right)^5}{t-1}\right] =$

e. $\frac{d}{d\theta}\left[\frac{\left(\theta^3+2\theta\right)^3}{(\theta+1)^2}\right] =$

f. $\frac{d}{dr}\left[r^2\left(r^2+3\right)^4\right] =$

g. $\frac{d}{du}\left[\left(u^2+4\right)^6\left(u^3-1\right)\right] =$

h. $\frac{d}{dt}\left[\left(t^3+2t^2+1\right)\left(t^2+t+1\right)^3\right] =$

i. $\frac{d}{dx}\left(\frac{x^3+3x}{1-x^2}\right)^8 =$

j. $\frac{d}{dx}\left(\frac{x+2}{x-2}\right)^2 =$

k. $\frac{d}{dx}\left[\frac{(2x+5)^3}{(1+x)^2}\right] =$

l. $\frac{d}{dx}\left[\frac{(1-x)^2}{x^3+2x}\right] =$

m. $\frac{d}{dx}\left(x^2+\frac{1}{x^3+5}\right)^4 =$

n. $\frac{d}{dx}\left[\left(2x^3+1\right)^3\left(x^2+1\right)^4\right] =$

Solutions:

a. $\boxed{\frac{d}{dx}\left(x^2+3\right)^5} = \boxed{5\left(x^2+3\right)^{5-1}\frac{d}{dx}\left(x^2+3\right)} = \boxed{5\left(x^2+3\right)^4\cdot 2x} = \boxed{\mathbf{10x\left(x^2+3\right)^4}}$

b. $\boxed{\frac{d}{dx}\left[\left(x^2+5\right)^3+1\right]^4} = \boxed{4\left[\left(x^2+5\right)^3+1\right]^{4-1}\frac{d}{dx}\left[\left(x^2+5\right)^3+1\right]} = \boxed{4\left[\left(x^2+5\right)^3+1\right]^3\cdot\left[3\left(x^2+5\right)^{3-1}\frac{d}{dx}\left(x^2+5\right)+0\right]}$

$$= \boxed{4\left[\left(x^2+5\right)^3+1\right]^3\cdot\left[3\left(x^2+5\right)^2\cdot 2x\right]} = \boxed{4\left[\left(x^2+5\right)^3+1\right]^3\cdot 6x\left(x^2+5\right)^2} = \boxed{\mathbf{24x\left[\left(x^2+5\right)^3+1\right]^3\left(x^2+5\right)^2}}$$

c. $$\boxed{\frac{d}{du}\left[\left(u^2+1\right)^3(u+5)\right]} = \boxed{\left[(u+5)\frac{d}{du}\left(u^2+1\right)^3\right]+\left[\left(u^2+1\right)^3\frac{d}{du}(u+5)\right]} = \boxed{\left\{(u+5)\cdot\left[3\left(u^2+1\right)^{3-1}\frac{d}{du}\left(u^2+1\right)\right]\right\}}$$

$$+\boxed{\left[\left(u^2+1\right)^3\cdot 1\right]} = \boxed{\left\{(u+5)\cdot\left[3\left(u^2+1\right)^2\cdot 2u\right]\right\}+\left(u^2+1\right)^3} = \boxed{\mathbf{6u(u+5)\left(u^2+1\right)^2+\left(u^2+1\right)^3}}$$

d. $$\boxed{\frac{d}{dt}\left[\frac{\left(t^2+3\right)^5}{t-1}\right]} = \boxed{\frac{\left[(t-1)\frac{d}{dt}\left(t^2+3\right)^5\right]-\left[\left(t^2+3\right)^5\frac{d}{dt}(t-1)\right]}{(t-1)^2}} = \boxed{\frac{\left[(t-1)\cdot 5\left(t^2+3\right)^{5-1}\cdot\frac{d}{dt}\left(t^2+3\right)\right]-\left[\left(t^2+3\right)^5\cdot 1\right]}{(t-1)^2}}$$

$$= \boxed{\frac{\left[(t-1)\cdot 5\left(t^2+3\right)^4\cdot 2t\right]-\left(t^2+3\right)^5}{(t-1)^2}} = \boxed{\frac{\mathbf{10t(t-1)\left(t^2+3\right)^4-\left(t^2+3\right)^5}}{\mathbf{(t-1)^2}}}$$

e. $$\boxed{\frac{d}{d\theta}\left[\frac{\left(\theta^3+2\theta\right)^3}{(\theta+1)^2}\right]} = \boxed{\frac{\left[(\theta+1)^2\frac{d}{d\theta}\left(\theta^3+2\theta\right)^3\right]-\left[\left(\theta^3+2\theta\right)^3\frac{d}{d\theta}(\theta+1)^2\right]}{(\theta+1)^4}} = \boxed{\frac{\left[(\theta+1)^2\cdot 3\left(\theta^3+2\theta\right)^{3-1}\cdot\frac{d}{d\theta}\left(\theta^3+2\theta\right)\right]}{(\theta+1)^4}}$$

$$= \boxed{\frac{-\left[\left(\theta^3+2\theta\right)^3\cdot 2(\theta+1)^{2-1}\cdot\frac{d}{d\theta}(\theta+1)\right]}{(\theta+1)^4}} = \boxed{\frac{\left[(\theta+1)^2\cdot 3\left(\theta^3+2\theta\right)^2\cdot\left(3\theta^2+2\right)\right]-\left[\left(\theta^3+2\theta\right)^3\cdot 2(\theta+1)\cdot 1\right]}{(\theta+1)^4}}$$

$$= \boxed{\frac{\mathbf{\left[3(\theta+1)^2\left(\theta^3+2\theta\right)^2\left(3\theta^2+2\right)\right]-\left[2\left(\theta^3+2\theta\right)^3(\theta+1)\right]}}{\mathbf{(\theta+1)^4}}}$$

f. $$\boxed{\frac{d}{dr}\left[r^2\left(r^2+3\right)^4\right]} = \boxed{\left[\left(r^2+3\right)^4\frac{d}{dr}r^2\right]+\left[r^2\frac{d}{dr}\left(r^2+3\right)^4\right]} = \boxed{\left[\left(r^2+3\right)^4\cdot 2r\right]+\left[r^2\cdot 4\left(r^2+3\right)^{4-1}\cdot\frac{d}{dr}\left(r^2+3\right)\right]}$$

$$= \boxed{2r\left(r^2+3\right)^4+\left[4r^2\left(r^2+3\right)^3\cdot 2r\right]} = \boxed{2r\left(r^2+3\right)^4+8r^3\left(r^2+3\right)^3} = \boxed{\mathbf{2r\left(r^2+3\right)^3\left[\left(r^2+3\right)+4r^2\right]}}$$

g. $$\boxed{\frac{d}{du}\left[\left(u^2+4\right)^6\left(u^3-1\right)\right]} = \boxed{\left[\left(u^3-1\right)\frac{d}{du}\left(u^2+4\right)^6\right]+\left[\left(u^2+4\right)^6\frac{d}{du}\left(u^3-1\right)\right]} = \boxed{\left[\left(u^3-1\right)\cdot 6\left(u^2+4\right)^{6-1}\cdot\frac{d}{du}\left(u^2+4\right)\right]}$$

$$+\boxed{\left[\left(u^2+4\right)^6\cdot 3u^2\right]} = \boxed{\left[\left(u^3-1\right)\cdot 6\left(u^2+4\right)^5\cdot 2u\right]+\left[3u^2\left(u^2+4\right)^6\right]} = \boxed{\left[12u\left(u^3-1\right)\left(u^2+4\right)^5\right]+\left[3u^2\left(u^2+4\right)^6\right]}$$

$$= \boxed{3u\left(u^2+4\right)^5\left[4\left(u^3-1\right)+u\left(u^2+4\right)\right]}$$

h. $\boxed{\frac{d}{dt}\left[\left(t^3+2t^2+1\right)\left(t^2+t+1\right)^3\right]} = \boxed{\left[\left(t^2+t+1\right)^3\frac{d}{dt}\left(t^3+2t^2+1\right)\right]+\left[\left(t^3+2t^2+1\right)\frac{d}{dt}\left(t^2+t+1\right)^3\right]}$

$$= \boxed{\left[\left(t^2+t+1\right)^3\cdot\left(3t^2+4t\right)\right]+\left[\left(t^3+2t^2+1\right)\cdot 3\left(t^2+t+1\right)^{3-1}\cdot\frac{d}{dt}\left(t^2+t+1\right)\right]} = \boxed{\left[\left(t^2+t+1\right)^3\cdot\left(3t^2+4t\right)\right]}$$

$$+ \boxed{\left[\left(t^3+2t^2+1\right)\cdot 3\left(t^2+t+1\right)^2\cdot(2t+1)\right]} = \boxed{\left(t^2+t+1\right)^2\left\{\left[t\left(t^2+t+1\right)(3t+4)\right]+\left[3\left(t^3+2t^2+1\right)(2t+1)\right]\right\}}$$

i. $\boxed{\frac{d}{dx}\left(\frac{x^3+3x}{1-x^2}\right)^8} = \boxed{8\left(\frac{x^3+3x}{1-x^2}\right)^{8-1}\cdot\frac{d}{dx}\left(\frac{x^3+3x}{1-x^2}\right)} = \boxed{8\left(\frac{x^3+3x}{1-x^2}\right)^7\cdot\frac{\left[\left(1-x^2\right)\frac{d}{dx}\left(x^3+3x\right)\right]-\left[\left(x^3+3x\right)\frac{d}{dx}\left(1-x^2\right)\right]}{\left(1-x^2\right)^2}}$

$$= \boxed{8\left(\frac{x^3+3x}{1-x^2}\right)^7\cdot\frac{\left[\left(1-x^2\right)\cdot\left(3x^2+3\right)\right]-\left[\left(x^3+3x\right)\cdot(-2x)\right]}{\left(1-x^2\right)^2}} = \boxed{8\left(\frac{x^3+3x}{1-x^2}\right)^7\cdot\frac{\left[\left(1-x^2\right)\left(3x^2+3\right)\right]+\left[2x^2\left(x^2+3\right)\right]}{\left(1-x^2\right)^2}}$$

j. $\boxed{\frac{d}{dx}\left(\frac{x+2}{x-2}\right)^2} = \boxed{2\left(\frac{x+2}{x-2}\right)^{2-1}\cdot\frac{d}{dx}\left(\frac{x+2}{x-2}\right)} = \boxed{2\left(\frac{x+2}{x-2}\right)\cdot\frac{\left[(x-2)\frac{d}{dx}(x+2)\right]-\left[(x+2)\frac{d}{dx}(x-2)\right]}{(x-2)^2}}$

$$= \boxed{2\left(\frac{x+2}{x-2}\right)\cdot\frac{[(x-2)\cdot 1]-[(x+2)\cdot 1]}{(x-2)^2}} = \boxed{2\left(\frac{x+2}{x-2}\right)\cdot\frac{\cancel{x}-2-\cancel{x}-2}{(x-2)^2}} = \boxed{2\left(\frac{x+2}{x-2}\right)\cdot\frac{-4}{(x-2)^2}} = \boxed{-\frac{8(x+2)}{(x-2)^3}}$$

k. $\boxed{\frac{d}{dx}\left[\frac{(2x+5)^3}{(1+x)^2}\right]} = \boxed{\frac{\left[(1+x)^2\frac{d}{dx}(2x+5)^3\right]-\left[(2x+5)^3\frac{d}{dx}(1+x)^2\right]}{(1+x)^4}} = \boxed{\frac{\left[(1+x)^2\cdot 3(2x+5)^{3-1}\frac{d}{dx}(2x+5)\right]}{(1+x)^4}}$

$$- \boxed{\frac{\left[(2x+5)^3\cdot 2(1+x)^{2-1}\frac{d}{dx}(1+x)\right]}{(1+x)^4}} = \boxed{\frac{\left[(1+x)^2\cdot 3(2x+5)^2\cdot 2\right]-\left[(2x+5)^3\cdot 2(1+x)\cdot 1\right]}{(1+x)^4}} = \boxed{\frac{\left[6(1+x)^2(2x+5)^2\right]}{(1+x)^4}}$$

$$- \boxed{\frac{\left[2(2x+5)^3(1+x)\right]}{(1+x)^4}} = \boxed{\frac{2\cancel{(1+x)}(2x+5)^2\left[3(1+x)-(2x+5)\right]}{(1+x)^{4-1=3}}} = \boxed{\frac{2(2x+5)^2(3+3x-2x-5)}{(1+x)^3}} = \boxed{\frac{2(2x+5)^2(x-2)}{(1+x)^3}}$$

l. $\boxed{\frac{d}{dx}\left[\frac{(1-x)^2}{x^3+2x}\right]} = \boxed{\frac{\left[\left(x^3+2x\right)\frac{d}{dx}(1-x)^2\right]-\left[(1-x)^2\frac{d}{dx}\left(x^3+2x\right)\right]}{\left(x^3+2x\right)^2}} = \boxed{\frac{\left[\left(x^3+2x\right)\cdot 2(1-x)^{2-1}\cdot\frac{d}{dx}(1-x)\right]}{\left(x^3+2x\right)^2}}$

$$-\frac{\left[(1-x)^2\cdot\left(3x^2+2\right)\right]}{\left(x^3+2x\right)^2} = \frac{\left[2\left(x^3+2x\right)(1-x)\cdot -1\right]-\left[(1-x)^2\left(3x^2+2\right)\right]}{\left(x^3+2x\right)^2} = \boxed{\frac{\left[-2\left(x^3+2x\right)(1-x)\right]-\left[(1-x)^2\left(3x^2+2\right)\right]}{\left(x^3+2x\right)^2}}$$

m. $\frac{d}{dx}\left(x^2+\frac{1}{x^3+5}\right)^4 = 4\left(x^2+\frac{1}{x^3+5}\right)^{4-1}\cdot\frac{d}{dx}\left(x^2+\frac{1}{x^3+5}\right) = 4\left(x^2+\frac{1}{x^3+5}\right)^3\cdot\left[\frac{d}{dx}\left(x^2\right)+\frac{d}{dx}\left(\frac{1}{x^3+5}\right)\right]$

$$= 4\left(x^2+\frac{1}{x^3+5}\right)^3\cdot\left[2x+\frac{\left[\left(x^3+5\right)\frac{d}{dx}(1)\right]-\left[1\cdot\frac{d}{dx}\left(x^3+5\right)\right]}{\left(x^3+5\right)^2}\right] = \boxed{4\left(x^2+\frac{1}{x^3+5}\right)^3\left[2x-\frac{3x^2}{\left(x^3+5\right)^2}\right]}$$

n. $\frac{d}{dx}\left[\left(2x^3+1\right)^3\left(x^2+1\right)^4\right] = \left[\left(x^2+1\right)^4\frac{d}{dx}\left(2x^3+1\right)^3\right]+\left[\left(2x^3+1\right)^3\frac{d}{dx}\left(x^2+1\right)^4\right]$

$$= \left[\left(x^2+1\right)^4\cdot 3\left(2x^3+1\right)^{3-1}\cdot\frac{d}{dx}\left(2x^3+1\right)\right]+\left[\left(2x^3+1\right)^3\cdot 4\left(x^2+1\right)^{4-1}\cdot\frac{d}{dx}\left(x^2+1\right)\right] = \left[\left(x^2+1\right)^4\cdot 3\left(2x^3+1\right)^2\cdot 6x^2\right]$$

$$+\left[\left(2x^3+1\right)^3\cdot 4\left(x^2+1\right)^3\cdot 2x\right] = \boxed{\left[18x^2\left(x^2+1\right)^4\left(2x^3+1\right)^2\right]+\left[8x\left(2x^3+1\right)^3\left(x^2+1\right)^3\right]}$$

In some instances students are asked to find the derivative of a function y, where y is a function of u and u is a function of x. We can solve this class of problems using one of two methods.

The first method, and perhaps the easiest one, is performed by substituting u into the y equation and taking the derivative of y with respect to x. The second method is to find the derivative of y by using the equation $\frac{dy}{dx}=\frac{dy}{du}\cdot\frac{du}{dx}$. This method is most often used in calculus books and can be time consuming. For example, let's find the derivative of the function $y=u^2+1$ where $u=x+1$ using each of these methods.

First Method: Given the function $y=u^2+1$ where $u=x+1$, substitute u with $x+1$ in the function y and simplify, i.e., $y = u^2+1 = (x+1)^2+1 = x^2+2x+1+1 = x^2+2x+2$. Next, take the derivative of y with respect to x, i.e., $\frac{dy}{dx} = \frac{dy}{dx}\left(x^2+2x+2\right) = 2x^{2-1}+2x^{1-1}+0 = \mathbf{2x+2}$.

Second Method: Given the function $y=u^2+1$ where $u=x+1$, find $\frac{dy}{du}$ and $\frac{du}{dx}$, i.e., $\frac{dy}{du} = 2u^{2-1}+0 = 2u$ and $\frac{du}{dx} = x^{1-1}+0 = x^0 = 1$. Next, substitute $\frac{dy}{du}$ and $\frac{du}{dx}$ in the equation $\frac{dy}{dx}=\frac{dy}{du}\cdot\frac{du}{dx}$ to find the derivative $\frac{dy}{dx}$, i.e., $\frac{dy}{dx} = 2u\cdot 1 = 2u$. Substituting $u=x+1$ in place of u we obtain $\frac{dy}{dx} = 2u = 2(x+1) = \mathbf{2x+2}$.

The second method is generally beyond the scope of this book, therefore the first method is used in order to solve this class of problems. Examples 5.4-4 and 5.4-5 below provide additional examples as to how these types of problems are solved.

Example 5.4-4: Find $\frac{dy}{dx}$ given:

a. $y=\frac{1}{1+u}$ and $u=3x+1$

b. $y=u^2+2u+1$ and $u=3x+2$

c. $y=\frac{u}{1+u^2}$ and $u=5x+1$

d. $y=u^2+1$ and $u=\frac{x+1}{x^2-1}$

e. $y=\frac{1+u}{u^3}$ and $u=x^2+1$

f. $y=u^3+1$ and $u=\left(x^2+1\right)^{-1}$

Solutions:

a. Given $y=\frac{1}{1+u}$ and $u=3x+1$, then $y=\frac{1}{1+(3x+1)}=\frac{1}{3x+2}$ and

$$\frac{dy}{dx}=\frac{\left[(3x+2)\frac{d}{dx}(1)\right]-\left[1\cdot\frac{d}{dx}(3x+2)\right]}{(3x+2)^2}=\frac{0-3}{(3x+2)^2}=-\frac{\mathbf{3}}{(\mathbf{3x+2})^{\mathbf{2}}}$$

b. Given $y=u^2+2u+1$ and $u=3x+2$, then $y=(3x+2)^2+2(3x+2)+1=(3x+2)^2+6x+5$ and

$$\frac{dy}{dx}=\frac{d}{dx}(3x+2)^2+\frac{d}{dx}6x+\frac{d}{dx}5=2(3x+2)^{2-1}\cdot\frac{d}{dx}(3x+2)+6+0=2(3x+2)\cdot3+6=\mathbf{18(x+1)}$$

c. Given $y=\frac{u}{1+u^2}$ and $u=5x+1$, then $y=\frac{(5x+1)}{1+(5x+1)^2}$ and

$$\frac{dy}{dx}=\frac{\left[\left[1+(5x+1)^2\right]\frac{d}{dx}(5x+1)\right]-\left[(5x+1)\cdot\frac{d}{dx}\left[1+(5x+1)^2\right]\right]}{\left[1+(5x+1)^2\right]^2}=\frac{\left\{\left[1+(5x+1)^2\right]\cdot5\right\}-\left[(5x+1)\cdot2(5x+1)\cdot5\right]}{\left[1+(5x+1)^2\right]^2}$$

$$=\frac{5\left[1+(5x+1)^2\right]-\left[10(5x+1)^2\right]}{\left[1+(5x+1)^2\right]^2}=\frac{5+5(5x+1)^2-10(5x+1)^2}{\left[1+(5x+1)^2\right]^2}=\frac{\mathbf{5-5(5x+1)^2}}{\left[\mathbf{1+(5x+1)^2}\right]^{\mathbf{2}}}$$

d. Given $y=u^2+1$ and $u=\frac{x+1}{x^2-1}$, then $y=\left(\frac{x+1}{x^2-1}\right)^2+1$ and

$$\frac{dy}{dx}=\left[2\left(\frac{x+1}{x^2-1}\right)^{2-1}\cdot\frac{d}{dx}\left(\frac{x+1}{x^2-1}\right)\right]+0=2\left(\frac{x+1}{x^2-1}\right)\cdot\frac{\left[\left(x^2-1\right)\frac{d}{dx}(x+1)\right]-\left[(x+1)\cdot\frac{d}{dx}\left(x^2-1\right)\right]}{\left(x^2-1\right)^2}$$

$$=2\left(\frac{x+1}{x^2-1}\right)\cdot\frac{\left[\left(x^2-1\right)\cdot1\right]-\left[(x+1)\cdot2x\right]}{\left(x^2-1\right)^2}=2\left(\frac{x+1}{x^2-1}\right)\cdot\left(\frac{x^2-1-2x^2-2x}{\left(x^2-1\right)^2}\right)=2\left(\frac{x+1}{x^2-1}\right)\cdot\left(-\frac{x^2+2x+1}{\left(x^2-1\right)^2}\right)$$

$$= \boxed{-2\left(\frac{x+1}{x^2-1}\right)\cdot\frac{(x+1)^2}{\left(x^2-1\right)^2}} = \boxed{-2\frac{(x+1)(x+1)^2}{\left(x^2-1\right)\left(x^2-1\right)^2}} = \boxed{-2\frac{(x+1)^3}{\left(x^2-1\right)^3}} = \boxed{\mathbf{-2\left(\frac{x+1}{x^2-1}\right)^3}}$$

e. Given $y=\frac{1+u}{u^3}$ and $u=x^2+1$, then $y=\frac{1+\left(x^2+1\right)}{\left(x^2+1\right)^3} = \frac{x^2+2}{\left(x^2+1\right)^3}$

$$\boxed{\frac{dy}{dx}} = \boxed{\frac{\left[\left(x^2+1\right)^3\cdot\frac{d}{dx}\left(x^2+2\right)\right]-\left[\left(x^2+2\right)\cdot\frac{d}{dx}\left(x^2+1\right)^3\right]}{\left(x^2+1\right)^6}} = \boxed{\frac{\left[\left(x^2+1\right)^3\cdot 2x\right]-\left[\left(x^2+2\right)\cdot 3\left(x^2+1\right)^2\frac{d}{dx}\left(x^2+1\right)\right]}{\left(x^2+1\right)^6}}$$

$$= \boxed{\frac{\left[2x\left(x^2+1\right)^3\right]-\left[\left(x^2+2\right)\cdot 3\left(x^2+1\right)^2\cdot 2x\right]}{\left(x^2+1\right)^6}} = \boxed{\mathbf{\frac{\left[2x\left(x^2+1\right)^3\right]-\left[6x\left(x^2+2\right)\left(x^2+1\right)^2\right]}{\left(x^2+1\right)^6}}}$$

f. Given $y=u^3+1$ and $u=\left(x^2+1\right)^{-1}$, then $y=\left[\left(x^2+1\right)^{-1}\right]^3+1 = \left(x^2+1\right)^{-3}+1$

$$\boxed{\frac{dy}{dx}} = \boxed{\left[-3\left(x^2+1\right)^{-3-1}\cdot\frac{d}{dx}\left(x^2+1\right)\right]+0} = \boxed{-3\left(x^2+1\right)^{-4}\cdot 2x} = \boxed{-6x\left(x^2+1\right)^{-4}} = \boxed{\mathbf{-\frac{6x}{\left(x^2+1\right)^4}}}$$

Example 5.4-5: Find y' given that:

a. $y=3u^3-1$ and $u=x^2+1$

b. $y=\frac{3u^2}{1+u}$ and $u=x^2$

c. $y=\frac{u+1}{u-1}$ and $u=x^2+3$

d. $y=\frac{1-5u}{u^2}$ and $u=1-x$

e. $y=\frac{u}{u^2+1}$ and $u=2x-1$

f. $y=u+\frac{1}{4}$ and $u=(2x-1)^4$

g. $y=\frac{2u}{(u-1)^2}$ and $u=x+2$

h. $y=u^4-1$ and $u=\frac{1+x}{1-x}$

Solutions:

a. Given $y=3u^3-1$ and $u=x^2+1$, then $y=3\left(x^2+1\right)^3-1$ and

$$\boxed{y'} = \boxed{3\cdot 3\left(x^2+1\right)^{3-1}\cdot 2x-0} = \boxed{9\left(x^2+1\right)^2\cdot 2x} = \boxed{\mathbf{18x\left(x^2+1\right)^2}}$$

b. Given $y=\frac{3u^2}{1+u}$ and $u=x^2$, then $y=\frac{3x^4}{1+x^2}$ and

$$\boxed{y'} = \boxed{\frac{\left[\left(3\cdot 4x^{4-1}\right)\cdot\left(1+x^2\right)\right]-\left[2x\cdot 3x^4\right]}{\left(1+x^2\right)^2}} = \boxed{\frac{12x^3\left(1+x^2\right)-6x^5}{\left(1+x^2\right)^2}} = \boxed{\frac{12x^3+6x^5}{\left(1+x^2\right)^2}} = \boxed{\mathbf{\frac{6x^3\left(2+x^2\right)}{\left(1+x^2\right)^2}}}$$

c. Given $y=\frac{u+1}{u-1}$ and $u=x^2+3$, then $y=\frac{(x^2+3)+1}{(x^2+3)-1}=\frac{x^2+4}{x^2+2}$ and

$$\boxed{y'}=\boxed{\frac{\left[2x^{2-1}\cdot(x^2+2)\right]-\left[2x^{2-1}\cdot(x^2+4)\right]}{(x^2+2)^2}}=\boxed{\frac{\left[2x(x^2+2)\right]-\left[2x(x^2+4)\right]}{(x^2+2)^2}}=\boxed{\frac{2x^3+4x-2x^3-8x}{(x^2+2)^2}}=\boxed{\mathbf{\frac{-4x}{(x^2+2)^2}}}$$

d. Given $y=\frac{1-5u}{u^2}$ and $u=1-x$, then $y=\frac{1-5(1-x)}{(1-x)^2}=\frac{1-5+5x}{(1-x)^2}=\frac{5x-4}{(1-x)^2}$

$$\boxed{y'}=\boxed{\frac{\left[5\cdot(1-x)^2\right]-\left[2(1-x)^{2-1}\cdot-1\cdot(5x-4)\right]}{(1-x)^4}}=\boxed{\frac{5(1-x)^2-\left[-2(1-x)(5x-4)\right]}{(1-x)^4}}=\boxed{\frac{5(1-x)^2+2(1-x)(5x-4)}{(1-x)^4}}$$

$$=\boxed{\frac{5(x^2-2x+1)+2(5x-4-5x^2+4x)}{(1-x)^4}}=\boxed{\frac{5x^2-10x+5-10x^2+18x-8}{(1-x)^4}}=\boxed{\mathbf{\frac{-5x^2+8x-3}{(1-x)^4}}}$$

e. Given $y=\frac{u}{u^2+1}$ and $u=2x-1$, then $y=\frac{2x-1}{(2x-1)^2+1}$ and

$$\boxed{y'}=\boxed{\frac{2\cdot\left[(2x-1)^2+1\right]-\left[2(2x-1)^{2-1}\cdot2\cdot(2x-1)\right]}{\left[(2x-1)^2+1\right]^2}}=\boxed{\frac{2\cdot\left[(2x-1)^2+1\right]-\left[4(2x-1)\cdot(2x-1)\right]}{\left[(2x-1)^2+1\right]^2}}$$

$$=\boxed{\frac{2(2x-1)^2+2-4(2x-1)^2}{\left[(2x-1)^2+1\right]^2}}=\boxed{\frac{2-2(2x-1)^2}{\left[(2x-1)^2+1\right]^2}}=\boxed{\frac{2\left[1-(2x-1)^2\right]}{\left[(2x-1)^2+1\right]^2}}=\boxed{\mathbf{\frac{-2\left[(2x-1)^2-1\right]}{\left[(2x-1)^2+1\right]^2}}}$$

f. Given $y=u+\frac{1}{4}$ and $u=(2x-1)^4$, then $y=(2x-1)^4+\frac{1}{4}$ and

$$\boxed{y'}=\boxed{4(2x-1)^{4-1}\cdot2+0}=\boxed{4(2x-1)^3\cdot2}=\boxed{\mathbf{8(2x-1)^3}}$$

g. Given $y=\frac{2u}{(u-1)^2}$ and $u=x+2$, then $y=\frac{2(x+2)}{(x+2-1)^2}=\frac{2x+4}{(x+1)^2}$

$$\boxed{y'}=\boxed{\frac{\left[2\cdot(x+1)^2\right]-\left[2(x+1)^{2-1}\cdot(2x+4)\right]}{(x+1)^4}}=\boxed{\frac{2(x+1)^2-2(x+1)(2x+4)}{(x+1)^4}}=\boxed{\frac{2(x^2+2x+1)-2(2x^2+6x+4)}{(x+1)^4}}$$

$$=\boxed{\frac{2x^2+4x+2-4x^2-12x-8}{(x+1)^4}}=\boxed{\frac{-2x^2-8x-6}{(x+1)^4}}=\boxed{\mathbf{\frac{-2(x^2+4x+3)}{(x+1)^4}}}$$

h. Given $y = u^4 - 1$ and $u = \frac{1+x}{1-x}$, then $y = \left(\frac{1+x}{1-x}\right)^4 - 1$ and

$$\boxed{y'} = \boxed{4\left(\frac{1+x}{1-x}\right)^{4-1} \cdot \frac{[1\cdot(1-x)]-[-1\cdot(1+x)]}{(1-x)^2} - 0} = \boxed{4\left(\frac{1+x}{1-x}\right)^3 \cdot \frac{(1-x)+(1+x)}{(1-x)^2}} = \boxed{4\left(\frac{1+x}{1-x}\right)^3 \cdot \frac{2}{(1-x)^2}}$$

$$= \boxed{4\left[\frac{(1+x)^3}{(1-x)^3} \cdot \frac{2}{(1-x)^2}\right]} = \boxed{\frac{8(1+x)^3}{(1-x)^{3+2}}} = \boxed{\mathbf{\frac{8(1+x)^3}{(1-x)^5}}}$$

Section 5.4 Practice Problems - The Chain Rule

1. Find the derivative of the following functions. Do not simplify the answer to its lowest term.

a. $y = (x^2+2)^3$ b. $y = (x^2+1)^{-2}$ c. $y = (x^3-1)^5$

d. $y = \left(1 - \frac{1}{x^2}\right)^2$ e. $y = 2x^3 + \frac{1}{3x^2}$ f. $y = \left(\frac{1+x^2}{r^3}\right)^4$

g. $y = x^2\left(\frac{x+1}{3}\right)^3$ h. $y = [x(x+1)^2 + 2x]^3$ i. $y = \left(\frac{x}{3} - 2x^3\right)^{-1}$

j. $y = (x^3+3x^2+1)^4$ k. $y = \left(\frac{t^2}{1+t^2}\right)^3$ l. $y = (1+x^{-2})^{-1}$

m. $y = \frac{(x+1)^{-2}}{x^3}$ n. $y = \left(\frac{1}{1-x^3}\right)^2 + \frac{1}{x}$ o. $y = \frac{x^3}{x^3+2} - x^2$

2. Find the derivative of the following functions at $x = 0$, $x = 1$, and $x = -1$.

a. $y = (x^3+1)^5$ b. $y = (x^3+3x^2-1)^4$ c. $y = \left(\frac{x}{x+1}\right)^2$

d. $y = x(x^2+1)^2$ e. $y = x^3 + 2(x^2+1)^3$ f. $y = \left(\frac{x^2}{1+x^2}\right)^3$

g. $y = \left(\frac{x}{x^2+1}\right)^5$ h. $y = (x^2+1)^3 \cdot \frac{1}{x^2}$ i. $y = \left(\frac{x^3}{x-1}\right)^2 + 5x$

3. Use the chain rule to differentiate the following functions.

a. $\frac{d}{dt}\left[\frac{(t+1)^3}{t^2}\right] =$ b. $\frac{d}{du}\left[\frac{(u^2+1)^3}{3u^4}\right] =$ c. $\frac{d}{dx}\left[\frac{(2x+1)^3}{(1-x)^2}\right] =$

d. $\frac{d}{dx}\left[(x^3-1)^2(2x+1)^3\right] =$ e. $\frac{d}{ds}\left[s^3 - \frac{1}{s^2+6}\right]^2 =$ f. $\frac{d}{dt}\left[\frac{(t^2-1)^3}{t^2+1}\right] =$

g. $\frac{d}{du}\left[(u^2+1)^3\left(\frac{1}{u+1}\right)^2\right] =$ h. $\frac{d}{d\theta}\left[\frac{\theta^2+3}{(\theta-1)^3}\right]^2 =$ i. $\frac{d}{dr}\left[\frac{r^7}{(r^2+2r)^3}\right] =$

4. Given the following y functions in terms of u, find y'.

a. $y = 2u^2 - 1$ and $u = x - 1$ b. $y = \frac{u}{u-1}$ and $u = x^3$ c. $y = \frac{u}{1+u^2}$ and $u = x^2 + 1$

d. $y = u^2 - \frac{1}{2}$ and $y = x^4$ e. $y = u^4$ and $u = \frac{1}{1-x^2}$ f. $y = \frac{u^2}{(u+1)^3}$ and $u = x - 1$

5.5 Implicit Differentiation

In many instances an equation is explicitly represented in the form where y is the only term in the left-hand side of the equation. In these instances y' is obtained by applying the differentiation rules to the right hand side of the equation. However, for cases where y is not explicitly given, we must either first solve for y (if y can be factored) and then differentiate or use the implicit differentiation method. For example, to differentiate the equation $xy = x^2 + y$ we can either solve for y by bringing the y terms to one side of the equation and then differentiate as follows:

$\boxed{xy = x^2 + y}$; $\boxed{xy - y = x^2}$; $\boxed{y(x-1) = x^2}$; $\boxed{y = \frac{x^2}{x-1}}$ therefore, y' is equal to:

$$\boxed{y'} = \boxed{\frac{\left[2x^{2-1}\cdot(x-1)\right]-\left[1\cdot x^2\right]}{(x-1)^2}} = \boxed{\frac{[2x(x-1)]-x^2}{(x-1)^2}} = \boxed{\frac{2x^2-2x-x^2}{(x-1)^2}} = \boxed{\mathbf{\frac{x^2-2x}{(x-1)^2}}}$$

or, we can use the implicit differentiation method as shown below.

$\boxed{xy = x^2 + y}$; $\boxed{(1\cdot y + y'\cdot x) = 2x^{2-1} + y'}$; $\boxed{y + y'x = 2x + y'}$; $\boxed{y'x - y' = 2x - y}$; $\boxed{y'(x-1) = 2x - y}$; $\boxed{y' = \frac{2x-y}{x-1}}$

Substituting $y = \frac{x^2}{x-1}$ into the y' equation we obtain:

$$\boxed{y'} = \boxed{\frac{2x-y}{x-1}} = \boxed{\frac{2x-\frac{x^2}{x-1}}{x-1}} = \boxed{\frac{\frac{2x(x-1)-x^2}{x-1}}{x-1}} = \boxed{\frac{\frac{2x(x-1)-x^2}{x-1}}{\frac{x-1}{1}}} = \boxed{\frac{\left[2x(x-1)-x^2\right]\cdot 1}{(x-1)\cdot(x-1)}} = \boxed{\mathbf{\frac{x^2-2x}{(x-1)^2}}}$$

Note that the key in using the implicit differentiation method is that the chain rule must be applied each time we come across a term with y in it. Following are additional examples showing the two methods of differentiation when y is not explicitly given:

Example 5.5-1: Given $xy + x = y + 3$, find y'.

Solution:

First Method: Let's solve for y by bringing the y terms to the left-hand side of the equation,

i.e., $\boxed{xy + x = y + 3}$; $\boxed{y(x-1) = -x + 3}$; $\boxed{y = \frac{-x+3}{x-1}}$

We can now solve for y' using the differentiation rule for division.

$$\boxed{y'} = \boxed{\frac{[-1\cdot(x-1)]-[1\cdot(-x+3)]}{(x-1)^2}} = \boxed{\frac{-\cancel{x}+1+\cancel{x}-3}{(x-1)^2}} = \boxed{\mathbf{\frac{-2}{(x-1)^2}}}$$

Second Method: Use the implicit differentiation method to solve for y', i.e., given

$\boxed{xy + x = y + 3}$ then, $\boxed{(1\cdot y + y'\cdot x) + 1 = y' + 0}$; $\boxed{y + 1 = y' - y'x}$; $\boxed{y'(1-x) = y + 1}$; $\boxed{y' = \frac{y+1}{1-x}}$

Substituting $y = \frac{-x+3}{x-1}$ into the y' equation we obtain:

$$\boxed{y'} = \boxed{y' = \frac{y+1}{1-x}} = \boxed{\frac{\frac{-x+3}{x-1}+1}{1-x}} = \boxed{\frac{\frac{-x+3}{x-1}+\frac{1}{1}}{1-x}} = \boxed{\frac{\frac{[(-x+3)\cdot 1]+[1\cdot(x-1)]}{(x-1)\cdot 1}}{1-x}} = \boxed{\frac{\frac{-\not{x}+3+\not{x}-1}{x-1}}{1-x}} = \boxed{\frac{\frac{2}{x-1}}{1-x}} = \boxed{\frac{\frac{2}{x-1}}{\frac{1-x}{1}}}$$

$$= \boxed{\frac{2\cdot 1}{(x-1)\cdot(1-x)}} = \boxed{\frac{2}{(x-1)\cdot -(x-1)}} = \boxed{-\frac{2}{(x-1)(x-1)}} = \boxed{\mathbf{\frac{-2}{(x-1)^2}}}$$

Example 5.5-2: Given $x^2y+5=y+2x$, find y'.

Solution:

First Method: Let's solve for y by bringing the y terms to the left-hand side of the equation,

i.e., $\boxed{x^2y-y=2x-5}$; $\boxed{y\left(x^2-1\right)=2x-5}$; $\boxed{y=\frac{2x-5}{x^2-1}}$

We can now solve for y' using the differentiation rule for division.

$$\boxed{y'} = \boxed{\frac{\left[2\cdot\left(x^2-1\right)\right]-\left[2x^{2-1}\cdot(2x-5)\right]}{\left(x^2-1\right)^2}} = \boxed{\frac{2x^2-2-4x^2+10x}{\left(x^2-1\right)^2}} = \boxed{\mathbf{\frac{-2x^2+10x-2}{\left(x^2-1\right)^2}}}$$

Second Method: Use the implicit differentiation method to solve for y', i.e., given

$\boxed{x^2y+5=y+2x}$ then, $\boxed{\left(2x\cdot y+y'\cdot x^2\right)+0=y'+2x^{1-1}}$; $\boxed{2xy+y'x^2=y'+2}$; $\boxed{y'x^2-y'=2-2xy}$

; $\boxed{y'\left(x^2-1\right)=2x-2xy}$; $\boxed{y'=\frac{2-2xy}{x^2-1}}$

Substituting $y = \frac{2x-5}{x^2-1}$ into the y' equation we obtain:

$$\boxed{y'} = \boxed{\frac{2-2xy}{x^2-1}} = \boxed{\frac{2-2x\cdot\frac{2x-5}{x^2-1}}{x^2-1}} = \boxed{\frac{2-\frac{4x^2-10x}{x^2-1}}{x^2-1}} = \boxed{\frac{\frac{2\left(x^2-1\right)-\left(4x^2-10x\right)}{x^2-1}}{x^2-1}} = \boxed{\frac{\frac{2x^2-2-4x^2+10x}{x^2-1}}{x^2-1}}$$

$$= \boxed{\frac{\frac{-2x^2+10x-2}{x^2-1}}{x^2-1}} = \boxed{\frac{\frac{-2x^2+10x-2}{x^2-1}}{\frac{x^2-1}{1}}} = \boxed{\frac{\left(-2x^2+10x-2\right)\cdot 1}{\left(x^2-1\right)\cdot\left(x^2-1\right)}} = \boxed{\mathbf{\frac{-2x^2+10x-2}{\left(x^2-1\right)^2}}}$$

In the previous examples, to find y' we could either first solve for y and then differentiate or use the implicit differentiation rule. However, sometimes we can not simply solve for y by bringing the y terms to the left-hand side of the equation. In these instances, as is shown in the following examples, we can only use implicit differentiation in order to differentiate y.

Example 5.5-3: Given $x^2y^2 + y = 3y^3 - 1$, find $y' = \frac{dy}{dx}$.

Solution:

$\boxed{\frac{d}{dx}\left(x^2y^2 + y\right) = \frac{d}{dx}\left(3y^3 - 1\right)}$; $\boxed{\left(2x \cdot y^2 + 2y^{2-1}y' \cdot x^2\right) + y' = (3 \cdot 3)y^{3-1} \cdot y' - 0}$; $\boxed{2xy^2 + 2yy'x^2 + y' = 9y^2y'}$

; $\boxed{2x^2yy' - 9y^2y' + y' = -2xy^2}$; $\boxed{y'\left(2x^2y - 9y^2 + 1\right) = -2xy^2}$; $\boxed{\mathbf{y' = -\frac{2xy^2}{2x^2y - 9y^2 + 1}}}$

Example 5.5-4: Given $xy + x^2y^2 + y^3 = 10x$, find $\frac{dy}{dx}$ in terms of x and y.

Solution:

$\boxed{\frac{d}{dx}\left(xy + x^2y^2 + y^3\right) = \frac{d}{dx}(10x)}$; $\boxed{\left(y + xy'\right) + \left(2xy^2 + 2x^2yy'\right) + 3y^2y' = 10}$

; $\boxed{xy' + +2x^2yy' + 3y^2y' = -2xy^2 - y + 10}$; $\boxed{y'\left(x + +2x^2y + 3y^2\right) = -2xy^2 - y + 10}$; $\boxed{\mathbf{y' = \frac{-2xy^2 - y + 10}{x + +2x^2y + 3y^2}}}$

Example 5.5-5: Given $3x^3y^3 + 2y^2 = y + 1$, find $\frac{dy}{dx}$ in terms of x and y.

Solution:

$\boxed{\frac{d}{dx}\left(3x^3y^3 + 2y^2\right) = \frac{d}{dx}(y + 1)}$; $\boxed{3\left(3x^2y^3 + 3y^2\,y'x^3\right) + \left(4yy'\right) = y' + 0}$; $\boxed{9x^2y^3 + 9x^3y^2\,y' + 4yy' = y'}$

; $\boxed{9x^3y^2\,y' + 4yy' - y' = -9x^2y^3}$; $\boxed{y'\left(9x^3y^2 + 4y - 1\right) = -9x^2y^3}$; $\boxed{\mathbf{y' = -\frac{9x^2y^3}{9x^3y^2 + 4y - 1}}}$

Example 5.5-6: Given $xy + x^3y^3 = 5$, find $\frac{dy}{dx}$ in terms of x and y.

Solution:

$\boxed{\frac{d}{dx}\left(xy + x^3y^3\right) = \frac{d}{dx}(5)}$; $\boxed{(1 \cdot y + y' \cdot x) + \left(3x^2 \cdot y^3 + 3y^2\,y' \cdot x^3\right) = 0}$; $\boxed{y + xy' + 3x^2y^3 + 3x^3y^2\,y' = 0}$

; $\boxed{xy' + 3x^3y^2\,y' = -3x^2y^3 - y}$; $\boxed{y'\left(x + 3x^3y^2\right) = -3x^2y^3 - y}$; $\boxed{\mathbf{y' = -\frac{3x^2y^3 + y}{x + 3x^3y^2}}}$

Example 5.5-7: Given $3xy + y = \left(x^2 + y^2\right)$, find $\frac{dy}{dx}$ in terms of x and y.

Solution:

$\boxed{\frac{d}{dx}(3xy + y) = \frac{d}{dx}\left(x^2 + y^2\right)}$; $\boxed{3(1 \cdot y + y' \cdot x) + y' = 2x + 2yy'}$; $\boxed{3y + 3xy' + y' = 2x + 2yy'}$

; $\boxed{3xy' + y' - 2yy' = 2x - 3y}$; $\boxed{y'(3x + 1 - 2y) = 2x - 3y}$; $\boxed{\mathbf{y' = \frac{2x - 3y}{3x - 2y + 1}}}$

Example 5.5-8: Given $\frac{1}{x} + \frac{1}{y^2} = 10x$, find $\frac{dy}{dx}$ in terms of x and y.

Solution:

$\boxed{\frac{d}{dx}\left(\frac{1}{x}+\frac{1}{y^2}\right)=\frac{d}{dx}(10x)}$; $\boxed{\frac{d}{dx}\left(x^{-1}+y^{-2}\right)=\frac{d}{dx}(10x)}$; $\boxed{-x^{-2}-2y^{-3}y'=10}$; $\boxed{-2y^{-3}y'=x^{-2}+10}$

; $\boxed{y'=\frac{x^{-2}+10}{-2y^{-3}}}$; $\boxed{y'=\frac{\frac{1}{x^2}+10}{\frac{-2}{y^3}}}$; $\boxed{y'=\frac{\frac{1+10x^2}{x^2}}{\frac{-2}{y^3}}}$; $\boxed{y'=\frac{y^3\left(1+10x^2\right)}{-2x^2}}$; $\boxed{\mathbf{y'=-\frac{y^3}{2}\left(\frac{1}{x^2}+10\right)}}$

Example 5.5-9: Given $xy^2+yx^2=x^2$, find $\frac{dy}{dx}$ in terms of x and y.

Solution:

$\boxed{\frac{d}{dx}\left(xy^2+yx^2\right)=\frac{d}{dx}\left(x^2\right)}$; $\boxed{\left(1\cdot y^2+2y\,y'\cdot x\right)+\left(y'\cdot x^2+2x\cdot y\right)=2x}$; $\boxed{y^2+2x\,y\,y'+x^2y'+2x\,y=2x}$

; $\boxed{2x\,y\,y'+x^2y'=-y^2-2x\,y+2x}$; $\boxed{y'\left(x^2+2x\,y\right)=-y^2-2x\,y+2x}$; $\boxed{\mathbf{y'=\frac{-y^2-2x\,y+2x}{x^2+2x\,y}}}$

Example 5.5-10: Given $y^{\frac{2}{3}}+x^3y=y$, find $\frac{dy}{dx}$ in terms of x and y.

Solution:

$\boxed{\frac{d}{dx}\left(y^{\frac{2}{3}}+x^3y\right)=\frac{d}{dx}(y)}$; $\boxed{\frac{2}{3}y^{\frac{2}{3}-1}\cdot y'+\left(3x^2\cdot y+y'\cdot x^3\right)=y'}$; $\boxed{\frac{2}{3}y^{-\frac{1}{3}}y'+3x^2y+x^3y'=y'}$

; $\boxed{\frac{2}{3}y^{-\frac{1}{3}}y'+x^3y'-y'=-3x^2y}$; $\boxed{y'\left(\frac{2}{3}y^{-\frac{1}{3}}+x^3-1\right)=-3x^2y}$; $\boxed{\mathbf{y'=\frac{-3x^2y}{\frac{2}{3}y^{-\frac{1}{3}}+x^3-1}}}$

Example 5.5-11: Given $xy+y^2=y^{\frac{1}{8}}$, find $\frac{dy}{dx}$ in terms of x and y.

Solution:

$\boxed{\frac{d}{dx}\left(xy+y^2\right)=\frac{d}{dx}\left(y^{\frac{1}{8}}\right)}$; $\boxed{\left[(1\cdot y)+(\,y'\cdot x)\right]+2yy'=\frac{1}{8}y^{\frac{1}{8}-1}\cdot y'}$; $\boxed{y+x\,y'+2y\,y'=\frac{1}{8}y^{-\frac{7}{8}}y'}$

; $\boxed{x\,y'+2y\,y'-\frac{1}{8}y^{-\frac{7}{8}}y'=-y}$; $\boxed{y'\left(x+2y-\frac{1}{8}y^{-\frac{7}{8}}\right)=-y}$; $\boxed{\mathbf{y'=-\frac{y}{x+2y-\frac{1}{8}y^{-\frac{7}{8}}}}}$

Example 5.5-12: Given $xy^2+y=x^2+3$, find $\frac{dy}{dx}$ in terms of x and y.

Solution:

$\boxed{\frac{d}{dx}\left(xy^2+y\right)=\frac{d}{dx}\left(x^2+3\right)}$; $\boxed{\left[\left(1\cdot y^2\right)+\left(2y\,y'\cdot x\right)\right]+y'=2x+0}$; $\boxed{y^2+2xy\,y'+y'=2x}$; $\boxed{2xy\,y'+y'=2x-y^2}$

; $\boxed{y'(2xy+1)=2x-y^2}$; $\boxed{\mathbf{y'=\frac{2x-y^2}{2xy+1}}}$

Example 5.5-13: Given $x^4y^3 + y^2 = x + 4$, find $\frac{dy}{dx}$ in terms of x and y.

Solution:

$\boxed{\frac{d}{dx}\left(x^4y^3 + y^2\right) = \frac{d}{dx}(x+4)}$; $\boxed{\left[\left(4x^3 \cdot y^3\right) + \left(3y^2y' \cdot x^4\right)\right] + 2y\,y' = 1 + 0}$; $\boxed{4x^3y^3 + 3x^4y^2y' + 2y\,y' = 1}$

; $\boxed{3x^4y^2y' + 2y\,y' = 1 - 4x^3y^3}$; $\boxed{y'\left(3x^4y^2 + 2y\right) = 1 - 4x^3y^3}$; $\boxed{\mathbf{y' = \frac{1 - 4x^3y^3}{3x^4y^2 + 2y}}}$

Example 5.5-14: Given $y^6 + x^3y^5 + x^2 = 5$, find $\frac{dy}{dx}$ in terms of x and y.

Solution:

$\boxed{\frac{d}{dx}\left(y^6 + x^3y^5 + x^2\right) = \frac{d}{dx}(5)}$; $\boxed{6y^5y' + \left[\left(3x^2 \cdot y^5\right) + \left(5y^4y' \cdot x^3\right)\right] + 2x = 0}$; $\boxed{6y^5y' + 3x^2y^5 + 5x^3y^4y' + 2x = 0}$

; $\boxed{6y^5y' + 5x^3y^4y' = -3x^2y^5 - 2x}$; $\boxed{y'\left(6y^5 + 5x^3y^4\right) = -3x^2y^5 - 2x}$; $\boxed{\mathbf{y' = -\frac{3x^2y^5 + 2x}{6y^5 + 5x^3y^4}}}$

Section 5.5 Practice Problems - Implicit Differentiation

Use implicit differentiation method to solve the following functions.

a. $x^2y + x = y$

b. $xy - 3x^2 + y = 0$

c. $x^2y^2 + y = 3y^3$

d. $xy + y^3 = 5x$

e. $4x^4y^4 + 2y^2 = y - 1$

f. $xy + x^2y^2 - 10 = 0$

g. $xy^2 + y = x^2$

h. $xy^3 + x^3y = x$

i. $y^{\frac{1}{2}} + x^2y = x$

j. $x^2y + y^2 = y^{\frac{1}{4}}$

k. $x + y^2 = x^2 - 3$

l. $x^4y^2 + y = -3$

m. $y^7 - x^2y^4 - x = 8$

n. $(x+3)^2 = y^2 - x$

o. $3x^2y^5 + y^2 = -x$

5.6 The Derivative of Functions with Fractional Exponents

The derivative of a function $f(x)$ with fractional exponent is obtained by applying the chain rule in the following way:

$$\frac{d}{dx}[f(x)]^{\frac{a}{b}} = \frac{a}{b}[f(x)]^{\frac{a}{b}-1} \cdot \frac{d}{dx}[f(x)]$$

For example, the derivative of $f(x) = x^{\frac{a}{b}}$ is equal to

$$\frac{d}{dx}x^{\frac{a}{b}} = \frac{a}{b}x^{\frac{a}{b}-1} \cdot \frac{d}{dx}x = \frac{a}{b}x^{\frac{a}{b}-1} \cdot 1 = \frac{a}{b}x^{\frac{a}{b}-1} = \frac{a}{b}x^{\frac{a-b}{b}}$$

Note that the steps in finding the derivative of a function with fractional exponent is similar to finding the derivative of a function that is raised to a power as discussed in Section 5.4. The following examples illustrate how to obtain the derivative of exponential functions:

Example 5.6-1: Find the derivative for the following exponential expressions.

a. $y = x^{\frac{2}{3}}$

b. $y = \left(3x^2\right)^{\frac{1}{3}}$

c. $y = \left(3x^3 + 2x\right)^{\frac{1}{4}}$

d. $y = \left(3x^2 + 6x\right)^{\frac{2}{5}}$

e. $y = (2x+1)^{\frac{3}{4}}$

f. $y = \left(3x^2 + 8\right)^{\frac{2}{7}}$

g. $y = \left(x^2\right)^{\frac{1}{3}} + (2x+1)^{\frac{3}{5}}$

h. $y = x\left(x^2 + 1\right)^{\frac{2}{3}}$

i. $y = (x+1)^{\frac{1}{2}}\left(x^2 + 3\right)^{\frac{1}{3}}$

j. $y = \dfrac{\left(x^2+1\right)^{\frac{1}{2}}}{x}$

k. $y = \dfrac{(x+3)^{\frac{1}{5}}}{x^{\frac{2}{3}}}$

l. $y = \dfrac{x^3}{(x+1)^{\frac{2}{3}}}$

m. $y = (x+1) \cdot \dfrac{1}{x^{\frac{1}{7}}}$

Solutions:

a. Given $\boxed{y = x^{\frac{2}{3}}}$ then $\boxed{y'} = \boxed{\frac{2}{3}x^{\frac{2}{3}-1}} = \boxed{\frac{2}{3}x^{\frac{2}{3}-\frac{1}{1}} \cdot 1} = \boxed{\frac{2}{3}x^{\frac{2-3}{3}}} = \boxed{\frac{2}{3}x^{-\frac{1}{3}}} = \boxed{\frac{2}{3} \cdot \frac{1}{x^{\frac{1}{3}}}} = \boxed{\frac{2}{3} \cdot \frac{1}{\sqrt[3]{x}}} = \boxed{\mathbf{\frac{2}{3\sqrt[3]{x}}}}$

Note that the answer does not necessarily need to be in radical form. We can simply stop when $y' = \frac{2}{3}x^{-\frac{1}{3}}$. However, for review purposes only, the answer to some of the problems are shown in radical form (see Sections 1.1 and 1.2 on the subjects of exponents and radicals).

b. Given $\boxed{y = \left(3x^2\right)^{\frac{1}{3}}}$ then $\boxed{y'} = \boxed{\frac{1}{3}\left(3x^2\right)^{\frac{1}{3}-1} \cdot 6x} = \boxed{\frac{6}{3}x\left(3x^2\right)^{\frac{1}{3}-\frac{1}{1}}} = \boxed{2x\left(3x^2\right)^{\frac{1-3}{3}}} = \boxed{2x\left(3x^2\right)^{-\frac{2}{3}}} = \boxed{2x\frac{1}{\left(3x^2\right)^{\frac{2}{3}}}}$

$= \boxed{2x\frac{1}{\sqrt[3]{\left(3x^2\right)^2}}} = \boxed{2x\frac{1}{\sqrt[3]{9x^4}}} = \boxed{\frac{2\cancel{x}}{\cancel{x}\sqrt[3]{9x}}} = \boxed{\mathbf{\frac{2}{\sqrt[3]{9x}}}}$

c. Given $y=(3x^3+2x)^{\frac{1}{4}}$ then $y'=\frac{1}{4}(3x^3+2x)^{\frac{1}{4}-1}(9x^2+2)=\frac{1}{4}(3x^3+2x)^{\frac{1}{4}-\frac{1}{1}}(9x^2+2)$

$=\frac{1}{4}(3x^3+2x)^{\frac{1-4}{4}}(9x^2+2)=\frac{1}{4}(3x^3+2x)^{-\frac{3}{4}}(9x^2+2)=\frac{9x^2+2}{4(3x^3+2x)^{\frac{3}{4}}}=\mathbf{\frac{9x^2+2}{4\sqrt[4]{(3x^3+2x)^3}}}$

d. Given $y=(3x^2+6x)^{\frac{2}{5}}$ then $y'=\frac{2}{5}(3x^2+6x)^{\frac{2}{5}-1}(6x+6)=\frac{12}{5}(3x^2+6x)^{\frac{2}{5}-\frac{1}{1}}(x+1)=\frac{12}{5}(3x^2+6x)^{\frac{2-5}{5}}(x+1)$

$=\frac{12}{5}(3x^2+6x)^{-\frac{3}{5}}(x+1)=\frac{12}{5}(x+1)\cdot\frac{1}{(3x^2+6x)^{\frac{3}{5}}}=\frac{12}{5}\cdot\frac{x+1}{\sqrt[5]{(3x^2+6x)^3}}=\mathbf{\frac{12(x+1)}{5\sqrt[5]{(3x^2+6x)^3}}}$

e. Given $y=(2x+1)^{\frac{3}{4}}$ then $y'=\frac{3}{4}(2x+1)^{\frac{3}{4}-1}\cdot 2=\frac{6}{4}(2x+1)^{\frac{3}{4}-\frac{1}{1}}=\frac{3}{2}(2x+1)^{\frac{3-4}{4}}=\frac{3}{2}(2x+1)^{-\frac{1}{4}}$

$=\frac{3}{2}\cdot\frac{1}{(2x+1)^{\frac{1}{4}}}=\frac{3}{2}\cdot\frac{1}{\sqrt[4]{2x+1}}=\mathbf{\frac{3}{2\sqrt[4]{2x+1}}}$

f. Given $y=(3x^2+8)^{\frac{2}{7}}$ then $y'=\frac{2}{7}(3x^2+8)^{\frac{2}{7}-1}\cdot 6x=\frac{12x}{7}(3x^2+8)^{\frac{2}{7}-\frac{1}{1}}=\frac{12x}{7}(3x^2+8)^{\frac{2-7}{7}}$

$=\frac{12x}{7}(3x^2+8)^{-\frac{5}{7}}=\frac{12x}{7}\cdot\frac{1}{(3x^2+8)^{\frac{5}{7}}}=\frac{12x}{7}\cdot\frac{1}{\sqrt[7]{(3x^2+8)^5}}=\mathbf{\frac{12x}{7\sqrt[7]{(3x^2+8)^5}}}$

g. Given $y=(x^2)^{\frac{1}{3}}+(2x+1)^{\frac{3}{5}}=x^{\frac{2}{3}}+(2x+1)^{\frac{3}{5}}$ then $y'=\frac{2}{3}x^{\frac{2}{3}-1}+\frac{3}{5}(2x+1)^{\frac{3}{5}-1}\cdot 2=\frac{2}{3}x^{\frac{2}{3}-\frac{1}{1}}+\frac{6}{5}(2x+1)^{\frac{3}{5}-\frac{1}{1}}$

$=\frac{2}{3}x^{\frac{2-3}{3}}+\frac{6}{5}(2x+1)^{\frac{3-5}{5}}=\frac{2}{3}x^{-\frac{1}{3}}+\frac{6}{5}(2x+1)^{-\frac{2}{5}}=\frac{2}{3}\cdot\frac{1}{x^{\frac{1}{3}}}+\frac{6}{5}\cdot\frac{1}{(2x+1)^{\frac{2}{5}}}=\mathbf{\frac{2}{3\sqrt[3]{x}}+\frac{6}{5\sqrt[5]{(2x+1)^2}}}$

h. Given $y=x(x^2+1)^{\frac{2}{3}}$ then $y'=\left[1\cdot(x^2+1)^{\frac{2}{3}}\right]+\left\{\left[\frac{2}{3}(x^2+1)^{\frac{2}{3}-1}\cdot 2x\right]\cdot x\right\}=(x^2+1)^{\frac{2}{3}}+\frac{4x^2}{3}(x^2+1)^{\frac{2}{3}-\frac{1}{1}}$

$=(x^2+1)^{\frac{2}{3}}+\frac{4x^2}{3}(x^2+1)^{\frac{2-3}{3}}=(x^2+1)^{\frac{2}{3}}+\frac{4x^2}{3}(x^2+1)^{-\frac{1}{3}}=(x^2+1)^{\frac{2}{3}}+\frac{4x^2}{3(x^2+1)^{\frac{1}{3}}}=\mathbf{\sqrt[3]{(x^2+1)^2}+\frac{4x^2}{3\sqrt[3]{x^2+1}}}$

i. Given $y=(x+1)^{\frac{1}{2}}(x^2+3)^{\frac{1}{3}}$ then $y'=\left[\frac{1}{2}(x+1)^{\frac{1}{2}-1}\cdot 1\cdot(x^2+3)^{\frac{1}{3}}\right]+\left[\frac{1}{3}(x^2+3)^{\frac{1}{3}-1}\cdot 2x\cdot(x+1)^{\frac{1}{2}}\right]$

$$= \boxed{\left[\frac{1}{2}(x+1)^{-\frac{1}{2}}\cdot\left(x^2+3\right)^{\frac{1}{3}}\right]+\left[\frac{2x}{3}\left(x^2+3\right)^{-\frac{2}{3}}\cdot(x+1)^{\frac{1}{2}}\right]} = \boxed{\left[\frac{1}{2}\cdot\left(x^2+3\right)^{\frac{1}{3}}\cdot\frac{1}{(x+1)^{\frac{1}{2}}}\right]+\left[\frac{2x}{3}\cdot(x+1)^{\frac{1}{2}}\cdot\frac{1}{\left(x^2+3\right)^{\frac{2}{3}}}\right]}$$

$$= \boxed{\frac{\left(x^2+3\right)^{\frac{1}{3}}}{2(x+1)^{\frac{1}{2}}}+\frac{2x(x+1)^{\frac{1}{2}}}{3\left(x^2+3\right)^{\frac{2}{3}}}} = \boxed{\mathbf{\frac{\sqrt[3]{x^2+3}}{2\sqrt{x+1}}+\frac{2x\sqrt{x+1}}{3\sqrt[3]{\left(x^2+3\right)^2}}}}$$

j. Given $\boxed{y=\frac{\left(x^2+1\right)^{\frac{1}{2}}}{x}}$ then $\boxed{y'}$ $= \boxed{\frac{\left\{\left[\frac{1}{2}\left(x^2+1\right)^{\frac{1}{2}-1}\cdot 2x\right]\cdot x\right\}-\left[1\cdot\left(x^2+1\right)^{\frac{1}{2}}\right]}{x^2}} = \boxed{\frac{\frac{2x^2}{2}\left(x^2+1\right)^{-\frac{1}{2}}-\left(x^2+1\right)^{\frac{1}{2}}}{x^2}}$

$$= \boxed{\frac{x^2\frac{1}{\left(x^2+1\right)^{\frac{1}{2}}}-\left(x^2+1\right)^{\frac{1}{2}}}{x^2}} = \boxed{\frac{\frac{x^2}{\sqrt{x^2+1}}-\sqrt{x^2+1}}{x^2}} = \boxed{\frac{\frac{x^2-\left(x^2+1\right)}{\sqrt{x^2+1}}}{x^2}} = \boxed{\frac{\frac{-1}{\sqrt{x^2+1}}}{x^2}} = \boxed{\mathbf{\frac{-1}{x^2\sqrt{x^2+1}}}}$$

k. Given $\boxed{y=\frac{(x+3)^{\frac{1}{5}}}{x^{\frac{2}{3}}}}$ then $\boxed{y'}$ $= \boxed{\frac{\left[\frac{1}{5}(x+3)^{\frac{1}{5}-1}\cdot x^{\frac{2}{3}}\right]-\left[\frac{2}{3}x^{\frac{2}{3}-1}\cdot(x+3)^{\frac{1}{5}}\right]}{x^{\frac{4}{3}}}} = \boxed{\frac{\left[\frac{1}{5}x^{\frac{2}{3}}(x+3)^{-\frac{4}{5}}\right]-\left[\frac{2}{3}x^{-\frac{1}{3}}(x+3)^{\frac{1}{5}}\right]}{x^{\frac{4}{3}}}}$

$$= \boxed{\frac{\frac{1}{5}\cdot\frac{x^{\frac{2}{3}}}{(x+3)^{\frac{4}{5}}}-\frac{2}{3}\cdot\frac{(x+3)^{\frac{1}{5}}}{x^{\frac{1}{3}}}}{x^{\frac{4}{3}}}} = \boxed{\frac{\frac{x^{\frac{2}{3}}}{5(x+3)^{\frac{4}{5}}}-\frac{2(x+3)^{\frac{1}{5}}}{3x^{\frac{1}{3}}}}{x^{\frac{4}{3}}}} = \boxed{\mathbf{\frac{\frac{\sqrt[3]{x^2}}{5\sqrt[5]{(x+3)^4}}-\frac{2\sqrt[5]{x+3}}{3\sqrt[3]{x}}}{\sqrt[3]{x^4}}}}$$

l. Given $\boxed{y=\frac{x^3}{(x+1)^{\frac{2}{3}}}}$ then $\boxed{y'}$ $= \boxed{\frac{\left[3x^{3-1}\cdot(x+1)^{\frac{2}{3}}\right]-\left[\frac{2}{3}(x+1)^{\frac{2}{3}-1}\cdot x^3\right]}{(x+1)^{\frac{4}{3}}}} = \boxed{\frac{\left[3x^2(x+1)^{\frac{2}{3}}\right]-\left[\frac{2x^3}{3}(x+1)^{-\frac{1}{3}}\right]}{(x+1)^{\frac{4}{3}}}}$

$$= \boxed{\frac{3x^2(x+1)^{\frac{2}{3}}-\frac{2x^3}{3(x+1)^{\frac{1}{3}}}}{(x+1)^{\frac{4}{3}}}} = \boxed{\frac{3x^2\sqrt[3]{(x+1)^2}-\frac{2x^3}{3\sqrt[3]{x+1}}}{\sqrt[3]{(x+1)^4}}} = \boxed{\mathbf{\frac{3x^2\sqrt[3]{(x+1)^2}-\frac{2x^3}{3\sqrt[3]{x+1}}}{(x+1)\sqrt[3]{x+1}}}}$$

m. Given $\boxed{y=(x+1)\cdot\frac{1}{x^{\frac{1}{7}}}=\frac{x+1}{x^{\frac{1}{7}}}}$ then $\boxed{y'}$ $= \boxed{\frac{\left[1\cdot x^{\frac{1}{7}}\right]-\left[\frac{1}{7}x^{\frac{1}{7}-1}\cdot(x+1)\right]}{x^{\frac{2}{7}}}} = \boxed{\frac{x^{\frac{1}{7}}-\frac{x+1}{7}x^{-\frac{6}{7}}}{x^{\frac{2}{7}}}} = \boxed{\frac{x^{\frac{1}{7}}-\frac{x+1}{7}\cdot\frac{1}{x^{\frac{6}{7}}}}{x^{\frac{2}{7}}}}$

$$= \boxed{\frac{\sqrt[7]{x}-\frac{x+1}{7}\cdot\frac{1}{\sqrt[7]{x^6}}}{\sqrt[7]{x^2}}} = \boxed{\mathbf{\frac{\sqrt[7]{x}-\frac{x+1}{7\sqrt[7]{x^6}}}{\sqrt[7]{x^2}}}}$$

Example 5.6-2: Find $\frac{d}{dx}$ of the following functions.

a. $\frac{d}{dx}\left(x^{-\frac{2}{3}}\right) =$

b. $\frac{d}{dx}(x+1)^{-\frac{1}{4}} =$

c. $\frac{d}{dx}\left(x^3+1\right)^{\frac{1}{8}} =$

d. $\frac{d}{dx}\left(3x^2+4x\right)^{\frac{7}{8}} =$

e. $\frac{d}{dx}x\left(x^2+1\right)^{-\frac{2}{3}} =$

f. $\frac{d}{dx}\left(3x^3+4\right)^{-\frac{1}{3}} =$

g. $\frac{d}{dx}\left[\frac{(x+1)^{-\frac{1}{2}}}{x}\right] =$

h. $\frac{d}{dx}\left(x^2+3x+5\right)^{\frac{1}{4}} =$

i. $\frac{d}{dx}\left[(x+1)(x-1)^{\frac{5}{4}}\right] =$

j. $\frac{d}{dx}\left(x^2+5\right)^{-\frac{1}{6}} =$

k. $\frac{d}{dx}\left[x^{\frac{3}{5}}\left(x^2+1\right)^{\frac{1}{4}}\right] =$

l. $\frac{d}{dx}\left[\frac{x^2}{(x+1)^{-\frac{1}{3}}}\right] =$

Solutions:

a. $\frac{d}{dx}\left(x^{-\frac{2}{3}}\right) = -\frac{2}{3}x^{-\frac{2}{3}-1}\cdot\frac{d}{dx}(x) = -\frac{2}{3}x^{-\frac{2}{3}-\frac{1}{1}}\cdot 1 = -\frac{2}{3}x^{\frac{-2-3}{3}} = -\frac{2}{3}x^{-\frac{5}{3}} = -\frac{2}{3}\cdot\frac{1}{x^{\frac{5}{3}}} = \frac{-2}{3\sqrt[3]{x^5}} = \mathbf{\frac{-2}{3x\sqrt[3]{x^2}}}$

b. $\frac{d}{dx}(x+1)^{-\frac{1}{4}} = -\frac{1}{4}(x+1)^{-\frac{1}{4}-1}\cdot\frac{d}{dx}(x+1) = -\frac{1}{4}(x+1)^{-\frac{1}{4}-\frac{1}{1}}\cdot 1 = -\frac{1}{4}(x+1)^{\frac{-1-4}{4}} = -\frac{1}{4}(x+1)^{-\frac{5}{4}}$

$= -\frac{1}{4}\cdot\frac{1}{(x+1)^{\frac{5}{4}}} = -\frac{1}{4\sqrt[4]{(x+1)^5}} = \mathbf{-\frac{1}{4(x+1)\sqrt[4]{x+1}}}$

c. $\frac{d}{dx}\left(x^3+1\right)^{\frac{1}{8}} = \frac{1}{8}\left(x^3+1\right)^{\frac{1}{8}-1}\cdot\frac{d}{dx}\left(x^3+1\right) = \frac{1}{8}\left(x^3+1\right)^{\frac{1}{8}-\frac{1}{1}}\cdot 3x^2 = \frac{3}{8}x^2\left(x^3+1\right)^{\frac{1-8}{8}} = \frac{3}{8}x^2\left(x^3+1\right)^{-\frac{7}{8}}$

$= \frac{3}{8}x^2\cdot\frac{1}{\left(x^3+1\right)^{\frac{7}{8}}} = \frac{3x^2}{8\left(x^3+1\right)^{\frac{7}{8}}} = \mathbf{\frac{3x^2}{8\sqrt[8]{\left(x^3+1\right)^7}}}$

d. $\frac{d}{dx}\left(3x^2+4x\right)^{\frac{7}{8}} = \frac{7}{8}\left(3x^2+4x\right)^{\frac{7}{8}-1}\cdot\frac{d}{dx}\left(3x^2+4x\right) = \frac{7}{8}\left(3x^2+4x\right)^{\frac{7}{8}-\frac{1}{1}}\cdot(6x+4) = \frac{7}{8}\left(3x^2+4x\right)^{\frac{7-8}{8}}\cdot(6x+4)$

$= \frac{7}{8}\left(3x^2+4x\right)^{-\frac{1}{8}}\cdot(6x+4) = \frac{7}{8}\cdot\frac{6x+4}{\left(3x^2+4x\right)^{\frac{1}{8}}} = \mathbf{\frac{7(6x+4)}{8\sqrt[8]{3x^2+4x}}}$

e. $\frac{d}{dx}x\left(x^2+1\right)^{-\frac{2}{3}} = \left[\left(x^2+1\right)^{-\frac{2}{3}}\cdot\frac{d}{dx}x\right]+\left[x\cdot\frac{d}{dx}\left(x^2+1\right)^{-\frac{2}{3}}\right] = \left[\left(x^2+1\right)^{-\frac{2}{3}}\cdot 1\right]+\left[-x\cdot\frac{2}{3}\left(x^2+1\right)^{-\frac{2}{3}-1}\cdot 2x\right]$

$= \left(x^2+1\right)^{-\frac{2}{3}}-\frac{4x^2}{3}\left(x^2+1\right)^{-\frac{2}{3}-\frac{1}{1}} = \left(x^2+1\right)^{-\frac{2}{3}}-\frac{4x^2}{3}\left(x^2+1\right)^{\frac{-2-3}{3}} = \left(x^2+1\right)^{-\frac{2}{3}}-\frac{4x^2}{3}\left(x^2+1\right)^{-\frac{5}{3}}$

$$= \boxed{\frac{1}{(x^2+1)^{\frac{2}{3}}} - \frac{4x^2}{3} \cdot \frac{1}{(x^2+1)^{\frac{5}{3}}}} = \boxed{\frac{1}{\sqrt[3]{(x^2+1)^2}} - \frac{4x^2}{3\sqrt[3]{(x^2+1)^5}}} = \boxed{\mathbf{\frac{1}{\sqrt[3]{(x^2+1)^2}} - \frac{4x^2}{3(x^2+1)\sqrt[3]{(x^2+1)^2}}}}$$

f. $$\boxed{\frac{d}{dx}(3x^3+4)^{-\frac{1}{3}}} = \boxed{-\frac{1}{3}(3x^3+4)^{-\frac{1}{3}-1} \cdot \frac{d}{dx}(3x^3+4)} = \boxed{-\frac{1}{3}(3x^3+4)^{-\frac{1}{3}-\frac{1}{1}} \cdot 9x^2} = \boxed{-\frac{9x^2}{3}(3x^3+4)^{\frac{-1-3}{3}}}$$

$$= \boxed{-3x^2(3x^3+4)^{-\frac{4}{3}}} = \boxed{-3x^2 \cdot \frac{1}{(3x^3+4)^{\frac{4}{3}}}} = \boxed{-\frac{3x^2}{\sqrt[3]{(3x^3+4)^4}}} = \boxed{\mathbf{-\frac{3x^2}{(3x^3+4)\sqrt[3]{3x^3+4}}}}$$

g. $$\boxed{\frac{d}{dx}\left[\frac{(x+1)^{-\frac{1}{2}}}{x}\right]} = \boxed{\frac{\left[x \cdot \frac{d}{dx}(x+1)^{-\frac{1}{2}}\right] - \left[(x+1)^{-\frac{1}{2}} \cdot \frac{d}{dx}(x)\right]}{x^2}} = \boxed{\frac{\left[x \cdot -\frac{1}{2}(x+1)^{-\frac{1}{2}-1} \cdot \frac{d}{dx}(x+1)\right] - \left[(x+1)^{-\frac{1}{2}} \cdot 1\right]}{x^2}}$$

$$= \boxed{\frac{\left[-\frac{x}{2}(x+1)^{-\frac{1}{2}-\frac{1}{1}} \cdot 1\right] - (x+1)^{-\frac{1}{2}}}{x^2}} = \boxed{\frac{-\frac{x}{2}(x+1)^{\frac{-1-2}{2}} - (x+1)^{-\frac{1}{2}}}{x^2}} = \boxed{\mathbf{\frac{-\frac{x}{2}(x+1)^{-\frac{3}{2}} - (x+1)^{-\frac{1}{2}}}{x^2}}}$$

h. $$\boxed{\frac{d}{dx}(x^2+3x+5)^{\frac{1}{4}}} = \boxed{\frac{1}{4}(x^2+3x+5)^{\frac{1}{4}-1} \cdot \frac{d}{dx}(x^2+3x+5)} = \boxed{\frac{1}{4}(x^2+3x+5)^{\frac{1}{4}-\frac{1}{1}} \cdot (2x+3)} = \boxed{\frac{1}{4}(x^2+3x+5)^{\frac{1-4}{4}} \cdot (2x+3)}$$

$$= \boxed{\frac{1}{4}(x^2+3x+5)^{-\frac{3}{4}}(2x+3)} = \boxed{\frac{2x+3}{4} \cdot \frac{1}{(x^2+3x+5)^{\frac{3}{4}}}} = \boxed{\frac{2x+3}{4} \cdot \frac{1}{\sqrt[4]{(x^2+3x+5)^3}}} = \boxed{\mathbf{\frac{2x+3}{4\sqrt[4]{(x^2+3x+5)^3}}}}$$

i. $$\boxed{\frac{d}{dx}\left[(x+1)(x-1)^{\frac{5}{4}}\right]} = \boxed{\left[(x-1)^{\frac{5}{4}} \frac{d}{dx}(x+1)\right] + \left[(x+1)\frac{d}{dx}(x-1)^{\frac{5}{4}}\right]} = \boxed{\left[(x-1)^{\frac{5}{4}} \cdot 1\right] + \left[(x+1) \cdot \frac{5}{4}(x-1)^{\frac{5}{4}-1} \cdot \frac{d}{dx}(x-1)\right]}$$

$$= \boxed{(x-1)^{\frac{5}{4}} + \left[(x+1) \cdot \frac{5}{4}(x-1)^{\frac{5}{4}-\frac{1}{1}} \cdot 1\right]} = \boxed{(x-1)^{\frac{5}{4}} + \left[(x+1) \cdot \frac{5}{4}(x-1)^{\frac{5-4}{4}}\right]} = \boxed{\mathbf{(x-1)^{\frac{5}{4}} + \frac{5}{4}(x+1)(x-1)^{\frac{1}{4}}}}$$

j. $$\boxed{\frac{d}{dx}(x^2+5)^{-\frac{1}{6}}} = \boxed{-\frac{1}{6}(x^2+5)^{-\frac{1}{6}-1} \cdot \frac{d}{dx}(x^2+5)} = \boxed{-\frac{1}{6}(x^2+5)^{-\frac{1}{6}-\frac{1}{1}} \cdot 2x} = \boxed{-\frac{2x}{6}(x^2+5)^{\frac{-1-6}{6}}} = \boxed{-\frac{x}{3}(x^2+5)^{-\frac{7}{6}}}$$

$$= \boxed{-\frac{x}{3} \cdot \frac{1}{(x^2+5)^{\frac{7}{6}}}} = \boxed{-\frac{x}{3} \cdot \frac{1}{\sqrt[6]{(x^2+5)^7}}} = \boxed{\mathbf{-\frac{x}{3(x^2+5)\sqrt[6]{x^2+5}}}}$$

k. $$\boxed{\frac{d}{dx}\left[x^{\frac{3}{5}}(x^2+1)^{\frac{1}{4}}\right]} = \boxed{\left[(x^2+1)^{\frac{1}{4}} \frac{d}{dx}x^{\frac{3}{5}}\right] + \left[x^{\frac{3}{5}} \frac{d}{dx}(x^2+1)^{\frac{1}{4}}\right]} = \boxed{\left[(x^2+1)^{\frac{1}{4}} \cdot \frac{3}{5}x^{\frac{3}{5}-1}\right] + \left[x^{\frac{3}{5}} \frac{1}{4}(x^2+1)^{\frac{1}{4}-1} \cdot \frac{d}{dx}(x^2+1)\right]}$$

$$= \boxed{\left[\frac{3}{5}(x^2+1)^{\frac{1}{4}} \cdot x^{\frac{3}{5}-\frac{1}{1}}\right] + \left[x^{\frac{3}{5}} \frac{1}{4}(x^2+1)^{\frac{1}{4}-\frac{1}{1}} \cdot 2x\right]} = \boxed{\left[\frac{3}{5}(x^2+1)^{\frac{1}{4}} \cdot x^{\frac{3-5}{5}}\right] + \left[\frac{2x \cdot x^{\frac{3}{5}}}{4} \cdot (x^2+1)^{\frac{1-4}{4}}\right]}$$

$$= \boxed{\left[\frac{3x^{-\frac{2}{5}}}{5}\left(x^2+1\right)^{\frac{1}{4}}\right]+\left[\frac{x^{\frac{8}{5}}}{2}\cdot\left(x^2+1\right)^{-\frac{3}{4}}\right]} = \boxed{\mathbf{\frac{3\sqrt[4]{x^2+1}}{5\sqrt[5]{x^2}}+\frac{x\sqrt[5]{x^3}}{2\sqrt[4]{\left(x^2+1\right)^3}}}}$$

1. $$\boxed{\frac{d}{dx}\left[\frac{x^2}{(x+1)^{-\frac{1}{3}}}\right]} = \boxed{\frac{d}{dx}\left[x^2(x+1)^{\frac{1}{3}}\right]} = \boxed{\left[(x+1)^{\frac{1}{3}}\frac{d}{dx}x^2\right]+\left[x^2\frac{d}{dx}(x+1)^{\frac{1}{3}}\right]}$$

$$= \boxed{\left[(x+1)^{\frac{1}{3}}\cdot 2x\right]+\left[x^2\cdot\frac{1}{3}(x+1)^{\frac{1}{3}-1}\cdot\frac{d}{dx}(x+1)\right]} = \boxed{2x(x+1)^{\frac{1}{3}}+\left[x^2\cdot\frac{1}{3}(x+1)^{\frac{1}{3}-\frac{1}{1}}\cdot 1\right]} = \boxed{\mathbf{2x(x+1)^{\frac{1}{3}}+\frac{x^2}{3}(x+1)^{-\frac{2}{3}}}}$$

Example 5.6-3: Find $y'(1)$ in example 5.6-1 for the given functions a through e.

Solutions:

In Example 5.6-1 we found the derivative of the exponential functions a through e to be equal to the following:

a. $y'=\frac{2}{3\sqrt[3]{x}}$ therefore $\boxed{y'(1)} = \boxed{\frac{2}{3\sqrt[3]{1}}} = \boxed{\frac{2}{3}} = \boxed{\mathbf{0.67}}$

b. $y'=\frac{2}{\sqrt[3]{9x}}$ therefore $\boxed{y'(1)} = \boxed{\frac{2}{\sqrt[3]{9\cdot 1}}} = \boxed{\frac{2}{\sqrt[3]{9}}} = \boxed{\frac{2}{9^{0.33}}} = \boxed{\frac{2}{2.08}} = \boxed{\mathbf{0.962}}$

c. $y'=\frac{9x^2+2}{4\sqrt[4]{\left(3x^3+2x\right)^3}}$ therefore $\boxed{y'(1)} = \boxed{\frac{9\cdot 1^2+2}{4\sqrt[4]{\left(3\cdot 1^3+2\cdot 1\right)^3}}} = \boxed{\frac{11}{4\sqrt[4]{5^3}}} = \boxed{\frac{11}{4\cdot 125^{0.25}}} = \boxed{\frac{11}{4\cdot 3.343}} = \boxed{\mathbf{0.823}}$

d. $y'=\frac{12(x+1)}{5\sqrt[5]{\left(3x^2+6x\right)^3}}$ therefore $\boxed{y'(1)} = \boxed{\frac{12(1+1)}{5\sqrt[5]{\left(3\cdot 1^2+6\cdot 1\right)^3}}} = \boxed{\frac{24}{5\sqrt[5]{729}}} = \boxed{\frac{24}{5\cdot 729^{0.2}}} = \boxed{\frac{24}{5\cdot 3.74}} = \boxed{\mathbf{1.28}}$

e. $y'=\frac{3}{2\sqrt[4]{2x+1}}$ therefore $\boxed{y'(1)} = \boxed{\frac{3}{2\sqrt[4]{2\cdot 1+1}}} = \boxed{\frac{3}{2\sqrt[4]{2+1}}} = \boxed{\frac{3}{2\sqrt[4]{3}}} = \boxed{\frac{3}{2\cdot 3^{0.25}}} = \boxed{\frac{3}{2\cdot 1.32}} = \boxed{\mathbf{1.14}}$

Section 5.6 Practice Problems - The Derivative of Functions with Fractional Exponents

1. Find the derivative of the following exponential expressions.

a. $y=x^{\frac{1}{5}}$

b. $y=\left(4x^3\right)^{\frac{1}{2}}$

c. $y=(2x+1)^{\frac{1}{3}}$

d. $y=\left(2x^2+1\right)^{\frac{1}{8}}$

e. $y=\left(2x^3+3x\right)^{\frac{3}{5}}$

f. $y=\left(x^3+8\right)^{\frac{2}{3}}$

g. $y=\left(x^3\right)^{\frac{1}{2}}-(3x-1)^{\frac{1}{3}}$

h. $y=x^2(x+1)^{\frac{1}{8}}$

i. $y=\left(x^3+1\right)^{\frac{2}{5}}+x^{\frac{1}{2}}$

j. $y=\frac{x+1}{x^{\frac{2}{3}}}$

k. $y=\frac{\left(x^2+1\right)^{\frac{1}{2}}}{x^2}$

l. $y=\frac{(x+1)^2}{x^{\frac{1}{3}}}$

2. Use the $\frac{d}{dx}$ notation to find the derivative of the following exponential expressions.

a. $\frac{d}{dx}\left(x^{\frac{1}{5}}\right)^2 =$ b. $\frac{d}{dx}(x-1)^{\frac{1}{2}} =$ c. $\frac{d}{dx}\left(x^2+1\right)^{\frac{1}{3}} =$

d. $\frac{d}{dx}\left(x^3+1\right)^{-\frac{1}{4}} =$ e. $\frac{d}{dx}\left[\frac{(x-1)^{\frac{1}{2}}}{x^2}\right] =$ f. $\frac{d}{dx}\left(x^3+2x\right)^{\frac{1}{8}} =$

g. $\frac{d}{dx}\left[\left(x^3+1\right)\left(x^2\right)^{\frac{1}{3}}\right] =$ h. $\frac{d}{dx}\left[x^3 \cdot \frac{1}{\left(x^2+1\right)^{\frac{1}{2}}}\right] =$ i. $\frac{d}{dx}\left[\frac{x^5}{\left(x^3+1\right)^{\frac{2}{3}}}\right] =$

j. $\frac{d}{dx}\left[(x-1)^{\frac{1}{2}}(x+1)^{\frac{1}{3}}\right] =$ k. $\frac{d}{dx}\left[x^3\left(x^2+1\right)^{\frac{1}{2}}\right]$ l. $\frac{d}{dx}\left[x^3\left(x^2+1\right)^{-\frac{1}{3}}\right] =$

5.7 The Derivative of Radical Functions

In this section finding the derivative of radical expressions and the steps as to how it is calculated is discussed. The derivative of radical functions is found by using the following steps:

First - Write the radical expression in its equivalent fraction form, i.e., write $\sqrt{x^3}$ as $x^{\frac{3}{2}}$.

Second - Apply the differentiation rules to find the derivative of the exponential expression.

Third - Change the answer from an expression with fractional exponent to an expression with radical expression (optional).

The following examples show the steps in solving functions containing radical terms. Students who have difficulty with simplifying radical expressions may want to review *radicals* addressed in Chapter 4 of the "*Mastering Algebra - An Introduction*".

Example 5.7-1: Find the derivative for the following Radical expressions.

a. $f(x)=\sqrt{x^3+1}$
b. $f(x)=\sqrt{x^2+3x+1}$
c. $f(x)=\sqrt{2x^5+1}$
d. $f(u)=\sqrt[5]{u^3}+3u$
e. $f(t)=t^2+\sqrt{t+1}$
f. $g(x)=x^2\sqrt{x^3+x-5}$
g. $h(w)=\sqrt[3]{w^2+1}$
h. $f(z)=\sqrt[4]{z^3-z^2}+z$
i. $f(x)=\dfrac{1}{\sqrt{x^2+1}}$
j. $f(x)=\dfrac{x}{\sqrt{x^2-1}}$
k. $r(\theta)=\dfrac{\theta^3+1}{\sqrt{\theta^2+1}}$
l. $p(r)=\dfrac{r^3}{\sqrt{r^3-1}}$
m. $g(u)=\dfrac{\sqrt{u-1}}{\sqrt{u+1}}$
n. $h(t)=\dfrac{\sqrt[3]{t^2+1}}{\sqrt{t^3}}$
o. $s(r)=\dfrac{r^2-1}{\sqrt{r-1}}$

Solutions:

a. Given $f(x) = \sqrt{x^3+1} = (x^3+1)^{\frac{1}{2}}$, then

$$\boxed{f'(x)} = \boxed{\frac{1}{2}(x^3+1)^{\frac{1}{2}-1}\cdot 3x^{3-1}} = \boxed{\frac{1}{2}(x^3+1)^{-\frac{1}{2}}\cdot 3x^2} = \boxed{\frac{3}{2}x^2(x^3+1)^{-\frac{1}{2}}} = \boxed{\frac{3}{2}\frac{x^2}{(x^3+1)^{\frac{1}{2}}}} = \boxed{\mathbf{\frac{3}{2}\frac{x^2}{\sqrt{x^3+1}}}}$$

b. Given $f(x) = \sqrt{x^2+3x+1} = (x^2+3x+1)^{\frac{1}{2}}$, then

$$\boxed{f'(x)} = \boxed{\frac{1}{2}(x^2+3x+1)^{\frac{1}{2}-1}\cdot(2x^{2-1}+3x^{1-1})} = \boxed{\frac{1}{2}(x^2+3x+1)^{-\frac{1}{2}}\cdot(2x+3)} = \boxed{\frac{2x+3}{2(x^2+3x+1)^{\frac{1}{2}}}} = \boxed{\mathbf{\frac{2x+3}{2\sqrt{x^2+3x+1}}}}$$

c. Given $f(x) = \sqrt{2x^5+1} = (2x^5+1)^{\frac{1}{2}}$, then

$$\boxed{f'(x)} = \boxed{\frac{1}{2}(2x^5+1)^{\frac{1}{2}-1}\cdot(2\times 5x^{5-1}+0)} = \boxed{\frac{1}{2}(2x^5+1)^{-\frac{1}{2}}\cdot 10x^4} = \boxed{\frac{\overset{5}{\cancel{10}}x^4}{\cancel{2}(2x^5+1)^{\frac{1}{2}}}} = \boxed{\mathbf{\frac{5x^4}{\sqrt{2x^5+1}}}}$$

d. Given $f(u) = \sqrt[5]{u^3}+3u = u^{\frac{3}{5}}+3u$, then

$$\boxed{f'(u)} = \boxed{\frac{3}{5}u^{\frac{3}{5}-1}+3u^{1-1}} = \boxed{\frac{3}{5}u^{-\frac{2}{5}}+3u^0} = \boxed{\frac{3}{5}u^{-\frac{2}{5}}+3} = \boxed{\frac{3}{5u^{\frac{2}{5}}}+3} = \boxed{\mathbf{\frac{3}{5\sqrt[5]{u^2}}+3}}$$

e. Given $f(t) = t^2+\sqrt{t+1} = t^2+(t+1)^{\frac{1}{2}}$, then

$$\boxed{f'(t)} = \boxed{2t^{2-1}+\frac{1}{2}(t+1)^{\frac{1}{2}-1}\cdot t^{1-1}} = \boxed{2t+\frac{1}{2}(t+1)^{-\frac{1}{2}}\cdot t^{0}} = \boxed{2t+\frac{1}{2(t+1)^{\frac{1}{2}}}} = \boxed{\mathbf{2t+\frac{1}{2\sqrt{t+1}}}}$$

f. Given $g(x) = x^2\sqrt{x^3+x-5} = x^2\left(x^3+x-5\right)^{\frac{1}{2}}$, then

$$\boxed{g'(x)} = \boxed{\left[2x^{2-1}\left(x^3+x-5\right)^{\frac{1}{2}}\right]+\left[\frac{1}{2}\left(x^3+x-5\right)^{\frac{1}{2}-1}\cdot\left(3x^{3-1}+x^{1-1}+0\right)\right]x^2} = \boxed{2x\left(x^3+x-5\right)^{\frac{1}{2}}+\frac{1}{2}\left(x^3+x-5\right)^{-\frac{1}{2}}}$$

$$\cdot\boxed{\left(3x^2+1\right)x^2} = \boxed{2x\left(x^3+x-5\right)^{\frac{1}{2}}+\frac{\left(3x^2+1\right)x^2}{2\left(x^3+x-5\right)^{\frac{1}{2}}}} = \boxed{\mathbf{2x\sqrt{x^3+x-5}+\frac{x^2\left(3x^2+1\right)}{2\sqrt{x^3+x-5}}}}$$

g. Given $h(w) = \sqrt[3]{w^2+1} = \left(w^2+1\right)^{\frac{1}{3}}$, then

$$\boxed{h'(w)} = \boxed{\frac{1}{3}\left(w^2+1\right)^{\frac{1}{3}-1}\left(2w^{2-1}+0\right)} = \boxed{\frac{1}{3}\left(w^2+1\right)^{-\frac{2}{3}}\cdot 2w} = \boxed{\frac{2w}{3\left(w^2+1\right)^{\frac{2}{3}}}} = \boxed{\mathbf{\frac{2w}{3\sqrt[3]{\left(w^2+1\right)^2}}}}$$

h. Given $f(z) = \sqrt[4]{z^3-z^2}+z = \left(z^3-z^2\right)^{\frac{1}{4}}+z$, then

$$\boxed{f'(z)} = \boxed{\frac{1}{4}\left(z^3-z^2\right)^{\frac{1}{4}-1}\left(3z^{3-1}-2z^{2-1}\right)+1} = \boxed{\frac{\left(z^3-z^2\right)^{-\frac{3}{4}}\left(3z^2-2z\right)}{4}+1} = \boxed{\frac{3\left(3z^2-2z\right)}{4\left(z^3-z^2\right)^{\frac{3}{4}}}+1} = \boxed{\mathbf{\frac{3z^2-2z}{4\sqrt[4]{\left(z^3-z^2\right)^3}}+1}}$$

i. Given $f(x) = \frac{1}{\sqrt{x^2+1}} = \frac{1}{\left(x^2+1\right)^{\frac{1}{2}}} = \left(x^2+1\right)^{-\frac{1}{2}}$, then

$$\boxed{f'(x)} = \boxed{-\frac{1}{2}\left(x^2+1\right)^{-\frac{1}{2}-1}\left(2x^{2-1}+0\right)} = \boxed{-\frac{1}{2}\left(x^2+1\right)^{-\frac{3}{2}}\cdot 2x} = \boxed{\frac{-x}{\left(x^2+1\right)^{\frac{3}{2}}}} = \boxed{\frac{-x}{\sqrt[2]{\left(x^2+1\right)^3}}} = \boxed{\mathbf{\frac{-x}{\left(x^2+1\right)\sqrt{x^2+1}}}}$$

j. Given $f(x) = \frac{x}{\sqrt{x^2-1}} = \frac{x}{\left(x^2-1\right)^{\frac{1}{2}}}$, then

$$\boxed{f'(x)} = \boxed{\frac{\left[1\cdot\left(x^2-1\right)^{\frac{1}{2}}\right]-\left[\frac{1}{2}\left(x^2-1\right)^{\frac{1}{2}-1}\left(2x^{2-1}-0\right)\cdot x\right]}{x^2-1}} = \boxed{\frac{\left(x^2-1\right)^{\frac{1}{2}}-\left[\frac{1}{2}\left(x^2-1\right)^{-\frac{1}{2}}\cdot 2x^2\right]}{x^2-1}} = \boxed{\frac{\left(x^2-1\right)^{\frac{1}{2}}-\frac{x^2}{\left(x^2-1\right)^{\frac{1}{2}}}}{x^2-1}}$$

$$= \boxed{\frac{\frac{\left(x^2-1\right)^{\frac{1}{2}}}{1}-\frac{x^2}{\left(x^2-1\right)^{\frac{1}{2}}}}{x^2-1}} = \boxed{\frac{\frac{\left(x^2-1\right)^{\frac{1}{2}}\cdot\left(x^2-1\right)^{\frac{1}{2}}-x^2}{\left(x^2-1\right)^{\frac{1}{2}}}}{x^2-1}} = \boxed{\frac{\frac{\left(x^2-1\right)^{\frac{1}{2}+\frac{1}{2}}-x^2}{\left(x^2-1\right)^{\frac{1}{2}}}}{x^2-1}} = \boxed{\frac{\frac{x^2-1-x^2}{\left(x^2-1\right)^{\frac{1}{2}}}}{x^2-1}}$$

$$= \boxed{-\frac{1}{\left(x^2-1\right)^{\frac{1}{2}}\left(x^2-1\right)}} = \boxed{-\frac{1}{\left(x^2-1\right)^{\frac{1}{2}+1}}} = \boxed{-\frac{1}{\left(x^2-1\right)^{\frac{3}{2}}}} = \boxed{-\frac{1}{\sqrt{\left(x^2-1\right)^3}}} = \boxed{\mathbf{-\frac{1}{\left(x^2-1\right)\sqrt{x^2-1}}}}$$

k. Given $r(\theta) = \dfrac{\theta^3+1}{\sqrt{\theta^2+1}} = \dfrac{\theta^3+1}{\left(\theta^2+1\right)^{\frac{1}{2}}}$, then

$$\boxed{r'(\theta)} = \boxed{\frac{\left[\left(3\theta^{3-1}+0\right)\left(\theta^2+1\right)^{\frac{1}{2}}\right]-\left[\frac{1}{2}\left(\theta^2+1\right)^{\frac{1}{2}-1}\left(2\theta^{2-1}+0\right)\left(\theta^3+1\right)\right]}{\theta^2+1}} = \boxed{\frac{\left[3\theta^2\left(\theta^2+1\right)^{\frac{1}{2}}\right]-\frac{1}{2}\left(\theta^2+1\right)^{-\frac{1}{2}}}{\theta^2+1}}$$

$$\cdot\boxed{\frac{2\theta\left(\theta^3+1\right)}{\theta^2+1}} = \boxed{\frac{3\theta^2\left(\theta^2+1\right)^{\frac{1}{2}}-\frac{2\theta\left(\theta^3+1\right)}{2\left(\theta^2+1\right)^{\frac{1}{2}}}}{\theta^2+1}} = \boxed{\frac{\frac{6\theta^2\left(\theta^2+1\right)-2\theta^4-2\theta}{2\left(\theta^2+1\right)^{\frac{1}{2}}}}{\theta^2+1}} = \boxed{\frac{6\theta^4+6\theta^2-2\theta^4-2\theta}{2\left(\theta^2+1\right)^{\frac{1}{2}}\left(\theta^2+1\right)}}$$

$$= \boxed{\frac{4\theta^4+6\theta^2-2\theta}{2\left(\theta^2+1\right)^{\frac{1}{2}+1}}} = \boxed{\frac{2\theta\left(2\theta^3+3\theta-1\right)}{2\left(\theta^2+1\right)^{\frac{3}{2}}}} = \boxed{\frac{\theta\left(2\theta^3+3\theta-1\right)}{\sqrt{\left(\theta^2+1\right)^3}}} = \boxed{\mathbf{\frac{\theta\left(2\theta^3+3\theta-1\right)}{\left(\theta^2+1\right)\sqrt{\theta^2+1}}}}$$

l. Given $p(r) = \dfrac{r^3}{\sqrt{r^3-1}} = \dfrac{r^3}{\left(r^3-1\right)^{\frac{1}{2}}}$, then

$$\boxed{p'(r)} = \boxed{\frac{\left[3r^{3-1}\left(r^3-1\right)^{\frac{1}{2}}\right]-\left[\frac{1}{2}\left(r^3-1\right)^{\frac{1}{2}-1}\left(3r^{3-1}-0\right)\right]}{r^3-1}} = \boxed{\frac{\left[3r^2\left(r^3-1\right)^{\frac{1}{2}}\right]-\left[\frac{1}{2}\left(r^3-1\right)^{-\frac{1}{2}}3r^2\right]\cdot r^3}{r^3-1}}$$

$$= \boxed{\frac{3r^2\left(r^3-1\right)^{\frac{1}{2}}-\frac{3r^5}{2\left(r^3-1\right)^{\frac{1}{2}}}}{r^3-1}} = \boxed{\frac{3r^2\left(r^3-1\right)^{\frac{1}{2}}-\frac{3r^5}{2\left(r^3-1\right)^{\frac{1}{2}}}}{r^3-1}} = \boxed{\frac{\frac{6r^2\left(r^3-1\right)-3r^5}{2\left(r^3-1\right)^{\frac{1}{2}}}}{r^3-1}} = \boxed{\frac{\frac{6r^5-6r^2-3r^5}{2\left(r^3-1\right)^{\frac{1}{2}}}}{r^3-1}}$$

$$= \boxed{\frac{3r^5-6r^2}{2\left(r^3-1\right)\left(r^3-1\right)^{\frac{1}{2}}}} = \boxed{\frac{3r^2\left(r^3-2\right)}{2\left(r^3-1\right)^{\frac{1}{2}+1}}} = \boxed{\frac{3r^2\left(r^3-2\right)}{2\left(r^3-1\right)^{\frac{3}{2}}}} = \boxed{\frac{3r^2\left(r^3-2\right)}{2\sqrt{\left(r^3-1\right)^3}}} = \boxed{\mathbf{\frac{3r^2\left(r^3-2\right)}{2\left(r^3-1\right)\sqrt{r^3-1}}}}$$

m. Given $g(u) = \dfrac{\sqrt{u-1}}{\sqrt{u+1}} = \dfrac{(u-1)^{\frac{1}{2}}}{(u+1)^{\frac{1}{2}}}$, then

$$\boxed{g'(u)} = \boxed{\frac{\left[\frac{1}{2}(u-1)^{\frac{1}{2}-1}(u+1)^{\frac{1}{2}}\right]-\left[\frac{1}{2}(u+1)^{\frac{1}{2}-1}(u-1)^{\frac{1}{2}}\right]}{u+1}} = \boxed{\frac{\left[\frac{1}{2}(u-1)^{-\frac{1}{2}}(u+1)^{\frac{1}{2}}\right]-\left[\frac{1}{2}(u+1)^{-\frac{1}{2}}(u-1)^{\frac{1}{2}}\right]}{u+1}}$$

$$= \boxed{\frac{\left[\frac{(u+1)^{\frac{1}{2}}}{2(u-1)^{\frac{1}{2}}}\right]-\left[\frac{(u-1)^{\frac{1}{2}}}{2(u+1)^{\frac{1}{2}}}\right]}{u+1}} = \boxed{\frac{\frac{2(u+1)-2(u-1)}{4(u-1)^{\frac{1}{2}}(u+1)^{\frac{1}{2}}}}{u+1}} = \boxed{\frac{\frac{2u+2-2u+2}{4(u-1)^{\frac{1}{2}}(u+1)^{\frac{1}{2}}}}{u+1}} = \boxed{\frac{\frac{4}{4(u-1)^{\frac{1}{2}}(u+1)^{\frac{1}{2}}}}{u+1}}$$

$$= \boxed{\frac{4}{4(u+1)(u-1)^{\frac{1}{2}}(u+1)^{\frac{1}{2}}}} = \boxed{\frac{1}{(u+1)(u+1)^{\frac{1}{2}}(u-1)^{\frac{1}{2}}}} = \boxed{\mathbf{\frac{1}{(u+1)\sqrt{u+1}\,\sqrt{u-1}}}}$$

n. Given $h(t) = \frac{\sqrt[3]{t^2+1}}{\sqrt{t^3}} = \frac{(t^2+1)^{\frac{1}{3}}}{t^{\frac{3}{2}}}$, then

$$\boxed{h'(t)} = \boxed{\frac{\left[\frac{1}{3}(t^2+1)^{\frac{1}{3}-1}\cdot(2t^{2-1}+0)\right]t^{\frac{3}{2}} - \left[\frac{3}{2}t^{\frac{3}{2}-1}\cdot(t^2+1)^{\frac{1}{3}}\right]}{t^3}} = \boxed{\frac{\left[\frac{1}{3}(t^2+1)^{-\frac{2}{3}}\cdot 2t\right]t^{\frac{3}{2}} - \left[\frac{3}{2}t^{\frac{1}{2}}(t^2+1)^{\frac{1}{3}}\right]}{t^3}}$$

$$= \boxed{\frac{\left[\frac{2}{3}t^{1+\frac{3}{2}}(t^2+1)^{-\frac{2}{3}}\right] - \left[\frac{3}{2}t^{\frac{1}{2}}(t^2+1)^{\frac{1}{3}}\right]}{t^3}} = \boxed{\frac{\left[\frac{2t^{\frac{5}{2}}}{3(t^2+1)^{\frac{2}{3}}}\right] - \left[\frac{3}{2}t^{\frac{1}{2}}(t^2+1)^{\frac{1}{3}}\right]}{t^3}} = \boxed{\frac{\frac{2t^{\frac{5}{2}}}{3(t^2+1)^{\frac{2}{3}}} - \frac{\frac{3}{2}t^{\frac{1}{2}}(t^2+1)^{\frac{1}{3}}}{1}}{t^3}}$$

$$= \boxed{\frac{\frac{2t^{\frac{5}{2}} - \frac{3}{2}t^{\frac{1}{2}}(t^2+1)^{\frac{1}{3}}\cdot 3(t^2+1)^{\frac{2}{3}}}{3(t^2+1)^{\frac{2}{3}}}}{t^3}} = \boxed{\frac{2t^{\frac{5}{2}} - \frac{9}{2}t^{\frac{1}{2}}(t^2+1)^{\frac{1}{3}+\frac{2}{3}}}{3t^3(t^2+1)^{\frac{2}{3}}}} = \boxed{\mathbf{\frac{2t^2\sqrt{t} - \frac{9\sqrt{t}(t^2+1)}{2}}{3t^3\sqrt[3]{(t^2+1)^2}}}}$$

o. Given $s(r) = \frac{r^2-1}{\sqrt{r-1}} = \frac{r^2-1}{(r-1)^{\frac{1}{2}}}$, then

$$\boxed{s'(r)} = \boxed{\frac{\left[(2r^{2-1}-0)(r-1)^{\frac{1}{2}}\right] - \left[\frac{1}{2}(r-1)^{\frac{1}{2}-1}(r^2-1)\right]}{r-1}} = \boxed{\frac{\left[2r(r-1)^{\frac{1}{2}}\right] - \left[\frac{1}{2}(r-1)^{-\frac{1}{2}}(r^2-1)\right]}{r-1}}$$

$$= \boxed{\frac{\left[2r(r-1)^{\frac{1}{2}}\right] - \left[\frac{(r^2-1)}{2(r-1)^{\frac{1}{2}}}\right]}{r-1}} = \boxed{\frac{\left[\frac{2r(r-1)^{\frac{1}{2}}}{1}\right] - \left[\frac{(r^2-1)}{2(r-1)^{\frac{1}{2}}}\right]}{r-1}} = \boxed{\frac{\frac{4r(r-1)-(r^2-1)}{2(r-1)^{\frac{1}{2}}}}{r-1}} = \boxed{\frac{\frac{4r^2-4r-r^2+1}{2(r-1)^{\frac{1}{2}}}}{r-1}}$$

$$= \boxed{\frac{3r^2-4r+1}{2(r-1)(r-1)^{\frac{1}{2}}}} = \boxed{\mathbf{\frac{3r^2-4r+1}{2(r-1)\sqrt{r-1}}}}$$

Example 5.7-2: Use the chain rule to differentiate the following radical expressions.

a. $\frac{d}{dx}\left(\sqrt{x} - \frac{1}{\sqrt{x}}\right) =$

b. $\frac{d}{dx}\left(\sqrt{\frac{x+2}{3x+1}}\right) =$

c. $\frac{d}{dx}\left(\frac{x^2}{\sqrt{x^2+1}}\right) =$

d. $\frac{d}{dt}\left(\frac{\sqrt{x^3+1}}{x}\right) =$

e. $\frac{d}{dx}\left(\sqrt[3]{x^2} + x^{-2}\right) =$

f. $\frac{d}{dx}\sqrt[5]{x^3+1} =$

Solutions:

a. $\boxed{\frac{d}{dx}\left(\sqrt{x} - \frac{1}{\sqrt{x}}\right)} = \boxed{\frac{d}{dx}\left(x^{\frac{1}{2}} - \frac{1}{x^{\frac{1}{2}}}\right)} = \boxed{\frac{d}{dx}\left(x^{\frac{1}{2}} - x^{-\frac{1}{2}}\right)} = \boxed{\frac{d}{dx}x^{\frac{1}{2}} - \frac{d}{dx}x^{-\frac{1}{2}}} = \boxed{\frac{1}{2}x^{\frac{1}{2}-1} + \frac{1}{2}x^{-\frac{1}{2}-1}}$

$$= \frac{1}{2}x^{-\frac{1}{2}} + \frac{1}{2}x^{-\frac{3}{2}} = \frac{1}{2x^{\frac{1}{2}}} + \frac{1}{2x^{\frac{3}{2}}} = \frac{1}{2\sqrt{x}} + \frac{1}{2\sqrt{x^3}} = \mathbf{\frac{1}{2\sqrt{x}} + \frac{1}{2x\sqrt{x}}}$$

b. $$\frac{d}{dx}\left(\sqrt{\frac{x+2}{3x+1}}\right) = \frac{d}{dx}\left(\frac{x+2}{3x+1}\right)^{\frac{1}{2}} = \frac{1}{2}\left(\frac{x+2}{3x+1}\right)^{\frac{1}{2}-1} \cdot \frac{d}{dx}\left(\frac{x+2}{3x+1}\right) = \frac{1}{2}\left(\frac{x+2}{3x+1}\right)^{-\frac{1}{2}}\left(\frac{[1\cdot(3x+1)]-[3\cdot(x+2)]}{(3x+1)^2}\right)$$

$$= \frac{1}{2\left(\frac{x+2}{3x+1}\right)^{\frac{1}{2}}}\left(\frac{3x+1-3x-6}{(3x+1)^2}\right) = \frac{1}{2\left(\frac{x+2}{3x+1}\right)^{\frac{1}{2}}}\left(\frac{-5}{(3x+1)^2}\right) = -\frac{5}{2\left(\frac{x+2}{3x+1}\right)^{\frac{1}{2}}(3x+1)^2} = -\frac{5(3x+1)^{\frac{1}{2}}}{2(x+2)^{\frac{1}{2}}(3x+1)^2}$$

$$= -\frac{5}{2(x+2)^{\frac{1}{2}}(3x+1)^2(3x+1)^{-\frac{1}{2}}} = -\frac{5}{2(x+2)^{\frac{1}{2}}(3x+1)^{2-\frac{1}{2}}} = -\frac{5}{2(x+2)^{\frac{1}{2}}(3x+1)^{\frac{4-1}{2}}} = -\frac{5}{2(x+2)^{\frac{1}{2}}(3x+1)^{\frac{3}{2}}}$$

$$= -\frac{5}{2\left(\sqrt{x+2}\right)\sqrt{(3x+1)^3}} = \mathbf{-\frac{5}{2\left(\sqrt{x+2}\right)(3x+1)\sqrt{3x+1}}}$$

c. $$\frac{d}{dx}\left(\frac{x^2}{\sqrt{x^2+1}}\right) = \frac{d}{dx}\frac{x^2}{(x^2+1)^{\frac{1}{2}}} = \frac{(x^2+1)^{\frac{1}{2}}\frac{d}{dx}x^2 - x^2\frac{d}{dx}(x^2+1)^{\frac{1}{2}}}{x^2+1} = \frac{2x(x^2+1)^{\frac{1}{2}} - \left[\frac{1}{2}(x^2+1)^{\frac{1}{2}-1}(2x^{2-1}+0)\right]x^2}{x^2+1}$$

$$= \frac{2x(x^2+1)^{\frac{1}{2}} - \left[\frac{1}{2}(x^2+1)^{-\frac{1}{2}}\cdot 2x\right]\cdot x^2}{x^2+1} = \frac{2x(x^2+1)^{\frac{1}{2}} - x^3(x^2+1)^{-\frac{1}{2}}}{x^2+1} = \frac{2x(x^2+1)^{\frac{1}{2}} - \frac{x^3}{(x^2+1)^{\frac{1}{2}}}}{x^2+1}$$

$$= \frac{\frac{2x(x^2+1)^{\frac{1}{2}}(x^2+1)^{\frac{1}{2}} - x^3}{(x^2+1)^{\frac{1}{2}}}}{x^2+1} = \frac{\frac{2x(x^2+1)-x^3}{(x^2+1)^{\frac{1}{2}}}}{x^2+1} = \frac{2x^3+2x-x^3}{(x^2+1)^{\frac{1}{2}}(x^2+1)} = \frac{x^3+2x}{(x^2+1)^{\frac{1}{2}+1}} = \frac{x^3+2x}{(x^2+1)^{\frac{3}{2}}}$$

$$= \frac{x^3+2x}{\sqrt{(x^2+1)^3}} = \mathbf{\frac{x(x^2+2)}{(x^2+1)\sqrt{x^2+1}}}$$

d. $$\frac{d}{dt}\left(\frac{\sqrt{x^3}}{x+1}\right) = \frac{d}{dt}\left(\frac{x^{\frac{3}{2}}}{x+1}\right) = \frac{\left[(x+1)\cdot\frac{d}{dx}x^{\frac{3}{2}}\right] - \left[x^{\frac{3}{2}}\frac{d}{dx}(x+1)\right]}{(x+1)^2} = \frac{\left[(x+1)\cdot\frac{3}{2}x^{\frac{3}{2}-1}\right] - \left[x^{\frac{3}{2}}\cdot 1\right]}{(x+1)^2}$$

$$= \frac{\frac{3}{2}x^{\frac{1}{2}}(x+1) - x^{\frac{3}{2}}}{(x+1)^2} = \frac{\frac{3}{2}x^{\frac{1}{2}+1} + \frac{3}{2}x^{\frac{1}{2}} - x^{\frac{3}{2}}}{(x+1)^2} = \frac{\frac{3}{2}x^{\frac{3}{2}} + \frac{3}{2}x^{\frac{1}{2}} - x^{\frac{3}{2}}}{(x+1)^2} = \frac{\left(\frac{3}{2}-1\right)x^{\frac{3}{2}} + \frac{3}{2}x^{\frac{1}{2}}}{(x+1)^2} = \frac{\frac{1}{2}x^{\frac{3}{2}} + \frac{3}{2}x^{\frac{1}{2}}}{(x+1)^2}$$

$$= \boxed{\frac{\frac{x^{\frac{3}{2}}+3x^{\frac{1}{2}}}{2}}{(x+1)^2}} = \boxed{\frac{x^{\frac{3}{2}}+3x^{\frac{1}{2}}}{2(x+1)^2}} = \boxed{\frac{\sqrt{x^3}+3\sqrt{x}}{2(x+1)^2}} = \boxed{\frac{x\sqrt{x}+3\sqrt{x}}{2(x+1)^2}} = \boxed{\mathbf{\frac{\sqrt{x}(x+3)}{2(x+1)^2}}}$$

e. $\boxed{\frac{d}{dx}\left(\sqrt[3]{x^2}+x^{-2}\right)} = \boxed{\frac{d}{dx}\left(x^{\frac{2}{3}}+x^{-2}\right)} = \boxed{\frac{d}{dx}x^{\frac{2}{3}}+\frac{d}{dx}x^{-2}} = \boxed{\frac{2}{3}x^{\frac{2}{3}-1}-2x^{-2-1}} = \boxed{\frac{2}{3}x^{\frac{2-3}{3}}-2x^{-3}} = \boxed{\frac{2}{3}x^{-\frac{1}{3}}-2x^{-3}}$

$$= \boxed{\frac{2}{3x^{\frac{1}{3}}}-\frac{2}{x^3}} = \boxed{\mathbf{\frac{2}{3\sqrt[3]{x}}-\frac{2}{x^3}}}$$

f. $\boxed{\frac{d}{dx}\sqrt[5]{x^3+1}} = \boxed{\frac{d}{dx}(x^3+1)^{\frac{1}{5}}} = \boxed{\frac{1}{5}(x^3+1)^{\frac{1}{5}-1}\cdot\frac{d}{dx}(x^3+1)} = \boxed{\frac{1}{5}(x^3+1)^{-\frac{4}{5}}(3x^{3-1}+0)} = \boxed{\frac{1}{5}(x^3+1)^{-\frac{4}{5}}\cdot 3x^2}$

$$= \boxed{\frac{3x^2(x^3+1)^{-\frac{4}{5}}}{5}} = \boxed{\frac{3x^2}{5(x^3+1)^{\frac{4}{5}}}} = \boxed{\mathbf{\frac{3x^2}{5\sqrt[5]{(x^3+1)^4}}}}$$

Example 5.7-3: Use the chain rule to differentiate the following radical expressions.

a. $y=x^3\sqrt{x^2-1}$ b. $y=3x^2+\sqrt{x+1}$ c. $y=\sqrt{x^3+x^2+1}$

d. $y=\sqrt[5]{x^5+x^2-1}$ e. $y=\sqrt{x^2-1}\cdot\sqrt{x+1}$ f. $y=\sqrt{x}(x+1)^3$

g. $y=\sqrt{(x^2+5x-1)^5}$ h. $y=x^2\sqrt[3]{x+1}$ i. $y=(x-1)^3\sqrt[3]{x^5}$

Solutions:

a. Given $y=x^3\sqrt{x^2-1} = x^3(x^2-1)^{\frac{1}{2}}$, then

$$\boxed{y'} = \boxed{\left[3x^2\cdot(x^2-1)^{\frac{1}{2}}\right]+\left[\frac{1}{2}(x^2-1)^{\frac{1}{2}-1}\cdot 2x\cdot x^3\right]} = \boxed{\left[3x^2(x^2-1)^{\frac{1}{2}}\right]+\left[\frac{2x^4}{2}(x^2-1)^{\frac{1}{2}-\frac{1}{1}}\right]} = \boxed{\left[3x^2(x^2-1)^{\frac{1}{2}}\right]}$$

$$+\boxed{\left[x^4(x^2-1)^{\frac{1-2}{2}}\right]} = \boxed{3x^2(x^2-1)^{\frac{1}{2}}+x^4(x^2-1)^{-\frac{1}{2}}} = \boxed{3x^2(x^2-1)^{\frac{1}{2}}+\frac{x^4}{(x^2-1)^{\frac{1}{2}}}} = \boxed{\mathbf{3x^2\sqrt{x^2-1}+\frac{x^4}{\sqrt{x^2-1}}}}$$

b. Given $y=3x^2+\sqrt{x+1} = 3x^2+(x+1)^{\frac{1}{2}}$, then

$$\boxed{y'} = \boxed{(3\cdot 2)x^{2-1}+\frac{1}{2}(x+1)^{\frac{1}{2}-1}} = \boxed{6x+\frac{1}{2}(x+1)^{\frac{1-2}{2}}} = \boxed{6x+\frac{1}{2}(x+1)^{-\frac{1}{2}}} = \boxed{6x+\frac{1}{2(x+1)^{\frac{1}{2}}}} = \boxed{\mathbf{6x+\frac{1}{2\sqrt{x+1}}}}$$

c. Given $y=\sqrt{x^3+x^2+1} = (x^3+x^2+1)^{\frac{1}{2}}$, then

$$\boxed{y'} = \boxed{\frac{1}{2}(x^3+x^2+1)^{\frac{1}{2}-1}\cdot(3x^2+2x)} = \boxed{\left(\frac{3x^2+2x}{2}\right)(x^3+x^2+1)^{\frac{1}{2}-\frac{1}{1}}} = \boxed{\left(\frac{3x^2+2x}{2}\right)(x^3+x^2+1)^{\frac{1-2}{2}}}$$

$$= \boxed{\left(\frac{3x^2+2x}{2}\right)\left(x^3+x^2+1\right)^{-\frac{1}{2}}} = \boxed{\left(\frac{3x^2+2x}{2}\right)\cdot\frac{1}{\left(x^3+x^2+1\right)^{\frac{1}{2}}}} = \boxed{\frac{\mathbf{3x^2+2x}}{\mathbf{2\sqrt{x^3+x^2+1}}}}$$

d. Given $y=\sqrt[5]{x^5+x^2-1} = \left(x^5+x^2-1\right)^{\frac{1}{5}}$, then

$$\boxed{y'} = \boxed{\frac{1}{5}\left(x^5+x^2-1\right)^{\frac{1}{5}-1}\cdot\left(5x^4+2x\right)} = \boxed{\left(\frac{5x^4+2x}{5}\right)\left(x^5+x^2-1\right)^{\frac{1}{5}-\frac{1}{1}}} = \boxed{\left(\frac{5x^4+2x}{5}\right)\left(x^5+x^2-1\right)^{\frac{1-5}{5}}}$$

$$= \boxed{\left(\frac{5x^4+2x}{5}\right)\left(x^5+x^2-1\right)^{-\frac{4}{5}}} = \boxed{\left(\frac{5x^4+2x}{5}\right)\cdot\frac{1}{\left(x^5+x^2-1\right)^{\frac{4}{5}}}} = \boxed{\frac{\mathbf{5x^4+2x}}{\mathbf{5\sqrt[5]{\left(x^5+x^2-1\right)^4}}}}$$

e. Given $y=\sqrt{x^2-1}\cdot\sqrt{x+1} = \left(x^2-1\right)^{\frac{1}{2}}\cdot(x+1)^{\frac{1}{2}}$, then

$$\boxed{y'} = \boxed{\left[\frac{1}{2}\left(x^2-1\right)^{\frac{1}{2}-1}\cdot 2x\cdot(x+1)^{\frac{1}{2}}\right]+\left[(x+1)^{\frac{1}{2}-1}\cdot\left(x^2-1\right)^{\frac{1}{2}}\right]} = \boxed{\left[\frac{2x}{2}\left(x^2-1\right)^{\frac{1}{2}-\frac{1}{1}}\cdot(x+1)^{\frac{1}{2}}\right]+\left[\frac{1}{2}(x+1)^{\frac{1}{2}-\frac{1}{1}}\cdot\left(x^2-1\right)^{\frac{1}{2}}\right]}$$

$$= \boxed{\left[x\left(x^2-1\right)^{-\frac{1}{2}}\cdot(x+1)^{\frac{1}{2}}\right]+\left[\frac{1}{2}(x+1)^{-\frac{1}{2}}\cdot\left(x^2-1\right)^{\frac{1}{2}}\right]} = \boxed{\left[x\cdot\frac{1}{\left(x^2-1\right)^{\frac{1}{2}}}\cdot(x+1)^{\frac{1}{2}}\right]+\left[\frac{1}{2(x+1)^{\frac{1}{2}}}\cdot\left(x^2-1\right)^{\frac{1}{2}}\right]}$$

$$= \boxed{\left[x\cdot\frac{1}{\sqrt{x^2-1}}\cdot\sqrt{x+1}\right]+\left[\frac{1}{2\sqrt{x+1}}\cdot\sqrt{x^2-1}\right]} = \boxed{\frac{\mathbf{x\sqrt{x+1}}}{\mathbf{\sqrt{x^2-1}}}+\frac{\mathbf{\sqrt{x^2-1}}}{\mathbf{2\sqrt{x+1}}}}$$

f. Given $y=\sqrt{x}(x+1)^3 = x^{\frac{1}{2}}(x+1)^3$, then

$$\boxed{y'} = \boxed{\left[\frac{1}{2}x^{\frac{1}{2}-1}\cdot(x+1)^3\right]+\left[3(x+1)^2\cdot x^{\frac{1}{2}}\right]} = \boxed{\left[\frac{(x+1)^3}{2}\cdot x^{\frac{1}{2}-\frac{1}{1}}\right]+\left[3(x+1)^2 x^{\frac{1}{2}}\right]} = \boxed{\left[\frac{(x+1)^3}{2}x^{-\frac{1}{2}}\right]+\left[3(x+1)^2 x^{\frac{1}{2}}\right]}$$

$$= \boxed{\left[\frac{(x+1)^3}{2}\cdot\frac{1}{x^{\frac{1}{2}}}\right]+\left[3(x+1)^2 x^{\frac{1}{2}}\right]} = \boxed{\frac{\mathbf{(x+1)^3}}{\mathbf{2\sqrt{x}}}+\mathbf{3\sqrt{x}\,(x+1)^2}}$$

g. Given $y=\sqrt{\left(x^2+5x-1\right)^5} = \left[\left(x^2+5x-1\right)^5\right]^{\frac{1}{2}} = \left(x^2+5x-1\right)^{\frac{5}{2}}$, then

$$\boxed{y'} = \boxed{\frac{5}{2}\left(x^2+5x-1\right)^{\frac{5}{2}-1}\cdot(2x+5)} = \boxed{\frac{5(2x+5)}{2}\cdot\left(x^2+5x-1\right)^{\frac{5}{2}-\frac{1}{1}}} = \boxed{\frac{5(2x+5)}{2}\cdot\left(x^2+5x-1\right)^{\frac{5-2}{2}}}$$

$$= \boxed{\frac{5(2x+5)}{2}\cdot\left(x^2+5x-1\right)^{\frac{3}{2}}} = \boxed{\frac{5(2x+5)}{2}\cdot\sqrt{\left(x^2+5x-1\right)^3}} = \boxed{\frac{\mathbf{5(2x+5)\left(x^2+5x-1\right)\sqrt{x^2+5x-1}}}{\mathbf{2}}}$$

h. Given $y=x^2\sqrt[3]{x+1} = x^2(x+1)^{\frac{1}{3}}$, then

$$\boxed{y'} = \boxed{\left[2x\cdot(x+1)^{\frac{1}{3}}\right]+\left[\frac{1}{3}(x+1)^{\frac{1}{3}-1}\cdot x^2\right]} = \boxed{\left[2x(x+1)^{\frac{1}{3}}\right]+\left[\frac{x^2}{3}(x+1)^{\frac{1}{3}-\frac{1}{1}}\right]} = \boxed{2x(x+1)^{\frac{1}{3}}+\frac{x^2}{3}(x+1)^{\frac{1-3}{3}}}$$

$$= \boxed{2x(x+1)^{\frac{1}{3}}+\frac{x^2}{3}(x+1)^{-\frac{2}{3}}} = \boxed{2x(x+1)^{\frac{1}{3}}+\frac{x^2}{3}\cdot\frac{1}{(x+1)^{\frac{2}{3}}}} = \boxed{\mathbf{2x\sqrt[3]{x+1}+\frac{x^2}{3\sqrt[3]{(x+1)^2}}}}$$

i. Given $y=(x-1)^3\sqrt[3]{x^5} = (x-1)^3\left(x^5\right)^{\frac{1}{3}} = (x-1)^3 x^{\frac{5}{3}}$, then

$$\boxed{y'} = \boxed{\left[3(x-1)^2\cdot x^{\frac{5}{3}}\right]+\left[\frac{5}{3}x^{\frac{5}{3}-1}\cdot(x-1)^3\right]} = \boxed{\left[3x^{\frac{5}{3}}(x-1)^2\right]+\left[\frac{5}{3}x^{\frac{5-3}{3}}(x-1)^3\right]} = \boxed{\left[3x^{\frac{5}{3}}(x-1)^2\right]+\left[\frac{5}{3}x^{\frac{2}{3}}(x-1)^3\right]}$$

$$= \boxed{\mathbf{3x\sqrt[3]{x^2}(x-1)^2+\frac{5}{3}\sqrt[3]{x^2}(x-1)^3}}$$

Example 5.7-4: Find $\frac{dy}{dx}$ by implicit differentiation.

a. $\sqrt{x}+\sqrt{y}=10$ b. $\sqrt{x^2+y^2}=x$ c. $\sqrt[3]{x^2}+\sqrt[5]{y^3}=2$

d. $\sqrt{2x+1}=y^2$ e. $\sqrt[5]{x^2+1}=y^{\frac{2}{3}}$ f. $\sqrt[4]{x^2y^2}=x$

g. $\sqrt[7]{xy^2}=3x^2$ h. $\sqrt{x^3+y^3}=2$ i. $\sqrt{x^2-1}=xy$

Solutions:

a. $\boxed{\frac{d}{dx}\left(\sqrt{x}+\sqrt{y}\right)=\frac{d}{dx}(10)}$; $\boxed{\frac{d}{dx}\left(x^{\frac{1}{2}}+y^{\frac{1}{2}}\right)=0}$; $\boxed{\frac{1}{2}x^{\frac{1}{2}-1}\cdot\frac{d}{dx}(x)+\frac{1}{2}y^{\frac{1}{2}-1}\cdot\frac{d}{dx}(y)=0}$; $\boxed{\frac{1}{2}x^{-\frac{1}{2}}\cdot 1+\frac{1}{2}y^{-\frac{1}{2}}\cdot y'=0}$

; $\boxed{\frac{1}{2}y^{-\frac{1}{2}}y'=-\frac{1}{2}x^{-\frac{1}{2}}}$; $\boxed{\frac{1}{2}\cdot\frac{y'}{y^{\frac{1}{2}}}=-\frac{1}{2}\cdot\frac{1}{x^{\frac{1}{2}}}}$; $\boxed{2y^{\frac{1}{2}}\cdot\frac{1}{2}\cdot\frac{y'}{y^{\frac{1}{2}}}=2y^{\frac{1}{2}}\cdot-\frac{1}{2}\cdot\frac{1}{x^{\frac{1}{2}}}}$; $\boxed{y'=\frac{-y^{\frac{1}{2}}}{x^{\frac{1}{2}}}}$; $\boxed{y'=\frac{-\sqrt{y}}{\sqrt{x}}}$; $\boxed{\mathbf{y'=-\sqrt{\frac{y}{x}}}}$

b. $\boxed{\frac{d}{dx}\left(\sqrt{x^2+y^2}\right)=\frac{d}{dx}(x)}$; $\boxed{\frac{d}{dx}\left(x^2+y^2\right)^{\frac{1}{2}}=1}$; $\boxed{\frac{1}{2}\left(x^2+y^2\right)^{\frac{1}{2}-1}\left(\frac{d}{dx}x^2+\frac{d}{dx}y^2\right)=1}$; $\boxed{\frac{1}{2}\left(x^2+y^2\right)^{-\frac{1}{2}}(2x+2yy')=1}$

; $\boxed{\frac{2x+2yy'}{2\left(x^2+y^2\right)^{\frac{1}{2}}}=1}$; $\boxed{2x+2yy'=2\left(x^2+y^2\right)^{\frac{1}{2}}}$; $\boxed{y'=\frac{2\left(x^2+y^2\right)^{\frac{1}{2}}-2x}{2y}}$; $\boxed{\mathbf{y'=\frac{\sqrt{x^2+y^2}-x}{y}}}$

c. $\boxed{\frac{d}{dx}\left(\sqrt[3]{x^2}+\sqrt[5]{y^3}\right)=\frac{d}{dx}(2)}$; $\boxed{\frac{d}{dx}\left[\left(x^2\right)^{\frac{1}{3}}+\left(y^3\right)^{\frac{1}{5}}\right]=0}$; $\boxed{\frac{d}{dx}\left[x^{\frac{2}{3}}+y^{\frac{3}{5}}\right]=0}$; $\boxed{\frac{2}{3}x^{\frac{2}{3}-1}\cdot\frac{d}{dx}(x)+\frac{3}{5}y^{\frac{3}{5}-1}\cdot\frac{d}{dx}(y)=0}$

; $\boxed{\frac{2}{3}x^{-\frac{1}{3}}\cdot 1+\frac{3}{5}y^{-\frac{2}{5}}\cdot y'=0}$; $\boxed{\frac{3}{5}y^{-\frac{2}{5}}y'=\frac{-2}{3}x^{-\frac{1}{3}}}$; $\boxed{\frac{3y'}{5y^{\frac{2}{5}}}=\frac{-2}{3x^{\frac{1}{3}}}}$; $\boxed{9x^{\frac{1}{3}}y'=-10y^{\frac{2}{5}}}$; $\boxed{y'=\frac{-10y^{\frac{2}{5}}}{9x^{\frac{1}{3}}}}$; $\boxed{\mathbf{y'=-\frac{10\sqrt[5]{y^2}}{9\sqrt[3]{x}}}}$

d. $\boxed{\frac{d}{dx}\left(\sqrt{2x+1}\right)=\frac{d}{dx}\left(y^2\right)}$; $\boxed{\frac{d}{dx}(2x+1)^{\frac{1}{2}}=2y\cdot\frac{d}{dx}(y)}$; $\boxed{6x+\frac{1}{2}(x+1)^{\frac{1-2}{2}}}$; $\boxed{\frac{1}{2}(2x+1)^{-\frac{1}{2}}\cdot 2x=2y\cdot y'}$

; $\boxed{2y\,y'=x(2x+1)^{-\frac{1}{2}}}$; $\boxed{y'=\frac{x(2x+1)^{-\frac{1}{2}}}{2y}}$; $\boxed{y'=\frac{x}{2y(2x+1)^{\frac{1}{2}}}}$; $\boxed{\mathbf{y'=\frac{x}{2y\sqrt{2x+1}}}}$

e. $\boxed{\frac{d}{dx}\left(\sqrt[5]{x^2+1}\right)=\frac{d}{dx}\left(y^{\frac{2}{3}}\right)}$; $\boxed{\frac{d}{dx}\left(x^2+1\right)^{\frac{1}{5}}=\frac{2}{3}y^{\frac{2}{3}-1}\cdot\frac{d}{dx}(y)}$; $\boxed{\frac{1}{5}\left(x^2+1\right)^{\frac{1}{5}-1}\cdot\left[\frac{d}{dx}\left(x^2\right)+\frac{d}{dx}(1)\right]=\frac{2}{3}y^{\frac{2}{3}-1}\cdot\frac{d}{dx}(y)}$

; $\boxed{\frac{1}{5}\left(x^2+1\right)^{-\frac{4}{5}}\cdot 2x=\frac{2}{3}y^{-\frac{1}{3}}\cdot y'}$; $\boxed{\frac{2}{3}y^{-\frac{1}{3}}y'=\frac{2x}{5}\left(x^2+1\right)^{-\frac{4}{5}}}$; $\boxed{\frac{2y'}{3y^{\frac{1}{3}}}=\frac{2x}{5\left(x^2+1\right)^{\frac{4}{5}}}}$; $\boxed{10\left(x^2+1\right)^{\frac{4}{5}}y'=6x\,y^{\frac{1}{3}}}$

; $\boxed{y'=\frac{6x\,y^{\frac{1}{3}}}{10\left(x^2+1\right)^{\frac{4}{5}}}}$; $\boxed{\mathbf{y'=\frac{3x\sqrt[3]{y}}{5\sqrt[5]{\left(x^2+1\right)^4}}}}$

f. $\boxed{\frac{d}{dx}\left(\sqrt[4]{x^2y^2}\right)=\frac{d}{dx}(x)}$; $\boxed{\frac{d}{dx}\left(x^2y^2\right)^{\frac{1}{4}}=1}$; $\boxed{\frac{1}{4}\left(x^2y^2\right)^{\frac{1}{4}-1}\left[\frac{d}{dx}\left(x^2y^2\right)\right]=1}$; $\boxed{\frac{1}{4}\left(x^2y^2-1\right)^{-\frac{3}{4}}\left(2xy^2+2x^2y\,y'\right)=1}$

; $\boxed{\frac{2xy^2+2x^2y\,y'}{4\left(x^2y^2-1\right)^{\frac{3}{4}}}=1}$; $\boxed{\frac{2\left(xy^2+x^2y\,y'\right)}{4\left(x^2y^2-1\right)^{\frac{3}{4}}}=1}$; $\boxed{xy^2+x^2y\,y'=2\left(x^2y^2-1\right)^{\frac{3}{4}}}$; $\boxed{x^2y\,y'=2\left(x^2y^2-1\right)^{\frac{3}{4}}-xy^2}$

; $\boxed{y'=\frac{2\left(x^2y^2-1\right)^{\frac{3}{4}}}{x^2y}-\frac{xy^2}{x^2y}}$; $\boxed{\mathbf{y'=\frac{2\left(x^2y^2-1\right)^{\frac{3}{4}}}{x^2y}-\frac{y}{x}}}$

g. $\boxed{\frac{d}{dx}\left(\sqrt[7]{xy^2}\right)=\frac{d}{dx}\left(3x^2\right)}$; $\boxed{\frac{d}{dx}\left(xy^2\right)^{\frac{1}{7}}=6x}$; $\boxed{\frac{1}{7}\left(xy^2\right)^{\frac{1}{7}-1}\cdot\frac{d}{dx}\left(xy^2\right)=6x}$; $\boxed{\frac{1}{7}\left(xy^2\right)^{-\frac{6}{7}}\left(1\cdot y^2+2y\,y'\cdot x\right)=6x}$

; $\boxed{\frac{y^2+2y\,y'x}{7\left(xy^2\right)^{\frac{6}{7}}}=6x}$; $\boxed{y^2+2y\,y'x=6x\cdot 7\left(xy^2\right)^{\frac{6}{7}}}$; $\boxed{2y\,y'x=42x\left(xy^2\right)^{\frac{6}{7}}-y^2}$; $\boxed{\mathbf{y'=\frac{42x^{\frac{13}{7}}y^{\frac{12}{7}}-y^2}{2x\,y}}}$

h. $\boxed{\frac{d}{dx}\left(\sqrt{x^3+y^3}\right)=\frac{d}{dx}(2)}$; $\boxed{\frac{d}{dx}\left(x^3+y^3\right)^{\frac{1}{2}}=0}$; $\boxed{\frac{1}{2}\left(x^3+y^3\right)^{\frac{1}{2}-1}\cdot\left(\frac{d}{dx}x^3+\frac{d}{dx}y^3\right)=0}$; $\boxed{\frac{1}{2}\left(x^3+y^3\right)^{-\frac{1}{2}}\cdot\left(3x^2+3y^2y'\right)=0}$

$\boxed{\frac{3x^2}{2}\left(x^3+y^3\right)^{-\frac{1}{2}}+\frac{3y^2y'}{2}\left(x^3+y^3\right)^{-\frac{1}{2}}=0}$; $\boxed{\frac{3y^2y'}{2}\left(x^3+y^3\right)^{-\frac{1}{2}}=-\frac{3x^2}{2}\left(x^3+y^3\right)^{-\frac{1}{2}}}$; $\boxed{y^2y'=-x^2}$; $\boxed{\mathbf{y'=-\frac{x^2}{y^2}}}$

i. $\boxed{\frac{d}{dx}\left(\sqrt{x^2-1}\right)=\frac{d}{dx}(xy)}$; $\boxed{\frac{d}{dx}\left(x^2-1\right)^{\frac{1}{2}}=1\cdot y+x\cdot\frac{d}{dx}(y)}$; $\boxed{\frac{1}{2}\left(x^2-1\right)^{\frac{1}{2}-1}\cdot 2x=y+x\,y'}$; $\boxed{x\left(x^2-1\right)^{-\frac{1}{2}}=y+x\,y'}$

$; \boxed{y + x y' = x\left(x^2 - 1\right)^{-\frac{1}{2}}}; \boxed{x y' = x\left(x^2 - 1\right)^{-\frac{1}{2}} - y}; \boxed{\mathbf{y' = \dfrac{x\left(x^2 - 1\right)^{-\frac{1}{2}} - y}{x}}}$

Example 5.7-5: Compute the derivative for the following Radical expressions.

a. $\dfrac{d}{dx}\left(2x + \sqrt{x}\right) =$ b. $\dfrac{d}{dx}\left(\sqrt{x+1}\right) =$ c. $\dfrac{d}{dx}\left(\sqrt[3]{x^2+1}\right) =$

d. $\dfrac{d}{dx}\left(x + \sqrt{x^3}\right) =$ e. $\dfrac{d}{dx}\left(\sqrt{x^5} - 3\right) =$ f. $\dfrac{d}{dx}\left(x^3\sqrt{x+1}\right) =$

g. $\dfrac{d}{dx}\left(\dfrac{x+1}{\sqrt{x}}\right) =$ h. $\dfrac{d}{dx}\left(\sqrt{\dfrac{1}{x+1}}\right) =$ i. $\dfrac{d}{dx}\left(\dfrac{x^2}{\sqrt[3]{x}}\right) =$

Solutions:

a. $\boxed{\dfrac{d}{dx}\left(2x + \sqrt{x}\right)} = \boxed{\dfrac{d}{dx}2x + \dfrac{d}{dx}\sqrt{x}} = \boxed{\dfrac{d}{dx}2x + \dfrac{d}{dx}x^{\frac{1}{2}}} = \boxed{2 + \dfrac{1}{2}x^{\frac{1}{2}-1}} = \boxed{2 + \dfrac{1}{2}x^{-\frac{1}{2}}} = \boxed{2 + \dfrac{1}{2x^{\frac{1}{2}}}} = \boxed{\mathbf{2 + \dfrac{1}{2\sqrt{x}}}}$

b. $\boxed{\dfrac{d}{dx}\left(\sqrt{x+1}\right)} = \boxed{\dfrac{d}{dx}(x+1)^{\frac{1}{2}}} = \boxed{\dfrac{1}{2}(x+1)^{\frac{1}{2}-1}\cdot 1} = \boxed{\dfrac{1}{2}(x+1)^{-\frac{1}{2}}} = \boxed{\dfrac{1}{2(x+1)^{\frac{1}{2}}}} = \boxed{\mathbf{\dfrac{1}{2\sqrt{x+1}}}}$

c. $\boxed{\dfrac{d}{dx}\left(\sqrt[3]{x^2+1}\right)} = \boxed{\dfrac{d}{dx}\left(x^2+1\right)^{\frac{1}{3}}} = \boxed{\dfrac{1}{3}\left(x^2+1\right)^{\frac{1}{3}-1}\cdot 2x} = \boxed{\dfrac{2x}{3}\left(x^2+1\right)^{-\frac{2}{3}}} = \boxed{\dfrac{2x}{3\left(x^2+1\right)^{\frac{2}{3}}}} = \boxed{\mathbf{\dfrac{2x}{3\sqrt[3]{\left(x^2+1\right)^2}}}}$

d. $\boxed{\dfrac{d}{dx}\left(x + \sqrt{x^3}\right)} = \boxed{\dfrac{d}{dx}\left(x + x^{\frac{3}{2}}\right)} = \boxed{\dfrac{d}{dx}x + \dfrac{d}{dx}x^{\frac{3}{2}}} = \boxed{1 + \dfrac{3}{2}x^{\frac{3}{2}-1}} = \boxed{1 + \dfrac{3}{2}x^{\frac{1}{2}}} = \boxed{\mathbf{1 + \dfrac{3}{2}\sqrt{x}}}$

e. $\boxed{\dfrac{d}{dx}\left(\sqrt{x^5} - 3\right)} = \boxed{\dfrac{d}{dx}\left(x^{\frac{5}{2}} - 3\right)} = \boxed{\dfrac{d}{dx}x^{\frac{5}{2}} + \dfrac{d}{dx}(-3)} = \boxed{\dfrac{5}{2}x^{\frac{5}{2}-1} + 0} = \boxed{\dfrac{5}{2}x^{\frac{3}{2}}} = \boxed{\dfrac{5}{2}\sqrt{x^3}} = \boxed{\mathbf{\dfrac{5x\sqrt{x}}{2}}}$

f. $\boxed{\dfrac{d}{dx}\left(x^3\sqrt{x+1}\right)} = \boxed{\dfrac{d}{dx}\left[x^3(x+1)^{\frac{1}{2}}\right]} = \boxed{(x+1)^{\frac{1}{2}}\dfrac{d}{dx}x^3 + x^3\dfrac{d}{dx}(x+1)^{\frac{1}{2}}} = \boxed{(x+1)^{\frac{1}{2}}\cdot 3x^2 + x^3\cdot\dfrac{1}{2}(x+1)^{\frac{1}{2}-1}}$

$= \boxed{3x^2(x+1)^{\frac{1}{2}} + \dfrac{x^3}{2}(x+1)^{-\frac{1}{2}}} = \boxed{3x^2(x+1)^{\frac{1}{2}} + \dfrac{x^3}{2(x+1)^{\frac{1}{2}}}} = \boxed{\dfrac{3x^2(x+1)^{\frac{1}{2}}\cdot 2(x+1)^{\frac{1}{2}} + x^3}{2(x+1)^{\frac{1}{2}}}} = \boxed{\dfrac{6x^2(x+1) + x^3}{2\sqrt{x+1}}}$

$= \boxed{\dfrac{6x^3 + 6x^2 + x^3}{2\sqrt{x+1}}} = \boxed{\dfrac{7x^3 + 6x^2}{2\sqrt{x+1}}} = \boxed{\mathbf{\dfrac{x^2(7x+6)}{2\sqrt{x+1}}}}$

g. $\boxed{\dfrac{d}{dx}\left(\dfrac{x+1}{\sqrt{x}}\right)} = \boxed{\dfrac{d}{dx}\left(\dfrac{x+1}{x^{\frac{1}{2}}}\right)} = \boxed{\dfrac{\left[x^{\frac{1}{2}}\cdot\dfrac{d}{dx}(x+1)\right] - \left[(x+1)\cdot\dfrac{d}{dx}x^{\frac{1}{2}}\right]}{x}} = \boxed{\dfrac{\left[x^{\frac{1}{2}}\cdot 1\right] - \left[(x+1)\cdot\dfrac{1}{2}x^{\frac{1}{2}-1}\right]}{x}}$

$$= \boxed{\frac{x^{\frac{1}{2}} - \left[(x+1)\cdot\frac{1}{2}x^{-\frac{1}{2}}\right]}{x}} = \boxed{\frac{x^{\frac{1}{2}} - \frac{x+1}{2x^{\frac{1}{2}}}}{x}} = \boxed{\frac{\frac{x^{\frac{1}{2}}\cdot 2x^{\frac{1}{2}} - (x+1)}{2x^{\frac{1}{2}}}}{x}} = \boxed{\frac{\frac{2x-x-1}{2\sqrt{x}}}{x}} = \boxed{\frac{\frac{x-1}{2\sqrt{x}}}{x}} = \boxed{\mathbf{\frac{x-1}{2x\sqrt{x}}}}$$

h. $$\boxed{\frac{d}{dx}\left(\sqrt{\frac{1}{x+1}}\right)} = \boxed{\frac{d}{dx}\left(\frac{1}{\sqrt{x+1}}\right)} = \boxed{\frac{d}{dx}\left(\frac{1}{(x+1)^{\frac{1}{2}}}\right)} = \boxed{\frac{\left[(x+1)^{\frac{1}{2}}\cdot\frac{d}{dx}(1)\right] - \left[1\cdot\frac{d}{dx}(x+1)^{\frac{1}{2}}\right]}{x+1}} = \boxed{\frac{0 - \frac{1}{2}(x+1)^{\frac{1}{2}-1}}{x+1}}$$

$$= \boxed{\frac{-\frac{1}{2}(x+1)^{-\frac{1}{2}}}{x+1}} = \boxed{\frac{-\frac{1}{2(x+1)^{\frac{1}{2}}}}{(x+1)}} = \boxed{-\frac{1}{2(x+1)^{\frac{1}{2}}(x+1)}} = \boxed{-\frac{1}{2(x+1)^{\frac{3}{2}}}} = \boxed{-\frac{1}{2\sqrt{(x+1)^3}}} = \boxed{\mathbf{-\frac{1}{2(x+1)\sqrt{x+1}}}}$$

i. $$\boxed{\frac{d}{dx}\left(\frac{x^2}{\sqrt[3]{x}}\right)} = \boxed{\frac{d}{dx}\left(\frac{x^2}{x^{\frac{1}{3}}}\right)} = \boxed{\frac{d}{dx}\left(x^2\cdot x^{-\frac{1}{3}}\right)} = \boxed{\frac{d}{dx}x^{2-\frac{1}{3}}} = \boxed{\frac{d}{dx}x^{\frac{6-1}{3}}} = \boxed{\frac{d}{dx}x^{\frac{5}{3}}} = \boxed{\frac{5}{3}x^{\frac{5}{3}-1}} = \boxed{\frac{5}{3}x^{\frac{2}{3}}} = \boxed{\mathbf{\frac{5\sqrt[3]{x^2}}{3}}}$$

Example 5.7-6: Evaluate the derivative of the following equations at the given values.

a. $y = \sqrt{(x^2+1)^3}$ at $x=2$ b. $y = \sqrt[3]{x}+\sqrt{x}+2$ at $x=10$ c. $y = x\sqrt{x^2+1}$ at $x=5$

d. $y = \sqrt[3]{2x}+\sqrt[3]{x^2}$ at $x=2$ e. $y = \frac{\sqrt{6x}+1}{x^2}$ at $x=3$ f. $y = \frac{1}{\sqrt{x^2+1}}$ at $x=-5$

g. $y = \frac{\sqrt{5x+1}}{2x+1}$ at $x=3$ h. $y = x\sqrt{x^2-10}+2x$ at $x=5$ i. $y = x^2\sqrt{(x+1)^3}$ at $x=1$

j. $y = \sqrt{\frac{x^2+1}{1-x^2}}$ at $x=0$ k. $y = \frac{\sqrt{x^5+1}}{2x}$ at $x=2$ l. $y = \frac{x+1}{\sqrt{x^3+1}}$ at $x=1$

Solutions:

a. Given $y = \sqrt{(x^2+1)^3} = (x^2+1)^{\frac{3}{2}}$, then

$$\boxed{y'} = \boxed{\frac{3}{2}(x^2+1)^{\frac{3}{2}-1}(2x^{2-1}+0)} = \boxed{\frac{3}{2}(x^2+1)^{\frac{3-2}{2}}\cdot 2x} = \boxed{3x(x^2+1)^{\frac{1}{2}}} = \boxed{3x\sqrt{x^2+1}}$$

Substituting $x=2$ in place of x in the y' equation we obtain the following value:

$$\boxed{y'} = (3\cdot 2)\sqrt{2^2+1} = 6\sqrt{5} = \mathbf{13.42}$$

b. Given $y = \sqrt[3]{x}+\sqrt{x}+2 = x^{\frac{1}{3}}+x^{\frac{1}{2}}+2$, then

$$\boxed{y'} = \boxed{\frac{1}{3}x^{\frac{1}{3}-1}+\frac{1}{2}x^{\frac{1}{2}-1}+0} = \boxed{\frac{1}{3}x^{-\frac{2}{3}}+\frac{1}{2}x^{-\frac{1}{2}}} = \boxed{\frac{1}{3x^{\frac{2}{3}}}+\frac{1}{2x^{\frac{1}{2}}}} = \boxed{\frac{1}{3\sqrt[3]{x^2}}+\frac{1}{2\sqrt{x}}}$$

Substituting $x=10$ in place of x in the y' equation we obtain the following value:

$$\boxed{y'}=\boxed{\frac{1}{3\cdot 10^{\frac{2}{3}}}+\frac{1}{2\cdot 10^{\frac{1}{2}}}}=\boxed{\frac{1}{3\cdot 4.634}+\frac{1}{2\cdot 3.162}}=\boxed{0.072+0.158}=\boxed{\mathbf{0.23}}$$

c. Given $y = x\sqrt{x^2+1} = x\left(x^2+1\right)^{\frac{1}{2}}$, then

$$\boxed{y'}=\boxed{\left[1\cdot\left(x^2+1\right)^{\frac{1}{2}}\right]+\left[\frac{1}{2}\left(x^2+1\right)^{\frac{1}{2}-1}\cdot\left(2x^{2-1}+0\right)\cdot x\right]}=\boxed{\left(x^2+1\right)^{\frac{1}{2}}+\left[\frac{1}{2}\left(x^2+1\right)^{-\frac{1}{2}}\cdot 2x^2\right]}=\boxed{\left(x^2+1\right)^{\frac{1}{2}}+x^2\left(x^2+1\right)^{-\frac{1}{2}}}$$

$$=\boxed{\left(x^2+1\right)^{\frac{1}{2}}+\frac{x^2}{\left(x^2+1\right)^{\frac{1}{2}}}}=\boxed{\frac{\left(x^2+1\right)^{\frac{1}{2}}\left(x^2+1\right)^{\frac{1}{2}}+x^2}{\left(x^2+1\right)^{\frac{1}{2}}}}=\boxed{\frac{\left(x^2+1\right)^{\frac{1}{2}+\frac{1}{2}}+x^2}{\left(x^2+1\right)^{\frac{1}{2}}}}=\boxed{\frac{x^2+1+x^2}{\left(x^2+1\right)^{\frac{1}{2}}}}=\boxed{\frac{2x^2+1}{\sqrt{x^2+1}}}$$

Substituting $x=5$ in place of x in the y' equation we obtain the following value:

$$\boxed{y'}=\boxed{\frac{2\cdot 5^2+1}{\left(5^2+1\right)^{\frac{1}{2}}}}=\boxed{\frac{2\cdot 25+1}{(25+1)^{\frac{1}{2}}}}=\boxed{\frac{50+1}{26^{\frac{1}{2}}}}=\boxed{\frac{51}{5.1}}=\boxed{\mathbf{10}}$$

d. Given $y = \sqrt[3]{2x}+\sqrt[3]{x^2} = (2x)^{\frac{1}{3}}+x^{\frac{2}{3}}$, then

$$\boxed{y'}=\boxed{\frac{1}{3}(2x)^{\frac{1}{3}-1}+\frac{2}{3}x^{\frac{2}{3}-1}}=\boxed{\frac{1}{3}(2x)^{-\frac{2}{3}}+\frac{2}{3}x^{-\frac{1}{3}}}=\boxed{\frac{1}{3(2x)^{\frac{2}{3}}}+\frac{2}{3x^{\frac{1}{3}}}}=\boxed{\frac{1}{3\sqrt[3]{(2x)^2}}+\frac{2}{3\sqrt[3]{x}}}$$

Substituting $x=2$ in place of x in the y' equation we obtain the following value:

$$\boxed{y'}=\boxed{\frac{1}{3\cdot(2\cdot 2)^{\frac{2}{3}}}+\frac{2}{3\cdot 2^{\frac{1}{3}}}}=\boxed{\frac{1}{3\cdot 4^{\frac{2}{3}}}+\frac{2}{3\cdot 2^{\frac{1}{3}}}}=\boxed{\frac{1}{3\cdot 2.53}+\frac{2}{3\cdot 1.26}}=\boxed{0.132+0.53}=\boxed{\mathbf{0.662}}$$

e. Given $y = \dfrac{\sqrt{6x}+1}{x^2} = \dfrac{(6x)^{\frac{1}{2}}+1}{x^2}$, then

$$\boxed{y'}=\boxed{\frac{\left\{\left[\frac{1}{2}(6x)^{\frac{1}{2}-1}\cdot 6+0\right]\cdot x^2\right\}-\left\{2x^{2-1}\cdot\left[(6x)^{\frac{1}{2}}+1\right]\right\}}{x^4}}=\boxed{\frac{\left\{\left[\frac{1}{2}(6x)^{-\frac{1}{2}}\cdot 6\right]\cdot x^2\right\}-\left\{2x\cdot\left[(6x)^{\frac{1}{2}}+1\right]\right\}}{x^4}}$$

$$=\boxed{\frac{3x^2(6x)^{-\frac{1}{2}}-2x(6x)^{\frac{1}{2}}-2x}{x^4}}$$

Substituting $x=3$ in place of x in the y' equation we obtain the following value:

$$\boxed{y'}=\boxed{\frac{3\cdot 3^2\cdot(6\cdot 3)^{-\frac{1}{2}}-2\cdot 3\cdot(6\cdot 3)^{\frac{1}{2}}-2\cdot 3}{3^4}}=\boxed{\frac{27\cdot(18)^{-\frac{1}{2}}-6\cdot(18)^{\frac{1}{2}}-6}{81}}=\boxed{\frac{(27\cdot 0.236)-(6\cdot 4.243)-6}{81}}$$

$$=\boxed{\frac{6.372-25.458-6}{81}}=\boxed{-\frac{25.086}{81}}=\boxed{\mathbf{-0.31}}$$

f. Given $y = \frac{1}{\sqrt{x^2+1}} = \frac{1}{(x^2+1)^{\frac{1}{2}}}$, then

$$\boxed{y'} = \boxed{\frac{\left[0\cdot(x^2+1)^{\frac{1}{2}}\right]-\left[\frac{1}{2}(x^2+1)^{\frac{1}{2}-1}\cdot(2x^{2-1}+0)\cdot 1\right]}{x^2+1}} = \boxed{\frac{0-\left[\frac{1}{2}(x^2+1)^{-\frac{1}{2}}\cdot 2x\right]}{x^2+1}} = \boxed{\frac{-x(x^2+1)^{-\frac{1}{2}}}{x^2+1}}$$

$$= \boxed{\frac{-x}{(x^2+1)(x^2+1)^{\frac{1}{2}}}} = \boxed{\frac{-x}{(x^2+1)^{1+\frac{1}{2}}}} = \boxed{\frac{-x}{(x^2+1)^{\frac{3}{2}}}} = \boxed{\frac{-x}{\sqrt{(x^2+1)^3}}} = \boxed{\frac{-x}{(x^2+1)\sqrt{x^2+1}}}$$

Substituting $x=-5$ in place of x in the y' equation we obtain the following value:

$$\boxed{y'} = \boxed{-\frac{-5}{\sqrt{\left[(-5)^2+1\right]^3}}} = \boxed{\frac{5}{\sqrt{(25+1)^3}}} = \boxed{\frac{5}{\sqrt{17576}}} = \boxed{\frac{5}{132.57}} = \boxed{\mathbf{0.038}}$$

g. Given $y = \frac{\sqrt{5x+1}}{2x+1} = \frac{(5x+1)^{\frac{1}{2}}}{2x+1}$, then

$$\boxed{y'} = \boxed{\frac{\left[\frac{1}{2}(5x+1)^{\frac{1}{2}-1}\cdot(5x^{1-1}+0)\cdot(2x+1)\right]-\left[(2x^{1-1}+0)(5x+1)^{\frac{1}{2}}\right]}{(2x+1)^2}} = \boxed{\frac{\left[2.5(5x+1)^{-\frac{1}{2}}\cdot(2x+1)\right]-\left[2(5x+1)^{\frac{1}{2}}\right]}{(2x+1)^2}}$$

Substituting $x=3$ in place of x in the y' equation we obtain the following value:

$$\boxed{y'} = \boxed{\frac{\left[2.5(15+1)^{-\frac{1}{2}}\cdot(6+1)\right]-\left[2(15+1)^{\frac{1}{2}}\right]}{(6+1)^2}} = \boxed{\frac{\left[17.5\cdot(16)^{-\frac{1}{2}}\right]-\left[2\cdot(16)^{\frac{1}{2}}\right]}{7^2}} = \boxed{\frac{[17.5\cdot 0.25]-[2\cdot 4]}{49}} = \boxed{\mathbf{-0.074}}$$

h. Given $y = x\sqrt{x^2-10}+2x = x(x^2-10)^{\frac{1}{2}}+2x$, then

$$\boxed{y'} = \boxed{\left[1\cdot(x^2-10)^{\frac{1}{2}}\right]+\left[0.5(x^2-10)^{\frac{1}{2}-1}\cdot(2x^{2-1}-0)\cdot x\right]+2x^{1-1}} = \boxed{(x^2-10)^{\frac{1}{2}}+\left[x^2(x^2-10)^{-\frac{1}{2}}\right]+2}$$

Substituting $x=5$ in place of x in the y' equation we obtain the following value:

$$\boxed{y'} = \boxed{(5^2-10)^{\frac{1}{2}}+\left[5^2(5^2-10)^{-\frac{1}{2}}\right]+2} = \boxed{15^{\frac{1}{2}}+\left[25\cdot 15^{-\frac{1}{2}}\right]+2} = \boxed{3.873+[25\cdot 0.258]+2} = \boxed{\mathbf{12.323}}$$

i. Given $y = x^2\sqrt{(x+1)^3} = x^2(x+1)^{\frac{3}{2}}$, then

$$\boxed{y'} = \boxed{\left[2x^{2-1}\cdot(x+1)^{\frac{3}{2}}\right]+\left[\frac{3}{2}(x+1)^{\frac{3}{2}-1}\cdot(x^{1-1}+0)\cdot x^2\right]} = \boxed{\left[2x(x+1)^{\frac{3}{2}}\right]+\left[\frac{3}{2}x^3(x+1)^{\frac{1}{2}}\right]}$$

Substituting $x=1$ in place of x in the y' equation we obtain the following value:

$$\boxed{y'} = \boxed{\left[2(1+1)^{\frac{3}{2}}\right]+\left[\frac{3}{2}(1+1)^{\frac{1}{2}}\right]} = \boxed{\left(2\cdot 2^{\frac{3}{2}}\right)+\left(\frac{3}{2}\cdot 2^{\frac{1}{2}}\right)} = \boxed{5.657+2.121} = \boxed{\mathbf{7.778}}$$

j. Given $y = \sqrt{\frac{x^2+1}{1-x^2}} = \left(\frac{x^2+1}{1-x^2}\right)^{\frac{1}{2}}$, then

$$\boxed{y'} = \boxed{\frac{1}{2}\left(\frac{x^2+1}{1-x^2}\right)^{\frac{1}{2}-1}\left[\frac{2x(1-x^2)+2x(x^2+1)}{(1-x^2)^2}\right]} = \boxed{\frac{1}{2}\left(\frac{x^2+1}{1-x^2}\right)^{-\frac{1}{2}}\left[\frac{2x-2x^3+2x^3+2x}{(1-x^2)^2}\right]} = \boxed{\frac{1}{2}\left(\frac{x^2+1}{1-x^2}\right)^{-\frac{1}{2}}\frac{4x}{(1-x^2)^2}}$$

Substituting $x=0$ in place of x in the y' equation we obtain the following value:

$$\boxed{y'} = \boxed{\frac{1}{2}\left(\frac{0+1}{1-0}\right)^{-\frac{1}{2}}\cdot\frac{4\cdot 0}{(1-0)^2}} = \boxed{\frac{1}{2}\cdot 1^{-\frac{1}{2}}\cdot 0} = \boxed{\mathbf{0}}$$

k. Given $y = \frac{\sqrt{x^5+1}}{2x} = \frac{(x^5+1)^{\frac{1}{2}}}{2x}$, then

$$\boxed{y'} = \boxed{\frac{\left[\frac{1}{2}(x^5+1)^{\frac{1}{2}-1}5x^4\cdot 2x\right]-\left[2x^{1-1}\cdot(x^5+1)^{\frac{1}{2}}\right]}{(2x)^2}} = \boxed{\frac{\left[5x^5(x^5+1)^{-\frac{1}{2}}\right]-\left[2\cdot(x^5+1)^{\frac{1}{2}}\right]}{4x^2}}$$

Substituting $x=2$ in place of x in the y' equation we obtain the following value:

$$\boxed{y'} = \boxed{\frac{\left[160(32+1)^{-\frac{1}{2}}\right]-\left[2\cdot(32+1)^{\frac{1}{2}}\right]}{4\cdot 4}} = \boxed{\frac{[160\cdot 0.174]-[2\cdot 5.744]}{16}} = \boxed{\frac{27.8-11.5}{16}} = \boxed{\mathbf{1.02}}$$

l. Given $y = \frac{x+1}{\sqrt{x^3+1}} = \frac{x+1}{(x^3+1)^{\frac{1}{2}}}$, then

$$\boxed{y'} = \boxed{\frac{\left[1\cdot(x^3+1)^{\frac{1}{2}}\right]-\left\{\left[\frac{1}{2}(x^3+1)^{\frac{1}{2}-1}(3x^2+0)\right](x+1)\right\}}{x^3+1}} = \boxed{\frac{(x^3+1)^{\frac{1}{2}}-\left\{\left[\frac{1}{2}(x^3+1)^{-\frac{1}{2}}3x^2\right](x+1)\right\}}{x^3+1}}$$

$$= \boxed{\frac{(x^3+1)^{\frac{1}{2}}-\frac{\frac{3}{2}x^2(x+1)}{(x^3+1)^{\frac{1}{2}}}}{x^3+1}} = \boxed{\frac{\frac{(x^3+1)^{\frac{1}{2}}}{1}-\frac{3x^3+3x^2}{2(x^3+1)^{\frac{1}{2}}}}{x^3+1}} = \boxed{\frac{\frac{(x^3+1)^{\frac{1}{2}}2(x^3+1)^{\frac{1}{2}}-3x^3-3x^2}{2(x^3+1)^{\frac{1}{2}}}}{x^3+1}} = \boxed{\frac{\frac{2(x^3+1)-3x^3-3x^2}{2(x^3+1)^{\frac{1}{2}}}}{x^3+1}}$$

$$= \boxed{\frac{2x^3+2-3x^3-3x^2}{2(x^3+1)^{\frac{1}{2}}(x^3+1)}} = \boxed{\frac{-x^3-3x^2+2}{2(x^3+1)^{\frac{1}{2}}(x^3+1)}} = \boxed{\frac{-x^3-3x^2+2}{2(x^3+1)^{\frac{1}{2}+1}}} = \boxed{\frac{-x^3-3x^2+2}{2(x^3+1)^{\frac{3}{2}}}}$$

Substituting $x=1$ in place of x in the y' equation we obtain the following value:

$$\boxed{y'} = \boxed{\frac{-1^3-(3\cdot 1^2)+2}{2(1^3+1)^{\frac{3}{2}}}} = \boxed{\frac{-1-3+2}{2\cdot 2^{\frac{3}{2}}}} = \boxed{\frac{-2}{2\cdot 2.828}} = \boxed{\frac{-1}{2.828}} = \boxed{\mathbf{-0.354}}$$

Section 5.7 Practice Problems - The Derivative of Radical Functions

1. Find the derivative of the following radical expressions. Do not simplify the answer to its lowest term.

a. $y=\sqrt{x^2+1}$ b. $y=\sqrt{x^3+3x-5}$ c. $y=x^2+\sqrt{x-1}$

d. $y=\dfrac{\sqrt{x+1}}{x}$ e. $y=\dfrac{x^2}{\sqrt{x^2-1}}$ f. $y=\sqrt{x^3}+3x^2$

g. $y=\dfrac{\sqrt{x^2+3}}{\sqrt{x+1}}$ h. $y=\dfrac{\sqrt[4]{x^3-1}}{\sqrt{x}}$ i. $y=\dfrac{x^3}{x^2\sqrt{x}}$

2. Use the $\dfrac{d}{dx}$ notation to find the derivative of the following radical expressions.

a. $\dfrac{d}{dx}\left(\sqrt{x^2}+\dfrac{1}{x}\right)=$ b. $\dfrac{d}{dx}\left(\sqrt{\dfrac{x}{x-1}}\right)=$ c. $\dfrac{d}{dx}\left(\dfrac{x^3}{\sqrt{x+1}}\right)=$

d. $\dfrac{d}{dx}\left(\dfrac{\sqrt{x+5}}{x}\right)=$ e. $\dfrac{d}{dx}\left(x^3+\dfrac{\sqrt{x}}{x}\right)=$ f. $\dfrac{d}{dx}\left(1+\dfrac{2\sqrt{x}}{x^3}\right)=$

3. Find the derivative of the following radical expressions.

a. $\dfrac{d}{dx}\left(\sqrt{x^3}+\sqrt{y}\right)=\dfrac{d}{dx}(x)$ b. $\dfrac{d}{dx}\left(\sqrt{x}+y^3\right)=\dfrac{d}{dx}(2)$ c. $\dfrac{d}{dx}(xy)=\dfrac{d}{dx}\left(\sqrt{x}\right)$

d. $\dfrac{d}{dx}\left(\sqrt{y}+x^3\right)=0$ e. $\dfrac{d}{dx}\left(\sqrt{x^4}+y^2\right)=\dfrac{d}{dx}(x)$ f. $\dfrac{d}{dx}\left(\sqrt{x+1}\right)=\dfrac{d}{dx}\left(y^3\right)$

g. $\dfrac{d}{dx}\left(xy^2+\sqrt{x}\right)=\dfrac{d}{dx}(2)$ h. $\dfrac{d}{dx}\left(\sqrt{x^3}\right)+\dfrac{d}{dx}(xy)=0$ i. $\dfrac{d}{dx}\left(\sqrt{x}+3y\right)=\dfrac{d}{dx}(y)$

4. Evaluate the derivative of the following radical expressions for the specified values of x.

a. $y=\sqrt{3x^3+x^2}$ at $x=1$ b. $y=\left(x^2+1\right)\sqrt{x}$ at $x=0$ c. $y=\dfrac{x^2-1}{\sqrt{4x^2}}$ at $x=2$

d. $y=\sqrt{\dfrac{x}{x^2+1}}$ at $x=1$ e. $y=\sqrt{x^3+1}+4x^3$ at $x=0$ f. $y=\dfrac{x^2+1}{\sqrt{x^3}}$ at $x=3$

5.8 Higher Order Derivatives

If the function $y = f(x)$ is differentiable, then we can form a new function $y' = f'(x)$ which is referred to as the first derivative of $y = f(x)$. Consequently, if $y' = f'(x)$ is differentiable, then we can form another new function $y'' = f''(x)$ called the second derivative of $y = f(x)$. This process of obtaining a new function can continue as long as we have differentiability. First, second, third, and higher order derivatives are denoted in various notations. In general, however, first, second, third, and nth derivatives are shown in the following forms:

$$\boxed{y'} = \boxed{f'(x)} = \boxed{\frac{dy}{dx}} = \boxed{\frac{d}{dx}f(x)} = \boxed{Dy} = \boxed{Df(x)}$$

$$\boxed{y''} = \boxed{f''(x)} = \boxed{\frac{d^2y}{dx^2}} = \boxed{\frac{d}{dx}\left(\frac{dy}{dx}\right)} = \boxed{\frac{d^2}{dx^2}f(x)} = \boxed{\frac{d}{dx}\left[\frac{d}{dx}f(x)\right]} = \boxed{D^2 y} = \boxed{D^2 f(x)}$$

$$\boxed{y'''} = \boxed{f'''(x)} = \boxed{\frac{d^3y}{dx^3}} = \boxed{\frac{d}{dx}\left(\frac{d^2y}{dx^2}\right)} = \boxed{\frac{d^3}{dx^3}f(x)} = \boxed{\frac{d}{dx}\left[\frac{d^2}{dx^2}f(x)\right]} = \boxed{D^3 y} = \boxed{D^3 f(x)}$$

$$\vdots$$

$$\boxed{y^n} = \boxed{f^n(x)} = \boxed{\frac{d^ny}{dx^n}} = \boxed{\frac{d}{dx}\left(\frac{d^{n-1}y}{dx^{n-1}}\right)} = \boxed{\frac{d^n}{dx^n}f(x)} = \boxed{\frac{d}{dx}\left[\frac{d^{n-1}}{dx^{n-1}}f(x)\right]} = \boxed{D^n y} = \boxed{D^n f(x)}$$

Students are encouraged to become familiar with these notations for finding the derivative of different functions. Note that the prime notation is not used beyond the third derivative. In general, the fourth or higher derivatives are shown as $y^4 = f^4(x)$; $y^5 = f^5(x)$; $y^6 = f^6(x)$; $\cdots$; $y^n = f^n(x)$ instead of $y^{iv} = f^{iv}(x)$; $y^v = f^v(x)$; $y^{vi} = f^{vi}(x)$, etc. For example, given $f(x) = x^6 + x^3 + 1$, then $f'(x) = 6x^5 + 3x^2$, $f''(x) = 30x^4 + 6x$, $f'''(x) = 120x^3 + 6$, $f^4(x) = 360x^2$, $f^5(x) = 720x$, $f^6(x) = 720$, and all derivatives higher than 7 are equal to zero. The following examples show in detail how higher order derivatives are obtained:

Example 5.8-1: Find the second derivative of the following functions.

a. $f(x) = 5x^8 - 3x^3 + 1$

b. $f(x) = x^3(x^2 + x + 5)$

c. $f(x) = x^2 + \frac{1}{x}$

d. $f(u) = \frac{u^3 - 1}{u + 1}$

e. $g(x) = x^2 + \frac{1}{x^3}$

f. $h(x) = (a^2 + x^3)^2$

g. $f(x) = (x^2 + 1)^{-1}$

h. $r(\theta) = \theta^2 + \frac{1}{(\theta + 1)^3}$

i. $s(r) = r^2(r^2 + 1)^3$

j. $f(t) = \frac{t^3 + t^2 + 1}{10}$

k. $p(r) = r^2 - \frac{1}{r}$

l. $f(x) = \frac{x^3}{x + 1}$

Solutions:

a. Given $f(x) = 5x^8 - 3x^3 + 1$, then

$\boxed{f'(x)} = \boxed{(5 \cdot 8)x^{8-1} - (3 \cdot 3)x^{3-1} + 0} = \boxed{\mathbf{40x^7 - 9x^2}}$ and

$\boxed{f''(x)} = \boxed{(40\cdot 7)x^{7-1} - (9\cdot 2)x^{2-1}} = \boxed{\mathbf{280x^6 - 18x}}$

b. Given $f(x) = x^3(x^2 + x + 5)$, then

$\boxed{f'(x)} = \boxed{\left[3x^{3-1}\cdot(x^2 + x + 5)\right] + \left[(2x^{2-1} + 1x^{1-1} + 0)\cdot x^3\right]} = \boxed{\left[3x^2(x^2 + x + 5)\right] + \left[(2x + 1)\cdot x^3\right]}$

$= \boxed{3x^4 + 3x^3 + 15x^2 + 2x^4 + x^3} = \boxed{\mathbf{5x^4 + 4x^3 + 15x^2}}$ and

$\boxed{f''(x)} = \boxed{(5\cdot 4)x^{4-1} + (4\cdot 3)x^{3-1} + (15\cdot 2)x^{2-1}} = \boxed{\mathbf{20x^3 + 12x^2 + 30x}}$

c. Given $f(x) = x^2 + \dfrac{1}{x}$, then

$\boxed{f'(x)} = \boxed{2x^{2-1} + \dfrac{[0\cdot x] - (1\cdot 1)}{x^2}} = \boxed{2x + \dfrac{0-1}{x^2}} = \boxed{\mathbf{2x - \dfrac{1}{x^2}}}$

A second way is to rewrite $f(x)$ as $f(x) = x^2 + x^{-1}$ and find its derivative, i.e.,

$\boxed{f'(x)} = \boxed{2x^{2-1} - 1\cdot x^{-1-1}} = \boxed{2x - x^{-2}} = \boxed{\mathbf{2x - \dfrac{1}{x^2}}}$ and

$\boxed{f''(x)} = \boxed{2x^{1-1} + (-1\cdot -2)x^{-2-1}} = \boxed{2x^0 + 2x^{-3}} = \boxed{2 + 2x^{-3}} = \boxed{\mathbf{2 + \dfrac{2}{x^3}}}$

d. Given $f(u) = \dfrac{u^3 - 1}{u + 1}$, then

$\boxed{f'(x)} = \boxed{\dfrac{\left[(3u^{3-1} - 0)\cdot(u+1)\right] - \left[(u^{1-1} + 0)\cdot(u^3 - 1)\right]}{(u+1)^2}} = \boxed{\dfrac{3u^2\cdot(u+1) - (u^3 - 1)}{(u+1)^2}} = \boxed{\dfrac{3u^3 + 3u^2 - u^3 + 1}{(u+1)^2}} = \boxed{\mathbf{\dfrac{2u^3 + 3u^2 + 1}{(u+1)^2}}}$

$\boxed{f''(x)} = \boxed{\dfrac{\left[(2\cdot 3u^{3-1} + 3\cdot 2u^{2-1} + 0)\cdot(u+1)^2\right] - \left[2(u+1)^{2-1}\cdot(2u^3 + 3u^2 + 1)\right]}{(u+1)^4}}$

$= \boxed{\dfrac{\left[(6u^2 + 6u)(u+1)^2\right] - \left[2(u+1)(2u^3 + 3u^2 + 1)\right]}{(u+1)^4}} = \boxed{\mathbf{\dfrac{\left[(6u^2 + 6u)(u+1)^2\right] + \left[(-2u - 2)(2u^3 + 3u^2 + 1)\right]}{(u+1)^4}}}$

e. Given $g(x) = x^2 + \dfrac{1}{x^3}$, then

$\boxed{g'(x)} = \boxed{2x^{2-1} + \dfrac{(0\cdot x^3) - (3x^{3-1}\cdot 1)}{x^6}} = \boxed{2x + \dfrac{-3x^2}{x^6}} = \boxed{2x - \dfrac{3x^2}{x^{6=4}}} = \boxed{\mathbf{2x - \dfrac{3}{x^4}}}$

A second way is to rewrite $g(x)$ as $g(x) = x^2 + x^{-3}$ and find its derivative, i.e.,

$\boxed{g'(x)} = \boxed{2x^{2-1} - 3 \cdot x^{-1-3}} = \boxed{2x - 3x^{-4}} = \boxed{\mathbf{2x - \frac{3}{x^4}}}$ and

$\boxed{g''(x)} = \boxed{2x^{1-1} + (-3 \cdot -4)x^{-4-1}} = \boxed{2x^0 + 12x^{-5}} = \boxed{2 + 12x^{-5}} = \boxed{\mathbf{2 + \frac{12}{x^5}}}$

f. Given $h(x) = \left(a^2 + x^3\right)^2$, then

$\boxed{h'(x)} = \boxed{2\left(a^2 + x^3\right)^{2-1} \cdot \left(0 + 3x^{3-1}\right)} = \boxed{2\left(a^2 + x^3\right) \cdot 3x^2} = \boxed{6x^2\left(a^2 + x^3\right)} = \boxed{6a^2x^2 + 6x^5} = \boxed{\mathbf{6x^5 + 6a^2x^2}}$

$\boxed{h''(x)} = \boxed{(6 \cdot 5)x^{5-1} + \left(6a^2 \cdot 2\right)x^{2-1}} = \boxed{\mathbf{30x^4 + 12a^2x}}$

g. Given $f(x) = \left(x^2 + 1\right)^{-1}$, then

$\boxed{f'(x)} = \boxed{\left[-1 \cdot \left(x^2 + 1\right)^{-1-1}\right] \cdot \left(2x^{2-1} + 0\right)} = \boxed{\left[-\left(x^2 + 1\right)^{-2}\right] \cdot 2x} = \boxed{\mathbf{-2x\left(x^2 + 1\right)^{-2}}}$

$\boxed{f''(x)} = \boxed{-2\left\{1 \cdot \left(x^2 + 1\right)^{-2} + \left[-2\left(x^2 + 1\right)^{-2-1} \cdot \left(2x^{2-1} + 0\right)\right]\right\}} = \boxed{-2\left\{\left(x^2 + 1\right)^{-2} + \left[-2\left(x^2 + 1\right)^{-3} \cdot 2x\right]\right\}}$

$= \boxed{-2\left\{\left(x^2 + 1\right)^{-2} - 4x\left(x^2 + 1\right)^{-3}\right\}} = \boxed{\mathbf{-2\left(x^2 + 1\right)^{-2} + 8x\left(x^2 + 1\right)^{-3}}}$

h. Given $r(\theta) = \theta^2 + \dfrac{1}{(\theta + 1)^3}$, then

$\boxed{r'(\theta)} = \boxed{2\theta^{2-1} + \dfrac{\left[0 \cdot (\theta + 1)^3\right] - \left[3(\theta + 1)^{3-1} \cdot 1\right]}{(\theta + 1)^6}} = \boxed{2\theta + \dfrac{0 - 3(\theta + 1)^2}{(\theta + 1)^6}} = \boxed{2\theta - \dfrac{3(\theta + 1)^2}{(\theta + 1)^{6=4}}} = \boxed{\mathbf{2\theta - \dfrac{3}{(\theta + 1)^2}}}$ and

$\boxed{r''(\theta)} = \boxed{2\theta^{1-1} - \dfrac{\left[0 \cdot (\theta + 1)^2\right] - \left[2(\theta + 1)^{2-1} \cdot 3\right]}{(\theta + 1)^4}} = \boxed{2 - \dfrac{0 - 6(\theta + 1)}{(\theta + 1)^4}} = \boxed{2 + \dfrac{6(\theta + 1)}{(\theta + 1)^{4=3}}} = \boxed{\mathbf{2 + \dfrac{6}{(\theta + 1)^3}}}$

i. Given $s(r) = r^2\left(r^2 + 1\right)^3$, then

$\boxed{s'(r)} = \boxed{\left[2r^{2-1} \cdot \left(r^2 + 1\right)^3\right] + \left[3\left(r^2 + 1\right)^{3-1} \cdot \left(2r^{2-1} + 0\right)\right] \cdot r^2} = \boxed{\left[2r\left(r^2 + 1\right)^3\right] + \left[3\left(r^2 + 1\right)^2 \cdot 2r\right] \cdot r^2}$

$= \boxed{\mathbf{2r\left(r^2 + 1\right)^3 + 6r^3\left(r^2 + 1\right)^2}}$ and

$\boxed{s''(r)} = \boxed{2\left\{\left[1 \cdot \left(r^2 + 1\right)^3\right] + \left[3\left(r^2 + 1\right)^{3-1} \cdot \left(2r^{2-1} + 0\right)\right] \cdot r\right\} + 6\left\{\left[3r^{3-1} \cdot \left(r^2 + 1\right)^2\right] + \left[2\left(r^2 + 1\right)^{2-1} \cdot \left(2r^{2-1} + 0\right)\right] \cdot r^3\right\}}$

$$= \boxed{2\left\{\left(r^2+1\right)^3 + \left[3\left(r^2+1\right)^{3-1}\cdot 2r\right]\cdot r\right\} + 6\left\{\left[3r^2\left(r^2+1\right)^2\right] + \left[2\left(r^2+1\right)\cdot 2r\right]\cdot r^3\right\}}$$

$$= \boxed{\mathbf{2\left\{\left(r^2+1\right)^3 + \left[6r^2\left(r^2+1\right)^2\right]\right\} + 6\left\{\left[3r^2\left(r^2+1\right)^2\right] + \left[4r^4\left(r^2+1\right)\right]\right\}}}$$

j. Given $f(t) = \frac{t^3+t^2+1}{10}$, then

$$\boxed{f'(t)} = \boxed{\frac{\left[\left(3t^{3-1}+2t^{2-1}+0\right)\cdot 10\right] - \left[0\cdot\left(t^3+t^2+1\right)\right]}{10^2}} = \boxed{\frac{10\left(3t^2+2t\right)-0}{100}} = \boxed{\frac{\not{10}\left(3t^2+2t\right)}{\not{100}=10}} = \boxed{\mathbf{\frac{3t^2+2t}{10}}} \text{ and}$$

$$\boxed{f''(t)} = \boxed{\frac{\left[\left(3\cdot 2t^{2-1}+2t^{1-1}\right)\cdot 10\right] - \left[0\cdot\left(3t^2+2t\right)\right]}{10^2}} = \boxed{\frac{10(6t+2)-0}{100}} = \boxed{\frac{\not{10}(6t+2)}{\not{100}=10}} = \boxed{\frac{2\,(3t+1)}{\not{10}=5}} = \boxed{\mathbf{\frac{3t+1}{5}}}$$

k. Given $p(r) = r^2 - \frac{1}{r}$ which is equal to $p(r) = r^2 - r^{-1}$, then

$$\boxed{p'(r)} = \boxed{2r^{2-1} - \left(-1\cdot r^{-1-1}\right)} = \boxed{\mathbf{2r+r^{-2}}} \text{ and}$$

$$\boxed{p''(r)} = \boxed{2r^{1-1} - 2r^{-2-1}} = \boxed{2r^0 - 2r^{-3}} = \boxed{\mathbf{2-2r^{-3}}}$$

l. Given $f(x) = \frac{x^3}{x+1}$, then

$$\boxed{f'(x)} = \boxed{\frac{\left[3x^{3-1}\cdot(x+1)\right] - \left[\left(1\cdot x^{1-1}+0\right)\cdot x^3\right]}{(x+1)^2}} = \boxed{\frac{3x^2(x+1)-x^3}{(x+1)^2}} = \boxed{\frac{3x^3+3x^2-x^3}{(x+1)^2}} = \boxed{\mathbf{\frac{2x^3+3x^2}{(x+1)^2}}} \text{ and}$$

$$\boxed{f''(x)} = \boxed{\frac{\left\{\left[(2\cdot 3)x^2+(3\cdot 2)x\right]\cdot(x+1)^2\right\} - \left\{2(x+1)\cdot\left(2x^3+3x^2\right)\right\}}{(x+1)^4}} = \boxed{\mathbf{\frac{\left[\left(6x^2+6x\right)(x+1)^2\right] - \left[2(x+1)\left(2x^3+3x^2\right)\right]}{(x+1)^4}}}$$

Example 5.8-2: Find $\frac{d^3y}{dx^3}$ for the following functions.

a. $y = (1-5x)^3$ b. $y = \left(a-bx^2\right)^{-2}$ c. $y = \frac{x^2+3x+1}{x+1}$

d. $y = \frac{1}{5}x^5 + \frac{1}{4}x^4 + x$ e. $y = \frac{ax^2+b}{c}$ f. $y = \frac{x^2+1}{x^3}$

Solutions:

a. Given $y = (1-5x)^3$, then

$$\boxed{y'} = \boxed{3(1-5x)^{3-1}\cdot\left(0-5x^{1-1}\right)} = \boxed{3(1-5x)^2\cdot(-5)} = \boxed{\mathbf{-15(1-5x)^2}}$$

$$\boxed{y''} = \boxed{(-15\cdot 2)(1-5x)^{2-1}\cdot\left(0-5x^{1-1}\right)} = \boxed{-30(1-5x)\cdot(-5)} = \boxed{\mathbf{150(1-5x)}}$$

$\boxed{y'''} = \boxed{150\left(0-5x^{1-1}\right)} = \boxed{150\cdot(-5)} = \boxed{\mathbf{-750}}$

b. Given $y=\left(a-bx^2\right)^{-2}$, then

$$\boxed{y'} = \boxed{-2\left(a-bx^2\right)^{-2-1}\cdot(0-2bx)} = \boxed{-2\left(a-bx^2\right)^{-3}\cdot(-2bx)} = \boxed{\mathbf{4bx\left(a-bx^2\right)^{-3}}}$$

$$\boxed{y''} = \boxed{\left[(4b\cdot 1)\left(a-bx^2\right)^{-3}\right]+\left\{\left[-3\left(a-bx^2\right)^{-3-1}\cdot(0-2bx)\right]\cdot(4bx)\right\}} = \boxed{\left[4b\left(a-bx^2\right)^{-3}\right]+\left[6bx\left(a-bx^2\right)^{-4}\cdot(4bx)\right]}$$

$$= \boxed{\left[\mathbf{4b\left(a-bx^2\right)^{-3}}\right]+\left[\mathbf{24b^2x^2\left(a-bx^2\right)^{-4}}\right]}$$

$$\boxed{y'''} = \boxed{\left[-12b\left(a-bx^2\right)^{-3-1}\cdot(-2bx)\right]+\left[\left(48b^2x\right)\cdot\left(a-bx^2\right)^{-4}+\left(-96b^2x^2\right)\left(a-bx^2\right)^{-4-1}\cdot(-2bx)\right]}$$

$$= \boxed{\left[\mathbf{24b^2x\left(a-bx^2\right)^{-4}}\right]+\left[\mathbf{48b^2x\left(a-bx^2\right)^{-4}+192b^3x^3\left(a-bx^2\right)^{-5}}\right]}$$

c. Given $y=\dfrac{x^2+3x+1}{x+1}$, then

$$\boxed{y'} = \boxed{\frac{\left[\left(2x^{2-1}+3x^{1-1}+0\right)\cdot(x+1)\right]-\left[\left(x^{1-1}+0\right)\cdot\left(x^2+3x+1\right)\right]}{(x+1)^2}} = \boxed{\frac{\left[(2x+3)\cdot(x+1)\right]-\left[1\cdot\left(x^2+3x+1\right)\right]}{(x+1)^2}}$$

$$= \boxed{\frac{\left(2x^2+2x+3x+3\right)-\left(x^2+3x+1\right)}{(x+1)^2}} = \boxed{\frac{2x^2+5x+3-x^2-3x-1}{(x+1)^2}} = \boxed{\mathbf{\frac{x^2+2x+2}{(x+1)^2}}}$$

$$\boxed{y''} = \boxed{\frac{\left[\left(2x^{2-1}+2x^{1-1}+0\right)\cdot(x+1)^2\right]-\left[2(x+1)^{2-1}\cdot\left(x^2+2x+2\right)\right]}{(x+1)^4}} = \boxed{\frac{\left[(2x+2)(x+1)^2\right]-\left[2(x+1)\left(x^2+2x+2\right)\right]}{(x+1)^4}}$$

$$= \boxed{\frac{\left[(2x+2)(x+1)^2\right]-\left[2(x+1)\left(x^2+2x+2\right)\right]}{(x+1)^4}} = \boxed{\frac{\left[2x^3+6x^2+6x+2\right]-\left[2x^3+6x^2+8x+4\right]}{(x+1)^4}}$$

$$= \boxed{\frac{2x^3+6x^2+6x+2-2x^3-6x^2-8x-4}{(x+1)^4}} = \boxed{\mathbf{\frac{-2x-2}{(x+1)^4}}}$$

$$\boxed{y'''} = \boxed{\frac{\left[\left(-2x^{1-1}+0\right)\cdot(x+1)^4\right]-\left[4(x+1)^{4-1}\cdot(-2x-2)\right]}{(x+1)^8}} = \boxed{\mathbf{\frac{\left[-2(x+1)^4\right]+\left[(x+1)^3(8x+8)\right]}{(x+1)^8}}}$$

d. Given $y=\frac{1}{5}x^5+\frac{1}{4}x^4+x$, then

$\boxed{y'}=\boxed{\frac{1}{5}\cdot 5x^{5-1}+\frac{1}{4}\cdot 4x^{4-1}+x^{1-1}}=\boxed{\mathbf{x^4+x^3+1}}$ $\boxed{y''}=\boxed{4x^{4-1}+3x^{3-1}+0}=\boxed{\mathbf{4x^3+3x^2}}$ and

$\boxed{y'''}=\boxed{(4\cdot 3)x^{3-1}+(3\cdot 2)x^{2-1}}=\boxed{\mathbf{12x^2+6x}}$

e. Given $y=\frac{ax^2+b}{c}$, then

$\boxed{y'}=\boxed{\frac{\left[\left(a\cdot 2x^{2-1}+0\right)\cdot c\right]-\left[0\cdot\left(ax^2+b\right)\right]}{c^2}}=\boxed{\frac{(2ax\cdot c)-0}{c^2}}=\boxed{\frac{2a\not{c}x}{c^{\not{2}=1}}}=\boxed{\mathbf{\frac{2ax}{c}}}$ $\boxed{y''}=\boxed{\mathbf{\frac{2a}{c}}}$ and $\boxed{y'''}=\boxed{\mathbf{0}}$

f. Given $y=\frac{x^2+1}{x^3}$, then

$\boxed{y'}=\boxed{\frac{\left[\left(2x^{2-1}+0\right)\cdot x^3\right]-\left[3x^{3-1}\cdot\left(x^2+1\right)\right]}{x^6}}=\boxed{\frac{\left[2x\cdot x^3\right]-\left[3x^2\cdot\left(x^2+1\right)\right]}{x^6}}=\boxed{\frac{2x^4-3x^4-3x^2}{x^6}}=\boxed{\frac{-x^4-3x^2}{x^6}}$

$=\boxed{\frac{-x^{\not{2}}\left(x^2+3\right)}{x^{\not{6}=4}}}=\boxed{\mathbf{-\frac{x^2+3}{x^4}}}$

$\boxed{y''}=-\boxed{\frac{\left[\left(2x^{2-1}+0\right)\cdot x^4\right]-\left[4x^{4-1}\cdot\left(x^2+3\right)\right]}{x^8}}=-\boxed{\frac{\left[2x\cdot x^4\right]-\left[4x^3\cdot\left(x^2+3\right)\right]}{x^8}}=-\boxed{\frac{2x^5-4x^5-12x^3}{x^8}}$

$=-\boxed{\frac{-2x^5-12x^3}{x^8}}=\boxed{\frac{x^{\not{3}}\left(2x^2+12\right)}{x^{\not{8}=5}}}=\boxed{\mathbf{\frac{2x^2+12}{x^5}}}$

$\boxed{y'''}=\boxed{\frac{\left[\left(4x^{2-1}+0\right)\cdot x^5\right]-\left[5x^{5-1}\cdot\left(2x^2+12\right)\right]}{x^{10}}}=\boxed{\frac{\left[4x\cdot x^5\right]-\left[5x^4\cdot\left(2x^2+12\right)\right]}{x^{10}}}=\boxed{\frac{4x^6-10x^6-60x^4}{x^{10}}}$

$=\boxed{\frac{-6x^6-60x^4}{x^{10}}}=\boxed{\frac{x^{\not{4}}\left(-6x^2-60\right)}{x^{\not{10}=6}}}=\boxed{\mathbf{-\frac{6x^2+60}{x^6}}}$

Example 5.8-3: Find $f''(0)$ and $f''(1)$ for the following functions.

a. $f(x)=6x^7+7x^2-2$ b. $f(x)=x^5(x-1)^2$ c. $f(x)=x-\frac{1}{x}$

d. $f(x)=\frac{x^3+1}{x}$ e. $f(x)=x^3-\frac{1}{x+1}$ f. $f(x)=(ax+b)^2$

g. $f(x)=(x-1)^{-2}$ h. $f(x)=\frac{(x+1)^2}{x}$ i. $f(x)=(x+1)\left(x^2+1\right)+5$

j. $f(x)=(1+5x)^3$ k. $f(x)=\frac{1+x}{x^3}$ l. $f(x)=\frac{1}{3}x^3+\frac{1}{2}x^2+x+10$

Solutions:

a. Given $f(x) = 6x^7 + 7x^2 - 2$, then

$$f'(x) = 6\cdot 7x^{7-1} + 7\cdot 2x^{2-1} - 0 = 42x^6 + 14x \text{ and } f''(x) = 42\cdot 6x^{6-1} + 14\cdot 1x^{1-1} = 252x^5 + 14$$

Therefore, $f''(0) = 252\cdot 0^5 + 14 = 0 + 14 = \mathbf{14}$ and $f''(1) = 252\cdot 1^5 + 14 = 252 + 14 = \mathbf{266}$

b. Given $f(x) = x^5(x-1)^2$, then

$$f'(x) = \left(5x^{5-1}\cdot 1\right)\cdot(x-1)^2 + [2(x-1)\cdot 1]\cdot x^5 = 5x^4(x-1)^2 + 2x^5(x-1) = 5x^4(x-1)^2 + 2x^6 - 2x^5 \text{ and}$$

$$f''(x) = \left[5\cdot 4x^{4-1}(x-1)^2 + 2(x-1)\cdot 5x^4\right] + 2\cdot 6x^{6-1} - 2\cdot 5x^{5-1} = 20x^3(x-1)^2 + 10x^4(x-1) + 12x^5 - 10x^4$$

$$= 20x^3\left(x^2 - 2x + 1\right) + 10x^5 - 10x^4 + 12x^5 - 10x^4 = 20x^5 - 40x^4 + 20x^3 + 22x^5 - 20x^4$$

$$= 42x^5 - 60x^4 + 20x^3$$ Therefore,

$$f''(0) = 42\cdot 0^5 - 60\cdot 0^4 + 20\cdot 0^3 = \mathbf{0} \text{ and } f''(1) = 42\cdot 1^5 - 60\cdot 1^4 + 20\cdot 1^3 = 42 - 60 + 20 = \mathbf{2}$$

c. Given $f(x) = x - \frac{1}{x}$, then

$$f'(x) = 1 - \frac{0\cdot x - 1\cdot 1}{x^2} = 1 + \frac{1}{x^2} \text{ and } f''(x) = 0 + \frac{0\cdot x - 2x\cdot 1}{x^4} = -\frac{2x}{x^4} = -\frac{2}{x^3}$$

Therefore, $f''(0) = -\frac{2}{0^3} = -\mathbf{\frac{2}{0}}$ which is not defined and $f''(1) = -\frac{2}{1^3} = -\frac{2}{1} = \mathbf{-2}$

d. Given $f(x) = \frac{x^3 + 1}{x}$, then

$$f'(x) = \frac{\left[3x^2\cdot x\right] - \left[1\cdot\left(x^3 + 1\right)\right]}{x^2} = \frac{3x^3 - x^3 - 1}{x^2} = \frac{2x^3 - 1}{x^2} \text{ and}$$

$$f''(x) = \frac{\left[6x^2\cdot x^2\right] - \left[2x\cdot\left(2x^3 - 1\right)\right]}{x^4} = \frac{6x^4 - 4x^4 + 2x}{x^4} = \frac{2x^4 + 2x}{x^4} = \frac{2\cancel{x}\left(x^3 + 1\right)}{x^{4-3}} = \frac{2\left(x^3 + 1\right)}{x^3}$$

Therefore, $f''(0) = \frac{2\left(0^3 + 1\right)}{0^3} = \mathbf{\frac{2}{0}}$ which is not defined and $f''(1) = \frac{2\left(1^3 + 1\right)}{1^3} = \frac{4}{1} = \mathbf{4}$

e. Given $f(x) = x^3 - \frac{1}{x+1}$, then

$$f'(x) = 3x^{3-1} - \frac{0\cdot(x+1) - 1\cdot 1}{(x+1)^2} = 3x^2 + \frac{1}{(x+1)^2} \text{ and}$$

$\boxed{f''(x)} = \boxed{3\cdot 2x^{2-1} + \frac{0\cdot(x+1)^2 - 2(x+1)^{2-1}\cdot 1}{(x+1)^4}} = \boxed{6x - \frac{2(x+1)}{(x+1)^4}} = \boxed{6x - \frac{2}{(x+1)^3}}$ Therefore,

$\boxed{f''(0)} = \boxed{6\cdot 0 - \frac{2}{(0+1)^3}} = \boxed{-\frac{2}{1}} = \boxed{-2}$ and $\boxed{f''(1)} = \boxed{6\cdot 1 - \frac{2}{(1+1)^3}} = \boxed{6 - \frac{2}{2^3}} = \boxed{6 - \frac{2}{8}} = \boxed{6 - \frac{1}{4}} = \boxed{\mathbf{5.75}}$

f. Given $f(x) = (ax+b)^2$, then

$\boxed{f'(x)} = \boxed{2(ax+b)^{2-1}\cdot a} = \boxed{2a(ax+b)} = \boxed{2a^2x + 2ab}$ and $\boxed{f''(x)} = \boxed{2a^2 + 0} = \boxed{2a^2}$

Therefore, $\boxed{f''(0)} = \boxed{\mathbf{2a^2}}$ and $\boxed{f''(1)} = \boxed{\mathbf{2a^2}}$

Note that since $f''(x)$ is independent of x, $f''(x)$ is equal to $2a^2$ for all values of x.

g. Given $f(x) = (x-1)^{-2}$, then

$\boxed{f'(x)} = \boxed{-2(x-1)^{-2-1}\cdot 1} = \boxed{-2(x-1)^{-3}}$ and $\boxed{f''(x)} = \boxed{-2\cdot\left[-3(x-1)^{-3-1}\cdot 1\right]} = \boxed{6(x-1)^{-4}}$ Therefore,

$\boxed{f''(0)} = \boxed{6(0-1)^{-4}} = \boxed{\frac{6}{(-1)^4}} = \boxed{\frac{6}{1}} = \boxed{\mathbf{6}}$ and $\boxed{f''(1)} = \boxed{6(1-1)^{-4}} = \boxed{\frac{6}{0^4}} = \boxed{\frac{\mathbf{6}}{\mathbf{0}}}$ which is undefined

h. Given $f(x) = \frac{(x+1)^2}{x}$, then

$\boxed{f'(x)} = \boxed{\frac{\left[2(x+1)^{2-1}\cdot 1\cdot x\right] - \left[1\cdot(x+1)^2\right]}{x^2}} = \boxed{\frac{2x(x+1) - (x+1)^2}{x^2}} = \boxed{\frac{2x^2 + 2x - x^2 - 2x - 1}{x^2}} = \boxed{\frac{x^2 - 1}{x^2}}$ and

$\boxed{f''(x)} = \boxed{\frac{\left[2x\cdot x^2\right] - \left[2x\cdot(x^2-1)\right]}{x^4}} = \boxed{\frac{2x^3 - 2x^3 + 2x}{x^4}} = \boxed{\frac{2x}{x^4}} = \boxed{\frac{2}{x^3}}$

Therefore, $\boxed{f''(0)} = \boxed{\frac{2}{0^3}} = \boxed{\frac{\mathbf{2}}{\mathbf{0}}}$ which is undefined and $\boxed{f''(1)} = \boxed{\frac{2}{1^3}} = \boxed{\frac{2}{1}} = \boxed{\mathbf{2}}$

i. Given $\boxed{f(x)} = \boxed{(x+1)(x^2+1)+5} = \boxed{x^3 + x + x^2 + 1 + 5} = \boxed{x^3 + x^2 + x + 6}$, then

$\boxed{f'(x)} = \boxed{3x^2 + 2x + 1}$ and $\boxed{f''(x)} = \boxed{6x + 2}$

Therefore, $\boxed{f''(0)} = \boxed{6\cdot 0 + 2} = \boxed{\mathbf{2}}$ and $\boxed{f''(1)} = \boxed{6\cdot 1 + 2} = \boxed{\mathbf{8}}$

j. Given $f(x) = (1+5x)^3$, then

$\boxed{f'(x)} = \boxed{3(1+5x)^{3-1}\cdot 5} = \boxed{15(1+5x)^2}$ and $\boxed{f''(x)} = \boxed{0\cdot(1+5x)^2 + 2(1+5x)\cdot 5\cdot 15} = \boxed{150(1+5x)}$

Therefore, $\boxed{f''(0)} = \boxed{150(1+5\cdot 0)} = \boxed{\mathbf{150}}$ and $\boxed{f''(1)} = \boxed{150(1+5\cdot 1)} = \boxed{150(1+5)} = \boxed{150\cdot 6} = \boxed{\mathbf{900}}$

k. Given $f(x) = \frac{1+x}{x^3}$, then

$$\boxed{f'(x)} = \boxed{\frac{1\cdot x^3 - 3x^2(1+x)}{x^6}} = \boxed{\frac{x^3 - 3x^2 - 3x^3}{x^6}} = \boxed{\frac{-2x^3 - 3x^2}{x^6}} = \boxed{\frac{-x^2(2x+3)}{x^{6-4}}} = \boxed{\frac{-2x-3}{x^4}}$$

$$\boxed{f''(x)} = \boxed{\frac{\left[-2\cdot x^4\right] - \left[4x^3(-2x-3)\right]}{x^8}} = \boxed{\frac{-2x^4 + 8x^4 + 12x^3}{x^8}} = \boxed{\frac{6x^4 + 12x^3}{x^8}} = \boxed{\frac{6x^3(x+2)}{x^{8-5}}} = \boxed{\frac{6x+12}{x^5}}$$

Therefore, $f''(0) = \boxed{\frac{6\cdot 0+12}{0^5}} = \boxed{\frac{\mathbf{12}}{\mathbf{0}}}$ which is not defined and $\boxed{f''(1)} = \boxed{\frac{6\cdot 1+12}{1^5}} = \boxed{\frac{18}{1}} = \boxed{\mathbf{18}}$

l. Given $f(x) = \frac{1}{3}x^3 + \frac{1}{2}x^2 + x + 10$, then

$$\boxed{f'(x)} = \boxed{\frac{3}{3}x^{3-1} + \frac{2}{2}x^{2-1} + x^{1-1} + 0} = \boxed{x^2 + x + 1} \text{ and } \boxed{f''(x)} = \boxed{2x^{2-1} + x^{1-1} + 0} = \boxed{2x+1}$$

Therefore, $\boxed{f''(0)} = \boxed{2\cdot 0+1} = \boxed{1}$ and $\boxed{f''(1)} = \boxed{2\cdot 1+1} = \boxed{3}$

Example 5.8-4: Find $\frac{dy}{dx}$, $\frac{d^2y}{dx^2}$, and $\frac{d^3y}{dx^3}$ for the following functions.

a. $y = x^4 + 5x^3 + 6x^2 + 1$

b. $y = x + \frac{1}{x}$

c. $y = x(x+1)^3$

d. $y = \left(x^2+1\right)^{-2}$

e. $y = x^3 + 3x^2 + 10$

f. $y = \frac{1}{1+x}$

g. $y = x - \frac{1}{x}$

h. $y = ax^3 + bx$

i. $y = \frac{x^3+1}{x^2}$

Solutions:

a. Given $y = x^4 + 5x^3 + 6x^2 + 1$, then

$$\boxed{\frac{dy}{dx}} = \boxed{\frac{d}{dx}x^4 + 5\frac{d}{dx}x^3 + 6\frac{d}{dx}x^2 + \frac{d}{dx}1} = \boxed{4x^{4-1} + (5\cdot 3)x^{3-1} + (6\cdot 2)x^{2-1} + 0} = \boxed{\mathbf{4x^3 + 15x^2 + 12x}}$$

$$\boxed{\frac{d^2y}{dx^2}} = \boxed{4\frac{d}{dx}x^3 + 15\frac{d}{dx}x^2 + 12\frac{d}{dx}x} = \boxed{(4\cdot 3)x^{3-1} + (15\cdot 2)x^{2-1} + (12\cdot 1)} = \boxed{\mathbf{12x^2 + 30x + 12}}$$

$$\boxed{\frac{d^3y}{dx^3}} = \boxed{12\frac{d}{dx}x^2 + 30\frac{d}{dx}x + \frac{d}{dx}12} = \boxed{(12\cdot 2)x^{2-1} + (30\cdot 1)x^{1-1} + 0} = \boxed{24x^1 + 30x^0} = \boxed{\mathbf{24x + 30}}$$

b. Given $y = x + \frac{1}{x}$ which is the same as $y = x + x^{-1}$, then

$$\boxed{\frac{dy}{dx}} = \boxed{\frac{d}{dx}x + \frac{d}{dx}x^{-1}} = \boxed{1 + \left(-1\cdot x^{-1-1}\right)} = \boxed{\mathbf{1 - x^{-2}}}$$

$$\boxed{\frac{d^2y}{dx^2}} = \boxed{\frac{d}{dx}1 - \frac{d}{dx}x^{-2}} = \boxed{0 - \left(-2\cdot x^{-2-1}\right)} = \boxed{\mathbf{2x^{-3}}}$$

$$\boxed{\frac{d^3y}{dx^3}} = \boxed{2\frac{d}{dx}x^{-3}} = \boxed{(2\cdot -3)x^{-3-1}} = \boxed{\mathbf{-6x^{-4}}}$$

c. Given $y = x(x+1)^3$, then

$$\boxed{\frac{dy}{dx}} = \boxed{\left[(x+1)^3\frac{d}{dx}x\right]+\left[x\frac{d}{dx}(x+1)^3\right]} = \boxed{\left[(x+1)^3\cdot 1\right]+\left[x\cdot 3(x+1)^{3-1}\cdot 1\right]} = \boxed{\mathbf{(x+1)^3+3x(x+1)^2}}$$

$$\boxed{\frac{d^2y}{dx^2}} = \boxed{\frac{d}{dx}(x+1)^3+\left[3(x+1)^2\frac{d}{dx}x\right]+\left[3x\frac{d}{dx}(x+1)^2\right]} = \boxed{\left[3(x+1)^{3-1}\cdot 1\right]+\left[3(x+1)^2\cdot 1\right]+\left[3x\cdot 2(x+1)^{2-1}\cdot 1\right]}$$

$$= \boxed{3(x+1)^2+3(x+1)^2+6x(x+1)} = \boxed{\mathbf{6(x+1)^2+6x(x+1)}}$$

$$\boxed{\frac{d^3y}{dx^3}} = \boxed{6\frac{d}{dx}(x+1)^2+\left\{\left[6(x+1)\frac{d}{dx}x\right]+\left[6x\frac{d}{dx}(x+1)\right]\right\}} = \boxed{\left[12(x+1)^{2-1}\cdot 1\right]+\left\{\left[6(x+1)\cdot 1\right]+\left[6x\cdot(x+1)^{1-1}\cdot 1\right]\right\}}$$

$$= \boxed{12(x+1)+6(x+1)+6x(x+1)^0} = \boxed{18(x+1)+6x} = \boxed{18x+18+6x} = \boxed{24x+18} = \boxed{\mathbf{6(4x+3)}}$$

d. Given $y = (x+1)^{-2}$, then

$$\boxed{\frac{dy}{dx}} = \boxed{\frac{d}{dx}(x+1)^{-2}} = \boxed{\left[-2(x+1)^{-2-1}\right]\cdot\frac{d}{dx}(x+1)} = \boxed{-2(x+1)^{-3}\cdot 1} = \boxed{\mathbf{-2(x+1)^{-3}}}$$

$$\boxed{\frac{d^2y}{dx^2}} = \boxed{-2\frac{d}{dx}(x+1)^{-3}} = \boxed{\left[(-2\cdot -3)(x+1)^{-3-1}\right]\cdot\frac{d}{dx}(x+1)} = \boxed{6(x+1)^{-4}\cdot 1} = \boxed{\mathbf{6(x+1)^{-4}}}$$

$$\boxed{\frac{d^3y}{dx^3}} = \boxed{6\frac{d}{dx}(x+1)^{-4}} = \boxed{\left[(6\cdot -4)(x+1)^{-4-1}\right]\cdot\frac{d}{dx}(x+1)} = \boxed{-24(x+1)^{-5}\cdot 1} = \boxed{\mathbf{-24(x+1)^{-5}}}$$

e. Given $y = x^3+3x^2+10$, then

$$\boxed{\frac{dy}{dx}} = \boxed{\frac{d}{dx}x^3+\frac{d}{dx}3x^2+\frac{d}{dx}10} = \boxed{\frac{d}{dx}x^3+3\frac{d}{dx}x^2+0} = \boxed{3x^{3-1}+(3\cdot 2)x^{2-1}} = \boxed{\mathbf{3x^2+6x}}$$

$$\boxed{\frac{d^2y}{dx^2}} = \boxed{\frac{d}{dx}3x^2+\frac{d}{dx}6x} = \boxed{3\frac{d}{dx}x^2+6\frac{d}{dx}x} = \boxed{(3\cdot 2)x^{2-1}+(6\cdot 1)} = \boxed{\mathbf{6x+6}}$$

$$\boxed{\frac{d^3y}{dx^3}} = \boxed{\frac{d}{dx}6x+\frac{d}{dx}6} = \boxed{6\frac{d}{dx}x+0} = \boxed{6\cdot 1} = \boxed{\mathbf{6}}$$

f. Given $y = \frac{1}{1+x}$, then

$$\boxed{\frac{dy}{dx}} = \boxed{\frac{\left[(1+x)\frac{d}{dx}1\right]-\left[1\cdot\frac{d}{dx}(1+x)\right]}{(1+x)^2}} = \boxed{\frac{0-(1\cdot 1)}{(1+x)^2}} = \boxed{\mathbf{-\frac{1}{(1+x)^2}}}$$

$$\frac{d^2y}{dx^2} = -\frac{\left[(1+x)^2\frac{d}{dx}1\right]-\left[1\cdot\frac{d}{dx}(1+x)^2\right]}{(1+x)^4} = -\frac{0-\left[(1\cdot 2)(1+x)^{2-1}\right]}{(1+x)^4} = \frac{2(1+x)}{(1+x)^{4=3}} = \mathbf{\frac{2}{(1+x)^3}}$$

$$\frac{d^3y}{dx^3} = \frac{\left[(1+x)^3\frac{d}{dx}2\right]-\left[2\cdot\frac{d}{dx}(1+x)^3\right]}{(1+x)^6} = \frac{0-\left[(2\cdot 3)(1+x)^{3-1}\right]}{(1+x)^6} = -\frac{6(1+x)^2}{(1+x)^{6=4}} = -\mathbf{\frac{6}{(1+x)^4}}$$

g. Given $y = x-\frac{1}{x}$ which is the same as $y = x - x^{-1}$, then

$$\frac{dy}{dx} = \frac{d}{dy}x - \frac{d}{dx}x^{-1} = 1+x^{-1-1} = \mathbf{1+x^{-2}}$$

$$\frac{d^2y}{dx^2} = \frac{d}{dy}1+\frac{d}{dx}x^{-2} = 0-2x^{-2-1} = \mathbf{-2x^{-3}}$$

$$\frac{d^3y}{dx^3} = -2\frac{d}{dy}x^{-3} = (-2\cdot -3)x^{-3-1} = \mathbf{6x^{-4}}$$

h. Given $y = ax^3+bx$, then

$$\frac{dy}{dx} = \frac{d}{dx}ax^3+\frac{d}{dx}bx = a\frac{d}{dx}x^3+b\frac{d}{dx}x = (a\cdot 3)x^{3-1}+(b\cdot 1)x^{1-1} = 3ax^2+bx^0 = \mathbf{3ax^2+b}$$

$$\frac{d^2y}{dx^2} = \frac{d}{dx}3ax^2+\frac{d}{dx}b = 3a\frac{d}{dx}x^2+0 = (3a\cdot 2)x^{2-1} = \mathbf{6ax}$$

$$\frac{d^3y}{dx^3} = \frac{d}{dx}6ax = 6a\frac{d}{dx}x = 6a\cdot 1 = \mathbf{6a}$$

i. Given $y = \frac{x^3+1}{x^2}$, then

$$\frac{dy}{dx} = \frac{\left[x^2\frac{d}{dx}\left(x^3+1\right)\right]-\left[\left(x^3+1\right)\frac{d}{dx}x^2\right]}{x^4} = \frac{x^2\left[\frac{d}{dx}x^3+\frac{d}{dx}1\right]-\left(x^3+1\right)\cdot 2x}{x^4} = \frac{x^2\left[3x^{3-1}+0\right]-2x\left(x^3+1\right)}{x^4}$$

$$= \frac{3x^4-2x^4-2x}{x^4} = \frac{x^4-2x}{x^4} = \frac{x\left(x^3-2\right)}{x^{4=3}} = \mathbf{\frac{x^3-2}{x^3}}$$

$$\frac{d^2y}{dx^2} = \frac{\left[x^3\frac{d}{dx}\left(x^3-2\right)\right]-\left[\left(x^3-2\right)\frac{d}{dx}x^3\right]}{x^6} = \frac{x^3\left[\frac{d}{dx}x^3-\frac{d}{dx}2\right]-\left(x^3-2\right)\cdot 3x^2}{x^6} = \frac{x^3\left[3x^{3-1}-0\right]-3x^2\left(x^3-2\right)}{x^6}$$

$$= \frac{3x^5-3x^5+6x^2}{x^6} = \frac{6x^2}{x^{6=4}} = \mathbf{\frac{6}{x^4}}$$

$$\boxed{\frac{d^3y}{dx^3}} = \boxed{\frac{x^4 \frac{d}{dx}6 - 6\frac{d}{dx}x^4}{x^8}} = \boxed{\frac{\left(x^4 \cdot 0\right) - (6 \cdot 4)x^{4-1}}{x^8}} = \boxed{\frac{0 - 24x^3}{x^8}} = \boxed{-\frac{24x^{\not 3}}{x^{\not 8 = 5}}} = \boxed{-\frac{\mathbf{24}}{\mathbf{x^5}}}$$

Example 5.8-5: Find y' and y'' for the following functions. Do not simplify the answer to its lowest term.

a. $x^2 + y^2 = 2$ b. $xy + y^2 = 1$ c. $1 + x^2y^2 = x$ d. $x^3y + y = 1$

Solutions:

a. Given $x^2 + y^2 = 2$, then y' is equal to $\boxed{2x^{2-1} + 2y \cdot y' = 0}$; $\boxed{2x + 2y\,y' = 0}$; $\boxed{2y\,y' = -2x}$; $\boxed{y' = \frac{-2x}{2y}}$

; $\boxed{y' = -\frac{x}{y}}$ and $\boxed{y'' = -\frac{(1 \cdot y) - (y' \cdot x)}{y^2}}$; $\boxed{y'' = -\frac{\mathbf{y - y'x}}{\mathbf{y^2}}}$

b. Given $xy + y^2 = 1$, then y' is equal to $\boxed{(1 \cdot y + y' \cdot x) + 2y \cdot y' = 0}$; $\boxed{y + y'x + 2y\,y' = 0}$; $\boxed{y'(x + 2y) = -y}$

; $\boxed{y' = -\frac{\mathbf{y}}{\mathbf{x + 2y}}}$ and $\boxed{y'' = -\frac{[y' \cdot (x + 2y)] - [(1 + 2y') \cdot y]}{(x + 2y)^2}}$; $\boxed{y'' = -\frac{xy' + 2y\,y' - y - 2y\,y'}{(x + 2y)^2}}$; $\boxed{y'' = -\frac{\mathbf{xy' - y}}{\mathbf{(x + 2y)^2}}}$

c. Given $1 + x^2y^2 = x$, then y' is equal to $\boxed{0 + \left(2x \cdot y^2 + 2yy' \cdot x^2\right) = 1}$; $\boxed{2x\,y^2 + 2y\,y'x^2 = 1}$

; $\boxed{2y\,y'x^2 = 1 - 2x\,y^2}$; $\boxed{y' = \frac{\mathbf{1 - 2x\,y^2}}{\mathbf{2x^2y}}}$ and $\boxed{y'' = \frac{\left\{\left[0 - 2\left(1 \cdot y^2 + 2yy' \cdot x\right)\right]\left(2x^2y\right)\right\} - \left\{2\left(2xy + x^2y'\right)\left(1 - 2xy^2\right)\right\}}{\left(2x^2y\right)^2}}$

; $\boxed{y'' = \frac{\mathbf{-4x^2y\left(y^2 + 2xyy'\right) - 2\left(2xy + x^2y'\right)\left(1 - 2xy^2\right)}}{\mathbf{4x^4y^2}}}$

d. Given $x^3y + y = 1$; $y\left(x^3 + 1\right) = 1$; $y = \frac{1}{x^3 + 1}$, then $\boxed{y'} = \boxed{\frac{\left[0 \cdot \left(x^3 + 1\right)\right] - \left[3x^2 \cdot 1\right]}{\left(x^3 + 1\right)^2}} = \boxed{\frac{\mathbf{-3x^2}}{\mathbf{\left(x^3 + 1\right)^2}}}$ and

$$\boxed{y''} = \boxed{-\frac{\left[-6x \cdot \left(x^3 + 1\right)^2\right] - \left[2\left(x^3 + 1\right) \cdot 3x^2 \cdot -3x^2\right]}{\left(x^3 + 1\right)^4}} = \boxed{\frac{-6x\left(x^3 + 1\right)^2 + 18x^4\left(x^3 + 1\right)}{\left(x^3 + 1\right)^4}} = \boxed{\frac{6x\left(x^3 + 1\right)\left[-\left(x^3 + 1\right) + 3x^3\right]}{\left(x^3 + 1\right)^4}}$$

$$= \boxed{\frac{6x\left(-x^3 - 1 + 3x^3\right)}{\left(x^3 + 1\right)^3}} = \boxed{\frac{\mathbf{6x\left(2x^3 - 1\right)}}{\mathbf{\left(x^3 + 1\right)^3}}}$$

Example 5.8-6: Find the first, second, and third derivative of the following functions.

a. $f(x)=6x^7+7x^2-2$

b. $f(x)=3x^4-2x^2+5x+9$

c. $f(x)=x^{-4}+3x^{-3}+x^{-2}+x$

d. $f(x)=x^7+6x^5+8x+3x^{-3}$

e. $f(x)=\frac{3}{x^3}+\frac{2}{x^2}+\frac{1}{x}$

f. $f(x)=x(2x+1)^3$

g. $f(x)=x^3+\frac{1}{x^3}$

h. $f(x)=(x+1)^2-x^3$

Solutions:

a. Given $f(x) = 6x^7+7x^2-2$, then

$$\boxed{f'(x)}=\boxed{(6\cdot 7)x^{7-1}+(7\cdot 2)x^{2-1}-0}=\boxed{\mathbf{42x^6+14x}}$$

$$\boxed{f''(x)}=\boxed{(42\cdot 6)x^{6-1}+14x^{1-1}}=\boxed{\mathbf{252x^5+14}}$$

$$\boxed{f'''(x)}=\boxed{(252\cdot 5)x^{5-1}+0}=\boxed{\mathbf{1260x^4}}$$

b. Given $f(x) = 3x^4-2x^2+5x+9$, then

$$\boxed{f'(x)}=\boxed{(3\cdot 4)x^{4-1}-(2\cdot 2)x^{2-1}+5x^{1-1}+0}=\boxed{\mathbf{12x^3-4x+5}}$$

$$\boxed{f''(x)}=\boxed{(12\cdot 3)x^{3-1}-4x^{1-1}}=\boxed{\mathbf{36x^2-4}}$$

$$\boxed{f'''(x)}=\boxed{(36\cdot 2)x^{2-1}-0}=\boxed{\mathbf{72x}}$$

c. Given $f(x) = x^{-4}+3x^{-3}+x^{-2}+x$, then

$$\boxed{f'(x)}=\boxed{-4x^{-4-1}+(3\cdot -3)x^{-3-1}-2x^{-2-1}+x^{1-1}}=\boxed{\mathbf{-4x^{-5}-9x^{-4}-2x^{-3}+1}}$$

$$\boxed{f''(x)}=\boxed{(-4\cdot -5)x^{-5-1}+(-9\cdot -4)x^{-4-1}+(-2\cdot -3)x^{-3-1}+0}=\boxed{\mathbf{20x^{-6}+36x^{-5}+6x^{-4}}}$$

$$\boxed{f'''(x)}=\boxed{(20\cdot -6)x^{-6-1}+(36\cdot -5)x^{-5-1}+(6\cdot -4)x^{-4-1}}=\boxed{\mathbf{-120x^{-7}-180x^{-6}-24x^{-5}}}$$

d. Given $f(x) = x^7+6x^5+8x+3x^{-3}$, then

$$\boxed{f'(x)}=\boxed{7x^{7-1}+(6\cdot 5)x^{5-1}+8x^{1-1}+(3\cdot -3)x^{-3-1}}=\boxed{\mathbf{7x^6+30x^4+8-9x^{-4}}}$$

$$\boxed{f''(x)}=\boxed{(7\cdot 6)x^{6-1}+(30\cdot 4)x^{4-1}+0+(-9\cdot -4)x^{-4-1}}=\boxed{\mathbf{42x^5+120x^3+36x^{-5}}}$$

$$\boxed{f'''(x)}=\boxed{(42\cdot 5)x^{5-1}+(120\cdot 3)x^{3-1}+(36\cdot -5)x^{-5-1}}=\boxed{\mathbf{210x^4+360x^2-180x^{-6}}}$$

e. Given $f(x) = \frac{3}{x^3} + \frac{2}{x^2} + \frac{1}{x}$ which is equal to $f(x) = 3x^{-3} + 2x^{-2} + x^{-1}$, then

$$\boxed{f'(x)} = \boxed{(3 \cdot -3)x^{-3-1} + (2 \cdot -2)x^{-2-1} - x^{-1-1}} = \boxed{\mathbf{-9x^{-4} - 4x^{-3} - x^{-2}}}$$

$$\boxed{f''(x)} = \boxed{(-9 \cdot -4)x^{-4-1} + (-4 \cdot -3)x^{-3-1} + (-1 \cdot -2)x^{-2-1}} = \boxed{\mathbf{36x^{-5} + 12x^{-4} + 2x^{-3}}}$$

$$\boxed{f'''(x)} = \boxed{(36 \cdot -5)x^{-5-1} + (12 \cdot -4)x^{-4-1} + (2 \cdot -3)x^{-3-1}} = \boxed{\mathbf{-180x^{-6} - 48x^{-5} - 6x^{-4}}}$$

f. Given $f(x) = x(2x+1)^3$, then

$$\boxed{f'(x)} = \boxed{\left[1 \cdot (2x+1)^3\right] + \left[3(2x+1)^{3-1} \cdot 2\right] \cdot x} = \boxed{\mathbf{(2x+1)^3 + 6x(2x+1)^2}}$$

$$\boxed{f''(x)} = \boxed{\left[3(2x+1)^{3-1} \cdot 2\right] + \left[6 \cdot (2x+1)^2 + 2(2x+1)^{2-1} \cdot 2 \cdot 6x\right]} = \boxed{6(2x+1)^2 + 6(2x+1)^2 + 24x(2x+1)}$$

$$= \boxed{\mathbf{12(2x+1)^2 + 24x(2x+1)}}$$

$$\boxed{f'''(x)} = \boxed{\left[(12 \cdot 2)(2x+1)^{2-1} \cdot 2\right] + \left[24 \cdot (2x+1) + (2x+1)^{1-1} \cdot 2 \cdot 24x\right]} = \boxed{48(2x+1) + \left[24(2x+1) + 48x\right]}$$

$$= \boxed{\mathbf{72(2x+1) + 48x}}$$

g. Given $f(x) = x^3 + \frac{1}{x^3}$ which is equal to $f(x) = x^3 + x^{-3}$, then

$$\boxed{f'(x)} = \boxed{3x^{3-1} - 3x^{-3-1}} = \boxed{\mathbf{3x^2 - 3x^{-4}}}$$

$$\boxed{f''(x)} = \boxed{(3 \cdot 2)x^{2-1} + (-3 \cdot -4)x^{-4-1}} = \boxed{\mathbf{6x + 12x^{-5}}}$$

$$\boxed{f'''(x)} = \boxed{6x^{1-1} + (12 \cdot -5)x^{-5-1}} = \boxed{6x^0 - 60x^{-6}} = \boxed{\mathbf{6 - 60x^{-6}}}$$

h. Given $f(x) = (x+1)^2 - x^3$, then

$$\boxed{f'(x)} = \boxed{\left[2(x+1)^{2-1} \cdot 1\right] - 3x^{3-1}} = \boxed{2(x+1) - 3x^2} = \boxed{2x + 2 - 3x^2} = \boxed{\mathbf{-3x^2 + 2x + 2}}$$

$$\boxed{f''(x)} = \boxed{(-3 \cdot 2)x^{2-1} + 2x^{1-1} + 0} = \boxed{-6x + 2x^0} = \boxed{\mathbf{-6x + 2}}$$

$$\boxed{f'''(x)} = \boxed{-6x^{1-1} + 0} = \boxed{-6x^0} = \boxed{\mathbf{-6}}$$

Section 5.8 Practice Problems - Higher Order Derivatives

1. Find the second derivative of the following functions.

a. $y = x^3 + 3x^2 + 5x - 1$

b. $y = x^2(x+1)^2$

c. $y = 3x^3 + 50x$

d. $y = x^5 + \frac{1}{x^2}$ e. $y = \frac{x^3}{x+1} - 5x^2$ f. $y = x^3\left(x^2 - 1\right)$

g. $y = x^4 + \frac{x^8 - 7x^5 + 5x}{10}$ h. $y = x^2 - \frac{1}{x+1}$ i. $y = \frac{1}{x^2} - 3x$

2. Find y''' for the following functions.

a. $y = x^5 + 6x^3 + 10$ b. $y = x^2 + \frac{1}{x}$ c. $y = 4x^3(x-1)^2$

d. $y = \frac{x}{x+1}$ e. $y = x^8 - 10x^5 + 5x - 10$ f. $y = \frac{x-1}{x^2} + 5x^3$

3. Find $f''(0)$ and $f''(1)$ for the following functions.

a. $f(x) = 6x^5 + 3x^3 + 5$ b. $f(x) = x^3(x+1)^2$ c. $f(x) = x + (x-1)^2$

d. $f(x) = (x-1)^{-3}$ e. $f(x) = (x-1)\left(x^2 + 1\right)$ f. $f(x) = \left(x^3 - 1\right)^2 + \sqrt{2x}$

Appendix - Exercise Solutions

Chapter 1 Solutions:

Section 1.1a Case I Solutions - Real Numbers Raised to Positive Integer Exponents

1. $4^3 = 4 \cdot 4 \cdot 4 = \mathbf{64}$

2. $(-10)^4 = -10 \cdot -10 \cdot -10 \cdot -10 = \mathbf{+10000}$

3. $0.25^3 = 0.25 \cdot 0.25 \cdot 0.25 = \mathbf{0.0156}$

4. $12^5 = 12 \cdot 12 \cdot 12 \cdot 12 \cdot 12 = \mathbf{248832}$

5. $-(3)^5 = -(3 \cdot 3 \cdot 3 \cdot 3 \cdot 3) = \mathbf{-243}$

6. $489^0 = \mathbf{1}$

Section 1.1a Case II Solutions - Real Numbers Raised to Negative Integer Exponents

1. $4^{-3} = \frac{1}{4^3} = \frac{1}{4 \cdot 4 \cdot 4} = \mathbf{\frac{1}{64}}$

2. $(-5)^{-4} = \frac{1}{(-5)^4} = \frac{1}{(-5) \cdot (-5) \cdot (-5) \cdot (-5)} = \mathbf{\frac{1}{625}}$

3. $0.25^{-3} = \frac{1}{0.25^3} = \frac{1}{(0.25) \cdot (0.25) \cdot (0.25)} = \mathbf{\frac{1}{0.0156}}$

4. $12^{-5} = \frac{1}{12^5} = \frac{1}{12 \cdot 12 \cdot 12 \cdot 12 \cdot 12} = \mathbf{\frac{1}{248832}}$

5. $-(3)^{-4} = -\frac{1}{3^4} = -\frac{1}{3 \cdot 3 \cdot 3 \cdot 3} = \mathbf{-\frac{1}{81}}$

6. $48^{-2} = \frac{1}{48^2} = \frac{1}{48 \cdot 48} = \mathbf{\frac{1}{2304}}$

Section 1.1b Case I Solutions - Multiplying Positive Integer Exponents

1. $x^2 \cdot x^3 \cdot x = x^2 \cdot x^3 \cdot x^1 = x^{2+3+1} = \boldsymbol{x^6}$

2. $2 \cdot a^2 \cdot b^0 \cdot a^3 \cdot b^2 = 2 \cdot \left(a^2 \cdot a^3\right) \cdot \left(b^0 \cdot b^2\right) = 2 \cdot \left(a^{2+3}\right) \cdot \left(b^{0+2}\right) = \boldsymbol{2a^5b^2}$

3. $\frac{4}{-6} a^2b^3ab^4b^5 = -\frac{\overset{2}{\cancel{4}}}{\underset{3}{\cancel{6}}} a^2b^3a^1b^4b^5 = -\frac{2}{3}\left(a^2a^1\right) \cdot \left(b^3b^4b^5\right) = -\frac{2}{3}\left(a^{2+1}\right) \cdot \left(b^{3+4+5}\right) = \boldsymbol{-\frac{2}{3}a^3b^{12}}$

4. $2^3 \cdot 2^2 \cdot x^{2a} \cdot x^{3a} \cdot x^a = \left(2^3 \cdot 2^2\right) \cdot \left(x^{2a} \cdot x^{3a} \cdot x^a\right) = \left(2^{3+2}\right) \cdot \left(x^{2a+3a+a}\right) = 2^5 \cdot x^{6a} = \boldsymbol{32x^{6a}}$

5. $\left(x \cdot y^2 \cdot z^3\right)^0 \cdot w^2z^3zw^4z^2 = 1 \cdot w^2z^3z^1w^4z^2 = \left(w^2w^4\right) \cdot \left(z^3z^1z^2\right) = \left(w^{2+4}\right) \cdot \left(z^{3+1+2}\right) = \boldsymbol{w^6z^6}$

6. $2^0 \cdot 4^2 \cdot 4^2 \cdot 2^2 \cdot 4^1 = \left(2^0 \cdot 2^2\right) \cdot \left(4^2 \cdot 4^2 \cdot 4^1\right) = \left(2^{0+2}\right) \cdot \left(4^{2+2+1}\right) = 2^2 \cdot 4^5 = 4 \cdot 1024 = \mathbf{4096}$

Section 1.1b Case II Solutions - Dividing Positive Integer Exponents

1. $\frac{x^5}{x^3} = \frac{x^5x^{-3}}{1} = \frac{x^{5-3}}{1} = \frac{x^2}{1} = \boldsymbol{x^2}$

2. $\frac{a^2b^3}{a} = \frac{a^2b^3}{a^1} = \frac{\left(a^2a^{-1}\right)b^3}{1} = \frac{a^{2-1}b^3}{1} = \frac{a^1b^3}{1} = \frac{ab^3}{1} = \boldsymbol{ab^3}$

3. $\frac{a^3b^3c^2}{a^2b^6c} = \frac{a^3b^3c^2}{a^2b^6c^1} = \frac{\left(a^3a^{-2}\right) \cdot \left(c^2c^{-1}\right)}{b^6b^{-3}} = \frac{\left(a^{3-2}\right) \cdot \left(c^{2-1}\right)}{b^{6-3}} = \frac{a^1 \cdot c^1}{b^3} = \boldsymbol{\frac{ac}{b^3}}$

4. $\frac{3^2\cdot\left(rs^2\right)}{(2rs)\cdot r^3} = \frac{9\cdot\left(rs^2\right)}{\left(2r^1s\right)\cdot r^3} = \frac{9rs^2}{2\left(r^3r^1\right)\cdot s} = \frac{9}{2}\frac{rs^2}{r^{3+1}s} = \frac{9}{2}\frac{r^1s^2}{r^4s^1} = \frac{9}{2}\frac{s^2s^{-1}}{r^4r^{-1}} = \frac{9}{2}\frac{s^{2-1}}{r^{4-1}} = \frac{9}{2}\frac{s^1}{r^3} = \mathbf{4\frac{1}{2}\left(\frac{s}{r^3}\right)}$

5. $\frac{2p^2q^3pr^4}{-6p^4q^2r} = -\frac{\overset{1}{\cancel{2}}}{\underset{3}{\cancel{6}}}\frac{p^2q^3p^1r^4}{p^4q^2r^1} = -\frac{1}{3}\frac{\left(q^3q^{-2}\right)\cdot\left(r^4r^{-1}\right)}{p^4p^{-2}p^{-1}} = -\frac{1}{3}\frac{\left(q^{3-2}\right)\cdot\left(r^{4-1}\right)}{p^{4-2-1}} = -\frac{1}{3}\frac{q^1r^3}{p^1} = -\mathbf{\frac{1}{3}\left(\frac{qr^3}{p}\right)}$

6. $\frac{\left(k^2l^3\right)\cdot\left(kl^2m^0\right)}{k^4l^3m^5} = \frac{\left(k^2l^3\right)\cdot\left(kl^2\cdot 1\right)}{k^4l^3m^5} = \frac{\left(k^2l^3\right)\cdot\left(kl^2\right)}{k^4l^3m^5} = \frac{k^2l^3k^1l^2}{k^4l^3m^5} = \frac{l^3l^{-3}l^2}{\left(k^4k^{-2}k^{-1}\right)m^5} = \frac{l^{3-3+2}}{\left(k^{4-2-1}\right)m^5} = \frac{l^2}{k^1m^5} = \mathbf{\frac{l^2}{km^5}}$

Section 1.1b Case III Solutions - Adding and Subtracting Positive Integer Exponents

1. $x^2+4xy-2x^2-2xy+z^3 = \left(x^2-2x^2\right)+(4xy-2xy)+z^3 = (1-2)x^2+(4-2)xy+z^3 = \mathbf{-x^2+2xy+z^3}$

2. $\left(a^3+2a^2+4^3\right)-\left(4a^3+20\right) = \left(a^3+2a^2+4^3\right)+\left(-4a^3-20\right) = a^3+2a^2+64-4a^3-20 = \left(a^3-4a^3\right)+2a^2+(64-20)$

$= (1-4)a^3+2a^2+44 = \mathbf{-3a^3+2a^2+44}$

3. $3x^4+2x^2+2x^4-\left(x^4-2x^2+3\right) = 3x^4+2x^2+2x^4+\left(-x^4+2x^2-3\right) = 3x^4+2x^2+2x^4-x^4+2x^2-3$

$= \left(3x^4+2x^4-x^4\right)+\left(2x^2+2x^2\right)-3 = (3+2-1)x^4+(2+2)x^2-3 = \mathbf{4x^4+4x^2-3}$

4. $-\left(-2l^3a^3+2l^2a^2-5^3\right)-\left(4l^3a^3-20\right) = \left(+2l^3a^3-2l^2a^2+5^3\right)+\left(-4l^3a^3+20\right) = 2l^3a^3-2l^2a^2+125-4l^3a^3+20$

$= \left(2l^3a^3-4l^3a^3\right)-2l^2a^2+(125+20) = (2-4)l^3a^3-2l^2a^2+145 = \mathbf{-2l^3a^3-2l^2a^2+145}$

5. $\left(m^{3n}-4m^{2n}\right)-\left(2m^{3n}+3m^{2n}\right)+5m = \left(m^{3n}-4m^{2n}\right)+\left(-2m^{3n}-3m^{2n}\right)+5m = m^{3n}-4m^{2n}-2m^{3n}-3m^{2n}+5m$

$= \left(m^{3n}-2m^{3n}\right)+\left(-4m^{2n}-3m^{2n}\right)+5m = (1-2)m^{3n}+(-4-3)m^{2n}+5m = \mathbf{-m^{3n}-7m^{2n}+5m}$

6. $\left(-7z^3+3z-5\right)-\left(-3z^3+z-4\right)+5z+20 = \left(-7z^3+3z-5\right)+\left(3z^3-z+4\right)+5z+20 = -7z^3+3z-5+3z^3-z+4+5z+20$

$= \left(-7z^3+3z^3\right)+(3z-z+5z)+(-5+4+20) = (-7+3)z^3+(3-1+5)z+19 = \mathbf{-4z^3+7z+19}$

Section 1.1c Case I Solutions - Multiplying Negative Integer Exponents

1. $\left(3^{-3}\cdot 2^{-1}\right)\cdot\left(2^{-3}\cdot 3^{-2}\cdot 2\right) = 3^{-3}\cdot 2^{-1}\cdot 2^{-3}\cdot 3^{-2}\cdot 2^1 = \left(3^{-3}\cdot 3^{-2}\right)\cdot\left(2^{-1}\cdot 2^{-3}\cdot 2^1\right) = \left(3^{-3-2}\right)\cdot\left(2^{-1-3+1}\right) = 3^{-5}\cdot 2^{-3}$

$= \frac{1}{3^5\cdot 2^3} = \frac{1}{243\cdot 8} = \mathbf{\frac{1}{1944}}$

2. $a^{-6}\cdot b^{-4}\cdot a^{-1}\cdot b^{-2}\cdot a^0 = \left(a^{-6}a^{-1}a^0\right)\cdot\left(b^{-2}b^{-4}\right) = \left(a^{-6-1+0}\right)\cdot\left(b^{-2-4}\right) = a^{-7}b^{-6} = \mathbf{\frac{1}{a^7b^6}}$

3. $\left(a^{-2}\cdot b^{-3}\right)^2\cdot\left(a\cdot b^{-2}\right) = \left(a^{-2\times 2}\cdot b^{-3\times 2}\right)\cdot\left(a\cdot b^{-2}\right) = \left(a^{-4}\cdot b^{-6}\right)\cdot\left(a\cdot b^{-2}\right) = a^{-4}\cdot b^{-6}\cdot a\cdot b^{-2} = \left(a^{-4}\cdot a^1\right)\cdot\left(b^{-6}\cdot b^{-2}\right)$

$= \left(a^{-4+1}\right)\cdot\left(b^{-6-2}\right) = a^{-3}\cdot b^{-8} = \mathbf{\frac{1}{a^3b^8}}$

4. $(-2)^{-4}\left(r^{-2}s^2t\right)\cdot\left(r^3st^{-2}s^{-1}\right) = \frac{1}{(-2)^4}r^{-2}s^2t^1r^3s^1t^{-2}s^{-1} = \frac{1}{(-2\cdot -2\cdot -2\cdot -2)}\left(r^{-2}r^3\right)\cdot\left(s^2s^1s^{-1}\right)\cdot\left(t^1t^{-2}\right)$

$= \frac{1}{+16}\left(r^{-2+3}\right)\cdot\left(s^{2+1-1}\right)\cdot\left(t^{1-2}\right) = \frac{1}{16}r^1\cdot s^2\cdot t^{-1} = \frac{1}{16}\cdot\frac{rs^2}{t^1} = \mathbf{\frac{1}{16}\left(\frac{rs^2}{t}\right)}$

5. $\left(\frac{4}{5}\right)^{-4}2^2v^{-5}2^{-4}v^3v^{-2} = \left(\frac{4^{1\times -4}}{5^{1\times -4}}\right)\cdot\left(2^22^{-4}\right)\cdot\left(v^{-5}v^3v^{-2}\right) = \left(\frac{4^{-4}}{5^{-4}}\right)\cdot\left(2^{2-4}\right)\cdot\left(v^{-5+3-2}\right) = \left(\frac{5^4}{4^4}\right)\cdot 2^{-2}\cdot v^{-4}$

$= \left(\frac{625}{256}\right)\cdot\frac{1}{2^2 v^4} = \left(\frac{625}{256}\right)\cdot\frac{1}{4\cdot v^4} = \left(\frac{625}{256\cdot 4}\right)\cdot\frac{1}{v^4} = \left(\frac{625}{1024}\right)\cdot\frac{1}{v^4} = \mathbf{\frac{625}{1024\,v^4}}$

6. $2^{-1}\cdot 3^2\cdot 3^{-5}\cdot 2^2\cdot 2^0 = \left(2^{-1}\cdot 2^2\cdot 2^0\right)\cdot\left(3^2\cdot 3^{-5}\right) = \left(2^{-1+2+0}\right)\cdot\left(3^{2-5}\right) = 2^1\cdot 3^{-3} = 2\cdot 3^{-3} = \frac{2}{3^3} = \mathbf{\frac{2}{27}}$

Section 1.1c Case II Solutions - Dividing Negative Integer Exponents

1. $\frac{x^{-2}x}{x^3x^0} = \frac{x^{-2}x^1}{x^3x^0} = \frac{x^{-2+1}}{x^{3+0}} = \frac{x^{-1}}{x^3} = \frac{1}{x^3x^1} = \frac{1}{x^{3+1}} = \mathbf{\frac{1}{x^4}}$

2. $\frac{-2a^{-2}b^3}{-6a^{-1}b^{-2}} = +\frac{\overset{1}{\cancel{2}}a^{-2}b^3}{\underset{3}{\cancel{6}}a^{-1}b^{-2}} = \frac{b^3b^2}{3a^2a^{-1}} = \frac{b^{3+2}}{3a^{2-1}} = \frac{b^5}{3a^1} = \mathbf{\frac{b^5}{3a}}$

3. $\frac{-(-3)^{-4}}{3\cdot(-3)^{-3}} = -\frac{(-3)^3}{3\cdot(-3)^4} = -\frac{-3\cdot -3\cdot -3}{3\cdot(-3\cdot -3\cdot -3\cdot -3)} = -\frac{-27}{3\cdot(+81)} = +\frac{27}{3\cdot 81} = \frac{\cancel{27}}{\underset{9}{\cancel{243}}} = \mathbf{\frac{1}{9}}$

4. $\frac{-3^3y^{-3}y\,w}{(-3)^{-2}y^2w^{-3}} = -\frac{3^3y^{-3}y^1w^1}{(-3)^{-2}y^2w^{-3}} = -\frac{27\cdot(-3)^2w^1w^3}{y^2y^3y^{-1}} = -\frac{27\cdot(-3\cdot -3)\,w^{1+3}}{y^{2+3-1}} = -\frac{(27\cdot 9)\,w^4}{y^4} = \mathbf{-\frac{243\,w^4}{y^4}}$

5. $\frac{a^{-2}b^2a^{-5}y^{-2}}{a^{-3}y} = \frac{a^{-2}b^2a^{-5}y^{-2}}{a^{-3}y^1} = \frac{b^2}{\left(a^{-3}a^2a^5\right)\cdot\left(y^2y^1\right)} = \frac{b^2}{a^{-3+2+5}\cdot y^{2+1}} = \frac{b^2}{a^4\cdot y^3} = \mathbf{\frac{b^2}{a^4y^3}}$

6. $\frac{(x\cdot y\cdot z)^0\cdot yx^{-2}}{x^{-4}y^{-1}} = \frac{1\cdot yx^{-2}}{x^{-4}y^{-1}} = \frac{y^1x^{-2}}{x^{-4}y^{-1}} = \frac{\left(x^4x^{-2}\right)\cdot\left(y^1y^1\right)}{1} = \frac{\left(x^{4-2}\right)\cdot\left(y^{1+1}\right)}{1} = \frac{x^2\cdot y^2}{1} = \mathbf{x^2y^2}$

Section 1.1c Case III Solutions - Adding and Subtracting Negative Integer Exponents

1. $x^{-1}+2x^{-2}+3x^{-1}-6x^{-2} = \left(2x^{-2}-6x^{-2}\right)+\left(x^{-1}+3x^{-1}\right) = (2-6)x^{-2}+(1+3)x^{-1} = -4x^{-2}+4x^{-1} = \frac{-4}{x^2}+\frac{4}{x^1}$

$= -\frac{4}{x^2}+\frac{4}{x} = \frac{(-4\cdot x)+\left(4\cdot x^2\right)}{x^2\cdot x} = \frac{4x^2-4x}{x^3} = \frac{4x(x-1)}{x^3} = \frac{4(x-1)}{x^3\cdot x^{-1}} = \frac{4(x-1)}{x^{3-1}} = \mathbf{\frac{4(x-1)}{x^2}}$

2. $\left(3a^{-4}-b^{-2}\right)+\left(-2a^{-4}+3b^{-2}\right) = 3a^{-4}-b^{-2}-2a^{-4}+3b^{-2} = \left(3a^{-4}-2a^{-4}\right)+\left(-b^{-2}+3b^{-2}\right) = (3-2)a^{-4}+(-1+3)b^{-2}$

$= a^{-4}+2b^{-2} = \frac{1}{a^4}+\frac{2}{b^2} = \frac{\left(b^2\cdot 1\right)+\left(2\cdot a^4\right)}{a^4\cdot b^2} = \mathbf{\frac{b^2+2a^4}{a^4b^2}}$

3. $(xy)^{-1}+y^{-2}+4(xy)^{-1}-3y^{-2}+2^{-3} = \left[(xy)^{-1}+4(xy)^{-1}\right]+\left(y^{-2}-3y^{-2}\right)+2^{-3} = [1+4](xy)^{-1}+(1-3)y^{-2}+2^{-3}$

$= 5(xy)^{-1}-2y^{-2}+2^{-3} = \frac{5}{xy}-\frac{2}{y^2}+\frac{1}{2^3} = \left(\frac{5}{xy}-\frac{2}{y^2}\right)+\frac{1}{8} = \left(\frac{\left(5\cdot y^2\right)-(2\cdot xy)}{xy\cdot y^2}\right)+\frac{1}{8} = \left(\frac{5y^2-2xy}{xy^3}\right)+\frac{1}{8}$

$= \frac{\left[8\cdot\left(5y^2-2xy\right)\right]+\left(1\cdot xy^3\right)}{8\cdot xy^3} = \frac{40y^2-16xy+xy^3}{8xy^3} = \frac{xy^3+40y^2-16xy}{8xy^3} = \frac{y\left(xy^2+40y-16x\right)}{8xy^3} = \frac{xy^2+40y-16x}{8xy^3y^{-1}}$

$= \frac{xy^2+40y-16x}{8xy^{3-1}} = \mathbf{\frac{xy^2+40y-16x}{8xy^2}}$

4. $4x^{-1}+y^{-3}+5y^{-3} = 4x^{-1}+\left(y^{-3}+5y^{-3}\right) = 4x^{-1}+(1+5)y^{-3} = 4x^{-1}+6y^{-3} = \frac{4}{x}+\frac{6}{y^3} = \frac{\left(4\cdot y^3\right)+(6\cdot x)}{x\cdot y^3} = \mathbf{\frac{4y^3+6x}{xy^3}}$

5. $m^{-5}-\left(m^{-2}-3m^{-5}+m^0\right)+3m^{-2} = m^{-5}-\left(m^{-2}-3m^{-5}+1\right)+3m^{-2} = m^{-5}+\left(-m^{-2}+3m^{-5}-1\right)+3m^{-2}$

$$= m^{-5} - m^{-2} + 3m^{-5} - 1 + 3m^{-2} = \left(m^{-5} + 3m^{-5}\right) + \left(-m^{-2} + 3m^{-2}\right) - 1 = (1+3)m^{-5} + (-1+3)m^{-2} - 1 = 4m^{-5} + 2m^{-2} - 1$$

$$= \frac{4}{m^5} + \frac{2}{m^2} - 1 = \left(\frac{4}{m^5} + \frac{2}{m^2}\right) - 1 = \left(\frac{\left(4 \cdot m^2\right) + \left(2 \cdot m^5\right)}{m^5 \cdot m^2}\right) - 1 = \left(\frac{4m^2 + 2m^5}{m^{5+2}}\right) - 1 = \frac{4m^2 + 2m^5}{m^7} - \frac{1}{1}$$

$$= \frac{\left[1 \cdot \left(4m^2 + 2m^5\right)\right] - \left(1 \cdot m^7\right)}{m^7 \cdot 1} = \frac{4m^2 + 2m^5 - m^7}{m^7} = \frac{m^2\left(4 + 2m^3 - m^5\right)}{m^7} = \frac{4 + 2m^3 - m^5}{m^7 m^{-2}} = \frac{4 + 2m^3 - m^5}{m^{7-2}} = \mathbf{\frac{-m^5 + 2m^3 + 4}{m^5}}$$

6. $\left(a^3\right)^{-2} + \left(a^{-2}b\right)^2 - 6a^{-6} + 3a^{-4}b^2 = \left(a^{3\times-2}\right) + \left(a^{-2\times2}b^{1\times2}\right) - 6a^{-6} + 3a^{-4}b^2 = a^{-6} + a^{-4}b^2 - 6a^{-6} + 3a^{-4}b^2$

$$= \left(a^{-6} - 6a^{-6}\right) + \left(a^{-4}b^2 + 3a^{-4}b^2\right) = (1-6)a^{-6} + (1+3)a^{-4}b^2 = -5a^{-6} + 4a^{-4}b^2 = -\frac{5}{a^6} + \frac{4b^2}{a^4} = \frac{-\left(5 \cdot a^4\right) + \left(a^6 \cdot 4b^2\right)}{a^6 \cdot a^4}$$

$$= \frac{-5a^4 + 4a^6b^2}{a^{6+4}} = \frac{a^4\left(-5 + 4a^2b^2\right)}{a^{10}} = \frac{-5 + 4a^2b^2}{a^{10}a^{-4}} = \frac{-5 + 4a^2b^2}{a^{10-4}} = \mathbf{\frac{4a^2b^2 - 5}{a^6}}$$

Section 1.2a Case I Solutions - Roots and Radical Expressions

1. $\sqrt[2]{98} = \sqrt{98} = \sqrt{49 \cdot 2} = \sqrt{7^2 \cdot 2} = \mathbf{7\sqrt{2}}$

2. $3\sqrt{75} = 3\sqrt{25 \cdot 3} = 3\sqrt{5^2 \cdot 3} = (3 \cdot 5)\sqrt{3} = \mathbf{15\sqrt{3}}$

3. $\sqrt[3]{125} = \sqrt[3]{5^3} = \mathbf{5}$

4. $\sqrt[5]{3125} = \sqrt[5]{5^5} = \mathbf{5}$

5. $\sqrt[4]{162} = \sqrt[4]{81 \cdot 2} = \sqrt[4]{3^4 \cdot 2} = \mathbf{3\sqrt[4]{2}}$

6. $\sqrt[2]{192} = \sqrt{192} = \sqrt{64 \cdot 3} = \sqrt{8^2 \cdot 3} = \mathbf{8\sqrt{3}}$

Section 1.2a Case II Solutions - Rational, Irrational, Real, and Imaginary Numbers

1. $\frac{5}{8}$; is a rational and real number

2. $\sqrt{45} = \sqrt{9 \cdot 5} = 3\sqrt{5}$; is an irrational and real number

3. 450; is a rational and real number

4. $-\frac{2}{\sqrt{10}}$; is an irrational and real number

5. $-\sqrt{-5}$; is not a real number

6. $\frac{\sqrt{5}}{-2}$; is an irrational and real number

Section 1.2a Case III Solutions - Simplifying Radical Expressions with Real Numbers as a Radicand

1. $-\sqrt{49} = -\sqrt{7 \cdot 7} = -\sqrt{7 \cdot 7} = -\sqrt{7^1 \cdot 7^1} = -\sqrt{7^{1+1}} = -\sqrt{7^2} = \mathbf{-7}$

2. $\sqrt{54} = \sqrt{9 \cdot 5} = \sqrt{(3 \cdot 3) \cdot 5} = \sqrt{\left(3^1 \cdot 3^1\right) \cdot 5} = \sqrt{3^{1+1} \cdot 5} = \sqrt{3^2 \cdot 5} = \mathbf{3\sqrt{5}}$

3. $-\sqrt{500} = -\sqrt{100 \cdot 5} = -\sqrt{(10 \cdot 10) \cdot 5} = -\sqrt{\left(10^1 \cdot 10^1\right) \cdot 5} = -\sqrt{\left(10^{1+1}\right) \cdot 5} = -\sqrt{10^2 \cdot 5} = \mathbf{-10\sqrt{5}}$

4. $\sqrt[5]{3^5 \cdot 5} = \mathbf{3\sqrt[5]{5}}$

5. $\sqrt[2]{216} = \sqrt{216} = \sqrt{36 \cdot 6} = \sqrt{(6 \cdot 6) \cdot 6} = \sqrt{\left(6^1 \cdot 6^1\right) \cdot 6} = \sqrt{6^{1+1} \cdot 6} = \sqrt{6^2 \cdot 6} = \mathbf{6\sqrt{6}}$

6. $-\frac{1}{4}\sqrt[4]{4^5\cdot 2} = -\frac{1}{4}\sqrt[4]{4^{4+1}\cdot 2} = -\frac{1}{4}\sqrt[4]{\left(4^4\cdot 4^1\right)\cdot 2} = -\frac{1}{4}\cdot 4\sqrt[4]{(4\cdot 2)} = -\frac{\overset{1}{\cancel{4}}}{\underset{1}{\cancel{4}}}\sqrt[4]{8} = \mathbf{-\sqrt[4]{8}}$

Section 1.2b Case I Solutions - Multiplying Monomial Expressions in Radical Form, with Real Numbers

1. $\sqrt{72}\cdot\sqrt{75} = \sqrt{36\cdot 2}\cdot\sqrt{25\cdot 3} = \sqrt{6^2\cdot 2}\cdot\sqrt{5^2\cdot 3} = 6\sqrt{2}\cdot 5\sqrt{3} = (6\cdot 5)\sqrt{2\cdot 3} = \mathbf{30\sqrt{6}}$

2. $-3\sqrt{20}\cdot 2\sqrt{32} = -3\sqrt{4\cdot 5}\cdot 2\sqrt{16\cdot 2} = -3\sqrt{2^2\cdot 5}\cdot 2\sqrt{4^2\cdot 2} = -(3\cdot 2)\sqrt{5}\cdot(2\cdot 4)\sqrt{2} = -6\sqrt{5}\cdot 8\sqrt{2} = -(6\cdot 8)\sqrt{5\cdot 2} = \mathbf{-48\sqrt{10}}$

3. $\sqrt[2]{16}\cdot\sqrt[2]{27} = \sqrt{16}\cdot\sqrt{27} = \sqrt{4^2}\cdot\sqrt{9\cdot 3} = 4\cdot\sqrt{3^2\cdot 3} = (4\cdot 3)\sqrt{3} = \mathbf{12\sqrt{3}}$

4. $\sqrt{64}\cdot\sqrt{100}\cdot\sqrt{54} = \sqrt{8^2}\cdot\sqrt{10^2}\cdot\sqrt{9\cdot 6} = (8\cdot 10)\cdot\sqrt{3^2\cdot 6} = (80\cdot 3)\sqrt{6} = \mathbf{240\sqrt{6}}$

5. $-\sqrt{125}\cdot -2\sqrt{98} = +2\sqrt{25\cdot 5}\cdot\sqrt{49\cdot 2} = 2\sqrt{5^2\cdot 5}\cdot\sqrt{7^2\cdot 2} = (2\cdot 5)\sqrt{5}\cdot 7\sqrt{2} = (10\cdot 7)\sqrt{5\cdot 2} = \mathbf{70\sqrt{10}}$

6. $\sqrt[4]{625}\cdot\sqrt[4]{324}\cdot\sqrt[4]{48} = \sqrt[4]{5^4}\cdot\sqrt[4]{81\cdot 4}\cdot\sqrt[4]{16\cdot 3} = 5\cdot\sqrt[4]{3^4\cdot 4}\cdot\sqrt[4]{2^4\cdot 3} = 5\cdot 3\sqrt[4]{4}\cdot 2\sqrt[4]{3} = (5\cdot 3\cdot 2)\sqrt[4]{4\cdot 3} = \mathbf{30\sqrt[4]{12}}$

Section 1.2b Case II Solutions - Multiplying Binomial Expressions in Radical Form, with Real Numbers

1. $\left(2\sqrt{3}+1\right)\cdot\left(2+\sqrt{2}\right) = (2\cdot 2)\sqrt{3}+\left(2\sqrt{3}\cdot\sqrt{2}\right)+(1\cdot 2)+\left(1\cdot\sqrt{2}\right) = 4\sqrt{3}+2\sqrt{3\cdot 2}+2+\sqrt{2} = \mathbf{4\sqrt{3}+2\sqrt{6}+\sqrt{2}+2}$

2. $\left(1+\sqrt{5}\right)\cdot\left(\sqrt{8}+\sqrt{5}\right) = \left(1+\sqrt{5}\right)\cdot\left(\sqrt{4\cdot 2}+\sqrt{5}\right) = \left(1+\sqrt{5}\right)\cdot\left(\sqrt{2^2\cdot 2}+\sqrt{5}\right) = \left(1+\sqrt{5}\right)\cdot\left(2\sqrt{2}+\sqrt{5}\right)$

$= \left(1\cdot 2\sqrt{2}\right)+\left(1\cdot\sqrt{5}\right)+\left(2\sqrt{2}\cdot\sqrt{5}\right)+\left(\sqrt{5}\cdot\sqrt{5}\right) = 2\sqrt{2}+\sqrt{5}+2\sqrt{2\cdot 5}+\sqrt{5\cdot 5} = 2\sqrt{2}+\sqrt{5}+2\sqrt{10}+\sqrt{5^2}$

$= \mathbf{2\sqrt{2}+\sqrt{5}+2\sqrt{10}+5}$

3. $\left(2-\sqrt{2}\right)\cdot\left(3+\sqrt{2}\right) = (2\cdot 3)+\left(2\cdot\sqrt{2}\right)-\left(3\cdot\sqrt{2}\right)-\left(\sqrt{2}\cdot\sqrt{2}\right) = 6+2\sqrt{2}-3\sqrt{2}-\sqrt{2\cdot 2} = 6+(2-3)\sqrt{2}-\sqrt{2^2}$

$= 6-\sqrt{2}-2 = (6-2)-\sqrt{2} = \mathbf{4-\sqrt{2}}$

4. $\left(5+\sqrt{5}\right)\cdot\left(5-\sqrt{5^3}\right) = \left(5+\sqrt{5}\right)\cdot\left(5-\sqrt{5^{2+1}}\right) = \left(5+\sqrt{5}\right)\cdot\left(5-\sqrt{5^2\cdot 5^1}\right) = \left(5+\sqrt{5}\right)\cdot\left(5-5\sqrt{5}\right)$

$= (5\cdot 5)-(5\cdot 5)\sqrt{5}+\left(5\cdot\sqrt{5}\right)-\left(5\sqrt{5}\cdot\sqrt{5}\right) = 25-25\sqrt{5}+5\sqrt{5}-5\sqrt{5\cdot 5} = 25+(-25+5)\sqrt{5}-5\sqrt{5^2}$

$= 25-20\sqrt{5}-5\cdot 5 = 25-20\sqrt{5}-25 = (25-25)-20\sqrt{5} = \mathbf{-20\sqrt{5}}$

5. $\left(2+\sqrt{6}\right)\cdot\left(\sqrt[4]{16}-\sqrt{18}\right) = \left(2+\sqrt{6}\right)\cdot\left(\sqrt[4]{2^4}-\sqrt{9\cdot 2}\right) = \left(2+\sqrt{6}\right)\cdot\left(2-\sqrt{3^2\cdot 2}\right) = \left(2+\sqrt{6}\right)\cdot\left(2-3\sqrt{2}\right)$

$= (2\cdot 2)-(2\cdot 3)\sqrt{2}+\left(2\cdot\sqrt{6}\right)-\left(3\sqrt{2}\cdot\sqrt{6}\right) = 4-6\sqrt{2}+2\sqrt{6}-3\sqrt{2\cdot 6} = 4-6\sqrt{2}+2\sqrt{6}-3\sqrt{12}$

$= 4-6\sqrt{2}+2\sqrt{6}-3\sqrt{4\cdot 3} = 4-6\sqrt{2}+2\sqrt{6}-3\sqrt{2^2\cdot 3} = 4-6\sqrt{2}+2\sqrt{6}-(3\cdot 2)\sqrt{3} = \mathbf{4-6\sqrt{2}+2\sqrt{6}-6\sqrt{3}}$

6. $\left(2-\sqrt{5}\right)\cdot\left(\sqrt{45}+\sqrt[4]{81}\right) = \left(2-\sqrt{5}\right)\cdot\left(\sqrt{9\cdot 5}+\sqrt[4]{3^4}\right) = \left(2-\sqrt{5}\right)\cdot\left(\sqrt{3^2\cdot 5}+3\right) = \left(2-\sqrt{5}\right)\cdot\left(3\sqrt{5}+3\right)$

$= (2\cdot 3)\sqrt{5}+(2\cdot 3)-\left(3\sqrt{5}\cdot\sqrt{5}\right)-\left(3\cdot\sqrt{5}\right) = 6\sqrt{5}+6-3\sqrt{5\cdot 5}-3\sqrt{5} = 6\sqrt{5}+6-3\sqrt{5^2}-3\sqrt{5} = 6\sqrt{5}+6-(3\cdot 5)-3\sqrt{5}$

$= 6\sqrt{5}-3\sqrt{5}+6-15 = (6-3)\sqrt{5}-9 = 3\sqrt{5}-9 = \mathbf{3\left(\sqrt{5}-3\right)}$

Section 1.2b Case III Solutions - Multiplying Monomial and Binomial Expressions in Radical Form, with Real Numbers

1. $2\sqrt{3}\cdot\left(2+\sqrt{2}\right) = (2\cdot 2)\sqrt{3}+\left(2\sqrt{3}\cdot\sqrt{2}\right) = 4\sqrt{3}+2\sqrt{3\cdot 2} = 4\sqrt{3}+2\sqrt{6} = \mathbf{2\left(2\sqrt{3}+\sqrt{6}\right)}$

2. $\sqrt{5}\cdot\left(\sqrt{8}+\sqrt{5}\right) = \left(\sqrt{5}\cdot\sqrt{8}\right)+\left(\sqrt{5}\cdot\sqrt{5}\right) = \left(\sqrt{5\cdot 8}\right)+\left(\sqrt{5\cdot 5}\right) = \sqrt{40}+\sqrt{5^2} = \sqrt{4\cdot 10}+5 = \sqrt{2^2\cdot 10}+5 = \mathbf{5+2\sqrt{10}}$

3. $-\sqrt{8}\cdot\left(3-\sqrt{3}\right) = -\sqrt{4\cdot 2}\cdot\left(3-\sqrt{3}\right) = -\sqrt{2^2\cdot 2}\cdot\left(3-\sqrt{3}\right) = -2\sqrt{2}\cdot\left(3-\sqrt{3}\right) = \left(-(2\cdot 3)\cdot\sqrt{2}\right)+\left(2\sqrt{2}\cdot\sqrt{3}\right)$

$= -6\sqrt{2}+2\sqrt{2\cdot 3} = -6\sqrt{2}+2\sqrt{6} = \mathbf{2\left(\sqrt{6}-3\sqrt{2}\right)}$

4. $4\sqrt{98}\cdot\left(3-\sqrt{2^3}\right) = 4\sqrt{49\cdot 2}\cdot\left(3-\sqrt{2^{2+1}}\right) = 4\sqrt{7^2\cdot 2}\cdot\left(3-\sqrt{2^2\cdot 2^1}\right) = (4\cdot 7)\sqrt{2}\cdot\left(3-2\sqrt{2}\right) = 28\sqrt{2}\cdot\left(3-2\sqrt{2}\right)$

$= (28\cdot 3)\sqrt{2}-(28\cdot 2)\cdot\left(\sqrt{2}\cdot\sqrt{2}\right) = 84\sqrt{2}-56\left(\sqrt{2\cdot 2}\right) = 84\sqrt{2}-56\sqrt{2^2} = 84\sqrt{2}-(56\cdot 2) = 84\sqrt{2}-112 = \mathbf{4\left(21\sqrt{2}-28\right)}$

5. $\sqrt[4]{48}\cdot\left(\sqrt[4]{324}+\sqrt[4]{32}\right) = \sqrt[4]{16\cdot 3}\cdot\left(\sqrt[4]{81\cdot 4}+\sqrt[4]{16\cdot 2}\right) = \sqrt[4]{2^4\cdot 3}\cdot\left(\sqrt[4]{3^4\cdot 4}+\sqrt[4]{2^4\cdot 2}\right) = 2\sqrt[4]{3}\cdot\left(3\sqrt[4]{4}+2\sqrt[4]{2}\right)$

$= (2\cdot 3)\cdot\left(\sqrt[4]{3}\cdot\sqrt[4]{4}\right)+(2\cdot 2)\cdot\left(\sqrt[4]{3}\cdot\sqrt[4]{2}\right) = 6\cdot\left(\sqrt[4]{3\cdot 4}\right)+4\cdot\left(\sqrt[4]{3\cdot 2}\right) = 6\sqrt[4]{12}+4\sqrt[4]{6} = \mathbf{2\left(3\sqrt[4]{12}+2\sqrt[4]{6}\right)}$

6. $2\sqrt{5}\cdot\left(\sqrt{45}+\sqrt[4]{81}\right) = 2\sqrt{5}\cdot\left(\sqrt{9\cdot 5}+\sqrt[4]{3^4}\right) = 2\sqrt{5}\cdot\left(\sqrt{3^2\cdot 5}+3\right) = 2\sqrt{5}\cdot\left(3\sqrt{5}+3\right) = (2\cdot 3)\left(\sqrt{5}\cdot\sqrt{5}\right)+(2\cdot 3)\sqrt{5}$

$= 6\left(\sqrt{5\cdot 5}\right)+6\sqrt{5} = 6\sqrt{5^2}+6\sqrt{5} = (6\cdot 5)+6\sqrt{5} = 30+6\sqrt{5} = \mathbf{6\left(5+\sqrt{5}\right)}$

Section 1.2b Case IV Solutions - Rationalizing Radical Expressions - Monomial Denominators with Real Numbers

1. $\sqrt{\frac{1}{8}} = \sqrt{\frac{1}{4\cdot 2}} = \sqrt{\frac{1}{2^2\cdot 2}} = \frac{1}{2}\sqrt{\frac{1}{2}} = \frac{1}{2}\cdot\frac{\sqrt{1}}{\sqrt{2}} = \frac{1}{2}\cdot\frac{1}{\sqrt{2}} = \frac{1}{2}\left(\frac{1}{\sqrt{2}}\times\frac{\sqrt{2}}{\sqrt{2}}\right) = \frac{1}{2}\left(\frac{1\times\sqrt{2}}{\sqrt{2}\times\sqrt{2}}\right) = \frac{1}{2}\left(\frac{\sqrt{2}}{\sqrt{2\cdot 2}}\right) = \frac{1}{2}\cdot\frac{\sqrt{2}}{\sqrt{2^2}}$

$= \frac{1}{2}\cdot\frac{\sqrt{2}}{2} = \frac{1\times\sqrt{2}}{2\cdot 2} = \mathbf{\frac{\sqrt{2}}{4}}$

2. $\sqrt[2]{\frac{50}{7}} = \sqrt{\frac{50}{7}} = \sqrt{\frac{25\cdot 2}{7}} = \sqrt{\frac{5^2\cdot 2}{7}} = 5\sqrt{\frac{2}{7}} = 5\frac{\sqrt{2}}{\sqrt{7}} = 5\frac{\sqrt{2}}{\sqrt{7}}\times\frac{\sqrt{7}}{\sqrt{7}} = 5\frac{\sqrt{2}\times\sqrt{7}}{\sqrt{7}\times\sqrt{7}} = 5\frac{\sqrt{2\cdot 7}}{\sqrt{7\cdot 7}} = 5\frac{\sqrt{14}}{\sqrt{7^2}} = \mathbf{5\frac{\sqrt{14}}{7}}$

3. $\frac{\sqrt{75}}{-5} = -\frac{\sqrt{25\cdot 3}}{5} = -\frac{\sqrt{5^2\cdot 3}}{5} = -\frac{\overset{1}{\cancel{5}}\sqrt{3}}{\underset{1}{\cancel{5}}} = -\frac{\sqrt{3}}{1} = \mathbf{-\sqrt{3}}$

4. $\sqrt[3]{\frac{25}{16}} = \sqrt[3]{\frac{25}{8\cdot 2}} = \sqrt[3]{\frac{25}{2^3\cdot 2}} = \frac{1}{2}\sqrt[3]{\frac{25}{2}} = \frac{1}{2}\frac{\sqrt[3]{25}}{\sqrt[3]{2^1}} = \frac{1}{2}\left(\frac{\sqrt[3]{25}}{\sqrt[3]{2^1}}\times\frac{\sqrt[3]{2^2}}{\sqrt[3]{2^2}}\right) = \frac{1}{2}\left(\frac{\sqrt[3]{25}\times\sqrt[3]{4}}{\sqrt[3]{2^1}\times\sqrt[3]{2^2}}\right) = \frac{1}{2}\left(\frac{\sqrt[3]{25\cdot 4}}{\sqrt[3]{2^1\cdot 2^2}}\right) = \frac{1}{2}\cdot\frac{\sqrt[3]{100}}{\sqrt[3]{2^{1+2}}}$

$= \frac{1}{2}\cdot\frac{\sqrt[3]{100}}{\sqrt[3]{2^3}} = \frac{1}{2}\cdot\frac{\sqrt[3]{100}}{2} = \frac{1\cdot\sqrt[3]{100}}{2\cdot 2} = \mathbf{\frac{\sqrt[3]{100}}{4}}$

5. $\sqrt[5]{\frac{32}{8}} = \sqrt[5]{\frac{2^5}{2^3}} = 2\sqrt[5]{\frac{1}{2^3}} = 2\frac{1}{\sqrt[5]{2^3}} = 2\frac{1}{\sqrt[5]{2^3}}\times\frac{\sqrt[5]{2^2}}{\sqrt[5]{2^2}} = 2\frac{1\times\sqrt[5]{2^2}}{\sqrt[5]{2^3}\times\sqrt[5]{2^2}} = 2\frac{\sqrt[5]{2^2}}{\sqrt[5]{2^3\cdot 2^2}} = 2\frac{\sqrt[5]{4}}{\sqrt[5]{2^{3+2}}} = 2\frac{\sqrt[5]{4}}{\sqrt[5]{2^5}} = \frac{\overset{1}{\cancel{2}}}{1}\cdot\frac{\sqrt[5]{4}}{\underset{1}{\cancel{2}}}$

$= \frac{1}{1}\cdot\frac{\sqrt[5]{4}}{1} = \frac{1\cdot\sqrt[5]{4}}{1\cdot 1} = \frac{\sqrt[5]{4}}{1} = \mathbf{\sqrt[5]{4}}$

The following are two other ways to solve this problem:

5. $\sqrt[5]{\frac{32}{8}} = \sqrt[5]{\frac{2^5}{2^3}} = \sqrt[5]{2^5\cdot 2^{-3}} = \sqrt[5]{2^{5-3}} = \sqrt[5]{2^2} = \mathbf{\sqrt[5]{4}}$ or, $\sqrt[5]{\frac{32}{8}} = \sqrt[5]{\frac{\overset{4}{\cancel{32}}}{\underset{1}{\cancel{8}}}} = \sqrt[5]{\frac{4}{1}} = \mathbf{\sqrt[5]{4}}$

6. $\dfrac{-3\sqrt{100}}{-5\sqrt{3000}} = +\dfrac{3\sqrt{10^2}}{5\sqrt{100\cdot 30}} = \dfrac{3\cdot 10}{5\sqrt{10^2\cdot 30}} = \dfrac{30}{(5\cdot 10)\sqrt{30}} = \dfrac{\overset{3}{\cancel{30}}}{\underset{5}{\cancel{50}}\sqrt{30}} = \dfrac{3}{5\sqrt{30}} = \dfrac{3}{5}\cdot\dfrac{1}{\sqrt{30}} = \dfrac{3}{5}\left(\dfrac{1}{\sqrt{30}}\times\dfrac{\sqrt{30}}{\sqrt{30}}\right)$

$= \dfrac{3}{5}\left(\dfrac{1\times\sqrt{30}}{\sqrt{30}\times\sqrt{30}}\right) = \dfrac{3}{5}\left(\dfrac{\sqrt{30}}{\sqrt{30\cdot 30}}\right) = \dfrac{3}{5}\cdot\dfrac{\sqrt{30}}{\sqrt{30^2}} = \dfrac{3}{5}\cdot\dfrac{\sqrt{30}}{30} = \dfrac{\overset{1}{\cancel{3}}\cdot\sqrt{30}}{5\cdot\underset{10}{\cancel{30}}} = \dfrac{1\cdot\sqrt{30}}{5\cdot 10} = \mathbf{\dfrac{\sqrt{30}}{50}}$

Section 1.2b Case V Solutions - Rationalizing Radical Expressions - Binomial Denominators with Real Numbers

1. $\dfrac{7}{1+\sqrt{7}} = \dfrac{7}{1+\sqrt{7}}\times\dfrac{1-\sqrt{7}}{1-\sqrt{7}} = \dfrac{7\times\left(1-\sqrt{7}\right)}{\left(1+\sqrt{7}\right)\times\left(1-\sqrt{7}\right)} = \dfrac{7\left(1-\sqrt{7}\right)}{(1\cdot 1)+\left(1\cdot\sqrt{7}\right)-\left(1\cdot\sqrt{7}\right)-\left(\sqrt{7}\cdot\sqrt{7}\right)} = \dfrac{7\left(1-\sqrt{7}\right)}{1+\sqrt{7}-\sqrt{7}-\sqrt{7\cdot 7}}$

$= \dfrac{7\left(1-\sqrt{7}\right)}{1-\sqrt{7^2}} = \dfrac{7\left(1-\sqrt{7}\right)}{1-7} = \dfrac{7\left(1-\sqrt{7}\right)}{-6} = -\mathbf{\dfrac{7\left(1-\sqrt{7}\right)}{6}}$

2. $\dfrac{1-\sqrt{18}}{2+\sqrt{18}} = \dfrac{1-\sqrt{9\cdot 2}}{2+\sqrt{9\cdot 2}} = \dfrac{1-\sqrt{3^2\cdot 2}}{2+\sqrt{3^2\cdot 2}} = \dfrac{1-3\sqrt{2}}{2+3\sqrt{2}} = \dfrac{1-3\sqrt{2}}{2+3\sqrt{2}}\times\dfrac{2-3\sqrt{2}}{2-3\sqrt{2}} = \dfrac{\left(1-3\sqrt{2}\right)\times\left(2-3\sqrt{2}\right)}{\left(2+3\sqrt{2}\right)\times\left(2-3\sqrt{2}\right)}$

$= \dfrac{(1\cdot 2)-(1\cdot 3)\sqrt{2}-(2\cdot 3)\sqrt{2}+(3\cdot 3)\cdot\left(\sqrt{2}\cdot\sqrt{2}\right)}{(2\cdot 2)-(2\cdot 3)\sqrt{2}+(2\cdot 3)\sqrt{2}-(3\cdot 3)\cdot\left(\sqrt{2}\cdot\sqrt{2}\right)} = \dfrac{2-3\sqrt{2}-6\sqrt{2}+9\sqrt{2\cdot 2}}{4-6\sqrt{2}+6\sqrt{2}-9\sqrt{2\cdot 2}} = \dfrac{2-(3+6)\sqrt{2}+9\sqrt{2^2}}{4-9\sqrt{2^2}} = \dfrac{2-9\sqrt{2}+(9\cdot 2)}{4-(9\cdot 2)}$

$= \dfrac{2-9\sqrt{2}+18}{4-18} = \dfrac{(2+18)-9\sqrt{2}}{-14} = -\mathbf{\dfrac{20-9\sqrt{2}}{14}}$

3. $\dfrac{\sqrt{5}}{\sqrt{5}+\sqrt{2}} = \dfrac{\sqrt{5}}{\sqrt{5}+\sqrt{2}}\times\dfrac{\sqrt{5}-\sqrt{2}}{\sqrt{5}-\sqrt{2}} = \dfrac{\sqrt{5}\times\left(\sqrt{5}-\sqrt{2}\right)}{\left(\sqrt{5}+\sqrt{2}\right)\times\left(\sqrt{5}-\sqrt{2}\right)} = \dfrac{\left(\sqrt{5}\cdot\sqrt{5}\right)-\left(\sqrt{5}\cdot\sqrt{2}\right)}{\left(\sqrt{5}\cdot\sqrt{5}\right)-\left(\sqrt{5}\cdot\sqrt{2}\right)+\left(\sqrt{2}\cdot\sqrt{5}\right)-\left(\sqrt{2}\cdot\sqrt{2}\right)}$

$= \dfrac{\sqrt{5\cdot 5}-\sqrt{5\cdot 2}}{\sqrt{5\cdot 5}-\sqrt{5\cdot 2}+\sqrt{2\cdot 5}-\sqrt{2\cdot 2}} = \dfrac{\sqrt{5^2}-\sqrt{10}}{\sqrt{5^2}-\sqrt{10}+\sqrt{10}-\sqrt{2^2}} = \dfrac{5-\sqrt{10}}{5-2} = \mathbf{\dfrac{5-\sqrt{10}}{3}}$

4. $\dfrac{3-\sqrt{5}}{\sqrt{7}-\sqrt{4}} = \dfrac{3-\sqrt{5}}{\sqrt{7}-\sqrt{2^2}} = \dfrac{3-\sqrt{5}}{\sqrt{7}-2} = \dfrac{3-\sqrt{5}}{\sqrt{7}-2}\times\dfrac{\sqrt{7}+2}{\sqrt{7}+2} = \dfrac{\left(3-\sqrt{5}\right)\times\left(\sqrt{7}+2\right)}{\left(\sqrt{7}-2\right)\times\left(\sqrt{7}+2\right)} = \dfrac{\left(3\cdot\sqrt{7}\right)+(3\cdot 2)-\left(\sqrt{5}\cdot\sqrt{7}\right)-\left(2\cdot\sqrt{5}\right)}{\left(\sqrt{7}\cdot\sqrt{7}\right)+\left(2\cdot\sqrt{7}\right)-\left(2\cdot\sqrt{7}\right)-(2\cdot 2)}$

$= \dfrac{3\sqrt{7}+6-\sqrt{5\cdot 7}-2\sqrt{5}}{\sqrt{7\cdot 7}+2\sqrt{7}-2\sqrt{7}-4} = \dfrac{3\sqrt{7}+6-\sqrt{35}-2\sqrt{5}}{\sqrt{7^2}-4} = \dfrac{3\sqrt{7}+6-\sqrt{35}-2\sqrt{5}}{7-4} = \mathbf{\dfrac{3\sqrt{7}-\sqrt{35}-2\sqrt{5}+6}{3}}$

5. $\dfrac{-3+\sqrt{3}}{4+\sqrt{5}} = \dfrac{-3+\sqrt{3}}{4+\sqrt{5}}\times\dfrac{4-\sqrt{5}}{4-\sqrt{5}} = \dfrac{\left(-3+\sqrt{3}\right)\times\left(4-\sqrt{5}\right)}{\left(4+\sqrt{5}\right)\times\left(4-\sqrt{5}\right)} = \dfrac{-(3\cdot 4)+\left(3\cdot\sqrt{5}\right)+\left(4\cdot\sqrt{3}\right)-\left(\sqrt{3}\cdot\sqrt{5}\right)}{(4\cdot 4)-\left(4\cdot\sqrt{5}\right)+\left(4\cdot\sqrt{5}\right)-\left(\sqrt{5}\cdot\sqrt{5}\right)}$

$= \dfrac{-12+3\sqrt{5}+4\sqrt{3}-\sqrt{3\cdot 5}}{16-4\sqrt{5}+4\sqrt{5}-\sqrt{5\cdot 5}} = \dfrac{-12+3\sqrt{5}+4\sqrt{3}-\sqrt{15}}{16-\sqrt{5^2}} = \dfrac{3\sqrt{5}+4\sqrt{3}-\sqrt{15}-12}{16-5} = \mathbf{\dfrac{3\sqrt{5}+4\sqrt{3}-\sqrt{15}-12}{11}}$

6. $\dfrac{3-\sqrt{3}}{3+\sqrt{3}} = \dfrac{3-\sqrt{3}}{3+\sqrt{3}}\times\dfrac{3-\sqrt{3}}{3-\sqrt{3}} = \dfrac{\left(3-\sqrt{3}\right)\times\left(3-\sqrt{3}\right)}{\left(3+\sqrt{3}\right)\times\left(3-\sqrt{3}\right)} = \dfrac{(3\cdot 3)-\left(3\cdot\sqrt{3}\right)-\left(3\cdot\sqrt{3}\right)+\left(\sqrt{3}\cdot\sqrt{3}\right)}{(3\cdot 3)-\left(3\cdot\sqrt{3}\right)+\left(3\cdot\sqrt{3}\right)-\left(\sqrt{3}\cdot\sqrt{3}\right)} = \dfrac{9-3\sqrt{3}-3\sqrt{3}+\sqrt{3\cdot 3}}{9-3\sqrt{3}+3\sqrt{3}-\sqrt{3\cdot 3}}$

$= \dfrac{9-(3+3)\sqrt{3}+\sqrt{3^2}}{9-\sqrt{3^2}} = \dfrac{9-6\sqrt{3}+3}{9-3} = \dfrac{(9+3)-6\sqrt{3}}{6} = \dfrac{12-6\sqrt{3}}{6} = \dfrac{\cancel{6}\left(2-\sqrt{3}\right)}{\cancel{6}} = \dfrac{2-\sqrt{3}}{1} = \mathbf{2-\sqrt{3}}$

Section 1.2b Case VI Solutions - Adding and Subtracting Radical Terms

1. $5\sqrt{3}+8\sqrt{3} = (5+8)\sqrt{3} = \mathbf{13\sqrt{3}}$
2. $2\sqrt[3]{3}-4\sqrt[3]{3} = (2-4)\sqrt[3]{3} = \mathbf{-2\sqrt[3]{3}}$
3. $12\sqrt[4]{5}+8\sqrt[4]{5}+2\sqrt[4]{5} = (12+8+2)\sqrt[4]{5} = \mathbf{22\sqrt[4]{5}}$
4. $a\sqrt{ab}-b\sqrt{ab}+c\sqrt{ab} = \mathbf{(a-b+c)\sqrt{ab}}$
5. $3x\sqrt[3]{x}-2x\sqrt[3]{x}+4x\sqrt[3]{x^2} = (3x-2x)\sqrt[3]{x}+4x\sqrt[3]{x^2} = \mathbf{x\sqrt[3]{x}+4x\sqrt[3]{x^2}}$
6. $5\sqrt[3]{2}+8\sqrt[3]{5}$ ***can not be simplified***

Section 1.3a Case I Solutions - Factoring the Greatest Common Factor to Monomial Terms

1. a. $5x^3 = 5\cdot x\cdot x^2 = 5\cdot x\cdot x\cdot x$
   b. $15x = 3\cdot 5\cdot x$
   Therefore, the common terms are 5 and x. Thus, G.C.F. = $5\cdot x$ = $\mathbf{5x}$
2. a. $18x^2y^3z^4 = 2\cdot 9\cdot x\cdot x\cdot y\cdot y^2\cdot z^2\cdot z^2 = 2\cdot 3\cdot 3\cdot x\cdot x\cdot y\cdot y\cdot y\cdot z\cdot z\cdot z\cdot z$
   b. $24xy^4z^5 = 8\cdot 3\cdot x\cdot y^2\cdot y^2\cdot z^2\cdot z^3 = 2\cdot 4\cdot 3\cdot x\cdot y\cdot y\cdot y\cdot y\cdot z\cdot z\cdot z\cdot z^2 = 2\cdot 2\cdot 2\cdot 3\cdot x\cdot y\cdot y\cdot y\cdot y\cdot z\cdot z\cdot z\cdot z\cdot z$
   Thus, the common terms are 2, 3, x, y, y, y, z, z, z, z, and z. Thus, G.C.F. = $2\cdot 3\cdot x\cdot y\cdot y\cdot y\cdot z\cdot z\cdot z\cdot z$ = $\mathbf{6xy^3z^4}$
3. a. $16a^2bc^3 = 2\cdot 8\cdot a\cdot a\cdot b\cdot c\cdot c^2 = 2\cdot 2\cdot 4\cdot a\cdot a\cdot b\cdot c\cdot c\cdot c = 2\cdot 2\cdot 2\cdot 2\cdot a\cdot a\cdot b\cdot c\cdot c\cdot c$
   b. $38ab^4c^2 = 2\cdot 19\cdot a\cdot b^2\cdot b^2\cdot c^2 = 2\cdot 19\cdot a\cdot b\cdot b\cdot b\cdot b\cdot c\cdot c$
   c. $6a^3bc = 2\cdot 3\cdot a\cdot a^2\cdot b\cdot c = 2\cdot 3\cdot a\cdot a\cdot a\cdot b\cdot c$
   Therefore, the common terms are 2, a, b, and c. Thus, G.C.F. = $2\cdot a\cdot b\cdot c$ = $\mathbf{2abc}$
4. a. $r^5s^4 = r^2\cdot r^3\cdot s^2\cdot s^2 = r\cdot r\cdot r\cdot r^2\cdot s\cdot s\cdot s\cdot s = r\cdot r\cdot r\cdot r\cdot r\cdot s\cdot s\cdot s\cdot s$
   b. $4r^3s^2 = 2\cdot 2\cdot r\cdot r^2\cdot s\cdot s = 2\cdot 2\cdot r\cdot r\cdot r\cdot s\cdot s$
   c. $3rs = 3\cdot r\cdot s$
   Therefore, the common terms are r and s. Thus, G.C.F. = $r\cdot s$ = $\mathbf{rs}$
5. a. $10u^2vw^3 = 2\cdot 5\cdot u\cdot u\cdot w\cdot w^2 = 2\cdot 5\cdot u\cdot u\cdot w\cdot w\cdot w$
   b. $2uv^3w^2 = 2\cdot u\cdot v\cdot v^2\cdot w\cdot w = 2\cdot u\cdot v\cdot v\cdot v\cdot w\cdot w$
   c. $uv^2 = u\cdot v\cdot v$
   Therefore, the common terms are u and v. Thus, G.C.F. = $u\cdot v$ = $\mathbf{uv}$
6. a. $19a^3b^3 = 19\cdot a\cdot a^2\cdot b\cdot b^2 = 19\cdot a\cdot a\cdot a\cdot b\cdot b\cdot b$
   b. $12ab^2 = 2\cdot 6\cdot a\cdot b\cdot b = 2\cdot 2\cdot 3\cdot a\cdot b\cdot b$
   c. $6ab = 2\cdot 3\cdot a\cdot b$
   Therefore, the common terms are a and b. Thus, G.C.F. = $a\cdot b$ = $\mathbf{ab}$

Section 1.3a Case II Solutions - Factoring the Greatest Common Factor to Binomial and Polynomial Terms

1. a. $18x^3y^3 = 2\cdot 9\cdot x\cdot x^2\cdot y\cdot y^2 = 2\cdot 3\cdot 3\cdot x\cdot x\cdot x\cdot y\cdot y\cdot y$
   b. $12x^2y = 2\cdot 6\cdot x\cdot x\cdot y = 2\cdot 2\cdot 3\cdot x\cdot x\cdot y$
   Therefore, the common terms are 2, 3, x, x, and y. This implies that G.C.F. = $2\cdot 3\cdot x\cdot x\cdot y = 6x^2y$. Thus,
   $18x^3y^3-12x^2y = \mathbf{6x^2y(3xy^2-2)}$
2. a. $3a^2b^3c = 3\cdot a\cdot a\cdot b\cdot b^2\cdot c = 3\cdot a\cdot a\cdot b\cdot b\cdot b\cdot c$
   b. $15ab^2c^3 = 3\cdot 5\cdot a\cdot b\cdot b\cdot c\cdot c^2 = 3\cdot 5\cdot a\cdot b\cdot b\cdot c\cdot c\cdot c$
   Therefore, the common terms are 3, a, b, b, and c. This implies that G.C.F. = $3\cdot a\cdot b\cdot b\cdot c = 3ab^2c$. Thus,
   $3a^2b^3c+15ab^2c^3 = \mathbf{3ab^2c(ab+5c^2)}$
3. a. $xyz^3 = x\cdot y\cdot z\cdot z^2 = x\cdot y\cdot z\cdot z\cdot z$
   b. $4x^2y^2z^5 = 2\cdot 2\cdot x\cdot x\cdot y\cdot y\cdot z^2\cdot z^3 = 2\cdot 2\cdot x\cdot x\cdot y\cdot y\cdot z\cdot z\cdot z\cdot z^2 = 2\cdot 2\cdot x\cdot x\cdot y\cdot y\cdot z\cdot z\cdot z\cdot z\cdot z$
   Therefore, the common terms are x, y, z, z, and z. This implies that G.C.F. = $x\cdot y\cdot z\cdot z\cdot z = xyz^3$. Thus,

$xyz^3 + 4x^2y^2z^5 = \mathbf{xyz^3\left(1 + 4xyz^2\right)}$

4. a. $25p^3 = 5 \cdot 5 \cdot p \cdot p^2 = 5 \cdot 5 \cdot p \cdot p \cdot p$

   b. $5p^2q^3 = 5 \cdot p \cdot p \cdot q \cdot q^2 = 5 \cdot p \cdot p \cdot q \cdot q \cdot q$

   c. $pq = p \cdot q$

   Therefore, the common term is p. This implies that G.C.F. = p. Thus, $25p^3 + 5p^2q^3 + pq = \mathbf{p\left(25p^2 + 5pq^3 + q\right)}$

5. a. $r^2s^2t = r \cdot r \cdot s \cdot s \cdot t$

   b. $5rst^2 = 5 \cdot r \cdot s \cdot t \cdot t$

   Therefore, the common terms are r, s, and t. This implies that G.C.F. = $r \cdot s \cdot t = rst$. Thus, $r^2s^2t - 5rst^2$
   $= \mathbf{rst(rs - 5t)}$

6. a. $36x^3yz^3 = 2 \cdot 18 \cdot x \cdot x^2 \cdot y \cdot z \cdot z^2 = 2 \cdot 2 \cdot 9 \cdot x \cdot x \cdot x \cdot y \cdot z \cdot z \cdot z = 2 \cdot 2 \cdot 3 \cdot 3 \cdot x \cdot x \cdot x \cdot y \cdot z \cdot z \cdot z$

   b. $4xy^2z^4 = 2 \cdot 2 \cdot x \cdot y \cdot y \cdot z^2 \cdot z^2 = 2 \cdot 2 \cdot x \cdot y \cdot y \cdot z \cdot z \cdot z \cdot z$

   c. $12x^3y^3z = 2 \cdot 6 \cdot x \cdot x^2 \cdot y \cdot y^2 \cdot z = 2 \cdot 2 \cdot 3 \cdot x \cdot x \cdot x \cdot y \cdot y \cdot y \cdot z$

   Therefore, the common terms are 2, 2, x, y, and z. This implies that G.C.F. = $2 \cdot 2 \cdot x \cdot y \cdot z = 4xyz$. Thus,
   $36x^3yz^3 + 4xy^2z^4 - 12x^3y^3z = \mathbf{4xyz\left(9x^2z^2 + yz^3 - 3x^2y^2\right)}$

Section 1.3b Solutions - Factoring Polynomials Using the Grouping Method

1. $2ab - 5b - 6a + 15 = b(2a - 5) - 3(2a - 5) = \mathbf{(2a - 5)(b - 3)}$
2. $y^3 + 4y^2 + y + 4 = y^2(y + 4) + (y + 4) = \mathbf{(y + 4)\left(y^2 + 1\right)}$
3. $42x^2y + 21xy - 70x - 35 = 21xy(2x + 1) - 35(2x + 1) = \mathbf{(2x + 1)(21xy - 35)}$
4. $(x + y)^3 + (x + y)^2 + x + y = (x + y)^3 + (x + y)^2 + (x + y) = (x + y)\left[(x + y)^2 + (x + y) + 1\right] = \mathbf{(x + y)\{(x + y)[(x + y) + 1] + 1\}}$
5. $4(a + b)^2 + 32a + 32b = 4(a + b)^2 + 32(a + b) = 4(a + b)[(a + b) + 8] = \mathbf{4(a + b)[a + b + 8]}$
6. $36r^3s - 6r^2s + 18r - 3 = 6r^2s(6r - 1) + 3(6r - 1) = (6r - 1)\left(6r^2s + 3\right) = \mathbf{3(6r - 1)\left(2r^2s + 1\right)}$

Section 1.3c Case I Solutions - Factoring Trinomials of the Form $ax^2 + bx + c$ where $a = 1$

1. $x^2 - 2x - 15 = \mathbf{(x + 3)(x - 5)}$
2. $y^2 - 9y + 8 = \mathbf{(y - 1)(y - 8)}$
3. $t^2 + 2t - 15 = \mathbf{(t - 3)(t + 5)}$
4. $y^2 - 2y + 11$ **is prime**
5. $x^2 + 10x + 21 = \mathbf{(x + 3)(x + 7)}$
6. $u^2 + 4u - 32 = \mathbf{(u - 4)(u + 8)}$

Section 1.3c Case II Solutions - Factoring Trinomials of the Form $ax^2 + bx + c$ where $a \rangle 1$

1. $10x^2 + 11x - 35 = 10x^2 + (25 - 14)x - 35 = 10x^2 + 25x - 14x - 35 = 5x(2x + 5) - 7(2x + 5) = \mathbf{(2x + 5)(5x - 7)}$
2. $6x^2 - x - 12 = 6x^2 + (-9 + 8)x - 12 = 6x^2 - 9x + 8x - 12 = 3x(2x - 3) + 4(2x - 3) = \mathbf{(2x - 3)(3x + 4)}$
3. $-7x^2 + 46x + 21 = -7x^2 + (49 - 3)x + 21 = -7x^2 + 49x - 3x + 21 = 7x(-x + 7) + 3(-x + 7) = \mathbf{(-x + 7)(7x + 3)}$
4. $6x^2 - 11xy + 3y^2 = 3y^2 + (-11x)y + 6x^2 = 3y^2 + (-9 - 2)xy + 6x^2 = 3y^2 - 9xy - 2xy + 6x^2 = 3y(y - 3x) - 2x(y - 3x)$
   $= \mathbf{(y - 3x)(3y - 2x)}$
5. $6x^2 + x - 40 = 6x^2 + (16 - 15)x - 40 = 6x^2 + 16x - 15x - 40 = 2x(3x + 8) - 5(3x + 8) = \mathbf{(3x + 8)(2x - 5)}$
6. $2x^2 + 3x - 27 = 2x^2 + (9 - 6)x - 27 = 2x^2 + 9x - 6x - 27 = x(2x + 9) - 3(2x + 9) = \mathbf{(2x + 9)(x - 3)}$

Section 1.3d Case I Solutions - Factoring Polynomials Using the Difference of Two Squares Method

1. $x^3-16x = x\left(x^2-16\right) = x\left(x^2-4^2\right) = \mathbf{x(x-4)(x+4)}$

2. $(x+1)^2-(y+3)^2 = [(x+1)-(y+3)][(x+1)+(y+3)] = (x+1-y-3)(x+1+y+3) = \mathbf{(x-y-2)(x+y+4)}$

3. $t^5-81t = t\left(t^4-81\right) = t\left(t^{2^2}-9^2\right) = t\left(t^2-9\right)\left(t^2+9\right) = t\left(t^2-3^2\right)\left(t^2+9\right) = \mathbf{t(t-3)(t+3)\left(t^2+9\right)}$

4. $\left(x^2+10x+25\right)-y^2 = (x+5)^2-y^2 = [(x+5)-y][(x+5)+y] = (x+5-y)\,(x+5+y) = \mathbf{(x-y+5)\,(x+y+5)}$

5. $c^4-9c^2 = c^2\left(c^2-9\right) = c^2\left(c^2-3^2\right) = \mathbf{c^2(c-3)(c+3)}$

6. $p^2-q^2-4q-4 = p^2-\left(q^2+4q+4\right) = p^2-(q+2)^2 = [p-(q+2)][p+(q+2)] = \mathbf{(p-q-2)(p+q+2)}$

Section 1.3d Case II Solutions - Factoring Polynomials Using the Sum and Difference of Two Cubes Method

1. $4x^6+4 = 4\left(x^6+1\right) = 4\left(x^{2^3}+1^3\right) = 4\left[\left(x^2\right)^3+1^3\right] = 4\left(x^2+1\right)\left[\left(x^2\right)^2-x^2\cdot 1+1^2\right] = \mathbf{4\left(x^2+1\right)\left(x^4-x^2+1\right)}$

2. $x^6y^6+8 = x^{2^3}y^{2^3}+2^3 = \left(x^2y^2\right)^3+2^3 = \left(x^2y^2+2\right)\left[\left(x^2y^2\right)^2-2\cdot x^2y^2+2^2\right] = \mathbf{\left(x^2y^2+2\right)\left(x^4y^4-2x^2y^2+4\right)}$

3. $(x+2)^3-y^3 = (x+2)^3-y^3 = [(x+2)-y]\left[(x+2)^2+(x+2)\cdot y+y^2\right] = \mathbf{(x-y+2)\left[(x+2)^2+(x+2)y+y^2\right]}$

4. $2r^6-128 = 2\left(r^6-64\right) = 2\left(r^{2^3}-4^3\right) = 2\left[\left(r^2\right)^3-4^3\right] = 2\left(r^2-4\right)\left[\left(r^2\right)^2+4\cdot r^2+4^2\right] = \mathbf{2\left(r^2-4\right)\left(r^4+4r^2+16\right)}$

5. $(x-7)^3+y^3 = (x-7)^3+y^3 = [(x-7)+y]\left[(x-7)^2-(x-7)\cdot y+y^2\right] = \mathbf{(x+y-7)\left[(x-7)^2-(x-7)y+y^2\right]}$

6. $x^6y^5+x^3y^2 = x^3y^2\left(x^3y^3+1\right) = x^3y^2\left[(xy)^3+1^3\right] = x^3y^2\left\{[(xy)+1]\left[(xy)^2-(xy)\cdot 1+1^2\right]\right\}$

$= \mathbf{x^3y^2\left\{(xy+1)\left[(xy)^2-xy+1\right]\right\}}$

Section 1.3d Case III Solutions - Factoring Perfect Square Trinomials

1. $x^2+18x+81 = x^2+18x+9^2 = x^2+2\cdot(x\cdot 9)+9^2 = \mathbf{(x+9)^2}$

2. $9+64p^2-48p = 64p^2-48p+9 = 8^2p^2-48p+3^2 = (8p)^2-2\cdot(8p\cdot 3)+3^2 = \mathbf{(8p-3)^2}$

3. $9w^2+25+30w = 9w^2+30w+25 = 3^2w^2+30w+5^2 = (3w)^2+2\cdot(3w\cdot 5)+5^2 = \mathbf{(3w+5)^2}$

4. $25+k^2-10k = k^2-10k+25 = k^2-2\cdot(k\cdot 5)+5^2 = \mathbf{(k-5)^2}$

5. $49x^2-84x+36 = 7^2x^2-84x+6^2 = (7x)^2-2\cdot(7x\cdot 6)+6^2 = \mathbf{(7x-6)^2}$

6. $1+16z+64z^2 = 64z^2+16z+1 = 8^2z^2+16z+1^2 = (8z)^2+2\cdot(8z\cdot 1)+1^2 = \mathbf{(8z+1)^2}$

Section 1.4a Solutions - Quadratic Equations and the Quadratic Formula

1. First - Write the quadratic equation $3x=-5+2x^2$ in standard form $ax^2+bx+c=0$.

   $3x=-5+2x^2$; $-2x^2+3x=-5+2x^2-2x^2$; $-2x^2+3x=-5+0$; $-2x^2+3x=-5$; $-2x^2+3x+5=-5+5$; $-2x^2+3x+5=0$

Second - Equate the a, b, and c coefficients with the coefficients of the given quadratic equation.

Thus, $\boldsymbol{a=-2}$, $\boldsymbol{b=3}$, and $\boldsymbol{c=5}$

2. First - Write the quadratic equation $2x^2=5$ in standard form $ax^2+bx+c=0$.

$2x^2=5$; $2x^2-5=5-5$; $2x^2-5=0$ which is the same as $2x^2+0x-5=0$

Second - Equate the a, b, and c coefficients with the coefficients of the given quadratic equation.

Thus, $\boldsymbol{a=2}$, $\boldsymbol{b=0}$, and $\boldsymbol{c=-5}$

3. First - Write the quadratic equation $3w^2-5w=2$ in standard form $aw^2+bw+c=0$.

$3w^2-5w=2$; $3w^2-5w-2=2-2$; $3w^2-5w-2=0$

Second - Equate the a, b, and c coefficients with the coefficients of the given quadratic equation.

Thus, $\boldsymbol{a=3}$, $\boldsymbol{b=-5}$, and $\boldsymbol{c=-2}$

4. First - Write the quadratic equation $15=-y^2-3$ in standard form $ay^2+by+c=0$.

$15=-y^2-3$; $y^2+15=-y^2+y^2-3$; $y^2+15=0-3$; $y^2+15=-3$; $y^2+15+3=-3+3$; $y^2+18=0$

; which is the same as $y^2+0y+18=0$

Second - Equate the a, b, and c coefficients with the coefficients of the given quadratic equation.

Thus, $\boldsymbol{a=1}$, $\boldsymbol{b=0}$, and $\boldsymbol{c=18}$

5. First - Write the quadratic equation $x^2+3=5x$ in standard form $ax^2+bx+c=0$.

$x^2+3=5x$; $x^2-5x+3=5x-5x$; $x^2-5x+3=0$

Second - Equate the a, b, and c coefficients with the coefficients of the given quadratic equation.

Thus, $\boldsymbol{a=1}$, $\boldsymbol{b=-5}$, and $\boldsymbol{c=3}$

6. First - Write the quadratic equation $-u^2+2=3u$ in standard form $au^2+bu+c=0$.

$-u^2+2=3u$; $-u^2-3u+2=3u-3u$; $-u^2-3u+2=0$

Second - Equate the a, b, and c coefficients with the coefficients of the given quadratic equation.

Thus, $\boldsymbol{a=-1}$, $\boldsymbol{b=-3}$, and $\boldsymbol{c=2}$

Section 1.4b Case I Solutions - Solving Quadratic Equations of the Form ax^2+bx+c where $a=1$

1. $x^2=-5x-6$ Write the equation in standard form, i.e., $x^2+5x+6=0$.

Let: $a=4$, $b=6$, and $c=1$. Then,

Given: $u=\frac{-b\pm\sqrt{b^2-4ac}}{2a}$; $u=\frac{-6\pm\sqrt{6^2-4\times4\times1}}{2\times4}$; $u=\frac{-6\pm\sqrt{36-16}}{8}$; $u=\frac{-6\pm\sqrt{20}}{8}$; $u=\frac{-6\pm4.47}{8}$ therefore,

I. $u=\frac{-6+4.47}{8}$; $u=-\frac{1.53}{8}$; $x=-\frac{2}{1}$; $\boldsymbol{u=-0.19}$ and

II. $x=\frac{-5-1}{2}$; $x=-\frac{\overset{3}{\cancel{6}}}{\cancel{2}}$; $x=-\frac{3}{1}$; $\boldsymbol{x=-3}$

Check: I. *Let* $u=-0.19$ *in* $4u^2+6u+1=0$; $4\cdot(-0.19)^2+6\cdot-0.19+1\overset{?}{=}0$; $4\overset{?}{=}10-6$; $4=4$

II. *Let* $x=-3$ *in* $x^2=-5x-6$; $(-3)^2\overset{?}{=}(-5\times-3)-6$; $9\overset{?}{=}15-6$; $9=9$

Therefore, the equation $x^2+5x+6=0$ can be factored to $\boldsymbol{(u+0.19)(u+1.31)=0}$.

2. $y^2-40y=-300$ Write the equation in standard form, i.e., $4w^2+10w+3=0$.

Let: $a=4$, $b=10$, and $c=3$. Then,

Given: $w=\frac{-b\pm\sqrt{b^2-4ac}}{2a}$; $w=\frac{-10\pm\sqrt{10^2-4\times4\times3}}{2\times4}$; $w=\frac{-10\pm\sqrt{100-48}}{8}$; $w=\frac{-10\pm\sqrt{52}}{8}$

; $y=\frac{40\pm\sqrt{20^2}}{2}$; $y=\frac{40\pm20}{2}$ therefore, I. $w=\frac{-10+7.2}{8}$; $y=\frac{\overset{30}{\cancel{60}}}{\cancel{2}}$; $y=\frac{30}{1}$; $\mathbf{y=30}$ and

II. $y=\frac{40-20}{2}$; $y=\frac{\overset{10}{\cancel{20}}}{\cancel{2}}$; $y=\frac{10}{1}$; $\mathbf{y=10}$

Check: I. *Let* $w=-0.35$ *in* $4w^2+10w=-3$; $4\cdot(-0.35)^2+10\cdot-0.35\overset{?}{=}-3$; $900-1200\overset{?}{=}-300$; $-300=-300$

II. *Let* $y=10$ *in* $y^2-40y=-300$; $(10)^2-40\cdot10\overset{?}{=}-300$; $100-400\overset{?}{=}-300$; $-300=-300$

Therefore, the equation $y^2-40y+300=0$ can be factored to $\mathbf{(y-30)(y-10)=0}$.

3. $-x=-x^2+20$ Write the equation in standard form, i.e., $x^2-x-20=0$.

Let: $a=1$, $b=-1$, and $c=-20$. Then,

Given: $x=\frac{-b\pm\sqrt{b^2-4ac}}{2a}$; $x=\frac{-(-1)\pm\sqrt{(-1)^2-4\times1\times-20}}{2\times1}$; $x=\frac{1\pm\sqrt{1+80}}{2}$; $x=\frac{1\pm\sqrt{81}}{2}$; $x=\frac{1\pm\sqrt{9^2}}{2}$

; $x=\frac{1\pm9}{2}$ therefore, I. $x=\frac{1+9}{2}$; $x=\frac{\overset{5}{\cancel{10}}}{\cancel{2}}$; $x=\frac{5}{1}$; $\mathbf{x=5}$ and

II. $x=\frac{1-9}{2}$; $x=-\frac{\overset{4}{\cancel{8}}}{\cancel{2}}$; $x=-\frac{4}{1}$; $\mathbf{x=-4}$

Check: I. *Let* $x=5$ *in* $-x=-x^2+20$; $-5\overset{?}{=}-5^2+20$; $-5\overset{?}{=}-25+20$; $-5=-5$

II. *Let* $x=-4$ *in* $-x=-x^2+20$; $-(-4)\overset{?}{=}-(-4)^2+20$; $4\overset{?}{=}-16+20$; $4=4$

Therefore, the equation $x^2-x-20=0$ can be factored to $\mathbf{(x-5)(x+4)=0}$.

4. $x^2+3x+4=0$ The equation is already in standard form.

Let: $a=1$, $b=3$, and $c=4$. Then,

Given: $x=\frac{-b\pm\sqrt{b^2-4ac}}{2a}$; $x=\frac{-3\pm\sqrt{3^2-4\times1\times4}}{2\times1}$; $x=\frac{-3\pm\sqrt{9-16}}{2}$; $x=\frac{-3\pm\sqrt{-7}}{2}$

Since the number under the radical is negative (an imaginary number), the given equation is not factorable.

5. $x^2-80-2x=0$ Write the equation in standard form, i.e., $x^2-2x-80=0$.

Let: $a=1$, $b=-2$, and $c=-80$. Then,

Given: $x=\frac{-b\pm\sqrt{b^2-4ac}}{2a}$; $x=\frac{-(-2)\pm\sqrt{(-2)^2-4\times1\times-80}}{2\times1}$; $x=\frac{2\pm\sqrt{4+320}}{2}$; $x=\frac{2\pm\sqrt{324}}{2}$; $x=\frac{2\pm\sqrt{18^2}}{2}$

; $x=\frac{2\pm18}{2}$ therefore, I. $x=\frac{2+18}{2}$; $x=\frac{\overset{10}{\cancel{20}}}{\cancel{2}}$; $x=\frac{10}{1}$; $\mathbf{x=10}$ and

II. $x=\frac{2-18}{2}$; $x=-\frac{\overset{8}{\cancel{16}}}{\cancel{2}}$; $x=-\frac{8}{1}$; $\mathbf{x=-8}$

Check: I. *Let* $x=10$ *in* $x^2-80-2x=0$; $10^2-80-2\cdot10\overset{?}{=}0$; $100-80-20\overset{?}{=}0$; $100-100\overset{?}{=}0$; $0=0$

II. *Let* $x=-8$ *in* $x^2-80-2x=0$; $(-8)^2-80-2\cdot(-8)\overset{?}{=}0$; $64-80+16\overset{?}{=}0$; $80-80\overset{?}{=}0$; $0=0$

Therefore, the equation $x^2-2x-80=0$ can be factored to $\mathbf{(x-10)(x+8)=0}$.

6. $x^2+4x+4=0$ The equation is already in standard form.

Let: $a=1$, $b=4$, and $c=4$. Then,

Given: $x=\dfrac{-b\pm\sqrt{b^2-4ac}}{2a}$; $x=\dfrac{-4\pm\sqrt{4^2-4\times1\times4}}{2\times1}$; $x=\dfrac{-4\pm\sqrt{16-16}}{2}$; $x=\dfrac{-4\pm\sqrt{0}}{2}$; $x=\dfrac{-4\pm0}{2}$

; $x=-\dfrac{\overset{2}{\not 4}}{\not 2}$; $x=-\dfrac{2}{1}$; $x=-2$. In this case the equation has one repeated solution, i.e., $\mathbf{x=-2}$ and $\mathbf{x=-2}$

Check: *Let* $x=-2$ *in* $x^2+4x+4=0$; $(-2)^2+4\cdot(-2)+4\overset{?}{=}0$; $4-8+4\overset{?}{=}0$; $8-8\overset{?}{=}0$; $0=0$

Therefore, the equation $x^2+4x+4=0$ can be factored to $\mathbf{(x+2)(x+2)=0}$

Section 1.4b Case II Solutions - Solving Quadratic Equations of the Form ax^2+bx+c where $a\rangle1$

1. $4u^2+6u+1=0$ The quadratic equation is already in standard form.

Let: $a=4$, $b=6$, and $c=1$. Then,

Given: $u=\dfrac{-b\pm\sqrt{b^2-4ac}}{2a}$; $u=\dfrac{-6\pm\sqrt{6^2-4\times4\times1}}{2\times4}$; $u=\dfrac{-6\pm\sqrt{36-16}}{8}$; $u=\dfrac{-6\pm\sqrt{20}}{8}$; $u=\dfrac{-6\pm4.47}{8}$

therefore, I. $u=\dfrac{-6+4.47}{8}$; $u=-\dfrac{1.53}{8}$; $\mathbf{u=-0.19}$ and

II. $u=\dfrac{-6-4.47}{8}$; $u=-\dfrac{10.47}{8}$; $\mathbf{u=-1.31}$

The solution set is $\{\mathbf{-1.31,-0.9}\}$

Check: I. *Let* $u=-0.19$ *in* $4u^2+6u+1=0$; $4\cdot(-0.19)^2+6\cdot-0.19+1\overset{?}{=}0$; $4\cdot0.036-1.14+1\overset{?}{=}0$; $0.14-1.14+1\overset{?}{=}0$

; $1.14-1.14\overset{?}{=}0$; $0=0$

II. *Let* $u=-1.31$ *in* $4u^2+6u+1=0$; $4\cdot(-1.31)^2+6\cdot-1.31+1\overset{?}{=}0$; $4\cdot1.716-7.86+1\overset{?}{=}0$; $6.86-7.86+1\overset{?}{=}0$

; $7.86-7.86\overset{?}{=}0$; $0=0$

Therefore, the equation $4u^2+6u+1=0$ can be factored to $\mathbf{(u+0.19)(u+1.31)=0}$

2. $4w^2+10w=-3$ Write the equation in standard form, i.e., $4w^2+10w+3=0$.

Let: $a=4$, $b=10$, and $c=3$. Then,

Given: $w=\dfrac{-b\pm\sqrt{b^2-4ac}}{2a}$; $w=\dfrac{-10\pm\sqrt{10^2-4\times4\times3}}{2\times4}$; $w=\dfrac{-10\pm\sqrt{100-48}}{8}$; $w=\dfrac{-10\pm\sqrt{52}}{8}$

; $w=\dfrac{-10\pm7.2}{8}$ therefore, I. $w=\dfrac{-10+7.2}{8}$; $w=-\dfrac{2.8}{8}$; $\mathbf{w=-0.35}$ and

II. $w=\dfrac{-10-7.2}{8}$; $w=-\dfrac{17.2}{8}$; $\mathbf{w=-2.15}$

The solution set is $\{\mathbf{-2.15,-0.35}\}$

Check: I. *Let* $w=-0.35$ *in* $4w^2+10w=-3$; $4\cdot(-0.35)^2+10\cdot-0.35\overset{?}{=}-3$; $4\cdot0.123-3.5\overset{?}{=}-3$; $0.5-3.5\overset{?}{=}-3$

; $-3=-3$

II. *Let* $w=-2.15$ *in* $4w^2+10w=-3$; $4\cdot(-2.15)^2+10\cdot-2.15\stackrel{?}{=}-3$; $4\cdot4.62-21.5\stackrel{?}{=}-3$; $18.5-21.5\stackrel{?}{=}-3$; $-3=-3$

Therefore, the equation $4w^2+10w+3=0$ can be factored to $\mathbf{(w+0.35)(w+2.15)=0}$.

3. $6x^2+4x-2=0$ The quadratic equation is already in standard form.

Let: $a=6$, $b=4$, and $c=-2$. Then,

Given: $x=\frac{-b\pm\sqrt{b^2-4ac}}{2a}$; $x=\frac{-4\pm\sqrt{4^2-4\times6\times-2}}{2\times6}$; $x=\frac{-4\pm\sqrt{16+48}}{12}$; $x=\frac{-4\pm\sqrt{64}}{12}$; $x=\frac{-4\pm\sqrt{8^2}}{12}$; $x=\frac{-4\pm8}{12}$ therefore,

I. $x=\frac{-4+8}{12}$; $x=\frac{4}{12}$; $x=\frac{1}{3}$; $\mathbf{x=0.33}$ and

II. $x=\frac{-4-8}{12}$; $x=-\frac{12}{12}$; $x=-\frac{1}{1}$; $\mathbf{x=-1}$

The solution set is $\{\mathbf{-1, 0.33}\}$.

Check: I. *Let* $x=0.33$ *in* $6x^2+4x-2=0$; $6\cdot0.33^2+4\cdot0.33-2\stackrel{?}{=}0$; $6\cdot0.111+1.32-2\stackrel{?}{=}0$; $0.67+1.32-2\stackrel{?}{=}0$; $2-2\stackrel{?}{=}0$; $0=0$

II. *Let* $x=-1$ *in* $6x^2+4x-2=0$; $6\cdot(-1)^2+4\cdot-1-2\stackrel{?}{=}0$; $6\cdot1-4-2\stackrel{?}{=}0$; $6-6\stackrel{?}{=}0$; $0=0$

Therefore, the equation $6x^2+4x-2=0$ can be factored to $\mathbf{(x-0.33)(x+1)=0}$.

4. $15y^2+3=-14y$ Write the equation in standard form, i.e., $15y^2+14y+3=0$.

Let: $a=15$, $b=14$, and $c=3$. Then,

Given: $y=\frac{-b\pm\sqrt{b^2-4ac}}{2a}$; $y=\frac{-14\pm\sqrt{(-14)^2-4\times15\times3}}{2\times15}$; $y=\frac{-14\pm\sqrt{196-180}}{30}$; $y=\frac{-14\pm\sqrt{16}}{30}$; $y=\frac{-14\pm\sqrt{4^2}}{30}$; $y=\frac{-14\pm4}{30}$ therefore,

I. $y=\frac{-14+4}{30}$; $y=-\frac{10}{30}$; $y=-\frac{1}{3}$; $\mathbf{y=-0.33}$ and

II. $y=\frac{-14-4}{30}$; $y=-\frac{18}{30}$; $y=-\frac{3}{5}$; $\mathbf{y=-0.6}$

The solution set is $\{\mathbf{-0.6, -0.33}\}$.

Check I. *Let* $y=-0.33$ *in* $15y^2+3=-14y$; $15\cdot(-0.33)^2+3\stackrel{?}{=}-14\cdot-0.33$; $15\cdot0.108+3\stackrel{?}{=}4.62$; $1.62+3\stackrel{?}{=}4.62$; $4.62=4.62$

II. *Let* $y=-0.6$ *in* $15y^2+3=-14y$; $15\cdot(-0.6)^2+3\stackrel{?}{=}-14\cdot-0.6$; $15\cdot0.36+3\stackrel{?}{=}8.4$; $5.4+3\stackrel{?}{=}8.4$; $8.4=8.4$

Therefore, the equation $15y^2+3=-14y$ can be factored to $\mathbf{(y+0.6)(w+0.33)=0}$.

5. $2x^2-5x+3=0$ The equation is in standard form.

Let: $a=2$, $b=-5$, and $c=3$. Then,

Given: $x=\frac{-b\pm\sqrt{b^2-4ac}}{2a}$; $x=\frac{-(-5)\pm\sqrt{(-5)^2-4\times2\times3}}{2\times2}$; $x=\frac{5\pm\sqrt{25-24}}{4}$; $x=\frac{5\pm\sqrt{1}}{4}$; $x=\frac{5\pm1}{4}$

therefore, I. $x=\frac{5+1}{4}$; $x=\frac{6}{4}$; $x=\frac{3}{2}$; $\mathbf{x=1.5}$ and

II. $x=\frac{5-1}{4}$; $x=\frac{4}{4}$; $x=\frac{1}{1}$; $\mathbf{x=1}$

The solution set is $\{\mathbf{1,1.5}\}$.

Check I. *Let* $x=1$ *in* $2x^2-5x+3=0$; $2\cdot1^2-5\cdot1+3\overset{?}{=}0$; $2\cdot1-5+3\overset{?}{=}0$; $2-5+3\overset{?}{=}0$; $5-5\overset{?}{=}0$; $0=0$

II. *Let* $x=1.5$ *in* $2x^2-5x+3=0$; $2\cdot1.5^2-5\cdot1.5+3\overset{?}{=}0$; $2\cdot2.25-7.5+3\overset{?}{=}0$; $4.5-7.5+3\overset{?}{=}0$

; $7.5-7.5\overset{?}{=}0$; $0=0$

Therefore, the equation $2x^2-5x+3=0$ can be factored to $\mathbf{(x-1)(x-1.5)=0}$.

6. $2x^2+xy-y^2=0$ x is varaible Write the equation in standard form, i.e., $2x^2+yx-y^2=0$.

Let: $a=2$, $b=y$, and $c=-y^2$. Then,

Given: $x=\frac{-b\pm\sqrt{b^2-4ac}}{2a}$; $x=\frac{-y\pm\sqrt{y^2-4\times2\times-y^2}}{2\times2}$; $x=\frac{-y\pm\sqrt{y^2+8y^2}}{4}$; $x=\frac{-y\pm\sqrt{9y^2}}{4}$; $x=\frac{-y\pm3y}{4}$

therefore, I. $x=\frac{-y+3y}{4}$; $x=\frac{2y}{4}$; $x=\frac{1}{2}y$; $\mathbf{x=0.5y}$ and

II. $x=\frac{-y-3y}{4}$; $x=\frac{-4y}{4}$; $x=-\frac{4}{4}y$; $\mathbf{x=-y}$

The solution set is $\{\mathbf{-y,0.5y}\}$.

I. *Let* $x=0.5y$ *in* $2x^2+xy-y^2=0$; $2\cdot(0.5y)^2+(0.5y)\cdot y-y^2\overset{?}{=}0$; $2\cdot0.25y^2+0.5y^2-y^2\overset{?}{=}0$

; $0.5y^2+0.5y^2-y^2\overset{?}{=}0$; $y^2-y^2\overset{?}{=}0$; $0=0$

II. *Let* $x=-y$ *in* $2x^2+xy-y^2=0$; $2\cdot(-y)^2+(-y)\cdot y-y^2\overset{?}{=}0$; $2y^2-y^2-y^2\overset{?}{=}0$; $2y^2-2y^2\overset{?}{=}0$; $0=0$

Therefore, the equation $2x^2+xy-y^2=0$ can be factored to $\mathbf{(x+y)(x-0.5y)=0}$.

Section 1.4c Solutions - Solving Quadratic Equations Using the Square Root Property Method

1. First - Take the square root of both sides of the equation $(2y+5)^2=36$, i.e., $\sqrt{(2y+5)^2}=\pm\sqrt{36}$

Second - Simplify the terms on both sides to obtain the solutions, i.e., $\sqrt{(2y+5)^2}=\pm\sqrt{36}$; $2y+5=\pm6$

Therefore the two solutions are: I. $2y+5=-6$; $2y=-6-5$; $2y=-11$; $\frac{2y}{2}=-\frac{11}{2}$; $y=-\frac{11}{2}$; $\mathbf{y=-5.5}$ and

II. $2y+5=+6$; $2y=6-5$; $2y=1$; $\frac{2y}{2}=\frac{1}{2}$; $y=\frac{1}{2}$; $\mathbf{y=0.5}$

Thus, the solution set is $\{\mathbf{-5.5,0.5}\}$ and the equation $(2y+5)^2=36$ can be factored to $\mathbf{(y+5.5)(y-0.5)=0}$.

Check: I. *Let* $y=-5.5$ *in* $(2y+5)^2=36$; $(2\cdot-5.5+5)^2\overset{?}{=}36$; $(-11+5)^2\overset{?}{=}36$; $(-6)^2\overset{?}{=}36$; $6^2\overset{?}{=}36$; $36=36$

II. *Let* $y=0.5$ *in* $(2y+5)^2=36$; $(2\cdot0.5+5)^2\overset{?}{=}36$; $(1+5)^2\overset{?}{=}36$; $6^2\overset{?}{=}36$; $36=36$

2. First - Take the square root of both sides of the equation $(x+1)^2=7$, i.e., $\sqrt{(x+1)^2}=\pm\sqrt{7}$

Second - Simplify the terms on both sides to obtain the solutions, i.e., $\sqrt{(x+1)^2}=\pm\sqrt{7}$; $x+1=\pm2.65$

Therefore the two solutions are: I. $x+1=-2.65$; $x=-2.65-1$; $\mathbf{x=-3.65}$ and

II. $x+1=+2.65$; $x=2.65-1$; $\mathbf{x=1.65}$

Thus, the solution set is $\{-3.65, 1.65\}$ and the equation $(x+1)^2 = 7$ can be factored to $(x+3.65)(x-1.65)=0$.

Check: I. *Let* $x = -3.65$ *in* $(x+1)^2 = 7$; $(-3.65+1)^2 \stackrel{?}{=} 7$; $(-2.65)^2 \stackrel{?}{=} 7$; $7 = 7$

II. *Let* $x = 1.65$ *in* $(x+1)^2 = 7$; $(1.65+1)^2 \stackrel{?}{=} 7$; $2.65^2 \stackrel{?}{=} 7$; $7 = 7$

3. First - Take the square root of both sides of the equation $(2x-3)^2 = 1$, i.e., $\sqrt{(2x-3)^2} = \pm\sqrt{1}$

Second - Simplify the terms on both sides to obtain the solutions, i.e., $\sqrt{(2x-3)^2} = \pm\sqrt{1}$; $2x-3 = \pm 1$

Therefore the two solutions are: I. $2x-3=-1$; $2x=-1+3$; $2x=2$; $\frac{2x}{2}=\frac{2}{2}$; $x=\frac{1}{1}$; $\mathbf{x=1}$ and

II. $2x-3=+1$; $2x=1+3$; $2x=4$; $\frac{2x}{2}=\frac{\overset{2}{4}}{2}$; $x=\frac{2}{1}$; $\mathbf{x=2}$

Thus, the solution set is $\{1, 2\}$ and the equation $(2x-3)^2 = 1$ can be factored to $(x-1)(x-2)=0$.

Check: I. *Let* $x=1$ *in* $(2x-3)^2 = 1$; $(2\cdot1-3)^2 \stackrel{?}{=} 1$; $(2-3)^2 \stackrel{?}{=} 1$; $(-1)^2 \stackrel{?}{=} 1$; $1=1$

II. *Let* $x=2$ *in* $(2x-3)^2 = 1$; $(2\cdot2-3)^2 \stackrel{?}{=} 1$; $(4-3)^2 \stackrel{?}{=} 1$; $1^2 \stackrel{?}{=} 1$; $1=1$

4. First - Write the equation $x^2+3=0$ in the form of $x^2=b$, i.e., $x^2=-3$

Second - Take the square root of both sides of the equation, i.e., $\sqrt{x^2} = \pm\sqrt{-3}$

Since the number under the radical is a negative number (an imaginary number) therefore, **the equation $x^2+3=0$ has no real solutions.**

5. First - Take the square root of both sides of the equation $(y-5)^2 = 5$, i.e., $\sqrt{(y-5)^2} = \pm\sqrt{5}$

Second - Simplify the terms on both sides to obtain the solutions, i.e., $\sqrt{(y-5)^2} = \pm\sqrt{5}$; $y-5=\pm 2.24$

Therefore the two solutions are: I. $y-5=-2.24$; $y=-2.24+5$; $\mathbf{y=2.76}$ and

II. $y-5=+2.24$; $y=2.24+5$; $\mathbf{y=7.24}$

Thus, the solution set is $\{-2.76, 7.24\}$ and the equation $(y-5)^2 = 5$ can be factored to $(y-2.76)(y-7.24)=0$.

Check: I. *Let* $y=2.76$ *in* $(y-5)^2 = 5$; $(2.76-5)^2 \stackrel{?}{=} 5$; $(-2.24)^2 \stackrel{?}{=} 5$; $5=5$

II. *Let* $y=7.24$ *in* $(y-5)^2 = 5$; $(7.24-5)^2 \stackrel{?}{=} 5$; $(2.24)^2 \stackrel{?}{=} 5$; $5=5$

6. First - Write the equation $16x^2-25=0$ in the form of $ax^2=b$, i.e., $16x^2=25$

Second - Divide both sides of the equation $16x^2=25$ by the coefficient of x, i.e., $\frac{16x^2}{16}=\frac{25}{16}$; $x^2=\frac{25}{16}$

Third - Take the square root of both sides of the equation, i.e., $\sqrt{x^2}=\pm\sqrt{\frac{25}{16}}$

Fourth - Simplify the terms on both sides to obtain the solutions, i.e., $\mathbf{x=\pm\frac{5}{4}}$

Therefore, the solution set is $\left\{-\frac{5}{4}, \frac{5}{4}\right\}$ and the equation $16x^2-25=0$ can be factored to $\left(x-\frac{5}{4}\right)\left(x+\frac{5}{4}\right)=0$ which is the same as $(4x-5)(4x+5)=0$.

Check: I. *Let* $x=-\frac{5}{4}$ *in* $16x^2-25=0$; $16\cdot\left(-\frac{5}{4}\right)^2-25\stackrel{?}{=}0$; $16\cdot\frac{25}{16}-25\stackrel{?}{=}0$; $25-25\stackrel{?}{=}0$; $0=0$

II. *Let* $x=-\frac{5}{4}$ *in* $16x^2-25=0$; $16\cdot\left(\frac{5}{4}\right)^2-25\stackrel{?}{=}0$; $16\cdot\frac{25}{16}-25\stackrel{?}{=}0$; $25-25\stackrel{?}{=}0$; $0=0$

Section 1.4d Case I Solutions - Solving Quadratic Equations of the Form $ax^2+bx+c=0$, where $a=1$, by Completing the Square

1. First - Write the equation $x^2+10x-2=0$ in the form of $x^2+bx=-c$, i.e., $x^2+10x=2$.

   Second - Complete the square and simplify. $x^2+10x=2$; $x^2+10x+\left(\frac{\not{10}^{5}}{\not{2}}\right)^2=2+\left(\frac{\not{10}^{5}}{\not{2}}\right)^2$; $x^2+10x+5^2=2+5^2$

   ; $x^2+10x+25=2+25$; $x^2+10x+25=27$; $(x+5)^2=27$

   Third - Take the square root of both sides of the equation and solve for x.

   $(x+5)^2=27$; $\sqrt{(x+5)^2}=\pm\sqrt{27}$; $x+5=\pm 5.19$. Therefore,

   I. $x+5=+5.19$; $x=5.19-5$; $\mathbf{x=0.19}$ and II. $x+5=-5.19$; $x=-5.19-5$; $\mathbf{x=-10.19}$

   The solution set is $\{\mathbf{-10.19, 0.19}\}$.

   Fourth - Check the answers and write the quadratic equation in its factored form.

   I. *Let* $x=0.19$ *in* $x^2+10x-2=0$; $0.19^2+10\cdot 0.19-2\overset{?}{=}0$; $0.036+1.9-2\overset{?}{=}0$; $2-2\overset{?}{=}0$; $0=0$

   II. *Let* $x=-10.19$ *in* $x^2+10x-2=0$; $(-10.19)^2+10\cdot -10.19-2\overset{?}{=}0$; $103.8-101.9-2\overset{?}{=}0$; $101.9-101.9\overset{?}{=}0$; $0=0$

   Therefore, the equation $x^2+10x-2=0$ can be factored to $\mathbf{(x+10.19)(x-0.19)=0}$.

2. First - Write the equation $x^2-x-1=0$ in the form of $x^2+bx=-c$, i.e., $x^2-x=1$.

   Second - Complete the square and simplify. $x^2-x=1$; $x^2-x+\left(-\frac{1}{2}\right)^2=1+\left(-\frac{1}{2}\right)^2$; $x^2-x+\frac{1}{4}=1+\frac{1}{4}$

   ; $\left(x-\frac{1}{2}\right)^2=1.25$

   Third - Take the square root of both sides of the equation and solve for x.

   $\left(x-\frac{1}{2}\right)^2=1.25$; $\sqrt{\left(x-\frac{1}{2}\right)^2}=\pm\sqrt{1.25}$; $x-\frac{1}{2}=\pm 1.118$; $x-0.5=\pm 1.118$. Therefore,

   I. $x-0.5=+1.118$; $x=1.118+0.5$; $\mathbf{x=1.618}$ and II. $x-0.5=-1.118$; $x=-1.118+0.5$; $\mathbf{x=-0.618}$

   The solution set is $\{\mathbf{-0.618, 1.618}\}$.

   Fourth - Check the answers and write the quadratic equation in its factored form.

   I. *Let* $x=1.618$ *in* $x^2-x-1=0$; $1.618^2-1.618-1\overset{?}{=}0$; $2.618-1.618-1\overset{?}{=}0$; $2.618-2.618\overset{?}{=}0$; $0=0$

   II. *Let* $x=-0.618$ *in* $x^2-x-1=0$; $(-0.618)^2-(-0.618)-1\overset{?}{=}0$; $0.381+0.618-1\overset{?}{=}0$; $-0.618+0.618\overset{?}{=}0$; $0=0$

   Therefore, the equation $x^2-x-1=0$ can be factored to $\mathbf{(x+0.618)(x-1.618)=0}$.

3. First - Write the equation $x(x+2)=80$ in the form of $x^2+bx=-c$, i.e., $x^2+2x=80$.

   Second - Complete the square and simplify. $x^2+2x=80$; $x^2+2x+\left(\frac{2}{2}\right)^2=80+\left(\frac{2}{2}\right)^2$; $x^2+2x+1=80+1$

   ; $(x+1)^2=81$

   Third - Take the square root of both sides of the equation and solve for x.

   $(x+1)^2=81$; $\sqrt{(x+1)^2}=\pm\sqrt{81}$; $x+1=\pm 9$. Therefore,

   I. $x+1=+9$; $x=9-1$; $\mathbf{x=8}$ and II. $x+1=-9$; $x=-9-1$; $\mathbf{x=-10}$

   The solution set is $\{\mathbf{-10, 8}\}$.

   Fourth - Check the answers and write the quadratic equation in its factored form.

I. *Let* $x = 8$ *in* $x(x+2) = 80$; $8(8+2) \overset{?}{=} 80$; $8 \cdot 10 \overset{?}{=} 80$; $80 = 80$

II. *Let* $x = -10$ *in* $x(x+2) = 80$; $-10(-10+2) \overset{?}{=} 80$; $-10 \cdot -8 \overset{?}{=} 80$; $80 = 80$

Therefore, the equation $x(x+2) = 80$ can be factored to $\mathbf{(x+10)(x-8) = 0}$.

4. First - Write the equation $y^2 - 10y + 5 = 0$ in the form of $y^2 + by = -c$, i.e., $y^2 - 10y = -5$.

Second - Complete the square and simplify. $y^2 - 10y = -5$; $y^2 - 10y + \left(-\frac{\overset{5}{\cancel{10}}}{\cancel{2}}\right)^2 = -5 + \left(-\frac{\overset{5}{\cancel{10}}}{\cancel{2}}\right)^2$; $y^2 - 10y + 5^2 = -5 + 5^2$

; $y^2 - 10y + 25 = -5 + 25$; $y^2 - 10y + 25 = 20$; $(y-5)^2 = 20$

Third - Take the square root of both sides of the equation and solve for y.

$(y-5)^2 = 20$; $\sqrt{(y-5)^2} = \pm\sqrt{20}$; $y - 5 = \pm 4.47$. Therefore,

I. $y - 5 = +4.47$; $y = 4.47 + 5$; $\mathbf{y = 9.47}$ and II. $y - 5 = -4.47$; $y = -4.47 + 5$; $\mathbf{y = 0.53}$

The solution set is $\{\mathbf{0.53, 9.47}\}$.

Fourth - Check the answers and write the quadratic equation in its factored form.

I. *Let* $y = 0.53$ *in* $y^2 - 10y + 5 = 0$; $0.53^2 - 10 \cdot 0.53 + 5 \overset{?}{=} 0$; $0.3 - 5.3 + 5 \overset{?}{=} 0$; $5.3 - 5.3 \overset{?}{=} 0$; $0 = 0$

II. *Let* $y = 9.47$ *in* $y^2 - 10y + 5 = 0$; $9.47^2 - 10 \cdot 9.47 + 5 \overset{?}{=} 0$; $89.7 - 94.7 + 5 \overset{?}{=} 0$; $94.7 - 94.7 \overset{?}{=} 0$; $0 = 0$

Therefore, the equation $y^2 - 10y + 5 = 0$ can be factored to $\mathbf{(y-0.53)(y-9.47) = 0}$.

5. First - Write the equation $x^2 + 4x - 5 = 0$ in the form of $x^2 + bx = -c$, i.e., $x^2 + 4x = 5$.

Second - Complete the square and simplify. $x^2 + 4x = 5$; $x^2 + 4x + \left(\frac{\overset{2}{\cancel{4}}}{\cancel{2}}\right)^2 = 5 + \left(\frac{\overset{2}{\cancel{4}}}{\cancel{2}}\right)^2$; $x^2 + 4x + 2^2 = 5 + 2^2$

; $x^2 + 4x + 4 = 5 + 4$; $x^2 + 4x + 4 = 9$; $(x+2)^2 = 9$

Third - Take the square root of both sides of the equation and solve for x.

$(x+2)^2 = 9$; $\sqrt{(x+2)^2} = \pm\sqrt{9}$; $x + 2 = \pm 3$. Therefore,

I. $x + 2 = +3$; $x = 3 - 2$; $\mathbf{x = 1}$ and II. $x + 2 = -3$; $x = -2 - 3$; $\mathbf{x = -5}$

The solution set is $\{\mathbf{-5, 1}\}$.

Fourth - Check the answers and write the quadratic equation in its factored form.

I. *Let* $x = 1$ *in* $x^2 + 4x - 5 = 0$; $1^2 + 4 \cdot 1 - 5 \overset{?}{=} 0$; $1 + 4 - 5 \overset{?}{=} 0$; $5 - 5 \overset{?}{=} 0$; $0 = 0$

II. *Let* $x = -5$ *in* $x^2 + 4x - 5 = 0$; $(-5)^2 + 4 \cdot (-5) - 5 \overset{?}{=} 0$; $25 - 20 - 5 \overset{?}{=} 0$; $25 - 25 \overset{?}{=} 0$; $0 = 0$

Therefore, the equation $x^2 + 4x - 5 = 0$ can be factored to $\mathbf{(x+5)(x-1) = 0}$.

6. The equation $y^2 + 4y = 14$ is already in the form of $y^2 + by = -c$.

First - Complete the square and simplify. $y^2 + 4y = 14$; $y^2 + 4y + \left(\frac{\overset{2}{\cancel{4}}}{\cancel{2}}\right)^2 = 14 + \left(\frac{\overset{2}{\cancel{4}}}{\cancel{2}}\right)^2$; $y^2 + 4y + 2^2 = 14 + 2^2$

; $y^2 + 4y + 4 = 14 + 4$; $y^2 + 4y + 4 = 18$; $(y+2)^2 = 18$

Second - Take the square root of both sides of the equation and solve for y.

$(y+2)^2 = 8$; $\sqrt{(y+2)^2} = \pm\sqrt{18}$; $y + 2 = \pm 4.24$. Therefore,

I. $y+2=+4.24$; $y=4.24-2$; $\mathbf{y=2.24}$ and II. $y+2=-4.24$; $y=-4.24-2$; $\mathbf{y=-6.24}$

The solution set is $\{\mathbf{-6.24, 2.24}\}$.

Third - Check the answers and write the quadratic equation in its factored form.

I. *Let* $y=2.24$ *in* $y^2+4y=14$; $2.24^2+4\cdot 2.24\overset{?}{=}14$; $5+9\overset{?}{=}14$; $14=14$

II. *Let* $y=-6.24$ *in* $y^2+4y=14$; $(-6.24)^2+4\cdot -6.24\overset{?}{=}14$; $39-25\overset{?}{=}14$; $14=14$

Therefore, the equation $y^2+4y=14$ can be factored to $\mathbf{(y+6.24)(y-2.24)=0}$.

Section 1.4d Case II Solutions - Solving Quadratic Equations of the Form $ax^2+bx+c=0$, where $a>1$, by Completing the Square

1. First - Write the equation $4u^2+6u+1=0$ in the form of $au^2+bu=-c$, i.e., $4u^2+6u=-1$.

Second - Divide both sides of the equation by the coefficient of u^2, i.e., $\frac{\cancel{4}}{\cancel{4}}u^2+\frac{\overset{3}{\cancel{6}}}{\underset{2}{\cancel{4}}}u=-\frac{1}{4}$; $u^2+\frac{3}{2}u=-\frac{1}{4}$

Third - Complete the square and simplify. $u^2+\frac{3}{2}u=-\frac{1}{4}$; $u^2+\frac{3}{2}u+\left(\frac{3}{4}\right)^2=-\frac{1}{4}+\left(\frac{3}{4}\right)^2$; $u^2+\frac{3}{2}u+\frac{9}{16}=-\frac{1}{4}+\frac{9}{16}$

; $\left(u+\frac{3}{4}\right)^2=\frac{(-1\cdot 16)+(9\cdot 4)}{4\cdot 16}$; $\left(u+\frac{3}{4}\right)^2=\frac{-16+36}{64}$; $\left(u+\frac{3}{4}\right)^2=\frac{\overset{5}{\cancel{20}}}{\underset{16}{\cancel{64}}}$; $\left(u+\frac{3}{4}\right)^2=\frac{5}{16}$

Fourth - Take the square root of both sides of the equation and solve for u.

$\left(u+\frac{3}{4}\right)^2=\frac{5}{16}$; $\sqrt{\left(u+\frac{3}{4}\right)^2}=\pm\sqrt{\frac{5}{16}}$; $u+\frac{3}{4}=\pm\sqrt{0.313}$; $u+0.75=\pm 0.56$. Therefore,

I. $u+0.75=+0.56$; $u=0.56-0.75$; $\mathbf{u=-0.19}$ and II. $u+0.75=-0.56$; $u=-0.56-0.75$; $\mathbf{u=-1.31}$

The solution set is $\{\mathbf{-1.31, -0.19}\}$.

Fifth - Check the answers and write the quadratic equation in its factored form.

I. *Let* $u=-0.19$ *in* $4u^2+6u+1=0$; $4\cdot(-0.19)^2+6\cdot -0.19+1\overset{?}{=}0$; $4\cdot 0.036-1.14+1\overset{?}{=}0$; $0.14-1.14+1\overset{?}{=}0$

; $1.14-1.14\overset{?}{=}0$; $0=0$

II. *Let* $u=-1.31$ *in* $4u^2+6u+1=0$; $4\cdot(-1.31)^2+6\cdot -1.31+1\overset{?}{=}0$; $4\cdot 1.716-7.86+1\overset{?}{=}0$; $6.86-7.86+1\overset{?}{=}0$

; $7.86-7.86\overset{?}{=}0$; $0=0$

Therefore, the equation $4u^2+6u+1=0$ can be factored to $\mathbf{(u+1.31)(u+0.19)=0}$.

2. The equation $4w^2+10w=-3$ is already in standard form of $aw^2+bw=-c$.

First - Divide both sides of the equation by the coefficient of w^2, i.e., $\frac{\cancel{4}}{\cancel{4}}w^2+\frac{\overset{5}{\cancel{10}}}{\underset{2}{\cancel{4}}}w=-\frac{3}{4}$; $w^2+\frac{5}{2}w=-\frac{3}{4}$

Second - Complete the square and simplify. $w^2+\frac{5}{2}w=-\frac{3}{4}$; $w^2+\frac{5}{2}w+\left(\frac{5}{4}\right)^2=-\frac{3}{4}+\left(\frac{5}{4}\right)^2$; $w^2+\frac{5}{2}w+\frac{25}{16}=-\frac{3}{4}+\frac{25}{16}$

; $\left(w+\frac{5}{4}\right)^2=\frac{(-3\cdot 16)+(25\cdot 4)}{4\cdot 16}$; $\left(w+\frac{5}{4}\right)^2=\frac{-48+100}{64}$; $\left(w+\frac{5}{4}\right)^2=\frac{\overset{13}{\cancel{52}}}{\underset{16}{\cancel{64}}}$; $\left(w+\frac{5}{4}\right)^2=\frac{13}{16}$

Third - Take the square root of both sides of the equation and solve for w.

$\left(w+\frac{5}{4}\right)^2 = \frac{13}{16}$; $\sqrt{\left(w+\frac{5}{4}\right)^2} = \pm\sqrt{\frac{13}{16}}$; $w+\frac{5}{4} = \pm\sqrt{0.813}$; $w+1.25 = \pm 0.9$. Therefore,

I. $w+1.25 = +0.9$; $w = 0.9 - 1.25$; $\mathbf{w = -0.35}$ and II. $w+1.25 = -0.9$; $w = -0.9 - 1.25$; $\mathbf{w = -2.15}$

The solution set is $\{\mathbf{-2.15, -0.35}\}$.

Fourth - Check the answers and write the quadratic equation in its factored form.

I. *Let* $w = -0.35$ *in* $4w^2 + 10w = -3$; $4\cdot(-0.35)^2 + 10\cdot -0.35 \stackrel{?}{=} -3$; $4\cdot 0.123 - 3.5 \stackrel{?}{=} -3$; $0.5 - 3.5 \stackrel{?}{=} -3$; $-3 = -3$

II. *Let* $w = -2.15$ *in* $4w^2 + 10w = -3$; $4\cdot(-2.15)^2 + 10\cdot -2.15 \stackrel{?}{=} -3$; $4\cdot 4.62 - 21.5 \stackrel{?}{=} -3$; $18.5 - 21.5 \stackrel{?}{=} -3$; $-3 = -3$

Therefore, the equation $4w^2 + 10w = -3$ can be factored to $\mathbf{(w+2.15)(w+0.35) = 0}$.

3. First - Write the equation $6x^2 + 4x - 2 = 0$ in the form of $ax^2 + bx = -c$, i.e., $6x^2 + 4x = 2$.

Second - Divide both sides of the equation by the coefficient of x^2, i.e., $\frac{\cancel{6}}{\cancel{6}}x^2 + \frac{\overset{2}{\cancel{4}}}{\underset{3}{\cancel{6}}}x = \frac{\cancel{2}}{\underset{3}{\cancel{6}}}$; $x^2 + \frac{2}{3}x = \frac{1}{3}$

Third - Complete the square and simplify. $x^2 + \frac{2}{3}x = \frac{1}{3}$; $x^2 + \frac{2}{3}x + \left(\frac{2}{\underset{3}{\cancel{6}}}\right)^2 = \frac{1}{3} + \left(\frac{\cancel{2}}{\underset{3}{\cancel{6}}}\right)^2$; $x^2 + \frac{2}{3}x + \left(\frac{1}{3}\right)^2 = \frac{1}{3} + \left(\frac{1}{3}\right)^2$

; $x^2 + \frac{2}{3}x + \frac{1}{9} = \frac{1}{3} + \frac{1}{9}$; $\left(x+\frac{1}{3}\right)^2 = \frac{(1\cdot 9)+(1\cdot 3)}{3\cdot 9}$; $\left(x+\frac{1}{3}\right)^2 = \frac{9+3}{27}$; $\left(x+\frac{1}{3}\right)^2 = \frac{12}{27}$

Fourth - Take the square root of both sides of the equation and solve for x.

$\left(x+\frac{1}{3}\right)^2 = \frac{12}{27}$; $\sqrt{\left(x+\frac{1}{3}\right)^2} = \pm\sqrt{\frac{12}{27}}$; $x+\frac{1}{3} = \pm\sqrt{0.44}$; $x + 0.33 = \pm 0.66$. Therefore,

I. $x + 0.33 = +0.66$; $x = 0.66 - 0.33$; $\mathbf{x = 0.33}$ and II. $x + 0.33 = -0.66$; $x = -0.66 - 0.33$; $\mathbf{x = -1}$

The solution set is $\{\mathbf{-1, 0.33}\}$.

Fifth - Check the answers and write the quadratic equation in its factored form.

I. *Let* $x = 0.33$ *in* $6x^2 + 4x - 2 = 0$; $6\cdot(0.33)^2 + 4\cdot 0.33 - 2 \stackrel{?}{=} 0$; $6\cdot 0.11 + 1.32 - 2 \stackrel{?}{=} 0$; $0.66 + 1.32 - 2 \stackrel{?}{=} 0$

; $2 - 2 \stackrel{?}{=} 0$; $0 = 0$

II. *Let* $x = -1$ *in* $6x^2 + 4x - 2 = 0$; $6\cdot(-1)^2 + 4\cdot -1 - 2 \stackrel{?}{=} 0$; $6\cdot 1 - 4 - 2 \stackrel{?}{=} 0$; $6 - 4 - 2 \stackrel{?}{=} 0$; $6 - 6 \stackrel{?}{=} 0$; $0 = 0$

Therefore, the equation $6x^2 + 4x - 2 = 0$ can be factored to $\mathbf{(x-0.33)(x+1) = 0}$.

4. First - Write the equation $15y^2 + 3 = -14y$ in the form of $ay^2 + by = -c$, i.e., $15y^2 + 14y = -3$.

Second - Divide both sides of the equation by the coefficient of y^2, i.e., $\frac{\cancel{15}}{\cancel{15}}y^2 + \frac{14}{15}y = -\frac{\cancel{3}}{\underset{5}{\cancel{15}}}$; $y^2 + \frac{14}{15}y = -\frac{1}{5}$

Third - Complete the square. $y^2 + \frac{14}{15}y = -\frac{1}{5}$; $y^2 + \frac{14}{15}y + \left(\frac{14}{30}\right)^2 = -\frac{1}{5} + \left(\frac{14}{30}\right)^2$; $y^2 + \frac{14}{15}y + \frac{196}{900} = -\frac{1}{5} + \frac{196}{900}$

; $\left(y + \frac{\overset{7}{\cancel{14}}}{\underset{15}{\cancel{30}}}\right)^2 = \frac{(-1\cdot 900)+(196\cdot 5)}{5\cdot 900}$; $\left(y+\frac{7}{15}\right)^2 = \frac{-900+980}{4500}$; $\left(y+\frac{7}{15}\right)^2 = \frac{\overset{4}{\cancel{80}}}{\underset{225}{\cancel{4500}}}$; $\left(y+\frac{7}{15}\right)^2 = \frac{4}{225}$

Fourth - Take the square root of both sides of the equation and solve for y.

$\left(y+\frac{7}{15}\right)^2 = \frac{4}{225}$; $\sqrt{\left(y+\frac{7}{15}\right)^2} = \pm\sqrt{\frac{4}{225}}$; $y + \frac{7}{15} = \pm\sqrt{0.02}$; $y + 0.46 = \pm 0.13$. Therefore,

I. $y + 0.46 = +0.13$; $y = 0.13 - 0.46$; $\mathbf{y = -0.33}$ and II. $y + 0.46 = -0.13$; $y = -0.13 - 0.46$; $\mathbf{y = -0.59}$

The solution set is $\{\mathbf{-0.59, -0.33}\}$.

Fifth - Check the answers and write the quadratic equation in its factored form.

I. *Let* $y=-0.33$ *in* $15y^2+3=-14y$; $15\cdot(-0.33)^2+3\overset{?}{=}-14\cdot-0.33$; $15\cdot 0.108+3\overset{?}{=}4.62$; $1.62+3\overset{?}{=}4.62$; $4.62=4.62$

II. *Let* $y=-0.59$ *in* $15y^2+3=-14y$; $15\cdot(-0.59)^2+3\overset{?}{=}-14\cdot-0.59$; $15\cdot 0.348+3\overset{?}{=}8.26$; $5.23+3\overset{?}{=}8.26$; $8.26=8.26$

Therefore, the equation $15y^2+3=-14y$ can be factored to $\mathbf{(y+0.59)(w+0.33)=0}$.

5. First - Write the equation $2x^2-5x+3=0$ in the form of $ax^2+bx=-c$, i.e., $2x^2-5x=-3$.

Second - Divide both sides of the equation by the coefficient of x^2, i.e., $\frac{\cancel{2}}{\cancel{2}}x^2-\frac{5}{2}x=-\frac{3}{2}$; $x^2-\frac{5}{2}x=-\frac{3}{2}$

Third - Complete the square and simplify. $x^2-\frac{5}{2}x=-\frac{3}{2}$; $x^2-\frac{5}{2}x+\left(-\frac{5}{4}\right)^2=-\frac{3}{2}+\left(-\frac{5}{4}\right)^2$; $x^2-\frac{5}{2}x+\frac{25}{16}=-\frac{3}{2}+\frac{25}{16}$

; $\left(x-\frac{5}{4}\right)^2=\frac{(-3\cdot16)+(2\cdot25)}{2\cdot16}$; $\left(x-\frac{5}{4}\right)^2=\frac{-48+50}{32}$; $\left(x-\frac{5}{4}\right)^2=\frac{\cancel{2}}{\cancel{32}_{16}}$; $\left(x-\frac{5}{4}\right)^2=\frac{1}{16}$

Fourth - Take the square root of both sides of the equation and solve for x.

$\left(x-\frac{5}{4}\right)^2=\frac{1}{16}$; $\sqrt{\left(x-\frac{5}{4}\right)^2}=\pm\sqrt{\frac{1}{16}}$; $x-\frac{5}{4}=\pm\frac{1}{4}$; $x-1.25=\pm0.25$. Therefore,

I. $x-1.25=+0.25$; $x=0.25+1.25$; $\mathbf{x=1.5}$ and II. $x-1.25=-0.25$; $x=-0.25+1.25$; $\mathbf{x=1}$

The solution set is $\mathbf{\{1, 1.5\}}$.

Fifth - Check the answers and write the quadratic equation in its factored form.

I. *Let* $x=1$ *in* $2x^2-5x+3=0$; $2\cdot1^2-5\cdot1+3\overset{?}{=}0$; $2\cdot1-5+3\overset{?}{=}0$; $2-5+3\overset{?}{=}0$; $5-5\overset{?}{=}0$; $0=0$

II. *Let* $x=1.5$ *in* $2x^2-5x+3=0$; $2\cdot1.5^2-5\cdot1.5+3\overset{?}{=}0$; $2\cdot2.25-7.5+3\overset{?}{=}0$; $4.5-7.5+3\overset{?}{=}0$; $7.5-7.5\overset{?}{=}0$; $0=0$

Therefore, the equation $2x^2-5x+3=0$ can be factored to $\mathbf{(x-1)(x-1.5)=0}$.

6. First - Write the equation $2x^2+xy-y^2=0$, where x is variable, in the form of $ax^2+bx=-c$, i.e., $2x^2+yx=y^2$.

Second - Divide both sides of the equation by the coefficient of x^2, i.e., $\frac{\cancel{2}}{\cancel{2}}x^2+\frac{y}{2}x=\frac{y^2}{2}$; $x^2+\frac{y}{2}x=\frac{y^2}{2}$

Third - Complete the square and simplify. $x^2+\frac{y}{2}x=\frac{y^2}{2}$; $x^2+\frac{y}{2}x+\left(\frac{y}{4}\right)^2=\frac{y^2}{2}+\left(\frac{y}{4}\right)^2$; $x^2+\frac{y}{2}x+\left(\frac{y}{4}\right)^2=\frac{y^2}{2}+\frac{y^2}{16}$

; $\left(x+\frac{y}{4}\right)^2=\frac{(y^2\cdot16)+(y^2\cdot2)}{2\cdot16}$; $\left(x+\frac{y}{4}\right)^2=\frac{16y^2+2y^2}{32}$; $\left(x+\frac{y}{4}\right)^2=\frac{\cancel{18}^{9}}{\cancel{32}_{16}}y^2$; $\left(x+\frac{y}{4}\right)^2=\frac{9}{16}y^2$

Fourth - Take the square root of both sides of the equation and solve for x.

$\left(x+\frac{y}{4}\right)^2=\frac{9}{16}y^2$; $\sqrt{\left(x+\frac{y}{4}\right)^2}=\pm\sqrt{\frac{9}{16}y^2}$; $x+\frac{y}{4}=\pm\frac{3}{4}y$; $x+0.25y=\pm0.75y$. Therefore,

I. $x+0.25y=+0.75y$; $x=0.75y-0.25y$; $\mathbf{x=0.5y}$ and II. $x+0.25y=-0.75y$; $x=-0.75y-0.25y$; $\mathbf{x=-y}$

The solution set is $\mathbf{\{-y, 0.5y\}}$.

Fifth - Check the answers and write the quadratic equation in its factored form.

I. *Let* $x=0.5y$ *in* $2x^2+xy-y^2=0$; $2\cdot(0.5y)^2+(0.5y)\cdot y-y^2\overset{?}{=}0$; $2\cdot0.25y^2+0.5y^2-y^2\overset{?}{=}0$; $0.5y^2+0.5y^2-y^2\overset{?}{=}0$; $y^2-y^2\overset{?}{=}0$; $0=0$

II. *Let* $x=-y$ *in* $2x^2+xy-y^2=0$; $2\cdot(-y)^2+(-y)\cdot y-y^2\overset{?}{=}0$; $2y^2-y^2-y^2\overset{?}{=}0$; $2y^2-2y^2\overset{?}{=}0$; $0=0$

Therefore, the equation $2x^2+xy-y^2=0$ can be factored to $\mathbf{(x+y)(x-0.5y)=0}$.

Section 1.4e Solutions - How to Choose the Best Factoring or Solution Method

1. **First Method:** (The Trial and Error Method)

   Write the equation $x^2 = 16$ in the standard quadratic equation form $ax^2 + bx + c = 0$, i.e., write $x^2 = 16$ as $x^2 + 0x - 16 = 0$. Consider the left hand side of the equation which is a polynomial. To factor the given polynomial we need to obtain two numbers whose sum is 0 and whose product is -16. Let's construct a table as follows:

Sum	***Product***
$1-1=0$	$1\cdot(-1)=-1$
$2-2=0$	$2\cdot(-2)=-4$
$3-3=0$	$3\cdot(-3)=-9$
$\mathbf{4-4=0}$	$\mathbf{4\cdot(-4)=-16}$

   The last line contains the sum and the product of the two numbers that we need. Thus, $x^2 = 16$ or $x^2 + 0x - 16 = 0$ can be factored to $\mathbf{(x-4)(x+4)=0}$

   Second Method: (The Quadratic Formula Method)

   First, write the equation in the standard quadratic equation form $ax^2 + bx + c = 0$, i.e., write $x^2 = 16$ as $x^2 + 0x - 16 = 0$. Second, equate the coefficients of $x^2 + 0x - 16 = 0$ with the standard quadratic equation by letting $a = 1$, $b = 0$, and $c = -16$. Then,

   Given: $x = \dfrac{-b \pm \sqrt{b^2 - 4ac}}{2a}$; $x = \dfrac{-0 \pm \sqrt{0^2 - (4 \times 1 \times -16)}}{2 \times 1}$; $x = \dfrac{\pm\sqrt{0+64}}{2}$; $x = \dfrac{\pm\sqrt{64}}{2}$; $x = \pm\dfrac{\sqrt{8^2}}{2}$; $x = \pm\dfrac{8}{2}$.

   Therefore, the two solutions are $\mathbf{x = -4}$ and $\mathbf{x = 4}$ and the equation $x^2 + 0x - 16 = 0$ can be factored to $\mathbf{(x+4)(x-4)=0}$.

   Third Method: (The Square Root Property Method)

   Take the square root of both sides of the equation, i.e., write $x^2 = 16$ as $\sqrt{x^2} = \pm\sqrt{16}$; $x = \pm\sqrt{4^2}$; $x = \pm 4$. Thus, $\mathbf{x = +4}$ and $\mathbf{x = -4}$ are the solution sets to the equation $x^2 = 16$ which can be represented in its factorable form as $\mathbf{(x+4)(x-4)=0}$.

   Check: $(x-4)(x+4) = 0$; $x \cdot x + 4 \cdot x - 4 \cdot x + 4 \cdot (-4) = 0$; $x^2 + 4x - 4x - 16 = 0$; $x^2 + (4-4)x - 16 = 0$

   ; $x^2 + 0x - 16 = 0$

   From the above three methods using the Square Root Property method is the easiest method to use. The Trial and Error method is the second easiest method to use. Followed by the Quadratic Formula method which is the longest and somewhat a more difficult way of obtaining the factored terms.

2. **First Method:** (The Trial and Error Method)

   Consider the left hand side of the equation which is a polynomial. To factor the polynomial $x^2 + 7x + 3$ we need to obtain two numbers whose sum is 7 and whose product is 3. However, after few trials, it becomes clear that such a combination of integer numbers is not possible to obtain. Therefore, **the given equation is not factorable and is referred to as PRIME.**

   Second Method: (The Quadratic Formula Method)

   Given the standard quadratic equation $ax^2 + bx + c = 0$, equate the coefficients of $x^2 + 7x + 3 = 0$ with the standard quadratic equation by letting $a = 1$, $b = 7$, and $c = 3$. Then,

   Given: $x = \dfrac{-b \pm \sqrt{b^2 - 4ac}}{2a}$; $x = \dfrac{-7 \pm \sqrt{7^2 - (4 \times 1 \times 3)}}{2 \times 1}$; $x = \dfrac{-7 \pm \sqrt{49 - 12}}{2}$; $x = \dfrac{-7 \pm \sqrt{37}}{2}$; $x = \dfrac{-7 \pm 6.08}{2}$.

   Therefore, the two solutions are $\mathbf{x = -6.54}$ and $\mathbf{x = -0.46}$ and the equation $x^2 + 7x + 3 = 0$ can be factored to $\mathbf{(x+6.54)(x+0.46)=0}$.

   Third Method: (Completing the Square Method)

$x^2+7x+3=0$; $x^2+7x=-3$; $x^2+7x+\left(\frac{7}{2}\right)^2=-3+\left(\frac{7}{2}\right)^2$; $x^2+7x+\frac{49}{4}=-3+\frac{49}{4}$; $\left(x+\frac{7}{2}\right)^2=-\frac{3}{1}+\frac{49}{4}$; $\left(x+\frac{7}{2}\right)^2=\frac{(-3\cdot 4)+(1\cdot 49)}{1\cdot 4}$; $\left(x+\frac{7}{2}\right)^2=\frac{-12+49}{4}$; $\left(x+\frac{7}{2}\right)^2=\frac{37}{4}$; $x+\frac{7}{2}=\pm\sqrt{\frac{37}{4}}$; $x+\frac{7}{2}=\pm\frac{\sqrt{37}}{2}$. Therefore, the two solutions are $\mathbf{x=-6.54}$ and $\mathbf{x=-0.46}$ and the equation $x^2+7x+3=0$ can be factored to $\mathbf{(x+6.54)(x+0.46)=0}$.

Check: I. *Let* $x=-0.46$ *in* $x^2+7x+3=0$; $(-0.46)^2+7\cdot(-0.46)+3\overset{?}{=}0$; $0.2-3.2+3\overset{?}{=}0$; $-3+3\overset{?}{=}0$; $0=0$

II. *Let* $x=-6.54$ *in* $x^2+7x+3=0$; $(-6.54)^2+7\cdot(-6.54)+3\overset{?}{=}0$; $42.8-45.8+3\overset{?}{=}0$; $42.8-42.8\overset{?}{=}0$; $0=0$

Therefore, the equation $x^2+7x+3=0$ can be factored to $(x+0.46)(x+6.54)=0$.

From the above three methods using the Quadratic Formula method may be the faster method than Completing the Square method.

3. **First Method:** (The Square Root Property Method)

$(3x+4)^2=36$; $\sqrt{(3x+4)^2}=\pm\sqrt{36}$; $3x+4=\pm 6$; $3x=\pm 6-4$; $x=\frac{\pm 6-4}{3}$. Thus, the two solutions are $x=\frac{6-4}{3}$; $\mathbf{x=\frac{2}{3}}$; and $x=\frac{-6-4}{3}$; $\mathbf{x=-\frac{10}{3}}$ and the equation $(3x+4)^2=36$ can be factored to $\left(x-\frac{2}{3}\right)\left(x+\frac{10}{3}\right)=0$ which is the same as $\mathbf{(3x-2)(3x+10)=0}$.

Second Method: (The Quadratic Formula Method)

Complete the square term on the left hand side and write the equation in standard form, i.e., $(3x+4)^2=36$; $9x^2+24x+16=36$; $9x^2+24x+16-36=36-36$; $9x^2+24x-20=0$.

Given the standard quadratic equation $ax^2+bx+c=0$, equate the coefficients of $9x^2+24x-20=0$ with the standard quadratic equation by letting $a=9$, $b=24$, and $c=-20$. Then,

Given: $x=\frac{-b\pm\sqrt{b^2-4ac}}{2a}$; $x=\frac{-24\pm\sqrt{24^2-(4\times 9\times -20)}}{2\times 9}$; $x=\frac{-24\pm\sqrt{576+720}}{18}$; $x=\frac{-24\pm\sqrt{1296}}{18}$; $x=\frac{-24\pm\sqrt{36^2}}{18}$; $x=\frac{-24\pm 36}{18}$. Therefore, the two solutions are $x=\frac{-24-36}{18}$; $x=-\frac{\overset{10}{\cancel{60}}}{\underset{3}{\cancel{18}}}$; $\mathbf{x=-\frac{10}{3}}$; and $x=\frac{-24+36}{18}$; $x=\frac{\overset{2}{\cancel{12}}}{\underset{3}{\cancel{18}}}$; $\mathbf{x=\frac{2}{3}}$ and the equation $(3x+4)^2=36$ can be factored to $\left(x-\frac{2}{3}\right)\left(x+\frac{10}{3}\right)=0$ which is the same as $\mathbf{(3x-2)(3x+10)=0}$.

Third Method: (Completing-the-Square Method)

First complete the square term on the left hand side and simplify the equation, i.e., $(3x+4)^2=36$; $9x^2+24x+16=36$; $9x^2+24x+16-16=36--16$; $9x^2+24x=20$; $\frac{\cancel{9}}{\cancel{9}}x^2+\frac{24}{9}x=\frac{20}{9}$; $x^2+\frac{24}{9}x=\frac{20}{9}$.

Then, complete the square in the following way:

$x^2+\frac{24}{9}x=\frac{20}{9}$; $x^2+\frac{24}{9}x+\left(\frac{\overset{4}{\cancel{24}}}{\underset{3}{\cancel{18}}}\right)^2=\frac{20}{9}+\left(\frac{\overset{4}{\cancel{24}}}{\underset{3}{\cancel{18}}}\right)^2$; $x^2+\frac{24}{9}x+\left(\frac{4}{3}\right)^2=\frac{20}{9}+\left(\frac{4}{3}\right)^2$; $x^2+\frac{24}{9}x+\frac{16}{9}=\frac{20}{9}+\frac{16}{9}$; $\left(x+\frac{4}{3}\right)^2=\frac{20+16}{9}$; $\left(x+\frac{4}{3}\right)^2=\frac{36}{9}$; $\sqrt{\left(x+\frac{4}{3}\right)^2}=\pm\sqrt{\frac{36}{9}}$; $x+\frac{4}{3}=\pm\sqrt{\frac{6^2}{3^2}}$; $x+\frac{4}{3}=\pm\frac{6}{3}$; $x=-\frac{4}{3}\pm\frac{6}{3}$; $x=-\frac{4\pm 6}{3}$.

Therefore, the two solutions are $x=\frac{-4-6}{3}$; $\mathbf{x=-\frac{10}{3}}$; and $x=\frac{-4+6}{3}$; $\mathbf{x=\frac{2}{3}}$. In addition, the equation $(3x+4)^2=36$

can be factored to $\left(x-\frac{2}{3}\right)\left(x+\frac{10}{3}\right)=0$ which is the same as $\mathbf{(3x-2)(3x+10)=0}$.

Check: $(3x-2)(3x+10)=0$; $3x\cdot 3x+10\cdot 3x-2\cdot 3x-2\cdot 10=0$; $9x^2+30x-6x-20=0$; $9x^2+(30-6)x-20=0$

; $9x^2+24x-20=0$ which is the same as $(3x+4)^2=36$.

From the above three methods the Square Root Property method is the easiest method in factoring the quadratic equation, followed by the Quadratic Formula method and Completing the Square method.

4. **First Method:** (The Trial and Error Method)

Consider the left hand side of the equation which is a polynomial. To factor the polynomial $x^2+11x+30$ we need to obtain two numbers whose sum is 11 and whose product is 30. Let's construct a table as follows:

Sum	***Product***
$1+10=11$	$1\cdot 10=10$
$2+9=11$	$2\cdot 9=18$
$3+8=11$	$3\cdot 8=24$
$4+7=11$	$4\cdot 7=28$
$\mathbf{5+6=11}$	$\mathbf{5\cdot 6=30}$

The last line contains the sum and the product of the two numbers that we need. Thus, $x^2+11x+30=0$ can be factored to $\mathbf{(x+5)(x+6)=0}$

Second Method: (The Quadratic Formula Method)

Given the standard quadratic equation $ax^2+bx+c=0$, equate the coefficients of $x^2+11x+30=0$ with the standard quadratic equation by letting $a=1$, $b=11$, and $c=30$. Then,

Given: $x=\frac{-b\pm\sqrt{b^2-4ac}}{2a}$; $x=\frac{-11\pm\sqrt{11^2-(4\times 1\times 30)}}{2\times 1}$; $x=\frac{-11\pm\sqrt{121-120}}{2}$; $x=\frac{-11\pm\sqrt{1}}{2}$; $x=\frac{-11\pm 1}{2}$.

Therefore, the two solutions are $x=\frac{-11+1}{2}$; $x=-\frac{\overset{5}{\cancel{10}}}{\cancel{2}}$; $\mathbf{x=-5}$ and $x=\frac{-11-1}{2}$; $x=-\frac{\overset{6}{\cancel{12}}}{\cancel{2}}$; $\mathbf{x=-6}$ and the equation $x^2+11x+30=0$ can be factored to $\mathbf{(x+5)(x+6)=0}$.

Third Method: (Completing-the-Square Method)

$x^2+11x+30=0$; $x^2+11x=-30$; $x^2+11x+\left(\frac{11}{2}\right)^2=-30+\left(\frac{11}{2}\right)^2$; $x^2+11x+\frac{121}{4}=-30+\frac{121}{4}$

; $\left(x+\frac{11}{2}\right)^2=-\frac{30}{1}+\frac{121}{4}$; $\left(x+\frac{11}{2}\right)^2=\frac{(-30\cdot 4)+(1\cdot 121)}{1\cdot 4}$; $\left(x+\frac{11}{2}\right)^2=\frac{-120+121}{4}$; $\left(x+\frac{11}{2}\right)^2=\frac{1}{4}$; $x+\frac{11}{2}=\pm\sqrt{\frac{1}{4}}$

; $x+\frac{11}{2}=\pm\frac{1}{2}$. Therefore, the two solutions are $\mathbf{x=-5}$ and $\mathbf{x=-6}$ and the equation $x^2+11x+30=0$ can be factored to $\mathbf{(x+5)(x+6)=0}$.

Check: $(x+5)(x+6)=0$; $x\cdot x+6\cdot x+5\cdot x+5\cdot 6=0$; $x^2+6x+5x+30=0$; $x^2+(6+5)x+30=0$; $x^2+11x+30=0$

From the above three methods using the Trial and Error method is the easiest method to obtain the factored terms. Completing the Square method is the second easiest method to use, followed by the Quadratic Formula method which is the longest and perhaps the most difficult way of obtaining the factored terms.

Section 1.5a Solutions - Introduction to Algebraic Fractions

A. Write the correct sign for the following fractions:

1. $-\frac{2}{-5}=+\mathbf{\frac{2}{5}}$
2. $+\frac{-3}{-6}=+\frac{3}{6}=+\mathbf{\frac{1}{2}}$
3. $-\frac{-8}{-4}=-\frac{8}{4}=-\frac{2}{1}=\mathbf{-2}$
4. $\frac{10}{-2}=-\frac{10}{2}=-\frac{5}{1}=\mathbf{-5}$
5. $-\frac{5}{-15}=+\frac{5}{15}=+\mathbf{\frac{1}{3}}$
6. $+\frac{-8}{6}=-\frac{8}{6}=-\mathbf{\frac{4}{3}}$

B. State the value(s) of the variable for which the following fractions are not defined.

1. $\frac{3}{x-1}$ is not defined when $\mathbf{x=1}$

2. $\frac{x-5}{5-x}$ is not defined when $\mathbf{x=5}$

3. $-\frac{1}{x}$ is not defined when $\mathbf{x=0}$

4. $\frac{x}{x+10}$ is not defined when $\mathbf{x=-10}$

5. $\frac{2x}{3x-5}$ is not defined when $\mathbf{x=\frac{5}{3}}$

6. $\frac{5x-2}{x-7}$ is not defined when $\mathbf{x=7}$

C. State which of the following algebraic fractions are equivalent fractions.

1. $\frac{2x}{3y}$ and $\frac{4x}{9y}$ are **not equivalent** fractions.

2. $\frac{3x+1}{2x}$ and $\frac{9x+3}{6x}$ are **equivalent** fractions.

3. $\frac{2}{a-b}$ and $-\frac{2}{b-a}$ are **equivalent** fractions.

4. $\frac{x-5}{x+1}$ and $\frac{5-x}{x-1}$ are **not equivalent** fractions.

5. $-\frac{a}{a-1}$ and $\frac{a}{1-a}$ are **equivalent** fractions.

6. $\frac{3-x}{-x}$ and $\frac{x+3}{x}$ are **not equivalent** fractions.

Section 1.5b Solutions - Simplifying Algebraic Fractions to Lower Terms

1. $\frac{x^2y^2z^5}{-xy^3z^2} = -\frac{\overset{x}{\cancel{x^2}}\,\cancel{y^2}\,\overset{z^3}{\cancel{z^5}}}{\cancel{x}\,\underset{y}{\cancel{y^3}}\,\cancel{z^2}} = -\frac{\mathbf{xz^3}}{\mathbf{y}}$

2. $-\frac{3a^2bc^3}{-9ab^2c} = +\frac{\cancel{3}\overset{a}{\cancel{a^2}}\,\cancel{b}\,\overset{c^2}{\cancel{c^3}}}{\underset{3}{\cancel{9}}\,\cancel{a}\,\underset{b}{\cancel{b^2}}\,\cancel{c}} = \frac{\mathbf{ac^2}}{\mathbf{3b}}$

3. $\frac{1+2m}{1-2m}$ **Can not be simplified.**

4. $\frac{2uvw^3}{10u^2v} = \frac{\cancel{2}\cancel{u}\cancel{v}w^3}{\underset{5}{\cancel{10}}\,\underset{u}{\cancel{u^2}}\,\cancel{v}} = \frac{\mathbf{w^3}}{\mathbf{5u}}$

5. $\frac{y^2-4}{y^2-y-6} = \frac{y^2-2^2}{(y+2)(y-3)} = \frac{(y-2)(y+2)}{(y+2)(y-3)} = \frac{(y-2)\cancel{(y+2)}}{\cancel{(y+2)}(y-3)} = \frac{\mathbf{y-2}}{\mathbf{y-3}}$

6. $\frac{x^3-3x^2}{x^2-9} = \frac{x^2(x-3)}{x^2-3^2} = \frac{x^2(x-3)}{(x-3)(x+3)} = \frac{x^2\cancel{(x-3)}}{\cancel{(x-3)}(x+3)} = \frac{\mathbf{x^2}}{\mathbf{x+3}}$

Section 1.5c Case I Solutions - Addition amd Subtraction of Algebraic Fractions with Common Denominators

1. $\frac{x}{7}+\frac{3}{7} = \frac{\mathbf{x+3}}{\mathbf{7}}$

2. $\frac{8}{a+b}-\frac{7}{a+b} = \frac{8-7}{a+b} = \frac{\mathbf{1}}{\mathbf{a+b}}$

3. $\frac{3x+1}{2y}+\frac{4x+1}{2y}+\frac{3}{2y} = \frac{3x+1+4x+1+3}{2y} = \frac{(3x+4x)+(1+1+3)}{2y} = \frac{(3+4)x+5}{2y} = \frac{\mathbf{7x+5}}{\mathbf{2y}}$

4. $\frac{4x}{x-2}-\frac{8}{x-2} = \frac{4x-8}{x-2} = \frac{4\cancel{(x-2)}}{\cancel{(x-2)}} = \frac{4}{1} = \mathbf{4}$

5. $\frac{15a}{5a+b}-\frac{-5b}{5a+b} = \frac{15a-(-5b)}{5a+b} = \frac{15a+5b}{5a+b} = \frac{\mathbf{5(3a+b)}}{\mathbf{5a+b}}$ The answer is in its lowest terms.

6. $\frac{6x}{3x^2y^2}+\frac{5y}{3x^2y^2} = \frac{\mathbf{6x+5y}}{\mathbf{3x^2y^2}}$ The answer is in its lowest terms.

Section 1.5c Case II Solutions - Addition and Subtraction of Algebraic Fractions without Common Denominators

1. $\frac{3}{4x^2}+\frac{5}{2x^3} = \frac{(3\cdot 2x^3)+(5\cdot 4x^2)}{4x^2\cdot 2x^3} = \frac{6x^3+20x^2}{8x^5} = \frac{\cancel{2x^2}(3x+10)}{\underset{4\,x^3}{\cancel{8x^5}}} = \frac{\mathbf{3x+10}}{\mathbf{4x^3}}$

2. $\frac{x}{x+4}-\frac{2}{x-1}=\frac{[x\cdot(x-1)]-[2\cdot(x+4)]}{(x+4)\cdot(x-1)}=\frac{x^2-x-(2x+8)}{(x+4)(x-1)}=\frac{x^2-x-2x-8}{(x+4)(x-1)}=\mathbf{\frac{x^2-3x-8}{(x+4)(x-1)}}$

3. $\frac{a-b}{a+b}-\frac{a}{b}=\frac{[b\cdot(a-b)]-[a\cdot(a+b)]}{b\cdot(a+b)}=\frac{(ab-b^2)-(a^2+ab)}{b(a+b)}=\frac{ab-b^2-a^2-ab}{b(a+b)}=\frac{-b^2-a^2}{b(a+b)}=\mathbf{\frac{-(a^2+b^2)}{b(a+b)}}$

4. $\frac{x^2}{x+3}+\frac{5x}{x-5}=\frac{[x^2\cdot(x-5)]+[5x\cdot(x+3)]}{(x+3)\cdot(x-5)}=\frac{(x^3-5x^2)+(5x^2+15x)}{(x+3)(x-5)}=\frac{x^3-5x^2+5x^2+15x}{(x+3)(x-5)}=\frac{x^3+15x}{(x+3)(x-5)}$

$=\mathbf{\frac{x(x^2+15)}{(x+3)(x-5)}}$

5. $\frac{1}{4x^2y^2z}-\frac{2}{xy^2z}=\frac{(1\cdot xy^2z)-(2\cdot 4x^2y^2z)}{4x^2y^2z\cdot xy^2z}=\frac{xy^2z-8x^2y^2z}{4x^3y^4z^2}=\frac{xy^2z(y-8x)}{4x^3y^4z^2}=\mathbf{\frac{1-8x}{4x^2y^2z}}$

6. $\frac{3}{x+1}+\frac{2}{x-1}-\frac{5}{x}=\left(\frac{3}{x+1}+\frac{2}{x-1}\right)-\frac{5}{x}=\left(\frac{[3\cdot(x-1)]+[2\cdot(x+1)]}{(x+1)\cdot(x-1)}\right)-\frac{5}{x}=\left(\frac{3x-3+2x+2}{(x+1)(x-1)}\right)-\frac{5}{x}=\left(\frac{5x-1}{(x+1)(x-1)}\right)-\frac{5}{x}$

$=\frac{[x\cdot(5x-1)]-\{5\cdot[(x+1)(x-1)]\}}{x\cdot[(x+1)(x-1)]}=\frac{(5x^2-x)-[5\cdot(x^2-x+x-1)]}{x(x+1)(x-1)}=\frac{(5x^2-x)-[5\cdot(x^2-1)]}{x(x+1)(x-1)}=\frac{(5x^2-x)-(5x^2-5)}{x(x+1)(x-1)}$

$=\frac{5x^2-x-5x^2+5}{x(x+1)(x-1)}=\frac{-x+5}{x(x+1)(x-1)}=-\mathbf{\frac{(x-5)}{x(x+1)(x-1)}}$

Section 1.5c Case III Solutions - Multiplication of Algebraic Fractions

1. $\frac{1}{xy}\cdot\frac{x^2y^2}{2}=\frac{1\cdot x^2y^2}{xy\cdot 2}=\frac{x^2y^2}{2xy}=\frac{x^2y^2}{2xy}=\mathbf{\frac{xy}{2}}$

2. $\frac{2a^2}{a^3}\cdot\frac{1}{8a}=\frac{2a^2\cdot 1}{a^3\cdot 8a}=\frac{2a^2}{8a^4}=\frac{2a^2}{8a^4}=\mathbf{\frac{1}{4a^2}}$

3. $xyz\cdot\frac{1}{x^2y^2z^2}\cdot\frac{x^2}{y}=\frac{xyz}{1}\cdot\frac{1}{x^2y^2z^2}\cdot\frac{x^2}{y}=\frac{xyz\cdot 1\cdot x^2}{1\cdot x^2y^2z^2\cdot y}=\frac{x^3yz}{x^2y^3z^2}=\frac{x^3yz}{x^2y^3z^2}=\mathbf{\frac{x}{y^2z}}$

4. $\frac{5u^2v^2}{uv}\cdot\frac{uv^3}{15v^2}\cdot\frac{1}{u^4}=\frac{5u^2v^2\cdot uv^3\cdot 1}{uv\cdot 15v^2\cdot u^4}=\frac{5u^3v^5}{15u^5v^3}=\frac{5u^3v^5}{15u^5v^3}=\mathbf{\frac{v^2}{3u^2}}$

5. $8x\cdot\frac{2}{x^3}\cdot\frac{1}{4x^2}=\frac{8x}{1}\cdot\frac{2}{x^3}\cdot\frac{1}{4x^2}=\frac{8x\cdot 2\cdot 1}{1\cdot x^3\cdot 4x^2}=\frac{16x}{4x^5}=\frac{16x}{4x^5}=\mathbf{\frac{4}{x^4}}$

6. $\frac{x-4}{x+2}\cdot\frac{x^2-4}{2}\cdot\frac{1}{x-4}=\frac{(x-4)\cdot(x^2-4)\cdot 1}{(x+2)\cdot 2\cdot(x-4)}=\frac{(x-4)(x-2)(x+2)}{2(x+2)(x-4)}=\frac{(x-4)(x-2)(x+2)}{2(x+2)(x-4)}=\mathbf{\frac{x-2}{2}}$

Section 1.5c Case IV Solutions - Division of Algebraic Fractions

1. $\frac{x^2y}{x^3y^2}\div xy=\frac{x^2y}{x^3y^2}\div\frac{xy}{1}=\frac{x^2y}{x^3y^2}\cdot\frac{1}{xy}=\frac{x^2y\cdot 1}{x^3y^2\cdot xy}=\frac{x^2y}{x^4y^3}=\frac{x^2y}{x^4y^3}=\mathbf{\frac{1}{x^2y^2}}$

2. $\frac{uv^2w}{vw^2}\div\frac{uv^3}{w} = \frac{uv^2w}{vw^2}\cdot\frac{w}{uv^3} = \frac{uv^2w\cdot w}{vw^2\cdot uv^3} = \frac{uv^2w^2}{uv^4w^2} = \frac{\cancel{u}\cancel{v^2}\cancel{w^2}}{\cancel{u}\underset{v^2}{\cancel{v^4}}\cancel{w^2}} = \mathbf{\frac{1}{v^2}}$

3. $a^2b^2c^4\div\frac{a^2b}{2ac} = \frac{a^2b^2c^4}{1}\div\frac{a^2b}{2ac} = \frac{a^2b^2c^4}{1}\cdot\frac{2ac}{a^2b} = \frac{a^2b^2c^4\cdot 2ac}{1\cdot a^2b} = \frac{2a^3b^2c^5}{a^2b} = \frac{2\overset{a}{\cancel{a^3}}\overset{b}{\cancel{b^2}}c^5}{\cancel{a^2}\cancel{b}} = \frac{2abc^5}{1} = \mathbf{2abc^5}$

4. $\frac{xyz}{x^2z^3}\div\frac{x^2z^2}{yz} = \frac{xyz}{x^2z^3}\cdot\frac{yz}{x^2z^2} = \frac{xyz\cdot yz}{x^2z^3\cdot x^2z^2} = \frac{xy^2z^2}{x^4z^5} = \frac{\cancel{x}y^2\cancel{z^2}}{\underset{x^3}{\cancel{x^4}}\underset{z^3}{\cancel{z^5}}} = \mathbf{\frac{y^2}{x^3z^3}}$

5. $\left(\frac{uv^2}{v^3}\div 2u^2\right)\div\frac{uv}{3} = \left(\frac{uv^2}{v^3}\div\frac{2u^2}{1}\right)\div\frac{uv}{3} = \left(\frac{uv^2}{v^3}\cdot\frac{1}{2u^2}\right)\div\frac{uv}{3} = \left(\frac{uv^2\cdot 1}{v^3\cdot 2u^2}\right)\div\frac{uv}{3} = \frac{uv^2}{2u^2v^3}\div\frac{uv}{3} = \frac{uv^2}{2u^2v^3}\cdot\frac{3}{uv}$

$= \frac{uv^2\cdot 3}{2u^2v^3\cdot uv} = \frac{3uv^2}{2u^3v^4} = \frac{3\cancel{u}\cancel{v^2}}{2\underset{u^2}{\cancel{u^3}}\underset{v^2}{\cancel{v^4}}} = \mathbf{\frac{3}{2u^2v^2}}$

6. $\frac{x^2y^2z}{xz}\div\left(x^2y\div\frac{4}{yz^3}\right) = \frac{x^2y^2z}{xz}\div\left(\frac{x^2y}{1}\div\frac{4}{yz^3}\right) = \frac{x^2y^2z}{xz}\div\left(\frac{x^2y}{1}\cdot\frac{yz^3}{4}\right) = \frac{x^2y^2z}{xz}\div\left(\frac{x^2y\cdot yz^3}{1\cdot 4}\right) = \frac{x^2y^2z}{xz}\div\frac{x^2y^2z^3}{4}$

$= \frac{x^2y^2z}{xz}\cdot\frac{4}{x^2y^2z^3} = \frac{x^2y^2z\cdot 4}{xz\cdot x^2y^2z^3} = \frac{4x^2y^2z}{x^3y^2z^4} = \frac{4\cancel{x^2}\cancel{y^2}\cancel{z}}{\underset{x}{\cancel{x^3}}\cancel{y^2}\underset{z^3}{\cancel{z^4}}} = \mathbf{\frac{4}{xz^3}}$

Section 1.5d Case I Solutions - Addition and Subtraction of Complex Algebraic Fractions

1. $\frac{2-\frac{1}{5a}}{3-\frac{2}{15a}} = \frac{\frac{2}{1}-\frac{1}{5a}}{\frac{3}{1}-\frac{2}{15a}} = \frac{\frac{(2\cdot 5a)-(1\cdot 1)}{1\cdot 5a}}{\frac{(3\cdot 15a)-(2\cdot 1)}{1\cdot 15a}} = \frac{\frac{10a-1}{5a}}{\frac{45a-2}{15a}} = \frac{(10a-1)\cdot\overset{3}{\cancel{15a}}}{\cancel{5a}\cdot(45a-2)} = \frac{3(10a-1)}{45a-2} = \mathbf{\frac{30a-3}{45a-2}}$

2. $\frac{\frac{2x^3y^2z}{4x^2z}}{\frac{2x}{xy^2}}-1 = \frac{(2x^3y^2z)\cdot(xy^2)}{(4x^2z)\cdot(2x)}-1 = \frac{2x^4y^4z}{8x^3z}-1 = \frac{\cancel{2}\overset{x}{\cancel{x^4}}y^4\cancel{z}}{\underset{4}{\cancel{8}}\cancel{x^3}\cancel{z}}-1 = \frac{xy^4}{4}-1 = \frac{xy^4}{4}-\frac{1}{1} = \frac{(xy^4\cdot 1)-(1\cdot 4)}{4\cdot 1} = \mathbf{\frac{xy^4-4}{4}}$

3. $\frac{\frac{2}{a}+\frac{1}{a^3}}{\frac{2}{a^3}+2} = \frac{\frac{2}{a}+\frac{1}{a^3}}{\frac{2}{a^3}+\frac{2}{1}} = \frac{\frac{(2\cdot a^3)+(1\cdot a)}{a\cdot a^3}}{\frac{(2\cdot 1)+(2\cdot a^3)}{a^3\cdot 1}} = \frac{\frac{2a^3+a}{a^4}}{\frac{2+2a^3}{a^3}} = \frac{(2a^3+a)\cdot\cancel{a^3}}{a^4\cdot(2+2a^3)} = \frac{\cancel{a}(2a^2+1)}{2\cancel{a}(1+a^3)} = \frac{2a^2+1}{2(1+a^3)} = \mathbf{\frac{2a^2+1}{2(1+a)(1-a+a^2)}}$

4. $\frac{a}{a+\frac{1}{a^2}} = \frac{\frac{a}{1}}{\frac{a}{1}+\frac{1}{a^2}} = \frac{\frac{a}{1}}{\frac{(a\cdot a^2)+(1\cdot 1)}{1\cdot a^2}} = \frac{\frac{a}{1}}{\frac{a^3+1}{a^2}} = \frac{a\cdot a^2}{(a^3+1)\cdot 1} = \frac{a^3}{a^3+1} = \mathbf{\frac{a^3}{(a+1)(a^2-a+1)}}$

5. $\frac{\frac{x}{y}-3}{3-\frac{y}{x}} = \frac{\frac{x}{y}-\frac{3}{1}}{\frac{3}{1}-\frac{y}{x}} = \frac{\frac{(x\cdot 1)-(3\cdot y)}{y\cdot 1}}{\frac{(3\cdot x)-(y\cdot 1)}{1\cdot x}} = \frac{\frac{x-3y}{y}}{\frac{3x-y}{x}} = \frac{x\cdot(x-3y)}{y\cdot(3x-y)} = \mathbf{\frac{x(x-3y)}{y(3x-y)}}$

6. $\frac{\frac{1}{x}-\frac{1}{x+4}}{\frac{3}{x+4}+1} = \frac{\frac{1}{x}-\frac{1}{x+4}}{\frac{3}{x+4}+\frac{1}{1}} = \frac{\frac{[1\cdot(x+4)]-(1\cdot x)}{x\cdot(x+4)}}{\frac{(3\cdot 1)+[1\cdot(x+4)]}{(x+4)\cdot 1}} = \frac{\frac{\cancel{x}+4-\cancel{x}}{x\cdot(x+4)}}{\frac{3+x+4}{x+4}} = \frac{\frac{4}{x\cdot(x+4)}}{\frac{x+7}{x+4}} = \frac{4\cdot\cancel{(x+4)}}{x\cdot\cancel{(x+4)}\cdot(x+7)} = \mathbf{\frac{4}{x(x+7)}}$

Section 1.5d Case II Solutions - Multiplication of Complex Algebraic Fractions

1. $$\frac{y^2}{\frac{x^2y^2}{xy}\cdot\frac{2x^3y}{x^4}} = \frac{y^2}{\frac{x^2y^2\cdot 2x^3y}{xy\cdot x^4}} = \frac{y^2}{\frac{2x^5y^3}{x^5y}} = \frac{\frac{y^2}{1}}{\frac{2x^5y^3}{x^5y}} = \frac{y^2\cdot x^5y}{1\cdot 2x^5y^3} = \frac{x^5y^3}{2x^5y^3} = \frac{\cancel{x^5}\cancel{y^3}}{2\cancel{x^5}\cancel{y^3}} = \mathbf{\frac{1}{2}}$$

2. $$\frac{\frac{1}{2x^3y^2}\cdot\frac{4x}{y}}{\frac{1}{xy^2}} = \frac{\frac{1\cdot 4x}{2x^3y^2\cdot y}}{\frac{1}{xy^2}} = \frac{\frac{4x}{2x^3y^3}}{\frac{1}{xy^2}} = \frac{4x\cdot xy^2}{2x^3y^3\cdot 1} = \frac{4x^2y^2}{2x^3y^3} = \frac{\overset{2}{\cancel{4}}\cancel{x^2}\cancel{y^2}}{\cancel{2}\underset{x}{\cancel{x^3}}\underset{y}{\cancel{y^3}}} = \mathbf{\frac{2}{xy}}$$

3. $$\frac{\frac{x^2}{x^3}}{2x\cdot\frac{3x^2}{18x^5}} = \frac{\frac{x^2}{x^3}}{\frac{2x}{1}\cdot\frac{3x^2}{18x^5}} = \frac{\frac{x^2}{x^3}}{\frac{2x\cdot 3x^2}{1\cdot 18x^5}} = \frac{\frac{x^2}{x^3}}{\frac{6x^3}{18x^5}} = \frac{x^2\cdot 18x^5}{x^3\cdot 6x^3} = \frac{18x^7}{6x^6} = \frac{\overset{3}{\cancel{18}}\overset{x}{\cancel{x^7}}}{\cancel{6}\cancel{x^6}} = \frac{3x}{1} = \mathbf{3x}$$

4. $$\frac{\left(x^2yz^2\cdot\frac{2x}{y^2}\right)\cdot\frac{3x}{x^3z^4}}{\frac{1}{y^2z^3}} = \frac{\left(\frac{x^2yz^2}{1}\cdot\frac{2x}{y^2}\right)\cdot\frac{3x}{x^3z^4}}{\frac{1}{y^2z^3}} = \frac{\left(\frac{x^2yz^2\cdot 2x}{1\cdot y^2}\right)\cdot\frac{3x}{x^3z^4}}{\frac{1}{y^2z^3}} = \frac{\frac{2x^3yz^2}{y^2}\cdot\frac{3x}{x^3z^4}}{\frac{1}{y^2z^3}} = \frac{\frac{2x^3yz^2\cdot 3x}{y^2\cdot x^3z^4}}{\frac{1}{y^2z^3}} = \frac{\frac{6x^4yz^2}{x^3y^2z^4}}{\frac{1}{y^2z^3}}$$

$$= \frac{6x^4yz^2\cdot y^2z^3}{x^3y^2z^4\cdot 1} = \frac{6x^4y^3z^5}{x^3y^2z^4} = \frac{6\overset{x}{\cancel{x^4}}\overset{y}{\cancel{y^3}}\overset{z}{\cancel{z^5}}}{\cancel{x^3}\cancel{y^2}\cancel{z^4}} = \frac{6xyz}{1} = \mathbf{6xyz}$$

5. $$\frac{\left(\frac{2a^2b^2}{a^3}\cdot 4ab^3\right)\cdot\frac{1}{3ab^4}}{\frac{8}{3}} = \frac{\left(\frac{2a^2b^2}{a^3}\cdot\frac{4ab^3}{1}\right)\cdot\frac{1}{3ab^4}}{\frac{8}{3}} = \frac{\left(\frac{2a^2b^2\cdot 4ab^3}{a^3\cdot 1}\right)\cdot\frac{1}{3ab^4}}{\frac{8}{3}} = \frac{\frac{8a^3b^5}{a^3}\cdot\frac{1}{3ab^4}}{\frac{8}{3}} = \frac{\frac{8a^3b^5\cdot 1}{a^3\cdot 3ab^4}}{\frac{8}{3}} = \frac{\frac{8a^3b^5}{3a^4b^4}}{\frac{8}{3}}$$

$$= \frac{8a^3b^5\cdot 3}{3a^4b^4\cdot 8} = \frac{24a^3b^5}{24a^4b^4} = \frac{\cancel{24}\cancel{a^3}\overset{b}{\cancel{b^5}}}{\cancel{24}\underset{a}{\cancel{a^4}}\cancel{b^4}} = \mathbf{\frac{b}{a}}$$

6. $$\frac{\frac{u^3v^3w^2}{w^3}\cdot\frac{1}{3w}}{\frac{1}{6uv^2}} = \frac{\frac{u^3v^3w^2\cdot 1}{w^3\cdot 3w}}{\frac{1}{6uv^2}} = \frac{\frac{u^3v^3w^2}{3w^4}}{\frac{1}{6uv^2}} = \frac{u^3v^3w^2\cdot 6uv^2}{3w^4\cdot 1} = \frac{6u^4v^5w^2}{3w^4} = \frac{\overset{2}{\cancel{6}}u^4v^5\cancel{w^2}}{\cancel{3}\underset{w^2}{\cancel{w^4}}} = \mathbf{\frac{2u^4v^5}{w^2}}$$

Section 1.5d Case III Solutions - Division of Complex Algebraic Fractions

1. $$\frac{\frac{xy}{x^3}\div\frac{x^2}{xy^2}}{2x} = \frac{\frac{xy}{x^3}\cdot\frac{xy^2}{x^2}}{2x} = \frac{\frac{xy\cdot xy^2}{x^3\cdot x^2}}{2x} = \frac{\frac{x^2y^3}{x^5}}{2x} = \frac{\frac{x^2y^3}{x^5}}{\frac{2x}{1}} = \frac{x^2y^3\cdot 1}{x^5\cdot 2x} = \frac{x^2y^3}{2x^6} = \frac{\cancel{x^2}y^3}{2\underset{x^4}{\cancel{x^6}}} = \mathbf{\frac{y^3}{2x^4}}$$

2. $$\frac{\frac{1}{a}\div\frac{1}{b}}{\frac{a}{ab^2}\div b} = \frac{\frac{1}{a}\div\frac{1}{b}}{\frac{a}{ab^2}\div\frac{b}{1}} = \frac{\frac{1}{a}\cdot\frac{b}{1}}{\frac{a}{ab^2}\cdot\frac{1}{b}} = \frac{\frac{1\cdot b}{a\cdot 1}}{\frac{a\cdot 1}{ab^2\cdot b}} = \frac{\frac{b}{a}}{\frac{a}{ab^3}} = \frac{b\cdot ab^3}{a\cdot a} = \frac{ab^4}{a^2} = \frac{\cancel{a}b^4}{\underset{a}{\cancel{a^2}}} = \mathbf{\frac{b^4}{a}}$$

3. $$xy\div\left(\frac{\frac{1}{xy}\div x}{x^3}\right) = xy\div\left(\frac{\frac{1}{xy}\div\frac{x}{1}}{x^3}\right) = xy\div\left(\frac{\frac{1}{xy}\cdot\frac{1}{x}}{x^3}\right) = xy\div\left(\frac{\frac{1\cdot 1}{xy\cdot x}}{x^3}\right) = xy\div\frac{\frac{1}{x^2y}}{x^3} = xy\div\frac{\frac{1}{x^2y}}{\frac{x^3}{1}} = xy\div\frac{1\cdot 1}{x^2y\cdot x^3}$$

$$= xy \div \frac{1}{x^5 y} = xy \cdot \frac{x^5 y}{1} = \frac{xy}{1} \cdot \frac{x^5 y}{1} = \frac{xy \cdot x^5 y}{1 \cdot 1} = \frac{x^6 y^2}{1} = \mathbf{x^6 y^2}$$

4. $$\frac{\frac{u^2 vw}{w^2} \div \frac{u^3 v}{w}}{w^3} = \frac{\frac{u^2 vw}{w^2} \cdot \frac{w}{u^3 v}}{w^3} = \frac{\frac{u^2 vw \cdot w}{w^2 \cdot u^3 v}}{w^3} = \frac{\frac{u^2 vw^2}{u^3 vw^2}}{w^3} = \frac{\frac{u^2 vw^2}{u^3 vw^2}}{\frac{w^3}{1}} = \frac{u^2 vw^2 \cdot 1}{u^3 vw^2 \cdot w^3} = \frac{u^2 vw^2}{u^3 vw^5} = \frac{\cancel{u^2}\cancel{v}\cancel{w^2}}{\underset{u}{\cancel{u^3}}\cancel{v}\underset{w^3}{\cancel{w^5}}} = \mathbf{\frac{1}{uw^3}}$$

5. $$\frac{ab^3 \div \frac{a^2 b^2}{b}}{a \div \frac{a}{b}} = \frac{\frac{ab^3}{1} \div \frac{a^2 b^2}{b}}{\frac{a}{1} \div \frac{a}{b}} = \frac{\frac{ab^3}{1} \cdot \frac{b}{a^2 b^2}}{\frac{a}{1} \cdot \frac{b}{a}} = \frac{\frac{ab^3 \cdot b}{1 \cdot a^2 b^2}}{\frac{a \cdot b}{1 \cdot a}} = \frac{\frac{ab^4}{a^2 b^2}}{\frac{ab}{a}} = \frac{ab^4 \cdot a}{a^2 b^2 \cdot ab} = \frac{a^2 b^4}{a^3 b^3} = \frac{\cancel{a^2}\overset{b}{\cancel{b^4}}}{\underset{a}{\cancel{a^3}}\cancel{b^3}} = \mathbf{\frac{b}{a}}$$

6. $$\frac{\frac{z^3}{z^2} \div 2z}{z^3 \div \frac{z^2}{2}} = \frac{\frac{z^3}{z^2} \div \frac{2z}{1}}{\frac{z^3}{1} \div \frac{z^2}{2}} = \frac{\frac{z^3}{z^2} \cdot \frac{1}{2z}}{\frac{z^3}{1} \cdot \frac{2}{z^2}} = \frac{\frac{z^3 \cdot 1}{z^2 \cdot 2z}}{\frac{z^3 \cdot 2}{1 \cdot z^2}} = \frac{\frac{z^3}{2z^3}}{\frac{2z^3}{z^2}} = \frac{z^3 \cdot z^2}{2z^3 \cdot 2z^3} = \frac{z^5}{4z^6} = \frac{\cancel{z^5}}{4\underset{z}{\cancel{z^6}}} = \mathbf{\frac{1}{4z}}$$

Chapter 2 Solutions:

Section 2.1 Solutions – Introduction to Functions of Real Variables

1. Find the corresponding y values.

a. Given $x-4y=0$, at $x=0$ the y value is equal to $0-4y=0$; $-4y=0$; $\mathbf{y=0}$

at $x=-1$ the y value is equal to $-1-4y=0$; $-4y=1$; $y=-\frac{1}{4}$; $\mathbf{y=-0.25}$

at $x=3$ the y value is equal to $3-4y=0$; $-4y=-3$; $y=\frac{3}{4}$; $\mathbf{y=0.75}$

Therefore, the ordered pairs are $\{(\mathbf{0, 0}), (\mathbf{-1, -0.25}), (\mathbf{3, 0.75})\}$

b. Given $y-x^2+1=0$, at $x=-1$ the y value is equal to $y-(-1)^2+1=0$; $y-1+1=0$; $\mathbf{y=0}$

at $x=3$ the y value is equal to $y-3^2+1=0$; $y-9+1=0$; $y-8=0$; $\mathbf{y=8}$

at $x=-3$ the y value is equal to $y-(-3)^2+1=0$; $y-9+1=0$; $y-8=0$; $\mathbf{y=8}$

Therefore, the ordered pairs are $\{(\mathbf{0, 0}), (\mathbf{3, 8}), (\mathbf{-3, 8})\}$

c. Given $y-\sqrt{x^2+1}=0$, at $x=2$ the y value is equal to $y-\sqrt{2^2+1}=0$; $y-\sqrt{4+1}=0$; $y-\sqrt{5}=0$; $y=\sqrt{5}$; $\mathbf{y=2.24}$

at $x=-2$ the y value is equal to $y-\sqrt{(-2)^2+1}=0$; $y-\sqrt{4+1}=0$; $y-\sqrt{5}=0$; $y=\sqrt{5}$; $\mathbf{y=2.24}$

at $x=-5$ the y value is equal to $y-\sqrt{(-5)^2+1}=0$; $y-\sqrt{25+1}=0$; $y-\sqrt{26}=0$; $y=\sqrt{26}$; $\mathbf{y=5.09}$

Therefore, the ordered pairs are $\{(\mathbf{2, 2.24}), (\mathbf{-2, 2.24}), (\mathbf{-5, 5.09})\}$

d. Given $x+4y=-5$, at $x=0$ the y value is equal to $0+4y=-5$; $4y=-5$; $y=-\frac{5}{4}$; $\mathbf{y=-1.25}$

at $x=-2$ the y value is equal to $-2+4y=-5$; $4y=-5+2$; $4y=-3$; $y=-\frac{3}{4}$; $\mathbf{y=-0.75}$

at $x=4$ the y value is equal to $4+4y=-5$; $4y=-5-4$; $4y=-9$; $y=-\frac{9}{4}$; $\mathbf{y=-2.25}$

Therefore, the ordered pairs are $\{(\mathbf{0, -1.25}), (\mathbf{-2, -0.75}), (\mathbf{4, -2.25})\}$

e. Given $y=2x^2-6$, at $x=0$ the y value is equal to $y=2\cdot 0^2-6$; $y=0-6$; $\mathbf{y=-6}$

at $x=-2$ the y value is equal to $y=2\cdot(-2)^2-6$; $y=(2\cdot 4)-6$; $y=8-6$; $\mathbf{y=2}$

at $x=-3$ the y value is equal to $y=2\cdot(-3)^2-6$; $y=(2\cdot 9)-6$; $y=18-6$; $\mathbf{y=12}$

Therefore, the ordered pairs are $\{(\mathbf{0, -6}), (\mathbf{-2, 2}), (\mathbf{-3, 12})\}$

2. a. Specify the domain and the range for each of the following ordered pairs. b. State which set constitute a relation or a function.

a. The domain and the range for the ordered pairs $\{(1,4), (2,5), (3,6), (6,9), (8,12)\}$ are: $\{1, 2, 3, 6, 8\}$ and $\{4, 5, 6, 9, 12\}$, respectively. Since each domain value corresponds with only one range value **it is a function.**

b. The domain and the range for the ordered pairs $\{(1,-1), (2,3), (4,6), (7,9), (10,11)\}$ are: $\{1, 2, 4, 7, 10\}$ and $\{-1, 3, 6, 9, 11\}$, respectively. Since each domain value corresponds with only one range value **it is a function.**

c. The domain and the range for the ordered pairs $\{(1,2), (1,-2), (2,5), (6,8), (9,12)\}$are: $\{1, 2, 6, 9\}$ and $\{2, -2, 5, 8, 12\}$, respectively. Since each domain value does not corresponds with only one range value **it is a relation.**

d. The domain and the range for the ordered pairs $\{(0,0), (1,6), (2,5), (2,7), (5,8)\}$are: $\{0, 1, 2, 5\}$ and $\{0, 5, 6, 7, 8\}$, respectively. Since each domain value does not corresponds with only one range value **it is a relation**.

e. The domain and the range for the ordered pairs $\{(-1,3),(-1,6),(2,5),(8,10),(10,12)\}$are: $\{-1,2,8,10\}$ and $\{3,5,6,10,12\}$, respectively. Since each domain value does not corresponds with only one range value **it is a relation**.

f. The domain and the range for the ordered pairs $\{(1,3),(2,5),(5,6),(7,10),(8,13)\}$ are: $\{1,2,5,7,8\}$ and $\{3,5,6,10,13\}$, respectively. Since each domain value corresponds with only one range value **it is a function**.

3. State which of the following equations defines a function.

a. The equation $x+y=12$ defines a **function** because there is only one value of y associated with each value of x.

b. The equation $x^2+y^2=81$; $y^2=81-x^2$; $y=\pm\sqrt{81-x^2}$ defines a **relation** because for each value of x there are two values of y.

c. The equation $y^2=15-x^2$; $y=\pm\sqrt{15-x^2}$ defines a **relation** because for each value of x there are two values of y.

d. The equation $y^2=x^5$; $y=\pm\sqrt{x^5}$ defines a **relation** because for each value of x there are two values of y.

e. The equation $y=x^2+9$ defines a **function** because there is only one value of y associated with each value of x.

f. The equation $2y-6x=10$; $2y=10+6x$; $y=\frac{10+6x}{2}$; $y=5+3x$ defines a **function** because there is only one value of y associated with each value of x.

g. The equation $y = \sqrt{x^2}-4 = x-4$ defines a **function** because there is only one value of y associated with each value of x.

h. The equation $x-4y=8$; $-4y=8-x$; $y=\frac{8-x}{-4}$; $y=\frac{x}{4}-2$ defines a **function** because there is only one value of y associated with each value of x.

i. The equation $x+y^2=9$; $y^2=9-x$; $y=\pm\sqrt{9-x}$ defines a **relation** because for each value of x there are two values of y.

4. Find the corresponding range values for each of the following functions.

a. Given $f(x)=-x^2+2x$, then

$f(-4) = -(-4)^2+(2\cdot-4) = -16-8 = \mathbf{-24}$

$f(-1) = -(-1)^2+(2\cdot-1) = -1-2 = \mathbf{-3}$

$f(0) = -0^2+(2\cdot0) = 0+0 = \mathbf{0}$

$f(1) = -1^2+(2\cdot1) = -1+2 = \mathbf{1}$

Therefore, the ordered pairs are $\{(\mathbf{-4,-24}),(\mathbf{-1,-3}),(\mathbf{0,0}),(\mathbf{1,1})\}$

b. Given $f(x)=x^3-2x^2+1$, then

$f(-2) = (-2)^3-2\cdot(-2)^2+1 = -8-(2\cdot4)+1 = \mathbf{-15}$

$f(0) = 0^3-2\cdot0^2+1 = 0-0+1 = \mathbf{1}$

$f(2) = 2^3-2\cdot2^2+1 = 8-(2\cdot4)+1 = 8-8+1 = \mathbf{1}$

$f(-a) = (-a)^3-2\cdot(-a)^2+1 = \mathbf{-a^3-2a^2+1}$

Therefore, the ordered pairs are $\{(\mathbf{-2,-15}),(\mathbf{0,1}),(\mathbf{2,1}),(\mathbf{-a,-a^3-2a^2+1})\}$

c. Given $f(x)=\sqrt{x^2+1}$, then

$f(-2) = \sqrt{(-2)^2+1} = \sqrt{4+1} = \sqrt{5} = \mathbf{2.24}$

$f(-1) = \sqrt{(-1)^2+1} = \sqrt{1+1} = \sqrt{2} = \mathbf{1.414}$

$f(0) = \sqrt{0^2+1} = \sqrt{0+1} = \sqrt{1} = \mathbf{1}$

$f(2) = \sqrt{2^2+1} = \sqrt{4+1} = \sqrt{5} = \mathbf{2.24}$

Therefore, the ordered pairs are $\{(\mathbf{-2,2.24}),(\mathbf{-1,1.414}),(\mathbf{0,1}),(\mathbf{2,2.24})\}$

d. Given $f(x)=\frac{1}{x^2+2}$, then

$f(-3) = \frac{1}{(-3)^2+2} = \frac{1}{9+2} = \frac{1}{11} = \mathbf{0.091}$

$f(-1) = \frac{1}{(-1)^2+2} = \frac{1}{1+2} = \frac{1}{3} = \mathbf{0.333}$

$f(0) = \frac{1}{0^2+2} = \frac{1}{2} = \mathbf{0.5}$ $\quad f(3) = \frac{1}{3^2+2} = \frac{1}{9+2} = \frac{1}{11} = \mathbf{0.091}$

Therefore, the ordered pairs are $\{(-3, \mathbf{0.091}), (-1, \mathbf{0.333}), (0, \mathbf{0.5}), (3, \mathbf{0.091})\}$

5. Given the following functions, find $f(a+h)-f(a)$.

a. Given $f(x)=2x-1$, then $f(a+h) = 2(a+h)-1 = 2a+2h-1$ and $f(a) = 2a-1$. Therefore, $f(a+h)-f(a)$ $= (2a+2h-1)-(2a-1) = 2a+2h-1-2a+1 = \mathbf{2h}$

b. Given $f(x)=2x^2-3$, then $f(a+h) = 2(a+h)^2-3 = 2(a^2+h^2+2ah)-3 = 2a^2+2h^2+4ah-3$ and $f(a) = 2a^2-3$. Thus, $f(a+h)-f(a) = (2a^2+2h^2+4ah-3)-(2a^2-3) = 2a^2+2h^2+4ah-3-2a^2+3 = \mathbf{2h^2+4ah}$

c. Given $f(x)=\frac{x^2-a^2}{x}$, then $f(a+h) = \frac{(a+h)^2-a^2}{a+h} = \frac{a^2+h^2+2ah-a^2}{a+h} = \frac{h^2+2ah}{a+h}$ and $f(a) = \frac{a^2-a^2}{a} = \frac{0}{a}$ $= 0$. Therefore, $f(a+h)-f(a) = \frac{h^2+2ah}{a+h}-0 = \mathbf{\frac{h^2+2ah}{a+h}}$

d. Given $f(x)=(x-3)(x+1) = x^2-2x-3$, then $f(a+h) = (a+h)^2-2(a+h)-3 = a^2+h^2+2ah-2a-2h-3$ and $f(a) = a^2-2a-3$. Thus, $f(a+h)-f(a) = (a^2+h^2+2ah-2a-2h-3)-(a^2-2a-3) = a^2+h^2+2ah-2a-2h$ $-3-a^2+2a+3 = \mathbf{h^2+2ah-2h}$

Section 2.2 Solutions – Math Operations Involving Functions of Real Variables

1. Given $f(x) = x^2-3x+5$ and $g(x) = 2x^2$, find

a. $2f(x)+g(x) = 2(x^2-3x+5)+2x^2 = 2x^2-6x+10+2x^2 = (2x^2+2x^2)-6x+10 = \mathbf{4x^2-6x+10}$

b. $f(x)-3g(x) = (x^2-3x+5)-3\cdot 2x^2 = x^2-3x+5-6x^2 = (x^2-6x^2)-3x+5 = \mathbf{-5x^2-3x+5}$

c. $\frac{f(x)}{x}+5g(x) = \frac{x^2-3x+5}{x}+5\cdot 2x^2 = \frac{x^2-3x+5}{x}+\frac{10x^2}{1} = \frac{[(x^2-3x+5)\cdot 1]+(10x^2\cdot x)}{x\cdot 1} = \frac{x^2-3x+5+10x^3}{x}$ $= \frac{10x^3+x^2-3x+5}{x} = \frac{10x^3}{x}+\frac{x^2}{x}+\frac{-3x}{x}+\frac{5}{x} = \mathbf{10x^2+x-3+\frac{5}{x}}$

d. $3f(x)\cdot g(x) = 3(x^2-3x+5)\cdot(2x^2) = 6x^2(x^2-3x+5) = \mathbf{6x^4-18x^3+30x^2}$

e. $3f(x)-5g(x) = 3(x^2-3x+5)-5\cdot 2x^2 = 3x^2-9x+15-10x^2 = (3x^2-10x^2)-9x+15 = \mathbf{-7x^2-9x+15}$

f. $\frac{3}{g(x)}+2f(x) = \frac{3}{2x^2}+2(x^2-3x+5) = \frac{3}{2x^2}+\frac{2x^2-6x+10}{1} = \frac{(3\cdot 1)+[(2x^2-6x+10)\cdot 2x^2]}{2x^2\cdot 1} = \frac{3+4x^4-12x^3+20x^2}{2x^2}$ $= \frac{4x^4-12x^3+20x^2+3}{2x^2} = \frac{4x^4}{2x^2}-\frac{12x^3}{2x^2}+\frac{20x^2}{2x^2}+\frac{3}{2x^2} = \mathbf{2x^2-6x+10+\frac{3}{2x^2}}$

g. $(3f-2g)(x) = 3(x^2-3x+5)-2\cdot 2x^2 = 3x^2-9x+15-4x^2 = (3x^2-4x^2)-9x+15 = \mathbf{-x^2-9x+15}$

h. $\left(\frac{f}{4g}\right)(x)-2 = \frac{x^2-3x+5}{4\cdot 2x^2}-2 = \frac{x^2-3x+5}{8x^2}-\frac{2}{1} = \frac{[(x^2-3x+5)\cdot 1]-(2\cdot 8x^2)}{8x^2\cdot 1} = \frac{x^2-3x+5-16x^2}{8x^2} = \frac{-15x^2-3x+5}{8x^2}$ $= -\frac{15x^2}{8x^2}-\frac{3x}{8x^2}+\frac{5}{8x^2} = \mathbf{-\frac{15}{8}-\frac{3}{8x}+\frac{5}{8x^2}}$

i. $\frac{f(x)-20}{g(x)}+x = \frac{(x^2-3x+5)-20}{2x^2}+x = \frac{x^2-3x-15}{2x^2}+\frac{x}{1} = \frac{[(x^2-3x-15)\cdot 1]+(x\cdot 2x^2)}{2x^2\cdot 1} = \frac{x^2-3x-15+2x^3}{2x^2}$

$= \dfrac{2x^3+x^2-3x-15}{2x^2} = \dfrac{2x^3}{2x^2}+\dfrac{x^2}{2x^2}-\dfrac{3x}{2x^2}-\dfrac{15}{2x^2} = \mathbf{x+\dfrac{1}{2}-\dfrac{3}{2x}-\dfrac{15}{2x^2}}$

2. Let $f(x)=x^2+2$ and $g(x)=2x+5$. Find and simplify the following expressions.

a. $(f+g)(x) = (x^2+2)+(2x+5) = x^2+2x+7$ therefore $(f+g)(-2) = (-2)^2+(2\cdot-2)+7 = 4-4+7 = \mathbf{7}$

b. $(g-f)(x) = (2x+5)-(x^2+2) = 2x+5-x^2-2 = -x^2+2x+3$ therefore $(g-f)(0) = -0^2+(2\cdot 0)+3 = \mathbf{3}$

c. $\left(\dfrac{f}{g}\right)(x) = \dfrac{x^2+2}{2x+5}$ therefore $\left(\dfrac{f}{g}\right)(-1) = \dfrac{(-1)^2+2}{(2\cdot-1)+5} = \dfrac{1+2}{-2+5} = \dfrac{3}{3} = \mathbf{1}$

d. $(f+2g)(x) = (x^2+2)+2(2x+5) = x^2+2+4x+10 = x^2+4x+12$ therefore $(f+2g)(0) = 0^2+(4\cdot 0)+12 = \mathbf{12}$

e. $\left(\dfrac{g}{f}\right)(x) = \dfrac{2x+5}{x^2+2}$ therefore $\left(\dfrac{g}{f}\right)(-2) = \dfrac{(2\cdot-2)+5}{(-2)^2+2} = \dfrac{-4+5}{4+2} = \mathbf{\dfrac{1}{6}}$

f. $(f\cdot g)(x) = (x^2+2)(2x+5) = 2x^3+5x^2+4x+10$ therefore $(f\cdot g)(2) = 2\cdot 2^3+5\cdot 2^2+4\cdot 2+10 = 16+20+8+10 = \mathbf{54}$

3. State which of the following functions are odd or even.

a. Given $f(x) = x-1$ compute $f(-x)$ and $-f(x)$, i.e., $f(-x) = -x-1$ and $-f(x) = -(x-1) = -x+1$. Since $f(-x) \neq -f(x)$ and $f(-x) \neq f(x)$ the function $f(x) = x-1$ **is neither an odd nor an even function.**

b. Given $f(x) = x^6+1$ compute $f(-x)$ and $-f(x)$, i.e., $f(-x) = (-x)^6+1 = x^6+1$ and $-f(x) = -(x^6+1) = -x^6-1$. Since $f(-x) = f(x)$ the function $f(x) = x^6+1$ is an **even function.**

c. Given $f(x) = x^2(x-1) = x^3-x^2$ compute $f(-x)$ and $-f(x)$, i.e., $f(-x) = (-x)^2(-x-1) = x^2(-x-1) = -x^3-x^2$ and $-f(x) = -x^2(x-1) = -x^3+x^2$. Since $f(-x) \neq -f(x)$ and $f(-x) \neq f(x)$ the function $f(x) = x^2(x-1) = x^3-x^2$ is **neither an odd nor an even function.**

d. Given $f(x) = x^2+\dfrac{1}{x}$ compute $f(-x)$ and $-f(x)$, i.e., $f(-x) = (-x)^2+\dfrac{1}{-x} = x^2-\dfrac{1}{x}$ and $-f(x) = -\left(x^2+\dfrac{1}{x}\right)$ $= -x^2-\dfrac{1}{x}$. Since $f(-x) \neq -f(x)$ and $f(-x) \neq f(x)$ the function $f(x) = x^2+\dfrac{1}{x}$ **is neither an odd nor an even function.**

e. Given $f(x) = 1+x^3$ compute $f(-x)$ and $-f(x)$, i.e., $f(-x) = 1+(-x)^3 = 1-x^3$ and $-f(x) = -(1+x^3) = -1-x^3$. Since $f(-x) \neq -f(x)$ and $f(-x) \neq f(x)$ the function $f(x) = 1+x^3$ **is neither an odd nor an even function.**

f. Given $f(x) = |x|+3$ compute $f(-x)$ and $-f(x)$, i.e., $f(-x) = |-x|+3 = |x|+3$ and $-f(x) = -(|x|+3) = -|x|-3$. Since $f(-x) = f(x)$ the function $f(x) = |x|+3$ is an **even function.**

g. Given $f(x) = \dfrac{x^2}{1-2x}$ compute $f(-x)$ and $-f(x)$, i.e., $f(-x) = \dfrac{(-x)^2}{1-2\cdot(-x)} = \dfrac{x^2}{1+2x}$ and $-f(x) = -\left(\dfrac{x^2}{1-2x}\right) = -\dfrac{x^2}{1-2x}$

Since $f(-x) \neq -f(x)$ and $f(-x) \neq f(x)$ the function $f(x) = \dfrac{x^2}{1-2x}$ **is neither an odd nor an even function.**

h. Given $f(x) = x-\dfrac{1}{1+x}$ compute $f(-x)$ and $-f(x)$, i.e., $f(-x) = -x-\dfrac{1}{1-x}$ and $-f(x) = -\left(x-\dfrac{1}{1+x}\right) = -x+\dfrac{1}{1+x}$

Since $f(-x) \neq -f(x)$ and $f(-x) \neq f(x)$ the function $f(x) = x-\dfrac{1}{1+x}$ **is neither an odd nor an even function.**

i. Given $f(x) = x^4 + x^2 - 2$ compute $f(-x)$ and $-f(x)$, i.e., $f(-x) = (-x)^4 + (-x)^2 - 2 = x^4 + x^2 - 2$ and $-f(x)$ $= -\left(x^4 + x^2 - 2\right) = -x^4 - x^2 + 2$. Since $f(-x) = f(x)$ the function $f(x) = x^4 + x^2 - 2$ is an **even function.**

Section 2.3 Solutions – Composite Functions of Real Variables

1. Find the composition $f(g(x)) = (f \circ g)(x)$ for the following $f(x)$ and $g(x)$ functions.

a. Given $f(x) = 2x - 1$ and $g(x) = -x^2$, then $f(g(x)) = f\left(-x^2\right) = 2 \cdot \left(-x^2\right) - 1 = \mathbf{-2x^2 - 1}$

b. Given $f(x) = 2x + 5$; $g(x) = x + 10$, then $f(g(x)) = f(x+10) = 2(x+10) + 5 = 2x + 20 + 5 = \mathbf{2x + 25}$

c. Given $f(x) = \frac{1}{x+1}$ and $g(x) = x^3$, then $f(g(x)) = f\left(x^3\right) = \mathbf{\frac{1}{x^3 + 1}}$

d. Given $f(x) = x - 3$ and $g(x) = -x^2$, then $f(g(x)) = f\left(-x^2\right) = \mathbf{-x^2 - 3}$

e. Given $f(x) = x^2 + 1$ and $g(x) = \sqrt{x}$, then $f(g(x)) = f\left(\sqrt{x}\right) = \left(\sqrt{x}\right)^2 + 1 = x^{\frac{2}{2}} + 1 = \mathbf{x + 1}$

f. Given $f(x) = \sqrt{x} + 10x$ and $g(x) = x - 3$, then $f(g(x)) = f(x-3) = \sqrt{x-3} + 10(x-3) = \mathbf{\sqrt{x-3} + 10x - 30}$

2. Find the composition $g(f(x)) = (g \circ f)(x)$ for the following $g(x)$ and $f(x)$ functions.

a. Given $g(x) = x^2$ and $f(x) = -\frac{1}{x}$, then $g(f(x)) = g\left(-\frac{1}{x}\right) = \left(-\frac{1}{x}\right)^2 = \mathbf{\frac{1}{x^2}}$

b. Given $g(x) = x^2 - 1$ and $f(x) = 3x$, then $g(f(x)) = g(3x) = (3x)^2 - 1 = \mathbf{9x^2 - 1}$

c. Given $g(x) = x + 2$ and $f(x) = \sqrt{x+5}$, then $g(f(x)) = g\left(\sqrt{x+5}\right) = \mathbf{\sqrt{x+5} + 2}$

d. Given $g(x) = 2x + 1$ and $f(x) = \frac{1}{x^2}$, then $g(f(x)) = g\left(\frac{1}{x^2}\right) = 2 \cdot \frac{1}{x^2} + 1 = \mathbf{\frac{2}{x^2} + 1}$

e. Given $g(x) = \frac{1}{x+1}$ and $f(x) = -\sqrt{x}$, then $g(f(x)) = g\left(-\sqrt{x}\right) = \mathbf{\frac{1}{-\sqrt{x} + 1}}$

f. Given $g(x) = \frac{1}{2x-3} + x$ and $f(x) = 3x$, then $g(f(x)) = g(3x) = \frac{1}{(2 \cdot 3x) - 3} + 3x = \mathbf{\frac{1}{6x - 3} + 3x}$

3. Given $f(x) = -x^2 + 5$ and $g(x) = x^3 + 2x - 1$, find

a. $f(0) = -0^2 + 5 = \mathbf{5}$ and $g(1) = 1^3 + (2 \cdot 1) - 1 = 1 + 2 - 1 = \mathbf{2}$

b. $f(1) = -1^2 + 5 = -1 + 5 = \mathbf{4}$ and $g(2) = 2^3 + (2 \cdot 2) - 1 = 8 + 4 - 1 = \mathbf{11}$

c. $f(k) = \mathbf{-k^2 + 5}$ and $g(-k) = (-k)^3 + (2 \cdot -k) - 1 = \mathbf{-k^3 - 2k - 1}$

d. $f(3) = -3^2 + 5 = -9 + 5 = \mathbf{-4}$ and $g(-3) = (-3)^3 + (2 \cdot -3) - 1 = -27 - 6 - 1 = \mathbf{-34}$

e. $f(-n) = -(-n)^2 + 5 = \mathbf{-n^2 + 5}$ and $g(2n) = (2n)^3 + (2 \cdot 2n) - 1 = \mathbf{8n^3 + 4n - 1}$

f. $f(n+1) = -(n+1)^2 + 5 = -\left(n^2 + 1 + 2n\right) + 5 = -n^2 - 1 - 2n + 5 = \mathbf{-n^2 - 2n + 4}$ and $g(0) = 0^3 + (2 \cdot 0) - 1 = \mathbf{-1}$

4. Given $f(x) = x - 2$ and $g(x) = x^2 - 1$ let's find $f(g(x))$ and $g(f(x))$, i.e.,

$f(g(x)) = g(x) - 2 = \left(x^2 - 1\right) - 2 = x^2 - 3$

$g(f(x)) = [f(x)]^2 - 1 = (x-2)^2 - 1 = \left(x^2 + 4 - 4x\right) - 1 = x^2 - 4x + 3$

a. $f(g(-1)) = (-1)^2 - 3 = 1 - 3 = \mathbf{-2}$

b. $g(f(0)) = 0^2 - (4 \cdot 0) + 3 = \mathbf{3}$

c. $f(g(-3)) = (-3)^2 - 3 = 9 - 3 = \mathbf{6}$

d. $g(f(2)) = 2^2 - (4 \cdot 2) + 3 = 4 - 8 + 3 = \mathbf{-1}$

e. $f(g(n-1)) = (n-1)^2 - 3 = (n^2+1-2n)-3 = \mathbf{n^2-2n-2}$ f. $g(f(2n)) = (2n)^2-(4\cdot 2n)+3 = \mathbf{4n^2-8n+3}$

5. Given the functions $f(g(h(x)))=x^3-2x^2+1$, $g(f(h(x)))=x-1$, and $h(g(f(x)))=-x^2+2$, find

a. $f(g(h(2))) = 2^3-(2\cdot 2^2)+1 = 8-(2\cdot 4)+1 = 8-8+1 = \mathbf{1}$

b. $f(g(h(-2))) = (-2)^3-[2\cdot(-2)^2]+1 = -8-(2\cdot 4)+1 = -8-8+1 = \mathbf{-15}$

c. $h(g(f(a-1))) = -(a-1)^2+2 = -(a^2+1-2a)+2 = -a^2-1+2a+2 = \mathbf{-a^2+2a+1}$

d. $g(f(h(-1))) = -1-1 = \mathbf{-2}$ e. $g(f(h(2n+1))) = (2n+1)-1 = \mathbf{2n}$ f. $h(g(f(0))) = -0^2+2 = \mathbf{2}$

Section 2.4 Solutions – One-to-One and Inverse Functions of Real Variables

1. State which of the following functions are one-to-one.

A function is a one-to-one function because for each value of x there is only one corresponding value of y.

A function is not a one-to-one function because each x value does not correspond to only one y value.

a. The function $f(x)=2x+5$ is a **one-to-one function.** b. The function $f(x)=-5+\frac{2}{3}x$ is a **one-to-one function.**

c. The function $f(x)=\frac{1}{5}x-1$ is a **one-to-one function.** d. The function $f(x)=x^2-25$ is **not a one-to-one function.**

e. The function $f(x)=\sqrt{6x-5}$ is a **one-to-one function.** f. The function $f(x)=\frac{1}{4}(16-3x)$ is a **one-to-one function.**

g. The function $f(x)=x^3-2$ is a **one-to-one function.** h. The function $f(x)=2|x|$ is **not a one-to-one function.**

i. The function $f(x)=x^4$ is **not a one-to-one function.** j. The function $f(x)=1+e^{2x}$ is a **one-to-one function.**

k. The function $f(x)=x^8+1$ is **not a one-to-one function.** l. The function $f(x)=x^2+4$ is **not a one-to-one function.**

2. Given the following functions are one-to-one, use the first method to find their inverse.

Interchange the x variable with y and the y variable with x in the given equation and solve for y. Next, replace y with $f^{-1}(x)$ to obtain:

a. Given $y=x+3$, then $x=y+3$; $y=x-3$; $\mathbf{f^{-1}(x)=x-3}$

b. Given $y=5x$, then $x=5y$; $y=\frac{x}{5}$; $\mathbf{f^{-1}(x)=\frac{x}{5}}$

c. Given $y=\sqrt{5x-1}$, then $x=\sqrt{5y-1}$; $x=(5y-1)^{\frac{1}{2}}$; $x^2=5y-1$; $x^2+1=5y$; $y=\frac{x^2+1}{5}$; $\mathbf{f^{-1}(x)=\frac{x^2+1}{5}}$

d. Given $y=1-2x^3$, then $x=1-2y^3$; $x-1=-2y^3$; $\frac{x-1}{-2}=y^3$; $\frac{1-x}{2}=y^3$; $\left(\frac{1-x}{2}\right)^{\frac{1}{3}}=\left(y^3\right)^{\frac{1}{3}}$; $\left(\frac{1-x}{2}\right)^{\frac{1}{3}}=y$

; $y=\sqrt[3]{\frac{1-x}{2}}$; $\mathbf{f^{-1}(x)=\sqrt[3]{\frac{1-x}{2}}}$

e. Given $y=\sqrt{2x}+1$, then $x=\sqrt{2y}+1$; $x-1=\sqrt{2y}$; $x-1=(2y)^{\frac{1}{2}}$; $(x-1)^2=2y$; $y=\frac{(x-1)^2}{2}$; $\mathbf{f^{-1}(x)=\frac{(x-1)^2}{2}}$

f. Given $y=0.2x+10$, then $x=0.2y+10$; $x-10=0.2y$; $\frac{x-10}{0.2}=y$; $y=\frac{10x-100}{2}$; $y=5x-50$; $\mathbf{f^{-1}(x)=5x-50}$

g. Given $y=\sqrt[3]{x-2}+1$, $x\ge 2$, then $x=\sqrt[3]{y-2}+1$; $x-1=\sqrt[3]{y-2}$; $x-1=(y-2)^{\frac{1}{3}}$; $(x-1)^3=y-2$; $(x-1)^3+2=y$

; $y=(x-1)^3+2$; $\mathbf{f^{-1}(x)=(x-1)^3+2}$

h. Given $y=\frac{2x-3}{x}$, $x\ne 0$, then $x=\frac{2y-3}{y}$; $xy=2y-3$; $xy-2y=-3$; $y(x-2)=-3$; $y=\frac{-3}{x-2}$; $\mathbf{f^{-1}(x)=-\frac{3}{x-2}}$

i. Given, $y = 2-5x$, then $x = 2-5y$; $x-2 = -5y$; $\frac{x-2}{-5} = y$; $y = \frac{2-x}{5}$; $\boldsymbol{f^{-1}(x) = \frac{2-x}{5}}$

j. Given, $y = \frac{2x+5}{x+1}$, $x \neq -1$, then $x = \frac{2y+5}{y+1}$; $x(y+1) = 2y+5$; $xy+x = 2y+5$; $xy-2y = 5-x$; $y(x-2) = 5-x$; $y = \frac{5-x}{x-2}$; $\boldsymbol{f^{-1}(x) = \frac{5-x}{x-2}}$

k. Given, $y = \frac{2x-3}{x-5}$, $x \neq 5$, then $x = \frac{2y-3}{y-5}$; $x(y-5) = 2y-3$; $xy-5x = 2y-3$; $xy-2y = 5x-3$; $y(x-2) = 5x-3$; $y = \frac{5x-3}{x-2}$; $\boldsymbol{f^{-1}(x) = \frac{5x-3}{x-2}}$

l. Given, $y = 2x^3-9$, then $x = 2y^3-9$; $x+9 = 2y^3$; $y^3 = \frac{x+9}{2}$; $\left(y^3\right)^{\frac{1}{3}} = \left(\frac{x+9}{2}\right)^{\frac{1}{3}}$; $y = \left(\frac{x+9}{2}\right)^{\frac{1}{3}}$; $\boldsymbol{f^{-1}(x) = \sqrt[3]{\frac{x+9}{2}}}$

Section 2.5 Solutions – Complex Numbers and Functions of Complex Variables

1. Simplify the following imaginary numbers.

a. $i^{10} = \left(i^2\right)^5 = (-1)^5 = \mathbf{-1}$

b. $i^{13} = i^{12+1} = i^{12} \cdot i^1 = \left(i^2\right)^6 \cdot i = (-1)^6 \cdot i = 1 \cdot i = \boldsymbol{i}$

c. $i^{17} = i^{16+1} = i^{16} \cdot i^1 = \left(i^2\right)^8 \cdot i = (-1)^8 \cdot i = 1 \cdot i = \boldsymbol{i}$

d. $i^{21} = i^{20+1} = i^{20} \cdot i^1 = \left(i^2\right)^{10} \cdot i = (-1)^{10} \cdot i = 1 \cdot i = \boldsymbol{i}$

e. $i^{50} = \left(i^2\right)^{25} = (-1)^{25} = \mathbf{-1}$

f. $i^{100} = \left(i^2\right)^{50} = (-1)^{50} = \mathbf{1}$

g. $i^{47} = i^{46+1} = i^{46} \cdot i^1 = \left(i^2\right)^{23} \cdot i = (-1)^{23} \cdot i = -1 \cdot i = \boldsymbol{-i}$

h. $i^{29} = i^{28+1} = i^{28} \cdot i^1 = \left(i^2\right)^{14} \cdot i = (-1)^{14} \cdot i = 1 \cdot i = \boldsymbol{i}$

2. Write the following expressions in the standard form $a+bi$.

a. $\sqrt{-6}+\sqrt{-12} = \left(\sqrt{6} \cdot \sqrt{-1}\right)+\left(\sqrt{12} \cdot \sqrt{-1}\right) = \left(\sqrt{6} \cdot i\right)+\left(\sqrt{12} \cdot i\right) = \sqrt{6}\,i+\sqrt{12}\,i = 2.45i+3.46i = 5.91i = \mathbf{0+5.91}\boldsymbol{i}$

b. $\left(\sqrt{-2} \cdot \sqrt{-3}\right)^4 = \left[\left(\sqrt{2} \cdot \sqrt{-1} \times \sqrt{3} \cdot \sqrt{-1}\right)\right]^4 = \left(\sqrt{2} \cdot i \times \sqrt{3} \cdot i\right)^4 = \left(\sqrt{2 \cdot 3} \cdot i^2\right)^4 = \left(\sqrt{6} \cdot -1\right)^4 = \left(-\sqrt{6}\right)^4 = 6^{\frac{4}{2}} = 6^2 = 36$ $= \mathbf{36+0}\boldsymbol{i}$

c. $\sqrt{-3} \cdot \sqrt{-5} = \left(\sqrt{3} \cdot \sqrt{-1}\right) \times \left(\sqrt{5} \cdot \sqrt{-1}\right) = \left(\sqrt{3} \cdot i\right) \times \left(\sqrt{5} \cdot i\right) = \sqrt{3 \cdot 5} \cdot i^2 = \sqrt{15} \cdot -1 = -\sqrt{15} = -3.873 = \mathbf{-3.873+0}\boldsymbol{i}$

d. $\left(\sqrt{-2}-\sqrt{-25}\right)^3 = \left[\left(\sqrt{2} \cdot \sqrt{-1}\right)-\left(\sqrt{25} \cdot \sqrt{-1}\right)\right]^3 = \left[\left(1.414 \cdot \sqrt{-1}\right)-\left(5 \cdot \sqrt{-1}\right)\right]^3 = \left[(1.414 \cdot i)-(5 \cdot i)\right]^3 = (1.414i-5i)^3$ $= (-3.586i)^3 = (-3.586)^3 \cdot i^3 = -46.11 \cdot -i = 46.11i = \mathbf{0+46.11}\boldsymbol{i}$

e. $\sqrt{-1}\left(\sqrt{-2}+\sqrt{-3}\right) = \sqrt{-1}\left[\left(\sqrt{2} \cdot \sqrt{-1}\right)+\left(\sqrt{3} \cdot \sqrt{-1}\right)\right] = i \cdot \left[\left(\sqrt{2} \cdot i\right)+\left(\sqrt{3} \cdot i\right)\right] = i \cdot \left(\sqrt{2}\,i+\sqrt{3}\,i\right) = i \cdot (1.414i+1.732i)$ $= i \cdot 3.146i = 3.146i^2 = 3.146 \times -1 = -3.146 = \mathbf{-3.146+0}\boldsymbol{i}$

f. $5-\sqrt{(-1)^3-2} = 5-\sqrt{-1-2} = 5-\sqrt{-3} = 5-\left(\sqrt{3} \cdot \sqrt{-1}\right) = 5-\left(\sqrt{3} \cdot i\right) = 5-\sqrt{3}\,i = \mathbf{5-1.732}\boldsymbol{i}$

3. Given $f(x) = x^2+1$ and $g(x) = x^2-2x+1$, find

a. $f(1-i) = (1-i)^2+1 = \left(1+i^2-2i\right)+1 = 1-1-2i+1 = -2i+1 = \mathbf{1-2}\boldsymbol{i}$

b. $f(-i) = (-i)^2+1 = i^2+1 = -1+1 = 0 = \mathbf{0+0}\boldsymbol{i}$

c. $g\left(1-\sqrt{2}\,i\right) = \left(1-\sqrt{2}\,i\right)^2-2\left(1-\sqrt{2}\,i\right)+1 = \left[1+\left(\sqrt{2}\,i\right)^2-2\sqrt{2}\,i\right]-2+2\sqrt{2}\,i+1 = 1+2i^2-2\sqrt{2}\,i-2+2\sqrt{2}\,i+1 = 1-2-2+1$ $= 2-4 = \mathbf{-2+0}\boldsymbol{i}$

d. $g\left(1+i^3\right) = g(1-i) = (1-i)^2-2(1-i)+1 = 1+i^2-2i-2+2i+1 = 1-1-2+1 = -1 = \mathbf{-1+0}\boldsymbol{i}$

e. $f(1+i) = (1+i)^2+1 = (1^2+i^2+2i)+1 = (1-1+2i)+1 = 2i+1 = \mathbf{1+2i}$

f. $f(2+3i) = (2+3i)^2+1 = (4+9i^2+12i)+1 = (4-9+12i)+1 = (4-9+1)+12i = \mathbf{-4+12i}$

g. $g(-\sqrt{-1}) = (-\sqrt{-1})^2 - 2(-\sqrt{-1})+1 = (-i)^2 - 2(-i)+1 = i^2+2i+1 = -1+2i+1 = 2i = \mathbf{0+2i}$

h. $g(2+i) = (2+i)^2 - 2(2+i)+1 = (4+i^2+4i)+(-4-2i)+1 = (4-1+4i)+(-4-2i)+1 = (4-4-1+1)+(4i-2i) = \mathbf{0+2i}$

i. $f(3+\sqrt{-3}) = (3+\sqrt{-3})^2+1 = (3+\sqrt{3}i)^2+1 = 9+(\sqrt{3}i)^2+6\sqrt{3}i+1 = 9+3i^2+10.39i+1 = 9-3+10.39i+1$
$= (9-3+1)+10.39i = \mathbf{7+10.39i}$

4. Given $f(x)=x^2-1$ and $g(x)=x-5$ let's find $f(g(x))$ and $g(f(x))$, i.e.,

$f(g(x)) = [g(x)]^2-1 = (x-5)^2-1 = (x^2+25-10x)-1 = x^2-10x+24$

$g(f(x)) = f(x)-5 = (x^2-1)-5 = x^2-6$

a. $f(g(1-\sqrt{-1})) = f(g(1-i)) = (1-i)^2-10(1-i)+24 = (1+i^2-2i)-10+10i+24 = (1-1-2i)+10i+14 = -2i+10i+14$
$= 8i+14 = \mathbf{14+8i}$

b. $g(f(1-i)) = (1-i)^2-6 = (1+i^2-2i)-6 = (1-1-2i)-6 = -2i-6 = \mathbf{-6-2i}$

c. $g(f(-i^4)) = g(f(-1)) = (-1)^2-6 = 1-6 = -5 = \mathbf{-5+0i}$

d. $f(g(2+5i)) = (2+5i)^2-10(2+5i)+24 = 4+25i^2+20i-20-50i+24 = (4-25-20+24)+(20i-50i) = \mathbf{-17-30i}$

e. $f(g(i^6)) = f(g(-1)) = (-1)^2-10\cdot(-1)+24 = 1+10+24 = 35 = \mathbf{35+0i}$

f. $g(f(1+i^5)) = g(f(1+i)) = (1+i)^2-6 = 1+i^2+2i-6 = 1-1+2i-6 = 2i-6 = \mathbf{-6+2i}$

5. Given the quadratic formula $x=\dfrac{-b\pm\sqrt{b^2-4ac}}{2a}$, solve for x for the following values of a, b, and c.

a. Substituting $a=2$, $b=3$, and $c=5$ into $x_{1,2}=\dfrac{-b\pm\sqrt{b^2-4ac}}{2a}$ we obtain:

$$x_1 = \frac{-3+\sqrt{3^2-(4\cdot2\cdot5)}}{2\cdot2} = \frac{-3+\sqrt{9-40}}{4} = \frac{-3+\sqrt{-31}}{4} = \frac{-3+\sqrt{31}\cdot\sqrt{-1}}{4} = \frac{-3+\sqrt{31}\cdot i}{4} = -\frac{3}{4}+\frac{\sqrt{31}}{4}i$$
$$= -0.75+\frac{5.57}{4}i = \mathbf{-0.75+1.39i} \text{ and}$$

$$x_2 = \frac{-3-\sqrt{3^2-(4\cdot2\cdot5)}}{2\cdot2} = \frac{-3-\sqrt{9-40}}{4} = \frac{-3-\sqrt{-31}}{4} = \frac{-3-\sqrt{31}\cdot\sqrt{-1}}{4} = \frac{-3-\sqrt{31}\cdot i}{4} = -\frac{3}{4}-\frac{\sqrt{31}}{4}i$$
$$= -0.75-\frac{5.57}{4}i = \mathbf{-0.75-1.39i}$$

b. Substituting $a=3$, $b=4$, and $c=6$ into $x_{1,2}=\dfrac{-b\pm\sqrt{b^2-4ac}}{2a}$ we obtain:

$$x_1 = \frac{-4+\sqrt{4^2-(4\cdot3\cdot6)}}{2\cdot3} = \frac{-4+\sqrt{16-72}}{6} = \frac{-4+\sqrt{-56}}{6} = \frac{-4+\sqrt{56}\cdot\sqrt{-1}}{6} = \frac{-4+\sqrt{56}\cdot i}{6} = -\frac{4}{6}+\frac{\sqrt{56}}{6}i$$
$$= -0.67+\frac{7.48}{6}i = \mathbf{-0.67+1.25i} \text{ and}$$

$$x_2 = \frac{-4-\sqrt{4^2-(4\cdot3\cdot6)}}{2\cdot3} = \frac{-4-\sqrt{16-72}}{6} = \frac{-4-\sqrt{-56}}{6} = \frac{-4-\sqrt{56}\cdot\sqrt{-1}}{6} = \frac{-4-\sqrt{56}\cdot i}{6} = -\frac{4}{6}-\frac{\sqrt{56}}{6}i$$
$$= -0.67-\frac{7.48}{6}i = \mathbf{-0.67-1.25i}$$

c. Substituting $a=4$, $b=1$, and $c=10$ into $x_{1,2}=\frac{-b\pm\sqrt{b^2-4ac}}{2a}$ we obtain:

$$x_1 = \frac{-1+\sqrt{1^2-(4\cdot4\cdot10)}}{2\cdot4} = \frac{-1+\sqrt{1-160}}{8} = \frac{-1+\sqrt{-159}}{8} = \frac{-1+\sqrt{159}\cdot\sqrt{-1}}{8} = \frac{-1+\sqrt{159}\cdot i}{8} = -\frac{1}{8}+\frac{\sqrt{159}}{8}i$$

$$= -0.125+\frac{12.61}{8}i = \mathbf{-0.125+1.58}\boldsymbol{i} \text{ and}$$

$$x_2 = \frac{-1-\sqrt{1^2-(4\cdot4\cdot10)}}{2\cdot4} = \frac{-1-\sqrt{1-160}}{8} = \frac{-1-\sqrt{-159}}{8} = \frac{-1-\sqrt{159}\cdot\sqrt{-1}}{8} = \frac{-1-\sqrt{159}\cdot i}{8} = -\frac{1}{8}-\frac{\sqrt{159}}{8}i$$

$$= -0.125-\frac{12.61}{8}i = \mathbf{-0.125-1.58}\boldsymbol{i}$$

Section 2.6 Solutions – Math Operations Involving Complex Numbers

Section 2.6 Case I Practice Problems – Add or subtract the following complex numbers:

a. $(4+2i)+(8-5i) = (4+8)+(2i-5i) = 12+(2-5)i = \mathbf{12-3}\boldsymbol{i}$

b. $(7-3i)-(-5+4i) = (7-3i)+(5-4i) = (7+5)+(-3i-4i) = 12+(-3-4)i = \mathbf{12-7}\boldsymbol{i}$

c. $(4+7i)+[(2-5i)-(6-i)] = (4+7i)+[(2-5i)+(-6+i)] = (4+7i)+[(2-6)+(-5i+i)] = (4+7i)+(-4-4i)$

$= (4-4)+(7i-4i) = 0+(7-4)i = \mathbf{0+3}\boldsymbol{i}$

d. $(2\sqrt{5}+\sqrt{-3})-(4-\sqrt{-25}) = (2\sqrt{5}+\sqrt{-3})+(-4+\sqrt{-25}) = (2\sqrt{5}+\sqrt{3}\,i)+(-4+5i) = (2\sqrt{5}-4)+(\sqrt{3}\,i+5i)$

$= (4.472-4)+(1.732i+5i) = 0.472+(1.732+5)i = \mathbf{0.472+6.732}\boldsymbol{i}$

e. $[(2+5i)+\sqrt{5}\,i^3]-(1+3\sqrt{5}\,i) = [(2+5i)-\sqrt{5}\,i]-(1+3\sqrt{5}\,i) = (2+5i-2.236i)-(1+3\sqrt{5}\,i) = (2+2.764i)+(-1-3\sqrt{5}\,i)$

$= (2-1)+(2.764i-3\sqrt{5}\,i) = 1+(2.764i-6.708i) = 1+(2.764-6.708)i = \mathbf{1-3.944}\boldsymbol{i}$

f. $(3-\sqrt{6}\,i)-(1+\sqrt{2}\,i) = (3-\sqrt{6}\,i)+(-1-\sqrt{2}\,i) = (3-2.449i)+(-1-1.414i) = (3-1)+(-2.449i-1.414i) = \mathbf{2-3.863}\boldsymbol{i}$

Section 2.6 Case II Practice Problems – Multiply the following complex numbers by one another:

a. $(5+2i)(3-6i) = (5\times3)+(5\times-6i)+(2i\times3)+(2i\times-6i) = 15-30i+6i-12i^2 = 15-30i+6i+(-12\times-1) = 15-30i+6i+12$

$= (15+12)+(-30i+6i) = \mathbf{27-24}\boldsymbol{i}$

b. $(-6-2i)(-7+i) = (-6-2i)(-7+i) = (-6\times-7)+(-6\times i)+(-2i\times-7)+(-2i\times i) = 42-6i+14i-2i^2 = 42-6i+14i+(-2\times-1)$

$= 42-6i+14i+2 = (42+2)+(-6i+14i) = \mathbf{44+8}\boldsymbol{i}$

c. $[3i^6(5-i)](3+i) = [(3\times-1)(5-i)](3+i) = [-3\times(5-i)](3+i) = (-15+3i)(3+i) = (-15\times3)+(-15\times i)+(3i\times3)+(3i\times i)$

$= -45-15i+9i+3i^2 = -45-15i+9i+(3\times-1) = -45-15i+9i-3 = (-45-3)+(-15i+9i) = \mathbf{-48-6}\boldsymbol{i}$

d. $(3-\sqrt{9}\,i)(5+\sqrt{3}\,i) = (3-3i)(5+1.732i) = (3\times5)+(3\times1.732i)+(-3i\times5)+(-3i\times1.732i) = 15+5.196i-15i-5.196i^2$

$= 15+5.196i-15i+(-5.196\times-1) = 15+5.196i-15i+5.196 = (15+5.196)+(5.196i-15i) = \mathbf{20.196-9.804}\boldsymbol{i}$

e. $(\sqrt{5}-i)(-3\sqrt{5}+i) = (2.236-i)(-6.708+i) = (2.236\times-6.708)+(2.236\times i)+(-i\times-6.708)+(-i\times i) = -15+2.236i+6.708i-i^2$

$= -15+2.236i+6.708i+1 = (-15+1)+(2.236i+6.708i) = \mathbf{-14+8.944}\boldsymbol{i}$

f. $i^3[(2+4i)(4-2i)] = -i[(2+4i)(4-2i)] = -i[(2\times4)+(2\times-2i)+(4i\times4)+(4i\times-2i)] = -i(8-4i+16i-8i^2) = -i(8-4i+16i+8)$

$= -8i+4i^2-16i^2-8i = -8i+(4\times-1)+(-16\times-1)-8i = -8i-4+16-8i = (-4+16)+(-8i-8i) = \mathbf{12-16}\boldsymbol{i}$

Section 2.6 Case III Practice Problems – Divide the following complex numbers by one another:

a. $\frac{1-4i}{5+3i} = \frac{1-4i}{5+3i}\times\frac{5-3i}{5-3i} = \frac{(1-4i)\times(5-3i)}{(5+3i)\times(5-3i)} = \frac{5-3i-20i+12i^2}{25-15i+15i-9i^2} = \frac{5-3i-20i-12}{25+9} = \frac{(5-12)+(-3i-20i)}{25+9} = \frac{-7-23i}{34}$

$= -\frac{7}{34}-\frac{23}{34}i = \mathbf{-0.21-0.68}\boldsymbol{i}$

b. $\frac{i^3}{1-8i} = \frac{-i}{1-8i} = \frac{-i}{1-8i}\times\frac{1+8i}{1+8i} = \frac{-i\times(1+8i)}{(1-8i)\times(1+8i)} = \frac{-i-8i^2}{1+8i-8i-64i^2} = \frac{-i+8}{1+64} = \frac{8-i}{65} = \frac{8}{65}-\frac{1}{65}i = \mathbf{0.123-0.015}\boldsymbol{i}$

c. $\frac{2-i}{\sqrt{3}+4i} = \frac{2-i}{\sqrt{3}+4i}\times\frac{\sqrt{3}-4i}{\sqrt{3}-4i} = \frac{(2-i)\times(\sqrt{3}-4i)}{(\sqrt{3}+4i)\times(\sqrt{3}-4i)} = \frac{2\sqrt{3}-8i-\sqrt{3}\,i+4i^2}{\sqrt{3}\cdot\sqrt{3}-4\sqrt{3}\,i+4\sqrt{3}\,i-16i^2} = \frac{2\sqrt{3}-8i-\sqrt{3}\,i-4}{\sqrt{3\cdot 3}+16}$

$= \frac{(2\sqrt{3}-4)+(-8i-\sqrt{3}\,i)}{\sqrt{9}+16} = \frac{(3.464-4)+(-8-1.732)i}{3+16} = \frac{-0.536-9.732i}{19} = -\frac{0.536}{19}-\frac{9.732}{19}i = \mathbf{-0.028-0.512}\boldsymbol{i}$

d. $\frac{-2+\sqrt{2}\,i}{\sqrt{5}-\sqrt{36}\,i} = \frac{-2+\sqrt{2}\,i}{\sqrt{5}-6i} = \frac{-2+\sqrt{2}\,i}{\sqrt{5}-6i}\times\frac{\sqrt{5}+6i}{\sqrt{5}+6i} = \frac{(-2+\sqrt{2}\,i)\times(\sqrt{5}+6i)}{(\sqrt{5}-6i)\times(\sqrt{5}+6i)} = \frac{-2\sqrt{5}-12i+\sqrt{10}\,i+6\sqrt{2}\,i^2}{\sqrt{5}\cdot\sqrt{5}+6\sqrt{5}\,i-6\sqrt{5}\,i-36i^2}$

$= \frac{-2\sqrt{5}-12i+\sqrt{10}\,i-6\sqrt{2}}{\sqrt{25}+36} = \frac{(-2\sqrt{5}-6\sqrt{2})+(-12i+\sqrt{10}\,i)}{5+36} = \frac{(-4.472-8.485)+(-12+3.162)i}{41} = \frac{-12.957-8.838i}{41}$

$= -\frac{12.957}{41}-\frac{8.838}{41}i = \mathbf{-0.316-0.216}\boldsymbol{i}$

e. $\frac{3-\sqrt{3}\,i}{1-8i^5} = \frac{3-\sqrt{3}\,i}{1-8i} = \frac{3-\sqrt{3}\,i}{1-8i}\times\frac{1+8i}{1+8i} = \frac{(3-\sqrt{3}\,i)\times(1+8i)}{(1-8i)\times(1+8i)} = \frac{3+24i-\sqrt{3}\,i-8\sqrt{3}\,i^2}{1+8i-8i-64i^2} = \frac{3+24i-\sqrt{3}\,i+8\sqrt{3}}{1+64}$

$= \frac{(3+8\sqrt{3})+(24i-\sqrt{3}\,i)}{65} = \frac{(3+13.856)+(24-1.732)i}{65} = \frac{16.856+22.27i}{65} = \frac{16.856}{65}+\frac{22.27}{65}i = \mathbf{0.259+0.343}\boldsymbol{i}$

f. $\frac{3+2i}{1-i^7} = \frac{3+2i}{1-(-i)} = \frac{3+2i}{1+i} = \frac{3+2i}{1+i}\times\frac{1-i}{1-i} = \frac{(3+2i)\times(1-i)}{(1+i)\times(1-i)} = \frac{3-3i+2i-2i^2}{1-i+i-i^2} = \frac{3-3i+2i+2}{1+1} = \frac{(3+2)+(-3i+2i)}{2}$

$= \frac{5-i}{2} = \frac{5}{2}-\frac{1}{2}i = \mathbf{2.5-0.5}\boldsymbol{i}$

Section 2.6 Case IV Practice Problems – Simplify the following expressions involving complex numbers:

a. $i^3(1-2i)+i^4(2-5i) = -i(1-2i)+(2-5i) = -i+2i^2+(2-5i) = -i-2+2-5i = (-2+2)+(-i-5i) = \mathbf{0-6}\boldsymbol{i}$

b. $\frac{4+5i}{1-i}\div\frac{1}{1+3i} = \frac{4+5i}{1-i}\times\frac{1+3i}{1} = \frac{(4+5i)\times(1+3i)}{(1-i)\times 1} = \frac{4+12i+5i+15i^2}{1-i} = \frac{4+12i+5i-15}{1-i} = \frac{(4-15)+(12i+5i)}{1-i}$

$= \frac{-11+17i}{1-i} = \frac{-11+17i}{1-i}\times\frac{1+i}{1+i} = \frac{(-11+17i)\times(1+i)}{(1-i)\times(1+i)} = \frac{-11+11i+17i+17i^2}{1+i-i-i^2} = \frac{-11+11i+17i-17}{1+1} = \frac{(-11-17)+(11i+17i)}{2}$

$= \frac{-28+28i}{2} = -\frac{28}{2}+\frac{28}{2}i = \mathbf{-14+14}\boldsymbol{i}$

c. $\frac{1-i}{2+3i}\times\frac{1}{1+i} = \frac{(1-i)\times 1}{(2+3i)\times(1+i)} = \frac{1-i}{2+2i+3i+3i^2} = \frac{1-i}{2+2i+3i-3} = \frac{1-i}{(2-3)+(2i+3i)} = \frac{1-i}{-1+5i} = \frac{1-i}{-1+5i}\times\frac{-1-5i}{-1-5i}$

$= \frac{(1-i)\times(-1-5i)}{(-1+5i)\times(-1-5i)} = \frac{-1-5i+i+5i^2}{1+5i-5i-25i^2} = \frac{-1-5i+i-5}{1+25} = \frac{(-1-5)+(-5i+i)}{26} = \frac{-6-4i}{26} = -\frac{6}{26}-\frac{4i}{26} = \mathbf{-0.231-0.154}\boldsymbol{i}$

d. $i^8\div\frac{2-4i}{1+2i} = 1\div\frac{2-4i}{1+2i} = \frac{1}{1}\times\frac{1+2i}{2-4i} = \frac{1+2i}{2-4i} = \frac{1+2i}{2-4i}\times\frac{2+4i}{2+4i} = \frac{(1+2i)\times(2+4i)}{(2-4i)\times(2+4i)} = \frac{2+4i+4i+8i^2}{4+8i-8i-16i^2} = \frac{2+4i+4i-8}{4+16}$

$= \frac{(2-8)+(4+4)i}{20} = \frac{-6+8i}{20} = -\frac{6}{20}+\frac{8}{20}i = \mathbf{-0.3+0.4i}$

e. $\frac{(2+5i)(1-i)}{(1+4i)(2+3i)} = \frac{2-2i+5i-5i^2}{2+3i+8i+12i^2} = \frac{2-2i+5i+5}{2+3i+8i-12} = \frac{(2+5)+(-2i+5i)}{(2-12)+(3i+8i)} = \frac{7+3i}{-10+11i} = \frac{7+3i}{-10+11i}\times\frac{-10-11i}{-10-11i}$

$= \frac{(7+3i)\times(-10-11i)}{(-10+11i)\times(-10-11i)} = \frac{-70-77i-30i-33i^2}{100+110i-110i-121i^2} = \frac{-70-77i-30i+33}{100+121} = \frac{(-70+33)+(-77i-30i)}{221} = \frac{-37-107i}{221}$

$= -\frac{37}{221}-\frac{107}{221}i = \mathbf{-0.167-0.484i}$

f. $(5+8i)-\frac{1}{1-i} = \frac{5+8i}{1}-\frac{1}{1-i} = \frac{[(5+8i)(1-i)]-(1\times1)}{1\times(1-i)} = \frac{5-5i+8i-8i^2-1}{1-i} = \frac{5-5i+8i+8-1}{1-i} = \frac{(5+8-1)+(-5i+8i)}{1-i}$

$= \frac{12+3i}{1-i} = \frac{12+3i}{1-i}\times\frac{1+i}{1+i} = \frac{(12+3i)\times(1+i)}{(1-i)\times(1+i)} = \frac{12+12i+3i+3i^2}{1+i-i-i^2} = \frac{12+12i+3i-3}{1+1} = \frac{(12-3)+(12i+3i)}{2} = \frac{9+15i}{2}$

$= \frac{9}{2}+\frac{15}{2}i = \mathbf{4.5+7.5i}$

g. $(4+2i)\div\frac{2}{1-3i} = \frac{4+2i}{1}\div\frac{2}{1-3i} = \frac{4+2i}{1}\times\frac{1-3i}{2} = \frac{(4+2i)\times(1-3i)}{1\times2} = \frac{4-12i+2i-6i^2}{2} = \frac{4-12i+2i+6}{2} = \frac{10-10i}{2}$

$= \frac{10}{2}-\frac{10}{2}i = \mathbf{5-5i}$

h. $(2-5i)+\frac{1-3i}{1+i} = \frac{2-5i}{1}+\frac{1-3i}{1+i} = \frac{[(2-5i)(1+i)]+[1\times(1-3i)]}{1\times(1+i)} = \frac{2+2i-5i-5i^2+1-3i}{1+i} = \frac{(2+5+1)+(2i-5i-3i)}{1+i}$

$= \frac{8-6i}{1+i} = \frac{8-6i}{1+i}\times\frac{1-i}{1-i} = \frac{(8-6i)\times(1-i)}{(1+i)\times(1-i)} = \frac{8-8i-6i+6i^2}{1-i+i-i^2} = \frac{8-8i-6i-6}{1+1} = \frac{(8-6)+(-8i-6i)}{2} = \frac{2-14i}{2} = \mathbf{1-7i}$

i. $\frac{2+3i}{i}\div\frac{2-4i}{1+i} = \frac{2+3i}{i}\times\frac{1+i}{2-4i} = \frac{(2+3i)\times(1+i)}{i\times(2-4i)} = \frac{2+2i+3i+3i^2}{2i-4i^2} = \frac{2+2i+3i-3}{2i+4} = \frac{(2-3)+(2i+3i)}{4+2i} = \frac{-1+5i}{4+2i}$

$= \frac{-1+5i}{4+2i}\times\frac{4-2i}{4-2i} = \frac{(-1+5i)\times(4-2i)}{(4+2i)\times(4-2i)} = \frac{-4+2i+20i-10i^2}{16-8i+8i-4i^2} = \frac{-4+2i+20i+10}{16+4} = \frac{(-4+10)+(2i+20i)}{20} = \frac{6+22i}{20}$

$= \frac{6}{20}+\frac{22}{20}i = \mathbf{0.3+1.1i}$

Chapter 3 Solutions:

Section 3.1 Solutions - Introduction to Matrices

1. State the order and find the transpose of each matrix.

a. $\begin{bmatrix} 1 & 0 \\ 2 & 3 \end{bmatrix}$ is a 2×2 matrix. The transpose of the $\begin{bmatrix} 1 & 0 \\ 2 & 3 \end{bmatrix}$ matrix is equal to $\begin{bmatrix} 1 & 0 \\ 2 & 3 \end{bmatrix}^t = \begin{bmatrix} \mathbf{1} & \mathbf{2} \\ \mathbf{0} & \mathbf{3} \end{bmatrix}$

b. $\begin{bmatrix} 1 & 4 \\ 0 & 2 \end{bmatrix}$ is a 2×2 matrix. The transpose of the $\begin{bmatrix} 1 & 4 \\ 0 & 2 \end{bmatrix}$ matrix is equal to $\begin{bmatrix} 1 & 4 \\ 0 & 2 \end{bmatrix}^t = \begin{bmatrix} \mathbf{1} & \mathbf{0} \\ \mathbf{4} & \mathbf{2} \end{bmatrix}$

c. $\begin{bmatrix} 1 & 2 & -3 \\ -1 & 0 & 1 \end{bmatrix}$ is a 2×3 matrix. The transpose of the $\begin{bmatrix} 1 & 2 & -3 \\ -1 & 0 & 1 \end{bmatrix}$ matrix is equal to $\begin{bmatrix} 1 & 2 & -3 \\ -1 & 0 & 1 \end{bmatrix}^t = \begin{bmatrix} \mathbf{1} & \mathbf{-1} \\ \mathbf{2} & \mathbf{0} \\ \mathbf{-3} & \mathbf{1} \end{bmatrix}$

d. $\begin{bmatrix} 1 & 0 \\ 2 & 1 \\ -1 & -3 \end{bmatrix}$ is a 3×2 matrix. The transpose of the $\begin{bmatrix} 1 & 0 \\ 2 & 1 \\ -1 & -3 \end{bmatrix}$ matrix is equal to $\begin{bmatrix} 1 & 0 \\ 2 & 1 \\ -1 & -3 \end{bmatrix}^t = \begin{bmatrix} \mathbf{1} & \mathbf{2} & \mathbf{-1} \\ \mathbf{0} & \mathbf{1} & \mathbf{-3} \end{bmatrix}$

e. $\begin{bmatrix} 2 & 3 & 5 \\ 1 & 4 & -1 \\ 3 & 0 & 2 \end{bmatrix}$ is a 3×3 matrix. The transpose of the $\begin{bmatrix} 2 & 3 & 5 \\ 1 & 4 & -1 \\ 3 & 0 & 2 \end{bmatrix}$ matrix is equal to $\begin{bmatrix} 2 & 3 & 5 \\ 1 & 4 & -1 \\ 3 & 0 & 2 \end{bmatrix}^t = \begin{bmatrix} \mathbf{2} & \mathbf{1} & \mathbf{3} \\ \mathbf{3} & \mathbf{4} & \mathbf{0} \\ \mathbf{5} & \mathbf{-1} & \mathbf{2} \end{bmatrix}$

f. $\begin{bmatrix} 1 \\ 2 \\ 3 \end{bmatrix}$ is a 3×1 matrix. The transpose of the $\begin{bmatrix} 1 \\ 2 \\ 3 \end{bmatrix}$ matrix is equal to $\begin{bmatrix} 1 \\ 2 \\ 3 \end{bmatrix}^t = \begin{bmatrix} \mathbf{1} & \mathbf{2} & \mathbf{3} \end{bmatrix}$

g. $\begin{bmatrix} 1 & -1 & 2 \end{bmatrix}$ is a 1×3 matrix. The transpose of the $\begin{bmatrix} 1 & -1 & 2 \end{bmatrix}$ matrix is equal to $\begin{bmatrix} 1 & -1 & 2 \end{bmatrix}^t = \begin{bmatrix} \mathbf{1} \\ \mathbf{-1} \\ \mathbf{2} \end{bmatrix}$

h. $\begin{bmatrix} 1 & 0 & -1 & 2 \\ 2 & 3 & 1 & 0 \\ -1 & 0 & -2 & 3 \end{bmatrix}$ is a 3×4 matrix. The transpose of the given matrix is equal to $\begin{bmatrix} 1 & 0 & -1 & 2 \\ 2 & 3 & 1 & 0 \\ -1 & 0 & -2 & 3 \end{bmatrix}^t = \begin{bmatrix} \mathbf{1} & \mathbf{2} & \mathbf{-1} \\ \mathbf{0} & \mathbf{3} & \mathbf{0} \\ \mathbf{-1} & \mathbf{1} & \mathbf{-2} \\ \mathbf{2} & \mathbf{0} & \mathbf{3} \end{bmatrix}$

2. Given $A = \begin{bmatrix} 1 & 2 & -3 \\ 0 & 1 & 1 \\ -2 & 3 & 0 \end{bmatrix}$, $\boldsymbol{a_{12} = 2}$, $\boldsymbol{a_{23} = 1}$, $\boldsymbol{a_{33} = 0}$, $\boldsymbol{a_{22} = 1}$, $\boldsymbol{a_{31} = -2}$, $\boldsymbol{a_{21} = 0}$, and $\boldsymbol{a_{11} = 1}$.

3. Given $B = \begin{bmatrix} 1 & -1 & 2 & 3 \\ 0 & 1 & 1 & 3 \\ 3 & -1 & -2 & 5 \end{bmatrix}$, $\boldsymbol{b_{12} = -1}$, $\boldsymbol{b_{21} = 0}$, $\boldsymbol{b_{32} = -1}$, $\boldsymbol{b_{33} = -2}$, $\boldsymbol{b_{23} = 1}$, $\boldsymbol{b_{24} = 3}$, $\boldsymbol{b_{34} = 5}$, and $\boldsymbol{b_{14} = 3}$.

4. Given $A_{2\times3}$, $B_{3\times1}$, $A_{1\times2}$, $A_{3\times3}$, $B_{2\times4}$, $B_{1\times4}$, $B_{4\times1}$, and $A_{3\times4}$ write a matrix that corresponds to the order given.

$A_{2\times3} = \begin{bmatrix} \mathbf{1} & \mathbf{0} & \mathbf{-1} \\ \mathbf{-1} & \mathbf{3} & \mathbf{1} \end{bmatrix}$, $B_{3\times1} = \begin{bmatrix} \mathbf{1} \\ \mathbf{0} \\ \mathbf{2} \end{bmatrix}$, $A_{1\times2} = \begin{bmatrix} \mathbf{1} & \mathbf{3} \end{bmatrix}$, $A_{3\times3} = \begin{bmatrix} \mathbf{1} & \mathbf{0} & \mathbf{-1} \\ \mathbf{0} & \mathbf{0} & \mathbf{1} \\ \mathbf{-1} & \mathbf{1} & \mathbf{0} \end{bmatrix}$, $B_{2\times4} = \begin{bmatrix} \mathbf{1} & \mathbf{2} & \mathbf{3} & \mathbf{0} \\ \mathbf{-1} & \mathbf{0} & \mathbf{2} & \mathbf{1} \end{bmatrix}$,

$B_{1\times4} = \begin{bmatrix} \mathbf{-1} & \mathbf{2} & \mathbf{0} & \mathbf{-3} \end{bmatrix}$, $B_{4\times1} = \begin{bmatrix} \mathbf{1} \\ \mathbf{2} \\ \mathbf{0} \\ \mathbf{-1} \end{bmatrix}$, $A_{3\times4} = \begin{bmatrix} \mathbf{1} & \mathbf{-1} & \mathbf{2} & \mathbf{3} \\ \mathbf{0} & \mathbf{2} & \mathbf{-2} & \mathbf{1} \\ \mathbf{0} & \mathbf{1} & \mathbf{0} & \mathbf{2} \end{bmatrix}$

5. State if the given paired matrices are equal to each other.

a. $\begin{bmatrix} 1 & 0 \\ -1 & 3 \end{bmatrix} \neq \begin{bmatrix} 2 & 0 \\ -1 & 3 \end{bmatrix}$ The two matrices **are not equal** to each other because not all the entry elements are the same.

b. $\begin{bmatrix} -1 & 2 \\ 3 & 6 \end{bmatrix} \neq \begin{bmatrix} -1 & \frac{1}{2} \\ 6 & 12 \end{bmatrix}$ The two matrices **are not equal** to each other because not all the entry elements are the same.

c. $\begin{bmatrix} -1 & 0 & 2 \\ 1 & 2 & 3 \end{bmatrix} = \begin{bmatrix} -1 & 0 & 2 \\ 1 & 2 & 3 \end{bmatrix}$ The two matrices **are equal** to each other because all the entry elements are the same.

d. $\begin{bmatrix} \frac{3}{4} & -4 \\ -0.8 & 1 \end{bmatrix} \neq \begin{bmatrix} \frac{3}{4} & -4 \\ 0.8 & 1 \end{bmatrix}$ The two matrices **are not equal** to each other because not all the entry elements are the same.

e. $\begin{bmatrix} 1 & 0 \\ -2 & 3 \end{bmatrix} = \begin{bmatrix} 1 & 0 \\ -2 & 3 \end{bmatrix}$ The two matrices **are equal** to each other because all the entry elements are the same.

f. $\begin{bmatrix} 1 & 4 \\ 3 & 5 \\ -1 & 2 \end{bmatrix} \neq \begin{bmatrix} 1 & 4 \\ 3 & 5 \\ -2 & 2 \end{bmatrix}$ The two matrices **are not equal** to each other because not all the entry elements are the same.

Section 3.2 Case I Solutions - Matrix Addition and Subtraction

1. Add or subtract the following matrices.

a. $\begin{bmatrix} 1 & 3 & 5 \\ 3 & -1 & 2 \end{bmatrix} + \begin{bmatrix} -1 & -5 & 2 \\ 0 & 6 & -3 \end{bmatrix} = \begin{bmatrix} 1-1 & 3-5 & 5+2 \\ 3+0 & -1+6 & 2-3 \end{bmatrix} = \begin{bmatrix} \mathbf{0} & \mathbf{-2} & \mathbf{7} \\ \mathbf{3} & \mathbf{5} & \mathbf{-1} \end{bmatrix}$

b. $\begin{bmatrix} 1 & -8 \\ -3 & 5 \end{bmatrix} - \begin{bmatrix} 1 & -3 \\ -2 & 4 \end{bmatrix} = \begin{bmatrix} 1 & -8 \\ -3 & 5 \end{bmatrix} + \begin{bmatrix} -1 & 3 \\ 2 & -4 \end{bmatrix} = \begin{bmatrix} 1-1 & -8+3 \\ -3+2 & 5-4 \end{bmatrix} = \begin{bmatrix} \mathbf{0} & \mathbf{-5} \\ \mathbf{-1} & \mathbf{1} \end{bmatrix}$

c. $\begin{bmatrix} 10 & 3 & -1 \\ 1 & 5 & 6 \\ 2 & 3 & 8 \end{bmatrix} + \begin{bmatrix} 1 & 7 & 9 \\ 2 & -3 & 0 \\ 1 & 5 & -3 \end{bmatrix} = \begin{bmatrix} 10+1 & 3+7 & -1+9 \\ 1+2 & 5-3 & 6+0 \\ 2+1 & 3+5 & 8-3 \end{bmatrix} = \begin{bmatrix} \mathbf{11} & \mathbf{10} & \mathbf{8} \\ \mathbf{3} & \mathbf{2} & \mathbf{6} \\ \mathbf{3} & \mathbf{8} & \mathbf{5} \end{bmatrix}$

d. $\begin{bmatrix} 2 & 3 \\ 1 & 5 \\ -3 & 1 \end{bmatrix} - \begin{bmatrix} 1 & 8 \\ 3 & 0 \\ 1 & -2 \end{bmatrix} = \begin{bmatrix} 2 & 3 \\ 1 & 5 \\ -3 & 1 \end{bmatrix} + \begin{bmatrix} -1 & -8 \\ -3 & 0 \\ -1 & 2 \end{bmatrix} = \begin{bmatrix} 2-1 & 3-8 \\ 1-3 & 5+0 \\ -3-1 & 1+2 \end{bmatrix} = \begin{bmatrix} \mathbf{1} & \mathbf{-5} \\ \mathbf{-2} & \mathbf{5} \\ \mathbf{-4} & \mathbf{3} \end{bmatrix}$

e. $\begin{bmatrix} 1 & 3 & 6 \end{bmatrix} - \begin{bmatrix} 0 & 5 & -3 \end{bmatrix} = \begin{bmatrix} 1 & 3 & 6 \end{bmatrix} + \begin{bmatrix} 0 & -5 & 3 \end{bmatrix} = \begin{bmatrix} 1+0 & 3-5 & 6+3 \end{bmatrix} = \begin{bmatrix} \mathbf{1} & \mathbf{-2} & \mathbf{9} \end{bmatrix}$

f. $\begin{bmatrix} 1 & -5 \\ -3 & 8 \end{bmatrix} + \begin{bmatrix} 0 & 0 \\ 0 & 0 \end{bmatrix} = \begin{bmatrix} 1+0 & -5+0 \\ -3+0 & 8+0 \end{bmatrix} = \begin{bmatrix} \mathbf{1} & \mathbf{-5} \\ \mathbf{-3} & \mathbf{8} \end{bmatrix}$

g. $\begin{bmatrix} 6 & 3 & -1 \\ 1 & 5 & 0 \\ 0 & 0 & 1 \end{bmatrix} + \begin{bmatrix} 1 & 0 & 0 \\ 0 & 1 & 0 \\ 0 & 0 & 1 \end{bmatrix} = \begin{bmatrix} 6+1 & 3+0 & -1+0 \\ 1+0 & 5+1 & 0+0 \\ 0+0 & 0+0 & 1+1 \end{bmatrix} = \begin{bmatrix} \mathbf{7} & \mathbf{3} & \mathbf{-1} \\ \mathbf{1} & \mathbf{6} & \mathbf{0} \\ \mathbf{0} & \mathbf{0} & \mathbf{2} \end{bmatrix}$

h. $\begin{bmatrix} 1 & 5 \\ 2 & -3 \\ -1 & 0 \end{bmatrix} + \begin{bmatrix} -1 & -5 \\ -2 & 3 \\ 1 & 0 \end{bmatrix} = \begin{bmatrix} 1-1 & 5-5 \\ 2-2 & -3+3 \\ -1+1 & 0+0 \end{bmatrix} = \begin{bmatrix} \mathbf{0} & \mathbf{0} \\ \mathbf{0} & \mathbf{0} \\ \mathbf{0} & \mathbf{0} \end{bmatrix}$

2. Given $A = \begin{bmatrix} 1 & 3 & 5 \\ 2 & -1 & 0 \\ -1 & 2 & 4 \end{bmatrix}$ and $B = \begin{bmatrix} -1 & 0 & 1 \\ 2 & 3 & 0 \\ 1 & 5 & -3 \end{bmatrix}$ find.

a. $A+B^t = \begin{bmatrix} 1 & 3 & 5 \\ 2 & -1 & 0 \\ -1 & 2 & 4 \end{bmatrix} + \begin{bmatrix} -1 & 0 & 1 \\ 2 & 3 & 0 \\ 1 & 5 & -3 \end{bmatrix}^t = \begin{bmatrix} 1 & 3 & 5 \\ 2 & -1 & 0 \\ -1 & 2 & 4 \end{bmatrix} + \begin{bmatrix} -1 & 2 & 1 \\ 0 & 3 & 5 \\ 1 & 0 & -3 \end{bmatrix} = \begin{bmatrix} 1-1 & 3+2 & 5+1 \\ 2+0 & -1+3 & 0+5 \\ -1+1 & 2+0 & 4-3 \end{bmatrix} = \begin{bmatrix} \mathbf{0} & \mathbf{5} & \mathbf{6} \\ \mathbf{2} & \mathbf{2} & \mathbf{5} \\ \mathbf{0} & \mathbf{2} & \mathbf{1} \end{bmatrix}$

b. $A^t + B^t = \begin{bmatrix} 1 & 3 & 5 \\ 2 & -1 & 0 \\ -1 & 2 & 4 \end{bmatrix}^t + \begin{bmatrix} -1 & 0 & 1 \\ 2 & 3 & 0 \\ 1 & 5 & -3 \end{bmatrix}^t = \begin{bmatrix} 1 & 2 & -1 \\ 3 & -1 & 2 \\ 5 & 0 & 4 \end{bmatrix} + \begin{bmatrix} -1 & 2 & 1 \\ 0 & 3 & 5 \\ 1 & 0 & -3 \end{bmatrix} = \begin{bmatrix} 1-1 & 2+2 & -1+1 \\ 3+0 & -1+3 & 2+5 \\ 5+1 & 0+0 & 4-3 \end{bmatrix} = \begin{bmatrix} \mathbf{0} & \mathbf{4} & \mathbf{0} \\ \mathbf{3} & \mathbf{2} & \mathbf{7} \\ \mathbf{6} & \mathbf{0} & \mathbf{1} \end{bmatrix}$

c. $(A+B)-B^t = \left(\begin{bmatrix} 1 & 3 & 5 \\ 2 & -1 & 0 \\ -1 & 2 & 4 \end{bmatrix} + \begin{bmatrix} -1 & 0 & 1 \\ 2 & 3 & 0 \\ 1 & 5 & -3 \end{bmatrix}\right) - \begin{bmatrix} -1 & 0 & 1 \\ 2 & 3 & 0 \\ 1 & 5 & -3 \end{bmatrix}^t = \begin{bmatrix} 1-1 & 3+0 & 5+1 \\ 2+2 & -1+3 & 0+0 \\ -1+1 & 2+5 & 4-3 \end{bmatrix} - \begin{bmatrix} -1 & 2 & 1 \\ 0 & 3 & 5 \\ 1 & 0 & -3 \end{bmatrix}$

$= \begin{bmatrix} 0 & 3 & 6 \\ 4 & 2 & 0 \\ 0 & 7 & 1 \end{bmatrix} + \begin{bmatrix} 1 & -2 & -1 \\ 0 & -3 & -5 \\ -1 & 0 & 3 \end{bmatrix} = \begin{bmatrix} 0+1 & 3-2 & 6-1 \\ 4+0 & 2-3 & 0-5 \\ 0-1 & 7+0 & 1+3 \end{bmatrix} = \begin{bmatrix} \mathbf{1} & \mathbf{1} & \mathbf{5} \\ \mathbf{4} & \mathbf{-1} & \mathbf{-5} \\ \mathbf{-1} & \mathbf{7} & \mathbf{4} \end{bmatrix}$

d. $(A-B)+B^t = \left(\begin{bmatrix} 1 & 3 & 5 \\ 2 & -1 & 0 \\ -1 & 2 & 4 \end{bmatrix} - \begin{bmatrix} -1 & 0 & 1 \\ 2 & 3 & 0 \\ 1 & 5 & -3 \end{bmatrix}\right) + \begin{bmatrix} -1 & 0 & 1 \\ 2 & 3 & 0 \\ 1 & 5 & -3 \end{bmatrix}^t = \left(\begin{bmatrix} 1 & 3 & 5 \\ 2 & -1 & 0 \\ -1 & 2 & 4 \end{bmatrix} + \begin{bmatrix} 1 & 0 & -1 \\ -2 & -3 & 0 \\ -1 & -5 & 3 \end{bmatrix}\right) + \begin{bmatrix} -1 & 2 & 1 \\ 0 & 3 & 5 \\ 1 & 0 & -3 \end{bmatrix}$

$= \begin{bmatrix} 1+1 & 3+0 & 5-1 \\ 2-2 & -1-3 & 0+0 \\ -1-1 & 2-5 & 4+3 \end{bmatrix} + \begin{bmatrix} -1 & 2 & 1 \\ 0 & 3 & 5 \\ 1 & 0 & -3 \end{bmatrix} = \begin{bmatrix} 2 & 3 & 4 \\ 0 & -4 & 0 \\ -2 & -3 & 7 \end{bmatrix} + \begin{bmatrix} -1 & 2 & 1 \\ 0 & 3 & 5 \\ 1 & 0 & -3 \end{bmatrix} = \begin{bmatrix} 2-1 & 3+2 & 4+1 \\ 0+0 & -4+3 & 0+5 \\ -2+1 & -3+0 & 7-3 \end{bmatrix} = \begin{bmatrix} \mathbf{1} & \mathbf{5} & \mathbf{5} \\ \mathbf{0} & \mathbf{-1} & \mathbf{5} \\ \mathbf{-1} & \mathbf{-3} & \mathbf{4} \end{bmatrix}$

e. $(A^t + B^t) - (A+B) = \left(\begin{bmatrix} 1 & 3 & 5 \\ 2 & -1 & 0 \\ -1 & 2 & 4 \end{bmatrix}^t + \begin{bmatrix} -1 & 0 & 1 \\ 2 & 3 & 0 \\ 1 & 5 & -3 \end{bmatrix}^t\right) - \left(\begin{bmatrix} 1 & 3 & 5 \\ 2 & -1 & 0 \\ -1 & 2 & 4 \end{bmatrix} + \begin{bmatrix} -1 & 0 & 1 \\ 2 & 3 & 0 \\ 1 & 5 & -3 \end{bmatrix}\right) = \left(\begin{bmatrix} 1 & 2 & -1 \\ 3 & -1 & 2 \\ 5 & 0 & 4 \end{bmatrix} + \begin{bmatrix} -1 & 2 & 1 \\ 0 & 3 & 5 \\ 1 & 0 & -3 \end{bmatrix}\right)$

$= - \begin{bmatrix} 1-1 & 3+0 & 5+1 \\ 2+2 & -1+3 & 0+0 \\ -1+1 & 2+5 & 4-3 \end{bmatrix} = \begin{bmatrix} 1-1 & 2+2 & -1+1 \\ 3+0 & -1+3 & 2+5 \\ 5+1 & 0+0 & 4-3 \end{bmatrix} - \begin{bmatrix} 0 & 3 & 6 \\ 4 & 2 & 0 \\ 0 & 7 & 1 \end{bmatrix} = \begin{bmatrix} 0 & 4 & 0 \\ 3 & 2 & 7 \\ 6 & 0 & 1 \end{bmatrix} + \begin{bmatrix} 0 & -3 & -6 \\ -4 & -2 & 0 \\ 0 & -7 & -1 \end{bmatrix} = \begin{bmatrix} \mathbf{0} & \mathbf{1} & \mathbf{-6} \\ \mathbf{-1} & \mathbf{0} & \mathbf{7} \\ \mathbf{6} & \mathbf{-7} & \mathbf{0} \end{bmatrix}$

f. $(A-B)-A^t = \left(\begin{bmatrix} 1 & 3 & 5 \\ 2 & -1 & 0 \\ -1 & 2 & 4 \end{bmatrix} - \begin{bmatrix} -1 & 0 & 1 \\ 2 & 3 & 0 \\ 1 & 5 & -3 \end{bmatrix}\right) - \begin{bmatrix} 1 & 3 & 5 \\ 2 & -1 & 0 \\ -1 & 2 & 4 \end{bmatrix}^t = \left(\begin{bmatrix} 1 & 3 & 5 \\ 2 & -1 & 0 \\ -1 & 2 & 4 \end{bmatrix} + \begin{bmatrix} 1 & 0 & -1 \\ -2 & -3 & 0 \\ -1 & -5 & 3 \end{bmatrix}\right) - \begin{bmatrix} 1 & 2 & -1 \\ 3 & -1 & 2 \\ 5 & 0 & 4 \end{bmatrix}$

$= \begin{bmatrix} 1+1 & 3+0 & 5-1 \\ 2-2 & -1-3 & 0+0 \\ -1-1 & 2-5 & 4+3 \end{bmatrix} + \begin{bmatrix} -1 & -2 & 1 \\ -3 & 1 & -2 \\ -5 & 0 & -4 \end{bmatrix} = \begin{bmatrix} 2 & 3 & 4 \\ 0 & -4 & 0 \\ -2 & -3 & 7 \end{bmatrix} + \begin{bmatrix} -1 & -2 & 1 \\ -3 & 1 & -2 \\ -5 & 0 & -4 \end{bmatrix} = \begin{bmatrix} 2-1 & 3-2 & 4+1 \\ 0-3 & -4+1 & 0-2 \\ -2-5 & -3+0 & 7-4 \end{bmatrix} = \begin{bmatrix} \mathbf{1} & \mathbf{1} & \mathbf{5} \\ \mathbf{-3} & \mathbf{-3} & \mathbf{-2} \\ \mathbf{-7} & \mathbf{-3} & \mathbf{3} \end{bmatrix}$

g. $2A-3B = 2\begin{bmatrix} 1 & 3 & 5 \\ 2 & -1 & 0 \\ -1 & 2 & 4 \end{bmatrix} - 3\begin{bmatrix} -1 & 0 & 1 \\ 2 & 3 & 0 \\ 1 & 5 & -3 \end{bmatrix} = \begin{bmatrix} 2 & 6 & 10 \\ 4 & -2 & 0 \\ -2 & 4 & 8 \end{bmatrix} + \begin{bmatrix} 3 & 0 & -3 \\ -6 & -9 & 0 \\ -3 & -15 & 9 \end{bmatrix} = \begin{bmatrix} 2+3 & 6+0 & 10-3 \\ 4-6 & -2-9 & 0+0 \\ -2-3 & 4-15 & 8+9 \end{bmatrix} = \begin{bmatrix} \mathbf{5} & \mathbf{6} & \mathbf{7} \\ \mathbf{-2} & \mathbf{-11} & \mathbf{0} \\ \mathbf{-5} & \mathbf{-11} & \mathbf{17} \end{bmatrix}$

3. Given $A = \begin{bmatrix} 1 & 0 & -3 \end{bmatrix}$, $B = \begin{bmatrix} 3 & 1 & 5 \end{bmatrix}$, $C = \begin{bmatrix} 1 & 5 & -1 \\ 2 & 0 & 3 \end{bmatrix}$, and $D = \begin{bmatrix} 10 & 8 & 7 \\ 3 & -5 & 2 \end{bmatrix}$ perform the following operations, if possible.

a. $A+C = \begin{bmatrix} 1 & 0 & -3 \end{bmatrix} + \begin{bmatrix} 1 & 5 & -1 \\ 2 & 0 & 3 \end{bmatrix}$ **The two matrices have different order.** Therefore, they can not be added.

b. $A-B = \begin{bmatrix} 1 & 0 & -3 \end{bmatrix} - \begin{bmatrix} 3 & 1 & 5 \end{bmatrix} = \begin{bmatrix} 1 & 0 & -3 \end{bmatrix} + \begin{bmatrix} -3 & -1 & -5 \end{bmatrix} = \begin{bmatrix} 1-3 & 0-1 & -3-5 \end{bmatrix} = \begin{bmatrix} \mathbf{-2} & \mathbf{-1} & \mathbf{-8} \end{bmatrix}$

c. $(C+D)-D^t = \begin{bmatrix} 1 & 5 & -1 \\ 2 & 0 & 3 \end{bmatrix} + \begin{bmatrix} 10 & 8 & 7 \\ 3 & -5 & 2 \end{bmatrix} - \begin{bmatrix} 10 & 8 & 7 \\ 3 & -5 & 2 \end{bmatrix}^t = \begin{bmatrix} 1+10 & 5+8 & -1+7 \\ 2+3 & 0-5 & 3+2 \end{bmatrix} - \begin{bmatrix} 10 & 3 \\ 8 & -5 \\ 7 & 2 \end{bmatrix} = \begin{bmatrix} 11 & 13 & 6 \\ 5 & -5 & 5 \end{bmatrix}$

$+\begin{bmatrix} -10 & -3 \\ -8 & 5 \\ -7 & -2 \end{bmatrix}$ **The two matrices have different order.** Therefore, they can not be added.

4. Add or subtract the following matrices.

a. $\left(\begin{bmatrix} 1 & 2 \\ 3 & 0 \end{bmatrix} - \begin{bmatrix} 1 & 3 \\ -5 & 2 \end{bmatrix}\right) + \begin{bmatrix} 2 & 3 \\ -1 & 6 \end{bmatrix} = \begin{bmatrix} 1-1 & 2-3 \\ 3+5 & 0-2 \end{bmatrix} + \begin{bmatrix} 2 & 3 \\ -1 & 6 \end{bmatrix} = \begin{bmatrix} 0 & -1 \\ 8 & -2 \end{bmatrix} + \begin{bmatrix} 2 & 3 \\ -1 & 6 \end{bmatrix} = \begin{bmatrix} 0+2 & -1+3 \\ 8-1 & -2+6 \end{bmatrix} = \begin{bmatrix} \mathbf{2} & \mathbf{2} \\ \mathbf{7} & \mathbf{4} \end{bmatrix}$

b. $\begin{bmatrix} 1 & 3 \\ 5 & 0 \\ 1 & -1 \end{bmatrix} - \left(\begin{bmatrix} 2 & 3 \\ -1 & 0 \\ 1 & 3 \end{bmatrix} - \begin{bmatrix} 1 & 5 \\ -1 & 2 \\ 4 & -3 \end{bmatrix}\right) = \begin{bmatrix} 1 & 3 \\ 5 & 0 \\ 1 & -1 \end{bmatrix} - \left(\begin{bmatrix} 2 & 3 \\ -1 & 0 \\ 1 & 3 \end{bmatrix} + \begin{bmatrix} -1 & -5 \\ 1 & -2 \\ -4 & 3 \end{bmatrix}\right) = \begin{bmatrix} 1 & 3 \\ 5 & 0 \\ 1 & -1 \end{bmatrix} - \begin{bmatrix} 2-1 & 3-5 \\ -1+1 & 0-2 \\ 1-4 & 3+3 \end{bmatrix} = \begin{bmatrix} 1 & 3 \\ 5 & 0 \\ 1 & -1 \end{bmatrix} - \begin{bmatrix} 1 & -2 \\ 0 & -2 \\ -3 & 6 \end{bmatrix}$

$= \begin{bmatrix} 1 & 3 \\ 5 & 0 \\ 1 & -1 \end{bmatrix} + \begin{bmatrix} -1 & 2 \\ 0 & 2 \\ 3 & -6 \end{bmatrix} = \begin{bmatrix} 1-1 & 3+2 \\ 5+0 & 0+2 \\ 1+3 & -1-6 \end{bmatrix} = \begin{bmatrix} \mathbf{0} & \mathbf{5} \\ \mathbf{5} & \mathbf{2} \\ \mathbf{4} & \mathbf{-7} \end{bmatrix}$

c. $\begin{bmatrix} 1 & 3 & 5 \\ 2 & -1 & 3 \\ -1 & 4 & 0 \end{bmatrix} + \left(\begin{bmatrix} 1 & 5 & 0 \\ 0 & -1 & 1 \\ 1 & 2 & 3 \end{bmatrix} + \begin{bmatrix} 1 & 0 & 0 \\ 0 & 1 & 0 \\ 0 & 0 & 1 \end{bmatrix}\right) = \begin{bmatrix} 1 & 3 & 5 \\ 2 & -1 & 3 \\ -1 & 4 & 0 \end{bmatrix} + \begin{bmatrix} 1+1 & 5+0 & 0+0 \\ 0+0 & -1+1 & 1+0 \\ 1+0 & 2+0 & 3+1 \end{bmatrix} = \begin{bmatrix} 1 & 3 & 5 \\ 2 & -1 & 3 \\ -1 & 4 & 0 \end{bmatrix} + \begin{bmatrix} 2 & 5 & 0 \\ 0 & 0 & 1 \\ 1 & 2 & 4 \end{bmatrix}$

$= \begin{bmatrix} 1+2 & 3+5 & 5+0 \\ 2+0 & -1+0 & 3+1 \\ -1+1 & 4+2 & 0+4 \end{bmatrix} = \begin{bmatrix} \mathbf{3} & \mathbf{8} & \mathbf{5} \\ \mathbf{2} & \mathbf{-1} & \mathbf{4} \\ \mathbf{0} & \mathbf{6} & \mathbf{4} \end{bmatrix}$

d. $\left(\begin{bmatrix} 2 & 3 \\ -1 & 0 \end{bmatrix} + \begin{bmatrix} -2 & -3 \\ 1 & 0 \end{bmatrix}\right) - \begin{bmatrix} 3 & 5 \\ 1 & -2 \end{bmatrix} = \begin{bmatrix} 2-2 & 3-3 \\ -1+1 & 0+0 \end{bmatrix} - \begin{bmatrix} 3 & 5 \\ 1 & -2 \end{bmatrix} = \begin{bmatrix} 0 & 0 \\ 0 & 0 \end{bmatrix} - \begin{bmatrix} 3 & 5 \\ 1 & -2 \end{bmatrix} = \begin{bmatrix} 0 & 0 \\ 0 & 0 \end{bmatrix} + \begin{bmatrix} -3 & -5 \\ -1 & 2 \end{bmatrix}$

$= \begin{bmatrix} 0-3 & 0-5 \\ 0-1 & 0+2 \end{bmatrix} = \begin{bmatrix} \mathbf{-3} & \mathbf{-5} \\ \mathbf{-1} & \mathbf{2} \end{bmatrix}$

5. Given $A = \begin{bmatrix} \mathbf{2} & \mathbf{3} & \mathbf{-1} \\ \mathbf{5} & \mathbf{0} & \mathbf{2} \\ \mathbf{8} & \mathbf{-3} & \mathbf{1} \end{bmatrix}$, then $A^t = \begin{bmatrix} 2 & 3 & -1 \\ 5 & 0 & 2 \\ 8 & -3 & 1 \end{bmatrix}^t = \begin{bmatrix} 2 & 5 & 8 \\ 3 & 0 & -3 \\ -1 & 2 & 1 \end{bmatrix}$ and $[A^t]^t = \begin{bmatrix} 2 & 5 & 8 \\ 3 & 0 & -3 \\ -1 & 2 & 1 \end{bmatrix}^t = \begin{bmatrix} \mathbf{2} & \mathbf{3} & \mathbf{-1} \\ \mathbf{5} & \mathbf{0} & \mathbf{2} \\ \mathbf{8} & \mathbf{-3} & \mathbf{1} \end{bmatrix}$

Therefore, $[A^t]^t = A$.

6. Given $A = \begin{bmatrix} 1 & 2 \\ 0 & 3 \end{bmatrix}$ and $B = \begin{bmatrix} 3 & -5 \\ -3 & 2 \end{bmatrix}$ show that $[A+B]^t = A^t + B^t$.

$A+B = \begin{bmatrix} 1 & 2 \\ 0 & 3 \end{bmatrix} + \begin{bmatrix} 3 & -5 \\ -3 & 2 \end{bmatrix} = \begin{bmatrix} 1+3 & 2-5 \\ 0-3 & 3+2 \end{bmatrix} = \begin{bmatrix} 4 & -3 \\ -3 & 5 \end{bmatrix}$. Then, $[A+B]^t = \begin{bmatrix} 4 & -3 \\ -3 & 5 \end{bmatrix}^t = \begin{bmatrix} \mathbf{4} & \mathbf{-3} \\ \mathbf{-3} & \mathbf{5} \end{bmatrix}$

$A^t = \begin{bmatrix} 1 & 2 \\ 0 & 3 \end{bmatrix}^t = \begin{bmatrix} 1 & 0 \\ 2 & 3 \end{bmatrix}$ and $B^t = \begin{bmatrix} 3 & -5 \\ -3 & 2 \end{bmatrix}^t = \begin{bmatrix} 3 & -3 \\ -5 & 2 \end{bmatrix}$. Therefore, $A^t + B^t = \begin{bmatrix} 1 & 0 \\ 2 & 3 \end{bmatrix} + \begin{bmatrix} 3 & -3 \\ -5 & 2 \end{bmatrix} = \begin{bmatrix} \mathbf{4} & \mathbf{-3} \\ \mathbf{-3} & \mathbf{5} \end{bmatrix}$

7. Given the equal matrices solve for the unknowns.

a. Given $\begin{bmatrix} u-1 & v & 3 \\ 8 & 6 & 16 \\ 0 & 6 & \frac{z}{3} \end{bmatrix} = \begin{bmatrix} 5 & 2v+3 & \sqrt{9} \\ w+1 & 6 & 2x \\ 0 & 2y-3 & 3 \end{bmatrix}$, by equating the matrix entries we obtain:

$u-1=5$; $u=5+1$; $\boldsymbol{u=6}$ $\qquad$ $v=2v+3$; $v-2v=3$; $-v=3$; $\boldsymbol{v=-3}$

$8 = w+1$; $8-1 = w$; $\mathbf{w = 7}$ $16 = 2x$; $\frac{16}{2} = x$; $\mathbf{x = 8}$

$6 = 2y-3$; $6+3 = 2y$; $9 = 2y$; $y = \frac{9}{2}$; $\mathbf{y = 4.5}$ $\frac{z}{3} = 3$; $z = 3 \cdot 3$; $\mathbf{z = 9}$

b. Given $\begin{bmatrix} 3x & 5 \\ 3 & y-8 \end{bmatrix} + \begin{bmatrix} 3 & 6 \\ -5 & 4 \end{bmatrix} = \begin{bmatrix} 6 & 11 \\ -2 & -4 \end{bmatrix}$ which is equal to $\begin{bmatrix} 3x+3 & 11 \\ -2 & y-4 \end{bmatrix} = \begin{bmatrix} 6 & 11 \\ -2 & -4 \end{bmatrix}$ we can solve for the x and y values by equating the entries on both sides of the matrix. Therefore, $3x+3 = 6$; $3x = 6-3$; $3x = 3$; $\mathbf{x = 1}$ and $y-4 = -4$; $y = -4+4$; $\mathbf{y = 0}$

c. Given $\begin{bmatrix} 3 & 5 \\ -3 & 8 \end{bmatrix} - \begin{bmatrix} 7 & 1 \\ 2 & -2x \end{bmatrix} = \begin{bmatrix} -4 & 4 \\ -5 & 10 \end{bmatrix}$ which is equal to $\begin{bmatrix} -4 & 4 \\ -5 & 8+2x \end{bmatrix} = \begin{bmatrix} -4 & 4 \\ -5 & 10 \end{bmatrix}$ we can solve for x by equating the entries on both sides of the matrix. Therefore, $8+2x = 10$; $2x = 10-8$; $2x = 2$; $\mathbf{x = 1}$

d. Given $\begin{bmatrix} x \\ 0 \\ 0 \end{bmatrix} + \begin{bmatrix} 3y \\ y \\ 0 \end{bmatrix} + \begin{bmatrix} z \\ -2z \\ 3z \end{bmatrix} = \begin{bmatrix} 6 \\ 3 \\ 9 \end{bmatrix}$ which is equal to $\begin{bmatrix} x+3y+z \\ y-2z \\ 3z \end{bmatrix} = \begin{bmatrix} 6 \\ 3 \\ 9 \end{bmatrix}$ we can solve for x, y, and z by equating the entries on both sides of the matrix. Therefore, $x+3y+z = 6$, $y-2z = 3$, and $3z = 9$. Solving for the unknowns we obtain $\mathbf{z = 3}$, $y-(2 \cdot 3) = 3$; $y-6 = 3$; $\mathbf{y = 9}$, and $x+(3 \cdot 9)+3 = 6$; $x+30 = 6$; $\mathbf{x = -24}$

Section 3.2 Case II Solutions - Matrix Multiplication

1. Find the product of the following matrix operations.

a. $$\begin{bmatrix} 1 & 5 & 3 \\ 0 & -3 & 1 \end{bmatrix} \begin{bmatrix} 1 & -1 \\ 2 & 0 \\ 1 & 3 \end{bmatrix} = \begin{bmatrix} (1\times1)+(5\times2)+(3\times1) & (1\times-1)+(5\times0)+(3\times3) \\ (0\times1)+(-3\times2)+(1\times1) & (0\times-1)+(-3\times0)+(1\times3) \end{bmatrix} = \begin{bmatrix} 1+10+3 & -1+0+9 \\ 0-6+1 & 0+0+3 \end{bmatrix} = \begin{bmatrix} \mathbf{14} & \mathbf{8} \\ \mathbf{-5} & \mathbf{3} \end{bmatrix}$$

b. $$\begin{bmatrix} 1 & -2 & 2 \end{bmatrix} \begin{bmatrix} 1 \\ 3 \\ -1 \end{bmatrix} = (1\times1)+(-2\times3)+(2\times-1) = 1-6-2 = \mathbf{-7}$$

c. $$\begin{bmatrix} 1 & -1 \\ 2 & 3 \end{bmatrix} \begin{bmatrix} 1 & -5 \\ 3 & 1 \end{bmatrix} = \begin{bmatrix} (1\times1)+(-1\times3) & (1\times-5)+(-1\times1) \\ (2\times1)+(3\times3) & (2\times-5)+(3\times1) \end{bmatrix} = \begin{bmatrix} 1-3 & -5-1 \\ 2+9 & -10+3 \end{bmatrix} = \begin{bmatrix} \mathbf{-2} & \mathbf{-6} \\ \mathbf{11} & \mathbf{-7} \end{bmatrix}$$

d. $$\begin{bmatrix} 4 & -3 \\ 7 & 3 \end{bmatrix} \begin{bmatrix} 5 & 1 \\ -2 & 3 \end{bmatrix} = \begin{bmatrix} (4\times5)+(-3\times-2) & (4\times1)+(-3\times3) \\ (7\times5)+(3\times-2) & (7\times1)+(3\times3) \end{bmatrix} = \begin{bmatrix} 20+6 & 4-9 \\ 35-6 & 7+9 \end{bmatrix} = \begin{bmatrix} \mathbf{26} & \mathbf{-5} \\ \mathbf{29} & \mathbf{16} \end{bmatrix}$$

e. $$\begin{bmatrix} -1 & 0 & 1 \\ 3 & 1 & 0 \\ 1 & 0 & 2 \end{bmatrix} \begin{bmatrix} 0 & 1 & 2 \\ 3 & 0 & 1 \\ -1 & 1 & 0 \end{bmatrix} = \begin{bmatrix} (-1\times0)+(0\times3)+(1\times-1) & (-1\times1)+(0\times0)+(1\times1) & (-1\times2)+(0\times1)+(1\times0) \\ (3\times0)+(1\times3)+(0\times-1) & (3\times1)+(1\times0)+(0\times1) & (3\times2)+(1\times1)+(0\times0) \\ (1\times0)+(0\times3)+(2\times-1) & (1\times1)+(0\times0)+(2\times1) & (1\times2)+(0\times1)+(2\times0) \end{bmatrix}$$

$$= \begin{bmatrix} 0+0-1 & -1+0+1 & -2+0+0 \\ 0+3+0 & 3+0+0 & 6+1+0 \\ 0+0-2 & 1+0+2 & 2+0+0 \end{bmatrix} = \begin{bmatrix} \mathbf{-1} & \mathbf{0} & \mathbf{-2} \\ \mathbf{3} & \mathbf{3} & \mathbf{7} \\ \mathbf{-2} & \mathbf{3} & \mathbf{2} \end{bmatrix}$$

f. $$\begin{bmatrix} 2 & -3 & -1 \\ 0 & 1 & -2 \\ 0 & 1 & 0 \end{bmatrix} \begin{bmatrix} 1 & 0 & 0 \\ 0 & 1 & 0 \\ 0 & 0 & 1 \end{bmatrix} = \begin{bmatrix} (2\times1)+(-3\times0)+(-1\times0) & (2\times0)+(-3\times1)+(-1\times0) & (2\times0)+(-3\times0)+(-1\times1) \\ (0\times1)+(1\times0)+(-2\times0) & (0\times0)+(1\times1)+(-2\times0) & (0\times0)+(1\times0)+(-2\times1) \\ (0\times1)+(1\times0)+(0\times0) & (0\times0)+(1\times1)+(0\times0) & (0\times0)+(1\times0)+(0\times1) \end{bmatrix}$$

$$= \begin{bmatrix} 2+0+0 & 0-3+0 & 0+0-1 \\ 0+0+0 & 0+1+0 & 0+0-2 \\ 0+0+0 & 0+1+0 & 0+0+0 \end{bmatrix} = \begin{bmatrix} \mathbf{2} & \mathbf{-3} & \mathbf{-1} \\ \mathbf{0} & \mathbf{1} & \mathbf{-2} \\ \mathbf{0} & \mathbf{1} & \mathbf{0} \end{bmatrix}$$

g. $3\begin{bmatrix}1&2&3&5\\0&1&-1&3\\1&0&1&2\end{bmatrix}=\begin{bmatrix}1\times3&2\times3&3\times3&5\times3\\0\times3&1\times3&-1\times3&3\times3\\1\times3&0\times3&1\times3&2\times3\end{bmatrix}=\begin{bmatrix}\mathbf{3}&\mathbf{6}&\mathbf{9}&\mathbf{15}\\\mathbf{0}&\mathbf{3}&\mathbf{-3}&\mathbf{9}\\\mathbf{3}&\mathbf{0}&\mathbf{3}&\mathbf{6}\end{bmatrix}$

h. $-5\begin{bmatrix}0&1&-2\\-3&-1&2\end{bmatrix}=\begin{bmatrix}0\times-5&1\times-5&-2\times-5\\-3\times-5&-1\times-5&2\times-5\end{bmatrix}=\begin{bmatrix}\mathbf{0}&\mathbf{-5}&\mathbf{10}\\\mathbf{15}&\mathbf{5}&\mathbf{-10}\end{bmatrix}$

2. Given $A=\begin{bmatrix}1&2\\-1&0\end{bmatrix}$ and $B=\begin{bmatrix}-1&2\\0&1\end{bmatrix}$ find.

a. $AB=\begin{bmatrix}1&2\\-1&0\end{bmatrix}\begin{bmatrix}-1&2\\0&1\end{bmatrix}=\begin{bmatrix}(1\times-1)+(2\times0)&(1\times2)+(2\times1)\\(-1\times-1)+(0\times0)&(-1\times2)+(0\times1)\end{bmatrix}=\begin{bmatrix}-1+0&2+2\\1+0&-2+0\end{bmatrix}=\begin{bmatrix}\mathbf{-1}&\mathbf{4}\\\mathbf{1}&\mathbf{-2}\end{bmatrix}$

b. $BA=\begin{bmatrix}-1&2\\0&1\end{bmatrix}\begin{bmatrix}1&2\\-1&0\end{bmatrix}=\begin{bmatrix}(-1\times1)+(2\times-1)&(-1\times2)+(2\times0)\\(0\times-1)+(1\times-1)&(0\times2)+(1\times0)\end{bmatrix}=\begin{bmatrix}-1-2&-2+0\\0-1&0+0\end{bmatrix}=\begin{bmatrix}\mathbf{-3}&\mathbf{-2}\\\mathbf{-1}&\mathbf{0}\end{bmatrix}$

c. $(AB)^tA=\left(\begin{bmatrix}1&2\\-1&0\end{bmatrix}\begin{bmatrix}-1&2\\0&1\end{bmatrix}\right)^t\begin{bmatrix}1&2\\-1&0\end{bmatrix}=\begin{bmatrix}-1&4\\1&-2\end{bmatrix}^t\begin{bmatrix}1&2\\-1&0\end{bmatrix}=\begin{bmatrix}-1&1\\4&-2\end{bmatrix}\begin{bmatrix}1&2\\-1&0\end{bmatrix}=\begin{bmatrix}(-1\times1)+(1\times-1)&(-1\times2)+(1\times0)\\(4\times1)+(-2\times-1)&(4\times2)+(-2\times0)\end{bmatrix}$

$=\begin{bmatrix}-1-1&-2+0\\4+2&8+0\end{bmatrix}=\begin{bmatrix}\mathbf{-2}&\mathbf{-2}\\\mathbf{6}&\mathbf{8}\end{bmatrix}$

d. $(BA)^tB=\left(\begin{bmatrix}-1&2\\0&1\end{bmatrix}\begin{bmatrix}1&2\\-1&0\end{bmatrix}\right)^t\begin{bmatrix}-1&2\\0&1\end{bmatrix}=\begin{bmatrix}-3&-2\\-1&0\end{bmatrix}^t\begin{bmatrix}-1&2\\0&1\end{bmatrix}=\begin{bmatrix}-3&-1\\-2&0\end{bmatrix}\begin{bmatrix}-1&2\\0&1\end{bmatrix}=\begin{bmatrix}(-3\times-1)+(-1\times0)&(-3\times2)+(-1\times1)\\(-2\times-1)+(0\times0)&(-2\times2)+(0\times1)\end{bmatrix}$

$=\begin{bmatrix}3+0&-6-1\\2+0&-4+0\end{bmatrix}=\begin{bmatrix}\mathbf{3}&\mathbf{-7}\\\mathbf{2}&\mathbf{-4}\end{bmatrix}$

e. $A^t(BA)=\begin{bmatrix}1&2\\-1&0\end{bmatrix}^t\left(\begin{bmatrix}-1&2\\0&1\end{bmatrix}\begin{bmatrix}1&2\\-1&0\end{bmatrix}\right)=\begin{bmatrix}1&2\\-1&0\end{bmatrix}^t\begin{bmatrix}-3&-2\\-1&0\end{bmatrix}=\begin{bmatrix}1&-1\\2&0\end{bmatrix}\begin{bmatrix}-3&-2\\-1&0\end{bmatrix}=\begin{bmatrix}(1\times-3)+(-1\times-1)&(1\times-2)+(-1\times0)\\(2\times-3)+(0\times-1)&(2\times-2)+(0\times0)\end{bmatrix}$

$=\begin{bmatrix}-3+1&-2+0\\-6+0&-4+0\end{bmatrix}=\begin{bmatrix}\mathbf{-2}&\mathbf{-2}\\\mathbf{-6}&\mathbf{-4}\end{bmatrix}$

f. $AB^t=\begin{bmatrix}1&2\\-1&0\end{bmatrix}\begin{bmatrix}-1&2\\0&1\end{bmatrix}^t=\begin{bmatrix}1&2\\-1&0\end{bmatrix}\begin{bmatrix}-1&0\\2&1\end{bmatrix}=\begin{bmatrix}(1\times-1)+(2\times2)&(1\times0)+(2\times1)\\(-1\times-1)+(0\times2)&(-1\times0)+(0\times1)\end{bmatrix}=\begin{bmatrix}-1+4&0+2\\1+0&0+0\end{bmatrix}=\begin{bmatrix}\mathbf{3}&\mathbf{2}\\\mathbf{1}&\mathbf{0}\end{bmatrix}$

g. $A^tB^t=\begin{bmatrix}1&2\\-1&0\end{bmatrix}^t\begin{bmatrix}-1&2\\0&1\end{bmatrix}^t=\begin{bmatrix}1&-1\\2&0\end{bmatrix}\begin{bmatrix}-1&0\\2&1\end{bmatrix}=\begin{bmatrix}(1\times-1)+(-1\times2)&(1\times0)+(-1\times1)\\(2\times-1)+(0\times2)&(2\times0)+(0\times1)\end{bmatrix}=\begin{bmatrix}-1-2&0-1\\-2+0&0+0\end{bmatrix}=\begin{bmatrix}\mathbf{-3}&\mathbf{-1}\\\mathbf{-2}&\mathbf{0}\end{bmatrix}$

h. $(A^tB^t)A=\left(\begin{bmatrix}1&2\\-1&0\end{bmatrix}^t\begin{bmatrix}-1&2\\0&1\end{bmatrix}^t\right)\begin{bmatrix}1&2\\-1&0\end{bmatrix}=\begin{bmatrix}-3&-1\\-2&0\end{bmatrix}\begin{bmatrix}1&2\\-1&0\end{bmatrix}=\begin{bmatrix}(-3\times1)+(-1\times-1)&(-3\times2)+(-1\times0)\\(-2\times1)+(0\times-1)&(-2\times2)+(0\times0)\end{bmatrix}=\begin{bmatrix}\mathbf{-2}&\mathbf{-6}\\\mathbf{-2}&\mathbf{-4}\end{bmatrix}$

3. Given the following equations in matrix form find the matrix Y.

a. Let the matrix Y be equal to $Y=\begin{bmatrix}a&b\\c&d\end{bmatrix}$. Then, $2\begin{bmatrix}a&b\\c&d\end{bmatrix}+\begin{bmatrix}1&0\\0&3\end{bmatrix}=\begin{bmatrix}-3&2\\1&3\end{bmatrix}$; $\begin{bmatrix}2a+1&2b\\2c&2d+3\end{bmatrix}=\begin{bmatrix}-3&2\\1&3\end{bmatrix}$. Equating the entry terms on both sides of the equality we obtain: $2a+1=-3$; $a=-2$ $\quad 2b=2$; $b=1$

$2c=1$; $c=\frac{1}{2}$ $\quad 2d+3=3$; $d=0$. Therefore, $\mathbf{Y}=\begin{bmatrix}\mathbf{-2}&\mathbf{1}\\\mathbf{\frac{1}{2}}&\mathbf{0}\end{bmatrix}$

b. Let the matrix Y be equal to $Y=\begin{bmatrix}a&b\\c&d\end{bmatrix}$. Then, $2\begin{bmatrix}a&b\\c&d\end{bmatrix}+3\begin{bmatrix}1&1\\-1&2\end{bmatrix}=\begin{bmatrix}7&-1\\5&0\end{bmatrix}$; $\begin{bmatrix}2a+3&2b+3\\2c-3&2d+6\end{bmatrix}=\begin{bmatrix}7&-1\\5&0\end{bmatrix}$

Equating the entry terms on both sides of the equality we obtain: $2a+3=7$; $a=2$ $2b+3=-1$; $b=-2$

$2c-3=5$; $c=4$ $2d+6=0$; $d=-3$. Therefore, $\boldsymbol{Y}=\begin{bmatrix}\mathbf{2} & \mathbf{-2}\\ \mathbf{4} & \mathbf{-3}\end{bmatrix}$

c. Let the matrix Y be equal to $Y=\begin{bmatrix}a & b\\ c & d\end{bmatrix}$. Then, $3\begin{bmatrix}a & b\\ c & d\end{bmatrix}-2\begin{bmatrix}1 & 0\\ 0 & 1\end{bmatrix}=2\begin{bmatrix}5 & 3\\ 1 & -2\end{bmatrix}$; $\begin{bmatrix}3a-2 & 3b\\ 3c & 3d-2\end{bmatrix}=\begin{bmatrix}10 & 6\\ 2 & -4\end{bmatrix}$

Equating the entry terms on both sides of the equality we obtain: $3a-2=10$; $a=4$ $3b=6$; $b=2$

$3c=2$; $c=\frac{2}{3}$ $3d-2=-4$; $d=-\frac{2}{3}$. Therefore, $\boldsymbol{Y}=\begin{bmatrix}\mathbf{4} & \mathbf{2}\\ \mathbf{\frac{2}{3}} & \mathbf{-\frac{2}{3}}\end{bmatrix}$

d. Let the matrix Y be equal to $Y=\begin{bmatrix}a & b & c\\ d & e & f\\ g & h & i\end{bmatrix}$. Then, $\begin{bmatrix}a & b & c\\ d & e & f\\ g & h & i\end{bmatrix}+4\begin{bmatrix}1 & 0 & 0\\ 0 & 1 & 0\\ 0 & 0 & 1\end{bmatrix}=\begin{bmatrix}8 & 0 & 0\\ 10 & 0 & 1\\ -1 & 2 & 3\end{bmatrix}$; $\begin{bmatrix}a+4 & b & c\\ d & e+4 & f\\ g & h & i+4\end{bmatrix}$

$=\begin{bmatrix}8 & 0 & 0\\ 10 & 0 & 1\\ -1 & 2 & 3\end{bmatrix}$. Equating the entry terms on both sides of the equality we obtain: $a+4=8$; $a=4$

$b=0$ $c=0$ $d=10$ $e+4=0$; $e=-4$ $f=1$ $g=-1$

$h=2$ $i+4=3$; $i=-1$. Therefore, $\boldsymbol{Y}=\begin{bmatrix}\mathbf{4} & \mathbf{0} & \mathbf{0}\\ \mathbf{10} & \mathbf{-4} & \mathbf{1}\\ \mathbf{-1} & \mathbf{2} & \mathbf{-1}\end{bmatrix}$

4. Given $A=\begin{bmatrix}1 & 2\\ 0 & -1\end{bmatrix}$, $B=\begin{bmatrix}1 & 0\\ -1 & 2\end{bmatrix}$, and $C=\begin{bmatrix}2 & -1\\ 0 & 1\end{bmatrix}$ show that: a. $(AB)C = A(BC)$ b. $A(B+C) = AB+AC$

c. $(B+C)A = BA+CA$ d. $3(A+B) = 3A+3B$

a. $(AB)C=\left(\begin{bmatrix}1 & 2\\ 0 & -1\end{bmatrix}\begin{bmatrix}1 & 0\\ -1 & 2\end{bmatrix}\right)\begin{bmatrix}2 & -1\\ 0 & 1\end{bmatrix}=\begin{bmatrix}(1\times1)+(2\times-1) & (1\times0)+(2\times2)\\ (0\times1)+(-1\times-1) & (0\times0)+(-1\times2)\end{bmatrix}\begin{bmatrix}2 & -1\\ 0 & 1\end{bmatrix}=\begin{bmatrix}1-2 & 0+4\\ 0+1 & 0-2\end{bmatrix}\begin{bmatrix}2 & -1\\ 0 & 1\end{bmatrix}$

$=\begin{bmatrix}-1 & 4\\ 1 & -2\end{bmatrix}\begin{bmatrix}2 & -1\\ 0 & 1\end{bmatrix}=\begin{bmatrix}(-1\times2)+(4\times0) & (-1\times-1)+(4\times1)\\ (1\times2)+(-2\times0) & (1\times-1)+(-2\times1)\end{bmatrix}=\begin{bmatrix}-2+0 & 1+4\\ 2+0 & -1-2\end{bmatrix}=\begin{bmatrix}\mathbf{-2} & \mathbf{5}\\ \mathbf{2} & \mathbf{-3}\end{bmatrix}$

$A(BC)=\begin{bmatrix}1 & 2\\ 0 & -1\end{bmatrix}\left(\begin{bmatrix}1 & 0\\ -1 & 2\end{bmatrix}\begin{bmatrix}2 & -1\\ 0 & 1\end{bmatrix}\right)=\begin{bmatrix}1 & 2\\ 0 & -1\end{bmatrix}\begin{bmatrix}(1\times2)+(0\times0) & (1\times-1)+(0\times1)\\ (-1\times2)+(2\times0) & (-1\times-1)+(2\times1)\end{bmatrix}=\begin{bmatrix}1 & 2\\ 0 & -1\end{bmatrix}\begin{bmatrix}2+0 & -1+0\\ -2+0 & 1+2\end{bmatrix}$

$=\begin{bmatrix}1 & 2\\ 0 & -1\end{bmatrix}\begin{bmatrix}2 & -1\\ -2 & 3\end{bmatrix}=\begin{bmatrix}(1\times2)+(2\times-2) & (1\times-1)+(2\times3)\\ (0\times2)+(-1\times-2) & (0\times-1)+(-1\times3)\end{bmatrix}=\begin{bmatrix}2-4 & -1+6\\ 0+2 & 0-3\end{bmatrix}=\begin{bmatrix}\mathbf{-2} & \mathbf{5}\\ \mathbf{2} & \mathbf{-3}\end{bmatrix}$

b. $A(B+C)=\begin{bmatrix}1 & 2\\ 0 & -1\end{bmatrix}\left(\begin{bmatrix}1 & 0\\ -1 & 2\end{bmatrix}+\begin{bmatrix}2 & -1\\ 0 & 1\end{bmatrix}\right)=\begin{bmatrix}1 & 2\\ 0 & -1\end{bmatrix}\begin{bmatrix}1+2 & 0-1\\ -1+0 & 2+1\end{bmatrix}=\begin{bmatrix}1 & 2\\ 0 & -1\end{bmatrix}\begin{bmatrix}3 & -1\\ -1 & 3\end{bmatrix}=\begin{bmatrix}(1\times3)+(2\times-1) & (1\times-1)+(2\times3)\\ (0\times3)+(-1\times-1) & (0\times-1)+(-1\times3)\end{bmatrix}$

$=\begin{bmatrix}3-2 & -1+6\\ 0+1 & 0-3\end{bmatrix}=\begin{bmatrix}\mathbf{1} & \mathbf{5}\\ \mathbf{1} & \mathbf{-3}\end{bmatrix}$

$AB+AC=\begin{bmatrix}1 & 2\\ 0 & -1\end{bmatrix}\begin{bmatrix}1 & 0\\ -1 & 2\end{bmatrix}+\begin{bmatrix}1 & 2\\ 0 & -1\end{bmatrix}\begin{bmatrix}2 & -1\\ 0 & 1\end{bmatrix}=\begin{bmatrix}(1\times1)+(2\times-1) & (1\times0)+(2\times2)\\ (0\times1)+(-1\times-1) & (0\times0)+(-1\times2)\end{bmatrix}+\begin{bmatrix}1 & 2\\ 0 & -1\end{bmatrix}\begin{bmatrix}2 & -1\\ 0 & 1\end{bmatrix}$

$=\begin{bmatrix}-1 & 4\\ 1 & -2\end{bmatrix}+\begin{bmatrix}(1\times2)+(2\times0) & (1\times-1)+(2\times1)\\ (0\times2)+(-1\times0) & (0\times-1)+(-1\times1)\end{bmatrix}=\begin{bmatrix}-1 & 4\\ 1 & -2\end{bmatrix}+\begin{bmatrix}2 & 1\\ 0 & -1\end{bmatrix}=\begin{bmatrix}-1+2 & 4+1\\ 1+0 & -2-1\end{bmatrix}=\begin{bmatrix}\mathbf{1} & \mathbf{5}\\ \mathbf{1} & \mathbf{-3}\end{bmatrix}$

c. $(B+C)A = \left(\begin{bmatrix}1 & 0\\-1 & 2\end{bmatrix}+\begin{bmatrix}2 & -1\\0 & 1\end{bmatrix}\right)\begin{bmatrix}1 & 2\\0 & -1\end{bmatrix} = \begin{bmatrix}1+2 & 0-1\\-1+0 & 2+1\end{bmatrix}\begin{bmatrix}1 & 2\\0 & -1\end{bmatrix} = \begin{bmatrix}3 & -1\\-1 & 3\end{bmatrix}\begin{bmatrix}1 & 2\\0 & -1\end{bmatrix} = \begin{bmatrix}(3\times1)+(-1\times0) & (3\times2)+(-1\times-1)\\(-1\times1)+(3\times0) & (-1\times2)+(3\times-1)\end{bmatrix}$

$= \begin{bmatrix}3+0 & 6+1\\-1+0 & -2-3\end{bmatrix} = \begin{bmatrix}\mathbf{3} & \mathbf{7}\\\mathbf{-1} & \mathbf{-5}\end{bmatrix}$

$BA+CA = \begin{bmatrix}1 & 0\\-1 & 2\end{bmatrix}\begin{bmatrix}1 & 2\\0 & -1\end{bmatrix}+\begin{bmatrix}2 & -1\\0 & 1\end{bmatrix}\begin{bmatrix}1 & 2\\0 & -1\end{bmatrix} = \begin{bmatrix}(1\times1)+(0\times0) & (1\times2)+(0\times1)\\(-1\times1)+(2\times0) & (-1\times2)+(2\times-1)\end{bmatrix} + \begin{bmatrix}(2\times1)+(-1\times0) & (2\times2)+(-1\times-1)\\(0\times1)+(1\times0) & (0\times2)+(1\times-1)\end{bmatrix}$

$= \begin{bmatrix}1 & 2\\-1 & -4\end{bmatrix} + \begin{bmatrix}2 & 5\\0 & -1\end{bmatrix} = \begin{bmatrix}1+2 & 2+5\\-1+0 & -4-1\end{bmatrix} = \begin{bmatrix}\mathbf{3} & \mathbf{7}\\\mathbf{-1} & \mathbf{-5}\end{bmatrix}$

d. $3(A+B) = 3\left(\begin{bmatrix}1 & 2\\0 & -1\end{bmatrix}+\begin{bmatrix}1 & 0\\-1 & 2\end{bmatrix}\right) = 3\begin{bmatrix}1+1 & 2+0\\0-1 & -1+2\end{bmatrix} = 3\begin{bmatrix}2 & 2\\-1 & 1\end{bmatrix} = \begin{bmatrix}3\times2 & 3\times2\\3\times-1 & 3\times1\end{bmatrix} = \begin{bmatrix}\mathbf{6} & \mathbf{6}\\\mathbf{-3} & \mathbf{3}\end{bmatrix}$

$3A+3B = 3\begin{bmatrix}1 & 2\\0 & -1\end{bmatrix}+3\begin{bmatrix}1 & 0\\-1 & 2\end{bmatrix} = \begin{bmatrix}3 & 6\\0 & -3\end{bmatrix}+\begin{bmatrix}3 & 0\\-3 & 6\end{bmatrix} = \begin{bmatrix}3+3 & 6+0\\0-3 & -3+6\end{bmatrix} = \begin{bmatrix}\mathbf{6} & \mathbf{6}\\\mathbf{-3} & \mathbf{3}\end{bmatrix}$

5. Given $A=\begin{bmatrix}1 & 2\\0 & 1\end{bmatrix}$ and $B=\begin{bmatrix}2 & -1\\3 & 0\end{bmatrix}$ show that:

a. $(AB)^t = \left(\begin{bmatrix}1 & 2\\0 & 1\end{bmatrix}\begin{bmatrix}2 & -1\\3 & 0\end{bmatrix}\right)^t = \begin{bmatrix}(1\times2)+(2\times3) & (1\times-1)+(2\times0)\\(0\times2)+(1\times3) & (0\times-1)+(1\times0)\end{bmatrix}^t = \begin{bmatrix}2+6 & -1+0\\0+3 & 0+0\end{bmatrix}^t = \begin{bmatrix}8 & -1\\3 & 0\end{bmatrix}^t = \begin{bmatrix}\mathbf{8} & \mathbf{3}\\\mathbf{-1} & \mathbf{0}\end{bmatrix}$

$B^tA^t = \begin{bmatrix}2 & -1\\3 & 0\end{bmatrix}^t\begin{bmatrix}1 & 2\\0 & 1\end{bmatrix}^t = \begin{bmatrix}2 & 3\\-1 & 0\end{bmatrix}\begin{bmatrix}1 & 0\\2 & 1\end{bmatrix} = \begin{bmatrix}(2\times1)+(3\times2) & (2\times0)+(3\times1)\\(-1\times1)+(0\times2) & (-1\times0)+(0\times1)\end{bmatrix} = \begin{bmatrix}2+6 & 0+3\\-1+0 & 0+0\end{bmatrix} = \begin{bmatrix}\mathbf{8} & \mathbf{3}\\\mathbf{-1} & \mathbf{0}\end{bmatrix}$

Therefore, $\boldsymbol{(AB)^t = B^tA^t}$

b. $(A+B)(A+B) = \left(\begin{bmatrix}1 & 2\\0 & 1\end{bmatrix}+\begin{bmatrix}2 & -1\\3 & 0\end{bmatrix}\right)\left(\begin{bmatrix}1 & 2\\0 & 1\end{bmatrix}+\begin{bmatrix}2 & -1\\3 & 0\end{bmatrix}\right) = \begin{bmatrix}1+2 & 2-1\\0+3 & 1+0\end{bmatrix}\begin{bmatrix}1+2 & 2-1\\0+3 & 1+0\end{bmatrix} = \begin{bmatrix}3 & 1\\3 & 1\end{bmatrix}\begin{bmatrix}3 & 1\\3 & 1\end{bmatrix}$

$= \begin{bmatrix}(3\times3)+(1\times3) & (3\times1)+(1\times1)\\(3\times3)+(1\times3) & (3\times1)+(1\times1)\end{bmatrix} = \begin{bmatrix}9+3 & 3+1\\9+3 & 3+1\end{bmatrix} = \begin{bmatrix}\mathbf{12} & \mathbf{4}\\\mathbf{12} & \mathbf{4}\end{bmatrix}$

$A^2+2AB+B^2 = AA+2AB+BB = \begin{bmatrix}1 & 2\\0 & 1\end{bmatrix}\begin{bmatrix}1 & 2\\0 & 1\end{bmatrix}+2\begin{bmatrix}8 & -1\\3 & 0\end{bmatrix}+\begin{bmatrix}2 & -1\\3 & 0\end{bmatrix}\begin{bmatrix}2 & -1\\3 & 0\end{bmatrix} = \begin{bmatrix}1 & 2\\0 & 1\end{bmatrix}\begin{bmatrix}1 & 2\\0 & 1\end{bmatrix}+\begin{bmatrix}16 & -2\\6 & 0\end{bmatrix}+\begin{bmatrix}2 & -1\\3 & 0\end{bmatrix}\begin{bmatrix}2 & -1\\3 & 0\end{bmatrix}$

$= \begin{bmatrix}(1\times1)+(2\times0) & (1\times2)+(2\times1)\\(0\times1)+(1\times0) & (0\times2)+(1\times1)\end{bmatrix}+\begin{bmatrix}16 & -2\\6 & 0\end{bmatrix}+\begin{bmatrix}(2\times2)+(-1\times3) & (2\times-1)+(-1\times0)\\(3\times2)+(0\times3) & (3\times-1)+(0\times0)\end{bmatrix} = \begin{bmatrix}1+0 & 2+2\\0+0 & 0+1\end{bmatrix}+\begin{bmatrix}16 & -2\\6 & 0\end{bmatrix}+\begin{bmatrix}4-3 & -2+0\\6+0 & -3+0\end{bmatrix}$

$= \begin{bmatrix}1 & 4\\0 & 1\end{bmatrix}+\begin{bmatrix}16 & -2\\6 & 0\end{bmatrix}+\begin{bmatrix}1 & -2\\6 & -3\end{bmatrix} = \begin{bmatrix}1+16+1 & 4-2-2\\0+6+6 & 1+0-3\end{bmatrix} = \begin{bmatrix}\mathbf{18} & \mathbf{0}\\\mathbf{12} & \mathbf{-2}\end{bmatrix}$ Therefore, $\boldsymbol{(A+B)(A+B) \neq A^2+2AB+B^2}$

c. $(A+B)(A-B) = \left(\begin{bmatrix}1 & 2\\0 & 1\end{bmatrix}+\begin{bmatrix}2 & -1\\3 & 0\end{bmatrix}\right)\left(\begin{bmatrix}1 & 2\\0 & 1\end{bmatrix}-\begin{bmatrix}2 & -1\\3 & 0\end{bmatrix}\right) = \begin{bmatrix}1+2 & 2-1\\0+3 & 1+0\end{bmatrix}\begin{bmatrix}1-2 & 2+1\\0-3 & 1-0\end{bmatrix} = \begin{bmatrix}3 & 1\\3 & 1\end{bmatrix}\begin{bmatrix}-1 & 3\\-3 & 1\end{bmatrix}$

$= \begin{bmatrix}(3\times-1)+(1\times-3) & (3\times3)+(1\times1)\\(3\times-1)+(1\times-3) & (3\times3)+(1\times1)\end{bmatrix} = \begin{bmatrix}-3-3 & 9+1\\-3-3 & 9+1\end{bmatrix} = \begin{bmatrix}\mathbf{-6} & \mathbf{10}\\\mathbf{-6} & \mathbf{10}\end{bmatrix}$

$A^2-B^2 = A\cdot A-B\cdot B = \begin{bmatrix}1 & 2\\0 & 1\end{bmatrix}\begin{bmatrix}1 & 2\\0 & 1\end{bmatrix}-\begin{bmatrix}2 & -1\\3 & 0\end{bmatrix}\begin{bmatrix}2 & -1\\3 & 0\end{bmatrix} = \begin{bmatrix}(1\times1)+(2\times0) & (1\times2)+(2\times1)\\(0\times1)+(1\times0) & (0\times2)+(1\times1)\end{bmatrix}-\begin{bmatrix}(2\times2)+(-1\times3) & (2\times-1)+(-1\times0)\\(3\times2)+(0\times3) & (3\times-1)+(0\times0)\end{bmatrix}$

$$= \begin{bmatrix} 1+0 & 2+2 \\ 0+0 & 0+1 \end{bmatrix} - \begin{bmatrix} 4-3 & -2+0 \\ 6+0 & -3+0 \end{bmatrix} = \begin{bmatrix} 1 & 4 \\ 0 & 1 \end{bmatrix} - \begin{bmatrix} 1 & -2 \\ 6 & -3 \end{bmatrix} = \begin{bmatrix} 1 & 4 \\ 0 & 1 \end{bmatrix} + \begin{bmatrix} -1 & 2 \\ -6 & 3 \end{bmatrix} = \begin{bmatrix} 1-1 & 4+2 \\ 0-6 & 1+3 \end{bmatrix} = \begin{bmatrix} \mathbf{0} & \mathbf{6} \\ \mathbf{-6} & \mathbf{4} \end{bmatrix}$$

Therefore, $(\boldsymbol{A}+\boldsymbol{B})(\boldsymbol{A}-\boldsymbol{B}) \neq \boldsymbol{A}^2 - \boldsymbol{B}^2$

d. $$AB = \begin{bmatrix} 1 & 2 \\ 0 & 1 \end{bmatrix}\begin{bmatrix} 2 & -1 \\ 3 & 0 \end{bmatrix} = \begin{bmatrix} (1\times 2)+(2\times 3) & (1\times -1)+(2\times 0) \\ (0\times 2)+(1\times 3) & (0\times -1)+(1\times 0) \end{bmatrix} = \begin{bmatrix} 2+6 & -1+0 \\ 0+3 & 0+0 \end{bmatrix} = \begin{bmatrix} \mathbf{8} & \mathbf{-1} \\ \mathbf{3} & \mathbf{0} \end{bmatrix}$$

$$BA = \begin{bmatrix} 2 & -1 \\ 3 & 0 \end{bmatrix}\begin{bmatrix} 1 & 2 \\ 0 & 1 \end{bmatrix} = \begin{bmatrix} (2\times 1)+(-1\times 0) & (2\times 2)+(-1\times 1) \\ (3\times 1)+(0\times 0) & (3\times 2)+(0\times 1) \end{bmatrix} = \begin{bmatrix} 2+0 & 4-1 \\ 3+0 & 6+0 \end{bmatrix} = \begin{bmatrix} \mathbf{2} & \mathbf{3} \\ \mathbf{3} & \mathbf{6} \end{bmatrix}$$ Therefore, $\boldsymbol{AB} \neq \boldsymbol{BA}$

6. Multiply the following matrices.

a. $$3\begin{bmatrix} 1 & 2 \\ 0 & 1 \end{bmatrix}\begin{bmatrix} -1 & 0 \\ 2 & -2 \end{bmatrix} = \begin{bmatrix} 1\times 3 & 2\times 3 \\ 0\times 3 & 1\times 3 \end{bmatrix}\begin{bmatrix} -1 & 0 \\ 2 & -2 \end{bmatrix} = \begin{bmatrix} 3 & 6 \\ 0 & 3 \end{bmatrix}\begin{bmatrix} -1 & 0 \\ 2 & -2 \end{bmatrix} = \begin{bmatrix} (3\times -1)+(6\times 2) & (3\times 0)+(6\times -2) \\ (0\times -1)+(3\times 2) & (0\times 0)+(3\times -2) \end{bmatrix} = \begin{bmatrix} -3+12 & 0-12 \\ 0+6 & 0-6 \end{bmatrix}$$

$$= \begin{bmatrix} \mathbf{9} & \mathbf{-12} \\ \mathbf{6} & \mathbf{-6} \end{bmatrix}$$ or we can first multiply the entries of the two matrices by one another and then multily the result by 3 as follows:

$$3\begin{bmatrix} 1 & 2 \\ 0 & 1 \end{bmatrix}\begin{bmatrix} -1 & 0 \\ 2 & -2 \end{bmatrix} = 3\begin{bmatrix} (1\times -1)+(2\times 2) & (1\times 0)+(2\times -2) \\ (0\times -1)+(1\times 2) & (0\times 0)+(1\times -2) \end{bmatrix} = 3\begin{bmatrix} -1+4 & 0-4 \\ 0+2 & 0-2 \end{bmatrix} = 3\begin{bmatrix} 3 & -4 \\ 2 & -2 \end{bmatrix} = \begin{bmatrix} 3\times 3 & -4\times 3 \\ 2\times 3 & -2\times 3 \end{bmatrix} = \begin{bmatrix} \mathbf{9} & \mathbf{-12} \\ \mathbf{6} & \mathbf{-6} \end{bmatrix}$$

b. $$\begin{bmatrix} 1 & 2 \\ -1 & 0 \end{bmatrix}\begin{bmatrix} 1 & -1 \\ 2 & -3 \end{bmatrix} + \begin{bmatrix} 1 & 0 \\ 0 & -3 \end{bmatrix}\begin{bmatrix} -1 & 2 \\ 0 & 3 \end{bmatrix} = \begin{bmatrix} (1\times 1)+(2\times 2) & (1\times -1)+(2\times -3) \\ (-1\times 1)+(0\times 2) & (-1\times -1)+(0\times -3) \end{bmatrix} + \begin{bmatrix} (1\times -1)+(0\times 0) & (1\times 2)+(0\times 3) \\ (0\times -1)+(-3\times 0) & (0\times 2)+(-3\times 3) \end{bmatrix}$$

$$= \begin{bmatrix} 1+4 & -1-6 \\ -1+0 & 1+0 \end{bmatrix} + \begin{bmatrix} -1+0 & 2+0 \\ 0+0 & 0-9 \end{bmatrix} = \begin{bmatrix} 5 & -7 \\ -1 & 1 \end{bmatrix} + \begin{bmatrix} -1 & 2 \\ 0 & -9 \end{bmatrix} = \begin{bmatrix} 5-1 & -7+2 \\ -1+0 & 1-9 \end{bmatrix} = \begin{bmatrix} \mathbf{4} & \mathbf{-5} \\ \mathbf{-1} & \mathbf{-8} \end{bmatrix}$$

c. $$-2\begin{bmatrix} -1 & 0 \\ -2 & 3 \end{bmatrix}\begin{bmatrix} 2 & 0 \\ 3 & -1 \end{bmatrix} + 2I = -2\begin{bmatrix} (-1\times 2)+(0\times 3) & (-1\times 0)+(0\times -1) \\ (-2\times 2)+(3\times 3) & (-2\times 0)+(3\times -1) \end{bmatrix} + 2I = -2\begin{bmatrix} -2+0 & 0+0 \\ -4+9 & 0-3 \end{bmatrix} + 2I = -2\begin{bmatrix} -2 & 0 \\ 5 & -3 \end{bmatrix} + 2I$$

$$= \begin{bmatrix} -2\times -2 & 0\times -2 \\ 5\times -2 & -3\times -2 \end{bmatrix} + 2I = \begin{bmatrix} 4 & 0 \\ -10 & 6 \end{bmatrix} + 2\begin{bmatrix} 1 & 0 \\ 0 & 1 \end{bmatrix} = \begin{bmatrix} 4 & 0 \\ -10 & 6 \end{bmatrix} + \begin{bmatrix} 2 & 0 \\ 0 & 2 \end{bmatrix} = \begin{bmatrix} 4+2 & 0+0 \\ -10+0 & 6+2 \end{bmatrix} = \begin{bmatrix} \mathbf{6} & \mathbf{0} \\ \mathbf{-10} & \mathbf{8} \end{bmatrix}$$

d. $$2\begin{bmatrix} 1 & 0 \\ -2 & 3 \end{bmatrix} - \left(\begin{bmatrix} 1 & 0 \\ -1 & 2 \end{bmatrix}\begin{bmatrix} -2 & 3 \\ -1 & 1 \end{bmatrix}\right) = 2\begin{bmatrix} 1 & 0 \\ -2 & 3 \end{bmatrix} - \begin{bmatrix} (1\times -2)+(0\times -1) & (1\times 3)+(0\times 1) \\ (-1\times -2)+(2\times -1) & (-1\times 3)+(2\times 1) \end{bmatrix} = 2\begin{bmatrix} 1 & 0 \\ -2 & 3 \end{bmatrix} - \begin{bmatrix} -2+0 & 3+0 \\ 2-2 & -3+2 \end{bmatrix}$$

$$= 2\begin{bmatrix} 1 & 0 \\ -2 & 3 \end{bmatrix} - \begin{bmatrix} -2 & 3 \\ 0 & -1 \end{bmatrix} = \begin{bmatrix} 1\times 2 & 0\times 2 \\ -2\times 2 & 3\times 2 \end{bmatrix} + \begin{bmatrix} 2 & -3 \\ 0 & 1 \end{bmatrix} = \begin{bmatrix} 2 & 0 \\ -4 & 6 \end{bmatrix} + \begin{bmatrix} 2 & -3 \\ 0 & 1 \end{bmatrix} = \begin{bmatrix} 2+2 & 0-3 \\ -4+0 & 6+1 \end{bmatrix} = \begin{bmatrix} \mathbf{4} & \mathbf{-3} \\ \mathbf{-4} & \mathbf{7} \end{bmatrix}$$

e. $$\begin{bmatrix} 1 & 0 & 0 \\ -1 & 0 & -2 \\ 0 & 1 & 2 \end{bmatrix}\left(\begin{bmatrix} 1 & 0 & -1 \\ 1 & 1 & 0 \\ 0 & 0 & 1 \end{bmatrix} + \begin{bmatrix} -3 & 1 & 0 \\ 0 & -2 & 1 \\ 1 & -1 & 0 \end{bmatrix}\right) = \begin{bmatrix} 1 & 0 & 0 \\ -1 & 0 & -2 \\ 0 & 1 & 2 \end{bmatrix}\begin{bmatrix} 1-3 & 0+1 & -1+0 \\ 1+0 & 1-2 & 0+1 \\ 0+1 & 0-1 & 1+0 \end{bmatrix} = \begin{bmatrix} 1 & 0 & 0 \\ -1 & 0 & -2 \\ 0 & 1 & 2 \end{bmatrix}\begin{bmatrix} -2 & 1 & -1 \\ 1 & -1 & 1 \\ 1 & -1 & 1 \end{bmatrix}$$

$$= \begin{bmatrix} (1\times -2)+(0\times 1)+(0\times 1) & (1\times 1)+(0\times -1)+(0\times -1) & (1\times -1)+(0\times 1)+(0\times 1) \\ (-1\times -2)+(0\times 1)+(-2\times 1) & (-1\times 1)+(0\times -1)+(-2\times -1) & (-1\times -1)+(0\times 1)+(-2\times 1) \\ (0\times -2)+(1\times 1)+(2\times 1) & (0\times 1)+(1\times -1)+(2\times -1) & (0\times -1)+(1\times 1)+(2\times 1) \end{bmatrix} = \begin{bmatrix} -2+0+0 & 1+0+0 & -1+0+0 \\ 2+0-2 & -1+0+2 & 1+0-2 \\ 0+1+2 & 0-1-2 & 0+1+2 \end{bmatrix}$$

$$= \begin{bmatrix} \mathbf{-2} & \mathbf{1} & \mathbf{-1} \\ \mathbf{0} & \mathbf{1} & \mathbf{-1} \\ \mathbf{3} & \mathbf{-3} & \mathbf{3} \end{bmatrix}$$

f. $$\left(\begin{bmatrix} 1 & 2 & -1 \\ 0 & 1 & 0 \\ -1 & 0 & 2 \end{bmatrix} - \begin{bmatrix} 1 & 2 & -3 \\ 1 & 0 & 1 \\ 0 & -2 & -1 \end{bmatrix}\right)\begin{bmatrix} 0 & -1 & 2 \\ -3 & 2 & -1 \\ 0 & 1 & -3 \end{bmatrix} = \left(\begin{bmatrix} 1 & 2 & -1 \\ 0 & 1 & 0 \\ -1 & 0 & 2 \end{bmatrix} + \begin{bmatrix} -1 & -2 & 3 \\ -1 & 0 & -1 \\ 0 & 2 & 1 \end{bmatrix}\right)\begin{bmatrix} 0 & -1 & 2 \\ -3 & 2 & -1 \\ 0 & 1 & -3 \end{bmatrix} = \begin{bmatrix} 1-1 & 2-2 & -1+3 \\ 0-1 & 1+0 & 0-1 \\ -1+0 & 0+2 & 2+1 \end{bmatrix}$$

$$\begin{bmatrix} 0 & -1 & 2 \\ -3 & 2 & -1 \\ 0 & 1 & -3 \end{bmatrix} = \begin{bmatrix} 0 & 0 & 2 \\ -1 & 1 & -1 \\ -1 & 2 & 3 \end{bmatrix}\begin{bmatrix} 0 & -1 & 2 \\ -3 & 2 & -1 \\ 0 & 1 & -3 \end{bmatrix} = \begin{bmatrix} (0\times 0)+(0\times -3)+(2\times 0) & (0\times -1)+(0\times 2)+(2\times 1) & (0\times 2)+(0\times -1)+(2\times -3) \\ (-1\times 0)+(1\times -3)+(-1\times 0) & (-1\times -1)+(1\times 2)+(-1\times 1) & (-1\times 2)+(1\times -1)+(-1\times -3) \\ (-1\times 0)+(2\times -3)+(3\times 0) & (-1\times -1)+(2\times 2)+(3\times 1) & (-1\times 2)+(2\times -1)+(3\times -3) \end{bmatrix}$$

$$= \begin{bmatrix} 0+0+0 & 0+0+2 & 0+0-6 \\ 0-3+0 & 1+2-1 & -2-1+3 \\ 0-6+0 & 1+4+3 & -2-2-9 \end{bmatrix} = \begin{bmatrix} \mathbf{0} & \mathbf{2} & \mathbf{-6} \\ \mathbf{-3} & \mathbf{2} & \mathbf{0} \\ \mathbf{-6} & \mathbf{8} & \mathbf{-13} \end{bmatrix}$$

7. Given $B = \begin{bmatrix} 1 & -2 & 3 \\ 5 & -3 & 4 \\ 1 & 2 & 3 \end{bmatrix}$, $U = \begin{bmatrix} u \\ v \\ w \end{bmatrix}$, and $C = \begin{bmatrix} 1 \\ -2 \\ 4 \end{bmatrix}$ write the linear system $BU = C$.

$$\begin{bmatrix} 1 & -2 & 3 \\ 5 & -3 & 4 \\ 1 & 2 & 3 \end{bmatrix}\begin{bmatrix} u \\ v \\ w \end{bmatrix} = \begin{bmatrix} 1 \\ -2 \\ 4 \end{bmatrix} ; \begin{bmatrix} (1\times u)+(-2\times v)+(3\times w) \\ (5\times u)+(-3\times v)+(4\times w) \\ (1\times u)+(2\times v)+(3\times w) \end{bmatrix} = \begin{bmatrix} 1 \\ -2 \\ 4 \end{bmatrix} ; \begin{bmatrix} u-2v+3w \\ 5u-3v+4w \\ u+2v+3w \end{bmatrix} = \begin{bmatrix} 1 \\ -2 \\ 4 \end{bmatrix} ; \begin{matrix} \mathbf{u-2v+3w=1} \\ \mathbf{5u-3v+4w=-2} \\ \mathbf{u+2v+3w=4} \end{matrix}$$

Section 3.3 Solutions - Determinants

1. Given $A = \begin{bmatrix} 1 & 3 & 4 & -1 \\ 0 & 1 & -2 & 0 \\ 1 & -1 & -3 & 5 \\ -3 & 0 & 6 & 8 \end{bmatrix}$, the entry elements are: a. $a_{23} = \mathbf{-2}$ b. $a_{11} = \mathbf{1}$ c. $a_{33} = \mathbf{-3}$ d. $a_{41} = \mathbf{-3}$

e. $a_{43} = \mathbf{6}$ f. $a_{44} = \mathbf{8}$ g. $a_{34} = \mathbf{5}$ h. $a_{32} = \mathbf{-1}$ i. $a_{24} = \mathbf{0}$ j. $a_{31} = \mathbf{1}$ k. $a_{13} = \mathbf{4}$

l. $a_{42} = \mathbf{0}$

2. Let $A = \begin{bmatrix} 1 & 2 \\ 3 & -1 \end{bmatrix}$. Compute the following minors and cofactors.

a. $M_{11} = \mathbf{-1}$ b. $M_{12} = \mathbf{3}$ c. $M_{21} = \mathbf{2}$ d. $M_{22} = \mathbf{1}$

e. $A_{11} = (-1)^{1+1}M_{11} = (-1)^2 \times -1 = 1\times -1 = \mathbf{-1}$

f. $A_{12} = (-1)^{1+2}M_{12} = (-1)^3 \times 3 = -1\times 3 = \mathbf{-3}$

g. $A_{21} = (-1)^{2+1}M_{21} = (-1)^3 \times 2 = -1\times 2 = \mathbf{-2}$

h. $A_{22} = (-1)^{2+2}M_{22} = (-1)^4 \times 1 = 1\times 1 = \mathbf{1}$

3. Let $A = \begin{bmatrix} 1 & 3 & -1 \\ 2 & 3 & -2 \\ 1 & 0 & 4 \end{bmatrix}$. Compute the indicated minors and cofactors.

a. $M_{23} = \begin{vmatrix} 1 & 3 \\ 1 & 0 \end{vmatrix} = (1\times 0)-(3\times 1) = 0-3 = \mathbf{-3}$

b. $M_{13} = \begin{vmatrix} 2 & 3 \\ 1 & 0 \end{vmatrix} = (2\times 0)-(3\times 1) = 0-3 = \mathbf{-3}$

c. $M_{32} = \begin{vmatrix} 1 & -1 \\ 2 & -2 \end{vmatrix} = (1\times -2)-(-1\times 2) = -2+2 = \mathbf{0}$

d. $M_{11} = \begin{vmatrix} 3 & -2 \\ 0 & 4 \end{vmatrix} = (3\times 4)-(0\times -2) = 12-0 = \mathbf{12}$

e. $A_{23} = (-1)^{2+3}M_{23} = (-1)^5 \times -3 = -1\times -3 = \mathbf{3}$

f. $A_{13} = (-1)^{1+3}M_{13} = (-1)^4 \times -3 = 1\times -3 = \mathbf{-3}$

g. $A_{32} = (-1)^{3+2}M_{32} = (-1)^5 \times 0 = -1\times 0 = \mathbf{0}$

h. $A_{11} = (-1)^{1+1}M_{11} = (-1)^2 \times 12 = 1\times 12 = \mathbf{12}$

4. Let $A = \begin{bmatrix} 1 & 0 & 0 \\ 1 & -1 & 0 \\ 2 & 1 & 1 \end{bmatrix}$. Compute the indicated minors and cofactors.

a. $M_{31} = \begin{vmatrix} 0 & 0 \\ -1 & 0 \end{vmatrix} = (0\times 0)-(0\times -1) = 0-0 = \mathbf{0}$

b. $M_{33} = \begin{vmatrix} 1 & 0 \\ 1 & -1 \end{vmatrix} = (1\times -1)-(0\times 1) = -1-0 = \mathbf{-1}$

c. $M_{12} = \begin{vmatrix} 1 & 0 \\ 2 & 1 \end{vmatrix} = (1\times 1)-(0\times 2) = 1-0 = \mathbf{1}$

d. $M_{22} = \begin{vmatrix} 1 & 0 \\ 2 & 1 \end{vmatrix} = (1\times 1)-(0\times 2) = 1-0 = \mathbf{1}$

e. $A_{31} = (-1)^{3+1} M_{31} = (-1)^4 \times 0 = 1 \times 0 = \mathbf{0}$

f. $A_{33} = (-1)^{3+3} M_{33} = (-1)^6 \times -1 = 1 \times -1 = \mathbf{-1}$

g. $A_{12} = (-1)^{1+2} M_{12} = (-1)^3 \times 1 = -1 \times 1 = \mathbf{-1}$

h. $A_{22} = (-1)^{2+2} M_{22} = (-1)^4 \times 1 = 1 \times 1 = \mathbf{1}$

5. Find the determinant of the following matrices.

a. $\delta(A) = \delta\begin{bmatrix} -1 & 2 \\ 3 & 0 \end{bmatrix} = \begin{vmatrix} -1 & 2 \\ 3 & 0 \end{vmatrix} = (-1\times 0) - (2\times 3) = 0 - 6 = \mathbf{-6}$

b. $\delta(B) = \delta\begin{bmatrix} 4 & 0 \\ 0 & 3 \end{bmatrix} = \begin{vmatrix} 4 & 0 \\ 0 & 3 \end{vmatrix} = (4\times 3) - (0\times 0) = 12 - 0 = \mathbf{12}$

c. $\delta(C) = \delta\begin{bmatrix} 1 & -3 \\ 0 & 2 \end{bmatrix} = \begin{vmatrix} 1 & -3 \\ 0 & 2 \end{vmatrix} = (1\times 2) - (-3\times 0) = 2 - 0 = \mathbf{2}$

d. $\delta(D) = \delta\begin{bmatrix} 3 & -2 \\ 5 & -3 \end{bmatrix} = \begin{vmatrix} 3 & -2 \\ 5 & -3 \end{vmatrix} = (3\times -3) - (-2\times 5) = -9 + 10 = \mathbf{1}$

e. $\delta(E) = \delta\begin{bmatrix} 1 & 0 \\ 0 & 1 \end{bmatrix} = \begin{vmatrix} 1 & 0 \\ 0 & 1 \end{vmatrix} = (1\times 1) - (0\times 0) = 1 - 0 = \mathbf{1}$

6. Given $A = \begin{bmatrix} 1 & 0 & -1 \\ 2 & 3 & 1 \\ 0 & 0 & 4 \end{bmatrix}$, the entry elements are: $a_{11} = 1$, $a_{12} = 0$, $a_{13} = -1$, $a_{21} = 2$, $a_{22} = 3$, $a_{23} = 1$, $a_{31} = 0$, $a_{32} = 0$, and $a_{33} = 4$. Thus,

a. Expanding about the second row we obtain:

$\delta(A) = a_{21}\cdot A_{21} + a_{22}\cdot A_{22} + a_{23}\cdot A_{23} = 2\cdot A_{21} + 3\cdot A_{22} + 1\cdot A_{23} = 2\cdot(-1)^{2+1} M_{21} + 3\cdot(-1)^{2+2} M_{22} + 1\cdot(-1)^{2+3} M_{23}$

$= -2M_{21} + 3M_{22} - M_{23} = -2\begin{vmatrix} 0 & -1 \\ 0 & 4 \end{vmatrix} + 3\begin{vmatrix} 1 & -1 \\ 0 & 4 \end{vmatrix} - \begin{vmatrix} 1 & 0 \\ 0 & 0 \end{vmatrix} = -2[(0\times 4) - (-1\times 0)] + 3[(1\times 4) - (-1\times 0)] - [(1\times 0) - (0\times 0)]$

$= -2(0-0) + 3(4-0) - (0-0) = 0 + 12 + 0 = \mathbf{12}$

b. Expanding about the third row we obtain:

$\delta(A) = a_{31}\cdot A_{31} + a_{32}\cdot A_{32} + a_{33}\cdot A_{33} = 0\cdot A_{31} + 0\cdot A_{32} + 4\cdot A_{33} = 4\cdot A_{33} = 4\cdot(-1)^{3+3} M_{33} = 4M_{33} = 4\begin{vmatrix} 1 & 0 \\ 2 & 3 \end{vmatrix}$

$= 4[(1\times 3) - (0\times 2)] = 12 - 0 = \mathbf{12}$

c. Expanding about the first column we obtain:

$\delta(A) = a_{11}\cdot A_{11} + a_{21}\cdot A_{21} + a_{31}\cdot A_{31} = 1\cdot A_{11} + 2\cdot A_{21} + 0\cdot A_{31} = A_{11} + 2A_{21} = (-1)^{1+1} M_{11} + 2(-1)^{2+1} M_{21} = M_{11} - 2M_{21}$

$= \begin{vmatrix} 3 & 1 \\ 0 & 4 \end{vmatrix} - 2\begin{vmatrix} 0 & -1 \\ 0 & 4 \end{vmatrix} = [(3\times 4) - (1\times 0)] - 2[(0\times 4) - (-1\times 0)] = 12 - 0 = \mathbf{12}$

d. Expanding about the second column we obtain:

$\delta(A) = a_{12}\cdot A_{12} + a_{22}\cdot A_{22} + a_{32}\cdot A_{32} = 0\cdot A_{12} + 3\cdot A_{22} + 0\cdot A_{32} = 3A_{22} = 3\cdot(-1)^{2+2} M_{22} = 3M_{22} = 3\begin{vmatrix} 1 & -1 \\ 0 & 4 \end{vmatrix}$

$= 3[(1\times 4) - (-1\times 0)] = 12 - 0 = \mathbf{12}$

7. Find the determinant of the following A matrices. Expand about the indicated rows and columns.

a. Given $\begin{bmatrix} 2 & 1 & 4 \\ 0 & -1 & 3 \\ 0 & 0 & 5 \end{bmatrix}$, 1. Expanding about the third column we obtain:

$\delta(A) = a_{13}\cdot A_{13} + a_{23}\cdot A_{23} + a_{33}\cdot A_{33} = 4\cdot A_{13} + 3\cdot A_{23} + 5\cdot A_{33} = 4\cdot(-1)^{1+3} M_{13} + 3\cdot(-1)^{2+3} M_{23} + 5\cdot(-1)^{3+3} M_{33}$

$= 4\cdot M_{13} - 3\cdot M_{23} + 5\cdot M_{33} = 4\begin{vmatrix} 0 & -1 \\ 0 & 0 \end{vmatrix} - 3\begin{vmatrix} 2 & 1 \\ 0 & 0 \end{vmatrix} + 5\begin{vmatrix} 2 & 1 \\ 0 & -1 \end{vmatrix} = 4[(0\times 0) - (-1\times 0)] - 3[(2\times 0) - (1\times 0)] + 5[(2\times -1) - (1\times 0)]$

$= 0+0-10 = \mathbf{-10}$

2. Expanding about the third row we obtain:

$$\delta(A) = a_{31}\cdot A_{31}+a_{32}\cdot A_{32}+a_{33}\cdot A_{33} = 0\cdot A_{31}+0\cdot A_{32}+5\cdot A_{33} = 5\cdot A_{33} = 5\cdot(-1)^{3+3}M_{33} = 5M_{33} = 5\begin{vmatrix}2 & 1\\ 0 & -1\end{vmatrix}$$

$= 5[(2\times -1)-(1\times 0)] = \mathbf{-10}$

b. Given $\begin{bmatrix}2 & -1 & 1\\ -1 & 3 & 0\\ 0 & 4 & 0\end{bmatrix}$, 1. Expanding about the third column we obtain:

$$\delta(A) = a_{13}\cdot A_{13}+a_{23}\cdot A_{23}+a_{33}\cdot A_{33} = 1\cdot A_{13}+0\cdot A_{23}+0\cdot A_{33} = A_{13} = (-1)^{1+3}M_{13} = M_{13} = \begin{vmatrix}-1 & 3\\ 0 & 4\end{vmatrix} = (-1\times 4)-(3\times 0) = \mathbf{-4}$$

2. Expanding about the first row we obtain:

$$\delta(A) = a_{11}\cdot A_{11}+a_{12}\cdot A_{12}+a_{13}\cdot A_{13} = 2\cdot A_{11}+(-1)\cdot A_{12}+1\cdot A_{13} = 2\cdot(-1)^{1+1}M_{11}-1\cdot(-1)^{1+2}M_{12}+1\cdot(-1)^{1+3}M_{13}$$

$$= 2M_{11}+M_{12}+M_{13} = 2\begin{vmatrix}3 & 0\\ 4 & 0\end{vmatrix}+\begin{vmatrix}-1 & 0\\ 0 & 0\end{vmatrix}+\begin{vmatrix}-1 & 3\\ 0 & 4\end{vmatrix} = -2[(3\times 0)-(0\times 4)]+[(-1\times 0)-(0\times 0)]+[(-1\times 4)-(3\times 0)]$$

$= -2(0-0)+(0-0)+(-4-0) = 0+0-4 = \mathbf{-4}$

c. Given $\begin{bmatrix}3 & 1 & 1\\ 2 & -3 & 2\\ 9 & 3 & 3\end{bmatrix}$, 1. Expanding about the second row we obtain:

$$\delta(A) = a_{21}\cdot A_{21}+a_{22}\cdot A_{22}+a_{23}\cdot A_{23} = 2\cdot A_{21}-3\cdot A_{22}+2\cdot A_{23} = 2\cdot(-1)^{2+1}M_{21}-3\cdot(-1)^{2+2}M_{22}+2\cdot(-1)^{2+3}M_{23}$$

$$= -2M_{21}-3M_{22}-2M_{23} = -2\begin{vmatrix}1 & 1\\ 3 & 3\end{vmatrix}-3\begin{vmatrix}3 & 1\\ 9 & 3\end{vmatrix}-2\begin{vmatrix}3 & 1\\ 9 & 3\end{vmatrix} = -2[(1\times 3)-(1\times 3)]-3[(3\times 3)-(1\times 9)]-2[(3\times 3)-(1\times 9)]$$

$= -2(0-0)-3(9-9)-2(9-9) = 0+0+0 = \mathbf{0}$

2. Expanding about the second column we obtain:

$$\delta(A) = a_{12}\cdot A_{12}+a_{22}\cdot A_{22}+a_{32}\cdot A_{32} = 1\cdot A_{12}-3\cdot A_{22}+3\cdot A_{32} = 1\cdot(-1)^{1+2}M_{12}-3\cdot(-1)^{2+2}M_{22}+3\cdot(-1)^{3+2}M_{32}$$

$$= -M_{12}-3M_{22}-3M_{32} = -\begin{vmatrix}2 & 2\\ 9 & 3\end{vmatrix}-3\begin{vmatrix}3 & 1\\ 9 & 3\end{vmatrix}-3\begin{vmatrix}3 & 1\\ 2 & 2\end{vmatrix} = -[(2\times 3)-(2\times 9)]-3[(3\times 3)-(1\times 9)]-3[(3\times 2)-(1\times 2)]$$

$= -(6-18)-3(9-9)-3(6-2) = 12-0-12 = \mathbf{0}$

d. Given $\begin{bmatrix}-2 & 0 & 0\\ -1 & 2 & 0\\ 5 & 7 & 0\end{bmatrix}$, 1. Expanding about the third column we obtain:

$$\delta(A) = a_{13}\cdot A_{13}+a_{23}\cdot A_{23}+a_{33}\cdot A_{33} = 0\cdot A_{13}+0\cdot A_{23}+0\cdot A_{33} = 0+0+0 = \mathbf{0}$$

2. Expanding about the first column we obtain:

$$\delta(A) = a_{11}\cdot A_{11}+a_{21}\cdot A_{21}+a_{31}\cdot A_{31} = -2\cdot A_{11}-1\cdot A_{21}+5\cdot A_{31} = -2\cdot(-1)^{1+1}M_{11}-(-1)^{2+1}M_{21}+5\cdot(-1)^{3+1}M_{31}$$

$$= -2M_{11}+M_{21}+5M_{31} = -2\begin{vmatrix}2 & 0\\ 7 & 0\end{vmatrix}+\begin{vmatrix}0 & 0\\ 7 & 0\end{vmatrix}+5\begin{vmatrix}0 & 0\\ 2 & 0\end{vmatrix} = 4[(2\times 0)-(0\times 7)]+[(0\times 0)-(0\times 7)]+5[(0\times 0)-(0\times 2)] = 0+0+0 = \mathbf{0}$$

8. Determine the determinant of the following matrices by observation only.

a. $\delta(A) = \begin{vmatrix}1 & 3 & -1\\ 0 & 0 & 0\\ -3 & 5 & 8\end{vmatrix} = \mathbf{0}$ b. $\delta(A) = \begin{vmatrix}5 & 1 & 5\\ 0 & -1 & 0\\ 2 & 1 & 2\end{vmatrix} = \mathbf{0}$ c. $\delta(A) = \begin{vmatrix}1 & 5 & 6 & 8\\ 2 & 3 & -1 & 0\\ 1 & 5 & 6 & 8\\ 0 & 1 & 2 & -3\end{vmatrix} = \mathbf{0}$ d. $\delta(A) = \begin{vmatrix}1 & 0 & 1 & 2\\ 2 & 0 & -1 & -3\\ -3 & 0 & 3 & 5\\ 4 & 0 & 5 & 6\end{vmatrix} = \mathbf{0}$

9. Solve for the unknown.

a. $\begin{vmatrix} x & 2 \\ 6 & 4 \end{vmatrix} = 4$; $(x \times 4) - (2 \times 6) = 4$; $4x - 12 = 4$; $4x = 4 + 12$; $4x = 16$; $x = \frac{16}{4}$; $\mathbf{x = 4}$

b. $\begin{vmatrix} -2 & 3 \\ 4 & x \end{vmatrix} = -4$; $(-2 \times x) - (3 \times 4) = -4$; $-2x - 12 = -4$; $-2x = -4 + 12$; $-2x = 8$; $x = -\frac{8}{2}$; $\mathbf{x = -4}$

c. $\begin{vmatrix} 3 & y \\ 2 & 5 \end{vmatrix} = 1$; $(3 \times 5) - (y \times 2) = 1$; $15 - 2y = 1$; $-2y = 1 - 15$; $-2y = -14$; $y = \frac{14}{2}$; $\mathbf{y = 7}$

d. $\begin{vmatrix} 1 & 3 \\ w & -9 \end{vmatrix} = 12$; $(1 \times -9) - (3 \times w) = 12$; $-9 - 3w = 12$; $-3w = 12 + 9$; $-3w = 21$; $w = -\frac{21}{3}$; $\mathbf{w = -7}$

Section 3.4 Solutions - Inverse Matrices

1. Use the minor and cofactor method to find the inverse of the following 2×2 matrices, if it exists. Verify each answer by showing that $A \times A^{-1} = I$.

a. Given $A = \begin{bmatrix} 1 & 2 \\ 3 & 6 \end{bmatrix}$, first find the determinant of A, i.e.,

$$\delta(A) = a_{11} \cdot A_{11} + a_{12} \cdot A_{12} = 1 \cdot A_{11} + 2 \cdot A_{12} = 1 \cdot (-1)^{1+1} \cdot M_{11} + 2 \cdot (-1)^{1+2} \cdot M_{12} = 1 \cdot (-1)^2 \cdot M_{11} + 2 \cdot (-1)^3 \cdot M_{12}$$

$= 1 \cdot M_{11} - 2 \cdot M_{12} = (1 \cdot 6) - (2 \cdot 3) = 6 - 6 = 0$. **Since $\delta(A) = 0$ matrix A does not have an inverse.**

b. Given $A = \begin{bmatrix} -3 & -1 \\ 9 & 3 \end{bmatrix}$, first find the determinant of A, i.e.,

$$\delta(A) = a_{11} \cdot A_{11} + a_{12} \cdot A_{12} = -3 \cdot A_{11} + (-1) \cdot A_{12} = -3 \cdot (-1)^{1+1} \cdot M_{11} - 1 \cdot (-1)^{1+2} \cdot M_{12} = -3 \cdot (-1)^2 \cdot M_{11} - (-1)^3 \cdot M_{12}$$

$= -3 \cdot M_{11} + M_{12} = (-3 \cdot 3) + 9 = -9 + 9 = 0$. **Since $\delta(A) = 0$ matrix A does not have an inverse.**

c. Given $A = \begin{bmatrix} 2 & 3 \\ 2 & 4 \end{bmatrix}$ first find the determinant of A, i.e.,

$$\delta(A) = a_{11} \cdot A_{11} + a_{12} \cdot A_{12} = 2 \cdot A_{11} + 3 \cdot A_{12} = 2 \cdot (-1)^{1+1} \cdot M_{11} + 3 \cdot (-1)^{1+2} \cdot M_{12} = 2 \cdot (-1)^2 \cdot M_{11} + 3 \cdot (-1)^3 \cdot M_{12}$$

$= 2 \cdot M_{11} - 3 \cdot M_{12} = (2 \cdot 4) - (3 \cdot 2) = 8 - 6 = 2$. Since $\delta(A) \neq 0$ matrix A has an inverse.

Next, replace each entry in A with its cofactor.

$A_{11} = (-1)^{1+1} M_{11} = (-1)^2 \times 4 = 4$, $A_{12} = (-1)^{1+2} M_{12} = (-1)^3 \times 2 = -2$, $A_{21} = (-1)^{2+1} M_{21} = (-1)^3 \times 3 = -3$, and $A_{22} = (-1)^{2+2} M_{22} = (-1)^4 \times 2 = 2$. Therefore, the cofactor matrix is equal to $C = \begin{bmatrix} A_{11} & A_{12} \\ A_{21} & A_{22} \end{bmatrix} = \begin{bmatrix} 4 & -2 \\ -3 & 2 \end{bmatrix}$ and

$C^t = \begin{bmatrix} 4 & -3 \\ -2 & 2 \end{bmatrix}$. Thus, $A^{-1} = \frac{1}{\delta(A)} C^t = \frac{1}{2} \cdot \begin{bmatrix} 4 & -3 \\ -2 & 2 \end{bmatrix} = \begin{bmatrix} 4 \times \frac{1}{2} & -3 \times \frac{1}{2} \\ -2 \times \frac{1}{2} & 2 \times \frac{1}{2} \end{bmatrix} = \begin{bmatrix} \frac{4}{2} & -\frac{3}{2} \\ -\frac{2}{2} & \frac{2}{2} \end{bmatrix} = \begin{bmatrix} \mathbf{2} & -\frac{\mathbf{3}}{\mathbf{2}} \\ \mathbf{-1} & \mathbf{1} \end{bmatrix}$

Check the answer by multiplying the A matrix with A^{-1}. The result should be equal to the identity matrix.

$$A \times A^{-1} = \begin{bmatrix} 2 & 3 \\ 2 & 4 \end{bmatrix} \times \begin{bmatrix} 2 & -\frac{3}{2} \\ -1 & 1 \end{bmatrix} = \begin{bmatrix} (2 \times 2) + (3 \times -1) & \left(2 \times -\frac{3}{2}\right) + (3 \times 1) \\ (2 \times 2) + (4 \times -1) & \left(2 \times -\frac{3}{2}\right) + (4 \times 1) \end{bmatrix} = \begin{bmatrix} 4-3 & -3+3 \\ 4-4 & -3+4 \end{bmatrix} = \begin{bmatrix} 1 & 0 \\ 0 & 1 \end{bmatrix}$$

d. Given $A = \begin{bmatrix} 3 & 1 \\ -1 & 2 \end{bmatrix}$ first find the determinant of A, i.e.,

$$\delta(A) = a_{11} \cdot A_{11} + a_{12} \cdot A_{12} = 3 \cdot A_{11} + 1 \cdot A_{12} = 3 \cdot (-1)^{1+1} \cdot M_{11} + 1 \cdot (-1)^{1+2} \cdot M_{12} = 3 \cdot (-1)^2 \cdot M_{11} + 1 \cdot (-1)^3 \cdot M_{12}$$

$= 3 \cdot M_{11} - 1 \cdot M_{12} = (3 \cdot 2) - (1 \cdot -1) = 6 + 1 = 7$. Since $\delta(A) \neq 0$ matrix A has an inverse.

Next, replace each entry in A with its cofactor.

$A_{11} = (-1)^{1+1} M_{11} = (-1)^2 \times 2 = 2$, $A_{12} = (-1)^{1+2} M_{12} = (-1)^3 \times -1 = 1$, $A_{21} = (-1)^{2+1} M_{21} = (-1)^3 \times 1 = -1$, and

$A_{22} = (-1)^{2+2} M_{22} = (-1)^4 \times 3 = 3$. Therefore, the cofactor matrix is equal to $C = \begin{bmatrix} A_{11} & A_{12} \\ A_{21} & A_{22} \end{bmatrix} = \begin{bmatrix} 2 & 1 \\ -1 & 3 \end{bmatrix}$ and

$C^t = \begin{bmatrix} 2 & -1 \\ 1 & 3 \end{bmatrix}$. Thus, $A^{-1} = \frac{1}{\delta(A)} C^t = \frac{1}{7} \cdot \begin{bmatrix} 2 & -1 \\ 1 & 3 \end{bmatrix} = \begin{bmatrix} 2 \times \frac{1}{7} & -1 \times \frac{1}{7} \\ 1 \times \frac{1}{7} & 3 \times \frac{1}{7} \end{bmatrix} = \begin{bmatrix} \mathbf{\frac{2}{7}} & \mathbf{-\frac{1}{7}} \\ \mathbf{\frac{1}{7}} & \mathbf{\frac{3}{7}} \end{bmatrix}$

Check the answer by multiplying the A matrix with A^{-1}. The result should be equal to the identity matrix.

$$A \times A^{-1} = \begin{bmatrix} 3 & 1 \\ -1 & 2 \end{bmatrix} \times \begin{bmatrix} \frac{2}{7} & -\frac{1}{7} \\ \frac{1}{7} & \frac{3}{7} \end{bmatrix} = \begin{bmatrix} \left(3 \times \frac{2}{7}\right) + \left(1 \times \frac{1}{7}\right) & \left(3 \times -\frac{1}{7}\right) + \left(1 \times \frac{3}{7}\right) \\ \left(-1 \times \frac{2}{7}\right) + \left(2 \times \frac{1}{7}\right) & \left(-1 \times -\frac{1}{7}\right) + \left(2 \times \frac{3}{7}\right) \end{bmatrix} = \begin{bmatrix} \frac{6}{7} + \frac{1}{7} & -\frac{3}{7} + \frac{3}{7} \\ -\frac{2}{7} + \frac{2}{7} & \frac{1}{7} + \frac{6}{7} \end{bmatrix} = \begin{bmatrix} 1 & 0 \\ 0 & 1 \end{bmatrix}$$

e. Given $A = \begin{bmatrix} -2 & -1 \\ 2 & 3 \end{bmatrix}$ first find the determinant of A, i.e.,

$\delta(A) = a_{11} \cdot A_{11} + a_{12} \cdot A_{12} = -2 \cdot A_{11} + (-1) \cdot A_{12} = -2 \cdot (-1)^{1+1} \cdot M_{11} + (-1) \cdot (-1)^{1+2} \cdot M_{12} = -2 \cdot (-1)^2 \cdot M_{11} - (-1)^3 \cdot M_{12}$

$= -2 \cdot M_{11} + 1 \cdot M_{12} = (-2 \cdot 3) + (1 \cdot 2) = -6 + 2 = -4$. Since $\delta(A) \neq 0$ matrix A has an inverse.

Next, replace each entry in A with its cofactor.

$A_{11} = (-1)^{1+1} M_{11} = (-1)^2 \times 3 = 3$, $A_{12} = (-1)^{1+2} M_{12} = (-1)^3 \times 2 = -2$, $A_{21} = (-1)^{2+1} M_{21} = (-1)^3 \times -1 = 1$, and

$A_{22} = (-1)^{2+2} M_{22} = (-1)^4 \times -2 = -2$. Therefore, the cofactor matrix is equal to $C = \begin{bmatrix} A_{11} & A_{12} \\ A_{21} & A_{22} \end{bmatrix} = \begin{bmatrix} 3 & -2 \\ 1 & -2 \end{bmatrix}$ and

$C^t = \begin{bmatrix} 3 & 1 \\ -2 & -2 \end{bmatrix}$. Thus, $A^{-1} = \frac{1}{\delta(A)} C^t = -\frac{1}{4} \cdot \begin{bmatrix} 3 & 1 \\ -2 & -2 \end{bmatrix} = \begin{bmatrix} 3 \times -\frac{1}{4} & 1 \times -\frac{1}{4} \\ -2 \times -\frac{1}{4} & -2 \times -\frac{1}{4} \end{bmatrix} = \begin{bmatrix} \mathbf{-\frac{3}{4}} & \mathbf{-\frac{1}{4}} \\ \mathbf{\frac{1}{2}} & \mathbf{\frac{1}{2}} \end{bmatrix}$

Check the answer by multiplying the A matrix with A^{-1}. The result should be equal to the identity matrix.

$$A \times A^{-1} = \begin{bmatrix} -2 & -1 \\ 2 & 3 \end{bmatrix} \times \begin{bmatrix} -\frac{3}{4} & -\frac{1}{4} \\ \frac{1}{2} & \frac{1}{2} \end{bmatrix} = \begin{bmatrix} \left(-2 \times -\frac{3}{4}\right) + \left(-1 \times \frac{1}{2}\right) & \left(-2 \times -\frac{1}{4}\right) + \left(-1 \times \frac{1}{2}\right) \\ \left(2 \times -\frac{3}{4}\right) + \left(3 \times \frac{1}{2}\right) & \left(2 \times -\frac{1}{4}\right) + \left(3 \times \frac{1}{2}\right) \end{bmatrix} = \begin{bmatrix} \frac{3}{2} - \frac{1}{2} & \frac{1}{2} - \frac{1}{2} \\ -\frac{3}{2} + \frac{3}{2} & -\frac{1}{2} + \frac{3}{2} \end{bmatrix} = \begin{bmatrix} 1 & 0 \\ 0 & 1 \end{bmatrix}$$

f. Given $A = \begin{bmatrix} 3 & 15 \\ 1 & 5 \end{bmatrix}$ first find the determinant of A, i.e.,

$\delta(A) = a_{11} \cdot A_{11} + a_{12} \cdot A_{12} = 3 \cdot A_{11} + 15 \cdot A_{12} = 3 \cdot (-1)^{1+1} \cdot M_{11} + 15 \cdot (-1)^{1+2} \cdot M_{12} = 3 \cdot (-1)^2 \cdot M_{11} + 15 \cdot (-1)^3 \cdot M_{12}$

$= 3 \cdot M_{11} - 15 \cdot M_{12} = (3 \cdot 5) - (15 \cdot 1) = 15 - 15 = 0$. **Since $\delta(A) = 0$ matrix A does not have an inverse.**

2. Use the minor and cofactor method to find the inverse of the following 3×3 matrices, if it exists. Verify each answer by showing that $A \times A^{-1} = I$.

a. Given $A = \begin{bmatrix} 1 & 0 & 2 \\ 2 & -3 & 0 \\ 3 & 4 & 0 \end{bmatrix}$ first find the determinant of A, i.e., obtain $\delta(A)$ by expanding about the third column. Note that $a_{13} = 2$, $a_{23} = 0$, and $a_{33} = 0$.

$\delta(A) = a_{13} \cdot A_{13} + a_{23} \cdot A_{23} + a_{33} \cdot A_{33} = 2 \cdot A_{13} + 0 \cdot A_{23} + 0 \cdot A_{33} = 2A_{13} + 0 + 0 = 2A_{13} = 2 \cdot (-1)^{1+3} \cdot M_{13} = 2 \cdot 1 \cdot M_{13}$

$= 2M_{13} = 2 \cdot \begin{vmatrix} 2 & -3 \\ 3 & 4 \end{vmatrix} = 2 \cdot [(2 \times 4) - (-3 \times 3)] = 2 \cdot (8 + 9) = 2 \cdot 17 = 34$. Since $\delta(A) \neq 0$ matrix A has an inverse.

First, replace each entry in A with its cofactor.

$$A_{11} = (-1)^{1+1} \cdot M_{11} = (-1)^2 \cdot M_{11} = M_{11} = \begin{vmatrix} -3 & 0 \\ 4 & 0 \end{vmatrix} = (-3\times 0)-(0\times 4) = 0-0 = 0$$

$$A_{12} = (-1)^{1+2} \cdot M_{12} = (-1)^3 \cdot M_{12} = -M_{12} = -\begin{vmatrix} 2 & 0 \\ 3 & 0 \end{vmatrix} = -[(2\times 0)-(0\times 3)] = -(0-0) = 0$$

$$A_{13} = (-1)^{1+3} \cdot M_{13} = (-1)^4 \cdot M_{13} = M_{13} = \begin{vmatrix} 2 & -3 \\ 3 & 4 \end{vmatrix} = (2\times 4)-(-3\times 3) = 8+9 = 17$$

$$A_{21} = (-1)^{2+1} \cdot M_{21} = (-1)^3 \cdot M_{21} = -M_{21} = -\begin{vmatrix} 0 & 2 \\ 4 & 0 \end{vmatrix} = -[(0\times 0)-(2\times 4)] = -(0-8) = 8$$

$$A_{22} = (-1)^{2+2} \cdot M_{22} = (-1)^4 \cdot M_{22} = M_{22} = \begin{vmatrix} 1 & 2 \\ 3 & 0 \end{vmatrix} = (1\times 0)-(2\times 3) = 0-6 = -6$$

$$A_{23} = (-1)^{2+3} \cdot M_{23} = (-1)^5 \cdot M_{23} = -M_{23} = -\begin{vmatrix} 1 & 0 \\ 3 & 4 \end{vmatrix} = -[(1\times 4)-(0\times 3)] = -(4-0) = -4$$

$$A_{31} = (-1)^{3+1} \cdot M_{31} = (-1)^4 \cdot M_{31} = M_{31} = \begin{vmatrix} 0 & 2 \\ -3 & 0 \end{vmatrix} = (0\times 0)-(2\times -3) = 0+6 = 6$$

$$A_{32} = (-1)^{3+2} \cdot M_{32} = (-1)^5 \cdot M_{32} = -M_{32} = -\begin{vmatrix} 1 & 2 \\ 2 & 0 \end{vmatrix} = -[(1\times 0)-(2\times 2)] = -(0-4) = 4$$

$$A_{33} = (-1)^{3+3} \cdot M_{33} = (-1)^6 \cdot M_{33} = M_{33} = \begin{vmatrix} 1 & 0 \\ 2 & -3 \end{vmatrix} = (1\times -3)-(0\times -2) = -3+0 = -3$$

Therefore, the cofactor matrix is equal to: $C = \begin{bmatrix} A_{11} & A_{12} & A_{13} \\ A_{21} & A_{22} & A_{23} \\ A_{31} & A_{32} & A_{33} \end{bmatrix} = \begin{bmatrix} 0 & 0 & 17 \\ 8 & -6 & -4 \\ 6 & 4 & -3 \end{bmatrix}$ and $C^t = \begin{bmatrix} 0 & 8 & 6 \\ 0 & -6 & 4 \\ 17 & -4 & -3 \end{bmatrix}$

Next, compute $A^{-1} = \dfrac{1}{\delta(A)} C^t = \dfrac{1}{34} \cdot \begin{bmatrix} 0 & 8 & 6 \\ 0 & -6 & 4 \\ 17 & -4 & -3 \end{bmatrix} = \begin{bmatrix} 0\times\frac{1}{34} & 8\times\frac{1}{34} & 6\times\frac{1}{34} \\ 0\times\frac{1}{34} & -6\times\frac{1}{34} & 4\times\frac{1}{34} \\ 17\times\frac{1}{34} & -4\times\frac{1}{34} & -3\times\frac{1}{34} \end{bmatrix} = \begin{bmatrix} \mathbf{0} & \mathbf{\frac{4}{17}} & \mathbf{\frac{3}{17}} \\ \mathbf{0} & \mathbf{-\frac{3}{17}} & \mathbf{\frac{2}{17}} \\ \mathbf{\frac{1}{2}} & \mathbf{-\frac{2}{17}} & \mathbf{-\frac{3}{34}} \end{bmatrix}$

Finally, check the answer by multiplying the A matrix with A^{-1}. The result should be equal to the identity matrix.

$$A\times A^{-1} = \begin{bmatrix} 1 & 0 & 2 \\ 2 & -3 & 0 \\ 3 & 4 & 0 \end{bmatrix} \times \begin{bmatrix} 0 & \frac{4}{17} & \frac{3}{17} \\ 0 & -\frac{3}{17} & \frac{2}{17} \\ \frac{1}{2} & -\frac{2}{17} & -\frac{3}{34} \end{bmatrix} = \begin{bmatrix} \frac{2}{2} & \frac{4}{17}-\frac{4}{17} & \frac{3}{17}-\frac{6}{34} \\ 0 & \frac{8}{17}+\frac{9}{17} & \frac{6}{17}-\frac{6}{17} \\ 0 & \frac{12}{17}-\frac{12}{17} & \frac{9}{17}+\frac{8}{17} \end{bmatrix} = \begin{bmatrix} 1 & \frac{4-4}{17} & \frac{3-3}{17} \\ 0 & \frac{8+9}{17} & \frac{-6+6}{17} \\ 0 & \frac{12-12}{17} & \frac{9+8}{17} \end{bmatrix} = \begin{bmatrix} 1 & 0 & 0 \\ 0 & 1 & 0 \\ 0 & 0 & 1 \end{bmatrix}$$

Since $A\times A^{-1}$ is equal to the identity matrix A^{-1} was computed correctly.

b. Given $A = \begin{bmatrix} 0 & 4 & 2 \\ 1 & 0 & 3 \\ 1 & -1 & 2 \end{bmatrix}$ first find the determinant of A, i.e., obtain $\delta(A)$ by expanding about the first column. Note that $a_{11} = 0$, $a_{21} = 1$, and $a_{31} = 1$.

$$\delta(A) = a_{11} \cdot A_{11} + a_{21} \cdot A_{21} + a_{31} \cdot A_{31} = 0 \cdot A_{11} + 1 \cdot A_{21} + 1 \cdot A_{31} = A_{21} + A_{31} = (-1)^{2+1} \cdot M_{21} + (-1)^{3+1} \cdot M_{31} = -M_{21} + M_{31}$$

$= -\begin{vmatrix} 4 & 2 \\ -1 & 2 \end{vmatrix} + \begin{vmatrix} 4 & 2 \\ 0 & 3 \end{vmatrix} = -[(4\times 2)-(2\times -1)]+[(4\times 3)-(2\times 0)] = -10+12 = 2$. Since $\delta(A) \neq 0$ matrix A has an inverse.

First, replace each entry in A with its cofactor.

$$A_{11} = (-1)^{1+1} \cdot M_{11} = (-1)^2 \cdot M_{11} = M_{11} = \begin{vmatrix} 0 & 3 \\ -1 & 2 \end{vmatrix} = (0\times 2)-(3\times -1) = 0+3 = 3$$

$$A_{12} = (-1)^{1+2} \cdot M_{12} = (-1)^3 \cdot M_{12} = -M_{12} = -\begin{vmatrix} 1 & 3 \\ 1 & 2 \end{vmatrix} = -[(1\times 2)-(3\times 1)] = -(2-3) = 1$$

$$A_{13} = (-1)^{1+3} \cdot M_{13} = (-1)^4 \cdot M_{13} = M_{13} = \begin{vmatrix} 1 & 0 \\ 1 & -1 \end{vmatrix} = (1\times -1)-(0\times 1) = -1+0 = -1$$

$$A_{21} = (-1)^{2+1} \cdot M_{21} = (-1)^3 \cdot M_{21} = -M_{21} = -\begin{vmatrix} 4 & 2 \\ -1 & 2 \end{vmatrix} = -[(4\times 2)-(2\times -1)] = -(8+2) = -10$$

$$A_{22} = (-1)^{2+2} \cdot M_{22} = (-1)^4 \cdot M_{22} = M_{22} = \begin{vmatrix} 0 & 2 \\ 1 & 2 \end{vmatrix} = (0\times 2)-(2\times 1) = 0-2 = -2$$

$$A_{23} = (-1)^{2+3} \cdot M_{23} = (-1)^5 \cdot M_{23} = -M_{23} = -\begin{vmatrix} 0 & 4 \\ 1 & -1 \end{vmatrix} = -[(0\times -1)-(4\times 1)] = -(0-4) = 4$$

$$A_{31} = (-1)^{3+1} \cdot M_{31} = (-1)^4 \cdot M_{31} = M_{31} = \begin{vmatrix} 4 & 2 \\ 0 & 3 \end{vmatrix} = (4\times 3)-(2\times 0) = 12+0 = 12$$

$$A_{32} = (-1)^{3+2} \cdot M_{32} = (-1)^5 \cdot M_{32} = -M_{32} = -\begin{vmatrix} 0 & 2 \\ 1 & 3 \end{vmatrix} = -[(0\times 3)-(2\times 1)] = -(0-2) = 2$$

$$A_{33} = (-1)^{3+3} \cdot M_{33} = (-1)^6 \cdot M_{33} = M_{33} = \begin{vmatrix} 0 & 4 \\ 1 & 0 \end{vmatrix} = (0\times 0)-(4\times 1) = 0-4 = -4$$

Therefore, the cofactor matrix is equal to: $C = \begin{bmatrix} A_{11} & A_{12} & A_{13} \\ A_{21} & A_{22} & A_{23} \\ A_{31} & A_{32} & A_{33} \end{bmatrix} = \begin{bmatrix} 3 & 1 & -1 \\ -10 & -2 & 4 \\ 12 & 2 & -4 \end{bmatrix}$ and $C^t = \begin{bmatrix} 3 & -10 & 12 \\ 1 & -2 & 2 \\ -1 & 4 & -4 \end{bmatrix}$

Next, compute $A^{-1} = \frac{1}{\delta(A)} C^t = \frac{1}{2} \cdot \begin{bmatrix} 3 & -10 & 12 \\ 1 & -2 & 2 \\ -1 & 4 & -4 \end{bmatrix} = \begin{bmatrix} 3\times\frac{1}{2} & -10\times\frac{1}{2} & 12\times\frac{1}{2} \\ 1\times\frac{1}{2} & -2\times\frac{1}{2} & 2\times\frac{1}{2} \\ -1\times\frac{1}{2} & 4\times\frac{1}{2} & -4\times\frac{1}{2} \end{bmatrix} = \begin{bmatrix} \mathbf{\frac{3}{2}} & \mathbf{-5} & \mathbf{6} \\ \mathbf{\frac{1}{2}} & \mathbf{-1} & \mathbf{1} \\ \mathbf{-\frac{1}{2}} & \mathbf{2} & \mathbf{-2} \end{bmatrix}$

Finally, check the answer by multiplying the A matrix with A^{-1}. The result should be equal to the identity matrix.

$$A\times A^{-1} = \begin{bmatrix} 0 & 4 & 2 \\ 1 & 0 & 3 \\ 1 & -1 & 2 \end{bmatrix} \times \begin{bmatrix} \frac{3}{2} & -5 & 6 \\ \frac{1}{2} & -1 & 1 \\ -\frac{1}{2} & 2 & -2 \end{bmatrix} = \begin{bmatrix} 2-1 & -4+4 & 4-4 \\ \frac{3}{2}-\frac{3}{2} & -5+6 & 6-6 \\ \frac{3}{2}-\frac{1}{2}-1 & -5+1+4 & 6-1-4 \end{bmatrix} = \begin{bmatrix} 1 & 0 & 0 \\ 0 & 1 & 0 \\ 0 & 0 & 1 \end{bmatrix}$$

Since $A\times A^{-1}$ is equal to the identity matrix A^{-1} was computed correctly.

c. Given $A = \begin{bmatrix} 3 & 1 & 1 \\ 1 & 0 & 3 \\ 6 & 2 & 2 \end{bmatrix}$ first find the determinant of A, i.e., obtain $\delta(A)$ by expanding about the second column. Note that $a_{12} = 1$, $a_{22} = 0$, and $a_{32} = 2$.

$$\delta(A) = a_{12} \cdot A_{12} + a_{22} \cdot A_{22} + a_{32} \cdot A_{32} = 1 \cdot A_{12} + 0 \cdot A_{22} + 2 \cdot A_{32} = A_{12} + 2A_{32} = (-1)^{1+2} \cdot M_{12} + 2 \cdot (-1)^{3+2} \cdot M_{32}$$

$= -M_{12} - 2M_{32} = -\begin{vmatrix} 1 & 3 \\ 6 & 2 \end{vmatrix} - 2\begin{vmatrix} 3 & 1 \\ 1 & 3 \end{vmatrix} = -[(1\times 2)-(3\times 6)] - 2[(3\times 3)-(1\times 1)] = 16-16 = 0$. **Since $\delta(A) = 0$ matrix A does not have an inverse.**

d. Given $A=\begin{bmatrix}0&0&1\\0&1&0\\1&0&0\end{bmatrix}$ first find the determinant of A, i.e.,

Obtain $\delta(A)$ by expanding about the third row. Note that $a_{31}=1$, $a_{32}=0$, and $a_{33}=0$.

$$\delta(A) = a_{31}\cdot A_{31}+a_{32}\cdot A_{32}+a_{33}\cdot A_{33} = 1\cdot A_{31}+0\cdot A_{32}+0\cdot A_{33} = A_{31}+0+0 = A_{31} = (-1)^{3+1}\cdot M_{31} = \begin{vmatrix}0&1\\1&0\end{vmatrix}$$

$= (0\times 0)-(1\times 1) = -1$. Since $\delta(A)\neq 0$ matrix A has an inverse.

First, replace each entry in A with its cofactor.

$$A_{11} = (-1)^{1+1}\cdot M_{11} = (-1)^2\cdot M_{11} = M_{11} = \begin{vmatrix}1&0\\0&0\end{vmatrix} = (1\times 0)-(0\times 0) = 0-0 = 0$$

$$A_{12} = (-1)^{1+2}\cdot M_{12} = (-1)^3\cdot M_{12} = -M_{12} = -\begin{vmatrix}0&0\\1&0\end{vmatrix} = -[(0\times 0)-(0\times 1)] = -(0-0) = 0$$

$$A_{13} = (-1)^{1+3}\cdot M_{13} = (-1)^4\cdot M_{13} = M_{13} = \begin{vmatrix}0&1\\1&0\end{vmatrix} = (0\times 0)-(1\times 1) = 0-1 = -1$$

$$A_{21} = (-1)^{2+1}\cdot M_{21} = (-1)^3\cdot M_{21} = -M_{21} = -\begin{vmatrix}0&1\\0&0\end{vmatrix} = -[(0\times 0)-(1\times 0)] = -(0-0) = 0$$

$$A_{22} = (-1)^{2+2}\cdot M_{22} = (-1)^4\cdot M_{22} = M_{22} = \begin{vmatrix}0&1\\1&0\end{vmatrix} = (0\times 0)-(1\times 1) = 0-1 = -1$$

$$A_{23} = (-1)^{2+3}\cdot M_{23} = (-1)^5\cdot M_{23} = -M_{23} = -\begin{vmatrix}0&0\\1&0\end{vmatrix} = -[(0\times 0)-(0\times 1)] = -(0-0) = 0$$

$$A_{31} = (-1)^{3+1}\cdot M_{31} = (-1)^4\cdot M_{31} = M_{31} = \begin{vmatrix}0&1\\1&0\end{vmatrix} = (0\times 0)-(1\times 1) = 0-1 = -1$$

$$A_{32} = (-1)^{3+2}\cdot M_{32} = (-1)^5\cdot M_{32} = -M_{32} = -\begin{vmatrix}0&1\\0&0\end{vmatrix} = -[(0\times 0)-(1\times 0)] = -(0-0) = 0$$

$$A_{33} = (-1)^{3+3}\cdot M_{33} = (-1)^6\cdot M_{33} = M_{33} = \begin{vmatrix}0&0\\0&1\end{vmatrix} = (0\times 1)-(0\times 0) = 0+0 = 0$$

Therefore, the cofactor matrix is equal to: $C=\begin{bmatrix}A_{11}&A_{12}&A_{13}\\A_{21}&A_{22}&A_{23}\\A_{31}&A_{32}&A_{33}\end{bmatrix}=\begin{bmatrix}0&0&-1\\0&-1&0\\-1&0&0\end{bmatrix}$ and $C^t=\begin{bmatrix}0&0&-1\\0&-1&0\\-1&0&0\end{bmatrix}$

Next, compute $A^{-1} = \frac{1}{\delta(A)}C^t = \frac{1}{-1}\cdot\begin{bmatrix}0&0&-1\\0&-1&0\\-1&0&0\end{bmatrix} = \begin{bmatrix}\mathbf{0}&\mathbf{0}&\mathbf{1}\\\mathbf{0}&\mathbf{1}&\mathbf{0}\\\mathbf{1}&\mathbf{0}&\mathbf{0}\end{bmatrix}$

Finally, check the answer by multiplying the A matrix with A^{-1}. The result should be equal to the identity matrix.

$$A\times A^{-1} = \begin{bmatrix}0&0&1\\0&1&0\\1&0&0\end{bmatrix}\times\begin{bmatrix}0&0&1\\0&1&0\\1&0&0\end{bmatrix} = \begin{bmatrix}0+0+1&0+0+0&0+0+0\\0+0+0&0+1+0&0+0+0\\0+0+0&0+0+0&1+0+0\end{bmatrix} = \begin{bmatrix}1&0&0\\0&1&0\\0&0&1\end{bmatrix}$$

Since $A\times A^{-1}$ is equal to the identity matrix A^{-1} was computed correctly.

e. Given $A=\begin{bmatrix}1&0&0\\0&0&1\\1&0&0\end{bmatrix}$ first find the determinant of A, i.e., obtain $\delta(A)$ by expanding about the second column. Note that $a_{12}=0$, $a_{22}=0$, and $a_{32}=0$.

$\delta(A) = a_{12}\cdot A_{12}+a_{22}\cdot A_{22}+a_{32}\cdot A_{32} = 0\cdot A_{12}+0\cdot A_{22}+0\cdot A_{32} = 0+0+0 = 0$. **Since $\delta(A)=0$ matrix A does not have an inverse.**

f. Given $A = \begin{bmatrix} 1 & 2 & -3 \\ 3 & -1 & 2 \\ 5 & 2 & 0 \end{bmatrix}$ first find the determinant of A, i.e., obtain $\delta(A)$ by expanding about the third row. Note that $a_{31} = 5$, $a_{32} = 2$, and $a_{33} = 0$.

$$\delta(A) = a_{31} \cdot A_{31} + a_{32} \cdot A_{32} + a_{33} \cdot A_{33} = 5 \cdot A_{31} + 2 \cdot A_{32} + 0 \cdot A_{33} = 5A_{31} + 2A_{32} = 5 \cdot (-1)^{3+1} \cdot M_{31} + 2 \cdot (-1)^{3+2} \cdot M_{32}$$

$$= 5M_{31} - 2M_{32} = 5 \cdot \begin{vmatrix} 2 & -3 \\ -1 & 2 \end{vmatrix} - 2 \cdot \begin{vmatrix} 1 & -3 \\ 3 & 2 \end{vmatrix} = 5 \cdot [(2 \times 2) - (-3 \times -1)] - 2 \cdot [(1 \times 2) - (-3 \times 3)] = 5 \cdot (4-3) - 2 \cdot (2+9) = 5 - 22$$

$= -17$. Since $\delta(A) \neq 0$ matrix A has an inverse.

First, replace each entry in A with its cofactor.

$$A_{11} = (-1)^{1+1} \cdot M_{11} = (-1)^2 \cdot M_{11} = M_{11} = \begin{vmatrix} -1 & 2 \\ 2 & 0 \end{vmatrix} = (-1 \times 0) - (2 \times 2) = 0 - 4 = -4$$

$$A_{12} = (-1)^{1+2} \cdot M_{12} = (-1)^3 \cdot M_{12} = -M_{12} = -\begin{vmatrix} 3 & 2 \\ 5 & 0 \end{vmatrix} = -[(3 \times 0) - (2 \times 5)] = -(0 - 10) = 10$$

$$A_{13} = (-1)^{1+3} \cdot M_{13} = (-1)^4 \cdot M_{13} = M_{13} = \begin{vmatrix} 3 & -1 \\ 5 & 2 \end{vmatrix} = (3 \times 2) - (-1 \times 5) = 6 + 5 = 11$$

$$A_{21} = (-1)^{2+1} \cdot M_{21} = (-1)^3 \cdot M_{21} = -M_{21} = -\begin{vmatrix} 2 & -3 \\ 2 & 0 \end{vmatrix} = -[(2 \times 0) - (-3 \times 2)] = -(0 + 6) = -6$$

$$A_{22} = (-1)^{2+2} \cdot M_{22} = (-1)^4 \cdot M_{22} = M_{22} = \begin{vmatrix} 1 & -3 \\ 5 & 0 \end{vmatrix} = (1 \times 0) - (-3 \times 5) = 0 + 15 = 15$$

$$A_{23} = (-1)^{2+3} \cdot M_{23} = (-1)^5 \cdot M_{23} = -M_{23} = -\begin{vmatrix} 1 & 2 \\ 5 & 2 \end{vmatrix} = -[(1 \times 2) - (2 \times 5)] = -(2 - 10) = 8$$

$$A_{31} = (-1)^{3+1} \cdot M_{31} = (-1)^4 \cdot M_{31} = M_{31} = \begin{vmatrix} 2 & -3 \\ -1 & 2 \end{vmatrix} = (2 \times 2) - (-3 \times -1) = 4 - 3 = 1$$

$$A_{32} = (-1)^{3+2} \cdot M_{32} = (-1)^5 \cdot M_{32} = -M_{32} = -\begin{vmatrix} 1 & -3 \\ 3 & 2 \end{vmatrix} = -[(1 \times 2) - (-3 \times 3)] = -(2 + 9) = -11$$

$$A_{33} = (-1)^{3+3} \cdot M_{33} = (-1)^6 \cdot M_{33} = M_{33} = \begin{vmatrix} 1 & 2 \\ 3 & -1 \end{vmatrix} = (1 \times -1) - (2 \times 3) = -1 - 6 = -7$$

Therefore, the cofactor matrix is equal to: $C = \begin{bmatrix} A_{11} & A_{12} & A_{13} \\ A_{21} & A_{22} & A_{23} \\ A_{31} & A_{32} & A_{33} \end{bmatrix} = \begin{bmatrix} -4 & 10 & 11 \\ -6 & 15 & 8 \\ 1 & -11 & -7 \end{bmatrix}$ and $C^t = \begin{bmatrix} -4 & -6 & 1 \\ 10 & 15 & -11 \\ 11 & 8 & -7 \end{bmatrix}$

Next, compute $A^{-1} = \dfrac{1}{\delta(A)} C^t = \dfrac{1}{-17} \cdot \begin{bmatrix} -4 & -6 & 1 \\ 10 & 15 & -11 \\ 11 & 8 & -7 \end{bmatrix} = \begin{bmatrix} -4 \times -\frac{1}{17} & -6 \times -\frac{1}{17} & 1 \times -\frac{1}{17} \\ 10 \times -\frac{1}{17} & 15 \times -\frac{1}{17} & -11 \times -\frac{1}{17} \\ 11 \times -\frac{1}{17} & 8 \times -\frac{1}{17} & -7 \times -\frac{1}{17} \end{bmatrix} = \begin{bmatrix} \mathbf{\frac{4}{17}} & \mathbf{\frac{6}{17}} & -\mathbf{\frac{1}{17}} \\ -\mathbf{\frac{10}{17}} & -\mathbf{\frac{15}{17}} & \mathbf{\frac{11}{17}} \\ -\mathbf{\frac{11}{17}} & -\mathbf{\frac{8}{17}} & \mathbf{\frac{7}{17}} \end{bmatrix}$

Finally, check the answer by multiplying the A matrix with A^{-1}. The result should be equal to the identity matrix.

$$A \times A^{-1} = \begin{bmatrix} 1 & 2 & -3 \\ 3 & -1 & 2 \\ 5 & 2 & 0 \end{bmatrix} \times \frac{1}{17} \begin{bmatrix} 4 & 6 & -1 \\ -10 & -15 & 11 \\ -11 & -8 & 7 \end{bmatrix} = \frac{1}{17} \begin{bmatrix} 4-20+33 & 6-30+24 & -1+22-21 \\ 12+10-22 & 18+15-16 & -3-11+14 \\ 20-20+0 & 30-30+0 & -5+22+0 \end{bmatrix} = \frac{1}{17} \begin{bmatrix} 17 & 0 & 0 \\ 0 & 17 & 0 \\ 0 & 0 & 17 \end{bmatrix}$$

$= \begin{bmatrix} 1 & 0 & 0 \\ 0 & 1 & 0 \\ 0 & 0 & 1 \end{bmatrix}$. Since $A \times A^{-1}$ is equal to the identity matrix A^{-1} was computed correctly.

3. Find the inverse of the following matrices, if it exists, using $A \times A^{-1} = I$.

a. Given $A = \begin{bmatrix} -3 & \frac{1}{3} \\ 2 & 1 \end{bmatrix}$ first find the determinant of A, i.e., $\delta(A) = (-3 \times 1) - \left(\frac{1}{3} \times 2\right) = -3.67$. Since $\delta(A) \neq 0$ the matrix A has an inverse. Let $A^{-1} = \begin{bmatrix} a & b \\ c & d \end{bmatrix}$. Then, $A \times A^{-1} = I$; $\begin{bmatrix} -3 & \frac{1}{3} \\ 2 & 1 \end{bmatrix} \times \begin{bmatrix} a & b \\ c & d \end{bmatrix} = \begin{bmatrix} 1 & 0 \\ 0 & 1 \end{bmatrix}$; $\begin{bmatrix} -3a + \frac{1}{3}c & -3b + \frac{1}{3}d \\ 2a + c & 2b + d \end{bmatrix} = \begin{bmatrix} 1 & 0 \\ 0 & 1 \end{bmatrix}$.

Equating the entries on both sides of the equality we obtain:

1. $-3a + \frac{1}{3}c = 1$; $a = -\frac{1}{3} + \frac{1}{9}c$
2. $-3b + \frac{1}{3}d = 0$; $b = \frac{1}{9}d$
3. $2a + c = 0$; $c = -2a$
4. $2b + d = 1$; $d = 1 - 2b$

Using the back substitution method we obtain: $a = -\frac{3}{11}$, $b = \frac{1}{11}$, $c = \frac{6}{11}$, and $d = \frac{9}{11}$. Therefore,

$A^{-1} = \begin{bmatrix} a & b \\ c & d \end{bmatrix} = \begin{bmatrix} -\frac{\mathbf{3}}{\mathbf{11}} & \frac{\mathbf{1}}{\mathbf{11}} \\ \frac{\mathbf{6}}{\mathbf{11}} & \frac{\mathbf{9}}{\mathbf{11}} \end{bmatrix}$. Finally, check the result by multiplying A by A^{-1} to obtain the identity matrix, i.e.,

$$A \times A^{-1} = \begin{bmatrix} -3 & \frac{1}{3} \\ 2 & 1 \end{bmatrix} \times \begin{bmatrix} -\frac{3}{11} & \frac{1}{11} \\ \frac{6}{11} & \frac{9}{11} \end{bmatrix} = \begin{bmatrix} \left(-3 \times -\frac{3}{11}\right) + \left(\frac{1}{3} \times \frac{6}{11}\right) & \left(-3 \times \frac{1}{11}\right) + \left(\frac{1}{3} \times \frac{9}{11}\right) \\ \left(2 \times -\frac{3}{11}\right) + \left(1 \times \frac{6}{11}\right) & \left(2 \times \frac{1}{11}\right) + \left(1 \times \frac{9}{11}\right) \end{bmatrix} = \begin{bmatrix} \frac{9}{11} + \frac{2}{11} & -\frac{3}{11} + \frac{3}{11} \\ -\frac{6}{11} + \frac{6}{11} & \frac{2}{11} + \frac{9}{11} \end{bmatrix} = \begin{bmatrix} 1 & 0 \\ 0 & 1 \end{bmatrix}$$

b. Given $A = \begin{bmatrix} 2 & 0 \\ 1 & 0 \end{bmatrix}$ first find the determinant of A, i.e., $\delta(A) = (2 \times 0) - (0 \times 1) = 0 - 0 = 0$. **Since $\delta(A) = 0$ matrix A does not have an inverse.**

c. Given $A = \begin{bmatrix} 1 & 0 \\ -2 & 3 \end{bmatrix}$ first find the determinant of A, i.e., $\delta(A) = (1 \times 3) - (0 \times -2) = 3$. Since $\delta(A) \neq 0$ the matrix A has an inverse. Let $A^{-1} = \begin{bmatrix} a & b \\ c & d \end{bmatrix}$. Then, $A \times A^{-1} = I$; $\begin{bmatrix} 1 & 0 \\ -2 & 3 \end{bmatrix} \times \begin{bmatrix} a & b \\ c & d \end{bmatrix} = \begin{bmatrix} 1 & 0 \\ 0 & 1 \end{bmatrix}$; $\begin{bmatrix} a & b \\ -2a + 3c & -2b + 3d \end{bmatrix} = \begin{bmatrix} 1 & 0 \\ 0 & 1 \end{bmatrix}$.

Equating the entries on both sides of the equality we obtain:

1. $a = 1$
2. $b = 0$
3. $-2a + 3c = 0$; $c = \frac{2}{3}$
4. $-2b + 3d = 1$; $d = \frac{1}{3}$

Using the back substitution method we obtain: $a = 1$, $b = 0$, $c = \frac{2}{3}$, and $d = \frac{1}{3}$. Therefore,

$A^{-1} = \begin{bmatrix} a & b \\ c & d \end{bmatrix} = \begin{bmatrix} \mathbf{1} & \mathbf{0} \\ \frac{\mathbf{2}}{\mathbf{3}} & \frac{\mathbf{1}}{\mathbf{3}} \end{bmatrix}$. Finally, check the result by multiplying A by A^{-1} to obtain the identity matrix, i.e.,

$$A \times A^{-1} = \begin{bmatrix} 1 & 0 \\ -2 & 3 \end{bmatrix} \times \begin{bmatrix} 1 & 0 \\ \frac{2}{3} & \frac{1}{3} \end{bmatrix} = \begin{bmatrix} (1 \times 1) + \left(0 \times \frac{2}{3}\right) & (1 \times 0) + \left(0 \times \frac{1}{3}\right) \\ (-2 \times 1) + \left(3 \times \frac{2}{3}\right) & (-2 \times 0) + \left(3 \times \frac{1}{3}\right) \end{bmatrix} = \begin{bmatrix} 1+0 & 0+0 \\ -2+2 & 0+1 \end{bmatrix} = \begin{bmatrix} 1 & 0 \\ 0 & 1 \end{bmatrix}$$

d. Given $A = \begin{bmatrix} 6 & 0 \\ 0 & -\frac{1}{2} \end{bmatrix}$ first find the determinant of A, i.e., $\delta(A) = \left(6 \times -\frac{1}{2}\right) - (0 \times 0) = -3$. Since $\delta(A) \neq 0$ the matrix A has an inverse. Let $A^{-1} = \begin{bmatrix} a & b \\ c & d \end{bmatrix}$. Then, $A \times A^{-1} = I$; $\begin{bmatrix} 6 & 0 \\ 0 & -\frac{1}{2} \end{bmatrix} \times \begin{bmatrix} a & b \\ c & d \end{bmatrix} = \begin{bmatrix} 1 & 0 \\ 0 & 1 \end{bmatrix}$; $\begin{bmatrix} 6a & 6b \\ -\frac{1}{2}c & -\frac{1}{2}d \end{bmatrix} = \begin{bmatrix} 1 & 0 \\ 0 & 1 \end{bmatrix}$.

Equating the entries on both sides of the equality we obtain:

1. $6a = 1$; $a = \frac{1}{6}$
2. $6b = 0$; $b = 0$
3. $-\frac{1}{2}c = 0$; $c = 0$
4. $-\frac{1}{2}d = 1$; $d = -2$

Using the back substitution method we obtain: $a = \frac{1}{6}$, $b = 0$, $c = 0$, and $d = -2$. Therefore,

$A^{-1} = \begin{bmatrix} a & b \\ c & d \end{bmatrix} = \begin{bmatrix} \mathbf{\frac{1}{6}} & \mathbf{0} \\ \mathbf{0} & \mathbf{-2} \end{bmatrix}$. Finally, check the result by multiplying A by A^{-1} to obtain the identity matrix, i.e.,

$$A \times A^{-1} = \begin{bmatrix} 6 & 0 \\ 0 & -\frac{1}{2} \end{bmatrix} \times \begin{bmatrix} \frac{1}{6} & 0 \\ 0 & -2 \end{bmatrix} = \begin{bmatrix} \left(6 \times \frac{1}{6}\right) + (0 \times 0) & (6 \times 0) + (0 \times -2) \\ \left(0 \times \frac{1}{6}\right) + \left(-\frac{1}{2} \times 0\right) & (0 \times 0) + \left(-\frac{1}{2} \times -2\right) \end{bmatrix} = \begin{bmatrix} 1+0 & 0+0 \\ 0+0 & 0+1 \end{bmatrix} = \begin{bmatrix} 1 & 0 \\ 0 & 1 \end{bmatrix}$$

e. Given $A = \begin{bmatrix} 1 & 0 & 0 \\ 0 & 0 & 1 \\ 1 & 0 & 0 \end{bmatrix}$ first find the determinant of A by expanding about the second column. Note that $a_{12} = 0$, $a_{22} = 0$, and $a_{32} = 0$. Therefore, $\delta(A) = a_{12} \cdot A_{12} + a_{22} \cdot A_{22} + a_{32} \cdot A_{32} = 0 \cdot A_{12} + 0 \cdot A_{22} + 0 \cdot A_{32} = 0$. **Since $\delta(A) = 0$ matrix A does not have an inverse.**

f. Given $A = \begin{bmatrix} 1 & 0 & 2 \\ 2 & 1 & 4 \\ 2 & 0 & 4 \end{bmatrix}$ first find the determinant of A by expanding about the second column. Note that $a_{12} = 0$, $a_{22} = 1$, and $a_{32} = 0$. Therefore, $\delta(A) = a_{12} \cdot A_{12} + a_{22} \cdot A_{22} + a_{32} \cdot A_{32} = 0 \cdot A_{12} + 1 \cdot A_{22} + 0 \cdot A_{32} = A_{22} = (-1)^{2+2} \cdot M_{22} = M_{22}$ $= \begin{vmatrix} 1 & 2 \\ 2 & 4 \end{vmatrix} = (1 \times 4) - (2 \times 2) = 4 - 4 = 0$. **Since $\delta(A) = 0$ matrix A does not have an inverse.**

4. Show that $\left(\begin{bmatrix} 1 & 3 \\ 0 & -1 \end{bmatrix} \cdot \begin{bmatrix} 1 & 0 \\ 2 & 3 \end{bmatrix}\right)^{-1} = \begin{bmatrix} 1 & 0 \\ 2 & 3 \end{bmatrix}^{-1} \cdot \begin{bmatrix} 1 & 3 \\ 0 & -1 \end{bmatrix}^{-1}$. Let $A = \begin{bmatrix} 1 & 3 \\ 0 & -1 \end{bmatrix} \cdot \begin{bmatrix} 1 & 0 \\ 2 & 3 \end{bmatrix}$, $B = \begin{bmatrix} 1 & 0 \\ 2 & 3 \end{bmatrix}$ and $D = \begin{bmatrix} 1 & 3 \\ 0 & -1 \end{bmatrix}$

First – Find the inverse matrix for $A = \begin{bmatrix} 1 & 3 \\ 0 & -1 \end{bmatrix} \cdot \begin{bmatrix} 1 & 0 \\ 2 & 3 \end{bmatrix}$ by finding its determinant and cofactors, i.e.,

$$\begin{bmatrix} 1 & 3 \\ 0 & -1 \end{bmatrix} \cdot \begin{bmatrix} 1 & 0 \\ 2 & 3 \end{bmatrix} = \begin{bmatrix} (1 \times 1) + (3 \times 2) & (1 \times 0) + (3 \times 3) \\ (0 \times 1) + (-1 \times 2) & (0 \times 0) + (-1 \times 3) \end{bmatrix} = \begin{bmatrix} 1+6 & 0+9 \\ 0-2 & 0-3 \end{bmatrix} = \begin{bmatrix} 7 & 9 \\ -2 & -3 \end{bmatrix}$$

$\delta(A) = (7 \times -3) - (9 \times -2) = -21 + 18 = -3$. Since $\delta(A) \neq 0$ matrix A has an inverse. Next, replace each entry in A with its cofactor. $A_{11} = (-1)^{1+1} M_{11} = (-1)^2 \times -3 = -3$, $A_{12} = (-1)^{1+2} M_{12} = (-1)^3 \times -2 = 2$, $A_{21} = (-1)^{2+1} M_{21} = (-1)^3 \times 9$ $= -9$, and $A_{22} = (-1)^{2+2} M_{22} = (-1)^4 \times 7 = 7$. Therefore, the cofactor matrix is equal to $C = \begin{bmatrix} A_{11} & A_{12} \\ A_{21} & A_{22} \end{bmatrix} = \begin{bmatrix} -3 & 2 \\ -9 & 7 \end{bmatrix}$ and

$C^t = \begin{bmatrix} -3 & -9 \\ 2 & 7 \end{bmatrix}$. Thus, $A^{-1} = \frac{1}{\delta(A)} C^t = \frac{1}{-3} \cdot \begin{bmatrix} -3 & -9 \\ 2 & 7 \end{bmatrix} = \begin{bmatrix} -3 \times -\frac{1}{3} & -9 \times -\frac{1}{3} \\ 2 \times -\frac{1}{3} & 7 \times -\frac{1}{3} \end{bmatrix} = \begin{bmatrix} 1 & 3 \\ -\frac{2}{3} & -\frac{7}{3} \end{bmatrix}$.

Therefore, $\left(\begin{bmatrix} 1 & 3 \\ 0 & -1 \end{bmatrix} \cdot \begin{bmatrix} 1 & 0 \\ 2 & 3 \end{bmatrix}\right)^{-1} = \begin{bmatrix} \mathbf{1} & \mathbf{3} \\ \mathbf{-\frac{2}{3}} & \mathbf{-\frac{7}{3}} \end{bmatrix}$

Second – Find the inverse matrix for $B = \begin{bmatrix} 1 & 0 \\ 2 & 3 \end{bmatrix}$ by finding its determinant and cofactors, i.e.,

$\delta(B) = (1 \times 3) - (0 \times 2) = 3 - 0 = 3$. Since $\delta(B) \neq 0$ matrix B has an inverse. Next, replace each entry in A with its cofactor. $B_{11} = (-1)^{1+1} M_{11} = (-1)^2 \times 3 = 3$, $B_{12} = (-1)^{1+2} M_{12} = (-1)^3 \times 2 = -2$, $B_{21} = (-1)^{2+1} M_{21} = (-1)^3 \times 0 = 0$, and $B_{22} = (-1)^{2+2} M_{22} = (-1)^4 \times 1 = 1$. Therefore, the cofactor matrix is equal to $C = \begin{bmatrix} B_{11} & B_{12} \\ B_{21} & B_{22} \end{bmatrix} = \begin{bmatrix} 3 & -2 \\ 0 & 1 \end{bmatrix}$ and

$C^t = \begin{bmatrix} 3 & 0 \\ -2 & 1 \end{bmatrix}$. Thus, $B^{-1} = \frac{1}{\delta(B)} C^t = \frac{1}{3} \cdot \begin{bmatrix} 3 & 0 \\ -2 & 1 \end{bmatrix} = \begin{bmatrix} 1 & 0 \\ -\frac{2}{3} & \frac{1}{3} \end{bmatrix}$ and the inverse of $B = \begin{bmatrix} 1 & 0 \\ 2 & 3 \end{bmatrix}$ is $\begin{bmatrix} 1 & 0 \\ -\frac{2}{3} & \frac{1}{3} \end{bmatrix}$.

Third – Find the inverse matrix for $D = \begin{bmatrix} 1 & 3 \\ 0 & -1 \end{bmatrix}$ by finding its determinant and cofactors, i.e.,

$\delta(D) = (1 \times -1) - (0 \times 3) = -1 + 0 = -1$. Since $\delta(D) \neq 0$ matrix D has an inverse. Next, replace each entry in A with its cofactor. $D_{11} = (-1)^{1+1} M_{11} = (-1)^2 \times -1 = -1$, $D_{12} = (-1)^{1+2} M_{12} = (-1)^3 \times 0 = 0$, $D_{21} = (-1)^{2+1} M_{21} = (-1)^3 \times 3$ $= -3$, and $D_{22} = (-1)^{2+2} M_{22} = (-1)^4 \times 1 = 1$. Therefore, the cofactor matrix is equal to $C = \begin{bmatrix} D_{11} & D_{12} \\ D_{21} & D_{22} \end{bmatrix} = \begin{bmatrix} -1 & 0 \\ -3 & 1 \end{bmatrix}$ and

$C^t = \begin{bmatrix} -1 & -3 \\ 0 & 1 \end{bmatrix}$. Thus, $D^{-1} = \frac{1}{\delta(D)} C^t = \frac{1}{-1} \cdot \begin{bmatrix} -1 & -3 \\ 0 & 1 \end{bmatrix} = \begin{bmatrix} 1 & 3 \\ 0 & -1 \end{bmatrix}$ and the inverse of $D = \begin{bmatrix} 1 & 3 \\ 0 & -1 \end{bmatrix}$ is $\begin{bmatrix} 1 & 3 \\ 0 & -1 \end{bmatrix}$.

Fourth – Multiply the inverse of the matrices B and D together. The result should be the same as the inverse of the A matrix.

$$\begin{bmatrix} 1 & 0 \\ -\frac{2}{3} & \frac{1}{3} \end{bmatrix} \times \begin{bmatrix} 1 & 3 \\ 0 & -1 \end{bmatrix} = \begin{bmatrix} (1 \times 1) + (0 \times 0) & (1 \times 3) + (0 \times -1) \\ \left(-\frac{2}{3} \times 1\right) + \left(\frac{1}{3} \times 0\right) & \left(-\frac{2}{3} \times 3\right) + \left(\frac{1}{3} \times -1\right) \end{bmatrix} = \begin{bmatrix} 1+0 & 3+0 \\ -\frac{2}{3}+0 & -2-\frac{1}{3} \end{bmatrix} = \begin{bmatrix} 1 & 3 \\ -\frac{2}{3} & -\frac{7}{3} \end{bmatrix}.$$

Therefore, $\begin{bmatrix} 1 & 0 \\ 2 & 3 \end{bmatrix}^{-1} \times \begin{bmatrix} 1 & 3 \\ 0 & -1 \end{bmatrix}^{-1}$ is also equal to $\begin{bmatrix} \mathbf{1} & \mathbf{3} \\ -\frac{\mathbf{2}}{\mathbf{3}} & -\frac{\mathbf{7}}{\mathbf{3}} \end{bmatrix}$.

5. Find the determinant of $A = \begin{bmatrix} 1 & 2 & 3 \\ 0 & 0 & 0 \\ -3 & 4 & 5 \end{bmatrix}$ by expanding about the second row. Note that $a_{21} = 0$, $a_{22} = 0$, and $a_{23} = 0$.

Therefore, $\delta(A) = a_{21} \cdot A_{21} + a_{22} \cdot A_{22} + a_{23} \cdot A_{23} = 0 \cdot A_{21} + 0 \cdot A_{22} + 0 \cdot A_{23} = 0$. **Since $\delta(A) = 0$ matrix A does not have an inverse.**

Section 3.5 Case I Solutions - Solving Linear Systems Using the Addition Method

1. Find the solution set of the given systems by using the addition method.

a. **First** - Given $\begin{matrix} x + 3y = 2 \\ 2x + 2y = -1 \end{matrix}$, let's eliminate x from the two equations by multiplying the first equation by -2, i.e.,

$$\begin{matrix} -2 \cdot (x + 3y) = -2 \cdot 2 \\ 2x + 2y = -1 \end{matrix} ; \quad \begin{matrix} -2x - 6y = -4 \\ \underline{2x + 2y = -1} \\ -4y = -5 \end{matrix} \qquad \text{therefore, } y = \frac{5}{4}$$

Second - Substitute the y value into $x + 3y = 2$ and solve for x, i.e., $x + 3 \cdot \frac{5}{4} = 2$; $x = 2 - \frac{15}{4}$; $x = \frac{(2 \cdot 4) - (15 \cdot 1)}{4}$; $x = \frac{8 - 15}{4}$; $x = -\frac{7}{4}$ Therefore, $\mathbf{x = -\frac{7}{4}}$ and $\mathbf{y = \frac{5}{4}}$ and the solution set is $\left\{\left(-\frac{7}{4}, \frac{5}{4}\right)\right\}$

Third - Check the answers by substituting the x and y values into one of the original equations. Let's substitute the x and y values into $2x + 2y = -1$.

$$\left(2 \times -\frac{7}{4}\right) + \left(2 \times \frac{5}{4}\right) \stackrel{?}{=} -1 \; ; \; -\frac{14}{4} + \frac{10}{4} \stackrel{?}{=} -1 \; ; \; \frac{-14 + 10}{4} \stackrel{?}{=} -1 \; ; \; -\frac{4}{4} \stackrel{?}{=} -1 \; ; \; -1 = -1$$

b. **First** - Given $\begin{matrix} x + 2y = 0 \\ -x + y = 2 \end{matrix}$, since the coefficient of x is opposite of each other we can eliminate x from both equations by adding the two with each other, i.e., $\begin{array}{r} x+2y=0 \\ -x+y=2 \\ \hline 3y=2 \end{array}$ therefore, $y=\frac{2}{3}$

Second - Substitute the y value into $-x+y=2$ and solve for x, i.e., $-x+\frac{2}{3}=2$; $-x=2-\frac{2}{3}$; $-x=\frac{(2\cdot3)-(2\cdot1)}{3}$; $x=-\frac{6-2}{3}$; $x=-\frac{4}{3}$ Therefore, $\mathbf{x=-\frac{4}{3}}$ and $\mathbf{y=\frac{2}{3}}$ and the solution set is $\left\{\left(-\frac{4}{3},\frac{2}{3}\right)\right\}$.

Third - Check the answers by substituting the x and y values into one of the original equations. Let's substitute the x and y values into $x+2y=0$, i.e., $-\frac{4}{3}+\left(2\times\frac{2}{3}\right)\overset{?}{=}0$; $-\frac{4}{3}+\frac{4}{3}\overset{?}{=}0$; $0=0$

c. Given $\begin{array}{l} 2x+4y=-1 \\ 4x+8y=5 \end{array}$, let's eliminate x from the two equations by multiplying the first equation by -2, i.e.,

$\begin{array}{r} -2\cdot(2x+4y)=-2\cdot-1 \\ 4x+8y=5 \end{array}$; $\begin{array}{r} -4x-8y=2 \\ 4x+8y=5 \\ \hline 0x+0y=7 \end{array}$; $0\neq7$

since 0 can not be equal to 7 **the linear system has no solution (is an inconsistent system).**

d. **First** - Given $\begin{array}{l} 4x-2y=-3 \\ x-2y=1 \end{array}$, let's eliminate y from the two equations by multiplying the first equation by -1, i.e.,

$\begin{array}{r} -1\cdot(4x-2y)=-1\cdot-3 \\ x-2y=1 \end{array}$; $\begin{array}{r} -4x+2y=3 \\ x-2y=1 \\ \hline -3x+0y=4 \end{array}$ therefore, $-3x=4$; $x=-\frac{4}{3}$

Second - Substitute the x value into $4x-2y=-3$ and solve for y, i.e., $\left(4\cdot-\frac{4}{3}\right)-2y=-3$; $-\frac{16}{3}-2y=-3$; $-2y=-3+\frac{16}{3}$; $-2y=\frac{-9+16}{3}$; $y=-\frac{7}{6}$ Thus, $\mathbf{x=-\frac{4}{3}}$ and $\mathbf{y=-\frac{7}{6}}$ and the solution set is $\left\{\left(-\frac{4}{3},-\frac{7}{6}\right)\right\}$

Third - Check the answers by substituting the x and y values into one of the original equations. Let's substitute the x and y values into $x-2y=1$, i.e., $-\frac{4}{3}+\left(-2\times-\frac{7}{6}\right)\overset{?}{=}1$; $-\frac{4}{3}+\frac{7}{3}\overset{?}{=}1$; $\frac{-4+7}{3}\overset{?}{=}1$; $\frac{3}{3}\overset{?}{=}1$; $1=1$

e. Given $\begin{array}{l} 2x+y=3 \\ 4x+2y=6 \end{array}$, let's eliminate x from the two equations by multiplying the first equation by -2, i.e.,

$\begin{array}{r} -2\cdot(2x+y)=-2\cdot3 \\ 4x+2y=6 \end{array}$; $\begin{array}{r} -4x-2y=-6 \\ 4x+2y=6 \\ \hline 0x+0y=0 \end{array}$; $0=0$

since both sides of the resulting equation are equal **the linear system has an infinite number of solutions (no unique solutions). This class of systems is referred to as dependent systems.**

f. **First** - Given $\begin{array}{l} x+y=2 \\ 2x-z=-1 \\ 2y+2z=3 \end{array}$ let's eliminate z from the second and third equations by multiplying the second equation by 2 i.e.,

$\begin{array}{r} 2\cdot(2x-z)=2\cdot-1 \\ 2y+2z=3 \end{array}$; $\begin{array}{r} 4x-2z=-2 \\ 2y+2z=3 \\ \hline 4x+2y=1 \end{array}$ (1) then, eliminate y from the second and third equation in the following way

Since the first equation is already in terms of x and y, let's label $x+y=2$ as equation no. (2)

Second - Solve for x and y using the two reduced equations (1) and (2), i.e.,

$$\begin{array}{l} 4x+2y=1 \\ -4\cdot(x+y)=-4\cdot 2 \end{array} ; \begin{array}{r} 4x+2y=1 \\ \underline{-4x-4y=-8} \\ -2y=-7 \end{array}$$

thus, $y=\frac{7}{2}$ and x is equal to $x+\frac{7}{2}=2$; $x=2-\frac{7}{2}$; $x=\frac{4-7}{2}$; $x=-\frac{3}{2}$

Third - Substitute the x and y values into $2y+2z=3$ and solve for z, i.e.,

$2\times\frac{7}{2}+2z=3$; $7+2z=3$; $2z=3-7$; $2z=-4$; $z=-2$

Therefore, $\mathbf{x=-\frac{3}{2}}$, $\mathbf{y=\frac{7}{2}}$, $\mathbf{z=-2}$ and the solution set is $\left\{\left(-\frac{3}{2}, \frac{7}{2}, -2\right)\right\}$.

Fourth - Check the answers by substituting the x, y, and z values into one of the original equation. Let's substitute the y, and z values into $2y+2z=3$, i.e., $\left(2\times\frac{7}{2}\right)+2\times-2\overset{?}{=}3$; $7-4\overset{?}{=}3$; $3=3$

g. Given $\begin{array}{l} x-y+3z=2 \\ x-z=-3 \\ 2x-2y+6z=-1 \end{array}$ let's eliminate $\boldsymbol{x}$ from the first and third equations by multiplying the first equation by -2 i.e.,

$$\begin{array}{r} -2\cdot(x-y+3z)=-2\cdot 2 \\ 2x-2y+6z=-1 \end{array} ; \begin{array}{r} -2x+2y-6z=-4 \\ \underline{2x-2y+6z=-1} \\ 0x+0y+0z=-5 \end{array} ; \ 0\neq-5$$

since 0 can not be equal to -5 **the linear system has no solution (is an inconsistent system).**

h. **First** - Given $\begin{array}{l} x+3y-z=-2 \\ -x+2y+3z=1 \\ x+y-2z=0 \end{array}$ let's eliminate x from the first and second equations i.e.,

$$\begin{array}{r} x+3y-z=-2 \\ \underline{-x+2y+3z=1} \\ 5y+2z=-1 \end{array} \quad (1)$$

then, eliminate x from the second and third equation

$$\begin{array}{r} -x+2y+3z=1 \\ \underline{x+y-2z=0} \\ 3y+z=1 \end{array} \quad (2)$$

Second - Solve for y and z using the two reduced equations (1) and (2), i.e.,

$$\begin{array}{l} 5y+2z=-1 \\ -2\cdot(3y+z)=-2\cdot 1 \end{array} ; \begin{array}{r} 5y+2z=-1 \\ \underline{-6y-2z=-2} \\ -y=-3 \end{array}$$

thus, $y=3$ and z is equal to $(5\cdot 3)+2z=-1$; $15+2z=-1$; $z=-\frac{16}{2}$; $z=-8$

Third - Substitute the y and z values into $x+3y-z=-2$ and solve for x, i.e.,

$x+(3\times 3)+8=-2$; $x+9+8=-2$; $x+17=-2$; $x=-17-2$; $x=-19$

Therefore, $\mathbf{x=-19}$, $\mathbf{y=3}$, $\mathbf{z=-8}$ and the solution set is $\{(\mathbf{-19, 3, -8})\}$.

Fourth - Check the answers by substituting the x, y, and z values into one of the original equation. Let's substitute the x, y, and z values into $x+y-2z=0$.

$-19+3+(-2\times-8)\overset{?}{=}0$; $-19+3+16\overset{?}{=}0$; $-19+19\overset{?}{=}0$; $0=0$

Section 3.5 Case II Solutions - Solving Linear Systems Using the Substitution Method

1. Find the solution set of the given systems by using the substitution method.

a. **First** - Given $\begin{array}{l} 3x-4y=2 \\ 5x-3y=1 \end{array}$, let's solve the second equation for y in terms of x, i.e., $5x-3y=1$, then $y=\frac{5}{3}x-\frac{1}{3}$

Second - Substitute the y value into the first equation $3x-4y=2$ and solve for x

$3x-4y=2$; $3x-4\left(\frac{5}{3}x-\frac{1}{3}\right)=2$; $3x-\frac{20}{3}x+\frac{4}{3}=2$; $\left(3-\frac{20}{3}\right)x=2-\frac{4}{3}$; $-\frac{11}{3}x=\frac{2}{3}$; $-11x=2$; $x=-\frac{2}{11}$

Third - Substitute the x value into the second equation $5x-3y=1$ and solve for y

$5x-3y=1$; $\left(5\times-\frac{2}{11}\right)-3y=1$; $-\frac{10}{11}-3y=1$; $-3y=1+\frac{10}{11}$; $-3y=\frac{21}{11}$; $-3y=-\frac{21}{33}$; $y=-\frac{7}{11}$

Therefore, $\boldsymbol{x=-\frac{2}{11}}$ and $\boldsymbol{y=-\frac{7}{11}}$ and the solution set is $\left\{\left(-\frac{2}{11},-\frac{7}{11}\right)\right\}$

Fourth - Check the answers by substituting the x and y values into one of the original equations. Let's substitute the x and y values into $3x-4y=2$.

$3\times-\frac{2}{11}+\left(-4\times-\frac{7}{11}\right)\stackrel{?}{=}2$; $-\frac{6}{11}+\frac{28}{11}\stackrel{?}{=}2$; $\frac{-6+28}{11}\stackrel{?}{=}2$; $\frac{22}{11}\stackrel{?}{=}2$; $2=2$

b. **First** - Given $\begin{matrix}x-3y=2\\y=3x-5\end{matrix}$, the second equation is already solved in terms of x, i.e., $y=3x-5$

Second - Substitute the y value into the first equation $x-3y=2$ and solve for x

$x-3y=2$; $x-3(3x-5)=2$; $x-9x+15=2$; $-8x=2-15$; $-8x=-13$; $x=\frac{13}{8}$

Third - Substitute the x value into the second equation $y=3x-5$ and solve for y

$y=3x-5$; $y=\left(3\times\frac{13}{8}\right)-5$; $y=\frac{39}{8}-5$; $y=\frac{39-40}{8}$; $y=-\frac{1}{8}$

Therefore, $\boldsymbol{x=\frac{13}{8}}$ and $\boldsymbol{y=-\frac{1}{8}}$ and the solution set is $\left\{\left(\frac{13}{8},-\frac{1}{8}\right)\right\}$

Fourth - Check the answers by substituting the x and y values into one of the original equations. Let's substitute the x and y values into $x-3y=2$.

$\frac{13}{8}+\left(-3\times-\frac{1}{8}\right)\stackrel{?}{=}2$; $\frac{13}{8}+\frac{3}{8}\stackrel{?}{=}2$; $\frac{13+3}{8}\stackrel{?}{=}2$; $\frac{16}{8}\stackrel{?}{=}2$; $2=2$

c. **First** - Given $\begin{matrix}x+4y=-3\\2x-3y=1\end{matrix}$, let's solve the first equation in terms of y, i.e., $x+4y=-3$, then $x=-4y-3$

Second - Substitute the x value into the second equation $2x-3y=1$ and solve for y

$2x-3y=1$; $2(-4y-3)-3y=1$; $-8y-6-3y=1$; $-11y=6+1$; $-11y=7$; $y=-\frac{7}{11}$

Third - Substitute the y value into the first equation $x+4y=-3$ and solve for x

$x+4y=-3$; $x+\left(4\times-\frac{7}{11}\right)=-3$; $x-\frac{28}{11}=-3$; $x=-3+\frac{28}{11}$; $x=\frac{-33+28}{11}$; $x=-\frac{5}{11}$

Therefore, $\boldsymbol{x=-\frac{5}{11}}$ and $\boldsymbol{y=-\frac{7}{11}}$ and the solution set is $\left\{\left(-\frac{5}{11},-\frac{7}{11}\right)\right\}$

Fourth - Check the answers by substituting the x and y values into one of the original equations. Let's substitute the x and y values into $x+4y=-3$.

$-\frac{5}{11}+\left(4\times-\frac{7}{11}\right)\stackrel{?}{=}-3$; $-\frac{5}{11}-\frac{28}{11}\stackrel{?}{=}-3$; $\frac{-5-28}{11}\stackrel{?}{=}-3$; $-\frac{33}{11}\stackrel{?}{=}-3$; $-3=-3$

d. **First** - Given $\begin{matrix}x+y=-5\\2x-5y=1\end{matrix}$, let's solve the first equation in terms of y, i.e., $x+y=-5$, then $x=-y-5$

Second - Substitute the x value into the second equation $2x-5y=1$ and solve for y

$2x-5y=1$; $2(-y-5)-5y=1$; $-2y-10-5y=1$; $-7y=10+1$; $-7y=11$; $y=-\frac{11}{7}$

Third - Substitute the y value into the first equation $x+y=-5$ and solve for x

$x+y=-5$; $x-\frac{11}{7}=-5$; $x=-5+\frac{11}{7}$; $x=\frac{-35+11}{7}$; $x=-\frac{24}{7}$

Therefore, $\mathbf{x=-\frac{24}{7}}$ and $\mathbf{y=-\frac{11}{7}}$ and the solution set is $\left\{\left(-\frac{24}{7},-\frac{11}{7}\right)\right\}$

Fourth - Check the answers by substituting the x and y values into one of the original equations. Let's substitute the x and y values into $2x-5y=1$.

$\left(2\times-\frac{24}{7}\right)+\left(-5\times-\frac{11}{7}\right)\stackrel{?}{=}1$; $-\frac{48}{7}+\frac{55}{7}\stackrel{?}{=}1$; $\frac{-48+55}{7}\stackrel{?}{=}1$; $\frac{7}{7}\stackrel{?}{=}1$; $1=1$

e. **First** - Given $\begin{matrix}x-2y=3\\2x-4y=5\end{matrix}$, let's solve the first equation in terms of y, i.e., $x-2y=3$, then $x=2y+3$

Second - Substitute the x value into the second equation $2x-4y=5$ and solve for y

$2x-4y=5$; $2(2y+3)-4y=5$; $4y+6-4y=5$; $\cancel{4y}+6-\cancel{4y}=5$; $6\neq5$

since 6 can not be equal to 5 **the linear system has no solution (is an inconsistent system).**

f. **First** - Given $\begin{matrix}2x+3y=3\\6x+9y=9\end{matrix}$, let's solve the second equation for y in terms of x, i.e., $6x+9y=9$, then $y=-\frac{2}{3}x+1$

Second - Substitute the y value into the first equation $2x+3y=3$ and solve for x

$2x+3y=3$; $2x+3\left(-\frac{2}{3}x+1\right)=3$; $2x-2x+3=3$; $\cancel{2x}-\cancel{2x}+3=3$; $3=3$

since both sides are equal to 3 **the linear system has an infinite number of solutions (no unique solutions) - dependent system.**

g. **First** - Given $\begin{matrix}2x+3y-z=3\\x-y+2z=1\\x-y+z=-2\end{matrix}$, let's solve the second equation in terms of y and z, i.e., $x-y+2z=1$, then $x=y-2z+1$

Second - Substitute the x value into the first and third equations

$2x+3y-z=3$; $2(y-2z+1)+3y-z=3$; $2y-4z+2+3y-z=3$; $5y-5z=1$ (1)

$x-y+z=-2$; $(y-2z+1)-y+z=-2$; $\cancel{y}-2z+1-\cancel{y}+z=-2$; $-z+1=-2$; $z=3$ (2)

Third - Substitute the z value in (2) into the first equation (1) and solve for y

$5y-5z=1$; $5y-(5\times3)=1$; $5y-15=1$; $5y=16$; $y=\frac{16}{5}$

Fourth - Substitute the y and z values into the first equation and solve for x

$x=y-2z+1$; $x=\frac{16}{5}+(-2\times3)+1$; $x=\frac{16}{5}-5$; $x=\frac{16-25}{5}$; $x=-\frac{9}{5}$

Therefore, $\mathbf{x=-\frac{9}{5}}$, $\mathbf{y=\frac{16}{5}}$, and $\mathbf{z=3}$ and the solution set is $\left\{\left(-\frac{9}{5},\frac{16}{5},3\right)\right\}$

Fifth - Check the answers by substituting the x, y, and z values into one of the original equations. Let's substitute the x, y, and z values into $2x+3y-z=3$.

$\left(2\times-\frac{9}{5}\right)+\left(3\times\frac{16}{5}\right)-3\stackrel{?}{=}3$; $-\frac{18}{5}+\frac{48}{5}-3\stackrel{?}{=}3$; $\frac{-18+48}{5}-3\stackrel{?}{=}3$; $\frac{30}{5}-3\stackrel{?}{=}3$; $6-3\stackrel{?}{=}3$; $3=3$

h. **First** - Given $\begin{matrix}x=2y-z\\x+3y+z=1\\2x-y+z=3\end{matrix}$, the first equation is already solved in terms of y and z, i.e., $x=2y-z$

Second - Substitute the x value into the second and third equations

$x+3y+z=1$; $(2y-z)+3y+z=1$; $2y-\not{z}+3y+\not{z}=1$; $5y=1$; $y=\frac{1}{5}$ (1)

$2x-y+z=3$; $2(2y-z)-y+z=3$; $4y-2z-y+z=3$; $3y-z=3$ (2)

Third - Substitute the y value in (1) into the second equation (2) and solve for z

$3y-z=3$; $3\times\frac{1}{5}-z=3$; $-z=3-\frac{3}{5}$; $-z=\frac{15-3}{5}$; $z=-\frac{12}{5}$

Fourth - Substitute the y and z values into the first equation and solve for x

$x=2y-z$; $x=2\times\frac{1}{5}+\frac{12}{5}$; $x=\frac{2}{5}+\frac{12}{5}$; $x=\frac{2+12}{5}$; $x=\frac{14}{5}$

Therefore, $\mathbf{x=\frac{14}{5}}$, $\mathbf{y=\frac{1}{5}}$, and $\mathbf{z=-\frac{12}{5}}$ and the solution set is $\left\{\left(\frac{14}{5},\frac{1}{5},-\frac{12}{5}\right)\right\}$

Fifth - Check the answers by substituting the x, y, and z values into one of the original equations. Let's substitute the x, y, and z values into $2x-y+z=3$.

$\left(2\times\frac{14}{5}\right)-\frac{1}{5}-\frac{12}{5}\overset{?}{=}3$; $\frac{28}{5}-\frac{1}{5}-\frac{12}{5}\overset{?}{=}3$; $\frac{28-1-12}{5}\overset{?}{=}3$; $\frac{15}{5}\overset{?}{=}3$; $3=3$

Section 3.5 Case III Solutions - Solving Linear Systems Using Inverse Matrices

1. Find the solution set of the given systems by using inverse matrices method.

a. **First** - Write the given system $\begin{matrix} x+2y=3 \\ -2x+5y=-1 \end{matrix}$ in the form of $AX=B$, i.e., $\begin{bmatrix}1 & 2\\-2 & 5\end{bmatrix}\begin{bmatrix}x\\y\end{bmatrix}=\begin{bmatrix}3\\-1\end{bmatrix}$

Second - Find determinant of the coefficient matrix A, i.e., $\delta(A)=a_{11}A_{11}+a_{12}A_{12}=1\cdot A_{11}+2\cdot A_{12}=1\cdot(-1)^{1+1}M_{11}$ $+2\cdot(-1)^{1+2}M_{12}=1\cdot M_{11}-2\cdot M_{12}=(1\cdot 5)+(-2\cdot -2)=9$. Since $\delta(A)\neq 0$ the inverse matrix exist.

Third - Find inverse of the A matrix. Note that $A_{11}=(-1)^{1+1}M_{11}=M_{11}=5$ $A_{12}=(-1)^{1+2}M_{12}=-M_{12}=2$ $A_{21}=(-1)^{2+1}M_{21}=-M_{21}=-2$, and $A_{22}=(-1)^{22}M_{22}=M_{22}=1$. Thus, $C=\begin{bmatrix}A_{11} & A_{12}\\A_{21} & A_{22}\end{bmatrix}=\begin{bmatrix}5 & 2\\-2 & 1\end{bmatrix}$ and $C^t=\begin{bmatrix}5 & -2\\2 & 1\end{bmatrix}$

Therefore, $A^{-1}=\frac{1}{\delta(A)}C^t=\frac{1}{9}\cdot\begin{bmatrix}5 & -2\\2 & 1\end{bmatrix}=\begin{bmatrix}\frac{5}{9} & -\frac{2}{9}\\\frac{2}{9} & \frac{1}{9}\end{bmatrix}$

Check the answer by multiplying the A matrix with A^{-1}.

$$A\times A^{-1}=\begin{bmatrix}1 & 2\\-2 & 5\end{bmatrix}\times\begin{bmatrix}\frac{5}{9} & -\frac{2}{9}\\\frac{2}{9} & \frac{1}{9}\end{bmatrix}=\begin{bmatrix}\left(1\times\frac{5}{9}\right)+\left(2\times\frac{2}{9}\right) & \left(1\times-\frac{2}{9}\right)+\left(2\times\frac{1}{9}\right)\\\left(-2\times\frac{5}{9}\right)+\left(5\times\frac{2}{9}\right) & \left(-2\times-\frac{2}{9}\right)+\left(5\times\frac{1}{9}\right)\end{bmatrix}=\begin{bmatrix}\frac{5}{9}+\frac{4}{9} & -\frac{2}{9}+\frac{2}{9}\\-\frac{10}{9}+\frac{10}{9} & \frac{4}{9}+\frac{5}{9}\end{bmatrix}=\begin{bmatrix}1 & 0\\0 & 1\end{bmatrix}$$

Since $A\times A^{-1}$ is equal to the identity matrix A^{-1} was computed correctly.

Fourth - Use $X=A^{-1}B$ to solve for x and y, i.e., $\begin{bmatrix}x\\y\end{bmatrix}=\begin{bmatrix}\frac{5}{9} & -\frac{2}{9}\\\frac{2}{9} & \frac{1}{9}\end{bmatrix}\begin{bmatrix}3\\-1\end{bmatrix}=\begin{bmatrix}\frac{15}{9}+\frac{2}{9}\\\frac{6}{9}-\frac{1}{9}\end{bmatrix}=\begin{bmatrix}\frac{17}{9}\\\frac{5}{9}\end{bmatrix}$

Therefore, $\mathbf{x=\frac{17}{9}}$ and $\mathbf{y=\frac{5}{9}}$ and the solution set is $\left\{\left(\frac{17}{9},\frac{5}{9}\right)\right\}$

Fifth - Check the answers by substituting the x and y values into one of the original equations.

Let's substitute the x and y values into $x+2y=3$, i.e., $\frac{17}{9}+2\times\frac{5}{9}\stackrel{?}{=}3$; $\frac{17}{9}+\frac{10}{9}\stackrel{?}{=}3$; $\frac{17+10}{9}\stackrel{?}{=}3$; $\frac{27}{9}\stackrel{?}{=}3$; $3=3$

b. **First** - Write the given system $\begin{matrix} x+3y=0 \\ -x+4y=-1 \end{matrix}$ in the form of $AX=B$, i.e., $\begin{bmatrix} 1 & 3 \\ -1 & 4 \end{bmatrix}\begin{bmatrix} x \\ y \end{bmatrix}=\begin{bmatrix} 0 \\ -1 \end{bmatrix}$

Second - Find determinant of the coefficient matrix A, i.e., $\delta(A) = a_{11}A_{11}+a_{12}A_{12} = 1\cdot A_{11}+3\cdot A_{12} = 1\cdot(-1)^{1+1}M_{11}$ $+3\cdot(-1)^{1+2}M_{12} = 1\cdot M_{11}-3\cdot M_{12} = (1\cdot 4)+(-3\cdot -1) = 7$. Since $\delta(A)\neq 0$ the inverse matrix exist.

Third - Find inverse of the A matrix. Note that $A_{11}=(-1)^{1+1}M_{11}=M_{11}=4$ $\quad A_{12}=(-1)^{1+2}M_{12}=-M_{12}=1$ $A_{21}=(-1)^{2+1}M_{21}=-M_{21}=-3$, and $A_{22}=(-1)^{22}M_{22}=M_{22}=1$. Thus, $C=\begin{bmatrix} A_{11} & A_{12} \\ A_{21} & A_{22} \end{bmatrix}=\begin{bmatrix} 4 & 1 \\ -3 & 1 \end{bmatrix}$ and $C^t=\begin{bmatrix} 4 & -3 \\ 1 & 1 \end{bmatrix}$.

Therefore, $A^{-1} = \frac{1}{\delta(A)}C^t = \frac{1}{7}\cdot\begin{bmatrix} 4 & -3 \\ 1 & 1 \end{bmatrix} = \begin{bmatrix} \frac{4}{7} & -\frac{3}{7} \\ \frac{1}{7} & \frac{1}{7} \end{bmatrix}$

Check the answer by multiplying the A matrix with A^{-1}.

$$A\times A^{-1} = \begin{bmatrix} 1 & 3 \\ -1 & 4 \end{bmatrix}\times\begin{bmatrix} \frac{4}{7} & -\frac{3}{7} \\ \frac{1}{7} & \frac{1}{7} \end{bmatrix} = \begin{bmatrix} \left(1\times\frac{4}{7}\right)+\left(3\times\frac{1}{7}\right) & \left(1\times-\frac{3}{7}\right)+\left(3\times\frac{1}{7}\right) \\ \left(-1\times\frac{4}{7}\right)+\left(4\times\frac{1}{7}\right) & \left(-1\times-\frac{3}{7}\right)+\left(4\times\frac{1}{7}\right) \end{bmatrix} = \begin{bmatrix} \frac{4}{7}+\frac{3}{7} & -\frac{3}{7}+\frac{3}{7} \\ -\frac{4}{7}+\frac{4}{7} & \frac{3}{7}+\frac{4}{7} \end{bmatrix} = \begin{bmatrix} 1 & 0 \\ 0 & 1 \end{bmatrix}$$

Since $A\times A^{-1}$ is equal to the identity matrix A^{-1} was computed correctly.

Fourth - Use $X=A^{-1}B$ to solve for x and y, i.e., $\begin{bmatrix} x \\ y \end{bmatrix} = \begin{bmatrix} \frac{4}{7} & -\frac{3}{7} \\ \frac{1}{7} & \frac{1}{7} \end{bmatrix}\begin{bmatrix} 0 \\ -1 \end{bmatrix} = \begin{bmatrix} 0+\frac{3}{7} \\ 0-\frac{1}{7} \end{bmatrix} = \begin{bmatrix} \frac{3}{7} \\ -\frac{1}{7} \end{bmatrix}$

Therefore, $\boldsymbol{x=\frac{3}{7}}$ and $\boldsymbol{y=-\frac{1}{7}}$ and the solution set is $\left\{\left(\frac{3}{7},-\frac{1}{7}\right)\right\}$

Fifth - Check the answers by substituting the x and y values into one of the original equations.

Let's substitute the x and y values into $x+3y=0$, i.e., $\frac{3}{7}+3\times-\frac{1}{7}\stackrel{?}{=}0$; $\frac{3}{7}-\frac{3}{7}\stackrel{?}{=}0$; $0=0$

c. **First** - Write the given system $\begin{matrix} 3x+4y=-2 \\ 6x+8y=10 \end{matrix}$ in the form of $AX=B$, i.e., $\begin{bmatrix} 3 & 4 \\ 6 & 8 \end{bmatrix}\begin{bmatrix} x \\ y \end{bmatrix}=\begin{bmatrix} -2 \\ 10 \end{bmatrix}$

Second - Find determinant of the coefficient matrix A, i.e., $\delta(A) = a_{11}A_{11}+a_{12}A_{12} = 3\cdot A_{11}+4\cdot A_{12} = 3\cdot(-1)^{1+1}M_{11}$ $+4\cdot(-1)^{1+2}M_{12} = 3\cdot M_{11}-4\cdot M_{12} = (3\cdot 8)+(-4\cdot 6) = 24-24 = 0$. **Since $\delta(A)=0$ the inverse matrix does not exist.**

Thus, the linear system does not have a solution.

d. **First** - Write the given system $\begin{matrix} 3x-2y=-3 \\ x-2y=0 \end{matrix}$ in the form of $AX=B$, i.e., $\begin{bmatrix} 3 & -2 \\ 1 & -2 \end{bmatrix}\begin{bmatrix} x \\ y \end{bmatrix}=\begin{bmatrix} -3 \\ 0 \end{bmatrix}$

Second - Find determinant of the coefficient matrix A, i.e., $\delta(A) = a_{11}A_{11}+a_{12}A_{12} = 3\cdot A_{11}+(-2)\cdot A_{12} = 3\cdot(-1)^{1+1}M_{11}$ $+(-2)\cdot(-1)^{1+2}M_{12} = 3\cdot M_{11}+2\cdot M_{12} = (3\cdot -2)+(2\cdot 1) = -4$. Since $\delta(A)\neq 0$ the inverse matrix exist.

Third - Find inverse of the A matrix. Note that $A_{11}=(-1)^{1+1}M_{11}=M_{11}=-2$ $\quad A_{12}=(-1)^{1+2}M_{12}=-M_{12}=-1$

$A_{21} = (-1)^{2+1} M_{21} = -M_{21} = 2$, and $A_{22} = (-1)^{2+2} M_{22} = M_{22} = 3$. Thus, $C = \begin{bmatrix} A_{11} & A_{12} \\ A_{21} & A_{22} \end{bmatrix} = \begin{bmatrix} -2 & -1 \\ 2 & 3 \end{bmatrix}$ and $C^t = \begin{bmatrix} -2 & 2 \\ -1 & 3 \end{bmatrix}$.

Therefore, $A^{-1} = \frac{1}{\delta(A)} C^t = \frac{1}{-4} \cdot \begin{bmatrix} -2 & 2 \\ -1 & 3 \end{bmatrix} = \begin{bmatrix} \frac{1}{2} & -\frac{1}{2} \\ \frac{1}{4} & -\frac{3}{4} \end{bmatrix}$

Check the answer by multiplying the A matrix with A^{-1}.

$$A \times A^{-1} = \begin{bmatrix} 3 & -2 \\ 1 & -2 \end{bmatrix} \times \begin{bmatrix} \frac{1}{2} & -\frac{1}{2} \\ \frac{1}{4} & -\frac{3}{4} \end{bmatrix} = \begin{bmatrix} \left(3 \times \frac{1}{2}\right) + \left(-2 \times \frac{1}{4}\right) & \left(3 \times -\frac{1}{2}\right) + \left(-2 \times -\frac{3}{4}\right) \\ \left(1 \times \frac{1}{2}\right) + \left(-2 \times \frac{1}{4}\right) & \left(1 \times -\frac{1}{2}\right) + \left(-2 \times -\frac{3}{4}\right) \end{bmatrix} = \begin{bmatrix} \frac{3}{2} - \frac{1}{2} & -\frac{3}{2} + \frac{3}{2} \\ \frac{1}{2} - \frac{1}{2} & -\frac{1}{2} + \frac{3}{2} \end{bmatrix} = \begin{bmatrix} 1 & 0 \\ 0 & 1 \end{bmatrix}$$

Since $A \times A^{-1}$ is equal to the identity matrix A^{-1} was computed correctly.

Fourth - Use $X = A^{-1} B$ to solve for x and y, i.e., $\begin{bmatrix} x \\ y \end{bmatrix} = \begin{bmatrix} \frac{1}{2} & -\frac{1}{2} \\ \frac{1}{4} & -\frac{3}{4} \end{bmatrix} \begin{bmatrix} -3 \\ 0 \end{bmatrix} = \begin{bmatrix} -\frac{3}{2} + 0 \\ -\frac{3}{4} + 0 \end{bmatrix} = \begin{bmatrix} -\frac{3}{2} \\ -\frac{3}{4} \end{bmatrix}$

Therefore, $\mathbf{x = -\frac{3}{2}}$ and $\mathbf{y = -\frac{3}{4}}$ and the solution set is $\left\{\left(-\frac{3}{2}, -\frac{3}{4}\right)\right\}$

Fifth - Check the answers by substituting the x and y values into one of the original equations.

Let's substitute the x and y values into $x - 2y = 0$, i.e., $-\frac{3}{2} + \left(-2 \times -\frac{3}{4}\right) \stackrel{?}{=} 0$; $-\frac{3}{2} + \frac{3}{2} \stackrel{?}{=} 0$; $0 = 0$

e. **First** - Write the given system $\begin{matrix} 2x + y = 1 \\ 4x + 2y = 8 \end{matrix}$ in the form of $AX = B$, i.e., $\begin{bmatrix} 2 & 1 \\ 4 & 2 \end{bmatrix} \begin{bmatrix} x \\ y \end{bmatrix} = \begin{bmatrix} 1 \\ 8 \end{bmatrix}$

Second - Find determinant of the coefficient matrix A, i.e., $\delta(A) = a_{11} A_{11} + a_{12} A_{12} = 2 \cdot A_{11} + 1 \cdot A_{12} = 2 \cdot (-1)^{1+1} M_{11}$ $+ 1 \cdot (-1)^{1+2} M_{12} = 2 \cdot M_{11} - 1 \cdot M_{12} = (2 \cdot 2) + (-1 \cdot 4) = 4 - 4 = 0$. **Since $\delta(A) = 0$ the inverse matrix does not exist. Thus, the linear system does not have a solution.**

f. **First** - Write the given system $\begin{matrix} 2x - y = -5 \\ 3x - 4y = -4 \end{matrix}$ in the form of $AX = B$, i.e., $\begin{bmatrix} 2 & -1 \\ 3 & -4 \end{bmatrix} \begin{bmatrix} x \\ y \end{bmatrix} = \begin{bmatrix} -5 \\ -4 \end{bmatrix}$

Second - Find determinant of the coefficient matrix A, i.e., $\delta(A) = a_{11} A_{11} + a_{12} A_{12} = 2 \cdot A_{11} + (-1) \cdot A_{12} = 2 \cdot (-1)^{1+1} M_{11}$ $+ (-1) \cdot (-1)^{1+2} M_{12} = 2 \cdot M_{11} + 1 \cdot M_{12} = (2 \cdot -4) + (1 \cdot 3) = -5$. Since $\delta(A) \neq 0$ the inverse matrix exist.

Third - Find inverse of the A matrix. Note that $A_{11} = (-1)^{1+1} M_{11} = M_{11} = -4$ $\quad A_{12} = (-1)^{1+2} M_{12} = -M_{12} = -3$

$A_{21} = (-1)^{2+1} M_{21} = -M_{21} = 1$, and $A_{22} = (-1)^{2+2} M_{22} = M_{22} = 2$. Thus, $C = \begin{bmatrix} A_{11} & A_{12} \\ A_{21} & A_{22} \end{bmatrix} = \begin{bmatrix} -4 & -3 \\ 1 & 2 \end{bmatrix}$ and $C^t = \begin{bmatrix} -4 & 1 \\ -3 & 2 \end{bmatrix}$

Therefore, $A^{-1} = \frac{1}{\delta(A)} C^t = \frac{1}{-5} \cdot \begin{bmatrix} -4 & 1 \\ -3 & 2 \end{bmatrix} = \begin{bmatrix} \frac{4}{5} & -\frac{1}{5} \\ \frac{3}{5} & -\frac{2}{5} \end{bmatrix}$

Check the answer by multiplying the A matrix with A^{-1}.

$$A \times A^{-1} = \begin{bmatrix} 2 & -1 \\ 3 & -4 \end{bmatrix} \times \begin{bmatrix} \frac{4}{5} & -\frac{1}{5} \\ \frac{3}{5} & -\frac{2}{5} \end{bmatrix} = \begin{bmatrix} \left(2 \times \frac{4}{5}\right) + \left(-1 \times \frac{3}{5}\right) & \left(2 \times -\frac{1}{5}\right) + \left(-1 \times -\frac{2}{5}\right) \\ \left(3 \times \frac{4}{5}\right) + \left(-4 \times \frac{3}{5}\right) & \left(3 \times -\frac{1}{5}\right) + \left(-4 \times -\frac{2}{5}\right) \end{bmatrix} = \begin{bmatrix} \frac{8}{5} - \frac{3}{5} & -\frac{2}{5} + \frac{2}{5} \\ \frac{12}{5} - \frac{12}{5} & -\frac{3}{5} + \frac{8}{5} \end{bmatrix} = \begin{bmatrix} 1 & 0 \\ 0 & 1 \end{bmatrix}$$

Since $A \times A^{-1}$ is equal to the identity matrix A^{-1} was computed correctly.

Fourth - Use $X = A^{-1}B$ to solve for x and y, i.e., $\begin{bmatrix} x \\ y \end{bmatrix} = \begin{bmatrix} \frac{4}{5} & -\frac{1}{5} \\ \frac{3}{5} & -\frac{2}{5} \end{bmatrix} \begin{bmatrix} -5 \\ -4 \end{bmatrix} = \begin{bmatrix} -\frac{20}{5} + \frac{4}{5} \\ -\frac{15}{5} + \frac{8}{5} \end{bmatrix} = \begin{bmatrix} -\frac{16}{5} \\ -\frac{7}{5} \end{bmatrix}$

Therefore, $\boldsymbol{x = -\frac{16}{5}}$ and $\boldsymbol{y = -\frac{7}{5}}$ and the solution set is $\left\{\left(-\frac{\mathbf{16}}{\mathbf{5}}, -\frac{\mathbf{7}}{\mathbf{5}}\right)\right\}$

Fifth - Check the answers by substituting the x and y values into one of the original equations.

Let's substitute the x and y values into $2x - y = -5$, i.e., $2 \times -\frac{16}{5} + \frac{7}{5} \overset{?}{=} -5$; $-\frac{32}{5} + \frac{7}{5} \overset{?}{=} -5$; $-\frac{25}{5} \overset{?}{=} -5$; $-5 = -5$

g. **First** - Write the given system $\begin{matrix} x + y = 2 \\ 2x - z = -1 \\ 2y + 2z = 3 \end{matrix}$ in the form of $AX = B$, i.e., $\begin{bmatrix} 1 & 1 & 0 \\ 2 & 0 & -1 \\ 0 & 2 & 2 \end{bmatrix} \begin{bmatrix} x \\ y \\ z \end{bmatrix} = \begin{bmatrix} 2 \\ -1 \\ 3 \end{bmatrix}$

Second - Find determinant of the coefficient matrix A.

$\delta(A) = a_{11}A_{11} + a_{12}A_{12} + a_{13}A_{13} = 1 \cdot A_{11} + 1 \cdot A_{12} + 0 \cdot A_{13} = A_{11} + A_{12} = (-1)^{1+1}M_{11} + (-1)^{1+2}M_{12} = M_{11} - M_{12}$

$= \begin{vmatrix} 0 & -1 \\ 2 & 2 \end{vmatrix} - \begin{vmatrix} 2 & -1 \\ 0 & 2 \end{vmatrix} = (0+2) - (4-0) = 2 - 4 = -2$. Since $\delta(A) \neq 0$ matrix A has an inverse.

Third - Find inverse of the A matrix. Note that

$A_{11} = (-1)^{1+1}M_{11} = M_{11} = \begin{vmatrix} 0 & -1 \\ 2 & 2 \end{vmatrix} = 0 + 2 = 2$

$A_{12} = (-1)^{1+2}M_{12} = -M_{12} = -\begin{vmatrix} 2 & -1 \\ 0 & 2 \end{vmatrix} = -(4+0) = -4$

$A_{13} = (-1)^{1+3}M_{13} = M_{13} = \begin{vmatrix} 2 & 0 \\ 0 & 2 \end{vmatrix} = 4 + 0 = 4$

$A_{21} = (-1)^{2+1}M_{21} = -M_{21} = -\begin{vmatrix} 1 & 0 \\ 2 & 2 \end{vmatrix} = -(2+0) = -2$

$A_{22} = (-1)^{2+2}M_{22} = M_{22} = \begin{vmatrix} 1 & 0 \\ 0 & 2 \end{vmatrix} = 2 + 0 = 2$

$A_{23} = (-1)^{2+3}M_{23} = -M_{23} = -\begin{vmatrix} 1 & 1 \\ 0 & 2 \end{vmatrix} = -(2+0) = -2$

$A_{31} = (-1)^{3+1}M_{31} = M_{31} = \begin{vmatrix} 1 & 0 \\ 0 & -1 \end{vmatrix} = -1 + 0 = -1$

$A_{32} = (-1)^{3+2}M_{32} = -M_{32} = -\begin{vmatrix} 1 & 0 \\ 2 & -1 \end{vmatrix} = -(-1+0) = 1$

$A_{33} = (-1)^{3+3}M_{33} = M_{33} = \begin{vmatrix} 1 & 1 \\ 2 & 0 \end{vmatrix} = 0 - 2 = -2$

Thus, $C = \begin{bmatrix} A_{11} & A_{12} & A_{13} \\ A_{21} & A_{22} & A_{23} \\ A_{31} & A_{32} & A_{33} \end{bmatrix} = \begin{bmatrix} 2 & -4 & 4 \\ -2 & 2 & -2 \\ -1 & 1 & -2 \end{bmatrix}$ and $C^t = \begin{bmatrix} 2 & -2 & -1 \\ -4 & 2 & 1 \\ 4 & -2 & -2 \end{bmatrix}$. Therefore,

$$A^{-1} = \frac{1}{\delta(A)}C^t = \frac{1}{-2}\begin{bmatrix} 2 & -2 & -1 \\ -4 & 2 & 1 \\ 4 & -2 & -2 \end{bmatrix} = \begin{bmatrix} -\frac{2}{2} & \frac{2}{2} & \frac{1}{2} \\ \frac{4}{2} & -\frac{2}{2} & -\frac{1}{2} \\ -\frac{4}{2} & \frac{2}{2} & \frac{2}{2} \end{bmatrix} = \begin{bmatrix} -1 & 1 & \frac{1}{2} \\ 2 & -1 & -\frac{1}{2} \\ -2 & 1 & 1 \end{bmatrix}$$

Check the answer by multiplying the A matrix with A^{-1}.

$$A \times A^{-1} = \begin{bmatrix} 1 & 1 & 0 \\ 2 & 0 & -1 \\ 0 & 2 & 2 \end{bmatrix} \times \begin{bmatrix} -1 & 1 & \frac{1}{2} \\ 2 & -1 & -\frac{1}{2} \\ -2 & 1 & 1 \end{bmatrix} == \begin{bmatrix} -1+2+0 & 1-1+0 & \frac{1}{2}-\frac{1}{2}+0 \\ 1-1+0 & 2+0-1 & 1+0-1 \\ 0+4-4 & 0-2+2 & 0-1+2 \end{bmatrix} = \begin{bmatrix} 1 & 0 & 0 \\ 0 & 1 & 0 \\ 0 & 0 & 1 \end{bmatrix}$$

Since $A \times A^{-1}$ is equal to the identity matrix A^{-1} was computed correctly.

Fourth - Use $X = A^{-1}B$ to solve for x, y, and z.

$$\begin{bmatrix}x\\y\\z\end{bmatrix} = \begin{bmatrix}-1 & 1 & \frac{1}{2}\\ 2 & -1 & -\frac{1}{2}\\ -2 & 1 & 1\end{bmatrix}\begin{bmatrix}2\\-1\\3\end{bmatrix} = \begin{bmatrix}-2-1+\frac{3}{2}\\ 4+1-\frac{3}{2}\\ -4-1+3\end{bmatrix} = \begin{bmatrix}-\frac{3}{2}\\ \frac{7}{2}\\ -2\end{bmatrix}$$

Therefore, $\boldsymbol{x=-\frac{3}{2}}$, $\boldsymbol{y=\frac{7}{2}}$, $\boldsymbol{z=-2}$ and the solution set is $\left\{\left(-\frac{3}{2}, \frac{7}{2}, -2\right)\right\}$

Fifth - Check the answers by substituting the x, y, and z values into one of the original equation.

Let's substitute the x, y, and z values into $x+y=2$. $-\frac{3}{2}+\frac{7}{2}\overset{?}{=}2$; $\frac{-3+7}{2}\overset{?}{=}2$; $\frac{4}{2}\overset{?}{=}2$; $2=2$

h. **First** - Write the given system $\begin{aligned}x-y+3z&=2\\ x-z&=-3\\ 2x-2y+6z&=-1\end{aligned}$ in the form of $AX=B$, i.e., $\begin{bmatrix}1 & -1 & 3\\ 1 & 0 & -1\\ 2 & -2 & 6\end{bmatrix}\begin{bmatrix}x\\y\\z\end{bmatrix} = \begin{bmatrix}2\\-3\\-1\end{bmatrix}$

Second - Find determinant of the coefficient matrix A by expanding about the second row.

$$\delta(A) = a_{21}A_{21}+a_{22}A_{22}+a_{23}A_{23} = 1\cdot A_{21}+0\cdot A_{22}-1\cdot A_{23} = A_{21}-A_{23} = (-1)^{2+1}M_{21}-(-1)^{2+3}M_{23} = -M_{21}+M_{23}$$

$= -\begin{vmatrix}-1 & 3\\ -2 & 6\end{vmatrix}+\begin{vmatrix}1 & -1\\ 2 & -2\end{vmatrix} = -(-6+6)+(-2+2) = 0+0 = 0$. **Since $\boldsymbol{\delta(A)=0}$ matrix $\boldsymbol{A}$ does not have an inverse.**

i. **First** - Write the given system $\begin{aligned}x+3y-z&=-2\\ -x+2y+3z&=1\\ x+y-2z&=0\end{aligned}$ in the form of $AX=B$, i.e., $\begin{bmatrix}1 & 3 & -1\\ -1 & 2 & 3\\ 1 & 1 & -2\end{bmatrix}\begin{bmatrix}x\\y\\z\end{bmatrix} = \begin{bmatrix}-2\\1\\0\end{bmatrix}$

Second - Find determinant of the coefficient matrix A.

$$\delta(A) = a_{11}A_{11}+a_{12}A_{12}+a_{13}A_{13} = 1\cdot A_{11}+3\cdot A_{12}+(-1)\cdot A_{13} = A_{11}+3A_{12}-A_{13} = (-1)^{1+1}M_{11}+3(-1)^{1+2}M_{12}-(-1)^{1+3}M_{13}$$

$= M_{11}-3M_{12}-M_{13} = \begin{vmatrix}2 & 3\\ 1 & -2\end{vmatrix}-3\begin{vmatrix}-1 & 3\\ 1 & -2\end{vmatrix}-\begin{vmatrix}-1 & 2\\ 1 & 1\end{vmatrix} = (-4-3)-3(2-3)-(-1-2) = -7+3+3 = -1$. Since $\delta(A)\neq 0$ matrix A has an inverse.

Third - Find inverse of the A matrix. Note that

$A_{11}=(-1)^{1+1}M_{11}=M_{11}=\begin{vmatrix}2 & 3\\ 1 & -2\end{vmatrix}=-4-3=-7$

$A_{12}=(-1)^{1+2}M_{12}=-M_{12}=-\begin{vmatrix}-1 & 3\\ 1 & -2\end{vmatrix}=-(2-3)=1$

$A_{13}=(-1)^{1+3}M_{13}=M_{13}=\begin{vmatrix}-1 & 2\\ 1 & 1\end{vmatrix}=-1-2=-3$

$A_{21}=(-1)^{2+1}M_{21}=-M_{21}=-\begin{vmatrix}3 & -1\\ 1 & -2\end{vmatrix}=-(-6+1)=5$

$A_{22}=(-1)^{2+2}M_{22}=M_{22}=\begin{vmatrix}1 & -1\\ 1 & -2\end{vmatrix}=-2+1=-1$

$A_{23}=(-1)^{2+3}M_{23}=-M_{23}=-\begin{vmatrix}1 & 3\\ 1 & 1\end{vmatrix}=-(1-3)=2$

$A_{31}=(-1)^{3+1}M_{31}=M_{31}=\begin{vmatrix}3 & -1\\ 2 & 3\end{vmatrix}=9+2=11$

$A_{32}=(-1)^{3+2}M_{32}=-M_{32}=-\begin{vmatrix}1 & -1\\ -1 & 3\end{vmatrix}=-(3-1)=-2$

$A_{33}=(-1)^{3+3}M_{33}=M_{33}=\begin{vmatrix}1 & 3\\ -1 & 2\end{vmatrix}=2+3=5$

Thus, $C=\begin{bmatrix}A_{11} & A_{12} & A_{13}\\ A_{21} & A_{22} & A_{23}\\ A_{31} & A_{32} & A_{33}\end{bmatrix}=\begin{bmatrix}-7 & 1 & -3\\ 5 & -1 & 2\\ 11 & -2 & 5\end{bmatrix}$ and $C^t=\begin{bmatrix}-7 & 5 & 11\\ 1 & -1 & -2\\ -3 & 2 & 5\end{bmatrix}$. Therefore,

$$A^{-1}=\frac{1}{\delta(A)}C^t=\frac{1}{-1}\begin{bmatrix}-7 & 5 & 11\\ 1 & -1 & -2\\ -3 & 2 & 5\end{bmatrix}=\begin{bmatrix}7 & -5 & -11\\ -1 & 1 & 2\\ 3 & -2 & -5\end{bmatrix}$$

Check the answer by multiplying the A matrix with A^{-1}.

$$A \times A^{-1} = \begin{bmatrix} 1 & 3 & -1 \\ -1 & 2 & 3 \\ 1 & 1 & -2 \end{bmatrix} \times \begin{bmatrix} 7 & -5 & -11 \\ -1 & 1 & 2 \\ 3 & -2 & -5 \end{bmatrix} = \begin{bmatrix} 7-3-3 & -5+3+2 & -11+6+5 \\ -7-2+9 & 5+2-6 & 11+4-15 \\ 7-1-6 & -5+1+4 & -11+2+10 \end{bmatrix} = \begin{bmatrix} 1 & 0 & 0 \\ 0 & 1 & 0 \\ 0 & 0 & 1 \end{bmatrix}$$

Since $A \times A^{-1}$ is equal to the identity matrix A^{-1} was computed correctly.

Fourth - Use $X = A^{-1}B$ to solve for x, y, and z.

$$\begin{bmatrix} x \\ y \\ z \end{bmatrix} = \begin{bmatrix} 7 & -5 & -11 \\ -1 & 1 & 2 \\ 3 & -2 & -5 \end{bmatrix} \begin{bmatrix} -2 \\ 1 \\ 0 \end{bmatrix} = \begin{bmatrix} -14-5+0 \\ 2+1+0 \\ -6-2+0 \end{bmatrix} = \begin{bmatrix} -19 \\ 3 \\ -8 \end{bmatrix}$$

Therefore, $\mathbf{x = -19}$, $\mathbf{y = 3}$, $\mathbf{z = -8}$ and the solution set is $\{(-19, 3, -8)\}$

Fifth - Check the answers by substituting the x, y, and z values into one of the original equation.

Let's substitute the x, y, and z values into $x + y - 2z = 0$. $-19+3+(-2\times -8) \overset{?}{=} 0$; $-19+3+16 \overset{?}{=} 0$; $0 = 0$

j. **First** - Write the given system $\begin{matrix} x+y=0 \\ 2x-z=1 \\ 4x+4y=-1 \end{matrix}$ in the form of $AX = B$, i.e., $\begin{bmatrix} 1 & 1 & 0 \\ 2 & 0 & -1 \\ 4 & 4 & 0 \end{bmatrix} \begin{bmatrix} x \\ y \\ z \end{bmatrix} = \begin{bmatrix} 0 \\ 1 \\ -1 \end{bmatrix}$

Second - Find determinant of the coefficient matrix A by expanding about the third column.

$$\delta(A) = a_{13}A_{13} + a_{23}A_{23} + a_{33}A_{33} = 0 \cdot A_{13} + (-1) \cdot A_{23} + 0 \cdot A_{33} = -A_{23} = -(-1)^{2+3} M_{23} = M_{23} = \begin{vmatrix} 1 & 1 \\ 4 & 4 \end{vmatrix} = 4 - 4 = 0.$$

Since $\delta(A) = 0$ matrix A does not have an inverse.

2. Use the result of exercise number 1-g (above) to find the solution set for the following linear equations.

a. **First** - Write the given system $\begin{matrix} x+y=-3 \\ 2x-z=-4 \\ 2y+2z=1 \end{matrix}$ in the form of $AX = B$, i.e., $\begin{bmatrix} 1 & 1 & 0 \\ 2 & 0 & -1 \\ 0 & 2 & 2 \end{bmatrix} \begin{bmatrix} x \\ y \\ z \end{bmatrix} = \begin{bmatrix} -3 \\ -4 \\ 1 \end{bmatrix}$

Second - From exercise number 1-g we have $A^{-1} = \begin{bmatrix} -1 & 1 & \frac{1}{2} \\ 2 & -1 & -\frac{1}{2} \\ -2 & 1 & 1 \end{bmatrix}$. Therefore, Using $X = A^{-1}B$ we obtain:

$$\begin{bmatrix} x \\ y \\ z \end{bmatrix} = \begin{bmatrix} -1 & 1 & \frac{1}{2} \\ 2 & -1 & -\frac{1}{2} \\ -2 & 1 & 1 \end{bmatrix} \begin{bmatrix} -3 \\ -4 \\ 1 \end{bmatrix} = \begin{bmatrix} 3-4+\frac{1}{2} \\ -6+4-\frac{1}{2} \\ 6-4+1 \end{bmatrix} = \begin{bmatrix} -\frac{1}{2} \\ -\frac{5}{2} \\ 3 \end{bmatrix}$$

Therefore, $\mathbf{x = -\frac{1}{2}}$, $\mathbf{y = -\frac{5}{2}}$, $\mathbf{z = 3}$ and the solution set is $\left\{\left(-\frac{1}{2}, -\frac{5}{2}, 3\right)\right\}$

Third - Check the answers by substituting the x, y, and z values into one of the original equation.

Let's substitute the x, y, and z values into $x + y = -3$. $-\frac{1}{2} - \frac{5}{2} \overset{?}{=} -3$; $\frac{-1-5}{2} \overset{?}{=} -3$; $-\frac{6}{2} \overset{?}{=} -3$; $-3 = -3$

b. **First** - Write the given system $\begin{matrix} x+y=0 \\ 2x-z=-1 \\ 2y+2z=1 \end{matrix}$ in the form of $AX = B$, i.e., $\begin{bmatrix} 1 & 1 & 0 \\ 2 & 0 & -1 \\ 0 & 2 & 2 \end{bmatrix} \begin{bmatrix} x \\ y \\ z \end{bmatrix} = \begin{bmatrix} 0 \\ -1 \\ 1 \end{bmatrix}$

Second - From exercise number 1-g we have $A^{-1} = \begin{bmatrix} -1 & 1 & \frac{1}{2} \\ 2 & -1 & -\frac{1}{2} \\ -2 & 1 & 1 \end{bmatrix}$. Therefore, Using $X = A^{-1}B$ we obtain:

$$\begin{bmatrix} x \\ y \\ z \end{bmatrix} = \begin{bmatrix} -1 & 1 & \frac{1}{2} \\ 2 & -1 & -\frac{1}{2} \\ -2 & 1 & 1 \end{bmatrix} \begin{bmatrix} 0 \\ -1 \\ 1 \end{bmatrix} = \begin{bmatrix} 0-1+\frac{1}{2} \\ 0+1-\frac{1}{2} \\ 0-1+1 \end{bmatrix} = \begin{bmatrix} -\frac{1}{2} \\ \frac{1}{2} \\ 0 \end{bmatrix}$$

Therefore, $x=-\frac{1}{2}$, $y=\frac{1}{2}$, $z=0$ and the solution set is $\left\{\left(-\frac{1}{2}, \frac{1}{2}, 0\right)\right\}$

Third - Check the answers by substituting the x, y, and z values into one of the original equation.

Let's substitute the x, y, and z values into $x+y=0$. $-\frac{1}{2}+\frac{1}{2}\overset{?}{=}0$; $\frac{-1+1}{2}\overset{?}{=}0$; $0=0$

3. Use the result of exercise number 1-i (above) to find the solution set for the following linear equations.

a. **First** - Write the given system $\begin{matrix} x+3y-z=1 \\ -x+2y+3z=3 \\ x+y-2z=-1 \end{matrix}$ in the form of $AX=B$, i.e., $\begin{bmatrix} 1 & 3 & -1 \\ -1 & 2 & 3 \\ 1 & 1 & -2 \end{bmatrix}\begin{bmatrix} x \\ y \\ z \end{bmatrix}=\begin{bmatrix} 1 \\ 3 \\ -1 \end{bmatrix}$

Second - From exercise number 1-i we have $A^{-1}=\begin{bmatrix} 7 & -5 & -11 \\ -1 & 1 & 2 \\ 3 & -2 & -5 \end{bmatrix}$. Therefore, Using $X=A^{-1}B$ we obtain:

$$\begin{bmatrix} x \\ y \\ z \end{bmatrix}=\begin{bmatrix} 7 & -5 & -11 \\ -1 & 1 & 2 \\ 3 & -2 & -5 \end{bmatrix}\begin{bmatrix} 1 \\ 3 \\ -1 \end{bmatrix}=\begin{bmatrix} 7-15+11 \\ -1+3-2 \\ 3-6+5 \end{bmatrix}=\begin{bmatrix} 3 \\ 0 \\ 2 \end{bmatrix}$$

Therefore, $x=3$, $y=0$, $z=2$ and the solution set is $\{(3, 0, 2)\}$

Third - Check the answers by substituting the x, y, and z values into one of the original equation.

Let's substitute the x, y, and z values into $x+3y-z=1$. $3+(3\times 0)-2\overset{?}{=}1$; $3-2\overset{?}{=}1$; $1=1$

b. **First** - Write the given system $\begin{matrix} x+3y-z=-1 \\ -x+2y+3z=0 \\ x+y-2z=2 \end{matrix}$ in the form of $AX=B$, i.e., $\begin{bmatrix} 1 & 3 & -1 \\ -1 & 2 & 3 \\ 1 & 1 & -2 \end{bmatrix}\begin{bmatrix} x \\ y \\ z \end{bmatrix}=\begin{bmatrix} -1 \\ 0 \\ 2 \end{bmatrix}$

Second - From exercise number 1-i we have $A^{-1}=\begin{bmatrix} 7 & -5 & -11 \\ -1 & 1 & 2 \\ 3 & -2 & -5 \end{bmatrix}$. Therefore, Using $X=A^{-1}B$ we obtain:

$$\begin{bmatrix} x \\ y \\ z \end{bmatrix}=\begin{bmatrix} 7 & -5 & -11 \\ -1 & 1 & 2 \\ 3 & -2 & -5 \end{bmatrix}\begin{bmatrix} -1 \\ 0 \\ 2 \end{bmatrix}=\begin{bmatrix} -7+0-22 \\ 1+0+4 \\ -3+0-10 \end{bmatrix}=\begin{bmatrix} -29 \\ 5 \\ -13 \end{bmatrix}$$

Therefore, $x=-29$, $y=5$, $z=-13$ and the solution set is $\{(-29, 5, -13)\}$

Third - Check the answers by substituting the x, y, and z values into one of the original equation.

Let's substitute the x, y, and z values into $x+3y-z=-1$. $-29+(3\times 5)+13\overset{?}{=}-1$; $-29+28\overset{?}{=}-1$; $-1=-1$

Section 3.5 Case IV Solutions - Solving Linear Systems Using Cramer's Rule

1. Find the solution set to each of the following linear systems using Cramer's rule. Note that Problems 1-e through 1-i are identical to the exercise 1-e through 1-i in Section 3.5 Case III.

a. **First** - Given the linear sytem $\begin{matrix} \frac{1}{3}x-\frac{1}{2}y=1 \\ \frac{1}{2}x-\frac{1}{3}y=0 \end{matrix}$ the coefficient matrix is equal to $A=\begin{bmatrix} \frac{1}{3} & -\frac{1}{2} \\ \frac{1}{2} & -\frac{1}{3} \end{bmatrix}$ and the augmented matrix is equal to $\begin{bmatrix} \frac{1}{3} & -\frac{1}{2} & \vdots & 1 \\ \frac{1}{2} & -\frac{1}{3} & \vdots & 0 \end{bmatrix}$

Second – Let's find the determininat of A, i.e., $\delta(A)=|A| = a_{11}\times a_{22}-a_{12}\times a_{21} = \frac{1}{3}\times -\frac{1}{3}+\frac{1}{2}\times\frac{1}{2} = -\frac{1}{9}+\frac{1}{4} = \frac{5}{36} = 0.14$.

Since $\delta(A)\neq 0$ we can proceed to the next step.

Third – Replace the entries in the first column of the augmented matrix with the entries in the right hand side of the augmented matrix to obtain matrix $A_1 = \begin{bmatrix} 1 & -\frac{1}{2} \\ 0 & -\frac{1}{3} \end{bmatrix}$. Next, find determinant of A_1, i.e., $\delta(A_1) = |A_1| = a_{11} \times a_{22} - a_{12} \times a_{21}$

$= 1 \times -\frac{1}{3} - 0 \times -\frac{1}{2} = -\frac{1}{3}$

Fourth – Replace the entries in the second column of the augmented matrix with the entries in the right hand side of the augmented matrix to obtain matrix $A_2 = \begin{bmatrix} \frac{1}{3} & 1 \\ \frac{1}{2} & 0 \end{bmatrix}$. Next, find the determinant of A_1, i.e., $\delta(A_2) = |A_2| = a_{11} \times a_{22} - a_{12} \times a_{21}$

$= \frac{1}{3} \times 0 - 1 \times \frac{1}{2} = -\frac{1}{2}$

Fifth – Solve for x and y using the Cramer's rule, i.e., $x = \frac{|A_1|}{|A|}$ and $y = \frac{|A_2|}{|A|}$. Therefore, $x = \frac{|A_1|}{|A|} = \frac{-0.333}{0.14} = \mathbf{-2.38}$, and $y = \frac{|A_2|}{|A|} = \frac{-0.5}{0.14} = \mathbf{-3.57}$. Thus, the solution set is equal to $\{(\mathbf{-2.38, -3.57})\}$. Let's check the answer by substituting the x and y values into $\frac{1}{3}x - \frac{1}{2}y = 1$, i.e., $\frac{1}{3} \times -2.38 - \frac{1}{2} \times -3.57 \stackrel{?}{=} 1$; $-0.79 + 1.79 \stackrel{?}{=} 1$; $1 = 1$

b. **First** - Given the linear system $\begin{matrix} x - 4y = 1 \\ y = 2 \end{matrix}$ the coefficient matrix is equal to $A = \begin{bmatrix} 1 & -4 \\ 0 & 1 \end{bmatrix}$ and the augmented matrix is equal to $\begin{bmatrix} 1 & -4 & \vdots & 1 \\ 0 & 1 & \vdots & 2 \end{bmatrix}$.

Second – Let's find the determinant of A, i.e., $\delta(A) = |A| = a_{11} \times a_{22} - a_{12} \times a_{21} = (1 \times 1) - (-4 \times 0) = 1$. Since $\delta(A) \neq 0$ we can proceed to the next step.

Third – Replace the entries in the first column of the augmented matrix with the entries in the right hand side of the augmented matrix to obtain matrix $A_1 = \begin{bmatrix} 1 & -4 \\ 2 & 1 \end{bmatrix}$. Next, find determinant of A_1, i.e., $\delta(A_1) = |A_1| = a_{11} \times a_{22} - a_{12} \times a_{21}$

$= (1 \times 1) - (-4 \times 2) = 1 + 8 = 9$

Fourth – Replace the entries in the second column of the augmented matrix with the entries in the right hand side of the augmented matrix to obtain matrix $A_2 = \begin{bmatrix} 1 & 1 \\ 0 & 2 \end{bmatrix}$. Next, find determinant of A_1, i.e., $\delta(A_2) = |A_2| = a_{11} \times a_{22} - a_{12} \times a_{21}$

$= (1 \times 2) - (1 \times 0) = 2 - 0 = 2$

Fifth – Solve for x and y using the Cramer's rule, i.e., $x = \frac{|A_1|}{|A|}$ and $y = \frac{|A_2|}{|A|}$. Therefore, $x = \frac{|A_1|}{|A|} = \frac{9}{1} = \mathbf{9}$, and $y = \frac{|A_2|}{|A|} = \frac{2}{1} = \mathbf{2}$. Thus, the solution set is equal to $\{(\mathbf{9, 2})\}$. Let's check the answer by substituting the x and y values into $x - 4y = 1$, i.e., $9 - 4 \times 2 \stackrel{?}{=} 1$; $9 - 8 \stackrel{?}{=} 1$; $1 = 1$.

Note that using direct substitution method would have been a much simpler and quicker method in solving this particular problem, i.e., since $y = 2$ substitution of the y value into $x - 4y = 1$ result in the x value to be equal to 9. However, the objective is to practice solution to linear systems using Cramer's rule method.

c. **First** - Given the linear system $\begin{matrix} x + y = -2a \\ x - y = 2b \end{matrix}$ the coefficient matrix is equal to $A = \begin{bmatrix} 1 & 1 \\ 1 & -1 \end{bmatrix}$ and the augmented matrix is equal to $\begin{bmatrix} 1 & 1 & \vdots & -2a \\ 1 & -1 & \vdots & 2b \end{bmatrix}$.

Second – Let's find the determinant of A, i.e., $\delta(A) = |A| = a_{11} \times a_{22} - a_{12} \times a_{21} = (1 \times -1) - (1 \times 1) = -1 - 1 = -2$. Since $\delta(A) \neq 0$ we can proceed to the next step.

Third – Replace the entries in the first column of the augmented matrix with the entries in the right hand side of the augmented matrix to obtain the matrix $A_1 = \begin{bmatrix} -2a & 1 \\ 2b & -1 \end{bmatrix}$. Next, find determinant of A_1, i.e., $\delta(A_1) = |A_1|$ = $a_{11} \times a_{22} - a_{12} \times a_{21}$ = $(-2a \times -1) - (1 \times 2b)$ = $2a - 2b$

Fourth – Replace the entries in the second column of the augmented matrix with the entries in the right hand side of the augmented matrix to obtain the matrix $A_2 = \begin{bmatrix} 1 & -2a \\ 1 & 2b \end{bmatrix}$. Next, find determinant of A_1, i.e., $\delta(A_2) = |A_2|$ = $a_{11} \times a_{22} - a_{12} \times a_{21}$ = $(1 \times 2b) - (-2a \times 1)$ = $2b + 2a$

Fifth – Solve for x and y using the Cramer's rule, i.e., $x = \frac{|A_1|}{|A|}$ and $y = \frac{|A_2|}{|A|}$. Therefore, x = $\frac{|A_1|}{|A|}$ = $\frac{2a-2b}{-2}$ = $\boldsymbol{b-a}$, and y = $\frac{|A_2|}{|A|}$ = $\frac{2b+2a}{-2}$ = $\boldsymbol{-b-a}$. Thus, the solution set is equal to $\{(\boldsymbol{b-a, -b-a})\}$. Let's check the answer by substituting the x and y values into $x - y = 2b$, i.e., $(b-a)-(-b-a) \stackrel{?}{=} 2b$; $b+b-a+a \stackrel{?}{=} 2b$; $2b = 2b$

d. **First** - Given the linear system $\begin{matrix} x-3y=1 \\ 2x-y=3 \end{matrix}$ the coefficient matrix is equal to $A = \begin{bmatrix} 1 & -3 \\ 2 & -1 \end{bmatrix}$ and the augmented matrix is equal to $\begin{bmatrix} 1 & -3 & \vdots & 1 \\ 2 & -1 & \vdots & 3 \end{bmatrix}$.

Second – Let's find the determinant of A, i.e., $\delta(A) = |A|$ = $a_{11} \times a_{22} - a_{12} \times a_{21}$ = $(1 \times -1) - (-3 \times 2)$ = $-1 + 6$ = 5. Since $\delta(A) \neq 0$ we can proceed to the next step.

Third – Replace the entries in the first column of the augmented matrix with the entries in the right hand side of the augmented matrix to obtain the matrix $A_1 = \begin{bmatrix} 1 & -3 \\ 3 & -1 \end{bmatrix}$. Next, find the determinant of A_1, i.e., $\delta(A_1) = |A_1|$ = $a_{11} \times a_{22} - a_{12} \times a_{21}$ = $(1 \times -1) - (-3 \times 3)$ = $-1 + 9$ = 8

Fourth – Replace the entries in the second column of the augmented matrix with the entries in the right hand side of the augmented matrix to obtain the matrix $A_2 = \begin{bmatrix} 1 & 1 \\ 2 & 3 \end{bmatrix}$. Next, find the determinant of A_1, i.e., $\delta(A_2) = |A_2|$ = $a_{11} \times a_{22} - a_{12} \times a_{21}$ = $(1 \times 3) - (1 \times 2)$ = $3 - 2$ = 1

Fifth – Solve for x and y using the Cramer's rule, i.e., $x = \frac{|A_1|}{|A|}$ and $y = \frac{|A_2|}{|A|}$. Therefore, x = $\frac{|A_1|}{|A|}$ = $\frac{8}{5}$ = **1.6**, and y = $\frac{|A_2|}{|A|}$ = $\frac{1}{5}$ = **0.2**. Thus, the solution set is equal to $\{(\mathbf{1.6, 0.2})\}$. Let's check the answer by substituting the x and y values into $2x - y = 3$, i.e., $2 \times 1.6 - 0.2 \stackrel{?}{=} 3$; $3.2 - 0.2 \stackrel{?}{=} 3$; $3 = 3$

e. **First** - Given the linear system $\begin{matrix} 2x+y=1 \\ 4x+2y=8 \end{matrix}$ the coefficient matrix is equal to $A = \begin{bmatrix} 2 & 1 \\ 4 & 2 \end{bmatrix}$ and the augmented matrix is equal to $\begin{bmatrix} 2 & 1 & \vdots & 1 \\ 4 & 2 & \vdots & 8 \end{bmatrix}$

Second – Let's find the determinant of A, i.e., $\delta(A) = |A|$ = $a_{11} \times a_{22} - a_{12} \times a_{21}$ = $(2 \times 2) - (1 \times 4)$ = $4 - 4$ = 0. **Since $\boldsymbol{\delta(A) = 0}$ the linear system is either a dependent or an inconsistent system. In this case, the linear system is inconsistent.**

f. **First** - Given the linear system $\begin{matrix} 2x-y=-5 \\ 3x-4y=-4 \end{matrix}$ the coefficient matrix is equal to $A = \begin{bmatrix} 2 & -1 \\ 3 & -4 \end{bmatrix}$ and the augmented matrix is equal to $\begin{bmatrix} 2 & -1 & \vdots & -5 \\ 3 & -4 & \vdots & -4 \end{bmatrix}$.

Second – Let's find the determinant of A, i.e., $\delta(A) = |A|$ = $a_{11} \times a_{22} - a_{12} \times a_{21}$ = $(2 \times -4) - (-1 \times 3)$ = $-8 + 3$ = -5. Since

$\delta(A) \neq 0$ we can proceed to the next step.

Third – Replace the entries in the first column of the augmented matrix with the entries in the right hand side of the augmented matrix to obtain matrix $A_1 = \begin{bmatrix} -5 & -1 \\ -4 & -4 \end{bmatrix}$. Next, find the determinant of A_1, i.e., $\delta(A_1) = |A_1| = a_{11} \times a_{22} - a_{12} \times a_{21}$ $= (-5 \times -4) - (-1 \times -4) = 20 - 4 = 16$

Fourth – Replace the entries in the second column of the augmented matrix with the entries in the right hand side of the augmented matrix to obtain matrix $A_2 = \begin{bmatrix} 2 & -5 \\ 3 & -4 \end{bmatrix}$. Next, find the determinant of A_1, i.e., $\delta(A_2) = |A_2| = a_{11} \times a_{22} - a_{12} \times a_{21}$ $= (2 \times -4) - (-5 \times 3) = -8 + 15 = 7$

Fifth – Solve for x and y using the Cramer's rule, i.e., $x = \frac{|A_1|}{|A|}$ and $y = \frac{|A_2|}{|A|}$. Therefore, $x = \frac{|A_1|}{|A|} = \frac{16}{-5} = \mathbf{-3.2}$, and $y = \frac{|A_2|}{|A|} = \frac{7}{-5} = \mathbf{-1.4}$. Thus, the solution set is equal to $\{(\mathbf{-3.2, -1.4})\}$. Let's check the answer by substituting the x and y values into $3x - 4y = -4$, i.e., $(3 \times -3.2) + (-4 \times -1.4) \overset{?}{=} -4$; $-9.6 + 5.6 \overset{?}{=} -4$; $-4 = -4$

g. **First** – Given the linear system $\begin{matrix} x + y = 2 \\ 2x - z = -1 \\ 2y + 2z = 3 \end{matrix}$ The coefficient matrix is equal to $A = \begin{bmatrix} 1 & 1 & 0 \\ 2 & 0 & -1 \\ 0 & 2 & 2 \end{bmatrix}$ and the augmented matrix is equal to $\begin{bmatrix} 1 & 1 & 0 & \vdots & 2 \\ 2 & 0 & -1 & \vdots & -1 \\ 0 & 2 & 2 & \vdots & 3 \end{bmatrix}$

Second - Let's find $\delta(A)$ by expanding about the first row, i.e., $\delta(A) = |A| = a_{11}A_{11} + a_{12}A_{12} + a_{13}A_{13} = 1 \cdot A_{11} + 1 \cdot A_{12}$ $+ 0 \cdot A_{13} = A_{11} + A_{12} = (-1)^{1+1}M_{11} + (-1)^{1+2}M_{12} = M_{11} - M_{12} = \begin{vmatrix} 0 & -1 \\ 2 & 2 \end{vmatrix} - \begin{vmatrix} 2 & -1 \\ 0 & 2 \end{vmatrix} = (0 + 2) - (4 + 0) = 2 - 4 = -2$. Since $\delta(A) \neq 0$ we can proceed to the next step.

Third - Replace the entries in the first column of the augmented matrix with the entries in the right hand side of the augmented matrix to obtain the matrix $A_1 = \begin{bmatrix} 2 & 1 & 0 \\ -1 & 0 & -1 \\ 3 & 2 & 2 \end{bmatrix}$. Next, find $\delta(A_1)$ by expanding about the first row, i.e., $\delta(A_1) = |A_1|$

$= a_{11}A_{11} + a_{12}A_{12} + a_{13}A_{13} = 2 \cdot A_{11} + 1 \cdot A_{12} + 0 \cdot A_{13} = 2A_{11} + A_{12} = 2 \cdot (-1)^{1+1}M_{11} + (-1)^{1+2}M_{12} = 2M_{11} - M_{12} = 2\begin{vmatrix} 0 & -1 \\ 2 & 2 \end{vmatrix}$ $- \begin{vmatrix} -1 & -1 \\ 3 & 2 \end{vmatrix} = 2[(0 \times 2) - (-1 \times 2)] - [(-1 \times 2) - (-1 \times 3)] = 2(0 + 2) - (-2 + 3) = 4 - 1 = 3$

Fourth - Replace the entries in the second column of the augmented matrix with the entries in the right hand side of the augmented matrix to obtain the matrix $A_2 = \begin{bmatrix} 1 & 2 & 0 \\ 2 & -1 & -1 \\ 0 & 3 & 2 \end{bmatrix}$. Next, find $\delta(A_2)$ by expanding about the first row, i.e.,.

$\delta(A_2) = |A_2| = a_{11}A_{11} + a_{12}A_{12} + a_{13}A_{13} = 1 \cdot A_{11} + 2 \cdot A_{12} + 0 \cdot A_{13} = A_{11} + 2A_{12} = (-1)^{1+1}M_{11} + 2(-1)^{1+2}M_{12} = M_{11} - 2M_{12}$ $= \begin{vmatrix} -1 & -1 \\ 3 & 2 \end{vmatrix} - 2\begin{vmatrix} 2 & -1 \\ 0 & 2 \end{vmatrix} = [(-1 \times 2) - (-1 \times 3)] - 2[(2 \times 2) - (-1 \times 0)] = (-2 + 3) - 2(4 + 0) = 1 - 8 = -7$

Fifth - Replace the entries in the third column of the augmented matrix with the entries in the right hand side of the augmented matrix to obtain the matrix $A_3 = \begin{bmatrix} 1 & 1 & 2 \\ 2 & 0 & -1 \\ 0 & 2 & 3 \end{bmatrix}$. Next, find $\delta(A_3)$ by expanding about the third row, i.e.,

$\delta(A_3) = |A_3| = a_{31}A_{31} + a_{32}A_{32} + a_{33}A_{33} = 0 \cdot A_{31} + 2 \cdot A_{32} + 3 \cdot A_{33} = 2A_{32} + 3A_{33} = 2(-1)^{3+2}M_{32} + 3(-1)^{3+3}M_{33}$

$= -2M_{32} + 3M_{33} = -2\begin{vmatrix} 1 & 2 \\ 2 & -1 \end{vmatrix} + 3\begin{vmatrix} 1 & 1 \\ 2 & 0 \end{vmatrix} = -2[(1\times -1)-(2\times 2)] + 3[(1\times 0)-(1\times 2)] = -2(-1-4)+3(0-2) = 10-6 = 4$

Sixth - Solve for x and y using the Cramer's rule, i.e., $x = \frac{|A_1|}{|A|}$, $y = \frac{|A_2|}{|A|}$, and $z = \frac{|A_3|}{|A|}$. Therefore,

$x = \frac{|A_1|}{|A|}$; $x = \frac{3}{-2}$; $\mathbf{x = -1.5}$ $\quad y = \frac{|A_2|}{|A|}$; $y = \frac{-7}{-2}$; $\mathbf{y = 3.5}$ $\quad z = \frac{|A_3|}{|A|}$; $z = \frac{4}{-2}$; $\mathbf{z = -2}$

and the solution set is equal to $\{(-\mathbf{1.5}, \mathbf{3.5}, -\mathbf{2})\}$. Let's check the answer by substituting the x, y, and z values into $2y + 2z = 3$, i.e., $(2\times 3.5)+(2\times -2)\overset{?}{=}3$; $7-4\overset{?}{=}3$; $3 = 3$

h. **First** – Given the linear system $\begin{matrix} x - y + 3z = 2 \\ x - z = -3 \\ 2x - 2y + 6z = -1 \end{matrix}$ The coefficient matrix is equal to $A = \begin{bmatrix} 1 & -1 & 3 \\ 1 & 0 & -1 \\ 2 & -2 & 6 \end{bmatrix}$ and the augmented matrix is equal to $\begin{bmatrix} 1 & -1 & 3 & \vdots & 2 \\ 1 & 0 & -1 & \vdots & -3 \\ 2 & -2 & 6 & \vdots & -1 \end{bmatrix}$.

Second - Let's find $\delta(A)$ by expanding about the first row, i.e., $\delta(A) = |A| = a_{11}A_{11} + a_{12}A_{12} + a_{13}A_{13} = 1\cdot A_{11} - 1\cdot A_{12}$ $+3\cdot A_{13} = A_{11} - A_{12} + 3A_{13} = (-1)^{1+1}M_{11} - (-1)^{1+2}M_{12} + 3(-1)^{1+3}M_{13} = M_{11} + M_{12} + 3M_{13} = \begin{vmatrix} 0 & -1 \\ -2 & 6 \end{vmatrix} + \begin{vmatrix} 1 & -1 \\ 2 & 6 \end{vmatrix} + 3\begin{vmatrix} 1 & 0 \\ 2 & -2 \end{vmatrix}$

$= [(0\times 6)-(-1\times -2)] + [(1\times 6)-(-1\times 2)] + 3[(1\times -2)-(0\times 2)] = (0-2)+(6+2)+3(-2+0) = -2+8-6 = 0$. **Since $\delta(A) = 0$ the linear system is either a dependent or an inconsistent system. In this case, the linear system is inconsistent.**

i. **First** – Given the linear system $\begin{matrix} x + 3y - z = -2 \\ -x + 2y + 3z = 1 \\ x + y - 2z = 0 \end{matrix}$ The coefficient matrix is equal to $A = \begin{bmatrix} 1 & 3 & -1 \\ -1 & 2 & 3 \\ 1 & 1 & -2 \end{bmatrix}$ and the augmented matrix is equal to $\begin{bmatrix} 1 & 3 & -1 & \vdots & -2 \\ -1 & 2 & 3 & \vdots & 1 \\ 1 & 1 & -2 & \vdots & 0 \end{bmatrix}$.

Second - Let's find $\delta(A)$ by expanding about the first row, i.e., $\delta(A) = |A| = a_{11}A_{11} + a_{12}A_{12} + a_{13}A_{13} = 1\cdot A_{11} + 3\cdot A_{12}$ $-1\cdot A_{13} = A_{11} + 3A_{12} - A_{13} = (-1)^{1+1}M_{11} + 3(-1)^{1+2}M_{12} - (-1)^{1+3}M_{13} = M_{11} - 3M_{12} - M_{13} = \begin{vmatrix} 2 & 3 \\ 1 & -2 \end{vmatrix} - 3\begin{vmatrix} -1 & 3 \\ 1 & -2 \end{vmatrix} - \begin{vmatrix} -1 & 2 \\ 1 & 1 \end{vmatrix}$

$= [(2\times -2)-(3\times 1)] - 3[(-1\times -2)-(3\times 1)] - [(-1\times 1)-(2\times 1)] = (-4-3)-3(2-3)-(-1-2) = -7+3+3 = -1$. Since $\delta(A) = 0$ $\delta(A) \neq 0$ we can proceed to the next step.

Third - Replace the entries in the first column of the augmented matrix with the entries in the right hand side of the augmented matrix to obtain the matrix $A_1 = \begin{bmatrix} -2 & 3 & -1 \\ 1 & 2 & 3 \\ 0 & 1 & -2 \end{bmatrix}$. Next, find $\delta(A_1)$ by expanding about the first row, i.e., $\delta(A_1) = |A_1|$

$= a_{11}A_{11} + a_{12}A_{12} + a_{13}A_{13} = -2\cdot A_{11} + 3\cdot A_{12} - 1\cdot A_{13} = -2A_{11} + 3A_{12} - A_{13} = -2(-1)^{1+1}M_{11} + 3(-1)^{1+2}M_{12} - (-1)^{1+3}M_{13}$

$= -2M_{11} - 3M_{12} - M_{13} = -2\begin{vmatrix} 2 & 3 \\ 1 & -2 \end{vmatrix} - 3\begin{vmatrix} 1 & 3 \\ 0 & -2 \end{vmatrix} - \begin{vmatrix} 1 & 2 \\ 0 & 1 \end{vmatrix} = -2[(2\times -2)-(3\times 1)] - 3[(1\times -2)-(3\times 0)] - [(1\times 1)-(2\times 0)]$

$= -2(-4-3)-3(-2+0)-(1+0) = 14+6-1 = 19$.

Fourth - Replace the entries in the second column of the augmented matrix with the entries in the right hand side of the augmented matrix to obtain $A_2 = \begin{bmatrix} 1 & -2 & -1 \\ -1 & 1 & 3 \\ 1 & 0 & -2 \end{bmatrix}$. Next, find $\delta(A_2)$ by expanding about the first row, i.e.,. $\delta(A_2) = |A_2|$

$= a_{11}A_{11} + a_{12}A_{12} + a_{13}A_{13} = 1 \cdot A_{11} - 2 \cdot A_{12} - 1 \cdot A_{13} = A_{11} - 2A_{12} - A_{13} = (-1)^{1+1}M_{11} - 2(-1)^{1+2}M_{12} - (-1)^{1+3}M_{13}$

$= M_{11} + 2M_{12} - M_{13} = \begin{vmatrix} 1 & 3 \\ 0 & -2 \end{vmatrix} + 2\begin{vmatrix} -1 & 3 \\ 1 & -2 \end{vmatrix} - \begin{vmatrix} -1 & 1 \\ 1 & 0 \end{vmatrix} = [(1 \times -2) - (3 \times 0)] + 2[(-1 \times -2) - (3 \times 1)] - [(-1 \times 0) - (1 \times 1)]$

$= (-2 - 0) + 2(2 - 3) - (0 - 1) = -2 - 2 + 1 = -3$.

Fifth - Replace the entries in the third column of the augmented matrix with the entries in the right hand side of the augmented matrix to obtain $A_3 = \begin{bmatrix} 1 & 3 & -2 \\ -1 & 2 & 1 \\ 1 & 1 & 0 \end{bmatrix}$. Next, find $\delta(A_3)$ by expanding about the first row, i.e., $\delta(A_3) = |A_3|$

$= a_{31}A_{31} + a_{32}A_{32} + a_{33}A_{33} = 1 \cdot A_{11} + 3 \cdot A_{12} - 2 \cdot A_{13} = A_{11} + 3A_{12} - 2A_{13} = (-1)^{1+1}M_{11} + 3(-1)^{1+2}M_{12} - 2(-1)^{1+3}M_{13}$

$= M_{11} - 3M_{12} - 2M_{13} = \begin{vmatrix} 2 & 1 \\ 1 & 0 \end{vmatrix} - 3\begin{vmatrix} -1 & 1 \\ 1 & 0 \end{vmatrix} - 2\begin{vmatrix} -1 & 2 \\ 1 & 1 \end{vmatrix} = [(2 \times 0) - (1 \times 1)] - 3[(-1 \times 0) - (1 \times 1)] - 2[(-1 \times 1) - (2 \times 1)]$

$= (0 - 1) - 3(0 - 1) - 2(-1 - 2) = -1 + 3 + 6 = 8$.

Sixth - Solve for x and y using the Cramer's rule, i.e., $x = \frac{|A_1|}{|A|}$, $y = \frac{|A_2|}{|A|}$, and $z = \frac{|A_3|}{|A|}$. Therefore,

$x = \frac{|A_1|}{|A|}$; $x = \frac{19}{-1}$; $\mathbf{x = -19}$ $y = \frac{|A_2|}{|A|}$; $y = \frac{-3}{-1}$; $\mathbf{y = 3}$ $z = \frac{|A_3|}{|A|}$; $z = \frac{8}{-1}$; $\mathbf{z = -8}$

and the solution set is equal to $\{(\mathbf{-19, 3, -8})\}$. Let's check the answer by substituting the x, y, and z values into $-x + 2y + 3z = 1$, i.e., $19 + (2 \times 3) + (3 \times -8) \stackrel{?}{=} 1$; $19 + 6 - 24 \stackrel{?}{=} 1$; $25 - 24 \stackrel{?}{=} 1$; $1 = 1$

2. Use the result of exercise number 1-g above to find the solution set for the following linear equations. (Note that the answers should agree with practice problems 2a and 2b in Section 3.5 Case III.)

a. **First** - Given the linear system $\begin{matrix} x + y = -3 \\ 2x - z = -4 \\ 2y + 2z = 1 \end{matrix}$ The coefficient matrix is equal to $A = \begin{bmatrix} 1 & 1 & 0 \\ 2 & 0 & -1 \\ 0 & 2 & 2 \end{bmatrix}$ and the augmented matrix is equal to $\begin{bmatrix} 1 & 1 & 0 & \vdots & -3 \\ 2 & 0 & -1 & \vdots & -4 \\ 0 & 2 & 2 & \vdots & 1 \end{bmatrix}$

Second - From exercise 1-g we know $\delta(A) = -2$. Since $\delta(A) \neq 0$ we can proceed to the next step.

Third - Replace the entries in the first column of the augmented matrix with the entries in the right hand side of the augmented matrix to obtain the matrix $A_1 = \begin{bmatrix} -3 & 1 & 0 \\ -4 & 0 & -1 \\ 1 & 2 & 2 \end{bmatrix}$. Next, find $\delta(A_1)$ by expanding about the first row, i.e., $\delta(A_1) = |A_1|$

$= a_{11}A_{11} + a_{12}A_{12} + a_{13}A_{13} = -3 \cdot A_{11} + 1 \cdot A_{12} + 0 \cdot A_{13} = -3A_{11} + A_{12} = -3 \cdot (-1)^{1+1}M_{11} + (-1)^{1+2}M_{12} = -3M_{11} - M_{12}$

$= -3\begin{vmatrix} 0 & -1 \\ 2 & 2 \end{vmatrix} - \begin{vmatrix} -4 & -1 \\ 1 & 2 \end{vmatrix} = -3[(0 \times 2) - (-1 \times 2)] - [(-4 \times 2) - (-1 \times 1)] = -3(0 + 2) - (-8 + 1) = -6 + 7 = 1$

Fourth - Replace the entries in the second column of the augmented matrix with the entries in the right hand side of the augmented matrix to obtain the matrix $A_2 = \begin{bmatrix} 1 & -3 & 0 \\ 2 & -4 & -1 \\ 0 & 1 & 2 \end{bmatrix}$. Next, find $\delta(A_2)$ by expanding about the first row, i.e.,.

$\delta(A_2) = |A_2| = a_{11}A_{11} + a_{12}A_{12} + a_{13}A_{13} = 1 \cdot A_{11} - 3 \cdot A_{12} + 0 \cdot A_{13} = A_{11} - 3A_{12} = (-1)^{1+1}M_{11} - 3(-1)^{1+2}M_{12} = M_{11} + 3M_{12}$

$= \begin{vmatrix} -4 & -1 \\ 1 & 2 \end{vmatrix} + 3\begin{vmatrix} 2 & -1 \\ 0 & 2 \end{vmatrix} = [(-4\times 2)-(-1\times 1)]+3[(2\times 2)-(-1\times 0)] = (-8+1)+3(4+0) = -7+12 = 5$

Fifth - Replace the entries in the third column of the augmented matrix with the entries in the right hand side of the augmented matrix to obtain the matrix $A_3 = \begin{bmatrix} 1 & 1 & -3 \\ 2 & 0 & -4 \\ 0 & 2 & 1 \end{bmatrix}$. Next, find $\delta(A_3)$ by expanding about the third row, i.e.,

$\delta(A_3) = |A_3| = a_{31}A_{31} + a_{32}A_{32} + a_{33}A_{33} = 0\cdot A_{31} + 2\cdot A_{32} + 1\cdot A_{33} = 2A_{32} + A_{33} = 2(-1)^{3+2}M_{32} + (-1)^{3+3}M_{33}$

$= -2M_{32} + M_{33} = -2\begin{vmatrix} 1 & -3 \\ 2 & -4 \end{vmatrix} + \begin{vmatrix} 1 & 1 \\ 2 & 0 \end{vmatrix} = -2[(1\times -4)-(-3\times 2)]+[(1\times 0)-(1\times 2)] = -2(-4+6)+(0-2) = -4-2 = -6$

Sixth - Solve for x and y using the Cramer's rule, i.e., $x = \frac{|A_1|}{|A|}$, $y = \frac{|A_2|}{|A|}$, and $z = \frac{|A_3|}{|A|}$. Therefore,

$x = \frac{|A_1|}{|A|}$; $x = \frac{1}{-2}$; $\mathbf{x = -0.5}$ $y = \frac{|A_2|}{|A|}$; $y = \frac{5}{-2}$; $\mathbf{y = -2.5}$ $z = \frac{|A_3|}{|A|}$; $z = \frac{-6}{-2}$; $\mathbf{z = 3}$

and the solution set is equal to $\{(\mathbf{-0.5, -2.5, 3})\}$. Let's check the answer by substituting the x, y, and z values into $2y+2z=1$, i.e., $(2\times -2.5)+(2\times 3) \overset{?}{=} 1$; $-5+6 \overset{?}{=} 3$; $1=1$

b. **First** - Given the linear system $\begin{matrix} x+y=0 \\ 2x-z=-1 \\ 2y+2z=1 \end{matrix}$ The coefficient matrix is equal to $A = \begin{bmatrix} 1 & 1 & 0 \\ 2 & 0 & -1 \\ 0 & 2 & 2 \end{bmatrix}$ and the augmented matrix is equal to $\begin{bmatrix} 1 & 1 & 0 & \vdots & 0 \\ 2 & 0 & -1 & \vdots & -1 \\ 0 & 2 & 2 & \vdots & 1 \end{bmatrix}$

Second - From exercise 1-g we know $\delta(A) = -2$. Since $\delta(A) \neq 0$ we can proceed to the next step.

Third - Replace the entries in the first column of the augmented matrix with the entries in the right hand side of the augmented matrix to obtain the matrix $A_1 = \begin{bmatrix} 0 & 1 & 0 \\ -1 & 0 & -1 \\ 1 & 2 & 2 \end{bmatrix}$. Next, find $\delta(A_1)$ by expanding about the first row, i.e., $\delta(A_1) = |A_1|$

$= a_{11}A_{11} + a_{12}A_{12} + a_{13}A_{13} = 0\cdot A_{11} + 1\cdot A_{12} + 0\cdot A_{13} = A_{12} = (-1)^{1+2}M_{12} = -M_{12} = -\begin{vmatrix} -1 & -1 \\ 1 & 2 \end{vmatrix} = -(-1\times 2)-(-1\times 1)$

$= -(-2+1) = 1$

Fourth - Replace the entries in the second column of the augmented matrix with the entries in the right hand side of the augmented matrix to obtain the matrix $A_2 = \begin{bmatrix} 1 & 0 & 0 \\ 2 & -1 & -1 \\ 0 & 1 & 2 \end{bmatrix}$. Next, find $\delta(A_2)$ by expanding about the first row, i.e.,.

$\delta(A_2) = |A_2| = a_{11}A_{11} + a_{12}A_{12} + a_{13}A_{13} = 1\cdot A_{11} + 0\cdot A_{12} + 0\cdot A_{13} = A_{11} = (-1)^{1+1}M_{11} = M_{11} = \begin{vmatrix} -1 & -1 \\ 1 & 2 \end{vmatrix} = (-1\times 2)$

$-(-1\times 1) = -1$

Fifth - Replace the entries in the third column of the augmented matrix with the entries in the right hand side of the augmented matrix to obtain the matrix $A_3 = \begin{bmatrix} 1 & 1 & 0 \\ 2 & 0 & -1 \\ 0 & 2 & 1 \end{bmatrix}$. Next, find $\delta(A_3)$ by expanding about the first row, i.e.,

$\delta(A_3) = |A_3| = a_{11}A_{11} + a_{12}A_{12} + a_{13}A_{13} = 1\cdot A_{11} + 1\cdot A_{12} + 0\cdot A_{13} = A_{11} + A_{12} = (-1)^{1+1}M_{11} + (-1)^{1+2}M_{12} = M_{11} - M_{12}$

$= \begin{vmatrix} 0 & -1 \\ 2 & 1 \end{vmatrix} - \begin{vmatrix} 2 & -1 \\ 0 & 1 \end{vmatrix} = [(0\times 1)-(-1\times 2)]-[(2\times 1)-(-1\times 0)] = (0+2)-(2+0) = 2-2 = 0$

Sixth - Solve for x and y using the cramer's rule, i.e., $x=\frac{|A_1|}{|A|}$, $y=\frac{|A_2|}{|A|}$, and $z=\frac{|A_3|}{|A|}$. Therefore,

$x=\frac{|A_1|}{|A|}$; $x=\frac{1}{-2}$; $\mathbf{x=-0.5}$ $\quad$ $y=\frac{|A_2|}{|A|}$; $y=\frac{-1}{-2}$; $\mathbf{y=0.5}$ $\quad$ $z=\frac{|A_3|}{|A|}$; $z=\frac{0}{-2}$; $\mathbf{z=0}$

and the solution set is equal to $\{(\mathbf{-0.5, 0.5, 0})\}$. Let's check the answer by substituting the x, y, and z values into $2x-z=-1$, i.e., $(2\times-0.5)-0\overset{?}{=}-1$; $-1=-1$

Section 3.5 Case V Solutions - Solving Linear Systems Using the Gaussian Elimination Method

1. Solve the following linear systems by applying the Gaussian Elimination method to each augmented matrix.

a. **First** - Write the linear system $\begin{matrix} x-2y=-3 \\ 2x+3y=4 \end{matrix}$ in the form of $AX=B$, i.e., $\begin{bmatrix} 1 & -2 \\ 2 & 3 \end{bmatrix}\begin{bmatrix} x \\ y \end{bmatrix}=\begin{bmatrix} -3 \\ 4 \end{bmatrix}$

Second - Write the augmented matrix $\begin{bmatrix} 1 & -2 & \vdots & -3 \\ 2 & 3 & \vdots & 4 \end{bmatrix}$

Third - Perform the elementary row operations, i.e., 1. Multiply each element of the first row by -2 and add the result to each element of the second row. $\begin{bmatrix} 1 & -2 & \vdots & -3 \\ 2-2 & 3+4 & \vdots & 4+6 \end{bmatrix}=\begin{bmatrix} 1 & -2 & \vdots & -3 \\ 0 & 7 & \vdots & 10 \end{bmatrix}$ and 2. Divide the elements of the second row by 7. $\begin{bmatrix} 1 & -2 & \vdots & -3 \\ 0 & \frac{7}{7} & \vdots & \frac{10}{7} \end{bmatrix}=\begin{bmatrix} 1 & -2 & \vdots & -3 \\ 0 & 1 & \vdots & \frac{10}{7} \end{bmatrix}$

Note that the augmented matrix has 1's in its main diagonal entries and zero in the lower triangle.

Fourth - Write the augmented matrix in its equivalent linear system form $\begin{matrix} x-2y=-3 \\ y=\frac{10}{7} \end{matrix}$. Since $y=\frac{10}{7}$, we can solve for x by back substitution. $x-2y=-3$; $x-2\times\frac{10}{7}=-3$; $x=-3+\frac{20}{7}$; $x=\frac{-21+20}{7}$; $x=-\frac{1}{7}$. Therefore, $\mathbf{x=-\frac{1}{7}}$ and $\mathbf{y=\frac{10}{7}}$ and the solution set is $\left\{\left(\mathbf{-\frac{1}{7}, \frac{10}{7}}\right)\right\}$

Fifth - Check the answers by substituting the x and y values into one of the original equations. Let's substitute the x and y values into $2x+3y=4$, i.e., $\left(2\times-\frac{1}{7}\right)+\left(3\times\frac{10}{7}\right)\overset{?}{=}4$; $-\frac{2}{7}+\frac{30}{7}\overset{?}{=}4$; $\frac{28}{7}\overset{?}{=}4$; $4=4$

b. **First** - Write the linear system $\begin{matrix} 2x+y=-2 \\ 3x-y=0 \end{matrix}$ in the form of $AX=B$, i.e., $\begin{bmatrix} 2 & 1 \\ 3 & -1 \end{bmatrix}\begin{bmatrix} x \\ y \end{bmatrix}=\begin{bmatrix} -2 \\ 0 \end{bmatrix}$

Second - Write the augmented matrix $\begin{bmatrix} 2 & 1 & \vdots & -2 \\ 3 & -1 & \vdots & 0 \end{bmatrix}$

Third - Perform the elementary row operations, i.e., 1. Divide each element of the first row by 2, i.e., $\begin{bmatrix} \frac{2}{2} & \frac{1}{2} & \vdots & -\frac{2}{2} \\ 3 & -1 & \vdots & 0 \end{bmatrix}$ $=\begin{bmatrix} 1 & \frac{1}{2} & \vdots & -1 \\ 3 & -1 & \vdots & 0 \end{bmatrix}$. 2. Multiply each element of the first row by -3 and adding the result to each element of the second row. $\begin{bmatrix} 1 & \frac{1}{2} & \vdots & -1 \\ 3-3 & -1-\frac{3}{2} & \vdots & 0+3 \end{bmatrix}=\begin{bmatrix} 1 & \frac{1}{2} & \vdots & -1 \\ 0 & -\frac{5}{2} & \vdots & 3 \end{bmatrix}$ 3. Multiply the elements of the second row by $-\frac{2}{5}$.

$\begin{bmatrix} 1 & \frac{1}{2} & \vdots & -1 \\ 0 & -\frac{5}{2}\times-\frac{2}{5} & \vdots & 3\times-\frac{2}{5} \end{bmatrix} = \begin{bmatrix} 1 & \frac{1}{2} & \vdots & -1 \\ 0 & 1 & \vdots & -\frac{6}{5} \end{bmatrix}$. Note that the augmented matrix has 1's in its main diagonal entries and zero in the lower triangle.

Fourth - Write the augmented matrix in its equivalent linear system form $\begin{matrix} x+\frac{1}{2}y=-1 \\ y=-\frac{6}{5} \end{matrix}$. Since $y=-\frac{6}{5}$, we can solve for x by back substitution. $x+\frac{1}{2}y=-1$; $x+\frac{1}{2}\times-\frac{6}{5}=-1$; $x-\frac{6}{10}=-1$; $x=\frac{6}{10}-1$; $x=\frac{6-10}{10}$; $x=-\frac{2}{5}$. Thus, $\boldsymbol{x=-\frac{2}{5}}$ and $\boldsymbol{y=-\frac{6}{5}}$ and the solution set is $\left\{\left(-\frac{2}{5},-\frac{6}{5}\right)\right\}$

Fifth - Check the answers by substituting the x and y values into one of the original equations. Let's substitute the x and y values into $2x+y=-2$, i.e., $2\times-\frac{2}{5}-\frac{6}{5}\overset{?}{=}-2$; $-\frac{4}{5}-\frac{6}{5}\overset{?}{=}-2$; $\frac{-4-6}{5}\overset{?}{=}-2$; $-\frac{10}{5}\overset{?}{=}-2$; $-2=-2$

c. **First** - Write the linear system $\begin{matrix} x-3y=1 \\ 2x+5y=0 \end{matrix}$ in the form of $AX=B$, i.e., $\begin{bmatrix} 1 & -3 \\ 2 & 5 \end{bmatrix}\begin{bmatrix} x \\ y \end{bmatrix}=\begin{bmatrix} 1 \\ 0 \end{bmatrix}$

Second - Write the augmented matrix $\begin{bmatrix} 1 & -3 & \vdots & 1 \\ 2 & 5 & \vdots & 0 \end{bmatrix}$

Third - Perform the elementary row operations, i.e., 1. Multiply each element of the first row by -2 and add the result to each element of the second row. $\begin{bmatrix} 1 & -3 & \vdots & 1 \\ 2-2 & 5+6 & \vdots & 0-2 \end{bmatrix} = \begin{bmatrix} 1 & -3 & \vdots & 1 \\ 0 & 11 & \vdots & -2 \end{bmatrix}$ and 2. Divide the elements of the second row by 11. $\begin{bmatrix} 1 & -3 & \vdots & 1 \\ 0 & \frac{11}{11} & \vdots & -\frac{2}{11} \end{bmatrix} = \begin{bmatrix} 1 & -3 & \vdots & 1 \\ 0 & 1 & \vdots & -\frac{2}{11} \end{bmatrix}$

Note that the augmented matrix has 1's in its main diagonal entries and zero in the lower triangle.

Fourth - Write the augmented matrix in its equivalent linear system form $\begin{matrix} x-3y=1 \\ y=-\frac{2}{11} \end{matrix}$. Since $y=-\frac{2}{11}$, we can solve for x by back substitution. $x-3y=1$; $x-3\times-\frac{2}{11}=1$; $x=1-\frac{6}{11}$; $x=\frac{11-6}{11}$; $x=\frac{5}{11}$. Therefore, $\boldsymbol{x=\frac{5}{11}}$ and $\boldsymbol{y=-\frac{2}{11}}$ and the solution set is $\left\{\left(\frac{5}{11},-\frac{2}{11}\right)\right\}$

Fifth - Check the answers by substituting the x and y values into one of the original equations. Let's substitute the x and y values into $2x+5y=0$, i.e., $\left(2\times\frac{5}{11}\right)+\left(5\times-\frac{2}{11}\right)\overset{?}{=}0$; $\frac{10}{11}-\frac{10}{11}\overset{?}{=}0$; $0=0$

d. **First** - Write the linear system $\begin{matrix} 4x-3y=1 \\ 3x+y=2 \end{matrix}$ in the form of $AX=B$, i.e., $\begin{bmatrix} 4 & -3 \\ 3 & 1 \end{bmatrix}\begin{bmatrix} x \\ y \end{bmatrix}=\begin{bmatrix} 1 \\ 2 \end{bmatrix}$

Second - Write the augmented matrix $\begin{bmatrix} 4 & -3 & \vdots & 1 \\ 3 & 1 & \vdots & 2 \end{bmatrix}$

Third - Perform the elementary row operations, i.e., 1. Divide the elements of the first row by 4 $\begin{bmatrix} \frac{4}{4} & -\frac{3}{4} & \vdots & \frac{1}{4} \\ 3 & 1 & \vdots & 2 \end{bmatrix}$

$= \begin{bmatrix} 1 & -\frac{3}{4} & \vdots & \frac{1}{4} \\ 3 & 1 & \vdots & 2 \end{bmatrix}$ 2. Multiply each element of the first row by -3 and add the result to each element of the second row. $\begin{bmatrix} 1 & -\frac{3}{4} & : & \frac{1}{4} \\ 3-3 & 1+\frac{9}{4} & : & 2-\frac{3}{4} \end{bmatrix} = \begin{bmatrix} 1 & -\frac{3}{4} & : & \frac{1}{4} \\ 0 & \frac{13}{4} & : & \frac{5}{4} \end{bmatrix}$ and 2. Multiply the elements of the second row by $\frac{4}{13}$.

$\begin{bmatrix} 1 & -\frac{3}{4} & : & \frac{1}{4} \\ 0 & \frac{13}{4}\times\frac{4}{13} & : & \frac{5}{4}\times\frac{4}{13} \end{bmatrix} = \begin{bmatrix} 1 & -\frac{3}{4} & : & \frac{1}{4} \\ 0 & 1 & : & \frac{5}{13} \end{bmatrix}$ Note that the augmented matrix has 1's in its main diagonal entries and zero in the lower triangle.

Fourth - Write the augmented matrix in its equivalent linear system form $\begin{matrix} x-\frac{3}{4}y=\frac{1}{4} \\ y=\frac{5}{13} \end{matrix}$. Since $y=\frac{5}{13}$, we can solve for x by back substitution. $x-\frac{3}{4}y=\frac{1}{4}$; $x-\frac{3}{4}\times\frac{5}{13}=\frac{1}{4}$; $x=\frac{1}{4}+\frac{15}{52}$; $x=\frac{52+60}{208}$; $x=\frac{112}{208}$ $= x=\frac{7}{13}$. Therefore, $\mathbf{x=\frac{7}{13}}$ and $\mathbf{y=\frac{5}{13}}$ and the solution set is $\left\{\left(\mathbf{\frac{7}{13},\frac{5}{13}}\right)\right\}$

Fifth - Check the answers by substituting the x and y values into one of the original equations. Let's substitute the x and y values into $3x+y=2$, i.e., $\left(3\times\frac{7}{13}\right)+\frac{5}{13}\stackrel{?}{=}2$; $\frac{21}{13}+\frac{5}{13}\stackrel{?}{=}2$; $\frac{21+5}{13}\stackrel{?}{=}2$; $\frac{26}{13}\stackrel{?}{=}2$; $2=2$

e. **First** - Write the linear system $\begin{matrix} 2x+3z=-1 \\ x+3y=0 \\ 2x-2y+3z=-2 \end{matrix}$ in the form of $AX=B$, i.e., $\begin{bmatrix} 2 & 0 & 3 \\ 1 & 3 & 0 \\ 2 & -2 & 3 \end{bmatrix}\begin{bmatrix} x \\ y \\ z \end{bmatrix} = \begin{bmatrix} -1 \\ 0 \\ -2 \end{bmatrix}$

Second - Write the augmented matrix $\begin{bmatrix} 2 & 0 & 3 & : & -1 \\ 1 & 3 & 0 & : & 0 \\ 2 & -2 & 3 & : & -2 \end{bmatrix}$

Third - Perform the elementary row operations, i.e., 1. Replace the second row with the first row. $\begin{bmatrix} 1 & 3 & 0 & : & 0 \\ 2 & 0 & 3 & : & -1 \\ 2 & -2 & 3 & : & -2 \end{bmatrix}$

2. Multiply each element of the first row by -2 and add the result to each element of the second row.

$\begin{bmatrix} 1 & 3 & 0 & : & 0 \\ 2-2 & 0-6 & 3+0 & : & -1+0 \\ 2 & -2 & 3 & : & -2 \end{bmatrix} = \begin{bmatrix} 1 & 3 & 0 & : & 0 \\ 0 & -6 & 3 & : & -1 \\ 2 & -2 & 3 & : & -2 \end{bmatrix}$ 3. Divide the second row by -6. $\begin{bmatrix} 1 & 3 & 0 & : & 0 \\ 0 & \frac{-6}{-6} & \frac{3}{-6} & : & \frac{-1}{-6} \\ 2 & -2 & 3 & : & -2 \end{bmatrix}$

$= \begin{bmatrix} 1 & 3 & 0 & : & 0 \\ 0 & 1 & -\frac{1}{2} & : & \frac{1}{6} \\ 2 & -2 & 3 & : & -2 \end{bmatrix}$ 4. Multiply each element of the first row by -2 and add the result to each element of the third row. $\begin{bmatrix} 1 & 3 & 0 & : & 0 \\ 0 & 1 & -\frac{1}{2} & : & \frac{1}{6} \\ 2-2 & -2-6 & 3+0 & : & -2+0 \end{bmatrix} = \begin{bmatrix} 1 & 3 & 0 & : & 0 \\ 0 & 1 & -\frac{1}{2} & : & \frac{1}{6} \\ 0 & -8 & 3 & : & -2 \end{bmatrix}$ 5. Multiply each element of the second row by 8 and add the result to each element of the third row. $\begin{bmatrix} 1 & 3 & 0 & : & 0 \\ 0 & 1 & -\frac{1}{2} & : & \frac{1}{6} \\ 0 & -8+8 & 3-4 & : & -2+\frac{4}{3} \end{bmatrix} = \begin{bmatrix} 1 & 3 & 0 & : & 0 \\ 0 & 1 & -\frac{1}{2} & : & \frac{1}{6} \\ 0 & 0 & -1 & : & -\frac{2}{3} \end{bmatrix}$ 6. Multiply each element of the third row by -1 $\begin{bmatrix} 1 & 3 & 0 & \vdots & 0 \\ 0 & 1 & -\frac{1}{2} & \vdots & \frac{1}{6} \\ 0 & 0 & -1\times -1 & \vdots & -\frac{2}{3}\times -1 \end{bmatrix} = \begin{bmatrix} 1 & 3 & 0 & \vdots & 0 \\ 0 & 1 & -\frac{1}{2} & \vdots & \frac{1}{6} \\ 0 & 0 & 1 & \vdots & \frac{2}{3} \end{bmatrix}$

Note that the augmented matrix has 1's in its main diagonal entries and zeros in the lower triangle.

Fourth - Write the augmented matrix in its equivalent linear system form $\begin{matrix} x+3y=0 \\ y-\frac{1}{2}z=\frac{1}{6} \\ z=\frac{2}{3} \end{matrix}$. Since $z=\frac{2}{3}$, we can solve for x and y by back substitution. $y-\frac{1}{2}z=\frac{1}{6}$; $y+\left(-\frac{1}{2}\times\frac{2}{3}\right)=\frac{1}{6}$; $y-\frac{1}{3}=\frac{1}{6}$; $y=\frac{1}{3}+\frac{1}{6}$; $y=\frac{(1\times 6)+(1\times 3)}{3\times 6}$; $y=\frac{9}{18}$; $y=\frac{1}{2}$ and since $x+3y=0$; $x+3\times\frac{1}{2}=0$; $x=-\frac{3}{2}$. Therefore, $\mathbf{x=-\frac{3}{2}}$, $\mathbf{y=\frac{1}{2}}$, and $\mathbf{z=\frac{2}{3}}$ and the solution set is $\left\{\left(-\frac{3}{2},\frac{1}{2},\frac{2}{3}\right)\right\}$

Fifth - Check the answers by substituting the x, y, and z values into one of the original equations. Let's substitute the x, y, and z values into $2x-2y+3z=-2$, i.e. $\left(2\times-\frac{3}{2}\right)+\left(-2\times\frac{1}{2}\right)+\left(3\times\frac{2}{3}\right)\stackrel{?}{=}-2$; $-3-1+2\stackrel{?}{=}-2$; $-4+2\stackrel{?}{=}-2$; $-2=-2$

f. **First** - Write the linear system $\begin{matrix} 3x-z=0 \\ 2x-y-z=1 \\ 3x+2y=-1 \end{matrix}$ in the form of $AX=B$, i.e., $\begin{bmatrix} 3 & 0 & -1 \\ 2 & -1 & -1 \\ 3 & 2 & 0 \end{bmatrix}\begin{bmatrix} x \\ y \\ z \end{bmatrix}=\begin{bmatrix} 0 \\ 1 \\ -1 \end{bmatrix}$

Second - Write the augmented matrix $\begin{bmatrix} 3 & 0 & -1 & \vdots & 0 \\ 2 & -1 & -1 & \vdots & 1 \\ 3 & 2 & 0 & \vdots & -1 \end{bmatrix}$

Third - Perform the elementary row operations. 1. Divide the first row by -2. $\begin{bmatrix} \frac{3}{3} & \frac{0}{3} & -\frac{1}{3} & \vdots & \frac{0}{3} \\ 2 & -1 & -1 & \vdots & 1 \\ 3 & 2 & 0 & \vdots & -1 \end{bmatrix} = \begin{bmatrix} 1 & 0 & -\frac{1}{3} & \vdots & 0 \\ 2 & -1 & -1 & \vdots & 1 \\ 3 & 2 & 0 & \vdots & -1 \end{bmatrix}$

2. Multiply each element of the first row by -2 and add the result to each element of the second row.

$\begin{bmatrix} 1 & 0 & -\frac{1}{3} & \vdots & 0 \\ 2-2 & -1+0 & -1+\frac{2}{3} & \vdots & 1+0 \\ 3 & 2 & 0 & \vdots & -1 \end{bmatrix} = \begin{bmatrix} 1 & 0 & -\frac{1}{3} & \vdots & 0 \\ 0 & -1 & -\frac{1}{3} & \vdots & 1 \\ 3 & 2 & 0 & \vdots & -1 \end{bmatrix}$ 3. Multiply the second row by -1 $\begin{bmatrix} 1 & 0 & -\frac{1}{3} & \vdots & 0 \\ 0 & 1 & \frac{1}{3} & \vdots & -1 \\ 3 & 2 & 0 & \vdots & -1 \end{bmatrix}$

4. Multiply each element of the first row by -3 and add the result to each element of the third row.

$\begin{bmatrix} 1 & 0 & -\frac{1}{3} & \vdots & 0 \\ 0 & 1 & \frac{1}{3} & \vdots & -1 \\ 3-3 & 2+0 & 0+1 & \vdots & -1+0 \end{bmatrix} = \begin{bmatrix} 1 & 0 & -\frac{1}{3} & \vdots & 0 \\ 0 & 1 & \frac{1}{3} & \vdots & -1 \\ 0 & 2 & 1 & \vdots & -1 \end{bmatrix}$ 5. Multiply each element of the second row by -2 and add the result to each element of the third row. $\begin{bmatrix} 1 & 0 & -\frac{1}{3} & \vdots & 0 \\ 0 & 1 & \frac{1}{3} & \vdots & -1 \\ 0+0 & 2-2 & 1-\frac{2}{3} & \vdots & -1+2 \end{bmatrix} = \begin{bmatrix} 1 & 0 & -\frac{1}{3} & \vdots & 0 \\ 0 & 1 & \frac{1}{3} & \vdots & -1 \\ 0 & 0 & \frac{1}{3} & \vdots & 1 \end{bmatrix}$ 6. Multiply each element of the third row by 3 $\begin{bmatrix} 1 & 0 & -\frac{1}{3} & \vdots & 0 \\ 0 & 1 & \frac{1}{3} & \vdots & -1 \\ 0 & 0 & \frac{1}{3}\times 3 & \vdots & 1\times 3 \end{bmatrix} = \begin{bmatrix} 1 & 0 & -\frac{1}{3} & \vdots & 0 \\ 0 & 1 & \frac{1}{3} & \vdots & -1 \\ 0 & 0 & 1 & \vdots & 3 \end{bmatrix}$

Note that the augmented matrix has 1's in its main diagonal entries and zeros in the lower triangle.

Fourth - Write the augmented matrix in its equivalent linear system form $\begin{matrix} x-\frac{1}{3}z=0 \\ y+\frac{1}{3}z=-1 \\ z=3 \end{matrix}$. Since $z=3$, we can solve for x and y by back substitution. $x-\frac{1}{3}z=0$; $x+\left(-\frac{1}{3}\times 3\right)=0$; $x=1$ and since $y+\frac{1}{3}z=-1$; $y+\frac{1}{3}\times 3=-1$; $y+1=-1$; $y=-2$. Thus, $\mathbf{x=1}$, $\mathbf{y=-2}$, and $\mathbf{z=3}$ and the solution set is $\{(\mathbf{1,\ -2,\ 3})\}$

Fifth - Check the answers by substituting the x, y, and z values into one of the original equations. Let's substitute the x, y, and z values into $3x+2y=-1$, i.e. $(3\times 1)+(2\times -2)\stackrel{?}{=}-1$; $3-4\stackrel{?}{=}-1$; $-1=-1$

2. In the following exercises write the linear system whose augmented matrix is given.

a. Given $\begin{bmatrix} \frac{1}{2} & 1 & \vdots & -1 \\ -2 & 3 & \vdots & 4 \end{bmatrix}$, the equivalent linear system is equal to $\begin{matrix} \mathbf{\frac{1}{2}x+y=-1} \\ \mathbf{-2x+3y=4} \end{matrix}$

b. Given $\begin{bmatrix} 2 & 0 & \vdots & 2 \\ -1 & 3 & \vdots & 5 \end{bmatrix}$, the equivalent linear system is equal to $\begin{matrix} \mathbf{2x=2} \\ \mathbf{-x+3y=5} \end{matrix}$

c. Given $\begin{bmatrix} 1 & 2 & -3 & \vdots & 1 \\ 0 & 1 & 2 & \vdots & 0 \\ -1 & 3 & 1 & \vdots & 2 \end{bmatrix}$, the equivalent linear system is equal to $\begin{matrix} \mathbf{x+2y-3z=1} \\ \mathbf{y+2z=0} \\ \mathbf{-x+3y+z=2} \end{matrix}$

d. Given $\begin{bmatrix} 2 & 3 & 5 & \vdots & 10 \\ 1 & -1 & 3 & \vdots & 11 \\ 0 & 2 & -1 & \vdots & -2 \end{bmatrix}$, the equivalent linear system is equal to $\begin{matrix} \mathbf{2x+3y+5z=10} \\ \mathbf{x-y+3z=11} \\ \mathbf{2y-z=-2} \end{matrix}$

3. Find the solution set to the following augmented matrices which have been transformed by elementary row operations.

a. Given the augmented matrix $\begin{bmatrix} 1 & 2 & 0 & \vdots & -5 \\ 0 & 1 & -2 & \vdots & 3 \\ 0 & 0 & 1 & \vdots & 2 \end{bmatrix}$ its equivalent linear system is $\begin{matrix} x+2y=-5 \\ y-2z=3 \\ z=2 \end{matrix}$. Since $z=2$, we can solve for x and y by back substitution. $y-2z=3$; $y+(-2\times 2)=3$; $y-4=3$; $y=7$ and since $x+2y=-5$; $x+2\times 7=-5$; $x=-19$. Therefore, $\mathbf{x=-19}$, $\mathbf{y=7}$, and $\mathbf{z=2}$ and the solution set is $\{(\mathbf{-19,\ 7,\ 2})\}$. Let's check the answers by substituting the x, y, and z values into one of the original equations. Let's substitute the x, y, and z values into $x+2y=-5$, i.e. $-19+(2\times 7)\stackrel{?}{=}-5$; $-19+14\stackrel{?}{=}-5$; $-5=-5$

b. Given the augmented matrix $\begin{bmatrix} 1 & -2 & 3 & \vdots & 4 \\ 0 & 1 & 3 & \vdots & 2 \\ 0 & 0 & 1 & \vdots & -3 \end{bmatrix}$ its equivalent linear system is. $\begin{matrix} x-2y+3z=4 \\ y+3z=2 \\ z=-3 \end{matrix}$. Since $z=-3$, we can solve for x and y by back substitution. $y+3z=2$; $y+(3\times -3)=2$; $y-9=2$; $y=11$ and since $x-2y+3z=4$; $x+(-2\times 11)+(3\times -3)=4$; $x-22-9=4$; $x=35$. Thus, $\mathbf{x=35}$, $\mathbf{y=11}$, and $\mathbf{z=-3}$ and the solution set is

$\{(35,\ 11,\ -3)\}$. Let's check the answers by substituting the x, y, and z values into one of the original equations. Let's substitute the x, y, and z values into $x-2y+3z=4$, i.e. $35+(-2\times 11)+(3\times -3)\overset{?}{=}4$; $35-22-9\overset{?}{=}4$; $4=4$

c. Given the augmented matrix $\begin{bmatrix} 1 & -2 & -1 & \vdots & 1 \\ 0 & 1 & 2 & \vdots & 2 \\ 0 & 0 & 1 & \vdots & 3 \end{bmatrix}$ its equivalent linear system is $\begin{matrix} x-2y-z=1 \\ y+2z=2 \\ z=3 \end{matrix}$. Since $z=3$, we can solve for x and y by back substitution. $y+2z=2$; $y+(2\times 3)=2$; $y+6=2$; $y=-4$ and since $x-2y-z=1$; $x+(-2\times -4)-3=1$; $x+8-3=1$; $x=-4$. Thus, $\mathbf{x=-4}$, $\mathbf{y=-4}$, and $\mathbf{z=3}$ and the solution set is $\{(\mathbf{-4,\ -4,\ 3})\}$

Let's check the answers by substituting the x, y, and z values into one of the original equations. Let's substitute the x, y, and z values into $x-2y-z=1$, i.e. $-4+(-2\times -4)-3\overset{?}{=}1$; $-4+8-3\overset{?}{=}1$; $8-7\overset{?}{=}-1$; $1=1$

Section 3.5 Case VI Solutions - Solving Linear Systems Using the Gauss-Jordan Elimination Method

1. Solve the following linear systems by applying Gauss-Jordan Elimination method to each augmented matrix.

a. **First** - Write the linear system $\begin{matrix} x-3y=-2 \\ 2x-y=-3 \end{matrix}$ in the form of $AX=B$, i.e., $\begin{bmatrix} 1 & -3 \\ 2 & -1 \end{bmatrix}\begin{bmatrix} x \\ y \end{bmatrix}=\begin{bmatrix} -2 \\ -3 \end{bmatrix}$

Second - Write the augmented matrix $\begin{bmatrix} 1 & -3 & \vdots & -2 \\ 2 & -1 & \vdots & -3 \end{bmatrix}$

Third - Perform the elementary row operations, i.e., 1. Multiply each element of the first row by -2 and add the result to each element of the second row.

$\begin{bmatrix} 1 & -3 & \vdots & -2 \\ 2-2 & -1+6 & \vdots & -3+4 \end{bmatrix}=\begin{bmatrix} 1 & -3 & \vdots & -2 \\ 0 & 5 & \vdots & 1 \end{bmatrix}$ 2. Divide the second row by 5. $\begin{bmatrix} 1 & -3 & \vdots & -2 \\ \frac{0}{5} & \frac{5}{5} & \vdots & \frac{1}{5} \end{bmatrix}=\begin{bmatrix} 1 & -3 & \vdots & -2 \\ 0 & 1 & \vdots & \frac{1}{5} \end{bmatrix}$

3. Multiply each element of the second row by 3 and add the result to each element of the first row.

$$\begin{bmatrix} 1+0 & -3+3 & \vdots & -2+\frac{3}{5} \\ 0 & 1 & \vdots & \frac{1}{5} \end{bmatrix}=\begin{bmatrix} 1 & 0 & \vdots & -\frac{7}{5} \\ 0 & 1 & \vdots & \frac{1}{5} \end{bmatrix}$$

Note that the augmented matrix has 1's in its main diagonal entries and zero in the lower and upper triangles. Therefore, $\mathbf{x=-\frac{7}{5}}$ and $\mathbf{y=\frac{1}{5}}$ and the solution set is $\left\{\left(\mathbf{-\frac{7}{5},\ \frac{1}{5}}\right)\right\}$

Fourth - Check the answers by substituting the x and y values into one of the original equations. Let's substitute the x and y values into $x-3y=-2$, i.e., $-\frac{7}{5}+\left(-3\times\frac{1}{5}\right)\overset{?}{=}-2$; $-\frac{7}{5}-\frac{3}{5}\overset{?}{=}-2$; $\frac{-7-3}{5}\overset{?}{=}-2$; $-\frac{10}{5}\overset{?}{=}-2$; $-2=-2$

b. **First** - Write the linear system $\begin{matrix} 2x-y=2 \\ 3x-\frac{2}{3}y=0 \end{matrix}$ in the form of $AX=B$, i.e., $\begin{bmatrix} 2 & -1 \\ 3 & -\frac{2}{3} \end{bmatrix}\begin{bmatrix} x \\ y \end{bmatrix}=\begin{bmatrix} 2 \\ 0 \end{bmatrix}$

Second - Write the augmented matrix $\begin{bmatrix} 2 & -1 & \vdots & 2 \\ 3 & -\frac{2}{3} & \vdots & 0 \end{bmatrix}$.

Third - Perform the elementary row operations, i.e., 1. Divide each element of the first row by -2 $\begin{bmatrix} \frac{2}{2} & -\frac{1}{2} & \vdots & \frac{2}{2} \\ 3 & -\frac{2}{3} & \vdots & 0 \end{bmatrix}$

$=\begin{bmatrix} 1 & -\frac{1}{2} & \vdots & 1 \\ 3 & -\frac{2}{3} & \vdots & 0 \end{bmatrix}$ 2. Multiply each element of the first row by -3 and add the result to each element of the second row.

$\begin{bmatrix} 1 & -\frac{1}{2} & \vdots & 1 \\ 3-3 & -\frac{2}{3}+\frac{3}{2} & \vdots & 0-3 \end{bmatrix} = \begin{bmatrix} 1 & -\frac{1}{2} & \vdots & 1 \\ 0 & \frac{5}{6} & \vdots & -3 \end{bmatrix}$ 3. Multiply the second row by $\frac{6}{5}$. $\begin{bmatrix} 1 & -\frac{1}{2} & \vdots & 1 \\ 0\times\frac{6}{5} & \frac{5}{6}\times\frac{6}{5} & \vdots & -3\times\frac{6}{5} \end{bmatrix}$

$= \begin{bmatrix} 1 & -\frac{1}{2} & \vdots & 1 \\ 0 & 1 & \vdots & -\frac{18}{5} \end{bmatrix}$ 4. Multiply each element of the second row by $\frac{1}{2}$ and add the result to each element of the first row. $\begin{bmatrix} 1+0 & -\frac{1}{2}+\frac{1}{2} & \vdots & 1-\frac{18}{10} \\ 0 & 1 & \vdots & -\frac{18}{5} \end{bmatrix} = \begin{bmatrix} 1 & 0 & \vdots & -\frac{4}{5} \\ 0 & 1 & \vdots & -\frac{18}{5} \end{bmatrix}$. Note that the augmented matrix has 1's in its main diagonal entries and zero in the lower and upper triangles. Therefore, $\mathbf{x = -\frac{4}{5}}$ and $\mathbf{y = -\frac{18}{5}}$ and the solution set is $\left\{\left(-\frac{4}{5}, -\frac{18}{5}\right)\right\}$

Fourth - Check the answers by substituting the x and y values into one of the original equations. Let's substitute the x and y values into $2x-y=2$, i.e., $\left(2\times-\frac{4}{5}\right)+\frac{18}{5}\stackrel{?}{=}2$; $-\frac{8}{5}+\frac{18}{5}\stackrel{?}{=}2$; $\frac{-8+18}{5}\stackrel{?}{=}2$; $\frac{10}{5}\stackrel{?}{=}2$; $2=2$

c. **First** - Write the linear system $\begin{matrix} -x+2y=-2 \\ \frac{1}{2}x-\frac{1}{3}y=-\frac{1}{2} \end{matrix}$ in the form of $AX=B$, i.e., $\begin{bmatrix} -1 & 2 \\ \frac{1}{2} & -\frac{1}{3} \end{bmatrix}\begin{bmatrix} x \\ y \end{bmatrix} = \begin{bmatrix} -2 \\ -\frac{1}{2} \end{bmatrix}$

Second - Write the augmented matrix $\begin{bmatrix} -1 & 2 & \vdots & -2 \\ \frac{1}{2} & -\frac{1}{3} & \vdots & -\frac{1}{2} \end{bmatrix}$

Third - Perform the elementary row operations, i.e., 1. Multiply the elements of the first row by -1. $\begin{bmatrix} 1 & -2 & \vdots & 2 \\ \frac{1}{2} & -\frac{1}{3} & \vdots & -\frac{1}{2} \end{bmatrix}$

2. Multiply each element of the first row by $-\frac{1}{2}$ and add the result to each element of the second row.

$\begin{bmatrix} 1 & -2 & \vdots & 2 \\ \frac{1}{2}-\frac{1}{2} & -\frac{1}{3}+1 & \vdots & -\frac{1}{2}-1 \end{bmatrix} = \begin{bmatrix} 1 & -2 & \vdots & 2 \\ 0 & \frac{2}{3} & \vdots & -\frac{3}{2} \end{bmatrix}$ 3. Multiply the second row by $\frac{3}{2}$. $\begin{bmatrix} 1 & -2 & \vdots & 2 \\ 0\times\frac{3}{2} & \frac{2}{3}\times\frac{3}{2} & \vdots & -\frac{3}{2}\times\frac{3}{2} \end{bmatrix}$

$= \begin{bmatrix} 1 & -2 & \vdots & 2 \\ 0 & 1 & \vdots & -\frac{9}{4} \end{bmatrix}$ 4. Multiply each element of the second row by 2 and add the result to each element of the first row.

$\begin{bmatrix} 1+0 & -2+2 & \vdots & 2-\frac{9}{2} \\ 0 & 1 & \vdots & -\frac{9}{4} \end{bmatrix} = \begin{bmatrix} 1 & 0 & \vdots & -\frac{5}{2} \\ 0 & 1 & \vdots & -\frac{9}{4} \end{bmatrix}$. Note that the augmented matrix has 1's in its main diagonal entries and zero in the lower and upper triangles. Therefore, $\mathbf{x = -\frac{5}{2}}$ and $\mathbf{y = -\frac{9}{4}}$ and the solution set is $\left\{\left(-\frac{5}{2}, -\frac{9}{4}\right)\right\}$

Fourth - Check the answers by substituting the x and y values into one of the original equations. Let's substitute the x and y values into $\frac{1}{2}x-\frac{1}{3}y=-\frac{1}{2}$, i.e., $\left(\frac{1}{2}\times-\frac{5}{2}\right)+\left(-\frac{1}{3}\times-\frac{9}{4}\right)\stackrel{?}{=}-\frac{1}{2}$; $-\frac{5}{4}+\frac{3}{4}\stackrel{?}{=}-\frac{1}{2}$; $\frac{-5+3}{4}\stackrel{?}{=}-\frac{1}{2}$; $-\frac{1}{2}=-\frac{1}{2}$

d. **First** - Write the linear system $\begin{matrix} x-y=0 \\ 2x-3y=-\frac{1}{4} \end{matrix}$ in the form of $AX=B$, i.e., $\begin{bmatrix} 1 & -1 \\ 2 & -3 \end{bmatrix}\begin{bmatrix} x \\ y \end{bmatrix} = \begin{bmatrix} 0 \\ -\frac{1}{4} \end{bmatrix}$

Second - Write the augmented matrix $\begin{bmatrix} 1 & -1 & \vdots & 0 \\ 2 & -3 & \vdots & -\frac{1}{4} \end{bmatrix}$

Third - Perform the elementary row operations, i.e., 1. Multiply each element of the first row by -2 and add the result to each element of the second row.

$\begin{bmatrix} 1 & -1 & \vdots & 0 \\ 2-2 & -3+2 & \vdots & -\frac{1}{4}+0 \end{bmatrix} = \begin{bmatrix} 1 & -1 & \vdots & 0 \\ 0 & -1 & \vdots & -\frac{1}{4} \end{bmatrix}$ 2. Multiply the second row by -1. $\begin{bmatrix} 1 & -1 & \vdots & 0 \\ 0 & -1\times -1 & \vdots & -\frac{1}{4}\times -1 \end{bmatrix}$

$= \begin{bmatrix} 1 & -1 & \vdots & 0 \\ 0 & 1 & \vdots & \frac{1}{4} \end{bmatrix}$ 3. Multiply each element of the second row by 1 and add the result to each element of the first row.

$\begin{bmatrix} 1+0 & -1+1 & \vdots & 0+\frac{1}{4} \\ 0 & 1 & \vdots & \frac{1}{4} \end{bmatrix} = \begin{bmatrix} 1 & 0 & \vdots & \frac{1}{4} \\ 0 & 1 & \vdots & \frac{1}{4} \end{bmatrix}$. Note that the augmented matrix has 1's in its main diagonal entries and zero in the lower and upper triangles. Therefore, $\boldsymbol{x = \frac{1}{4}}$ and $\boldsymbol{y = \frac{1}{4}}$ and the solution set is $\left\{\left(\boldsymbol{\frac{1}{4}, \frac{1}{4}}\right)\right\}$

Fourth - Check the answers by substituting the x and y values into one of the original equations. Let's substitute the x and y values into $2x-3y=-\frac{1}{4}$, i.e., $\left(2\times\frac{1}{4}\right)+\left(-3\times\frac{1}{4}\right) \stackrel{?}{=} -\frac{1}{4}$; $\frac{2}{4}-\frac{3}{4} \stackrel{?}{=} -\frac{1}{4}$; $\frac{2-3}{4} \stackrel{?}{=} -\frac{1}{4}$; $-\frac{1}{4}=-\frac{1}{4}$

e. **First** - Write the linear system $\begin{matrix} 3x-2z=-1 \\ x-y+z=0 \\ 2x+3y=-2 \end{matrix}$ in the form of $AX=B$, i.e., $\begin{bmatrix} 3 & 0 & -2 \\ 1 & -1 & 1 \\ 2 & 3 & 0 \end{bmatrix}\begin{bmatrix} x \\ y \\ z \end{bmatrix} = \begin{bmatrix} -1 \\ 0 \\ -2 \end{bmatrix}$

Second - Write the augmented matrix $\begin{bmatrix} 3 & 0 & -2 & \vdots & -1 \\ 1 & -1 & 1 & \vdots & 0 \\ 2 & 3 & 0 & \vdots & -2 \end{bmatrix}$

Third - Perform the elementary row operations, i.e., 1. Replace the second row with the first row. $\begin{bmatrix} 1 & -1 & 1 & \vdots & 0 \\ 3 & 0 & -2 & \vdots & -1 \\ 2 & 3 & 0 & \vdots & -2 \end{bmatrix}$

2. Multiply each element of the first row by -3 and add the result to each element of the second row.

$\begin{bmatrix} 1 & -1 & 1 & \vdots & 0 \\ 3-3 & 0+3 & -2-3 & \vdots & -1+0 \\ 2 & 3 & 0 & \vdots & -2 \end{bmatrix} = \begin{bmatrix} 1 & -1 & 1 & \vdots & 0 \\ 0 & 3 & -5 & \vdots & -1 \\ 2 & 3 & 0 & \vdots & -2 \end{bmatrix}$ 3. Multiply each element of the first row by -2 and add the result to the third row. $\begin{bmatrix} 1 & -1 & 1 & \vdots & 0 \\ 0 & 3 & -5 & \vdots & -1 \\ 2-2 & 3+2 & 0-2 & \vdots & -2+0 \end{bmatrix} = \begin{bmatrix} 1 & -1 & 1 & \vdots & 0 \\ 0 & 3 & -5 & \vdots & -1 \\ 0 & 5 & -2 & \vdots & -2 \end{bmatrix}$ 4. Divide the second row by 3.

$\begin{bmatrix} 1 & -1 & 1 & \vdots & 0 \\ \frac{0}{3} & \frac{3}{3} & -\frac{5}{3} & \vdots & -\frac{1}{3} \\ 0 & 5 & -2 & \vdots & -2 \end{bmatrix} = \begin{bmatrix} 1 & -1 & 1 & \vdots & 0 \\ 0 & 1 & -\frac{5}{3} & \vdots & -\frac{1}{3} \\ 0 & 5 & -2 & \vdots & -2 \end{bmatrix}$ 5. Multiply each element of the second row by 1 and add the result to each element of the first row. $\begin{bmatrix} 1+0 & -1+1 & 1-\frac{5}{3} & \vdots & 0-\frac{1}{3} \\ 0 & 1 & -\frac{5}{3} & \vdots & -\frac{1}{3} \\ 0 & 5 & -2 & \vdots & -2 \end{bmatrix} = \begin{bmatrix} 1 & 0 & -\frac{2}{3} & \vdots & -\frac{1}{3} \\ 0 & 1 & -\frac{5}{3} & \vdots & -\frac{1}{3} \\ 0 & 5 & -2 & \vdots & -2 \end{bmatrix}$ 6. Multiply each element of the second row by -5 and add the result to the third row $\begin{bmatrix} 1 & 0 & -\frac{2}{3} & \vdots & -\frac{1}{3} \\ 0 & 1 & -\frac{5}{3} & \vdots & -\frac{1}{3} \\ 0 & 5-5 & -2+\frac{25}{3} & \vdots & -2+\frac{5}{3} \end{bmatrix} = \begin{bmatrix} 1 & 0 & -\frac{2}{3} & \vdots & -\frac{1}{3} \\ 0 & 1 & -\frac{5}{3} & \vdots & -\frac{1}{3} \\ 0 & 0 & \frac{19}{3} & \vdots & -\frac{1}{3} \end{bmatrix}$

7. Multiply each element of the third row by $\frac{3}{19}$. $\begin{bmatrix} 1 & 0 & -\frac{2}{3} & \vdots & -\frac{1}{3} \\ 0 & 1 & -\frac{5}{3} & \vdots & -\frac{1}{3} \\ 0 & 0 & \frac{19}{3}\times\frac{3}{19} & \vdots & -\frac{1}{3}\times\frac{3}{19} \end{bmatrix} = \begin{bmatrix} 1 & 0 & -\frac{2}{3} & \vdots & -\frac{1}{3} \\ 0 & 1 & -\frac{5}{3} & \vdots & -\frac{1}{3} \\ 0 & 0 & 1 & \vdots & -\frac{1}{19} \end{bmatrix}$ 8. Multiply each element of the third row by $\frac{5}{3}$ and add the result to each element of the second row. $\begin{bmatrix} 1 & 0 & -\frac{2}{3} & \vdots & -\frac{1}{3} \\ 0 & 1 & -\frac{5}{3}+\frac{5}{3} & \vdots & -\frac{1}{3}-\frac{5}{57} \\ 0 & 0 & 1 & \vdots & -\frac{1}{19} \end{bmatrix}$

$= \begin{bmatrix} 1 & 0 & -\frac{2}{3} & \vdots & -\frac{1}{3} \\ 0 & 1 & 0 & \vdots & -\frac{72}{171} \\ 0 & 0 & 1 & \vdots & -\frac{1}{19} \end{bmatrix}$ 9. Multiply the third row by $\frac{2}{3}$ and add the results to the elements in the first row.

$\begin{bmatrix} 1 & 0 & -\frac{2}{3}+\frac{2}{3} & \vdots & -\frac{1}{3}-\frac{2}{57} \\ 0 & 1 & 0 & \vdots & -\frac{72}{171} \\ 0 & 0 & 1 & \vdots & -\frac{1}{19} \end{bmatrix} = \begin{bmatrix} 1 & 0 & 0 & \vdots & -\frac{63}{171} \\ 0 & 1 & 0 & \vdots & -\frac{72}{171} \\ 0 & 0 & 1 & \vdots & -\frac{1}{19} \end{bmatrix}$. Note that the augmented matrix has 1's in its main diagonal entries and zeros in the lower and upper triangles. Therefore, $\boldsymbol{x = -\frac{63}{171}}$, $\boldsymbol{y = -\frac{72}{171}}$, and $\boldsymbol{z = -\frac{1}{19}}$ and the solution set is $\boldsymbol{\left\{\left(-\frac{63}{171}, -\frac{72}{171}, -\frac{1}{19}\right)\right\}}$

Fourth - Check the answers by substituting the x, y, and z values into one of the original equations. Let's substitute the x, y, and z values into $x-y+z=0$, i.e., $-\frac{63}{171}+\frac{72}{171}-\frac{1}{19}\stackrel{?}{=}0$; $\frac{-63+72}{171}-\frac{1}{19}\stackrel{?}{=}0$; $\frac{9}{171}-\frac{1}{19}\stackrel{?}{=}0$; $0=0$

f. **First** - Write the linear system $\begin{matrix} x-z=0 \\ x-3y=-1 \\ x+y=0 \end{matrix}$ in the form of $AX=B$, i.e., $\begin{bmatrix} 1 & 0 & -1 \\ 1 & -3 & 0 \\ 1 & 1 & 0 \end{bmatrix}\begin{bmatrix} x \\ y \\ z \end{bmatrix} = \begin{bmatrix} 0 \\ -1 \\ 0 \end{bmatrix}$

Second - Write the augmented matrix $\begin{bmatrix} 1 & 0 & -1 & \vdots & 0 \\ 1 & -3 & 0 & \vdots & -1 \\ 1 & 1 & 0 & \vdots & 0 \end{bmatrix}$

Third - Perform the elementary row operations, i.e., 1. Multiply each element of the first row by -1 and add the result To each element of the second row.

$\begin{bmatrix} 1 & 0 & -1 & \vdots & 0 \\ 1-1 & -3+0 & 0+1 & \vdots & -1+0 \\ 1 & 1 & 0 & \vdots & 0 \end{bmatrix} = \begin{bmatrix} 1 & 0 & -1 & \vdots & 0 \\ 0 & -3 & 1 & \vdots & -1 \\ 1 & 1 & 0 & \vdots & 0 \end{bmatrix}$ 2. Multiply each element of the first row by -1 and add the result to the third row. $\begin{bmatrix} 1 & 0 & -1 & \vdots & 0 \\ 0 & -3 & 1 & \vdots & -1 \\ 1-1 & 1+0 & 0+1 & \vdots & 0+0 \end{bmatrix} = \begin{bmatrix} 1 & 0 & -1 & \vdots & 0 \\ 0 & -3 & 1 & \vdots & -1 \\ 0 & 1 & 1 & \vdots & 0 \end{bmatrix}$ 3. Divide the second row by -3.

$\begin{bmatrix} 1 & 0 & -1 & \vdots & 0 \\ \frac{0}{-3} & \frac{-3}{-3} & \frac{1}{-3} & \vdots & \frac{-1}{-3} \\ 0 & 1 & 1 & \vdots & 0 \end{bmatrix} = \begin{bmatrix} 1 & 0 & -1 & \vdots & 0 \\ 0 & 1 & -\frac{1}{3} & \vdots & \frac{1}{3} \\ 0 & 1 & 1 & \vdots & 0 \end{bmatrix}$ 4. Multiply each element of the second row by -1 and add the result to each element of the third row. $\begin{bmatrix} 1 & 0 & -1 & \vdots & 0 \\ 0 & 1 & -\frac{1}{3} & \vdots & \frac{1}{3} \\ 0+0 & 1-1 & 1+\frac{1}{3} & \vdots & 0-\frac{1}{3} \end{bmatrix} = \begin{bmatrix} 1 & 0 & -1 & \vdots & 0 \\ 0 & 1 & -\frac{1}{3} & \vdots & \frac{1}{3} \\ 0 & 0 & \frac{4}{3} & \vdots & -\frac{1}{3} \end{bmatrix}$ 5. Multiply each element of the third row by $\frac{3}{4}$ $\begin{bmatrix} 1 & 0 & -1 & \vdots & 0 \\ 0 & 1 & -\frac{1}{3} & \vdots & \frac{1}{3} \\ 0 & 0 & \frac{4}{3}\times\frac{3}{4} & \vdots & -\frac{1}{3}\times\frac{3}{4} \end{bmatrix} = \begin{bmatrix} 1 & 0 & -1 & \vdots & 0 \\ 0 & 1 & -\frac{1}{3} & \vdots & \frac{1}{3} \\ 0 & 0 & 1 & \vdots & -\frac{1}{4} \end{bmatrix}$ 6. Multiply each element of the third row by $\frac{1}{3}$ and add the result to the elements in the second row. $\begin{bmatrix} 1 & 0 & -1 & \vdots & 0 \\ 0+0 & 1+0 & -\frac{1}{3}+\frac{1}{3} & \vdots & \frac{1}{3}-\frac{1}{12} \\ 0 & 0 & 1 & \vdots & -\frac{1}{4} \end{bmatrix} = \begin{bmatrix} 1 & 0 & -1 & \vdots & 0 \\ 0 & 1 & 0 & \vdots & \frac{1}{4} \\ 0 & 0 & 1 & \vdots & -\frac{1}{4} \end{bmatrix}$ 7. Multiply each element of the third row by 1 and add the result to each element of the first row. $\begin{bmatrix} 1+0 & 0+0 & -1+1 & \vdots & 0-\frac{1}{4} \\ 0 & 1 & 0 & \vdots & \frac{1}{4} \\ 0 & 0 & 1 & \vdots & -\frac{1}{4} \end{bmatrix}$

$= \begin{bmatrix} 1 & 0 & 0 & \vdots & -\frac{1}{4} \\ 0 & 1 & 0 & \vdots & \frac{1}{4} \\ 0 & 0 & 1 & \vdots & -\frac{1}{4} \end{bmatrix}$ Note that the augmented matrix has 1's in its main diagonal entries and zeros in the lower and upper triangles. Therefore, $x = -\frac{1}{4}$, $y = \frac{1}{4}$, and $z = -\frac{1}{4}$ and the solution set is $\left\{\left(-\frac{1}{4}, \frac{1}{4}, -\frac{1}{4}\right)\right\}$

Fourth - Check the answers by substituting the x, y, and z values into one of the original equations. Let's substitute the x, y, and z values into $x-3y=-1$, i.e., $-\frac{1}{4}+\left(-3\times\frac{1}{4}\right)\overset{?}{=}-1$; $-\frac{1}{4}-\frac{3}{4}\overset{?}{=}-1$; $\frac{-1-3}{4}\overset{?}{=}-1$; $-\frac{4}{4}\overset{?}{=}-1$; $-1=-1$

2. Write the coefficient matrix and the augmented matrix for the following linear systems.

a. Given $\begin{matrix} x-2y=-1 \\ 2x+4y=-3 \end{matrix}$, the coefficient matrix is $C = \begin{bmatrix} 1 & -2 \\ 2 & 4 \end{bmatrix}$ and the augmented matrix is $A = \begin{bmatrix} 1 & -2 & \vdots & -1 \\ 2 & 4 & \vdots & -3 \end{bmatrix}$

b. Given $\begin{matrix} x+2y=-3 \\ x-y=0 \end{matrix}$, the coefficient matrix is $C = \begin{bmatrix} 1 & 2 \\ 1 & -1 \end{bmatrix}$ and the augmented matrix is $A = \begin{bmatrix} 1 & 2 & \vdots & -3 \\ 1 & -1 & \vdots & 0 \end{bmatrix}$

c. Given $\begin{matrix} x+2y-z=1 \\ y-3z=-3 \\ x-4z=-2 \end{matrix}$, the coefficient matrix is $C = \begin{bmatrix} 1 & 2 & -1 \\ 0 & 1 & -3 \\ 1 & 0 & -4 \end{bmatrix}$ and the augmented matrix is $A = \begin{bmatrix} 1 & 2 & -1 & \vdots & 1 \\ 0 & 1 & -3 & \vdots & -3 \\ 1 & 0 & -4 & \vdots & -2 \end{bmatrix}$

d. Given $\begin{matrix} x-2y=-4 \\ x+2y+z=-1 \\ y-3z=3 \end{matrix}$, the coefficient matrix is $C = \begin{bmatrix} 1 & -2 & 0 \\ 1 & 2 & 1 \\ 0 & 1 & -3 \end{bmatrix}$ and the augmented matrix is $A = \begin{bmatrix} 1 & -2 & 0 & \vdots & -4 \\ 1 & 2 & 1 & \vdots & -1 \\ 0 & 1 & -3 & \vdots & 3 \end{bmatrix}$

e. Given $\begin{array}{l} x+y-2z+w=-1 \\ 2y-4w=0 \\ x-2w=-1 \\ x+y-4w=0 \end{array}$, $C = \begin{bmatrix} 1 & 1 & -2 & 1 \\ 0 & 2 & 0 & -4 \\ 1 & 0 & 0 & -2 \\ 1 & 1 & 0 & -4 \end{bmatrix}$ and $A = \begin{bmatrix} 1 & 1 & -2 & 1 & \vdots & -1 \\ 0 & 2 & 0 & -4 & \vdots & 0 \\ 1 & 0 & 0 & -2 & \vdots & -1 \\ 1 & 1 & 0 & -4 & \vdots & 0 \end{bmatrix}$

f. Given $\begin{array}{l} x+y-2w=-1 \\ y-3w=-2 \\ x-y+2z-w=3 \\ y-3z=-2 \end{array}$, $C = \begin{bmatrix} 1 & 1 & 0 & -2 \\ 0 & 1 & 0 & -3 \\ 1 & -1 & 2 & -1 \\ 0 & 1 & -3 & 0 \end{bmatrix}$ and $A = \begin{bmatrix} 1 & 1 & 0 & -2 & \vdots & -1 \\ 0 & 1 & 0 & -3 & \vdots & -2 \\ 1 & -1 & 2 & -1 & \vdots & 3 \\ 0 & 1 & -3 & 0 & \vdots & -2 \end{bmatrix}$

Chapter 4 Solutions:

Section 4.1 Solutions - Sequences

1. List the first four and tenth terms of the given sequence.

a. Given $a_n = \frac{2n+1}{-2n}$, then $a_1 = \frac{2\cdot1+1}{-2\cdot1} = -\frac{3}{2} = \mathbf{-1.5}$ $\quad a_2 = \frac{2\cdot2+1}{-2\cdot2} = -\frac{5}{4} = \mathbf{-1.25}$

$a_3 = \frac{2\cdot3+1}{-2\cdot3} = -\frac{7}{6} = \mathbf{-1.17}$ $\quad a_4 = \frac{2\cdot4+1}{-2\cdot4} = -\frac{9}{8} = \mathbf{-1.13}$ $\quad a_{10} = \frac{2\cdot10+1}{-2\cdot10} = -\frac{21}{20} = \mathbf{-1.05}$

b. Given $b_k = \frac{k(k+1)}{k^2}$, then $b_1 = \frac{1\cdot(1+1)}{1^2} = \frac{1\cdot2}{1} = \frac{2}{1} = \mathbf{2}$ $\quad b_2 = \frac{2\cdot(2+1)}{2^2} = \frac{2\cdot3}{4} = \frac{6}{4} = \mathbf{1.5}$

$b_3 = \frac{3\cdot(3+1)}{3^2} = \frac{3\cdot4}{9} = \frac{12}{9} = \mathbf{1.33}$ $\quad b_4 = \frac{4\cdot(4+1)}{4^2} = \frac{4\cdot5}{16} = \frac{20}{16} = \mathbf{1.25}$ $\quad b_{10} = \frac{10\cdot(10+1)}{10^2} = \frac{110}{100} = \mathbf{1.1}$

c. Given $d_n = 3-(-2)^n$, then $d_1 = 3-(-2)^1 = 3-(-2) = 3+2 = \mathbf{5}$

$d_2 = 3-(-2)^2 = 3-(+4) = 3-4 = \mathbf{-1}$ $\quad d_3 = 3-(-2)^3 = 3-(-8) = 3+8 = \mathbf{11}$

$d_4 = 3-(-2)^4 = 3-(+16) = 3-16 = \mathbf{-13}$ $\quad d_{10} = 3-(-2)^{10} = 3-(+1024) = 3-1024 = \mathbf{-1021}$

d. Given $k_n = \left(-\frac{1}{2}\right)^n \frac{(-1)^{n+1}}{n+2}$, then $k_1 = \left(-\frac{1}{2}\right)^1 \frac{(-1)^{1+1}}{1+2} = -\frac{1}{2}\cdot\frac{(-1)^2}{3} = -\frac{1}{2}\cdot\frac{1}{3} = -\mathbf{\frac{1}{6}}$

$k_2 = \left(-\frac{1}{2}\right)^2 \frac{(-1)^{2+1}}{2+2} = \frac{1}{4}\cdot\frac{(-1)^3}{4} = \frac{1}{4}\cdot-\frac{1}{4} = -\mathbf{\frac{1}{16}}$ $\quad k_3 = \left(-\frac{1}{2}\right)^3 \frac{(-1)^{3+1}}{3+2} = -\frac{1}{8}\cdot\frac{(-1)^4}{5} = -\frac{1}{8}\cdot\frac{1}{5} = -\mathbf{\frac{1}{40}}$

$k_4 = \left(-\frac{1}{2}\right)^4 \frac{(-1)^{4+1}}{4+2} = \frac{1}{16}\cdot\frac{(-1)^5}{6} = \frac{1}{16}\cdot-\frac{1}{6} = -\mathbf{\frac{1}{96}}$ $\quad k_{10} = \left(-\frac{1}{2}\right)^{10} \frac{(-1)^{10+1}}{10+2} = \frac{1}{1024}\cdot\frac{(-1)^{11}}{12} = \mathbf{\frac{-1}{12,288}}$

2. Write s_3, s_4, s_5, s_8, and s_{10} for the following sequences.

a. Given $s_n = \frac{n(n+1)}{2n^{-1}}$, then $s_3 = \frac{3\cdot(3+1)}{2\cdot3^{-1}} = \frac{3\cdot4}{2\cdot3^{-1}} = \frac{3\cdot4\cdot3}{2} = \frac{36}{2} = \mathbf{18}$

$s_4 = \frac{4\cdot(4+1)}{2\cdot4^{-1}} = \frac{4\cdot5}{2\cdot4^{-1}} = \frac{4\cdot5\cdot4}{2} = \frac{80}{2} = \mathbf{40}$ $\quad s_5 = \frac{5\cdot(5+1)}{2\cdot5^{-1}} = \frac{5\cdot6}{2\cdot5^{-1}} = \frac{5\cdot6\cdot5}{2} = \frac{150}{2} = \mathbf{75}$

$s_8 = \frac{8\cdot(8+1)}{2\cdot8^{-1}} = \frac{8\cdot9}{2\cdot8^{-1}} = \frac{8\cdot9\cdot8}{2} = \frac{576}{2} = \mathbf{288}$ $\quad s_{10} = \frac{10\cdot(10+1)}{2\cdot10^{-1}} = \frac{10\cdot11}{2\cdot10^{-1}} = \frac{10\cdot11\cdot10}{2} = \frac{1100}{2} = \mathbf{550}$

b. Given $s_n = (-1)^{n+1}2^{n-2}$, then $s_3 = (-1)^{3+1}\cdot2^{3-2} = (-1)^4\cdot2^1 = 1\cdot2 = \mathbf{2}$

$s_4 = (-1)^{4+1}\cdot2^{4-2} = (-1)^5\cdot2^2 = -1\cdot4 = \mathbf{-4}$ $\quad s_5 = (-1)^{5+1}\cdot2^{5-2} = (-1)^6\cdot2^3 = 1\cdot8 = \mathbf{8}$

$s_8 = (-1)^{8+1}\cdot2^{8-2} = (-1)^9\cdot2^6 = -1\cdot64 = \mathbf{-64}$ $\quad s_{10} = (-1)^{10+1}\cdot2^{10-2} = (-1)^{11}\cdot2^8 = -1\cdot256 = \mathbf{-256}$

c. Given $s_n = \frac{(-2)^{n+1}(n-2)}{2n}$, then $s_3 = \frac{(-2)^{3+1}(3-2)}{2\cdot3} = \frac{(-2)^4\cdot1}{6} = \frac{16}{6} = \mathbf{2.67}$

$s_4 = \frac{(-2)^{4+1}(4-2)}{2\cdot4} = \frac{(-2)^5\cdot2}{8} = \frac{-32\cdot2}{8} = \mathbf{-8}$ $\quad s_5 = \frac{(-2)^{5+1}(5-2)}{2\cdot5} = \frac{(-2)^6\cdot3}{10} = \frac{64\cdot3}{10} = \mathbf{19.2}$

$s_8 = \frac{(-2)^{8+1}(8-2)}{2\cdot8} = \frac{(-2)^9\cdot6}{16} = \frac{-512\cdot6}{16} = \mathbf{-192}$ $\quad s_{10} = \frac{(-2)^{10+1}(10-2)}{2\cdot10} = \frac{(-2)^{11}\cdot8}{20} = \frac{-2048\cdot8}{20} = \mathbf{-819.2}$

3. Write the first five terms of the following sequences.

a. Given $a_n = (-1)^{n+1}(n+2)$, then

$a_1 = (-1)^{1+1}(1+2) = (-1)^2 \cdot 3 = 1 \cdot 3 = \mathbf{3}$

$a_2 = (-1)^{2+1}(2+2) = (-1)^3 \cdot 4 = -1 \cdot 4 = \mathbf{-4}$

$a_3 = (-1)^{3+1}(3+2) = (-1)^4 \cdot 5 = 1 \cdot 5 = \mathbf{5}$

$a_4 = (-1)^{4+1}(4+2) = (-1)^5 \cdot 6 = -1 \cdot 6 = \mathbf{-6}$

$a_5 = (-1)^{5+1}(5+2) = (-1)^6 \cdot 7 = 1 \cdot 7 = \mathbf{7}$

b. Given $a_i = 3\left(\frac{1}{100}\right)^{i-2}$, then

$a_1 = 3\left(\frac{1}{100}\right)^{1-2} = 3\left(\frac{1}{100}\right)^{-1} = 3 \cdot \frac{1}{\frac{1}{100}} = 3 \cdot \frac{100}{1} = \mathbf{300}$

$a_2 = 3\left(\frac{1}{100}\right)^{2-2} = 3\left(\frac{1}{100}\right)^{0} = 3 \cdot 1 = \mathbf{3}$

$a_3 = 3\left(\frac{1}{100}\right)^{3-2} = 3\left(\frac{1}{100}\right)^{1} = 3 \cdot \frac{1}{100^1} = \mathbf{\frac{3}{100}}$

$a_4 = 3\left(\frac{1}{100}\right)^{4-2} = 3\left(\frac{1}{100}\right)^{2} = 3 \cdot \frac{1}{100^2} = \mathbf{\frac{3}{10{,}000}}$

$a_5 = 3\left(\frac{1}{100}\right)^{5-2} = 3\left(\frac{1}{100}\right)^{3} = 3 \cdot \frac{1}{100^3} = \mathbf{\frac{3}{1{,}000{,}000}}$

c. Given $c_i = 3\left(-\frac{1}{5}\right)^{i-1}$, then

$c_1 = 3\left(-\frac{1}{5}\right)^{1-1} = 3\left(-\frac{1}{5}\right)^{0} = 3 \cdot 1 = \mathbf{3}$

$c_2 = 3\left(-\frac{1}{5}\right)^{2-1} = 3\left(-\frac{1}{5}\right)^{1} = 3 \cdot -\frac{1}{5} = -\frac{3}{5} = \mathbf{-0.6}$

$c_3 = 3\left(-\frac{1}{5}\right)^{3-1} = 3\left(-\frac{1}{5}\right)^{2} = 3 \cdot \frac{1}{5^2} = \frac{3}{25} = \mathbf{0.12}$

$c_4 = 3\left(-\frac{1}{5}\right)^{4-1} = 3\left(-\frac{1}{5}\right)^{3} = 3 \cdot -\frac{1}{5^3} = -\frac{3}{125} = \mathbf{-0.024}$

$c_5 = 3\left(-\frac{1}{5}\right)^{5-1} = 3\left(-\frac{1}{5}\right)^{4} = 3 \cdot \frac{1}{5^4} = \frac{3}{625} = \mathbf{0.0048}$

d. Given $a_n = (3n-5)^2$, then

$a_1 = (3 \cdot 1 - 5)^2 = (3-5)^2 = (-2)^2 = \mathbf{4}$

$a_2 = (3 \cdot 2 - 5)^2 = (6-5)^2 = 1^2 = \mathbf{1}$

$a_3 = (3 \cdot 3 - 5)^2 = (9-5)^2 = 4^2 = \mathbf{16}$

$a_4 = (3 \cdot 4 - 5)^2 = (12-5)^2 = 7^2 = \mathbf{49}$

$a_5 = (3 \cdot 5 - 5)^2 = (15-5)^2 = 10^2 = \mathbf{100}$

e. Given $u_k = ar^{k-2} + 2$, then

$u_1 = ar^{1-2} + 2 = ar^{-1} + 2 = \mathbf{\frac{a}{r} + 2}$

$u_2 = ar^{2-2} + 2 = ar^0 + 2 = \mathbf{a + 2}$

$u_3 = ar^{3-2} + 2 = ar^1 + 2 = \mathbf{ar + 2}$

$u_4 = ar^{4-2} + 2 = \mathbf{ar^2 + 2}$

$u_5 = ar^{5-2} + 2 = \mathbf{ar^3 + 2}$

f. Given $b_k = -3\left(\frac{2}{3}\right)^{k-2}$, then

$b_1 = -3 \cdot \left(\frac{2}{3}\right)^{1-2} = -3 \cdot \left(\frac{2}{3}\right)^{-1} = -3 \cdot \frac{1}{\frac{2}{3}} = -3 \cdot \frac{3}{2} = \mathbf{-\frac{9}{2}}$

$b_2 = -3 \cdot \left(\frac{2}{3}\right)^{2-2} = -3 \cdot \left(\frac{2}{3}\right)^{0} = -3 \cdot 1 = \mathbf{-3}$

$b_3 = -3 \cdot \left(\frac{2}{3}\right)^{3-2} = -3 \cdot \frac{2}{3} = -\frac{3 \cdot 2}{3} = \mathbf{-2}$

$b_4 = -3 \cdot \left(\frac{2}{3}\right)^{4-2} = -3 \cdot \left(\frac{2}{3}\right)^{2} = -3 \cdot \frac{4}{9} = \mathbf{-\frac{4}{3}}$

$b_5 = -3 \cdot \left(\frac{2}{3}\right)^{5-2} = -3 \cdot \left(\frac{2}{3}\right)^{3} = -3 \cdot \frac{8}{27} = \mathbf{-\frac{8}{9}}$

g. Given $c_j = \frac{j}{j+1} + j$, then

$c_1 = \frac{1}{1+1} + 1 = \frac{1}{2} + 1 = \frac{1+2}{2} = \mathbf{\frac{3}{2}}$

$c_2 = \frac{2}{2+1} + 2 = \frac{2}{3} + 2 = \frac{2+6}{3} = \mathbf{\frac{8}{3}}$

$c_3 = \frac{3}{3+1} + 3 = \frac{3}{4} + 3 = \frac{3+12}{4} = \mathbf{\frac{15}{4}}$

$c_4 = \frac{4}{4+1} + 4 = \frac{4}{5} + 4 = \frac{4+20}{5} = \mathbf{\frac{24}{5}}$

$c_5 = \frac{5}{5+1} + 5 = \frac{5}{6} + 5 = \frac{5+30}{6} = \mathbf{\frac{35}{6}}$

h. Given $y_n = \left(1 - \frac{1}{n+2}\right)^{n+1}$, then

$y_1 = \left(1 - \frac{1}{1+2}\right)^{1+1} = \left(1 - \frac{1}{3}\right)^2 = \left(\frac{3-1}{3}\right)^2 = \left(\frac{2}{3}\right)^2 = \mathbf{0.67^2}$

$y_2 = \left(1-\frac{1}{2+2}\right)^{2+1} = \left(1-\frac{1}{4}\right)^3 = \left(\frac{4-1}{4}\right)^3 = \left(\frac{3}{4}\right)^3 = \mathbf{0.75^3}$ $\quad y_3 = \left(1-\frac{1}{3+2}\right)^{3+1} = \left(1-\frac{1}{5}\right)^4 = \left(\frac{5-1}{5}\right)^4 = \left(\frac{4}{5}\right)^4 = \mathbf{0.8^4}$

$y_4 = \left(1-\frac{1}{4+2}\right)^{4+1} = \left(1-\frac{1}{6}\right)^5 = \left(\frac{6-1}{6}\right)^5 = \left(\frac{5}{6}\right)^5 = \mathbf{0.83^5}$ $\quad y_5 = \left(1-\frac{1}{5+2}\right)^{5+1} = \left(1-\frac{1}{7}\right)^6 = \left(\frac{7-1}{7}\right)^6 = \left(\frac{6}{7}\right)^6 = \mathbf{0.86^6}$

i. Given $u_k = 1-(-1)^{k+1}$, then $\quad u_1 = 1-(-1)^{1+1} = 1-(-1)^2 = 1-1 = \mathbf{0}$

$u_2 = 1-(-1)^{2+1} = 1-(-1)^3 = 1-(-1) = 1+1 = \mathbf{2}$ $\quad u_3 = 1-(-1)^{3+1} = 1-(-1)^4 = 1-1 = \mathbf{0}$

$u_4 = 1-(-1)^{4+1} = 1-(-1)^5 = 1-(-1) = 1+1 = \mathbf{2}$ $\quad u_5 = 1-(-1)^{5+1} = 1-(-1)^6 = 1-1 = \mathbf{0}$

j. Given $y_k = \frac{k}{2^{k-1}}$, then $\quad y_1 = \frac{1}{2^{1-1}} = \frac{1}{2^0} = \frac{1}{1} = \mathbf{1}$

$y_2 = \frac{2}{2^{2-1}} = \frac{2}{2^1} = \frac{2}{2} = \mathbf{1}$ $\quad y_3 = \frac{3}{2^{3-1}} = \frac{3}{2^2} = \frac{3}{4} = \mathbf{0.75}$

$y_4 = \frac{4}{2^{4-1}} = \frac{4}{2^3} = \frac{4}{8} = \mathbf{0.5}$ $\quad y_5 = \frac{5}{2^{5-1}} = \frac{5}{2^4} = \frac{5}{16} = \mathbf{0.313}$

k. Given $y_n = 9^{\frac{1}{k}}(k-2)$, then $\quad y_1 = 9^{\frac{1}{1}}(1-2) = 9\cdot -1 = \mathbf{-9}$

$y_2 = 9^{\frac{1}{2}}(2-2) = 9^{\frac{1}{2}}\cdot 0 = \mathbf{0}$ $\quad y_3 = 9^{\frac{1}{3}}(3-2) = 9^{\frac{1}{3}}\cdot 1 = 9^{\frac{1}{3}} = \mathbf{\sqrt[3]{9}}$

$y_4 = 9^{\frac{1}{4}}(4-2) = 9^{\frac{1}{4}}\cdot 2 = \mathbf{2\sqrt[4]{9}}$ $\quad y_5 = 9^{\frac{1}{5}}(5-2) = 9^{\frac{1}{5}}\cdot 3 = \mathbf{3\sqrt[5]{9}}$

l. Given $c_n = \frac{n^2-2}{n+1}$, then $\quad c_1 = \frac{1^2-2}{1+1} = \frac{1-2}{2} = \mathbf{-\frac{1}{2}}$

$c_2 = \frac{2^2-2}{2+1} = \frac{4-2}{3} = \mathbf{\frac{2}{3}}$ $\quad c_3 = \frac{3^2-2}{3+1} = \frac{9-2}{4} = \mathbf{\frac{7}{4}}$

$c_4 = \frac{4^2-2}{4+1} = \frac{16-2}{5} = \mathbf{\frac{14}{5}}$ $\quad c_5 = \frac{5^2-2}{5+1} = \frac{25-2}{6} = \mathbf{\frac{23}{6}}$

4. Given $n!$ read as "n factorial" which is defined as $n! = n(n-1)(n-2)(n-3)\cdots 5\cdot 4\cdot 3\cdot 2\cdot 1$, find

a. $8! = 8\cdot 7\cdot 6\cdot 5\cdot 4\cdot 3\cdot 2\cdot 1 = \mathbf{40{,}320}$

b. Given $a_n = \frac{2n+1}{n!}$, then $\quad a_1 = \frac{2\cdot 1+1}{1!} = \frac{2+1}{1!} = \frac{3}{1} = \mathbf{3}$

$a_2 = \frac{2\cdot 2+1}{2!} = \frac{4+1}{2!} = \frac{5}{2\cdot 1} = \frac{5}{2} = \mathbf{2.5}$ $\quad a_3 = \frac{2\cdot 3+1}{3!} = \frac{6+1}{3!} = \frac{7}{3\cdot 2\cdot 1} = \frac{7}{6} = \mathbf{1.17}$

$a_4 = \frac{2\cdot 4+1}{4!} = \frac{8+1}{4!} = \frac{9}{4\cdot 3\cdot 2\cdot 1} = \frac{3}{8} = \mathbf{0.375}$

c. Given $c_n = \frac{1+3^{n-1}}{(n!)^2}$, then $\quad c_{10} = \frac{1+3^{10-1}}{(10!)^2} = \frac{1+3^9}{10!10!} = \mathbf{\frac{19{,}684}{10!10!}}$ and $\quad c_{12} = \frac{1+3^{12-1}}{(12!)^2} = \frac{1+3^{11}}{12!12!} = \mathbf{\frac{177{,}148}{12!12!}}$

d. The first, fifth, tenth, and fifteenth terms of $y_n = \frac{n!(n-1)}{2+n!}$.

$y_1 = \frac{1!(1-1)}{2+1!} = \frac{1!\cdot 0}{2+1!} = \frac{0}{3} = \mathbf{0}$

$y_5 = \frac{5!(5-1)}{2+5!} = \frac{5!\cdot 4}{2+5!} = \frac{(5\cdot 4\cdot 3\cdot 2\cdot 1)\cdot 4}{2+(5\cdot 4\cdot 3\cdot 2\cdot 1)} = \frac{120\cdot 4}{2+120} = \frac{480}{122} = \mathbf{3.934}$

$y_{10} = \frac{10!(10-1)}{2+10!} = \frac{10!\cdot 9}{2+10!} = \frac{(10\cdot 9\cdot 8\cdot 7\cdot 6\cdot 5\cdot 4\cdot 3\cdot 2\cdot 1)\cdot 9}{2+(10\cdot 9\cdot 8\cdot 7\cdot 6\cdot 5\cdot 4\cdot 3\cdot 2\cdot 1)} = \frac{3{,}628{,}800\cdot 9}{2+3{,}628{,}800} = \frac{32{,}659{,}200}{3{,}628{,}802} = 8.9999 \approx \mathbf{9}$

or, a quicker way of solving this problem is as follows: $y_{10} = \frac{10!\,(10-1)}{2+10!} = \frac{10!\cdot 9}{2+10!} \approx \frac{10!\cdot 9}{10!} = \frac{\cancel{10}!\cdot 9}{\cancel{10}!} = \mathbf{9}$

$$y_{15} = \frac{15!\,(15-1)}{2+15!} = \frac{15!\cdot 14}{2+15!} = \frac{(15\cdot 14\cdot 13\cdot 12\cdot 11\cdot 10\cdot 9\cdot 8\cdot 7\cdot 6\cdot 5\cdot 4\cdot 3\cdot 2\cdot 1)\cdot 14}{2+(15\cdot 14\cdot 13\cdot 12\cdot 11\cdot 10\cdot 9\cdot 8\cdot 7\cdot 6\cdot 5\cdot 4\cdot 3\cdot 2\cdot 1)} \approx \mathbf{14} \quad \text{or}$$

$$y_{15} = \frac{15!\,(15-1)}{2+15!} = \frac{15!\cdot 14}{2+15!} \approx \frac{15!\cdot 14}{15!} = \frac{\cancel{15}!\cdot 14}{\cancel{15}!} = \mathbf{14}$$

5. Write the first three terms of the following sequences.

a. Given $c_n = \frac{(2n-3)(n+1)}{(n-4)n}$, then

$$c_1 = \frac{[(2\cdot 1)-3](1+1)}{(1-4)\cdot 1} = \frac{(2-3)\cdot 2}{-3\cdot 1} = \frac{-1\cdot 2}{-3} = \mathbf{\frac{2}{3}}$$

$$c_2 = \frac{[(2\cdot 2)-3](2+1)}{(2-4)\cdot 2} = \frac{(4-3)\cdot 3}{-2\cdot 2} = \frac{1\cdot 3}{-4} = \mathbf{-\frac{3}{4}}$$

$$c_3 = \frac{[(2\cdot 3)-3](3+1)}{(3-4)\cdot 3} = \frac{(6-3)\cdot 4}{-1\cdot 3} = \frac{3\cdot 4}{-3} = -\frac{12}{4} = \mathbf{-3}$$

b. Given, $a_n = \left(\frac{1}{n-1}\right)\left(\frac{n-2}{2+n}\right)$, then

$$a_1 = \left(\frac{1}{1-1}\right)\left(\frac{1-2}{2+1}\right) = \left(\frac{1}{0}\right)\left(\frac{-1}{3}\right) \ \textbf{which is undefined}$$

$$a_2 = \left(\frac{1}{2-1}\right)\left(\frac{2-2}{2+2}\right) = \left(\frac{1}{1}\right)\left(\frac{0}{4}\right) = \mathbf{0}$$

$$a_3 = \left(\frac{1}{3-1}\right)\left(\frac{3-2}{2+3}\right) = \left(\frac{1}{2}\right)\left(\frac{1}{5}\right) = \mathbf{\frac{1}{10}}$$

c. Given $s_n = (-1)^{n+1}2^{n+1}$, then

$$s_1 = (-1)^{1+1}\cdot 2^{1+1} = (-1)^2\cdot 2^2 = 1\cdot 4 = \mathbf{4}$$

$$s_2 = (-1)^{2+1}\cdot 2^{2+1} = (-1)^3\cdot 2^3 = -1\cdot 8 = \mathbf{-8}$$

$$s_3 = (-1)^{3+1}\cdot 2^{3+1} = (-1)^4\cdot 2^4 = 1\cdot 16 = \mathbf{16}$$

d. Given $y_k = (-1)^{k+1}\frac{k(k-1)}{2}$, then

$$y_1 = (-1)^{1+1}\cdot\frac{1\cdot(1-1)}{2} = (-1)^2\cdot\frac{1\cdot 0}{2} = \frac{0}{2} = \mathbf{0}$$

$$y_2 = (-1)^{2+1}\cdot\frac{2\cdot(2-1)}{2} = (-1)^3\cdot\frac{2\cdot 1}{2} = -\frac{2}{2} = \mathbf{1}$$

$$y_3 = (-1)^{3+1}\cdot\frac{3\cdot(3-1)}{2} = (-1)^4\cdot\frac{3\cdot 2}{2} = \frac{6}{2} = \mathbf{3}$$

e. Given $b_n = n^2\left(\frac{n-1}{2+n}\right)$, then

$$b1 = 1^2\cdot\left(\frac{1-1}{2+1}\right) = 1\cdot\left(\frac{0}{3}\right) = \mathbf{0}$$

$$b_2 = 2^2\cdot\left(\frac{2-1}{2+2}\right) = 4\cdot\left(\frac{1}{4}\right) = \frac{4}{4} = \mathbf{1}$$

$$b_3 = 3^2\cdot\left(\frac{3-1}{2+3}\right) = 9\cdot\left(\frac{2}{5}\right) = \frac{18}{5} = \mathbf{3.6}$$

f. Given $x_a = (5-a)^{a+1}2^a$, then

$$x_1 = (5-1)^{1+1}\cdot 2^1 = 4^2\cdot 2 = 16\cdot 2 = \mathbf{32}$$

$$x_2 = (5-2)^{2+1}\cdot 2^2 = 3^3\cdot 4 = 27\cdot 4 = \mathbf{108}$$

$$x_3 = (5-3)^{3+1}\cdot 2^3 = 2^4\cdot 8 = 16\cdot 8 = \mathbf{128}$$

Section 4.2 Solutions - Series

1. Given $\sum_{i=1}^{n} a_i = 10$ and $\sum_{i=1}^{n} b_i = 25$, find

a. $\sum_{i=1}^{n}(2a_i + 4b_i) = \sum_{i=1}^{n}2a_i + \sum_{i=1}^{n}4b_i = 2\sum_{i=1}^{n}a_i + 4\sum_{i=1}^{n}b_i = (2\cdot 10)+(4\cdot 25) = 20+100 = \mathbf{120}$

b. $\sum_{i=1}^{n}(-a_i + b_i) = \sum_{i=1}^{n}-a_i + \sum_{i=1}^{n}b_i = -\sum_{i=1}^{n}a_i + \sum_{i=1}^{n}b_i = -10+25 = \mathbf{15}$

c. $\sum_{i=1}^{n}(3a_i + 5b_i) = \sum_{i=1}^{n}3a_i + \sum_{i=1}^{n}5b_i = 3\sum_{i=1}^{n}a_i + 5\sum_{i=1}^{n}b_i = (3\cdot 10)+(5\cdot 25) = 30+125 = \mathbf{155}$

d. $\sum_{i=1}^{n}\left(\frac{1}{2}a_i + \frac{1}{5}b_i\right) = \sum_{i=1}^{n}\frac{1}{2}a_i + \sum_{i=1}^{n}\frac{1}{5}b_i = \frac{1}{2}\sum_{i=1}^{n}a_i + \frac{1}{5}\sum_{i=1}^{n}b_i = \left(\frac{1}{2}\cdot 10\right)+\left(\frac{1}{5}\cdot 25\right) = 5+5 = \mathbf{10}$

2. Evaluate each of the following series.

a. $\sum_{k=1}^{5} 2+k = (2+1)+(2+2)+(2+3)+(2+4)+(2+5) = 3+4+5+6+7 = \mathbf{25}$

b. $\sum_{n=0}^{6} \frac{1}{(-2)^{n+1}} = \frac{1}{(-2)^{0+1}}+\frac{1}{(-2)^{1+1}}+\frac{1}{(-2)^{2+1}}+\frac{1}{(-2)^{3+1}}+\frac{1}{(-2)^{4+1}}+\frac{1}{(-2)^{5+1}}+\frac{1}{(-2)^{6+1}} = \frac{1}{(-2)^1}+\frac{1}{(-2)^2}+\frac{1}{(-2)^3}+\frac{1}{(-2)^4}$

$+\frac{1}{(-2)^5}+\frac{1}{(-2)^6}+\frac{1}{(-2)^7} = -\frac{1}{2}+\frac{1}{4}-\frac{1}{8}+\frac{1}{16}-\frac{1}{32}+\frac{1}{64}-\frac{1}{128} = -0.5+0.25-0.125+0.0625-0.031+0.016-0.008$

$= (-0.5-0.125-0.031-0.008)+(0.25+0.0625+0.016) = -0.664+0.328 = \mathbf{-0.336}$

c. $\sum_{n=0}^{4} (-1)^{n+1} = (-1)^{0+1}+(-1)^{1+1}+(-1)^{2+1}+(-1)^{3+1}+(-1)^{4+1} = (-1)^1+(-1)^2+(-1)^3+(-1)^4+(-1)^5 = -1+1-1+1-1 = \mathbf{-1}$

d. $\sum_{j=-3}^{3} j-3j^2 = [-3-3\cdot(-3)^2]+[-2-3\cdot(-2)^2]+[-1-3\cdot(-1)^2]+[0-3\cdot 0^2]+[1-3\cdot 1^2]+[2-3\cdot 2^2]+[3-3\cdot 3^2]$

$= [-3-(3\cdot 9)]+[-2-(3\cdot 4)]+[-1-(3\cdot 1)]+0+[1-(3\cdot 1)]+[2-(3\cdot 4)]+[3-(3\cdot 9)] = [-3-27]+[-2-12]+[-1-3]+[1-3]$

$+[2-12]+[3-27] = -30-14-4-2-10-24 = \mathbf{-84}$

e. $\sum_{a=3}^{5} (a+2)^a = (3+2)^3+(4+2)^4+(5+2)^5 = 5^3+6^4+7^5 = 125+1296+16807 = \mathbf{18{,}228}$

f. $\sum_{i=0}^{5} \frac{(-1)^{i+1}}{2^i} = \frac{(-1)^{0+1}}{2^0}+\frac{(-1)^{1+1}}{2^1}+\frac{(-1)^{2+1}}{2^2}+\frac{(-1)^{3+1}}{2^3}+\frac{(-1)^{4+1}}{2^4}+\frac{(-1)^{5+1}}{2^5} = \frac{(-1)^1}{1}+\frac{(-1)^2}{2}+\frac{(-1)^3}{4}+\frac{(-1)^4}{8}+\frac{(-1)^5}{16}+\frac{(-1)^6}{32}$

$= -1+\frac{1}{2}-\frac{1}{4}+\frac{1}{8}-\frac{1}{16}+\frac{1}{32} = -1+0.5-0.25+0.125-0.063+0.031 = -1.313+0.656 = \mathbf{-0.657}$

g. $\sum_{k=-2}^{3} (2k-3)^{k+2} = [(2\cdot -2)-3]^{-2+2}+[(2\cdot -1)-3]^{-1+2}+[(2\cdot 0)-3]^{0+2}+[(2\cdot 1)-3]^{1+2}+[(2\cdot 2)-3]^{2+2}+[(2\cdot 3)-3]^{3+2}$

$= [-4-3]^0+[-2-3]^1+[0-3]^2+[2-3]^3+[4-3]^4+[6-3]^5 = (-7)^0+(-5)^1+(-3)^2+(-1)^3+1^4+3^5 = \mathbf{248}$

h. $\sum_{m=1}^{5}\left(\frac{1}{m}-1\right)^2 = \left(\frac{1}{1}-1\right)^2+\left(\frac{1}{2}-1\right)^2+\left(\frac{1}{3}-1\right)^2+\left(\frac{1}{4}-1\right)^2+\left(\frac{1}{5}-1\right)^2 = (1-1)^2+(0.5-1)^2+(0.333-1)^2+(0.25-1)^2+(0.2-1)^2$

$= 0^2+(-0.5)^2+(0.667)^2+(-0.75)^2+(-0.8)^2 = 0.25+0.445+0.563+0.64 = \mathbf{1.898}$

3. Find the sum of the following series within the specified range.

a. $\sum_{i=-3}^{3} 10^i = 10^{-3}+10^{-2}+10^{-1}+10^0+10^1+10^2+10^3 = 0.001+0.01+0.1+1+10+100+1000 = \mathbf{1111.111}$

b. $\sum_{n=0}^{6} \frac{n-1}{2^n} = \frac{0-1}{2^0}+\frac{1-1}{2^1}+\frac{2-1}{2^2}+\frac{3-1}{2^3}+\frac{4-1}{2^4}+\frac{5-1}{2^5}+\frac{6-1}{2^6} = -\frac{1}{1}+\frac{0}{2}+\frac{1}{4}+\frac{2}{8}+\frac{3}{16}+\frac{4}{32}+\frac{5}{64} = -1+0.25+0.25+0.1875$

$+0.125+0.08 = \mathbf{-0.108}$

c. $\sum_{a=0}^{4} \frac{1}{10^a} = \frac{1}{10^0}+\frac{1}{10^1}+\frac{1}{10^2}+\frac{1}{10^3}+\frac{1}{10^4} = \frac{1}{1}+\frac{1}{10}+\frac{1}{100}+\frac{1}{1000}+\frac{1}{10000} = 1+0.1+0.01+0.001+0.0001 = \mathbf{1.1111}$

d. $\sum_{n=1}^{5}(n^2-n) = (1^2-1)+(2^2-2)+(3^2-3)+(4^2-4)+(5^2-5) = 0+(4-2)+(9-3)+(16-4)+(25-5) = 2+6+12+20 = \mathbf{40}$

e. $\sum_{m=0}^{6}(-1)^{m+1} = (-1)^{0+1}+(-1)^{1+1}+(-1)^{2+1}+(-1)^{3+1}+(-1)^{4+1}+(-1)^{5+1}+(-1)^{6+1} = (-1)^1+(-1)^2+(-1)^3+(-1)^4+(-1)^5$
$+(-1)^6+(-1)^7 = -1+1-1+1-1+1-1 = \mathbf{-1}$

f. $\sum_{k=0}^{5}\frac{1+(-1)^k}{2^k} = \frac{1+(-1)^0}{2^0}+\frac{1+(-1)^1}{2^1}+\frac{1+(-1)^2}{2^2}+\frac{1+(-1)^3}{2^3}+\frac{1+(-1)^4}{2^4}+\frac{1+(-1)^5}{2^5} = \frac{1+1}{1}+\frac{1-1}{2}+\frac{1+1}{4}+\frac{1-1}{8}+\frac{1+1}{16}+\frac{1-1}{32}$
$= \frac{2}{1}+\frac{0}{2}+\frac{2}{4}+\frac{0}{8}+\frac{2}{16}+\frac{0}{32} = 2+0.5+0.125 = \mathbf{2.625}$

g. $\sum_{a=1}^{6}[5(a-1)+3] = [5(1-1)+3]+[5(2-1)+3]+[5(3-1)+3]+[5(4-1)+3]+[5(5-1)+3]+[5(6-1)+3] = [5\cdot 0+3]+[5\cdot 1+3]$
$+[5\cdot 2+3]+[5\cdot 3+3]+[5\cdot 4+3]+[5\cdot 5+3] = 3+8+13+18+23+28 = \mathbf{93}$

h. $\sum_{k=0}^{5}\left(-\frac{1}{3}\right)^{k-1} = \left(-\frac{1}{3}\right)^{0-1}+\left(-\frac{1}{3}\right)^{1-1}+\left(-\frac{1}{3}\right)^{2-1}+\left(-\frac{1}{3}\right)^{3-1}+\left(-\frac{1}{3}\right)^{4-1}+\left(-\frac{1}{3}\right)^{5-1} = \left(-\frac{1}{3}\right)^{-1}+\left(-\frac{1}{3}\right)^{0}+\left(-\frac{1}{3}\right)^{1}+\left(-\frac{1}{3}\right)^{2}$
$+\left(-\frac{1}{3}\right)^{3}+\left(-\frac{1}{3}\right)^{4} = -3+1-\frac{1}{3}+\frac{1}{9}-\frac{1}{27}+\frac{1}{81} = -3+1-0.33+0.11-0.04+0.01 = \mathbf{-2.25}$

i. $\sum_{j=1}^{5}\left(j-3j^2\right) = \left(1-3\cdot 1^2\right)+\left(2-3\cdot 2^2\right)+\left(3-3\cdot 3^2\right)+\left(4-3\cdot 4^2\right)+\left(5-3\cdot 5^2\right) = (1-3)+(2-12)+(3-27)+(4-48)+(5-75)$
$= -2-10-24-44-72 = \mathbf{-152}$

j. $\sum_{n=1}^{4}\frac{n+1}{n}-\sum_{n=1}^{4}\frac{n^2}{n+1} = \left(\frac{1+1}{1}+\frac{2+1}{2}+\frac{3+1}{3}+\frac{4+1}{4}\right)-\left(\frac{1^2}{1+1}+\frac{2^2}{2+1}+\frac{3^2}{3+1}+\frac{4^2}{4+1}\right) = \left(\frac{2}{1}+\frac{3}{2}+\frac{4}{3}+\frac{5}{4}\right)-\left(\frac{1}{2}+\frac{4}{3}+\frac{9}{4}+\frac{16}{5}\right)$
$= (2+1.5+1.33+1.25)-(0.5+1.33+2.25+3.2) = 6.08-7.28 = \mathbf{-1.2}$

k. $\sum_{k=1}^{5}5k^{-1} = 5\cdot 1^{-1}+5\cdot 2^{-1}+5\cdot 3^{-1}+5\cdot 4^{-1}+5\cdot 5^{-1} = \frac{5}{1}+\frac{5}{2}+\frac{5}{3}+\frac{5}{4}+\frac{5}{5} = 5+2.5+1.67+1.25+1 = \mathbf{11.42}$

l. $\sum_{i=1}^{4}(-0.1)^{2i-5} = (-0.1)^{2\cdot 1-5}+(-0.1)^{2\cdot 2-5}+(-0.1)^{2\cdot 3-5}+(-0.1)^{2\cdot 4-5} = (-0.1)^{-3}+(-0.1)^{-1}+(-0.1)^1+(-0.1)^3$
$= \frac{1}{(-0.1)^3}+\frac{1}{(-0.1)^1}-0.1-0.001 = \frac{1}{-0.001}+\frac{1}{-0.1}-0.1-0.001 = -1000-10-0.1-0.001 = \mathbf{-1010.101}$

4. Rewrite the following terms using the sigma notation.

a. $\frac{1}{2}+\frac{1}{3}+\frac{1}{4}+\frac{1}{5}+\frac{1}{6}+\frac{1}{7} = \mathbf{\sum_{n=0}^{5}\frac{1}{n+2}}$

b. $\frac{1}{2}+\frac{2}{3}+\frac{3}{4}+\frac{4}{5}+\frac{5}{6}+\frac{6}{7} = \mathbf{\sum_{n=1}^{6}\frac{n}{n+1}}$

c. $2+4+8+16+32+64 = \mathbf{\sum_{k=0}^{5}2^{k+1}}$

d. $1+\frac{1}{2}+\frac{1}{3}+\frac{1}{4}+\frac{1}{5}+\frac{1}{6} = \mathbf{\sum_{k=1}^{6}\frac{1}{k}}$

e. $0+\frac{1}{2}+\frac{2}{3}+\frac{3}{4}+\frac{4}{5}+\frac{5}{6} = \mathbf{\sum_{i=1}^{6}\frac{n-1}{n}}$

f. $1-\frac{1}{2}+\frac{1}{3}-\frac{1}{4}+\frac{1}{5}-\frac{1}{6} = \mathbf{\sum_{n=1}^{6}\frac{(-1)^{n+1}}{n}}$

Section 4.3 Solutions - Arithmetic Sequences and Arithmetic Series

1. Find the next seven terms of the following arithmetic sequences.

a. Substituting $s_1 = 3$ and $d = 2$ into the general arithmetic sequence $s_n = s_1+(n-1)d$ for $n = 2, 3, 4, 5, 6, and\ 7$ we obtain

$s_2 = s_1+(2-1)d = s_1+d = 3+2 = \mathbf{5}$
$s_3 = s_1+(3-1)d = s_1+2d = 3+(2\cdot 2) = 3+4 = \mathbf{7}$

$s_4 = s_1 + (4-1)d = s_1 + 3d = 3 + (3 \cdot 2) = 3 + 6 = \mathbf{9}$
$s_5 = s_1 + (5-1)d = s_1 + 4d = 3 + (4 \cdot 2) = 3 + 8 = \mathbf{11}$
$s_6 = s_1 + (6-1)d = s_1 + 5d = 3 + (5 \cdot 2) = 3 + 10 = \mathbf{13}$
$s_7 = s_1 + (7-1)d = s_1 + 6d = 3 + (6 \cdot 2) = 3 + 12 = \mathbf{15}$

Thus, the first seven terms of the arithmetic sequence are $(\mathbf{3, 5, 7, 9, 11, 13, 15})$

b. Substituting $s_1 = -3$ and $d = 2$ into the general arithmetic sequence $s_n = s_1 + (n-1)d$ for $n = 2, 3, 4, 5, 6, and\ 7$ we obtain

$s_2 = s_1 + (2-1)d = s_1 + d = -3 + 2 = \mathbf{-1}$
$s_3 = s_1 + (3-1)d = s_1 + 2d = -3 + (2 \cdot 2) = -3 + 4 = \mathbf{1}$
$s_4 = s_1 + (4-1)d = s_1 + 3d = -3 + (3 \cdot 2) = -3 + 6 = \mathbf{3}$
$s_5 = s_1 + (5-1)d = s_1 + 4d = -3 + (4 \cdot 2) = -3 + 8 = \mathbf{5}$
$s_6 = s_1 + (6-1)d = s_1 + 5d = -3 + (5 \cdot 2) = -3 + 10 = \mathbf{7}$
$s_7 = s_1 + (7-1)d = s_1 + 6d = -3 + (6 \cdot 2) = -3 + 12 = \mathbf{9}$

Thus, the first seven terms of the arithmetic sequence are $(\mathbf{-3, -1, 1, 3, 5, 7, 9})$

c. Substituting $s_1 = 10$ and $d = 0.8$ into the general arithmetic sequence $s_n = s_1 + (n-1)d$ for $n = 2, 3, 4, 5, 6, and\ 7$ we obtain

$s_2 = s_1 + (2-1)d = s_1 + d = 10 + 0.8 = \mathbf{10.8}$
$s_3 = s_1 + (3-1)d = s_1 + 2d = 10 + (2 \cdot 0.8) = 10 + 1.6 = \mathbf{11.6}$
$s_4 = s_1 + (4-1)d = s_1 + 3d = 10 + (3 \cdot 0.8) = 10 + 2.4 = \mathbf{12.4}$
$s_5 = s_1 + (5-1)d = s_1 + 4d = 10 + (4 \cdot 0.8) = 10 + 3.2 = \mathbf{13.2}$
$s_6 = s_1 + (6-1)d = s_1 + 5d = 10 + (5 \cdot 0.8) = 10 + 4 = \mathbf{14}$
$s_7 = s_1 + (7-1)d = s_1 + 6d = 10 + (6 \cdot 0.8) = 10 + 4.8 = \mathbf{14.8}$

Thus, the first seven terms of the arithmetic sequence are $(\mathbf{10, 10.8, 11.6, 12.4, 13.2, 14, 14.8})$

2. Find the general term and the eighth term of the following arithmetic sequences.

a. Given $s_1 = 3$ and $d = 4$, the nth term of the arithmetic sequence is equal to $s_n = s_1 + (n-1)d = 3 + (n-1) \cdot 4 = 3 + 4n - 4$

$= \mathbf{4n - 1}$. Substituting $n = 8$ into the general equation $s_n = 4n - 1$ we have $s_8 = 4 \cdot 8 - 1 = 32 - 1 = \mathbf{31}$

b. Given $s_1 = -3$ and $d = 5$, the nth term of the arithmetic sequence is equal to $s_n = s_1 + (n-1)d = -3 + (n-1) \cdot 5 = -3 + 5n - 5$

$= \mathbf{5n - 8}$. Substituting $n = 8$ into the general equation $s_n = 5n - 8$ we have $s_8 = 5 \cdot 8 - 8 = 40 - 8 = \mathbf{32}$

c. Given $s_1 = 8$ and $d = -1.2$, the nth term of the arithmetic sequence is equal to $s_n = s_1 + (n-1)d = 8 + (n-1) \cdot -1.2$

$= 8 - 1.2n + 1.2 = \mathbf{-1.2n + 9.2}$. Substituting $n = 8$ into the general equation $s_n = -1.2n + 9.2$ we have $s_8 = -1.2 \cdot 8 + 9.2$

$= -9.6 + 9.2 = \mathbf{-0.4}$

3. Find the next six terms in each of the following arithmetic sequences.

a. Given the arithmetic sequence $5, 8, \cdots$, $s_1 = 5$ and $d = 8 - 5 = 3$. Therefore, using the general arithmetic equation

$s_n = s_1 + (n-1)d$ or $s_{n+1} = s_n + d$ the next six terms are as follows:

$s_3 = s_2 + d = 8 + 3 = \mathbf{11}$ $s_4 = s_3 + d = 11 + 3 = \mathbf{14}$ $s_5 = s_4 + d = 14 + 3 = \mathbf{17}$

$s_6 = s_5 + d = 17 + 3 = \mathbf{20}$ $s_7 = s_6 + d = 20 + 3 = \mathbf{23}$ $s_8 = s_7 + d = 23 + 3 = \mathbf{26}$

Thus, the first eight terms of the arithmetic sequence are $(\mathbf{5, 8, 11, 14, 17, 20, 23, 26})$

b. Given the arithmetic sequence $x, x+4, \cdots$, $s_1 = x$ and $d = (x+4) - x = 4$. Therefore, using the general arithmetic equation

$s_n = s_1 + (n-1)d$ or $s_{n+1} = s_n + d$ the next six terms are as follows:

$s_3 = s_2 + d = (x+4) + 4 = \mathbf{x + 8}$ $s_4 = s_3 + d = (x+8) + 4 = \mathbf{x + 12}$

$s_5 = s_4 + d = (x+12)+4 = \mathbf{x+16}$ $\qquad$ $s_6 = s_5 + d = (x+16)+4 = \mathbf{x+20}$

$s_7 = s_6 + d = (x+20)+4 = \mathbf{x+24}$ $\qquad$ $s_8 = s_7 + d = (x+24)+4 = \mathbf{x+28}$

Thus, the first eight terms of the arithmetic sequence are $(\mathbf{x, x+4, x+8, x+12, x+16, x+20, x+24, x+28})$

c. Given the arithmetic sequence $3x+1, 3x+4, \cdots$, $s_1 = 3x+1$ and $d = (3x+4)-(3x+1) = 3$. Therefore, using the general arithmetic equation $s_n = s_1 + (n-1)d$ or $s_{n+1} = s_n + d$ the next six terms are as follows:

$s_3 = s_2 + d = (3x+4)+3 = \mathbf{3x+7}$ $\qquad$ $s_4 = s_3 + d = (3x+7)+3 = \mathbf{3x+10}$

$s_5 = s_4 + d = (3x+10)+3 = \mathbf{3x+13}$ $\qquad$ $s_6 = s_5 + d = (3x+13)+3 = \mathbf{3x+16}$

$s_7 = s_6 + d = (3x+16)+3 = \mathbf{3x+19}$ $\qquad$ $s_8 = s_7 + d = (3x+19)+3 = \mathbf{3x+22}$

Thus, the first eight terms of the arithmetic sequence are $(\mathbf{3x+1, 3x+4, 3x+7, 3x+10, 3x+13, 3x+16, 3x+19, 3x+22})$

d. Given the arithmetic sequence $w, w-10, \cdots$, $s_1 = w$ and $d = (w-10) - w = -10$. Using the general arithmetic equation $s_n = s_1 + (n-1)d$ or $s_{n+1} = s_n + d$ the next six terms are as follows:

$s_3 = s_2 + d = (w-10)-10 = \mathbf{w-20}$ $\qquad$ $s_4 = s_3 + d = (w-20)-10 = \mathbf{w-30}$

$s_5 = s_4 + d = (w-30)-10 = \mathbf{w-40}$ $\qquad$ $s_6 = s_5 + d = (w-40)-10 = \mathbf{w-50}$

$s_7 = s_6 + d = (w-50)-10 = \mathbf{w-60}$ $\qquad$ $s_8 = s_7 + d = (w-60)-10 = \mathbf{w-70}$

Thus, the first eight terms of the arithmetic sequence are $(\mathbf{w, w-10, w-20, w-30, w-40, w-50, w-60, w-70})$

4. Find the sum of the following arithmetic series.

a. The first three terms of the given series are $\sum_{i=10}^{20}(2i-4) = (2\cdot10-4)+(2\cdot11-4)+(2\cdot12-4)+\cdots = 16+18+20+\cdots$.

Therefore, $s_1 = 16$, $d = 18-16 = 2$, and $n = 11$. Substituting s_1, d, and n into the arithmetic series formula $S_n = \frac{n}{2}[2s_1 + (n-1)d]$ we can obtain $S_{11} = \frac{11}{2}[2\cdot16+(11-1)\cdot2] = 5.5\cdot(32+10\cdot2) = 5.5\cdot(32+20) = 5.5\cdot52 = \mathbf{286}$

b. The first three terms of the given series are $\sum_{k=1}^{1000} k = 1+2+3+\cdots$. Therefore, $s_1 = 1$, $d = 2-1 = 1$, and $n = 1000$.

Substituting s_1, d, and n into the arithmetic series formula $S_n = \frac{n}{2}[2s_1 + (n-1)d]$ we obtain

$S_{1000} = \frac{1000}{2}[2\cdot1+(1000-1)\cdot1] = 500\cdot[2+999] = 500\cdot1001 = \mathbf{500500}$

c. The first three terms of the given series are $\sum_{k=1}^{100}(2k-3) = (2\cdot1-3)+(2\cdot2-3)+(2\cdot3-3)+\cdots = -1+1+3+\cdots$.

Therefore, $s_1 = -1$, $d = 1-(-1) = 2$, and $n = 100$. Substituting s_1, d, and n into the arithmetic series formula $S_n = \frac{n}{2}[2s_1 + (n-1)d]$ we obtain $S_{100} = \frac{100}{2}[(2\cdot-1)+(100-1)\cdot2] = 50\cdot(-2+99\cdot2) = 50\cdot(-2+198) = 50\cdot196 = \mathbf{9800}$

d. The first three terms of the given series are $\sum_{i=1}^{15} 3i = (3\cdot1)+(3\cdot2)+(3\cdot3)+\cdots = 3+6+9+\cdots$.

Therefore, $s_1 = 3$, $d = 6-3 = 3$, and $n = 15$. Substituting s_1, d, and n into the arithmetic series formula $S_n = \frac{n}{2}[2s_1 + (n-1)d]$ we obtain $S_{15} = \frac{15}{2}[(2\cdot3)+(15-1)\cdot3] = 12.5\cdot(6+14\cdot3) = 12.5\cdot48 = \mathbf{600}$

e. The first three terms of the given series are $\sum_{i=1}^{10}(i+1) = (1+1)+(2+1)+(3+1)+\cdots = 2+3+4+\cdots$.

Therefore, $s_1 = 2$, $d = 3 - 2 = 1$, and $n = 10$. Substituting s_1, d, and n into the arithmetic series formula $S_n = \frac{n}{2}[2s_1 + (n-1)d]$ we obtain $S_{10} = \frac{10}{2}[(2 \cdot 2) + (10-1) \cdot 1] = 5 \cdot (4+9) = 5 \cdot 13 = \mathbf{65}$

f. The first three terms of the given series are $\sum_{k=5}^{15}(2k-1) = (2 \cdot 5 - 1) + (2 \cdot 6 - 1) + (2 \cdot 7 - 1) + \cdots = 9 + 11 + 13 + \cdots$.

Therefore, $s_1 = 9$, $d = 11 - 9 = 2$, and $n = 11$. Substituting s_1, d, and n into the arithmetic series formula $S_n = \frac{n}{2}[2s_1 + (n-1)d]$ we obtain $S_{11} = \frac{11}{2}[(2 \cdot 9) + (11-1) \cdot 2] = 5.5 \cdot (18+20) = 5.5 \cdot 38 = \mathbf{209}$

g. The first three terms of the given series are $\sum_{i=4}^{10}(3i+4) = (3 \cdot 4 + 4) + (3 \cdot 5 + 4) + (3 \cdot 6 + 4) + \cdots = 16 + 19 + 22 + \cdots$.

Therefore, $s_1 = 16$, $d = 19 - 16 = 3$, and $n = 7$. Substituting s_1, d, and n into the arithmetic series formula $S_n = \frac{n}{2}[2s_1 + (n-1)d]$ we obtain $S_7 = \frac{7}{2}[(2 \cdot 16) + (7-1) \cdot 3] = 3.5 \cdot (32+18) = 3.5 \cdot 50 = \mathbf{175}$

h. The first three terms of the given series are $\sum_{j=5}^{13}(3j+1) = (3 \cdot 5 + 1) + (3 \cdot 6 + 1) + (3 \cdot 7 + 1) + \cdots = 16 + 19 + 22 + \cdots$.

Therefore, $s_1 = 16$, $d = 19 - 16 = 3$, and $n = 9$. Substituting s_1, d, and n into the arithmetic series formula $S_n = \frac{n}{2}[2s_1 + (n-1)d]$ we obtain $S_9 = \frac{9}{2}[(2 \cdot 16) + (9-1) \cdot 3] = 4.5 \cdot (32+24) = 4.5 \cdot 56 = \mathbf{252}$

i. The first three terms of the given series are $\sum_{k=7}^{18}(4k-3) = (4 \cdot 7 - 3) + (4 \cdot 8 - 3) + (4 \cdot 9 - 3) + \cdots = 25 + 29 + 33 + \cdots$.

Therefore, $s_1 = 25$, $d = 29 - 25 = 4$, and $n = 12$. Substituting s_1, d, and n into the arithmetic series formula $S_n = \frac{n}{2}[2s_1 + (n-1)d]$ we obtain $S_{12} = \frac{12}{2}[(2 \cdot 25) + (12-1) \cdot 4] = 6 \cdot (50+44) = 6 \cdot 94 = \mathbf{564}$

5. The first term of an arithmetic sequence is 6 and the third term is 24. Find the tenth term.

Since $s_1 = 6$ and $s_3 = 24$ we use the general formula $s_n = s_1 + (n-1)d$ in order to solve for d. Therefore,

$s_3 = s_1 + (3-1)d$; $24 = 6 + 2d$; $24 - 6 = 2d$; $d = \frac{18}{2}$; $d = 9$. Then, $s_{10} = s_1 + (10-1)d = s_1 + 9 \cdot d = 6 + 9 \cdot 9 = \mathbf{87}$

6. Given the first term s_1 and d, find S_{50} for each of the following arithmetic sequences.

a. Given $s_1 = 2$ and $d = 5$, use the nth term for an arithmetic series $S_n = \frac{n}{2}[2s_1 + (n-1)d]$ to find S_{50}.

$S_{50} = \frac{50}{2}[(2 \cdot 2) + (50-1) \cdot 5] = \frac{50}{2}(4 + 245) = 25 \cdot 249 = \mathbf{6225}$

b. Given $s_1 = -5$ and $d = 6$, use the nth term for an arithmetic series $S_n = \frac{n}{2}[2s_1 + (n-1)d]$ to find S_{50}.

$S_{50} = \frac{50}{2}[(2 \cdot -5) + (50-1) \cdot 6] = \frac{50}{2}(-10 + 294) = 25 \cdot 284 = \mathbf{7100}$

c. Given $s_1 = 30$ and $d = 10$, use the nth term for an arithmetic series $S_n = \frac{n}{2}[2s_1 + (n-1)d]$ to find S_{50}.

$S_{50} = \frac{50}{2}[(2 \cdot 30) + (50-1) \cdot 10] = \frac{50}{2}(60 + 490) = 25 \cdot 550 = \mathbf{13750}$

7. Find the sum of the following sequences for the indicated values.

a. Given the sequence $-8, 6, \cdots$ the first term s_1 and the common difference d are equal to $s_1 = -8$ and $d = 6 - (-8) = 14$.

Thus, using the general arithmetic series $S_n = \frac{n}{2}[2s_1 + (n-1)d]$, S_{15} is equal to:

$$S_{15} = \frac{15}{2}[(2\cdot -8)+(15-1)\cdot 14] = \frac{15}{2}(-16+196) = 7.5\cdot 180 = \mathbf{1350}$$

b. Given the sequence $-20, 20, \cdots$ the first term s_1 and the common difference d are equal to $s_1 = -20$ and $d = 20-(-20)$ $= 40$. Thus, using the general arithmetic series $S_n = \frac{n}{2}[2s_1 + (n-1)d]$, S_{100} is equal to:

$$S_{100} = \frac{100}{2}[(2\cdot -20)+(100-1)\cdot 40] = 50(-40+3960) = 50\cdot 3920 = \mathbf{196{,}000}$$

Section 4.4 Solutions - Geometric Sequences and Geometric Series

1. Find the next four terms of the following geometric sequences.

a. Substituting $s_1 = 3$, $r = 0.5$ into $s_n = s_1 r^{n-1}$ we obtain

$$s_2 = 3\cdot r^{2-1} = 3r = 3\cdot 0.5 = \mathbf{1.5} \qquad s_3 = 3\cdot r^{3-1} = 3r^2 = 3\cdot 0.5^2 = 3\cdot 0.25 = \mathbf{0.75}$$

$$s_4 = 3\cdot r^{4-1} = 3r^3 = 3\cdot 0.5^3 = 3\cdot 0.125 = \mathbf{0.375} \qquad s_5 = 3\cdot r^{5-1} = 3r^4 = 3\cdot 0.5^4 = 3\cdot 0.0625 = \mathbf{0.1875}$$

Thus, the first five terms of the geometric sequence are $(\mathbf{3, 1.5, 0.75, 0.375, 0.1875})$

b. Substituting $s_1 = -5$, $r = 2$ into $s_n = s_1 r^{n-1}$ we obtain

$$s_2 = -5\cdot r^{2-1} = -5r = -5\cdot 2 = \mathbf{-10} \qquad s_3 = -5\cdot r^{3-1} = -5r^2 = -5\cdot 2^2 = -5\cdot 4 = \mathbf{-20}$$

$$s_4 = -5\cdot r^{4-1} = -5r^3 = -5\cdot 2^3 = -5\cdot 8 = \mathbf{-40} \qquad s_5 = -5\cdot r^{5-1} = -5r^4 = -5\cdot 2^4 = -5\cdot 16 = \mathbf{-80}$$

Thus, the first five terms of the geometric sequence are $(\mathbf{-5, -10, -20, -40, -80})$

c. Substituting $s_1 = 5$, $r = 0.75$ into $s_n = s_1 r^{n-1}$ we obtain

$$s_2 = 5\cdot r^{2-1} = 5r = 5\cdot 0.75 = \mathbf{3.75} \qquad s_3 = 5\cdot r^{3-1} = 5r^2 = 5\cdot 0.75^2 = 5\cdot 0.563 = \mathbf{2.81}$$

$$s_4 = 5\cdot r^{4-1} = 5r^3 = 5\cdot 0.75^3 = 5\cdot 0.42 = \mathbf{2.11} \qquad s_5 = 5\cdot r^{5-1} = 5r^4 = 5\cdot 0.75^4 = 5\cdot 0.316 = \mathbf{1.58}$$

Thus, the first five terms of the geometric sequence are $(\mathbf{5, 3.75, 2.81, 2.11, 1.58})$

2. Find the eighth and the general term of the following geometric sequences.

a. Substituting $s_1 = 2$, $r = \sqrt{3}$ into $s_n = s_1 r^{n-1}$ the eighth and the nth term are equal to:

$$s_8 = 2r^{8-1} = 2r^7 = 2\cdot\left(\sqrt{3}\right)^7 = 2\cdot 3^{\frac{7}{2}} = 2\cdot 46.76 = \mathbf{93.53} \text{ and } s_n = 2\cdot\left(\sqrt{3}\right)^{n-1} = \mathbf{2\cdot 3^{\frac{n-1}{2}}}$$

b. Substituting $s_1 = -4$, $r = 1.2$ into $s_n = s_1 r^{n-1}$ the eighth and the nth term are equal to:

$$s_8 = -4r^{8-1} = -4r^7 = -4\cdot(1.2)^7 = -4\cdot 1.2^7 = -4\cdot 3.583 = \mathbf{-14\cdot 33} \text{ and } s_n = \mathbf{-4\cdot(1.2)^{n-1}}$$

c. Substituting $s_1 = 4$, $r = -2.5$ into $s_n = s_1 r^{n-1}$ the eighth and the nth term are equal to:

$$s_8 = 4r^{8-1} = 4r^7 = 4\cdot(-2.5)^7 = -4\cdot 2.5^7 = -4\cdot 610.35 = \mathbf{-2441.4} \text{ and } s_n = \mathbf{4\cdot(-2.5)^{n-1}}$$

3. Find the next six terms and the nth term in each of the following geometric sequences.

a. Given $1, \frac{1}{4}, \cdots$, then $s_1 = 1$ and $r = \frac{\frac{1}{4}}{1} = \frac{1}{4}$. Using the general geometric equation $s_n = s_1 r^{n-1}$ the next six terms are:

$$s_3 = 1\cdot r^{3-1} = r^2 = \left(\frac{1}{4}\right)^2 = \mathbf{\frac{1}{4^2}} \qquad s_4 = 1\cdot r^{4-1} = r^3 = \left(\frac{1}{4}\right)^3 = \mathbf{\frac{1}{4^3}}$$

$$s_5 = 1\cdot r^{5-1} = r^4 = \left(\frac{1}{4}\right)^4 = \mathbf{\frac{1}{4^4}} \qquad s_6 = 1\cdot r^{6-1} = r^5 = \left(\frac{1}{4}\right)^5 = \mathbf{\frac{1}{4^5}}$$

$$s_7 = 1\cdot r^{7-1} = r^6 = \left(\frac{1}{4}\right)^6 = \mathbf{\frac{1}{4^6}} \qquad s_8 = 1\cdot r^{8-1} = r^7 = \left(\frac{1}{4}\right)^7 = \mathbf{\frac{1}{4^7}}$$

Thus, the first eight terms of the geometric sequence are $\left(\mathbf{1, \frac{1}{4}, \frac{1}{4^2}, \frac{1}{4^3}, \frac{1}{4^4}, \frac{1}{4^5}, \frac{1}{4^6}, \frac{1}{4^7}}\right)$ and the nth term is equal to

$$s_n = 1\cdot\left(\frac{1}{4}\right)^{n-1} = \frac{1^{n-1}}{4^{n-1}} = \mathbf{\frac{1}{4^{n-1}}}$$

b. Given $-\frac{1}{2}, \frac{1}{4}, \cdots$, then $s_1 = -\frac{1}{2}$ and $r = \frac{\frac{1}{4}}{-\frac{1}{2}} = -\frac{1\cdot 2}{4\cdot 1} = -\frac{2}{4} = -\frac{1}{2}$. Using the general geometric equation $s_n = s_1 r^{n-1}$ the next six terms are:

$s_3 = -\frac{1}{2}\cdot r^{3-1} = -\frac{1}{2}r^2 = -\frac{1}{2}\cdot\left(-\frac{1}{2}\right)^2 = -\mathbf{\frac{1}{2^3}}$

$s_4 = -\frac{1}{2}\cdot r^{4-1} = -\frac{1}{2}r^3 = -\frac{1}{2}\cdot\left(-\frac{1}{2}\right)^3 = \mathbf{\frac{1}{2^4}}$

$s_5 = -\frac{1}{2}\cdot r^{5-1} = -\frac{1}{2}r^4 = -\frac{1}{2}\cdot\left(-\frac{1}{2}\right)^4 = -\mathbf{\frac{1}{2^5}}$

$s_6 = -\frac{1}{2}\cdot r^{6-1} = r^5 = -\frac{1}{2}\cdot\left(-\frac{1}{2}\right)^5 = \mathbf{\frac{1}{2^6}}$

$s_7 = -\frac{1}{2}\cdot r^{7-1} = -\frac{1}{2}r^6 = -\frac{1}{2}\cdot\left(-\frac{1}{2}\right)^6 = -\mathbf{\frac{1}{2^7}}$

$s_8 = -\frac{1}{2}\cdot r^{8-1} = -\frac{1}{2}r^7 = -\frac{1}{2}\left(-\frac{1}{2}\right)^7 = \mathbf{\frac{1}{2^8}}$

Thus, the first eight terms of the geometric sequence are $\left(\mathbf{-\frac{1}{2}, \frac{1}{2^2}, -\frac{1}{2^3}, \frac{1}{2^4}, -\frac{1}{2^5}, \frac{1}{2^6}, -\frac{1}{2^7}, \frac{1}{2^8}}\right)$ and the nth term is equal to $s_n = -\frac{1}{2}\cdot\left(-\frac{1}{2}\right)^{n-1} = -\frac{1}{2}\cdot\frac{(-1)^{n-1}}{2^{n-1}} = -\frac{(-1)^{n-1}}{2\cdot 2^{n-1}} = -\frac{(-1)^{n-1}}{2^{n-1+1}} = -\mathbf{\frac{(-1)^{n-1}}{2^n}}$

c. Given $\frac{1}{3}p, -3p, \cdots$, then $s_1 = \frac{p}{3}$ and $r = \frac{-3p}{\frac{p}{3}} = \frac{\frac{-3p}{1}}{\frac{p}{3}} = \frac{-3p\cdot 3}{1\cdot p} = \frac{-9p}{p} = -9$. Using the general geometric equation $s_n = s_1 r^{n-1}$ the next six terms are:

$s_3 = \frac{p}{3}\cdot r^{3-1} = \frac{p}{3}\cdot r^2 = \frac{p}{3}\cdot(-9)^2 = \frac{p}{3}\cdot 9^2 = \mathbf{3^3 p}$

$s_4 = \frac{p}{3}\cdot r^{4-1} = \frac{p}{3}\cdot r^3 = \frac{p}{3}\cdot(-9)^3 = -\frac{p}{3}\cdot 9^3 = \mathbf{-3^5 p}$

$s_5 = \frac{p}{3}\cdot r^{5-1} = \frac{p}{3}\cdot r^4 = \frac{p}{3}\cdot(-9)^4 = \frac{p}{3}\cdot 9^4 = \mathbf{3^7 p}$

$s_6 = \frac{p}{3}\cdot r^{6-1} = \frac{p}{3}\cdot r^5 = \frac{p}{3}\cdot(-9)^5 = -\frac{p}{3}\cdot 9^5 = \mathbf{-3^9 p}$

$s_7 = \frac{p}{3}\cdot r^{7-1} = \frac{p}{3}\cdot r^6 = \frac{p}{3}\cdot(-9)^6 = \frac{p}{3}\cdot 9^6 = \mathbf{3^{11} p}$

$s_8 = \frac{p}{3}\cdot r^{8-1} = r^7 = \frac{p}{3}\cdot(-9)^7 = -\frac{p}{3}\cdot 9^7 = \mathbf{-3^{13} p}$

Thus, the first eight terms of the geometric sequence are $\left(\mathbf{\frac{1}{3p}, -3p, 3^3 p, -3^5 p, 3^7 p, -3^9 p, 3^{11} p, -3^{13} p}\right)$ and the nth term is equal to $s_n = \frac{p}{3}\cdot(-9)^{n-1} = \frac{p}{3}\cdot(-1)^{n-1}\left(3^2\right)^{n-1} = \frac{p}{3}\cdot(-1)^{n-1}\cdot 3^{2n-2} = p(-1)^{n-1}3^{2n-2-1} = \mathbf{p(-1)^{n-1}3^{2n-3}}$

4. Given the following terms of a geometric sequence, find the common ratio r.

a. Substitute $s_1 = 25$ and $s_4 = \frac{1}{5}$ into $s_n = s_1 r^{n-1}$ and solve for r, i.e., $s_4 = s_1 r^{4-1}$; $\frac{1}{5} = 25r^3$; $\frac{\frac{1}{5}}{25} = r^3$; $\frac{1}{125} = r^3$; $\frac{1}{5^3} = r^3$; $\left(\frac{1}{5^3}\right)^{\frac{1}{3}} = \left(r^3\right)^{\frac{1}{3}}$; $\frac{1}{5^{3\times\frac{1}{3}}} = r^{3\times\frac{1}{3}}$; $\frac{1}{5} = r$; $\mathbf{r = \frac{1}{5}}$

b. Substitute $s_1 = 4$ and $s_5 = \frac{1}{64}$ into $s_n = s_1 r^{n-1}$ and solve for r, i.e., $s_5 = s_1 r^{5-1}$; $\frac{1}{64} = 4r^{5-1}$; $\frac{\frac{1}{64}}{4} = r^4$; $\frac{\frac{1}{64}}{\frac{4}{1}} = r^4$; $\frac{1}{64\times 4} = r^4$; $\frac{1}{256} = r^4$; $\frac{1}{4^4} = r^4$; $\left(\frac{1}{4^4}\right)^{\frac{1}{4}} = \left(r^4\right)^{\frac{1}{4}}$; $\frac{1}{4^{4\times\frac{1}{4}}} = r^{4\times\frac{1}{4}}$; $\frac{1}{4} = r$; $\mathbf{r = \frac{1}{4}}$

c. Substitute $s_1 = 3$ and $s_8 = 1$ into $s_n = s_1 r^{n-1}$ and solve for r, i.e., $s_8 = s_1 r^{8-1}$; $1 = 3r^7$; $\frac{1}{3} = r^7$; $\left(\frac{1}{3}\right)^{\frac{1}{7}} = \left(r^7\right)^{\frac{1}{7}}$; $\frac{1}{3^{\frac{1}{7}}} = r^{7\times\frac{1}{7}}$; $\frac{1}{3^{\frac{1}{7}}} = r$; $\mathbf{r = \frac{1}{\sqrt[7]{3}}}$

5. Write the first five terms of the following geometric sequences.

a. Given $s_n = \left(-\frac{1}{3}\right)^{2n-1}$, then

$s_1 = \left(-\frac{1}{3}\right)^{2\cdot1-1} = \left(-\frac{1}{3}\right)^{2-1} = \mathbf{-\frac{1}{3}}$

$s_2 = \left(-\frac{1}{3}\right)^{2\cdot2-1} = \left(-\frac{1}{3}\right)^{4-1} = \left(-\frac{1}{3}\right)^{3} = \mathbf{-\frac{1}{3^3}}$

$s_3 = \left(-\frac{1}{3}\right)^{2\cdot3-1} = \left(-\frac{1}{3}\right)^{6-1} = \left(-\frac{1}{3}\right)^{5} = \mathbf{-\frac{1}{3^5}}$

$s_4 = \left(-\frac{1}{3}\right)^{2\cdot4-1} = \left(-\frac{1}{3}\right)^{8-1} = \left(-\frac{1}{3}\right)^{7} = \mathbf{-\frac{1}{3^7}}$

$s_5 = \left(-\frac{1}{3}\right)^{2\cdot5-1} = \left(-\frac{1}{3}\right)^{10-1} = \left(-\frac{1}{3}\right)^{9} = \mathbf{-\frac{1}{3^9}}$

Thus, the first five terms of the geometric sequence are $\left(\mathbf{-\frac{1}{3}, -\frac{1}{3^3}, -\frac{1}{3^5}, -\frac{1}{3^7}, -\frac{1}{3^9}}\right)$

b. Given $s_n = \left(\frac{1}{3}\right)^{2n+2}$, then

$s_1 = \left(\frac{1}{3}\right)^{2\cdot1+2} = \left(\frac{1}{3}\right)^{2+2} = \left(\frac{1}{3}\right)^{4} = \mathbf{\frac{1}{3^4}}$

$s_2 = \left(\frac{1}{3}\right)^{2\cdot2+2} = \left(\frac{1}{3}\right)^{4+2} = \left(\frac{1}{3}\right)^{6} = \mathbf{\frac{1}{3^6}}$

$s_3 = \left(\frac{1}{3}\right)^{2\cdot3+2} = \left(\frac{1}{3}\right)^{6+2} = \left(\frac{1}{3}\right)^{8} = \mathbf{\frac{1}{3^8}}$

$s_4 = \left(\frac{1}{3}\right)^{2\cdot4+2} = \left(\frac{1}{3}\right)^{8+2} = \left(\frac{1}{3}\right)^{10} = \mathbf{\frac{1}{3^{10}}}$

$s_5 = \left(\frac{1}{3}\right)^{2\cdot5+2} = \left(\frac{1}{3}\right)^{10+2} = \left(\frac{1}{3}\right)^{12} = \mathbf{\frac{1}{3^{12}}}$

Thus, the first five terms of the geometric sequence are $\left(\mathbf{\frac{1}{3^4}, \frac{1}{3^6}, \frac{1}{3^8}, \frac{1}{3^{10}}, \frac{1}{3^{12}}}\right)$

c. Given $s_n = \left(-\frac{1}{5}\right)^{2n-3}$, then

$s_1 = \left(-\frac{1}{5}\right)^{2\cdot1-3} = \left(-\frac{1}{5}\right)^{2-3} = \left(-\frac{1}{5}\right)^{-1} = \mathbf{-5}$

$s_2 = \left(-\frac{1}{5}\right)^{2\cdot2-3} = \left(-\frac{1}{5}\right)^{4-3} = \mathbf{-\frac{1}{5}}$

$s_3 = \left(-\frac{1}{5}\right)^{2\cdot3-3} = \left(-\frac{1}{5}\right)^{6-3} = \left(-\frac{1}{5}\right)^{3} = \mathbf{-\frac{1}{5^3}}$

$s_4 = \left(-\frac{1}{5}\right)^{2\cdot4-3} = \left(-\frac{1}{5}\right)^{8-3} = \left(-\frac{1}{5}\right)^{5} = \mathbf{-\frac{1}{5^5}}$

$s_5 = \left(-\frac{1}{5}\right)^{2\cdot5-3} = \left(-\frac{1}{5}\right)^{10-3} = \left(-\frac{1}{5}\right)^{7} = \mathbf{-\frac{1}{5^7}}$

Thus, the first five terms of the geometric sequence are $\left(\mathbf{-5, -\frac{1}{5}, -\frac{1}{5^3}, -\frac{1}{5^5}, -\frac{1}{5^7}}\right)$

d. Given $s_n = \left(-\frac{1}{2}\right)^{n}$, then

$s_1 = \left(-\frac{1}{2}\right)^{1} = -\frac{1}{2} = \mathbf{-0.5}$

$s_2 = \left(-\frac{1}{2}\right)^{2} = \frac{1}{2^2} = \frac{1}{4} = \mathbf{0.25}$

$s_3 = \left(-\frac{1}{2}\right)^{3} = -\frac{1}{2^3} = -\frac{1}{8} = \mathbf{-0.125}$

$s_4 = \left(-\frac{1}{2}\right)^{4} = \frac{1}{2^4} = \frac{1}{16} = \mathbf{0.063}$

$s_5 = \left(-\frac{1}{2}\right)^{5} = -\frac{1}{2^5} = -\frac{1}{32} = \mathbf{-0.031}$

Thus, the first five terms of the geometric sequence are $(\mathbf{-0.5, 0.25, -0.125, 0.063, -0.031})$

6. Evaluate the sum of the following geometric series.

a. $\sum_{k=1}^{6} 3^{k-1} = 3^{1-1}+3^{2-1}+3^{3-1}+3^{4-1}+3^{5-1}+3^{6-1} = 3^0+3^1+3^2+3^3+3^4+3^5 = 1+3+9+27+81+243 = \mathbf{364}$

or we can use the geometric series formula $S_n = \frac{s_1\left(1-r^n\right)}{1-r}$ where $s_1 = 1$, $r = \frac{3}{1} = 3$, and $n = 6$. Therefore,

$S_6 = \frac{1\cdot\left(1-3^6\right)}{1-3} = \frac{1-729}{-2} = \frac{728}{2} = \mathbf{364}$

b. $\sum_{k=3}^{10}(-2)^{k-3} = (-2)^{3-3}+(-2)^{4-3}+(-2)^{5-3}+(-2)^{6-3}+(-2)^{7-3}+(-2)^{8-3}+(-2)^{9-3}+(-2)^{10-3} = (-2)^0+(-2)^1+(-2)^2$
$+(-2)^3+(-2)^4+(-2)^5+(-2)^6+(-2)^7 = 1-2+4-8+16-32+64-128 = \mathbf{-85}$

or we can use the geometric series formula $S_n = \frac{s_1(1-r^n)}{1-r}$ where $s_1 = 1$, $r = \frac{-2}{1} = -2$, and $n = 8$. Therefore,

$$S_8 = \frac{1\cdot\left[1-(-2)^8\right]}{1-(-2)} = \frac{1-256}{1+2} = -\frac{255}{3} = \mathbf{-85}$$

c. $\sum_{j=4}^{8} 4\left(-\frac{1}{2}\right)^{j+1} = 4\sum_{j=4}^{8}\left(-\frac{1}{2}\right)^{j+1} = 4\left[\left(-\frac{1}{2}\right)^{4+1} + \left(-\frac{1}{2}\right)^{5+1} + \left(-\frac{1}{2}\right)^{6+1} + \left(-\frac{1}{2}\right)^{7+1} + \left(-\frac{1}{2}\right)^{8+1}\right] = 4\left[\left(-\frac{1}{2}\right)^{5} + \left(-\frac{1}{2}\right)^{6}\right.$

$\left. + 4\left[\left(-\frac{1}{2}\right)^{7} + \left(-\frac{1}{2}\right)^{8} + \left(-\frac{1}{2}\right)^{9}\right] = 4(-0.03+0.012-0.008+0.004-0.002) = 4(-0.024) = \mathbf{-0.096}$

or we can use the geometric series formula $S_n = \frac{s_1(1-r^n)}{1-r}$ where $s_1 = -0.12$, $r = \frac{0.012}{-0.03} = -0.4$, and $n = 5$. Therefore,

$$S_5 = \frac{-0.12\cdot\left[1-(-0.4)^5\right]}{1-(-0.4)} = \frac{-0.12\cdot(1+0.0102)}{1+0.4} = -\frac{0.1212}{1.4} = \mathbf{-0.09}$$

d. $\sum_{m=1}^{4}(-2)^{m-3} = (-2)^{1-3} + (-2)^{2-3} + (-2)^{3-3} + (-2)^{4-3} = (-2)^{-2} + (-2)^{-1} + (-2)^0 + (-2)^1 = \frac{1}{(-2)^2} + \frac{1}{-2} + 1 - 2$

$= \frac{1}{4} - \frac{1}{2} + 1 - 2 = 0.25 - 0.5 - 1 = \mathbf{-1.25}$

or we can use the geometric series formula $S_n = \frac{s_1(1-r^n)}{1-r}$ where $s_1 = 0.25$, $r = \frac{-0.5}{0.25} = -2$, and $n = 4$. Therefore,

$$S_4 = \frac{0.25\cdot\left[1-(-2)^4\right]}{1-(-2)} = \frac{0.25\cdot(1-16)}{1+2} = -\frac{3.75}{3} = \mathbf{-1.25}$$

e. $\sum_{n=5}^{10}(-3)^{n-4} = (-3)^{5-4} + (-3)^{6-4} + (-3)^{7-4} + (-3)^{8-4} + (-3)^{9-4} + (-3)^{10-4} = (-3)^1 + (-3)^2 + (-3)^3 + (-3)^4 + (-3)^5 + (-3)^6$

$= -3 + 9 - 27 + 81 - 243 + 729 = \mathbf{546}$

or we can use the geometric series formula $S_n = \frac{s_1(1-r^n)}{1-r}$ where $s_1 = -3$, $r = \frac{9}{-3} = -3$, and $n = 6$. Therefore,

$$S_6 = \frac{-3\cdot\left[1-(-3)^6\right]}{1-(-3)} = \frac{-3\cdot(1-729)}{1+3} = \frac{2184}{4} = \mathbf{546}$$

f. $\sum_{k=1}^{5}(-3)^{k-1} = (-3)^{1-1} + (-3)^{2-1} + (-3)^{3-1} + (-3)^{4-1} + (-3)^{5-1} = 1 + (-3)^1 + (-3)^2 + (-3)^3 + (-3)^4 = 1 - 3 + 9 - 27 + 81 = \mathbf{61}$

or we can use the geometric series formula $S_n = \frac{s_1(1-r^n)}{1-r}$ where $s_1 = 1$, $r = \frac{-3}{1} = -3$, and $n = 5$. Therefore,

$$S_5 = \frac{1\cdot\left[1-(-3)^5\right]}{1-(-3)} = \frac{1+243}{1+3} = \frac{244}{4} = \mathbf{61}$$

g. $\sum_{m=1}^{5}4^m = 4^1 + 4^2 + 4^3 + 4^4 + 4^5 = 4 + 16 + 64 + 256 + 1024 = \mathbf{1364}$

or we can use the geometric series formula $S_n = \frac{s_1(1-r^n)}{1-r}$ where $s_1 = 4$, $r = \frac{16}{4} = 4$, and $n = 5$. Therefore,

$$S_5 = \frac{4\cdot\left(1-4^5\right)}{1-4} = \frac{4\cdot(1-1024)}{-3} = \frac{4092}{3} = \mathbf{1364}$$

h. $\sum_{j=1}^{4}\frac{3^j}{27} = \frac{1}{27}\sum_{j=1}^{4}3^j = \frac{1}{27}\left(3^1 + 3^2 + 3^3 + 3^4\right) = \frac{1}{27}(3+9+27+81) = \frac{120}{27} = \mathbf{4.44}$

or we can use the geometric series formula $S_n = \frac{s_1(1-r^n)}{1-r}$ where, $s_1 = \frac{3}{27} = 0.111$, $r = \frac{9}{3} = 3$, and $n = 4$. Therefore,

$$S_4 = \frac{0.111\cdot(1-3^4)}{1-3} = \frac{0.111\cdot(1-81)}{-2} = \frac{8.88}{2} = \mathbf{4.44}$$

i. $\sum_{k=3}^{6} 6\left(\frac{1}{2}\right)^{k+1} = 6\sum_{k=3}^{6} 0.5^{k+1} = 6\left[0.5^{3+1}+0.5^{4+1}+0.5^{5+1}+0.5^{6+1}\right] = 6\left[0.5^4+0.5^5+0.5^6+0.5^7\right] = 6\left[0.063+0.031\right]$
$+6\left[0.016+0.008\right] = 6\,(0.118) = 0.708 \approx \mathbf{0.7}$

or we can use the geometric series formula $S_n = \frac{s_1(1-r^n)}{1-r}$ where, $s_1 = 6\cdot 0.5^4 = 0.375$, $r = \frac{0.031}{0.063} = 0.5$, and $n = 4$. Thus,

$$S_4 = \frac{0.375\cdot(1-0.5^4)}{1-0.5} = \frac{0.375\cdot(1-0.063)}{0.5} = \frac{0.351}{0.5} = 0.702 \approx \mathbf{0.7}$$

7. Given the first term s_1 and r, find S_8 for each of the following geometric sequences.

a. Given $s_1 = 3$ and $r = 3$ use the geometric series formula $S_n = \frac{s_1(1-r^n)}{1-r}$ to find S_8, i.e.,

$$S_8 = \frac{3\cdot(1-3^8)}{1-3} = \frac{3\cdot(1-6561)}{-2} = \frac{19680}{2} = \mathbf{9840}$$

b. Given $s_1 = -8$ and $r = 0.5$ use the geometric series formula $S_n = \frac{s_1(1-r^n)}{1-r}$ to find S_8, i.e.,

$$S_8 = \frac{-8\cdot(1-0.5^8)}{1-0.5} = \frac{-8\cdot(1-0.0039)}{0.5} = -\frac{7.968}{0.5} = \mathbf{-15.94}$$

c. Given $s_1 = 2$ and $r = -2.5$ use the geometric series formula $S_n = \frac{s_1(1-r^n)}{1-r}$ to find S_8, i.e.,

$$S_8 = \frac{2\cdot\left[1-(-2.5)^8\right]}{1-(-2.5)} = \frac{2\cdot\left[1-1525.88\right]}{1+2.5} = -\frac{2\cdot 1524.88}{3.5} = -\frac{3049.76}{3.5} = \mathbf{-871.36}$$

8. Solve for x and y.

a. Given $\sum_{i=3}^{7}(ix+2) = 30$, then $(3x+2)+(4x+2)+(5x+2)+(6x+2)+(7x+2) = 30$; $(3x+4x+5x+6x+7x)+10 = 30$; $25x+10 = 30$; $25x = 30-10$; $x = \frac{20}{25}$; $\mathbf{x = 0.8}$

b. Expanding $\sum_{i=1}^{4}(ix+y) = 20$ we obtain $(x+y)+(2x+y)+(3x+y)+(4x+y) = 20$; $10x+4y = 20$.

Expanding $\sum_{i=2}^{6}(ix+y) = 10$ we obtain $(2x+y)+(3x+y)+(4x+y)+(5x+y)+(6x+y) = 10$; $20x+5y = 10$. Using substitution method we obtain $\mathbf{x = -2}$ and $\mathbf{y = 10}$

Section 4.5 Solutions - Limits of Sequences and Series

1. State which of the following sequences are convergent.

To see if a sequence is *convergent* or *divergent* consider the nth term of the sequence and let it approach infinity.

a. $\lim_{n\to\infty}\frac{n+1}{2} \approx \lim_{n\to\infty}\frac{n}{2} = \frac{\infty}{2} = \infty$ **The sequence diverges**

b. $\lim_{n\to\infty}\frac{n^2-1}{n} \approx \lim_{n\to\infty}\frac{n^2}{n} = \lim_{n\to\infty}\frac{n^{2-1}}{\not n} = \lim_{n\to\infty} n = \infty$ **The sequence diverges**

c. $\lim_{n\to\infty}2^{n+1} \approx \lim_{n\to\infty}2^n = 2^\infty = \infty$ **The sequence diverges**

d. $\lim_{n\to\infty}\dfrac{1}{4^{n+1}} \approx \lim_{n\to\infty}\dfrac{1}{4^{n}} = \dfrac{1}{4^{\infty}} = \dfrac{1}{\infty} = \mathbf{0}$ **The sequence converges**

e. $\lim_{n\to\infty}\dfrac{n-1}{n^2} \approx \lim_{n\to\infty}\dfrac{n}{n^2} = \lim_{n\to\infty}\dfrac{\cancel{n}}{n^{\cancel{2}1}} = \lim_{n\to\infty}\dfrac{1}{n} = \dfrac{1}{\infty} = \mathbf{0}$ **The sequence converges**

f. $\lim_{n\to\infty}\left(\dfrac{1}{5}\right)^{n+1} \approx \lim_{n\to\infty}\left(\dfrac{1}{5}\right)^{n} = \lim_{n\to\infty}0.2^{n} = 0.2^{\infty} = \mathbf{0}$ **The sequence converges**

2. State which of the following geometric sequences are convergent.

   a. The sequence $\dfrac{1}{4}, \dfrac{1}{16}, \dfrac{1}{64}, \dfrac{1}{256}, \cdots, \dfrac{1}{4^n}, \cdots$ **converges to 0** since, for large values of n, the absolute value of the difference between $\dfrac{1}{4^n}$ and 0 is very small.

   b. The sequence $-5, 25, -125, 625, -3125, \cdots, (-5)^n, \cdots$ **diverges** since, as n increases, the nth term increases without bound.

   c. The sequence $2, -2, 2, -2, \cdots, 2(-1)^{n+1}, \cdots$ **diverges** since, as n increases, the nth term oscillates back and forth between $+2$ and -2.

   d. The sequence $1, \dfrac{1}{2}, \dfrac{1}{4}, \dfrac{1}{8}, \cdots, \left(\dfrac{1}{2}\right)^{n-1}, \cdots$ **converges to 0** since, for large values of n, the absolute value of the difference between $\left(\dfrac{1}{2}\right)^{n-1}$ and 0 is very small.

   e. The sequence $-9, 27, -81, 243, \cdots, (-1)^n 3^{n+1}, \cdots$ **diverges** since, as n increases, the nth term oscillates back and forth from a large positive number to a large negative number.

   f. The sequence $1, \dfrac{1}{3}, \dfrac{1}{9}, \dfrac{1}{27}, \dfrac{1}{81}, \cdots, \left(\dfrac{1}{3}\right)^{n-1}, \cdots$ **converges to 0** since, for large values of n, the absolute value of the difference between $\left(\dfrac{1}{3}\right)^{n-1}$ and 0 is very small.

   Again note that an easier way of knowing if a sequence is convergent or divergent is by considering the nth term and letting it approach to infinity as shown in practice problems 4.5-1 and 4.5-3.

3. State whether or not the following sequences converges or diverges as $n \to \infty$. If it does converge, find the limit.

   a. $\lim_{n\to\infty}\dfrac{n^2}{n^3-4} \approx \lim_{n\to\infty}\dfrac{n^2}{n^3} = \lim_{n\to\infty}\dfrac{n^{\cancel{2}}}{n^{\cancel{3}1}} = \lim_{n\to\infty}\dfrac{1}{n} = \dfrac{1}{\infty} = \mathbf{0}$ **converges to 0**

   b. $\lim_{n\to\infty}\dfrac{5n+1}{\sqrt{n^2+1}} \approx \lim_{n\to\infty}\dfrac{5n}{\sqrt{n^2}} = \lim_{n\to\infty}\dfrac{5n}{n} = \lim_{n\to\infty}\dfrac{5\cancel{n}}{\cancel{n}} = \lim_{n\to\infty}5 = \mathbf{5}$ **converges to 5**

   c. $\lim_{n\to\infty}\dfrac{25^n}{5^{n+1}} \approx \lim_{n\to\infty}\dfrac{5^{2n}}{5^n} = \lim_{n\to\infty}\dfrac{5^{2n}\cdot 5^{-n}}{1} = \lim_{n\to\infty}5^{2n-n} = \lim_{n\to\infty}5^n = 5^{\infty} = \boldsymbol{\infty}$ **diverges**

   d. $\lim_{n\to\infty}\dfrac{5^n+25}{125^n} \approx \lim_{n\to\infty}\dfrac{5^n}{5^{3n}} = \lim_{n\to\infty}\dfrac{1}{5^{3n}\cdot 5^{-n}} = \lim_{n\to\infty}\dfrac{1}{5^{3n-n}} = \lim_{n\to\infty}\dfrac{1}{5^{2n}} = \dfrac{1}{5^{2\cdot\infty}} = \dfrac{1}{5^{\infty}} = \dfrac{1}{\infty} = \mathbf{0}$ **converges to 0**

   e. $\lim_{n\to\infty}\dfrac{(n+2)^2}{n^2} \approx \lim_{n\to\infty}\dfrac{n^2}{n^2} = \lim_{n\to\infty}\dfrac{n^{\cancel{2}}}{n^{\cancel{2}}} = \lim_{n\to\infty}1 = \mathbf{1}$ **converges to 1**

   f. $\lim_{n\to\infty}\dfrac{2^n}{2^n+1} \approx \lim_{n\to\infty}\dfrac{2^n}{2^n} = \lim_{n\to\infty}\dfrac{2^{\cancel{n}}}{2^{\cancel{n}}} = \lim_{n\to\infty}1 = \mathbf{1}$ **converges to 1**

g. $\lim_{n\to\infty}\frac{\sqrt{n^2+2n}}{\sqrt{n^4+1}} \approx \lim_{n\to\infty}\frac{\sqrt{n^2}}{\sqrt{n^4}} = \lim_{n\to\infty}\frac{n}{n^2} = \lim_{n\to\infty}\frac{\not{n}}{n^{\not{2}}} = \lim_{n\to\infty}\frac{1}{n} = \frac{1}{\infty} = \mathbf{0}$ **converges to 0**

h. $\lim_{n\to\infty}\frac{5}{n^2+1} \approx \lim_{n\to\infty}\frac{5}{n^2} = \frac{5}{\infty^2} = \frac{5}{\infty} = \mathbf{0}$ **converges to 0**

i. $\lim_{n\to\infty}\frac{n+1}{n-1} \approx \lim_{n\to\infty}\frac{n}{n} = \lim_{n\to\infty}\frac{\not{n}}{\not{n}} = \lim_{n\to\infty}1 = \mathbf{1}$ **converges to 1**

j. $\lim_{n\to\infty}\frac{n}{n^3-1} \approx \lim_{n\to\infty}\frac{n}{n^3} = \lim_{n\to\infty}\frac{\not{n}}{n^{\not{3}=2}} = \lim_{n\to\infty}\frac{1}{n^2} = \frac{1}{\infty^2} = \frac{1}{\infty} = \mathbf{0}$ **converges to 0**

k. $\lim_{n\to\infty}10^{\frac{1}{n}} = 10^{\frac{1}{\infty}} = 10^0 = \mathbf{1}$ **converges to 1**

l. $\lim_{n\to\infty}\frac{(n-1)^2}{(1-n)(1+n)} \approx \lim_{n\to\infty}\frac{n^2}{-n\cdot n} = \lim_{n\to\infty}-\frac{n^2}{n^2} = \lim_{n\to\infty}-\frac{n^{\not{2}}}{n^{\not{2}}} = \lim_{n\to\infty}-1 = \mathbf{-1}$ **converges to –1**

m. $\lim_{n\to\infty}100^{-\frac{1}{n}} = \lim_{n\to\infty}\frac{1}{100^{\frac{1}{n}}} = \frac{1}{100^{\frac{1}{\infty}}} = \frac{1}{100^0} = \frac{1}{1} = \mathbf{1}$ **converges to 1**

n. $\lim_{n\to\infty}3^{\frac{3}{n}} = 3^{\frac{3}{\infty}} = 3^0 = \mathbf{1}$ **converges to 1**

o. $\lim_{n\to\infty}\frac{n+100}{n^3-10} \approx \lim_{n\to\infty}\frac{n}{n^3} = \lim_{n\to\infty}\frac{\not{n}}{n^{\not{3}=2}} = \lim_{n\to\infty}\frac{1}{n^2} = \frac{1}{\infty^2} = \frac{1}{\infty} = \mathbf{0}$ **converges to 0**

p. $\lim_{n\to\infty}\frac{100^{\frac{1}{n}}}{n^2+3} \approx \lim_{n\to\infty}\frac{100^{\frac{1}{n}}}{n^2} = \frac{100^{\frac{1}{\infty}}}{\infty^2} = \frac{100^0}{\infty} = \frac{1}{\infty} = \mathbf{0}$ **converges to 0**

q. $\lim_{n\to\infty}\frac{1}{n+1}-1 \approx \lim_{n\to\infty}\frac{1}{n}-1 = \frac{1}{\infty}-1 = 0-1 = \mathbf{-1}$ **converges to –1**

r. $\lim_{n\to\infty}(0.25)^{-n} = \lim_{n\to\infty}\frac{1}{0.25^n} = \frac{1}{0.25^\infty} = \frac{1}{0} = \infty$ **diverges**

s. $\lim_{n\to\infty}\frac{\sqrt{n+1}}{\sqrt{n}+1} \approx \lim_{n\to\infty}\frac{\sqrt{n}}{\sqrt{n}} = \lim_{n\to\infty}\frac{\sqrt{\not{n}}}{\sqrt{\not{n}}} = \lim_{n\to\infty}1 = \mathbf{1}$ **converges to 1**

t. $\lim_{n\to\infty}\frac{\sqrt{n}}{\sqrt{n+1}}+2 \approx \lim_{n\to\infty}\frac{\sqrt{n}}{\sqrt{n}}+2 = \lim_{n\to\infty}\frac{\sqrt{\not{n}}}{\sqrt{\not{n}}}+2 = \lim_{n\to\infty}(1+2) = 1+2 = \mathbf{3}$ **converges to 3**

4. Find the sum of the following geometric series.

a. Given $\sum_{j=0}^{\infty}3\left(\frac{1}{8}\right)^j$, then $s_1 = 3$ and $r = \frac{1}{8}$. Since $|r| \langle 1$ we can use the equation $S_\infty = \sum_{n=0}^{\infty}s_1r^n = \sum_{n=1}^{\infty}s_1r^{n-1} = \frac{s_1}{1-r}$

to obtain the sum, i.e., $\sum_{j=0}^{\infty}3\left(\frac{1}{8}\right)^j = \frac{3}{1-\frac{1}{8}} = \frac{3}{\frac{8-1}{8}} = \frac{3}{\frac{7}{8}} = \frac{\frac{3}{1}}{\frac{7}{8}} = \frac{3\times 8}{1\times 7} = \mathbf{\frac{24}{7}}$

b. Given $\sum_{j=0}^{\infty}3\left(-\frac{1}{4}\right)^j$, then $s_1 = 3$ and $r = -\frac{1}{4}$. Since $|r| \langle 1$ we can use the equation $S_\infty = \sum_{n=0}^{\infty}s_1r^n = \sum_{n=1}^{\infty}s_1r^{n-1} = \frac{s_1}{1-r}$

to obtain the sum, i.e., $\sum_{j=0}^{\infty}3\left(-\frac{1}{4}\right)^j = \frac{3}{1-\left(-\frac{1}{4}\right)} = \frac{3}{1+\frac{1}{4}} = \frac{3}{\frac{4+1}{4}} = \frac{3}{\frac{5}{4}} = \frac{\frac{3}{1}}{\frac{5}{4}} = \frac{3\times 4}{1\times 5} = \mathbf{\frac{12}{5}}$

c. Given $\sum_{k=1}^{\infty}3\left(\frac{3}{2}\right)^{k-1}$, then $s_1 = 3$ and $r = \frac{3}{2}$. Since $|r| \rangle 1$ the geometric series $\sum_{k=1}^{\infty}3\left(\frac{3}{2}\right)^{k-1}$ **has no finite sum.**

d. Given $\sum_{k=1}^{\infty}\frac{5}{100^{k+1}} = \sum_{k=1}^{\infty}\frac{5}{100^2 \cdot 100^{k-1}} = \sum_{k=1}^{\infty}\frac{5}{10000}\left(\frac{1}{100^{k-1}}\right) = \sum_{k=1}^{\infty}\frac{5}{10000}\left(\frac{1}{100}\right)^{k-1}$, then $s_1 = \frac{5}{10000}$ and $r = \frac{1}{100}$.

Since $|r| \langle 1$ we can use the equation $S_\infty = \sum_{n=0}^{\infty} s_1 r^n = \sum_{n=1}^{\infty} s_1 r^{n-1} = \frac{s_1}{1-r}$ to obtain the sum, i.e., $\sum_{k=1}^{\infty}\frac{5}{10000}\left(\frac{1}{100}\right)^{k-1}$

$$= \frac{\frac{5}{10000}}{1-\frac{1}{100}} = \frac{\frac{5}{10000}}{\frac{100-1}{100}} = \frac{\frac{5}{10000}}{\frac{99}{100}} = \frac{5\times 100}{10000\times 99} = \frac{500}{990000} = \frac{5}{9900} = \mathbf{\frac{1}{1980}}$$

e. Give $\sum_{j=0}^{\infty}\left(\frac{1}{3}\right)^j$, then $s_1 = 1$ and $r = \frac{1}{3}$. Since $|r| \langle 1$ we can use the equation $S_\infty = \sum_{n=0}^{\infty} s_1 r^n = \sum_{n=1}^{\infty} s_1 r^{n-1} = \frac{s_1}{1-r}$

to obtain the sum, i.e., $\sum_{j=0}^{\infty}\left(\frac{1}{3}\right)^j = \frac{1}{1-\frac{1}{3}} = \frac{1}{\frac{3-1}{3}} = \frac{1}{\frac{2}{3}} = \frac{\frac{1}{1}}{\frac{2}{3}} = \frac{1\times 3}{1\times 2} = \mathbf{\frac{3}{2}}$

f. Given $\sum_{j=0}^{\infty}\left(-\frac{1}{5}\right)^j$, then $s_1 = 1$ and $r = -\frac{1}{5}$. Since $|r| \langle 1$ we can use the equation $S_\infty = \sum_{n=0}^{\infty} s_1 r^n = \sum_{n=1}^{\infty} s_1 r^{n-1} = \frac{s_1}{1-r}$

to obtain the sum, i.e., $\sum_{j=0}^{\infty}\left(-\frac{1}{5}\right)^j = \frac{1}{1-\left(-\frac{1}{5}\right)} = \frac{1}{1+\frac{1}{5}} = \frac{1}{\frac{5+1}{5}} = \frac{1}{\frac{6}{5}} = \frac{\frac{1}{1}}{\frac{6}{5}} = \frac{1\times 5}{1\times 6} = \mathbf{\frac{5}{6}}$

5. Find the sum of the following infinite geometric series.

a. Given the series $5-1+\frac{1}{5}-\frac{1}{25}+\cdots$, $s_1 = 5$ and $r = -\frac{1}{5}$. Since $|r| = \left|-\frac{1}{5}\right| = \frac{1}{5} = 0.2 \langle 1$ we can use the equation

$S_\infty = \frac{s_1}{1-r}$ to obtain the sum, i.e., $5-1+\frac{1}{5}-\frac{1}{25}+\cdots = \frac{5}{1-\left(-\frac{1}{5}\right)} = \frac{5}{1+\frac{1}{5}} = \frac{5}{\frac{5+1}{5}} = \frac{5}{\frac{6}{5}} = \frac{\frac{5}{1}}{\frac{6}{5}} = \frac{5\times 5}{1\times 6} = \mathbf{\frac{25}{6}}$

b. Given the series $-\frac{1}{2}+2-8+32+\cdots$ $s_1 = -\frac{1}{2}$ and $r = \frac{2}{-\frac{1}{2}} = -4$. Since $|r| = |-4| = 4$ is greater than one the geometric

series $-\frac{1}{2}+2-8+32+\cdots$ **has no finite solution.**

c. Given the series $1+\frac{1}{6}+\frac{1}{36}+\frac{1}{216}+\cdots$ $s_1 = 1$ and $r = \frac{\frac{1}{6}}{1} = \frac{1}{6}$. Since $|r| = \left|\frac{1}{6}\right| = \frac{1}{6} = 0.17 \langle 1$ we can use the equation

$S_\infty = \frac{s_1}{1-r}$ to obtain the sum, i.e., $1+\frac{1}{6}+\frac{1}{36}+\frac{1}{216}+\cdots = \frac{1}{1-\frac{1}{6}} = \frac{1}{\frac{6-1}{6}} = \frac{1}{\frac{5}{6}} = \frac{\frac{1}{1}}{\frac{5}{6}} = \frac{1\times 6}{1\times 5} = \mathbf{\frac{6}{5}}$

d. Given the series $1+\frac{1}{10}+\frac{1}{100}+\frac{1}{1000}+\cdots$ $s_1 = 1$ and $r = \frac{\frac{1}{10}}{1} = \frac{1}{10}$. Since $|r| = \left|\frac{1}{10}\right| = \frac{1}{10} = 0.1 \langle 1$ we can use the equation

$S_\infty = \frac{s_1}{1-r}$ to obtain the sum, i.e., $1+\frac{1}{10}+\frac{1}{100}+\frac{1}{1000}+\cdots = \frac{1}{1-\frac{1}{10}} = \frac{1}{\frac{10-1}{10}} = \frac{1}{\frac{9}{10}} = \frac{\frac{1}{1}}{\frac{9}{10}} = \frac{1\times 10}{1\times 9} = \mathbf{\frac{10}{9}}$

6. Write the following repeating decimals as the quotient of two positive integers.

a. Given $0.66\overline{6666}\cdots = 0.66+0.0066+0.000066$, which is a gemetric series, then $s_1 = 0.66$ and $r = \frac{0.0066}{0.66} = 0.01$. Since th ratio r is less than one, we can use the infinite geometric series equation $s_\infty = \frac{s_1}{1-r}$ to obtain the sum of the infinite series $0.66+0.0066+0.000066$, i.e., $s_\infty = \frac{s_1}{1-r} = \frac{0.66}{1-0.01} = \frac{0.66}{0.99} = \frac{66}{99} = \frac{22}{33}$. Thus, $\mathbf{0.6666\overline{66}\cdots = \frac{22}{33}}$

b. Given $3.027027\overline{027}\cdots$, consider the decimal portion of the number $3.027027\overline{027}\cdots$ and write it in its equivalent form of $0.027027\overline{027}\cdots = 0.027+0.000027+0.000000027\cdots$. Since this is a geometric series, then $s_1 = 0.027$ and $r = \frac{0.000027}{0.027} = 0.001$. Since the ratio r is less than one, we can use the infinite geometric series equation $s_\infty = \frac{s_1}{1-r}$ to obtain the sum of the infinite series $0.027+0.000027+0.000000027\cdots$, i.e., $s_\infty = \frac{s_1}{1-r} = \frac{0.027}{1-0.001} = \frac{0.027}{0.999} = \frac{27}{999} = \frac{3}{111}$. Thus, $\mathbf{3.027027\overline{027}\cdots = 3\frac{3}{111}}$

c. $0.111\overline{111}\cdots = 0.11+0.0011+0.000011$, which is a gemetric series, then $s_1 = 0.11$ and $r = \frac{0.0011}{0.11} = 0.01$. Since the ratio r is less than one, we can use the infinite geometric series equation $s_\infty = \frac{s_1}{1-r}$ to obtain the sum of the infinite series $0.11+0.0011+0.000011$, i.e., $s_\infty = \frac{s_1}{1-r} = \frac{0.11}{1-0.01} = \frac{0.11}{0.99} = \frac{11}{99} = \frac{1}{9}$. Thus, $\mathbf{0.1111\overline{11}\cdots = \frac{1}{9}}$

Section 4.6 Solutions - The Factorial Notation

1. Expand and simplify the following factorial expressions.

a. $11! = 11\cdot10\cdot9\cdot8\cdot7\cdot6\cdot5\cdot4\cdot3\cdot2\cdot1 = \mathbf{39{,}916{,}800}$

b. $(10-3)! = 7! = 7\cdot6\cdot5\cdot4\cdot3\cdot2\cdot1 = \mathbf{5040}$

c. $\frac{12!}{5!} = \frac{12\cdot11\cdot10\cdot9\cdot8\cdot7\cdot6\cdot5!}{5!} = \frac{12\cdot11\cdot10\cdot9\cdot8\cdot7\cdot6\cdot\cancel{5!}}{\cancel{5!}} = \frac{12\cdot11\cdot10\cdot9\cdot8\cdot7\cdot6}{1} = \mathbf{3{,}991{,}680}$

d. $\frac{14!}{10!} = \frac{14\cdot13\cdot12\cdot11\cdot10!}{10!} = \frac{14\cdot13\cdot12\cdot11\cdot\cancel{10!}}{\cancel{10!}} = \frac{14\cdot13\cdot12\cdot11}{1} = \mathbf{24{,}024}$

e. $\frac{15!}{8!4!} = \frac{15\cdot14\cdot13\cdot12\cdot11\cdot10\cdot9\cdot8!}{8!4!} = \frac{15\cdot14\cdot13\cdot12\cdot11\cdot10\cdot9\cdot\cancel{8!}}{\cancel{8!}\cdot4\cdot3\cdot2\cdot1} = \frac{15\cdot14\cdot13\cdot\overset{3}{\cancel{12}}\cdot11\cdot\overset{5}{\cancel{10}}\cdot\overset{3}{\cancel{9}}}{\cancel{4}\cdot\cancel{3}\cdot\cancel{2}\cdot1} = \frac{15\cdot14\cdot13\cdot3\cdot11\cdot5\cdot3}{1} = \mathbf{1{,}351{,}350}$

f. $\frac{10!}{4!(10-2)!} = \frac{10!}{4!8!} = \frac{10\cdot9\cdot8!}{4!8!} = \frac{10\cdot9\cdot\cancel{8!}}{4!\cancel{8!}} = \frac{10\cdot9}{4\cdot3\cdot2\cdot1} = \frac{\overset{5}{\cancel{10}}\cdot\overset{3}{\cancel{9}}}{4\cdot\cancel{3}\cdot\cancel{2}\cdot1} = \frac{5\cdot3}{4} = \mathbf{\frac{15}{4}}$

g. $\frac{12!\,6!}{14!} = \frac{12!\,6!}{14\cdot13\cdot12!} = \frac{\cancel{12!}\,6!}{14\cdot13\cdot\cancel{12!}} = \frac{6!}{14\cdot13} = \frac{6\cdot5\cdot4\cdot3\cdot\cancel{2}\cdot1}{\underset{7}{\cancel{14}}\cdot13} = \frac{6\cdot5\cdot4\cdot3}{7\cdot13} = \mathbf{\frac{360}{91}}$

h. $\frac{(7-3)!\,9!}{12!(7-2)!} = \frac{4!\,9!}{12!\,5!} = \frac{4!\,9!}{12\cdot11\cdot10\cdot9!\cdot5\cdot4!} = \frac{\cancel{4!}\,\cancel{9!}}{12\cdot11\cdot10\cdot\cancel{9!}\cdot5\cdot\cancel{4!}} = \frac{1}{12\cdot11\cdot10\cdot5} = \mathbf{\frac{1}{6600}}$

2. Write the following products in factorial form.

a. $7\cdot6\cdot5\cdot4\cdot3\cdot2\cdot1 = \mathbf{7!}$

b. $10\cdot11\cdot12\cdot13\cdot14\cdot15 = \mathbf{\frac{15!}{9!}}$

c. $22\cdot23\cdot24\cdot25 = \mathbf{\frac{25!}{21!}}$

d. $8\cdot7\cdot6\cdot5\cdot4 = \mathbf{\frac{8!}{3!}}$

e. $4\cdot5\cdot6\cdot7\cdot8\cdot9 = \mathbf{\frac{9!}{3!}}$

f. $35 = \mathbf{\frac{35!}{34!}}$

3. Expand the following factorial expressions.

a. $5(n!) = \mathbf{5[n(n-1)(n-2)(n-3)(n-4)(n-5)(n-6)\cdots4\cdot3\cdot2\cdot1]}$

b. $(n-7)! = \mathbf{(n-7)(n-8)(n-9)(n-10)(n-11)(n-12)\cdots 4\cdot 3\cdot 2\cdot 1}$

c. $(n+10)! = \mathbf{(n+10)(n+9)(n+8)(n+7)(n+6)(n+5)(n+4)(n+3)(n+2)(n+1)\,n\,(n-1)(n-2)\cdots 4\cdot 3\cdot 2\cdot 1}$

d. $(5n-5)! = \mathbf{(5n-5)(5n-6)(5n-7)(5n-8)(5n-9)(5n-10)\cdots 4\cdot 3\cdot 2\cdot 1}$

e. $(2n-8)! = \mathbf{(2n-8)(2n-9)(2n-10)(2n-11)(2n-12)(2n-13)\cdots 4\cdot 3\cdot 2\cdot 1}$

f. $(2n+6)! = \mathbf{(2n+6)(2n+5)(2n+4)(2n+3)(2n+2)(2n+1)\,2n\,(2n-1)(2n-2)(2n-3)\cdots 4\cdot 3\cdot 2\cdot 1}$

g. $(2n-5)! = \mathbf{(2n-5)(2n-6)(2n-7)(2n-8)(2n-9)(2n-10)\cdots 4\cdot 3\cdot 2\cdot 1}$

h. $(3n+3)! = \mathbf{(3n+3)(3n+2)(3n+1)\,3n\,(3n-1)(3n-2)(3n-3)\cdots 4\cdot 3\cdot 2\cdot 1}$

4. Expand and simplify the following factorial expressions.

a. $\frac{(n-2)!}{(n-4)!} = \frac{(n-2)!}{(n-4)(n-3)(n-2)!} = \frac{\cancel{(n-2)!}}{(n-4)(n-3)\cancel{(n-2)!}} = \mathbf{\frac{1}{(n-4)(n-3)}}$

b. $\frac{(n+4)!}{n!} = \frac{(n+4)(n+3)(n+2)(n+1)n!}{n!} = \frac{(n+4)(n+3)(n+2)(n+1)\cancel{n!}}{\cancel{n!}} = \mathbf{(n+4)(n+3)(n+2)(n+1)}$

c. $\frac{(n+5)!}{(n-2)!} = \frac{(n+5)(n+4)(n+3)(n+2)(n+1)(n)(n-1)(n-2)!}{(n-2)!} = \frac{(n+5)(n+4)(n+3)(n+2)(n+1)(n)(n-1)\cancel{(n-2)!}}{\cancel{(n-2)!}}$

$= \mathbf{(n+5)(n+4)(n+3)(n+2)(n+1)(n)(n-1)}$

d. $\frac{(n-1)(n+1)!}{(n+2)!} = \frac{(n-1)(n+1)!}{(n+2)(n+1)!} = \frac{(n-1)\cancel{(n+1)!}}{(n+2)\cancel{(n+1)!}} = \mathbf{\frac{n-1}{n+2}}$

e. $\frac{(3n)!\,(3n-2)!}{(3n+1)!\,(3n-4)!} = \frac{(3n)!\,(3n-2)!}{(3n+1)(3n)!\,(3n-4)(3n-3)(3n-2)!} = \frac{\cancel{(3n)!}\,\cancel{(3n-2)!}}{(3n+1)\cancel{(3n)!}\,(3n-4)(3n-3)\cancel{(3n-2)!}} = \mathbf{\frac{1}{(3n+1)(3n-4)(3n-3)}}$

f. $\frac{(n-1)!}{(n+2)!\,(n!)^2} = \frac{(n-1)!}{(n+2)!\,(n!)(n!)} = \frac{(n-1)!}{(n+2)!\,(n!)(n)(n-1)!} = \frac{\cancel{(n-1)!}}{(n+2)!\,(n!)(n)\cancel{(n-1)!}} = \mathbf{\frac{1}{(n+2)!\,(n!)(n)}}$

g. $\frac{(2n-3)!\,2(n!)}{(2n)!\,(n-2)!} = \frac{(2n-3)!\,2[(n)(n-1)(n-2)!]}{(2n)(2n-1)(2n-2)(2n-3)!\,(n-2)!} = \frac{\cancel{(2n-3)!}\,2\,[(n)(n-1)\cancel{(n-2)!}]}{[(2n)(2n-1)(2n-2)\cancel{(2n-3)!}]\cancel{(n-2)!}} = \frac{2(n)(n-1)}{(2n)(2n-1)(2n-2)}$

$= \frac{\cancel{(2n)}(n-1)}{\cancel{(2n)}(2n-1)(2n-2)} = \mathbf{\frac{n-1}{(2n-1)(2n-2)}}$

5. Write the following expressions in factorial notation form. Simplify the answer.

a. $\binom{5}{3} = \frac{5!}{3!(5-3)!} = \frac{5!}{3!2!} = \frac{5\cdot 4\cdot \cancel{3!}}{\cancel{3!}2!} = \frac{5\cdot \overset{2}{\cancel{4}}}{1\cdot \cancel{2}} = \frac{5\cdot 2}{1} = \frac{10}{1} = \mathbf{10}$

b. $\binom{10}{6} = \frac{10!}{6!(10-6)!} = \frac{10!}{6!4!} = \frac{10\cdot 9\cdot 8\cdot 7\cdot \cancel{6!}}{\cancel{6!}4!} = \frac{\overset{5}{\cancel{10}}\cdot \overset{3}{\cancel{9}}\cdot \overset{2}{\cancel{8}}\cdot 7}{\cancel{4}\cdot \cancel{3}\cdot \cancel{2}\cdot 1} = \frac{5\cdot 3\cdot 2\cdot 7}{1} = \frac{210}{1} = \mathbf{210}$

c. $\binom{8}{0} = \frac{8!}{0!(8-0)!} = \frac{\cancel{8!}}{0!\cancel{8!}} = \frac{1}{1\cdot 1} = \mathbf{1}$

d. $\binom{8}{8} = \frac{8!}{8!(8-8)!} = \frac{\cancel{8!}}{\cancel{8!}0!} = \frac{1}{1\cdot 1} = \mathbf{1}$

e. $\binom{6}{3} = \frac{6!}{3!(6-3)!} = \frac{6!}{3!3!} = \frac{6\cdot 5\cdot 4\cdot \cancel{3!}}{3!\cancel{3!}} = \frac{\overset{2}{\cancel{6}}\cdot 5\cdot \overset{2}{\cancel{4}}}{1\cdot \cancel{2}\cdot \cancel{3}} = \frac{2\cdot 5\cdot 2}{1} = \frac{20}{1} = \mathbf{20}$

f. $\binom{5}{1} = \frac{5!}{1!(5-1)!} = \frac{5!}{1!4!} = \frac{5\cdot 4!}{1!4!} = \frac{5}{1} = \mathbf{5}$

g. $\binom{n}{n-5} = \frac{n!}{(n-5)![n-(n-5)]!} = \frac{n!}{(n-5)!(n-n+5)!} = \frac{n!}{(n-5)!5!} = \frac{n\cdot(n-1)\cdot(n-2)\cdot(n-3)\cdot(n-4)\cdot(n-5)!}{(n-5)!5!}$

$= \frac{n\cdot(n-1)\cdot(n-2)\cdot(n-3)\cdot(n-4)}{5!} = \frac{n\cdot(n-1)\cdot(n-2)\cdot(n-3)\cdot(n-4)}{5\cdot4\cdot3\cdot2\cdot1} = \mathbf{\frac{n\cdot(n-1)\cdot(n-2)\cdot(n-3)\cdot(n-4)}{120}}$

h. $\binom{2n}{2n-1} = \frac{2n!}{(2n-1)![2n-(2n-1)]!} = \frac{2n!}{(2n-1)!(2n-2n+1)!} = \frac{2n!}{(2n-1)!1!} = \frac{2n\cdot(2n-1)!}{(2n-1)!} = \mathbf{2n}$

i. $\binom{3n}{3n-3} = \frac{3n!}{(3n-3)![3n-(3n-3)]!} = \frac{3n!}{(3n-3)!(3n-3n+3)!} = \frac{3n!}{(3n-3)!3!} = \frac{3n\cdot(3n-1)\cdot(3n-2)\cdot(3n-3)!}{(3n-3)!\cdot1\cdot2\cdot3}$

$= \frac{3n\cdot(3n-1)\cdot(3n-2)}{1\cdot2\cdot3} = \mathbf{\frac{n\cdot(3n-1)\cdot(3n-2)}{2}}$

j. $\binom{n}{n-6} = \frac{n!}{(n-6)![n-(n-6)]!} = \frac{n!}{(n-6)!(n-n+6)!} = \frac{n!}{(n-6)!6!} = \frac{n\cdot(n-1)\cdot(n-2)\cdot(n-3)\cdot(n-4)\cdot(n-5)\cdot(n-6)!}{(n-6)!6!}$

$= \frac{n\cdot(n-1)\cdot(n-2)\cdot(n-3)\cdot(n-4)\cdot(n-5)}{6\cdot5\cdot4\cdot3\cdot2\cdot1} = \mathbf{\frac{n\cdot(n-1)\cdot(n-2)\cdot(n-3)\cdot(n-4)\cdot(n-5)}{720}}$

6. Expand the following binomial expressions.

a. $(x-2)^4 = \binom{4}{0}x^4 + \binom{4}{1}x^{4-1}\cdot(-2) + \binom{4}{2}x^{4-2}(-2)^2 + \binom{4}{3}x^{4-3}(-2)^3 + \binom{4}{4}x^{4-4}(-2)^4 = \binom{4}{0}x^4 - 2\binom{4}{1}x^3 + 4\binom{4}{2}x^2 - 8\binom{4}{3}x^1$

$+16\binom{4}{4}x^0 = \frac{4!}{0!(4-0)!}x^4 - \frac{2\cdot4!}{1!(4-1)!}x^3 + \frac{4\cdot4!}{2!(4-2)!}x^2 - \frac{8\cdot4!}{3!(4-3)!}x + \frac{16\cdot4!}{4!(4-4)!} = \frac{4!}{0!4!}x^4 - \frac{2\cdot4!}{3!}x^3 + \frac{4\cdot4!}{2!2!}x^2$

$-\frac{8\cdot4!}{3!1!}x + \frac{16\cdot4!}{4!0!} = \frac{4!}{0!4!}x^4 - \frac{2\cdot4\cdot3!}{3!}x^3 + \frac{4\cdot4\cdot3\cdot2!}{1\cdot2\cdot2!}x^2 - \frac{8\cdot4\cdot3!}{3!1!}x + \frac{16\cdot4!}{4!0!} = \mathbf{x^4 - 8x^3 + 24x^2 - 32x + 16}$

b. $(u+2)^7 = \binom{7}{0}u^7 + \binom{7}{1}u^{7-1}\cdot2 + \binom{7}{2}u^{7-2}\cdot2^2 + \binom{7}{3}u^{7-3}\cdot2^3 + \binom{7}{4}u^{7-4}\cdot2^4 + \binom{7}{5}u^{7-5}\cdot2^5 + \binom{7}{6}u^{7-6}\cdot2^6 + \binom{7}{7}u^{7-7}\cdot2^7$

$= \binom{7}{0}u^7 + 2\binom{7}{1}u^6 + 4\binom{7}{2}u^5 + 8\binom{7}{3}u^4 + 16\binom{7}{4}u^3 + 32\binom{7}{5}u^2 + 64\binom{7}{6}u + 128\binom{7}{7}u^0 = \frac{7!}{0!(7-0)!}u^7 + \frac{2\cdot7!}{1!(7-1)!}u^6 + \frac{4\cdot7!}{2!(7-2)!}u^5$

$+\frac{8\cdot7!}{3!(7-3)!}u^4 + \frac{16\cdot7!}{4!(7-4)!}u^3 + \frac{32\cdot7!}{5!(7-5)!}u^2 + \frac{64\cdot7!}{6!(7-6)!}u + \frac{128\cdot7!}{7!(7-7)!} = \frac{7!}{0!7!}u^7 + \frac{2\cdot7!}{1!6!}u^6 + \frac{4\cdot7!}{2!5!}u^5 + \frac{8\cdot7!}{3!4!}u^4$

$+\frac{16\cdot7!}{4!3!}u^3 + \frac{32\cdot7!}{5!2!}u^2 + \frac{64\cdot7!}{6!1!}u + \frac{128\cdot7!}{7!0!} = u^7 + (2\cdot7)u^6 + (7\cdot12)u^5 + (8\cdot35)u^4 + (16\cdot35)u^3 + (32\cdot21)u^2 + (64\cdot7)u + 128$

$= \mathbf{u^7 + 14u^6 + 84u^5 + 280u^4 + 560u^3 + 672u^2 + 448u + 128}$

c. $(y-3)^5 = \binom{5}{0}y^5 + \binom{5}{1}y^{5-1}\cdot(-3) + \binom{5}{2}y^{5-2}(-3)^2 + \binom{5}{3}y^{5-3}(-3)^3 + \binom{5}{4}y^{5-4}(-3)^4 + \binom{5}{5}y^{5-5}(-3)^5 = \binom{5}{0}y^5 - 3\binom{5}{1}y^4$

$+9\binom{5}{2}y^3 - 27\binom{5}{3}y^2 + 81\binom{5}{4}y^1 - 243\binom{5}{5}y^0 = \frac{5!}{0!(5-0)!}y^5 - \frac{3\cdot5!}{1!(5-1)!}y^4 + \frac{9\cdot5!}{2!(5-2)!}y^3 - \frac{27\cdot5!}{3!(5-3)!}y^2 + \frac{81\cdot5!}{4!(5-4)!}y$

$$-\frac{243\cdot 5!}{5!\,(5-5)!} = \frac{5!}{0!\,5!}y^5 - \frac{3\cdot 5!}{1!\,4!}y^4 + \frac{9\cdot 5!}{2!\,3!}y^3 - \frac{27\cdot 5!}{3!\,2!}y^2 + \frac{81\cdot 5!}{4!\,1!}y - \frac{243\cdot 5!}{5!\,0!} = y^5 - (3\cdot 5)y^4 + (9\cdot 10)y^3 - [27\cdot 10]y^2$$

$$+(81\cdot 5)y - 243 = \mathbf{y^5 - 15y^4 + 90y^3 - 270y^2 + 405y - 243}$$

7. Use the general equation for binomial expansion to solve the following exponential numbers to the nearest hundredth.

a. $(0.95)^5 = (1-0.05)^5$ therefore, $a=1$, $b=-0.05$, $n=5$. Using the general binomial expansion formula

$(a+b)^n = \binom{n}{0}a^n + \binom{n}{1}a^{n-1}b + \binom{n}{2}a^{n-2}b^2 + \cdots + \binom{n}{r-1}a^{n-r+1}b^{r-1} + \cdots + \binom{n}{n}b^n$ we obtain the following:

$$(1-0.05)^5 = \binom{5}{0}\cdot 1^5 + \binom{5}{1}\cdot 1^4\cdot(-0.05) + \binom{5}{2}\cdot 1^3\cdot(-0.05)^2 + \binom{5}{3}\cdot 1^2\cdot(-0.05)^3 + \binom{5}{4}\cdot 1^2\cdot(-0.05)^4 + \binom{5}{5}\cdot(-0.05)^5$$

$$= \binom{5}{0} - 0.05\binom{5}{1} + 0.0025\binom{5}{2} - \cdots = \frac{5!}{0!(5-0)!} - 0.05\frac{5!}{1!(5-1)!} + 0.0025\frac{5!}{2!(5-2)!} - \cdots = \frac{5!}{5!} - 0.05\frac{5!}{4!} + 0.0025\frac{5!}{2!3!} - \cdots$$

$$= \frac{\not{5}!}{\not{5}!} - 0.05\frac{5\cdot\not{4}!}{\not{4}!} + 0.0025\frac{5\cdot\overset{2}{\not{4}}\cdot\not{3}!}{\not{2}\cdot 1\cdot\not{3}!} - \cdots = 1 - 0.25 + 0.025 - \cdots \approx 0.775$$

Therefore, $(0.95)^5$ to the nearest hundredth is equal to **0.78**. (Note that this is an estimate.)

b. $(2.25)^7 = (2+0.25)^7$ therefore, $a=2$, $b=0.25$, $n=7$. Using the general binomial expansion formula

$(a+b)^n = \binom{n}{0}a^n + \binom{n}{1}a^{n-1}b + \binom{n}{2}a^{n-2}b^2 + \cdots + \binom{n}{r-1}a^{n-r+1}b^{r-1} + \cdots + \binom{n}{n}b^n$ we obtain the following:

$$(2+0.25)^7 = \binom{7}{0}2^7 + \binom{7}{1}2^6\cdot 0.25 + \binom{7}{2}2^5\cdot 0.25^2 + \binom{7}{3}2^4\cdot 0.25^3 + \binom{7}{4}2^3\cdot 0.25^4 + \binom{7}{5}2^2\cdot 0.25^5 + \binom{7}{6}2\cdot 0.25^6 + \binom{7}{7}0.25^7$$

$$= 128\binom{7}{0} + 16\binom{7}{1} + 2\binom{7}{2} + 0.25\binom{7}{3} + 0.0313\binom{7}{4} + \cdots = \frac{128\cdot 7!}{0!(7-0)!} + \frac{16\cdot 7!}{1!(7-1)!} + \frac{2\cdot 7!}{2!(7-2)!} + \frac{0.25\cdot 7!}{3!(7-3)!} + \frac{0.0313\cdot 7!}{4!(7-4)!} + \cdots$$

$$= \frac{128\cdot 7!}{7!} + \frac{16\cdot 7!}{6!} + \frac{2\cdot 7!}{2!\,5!} + \frac{0.25\cdot 7!}{3!\,4!} + \frac{0.0313\cdot 7!}{4!\,3!} + \cdots = \frac{128\cdot\not{7}!}{\not{7}!} + \frac{16\cdot 7\cdot\not{6}!}{\not{6}!} + \frac{\not{2}\cdot 7\cdot 6\cdot\not{5}!}{\not{2}\cdot 1\cdot\not{5}!} + \frac{0.25\cdot 7\cdot\not{6}\cdot 5\cdot\not{4}!}{\not{3}\cdot\not{2}\cdot 1\cdot\not{4}!} + \frac{0.0313\cdot 7\cdot\not{6}\cdot 5\cdot\not{4}!}{\not{4}!\cdot\not{3}\cdot\not{2}\cdot 1} + \cdots$$

$$= 128 + 112 + 42 + 8.75 + 1.095 + \cdots \approx 291.845$$

Therefore, $(2.25)^7$ to the nearest hundredth is equal to **291.85**. (Note that this is an estimate.)

c. $(1.05)^4 = (1+0.05)^4$ therefore, $a=1$, $b=0.05$, $n=4$. Using the general binomial expansion formula

$(a+b)^n = \binom{n}{0}a^n + \binom{n}{1}a^{n-1}b + \binom{n}{2}a^{n-2}b^2 + \cdots + \binom{n}{r-1}a^{n-r+1}b^{r-1} + \cdots + \binom{n}{n}b^n$ we obtain the following:

$$(1+0.05)^4 = \binom{4}{0}\cdot 1^4 + \binom{4}{1}\cdot 1^3\cdot 0.05 + \binom{4}{2}\cdot 1^2\cdot 0.05^2 + \binom{4}{3}\cdot 1^1\cdot 0.05^3 + \binom{4}{4}\cdot 0.05^4 = \binom{4}{0} + 0.05\binom{4}{1} + 0.0025\binom{4}{2} + 0.000125\binom{4}{3} + \cdots$$

$$= \frac{4!}{0!(4-0)!} + \frac{0.05\cdot 4!}{1!(4-1)!} + \frac{0.0025\cdot 4!}{2!(4-2)!} + \frac{0.000125\cdot 4!}{3!(4-3)!} + \cdots = \frac{4!}{4!} + \frac{0.05\cdot 4!}{3!} + \frac{0.0025\cdot 4!}{2!2!} + \frac{0.000125\cdot 4!}{3!\,1!} + \cdots$$

$$= \frac{\not{4}!}{\not{4}!} + \frac{0.05\cdot 4\cdot\not{3}!}{\not{3}!} + \frac{0.0025\cdot\overset{2}{\not{4}}\cdot 3\cdot\not{2}!}{\not{2}\cdot 1\cdot\not{2}!} + \frac{0.000125\cdot 4\cdot\not{3}!}{\not{3}!} + \cdots = 1 + 0.2 + 0.015 + 0.0005 + \cdots = 1.2155$$

Therefore, $(1.05)^4$ to the nearest hundredth is equal to **1.22**.

8. Find the stated term of the following binomial expressions.

a. To find the eighth term of $(x+3)^{12}$ first identify the a, b, r, and n terms, i.e., $a=x$, $b=3$, $r=8$, and $n=12$.

Then, use the equation $\binom{n}{r-1}a^{n-r+1}b^{r-1} = \frac{n!}{(r-1)!\,(n-r+1)!}a^{n-r+1}b^{r-1}$.

$$\binom{12}{7}x^5 \cdot 3^7 = \frac{12!}{7!(12-7)!}x^5 \cdot 3^7 = \frac{12!}{7!5!}x^5 \cdot 2187 = 2187x^5 \cdot \frac{\cancel{12} \cdot 11 \cdot \cancel{10} \cdot 9 \cdot 8 \cdot \cancel{7!}}{\cancel{7!} \cdot \cancel{5} \cdot \cancel{4} \cdot \cancel{3} \cdot \cancel{2} \cdot 1} = (2187 \cdot 792)x^5 = \mathbf{1{,}732{,}104x^5}$$

b. To find the ninth term of $(x-y)^{10}$ first identify the a, b, r, and n terms, i.e., $a=x$, $b=-y$, $r=9$, and $n=10$.

Then, use the equation $\binom{n}{r-1}a^{n-r+1}b^{r-1} = \frac{n!}{(r-1)!\,(n-r+1)!}a^{n-r+1}b^{r-1}$

$$\binom{10}{8}x^2 \cdot (-y)^8 = \frac{10!}{8!(10-8)!}x^2 \cdot y^8 = \frac{10!}{8!2!}x^2y^8 = \frac{10 \cdot 9 \cdot \cancel{8!}}{\cancel{8!} \cdot 2 \cdot 1}x^2y^8 = \frac{90}{2}x^2y^8 = \mathbf{45x^2y^8}$$

c. To find the seventh term of $(u-2a)^{11}$ first identify the a, b, r, and n terms, i.e., $a=u$, $b=-2a$, $r=7$, and $n=11$.

Then, use the equation $\binom{n}{r-1}a^{n-r+1}b^{r-1} = \frac{n!}{(r-1)!\,(n-r+1)!}a^{n-r+1}b^{r-1}$

$$\binom{11}{6}u^5 \cdot (-2a)^6 = \frac{11!}{6!(11-6)!}64a^6u^5 = \frac{11!}{6!5!}64a^6u^5 = 64a^6u^5\,\frac{11 \cdot \cancel{10} \cdot \overset{3}{\cancel{9}} \cdot \overset{2}{\cancel{8}} \cdot 7 \cdot \cancel{6!}}{\cancel{6!} \cdot \cancel{5} \cdot \cancel{4} \cdot \cancel{3} \cdot \cancel{2} \cdot 1} = (64 \cdot 462)a^6u^5 = \mathbf{29{,}568a^6u^5}$$

d. To find the twelfth term of $(x-1)^{18}$ first identify the a, b, r, and n terms, i.e., $a=x$, $b=-1$, $r=12$, and $n=18$.

Then, use the equation $\binom{n}{r-1}a^{n-r+1}b^{r-1} = \frac{n!}{(r-1)!\,(n-r+1)!}a^{n-r+1}b^{r-1}$

$$\binom{18}{11}x^7 \cdot (-1)^{11} = -x^7\frac{18!}{11!(18-11)!} = -x^7\frac{18!}{11!9!} = -x^7\frac{\cancel{18} \cdot 17 \cdot \overset{2}{\cancel{16}} \cdot \overset{3}{\cancel{15}} \cdot \overset{2}{\cancel{14}} \cdot 13 \cdot \cancel{12} \cdot \cancel{11!}}{\cancel{11!} \cdot \cancel{9} \cdot \cancel{8} \cdot \cancel{7} \cdot 6 \cdot \cancel{5} \cdot \cancel{4} \cdot \cancel{3} \cdot \cancel{2} \cdot 1} = -x^7\frac{17 \cdot 2 \cdot \cancel{3} \cdot \cancel{2} \cdot 13}{\cancel{6}} = -(17 \cdot 13 \cdot 2)x^7$$

$$= \mathbf{-442x^7}$$

Chapter 5 Solutions:

Section 5.1 Practice Problems – The Difference Quotient Method

1. Find the derivative of the following functions by using the difference quotient method.

a. Given $f(x)=x^2-1$, then $\frac{f(x+h)-f(x)}{h} = \frac{\left[(x+h)^2-1\right]-\left(x^2-1\right)}{h} = \frac{x^2+h^2+2hx-1-x^2+1}{h} = \frac{h^2+2hx}{h} = \frac{h(h+2x)}{h}$

$= h+2x$. Therefore, $f'(x) = \lim_{h\to 0}\frac{f(x+h)-f(x)}{h} = \lim_{h\to 0}(h+2x) = 0+2x = \mathbf{2x}$

b. Given $f(x)=x^3+2x-1$, then $\frac{f(x+h)-f(x)}{h} = \frac{\left[(x+h)^3+2(x+h)-1\right]-\left[x^3+2x-1\right]}{h}$

$= \frac{x^3+h^3+3x^2h+3xh^2+2x+2h-1-x^3-2x+1}{h} = \frac{h^3+3x^2h+3xh^2+2h}{h} = \frac{h\left(h^2+3x^2+3xh+2\right)}{h} = h^2+3x^2+3xh+2$.

Therefore, $f'(x) = \lim_{h\to 0}\frac{f(x+h)-f(x)}{h} = \lim_{h\to 0}\left(h^2+3x^2+3xh+2\right) = 0^2+3\cdot x^2+3\cdot 0\cdot h+2 = \mathbf{3x^2+2}$

c. Given $f(x)=\frac{x}{x-1}$, then $\frac{f(x+h)-f(x)}{h} = \frac{\frac{x+h}{x+h-1}-\frac{x}{x-1}}{h} = \frac{\frac{(x+h)(x-1)-x(x+h-1)}{(x+h-1)(x-1)}}{h} = \frac{(x+h)(x-1)-x(x+h-1)}{h\left[(x+h-1)(x-1)\right]}$

$= \frac{x^2-x+hx-h-x^2-hx+x}{h\left(x^2-x+hx-h-x+1\right)} = -\frac{h}{h\left(x^2-2x+hx-h+1\right)} = -\frac{1}{x^2-2x+hx-h+1}$. Therefore,

$f'(x) = \lim_{h\to 0}\frac{f(x+h)-f(x)}{h} = \lim_{h\to 0}\left(\frac{-1}{x^2-2x+hx-h+1}\right) = \frac{-1}{x^2-2x+0\cdot x-0+1} = \frac{-1}{x^2-2x+1} = -\frac{\mathbf{1}}{\mathbf{(x-1)^2}}$

d. Given $f(x)=-\frac{1}{x^2}$, then $\frac{f(x+h)-f(x)}{h} = \frac{-\frac{1}{(x+h)^2}+\frac{1}{x^2}}{h} = \frac{\frac{-x^2+(x+h)^2}{x^2(x+h)^2}}{h} = \frac{-x^2+(x+h)^2}{hx^2(x+h)^2} = \frac{-x^2+\left(x^2+h^2+2hx\right)}{hx^2\left(x^2+h^2+2hx\right)}$

$= \frac{-x^2+x^2+h^2+2hx}{hx^2\left(x^2+h^2+2hx\right)} = \frac{h(h+2x)}{hx^2\left(x^2+h^2+2hx\right)} = \frac{h+2x}{x^4+h^2x^2+2hx^3}$. Therefore,

$f'(x) = \lim_{h\to 0}\frac{f(x+h)-f(x)}{h} = \lim_{h\to 0}\left(\frac{h+2x}{x^4+h^2x^2+2hx^3}\right) = \frac{0+2x}{x^4+0^2\cdot x^2+2\cdot 0\cdot x^3} = \frac{2x}{x^{4-3}} = \frac{\mathbf{2}}{\mathbf{x^3}}$

e. Given $f(x)=20x^2-3$, then $\frac{f(x+h)-f(x)}{h} = \frac{\left[20(x+h)^2-3\right]-\left[20x^2-3\right]}{h} = \frac{\left[20\left(x^2+2hx+h^2\right)-3\right]-\left[20x^2-3\right]}{h}$

$= \frac{20x^2+40hx+20h^2-3-20x^2+3}{h} = \frac{40hx+20h^2}{h} = \frac{h(40x+20h)}{h} = 40x+20h$

Therefore, $f'(x) = \lim_{h\to 0}\frac{f(x+h)-f(x)}{h} = \lim_{h\to 0}(40x+20h) = 40x+(20\cdot 0) = 40x+0 = \mathbf{40x}$

f. Given $f(x)=\sqrt{x^3}$, then $\frac{f(x+h)-f(x)}{h} = \frac{\sqrt{(x+h)^3}-\sqrt{x^3}}{h} = \frac{\sqrt{(x+h)^3}-\sqrt{x^3}}{h}\cdot\frac{\sqrt{(x+h)^3}+\sqrt{x^3}}{\sqrt{(x+h)^3}+\sqrt{x^3}} = \frac{(x+h)^3-x^3}{h\sqrt{(x+h)^3}+\sqrt{x^3}}$

$= \frac{x^3+h^3+3x^2h+3xh^2-x^3}{h\sqrt{x^3+h^3+3x^2h+3xh^2}+\sqrt{x^3}} = \frac{h\left(h^2+3x^2+3xh\right)}{h\sqrt{x^3+h^3+3x^2h+3xh^2}+\sqrt{x^3}} = \frac{h^2+3x^2+3xh}{\sqrt{x^3+h^3+3x^2h+3xh^2}+\sqrt{x^3}}$. Therefore,

$$f'(x) = \lim_{h\to0}\frac{f(x+h)-f(x)}{h} = \lim_{h\to0}\left(\frac{h^2+3x^2+3xh}{\sqrt{x^3+h^3+3x^2h+3xh^2}+\sqrt{x^3}}\right) = \frac{0^2+3x^2+3x\cdot0}{\sqrt{x^3+0^3+3x^2\cdot0+3x\cdot0^2}+\sqrt{x^3}}$$

$$= \frac{3x^2}{\sqrt{x^3}+\sqrt{x^3}} = \frac{3x^2}{2\sqrt{x^3}} = \frac{3x^2}{2x^{\frac{3}{2}}} = \frac{3}{2}x^2\cdot x^{-\frac{3}{2}} = \frac{3}{2}x^{2-\frac{3}{2}} = \frac{3}{2}x^{\frac{4-3}{2}} = \frac{3}{2}x^{\frac{1}{2}} = \mathbf{\frac{3}{2}\sqrt{x}}$$

g. Given $f(x)=\frac{10}{\sqrt{x-5}}$, then $\frac{f(x+h)-f(x)}{h} = \frac{\frac{10}{\sqrt{x+h-5}}-\frac{10}{\sqrt{x-5}}}{h} = \frac{\frac{10\sqrt{x-5}-10\sqrt{x+h-5}}{\sqrt{x+h-5}\cdot\sqrt{x-5}}}{h} = \frac{10\sqrt{x-5}-10\sqrt{x+h-5}}{h\cdot\sqrt{x+h-5}\cdot\sqrt{x-5}}$

$$= 10\cdot\frac{\sqrt{x-5}-\sqrt{x+h-5}}{h\cdot\sqrt{x+h-5}\cdot\sqrt{x-5}}\cdot\frac{\sqrt{x-5}+\sqrt{x+h-5}}{\sqrt{x-5}+\sqrt{x+h-5}} = 10\cdot\frac{x-5-x-h+5}{\left(h\cdot\sqrt{x+h-5}\cdot\sqrt{x-5}\right)\cdot\left(\sqrt{x-5}+\sqrt{x+h-5}\right)}$$

$$= 10\cdot\frac{-h}{\left(h\cdot\sqrt{x+h-5}\cdot\sqrt{x-5}\right)\cdot\left(\sqrt{x-5}+\sqrt{x+h-5}\right)} = \frac{-10}{\left(\sqrt{x+h-5}\cdot\sqrt{x-5}\right)\cdot\left(\sqrt{x-5}+\sqrt{x+h-5}\right)}$$. Therefore,

$$f'(x) = \lim_{h\to0}\frac{f(x+h)-f(x)}{h} = \lim_{h\to0}\left(\frac{-10}{\left(\sqrt{x+h-5}\cdot\sqrt{x-5}\right)\cdot\left(\sqrt{x-5}+\sqrt{x+h-5}\right)}\right) = \frac{-10}{\left(\sqrt{x+0-5}\cdot\sqrt{x-5}\right)\cdot\left(\sqrt{x-5}+\sqrt{x+0-5}\right)}$$

$$= \frac{-10}{\left(\sqrt{x-5}\cdot\sqrt{x-5}\right)\cdot\left(\sqrt{x-5}+\sqrt{x-5}\right)} = \frac{-10}{(x-5)\cdot\left(2\sqrt{x-5}\right)} = \frac{-\overset{5}{10}}{2(x-5)\left(\sqrt{x-5}\right)} = \frac{-5}{(x-5)^{1+\frac{1}{2}}} = \frac{-5}{(x-5)^{\frac{3}{2}}} = -\mathbf{\frac{5}{\sqrt{(x-5)^3}}}$$

h. Given $f(x)=\frac{ax+b}{cx}$, then $\frac{f(x+h)-f(x)}{h} = \frac{\left[\frac{a(x+h)+b}{c(x+h)}\right]-\left[\frac{ax+b}{cx}\right]}{h} = \frac{\frac{cx(ax+ah+b)-(cx+ch)(ax+b)}{cx(cx+ch)}}{h}$

$$= \frac{acx^2+achx+bcx-acx^2-bcx-achx-bch}{chx(cx+ch)} = \frac{acx^2+achx+bcx-acx^2-bcx-achx-bch}{chx(cx+ch)} = \frac{-bch}{chx(cx+ch)} = \frac{-b}{cx^2+chx}.$$

Therefore, $f'(x) = \lim_{h\to0}\frac{f(x+h)-f(x)}{h} = \lim_{h\to0}\left(\frac{-b}{cx^2+chx}\right) = -\frac{b}{cx^2+c\cdot0\cdot x} = -\mathbf{\frac{b}{cx^2}}$

2. Compute $f'(x)$ for the specified values by using the difference quotient equation as the $\lim_{h\to0}$.

a. Given $f(x)=x^3$, then using $f'(x) = \lim_{h\to0}\frac{f(x+h)-f(x)}{h}$ we obtain $f'(x) = \lim_{h\to0}\frac{(x+h)^3-x^3}{h}$ at $x=1$

$$f'(1) = \lim_{h\to0}\frac{(1+h)^3-1^3}{h} = \lim_{h\to0}\frac{1+h^3+3h+3h^2-1}{h} = \lim_{h\to0}\frac{h(h^2+3+3h)}{h} = \lim_{h\to0}h^2+3+3h = 0^2+3+3\cdot0 = \mathbf{3}$$

b. Given $f(x)=1+2x$, then using $f'(x) = \lim_{h\to0}\frac{f(x+h)-f(x)}{h}$ we obtain $f'(x) = \lim_{h\to0}\frac{1+2(x+h)-(1+2x)}{h}$ at $x=0$

$$f'(0) = \lim_{h\to0}\frac{1+2(0+h)-(1+2\cdot0)}{h} = \lim_{h\to0}\frac{1+2h-1}{h} = \lim_{h\to0}\frac{2h}{h} = \lim_{h\to0}2 = \mathbf{2}$$

c. Given $f(x)=x^3+1$, then using $f'(x) = \lim_{h\to0}\frac{f(x+h)-f(x)}{h}$ we obtain $f'(x) = \lim_{h\to0}\frac{\left[(x+h)^3+1\right]-\left(x^3+1\right)}{h}$ at $x=-1$

$$f'(-1) = \lim_{h\to0}\frac{\left[(-1+h)^3+1\right]-\left[(-1)^3+1\right]}{h} = \lim_{h\to0}\frac{\left(-1+h^3+3h+3h^2+1\right)-(-1+1)}{h} = \lim_{h\to0}\frac{\left(h^3+3h+3h^2\right)-0}{h}$$

$$= \lim_{h\to0}\frac{h\left(h^2+3+3h\right)}{h} = \lim_{h\to0}\left(h^2+3+3h\right) = 0^2+3+3\cdot0 = \mathbf{3}$$

d. Given $f(x)=x^2(x+2)=x^3+2x^2$, then using $f'(x) = \lim_{h\to 0}\frac{f(x+h)-f(x)}{h}$ we obtain

$$f'(x) = \lim_{h\to 0}\frac{\left[(x+h)^3+2(x+h)^2\right]-\left(x^3+2x^2\right)}{h} \text{ at } x=2 \quad f'(2) = \lim_{h\to 0}\frac{\left[(2+h)^3+2(2+h)^2\right]-\left(2^3+2\cdot 2^2\right)}{h}$$

$$= \lim_{h\to 0}\frac{\left[8+h^3+12h+6h^2+8+2h^2+8h\right]-16}{h} = \lim_{h\to 0}\frac{16+h^3+20h+8h^2-16}{h} = \lim_{h\to 0}\frac{h\left(h^2+8h+20\right)}{h}$$

$$= \lim_{h\to 0}\left(h^2+8h+20\right) = 0^2+8\cdot 0+20 = \mathbf{20}$$

e. Given $f(x)=x^{-2}+x^{-1}+1 = \frac{1}{x^2}+\frac{1}{x}+1 = \frac{x+x^2}{x^3}+1 = \frac{x+x^2+x^3}{x^3} = \frac{x\left(1+x+x^2\right)}{x^{3-2}} = \frac{1+x+x^2}{x^2}$, then using

$$f'(x) = \lim_{h\to 0}\frac{f(x+h)-f(x)}{h} \text{ we obtain } f'(x) = \lim_{h\to 0}\frac{\frac{1+x+h+(x+h)^2}{(x+h)^2}-\frac{1+x+x^2}{x^2}}{h} \text{ at } x=1$$

$$f'(1) = \lim_{h\to 0}\frac{\frac{1+1+h+(1+h)^2}{(1+h)^2}-\frac{1+1+1^2}{1^2}}{h} = \lim_{h\to 0}\frac{\frac{2+h+1+h^2+2h}{1+h^2+2h}-3}{h} = \lim_{h\to 0}\frac{3+h^2+3h-3\left(1+h^2+2h\right)}{h\left(1+h^2+2h\right)}$$

$$= \lim_{h\to 0}\frac{3+h^2+3h-3-3h^2-6h}{h\left(1+h^2+2h\right)} = \lim_{h\to 0}\frac{-2h^2-3h}{h\left(1+h^2+2h\right)} = \lim_{h\to 0}\frac{h(-2h-3)}{h\left(1+h^2+2h\right)} = \lim_{h\to 0}\frac{-2h-3}{1+h^2+2h}$$

$$= \frac{(-2\cdot 0)-3}{1+0^2+2\cdot 0} = \frac{-3}{1} = \mathbf{-3}$$

f. Given $f(x)=\sqrt{x}+2$, then using $f'(x) = \lim_{h\to 0}\frac{f(x+h)-f(x)}{h}$ we obtain $f'(x) = \lim_{h\to 0}\frac{\left(\sqrt{x+h}+2\right)-\left(\sqrt{x}+2\right)}{h}$

$$\text{at } x=10 \quad f'(10) = \lim_{h\to 0}\frac{\left(\sqrt{10+h}+2\right)-\left(\sqrt{10}+2\right)}{h} = \lim_{h\to 0}\frac{\sqrt{10+h}+2-\sqrt{10}+2}{h} = \lim_{h\to 0}\frac{\sqrt{10+h}-\sqrt{10}}{h}$$

$$= \lim_{h\to 0}\frac{\left(\sqrt{10+h}\right)-\sqrt{10}}{h}\cdot\frac{\left(\sqrt{10+h}\right)+\sqrt{10}}{\left(\sqrt{10+h}\right)+\sqrt{10}} = \lim_{h\to 0}\frac{10+h-10}{h\cdot\left(\sqrt{10+h}+\sqrt{10}\right)} = \lim_{h\to 0}\frac{h}{h\cdot\left(\sqrt{10+h}+\sqrt{10}\right)} = \frac{1}{\sqrt{10+0}+\sqrt{10}}$$

$$= \frac{1}{\sqrt{10}+\sqrt{10}} = \frac{1}{2\sqrt{10}} = \frac{1}{2\cdot 3.16} = \frac{1}{6.32} = \mathbf{0.158}$$

Section 5.2 Solutions - Differentiation Rules Using the Prime Notation

1. Find the derivative of the following functions. Compare your answers with the practice problem number one in Section 5.1.

a. Given $f(x)=x^2-1$, then $f'(x) = 2x^{2-1}-0 = \mathbf{2x}$

b. Given $f(x)=x^3+2x-1$, then $f'(x) = 3x^{3-1}+2\cdot 1x^{1-1}-0 = 3x^2+2x^0 = \mathbf{3x^2+2}$

c. Given $f(x)=\frac{x}{x-1}$, then $f'(x) = \frac{[1\cdot(x-1)]-[1\cdot x]}{(x-1)^2} = \frac{x-1-x}{(x-1)^2} = -\frac{1}{(x-1)^2}$

d. Given $f(x)=-\frac{1}{x^2}$, then $f'(x) = -\frac{\left(0\cdot x^2\right)-(2x\cdot 1)}{x^4} = -\frac{0-2x}{x^4} = \frac{2x}{x^{4-3}} = \frac{\mathbf{2}}{\mathbf{x^3}}$

e. Given $f(x)=20x^2-3$, then $f'(x) = (20\cdot 2)x^{2-1}-0 = \mathbf{40x}$

f. Given $f(x)=\sqrt{x^3}=x^{\frac{3}{2}}$, then $f'(x) = \frac{3}{2}x^{\frac{3}{2}-1} = \frac{3}{2}x^{\frac{3-2}{2}} = \frac{3}{2}x^{\frac{1}{2}} = \mathbf{\frac{3}{2}\sqrt{x}}$

g. Given $f(x)=\frac{10}{\sqrt{x-5}}=\frac{10}{(x-5)^{\frac{1}{2}}}$, then $f'(x) = \frac{\left[0\cdot(x-5)^{\frac{1}{2}}\right]-\left[10\cdot\frac{1}{2}(x-5)^{-\frac{1}{2}}\right]}{x-5} = \frac{0-5(x-5)^{-\frac{1}{2}}}{x-5} = \frac{-5}{(x-5)^{\frac{1}{2}}\cdot(x-5)}$

$= \frac{-5}{(x-5)^{\frac{1}{2}+1}} = \frac{-5}{(x-5)^{\frac{3}{2}}} = \frac{-5}{\sqrt{(x-5)^3}} = -\frac{\mathbf{5}}{\mathbf{(x-5)\sqrt{x-5}}}$

h. Given $f(x)=\frac{ax+b}{cx}$, then $f'(x) = \frac{a\cdot cx-c\cdot(ax+b)}{(cx)^2} = \frac{acx-acx-bc}{(cx)^2} = -\frac{b\not{c}}{c^{2=1}x^2} = -\frac{\mathbf{b}}{\mathbf{cx^2}}$

2. Differentiate the following functions:

a. Given $f(x)=x^2+10x+1$, then $f'(x) = 2x^{2-1}+10x^{1-1}+0 = 2x+10x^0 = \mathbf{2x+10}$

b. Given $f(x)=x^8+3x^2-1$, then $f'(x) = 8x^{8-1}+(3\cdot2)x^{2-1}-0 = \mathbf{8x^7+6x}$

c. Given $f(x)=3x^4-2x^2+5$, then $f'(x) = (3\cdot4)x^{4-1}-(2\cdot2)x^{2-1}+0 = \mathbf{12x^3-4x}$

d. Given $f(x)=2(x^5+10x^4+5x)=2x^5+20x^4+10x$, then $f'(x) = (2\cdot5)x^{5-1}+(20\cdot4)x^{4-1}+(10\cdot1)x^{1-1} = 10x^4+80x^3+10x^0$

$= \mathbf{10x^4+80x^3+10}$

e. Given $f(x)=a^2x^3+b^2x+c^2$, then $f'(x) = (3\cdot a^2)x^{3-1}+(1\cdot b^2)x^{1-1}+0 = 3a^2x^2+b^2x^0 = \mathbf{3a^2x^2+b^2}$

f. Given $f(x)=x^2(x-1)+3x=x^3-x^2+3x$, then $f'(x) = 3x^{3-1}-2x^{2-1}+3x^{1-1} = 3x^2-2x+3x^0 = \mathbf{3x^2-2x+3}$

g. Given $f(x)=(x^3+1)(x^2-5)$, then $f'(x) = \left[3x^2(x^2-5)\right]+\left[2x(x^3+1)\right] = 3x^4-15x^2+2x^4+2x = \mathbf{5x^4-15x^2+2x}$

h. Given $f(x)=(3x^2+x-1)(x-1)$, then $f'(x) = [(6x+1)\cdot(x-1)]+\left[1\cdot(3x^2+x-1)\right] = 6x^2-6x+x-1+3x^2+x-1$

$= (6+3)x^2+(-6+1+1)x+(-1-1) = \mathbf{9x^2-4x-2}$

i. Given $f(x)=x(x^3+5x^2)-4x=x^4+5x^3-4x$, then $f'(x) = 4x^{4-1}+(5\cdot3)x^{3-1}-(4\cdot1)x^{1-1} = 4x^3+15x^2-4x^0$

$= \mathbf{4x^3+15x^2-4}$

j. Given $f(x)=\frac{x^3+1}{x}$, then $f'(x) = \frac{\left[(3x^2+0)\cdot x\right]-\left[1\cdot(x^3+1)\right]}{x^2} = \frac{3x^3-x^3-1}{x^2} = \frac{\mathbf{2x^3-1}}{\mathbf{x^2}}$

k. Given $f(x)=\frac{x^5+2x^2-1}{3x^2}$, then $f'(x) = \frac{\left[(5x^4+4x)\cdot3x^2\right]-\left[6x\cdot(x^5+2x^2-1)\right]}{(3x^2)^2} = \frac{15x^6+\not{12}\not{x^3}-6x^6-\not{12}\not{x^3}+6x}{9x^4}$

$= \frac{9x^6+6x}{9x^4} = \frac{\not{3}\not{x}(3x^5+2)}{3=\not{9}x^{4=3}} = \frac{\mathbf{3x^5+2}}{\mathbf{3x^3}}$

l. Given $f(x)=\frac{x^2}{(x-1)+3x}=\frac{x^2}{4x-1}$, then $f'(x) = \frac{[2x\cdot(4x-1)]-\left[4\cdot x^2\right]}{(4x-1)^2} = \frac{8x^2-2x-4x^2}{(4x-1)^2} = \frac{4x^2-2x}{(4x-1)^2} = \frac{\mathbf{2x(2x-1)}}{\mathbf{(4x-1)^2}}$

m. Given $f(x)=x^2\left(2+\frac{1}{x}\right)=2x^2+\frac{x^2}{x}=2x^2+x$, then $f'(x) = (2\cdot2)x^{2-1}+x^{1-1} = 4x+x^0 = \mathbf{4x+1}$

n. Given $f(x)=(x+1)\cdot\frac{2x}{x-1}=\frac{2x^2+2x}{x-1}$, then $f'(x) = \frac{[(4x+2)\cdot(x-1)]-\left[1\cdot(2x^2+2x)\right]}{(x-1)^2} = \frac{4x^2-4x+\not{2x}-2-2x^2-\not{2x}}{(x-1)^2}$

$= \frac{(4-2)x^2-4x-2}{(x-1)^2} = \frac{\mathbf{2x^2-4x-2}}{\mathbf{(x-1)^2}}$

o. Given $f(x)=\frac{x^3+3x-1}{x^4}$, then $f'(x) = \frac{\left[(3x^2+3)\cdot x^4\right]-\left[4x^3\cdot(x^3+3x-1)\right]}{x^8} = \frac{3x^6+3x^4-4x^6-12x^4+4x^3}{x^8}$

$= \frac{-x^6 - 9x^4 + 4x^3}{x^8} = \frac{x^3\left(-x^3 - 9x + 4\right)}{x^{8-5}} = -\frac{\mathbf{x^3 + 9x - 4}}{\mathbf{x^5}}$

p. Given $f(x) = \left(x^2 - 1\right)\left(\frac{2x^3 + 5}{x}\right)$, then $f'(x) = 2x \cdot \left(\frac{2x^3 + 5}{x}\right) + \left(\frac{\left(6x^2 \cdot x\right) - \left(2x^3 + 5\right)}{x^2}\right) \cdot \left(x^2 - 1\right) = 4x^3 + 10$

$+ \left(\frac{6x^3 - 2x^3 - 5}{x^2}\right) \cdot \left(x^2 - 1\right) = 4x^3 + 10 + \left(\frac{4x^3 - 5}{x^2}\right)\left(x^2 - 1\right) = 4x^3 + 10 + \frac{4x^5 - 4x^3 - 5x^2 + 5}{x^2} = \frac{\mathbf{8x^5 - 4x^3 + 5x^2 + 5}}{\mathbf{x^2}}$

q. Given $f(x) = \frac{3x^4 + x^2 + 2}{x - 1}$, then $f'(x) = \frac{\left[\left(12x^3 + 2x\right) \cdot (x - 1)\right] - \left[1 \cdot \left(3x^4 + x^2 + 2\right)\right]}{(x - 1)^2} = \frac{12x^4 - 12x^3 + 2x^2 - 2x - 3x^4 - x^2 - 2}{(x - 1)^2}$

$= \frac{\mathbf{9x^4 - 12x^3 + x^2 - 2x - 2}}{\mathbf{(x - 1)^2}}$

r. Given $f(x) = x^{-1} + \frac{1}{x^{-2}} = x^{-1} + x^2$, then $f'(x) = -x^{-1-1} + 2x^{2-1} = -x^{-2} + 2x = -\frac{\mathbf{1}}{\mathbf{x^2}} + \mathbf{2x}$

3. Compute $f'(x)$ at the specified value of x. Compare your answers with the practice problem number two in Section 5.1.

a. Given $f(x) = x^3$, then $f'(x) = 3x^{3-1} = 3x^2$ at $x = 1$ $f'(x) = 3 \cdot 1^2 = \mathbf{3}$

b. Given $f(x) = 1 + 2x$, then $f'(x) = 0 + (2 \cdot 1)x^{1-1} = 2x^0 = 2$ at $x = 0$ $f'(x) = \mathbf{2}$

c. Given $f(x) = x^3 + 1$, then $f'(x) = 3x^{3-1} + 0 = 3x^2$ at $x = -1$ $f'(x) = 3 \cdot (-1)^2 = \mathbf{3}$

d. Given $f(x) = x^2(x + 2) = x^3 + 2x^2$, then $f'(x) = 3x^{3-1} + (2 \cdot 2)x^{2-1} = 3x^2 + 4x$

at $x = 2$ $f'(x) = 3 \cdot 2^2 + 4 \cdot 2 = 3 \cdot 4 + 8 = 12 + 8 = \mathbf{20}$

e. Given $f(x) = x^{-2} + x^{-1} + 1$, then $f'(x) = -2x^{-2-1} - x^{-1-1} + 0 = -2x^{-3} - x^{-2} = -\frac{2}{x^3} - \frac{1}{x^2}$

at $x = 1$ $f'(x) = -\frac{2}{1^3} - \frac{1}{1^2} = -2 - 1 = \mathbf{-3}$

f. Given $f(x) = \sqrt{x} + 2 = x^{\frac{1}{2}} + 2$, then $f'(x) = \frac{1}{2}x^{\frac{1}{2}-1} + 0 = \frac{1}{2}x^{\frac{1-2}{2}} = \frac{1}{2}x^{-\frac{1}{2}} = \frac{1}{2x^{\frac{1}{2}}} = \frac{1}{2\sqrt{x}}$

at $x = 10$ $f'(x) = \frac{1}{2\sqrt{10}} = \frac{1}{2 \cdot 3.16} = \frac{1}{6.32} = \mathbf{0.158}$

4. Find $f'(0)$ and $f'(2)$ for the following functions:

a. Given $f(x) = x^3 - 3x^2 + 5$, then $f'(x) = 3x^{3-1} - (3 \cdot 2)x^{2-1} + 0 = 3x^2 - 6x$

Therefore, $f'(0) = \left(3 \cdot 0^2\right) - (6 \cdot 0) = 0 - 0 = \mathbf{0}$ and $f'(2) = \left(3 \cdot 2^2\right) - (6 \cdot 2) = 12 - 12 = \mathbf{0}$

b. Given $f(x) = \left(x^3 + 1\right)(x - 1)$, then $f'(x) = \left[3x^2 \cdot (x - 1)\right] + \left[1 \cdot \left(x^3 + 1\right)\right] = 3x^3 - 3x^2 + x^3 + 1 = 4x^3 - 3x^2 + 1$

Therefore, $f'(0) = \left(4 \cdot 0^3\right) - \left(3 \cdot 0^2\right) + 1 = 0 - 0 + 1 = \mathbf{1}$ and $f'(2) = \left(4 \cdot 2^3\right) - \left(3 \cdot 2^2\right) + 1 = 32 - 12 + 1 = \mathbf{21}$

c. Given $f(x) = x\left(x^2 + 1\right) = x^3 + x$, then $f'(x) = 3x^{2-1} + x^{1-1} = 3x + x^0 = 3x^2 + 1$

Therefore, $f'(0) = \left(3 \cdot 0^2\right) + 1 = 0 + 1 = \mathbf{1}$ and $f'(2) = \left(3 \cdot 2^2\right) + 1 = 12 + 1 = \mathbf{13}$

d. Given $f(x) = 2x^5 + 10x^4 - 4x$, then $f'(x) = (2 \cdot 5)x^{5-1} + (10 \cdot 4)x^{4-1} - 4x^{1-1} = 10x^4 + 40x^3 - 4x^0 = 10x^4 + 40x^3 - 4$

Therefore, $f'(0) = \left(10 \cdot 0^4\right) + \left(40 \cdot 0^3\right) - 4 = \mathbf{-4}$ and $f'(2) = \left(10 \cdot 2^4\right) + \left(40 \cdot 2^3\right) - 4 = 160 + 320 - 4 = \mathbf{476}$

e. Given $f(x)=2x^{-2}-3x^{-1}+5x$, then $f'(x) = (2\cdot -2)x^{-2-1}+(-3\cdot -1)x^{-1-1}+5x^{1-1} = -4x^{-3}+3x^{-2}+5x^{0}$

$= -4x^{-3}+3x^{-2}+5 = -\frac{4}{x^3}+\frac{3}{x^2}+5$. Therefore, $f'(0) = -\frac{4}{0^3}+\frac{3}{0^2}+5$ **which is undefined due to division by zero** and

$f'(2) = -\frac{4}{2^3}+\frac{3}{2^2}+5 = -\frac{4}{8}+\frac{3}{4}+5 = -0.5+0.75+5 = \mathbf{5.25}$

f. Given $f(x)=x^{-2}(x^5-x^3)+x=x^3-x+x=x^3$, then $f'(x) = 3x^{3-1} = 3x^2$

Therefore, $f'(0) = 3\cdot 0^2 = \mathbf{0}$ and $f'(2) = 3\cdot 2^2 = \mathbf{12}$

g. Given $f(x)=\frac{x}{1+x^2}$, then $f'(x) = \frac{[1\cdot(1+x^2)]-[2x\cdot x]}{(1+x^2)^2} = \frac{1+x^2-2x^2}{(1+x^2)^2} = \frac{1-x^2}{(1+x^2)^2}$

Therefore, $f'(0) = \frac{1-0^2}{(1+0^2)^2} = \mathbf{1}$ and $f'(2) = \frac{1-2^2}{(1+2^2)^2} = \frac{1-4}{(1+4)^2} = \frac{-3}{5^2} = -\frac{\mathbf{3}}{\mathbf{25}}$

h. Given $f(x)=\frac{1}{x}+x^3$, then $f'(x) = \frac{(0\cdot x)-(1\cdot 1)}{x^2}+3x^{3-1} = -\frac{1}{x^2}+3x^2$

Therefore, $f'(0) = -\frac{1}{0^2}+(3\cdot 0^2)$ **which is undefined due to division by zero** and $f'(2) = -\frac{1}{2^2}+(3\cdot 2^2) = -\frac{1}{4}+12 = \mathbf{11.75}$

i. Given $f(x)=\frac{ax^2+bx}{cx-d}$, then $f'(x) = \frac{[(2ax+b)\cdot(cx-d)]-[c\cdot(ax^2+bx)]}{(cx-d)^2} = \frac{2acx^2-2adx+\not{b}\not{c}\not{x}-bd-acx^2-\not{b}\not{c}\not{x}}{(cx-d)^2}$

$= \frac{acx^2-2adx-bd}{(cx-d)^2}$. Therefore, $f'(0) = \frac{(ac\cdot 0^2)-(2ad\cdot 0)-bd}{(c\cdot 0-d)^2} = \frac{0-0-bd}{(0-d)^2} = \frac{-b\not{d}}{d^{2-1}} = -\frac{\mathbf{b}}{\mathbf{d}}$ and

$f'(2) = \frac{(ac\cdot 2^2)-(2ad\cdot 2)-bd}{(c\cdot 2-d)^2} = \mathbf{\frac{4ac-4ad-bd}{(2c-d)^2}}$

5. Given $f(x)=x^2+1$ and $g(x)=2x-5$ find $h(x)$ and $h'(x)$.

a. Given $h(x)=x^3f(x)$ where $f(x)=x^2+1$, then $h(x) = x^3(x^2+1) = \mathbf{x^5+x^3}$ and $h'(x) = \mathbf{5x^4+3x^2}$

b. Given $f(x)=3+h(x)$ where $f(x)=x^2+1$, then $h(x) = f(x)-3 = (x^2+1)-3 = \mathbf{x^2-2}$ and $h'(x) = \mathbf{2x}$

c. Given $2g(x)=h(x)-1$ where $g(x)=2x-5$, then $h(x) = 2g(x)+1 = 2(2x-5)+1 = 4x-10+1 = \mathbf{4x-9}$ and $h'(x) = \mathbf{4}$

d. Given $3h(x)=2x\,g(x)-1$ where $g(x)=2x-5$, then $h(x) = \frac{2x\,g(x)-1}{3} = \frac{2x(2x-5)-1}{3} = \mathbf{\frac{4x^2-10x-1}{3}}$ and

$h'(x) = \frac{[(8x-10)\cdot 3]-[0\cdot(4x^2-10x-1)]}{3^2} = \frac{3(8x-10)}{9} = \mathbf{\frac{8x-10}{3}}$

e. Given $3[f(x)]^2-2h(x)=1$ where $f(x)=x^2+1$, then $h(x) = \frac{1}{2}(-1+3[f(x)]^2) = -\frac{1}{2}+\frac{3}{2}[f(x)]^2 = \mathbf{-\frac{1}{2}+\frac{3}{2}(x^2+1)^2}$ and

$h'(x) = 3(x^2+1)\cdot 2x = \mathbf{6x^3+6x}$

f. Given $h(x)=g(x)\cdot 3f(x)$ where $f(x)=x^2+1$ and $g(x)=2x-5$, then $h(x) = (2x-5)\cdot 3(x^2+1) = (2x-5)(3x^2+3)$

$= \mathbf{6x^3-15x^2+6x-15}$ and $h'(x) = \mathbf{18x^2-30x+6}$

g. Given $3h(x)-f(x)=0$ where $f(x)=x^2+1$, then $h(x) = \frac{f(x)}{3} = \mathbf{\frac{x^2+1}{3}}$ and $h'(x) = \mathbf{\frac{2}{3}x}$

h. Given $2g(x)+h(x)=f(x)$ where $f(x)=x^2+1$ and $g(x)=2x-5$, then $h(x) = f(x)-2g(x) = (x^2+1)-2(2x-5)$

$= x^2+1-4x+10 = \mathbf{x^2-4x+11}$ and $h'(x) = \mathbf{2x-4}$

i. Given $f(x)=x^3+5x^2+h(x)$ where $f(x)=x^2+1$, then $h(x) = f(x)-x^3-5x^2 = (x^2+1)-x^3-5x^2 = -x^3+(-5+1)x^2+1$

$= \mathbf{-x^3-4x^2+1}$ and $h'(x) = \mathbf{-3x^2-8x}$

j. Given $h(x)=\frac{x^3+1}{x}-f(x)$ where $f(x)=x^2+1$, then $h(x) = \mathbf{\frac{x^3+1}{x}-(x^2+1)}$ and $h'(x) = \frac{[3x^2\cdot x]-[1\cdot(x^3+1)]}{x^2}-2x$

$= \frac{3x^3-x^3-1}{x^2}-2x = \mathbf{\frac{2x^3-1}{x^2}-2x}$

k. Given $h(x)=2f(x)+g(x)$ where $f(x)=x^2+1$ and $g(x)=2x-5$, then $h(x) = 2(x^2+1)+(2x-5) = 2x^2+2+2x-5$

$= \mathbf{2x^2+2x-3}$ and $h'(x) = \mathbf{4x+2}$

l. Given $[h(x)]^2-f(x)=10$ where $f(x)=x^2+1$, then $[h(x)]^2=10+f(x)$; $h(x) = \sqrt{10+f(x)} = [10+(x^2+1)]^{\frac{1}{2}}$

$= \mathbf{(x^2+11)^{\frac{1}{2}}}$ and $h'(x) = \frac{1}{2}(x^2+11)^{\frac{1}{2}-1}\cdot 2x = \mathbf{x(x^2+11)^{-\frac{1}{2}}}$

m. Given $f(x)=\frac{2g(x)}{h(x)}$ where $f(x)=x^2+1$ and $g(x)=2x-5$, then $h(x) = \frac{2g(x)}{f(x)} = \frac{2(2x-5)}{x^2+1} = \mathbf{\frac{4x-10}{x^2+1}}$ and $h'(x)$

$= \frac{[4\cdot(x^2+1)]-[2x\cdot(4x-10)]}{(x^2+1)^2} = \frac{4x^2+4-8x^2+20x}{(x^2+1)^2}$ $\mathbf{\frac{-4x^2+20x+4}{(x^2+1)^2}}$

n. Given $\frac{3f(x)}{h(x)}=\frac{1}{x}$ where $f(x)=x^2+1$, then $h(x) = 3x\,f(x) = 3x(x^2+1) = \mathbf{3x^3+3x}$ and $h'(x) = \mathbf{9x^2+3}$

o. Given $f(x)=\frac{1}{h(x)+4}$, which is equivalent to $\frac{f(x)}{1}=\frac{1}{h(x)+4}$; $f(x)\cdot[h(x)+4]=1\cdot 1$; $f(x)h(x)+4f(x)=1$

; $f(x)h(x)=1-4f(x)$, and $f(x)=x^2+1$, then $h(x) = \frac{1-4f(x)}{f(x)} = \frac{1}{f(x)}-\frac{4f(x)}{f(x)} = \frac{1}{f(x)}-4 = \mathbf{\frac{1}{x^2+1}-4}$ and $h'(x)$

$= \frac{[0\cdot(x^2+1)]-[2x\cdot 1]}{(x^2+1)^2}-0 = \frac{0-2x}{(x^2+1)^2} = \mathbf{-\frac{2x}{(x^2+1)^2}}$

Section 5.3 Solutions - Differentiation Rules Using the $\frac{d}{dx}$ Notation

1. Find $\frac{dy}{dx}$ for the following functions:

a. Given $y=x^5+3x^2+1$, then $\frac{dy}{dx} = \frac{d}{dx}(x^5+3x^2+1) = \frac{d}{dx}x^5+\frac{d}{dx}3x^2+\frac{d}{dx}1 = 5x^4+(3\cdot 2)x+0 = \mathbf{5x^4+6x}$

b. Given $y=3x^2+5$, then $\frac{dy}{dx} = \frac{d}{dx}(3x^2+5) = \frac{d}{dx}3x^2+\frac{d}{dx}5 = (3\cdot 2)x+0 = \mathbf{6x}$

c. Given $y=x^3-\frac{1}{x}$, then $\frac{dy}{dx} = \frac{d}{dx}\left(x^3-\frac{1}{x}\right) = \frac{d}{dx}(x^3-x^{-1}) = \frac{d}{dx}x^3-\frac{d}{dx}x^{-1} = 3x^2+x^{-1-1} = 3x^2+x^{-2} = \mathbf{3x^2+\frac{1}{x^2}}$

d. Given $y=\frac{x^2}{1-x^3}$, then $\frac{dy}{dx} = \frac{d}{dx}\left(\frac{x^2}{1-x^3}\right) = \frac{\left[(1-x^3)\frac{d}{dx}x^2\right]-\left[x^2\frac{d}{dx}(1-x^3)\right]}{(1-x^3)^2} = \frac{\left[(1-x^3)\cdot 2x\right]-\left[x^2\cdot(-3x^2)\right]}{(1-x^3)^2}$

$= \frac{2x-2x^4+3x^4}{(1-x^3)^2} = \frac{x^4-2x}{(1-x^3)^2} = \mathbf{\frac{x(x^3-2)}{(1-x^3)^2}}$

e. Given $y = 4x^2 + \frac{1}{x-1}$, then $\frac{dy}{dx} = \frac{d}{dx}\left(4x^2 + \frac{1}{x-1}\right) = \frac{d}{dx}\left(4x^2\right) + \frac{d}{dx}\left(\frac{1}{x-1}\right) = 8x + \frac{\left[(x-1)\cdot\frac{d}{dx}1\right] - \left[1\cdot\frac{d}{dx}(x-1)\right]}{(x-1)^2}$

$= 8x + \frac{0-1}{(x-1)^2} = \mathbf{8x - \frac{1}{(x-1)^2}}$

f. Given $y = \frac{x^2+2x}{x^3+1}$, then $\frac{dy}{dx} = \frac{d}{dx}\left(\frac{x^2+2x}{x^3+1}\right) = \frac{\left[\left(x^3+1\right)\frac{d}{dx}\left(x^2+2x\right)\right] - \left[\left(x^2+2x\right)\frac{d}{dx}\left(x^3+1\right)\right]}{\left(x^3+1\right)^2} = \frac{\left[\left(x^3+1\right)\cdot(2x+2)\right]}{\left(x^3+1\right)^2}$

$\frac{-\left[\left(x^2+2x\right)\cdot 3x^2\right]}{\left(x^3+1\right)^2} = \frac{2x^4+2x^3+2x+2-3x^4-6x^3}{\left(x^3+1\right)^2} = \mathbf{\frac{-x^4-4x^3+2x+2}{\left(x^3+1\right)^2}}$

g. Given $y = x^3\left(x^2+5x-2\right) = x^5+5x^4-2x^3$, then $\frac{dy}{dx} = \frac{d}{dx}\left(x^5+5x^4-2x^3\right) = \frac{d}{dx}x^5 + \frac{d}{dx}5x^4 - \frac{d}{dx}2x^3 = \mathbf{5x^4+20x^3-6x^2}$

h. Given $y = x^2(x+3)(x-1) = \left(x^3+3x^2\right)(x-1)$, then $\frac{dy}{dx} = \frac{d}{dx}\left[\left(x^3+3x^2\right)(x-1)\right] = \left[(x-1)\frac{d}{dx}\left(x^3+3x^2\right)\right] + \left[\left(x^3+3x^2\right)\frac{d}{dx}(x-1)\right]$

$= \left[(x-1)\cdot\left(3x^2+6x\right)\right] + \left[\left(x^3+3x^2\right)\cdot 1\right] = 3x^3+6x^2-3x^2-6x+x^3+3x^2 = 4x^3+6x^2-6x = \mathbf{2x\left(2x^2+3x-3\right)}$

i. Given $y = 5x - \frac{1}{x^3}$, then $\frac{dy}{dx} = \frac{d}{dx}\left(5x - \frac{1}{x^3}\right) = \frac{d}{dx}\left(5x - x^{-3}\right) = \frac{d}{dx}5x - \frac{d}{dx}x^{-3} = 5 + 3x^{-3-1} = 5 + 3x^{-4} = \mathbf{5 + \frac{3}{x^4}}$

j. Given $y = \frac{(x-1)(x+3)}{x^2} = \frac{x^2+3x-x-3}{x^2} = \frac{x^2+(3-1)x-3}{x^2} = \frac{x^2+2x-3}{x^2}$, then $\frac{dy}{dx} = \frac{d}{dx}\left[\frac{x^2+2x-3}{x^2}\right]$

$= \frac{\left[x^2\frac{d}{dx}\left(x^2+2x-3\right)\right] - \left[\left(x^2+2x-3\right)\frac{d}{dx}x^2\right]}{x^4} = \frac{\left[x^2\cdot(2x+2)\right] - \left[\left(x^2+2x-3\right)\cdot 2x\right]}{x^4} = \frac{2x^3+2x^2-2x^3-4x^2+6x}{x^4}$

$= \frac{-2x^2+6x}{x^4} = \frac{-2\cancel{x}(x-3)}{x^{\cancel{4}=3}} = \mathbf{-\frac{2(x-3)}{x^3}}$

k. Given $y = x\left(\frac{x-1}{3}\right) = \frac{x^2-x}{3}$, then $\frac{dy}{dx} = \frac{d}{dx}\left(\frac{x^2-x}{3}\right) = \frac{\left[3\cdot\frac{d}{dx}\left(x^2-x\right)\right] - \left[\left(x^2-x\right)\cdot\frac{d}{dx}3\right]}{3^2} = \frac{\left[3\cdot(2x-1)\right] - \left[\left(x^2-x\right)\cdot 0\right]}{9}$

$= \frac{3(2x-1)-0}{9} = \frac{\cancel{3}(2x-1)}{\cancel{9}=3} = \mathbf{\frac{2x-1}{3}}$

l. Given $y = x^2(x+3)^{-1}$, then $\frac{dy}{dx} = \frac{d}{dx}\left[x^2(x+3)^{-1}\right] = \left[(x+3)^{-1}\frac{d}{dx}x^2\right] + \left[x^2\frac{d}{dx}(x+3)^{-1}\right] = \left[(x+3)^{-1}\cdot 2x\right] + \left[x^2\cdot\left(-(x+3)^{-2}\right)\right]$

$= 2x(x+3)^{-1} - x^2(x+3)^{-2} = \mathbf{\frac{2x}{x+3} - \frac{x^2}{(x+3)^2}}$

m. Given $y = \left(\frac{x}{1+x}\right)\left(\frac{x-3}{5}\right)$, then $\frac{dy}{dx} = \left(\frac{x-3}{5}\right)\frac{d}{dx}\left(\frac{x}{1+x}\right) + \left(\frac{x}{1+x}\right)\frac{d}{dy}\left(\frac{x-3}{5}\right) = \frac{x-3}{5}\cdot\frac{(1+x)\frac{d}{dx}x - x\frac{d}{dx}(1+x)}{(1+x)^2}$

$+ \frac{x}{1+x}\cdot\frac{5\frac{d}{dx}(x-3) - (x-3)\frac{d}{dx}5}{5^2} = \frac{x-3}{5}\cdot\frac{\left[(1+x)\cdot 1\right] - [x\cdot 1]}{(1+x)^2} + \frac{x}{1+x}\cdot\frac{[5\cdot 1] - \left[(x-3)\cdot 0\right]}{25} = \frac{x-3}{5}\cdot\frac{\cancel{x}+1-\cancel{x}}{(1+x)^2} + \frac{x}{1+x}\cdot\frac{\cancel{5}}{\cancel{25}_{5}}$

$= \frac{x-3}{5}\cdot\frac{1}{(1+x)^2} + \frac{x}{1+x}\cdot\frac{1}{5} = \mathbf{\frac{x-3}{5(1+x)^2} + \frac{x}{5(1+x)}}$

n. Given $y = x^3\left(1+\frac{1}{x-1}\right) = x^3 + \frac{x^3}{x-1}$, then $\frac{dy}{dx} = \frac{d}{dx}x^3 + \frac{d}{dx}\left(\frac{x^3}{x-1}\right) = 3x^2 + \frac{\left[(x-1)\frac{d}{dx}x^3\right]-\left[x^3\frac{d}{dx}(x-1)\right]}{(x-1)^2}$

$= 3x^2 + \frac{\left[(x-1)\cdot 3x^2\right]-\left[x^3\cdot 1\right]}{(x-1)^2} = 3x^2 + \frac{3x^3-3x^2-x^3}{(x-1)^2} = 3x^2 + \frac{2x^3-3x^2}{(x-1)^2} = \mathbf{3x^2 + \frac{x^2(2x-3)}{(x-1)^2}}$

o. Given $y = \frac{1}{x}\left(\frac{2x-1}{3x+1}\right) = \frac{2x-1}{3x^2+x}$, then $\frac{dy}{dx} = \frac{d}{dx}\left(\frac{2x-1}{3x^2+x}\right) = \frac{\left[(3x^2+x)\frac{d}{dx}(2x-1)\right]-\left[(2x-1)\frac{d}{dx}(3x^2+x)\right]}{(3x^2+x)^2}$

$= \frac{\left[(3x^2+x)\cdot 2\right]-\left[(2x-1)\cdot(6x+1)\right]}{(3x^2+x)^2} = \frac{6x^2+2x-(12x^2+2x-6x-1)}{(3x^2+x)^2} = \frac{6x^2+\cancel{2x}-12x^2-\cancel{2x}+6x+1}{(3x^2+x)^2} = \mathbf{\frac{-6x^2+6x+1}{(3x^2+x)^2}}$

p. Given $y = \frac{ax^2+bx+c}{bx}$, then $\frac{dy}{dx} = \frac{d}{dx}\left(\frac{ax^2+bx+c}{bx}\right) = \frac{\left[bx\frac{d}{dx}(ax^2+bx+c)\right]-\left[(ax^2+bx+c)\frac{d}{dx}bx\right]}{(bx)^2} = \frac{bx\cdot(2ax+b)}{(bx)^2}$

$\frac{-(ax^2+bx+c)\cdot b}{(bx)^2} = \frac{2abx^2+\cancel{b^2x}-abx^2-\cancel{b^2x}-bc}{b^2x^2} = \frac{abx^2-bc}{b^2x^2} = \frac{\cancel{b}(ax^2-c)}{b^{2-1}x^2} = \mathbf{\frac{ax^2-c}{bx^2}}$

q. Given $y = \frac{x^3-2}{x^4-3}$, then $\frac{dy}{dx} = \frac{d}{dx}\left(\frac{x^3-2}{x^4-3}\right) = \frac{\left[(x^4-3)\frac{d}{dx}(x^3-2)\right]-\left[(x^3-2)\frac{d}{dx}(x^4-3)\right]}{(x^4-3)^2} = \frac{\left[(x^4-3)\cdot 3x^2\right]-\left[(x^3-2)\cdot 4x^3\right]}{(x^4-3)^2}$

$= \frac{3x^6-9x^2-4x^6+8x^3}{(x^4-3)^2} = \mathbf{\frac{-x^6+8x^3-9x^2}{(x^4-3)^2}}$

r. Given $y = \frac{5x}{(1+x)^2}$, then $\frac{dy}{dx} = \frac{d}{dx}\left(\frac{5x}{(1+x)^2}\right) = \frac{\left[(1+x)^2\frac{d}{dx}5x\right]-\left[5x\frac{d}{dx}(1+x)^2\right]}{(1+x)^4} = \frac{\left[(1+x)^2\cdot 5\right]-\left[5x\cdot 2(1+x)\right]}{(1+x)^4}$

$= \frac{5(x^2+2x+1)-10x(1+x)}{(1+x)^4} = \frac{5x^2+\cancel{10x}+5-\cancel{10x}-10x^2}{(1+x)^4} = \mathbf{\frac{-5x^2+5}{(1+x)^4}}$

2. Find the derivative of the following functions:

a. $\frac{d}{dt}(3t^2+5t) = \frac{d}{dt}(3t^2)+\frac{d}{dt}(5t) = (3\cdot 2)t^{2-1}+(5\cdot 1)t^{1-1} = 6t+5t^0 = \mathbf{6t+5}$

b. $\frac{d}{dx}(6x^3+5x-2) = \frac{d}{dx}(6x^3)+\frac{d}{dx}(5x)+\frac{d}{dx}(-2) = (6\cdot 3)x^{3-1}+(5\cdot 1)x^{1-1}+0 = 18x^2+5x^0 = \mathbf{18x^2+5}$

c. $\frac{d}{du}(u^3+2u^2+5) = \frac{d}{du}(u^3)+\frac{d}{du}(2u^2)+\frac{d}{du}5 = 3u^{3-1}+(2\cdot 2)u^{2-1}+0 = \mathbf{3u^2+4u}$

d. $\frac{d}{dt}\left(\frac{t^2+2t}{5}\right) = \frac{\left[5\cdot\frac{d}{dt}(t^2+2t)\right]-\left[(t^2+2t)\cdot\frac{d}{dt}5\right]}{5^2} = \frac{\left[5\cdot(2t+2)\right]-\left[(t^2+2t)\cdot 0\right]}{25} = \frac{5(2t+2)-0}{25} = \frac{\cancel{5}(2t+2)}{\underset{5}{\cancel{25}}} = \mathbf{\frac{2t+2}{5}}$ or,

$\frac{d}{dt}\left(\frac{t^2+2t}{5}\right) = \frac{1}{5}\frac{d}{dt}(t^2+2t) = \frac{1}{5}\left(\frac{d}{dt}t^2+\frac{d}{dt}2t\right) = \frac{1}{5}(2t+2) = \mathbf{\frac{2t+2}{5}}$

e. $\frac{d}{ds}\left(\frac{s^3+3s-1}{s^2}\right) = \frac{\left[s^2\cdot\frac{d}{dt}(s^3+3s-1)\right]-\left[(s^3+3s-1)\cdot\frac{d}{dt}s^2\right]}{s^4} = \frac{\left[s^2\cdot(3s^2+3)\right]-\left[(s^3+3s-1)\cdot 2s\right]}{s^4} = \frac{3s^4+3s^2-2s^4-6s^2+2s}{s^4}$

$$\frac{s^4-3s^2+2s}{s^4} = \frac{\cancel{s}\left(s^3-3s+2\right)}{s^{4-3}} = \mathbf{\frac{s^3-3s+2}{s^3}}$$

f. $\frac{d}{dw}\left(w^3+\frac{w^2}{1+w}\right) = \frac{d}{dw}w^3+\frac{d}{dw}\left(\frac{w^2}{1+w}\right) = 3w^2+\frac{\left[(1+w)\cdot\frac{d}{dw}w^2\right]-\left[w^2\cdot\frac{d}{dt}(1+w)\right]}{(1+w)^2} = 3w^2+\frac{[(1+w)\cdot 2w]-\left[w^2\cdot 1\right]}{(1+w)^2}$

$= 3w^2+\frac{2w+2w^2-w^2}{(1+w)^2} = \mathbf{3w^2+\frac{w^2+2w}{(1+w)^2}}$

g. $\frac{d}{dt}\left[t^2(t+1)\left(t^2-3\right)\right] = \frac{d}{dt}\left[\left(t^3+t^2\right)\left(t^2-3\right)\right] = \left(t^2-3\right)\frac{d}{dt}\left(t^3+t^2\right)+\left(t^3+t^2\right)\frac{d}{dt}\left(t^2-3\right) = \left(t^2-3\right)\cdot\left(3t^2+2t\right)+\left(t^3+t^2\right)\cdot 2t$

$= 3t^4+2t^3-9t^2-6t+2t^4+2t^3 = \mathbf{5t^4+4t^3-9t^2-6t}$, or

$\frac{d}{dt}\left[t^2(t+1)\left(t^2-3\right)\right] = \frac{d}{dt}\left[\left(t^3+t^2\right)\left(t^2-3\right)\right] = \frac{d}{dt}\left(t^5+t^4-3t^3-3t^2\right) = \frac{d}{dt}t^5+\frac{d}{dt}t^4-3\frac{d}{dt}t^3-3\frac{d}{dt}t^2 = \mathbf{5t^4+4t^3-9t^2-6t}$

h. $\frac{d}{dx}\left[(x+1)\left(x^2+5\right)\right] = \left[\left(x^2+5\right)\frac{d}{dx}(x+1)\right]+\left[(x+1)\frac{d}{dx}\left(x^2+5\right)\right] = \left[\left(x^2+5\right)\cdot 1\right]+[(x+1)\cdot 2x] = x^2+5+2x^2+2x = \mathbf{3x^2+2x+5}$

or, $\frac{d}{dx}\left[(x+1)\left(x^2+5\right)\right] = \frac{d}{dx}\left(x^3+x^2+5x+5\right) = \frac{d}{dx}x^3+\frac{d}{dx}x^2+\frac{d}{dx}5x+\frac{d}{dx}5 = \mathbf{3x^2+2x+5}$

i. $\frac{d}{du}\left[\frac{u^2}{1-u}-\frac{u}{1+u}\right] = \frac{d}{du}\left(\frac{u^2}{1-u}\right)-\frac{d}{du}\left(\frac{u}{1+u}\right) = \frac{\left[(1-u)\cdot\frac{d}{du}u^2\right]-\left[u^2\cdot\frac{d}{du}(1-u)\right]}{(1-u)^2}-\frac{\left[(1+u)\cdot\frac{d}{du}u\right]-\left[u\cdot\frac{d}{du}(1+u)\right]}{(1+u)^2}$

$= \frac{[(1-u)\cdot 2u]-\left[u^2\cdot -1\right]}{(1-u)^2}-\frac{[(1+u)\cdot 1]-[u\cdot 1]}{(1+u)^2} = \frac{2u-2u^2+u^2}{(1-u)^2}-\frac{1+u-u}{(1+u)^2} = \frac{2u-u^2}{(1-u)^2}-\frac{1}{(1+u)^2} = \mathbf{\frac{u(2-u)}{(1-u)^2}-\frac{1}{(1+u)^2}}$

j. $\frac{d}{dr}\left(\frac{3r^3-2r^2+1}{r}\right) = \frac{\left[r\cdot\frac{d}{dr}\left(3r^3-2r^2+1\right)\right]-\left[\left(3r^3-2r^2+1\right)\cdot\frac{d}{dr}r\right]}{r^2} = \frac{\left[r\cdot\left(9r^2-4r\right)\right]-\left[\left(3r^3-2r^2+1\right)\cdot 1\right]}{r^2}$

$= \frac{9r^3-4r^2-3r^3+2r^2-1}{r^2} = \mathbf{\frac{6r^3-2r^2-1}{r^2}}$

k. $\frac{d}{ds}\left[\frac{3s^2}{s^3+1}-\frac{1}{s^2}\right] = \frac{d}{ds}\left(\frac{3s^2}{s^3+1}\right)-\frac{d}{ds}\left(\frac{1}{s^2}\right) = \frac{\left[\left(s^3+1\right)\cdot\frac{d}{ds}3s^2\right]-\left[3s^2\cdot\frac{d}{ds}\left(s^3+1\right)\right]}{\left(s^3+1\right)^2}-\frac{\left[s^2\cdot\frac{d}{ds}1\right]-\left[1\cdot\frac{d}{ds}s^2\right]}{s^4}$

$= \frac{\left[\left(s^3+1\right)\cdot 6s\right]-\left[3s^2\cdot 3s^2\right]}{\left(s^3+1\right)^2}-\frac{\left(s^2\cdot 0\right)-(1\cdot 2s)}{s^4} = \frac{6s^4+6s-9s^4}{\left(s^3+1\right)^2}-\frac{0-2s}{s^4} = \frac{-3s^4+6s}{\left(s^3+1\right)^2}+\frac{2\cancel{s}}{s^{4-3}} = \mathbf{-\frac{3s\left(s^3-2\right)}{\left(s^3+1\right)^2}+\frac{2}{s^3}}$

l. $\frac{d}{du}\left[\frac{u^3}{1-u}-\frac{u+1}{u^2}\right] = \frac{d}{du}\left(\frac{u^3}{1-u}\right)-\frac{d}{du}\left(\frac{u+1}{u^2}\right) = \frac{\left[(1-u)\cdot\frac{d}{du}u^3\right]-\left[u^3\cdot\frac{d}{du}(1-u)\right]}{(1-u)^2}-\frac{\left[u^2\cdot\frac{d}{du}(u+1)\right]-\left[(u+1)\cdot\frac{d}{du}u^2\right]}{u^4}$

$= \frac{\left[(1-u)\cdot 3u^2\right]-\left[u^3\cdot -1\right]}{(1-u)^2}-\frac{\left[u^2\cdot 1\right]-[(u+1)\cdot 2u]}{u^4} = \frac{3u^2-3u^3+u^3}{(1-u)^2}-\frac{u^2-2u^2-2u}{u^4} = \frac{-2u^3+3u^2}{(1-u)^2}-\frac{-u^2-2u}{u^4}$

$= \frac{u^2(-2u+3)}{(1-u)^2}+\frac{\cancel{u}(u+2)}{u^{4-3}} = \mathbf{-\frac{u^2(2u-3)}{(1-u)^2}+\frac{u+2}{u^3}}$

3. Find the derivative of the following functions at the specified value.

a. $\frac{d}{dx}\left(x^3+3x^2+1\right) = \frac{d}{dx}\left(x^3\right)+\frac{d}{dx}\left(3x^2\right)+\frac{d}{dx}(1) = 3x^{3-1}+(3\cdot 2)x^{2-1}+0 = 3x^2+6x$

at $x=2$ $\quad \frac{d}{dx}\left(x^3+3x^2+1\right) = 3\cdot 2^2+6\cdot 2 = 12+12 = \mathbf{24}$

b. $\frac{d}{dx}\left[(x+1)\left(x^2-1\right)\right] = \left(x^2-1\right)\frac{d}{dx}(x+1)+(x+1)\frac{d}{dx}\left(x^2-1\right) = \left(x^2-1\right)\cdot 1+(x+1)\cdot 2x = x^2-1+2x^2+2x = 3x^2+2x-1$

at $x=1$ $\quad \frac{d}{dx}\left[(x+1)\left(x^2-1\right)\right] = \left(3\cdot 1^2\right)+(2\cdot 1)-1 = 3+2-1 = \mathbf{4}$

c. $\frac{d}{ds}\left[3s^2(s-1)\right] = \frac{d}{ds}\left(3s^3-3s^2\right) = \frac{d}{ds}\left(3s^3\right)+\frac{d}{ds}\left(-3s^2\right) = 9s^2-6s$

at $s=0$ $\quad \frac{d}{ds}\left[3s^2(s-1)\right] = \left(9\cdot 0^2\right)-(6\cdot 0) = \mathbf{0}$

d. $\frac{d}{dt}\left[\frac{t^2+1}{t-1}\right] = \frac{\left[(t-1)\frac{d}{dt}\left(t^2+1\right)\right]-\left[\left(t^2+1\right)\frac{d}{dt}(t-1)\right]}{(t-1)^2} = \frac{[(t-1)\cdot 2t]-\left[\left(t^2+1\right)\cdot 1\right]}{(t-1)^2} = \frac{2t^2-2t-t^2-1}{(t-1)^2} = \frac{t^2-2t-1}{(t-1)^2}$

at $t=-1$ $\quad \frac{d}{dt}\left[\frac{t^2+1}{t-1}\right] = \frac{(-1)^2+(-2\cdot -1)-1}{(-1-1)^2} = \frac{1+2-1}{4} = \frac{2}{4} = \mathbf{\frac{1}{2}}$

e. $\frac{d}{du}\left[\frac{u^3}{(u+1)^2}\right] = \frac{\left[(u+1)^2\frac{d}{du}u^3\right]-\left[u^3\frac{d}{du}(u+1)^2\right]}{(u+1)^4} = \frac{\left[(u+1)^2\cdot 3u^2\right]-\left[u^3\cdot 2(u+1)\right]}{(u+1)^4} = \frac{\left[\left(u^2+2u+1\right)\cdot 3u^2\right]-\left[2u^4+2u^3\right]}{(u+1)^4}$

$= \frac{3u^4+6u^3+3u^2-2u^4-2u^3}{(u+1)^4} = \frac{u^4+4u^3+3u^2}{(u+1)^4}$ $\quad$ at $u=1$ $\quad \frac{d}{du}\left[\frac{u^3}{(u+1)^2}\right] = \frac{1^4+\left(4\cdot 1^3\right)+\left(3\cdot 1^2\right)}{(1+1)^4} = \frac{1+4+3}{2^4} = \frac{8}{16} = \mathbf{\frac{1}{2}}$

f. $\frac{d}{dw}\left[\frac{w\left(w^2+1\right)}{3w^2}\right] = \frac{d}{dw}\left[\frac{w\left(w^2+1\right)}{3w^{2-1}}\right] = \frac{d}{dw}\left[\frac{w^2+1}{3w}\right] = \frac{\left[3w\frac{d}{du}\left(w^2+1\right)\right]-\left[\left(w^2+1\right)\frac{d}{du}3w\right]}{(3w)^2} = \frac{[3w\cdot 2w]-\left[\left(w^2+1\right)\cdot 3\right]}{9w^2}$

$= \frac{6w^2-3w^2-3}{9w^2} = \frac{3\left(w^2-1\right)}{9w^2} = \frac{w^2-1}{3w^2}$ $\quad$ at $w=2$ $\quad \frac{d}{dw}\left[\frac{w\left(w^2+1\right)}{3w^2}\right] = \frac{2^2-1}{\left(3\cdot 2^2\right)} = \frac{3}{12} = \mathbf{\frac{1}{4}}$

g. $\frac{d}{dv}\left[\left(v^2+1\right)v^3\right] = \frac{d}{dv}\left(v^5+v^3\right) = \frac{d}{dv}v^5+\frac{d}{dv}v^3 = 5v^4+3v^2$

at $v=-2$ $\quad \frac{d}{dv}\left[\left(v^2+1\right)v^3\right] = 5\cdot(-2)^4+3\cdot(-2)^2 = (5\cdot 16)+(3\cdot 4) = 80+12 = \mathbf{92}$

h. $\frac{d}{dx}\left(\frac{x^3}{x^2+1}\right) = \frac{\left[\left(x^2+1\right)\frac{d}{dx}x^3\right]-\left[x^3\frac{d}{dx}\left(x^2+1\right)\right]}{\left(x^2+1\right)^2} = \frac{\left[\left(x^2+1\right)\cdot 3x^2\right]-\left[x^3\cdot 2x\right]}{\left(x^2+1\right)^2} = \frac{3x^4+3x^2-2x^4}{\left(x^2+1\right)^2} = \frac{x^4+3x^2}{\left(x^2+1\right)^2}$

at $x=0$ $\quad \frac{d}{dx}\left(\frac{x^3}{x^2+1}\right) = \frac{0^4+\left(3\cdot 0^2\right)}{\left(0^2+1\right)^2} = \frac{0+0}{1^2} = \frac{0}{1} = \mathbf{0}$

i. $\frac{d}{du}\left[u^3\left(\frac{u^2}{1-u}\right)\right] = \frac{d}{du}\left(\frac{u^5}{1-u}\right) = \frac{\left[(1-u)\frac{d}{du}u^5\right]-\left[u^5\frac{d}{du}(1-u)\right]}{(1-u)^2} = \frac{\left[(1-u)\cdot 5u^4\right]-\left[\left(u^5\cdot -1\right)\right]}{(1-u)^2} = \frac{5u^4-5u^5+u^5}{(1-u)^2} = \frac{-4u^5+5u^4}{(1-u)^2}$

at $u=0$ $\quad \frac{d}{du}\left[u^3\left(\frac{u^2}{1-u}\right)\right] = \frac{\left(-4\cdot 0^5\right)+\left(5\cdot 0^4\right)}{(1-0)^2} = \frac{0+0}{1} = \frac{0}{1} = \mathbf{0}$

4. Given the functions below find their derivatives at the specified value.

a. $\frac{ds}{dt}$ given $s=\left(t^2-1\right)+(3t+2)^2$, then $\frac{ds}{dt} = \frac{ds}{dt}\left(t^2-1\right)+\frac{ds}{dt}(3t+2)^2 = 2t+2(3t+2)^{2-1}\cdot\frac{ds}{dt}(3t+2) = 2t+2(3t+2)\cdot 3$

$= 2t + 18t + 12 = 20t + 12$ at $t = 2$ $\frac{ds}{dt} = (20 \cdot 2) + 12 = 40 + 12 = \mathbf{52}$

b. $\frac{dy}{dt}$ given $y = \frac{t^3 + 3t^2 + 1}{2t}$, then $\frac{dy}{dt} = \frac{2t\frac{d}{dt}(t^3 + 3t^2 + 1) - (t^3 + 3t^2 + 1)\frac{d}{dt}2t}{(2t)^2} = \frac{2t \cdot (3t^2 + 6t) - (t^3 + 3t^2 + 1) \cdot 2}{4t^2}$

$= \frac{6t^3 + 12t^2 - 2t^3 - 6t^2 - 2}{4t^2} = \frac{4t^3 + 6t^2 - 2}{4t^2} = \frac{2(2t^3 + 3t^2 - 1)}{\underset{2}{4}t^2} = \frac{2t^3 + 3t^2 - 1}{2t^2}$

at $t = 1$ $\frac{dy}{dt} = \frac{2 \cdot 1^3 + 3 \cdot 1^2 - 1}{2 \cdot 1^2} = \frac{2 + 3 - 1}{2} = \frac{4}{2} = \mathbf{2}$

c. $\frac{dw}{dx}$ given $w = (x^2 + 1)^2 + 3x$, then $\frac{dw}{dx} = \frac{d}{dx}(x^2 + 1)^2 + 3x = 2(x^2 + 1)^{2-1}\frac{d}{dx}(x^2 + 1) + \frac{d}{dx}3x = 2(x^2 + 1) \cdot 2x + 3$

$= 4x^3 + 4x + 3$ at $x = -1$ $\frac{dw}{dx} = 4 \cdot (-1)^3 + (4 \cdot -1) + 3 = -4 - 4 + 3 = \mathbf{-5}$

d. $\frac{dy}{dx}$ given $y = x^2(x^3 + 2x + 1)^2 + 3x$, then $\frac{dy}{dx} = \frac{d}{dx}\left[x^2(x^3 + 2x + 1)^2 + 3x\right] = \frac{d}{dx}\left[x^2(x^3 + 2x + 1)^2\right] + \frac{d}{dx}3x$

$= (x^3 + 2x + 1)^2\frac{d}{dx}x^2 + x^2\frac{d}{dx}(x^3 + 2x + 1)^2 + 3 = \left[(x^3 + 2x + 1)^2 \cdot 2x\right] + \left[x^2 \cdot 2(x^3 + 2x + 1)^{2-1} \cdot \frac{d}{dx}(x^3 + 2x + 1)\right] + 3$

$= 2x(x^3 + 2x + 1)^2 + \left[2x^2(x^3 + 2x + 1)(3x^2 + 2)\right] + 3$ at $x = 0$ $\frac{dy}{dx} = 0 + 0 + 3 = \mathbf{3}$

Section 5.4 Solutions - The Chain Rule

1. Find the derivative of the following functions. Do not simplify the answer to its lowest term.

a. Given $y = (x^2 + 2)^3$, then $y' = 3(x^2 + 2)^{3-1} \cdot 2x = \mathbf{6x(x^2 + 2)^2}$

b. Given $y = (x^2 + 1)^{-2}$, then $y' = -2(x^2 + 1)^{-2-1} \cdot 2x = \mathbf{-4x(x^2 + 1)^{-3}}$

c. Given $y = (x^3 - 1)^5$, then $y' = 5(x^3 - 1)^{5-1} \cdot 3x^2 = \mathbf{15x^2(x^3 - 1)^4}$

d. Given $y = \left(1 - \frac{1}{x^2}\right)^2$, then $y' = 2\left(1 - \frac{1}{x^2}\right)^{2-1} \cdot \frac{2x}{x^4} = 4\left(1 - \frac{1}{x^2}\right) \cdot \frac{\cancel{x}}{x^{4-3}} = \mathbf{\frac{4}{x^3}\left(1 - \frac{1}{x^2}\right)}$

e. Given $y = 2x^3 + \frac{1}{3x^2}$, then $y' = (2 \cdot 3)x^{3-1} + \frac{(0 \cdot 3x^2) - (6x \cdot 1)}{(3x^2)^2} = 6x^2 + \frac{0 - 6x}{9x^4} = 6x^2 - \frac{6\cancel{x}}{9x^{4-3}} = \mathbf{6x^2 - \frac{2}{3x^3}}$

f. Given $y = \left(\frac{1 + x^2}{x^3}\right)^4$, then $y' = 4\left(\frac{1 + x^2}{x^3}\right)^{4-1} \cdot \frac{\left[2x \cdot x^3\right] - \left[3x^2(1 + x^2)\right]}{x^6} = 4\left(\frac{1 + x^2}{x^3}\right)^3 \cdot \frac{2x^4 - 3x^2 - 3x^4}{x^6} = 4\left(\frac{1 + x^2}{x^3}\right)^3$

$\cdot \frac{-x^4 - 3x^2}{x^6} = 4\left(\frac{1 + x^2}{x^3}\right)^3 \cdot \frac{-\cancel{x^2}(x^2 + 3)}{x^{6-4}} = \mathbf{-4\left(\frac{1 + x^2}{x^3}\right)^3\left(\frac{x^2 + 3}{x^4}\right)}$

g. Given $y = x^2\left(\frac{x + 1}{3}\right)^3$, then $y' = \left[2x^{2-1} \cdot \left(\frac{x + 1}{3}\right)^3\right] + \left[3\left(\frac{x + 1}{3}\right)^{3-1} \cdot \frac{1}{3} \cdot x^2\right] = \mathbf{2x\left(\frac{x + 1}{3}\right)^3 + x^2\left(\frac{x + 1}{3}\right)^2}$

h. Given $y = \left[x(x + 1)^2 + 2x\right]^3 = \left[x(x^2 + 2x + 1) + 2x\right]^3 = (x^3 + 2x^2 + x + 2x)^3 = (x^3 + 2x^2 + 3x)^3$, then $y' = 3(x^3 + 2x^2 + 3x)^{3-1}$

$\cdot (3x^2 + 4x + 3) = \mathbf{3(x^3 + 2x^2 + 3x)^2(3x^2 + 4x + 3)}$

i. Given $y = \left(\frac{x}{3} - 2x^3\right)^{-1}$, then $y' = -\left(\frac{x}{3} - 2x^3\right)^{-1-1} \cdot \left(\frac{1}{3} - (2 \cdot 3)x^{3-1}\right) = \mathbf{-\left(\frac{x}{3} - 2x^3\right)^{-2}\left(\frac{1}{3} - 6x^2\right)}$

j. Given $y=\left(x^3+3x^2+1\right)^4$, then $y' = 4\left(x^3+3x^2+1\right)^{4-1}\cdot\left(3x^{3-1}+(3\cdot 2)x^{2-1}+0\right) = \mathbf{4\left(x^3+3x^2+1\right)^3\left(3x^2+6x\right)}$

k. Given $y=\left(\frac{t^2}{1+t^2}\right)^3$, then $y' = 3\left(\frac{t^2}{1+t^2}\right)^{3-1}\cdot\frac{\left[2t\cdot\left(1+t^2\right)\right]-\left[2t\cdot t^2\right]}{\left(1+t^2\right)^2} = 3\left(\frac{t^2}{1+t^2}\right)^2\cdot\frac{2t+2t^3-2t^3}{\left(1+t^2\right)^2} = \frac{3t^4}{\left(1+t^2\right)^2}\cdot\frac{2t}{\left(1+t^2\right)^2}$

$= \frac{3t^4\cdot 2t}{\left(1+t^2\right)^2\cdot\left(1+t^2\right)^2} = \frac{6t^{4+1}}{\left(1+t^2\right)^{2+2}} = \mathbf{\frac{6t^5}{\left(1+t^2\right)^4}}$

l. Given $y=\left(1+x^{-2}\right)^{-1}$, then $y' = -\left(1+x^{-2}\right)^{-1-1}\cdot -2x^{-2-1} = 2x^{-3}\left(1+x^{-2}\right)^{-2} = \mathbf{\frac{2}{x^3\left(1+x^{-2}\right)^2}}$

m. Given $y=\frac{(x+1)^{-2}}{x^3}$, then $y' = \frac{\left[-2(x+1)^{-2-1}\cdot 1\cdot x^3\right]-\left[3x^{3-1}\cdot(x+1)^{-2}\right]}{x^6} = \mathbf{\frac{\left[-2x^3(x+1)^{-3}\right]-\left[3x^2(x+1)^{-2}\right]}{x^6}}$

n. Given $y=\left(\frac{1}{1-x^3}\right)^2+\frac{1}{x}$, then $y' = \left[2\left(\frac{1}{1-x^3}\right)^{2-1}\cdot\frac{3x^2}{\left(1-x^3\right)^2}\right]-\frac{1}{x^2} = \left[\left(\frac{1}{1-x^3}\right)\cdot\frac{6x^2}{\left(1-x^3\right)^2}\right]-\frac{1}{x^2} = \mathbf{\frac{6x^2}{\left(1-x^3\right)^3}-\frac{1}{x^2}}$

o. Given $y=\frac{x^3}{x^3+2}-x^2$, then $y' = \frac{\left[3x^{3-1}\cdot\left(x^3+2\right)\right]-\left[3x^2\cdot x^3\right]}{\left(x^3+2\right)^2}-2x^{2-1} = \frac{3x^2\left(x^3+2\right)-3x^5}{\left(x^3+2\right)^2}-2x = \frac{3x^5+6x^2-3x^5}{\left(x^3+2\right)^2}-2x$

$= \frac{6x^2}{\left(x^3+2\right)^2}-\frac{2x}{1} = \mathbf{\frac{6x^2-2x\left(x^3+2\right)^2}{\left(x^3+2\right)^2}}$

2. Find the derivative of the following functions at $x=0$, $x=1$, and $x=-1$.

a. Given $y=\left(x^3+1\right)^5$, then $y' = 5\left(x^3+1\right)^{5-1}\cdot 3x^2 = 15x^2\left(x^3+1\right)^4$. Therefore,

$y'(0) = \left(15\cdot 0^2\right)\left(0^3+1\right)^4 = 0\cdot 1^4 = 0\cdot 1 = \mathbf{0}$

$y'(1) = \left(15\cdot 1^2\right)\left(1^3+1\right)^4 = 15\cdot 2^4 = 15\cdot 2^4 = 15\cdot 16 = \mathbf{240}$ and

$y'(-1) = \left[15\cdot(-1)^2\right]\left[(-1)^3+1\right]^4 = 15\cdot(-1+1)^4 = 15\cdot 0^4 = 15\cdot 0 = \mathbf{0}$

b. Given $y=\left(x^3+3x^2-1\right)^4$, then $y' = 4\left(x^3+3x^2-1\right)^{4-1}\left(3x^2+6x\right) = 12x\left(x^3+3x^2-1\right)^3(x+2)$. Therefore,

$y'(0) = (12\cdot 0)\left(0^3+3\cdot 0^2-1\right)^3(0+2) = 0.(-1)^3\cdot 2 = \mathbf{0}$

$y'(1) = (12\cdot 1)\left(1^3+3\cdot 1^2-1\right)^3(1+2) = 12\cdot(1+3-1)^3\cdot 3 = 12\cdot 27\cdot 3 = \mathbf{972}$ and

$y'(-1) = (12\cdot -1)\left[(-1)^3+3\cdot(-1)^2-1\right]^3(-1+2) = -12(-1+3-1)^3 = -12(-2+3)^3 = -12\cdot 1^3 = -12\cdot 1 = \mathbf{-12}$

c. Given $y=\left(\frac{x}{x+1}\right)^2$, then $y' = 2\left(\frac{x}{x+1}\right)^{2-1}\cdot\frac{1\cdot(x+1)-1\cdot x}{(x+1)^2} = 2\left(\frac{x}{x+1}\right)\cdot\frac{x+1-x}{(x+1)^2} = 2\left(\frac{x}{x+1}\right)\cdot\frac{1}{(x+1)^2} = \frac{2x}{(x+1)^3}$. Thus,

$y'(0) = \frac{2\cdot 0}{(0+1)^3} = \frac{0}{1^3} = \frac{0}{1} = \mathbf{0}$

$y'(1) = \frac{2\cdot 1}{(1+1)^3} = \frac{2}{2^3} = \frac{2}{8} = \frac{1}{4} = \mathbf{0.25}$ and

$y'(-1) = \frac{2\cdot -1}{(-1+1)^3} = \frac{-2}{0^3} = \mathbf{-\frac{2}{0}}$ **which is undefined due to division by zero**

d. Given $y=x\left(x^2+1\right)^2$, then $y' = \left[1\cdot\left(x^2+1\right)^2\right]+\left[2\left(x^2+1\right)^{2-1}\cdot 2x\right] = \left(x^2+1\right)^2+4x\left(x^2+1\right)$. Therefore,

$y'(0) = \left(0^2+1\right)^2+(4\cdot 0)\left(0^2+1\right) = 1^2+0 = \mathbf{1}$

$y'(1) = (1^2+1)^2 + (4\cdot 1)(1^2+1) = 2^2 + 4\cdot 2 = 4+8 = \mathbf{12}$ and

$y'(-1) = \left[(-1)^2+1\right]^2 + (4\cdot -1)\left[(-1)^2+1\right] = (1+1)^2 - 4\cdot(1+1) = 2^2 - 4\cdot 2 = 4-8 = \mathbf{-4}$

e. Given $y = x^3 + 2(x^2+1)^3$, then $y' = 3x^{3-1} + 2\cdot 3(x^2+1)^{3-1}\cdot 2x = 3x^2 + 12x(x^2+1)^2$. Therefore,

$y'(0) = 3\cdot 0^2 + (12\cdot 0)(0^2+1)^2 = 3\cdot 0 + 0\cdot 1^2 = 0+0 = \mathbf{0}$

$y'(1) = 3\cdot 1^2 + (12\cdot 1)(1^2+1)^2 = 3+12\cdot 2^2 = 3+12\cdot 4 = 3+48 = \mathbf{51}$ and

$y'(-1) = 3\cdot(-1)^2 + (12\cdot -1)\left[(-1)^2+1\right]^2 = 3\cdot 1 - 12\cdot(1+1)^2 = 3-12\cdot 2^2 = 3-12\cdot 4 = 3-48 = \mathbf{-45}$

f. Given $y = \left(\dfrac{x^2}{1+x^2}\right)^3$, then $y' = 3\left(\dfrac{x^2}{1+x^2}\right)^{3-1}\cdot\dfrac{\left[2x\cdot(1+x^2)\right]-\left[2x\cdot x^2\right]}{(1+x^2)^2} = 3\left(\dfrac{x^2}{1+x^2}\right)^2\cdot\dfrac{2x+2x^3-2x^3}{(1+x^2)^2}$

$= \dfrac{3x^4}{(1+x^2)^2}\cdot\dfrac{2x}{(1+x^2)^2} = \dfrac{(3\cdot 2)x^{4+1}}{(1+x^2)^{2+2}} = \dfrac{6x^5}{(1+x^2)^4}$. Therefore,

$y'(0) = \dfrac{6\cdot 0^5}{(1+0^2)^4} = \dfrac{0}{1^4} = \dfrac{0}{1} = \mathbf{0}$

$y'(1) = \dfrac{6\cdot 1^5}{(1+1^2)^4} = \dfrac{6}{2^4} = \dfrac{6}{16} = \dfrac{3}{8} = \mathbf{0.375}$ and

$y'(-1) = \dfrac{6\cdot(-1)^5}{\left[1+(-1)^2\right]^4} = \dfrac{6\cdot -1}{(1+1)^4} = -\dfrac{6}{2^4} = -\dfrac{6}{16} = -\dfrac{3}{8} = \mathbf{-0.375}$

g. Given $y = \left(\dfrac{x}{x^2+1}\right)^5$, then $y' = 5\left(\dfrac{x}{x^2+1}\right)^{5-1}\cdot\dfrac{\left[1\cdot(x^2+1)\right]-\left[2x\cdot x\right]}{(x^2+1)^2} = 5\left(\dfrac{x}{x^2+1}\right)^4\cdot\dfrac{x^2+1-2x^2}{(x^2+1)^2} = \dfrac{5x^4}{(x^2+1)^4}\cdot\dfrac{1-x^2}{(x^2+1)^2}$

$= \dfrac{5x^4(1-x^2)}{(x^2+1)^{4+2}} = \dfrac{5x^4(1-x^2)}{(x^2+1)^6}$. Therefore,

$y'(0) = \dfrac{5\cdot 0^4\cdot(1-0^2)}{(0^2+1)^6} = \dfrac{0\cdot 1}{1^6} = \dfrac{0}{1} = \mathbf{0}$

$y'(1) = \dfrac{5\cdot 1^4\cdot(1-1^2)}{(1^2+1)^6} = \dfrac{5\cdot(1-1)}{2^6} = \dfrac{5\cdot 0}{64} = \dfrac{0}{64} = \mathbf{0}$ and

$y'(-1) = \dfrac{5\cdot(-1)^4\cdot(1-(-1)^2)}{\left((-1)^2+1\right)^6} = \dfrac{5\cdot 1\cdot(1-1)}{(1+1)^6} = \dfrac{5\cdot 0}{2^6} = \dfrac{0}{64} = \mathbf{0}$

h. Given $y = (x^2+1)^3\cdot\dfrac{1}{x^2} = y = \dfrac{(x^2+1)^3}{x^2}$, then $y' = \dfrac{\left[3(x^2+1)^{3-1}\cdot 2x\cdot x^2\right]-\left[2x\cdot(x^2+1)^3\right]}{x^4} = \dfrac{6x^3(x^2+1)^2 - 2x(x^2+1)^3}{x^4}$

$= \dfrac{2x(x^2+1)^2\left[3x^2-(x^2+1)\right]}{x^{4=3}} = \dfrac{2(x^2+1)^2(3x^2-x^2-1)}{x^3} = \dfrac{2(x^2+1)^2(2x^2-1)}{x^3}$. Therefore,

$y'(0) = \dfrac{2(0^2+1)^2\cdot(2\cdot 0^2-1)}{0^3} = \dfrac{2\cdot 1^2\cdot(0-1)}{0} = \dfrac{2\cdot -1}{0} = -\dfrac{\mathbf{2}}{\mathbf{0}}$ **which is undefined due to division by zero**

$y'(1) = \dfrac{2(1^2+1)^2\cdot(2\cdot 1^2-1)}{1^3} = \dfrac{2\cdot 2^2\cdot(2-1)}{1} = \dfrac{2\cdot 4\cdot 1}{1} = \dfrac{8}{1} = \mathbf{8}$ and

$y'(-1) = \dfrac{2\left[(-1)^2+1\right]^2\cdot\left[2\cdot(-1)^2-1\right]}{(-1)^3} = \dfrac{2(1+1)^2\cdot(2\cdot 1-1)}{-1} = \dfrac{2\cdot 2^2\cdot(2-1)}{-1} = \dfrac{2\cdot 4\cdot 1}{-1} = -\dfrac{8}{1} = \mathbf{-8}$

i. Given $y=\left(\frac{x^3}{x-1}\right)^2+5x$, then $y' = 2\left(\frac{x^3}{x-1}\right)^{2-1}\cdot\frac{\left[3x^2\cdot(x-1)\right]-\left[1\cdot x^3\right]}{(x-1)^2}+5 = 2\left(\frac{x^3}{x-1}\right)\cdot\left(\frac{3x^3-3x^2-x^3}{(x-1)^2}\right)+5$

$= \frac{2x^3}{x-1}\cdot\frac{2x^3-3x^2}{(x-1)^2}+5 = \frac{2x^3}{x-1}\cdot\frac{x^2(2x-3)}{(x-1)^2}+5 = \frac{2x^3\cdot x^2\left(2x^3-3x^2\right)}{(x-1)(x-1)^2}+5 = \frac{2x^{3+2}(2x-3)}{(x-1)^{1+2}}+5 = \frac{2x^5(2x-3)}{(x-1)^3}+5$. Therefore,

$$y'(0) = \frac{2\cdot 0^5\cdot(2\cdot 0-3)}{(0-1)^3}+5 = \frac{0\cdot(0-3)}{1^3}+5 = \frac{0}{1}+5 = 0+5 = \mathbf{5}$$

$y'(1) = \frac{2\cdot 1^5\cdot(2\cdot 1-3)}{(1-1)^3}+5 = \frac{2\cdot(2-3)}{0^3}+5 = \frac{2\cdot -1}{0}+5 = -\frac{\mathbf{2}}{\mathbf{0}}+\mathbf{5}$ **which is undefined due to division by zero** and

$$y'(-1) = \frac{2\cdot(-1)^5\cdot[(2\cdot -1)-3]}{(-1-1)^3}+5 = \frac{2\cdot -1\cdot[-2-3]}{(-2)^3}+5 = \frac{-2\cdot -5}{-8}+5 = \frac{10}{8}+5 = \frac{5}{4}+5 = 1.25+5 = \mathbf{6.25}$$

3. Use the chain rule to differentiate the following functions.

a. $\frac{d}{dt}\left[\frac{(t+1)^3}{t^2}\right] = \frac{\left[t^2\cdot\frac{d}{dt}(t+1)^3\right]-\left[(t+1)^3\cdot\frac{d}{dt}t^2\right]}{t^4} = \frac{\left[t^2\cdot 3(t+1)^2\right]-\left[(t+1)^3\cdot 2t\right]}{t^4} = \frac{3t^2(t+1)^2-2t(t+1)^3}{t^4} = \frac{t(t+1)^2[3t-2(t+1)]}{t^{4=3}}$

$= \frac{(t+1)^2(3t-2t-2)}{t^3} = \frac{\mathbf{(t+1)^2(t-2)}}{\mathbf{t^3}}$

b. $\frac{d}{du}\left[\frac{\left(u^2+1\right)^3}{3u^4}\right] = \frac{\left[3u^4\frac{d}{du}\left(u^2+1\right)^3\right]-\left[\left(u^2+1\right)^3\frac{d}{du}3u^4\right]}{\left(3u^4\right)^2} = \frac{\left[3u^4\cdot 3\left(u^2+1\right)^{3-1}\cdot 2u\right]-\left[\left(u^2+1\right)^3\cdot(3\cdot 4)u^{4-1}\cdot 1\right]}{9u^8}$

$= \frac{18u^5\left(u^2+1\right)^2-12u^3\left(u^2+1\right)^3}{9u^8} = \frac{6u^3\left(u^2+1\right)^2\left[3u^2-2\left(u^2+1\right)\right]}{9u^{8=5}} = \frac{2\left(u^2+1\right)^2\left(3u^2-2u^2-2\right)}{3u^5} = \frac{\mathbf{2\left(u^2+1\right)^2\left(u^2-2\right)}}{\mathbf{3u^5}}$

c. $\frac{d}{dx}\left[\frac{(2x+1)^3}{(1-x)^2}\right] = \frac{\left[(1-x)^2\frac{d}{dx}(2x+1)^3\right]-\left[(2x+1)^3\frac{d}{dx}(1-x)^2\right]}{(1-x)^4} = \frac{\left[(1-x)^2\cdot 3(2x+1)^{3-1}\cdot 2\right]-\left[(2x+1)^3\cdot 2(1-x)^{2-1}\cdot -1\right]}{(1-x)^4}$

$= \frac{6(1-x)^2(2x+1)^2+2(2x+1)^3(1-x)}{(1-x)^4} = \frac{2(1-x)(2x+1)^2[3(1-x)+(2x+1)]}{(1-x)^{4=3}} = \frac{\mathbf{2(2x+1)^2(4-x)}}{\mathbf{(1-x)^3}}$

d. $\frac{d}{dx}\left[\left(x^3-1\right)^2(2x+1)^3\right] = \left[(2x+1)^3\frac{d}{dx}\left(x^3-1\right)^2\right]+\left[\left(x^3-1\right)^2\frac{d}{dx}(2x+1)^3\right] = \left[(2x+1)^3\cdot 2\left(x^3-1\right)^{2-1}\cdot 3x^2\right]$

$+\left[\left(x^3-1\right)^2\cdot 3(2x+1)^{3-1}\cdot 2\right] = \left[6x^2(2x+1)^3\left(x^3-1\right)\right]+\left[6\left(x^3-1\right)^2(2x+1)^2\right] = 6(2x+1)^2\left(x^3-1\right)\left[x^2(2x+1)+\left(x^3-1\right)\right]$

$= 6(2x+1)^2\left(x^3-1\right)\left(2x^3+x^2+x^3-1\right) = \mathbf{6(2x+1)^2\left(x^3-1\right)\left(3x^3+x^2-1\right)}$

e. $\frac{d}{ds}\left[s^3-\frac{1}{s^2+6}\right]^2 = 2\left[s^3-\frac{1}{s^2+6}\right]^{2-1}\cdot\frac{d}{ds}\left[s^3-\frac{1}{s^2+6}\right] = 2\left[s^3-\frac{1}{s^2+6}\right]\cdot\left[\frac{d}{ds}s^3-\frac{d}{ds}\frac{1}{s^2+6}\right] = 2\left[s^3-\frac{1}{s^2+6}\right]$

$\cdot\left[3s^{3-1}-\frac{\left(s^2+6\right)\cdot\frac{d}{ds}(1)-1\cdot\frac{d}{ds}\left(s^2+6\right)}{\left(s^2+6\right)^2}\right] = 2\left[s^3-\frac{1}{s^2+6}\right]\cdot\left[3s^2-\frac{0-2s}{\left(s^2+6\right)^2}\right] = \mathbf{2\left[s^3-\frac{1}{s^2+6}\right]\cdot\left[3s^2+\frac{2s}{\left(s^2+6\right)^2}\right]}$

f. $\frac{d}{dt}\left[\frac{\left(t^2-1\right)^3}{t^2+1}\right] = \frac{\left[\left(t^2+1\right)\frac{d}{dt}\left(t^2-1\right)^3\right]-\left[\left(t^2-1\right)^3\frac{d}{dt}\left(t^2+1\right)\right]}{\left(t^2+1\right)^2} = \frac{\left[\left(t^2+1\right)\cdot 3\left(t^2-1\right)^{3-1}\cdot 2t\right]-\left[\left(t^2-1\right)^3\cdot 2t\right]}{\left(t^2+1\right)^2}$

$$= \frac{\left[6t\left(t^2+1\right)\left(t^2-1\right)^2\right]-\left[2t\left(t^2-1\right)^3\right]}{\left(t^2+1\right)^2} = \frac{2t\left(t^2-1\right)^2\left[3\left(t^2+1\right)-\left(t^2-1\right)\right]}{\left(t^2+1\right)^2} = \frac{2t\left(t^2-1\right)^2\left[3t^2+3-t^2+1\right]}{\left(t^2+1\right)^2} = \frac{\mathbf{4t\left(t^2-1\right)^2\left(t^2+2\right)}}{\mathbf{\left(t^2+1\right)^2}}$$

g. $$\frac{d}{du}\left[\left(u^2+1\right)^3\left(\frac{1}{u+1}\right)^2\right] = \frac{d}{du}\left[\left(u^2+1\right)^3\frac{1}{(u+1)^2}\right] = \frac{d}{du}\left[\frac{\left(u^2+1\right)^3}{(u+1)^2}\right] = \frac{\left[(u+1)^2\frac{d}{du}\left(u^2+1\right)^3\right]-\left[\left(u^2+1\right)^3\frac{d}{du}(u+1)^2\right]}{(u+1)^4}$$

$$= \frac{\left[(u+1)^2\cdot 3\left(u^2+1\right)^{3-1}\cdot 2u\right]-\left[\left(u^2+1\right)^3\cdot 2(u+1)^{2-1}\right]}{(u+1)^4} = \frac{\left[6u(u+1)^2\left(u^2+1\right)^2\right]-\left[2\left(u^2+1\right)^3(u+1)\right]}{(u+1)^4}$$

$$= \frac{2(u+1)\left(u^2+1\right)^2\left[3u(u+1)-\left(u^2+1\right)\right]}{(u+1)^{4=3}} = \frac{2\left(u^2+1\right)^2\left(3u^2+3u-u^2-1\right)}{(u+1)^3} = \frac{\mathbf{2\left(u^2+1\right)^2\left(2u^2+3u-1\right)}}{\mathbf{(u+1)^3}}$$

h. $$\frac{d}{d\theta}\left[\frac{\theta^2+3}{(\theta-1)^3}\right]^2 = 2\left[\frac{\theta^2+3}{(\theta-1)^3}\right]^{2-1}\cdot\frac{d}{d\theta}\left[\frac{\theta^2+3}{(\theta-1)^3}\right] = 2\left[\frac{\theta^2+3}{(\theta-1)^3}\right]\cdot\frac{\left[(\theta-1)^3\frac{d}{d\theta}\left(\theta^2+3\right)\right]-\left[\left(\theta^2+3\right)\frac{d}{d\theta}(\theta-1)^3\right]}{(\theta-1)^6}$$

$$= \frac{2\left(\theta^2+3\right)}{(\theta-1)^3}\cdot\frac{\left[(\theta-1)^3\cdot 2\theta\right]-\left[\left(\theta^2+3\right)\cdot 3(\theta-1)^2\right]}{(\theta-1)^6} = \frac{2\left(\theta^2+3\right)}{(\theta-1)^3}\cdot\frac{2\theta(\theta-1)^3-3\left(\theta^2+3\right)(\theta-1)^2}{(\theta-1)^6} = \frac{2\left(\theta^2+3\right)}{(\theta-1)^3}\cdot\frac{(\theta-1)^2\left[2\theta(\theta-1)-3\left(\theta^2+3\right)\right]}{(\theta-1)^{6=4}}$$

$$= \frac{2\left(\theta^2+3\right)}{(\theta-1)^3}\cdot\frac{\left[2\theta(\theta-1)-3\left(\theta^2+3\right)\right]}{(\theta-1)^4} = \frac{2\left(\theta^2+3\right)\cdot\left(2\theta^2-2\theta-3\theta^2-9\right)}{(\theta-1)^{3+4}} = \frac{\mathbf{-2\left(\theta^2+3\right)\cdot\left(\theta^2+2\theta+9\right)}}{\mathbf{(\theta-1)^7}}$$

i. $$\frac{d}{dr}\left[\frac{r^7}{\left(r^2+2r\right)^3}\right] = \frac{\left(r^2+2r\right)^3\frac{d}{dr}r^7-r^7\frac{d}{dr}\left(r^2+2r\right)^3}{\left(r^2+2r\right)^6} = \frac{\left[\left(r^2+2r\right)^3\cdot 7r^{7-1}\right]-\left[r^7\cdot 3\left(r^2+2r\right)^{3-1}\cdot\frac{d}{dr}\left(r^2+2r\right)\right]}{\left(r^2+2r\right)^6}$$

$$= \frac{7r^6\left(r^2+2r\right)^3-\left[r^7\cdot 3\left(r^2+2r\right)\cdot(2r+2)\right]}{\left(r^2+2r\right)^6} = \frac{7r^6\left(r^2+2r\right)^3-\left[r^7\cdot 3r(r+2)\cdot 2(r+1)\right]}{\left(r^2+2r\right)^6} = \frac{\mathbf{7r^6\left(r^2+2r\right)^3-6r^8(r+2)(r+1)}}{\mathbf{\left(r^2+2r\right)^6}}$$

4. Given the following y functions in terms of u, find y'.

a. Given $y = 2u^2-1$ and $u = x-1$, then $y = 2(x-1)^2-1$ and $y' = 2\cdot 2(x-1)^{2-1}-0 = \mathbf{4(x-1)}$

b. Given $y = \frac{u}{u-1}$ and $u = x^3$, then $y = \frac{x^3}{x^3-1}$ and $y' = \frac{\left[3x^{3-1}\cdot\left(x^3-1\right)\right]-\left[3x^{3-1}\cdot x^3\right]}{\left(x^3-1\right)^2} = \frac{\left[3x^2\left(x^3-1\right)\right]-\left[3x^2\cdot x^3\right]}{\left(x^3-1\right)^2}$

$$= \frac{3x^5-3x^2-3x^5}{\left(x^3-1\right)^2} = -\frac{\mathbf{3x^2}}{\mathbf{\left(x^3-1\right)^2}}$$

c. Given $y = \frac{u}{1+u^2}$ and $u = x^2+1$, then $y = \frac{x^2+1}{1+\left(x^2+1\right)^2}$ and $y' = \frac{2x\cdot\left[1+\left(x^2+1\right)^2\right]-\left[2\left(x^2+1\right)^{2-1}\cdot 2x\cdot\left(x^2+1\right)\right]}{\left[1+\left(x^2+1\right)^2\right]^2}$

$$= \frac{\left[2x+2x\left(x^2+1\right)^2\right]-\left[4x\left(x^2+1\right)\cdot\left(x^2+1\right)\right]}{\left[1+\left(x^2+1\right)^2\right]^2} = \frac{2x+2x\left(x^2+1\right)^2-4x\left(x^2+1\right)^2}{\left[1+\left(x^2+1\right)^2\right]^2} = \frac{2x-2x\left(x^2+1\right)^2}{\left[1+\left(x^2+1\right)^2\right]^2} = \frac{\mathbf{2x\left[1-\left(x^2+1\right)^2\right]}}{\mathbf{\left[1+\left(x^2+1\right)^2\right]^2}}$$

d. Given $y = u^2-\frac{1}{2}$ and $u = x^4$, then $y = x^8-\frac{1}{2}$ and $y' = 8x^{8-1}-0 = \mathbf{8x^7}$

e. Given $y = u^4$ and $u = \frac{1}{1-x^2}$, then $y = \frac{1}{(1-x^2)^4}$ and $y' = \frac{\left[0\cdot(1-x^2)^4\right]-\left[4(1-x^2)^{4-1}\cdot -2x\cdot 1\right]}{(1-x^2)^8} = \frac{0+8x(1-x^2)^3}{(1-x^2)^8}$

$= \frac{8x(1-x^2)^3}{(1-x^2)^{8=5}} = \frac{\mathbf{8x}}{\mathbf{(1-x^2)^5}}$

f. Given $y = \frac{u^2}{(u+1)^3}$ and $u = x-1$, then $y = \frac{(x-1)^2}{(x-1+1)^3} = \frac{(x-1)^2}{x^3}$ and $y' = \frac{\left[2(x-1)^{2-1}\cdot x^3\right]-\left[3x^2\cdot(x-1)^2\right]}{x^6}$

$= \frac{2x^3(x-1)-3x^2(x-1)^2}{x^6} = \frac{x^2(x-1)\left[2x-3(x-1)\right]}{x^{6=4}} = \frac{(x-1)(2x-3x+3)}{x^4} = \frac{\mathbf{(x-1)(-x+3)}}{\mathbf{x^4}}$

Section 5.5 Solutions - Implicit Differentiation

Use implicit differentiation method to solve the following functions.

a. Given $x^2y+x=y$, then $\frac{d}{dx}(x^2y+x)=\frac{d}{dx}(y)$; $(2x\cdot y+x^2\cdot y')+1=y'$; $2xy+1=y'-x^2y'$; $2xy+1=y'(1-x^2)$

; $\mathbf{y' = \frac{2xy+1}{1-x^2}}$

b. Given $xy-3x^2+y=0$, then $\frac{d}{dx}(xy-3x^2+y)=\frac{d}{dx}(0)$; $(1\cdot y+x\cdot y')-6x+y'=0$; $y-6x=-xy'-y'$; $y-6x=-y'(x+1)$

; $\frac{y-6x}{x+1}=-y'$; $y'=-\frac{y-6x}{x+1}$; $\mathbf{y' = \frac{6x-y}{x+1}}$

c. Given $x^2y^2+y=3y^3$, then $\frac{d}{dx}(x^2y^2+y)=\frac{d}{dx}(3y^3)$; $(2x\cdot y^2+2yy'\cdot x^2)+y'=9y^2y'$; $2xy^2+2x^2yy'=9y^2y'-y'$

; $2xy^2=9y^2y'-y'-2x^2yy'$; $2xy^2=y'(9y^2-1-2x^2y)$; $\mathbf{y' = \frac{2xy^2}{9y^2-1-2x^2y}}$

d. Given $xy+y^3=5x$, then $\frac{d}{dx}(xy+y^3)=\frac{d}{dx}(5x)$; $(1\cdot y+y'\cdot x)+3y^2\cdot y'=5$; $y+y'x+3y^2y'=5$; $xy'+3y^2y'=5-y$

; $y'(x+3y^2)=5-y$; $\mathbf{y' = \frac{5-y}{x+3y^2}}$

e. Given $4x^4y^4+2y^2=y-1$, then $\frac{d}{dx}(4x^4y^4+2y^2)=\frac{d}{dx}(y-1)$; $4(4x^3\cdot y^4+4y^3y'\cdot x^4)+4yy'=y'$; $16x^3y^4+16x^4y^3y'$

$+4yy'=y'$; $16x^3y^4=y'-16x^4y^3y'-4yy'$; $16x^3y^4=y'(1-16x^4y^3-4y)$; $\mathbf{y' = \frac{16x^3y^4}{1-16x^4y^3-4y}}$

f. Given $xy+x^2y^2-10=0$, then $\frac{d}{dx}(xy+x^2y^2-10)=\frac{d}{dx}(0)$; $(1\cdot y+y'\cdot x)+(2x\cdot y^2+2yy'\cdot x^2)=0$; $y+y'x+2xy^2$

$+2x^2yy'=0$; $y+2xy^2=-y'x-2x^2yy'$; $y+2xy^2=-y'(x+2x^2y)$; $\mathbf{y' = -\frac{y+2xy^2}{x+2x^2y}}$

g. Given $xy^2+y=x^2$, then $\frac{d}{dx}(xy^2+y)=\frac{d}{dx}(x^2)$; $(1\cdot y^2+2yy'\cdot x)+y'=2x$; $2yy'x+y'=2x-y^2$; $\mathbf{y' = \frac{2x-y^2}{2xy+1}}$

h. Given $xy^3+x^3y=x$, then $\frac{d}{dx}(xy^3+x^3y)=\frac{d}{dx}(x)$; $(1\cdot y^3+3y^2y'\cdot x)+(3x^2\cdot y+y'\cdot x^3)=1$; $y^3+3xy^2y'+3x^2y+x^3y'=1$

; $3xy^2y'+x^3y'=1-y^3-3x^2y$; $y'(3xy^2+x^3)=1-y^3-3x^2y$; $\mathbf{y' = \frac{1-y^3-3x^2y}{3xy^2+x^3}}$

i. Given $y^{\frac{1}{2}}+x^2y=x$, then $\frac{d}{dx}\left(y^{\frac{1}{2}}+x^2y\right)=\frac{d}{dx}(x)$; $\frac{1}{2}y^{\frac{1}{2}-1}\cdot y'+(2x\cdot y+y'\cdot x^2)=1$; $\frac{1}{2}y^{-\frac{1}{2}}y'+2xy+x^2y'=1$

$; y'\left(\frac{1}{2}y^{-\frac{1}{2}}+x^2\right)=1-2xy \; ; \; y'\left(\frac{1}{2y^{\frac{1}{2}}}+x^2\right)=1-2xy \; ; \; \mathbf{y'=\frac{1-2xy}{\frac{1}{2\sqrt{y}}+x^2}}$

j. Given $x^2y+y^2=y^{\frac{1}{4}}$, then $\frac{d}{dx}\left(x^2y+y^2\right)=\frac{d}{dx}\left(y^{\frac{1}{4}}\right)$; $\left(2x\cdot y+y'\cdot x^2\right)+2y\,y'=\frac{1}{4}y^{\frac{1}{4}-1}y'$; $2x\,y+x^2y'+2y\,y'=\frac{1}{4}y^{-\frac{3}{4}}y'$

$; x^2y'+2y\,y'-\frac{1}{4}y^{-\frac{3}{4}}y'=-2x\,y \; ; \; y'\left(x^2+2y-\frac{1}{4}y^{-\frac{3}{4}}\right)=-2x\,y \; ; \; y'=\frac{-2x\,y}{x^2+2y-\frac{1}{4}y^{-\frac{3}{4}}} \; ; \; \mathbf{y'=\frac{-2x\,y}{x^2+2y-\frac{1}{4\sqrt[4]{y^3}}}}$

k. Given $x+y^2=x^2-3$, then $\frac{d}{dx}\left(x+y^2\right)=\frac{d}{dx}\left(x^2-3\right)$; $1+2y\,y'=2x$; $2y\,y'=2x-1$; $\mathbf{y'=\frac{2x-1}{2y}}$

l. Given $x^4y^2+y=-3$, then $\frac{d}{dx}\left(x^4y^2+y\right)=\frac{d}{dx}(-3)$; $\left(4x^3\cdot y^2+2y\,y'\cdot x^4\right)+y'=0$; $4x^3y^2+2x^4y\,y'+y'=0$

$; 2x^4y\,y'+y'=-4x^3y^2 \; ; \; y'\left(2x^4y+1\right)=-4x^3y^2 \; ; \; \mathbf{y'=\frac{-4x^3y^2}{2x^4y+1}}$

m. Given $y^7-x^2y^4-x=8$, then $\frac{d}{dx}\left(y^7-x^2y^4-x\right)=\frac{d}{dx}(8)$; $7y^6y'-\left(2x\cdot y^4+4y^3y'\cdot x^2\right)-1=0$; $7y^6y'-2xy^4-4x^2y^3y'=1$

$; 7y^6y'-4x^2y^3y'=1+2x\,y^4 \; ; \; y'\left(7y^6-4x^2y^3\right)=1+2x\,y^4 \; ; \; \mathbf{y'=\frac{1+2x\,y^4}{7y^6-4x^2y^3}}$

n. Given $(x+3)^2=y^2-x$, then $\frac{d}{dx}\left[(x+3)^2\right]=\frac{d}{dx}\left(y^2-x\right)$; $2(x+3)=2y\,y'-1$; $2x+6+1=2y\,y'$; $\mathbf{y'=\frac{2x+7}{2y}}$

o. Given $3x^2y^5+y^2=-x$, then $\frac{d}{dx}\left(3x^2y^5+y^2\right)=\frac{d}{dx}(-x)$; $3\left(2x\cdot y^5+5y^4y'\cdot x^2\right)+2y\,y'=-1$; $6x\,y^5+15x^2y^4y'+2y\,y'=-1$

$; 15x^2y^4y'+2y\,y'=-1-6x\,y^5 \; ; \; y'\left(15x^2y^4+2y\right)=-1-6x\,y^5 \; ; \; \mathbf{y'=-\frac{1+6x\,y^5}{15x^2y^4+2y}}$

Section 5.6 Solutions - The Derivative of Functions with Fractional Exponent

1. Find the derivative of the following exponential expressions.

a. Given $y=x^{\frac{1}{5}}$, then $y'=\frac{1}{5}x^{\frac{1}{5}-1}=\frac{1}{5}x^{\frac{1-5}{5}}=\mathbf{\frac{1}{5}x^{-\frac{4}{5}}}$

b. Given $y=\left(4x^3\right)^{\frac{1}{2}}=4x^{\frac{3}{2}}$, then $y'=\frac{3}{2}\cdot 4x^{\frac{3}{2}-1}=\frac{3}{\not{2}}\cdot\overset{2}{\not{4}}x^{\frac{3-2}{2}}=\mathbf{6x^{\frac{1}{2}}}$

c. Given $y=(2x+1)^{\frac{1}{3}}$, then $y'=\frac{1}{3}(2x+1)^{\frac{1}{3}-1}\cdot 2=\frac{2}{3}(2x+1)^{\frac{1-3}{3}}=\mathbf{\frac{2}{3}(2x+1)^{-\frac{2}{3}}}$

d. Given $y=\left(2x^2+1\right)^{\frac{1}{8}}$, then $y'=\frac{1}{8}\left(2x^2+1\right)^{\frac{1}{8}-1}\cdot 4x=\frac{4x}{8}\left(2x^2+1\right)^{\frac{1-8}{8}}=\mathbf{\frac{x}{2}\left(2x^2+1\right)^{-\frac{7}{8}}}$

e. Given $y=\left(2x^3+3x\right)^{\frac{3}{5}}$, then $y'=\frac{3}{5}\left(2x^3+3x\right)^{\frac{3}{5}-1}\cdot\left(6x^2+3\right)=\frac{3}{5}\left(2x^3+3x\right)^{-\frac{2}{5}}\cdot 3\left(2x^2+1\right)=\mathbf{\frac{9}{5}\left(2x^2+1\right)\left(2x^3+3x\right)^{-\frac{2}{5}}}$

f. Given $y=\left(x^3+8\right)^{\frac{2}{3}}$, then $y'=\frac{2}{3}\left(x^3+8\right)^{\frac{2}{3}-1}\cdot 3x^2=\frac{2}{\not{3}}\left(x^3+8\right)^{\frac{2-3}{3}}\cdot\not{3}x^2=\mathbf{2x^2\left(x^3+8\right)^{-\frac{1}{3}}}$

g. Given $y=\left(x^3\right)^{\frac{1}{2}}-(3x-1)^{\frac{1}{3}}=x^{\frac{3}{2}}-(3x-1)^{\frac{1}{3}}$, then $y'=\frac{3}{2}x^{\frac{3}{2}-1}-\frac{1}{3}(3x-1)^{\frac{1}{3}-1}\cdot 3=\frac{3}{2}x^{\frac{3-2}{2}}-\frac{1}{\not{3}}(3x-1)^{\frac{1-3}{3}}\cdot\not{3}$

$=\mathbf{\frac{3}{2}x^{\frac{1}{2}}-(3x-1)^{-\frac{2}{3}}}$

h. Given $y=x^2(x+1)^{\frac{1}{8}}$, then $y'=2x\cdot(x+1)^{\frac{1}{8}}+\frac{1}{8}(x+1)^{\frac{1}{8}-1}\cdot x^2=2x(x+1)^{\frac{1}{8}}+\frac{x^2}{8}(x+1)^{\frac{1-8}{8}}=\mathbf{2x(x+1)^{\frac{1}{8}}+\frac{x^2}{8}(x+1)^{-\frac{7}{8}}}$

i. Given $y=\left(x^3+1\right)^{\frac{2}{5}}+x^{\frac{1}{2}}$, then $y' = \frac{2}{5}\left(x^3+1\right)^{\frac{2}{5}-1}\cdot 3x^2+\frac{1}{2}x^{\frac{1}{2}-1} = \frac{6}{5}x^2\left(x^3+1\right)^{\frac{2-5}{5}}+\frac{1}{2}x^{\frac{1-2}{2}} = \mathbf{\frac{6}{5}x^2\left(x^3+1\right)^{-\frac{3}{5}}+\frac{1}{2}x^{-\frac{1}{2}}}$

j. Given $y=\frac{x+1}{x^{\frac{2}{3}}}$, then $y' = \frac{\left[1\cdot x^{\frac{2}{3}}\right]-\left[\frac{2}{3}x^{\frac{2}{3}-1}\cdot(x+1)\right]}{x^{\frac{4}{3}}} = \frac{x^{\frac{2}{3}}-\frac{2}{3}x^{\frac{2-3}{3}}(x+1)}{x^{\frac{4}{3}}} = \mathbf{\frac{x^{\frac{2}{3}}-\frac{2}{3}x^{-\frac{1}{3}}(x+1)}{x^{\frac{4}{3}}}}$

k. Given $y=\frac{\left(x^2+1\right)^{\frac{1}{2}}}{x^2}$, then $y' = \frac{\left[\frac{1}{2}\left(x^2+1\right)^{\frac{1}{2}-1}\cdot 2x\cdot x^2\right]-\left[2x\cdot\left(x^2+1\right)^{\frac{1}{2}}\right]}{x^4} = \frac{\left[\frac{2x^3}{2}\left(x^2+1\right)^{\frac{1-2}{2}}\right]-\left[2x\cdot\left(x^2+1\right)^{\frac{1}{2}}\right]}{x^4}$

$= \mathbf{\frac{x^3\left(x^2+1\right)^{-\frac{1}{2}}-2x\left(x^2+1\right)^{\frac{1}{2}}}{x^4}}$

l. Given $y=\frac{(x+1)^2}{x^{\frac{1}{3}}}$, then $y' = \frac{\left[2(x+1)^{2-1}\cdot x^{\frac{1}{3}}\right]-\left[\frac{1}{3}x^{\frac{1}{3}-1}\cdot(x+1)^2\right]}{x^{\frac{2}{3}}} = \mathbf{\frac{2x^{\frac{1}{3}}(x+1)-\frac{1}{3}x^{-\frac{2}{3}}(x+1)^2}{x^{\frac{2}{3}}}}$

2. Use the $\frac{d}{dx}$ notation to find the derivative of the following exponential expressions.

a. $\frac{d}{dx}\left(x^{\frac{1}{5}}\right)^2 = 2\left(x^{\frac{1}{5}}\right)^{2-1}\cdot\frac{d}{dx}x^{\frac{1}{5}} = 2x^{\frac{1}{5}}\cdot\frac{1}{5}x^{\frac{1}{5}-1} = \frac{2}{5}x^{\frac{1}{5}}\cdot x^{-\frac{4}{5}} = \frac{2}{5}x^{\frac{1}{5}-\frac{4}{5}} = \mathbf{\frac{2}{5}x^{-\frac{3}{5}}}$, or

$\frac{d}{dx}\left(x^{\frac{1}{5}}\right)^2 = \frac{d}{dx}x^{\frac{2}{5}} = \frac{2}{5}x^{\frac{2}{5}-1} = \frac{2}{5}x^{\frac{2-5}{5}} = \mathbf{\frac{2}{5}x^{-\frac{3}{5}}}$

b. $\frac{d}{dx}(x-1)^{\frac{1}{2}} = \frac{1}{2}(x-1)^{\frac{1}{2}-1}\cdot\frac{d}{dx}(x-1) = \frac{1}{2}(x-1)^{-\frac{1}{2}}\cdot 1 = \mathbf{\frac{1}{2}(x-1)^{-\frac{1}{2}}}$

c. $\frac{d}{dx}\left(x^2+1\right)^{\frac{1}{3}} = \frac{1}{3}\left(x^2+1\right)^{\frac{1}{3}-1}\cdot\frac{d}{dx}\left(x^2+1\right) = \frac{1}{3}\left(x^2+1\right)^{\frac{1-3}{3}}\cdot 2x = \mathbf{\frac{2x}{3}\left(x^2+1\right)^{-\frac{2}{3}}}$

d. $\frac{d}{dx}\left(x^3+1\right)^{-\frac{1}{4}} = -\frac{1}{4}\left(x^3+1\right)^{-\frac{1}{4}-1}\cdot\frac{d}{dx}\left(x^3+1\right) = -\frac{1}{4}\left(x^3+1\right)^{\frac{-1-4}{4}}\cdot 3x^2 = \mathbf{-\frac{3x^2}{4}\left(x^3+1\right)^{-\frac{5}{4}}}$

e. $\frac{d}{dx}\left[\frac{(x-1)^{\frac{1}{2}}}{x^2}\right] = \frac{\left[x^2\frac{d}{dx}(x-1)^{\frac{1}{2}}\right]-\left[(x-1)^{\frac{1}{2}}\frac{d}{dx}x^2\right]}{x^4} = \frac{\left[x^2\cdot\frac{1}{2}(x-1)^{\frac{1}{2}-1}\right]-\left[(x-1)^{\frac{1}{2}}\cdot 2x\right]}{x^4} = \mathbf{\frac{\frac{x^2}{2}(x-1)^{-\frac{1}{2}}-2x(x-1)^{\frac{1}{2}}}{x^4}}$

f. $\frac{d}{dx}\left(x^3+2x\right)^{\frac{1}{8}} = \frac{1}{8}\left(x^3+2x\right)^{\frac{1}{8}-1}\cdot\frac{d}{dx}\left(x^3+2x\right) = \frac{1}{8}\left(x^3+2x\right)^{-\frac{7}{8}}\cdot\left(3x^2+2\right) = \mathbf{\left(\frac{3x^2+2}{8}\right)\left(x^3+2x\right)^{-\frac{7}{8}}}$

g. $\frac{d}{dx}\left[\left(x^3+1\right)\left(x^2\right)^{\frac{1}{3}}\right] = \frac{d}{dx}\left[\left(x^3+1\right)x^{\frac{2}{3}}\right] = \left[x^{\frac{2}{3}}\cdot\frac{d}{dx}\left(x^3+1\right)\right]+\left[\left(x^3+1\right)\cdot\frac{d}{dx}x^{\frac{2}{3}}\right] = \left[x^{\frac{2}{3}}\cdot 3x^2\right]+\left[\left(x^3+1\right)\cdot\frac{2}{3}x^{\frac{2}{3}-1}\right]$

$= 3x^{2+\frac{2}{3}}+\frac{2}{3}\left(x^3+1\right)\cdot x^{-\frac{1}{3}} = \mathbf{3x^{\frac{8}{3}}+\frac{2x^{-\frac{1}{3}}}{3}\cdot\left(x^3+1\right)}$

h. $\frac{d}{dx}\left[x^3\cdot\frac{1}{\left(x^2+1\right)^{\frac{1}{2}}}\right] = \frac{d}{dx}\left[\frac{x^3}{\left(x^2+1\right)^{\frac{1}{2}}}\right] = \frac{\left[\left(x^2+1\right)^{\frac{1}{2}}\cdot\frac{d}{dx}x^3\right]-\left[x^3\cdot\frac{d}{dx}\left(x^2+1\right)^{\frac{1}{2}}\right]}{x^2+1} = \frac{\left[\left(x^2+1\right)^{\frac{1}{2}}\cdot 3x^2\right]-\left[x^3\cdot\frac{1}{2}\left(x^2+1\right)^{\frac{1}{2}-1}\cdot 2x\right]}{x^2+1}$

$= \frac{\left[3x^2\left(x^2+1\right)^{\frac{1}{2}}\right]-\left[x^4\left(x^2+1\right)^{\frac{1-2}{2}}\right]}{x^2+1} = \mathbf{\frac{\left[3x^2\left(x^2+1\right)^{\frac{1}{2}}\right]-\left[x^4\left(x^2+1\right)^{-\frac{1}{2}}\right]}{x^2+1}}$

i. $\frac{d}{dx}\left[\frac{x^5}{\left(x^3+1\right)^{\frac{2}{3}}}\right] = \frac{\left[\left(x^3+1\right)^{\frac{2}{3}}\cdot\frac{d}{dx}x^5\right]-\left[x^5\cdot\frac{d}{dx}\left(x^3+1\right)^{\frac{2}{3}}\right]}{\left(x^3+1\right)^{\frac{4}{3}}} = \frac{\left[\left(x^3+1\right)^{\frac{2}{3}}\cdot 5x^4\right]-\left[x^5\cdot\frac{2}{3}\left(x^3+1\right)^{\frac{2}{3}-1}\cdot 3x^2\right]}{\left(x^3+1\right)^{\frac{4}{3}}}$

$= \frac{\left[5x^4\left(x^3+1\right)^{\frac{2}{3}}\right]-\left[\frac{2x^7}{3}\left(x^3+1\right)^{\frac{2-3}{3}}\right]}{\left(x^3+1\right)^{\frac{4}{3}}} = \frac{\left[\mathbf{5x^4}\left(\mathbf{x^3+1}\right)^{\frac{2}{3}}\right]-\left[\mathbf{2x^7}\left(\mathbf{x^3+1}\right)^{-\frac{1}{3}}\right]}{\left(\mathbf{x^3+1}\right)^{\frac{4}{3}}}$

j. $\frac{d}{dx}\left[(x-1)^{\frac{1}{2}}(x+1)^{\frac{1}{3}}\right] = \left[(x+1)^{\frac{1}{3}}\frac{d}{dx}(x-1)^{\frac{1}{2}}\right]+\left[(x-1)^{\frac{1}{2}}\frac{d}{dx}(x+1)^{\frac{1}{3}}\right] = \left[(x+1)^{\frac{1}{3}}\cdot\frac{1}{2}(x-1)^{\frac{1}{2}-1}\right]+\left[(x-1)^{\frac{1}{2}}\cdot\frac{1}{3}(x+1)^{\frac{1}{3}-1}\right]$

$= \left[\frac{1}{2}(x+1)^{\frac{1}{3}}\cdot(x-1)^{\frac{1-2}{2}}\right]+\left[\frac{1}{3}(x-1)^{\frac{1}{2}}\cdot(x+1)^{\frac{1-3}{3}}\right] = \left[\frac{\mathbf{1}}{\mathbf{2}}(\mathbf{x+1})^{\frac{1}{3}}(\mathbf{x-1})^{-\frac{1}{2}}\right]+\left[\frac{\mathbf{1}}{\mathbf{3}}(\mathbf{x-1})^{\frac{1}{2}}(\mathbf{x+1})^{-\frac{2}{3}}\right]$

k. $\frac{d}{dx}\left[x^3\left(x^2+1\right)^{\frac{1}{2}}\right] = \left[\left(x^2+1\right)^{\frac{1}{2}}\frac{d}{dx}x^3\right]+\left[x^3\frac{d}{dx}\left(x^2+1\right)^{\frac{1}{2}}\right] = \left[\left(x^2+1\right)^{\frac{1}{2}}\cdot 3x^2\right]+\left[x^3\cdot\frac{1}{2}\left(x^2+1\right)^{\frac{1}{2}-1}\cdot 2x\right]$

$= \left[3x^2\left(x^2+1\right)^{\frac{1}{2}}\right]+\left[x^4\left(x^2+1\right)^{\frac{1-2}{2}}\right] = \mathbf{3x^2}\left(\mathbf{x^2+1}\right)^{\frac{1}{2}}+\mathbf{x^4}\left(\mathbf{x^2+1}\right)^{-\frac{1}{2}}$

l. $\frac{d}{dx}\left[x^3\left(x^2+1\right)^{-\frac{1}{3}}\right] = \left[\left(x^2+1\right)^{-\frac{1}{3}}\frac{d}{dx}x^3\right]+\left[x^3\frac{d}{dx}\left(x^2+1\right)^{-\frac{1}{3}}\right] = \left[\left(x^2+1\right)^{-\frac{1}{3}}\cdot 3x^2\right]+\left[x^3\cdot-\frac{1}{3}\left(x^2+1\right)^{-\frac{1}{3}-1}\cdot 2x\right]$

$= \left[3x^2\left(x^2+1\right)^{-\frac{1}{3}}\right]+\left[-\frac{2x^4}{3}\left(x^2+1\right)^{\frac{-1-3}{3}}\right] = \mathbf{3x^2}\left(\mathbf{x^2+1}\right)^{-\frac{1}{3}}-\frac{\mathbf{2x^4}}{\mathbf{3}}\left(\mathbf{x^2+1}\right)^{-\frac{4}{3}}$

Section 5.7 Solutions - The Derivative of Radical Functions

1. Find the derivative of the following radical expressions. Do not simplify the answer to its lowest term.

a. Given $y=\sqrt{x^2+1} = \left(x^2+1\right)^{\frac{1}{2}}$, then $y' = \frac{1}{2}\left(x^2+1\right)^{\frac{1}{2}-1}\cdot 2x = \frac{2x}{2}\left(x^2+1\right)^{-\frac{1}{2}} = \frac{\mathbf{x}}{\left(\mathbf{x^2+1}\right)^{\frac{1}{2}}}$

b. Given $y=\sqrt{x^3+3x-5} = \left(x^3+3x-5\right)^{\frac{1}{2}}$, then $y' = \frac{1}{2}\left(x^3+3x-5\right)^{\frac{1}{2}-1}\cdot\left(3x^2+3\right) = \frac{1}{2}\left(x^3+3x-5\right)^{-\frac{1}{2}}\left(3x^2+3\right) = \frac{\mathbf{3}\left(\mathbf{x^2+1}\right)}{\mathbf{2}\left(\mathbf{x^3+3x-5}\right)^{\frac{1}{2}}}$

c. Given $y=x^2+\sqrt{x-1} = x^2+(x-1)^{\frac{1}{2}}$, then $y' = 2x^{2-1}+(x-1)^{\frac{1}{2}-1} = 2x+(x-1)^{-\frac{1}{2}} = \mathbf{2x}+\frac{\mathbf{1}}{(\mathbf{x-1})^{\frac{1}{2}}}$

d. Given $y=\frac{\sqrt{x+1}}{x} = \frac{(x+1)^{\frac{1}{2}}}{x}$, then $y' = \frac{\left[\frac{1}{2}(x+1)^{\frac{1}{2}-1}\cdot x\right]-\left[1\cdot(x+1)^{\frac{1}{2}}\right]}{x^2} = \frac{\frac{\mathbf{x}}{\mathbf{2}}(\mathbf{x+1})^{-\frac{1}{2}}-(\mathbf{x+1})^{\frac{1}{2}}}{\mathbf{x^2}}$

e. Given $y=\frac{x^2}{\sqrt{x^2-1}} = \frac{x^2}{\left(x^2-1\right)^{\frac{1}{2}}}$, then $y' = \frac{\left[2x\cdot\left(x^2-1\right)^{\frac{1}{2}}\right]-\left[\frac{1}{2}\left(x^2-1\right)^{\frac{1}{2}-1}\cdot 2x\cdot x^2\right]}{x^2-1} = \frac{\mathbf{2x}\left(\mathbf{x^2-1}\right)^{\frac{1}{2}}-\mathbf{x^3}\left(\mathbf{x^2-1}\right)^{-\frac{1}{2}}}{\mathbf{x^2-1}}$

f. Given $y=\sqrt{x^3}+3x^2 = \left(x^3\right)^{\frac{1}{2}}+3x^2 = x^{\frac{3}{2}}+3x^2$, then $y' = \frac{3}{2}x^{\frac{3}{2}-1}+(3\cdot 2)x^{2-1} = \frac{3}{2}x^{\frac{3-2}{2}}+6x = \frac{\mathbf{3}}{\mathbf{2}}\mathbf{x}^{\frac{1}{2}}+\mathbf{6x}$

g. Given $y=\frac{\sqrt{x^2+3}}{\sqrt{x+1}} = \sqrt{\frac{x^2+3}{x+1}} = \left(\frac{x^2+3}{x+1}\right)^{\frac{1}{2}}$, then $y' = \frac{1}{2}\left(\frac{x^2+3}{x+1}\right)^{\frac{1}{2}-1}\cdot\frac{2x\cdot(x+1)-1\cdot\left(x^2+3\right)}{(x+1)^2} = \frac{\mathbf{1}}{\mathbf{2}}\left(\frac{\mathbf{x^2+3}}{\mathbf{x+1}}\right)^{-\frac{1}{2}}\cdot\frac{\mathbf{x^2+2x-3}}{(\mathbf{x+1})^2}$

h. Given $y = \frac{\sqrt[4]{x^3-1}}{\sqrt{x}} = \frac{\left(x^3-1\right)^{\frac{1}{4}}}{x^{\frac{1}{2}}}$, then $y' = \frac{\left[\frac{1}{4}\left(x^3-1\right)^{\frac{1}{4}-1}\cdot 3x^2\cdot x^{\frac{1}{2}}\right]-\left[\frac{1}{2}x^{\frac{1}{2}-1}\cdot\left(x^3-1\right)^{\frac{1}{4}}\right]}{x} = \mathbf{\frac{\frac{3x^{\frac{5}{2}}}{4}\left(x^3-1\right)^{-\frac{3}{4}}-\frac{1}{2}x^{-\frac{1}{2}}\left(x^3-1\right)^{\frac{1}{4}}}{x}}$

i. Given $y = \frac{x^3}{x^2\sqrt{x}} = \frac{x^{\cancel{3}-1}}{\cancel{x^2}\cdot x^{\frac{1}{2}}} = \frac{x}{x^{\frac{1}{2}}} = x\cdot x^{-\frac{1}{2}} = x^{1-\frac{1}{2}} = x^{\frac{1}{2}}$, then $y' = \frac{1}{2}x^{\frac{1}{2}-1} = \frac{1}{2}x^{\frac{1}{2}-\frac{1}{1}} = \frac{1}{2}x^{\frac{1-2}{2}} = \mathbf{\frac{1}{2}x^{-\frac{1}{2}}}$

2. Use the $\frac{d}{dx}$ notation to find the derivative of the following radical expressions.

a. $\frac{d}{dx}\left(\sqrt{x^2}+\frac{1}{x}\right) = \frac{d}{dx}\left(x+\frac{1}{x}\right) = \frac{d}{dx}x+\frac{d}{dx}\left(\frac{1}{x}\right) = 1+\frac{(0\cdot x)-(1\cdot 1)}{x^2} = \mathbf{1-\frac{1}{x^2}}$

b. $\frac{d}{dx}\left(\sqrt{\frac{x}{x-1}}\right) = \frac{d}{dx}\left(\frac{x}{x-1}\right)^{\frac{1}{2}} = \frac{1}{2}\left(\frac{x}{x-1}\right)^{\frac{1}{2}-1}\cdot\frac{\left[(x-1)\frac{d}{dx}x\right]-\left[x\frac{d}{dx}(x-1)\right]}{(x-1)^2} = \frac{1}{2}\left(\frac{x}{x-1}\right)^{\frac{1}{2}-1}\cdot\frac{[1\cdot(x-1)]-[1\cdot x]}{(x-1)^2}$

$= \frac{1}{2}\left(\frac{x}{x-1}\right)^{\frac{1-2}{2}}\cdot\frac{\cancel{x}-1-\cancel{x}}{(x-1)^2} = \mathbf{-\frac{1}{2}\left(\frac{x}{x-1}\right)^{-\frac{1}{2}}\cdot\frac{1}{(x-1)^2}}$

c. $\frac{d}{dx}\left(\frac{x^3}{\sqrt{x+1}}\right) = \frac{d}{dx}\left(\frac{x^3}{(x+1)^{\frac{1}{2}}}\right) = \frac{\left[(x+1)^{\frac{1}{2}}\frac{d}{dx}x^3\right]-\left[x^3\frac{d}{dx}(x+1)^{\frac{1}{2}}\right]}{x+1} = \frac{\left[(x+1)^{\frac{1}{2}}\cdot 3x^{3-1}\right]-\left[x^3\cdot\frac{1}{2}(x+1)^{\frac{1}{2}-1}\right]}{x+1}$

$= \frac{\left[3(x+1)^{\frac{1}{2}}x^2\right]-\left[\frac{x^3}{2}(x+1)^{\frac{1-2}{2}}\right]}{x+1} = \mathbf{\frac{\left[3x^2(x+1)^{\frac{1}{2}}\right]-\left[\frac{x^3}{2}(x+1)^{-\frac{1}{2}}\right]}{x+1}}$

d. $\frac{d}{dx}\left(\frac{\sqrt{x+5}}{x}\right) = \frac{d}{dx}\left(\frac{(x+5)^{\frac{1}{2}}}{x}\right) = \frac{\left[x\frac{d}{dx}(x+5)^{\frac{1}{2}}\right]-\left[(x+5)^{\frac{1}{2}}\frac{d}{dx}x\right]}{x^2} = \frac{\left[x\cdot\frac{1}{2}(x+5)^{\frac{1}{2}-1}\right]-\left[(x+5)^{\frac{1}{2}}\cdot 1\right]}{x^2} = \mathbf{\frac{\frac{x}{2}(x+5)^{-\frac{1}{2}}-(x+5)^{\frac{1}{2}}}{x^2}}$

e. $\frac{d}{dx}\left(x^3+\frac{\sqrt{x}}{x}\right) = \frac{d}{dx}\left(x^3+\frac{x^{\frac{1}{2}}}{x}\right) = \frac{d}{dx}\left(x^3+x^{\frac{1}{2}}\cdot x^{-1}\right) = \frac{d}{dx}\left(x^3+x^{\frac{1}{2}-1}\right) = \frac{d}{dx}\left(x^3+x^{-\frac{1}{2}}\right) = \frac{d}{dx}x^3+\frac{d}{dx}x^{-\frac{1}{2}}$

$= 3x^{3-1}-\frac{1}{2}x^{-\frac{1}{2}-1} = 3x^2-\frac{1}{2}x^{\frac{-1-2}{2}} = \mathbf{3x^2-\frac{1}{2}x^{-\frac{3}{2}}}$

f. $\frac{d}{dx}\left(1+\frac{2\sqrt{x}}{x^3}\right) = \frac{d}{dx}\left(1+\frac{2x^{\frac{1}{2}}}{x^3}\right) = \frac{d}{dx}\left(1+2x^{\frac{1}{2}}x^{-3}\right) = \frac{d}{dx}\left(1+2x^{\frac{1}{2}-3}\right) = \frac{d}{dx}\left(1+2x^{\frac{1-6}{2}}\right) = \frac{d}{dx}\left(1+2x^{-\frac{5}{2}}\right)$

$= \frac{d}{dx}(1)+\frac{d}{dx}\left(2x^{-\frac{5}{2}}\right) = 0-\frac{5}{2}x^{-\frac{5}{2}-1}\cdot 2 = -\frac{5}{2}x^{\frac{-5-2}{2}}\cdot 2 = \mathbf{-5x^{-\frac{7}{2}}}$

3. Find the derivative of the following radical expressions.

a. $\frac{d}{dx}\left(\sqrt{x^3}+\sqrt{y}\right)=\frac{d}{dx}(x)$; $\frac{d}{dx}\left(x^{\frac{3}{2}}+y^{\frac{1}{2}}\right)=1$; $\frac{d}{dx}x^{\frac{3}{2}}+\frac{d}{dx}y^{\frac{1}{2}}=1$; $\frac{3}{2}x^{\frac{3}{2}-1}+\frac{1}{2}y^{\frac{1}{2}-1}y'=1$; $\frac{3}{2}x^{\frac{3-2}{2}}+\frac{1}{2}y^{-\frac{1}{2}}y'=1$

; $\frac{3}{2}x^{\frac{1}{2}}+\frac{1}{2}y^{-\frac{1}{2}}y'=1$; $\frac{1}{2}y^{-\frac{1}{2}}y'=1-\frac{3}{2}x^{\frac{1}{2}}$; $\frac{y'}{2y^{\frac{1}{2}}}=1-\frac{3}{2}x^{\frac{1}{2}}$; $y'=2y^{\frac{1}{2}}\left(1-\frac{3}{2}x^{\frac{1}{2}}\right)$; $\mathbf{y'=2y^{\frac{1}{2}}-3(xy)^{\frac{1}{2}}}$

b. $\frac{d}{dx}\left(\sqrt{x}+y^3\right)=\frac{d}{dx}(2)$; $\frac{d}{dx}\left(x^{\frac{1}{2}}+y^3\right)=0$; $\frac{d}{dx}x^{\frac{1}{2}}+\frac{d}{dx}y^3=0$; $\frac{1}{2}x^{\frac{1}{2}-1}+3y^2y'=0$; $\frac{1}{2}x^{-\frac{1}{2}}+3y^2y'=0$; $3y^2y'=-\frac{1}{2}x^{-\frac{1}{2}}$

; $3y^2y' = -\frac{1}{2x^{\frac{1}{2}}}$; $\mathbf{y' = -\frac{1}{6x^{\frac{1}{2}}y^2}}$

c. $\frac{d}{dx}(xy)=\frac{d}{dx}(\sqrt{x})$; $\frac{d}{dx}(xy)=\frac{d}{dx}x^{\frac{1}{2}}$; $y\frac{d}{dx}x+x\frac{d}{dx}y=\frac{d}{dx}x^{\frac{1}{2}}$; $y+xy'=\frac{1}{2}x^{\frac{1}{2}-1}$; $xy'=\frac{1}{2}x^{-\frac{1}{2}}-y$; $\mathbf{y'=\frac{1}{x}\left(\frac{1}{2x^{\frac{1}{2}}}-y\right)}$

d. $\frac{d}{dx}\left(\sqrt{y}+x^3\right)=0$; $\frac{d}{dx}\left(y^{\frac{1}{2}}+x^3\right)=0$; $\frac{d}{dx}y^{\frac{1}{2}}+\frac{d}{dx}x^3=0$; $\frac{1}{2}y^{\frac{1}{2}-1}y'+3x^2=0$; $\frac{1}{2}y^{-\frac{1}{2}}y'=-3x^2$; $\frac{y'}{2y^{\frac{1}{2}}}=-3x^2$

; $\frac{y'}{2y^{\frac{1}{2}}}=\frac{-3x^2}{1}$; $\mathbf{y'=-6x^2y^{\frac{1}{2}}}$

e. $\frac{d}{dx}\left(\sqrt{x^4}+y^2\right)=\frac{d}{dx}(x)$; $\frac{d}{dx}\left(x^2+y^2\right)=1$; $\frac{d}{dx}x^2+\frac{d}{dx}y^2=1$; $2x+2yy'=1$; $2yy'=1-2x$; $\mathbf{y'=\frac{1-2x}{2y}}$

f. $\frac{d}{dx}\left(\sqrt{x+1}\right)=\frac{d}{dx}\left(y^3\right)$; $\frac{d}{dx}(x+1)^{\frac{1}{2}}=3y^2\frac{d}{dx}y$; $\frac{1}{2}(x+1)^{\frac{1}{2}-1}=3y^2y'$; $\frac{1}{2}(x+1)^{-\frac{1}{2}}=3y^2y'$; $\frac{1}{2(x+1)^{\frac{1}{2}}}=\frac{3y^2y'}{1}$

; $6y^2y'(x+1)^{\frac{1}{2}}=1$; $\mathbf{y'=\frac{1}{6y^2(x+1)^{\frac{1}{2}}}}$

g. $\frac{d}{dx}\left(xy^2+\sqrt{x}\right)=\frac{d}{dx}(2)$; $\frac{d}{dx}\left(xy^2+x^{\frac{1}{2}}\right)=0$; $y^2\frac{d}{dx}x+x\frac{d}{dx}y^2+\frac{1}{2}x^{\frac{1}{2}-1}=0$; $y^2\cdot1+x\cdot2yy'+\frac{1}{2}x^{-\frac{1}{2}}=0$

; $2xyy'=-\frac{1}{2}x^{-\frac{1}{2}}-y^2$; $y'=\frac{1}{2xy}\left(-\frac{1}{2}x^{-\frac{1}{2}}-y^2\right)$; $y'=-\frac{x^{-\frac{1}{2}}}{4xy}-\frac{y^2}{2xy}$; $y'=-\frac{1}{4x^{1+\frac{1}{2}}y}-\frac{y^2}{2xy}$; $\mathbf{y'=-\frac{1}{4x^{\frac{3}{2}}y}-\frac{y}{2x}}$

h. $\frac{d}{dx}\left(\sqrt{x^3}\right)+\frac{d}{dx}(xy)=0$; $\frac{d}{dx}x^{\frac{3}{2}}+\frac{d}{dx}(xy)=0$; $\frac{3}{2}x^{\frac{3}{2}-1}+\left(y\frac{d}{dx}x+x\frac{d}{dx}y\right)=0$; $\frac{3}{2}x^{\frac{1}{2}}+(y+xy')=0$; $xy'=-\frac{3}{2}x^{\frac{1}{2}}-y$

; $y'=\frac{1}{x}\left(-\frac{3}{2}x^{\frac{1}{2}}-y\right)$; $y'=-\frac{3x^{\frac{1}{2}}}{2x}-\frac{y}{x}$; $y'=-\frac{3}{2x\cdot x^{-\frac{1}{2}}}-\frac{y}{x}$; $y'=-\frac{3}{2x^{1-\frac{1}{2}}}-\frac{y}{x}$; $\mathbf{y'=-\frac{3}{2x^{\frac{1}{2}}}-\frac{y}{x}}$

i. $\frac{d}{dx}\left(\sqrt{x}+3y\right)=\frac{d}{dx}(y)$; $\frac{d}{dx}\left(x^{\frac{1}{2}}+3y\right)=y'$; $\frac{d}{dx}x^{\frac{1}{2}}+\frac{d}{dx}3y=y'$; $\frac{1}{2}x^{\frac{1}{2}-1}+3y'=y'$; $\frac{1}{2}x^{-\frac{1}{2}}+3y'-y'=0$; $\frac{1}{2}x^{-\frac{1}{2}}+2y'=0$

; $2y'=-\frac{1}{2}x^{-\frac{1}{2}}$; $2y'=-\frac{1}{2x^{\frac{1}{2}}}$; $\mathbf{y'=-\frac{1}{4x^{\frac{1}{2}}}}$

4. Evaluate the derivative of the following radical expressions for the specified value of x.

a. Given $y=\sqrt{3x^3}+x^2 = \left(3x^3\right)^{\frac{1}{2}}+x^2 = 3^{\frac{1}{2}}x^{\frac{3}{2}}+x^2$, then $y' = 3^{\frac{1}{2}}\cdot\frac{3}{2}x^{\frac{3}{2}-1}+2x^{2-1} = \frac{3^{\frac{3}{2}}}{2}x^{\frac{3-2}{2}}+2x = \frac{3^{\frac{3}{2}}}{2}x^{\frac{1}{2}}+2x$

at $x=1$ $y' = \frac{3^{\frac{3}{2}}}{2}\cdot1^{\frac{1}{2}}+(2\cdot1) = \frac{3^{\frac{3}{2}}}{2}+2 = \frac{5.196}{2}+2 = \mathbf{4.598}$

b. Given $y=\left(x^2+1\right)\sqrt{x} = \left(x^2+1\right)x^{\frac{1}{2}} = x^{2+\frac{1}{2}}+x^{\frac{1}{2}} = x^{\frac{4+1}{2}}+x^{\frac{1}{2}} = x^{\frac{5}{2}}+x^{\frac{1}{2}}$, then $y' = \frac{5}{2}x^{\frac{5}{2}-1}+\frac{1}{2}x^{\frac{1}{2}-1} = \frac{5}{2}x^{\frac{5-2}{2}}+\frac{1}{2}x^{\frac{1-2}{2}}$

$= \frac{5}{2}x^{\frac{3}{2}}+\frac{1}{2}x^{-\frac{1}{2}} = \frac{5}{2}x^{\frac{3}{2}}+\frac{1}{2x^{\frac{1}{2}}}$ at $x=0$ $y' = \frac{5}{2}\cdot0^{\frac{3}{2}}+\frac{1}{2\cdot0^{\frac{1}{2}}} = 0+\frac{1}{0}$ **is undefined due to division by zero**

c. Given $y=\frac{x^2-1}{\sqrt{4x^2}} = \frac{x^2-1}{2x}$, then $y' = \frac{[2x\cdot2x]-\left[2\cdot\left(x^2-1\right)\right]}{(2x)^2} = \frac{4x^2-2x^2+2}{4x^2} = \frac{2x^2+2}{4x^2} = \frac{2\left(x^2+1\right)}{4x^2} = \frac{x^2+1}{2x^2}$

at $x=2$ $\quad y' = \frac{2^2+1}{2\cdot 2^2} = \frac{4+1}{2\cdot 4} = \frac{5}{8} = \mathbf{0.625}$

d. Given $y=\sqrt{\frac{x}{x^2+1}} = \left(\frac{x}{x^2+1}\right)^{\frac{1}{2}}$, then $y' = \frac{1}{2}\left(\frac{x}{x^2+1}\right)^{\frac{1}{2}-1}\cdot\frac{\left[1\cdot\left(x^2+1\right)\right]-\left[2x\cdot x\right]}{\left(x^2+1\right)^2} = \frac{1}{2}\left(\frac{x}{x^2+1}\right)^{-\frac{1}{2}}\cdot\frac{-x^2+1}{\left(x^2+1\right)^2}$

at $x=1$ $\quad y' = \frac{1}{2}\left(\frac{1}{1^2+1}\right)^{-\frac{1}{2}}\cdot\frac{-1^2+1}{\left(1^2+1\right)^2} = \frac{1}{2}\left(\frac{1}{2}\right)^{-\frac{1}{2}}\cdot\frac{-1+1}{2^2} = \frac{1}{2}\left(\frac{1}{2}\right)^{-\frac{1}{2}}\cdot 0 = \mathbf{0}$

e. Given $y=\sqrt{x^3+1}+4x^3 = \left(x^3+1\right)^{\frac{1}{2}}+4x^3$, then $y' = \frac{1}{2}\left(x^3+1\right)^{\frac{1}{2}-1}\cdot 3x^2+(4\cdot 3)x^{3-1} = \frac{3x^2}{2}\left(x^3+1\right)^{-\frac{1}{2}}+12x^2$

at $x=0$ $\quad y' = \frac{3\cdot 0^2}{2}\cdot\left(0^3+1\right)^{-\frac{1}{2}}+12\cdot 0^2 = \frac{3\cdot 0^2}{2}\cdot\frac{1}{\left(0^3+1\right)^{\frac{1}{2}}}+12\cdot 0^2 = 0\cdot\frac{1}{1^{\frac{1}{2}}}+0 = 0+0 = \mathbf{0}$

f. Given $y=\frac{x^2+1}{\sqrt{x^3}} = \frac{x^2+1}{\left(x^3\right)^{\frac{1}{2}}} = \frac{x^2+1}{x^{\frac{3}{2}}}$, then $y' = \frac{\left[2x\cdot x^{\frac{3}{2}}\right]-\left[\frac{3}{2}x^{\frac{3}{2}-1}\cdot\left(x^2+1\right)\right]}{x^3} = \frac{2x^{\frac{5}{2}}-\frac{3}{2}x^{\frac{1}{2}}\left(x^2+1\right)}{x^3}$

at $x=3$ $\quad y' = \frac{2\cdot 3^{\frac{5}{2}}-\frac{3}{2}\cdot 3^{\frac{1}{2}}\left(3^2+1\right)}{3^3} = \frac{(2\cdot 15.58)-(1.5\cdot 1.732\cdot 10)}{27} = \frac{31.18-25.98}{27} = \mathbf{0.193}$

Section 5.8 Solutions - Higher Order Derivatives

1. Find the second derivative of the following functions.

a. Given $y=x^3+3x^2+5x-1$, then $y' = 3x^{3-1}+(3\cdot 2)x^{2-1}+5x^{1-1}-0 = 3x^2+6x+5x^0 = 3x^2+6x+5$ and

$y'' = (3\cdot 2)x^{2-1}+6x^{1-1}+0 = 6x+6x^0 = \mathbf{6x+6}$

b. Given $y=x^2(x+1)^2$, then $y' = 2x\cdot(x+1)^2+2(x+1)^{2-1}\cdot x^2 = 2x(x+1)^2+2x^2(x+1) = 2x(x+1)^2+2x^3+2x^2$ and

$y'' = \left[2\cdot(x+1)^2+2(x+1)^{2-1}\cdot 2x\right]+(2\cdot 3)x^{3-1}+(2\cdot 2)x^{2-1} = 2(x+1)^2+4x(x+1)+6x^2+4x = 2\left(x^2+2x+1\right)+4x^2+4xx$

$+6x^2+4x = 2x^2+4x+2+10x^2+8x = 12x^2+12x+2 = \mathbf{2\left(6x^2+6x+1\right)}$

c. Given $y=3x^3+50x$, then $y' = (3\cdot 3)x^{3-1}+50x^{1-1} = 9x^2+50$ and $y'' = (9\cdot 2)x^{2-1}+0 = \mathbf{18x}$

d. Given $y=x^5+\frac{1}{x^2} = x^5+x^{-2}$, then $y' = 5x^{5-1}-2x^{-2-1} = 5x^4-2x^{-3}$ and $y'' = (5\cdot 4)x^{4-1}+(-2\cdot -3)x^{-3-1} = \mathbf{20x^3+6x^{-4}}$

e. Given $y=\frac{x^3}{x+1}-5x^2$, then $y' = \frac{\left[3x^2\cdot(x+1)\right]-\left[1\cdot x^3\right]}{(x+1)^2}-(5\cdot 2)x^{2-1} = \frac{3x^3+3x^2-x^3}{(x+1)^2}-10x = \frac{2x^3+3x^2}{(x+1)^2}-10x$ and

$y'' = \frac{\left[\left(6x^2+6x\right)\cdot(x+1)^2\right]-\left[2(x+1)\cdot\left(2x^3+3x^2\right)\right]}{(x+1)^4}-10x^{1-1} = \frac{\left[6x(x+1)(x+1)^2\right]-\left[2(x+1)\left(2x^3+3x^2\right)\right]}{(x+1)^4}-10$

$= \frac{(x+1)\left[6x(x+1)^2-2\left(2x^3+3x^2\right)\right]}{(x+1)^{4=3}}-10 = \mathbf{\frac{6x(x+1)^2-2x^2(2x+3)}{(x+1)^3}-10}$

f. Given $y=x^3\left(x^2-1\right) = x^5-x^3$, then $y' = 5x^{5-1}-3x^{3-1} = 5x^4-3x^2$ and $y'' = (5\cdot 4)x^{4-1}-(3\cdot 2)x^{2-1} = \mathbf{20x^3-6x}$

g. Given $y=x^4+\frac{x^8-7x^5+5x}{10}$, then $y' = 4x^{4-1}+\frac{1}{10}\left(8x^{8-1}+(-7\cdot 5)x^{5-1}+5x^{1-1}\right) = 4x^3+\frac{1}{10}\left(8x^7-35x^4+5\right)$ and

$y'' = (4\cdot 3)x^{3-1}+\frac{1}{10}\left[(8\cdot 7)x^{7-1}+(-35\cdot 4)x^{4-1}+0\right] = 12x^2+\frac{1}{10}\left(56x^6-140x^3\right) = \mathbf{5.6x^6-14x^3+12x^2}$

h. Given $y = x^2 - \frac{1}{x+1} = x^2 - (x+1)^{-1}$, then $y' = 2x^{2-1} + (x+1)^{-1-1} = 2x + (x+1)^{-2}$ and $y'' = 2x^{1-1} - 2(x+1)^{-2-1}$ $= 2x^0 - 2(x+1)^{-3} = 2 - 2(x+1)^{-3} = \mathbf{2 - \frac{2}{(x+1)^3}}$

i. Given $y = \frac{1}{x^2} - 3x = x^{-2} - 3x$, then $y' = -2x^{-2-1} - 3x^{1-1} = -2x^{-3} - 3$ and $y'' = (-2 \cdot -3)x^{-3-1} - 0 = \mathbf{6x^{-4}}$

2. Find y''' for the following functions.

a. Given $y = x^5 + 6x^3 + 10$, then $y' = 5x^{5-1} + (6 \cdot 3)x^{3-1} + 0 = 5x^4 + 18x^2$, $y'' = (5 \cdot 4)x^{4-1} + (18 \cdot 2)x^{2-1} = 20x^3 + 36x$, and $y''' = (20 \cdot 3)x^{3-1} + 36x^{1-1} = \mathbf{60x^2 + 36}$

b. Given $y = x^2 + \frac{1}{x} = x^2 + x^{-1}$, then $y' = 2x^{2-1} - x^{-1-1} = 2x - x^{-2}$, $y'' = 2x^{1-1} + 2x^{-2-1} = 2 + 2x^{-3}$, and $y''' = \mathbf{-6x^{-4}}$

c. Given $y = 4x^3(x-1)^2$, then $y' = \left[(4 \cdot 3)x^{3-1} \cdot (x-1)^2\right] + \left[2(x-1)^{2-1} \cdot 4x^3\right] = 12x^2(x-1)^2 + 8x^3(x-1) = 12x^2(x^2 - 2x + 1)$ $+ 8x^4 - 8x^3 = 12x^4 - 24x^3 + 12x^2 + 8x^4 - 8x^3 = 20x^4 - 32x^3 + 12x^2$ $y'' = (20 \cdot 4)x^{4-1} - (32 \cdot 3)x^{3-1} + (12 \cdot 2)x^{2-1}$ $= 80x^3 - 96x^2 + 24x$ and $y''' = (80 \cdot 3)x^{3-1} - (96 \cdot 2)x^{2-1} + 24x^{1-1} = 240x^2 - 192x + 24x^0 = \mathbf{240x^2 - 192x + 24}$

d. Given $y = \frac{x}{x+1}$, then $y' = \frac{[1 \cdot (x+1)] - [1 \cdot x]}{(x+1)^2} = \frac{x+1-x}{(x+1)^2} = \frac{1}{(x+1)^2} = (x+1)^{-2}$, $y'' = -2(x+1)^{-2-1} = -2(x+1)^{-3}$ and $y''' = (-2 \cdot -3)(x+1)^{-3-1} = 6(x+1)^{-4} = \mathbf{\frac{6}{(x+1)^4}}$

e. Given $y = x^8 - 10x^5 + 5x - 10$, then $y' = 8x^{8-1} + (-10 \cdot 5)x^{5-1} + 5x^{1-1} - 0 = 8x^7 - 50x^4 + 5x^0 = 8x^7 - 50x^4 + 5$, $y'' = (8 \cdot 7)x^{7-1} - (50 \cdot 4)x^{4-1} + 0 = 56x^6 - 200x^3$ and $y''' = (56 \cdot 6)x^{6-1} - (200 \cdot 3)x^{3-1} = \mathbf{336x^5 - 600x^2}$

f. Given $y = \frac{x-1}{x^2} + 5x^3$, then $y' = \frac{[1 \cdot x^2] - [2x \cdot (x-1)]}{x^4} + (5 \cdot 3)x^{3-1} = \frac{x^2 - 2x^2 + 2x}{x^4} + 15x^2 = \frac{-x^2 + 2x}{x^4} + 15x^2$ $= \frac{\cancel{x}(-x+2)}{x^{4-3}} + 15x^2 = \frac{-x+2}{x^3} + 15x^2$, $y'' = \frac{[-1 \cdot x^3] - [3x^2 \cdot (-x+2)]}{x^6} + (15 \cdot 2)x^{2-1} = \frac{-x^3 + 3x^3 - 6x^2}{x^6} + 30x$ $= \frac{2x^3 - 6x^2}{x^6} + 30x = \frac{x^2(2x-6)}{x^{6-4}} + 30x = \frac{2x-6}{x^4} + 30x$ and $y''' = \frac{[2 \cdot x^4] - [4x^3 \cdot (2x-6)]}{x^8} + 30x^{1-1} = \frac{2x^4 - 8x^4 + 24x^3}{x^8} + 30$ $= \frac{-6x^4 + 24x^3}{x^8} + 30 = \frac{-6x^3(x-4)}{x^{8-5}} + 30 = \mathbf{-\frac{6(x-4)}{x^5} + 30}$

3. Find $f''(0)$ and $f''(1)$ for the following functions.

a. Given $f(x) = 6x^5 + 3x^3 + 5$, then $f'(x) = (6 \cdot 5)x^{5-1} + (3 \cdot 3)x^{3-1} + 0 = 30x^4 + 9x^2$ and $f''(x) = (30 \cdot 4)x^{4-1} + (9 \cdot 2)x^{2-1}$ $= 120x^3 + 18x$. Therefore, $f''(0) = 120 \cdot 0^3 + 18 \cdot 0 = \mathbf{0}$ and $f''(1) = 120 \cdot 1^3 + 18 \cdot 1 = 120 + 18 = \mathbf{138}$

b. Given $f(x) = x^3(x+1)^2$, then $f'(x) = \left[3x^{3-1} \cdot (x+1)^2\right] + \left[2(x+1)^{2-1} \cdot x^3\right] = 3x^2(x+1)^2 + 2x^3(x+1) = 3x^2(x^2 + 2x + 1)$ $+ 2x^4 + 2x^3 = 3x^4 + 6x^3 + 3x^2 + 2x^4 + 2x^3 = 5x^4 + 8x^3 + 3x^2$ and $f''(x) = (5 \cdot 4)x^{4-1} + (8 \cdot 3)x^{3-1} + (3 \cdot 2)x^{2-1}$ $= 20x^3 + 24x^2 + 6x$. Therefore, $f''(0) = (20 \cdot 0^3) + (24 \cdot 0^2) + (6 \cdot 0) = \mathbf{0}$ and $f''(1) = (20 \cdot 1^3) + (24 \cdot 1^2) + (6 \cdot 1) = \mathbf{50}$

c. Given $f(x)=x+(x-1)^2$, then $f'(x) = 1+2(x-1)^{2-1} = 1+2(x-1) = 1+2x-2 = 2x-1$ and $f''(x) = 2x^{1-1}-0 = 2$.
Therefore, $f''(0) = \mathbf{2}$ and $f''(1) = \mathbf{2}$

d. Given $f(x)=(x-1)^{-3}$, then $f'(x) = -3(x-1)^{-3-1} = -3(x-1)^{-4}$ and $f''(x) = (-3\cdot -4)(x-1)^{-4-1} = 12(x-1)^{-5} = \dfrac{12}{(x-1)^5}$

Therefore, $f''(0) = \dfrac{12}{(0-1)^5} = \dfrac{12}{-1} = \mathbf{-12}$ and $f''(1) = \dfrac{12}{(1-1)^5} = \mathbf{\dfrac{12}{0}}$ **which is undefined due to division by zero.**

e. Given $f(x)=(x-1)(x^2+1)$, then $f'(x) = \left[1\cdot(x^2+1)\right]+\left[2x\cdot(x-1)\right] = x^2+1+2x^2-2x = 3x^2-2x+1$ and

$f''(x) = (3\cdot 2)x^{2-1}-2x^{1-1}+0 = 6x-2$. Therefore, $f''(0) = (6\cdot 0)-2 = 0-2 = \mathbf{-2}$ and $f''(1) = (6\cdot 1)-2 = 6-2 = \mathbf{4}$

f. Given $f(x)=(x^3-1)^2+\sqrt{2x} = (x^3-1)^2+(2x)^{\frac{1}{2}}$, then $f'(x) = 2(x^3-1)^{2-1}\cdot 3x^2+\frac{1}{2}(2x)^{\frac{1}{2}-1}\cdot 2 = 6x^2(x^3-1)+(2x)^{-\frac{1}{2}}$

and $f''(x) = \left[12x\cdot(x^3-1)+3x^2\cdot 6x^2\right]-\frac{1}{2}(2x)^{-\frac{1}{2}-1}\cdot 2 = 12x^4-12x+18x^4-(2x)^{-\frac{3}{2}} = 30x^4-12x-(2x)^{-\frac{3}{2}}$. Thus, $f''(0)$
$= 30\cdot 0^4-12\cdot 0-\dfrac{1}{(2\cdot 0)^{\frac{3}{2}}} = \mathbf{-\dfrac{1}{0}}$ **which is undefined,** and $f''(1) = 30\cdot 1^4-12\cdot 1-\dfrac{1}{(2\cdot 1)^{\frac{3}{2}}} = 30-12-0.35 = \mathbf{17.65}$

Glossary

The following glossary terms are used throughout the Hamilton Education Guides math series:

Absolute value - The numerical value or magnitude of a quantity, as of a negative number, without regard to its sign. The symbol for absolute value is two parallel lines "| |". For instance, $|-2| = |2| = 2$, $|-35| = |35| = 35$, $|-0.23| = |0.23| = 0.23$, and $|-5.13| = |5.13| = 5.13$ are some examples of how absolute value is used.

Addend - Any of a set of numbers to be added.

Addition - The process of adding two or more numbers to get a number called the sum.

Adequate - To consider or treat as equal. To make or set equal.

Algebraic approach - An approach in which only numbers, letters, and arithmetic operations are used.

Algebraic expression - Designating an expression, equation, or function in which only numbers, letters, and arithmetic operations are contained or used.

Algebraic fractions - A fraction having variables in either the numerator or the denominator or both.

Apparent - Appearing to the eye or to the judgment; seeming, often in distinction to real; obvious.

Application - The act of applying or putting to use.

Apply - To put on. To put to or adapt for particular use. To use.

Approximation - An amount or estimate nearly exact or correct.

Arithmetic fractions - A fraction having positive or negative whole numbers in the numerator and the denominator; an integer fraction.

Associative - Pertaining to an operation in which the result is the same regardless of the way the elements are grouped, as, in addition, $2+(4+5)$ = $(2+4)+5$ = 11 and, in multiplication, $2\times(4\times5)$ = $(2\times4)\times5$ = 40.

Assumption - The act of assuming; supposition; the act of taking for granted.

Base - *a.* The number on which a system of numeration is based. For example, the base of the decimal system is 10. Computers use the binary system, which has the base 2. *b.* A number that is to be multiplied by itself the number of times indicated by an exponent or logarithm. For example, in 2^5, 2 is the base and 5 is the exponent.

Binomial - An expression consisting of two terms connected by a plus or minus sign. For example, $a+b$, $\sqrt{x^3}-\sqrt{y}$, x^3+3x, and a^2b^3-3ab are referred to as binomials.

Case - Supporting facts offered in justification of a statement.

Class - A group of persons or things that have something in common, a set, collection, group.

Coefficient - A number placed in front of an algebraic expression and multiplying it. For example, in the expression $3x^2 + 5x = 2$, 3 is the coefficient of x^2, and 5 is the coefficient of x.

Common denominator - A common multiple of the denominators of two or more fractions. For example, 10 is a common denominator of $\frac{1}{2}$ and $\frac{3}{5}$.

Common divisor - A number or quantity that can evenly divide two or more other numbers or quantities. For example, 4 is a common divisor of 12 and 20.

Common factor - Another name for common divisor.

Common fraction - A fraction whose numerator and denominator are both integers (whole numbers).

Commutative - Pertaining to an operation in which the order of the elements does not affect the result, as, in addition, $5+3 = 3+5$ and, in multiplication, $5 \times 3 = 3 \times 5$.

Complex fractions - A fraction in which either the numerator or the denominator or both contain a fraction.

Computation - The act or method of computing; calculation.

Conjugate - Inversely related to one of a group of otherwise identical properties.

Constant - Remaining the same; not changing. A number or other thing that never changes.

Conversion - A change in the form of a quantity or an expression without a change in the value.

Converge - To approach a limit.

Cube - The third power of a number or quantity.

Cube root $(\sqrt[3]{\ })$ - A number which, cubed, equals the number given. For example, the cube root of 216 is 6.

Decimal number - Any number written using base 10; a number containing a decimal point.

Decimal point - A period placed to the left of a decimal.

Degree - The greatest sum of the exponents of the variables in a term of a polynomial or polynomial equation. For example, the polynomial $w^3 + 3w + 5$ is a third degree polynomial.

Denominator - The term below the line in a fraction; the divisor of the numerator. For example, in the fraction $\frac{3}{5}$, 5 is the denominator.

Dependent variable - A variable restricted to one or more of a set of values for every value assumed by an independent variable.

Determinant - A square array of quantities, having a value determined by a rule of combining the elements of the array.

Difference - The amount by which one quantity differs from another; remainder left after subtraction.

Digit - Any of the numerals from 0 through 9 - in the base-ten system.

Distributive - Of the principle in multiplication that allows the multiplier to be used separately with each term of the multiplicand.

Diverge - To fail to approach a limit.

Dividend - A quantity to be divided. For example, in the problem $14 \div 2$, 14 is called the dividend.

Division - The process of finding how many times a number (the divisor) is contained in another number (the dividend). The number of times equals the quotient.

Divisor - The quantity by which another quantity, the dividend, is to be divided. For example, in the problem $14 \div 2$, 2 is called the divisor.

Domain - The set of possible values of an independent variable.

Equal - Exactly the same. Of the same quantity, size, number, value, degree, intensity, or quality.

Equality - The condition or quality of being equal.

Equate - To make or set equal. To put in the form of an equation.

Equation - A mathematical expression involving the use of an equal sign. For example, $x^3 + 3x^2 + 5x = 3$ is referred to as an equation.

Equivalent algebraic fractions - Algebraic fractions that are quantitatively the same.

Even number - A number which is exactly divisible by two; not odd. For example, $(0, 2, 4, 6, 8, 10, \ldots)$ are even numbers.

Expanded form - To write, a quantity, as a sum of terms, as a continued product, or as another extended form.

Exponent - A number placed as a superscript to show how many times another number is to be placed as a factor. For example, in the problem $5^3 = 5 \times 5 \times 5 = 125$, 3 is an exponent.

Exponential - Containing, involving, or expressed as an exponent.

Exponential notation - A way of expressing a number as the product of the factor and 10 raised to some power. The factor is either a whole number or a decimal number. For example, the exponential notation form of 0.0353, 0.048, 489, 3987 are 35.3×10^{-3}, 48×10^{-3}, 48.9×10^{1}, and 398.7×10^{1}, respectively.

Expression - A designation of any symbolic mathematical form, such as an equation. The means by which something is expressed.

Factor - One of two or more quantities having a designated product. For example, 3 and 5 are factors of 15.

Factorial - The product of all the positive integers from 1 to a given number. For example, 5 factorial, usually written as $5!$, is equal to $1 \cdot 2 \cdot 3 \cdot 4 \cdot 5 = 120$.

Factorize - Resolve into factors.

Finite - Bounded in an interval.

Formula - A statement of some fact or relationship in mathematical terms.

Fraction - A number which indicates the ratio between two quantities in the form of $\frac{a}{b}$ such that a is any real number and b is any real number not equal to zero.

Fractional - Having to do with or making up a fraction.

Greater than $(\rangle)$ - A symbol used to compare two numbers with the greater number given first. For example, $5\rangle 2$, $23\rangle 20$, $50\rangle 10$.

Greatest common factor - A greatest number that divides two or more numbers without a remainder. For example, 6 is the greatest common factor among 6, 12, and 36.

Group - An assemblage of objects or numbers.

Horizontal - Flat. Parallel to the horizon. Something that is horizontal, as a line, plane, or bar.

Imaginary number - The positive square root of a negative number. For example, $\sqrt{-5}$, $\sqrt{-3}$, and $\sqrt{-1}$ are imaginary numbers. Not real number.

Improper fraction - A fraction in which the numerator is larger than or equal to the denominator. For example, $\frac{6}{5}$, $\frac{10}{9}$, and $\frac{23}{7}$ are improper fractions.

Increment - An increase or addition. The amount by which a quantity increases.

Independent variable - Not dependant on other variables.

Index - A number or symbol, often written as a subscript or superscript to a mathematical expression, that indicates an operation to be performed on. For example, in the problem $\sqrt[3]{x^2}$, 3 is referred to as an index.

Inequality $(\neq)$ - A relation indicating that the two numbers are not the same.

Infinite - Existing beyond or being greater than any arbitrarily large value.

Integer fraction - A fraction having positive or negative whole numbers in the numerator and the denominator.

Integer number - Any member of the set of positive whole numbers $(1, 2, 3, 4, ...)$, negative whole numbers $(-1, -2, -3, -4, ...)$, and zero is an integer number.

Interval - A set consisting of all the numbers between a pair of given numbers.

Introduction - To inform of something for the first time. The act of introducing.

Invert - To turn upside down. To reverse the order of.

Irrational number - A number not capable of being expressed by an integer (a whole number) or an integer fraction (quotient of an integer). For example, $\sqrt{3}$, π, and $\sqrt[4]{7}$ are irrational numbers.

Law - A general principle or rule that is obeyed in all cases to which it is applicable.

Less than $(\langle)$ - A symbol used to compare two numbers with the lesser number given first. For example, $5 \langle 8$, $23 \langle 30$, $12 \langle 25$, and $125 \langle 258$.

Like terms - Similar terms.

Limit - A fixed value which a variable quantity may approach indefinitely but can never reach.

Linear - Of or having to do with a line or lines. Of the first degree, as an equation.

Linear equation - An algebraic equation in which variables are used as factors no more than once in each term. For example, $3x + 5y = 10$ is a linear equation.

Matrix - A rectangular array of numerical or algebraic quantities treated as an algebraic entity.

Mathematical operation - The process of performing addition, subtraction, multiplication, and division in a specified sequence.

Method - A way of doing or accomplishing something.

Mixed fraction - A fraction made up of a positive or negative whole number and an integer fraction.

Mixed operation - Combining addition, subtraction, multiplication, and division in a math process is defined as a mixed operation.

Monomial - An expression consisting of only one term. Being a simple algebraic term. For example, 5, $\sqrt{xy}$, x^3, and $2ab$ are referred to as monomials.

Multiplicand - The number that is or is to be multiplied by another.

Multiplication - The process of finding the number obtained by repeated additions of a number a specified number of times: Multiplication is symbolized in various ways, i.e., $3 \times 4 = 12$ or $3 \cdot 4 = 12$, which means $3 + 3 + 3 + 3 = 12$, to add the number three together four times.

Multiplier - The number by which the multiplicand is multiplied. For example, if 3 is multiplied by 4, 3 is the multiplicand, 4 is the multiplier, and 12 is the product.

Not Applicable - In this book *Not Applicable* pertains to a *step* that can not be put to a specific use. A *step* that is not relevant.

Notation - A set of symbols used in specialized fields to represent numbers, quantities, or words.

Non real number - Imaginary number.

Numerator - The term above the line in a fraction. For example, in the fraction $\frac{3}{5}$, 3 is the numerator.

Numerical coefficient - Coefficients represented by numbers rather than letters.

Odd number - A number having a remainder of one when divided by two; not even. For example, $(1, 3, 5, 7, 9, 11, \ldots)$ are even numbers.

Operation - A process or action, such as addition, subtraction, multiplication, or division, performed in a specified sequence and in accordance with specific rules of procedure.

Polynomial - An algebraic function of two or more summed terms, each term consisting of a constant multiplier and one or more variables raised to a power. For example, the general form of a polynomial of degree n in a single real variable x is $P(x) = a_n x^n + a_{n-1}x^{n-1} + a_{n-2}x^{n-2} + \cdots + a_0$.

Power - An exponent. The result of a number multiplied by itself a given number of times. For example, the third power of 3 is 27.

Primary - Something that is first in degree, quality, or importance. Occurring first in time or sequence. Original.

Prime factorization - A factorization that shows only prime factors. For example, $21 = 1 \times 3 \times 7$.

Prime number - A number that has itself and unity as its only factors. For example, 2, 3, 5, 7, and 11 are prime numbers since they have no common divisor except unity.

Principal - First, highest, or foremost in importance.

Process - A series of operations or a method for producing something. A series of actions, changes, or functions that bring about an end or result.

Product - The quantity obtained by multiplying two or more quantities together.

Proper fraction - A fraction in which the numerator is smaller than the denominator.

Quadratic equations - Indicating a mathematical expression or equation of the second degree. For example, $x^2 + 3x - 2 = 0$ is a quadratic equation.

Quality or Qualitative - That which makes something the way it is; distinctive feature or characteristic.

Quantity or Quantitative - An amount or number.

Quotient - The quantity resulting from division of one quantity by another.

Radical - The root of a quantity as indicated by the radical sign. Indicating or having to do with a square root or cube root.

Radical expression - A mathematical expression or form in which radical signs appear.

Radical sign $\left(\sqrt{\ }\right)$ - A sign that indicates a specified root of the number written under it. For example, $\sqrt[3]{27}$ = the cube root of 27, which is, 3.

Radicand - The quantity under a radical sign. For example, 27 is the radicand of $\sqrt[3]{27}$.

Range - The totality of points in a set established by mapping each number of the first set with a single number of the second.

Ratio - The relative size of two quantities expressed as the quotient of one divided by the other. For example, the ratio of 10 to 4 is written as $10 : 4$ or $\frac{10}{4}$.

Rational number - A number that can be represented as an integer (a whole number) or an integer fraction (quotient of integers). For example, $\frac{1}{5}$, $-\frac{2}{15}$, $12 = \frac{12}{1}$, $-230 = -\frac{230}{1} = \frac{230}{-1} = \ldots$,

$-10 = -\frac{10}{1} = -\frac{100}{10} = -\frac{50}{5} = \frac{350}{-35} = \ldots$, and $0.13 = \frac{13}{100} = \frac{130}{1000} = \frac{26}{200} = \ldots$ are rational numbers.

Rationalization - The act, process, or practice of rationalizing.

Rationalize - To remove radicals without changing the value of an expression or roots of an equation.

Real number - A number that is either a rational number or an irrational number. For example, $\frac{3}{5}$, $-\frac{4}{13}$, -23, 0.13, $\sqrt{5}$, and π are real numbers.

Reference - The directing of attention to a person or thing.

Remainder - *a.* What is left when a smaller number is subtracted from a larger number. *b.* What is left undivided when one number is divided by another that is not one of its factors.

Respectively - In their respective order; individually in their given order.

Result - To end in a particular way. The consequence of a particular action. An outcome.

Resultant - That which results. Consequence.

Root - A quantity that, multiplied by itself a specified number of times, produces a given quantity. For example, 5 is the square root (5×5) of 25 and the cube root $(5 \times 5 \times 5)$ of 125.

Round number - A number that is revised or rounded to the nearest unit, as ten, hundred or thousand. For example, 200 is a round number for 199 or 201.

Rounded off - To make into a round number.

Rule - A standard method or procedure prescribed for solving a class of mathematical problem.

Scientific notation - A way of expressing a number as the product of the factor and 10 raised to some power. The factor is always of the form where the decimal point is to the right of the first non-zero digit. For example, the scientific notation form of 0.0353, 0.048, 489, 3987 are 3.53×10^{-2}, 4.8×10^{-2}, 4.89×10^{2}, 3.987×10^{3}, respectively.

Sequence - An ordered set of quantities, as x, x^2, x^3, x^4. A number of things following each other.

Series - The indicated sum of a finite or of a sequentially ordered infinite set of terms.

Sign - A mark or symbol having an accepted and specific meaning. For example, the sign $+$ implies addition.

Signed number - A number which can have a positive or negative value as designated by $+$ or $-$ symbol. A signed number with no accompanying symbol is understood to be positive.

Similar radicals - Radical expressions with the same index and the same radicand. For example, $\sqrt[3]{x^2}$, $5\sqrt[3]{x^2}$, and $3\sqrt[3]{x^2}$ are referred to as similar radicals.

Simplify - To express in a less complex form; make easier.

Solution - The act, method, or process of solving a problem. The answer to a problem.

Solution set - The set of all the values that satisfy an equation or inequality.

Solve - To find a solution to; answer.

Square - To find the equivalent of in square measure; to multiply, as a number or quantity, by itself.

Square root $(\sqrt{\ })$ - The factor of a number which, multiplied by itself, gives the original number. For example, the square root of 36 is 6.

Standard - Any type, model, or example for comparison. Serving as a gauge or model.

Standard form - serving as a model.

Subscript - A number, letter, or a symbol, written below and to the right or left of a character. For example, 2 is the subscript in x_2.

Substitute - To put in the place of another; to put in exchange.

Subtraction - The mathematical process of finding the difference between two numbers.

Sum - The amount obtained as a result of adding two or more numbers together.

Summation - The act of summing or totaling; addition.

Superscript - A number, letter, or a symbol, written above a character. For example, 5 is the superscript in y^5.

Symbol - A sign used to represent a mathematical operation.

Term - The parts of a mathematical expression that are added or subtracted. For example, in the equation $ax^3 + bx^2 + cx - d$, ax^3, bx^2, cx, and d are referred to as terms.

Trinomial - An expression consisting of three terms connected by a plus or minus sign. For example, $a^2 + a + 3$, $\sqrt[3]{x^2} + \sqrt[3]{x} - 5$, and $x^3 + 3x^2 + 2$ are referred to as trinomials.

Undefined - Not defined or explained.

Variable - A quantity capable of assuming any of a set of values. Having no fixed quantitative value.

Vertical - Upright. At right angles to the horizon. Straight up and down.

Whole number - A whole number is defined as an integer number.

Zero - The symbol or numeral 0. The point, marked 0, from which positive or negative quantities are reckoned on a graduated scale.

The following references were used in developing this glossary:

1) *The Webster's New World Dictionary of American English, Victoria E. Neufeldt, editor in chief, third college edition, 1995.*

2) *The American Heritage Dictionary of the English Language, William Morris, editor, third edition, 1994.*

3) *HBJ School Dictionary, Harcourt Brace Jovanovich publishing, fourth edition, 1985.*

Index

D

E

F

G

I

S

About the Author

Said Hamilton received his B.S. degree in Electrical Engineering from Oklahoma State University and Master's degree, also in Electrical Engineering from the University of Texas at Austin. He has taught a number of math and engineering courses as a visiting lecturer at the University of Oklahoma, Department of Mathematics, and as a faculty member at Rose State College, Department of Engineering Technology, at Midwest City, Oklahoma. He is currently working in the field of aerospace technology and has published numerous technical papers.

About the Editor

Pat Eblen received his Bachelor of Science degree in Electrical Engineering from the University of Kentucky where he was a member of Eta Kappa Nu Electrical Engineering Honor Society. He has worked in the aerospace industry for nearly twenty years where he has received numerous awards for contributions to spacecraft technology programs. Mr. Eblen enjoys studying mathematical theories in probability and quantum mechanics and has developed several original concepts in these fields.